KT-216-440

Handbook of Data Communications and Networks

Handbook of Data Communications and Networks

by

Bill Buchanan
Napier University

KLUWER ACADEMIC PUBLISHERS
BOSTON/DORDRECHT/LONDON

Library of Congress Cataloging-in-Publication Data

ISBN 0-412-84060-X

Published by Kluwer Academic Publishers,
P.O. Box 17, 3300 AA Dordrecht, The Netherlands.

Sold and distributed in North, Central and South America
by Kluwer Academic Publishers,
101 Philip Drive, Norwell, MA 02061, U.S.A.

In all other countries, sold and distributed
by Kluwer Academic Publishers,
P.O. Box 322, 3300 AH Dordrecht, The Netherlands.

Printed on acid-free paper

Printed in the Netherlands

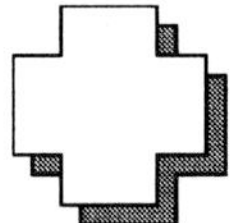

Table of Contents

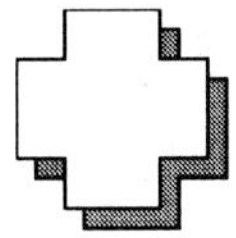

Preface

Data communications and networks are two of the fastest growing technological areas. This is because there is an almost unlimited demand for information transfer. Increased transfer rates give the opportunity for better applications and improved opportunities. These improvements lead to an increased demand for communications, and so on. Most people in business would now agree that networks and data communications makes their work easier and increases their productivity.

Communication systems now cover many important areas of modern life. Without them, there would be no ATM bank services, no Internet, no video conferencing and no electronic mail. Ask most organizations about what their most important resources were and many would say that their communications infrastructure was the most important. Without it, they could not effectively communicate, and business needs good and reliable communications. Within a few years, the most important component in a computer system will not be the processor or the amount of memory it is has, it will be its link to the network. Computers systems will become just another part of a local and a global communications network. It must also be said that communications is making the world a smaller place. It is reducing the influence that governments and organizations have on our life and makes it much easier for us search and retrieve information.

One thing that has changed in the modern working environment is that professionals need to be skilled in many different areas. The days of job demarcation are receding fast. Most technical people now require to be multi-skilled, in both hardware and software. They do not necessarily need to know exactly how to design an application program or how to design an electronic circuit, but they do need to know how systems fit together and how applications run. Thus a fault on a network could be caused by a faulty IC, or a cable break, or an electrical short-circuit, a network protocol error, or an application software error, and so on. Thus, professionals must be able to visualize communication systems from the hardware level right up to the application level. Failure to do this, results in badly designed systems or badly planned and maintained networks, or even inefficient working.

The object of this book is to cover most of the currently relevant areas of data communications and networks. These include:

- Communications protocols (especially TCP/IP).
- Networking (especially in Ethernet, Fast Ethernet, FDDI and ATM).
- Networking operating systems (especially in Windows NT, Novell NetWare and UNIX).
- Communications programs (especially in serial communications, parallel communications and TCP/IP).
- Computer hardware (especially in PC hardware, serial communications and parallel communication).

The book thus splits into 15 different areas, these are:

- General data compression (Chapters 2 and 3).
- Video, images and sound (Chapters 4–11).
- Error coding and encryption (Chapters 12–17).
- TCP/IP, WWW, Internets and Intranets (Chapters 18–20 and 23).
- Electronic Mail (Chapter 21).
- HTML (Chapters 25 and 26).
- Java (Chapters 27–29).
- Communication Programs (Chapters 20, 29 and 49).
- Network Operating Systems (Chapters 31–34).
- LANs/WANs (Chapters 35, 38–46).
- Serial Communications (Chapters 47 and 48).
- Parallel Communications (Chapters 50–52).
- Local Communications (Chapters 53–57).
- Routing and Protocols (Chapters 36 and 37).
- Cables and connectors (Chapters 58–60).

Many handbooks and reference guides on the market contain endless tables and mathematics, or are dry to read and contain very little insight in their subject area. I have tried to make this book readable, but also contain key information which can be used by professionals. I believe there are very few books which cover the amount of areas in this book, and many more areas are included on the WWW site and the associated CD-ROM.

Further information and source code can be found at:

`http://www.eece.napier.ac.uk/~bill_b/handb.html`

Help can be sought using the email address:

`w.buchanan@napier.ac.uk`

Dr. William Buchanan,
Senior Lecturer,
Napier University,
219 Colinton Road,
Edinburgh.

Trademarks

Intel™, Pentium™ and Pentium Pro™ are registered trademarks of Intel Corporation. CompuServe is a registered trademark of CompuServe Incorporated. IBM, PS/2 and PC/XT are trademarks of International Business Machines. Microsoft® is a registered trademark. Win32™, Win32s™, DOS™, MS-DOS™, Windows™, Windows 95™, Windows 98™, Windows NT™, Internet Explorer™, Visual Basic™ , and Visual C++™ are trademarks of Microsoft Corporation. Turbo Pascal™ , Turbo Debugger™ and Borland C++™ are trademarks of Borland International Incorporated. Unix™, Novell™ and NetWare™ are trademarks of Novell Incorporated. Java™ is a trademark of Sun Microsystems Incorporated. ANSI^M is a trademark of American National Standards Institute. Netscape Communicator™ is a trademark of Netscape Communications Incorporated.

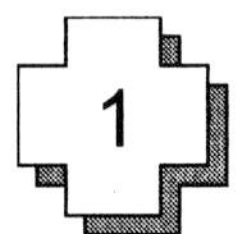

Introduction

1.1 Introduction

The usage of data communications and computer networks is ever increasing. It is one of the few technological areas which brings benefits to most of the countries and the peoples of the world. Without it, many industries could not exist. It is the objective of this book to discuss data communications in a readable form that both students and professionals all over the world can understand.

The book splits into eight main sections and some appendices, such as:

- General data compression.
- Video, images and sound.
- Error coding and encryption.
- TCP/IP and the Internet.
- Network operating systems.
- LANs/WANs.
- Local communications.
- Cables and connectors.

In the past, most electronic communication systems transmitted analogue signals. On an analogue telephone system, the voltage level from the telephone varies with the voice signal. Unwanted signals from external sources easily corrupt these signals. In a digital communication system, a series of digital codes represents the analogue signal, which are then transmitted as 1's and 0's. These digit forms are less likely to be affected by noise and thus have become the predominant form of communications.

Digital communication also offers a greater number of services, greater traffic and allows for high-speed communications between digital equipment. The usage of digital communications includes cable television, computer networks, facsimile, mobile digital radio, digital FM radio and so on.

1.2 A little bit of history

First, a little bit of history. This will lay the ground for the more technical coverage of the following chapters. Students and professionals in many other areas of work know of the important people of the past, so why not in data communications.

Communications, whether from smoke signals or in the form of pictures or the written word is as old as mankind. Before electrical communications, man has used fire, smoke and light to transmit messages over long distances. For example, Claude Chappe's developed the semaphore system in 1792, which has since been used to transmit messages with flags and light.

The history of communication can be traced to three main stages:

- The foundation of electrical engineering and radio wave transmission which owes a lot to the founding fathers of electrical engineering who were Coulomb, Ampère, Ohm, Gauss, Faraday, Henry and Maxwell, who laid down the basic principles of electrical engineering.
- The electronics revolution, which brought increased reliability, improved operations, improved sensitization and increased miniaturization.
- The desktop computer revolution, which has accelerated the usage of digital communication and has finally integrated all forms of electronic communications: text, speech, images and video.
- The usage of modern communications techniques, such as satellite communications, local area networks and digital networks.

1.2.1 History of electrical engineering

The Greek philosopher Thales appears to have been the first to document the observations of electrical force. He noted that on rubbing a piece of amber with fur caused it to attracted feathers. It is interesting that the Greek name for amber was *elektron* and the name has since been used in electrical engineering.

An important concept in electrical devices is that electrical energy is undoubtedly tied to magnetic energy. When there is an electric force, there is an associated magnetic force. The growth in understanding of electrics and magnetics began during the 1600s when the court physician of Queen Elizabeth I, William Gilbert, investigated magnets and found that the Earth had a magnet field. From this he found that a freely suspended magnetic tends to align itself with the magnetic field lines of the Earth. From then on, travelers around the world could easily plot their course because they knew which way was North.

Much of the early research in magnetics and electrics was conducted in the Old World, mainly in England, France and Germany. However, in 1752, Benjamin Franklin put the USA on the scientific map when he flew a kite in an electrical storm and discovered the flow of electrical current. This experiment is not recommended and resulted in the untimely deaths of several scientists.

In 1785, the French scientist Charles Coulomb showed that the force of attraction and repulsion of electrical charges varied inversely with the square of the distance between them. He also showed that two similar charges repel each other, while two dissimilar charges attract.

In 1820, the French scientist André Ampère studied electrical current in wires and the forces between them. Then, in 1827, the German scientist Georg Ohm studied the resistance to electrical flow. From this, he determined that resistance in a conductor was equal to the voltage across the material divided by current through it. Soon after this, the English scientist Michael Faraday produced an electric generator when he found that the motion of a wire through an electric field generated electricity. From this, he mathematically expressed the link between magnetism and electricity.

The root of modern communication can be traced back to the work of Henry, Maxwell, Hertz, Bell, Marconi and Watt. American Joseph Henry produced the first electromagnet when he wrapped a coil of insulated electrical wire around a metal inner. Henry, unfortunately, like many other great scientists, did not patent his discovery and its first application was in telegraphy, which was the beginning of the communications industry. Henry sent coded electrical pulses over telegraph wires to an electromagnet at the other end. It was a great success, but it was left to the artist Samuel Morse (the American Leonardo, according to one of his biographers) to take much of the credit. Morse, of-course developed Morse Code, which is a code of dots and dashes. Using Henry's system, he installed a telegraph system from Washington to Baltimore. The first transmitted message was "What hath God wrought." It received excellent publicity and after eight years there were over 23 000 miles (37 000 km) of telegraph wires in the USA. Several of the first companies to develop telegraph systems went on to become very large corporations, such as the Mississippi Valley Printing Telegraph Company which later became the Western Union. One of the first non-commercial uses of telegraph was the Crimean War and the American Civil War. A communications line from New York to San Francisco was an important mechanism for transmitting information to and from troops.

Other important developers of telegraph systems around the world were P.L. Shilling in Russia, Gauss and Weber in Germany, and Cooke and Wheatstone in Britain.

One of the all time greats was the James Clerk Maxwell, who was born in Edinburgh in 1831. He rates amongst the greatest of all the human beings who have walked on this planet and his importance to science puts him on par with Isaac Newton, Albert Einstein, James Watt and Michael Faraday. Maxwell's most famous formulation was a set of four equations that define the basic laws of electricity and magnetism (Maxwell's equations). Before Maxwell's work, many scientists had observed the relationship between electricity and magnetism. However, it was Maxwell, who finally derived the mathematical link between these forces. His four short equations described exactly the behavior and interaction of electric and magnetic fields. From this work, he also proved that all electromagnetic waves, in a vacuum, travel at 300 000 km per second (or 186 000 miles per second). This, Maxwell recognized, was equal to the speed of light and from this, he deduced that light was also an electromagnetic wave. He then reasoned that the electromagnetic wave spectrum must contain many invisible waves, each with its own wavelength and characteristic. Other practical scientists, such as Hertz and Marconi soon discovered these 'unseen' waves. The electromagnetic spectrum was soon filled with infrared waves, ultraviolet, gamma ray, X-rays and radio waves.

Another Scot, Alexander Graham Bell, had a great interest in the study of speech and elocution. In the USA, he opened the Boston School for the Deaf. His other interest was in multiple telegraphy and he worked on a device which he called a harmonic telegraph. This he used to aid the teaching of speech to deaf people. In 1876, out of this research he produced the first telephone with an electromagnet for the mouthpiece and the receiver. "It talks" was the headline (it has not stopped since). Even the great Maxwell was even amazed that anything so simple could reproduce the human voice and, in 1877, Queen Victoria acquired a telephone. Edison then enhanced it by using carbon powder in the diaphragm that produced an increased amount of electrical current. Bell along with several others formed the Bell Telephone Company which then developed the telephone so that, by 1915, long-distance telephone calls were possible. Bell's patent number 174 465 is the most lucrative ever issued. At the time, a reporter wrote, about the telephone, "It is an interesting toy ... but it can never be of any practical value."

The great inventor Edison, along with D.E. Hughes, produced a carbon transmitter, which was a basic microphone. Edison also first patented the phonograph which consisted of tinfoil wrapped around a rotating cylinder with grooves in it.

Around 1851, the brothers Jacob and John Watkins Brett laid a cable across the English Channel between Dover and Cape Griz Nez. It was the first use of electrical communications between England and France (unfortunately a French fisherman mistook it for a sea monster and trawled it up). The British maintained a monopoly on submarine cables and laid cables across the Thames, Scotland to Ireland, England to Holland, as well as cables under the Black Sea, the Mississippi River and the Gulf of St Lawrence. Submarine cables have since been placed under most of the major seas and oceans around the world.

Around 1888, German Heinrich Hertz detected radio waves (as predicted by Maxwell) when he found that a spark produced an electrical current in a wire on the other side of the room. Then, Marconi, in 1896, succeeded in transmitting radio waves over a distance of two miles. From this humble start, he soon managed to transmit a radio wave across the Atlantic Ocean.

Scot Robert Watson-Watt made RADAR (radio detection and ranging) practicable in 1935. Today it is used in many applications from detect missiles and planes, to detecting rain clouds and detecting the speed of motor cars. The next great revolution occurred with electronics.

1.2.2 History of electronics

The science of electronics began in 1895 when H. Lorentz postulated the existence of charges called electrons. A few years later, Braun built the first electron valve, which was a simple cathode-ray tube. Then, at the beginning of the century, Fleming invented a diode, called a valve, which used a heated cylindrical plate in a vacuum. A positively charged heater plate caused a current to flow, but when negatively charged it had virtually no current flow. Lee De Forest further enhanced it by adding a grid to make a triode. This allowed a small control voltage to control a large current. Their main application was in the amplification of electrical signals and then later as an electrical switch (to create digital levels). By the 1940s, several scientists at the Bell Laboratories were investigating materials called semiconductors, such as silicon and germanium. These substances only conducted electricity moderately well, but when they where doped with impurities their resistance changed. From this work, they made a crystal called a diode, which worked just like a valve, but had many advantages, including the fact that it did not require a vacuum and was much smaller. It also worked well at room temperatures, required little current and had no warm-up time. This was the start of microelectronics.

In 1948, William Shockley at the Bell Labs produced a transistor that could act as a triode. It was made from a germanium crystal with a thin p-type section sandwiched between two n-type materials. Rather that release its details to the world, Bell Laboratories kept its invention secret for over seven months so that they could fully understand its operation. Then, on 30 June 1948, Shockley finally revealed the transistor to the world. Unfortunately, as with many other great inventions, it received little public attention and even less press coverage (the *New York Times* gave it 4½ inches on page 46). It must be said that few men have made such a profound change on the world and Shockley deservedly received a Nobel Prize in 1956.

Transistors had initially been made from germanium, which is not a robust material and cannot withstand high temperatures. The first company to propose a method of using silicon transistors was a geological research company named Texas Instruments (which had diversified into transistors). Soon many companies were producing silicon transistors, and, by 1955, the electronic valve market had peaked, while the market for transistors was rocketing. The larger electronic valve manufacturers, such as Western Electric, CBS, Raytheon and Westinghouse failed to adapt to the changing market and quickly lost their market share to the new transistor manufacturing companies, such as Texas Instruments, Motorola, Hughes and RCA.

Within a few years, transistors were small enough to make hearing aids that fitted into the ear. They soon, with the help of companies such as Sony, operated over higher frequencies and within larger temperature ranges. Eventually they became so small that many of them could be placed on a single piece of silicon. These were referred to as microchips and they started the microelectronics industry.

1.2.3 History of computing

In 1959, IBM built the first commercial transistorized computer named the IBM 7090/7094 series. It was so successful that it dominated the computer market for many years. In 1965, they produced the famous IBM system 360 which was built with integrated circuits. Then, in 1970, they introduced the 370 system, which included semiconductor memories. These were the great computing workhorses of the time, but unfortunately, were extremely expensive to purchase and maintain. Most companies had to lease their computer systems, as they could not afford to purchase them.

Around the same time, the electronics industry was producing cheap pocket calculators. These led to the development of affordable computers, when the Japanese company Busicon commissioned a small company named Intel to produce a set of between eight and twelve ICs for a calculator. Then instead of designing a complete set of ICs, Intel produced a set of ICs which could be programmed to perform different tasks. These were the first ever microprocessors and soon Intel (short for *Int*egrated *El*ectronics) produced a general-purpose 4-bit microprocessor, named the 4004 and a more powerful 8-bit version, named the 8080. Other companies, such as Motorola, MOS Technologies and Zilog were soon also making microprocessors.

IBM's virtual monopoly on computer systems soon started to slip as many companies developed computers based around the newly available 8-bit microprocessors, namely MOS Technologies 6502 and Zilog's Z-80. IBM's main contenders were Apple and Commodore who introduced a new type of computer – the personal computer (PC). The leading systems, at the time, where the Apple I and the Commodore PET. These captured the interest of the home user and for the first time individuals had access to cheap computing power. These flagship computers spawned many others, such as the Sinclair ZX80/ZX81, the BBC microcomputer, the Sinclair Spectrum, the Commodore Vic-20 and the classic Apple II (all of which where based on the 6502 or Z-80). Most of these computers were aimed at the lower end of the market and were mainly used for game playing and not for business applications.

IBM realized the potential of the PC and microprocessor. Unlike many of their previous computer systems, they developed their version of the PC using standard components, such as Intel's 16-bit 8086 microprocessor. They released it as a business computer, which could run word processors, spread sheets and databases and was named the IBM PC. It has since become the parent of all the PCs ever produced. To increase the production of this software for the PC they made information on the hardware freely available. This resulted in many software packages being developed and also helped clone manufacturers to copy the original design. So the term 'IBM compatible' was born and it quickly became an industry standard by sheer market dominance.

On previous computers IBM had written most of their programs for their systems. For the PC they had a strict time limit, so they went to a small computer company called Microsoft to develop the operating system program. The developed program was hardly earth shattering, but has since gone on to make billions of dollars. It was named the Disk Operating System (DOS) because of its original purpose of controlling the disk drives. It accepted commands from the keyboard and displayed them to the monitor. The language of DOS consisted of a set of commands which were entered directly by the user and interpreted to perform file management tasks, program execution and system configuration. Its function was to run programs, copy and remove files, create directories, move within a directory structure and to list files. Microsoft have since gone on to develop industry-standard software such as Microsoft Windows Version 3, Microsoft Office, Microsoft Windows 95/98 and Microsoft NT. Intel has also benefited greatly from the development of the PC and has developed a large market share for their industry-standard microprocessors, such as the 80286, 80386, 80486 and Pentium processors. Microsoft's operating system Windows NT has networking built into its core.

The power of modern desktop computers has allowed great advances in electronic communications. For many computers, networking is now a standard part of the computer, whether it is through a modem or directly onto a digital network.

1.2.4 History of modern communications

After the telephone's initial development, call switching was achieved by using operators. However, in 1889, Strowger, a Kansas City undertaker, patented an automatic switching system. In one of the least catchy advertising slogans, it was advertised as "a girl-less, cuss-less, out-of-orderless, wait-less telephone system". It used a pawl-and-ratchet system to move a wiper over a set of electrical contacts. This led to the development of the Strowger exchange, which was used extensively until the 1970s. Another important improvement came with the crossbar, which allowed many inputs to connect to many outputs, simply by addressing the required connection. The first inventor is claimed to be J.N Reynolds of Bell Systems, but it is normally given to G.A. Betulander.

One of the few benefits of war (whether it be a real war or a cold war) is the rapid development of science and technology. Radio transmission benefited from this over World War I. A by-product of this work was frequency modulation (FM) and amplitude modulation (AM). In these, signals to be carried on (modulated) high frequency carrier waves which traveled through the air better than unmodulated waves. Another by-product of the war effort was frequency division multiplexing (FDM) which allowed many signals to be transmitted over the same channel, but with a different carrier frequency.

After the Second World War, the first telephone cable across the Atlantic was laid from Oban, in Scotland to Clarenville in Newfoundland. Previously, in 1902, the first Pacific Ocean cable was laid. A cable, laid in 1963, stretches from Australia to Canada.

The ATT-owned Telstar satellite started a great revolution as it allowed communications over large distances using microwave signals. These electromagnetic waves can propagate through rain and clouds and bounce off the satellite. The amount of information that can be transmitted varies with the bandwidth of the system. This is normally limited by the transmission system. A satellite system can carry as much as 10 times the amount of information that a radio wave can carry. This allows several TV channels to be transmitted simultaneously and/or thousands of telephone calls. Light waves have an even greater capacity for transmitting information (almost limitless). Thus, a great leap forward has been with the transmission of data using fiber-optic cables.

Most information transmitted is in the form of digital pulses. A standard code for this transmission, called Pulse code modulation (PCM), was invented by A.H. Reeves in the 1930s, but not used until the 1960s.

A major problem in the past with computers systems was that they used different codes to transmit text. Baudot developed a 5-unit standard code for telegraph systems. Unfortunately, it had a limited alphabet of upper-case letters and had only a few punctuation symbols. In 1966, ANSI defined a standard code called ASCII. This has since become the standard coding system for text-based communications. In its standard form it uses 7 bits and can thus only represent up to 128 characters. It has since been modified to support an 8-bit code (called Extended ASCII).

1.3 Information

Information is available in an analogue form or in a digital form, as illustrated in Figure 1.1. Computer-generated data can be easily stored in a digital format, but analogue signals, such as speech and video, must first be sampled at regular intervals and then converted into a digital form. This process is known as digitization and has the following advantages:

- Digital data is less affected by noise, as illustrated in Figure 1.2.
- Extra information can be added to digital signals so that errors can either be detected or corrected.
- Digital data tends not degrade over time.
- Processing of digital information is relatively easy, either in real-time or non real-time.
- A single type of media can be used to store many different types of information (such video, speech, audio and computer data can be stored on tape, hard-disk or CD-ROM).
- A digital system has a more dependable response, whereas an analogue system's accuracy depends on parameters such as component tolerance, temperature, power supply variations, and so on. Analogue systems thus produce a variable response and no two analogue systems are identical.
- Digital systems are more adaptable and can be reprogrammed with software. Analogue systems normally require a change of hardware for any functional changes (although programmable analogue devices are now available).

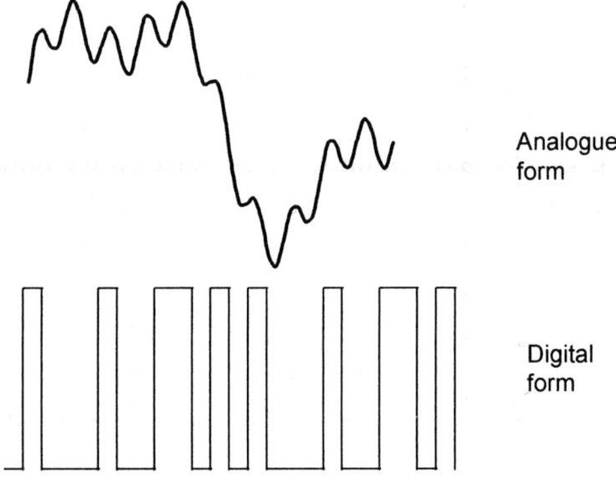

Figure 1.1 Analogue and digital format.

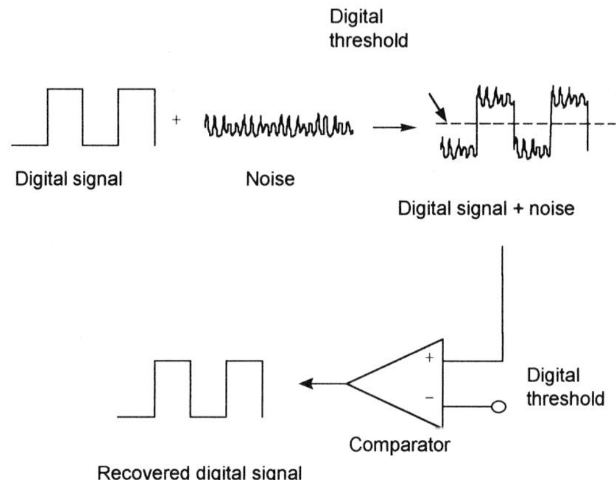

Figure 1.2 Recovery of a digital signal with noise added to it.

The main disadvantage with digital conversion is:

- Data samples must be quantized to given levels; this adds an error called quantization error. The larger the number of bits used to represent each sample, the smaller the quantization error.

As the analogue signal must be sampled at regular intervals, digital representations of analogue waveforms require large storage space. For example, 70 minutes of hi-fi quality music requires over 600 MB of data storage. The data once stored tends to be reliable and will not degrade over time. Typically, digital data is stored as either magnetic fields on a magnetic disk or as pits on an optical disk. A great advantage of digital technology is that once the analogue data has been converted to digital then it is relatively easy to store it with other purely digital data. Once stored in digital form it is relatively easy to process the data before it is converted back into analogue form. Analogue signals are relatively easy to store, such as video and audio signals are stored as magnetic fields on tape and a picture is stored on photographic paper. These media, though, tend to add noise (such as tape hiss) during storage and recovery.

As has been said, the accuracy of a digital systems depends on the number of bits used for each sample, whereas an analogue system's accuracy depends on component tolerance. Analogue systems also produce a differing response for different systems whereas a digital system has a dependable response. It is very difficult (if not impossible) to recover the original analogue signal after it is affected by noise (especially if it is affected by random noise). Most methods of reducing this noise involves some form of filtering or smoothing of the signal.

1.4 Conversion to digital

Figure 1.3 outlines the conversion process for digital data (the upper diagram) and for analogue data (the lower diagram). The lower diagram shows how an analogue signal (such as speech or video) is first sampled at regular intervals of time. These samples are then converted into a digital form with an ADC (analogue-to-digital converter). It can then be compressed and/or stored in a defined digital format (such as WAV, JPG, and so on). This digital form is then converted back into an analogue form with a DAC (digital-to-analogue converter). When data is already in a digital form (such as text or animation) it is converted into a given data format (such as BMP, GIF, JPG, and so on). It can be further compressed before it is stored, transmitted or processed.

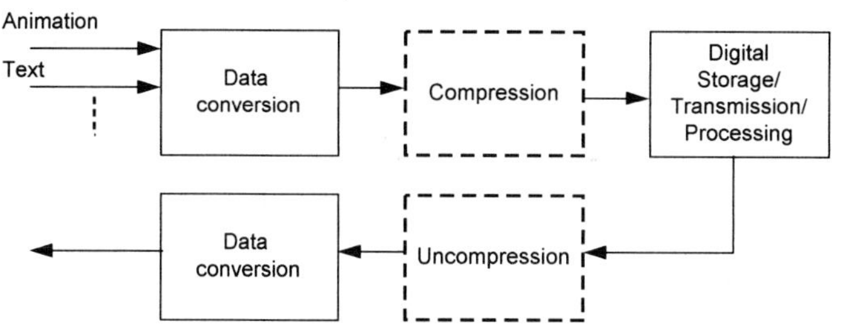

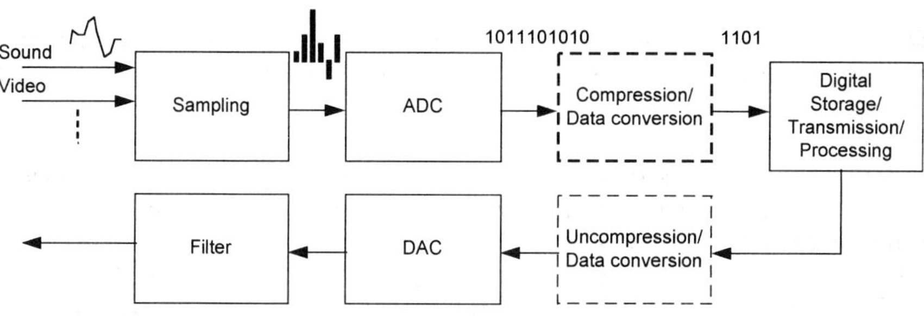

Figure 1.3 Information conversion into a digital form.

1.5 Sampling theory

As a signal may be continually changing, a sample of it must be taken at given time intervals, where the rate of sampling depends on its rate of change. For example, the temperature of the sea will not vary much over a short time but a video image of a sports match will. Digital encoding normally involves sampling a signal at fixed time intervals. Sufficient information is then extracted to allow the signal to be processed or reconstructed. Figure 1.4 shows a signal sampled every T_S seconds.

If a signal is to be reconstructed as the original signal it must be sampled at a rate defined by the Nyquist criterion. This states:

the sampling rate must be twice the highest frequency of the signal

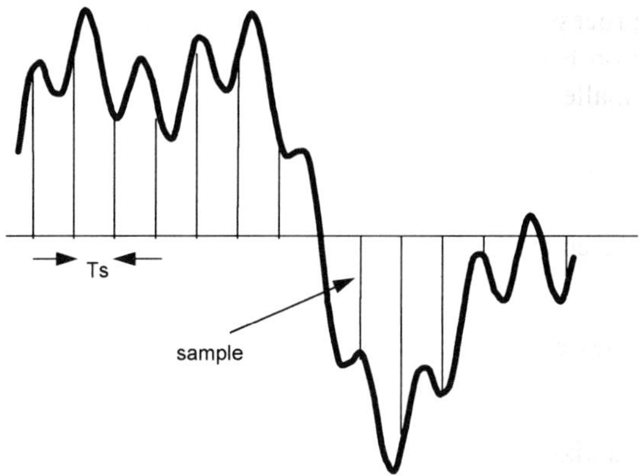

Figure 1.4 The sampling process.

For telephone speech channels, the maximum signal frequency is limited to 4 kHz and must thus be sampled at least 8 000 times per second (8 kHz). This gives one sample every 125 μs. Hi-fi quality audio has a maximum signal frequency of 20 kHz and must be sampled at least 40 000 times per second (many professional hi-fi sampling systems sample at 44.1 kHz). Video signals have a maximum frequency of 6 MHz, thus a video signal must be sampled at 12 MHz (or once every 83.3 ns).

1.6 Quantization

Quantization involves converting an analogue level into a discrete quantized level. Figure 1.5 shows the conversion of a waveform into a 4-bit digital code. In this case, there are 16 discrete levels which are represented with the binary values 0000 to 1111. Value 1111 represents the maximum voltage level and value 0000 the minimum. It can be seen, in this case, that the digital codes for the four samples are 1011, 1011, 1001, 0111.

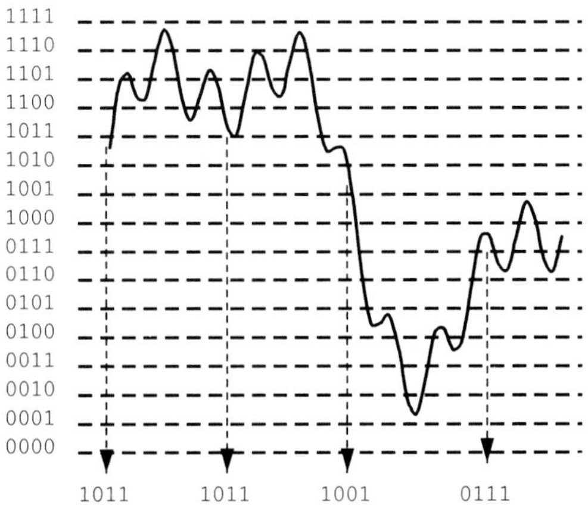

Figure 1.5 Converting an analogue waveform into a 4-bit digital form.

The quantization process approximates the level of the analogue to the nearest quantized level, this approximation leads to an error known as the quantization error. The greater the number of levels the smaller the quantization error. Table 1.1 gives the number of levels for a given number of bits.

The maximum error between the original level and the quantized level occurs when the original level falls exactly halfway between two quantized levels. The maximum error will be a half of the smallest increment or

$$\text{Max error} = \pm\frac{1}{2}\frac{\text{Maximum range}}{2^N}$$

Table 1.1 states the quantization error (as a percentage) of a given number of bits. For example, the maximum error with 8 bits is 0.2%, while for 16 bits it is only 0.000 76%.

Table 1.1 Number of quantization levels as a function of bits.

Bits (N)	Quantization levels	Accuracy (%)
1	2	50
2	4	25
3	8	12.5
4	16	6.25
8	256	0.2
12	4 096	0.012
14	16 384	0.003
16	65 536	0.000 76

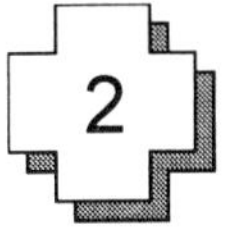

Compression Techniques

2.1 Introduction

Most sources of information contain redundant information or information that adds little to the stored data. An example of this might be the storage of a rectangular image. As a single color bitmapped image it could be stored as:

```
000000000000000000000000000000000000000000000
000000000000000000000000000000000000000000000
000000000000000000000000000000000000000000000
000000000111111111111111111111111110000000000
000000000100000000000000000000000010000000000
000000000100000000000000000000000010000000000
000000000100000000000000000000000010000000000
000000000100000000000000000000000010000000000
000000000100000000000000000000000010000000000
000000000111111111111111111111111110000000000
000000000000000000000000000000000000000000000
000000000000000000000000000000000000000000000
```

An improved method might be to store the image as a graphics metafile, such as:

```
Rectangle 11, 4, 26, 7
;     rectangle at start co-ordinate (11,4) of width 26 and length 7
;     pixels
```

or to compress long sequences of identical bit states, such as:

```
0,45     ; 45 bits of a '0'
0,45     ; 45 bits of a '0'
0,45     ; 45 bits of a '0'
0,10 1,26 0,9  ; 10 bits of a '0', 26 1's and 9 0's
0,10 1,1 0,24 1,1 0,9
0,10 1,1 0,24 1,1 0,9
0,10 1,1 0,24 1,1 0,9
0,10 1,1 0,24 1,1 0,9
0,10 1,1 0,24 1,1 0,9
0,10 1,26 0,9
0,45
0,45
```

Data compression is becoming an important subject as increasingly digital information is required to be stored and transmitted. Many data transmission systems have a limited data rate and data storage is limited. For example, an uncompressed digitized video stored on CD-ROM could give only 30 seconds of video (approximately 648 MB of data).

Data compression normally involves analyzing the source of the information and determining if there is redundant data in it.

2.2 Compression methods

When compressing data it is important to take into account the type of data and how it is interpreted. For example pixels in an image may be distorted but it would still contain the required information. Whereas, a single erroneous bit in a computer data file could cause severe problems.

Video and sound images are normally compressed with a lossy compression whereas computer-type data has a lossless compression. The basic definitions are:

- Lossless compression – where the information, once uncompressed, will be identical to the original uncompressed data. This will obviously be the case with computer-type data, such as data files, computer programs, and so on. Any loss of data will cause the file to be corrupted.

- Lossy compression – where the information, once uncompressed, cannot be fully recovered. Lossy compression normally involves analyzing the data and determining which information has little effect on the resulting compressed data. For example, there is little difference, to the eye, between an image with 16.7 million colors (24-bit color information) and an image stored with 1024 colors (10-bit color information), but the storage will be reduced to 41.67% (many computer systems cannot even display 24-color information in certain resolutions). Compression of an image might also be used to reduce the resolution of the image. Again, the human eye might compensate for the loss of resolution.

2.3 Letter probabilities

The English language has a great deal of redundancy in it, thus common occurrences in text can be coded with short bit sequences. The probability of each letter also varies. For example a letter '*e*' occurs many more times than the letter '*z*'. Program 2.1 determines the number of occurrences of the characters from '*a*' to '*z*' in an example text input. Sample run 2.1 shows a sample run with an example piece of text. It can be seen that the most common character in the text is an '*e*' (30 occurrences), followed by a '*t*' (22 occurrences), followed by '*a*' (17 occurrences).

Program 2.1

```
#include    <stdio.h>
#include    <string.h>

int   get_occurances(char c, char txt[]);

int   main(void)
{
```

```
char  ch, text[BUFSIZ];
int   occ;

      printf("Enter text >>");
      gets(text);

      for (ch='a';ch<='z';ch++)
      {
         occ=get_occurances(ch,text);
         printf("%c %d\n",ch,occ);
      }

      return(0);
}

int   get_occurances(char c, char txt[])
{
int   occ=0,i;

      for (i=0;i<strlen(txt);i++)
         if (c==txt[i]) occ++;
      return(occ);
}
```

Sample run 2.1

```
Enter text >>  this is an example text input to determine the number of occur-
rences of a character within a piece of text. it shows how the characters used
in the english language vary in their occurrences. the e character is the most
common character in the english language.
a 17
b 1
c 16
d 2
e 30
f 3
g 6
h 17
i 15
j 0
k 0
l 5
m 5
n 15
o 11
p 3
q 0
r 16
s 12
t 22
u 7
v 1
w 3
x 3
y 1
z 0
```

Program 2.6 in Section 2.9 gives a simple C program which determines the probability of letters within a text file. This program can be used to determine typical letter probabilities. Sample run 2.2 shows a sample run using the text from Chapter 1 as the input to the program. It can be seen that the highest probability is with the letter '*e*', which occurs, on average, 94.3 times every 1000 letters. Table 2.1 lists the letters in order of their probability. Notice that the letters which are worth the least in the popular board game Scrabble (such as, '*e*', '*t*', '*a*', and so on) are the most probable and the letters with the highest scores (such as '*x*', '*z*' and '*q*') are the least probable.

Sample run 2.2

```
Char. Occur. Prob.
  a   1963   0.0672
  b    284   0.0097
  c    914   0.0313
  d    920   0.0315
  e   2752   0.0943
  f    471   0.0161
  g    473   0.0162
  h    934   0.0320
  i   1680   0.0576
  j     13   0.0004
  k     96   0.0033
  l    968   0.0332
  m    724   0.0248
  n   1541   0.0528
  o   1599   0.0548
  p    443   0.0152
  q     49   0.0017
  r   1410   0.0483
  s   1521   0.0521
  t   2079   0.0712
  u    552   0.0189
  v    264   0.0090
  w    383   0.0131
  x     57   0.0020
  y    278   0.0095
  z     44   0.0015
  .    292   0.0100
  SP  4474   0.1533
  ,    189   0.0065
```

Table 2.1 Letters and their occurrence in a sample text file.

Character	*Occurrences*	*Probability*	*Character*	*Occurrences*	*Probability*
SPACE	4474	0.1533	g	473	0.0162
e	2752	0.0943	f	471	0.0161
t	2079	0.0712	p	443	0.0152
a	1963	0.0672	w	383	0.0131
i	1680	0.0576	.	292	0.0100
o	1599	0.0548	b	284	0.0097
n	1541	0.0528	y	278	0.0095
s	1521	0.0521	v	264	0.0090
r	1410	0.0483	,	189	0.0065
l	968	0.0332	k	96	0.0033
h	934	0.0320	x	57	0.0020
d	920	0.0315	q	49	0.0017
c	914	0.0313	z	44	0.0015
m	724	0.0248	j	13	0.0004
u	552	0.0189			

2.4 Coding methods

Apart from lossy and lossless compression, data encoding is normally classified into two main areas: entropy encoding and source encoding.

2.4.1 Entropy coding

Entropy coding does not take into account the characteristics of the data and treats all the bits in the same way; it produces lossless coding. Typically, it uses:

- Statistical encoding – where the coding analyses the statistical pattern of the data. For example if a source of text contains many more 'e' characters than 'z' characters then the character 'e' could be coded with very few bits and the character 'z' with many bits.

- Suppressing repetitive sequences – many sources of information contain large amount of receptive data. For example, this page contains large amounts of 'white space'. If the image of this page were to be stored, a special character sequence could represent long runs of 'white space'.

2.4.2 Source encoding

Source encoding normally takes into account characteristics of the information. For example, images normally contain many repetitive sequences, such as common pixel colors in neighboring pixels. This can be encoded as a special coding sequence. In video pictures, also, there are very few changes between one frame and the next. Thus, typically the data encoded only stores the changes from one frame to the next.

2.5 Statistical encoding

Statistical encoding is an entropy technique which identifies certain sequences within the data. Frequently used patterns are coded with fewer bits than less common patterns. For example, text files normally contain many more '*e*' characters than '*z*' characters. Thus, the '*e*' character could be encoded with a few bits and the '*z*' with many bits. Statistical encoding is also known as arithmetic compression.

A typical statistical coding scheme is Huffman encoding. Initially the encoder scans through the file and generates a table of occurrences of each character. The codes are assigned to minimize the number of encoded bits, then stored in a codebook which must be transmitted with the data.

Table 2.2 shows a typical coding scheme for the characters '*a*' to '*z*'. It uses the same number of bits for each character. Morse code is an example of statistical encoding. It uses dots (a zero) and dashes (a one) to code characters, where a short space in time delimits each character. It uses short codes for the most probable letters and longer codes for less probable letters. In the form of zeros and ones it is stated in Table 2.3.

Thus the message:

```
this an
```

would be encoded as:

```
Message:       t      h      i      s             a      n
Simple code: 10011  00111  01000  10010  11010  00000  01101
Morse code:  1      0000   00     000    0011   01     10
```

This has reduced the number of bits used to represent the message from 35 (7×5) to 18.

Table 2.2 Simple coding scheme.

a	00000	b	00001	c	00010	d	00011	e	00100
f	00101	g	00110	h	00111	i	01000	j	01001
k	01010	l	01011	m	01100	n	01101	o	01110
p	01111	q	10000	r	10001	s	10010	t	10011
u	10100	v	10101	w	10110	x	10111	y	11000
z	11001	SP	11010						

Table 2.3 Morse coding scheme.

a	01	b	1000	c	1010	d	100	e	0
f	0010	g	110	h	0000	i	00	j	0111
k	101	l	0100	m	11	n	10	o	111
p	0110	q	1101	r	010	s	000	t	1
u	001	v	0001	w	011	x	1001	y	1011
z	1100	SP	0011						

2.6 Repetitive sequence suppression

Repetitive sequence suppression involves representing long runs of a certain bit sequence with a special character. A special bit sequence is then used to represent that character, followed by the number of times it appears in sequence. Typically 0s (zero) and ' ' (spaces) occur repetitively in text files. For example the data:

```
8.320000000000O
```

could be coded as:

```
8.32F11
```

where F represents the flag. In this case, the number of stored characters has been reduced from 16 to 7. Many text sources have other characters which occur repetitively. Run-length encoding (RLE) uses this to encode any character sequence with a special flag followed by the number of characters and finally the character which is repeated. For example

```
Fred        has     when.........
```

could be coded as:

```
FredF7 hasF7 whenF9.
```

where F represents the flag. In this case, the number of stored characters has been reduced from 32 to 20. The 'F7 ' character code represents seven ' ' (spaces) and 'F9.' represents nine '.' characters.

Program 2.2 is a very simple program which scans a file `IN.DAT` and, using RLE, stores to a file `OUT.DAT`. The special character sequence is:

ZZ*cxx*

where ZZ is the flag sequence, *c* is the repeating character and *xx* the number of times the character occurs. The ZZ flag sequence is chosen because, in a text file, it is unlikely to occur within the file. File listing 2.1 shows a sample `IN.DAT` and File listing 2.2 shows the RLE encoded file (`OUT.DAT`).

Program 2.2

```
/*    ENCODE.C     */
#include <stdio.h>

int   main(void)
{
FILE  *in,*out;
char  previous,current;
int      count;

      if ((in=fopen("in.dat","r"))==NULL)
      {
         printf("Cannot open <in.dat>");
         return(1);
      }
      if ((out=fopen("out.dat","w"))==NULL)
      {
         printf("Cannot open <out.dat>");
         return(1);
      }

      do
      {
         count=1;
         previous=current;
         current=fgetc(in);
         do
         {
            previous=current;
            current=fgetc(in);
            if (previous!=current) ungetc(current,in);
            else count++;
         } while (previous==current);

         if (count>1) fprintf(out,"ZZ%c%02d",previous,count);
         else fprintf(out,"%c",previous);
      } while (!feof(in));
      fclose(in);
      fclose(out);
      return(0);
}
```

File list 2.1

```
The      bbbbbbboy stood onnnnn the burning
deck          and still did.
1.000000000
3.000000010
5.000000000
```

File list 2.2

```
TheZZ 05ZZb07oy stZZo02d oZZn05 the burning
deckZZ 09and stiZZl02 did.
1.ZZ009
3.ZZ00710
5.ZZ009
```

Program 2.3 is a simple C program which decodes the RLE file produced by the previous program.

Program 2.3

```
/*    UNENCODE.C     */
#include <stdio.h>

int   main(void)
{
FILE  *in,*out;
char  ch;
int   count,i;
      if ((in=fopen("out.dat","r"))==NULL)
      {
         printf("Cannot open <out.dat>");
         return(1);
      }
      if ((out=fopen("in1.dat","w"))==NULL)
      {
         printf("Cannot open <in1.dat>");
         return(1);
      }

      do
      {
         ch=fgetc(in);
         if (ch=='Z')
         {
            ch=fgetc(in);
            if (ch=='Z')
            {
               fscanf(in,"%c%02d",&ch,&count);
               for (i=0;i<count;i++)
                  fprintf(out,"%c",ch);
            }
            else
            {
               ungetc(ch,in); fprintf(out,"Z");
            }
         }
         else fprintf(out,"%c",ch);
      } while (!feof(in));
      fclose(in); fclose(out);
      return(0);
}
```

The ZZ flag sequence is inefficient as it uses two characters to store the flag, a better flag could be an 8-bit character which cannot occur, such as 11111111b, or ffh. Program 2.4 is an example of this and Program 2.5 shows the decoding program.

Program 2.4

```
#include <stdio.h>

#define  FLAG  0xff  /* 1111 1111b  */

int   main(void)
{
FILE  *in,*out;
char  previous,current;
int   count;

 ;;; ;;;;;
        if (count>1) fprintf(out,"%c%c%02d",FLAG,previous,count);
        else fprintf(out,"%c",previous);
     }  while (!feof(in));
     fclose(in);
     fclose(out);
     return(0);
}
```

Program 2.5

```
/*    UNENCODE.C      */
#include <stdio.h>

#define  FLAG  0xff  /* 1111 1111b  */

int   main(void)
{
FILE  *in,*out;
char  ch;
int   count,i;

      ;;; ;;;;
      ;;; ;;;;

      do
      {
         ch=fgetc(in);

         if (ch==FLAG)
         {
            ch=fgetc(in);
            fscanf(in,"%c%02d",&ch,&count);
            for (i=0;i<count;i++)
                  fprintf(out,"%c",ch);
         }
         else fprintf(out,"%c",ch);

      }  while (!feof(in));

      fclose(in);

      fclose(out);

      return(0);
}
```

In a binary file any bit sequence can occur. To overcome this a flag sequence, such as 10101010, can be used to identify the flag. If this sequence occurs within the data then it will be coded with two flags, two consecutive flags within the data are coded with three flags, and so on. For example:

```
00000000 10101010 10101010 00011100 01001100
```

would be encoded as:

```
00000000 10101010 10101010 10101010 00011100 01001100
```

thus when the three flags are detected then one of them is deleted.

Repetitive sequence suppression is an excellent general-purpose compression technique for images, as most images tend to have long sequences of the same pixel intensity or color.

2.7 Differential encoding

Differential coding is a source coding method which is used when there is a limited change from one value to the next. It is well suited to video and audio signals, especially audio, where the sampled values can only change within a given range. It is typically used in PCM (pulse code modulation) schemes to encode audio and video signals. PCM converts analogue samples into a digital code. Examples are:

- Delta modulation PCM – delta PCM uses a single-bit code to represent an analogue signal. With delta modulation a '1' is transmitted (or stored) if the analogue input is higher than the previous sample or a '0' if it is lower. It must obviously work at a higher rate than the Nyquist frequency but because it only uses 1 bit it normally results in a lower output bit rate (because the factor of increasing the sampling rate is normally less than the factor of reducing the number of encoded bits).
- Adaptive delta modulation PCM – unfortunately delta modulation cannot react to rapidly changing signals and will thus take a relatively long time to catch up (known as slope overload). It also suffers when the signal does not change much as this produces a square wave signal (known as granular noise). One method of reducing granular noise and slope overload is to use adaptive delta PCM. With this method, the slope of the input signal varies the step size. The larger the slope, the larger the step size. Algorithms usually depend on the system and the characteristics of the signal. A typical algorithm is to start with a small step and increase it by a multiple until the required level is reached. The number of slopes will depend on the number of coded bits, such as four step sizes for 2 bits, eight for 3 bits, and so on.
- Differential PCM (DPCM) – speech signals tend not to change much between two samples. Thus, similar codes are sent, which leads to a degree of redundancy. Certain signals, such as speech, have a limited range for a sample amplitude for a given sample time. DPCM reduces the redundancy by transmitting the difference in the amplitude of two consecutive samples. Since the range of sample differences is typically less than the range of individual samples, fewer bits are required for DPCM than for conventional PCM.

2.8 Transform encoding

Transform encoding is a source-encoding scheme where the data is transformed by a mathematical transform in order to reduce the transmitted (or stored) data. A typical method is to conduct a Fourier transform to determine the frequency information from the data. The coefficients of the transform can then be stored. The strongest coefficients are accurately coded and the less significant coefficients can be coded less accurately (thus fewer bits are used to code these coefficients).

Transform encoding is suitable for compressing images and a typical transform is the discrete cosine transform (DCT).

2.9 Letter probability program

Program 2.6

```
#include <stdio.h>
#include <string.h>
#include <ctype.h>

#define  NUM_LETTERS 29

int   get_occurances(char c, char txt[]);

int   main(void)
{
char  ch, fname[BUFSIZ];
int   occ[NUM_LETTERS]={0,0,0,0,0,0,0,0,0,0,0,0,0,0,0,
                        0,0,0,0,0,0,0,0,0,0,0,0,0,0};
unsigned int   total,i;
FILE           *in;

      printf("Enter text file>>");
      gets(fname);

      if ((in=fopen(fname,"r"))==NULL)
      {
         printf("Can't find file %s\n",fname);
         return(1);
      }

      do
      {
         ch=tolower(getc(in));

         if (isalpha(ch))
         {
            (occ[ch-'a'])++;
            total++;
         }
         else if (ch=='.') { occ[NUM_LETTERS-3]++; total++; }
         else if (ch==' ') { occ[NUM_LETTERS-2]++; total++; }
         else if (ch==',') { occ[NUM_LETTERS-1]++; total++; }
      } while (!feof(in));

      fclose(in);

      puts("Char. Occur. Prob.");
```

```
       for (i=0;i<NUM_LETTERS;i++)
       {
          printf("  %c  %5d %5.4f\n",'a'+i,occ[i],(float)occ[i]/(float)total);
       }

       return(0);
}

int    get_occurances(char c, char txt[])
{
int    occ=0,i;

       for (i=0;i<strlen(txt);i++)
          if (c==txt[i]) occ++;

       return(occ);
}
```

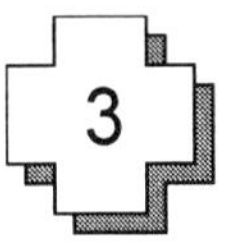

Huffman/Lempel-Ziv Compression Methods

3.1 Introduction

Normally, general data compression does not take into account the type of data which is being compressed and is lossless. It can be applied to computer data files, documents, images, and so on. The two main techniques are statistical coding and repetitive sequence suppression. This chapter discusses two of the most widely used methods for general data compression: Huffman coding and Lempel-Ziv coding.

3.2 Huffman coding

Huffman coding uses a variable length code for each of the elements within the information. This normally involves analyzing the information to determine the probability of elements within the information. The most probable elements are coded with a few bits and the least probable coded with a greater number of bits.

The following example relates to characters. First, the textual information is scanned to determine the number of occurrences of a given letter. For example:

'b'	'c'	'e'	'i'	'o'	'p'
12	3	57	51	33	20

Next the characters are arranged in order of their number of occurrences, such as:

'e'	'i'	'o'	'p'	'b'	'c'
57	51	33	20	12	3

Next the two least probable characters are assigned either a 0 or a 1. Figure 3.1 shows that the least probable ('c') has been assigned a 0 and the next least probable ('b') has been assigned a 1. The summation of the two occurrences is then taken to the next column and the occurrence values are again arranged in descending order (that is, 57, 51, 33, 20 and 15). As with the first column, the least probable occurrence is assigned a 0 and the next least probable occurrence is assigned a 1. This continues until the last column. The Huffman-coded values are then read from left to right and the bits are listed from right to left.

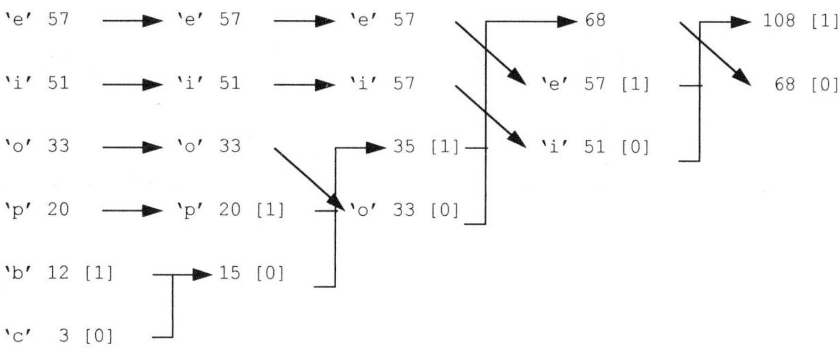

Figure 3.1 Huffman coding example.

The final coding will be:

```
'e'    11
'i'    10
'o'    00
'p'    011
'b'    0101
'c'    0100
```

The great advantage of Huffman coding is that, although each character is coded with a different number of bits, the receiver will automatically determine the character whatever their order. For example if a 1 is followed by a 1 then the received character is an 'e'. If it is then followed by two 0s then it is an 'o'. Here is an example:

```
11000110100100110100
```

will be decoded as:

'e' 'o' 'p' 'c' 'i' 'p' 'c'

When transmitting or storing Huffman-coded data, the coding table needs to be stored with the data (if the table is generated dynamically). It is generally a good compression technique but it does not take into account higher order associations between characters. For example, the character 'q' is normally followed by the character 'u' (apart from words such as Iraq and in many old Scots words which begin with *qw*). An efficient coding scheme for text would be to encode a single character 'q' with a longer bit sequence than a 'qu' sequence.

3.3 Adaptive Huffman coding

Adaptive Huffman coding was first conceived by Faller and Gallager and then further refined by Knuth (so it is often called the FGK algorithm). It uses defined word schemes which determine the mapping from source messages to code words based upon a running estimate of the source message probabilities. The code is adaptive and changes so as to remain optimal for the current estimates. In this way, the adaptive Huffman codes respond to locality and the encoder thus learns the characteristics of the source data. The decoder must then learn along with the encoder by continually updating the Huffman tree so as to stay in synchronization with the encoder.

A second advantage of adaptive Huffman coding is that it only requires a single pass over the data. In many cases, the adaptive Huffman method actually gives a better performance, in terms of number of bits transmitted, than static Huffman coding. This does not contradict the optimality of the static method, as the static method is optimal only over all methods, which assumes a time-invariant mapping. The performance of the adaptive methods can also be worse than that of the static method. Upper bounds on the redundancy of these methods are presented in this section. As discussed in the introduction, the FGK method is the basis of the UNIX compact program.

3.3.1 FGK algorithm

The basis for the FGK algorithm is the Sibling Property: a binary code tree has the sibling property if each node (except the root) has a sibling and if the nodes can be listed in order of non-increasing weight with each node adjacent to its sibling.

3.4 Lempel-Ziv coding

Around 1977, Abraham Lempel and Jacob Ziv developed the Lempel-Ziv class of adaptive dictionary data compression techniques. Also known as LZ-77 coding, they are now some of the most popular compression techniques.

The LZ coding scheme takes into account repetition in phases, words or parts of words. These repeated parts can either be text or binary. A flag is normally used to identify coded and unencoded parts. An example piece of text could be:

'The receiver requires a receipt which is automatically sent when it is received.'

This has the repetitive sequence 'recei'. The encoded sequence could be modified with the flag sequence *#m#n* where *m* represents the number of characters to trace back to find the character sequence and *n* the number of replaced characters. Thus the encoded message could become:

'The receiver requires a #20#5pt which is automatically sent wh#6#2 it #30#2 #47#5ved.'

Normally a long sequence of text has many repeated words and phases, such as 'and', 'there', and so on. Note that in some cases this could lead to longer files if short sequences were replaced with codes that were longer than the actual sequence itself.

3.5 Lempel-Ziv-Welsh coding

The Lempel-Ziv-Welsh (LZW) algorithm (also known LZ-78) builds a dictionary of frequently used groups of characters (or 8-bit binary values). Before the file is decoded, the compression dictionary must be sent (if transmitting data) or stored (if data is being stored). This method is good at compressing text files because text files contain ASCII characters (which are stored as 8-bit binary values) but not so good for graphics files, which may have repeating patterns of binary digits that might not be multiples of 8 bits.

A simple example is to use a 6 character alphabet and a 16 entry dictionary, thus the resulting code word will have 4 bits. If the transmitted message is:

```
ababacdcdaaaaaaef
```

Then the transmitter and receiver would initially add the following to its dictionary:

0000	'a'
0001	'b'
0010	'c'
0011	'd'
0100	'e'
0101	'f'
0110–1111	empty

First the 'a' character is sent with 0000, next the 'b' character is sent and the transmitter checks to see that the 'ab' sequence has been stored in the dictionary. As it has not, it adds 'ab' to the dictionary, to give:

0000	'a'
0001	'b'
0010	'c'
0011	'd'
0100	'e'
0101	'f'
0110	'ab'
0111–1111	empty

The receiver will also add this to its table (thus the transmitter and receiver will always have the same tables). Next the transmitter reads the 'a' character and checks to see if the 'ba' sequence is in the code table. As it is not, it transmits the 'a' character as 0000, adds the 'ba' sequence to the dictionary, which will now contain:

0000	'a'
0001	'b'
0010	'c'
0011	'd'
0100	'e'
0101	'f'
0110	'ab'
0111	'ba'
1000–1111	empty

0000	0001	0000	0110	0010
↑	↑	↑	↑	↑
'a'	'b'	'a'	'ba'	'c'

Next the transmitter reads the 'b' character and checks to see if the 'ba' sequence is in the table. As it is, it will transmit the code table address which identifies it, i.e. 0111. When this is received, the receiver detects that it is in its dictionary and it knows that the addressed sequence is 'ba'.

Next the transmitter reads an 'c' and checks for the character in its dictionary. As it is included, it transmits its address, i.e. 0010. When this is received, the receiver checks its dictionary and locates the character 'c'. This then continues with the transmitter and receiver maintaining identical copies of their dictionaries. A great deal of compression occurs when sending a sequence of one character, such as a long sequence of 'a'.

Typically, in a practical implementation of LZW, the dictionary size for LZW starts at 4K (4096). The dictionary then stores bytes from 0 to 255 and the addresses 256 to 4095 are used for strings (which can contain two or more characters). As there are 4096 entries then it is a 12-bit coding scheme (0 to 4096 gives 0 to $2^{12}-1$ different addresses).

3.6 Variable-length-code LZW compression

The Variable-length-code LZW (VLC-LZW) uses a variation of the LZW algorithm where variable-length codes are used to replace patterns detected in the original data. It uses a dictionary constructed from the patterns encountered in the original data. Each new pattern is entered into it and its indexed address is used to replace it in the compressed stream. The transmitter and receiver maintain the same dictionary.

The VLC part of the algorithm is based on an initial code size (the LZW initial code size), which specifies the initial number of bits used for the compression codes. When the number of patterns detected by the compressor in the input stream exceeds the number of patterns encodable with the current number of bits then the number of bits per LZW code is increased by one. The code size is initially transmitted (or stored) so that the receiver (or uncompressor) knows the size of the dictionary and the length of the codewords.

In 1985, the LZW algorithm was patented by the Sperry Corp. It is used by the GIF file format and is similar to the technique used to compress data in V.42bis modems.

3.7 Disadvantages with LZ compression

LZ compression substitutes the detected repeated patterns with references to a dictionary. Unfortunately the larger the dictionary, the greater the number of bits that are necessary for the references. The optimal size of the dictionary also varies for different types of data; the more variable the data, the smaller the optimal size of the directory.

3.8 Practical Lempel-Ziv/Huffman coding

This section contains practical examples of programs which use Lempel-Ziv and/or Huffman coding. Most compression programs use either one or both of these techniques. As previously mentioned, both techniques are lossless. In general, Huffman is the most efficient but requires two passes over the data, while Lempel-Ziv uses just one pass. This feature of a single pass is obviously important when saving to a hard disk drive or when encoding and decoding data in real-time communications. One of the most widely used variants is LZS, owned by Stac Electronics (who were the first commercial company to produce a compressed drive, named Stacker). Microsoft have included a variation of this program, called DoubleSpace in DOS Version 6 and DriveSpace in Windows 95.

The LZS technique is typically used in mass backup devices, such as tape drives, where the compression can either be implemented in hardware or in software. This typically allows the tape to store at least twice the quoted physical capacity of the tape.

The amount of compression, of course, depends on the type of file being compressed. Random data, such as executable programs or object code files, typically has low compression (resulting in a file which is 50 to 95% of the original file size). Still images and animation files tend to have high compression and typically result in a file which is between only 2 and 20% of the original file size. It should be noted that once a file has been compressed there is virtually no gain in compressing it again (unless a differential method is used). Thus storing or transmitting compressed files over a system which has further compression will not increase the compression ratio (unless another algorithm is used).

Typical files produced from LZ77/LZ78 compression methods are ZIP, ARJ, LZH, Z, and so on. Huffman is used in ARC and PKARC utilities, and in the UNIX compact command.

3.8.1 Lempel-Ziv/Huffman practical compression

DriveSpace and DoubleSpace are programs used in PC systems to compress files on hard disk drives. They use a mixture of Huffman and Lempel-Ziv coding, where Huffman codes are used to differentiate between data (literal values) and back references and LZ coding is used for back references.

A DriveSpace disk starts with a 4 byte magic number (52 B2 00 00 08h). This identifies that the disk is using DriveSpace. Following this there are either literal values or back references which are preceded by control bits of either `0`, `11`, `100`, `1010` or `1011` (these are Huffman values coded to differentiate them). If the control bit is 0 then the following 7 bits of `abcdefg` correspond to data of `0gfedcba`, else if it is 11 then the following 7 bits of `abcdefg` correspond to data `1gfedcba`. Thus the data:

```
10110101 01111110 11100000 11111111
```

would be encoded as:

```
11 1010110 0 0111111 11 0000011 11 1111111
```

The back-reference values are preceded by `100`, `1010` or `1011`. If preceded by `100` then followed by `abcdef`*X*, which is a 6-bit back reference of a length given by *X*. A `1010` followed by `abcdefgh`*X* is an 8-bit back reference of 64+`hgfedcba` with length given by *X*. A `1011` followed by `abcdefghijkl`*X* is a 12-bit back reference of 64+256+ `lkjihgfed-cba` with length given by *X*.

The back reference consists of a code indicating the number of bits back to find the start of the referenced data, followed by the length of the data itself. This code consists of *N* zeros followed by a 1. The number of zeros, *N*, indicates the number of bits of length data and the length of the back reference is $M+2^N+2$, where *M* is the *N*-bit unsigned number comprising the data length. Thus the minimum length of a back reference will be when $M=0$ and $N=0$ giving a value of 3. An example format of a back pointer is:

```
100  abcdef 000001 ghijk
```

where N will be 5 since there are five zeros after the 5-bit back reference and `fedcba` corresponds to the back reference `fedcba`. The length of the reference values will be $M+2^5+2$, where M is the 5-digit unsigned binary number `kjihg`. For example if the stored bit field were:

```
0101001010110010000000000000000000000010010000100110010000010
0001001100101000010
```

It would be decoded as:

01010010101100100000000000000000000001001
MAGIC NUMBER

0 0001001	**100** 100000 1
'H' (100 1000)	As the control bit field is **100** then it has a 6-bit back reference of `000001` (one place back) followed by `1` which shows that the back reference length of bits is 0. Thus, using the formula $M+2^N+2$ gives $0+2^0+2=3$. The back reference has a length of 3 bytes, giving the output 'HHH'

0 0001001	**100** 1010000 10
'E' (01010001)	As the control bit field is **100**, it has a 6-bit back reference of `000101` (five places back) followed by `01` which shows that the back-reference length of bits is `1`. Thus, using the formula $M+2^N+2$ gives $0+2^1+2=4$. The character five places back is an 'H', thus 'H' is repeated four times.

This then gives the sequence 'HHHHEHHHH'.

In DriveSpace, each of the fields after the magic number is a group. A group consist of a control part (the Huffman code) and an item. An item may be either a literal item or a copy item (i.e. a 6, 8 or 12 bit back reference). The end of a file in DriveSpace is identified with a special 12-bit back-reference value of 1111 1111 1111 1111 (FFFFh).

3.8.2 GIF files

The graphic interface format (GIF) uses a compression algorithm based on the Lempel-Ziv-Welsh (LZW) compression scheme. When compressing an image the compression program maintains a list of substrings that have been found previously. When a repeated string is found, the referred item is replaced with a pointer to the original. Since images tends to contain many repeated values, the GIF format is a good compression technique. The format of the data file will be discussed in Chapter 5.

3.8.3 UNIX compress/uncompress

The UNIX programs `compress` and `uncompress` use adaptive Lempel-Ziv coding. They are generally better than `pack` and `unpack` which are based on Huffman coding. Where possible, the `compress` program adds a `.Z` onto a file when compressed. Compressed files can be restored using the `uncompress` or `zcat` programs.

3.8.4 UNIX archive/zoo

The UNIX-based `zoo` freeware file compression utility employs the Lempel-Ziv algorithm. It can store and selectively extract multiple generations of the same file. Data can thus be recovered from damaged archives by skipping the damaged portion and locating undamaged data (using the `fiz` program).

4 Image Compression (GIF/TIFF/PCX)

4.1 Introduction

Data communication increasingly involves the transmission of still and moving images. Compressing images into a standard form can give great savings in transmission times and storage lengths. Some of these forms are outlined in Table 4.1. The main parameters in a graphics file are:

- The picture resolution. This is defined by the number of pixels in the *x*- and *y*-directions.
- The number of colors per pixel. If N bits are used for the bit color then the total number of displayable colors will be 2^N. For example an 8-bit color field defines 256 colors, a 24-bit color field gives 2^{24} or 16.7M colors. Many new computer systems allow for 32-bit color which gives over 4 billion colors.
- Palette size. Some systems reduce the number of bits used to display a color by reducing the number of displayable colors for a given palette size.

Table 4.1 Typical standard compressed graphics formats.

File	*Compression type*	*Max. resolution or colors*	
TIFF	Huffman RLE and/or LZW	48-bit color	TIFF (tagged image file format) is typically used to transfer graphics from one computer system to another. It allows high resolutions and colors of up to 48 bits (16 bits for red, green and blue).
PCX	RLE	65 536 × 65 536 (24-bit color)	Graphics file format which uses RLE to compress the image. Unfortunately, it make no provision for storing gray scale or color-correcting tables.
GIF	LZW	65 536 × 65 536 (24-bit color, but only 256 display-able colors)	Standardized graphics file format which can be read by most graphics packages. It has similar graphics characteristics to PCX files and allows multiple images in a single file and interlaced graphics.
JPG	JPEG compression (DCT, Quantization and Huffman)	Depends on the compression	Excellent compression technique which produces lossy compression. It normally results in much greater compression than the methods outlined above.

4.2 Comparison of the different methods

This section uses example bitmapped images and shows how much the different techniques manage to compress them. Figure 4.1 shows an image and Table 4.2 shows the resultant file size when it is saved in different formats. It can be seen that the BMP file format has the largest storage. The two main forms of BMP files are RGB (red, green, blue) encoded and RLE encoded. RGB coding saves the bit-map in an uncompressed form, whereas the RLE coding will reduce the total storage by compressing repetitive sequences. Next is the PCX file which has limited compression abilities (the format used in this case is version 5). The GIF format manages to compress the file to around 40% of its original size and the TIF file achieves similar compression (mainly because both techniques use LZH compression). It can be seen that by far the best compression is achieved with JPEG, which in both forms has compressed the file to under 10% of its original size.

The reason that the compression ratios for GIF, TIF and BMP RLE are relatively high is that the image in Figure 4.1 contains a lot of changing data. Most images will compress to less than 10% because they have large areas which do not change much.

Figure 4.1 Sample graphics image.

Table 4.2 Compression on a graphics file.

Type	*Size(B)*	*Compression(%)*	
BMP	308 278	100.0	BMP, RBG encoded (640×480, 256 colors)
BMP	301 584	97.8	BMP, RLE encoded
PCX	274 050	88.9	PCX, Version 5
GIF	124 304	40.3	GIF, Version 89a, non-interlaced
GIF	127 849	41.5	GIF, Version 89a, interlaced
TIF	136 276	44.2	TIF, LZW compressed
TIF	81 106	26.3	TIF, CCITT Group 3, MONOCHROME
JPG	28 271	9.2	JPEG - JFIF Complaint (Standard coding)
JPG	26 511	8.6	JPEG - JFIF Complaint (Progressive coding)

Figure 4.2 shows a simple graphic of 500 × 500, 24-bit, which has large areas with identical colors. Table 4.3 shows that, in this case, the compression ratio is low. The RLE encoded BMP file is only 1% of the original as the graphic contains long runs of the same color. The GIF file has compressed to less than 1%. Note that the PCX, GIF and BMP RLE files have saved the image with only 256 colors. The JPG formats have the advantage that they have saved the image with the full 16.7M colors and give compression rates of around 2%.

Table 4.3 Compression on a graphics file with highly redundant data.

Type	*Size (B)*	*Compression (%)*	
BMP	750054	100.0	BMP, RBG encoded (500 × 500, 16.7M colors)
BMP	7832	1.0	BMP, RLE encoded (256 colors)
PCX	31983	4.3	PCX, Version 5 (256 colors)
GIF	4585	0.6	GIF, Version 89a, non-interlaced (256 colors)
TIF	26072	3.5	TIF, LZW compressed (16.7M colors)
JPG	15800	2.1	JPEG (Standard coding, 16.7M colors)
JPG	12600	1.7	JPEG (Progressive coding 16.7M colors)

4.3 GIF coding

The graphics interchange format (GIF) is the copyright of CompuServe Incorporated. Its popularity has increased mainly because of its wide usage on the Internet. CompuServe Incorporated, luckily, has granted a limited, non-exclusive, royalty-free license for the use of GIF (but any software using the GIF format must acknowledge the ownership of the GIF format).

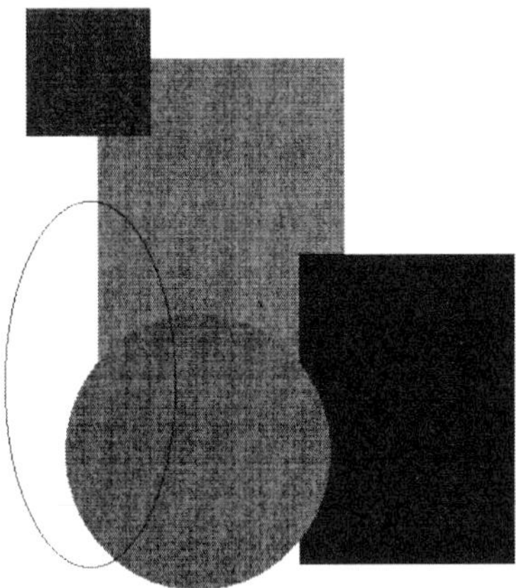

Figure 4.2 Sample graphics image.

Most graphics software supports the Version 87a or 89a format (the 89a format is an update the 87a format). Both have basic specification:

- A header with GIF identification.
- A logical screen descriptor block which defines the size, aspect ratio and color depth of the image place.
- A global color table.
- Data blocks with bitmapped images and the possibility of text overlay.
- Multiple images, with image sequencing or interlacing. This process is defined in a graphic-rendering block.
- LZW compressed bitmapped images.

4.3.1 Color tables

Color tables store the color information of part of an image (a local color table) or they can be global (a global table).

4.3.2 Blocks, extensions and scope

Blocks can be specified into three groups: control, graphic-rendering and special purpose. Control blocks contain information used to control the process of the data stream or information used in setting hardware parameters. They include:

- GIF Header – which contains basic information on the GIF file, such as the version number and the GIF file signature.
- Logical screen descriptor – which contains information about the active screen display, such as screen width and height, and the aspect ratio.
- Global color table – which contains up to 256 colors from a palette of 16.7M colors (i.e. 256 colors with 24-bit color information).
- Data subblocks – which contain the compressed image data.
- Image description – which contains, possibly, a local color table and defines the image width and height, and its top left coordinate.
- Local color table – an optional block which contains local color information for an image as with the global color table, it has a maximum of 256 colors from a palette of 16.7M.
- Table-based image data – which contains compressed image data.
- Graphic control extension – an optional block which has extra graphic-rendering information, such as timing information and transparency.
- Comment extension – an optional block which contains comments ignored by the decoder.
- Plain text extension – an optional block which contains textual data.
- Application extension – which contains application-specific data. This block can be used by a software package to add extra information to the file.
- Trailer – which defines the end of a block of data.

4.3.3 GIF header

The header is 6 bytes long and identifies the GIF signature and the version number of the chosen GIF specification. Its format is:

- 3 bytes with the characters 'G', 'I' and 'F'.
- 3 bytes with the version number (such as 87a or 89a). Version numbers are ordered with

two digits for the year, followed by a letter ('a', 'b', and so on).

Program 4.1 is a C program for reading the 6-byte header and Sample run 4.1 shows a sample run with a GIF file. It can be seen that the file in the test run has the required signature and has been stored with Version 89a.

Program 4.1

```
#include    <stdio.h>

int   main(void)
{
FILE  *in;
char  fname[BUFSIZ], str[BUFSIZ];

      printf("Enter GIF file>>");
      gets(fname);

      if ((in=fopen(fname,"r"))==NULL)
      {
         printf("Can't find file %s\n",fname);
         return(1);
      }

      fread(str,3,1,in);
      str[3]=NULL; /* terminate string */
      printf("Signature: %s\n",str);
      fread(str,3,1,in);
      str[3]=NULL; /* terminate string */
      printf("Version: %s\n",str);

      fclose(in);
      return(0);
}
```

Sample run 4.1

```
Enter GIF file>> clouds.gif
Signature: GIF
Version: 89a
```

4.3.4 Logical screen descriptor

The logical screen descriptor appears after the header. Its format is:

- 2 bytes with the logical screen width (unsigned integer).
- 2 bytes with the logical screen height (unsigned integer).
- 1 byte of a packed bit field, with 1 bit for global color table flag, 3 bits for color resolution, 1 bit for sort flag and 3 bits to give an indication of the number of colors in the global color table
- 1 byte for the background color index.
- 1 byte for the pixel aspect ratio.

Program 4.2 is a C program which reads the header and the logical descriptor field, and Sample run 4.2 shows a sample run. It can be seen, in this case, that the logic screen size is 640×480. The packed field, in this case, has a hexadecimal value of F7h, which is 1111 0111b in binary. Thus, all the bits of the packed bit field are set, apart from the sort flag. If this is set then the global color table is sorted in order of decreasing importance (the most frequent color appearing first and the least frequent color last). The total number of colors in the global color table is found by raising 2 to the power of 1 + the color value in the packed bit field:

$$\text{Number of colors} = 2^{\text{Color value in packed bit field}+1}$$

In this case, there is a bit field of seven colors, thus the total number of colors is 2^8, or 256.

It can be seen that the aspect ratio in Sample run 4.2 is zero. If it is zero then no aspect ratio is given. If it is not equal to zero then the aspect ratio of the image is computed by:

$$\text{Aspect ratio} = \frac{\text{Pixel aspect ratio} + 15}{64}$$

where the pixel ratio is the pixel's width divided by its height.

Program 4.2

```
#include    <stdio.h>

int    main(void)
{
FILE   *in;
char   fname[BUFSIZ], str[BUFSIZ];
int    x,y;
char   color_index, aspect, packed;

       printf("Enter GIF file>>");
       gets(fname);

       if ((in=fopen(fname,"r"))==NULL)
       {
          printf("Can't find file %s\n",fname);
          return(1);
       }

       fread(str,3,1,in);   str[3]=NULL; /* terminate string */
       printf("Signature: %s\n",str);
       fread(str,3,1,in);   str[3]=NULL; /* terminate string */
       printf("Version: %s\n",str);

       fread(&x,2,1,in); str[3]=NULL; /* terminate string */
       printf("Screen width: %d\n",x);
       fread(&y,2,1,in); str[3]=NULL; /* terminate string */
       printf("Screen height: %d\n",y);

       fread(&packed,1,1,in);
       printf("Packed: %x\n",packed & 0xff); /* mask-off the bottom 8 bits */
       fread(&color_index,1,1,in);
       printf("Color index: %d\n",color_index);
       fread(&aspect,1,1,in);
       printf("Aspect ratio: %d\n",aspect);

       fclose(in);
       return(0);
}
```

Sample run 4.2

```
Enter GIF file>> clouds.gif
Signature: GIF
Version: 89a
Screen width: 640
Screen height: 480
Packed: f7
Color index: 0
Aspect ratio: 0
```

4.3.5 Global color table

After the header and the logical display descriptor comes the global color table. It contains up to 256 colors from a palette of 16.7M colors. Each of the colors is defined as a 24-bit color of red (8 bits), green (8 bits) and blue (8 bits). The format in memory is:

```
RRRRRRRR
GGGGGGGG
BBBBBBBB
RRRRRRRR
GGGGGGGG
BBBBBBBB
  :    :
RRRRRRRR
GGGGGGGG
BBBBBBBB
```

Thus the number of bytes that the table will contain will be:

$$\text{Number of bytes} = 3 \times 2^{\text{Size of global color table}+1}$$

The 24-bit color scheme allows a total of 16777216 (2^{24}) different colors to be displayed. Table 4.4 defines some colors in the RGB (red/green/blue) strength. The format is `rrggbbh`, where `rr` is the hexadecimal equivalent for the red component, `gg` the hexadecimal equivalent for the green component and `bb` the hexadecimal equivalent for the blue component. For example, in binary:

`000000000000000000000000`	represents black (000000h)
`111111111111111111111111`	represents white (FFFFFFh)
`011101110111011101110111`	represents gray (777777h)
`111110101110010100000011`	represents yellow (FCE503h)
`001110100000101101011001`	represents purple (3A0B59h)

Program 4.3 is a C program which reads the header, the image descriptor and the color table. Sample run 4.3 shows a truncated color table. The first three are:

0111 1011 1010 1101 1101 0110 (7BADD6h)
1000 0100 1011 0101 1101 1110 (84B5DEh)
0111 0011 1010 1101 1101 0110 (73ADD6h)

Table 4.4 Hexadecimal colors for 24-bit color representation.

Color	*Code*	*Color*	*Code*
White	FFFFFFh	Dark red	C91F16h
Light red	DC640Dh	Orange	F1A60Ah
Yellow	FCE503h	Light green	BED20Fh
Dark green	088343h	Light blue	009DBEh
Dark blue	0D3981h	Purple	3A0B59h
Pink	F3D7E3h	Nearly black	434343h
Dark gray	777777h	Gray	A7A7A7h
Light gray	D4D4D4h	Black	000000h

These colors have a strong blue component (D6h and DEh) and reduced strength red and green components. The image itself is a picture of clouds on a blue sky, thus the image is likely to have strong blue colors.

Program 4.3

```
#include    <stdio.h>

int   main(void)
{
FILE  *in;
char  fname[BUFSIZ], str[BUFSIZ];
int   x,y,i;
char  color_index, aspect, packed,red,blue,green;

      printf("Enter GIF file>>");
      gets(fname);
      if ((in=fopen(fname,"r"))==NULL)
      {
         printf("Can't find file %s\n",fname);
         return(1);
      }

      fread(str,3,1,in);   str[3]=NULL; /* terminate string */
      printf("Signature: %s\n",str);
      fread(str,3,1,in);   str[3]=NULL; /* terminate string */
      printf("Version: %s\n",str);

      fread(&x,2,1,in); str[3]=NULL; /* terminate string */
      printf("Screen width: %d\n",x);
      fread(&y,2,1,in); str[3]=NULL; /* terminate string */
      printf("Screen height: %d\n",y);

      fread(&packed,1,1,in);
      printf("Packed: %x\n",packed & 0xff); /* mask-off the bottom 8 bits */
      fread(&color_index,1,1,in);
      printf("Color index: %d\n",color_index);
      fread(&aspect,1,1,in);
      printf("Aspect ratio: %d\n",aspect);

      for (i=0;i<64;i++)
      {
         fread(&red,1,1,in);
         printf("Red: %x ",red & 0xff);       /* display 8 bits */
         fread(&green,1,1,in);
         printf("Green: %x ",green & 0xff);   /* display 8 bits */
         fread(&blue,1,1,in);
         printf("Blue: %x\n",blue & 0xff);    /* display 8 bits */
```

```
    }

    fclose(in);
    return(0);
}
```

Sample run 4.3

```
Enter GIF file>> clouds.gif
Signature: GIF
Version: 89a
Screen width: 640
Screen height: 480
Packed: f7
Color index: 0
Aspect ratio: 0
Red: 7b Green: ad Blue: d6
Red: 84 Green: b5 Blue: de
Red: 73 Green: ad Blue: d6
Red: 7b Green: ad Blue: de
Red: 94 Green: bd Blue: de
Red: 7b Green: b5 Blue: de
Red: 8c Green: b5 Blue: de
Red: 8c Green: bd Blue: de
Red: 9c Green: c6 Blue: de
Red: ce Green: de Blue: ef
Red: de Green: e7 Blue: ef
Red: a5 Green: c6 Blue: e7
      ::::::
Red: 8c Green: bd Blue: e7
Red: ff Green: ff Blue: f7
Red: ad Green: d6 Blue: ef
Red: 8c Green: b5 Blue: e7
Red: 84 Green: b5 Blue: e7
```

4.3.6 Image descriptor

After the global color table is the image descriptor. Its format is:

- 1 byte for the image separator (always 2Ch).
- 2 bytes for the image left position (unsigned integer).
- 2 bytes for the image top position (unsigned integer).
- 2 bytes for the image width (unsigned integer).
- 2 bytes for the image height (unsigned integer).
- 1 byte of a packed bit field, with 1 bit for local color table flag, 1 bit for interlace flag, 1 bit for sort flag, 2 bits are reserved and 3 bits for the size of the local color table.

Program 4.4 is a C program which searches for the image separator (2Ch) and displays the image descriptor data that follows. Sample run 4.4 shows a sample run. It can be seen from this sample run that the image is to be displayed at (0, 0), its width is 640 pixels and its height is 480 pixels. The packed bit field contains all zeros, thus there is no local color table (and the global color table should be used).

Program 4.4

```
#include    <stdio.h>

int   main(void)
{
```

```
FILE   *in;
char   fname[BUFSIZ];
int    i,left,top,width,height;
char   ch,packed;

       printf("Enter GIF file>>");
       gets(fname);

       if ((in=fopen(fname,"r"))==NULL)
       {
          printf("Can't find file %s\n",fname);
          return(1);
       }

       do
       {
          fread(&ch,1,1,in);
       } while (ch!=0x2C); /* find image seperator */

       fread(&left,2,1,in);
       printf("Image left position: %d\n",left);
       fread(&top,2,1,in);
       printf("Image top position: %d\n",top);
       fread(&width,2,1,in);
       printf("Image width: %d\n",width);
       fread(&height,2,1,in);
       printf("Image height: %d\n",height);
       fread(&packed,1,1,in);
       printf("Packed: %x\n",packed & 0xff);
       fclose(in);
       return(0);
}
```

Sample run 4.4

```
Enter GIF file>> clouds.gif
Image left position: 0
Image top position: 0
Image width: 640
Image height: 480
Packed: 0
```

4.3.7 Local color table

The local color table is an optional block which defines the color map for the image that precedes it. The format is identical to the global color map, i.e. 3 bytes for each of the colors.

4.3.8 Table-based image data

The table-based image data follows the local color table. This table contains compressed image data. It consists of a series of subblocks of up to 255 bytes. The data consists of an index to the color table (either global or local) for each pixel in the image. As the global (or local) color table has 256 entries, the data value (in its uncompressed form) will range from 0 to 255 (8 bits). The tables format is:

- 1 byte for the LZW minimum code size, which is the initial number of bits used in the LZW coding.
- *N* bytes for the LZW compressed image data. The first block is preceded by the data size.

To recap from Chapter 3, GIF coding uses the variable-length-code LZW technique where a variable-length code replaces image data (pixel color references). These variable-length codes are specified in a Huffman code table. The encoder replaces the data from the input and builds a dictionary with the patterns in the data. Every new pattern is entered into the dictionary and the index value of the table is added to coded data. When a previously stored pattern is encountered, its dictionary index value is added to the coded data. The decoder takes the compressed data and builds the dictionary which is identical to the encoder. It then replaces indexed terms from the dictionary.

The VLC algorithm uses an initial code size to specify the initial number of bits used for the compression codes. When the number of patterns detected by the encoder exceeds the number of patterns encodable with the current number of bits then the number of bits per LZW is increased by 1.

Program 4.5 reads the LZW code size byte. The byte after this is the block size, followed by the number of bytes of data as defined in the block size byte. Sample run 4.5 gives a sample run. It can be seen that the initial LZW code size is 8 and that the block size of the first block is 254 bytes. The dictionary entries will thus start at entry 256 (2^8).

Program 4.5

```
#include    <stdio.h>

int   main(void)
{
FILE  *in;
char  fname[BUFSIZ];
int   i,left,top,width,height;
char  ch,packed,code,block;

      printf("Enter GIF file>>");
      gets(fname);

      if ((in=fopen(fname,"r"))==NULL)
      {
         printf("Can't find file %s\n",fname);
         return(1);
      }

      do
      {
         fread(&ch,1,1,in);
      } while (ch!=0x2C);
      fread(&left,2,1,in);
      printf("Image left position: %d\n",left);
      fread(&top,2,1,in);
      printf("Image top position: %d\n",top);
      fread(&width,2,1,in);
      printf("Image width: %d\n",width);
      fread(&height,2,1,in);
      printf("Image height: %d\n",height);
      fread(&packed,1,1,in);
      printf("Packed: %x\n",packed & 0xff);
      fread(&code,1,1,in);
      printf("LZW code size: %d\n",code & 0xff);
      fread(&block,1,1,in);
      printf("Block size: %d\n",block & 0xff);
      fclose(in);
      return(0);
}
```

Sample run 4.5

```
Enter GIF file>> clouds.gif
Image left position: 0
Image top position: 0
Image width: 640
Image height: 480
Packed: 0
LZW code size: 8
Block size: 254
```

4.3.9 Graphic control extension

The graphic control extension is optional and contains information on the rendering of the image that follows. Its format is:

- 1 byte with the extension identifier (21h).
- 1 byte with the graphic control label (F9h).
- 1 byte with the block size following this field and up to but not including, the end terminator. It always has a fixed value of 4.
- 1 byte with a packed array of which the first 3 bits are reserved, 3 bits define the disposal method, 1 bit defines the user input flag and 1 bit defines the transparent color flag.
- 2 bytes with the delay time for the encode wait, in hundreds of a seconds, before encoding the image data.
- 1 byte with the transparent color index.
- 1 byte for the block terminator (00h).

4.3.10 Comment extension

The comment extension is optional and contains information which is ignored by the encoder. Its format is:

- 1 byte with the extension identifier (21h).
- 1 byte with the comment extension label (FEh).
- *N* bytes, with comment data.
- 1 byte for the block terminator (00h).

4.3.11 Plain text extension

The plain text extension is optional and contains text information. Its format is:

- 1 byte with the extension identifier (21h).
- 1 byte with the plain text label (01h).
- 1 byte with the block size. This is the number of bytes after the block size field up to but not including the beginning of the plain text data block. It always contains the value 12.
- 2 bytes for the text grid left position.
- 2 bytes for the text grid top position.
- 2 bytes for the text width.
- 2 bytes for the text height.
- 1 byte for the character cell width.
- 1 byte for the character cell height.
- 1 byte for the text foreground color.

- 1 byte for the text background color.
- *N* bytes for the plain text data.
- 1 byte for the block terminator (00h).

4.3.12 Application extension

The application extension is optional and contains information for application programs. Its format is:

- 1 byte with the extension identifier (21h).
- 1 byte with the application extension label (FFh).
- 1 byte for the block size. This is the number of bytes after the block size field up to but not including the beginning of the application data. It always contains the value 11.
- 8 bytes for the application identifier.
- 3 bytes for the application authentication code.
- *N* bytes, for the application data.
- 1 byte for the block terminator (00h).

4.3.13 Trailer

The trailer indicates the end of the GIF file. Its format is:

- 1 byte identifying the trailer (3Bh).

4.4 TIFF coding

Tag image file format (TIFF) is an excellent method of transporting images between file systems and software packages. It is supported by most graphics import packages and has a high resolution that is typically used when scanning images. There are two main types of TIFF coding, baseline TIFF and extended TIFF. It can also use different compression methods and different file formats, depending on the type of data stored.

In TIFF 6.0, defined in June 1992, the pixel data can be stored in several different compression formats, such as:

- Code number 1, no compression.
- Code number 2, CCITT Group 3 modified Huffman RLE encoding (see Section 33.9.1).
- Code number 3, Fax-compatible CCITT Group 3 (see Section 33.9).
- Code number 4, Fax-compatible CCITT Group 4 (see Section 33.9).
- Code number 5, LZW compression (see Chapter 3).

Codes 1 and 2 are baseline TIFF files whereas the others are extended.

4.4.1 File structure

TIFF files have a three-level hierarchy:

- A file header.
- One or more IFDs (image file directories). These contain codes and their data (or pointers to the data).

- Data.

The file header contains 8 bytes: a byte order field (2 bytes), the version number field (2 bytes) and the pointer to the first IFD (4 bytes). Figure 4.3 shows the file header format. The byte order field defines whether Motorola architecture is used (the character sequence is 'MM', or 4D4Dh) or Intel architecture (the character sequence is 'II', or 4949h). The Motorola format defines that the bytes are ordered from the most significant to the least significant, the Intel format defines that the bytes are organized from least significant to the most significant.

The version number field always contains the decimal number 42 (maybe related to Douglas Adam's *Hitchhikers Guide to the Galaxy*, where 42 is described as the answer to the life, the universe and everything). It is used to identify that the file is TIFF format.

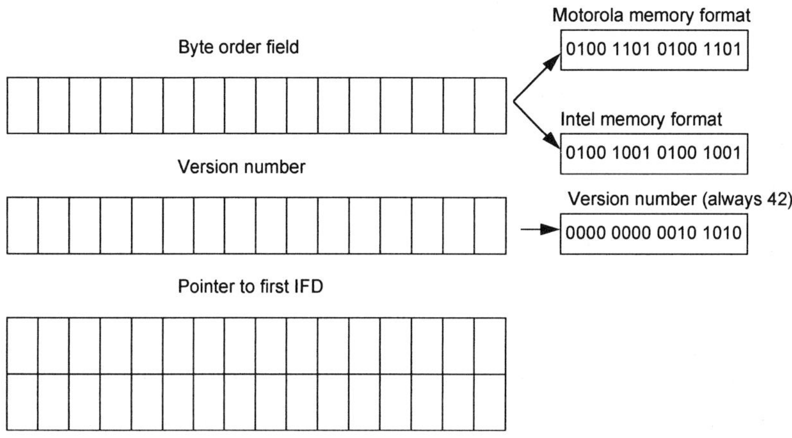

Figure 4.3 TIFF file header.

The first IFD offset pointer is a 4-byte pointer to the first IFD. If the format is Intel then the bytes are arranged from least significant to most significant else they are arranged from most significant to least significant.

Program 4.6 is a C program which reads the header of a TIFF file and Sample run 4.6 shows that, in this case, it uses the Intel format and the second byte field contains 2Ah (or 42 decimal).

Program 4.6

```
#include    <stdio.h>

int   main(void)
{
FILE  *in;
char  ch1,ch2,fname[BUFSIZ];

      printf("Enter TIFF file>>");
      gets(fname);

      if ((in=fopen(fname,"r"))==NULL)
      {
         printf("Can't find file %s\n",fname);
```

```
        return(1);
    }

    ch1=fgetc(in); ch2=getc(in);
    printf("Memory model %c%c\n",ch1,ch2);
    ch1=fgetc(in); ch2=getc(in);
    printf("Version %x%x\n",ch2,ch1);

    fclose(in);
    return(0);
}
```

Sample run 4.6

```
Enter TIFF file>> image1.tif
Memory model II
Version 02a
```

4.4.2 IFD

Typically, the first IFD will be the only IFD, which is pointed to by the first IFD in the header field.

4.4.3 Compression code 2: Huffman RLE coding

TIFF compression code 2 uses the CCITT Group 3 type compression which is a modified Huffman coding and is used in many fax transmissions. It specifies a 1-bit monochrome code with alternate black and white sequences of pixels. Tables 4.5 and 4.6 give the predefined coding table for white and black sequence runs. These tables contain codes in which the most frequent run lengths are coded with a short code. The compressed code always starts on white code. Codes themselves range from 0 to 63. Values from 64 to 2560 use two codes. The first gives the multiple of 64 followed by the normally coded remainder. There is no special end-of-line identifier because the size of the image is known by the defined ImageWidth tag field. There are thus `ImageWidth` pixels on a line.

Table 4.5 White run-length coding

Run length	*Coding*	*Run length*	*Coding*	*Run length*	*Coding*	*Run length*	*Coding*
0	00110101	1	000111	2	0111	3	1000
4	1011	5	1100	6	1110	7	1111
8	10011	9	10100	10	00111	11	01000
12	001000	13	000011	14	110100	15	110101
16	101010	17	101011	18	0100111	19	0001100
61	00110010	62	00110011	63	00110100	64	110011

Table 4.6 Black run-length coding

Run length	*Coding*	*Run length*	*Coding*	*Run length*	*Coding*	*Run length*	*Coding*
0	0000110111	1	010	2	11	3	10
4	011	5	0011	6	0010	7	00011
8	000101	9	000100	10	0000100	11	0000101
12	0000111	13	00000100	14	00000111	15	000011000
16	0000010111	17	0000011000	18	0000001000	19	00001100111
61	000001011010	62	0000001100110	63	000001100111	64	0000001111

For example, if the data were:

16 white   4 black   16 white   2 black   63 white   10 black   63 white

it would be coded as:

```
101010  011  101010  11 00110100  0000100      00110100
```

This would take 40 bits to code, whereas it would take 184 bits if coded with pixel colors (i.e., 16+4+16+2+63+10+63). This results in a compression ratio of 4.6:1.

4.4.4 Compression code 5: LZW compression

The compression technique used by TIFF code 5 is the same as is used in GIF files, but has a fixed code size of 8 (refer to Chapter 2 for more information on LZW code sizes). The dictionary starts with the values 0 to 255 stored in the entries 0 to 255. There are two codes for Clear (at 256) and EndOfInformation (at 257) and the dictionary is then built up from 258 to 4095. The Clear code is a special code which resets the dictionary entries to the original entries from 0 to 255.

A basic encoding algorithm could be:

```
Byte: byte;
Buffer, Test, String: string;
Table: array[1..4096] of string;

begin
    clear Table;  clear Buffer; clear Test; clear String;

    write ClearCode code;

    while (valid data)
    begin
        read Byte;
        Test=String+Byte;
        if (Test in Table) then String=String+Byte;
        else
        begin
            write String code;
            add Test to Table;
            String=Byte;
        end;
    end;

    write String code;
    write EndOfInformation code.
end.
```

4.5 GIF interlaced images

GIF images can be stored in an interlaced manner. This facility is useful when receiving information over a relatively slow transmission line, as it allows an outline of an image to be displayed before the entire image has been encoded (or received). The images stored are:

Group 1: Starting at row 0, every 8th row.
Group 2: Starting at row 4, every 8th row.
Group 3: Starting at row 2, every 4th row.
Group 4: Starting at row 1, every 2nd row.

For example if the image has 16 rows (0–15) then the following would be stored:

	Scanned line displayed			
	1	2	3	4
Row 0	X			
Row 1				X
Row 2			X	
Row 3				X
Row 4		X		
Row 5				X
Row 6			X	
Row 7				X
Row 8	X			
Row 9				X
Row 10			X	
Row 11				X
Row 12		X		
Row 13				X
Row 14			X	
Row 15				X

It can be seen that the first 1/8 of the data displays an outline of the image, the next 1/8 then improves the quality. After this, the next 1/4 further improves the quality and then the final 1/2 gives the completed image.

4.6 PCX coding

PCX is a well-supported graphics file format. It uses RLE coding to compress the data. As it uses RLE coding the amount of compression normally depends on the amount of data which is repeated. The less the image changes, the more compression can be achieved. This can be seen from Tables 4.2 and 4.3. The image in Figure 4.1 has many color changes, whereas Figure 4.2 has very few changes. Table 4.2 shows that the compression for the image in figure is only 88.9% of the original file size, while Table 4.3 shows that the compression for the image in Figure 4.2 is 4.3%.

PCX Version 0 only allows for a basic 2-color or 4-color image, Version 2 supports 16-color images and Version 5 supports 256 colors from 24-bit palettes.

4.6.1 File structure

A PCX file consists of three main parts:

- A file header.
- Bitmapped data.
- A color palette of 256 colors.

4.6.2 Header file

Figure 4.4 defines the header, which always contains 256 bytes. Program 4.7 outlines a C program which can be used to read the header. It can be seen from Sample run 4.7 that a sample file IMAGE1.PCX has the flag set to 0Ah (it is thus a PCX file), the version number is set to 5, the encoding field is set to 1, the image size ranges from (0,0) to (639,479), the bits per pixel is 8 and the printable dots per inch in the x- and y-direction is 150. In this case, the number of displayable colors will be 256 as there are 8 bits per pixel. Also, the screen dimensions are set for 640×480 (to fit a standard VGA screen on a PC).

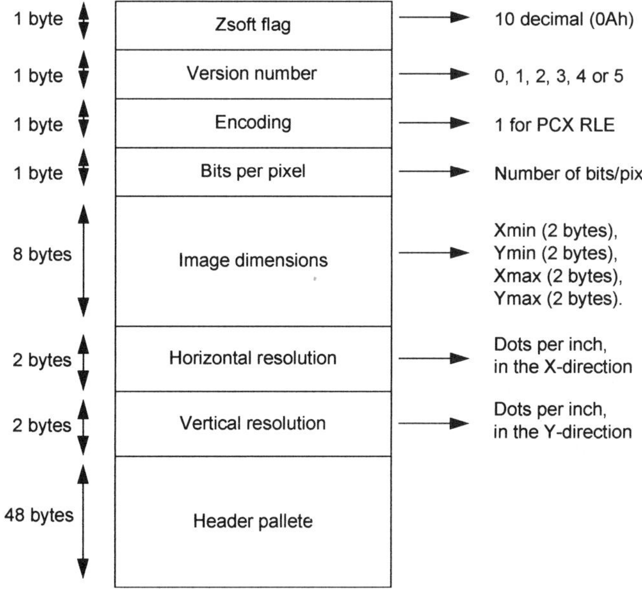

Figure 4.4 PCX file header.

Program 4.7

```
#include    <stdio.h>

int   main(void)
{
FILE  *in;
char  flag,version,coding,bits,fname[BUFSIZ];
```

```
int     xmin,ymin,xmax,ymax, xres, yres;

        printf("Enter PCX file>>");
        gets(fname);

        if ((in=fopen(fname,"r"))==NULL)
        {
           printf("Can't find file %s\n",fname);
           return(1);
        }

        fread(&flag,1,1,in);    /* read 1 byte from file and put into flag */
        fread(&version,1,1,in); /* read 1 byte from file and put in version */

        fread(&coding,1,1,in);  /* read 1 byte from file and put into coding */
        fread(&bits,1,1,in); /* read 1 byte from file and put into bits */

        fread(&xmin,2,1,in); /* read 2 bytes from file and put into xmin */
        fread(&ymin,2,1,in); /* read 2 bytes from file and put into ymin */

        fread(&xmax,2,1,in); /* read 2 bytes from file and put into xmax */
        fread(&ymax,2,1,in); /* read 2 bytes from file and put into ymax */

        printf("Flag %X\n",flag);
        printf("Version %X\n",version);

        printf("Coding %X\n",coding);

        printf("Bits per pixel %X (%d decimal)\n",bits,bits);
        printf("Min (%d,%d) Max (%d,%d)\n",xmin,ymin,xmax,ymax);

        fread(&xres,2,1,in); /* read 2 bytes from file and put into xres */
        fread(&yres,2,1,in); /* read 2 bytes from file and put into yres */
        printf("Resolution (%d,%d)\n",xres,yres);

        fclose(in);
        return(0);
}
```

Sample run 4.7

```
Enter PCX file>> image1.pcx
Flag A
Version 5
Coding 1
Bits per pixel 8 (8 decimal)
Min (0,0) Max (639,479)
Resolution (150,150)
```

4.6.3 Bitmapped data

If the file does not use a palette then the data contains actual pixel colors, whereas if a palette is used then the data relates to pointers for the color palette.

5 Image Compression (JPEG)

5.1 Introduction

JPEG is an excellent compression technique which produces lossy compression (although in one mode it is lossless). As seen from the previous chapter it has excellent compression ratio when applied to a color image. This chapter introduces the JPEG standard and the method used to compress an image. It also discusses the JFIF file standard which defines the file format for JPEG encoded images. Along with GIF files, JPEG is now one of the most widely used standards for image compression.

5.2 JPEG coding

A typical standard for image compression has been devised by the Joint Photographic Expert Group (JPEG), a subcommittee of the ISO/IEC, and the standards produced can be summarized as follows:

It is a compression technique for gray-scale or color images and uses a combination of discrete cosine transform, quantization, run-length and Huffman coding.

It has resulted from research into compression ratios and the resultant image quality. The main steps are:

- Data blocks — Generation of data blocks
- Source-encoding — Discrete cosine transform and quantization
- Entropy-encoding — Run-length encoding and Huffman encoding

Unfortunately, compared with GIF, TIFF and PCX, the compression process is relatively slow. It is also lossy in that some information is lost in the compression process. This information is perceived to have little effect on the decoded image.

GIF files typically take 24-bit color information (8 bits for red, 8 bits for green and 8 bits for blue) and convert it into an 8-bit color palette (thus reducing the number of bits stored to approximately one-third of the original). It then uses LZW compression to further reduce the storage. JPEG operates differently in that it stores changes in color. As the eye is very sensitive to brightness changes and less on color changes, then if these changes are similar to the original then the eye perceives the recovered image as very similar to the original.

5.2.1 Color conversion and subsampling

The first part of the JPEG compression separates each color component (red, green and blue) in terms of luminance (brightness) and chrominance (color information). JPEG allows more losses on the chrominance and less on the luminance as the human eye is less sensitive to color changes than to brightness changes. In an RGB image, all three colors carry some brightness information but the green component has a stronger effect on brightness than the blue component.

A typical scheme for converting RGB into luminance and color is CCIR 601, which converts the components into Y (can be equated to brightness), C_b (blueness) and C_r (redness). The Y component can be used as a black and white version of the image.

Each component is computed from the RGB components as:

$$Y=0.299R+0.587G+0.114B$$
$$C_b=0.1687\text{R}-0.3313G+0.5B$$
$$C_r=0.5R-0.4187G+0.0813B$$

For the brightness (Y) it can be seen that green has the most effect and blue has the least effect. For the redness (C_r), the red color (of course) has the most effect and green the least. For blueness (C_b), the blue color has the most effect and green the least. Note that the YC_bC_r components are often known as YUV (especially in TV systems).

A subsampling process is then samples the C_b and C_r components at a lower rate than the Y component. A typical sampling rate is four samples of the Y component for a single sample of the C_b and C_r components (4:1:1). This sampling rate is normally set with the compression parameters, the lower the sampling, the smaller the compressed data and the shorter the compression time. The JPEG header contains all the information necessary to properly decode the JPEG data.

5.2.2 DCT coding

The DCT (discrete cosine transform) converts intensity data into frequency data, which can be used to tell how fast the intensities vary. In JPEG coding the image is segmented into 8×8 pixel rectangles, as illustrated in Figure 5.1. If the image contains several components (such as Y,C_b,C_r or R,G,B), then each of the components in the pixel blocks is operated on separately. If an image is subsampled, there will be more blocks of some components than of others. For example, for 2×2 sampling there will be four blocks of Y data for each block of C_b or C_r data.

The data points in the 8×8 pixel array starts at the upper right at $(0,0)$ and finish at the lower right at $(7,7)$. At the point (x,y) the data value is $f(x,y)$. The DCT produces a new 8×8 block ($u\times v$) of transformed data using the formula:

$$F(u,v)=\frac{1}{4}C(u)C(v)\left[\sum_{x=0}^{7}\sum_{y=0}^{7}f(x,y)\cos\frac{(2x+1)u\pi}{16}\cos\frac{(2y+1)v\pi}{16}\right]$$

$$\text{where}\quad C(z)=\frac{1}{\sqrt{2}}\quad \text{if } z=0$$

$$\text{or}\quad =1\quad \text{if } z\neq 0$$

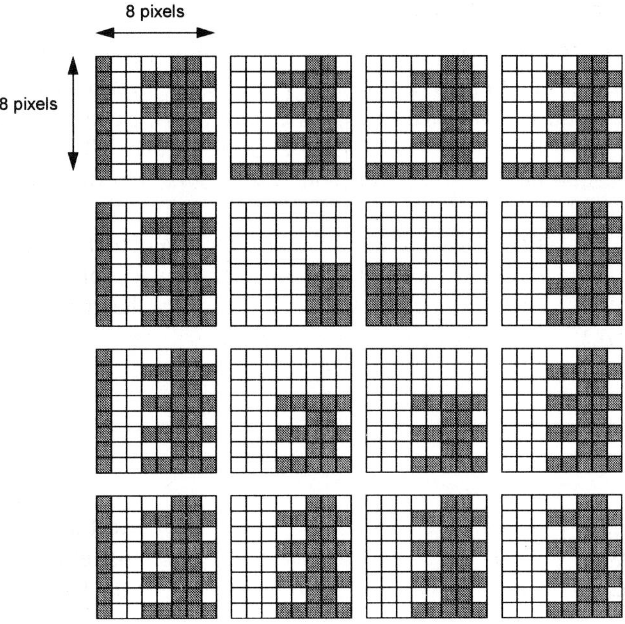

Figure 5.1 Segment of an image in 8x8 pixel blocks.

This results in an array of space frequency $F(u, v)$ which gives the rate of change at a given point. These are normally 12-bit values which give a range of 0 to 1024. Each component specifies the degree to which the image changes over the sampled block. For example:

- F(0,0) gives the average value of the 8×8 array.
- $F(1,0)$ gives the degree to which the values change slowly (low frequency).
- $F(7,7)$ gives indicates the degree to which the values change most quickly in both directions (high frequency).

The coefficients are equivalent to representing changes of frequency within the data block. The value in the upper left block $(0,0)$ is the DC or average value. The values to the right of a row have increasing horizontal frequency and the values to the bottom of a column have increasing vertical frequency. Many of the bands end up having zero or almost zero terms.

Program 5.1 gives a C program which determines the DCT of an 8×8 block and Sample run 5.1 shows a sample run with the resultant coefficients.

Program 5.1

```
#include <stdio.h>
#include <math.h>
#define  PI 3.1415926535897

int   main(void)
{
int   x,y,u,v;
float in[8][8]=   {{144,139,149,155,153,155,155,155},
```

```
                {151,151,151,159,156,156,156,158},
                {151,156,160,162,159,151,151,151},
                {158,163,161,160,160,160,160,161},
                {158,160,161,162,160,155,155,156},
                {161,161,161,161,160,157,157,157},
                {162,162,161,160,161,157,157,157},
                {162,162,161,160,163,157,158,154}};
float out[8][8],sum,Cu,Cv;

      for (u=0;u<8;u++)
      {
         for (v=0;v<8;v++)
         {
            sum=0;
            for (x=0;x<8;x++)
               for (y=0;y<8;y++)
               {
                  sum=sum+in[x][y]*cos(((2.0*x+1)*u*PI)/16.0)*
                     cos(((2.0*y+1)*v*PI)/16.0);
               }
            if (u==0) Cu=1/sqrt(2); else Cu=1;
            if (v==0) Cv=1/sqrt(2); else Cv=1;

            out[u][v]=1/4.0*Cu*Cv*sum;
            printf("%8.1f ",out[u][v]);
         }
         printf("\n");
      }
      printf("\n");
      return(0);
}
```

The program uses a fixed 8×8 block of:

```
144   139   149   155   153   155   155   155
151   151   151   159   156   156   156   158
151   156   160   162   159   151   151   151
158   163   161   160   160   160   160   161
158   160   161   162   160   155   155   156
161   161   161   161   160   157   157   157
162   162   161   160   161   157   157   157
162   162   161   160   163   157   158   154
```

Sample run 5.1

```
  1257.9       2.3      -9.7      -4.1       3.9       0.6      -2.1       0.7
   -21.0     -15.3      -4.3      -2.7       2.3       3.5       2.1      -3.1
   -11.2      -7.6      -0.9       4.1       2.0       3.4       1.4       0.9
    -4.9      -5.8       1.8       1.1       1.6       2.7       2.8      -0.7
     0.1      -3.8       0.5       1.3      -1.4       0.7       1.0       0.9
     0.9      -1.6       0.9      -0.3      -1.8      -0.3       1.4       0.8
    -4.4       2.7      -4.4      -1.5      -0.1       1.1       0.4       1.9
    -6.4       3.8      -5.0      -2.6       1.6       0.6       0.1       1.5
```

Notice that the values of the most significant values are in the top left-hand corner and that many terms are near to zero. It is this property which allows many values to become zeros when quantized. These zeros can then be compressed using run-length coding and Huffman codes.

5.2.3 Quantization

The next stage of the JPEG compression is quantization where bias is given to lower-frequency components. JPEG divides each of the DCT values by a quantization factor, which is then rounded to the nearest integer. As the DCT factors are 8×8 then a table of 8×8 of quantization factors are used, corresponding to each term of the DCT output. The JPEG file then stores this table so that the decoding process may use this table or a standard quantization table. Note that files with multiple components must have multiple tables, such as one each for the Y, C_b and C_r components.

For example the values of the quantized high-frequency term (such as $F(7,7)$) could have a term of around 100, while the low-frequency term could have a factor of 16. These values define the accuracy of the final value. When decoding, the original values are (approximately) recovered by multiplying by the quantization factor.

Figure 5.2 shows that, for a factor of 100, the values between 50 and 150 would be quantized as a 1, thus the maximum error would be ±50. The maximum error for the factor of 16 is ±8. Thus the maximum error of the final unquantized value for a scale factor of 100 is 1.22% (5000/4096), while a factor of 16 gives a maximum error of 0.20% (800/4096). So, using the factors of 100 for $F(7,7)$ and 16 for $F(0,0)$, and a 12-bit DCT, the $F(0,0)$ term would range from 0 to 256 and the $F(7,7)$ term would range from 0 to 41. The $F(0,0)$ term could be coded with 8 bits (0 to 255) and the $F(7,7)$ term with 6 bits (0 to 63).

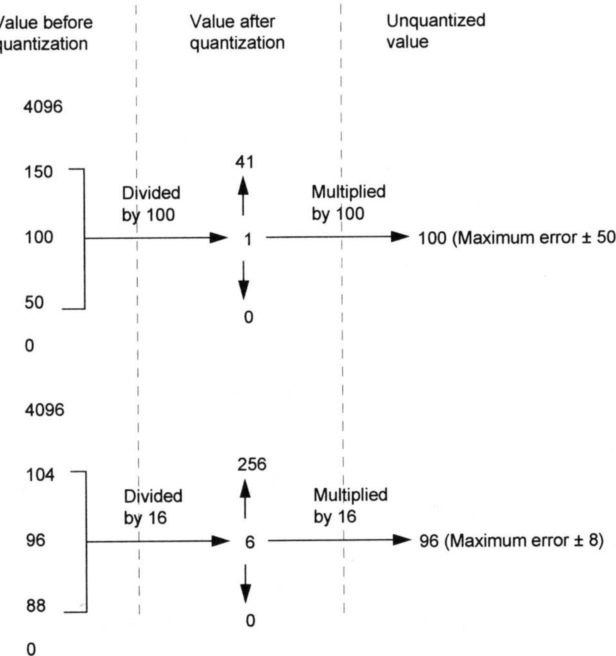

Figure 5.2 Example of quantization.

Program 5.2 normalizes and quantizes (to the nearest integer) the example given previously. To recap, the input 8×8 block is:

144	139	149	155	153	155	155	155
151	151	151	159	156	156	156	158
151	156	160	162	159	151	151	151
158	163	161	160	160	160	160	161
158	160	161	162	160	155	155	156
161	161	161	161	160	157	157	157
162	162	161	160	161	157	157	157
162	162	161	160	163	157	158	154

The applied normalization matrix is:

5	3	4	4	4	3	5	4
4	4	5	5	5	6	7	12
8	7	7	7	7	15	11	11
9	12	13	15	18	18	17	15
20	20	20	20	20	20	20	20
20	20	20	20	20	20	20	20
20	20	20	20	20	20	20	20
20	20	20	20	20	20	20	20

Program 5.2

```
#include <stdio.h>
#include <math.h>
#define  PI 3.1415926535897

int   main(void)
{
int   x,y,u,v;
float in[8][8]=   {
         {144,139,149,155,153,155,155,155},
         {151,151,151,159,156,156,156,158},
         {151,156,160,162,159,151,151,151},
         {158,163,161,160,160,160,160,161},
         {158,160,161,162,160,155,155,156},
         {161,161,161,161,160,157,157,157},
         {162,162,161,160,161,157,157,157},
         {162,162,161,160,163,157,158,154}};

float norm[8][8]= {
      {5,3,4,4,4,3,5,4},
      {4,4,5,5,5,6,7,12},
      {8,7,7,7,7,15,11,11},
      {9,12,13,15,18,18,17,15},
      {20,20,20,20,20,20,20,20},
      {20,20,20,20,20,20,20,20},
      {20,20,20,20,20,20,20,20},
      {20,20,20,20,20,20,20,20}};

int   out[8][8];
float sum,Cu,Cv;

      for (u=0;u<8;u++)
      {
         for (v=0;v<8;v++)
         {
            sum=0;
            for (x=0;x<8;x++)
               for (y=0;y<8;y++)
               {
                  sum=sum+in[x][y]*cos(((2.0*x+1)*u*PI)/16.0)*
                     cos(((2.0*y+1)*v*PI)/16.0);
```

```
            }
            if (u==0) Cu=1/sqrt(2); else Cu=1;
            if (v==0) Cv=1/sqrt(2); else Cv=1;

            out[u][v]=(int)1/4.0*Cu*Cv*sum/norm[u][v];
            printf("%8d ",out[u][v]);
        }
        printf("\n");
    }

    printf("\n");
    return(0);
}
```

It can be seen from Sample run 5.2 that most of the normalized and quantized components are zero. This helps in the next stages of the compression, which involve either LZW or RLE. Thus, a scheme which stores similar values results in a larger compression than non-arranged values. It can also be seen from Sample run 5.2 that most of the non-zero values are in the top left-hand corner.

To achieve the compression, the DC components are stored as the difference in the DC value from one block to the next. This is because DC components, from block to block tend to be similar. The AC components are then stored logically in a zigzag order, as illustrated in Figure 5.3. This puts the lowest-frequency components first. Typically the quantized high-frequency components will be zero, and the final compression stage will compress these to a high degree.

Sample run 5.2 would be stored as:

```
251, 0, -5, -1, -3, -2, 0, -1, 0, 0, 0, 0, -1, 0, 0, 0, 0,...,0
```

which has a run of 51 zero (as well as an earlier run of 4 zeros).

Sample run 5.2

251	0	-2	-1	0	0	0	0
-5	-3	0	0	0	0	0	0
-1	-1	0	0	0	0	0	0
0	0	0	0	0	0	0	0
0	0	0	0	0	0	0	0
0	0	0	0	0	0	0	0
0	0	0	0	0	0	0	0
0	0	0	0	0	0	0	0

5.2.4 Final compression

The final part of JPEG compression uses either a modified Huffman coding or arithmetic coding. Huffman coding is by far the most popular technique, but tends to lead to a larger compressed file.

Data values are coded with a modified Huffman code called the variable-length code (VLC). This encodes values as the difference between consecutive values. A positive value is stored with its binary equivalent and a negative value as the one's complement (all the bits inverted) equivalent, such as:

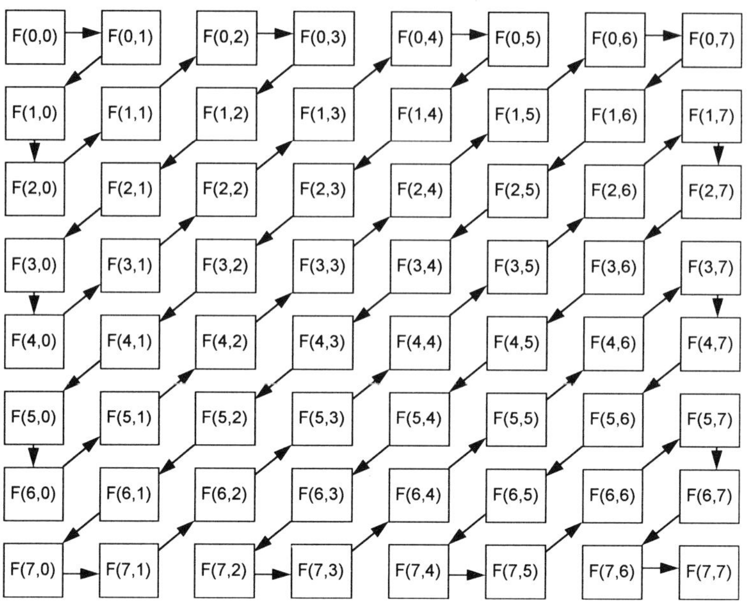

Figure 5.3 Zigzag storage.

5	101	–5	010
10	1010	–10	0101
1	1	–1	0
23	10111	–23	01000

This difference value is preceded by a 4-bit binary value which defines the number of bits in the data values. Figure 5.4 gives an example. In this case, the data is 12, 10, 11, 11 and 11. The initial value for the difference encoding is taken as zero, thus the difference values will be 12, –2, 1, 0, 0. The first four bits will be 0100 (4) as the value 12 requires four bits. Next the value of 12 is stored (1010). The next difference is –2, which is 01 in 1's complement. This requires two bits, thus the next four bits will be 0010 (to define two bits), followed by the –2 value (01). This then continues. Note that a zero value is stored as a single 4-bit value of 0000, so no other bits follow it.

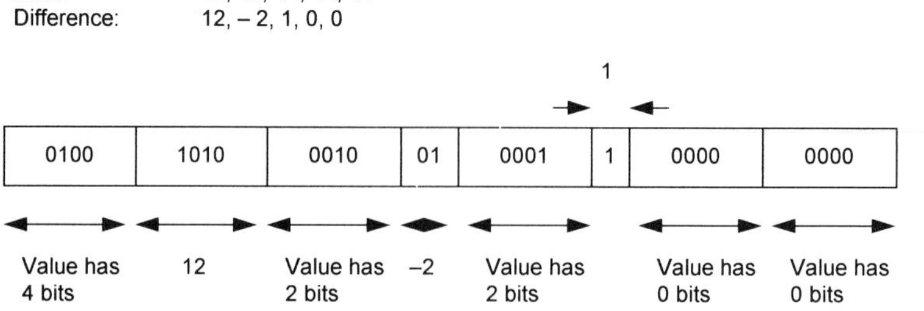

Figure 5.4 VLC storage.

AC components (such as $F(0,1)$, $F(1,0)$, and so on) are stored as an 8-bit Huffman code followed by a variable-length integer. The Huffman code is made up of high 4 bits which give the number of zero values preceding this value, and the low 4 bits which give the length of the variable-length integer. Figure 5.5 shows an example of AC coding with the data 0, 0, 0, 0, 2, 0, 0, 6. In this case, the first four bits of the 8-bit Huffman code (before it is converted to a Huffman code) is 0101, because there are four consecutive zeros. The value after these zeros is 2, which requires only two bits. Thus, the second part of the Huffman code (before coding) will be 0010 to specify two bits. Next, the data contains two zeros so the Huffman code (before coding) will be 0010. The data after the two zeros is a 6 which requires four bits, thus the second part of the Huffman code (before coding) is 0100. After this, the data value for 6 is represented in binary (1010). The AC components contain many runs of zeros so the code produced will tend to be extremely compressed.

The binary value of 0000 0000 (00h) can never occur in the AC coding scheme. This code is used as a special code to identify that all of the values until the end of a block are zero. This is a common occurrence and thus saves coding bits.

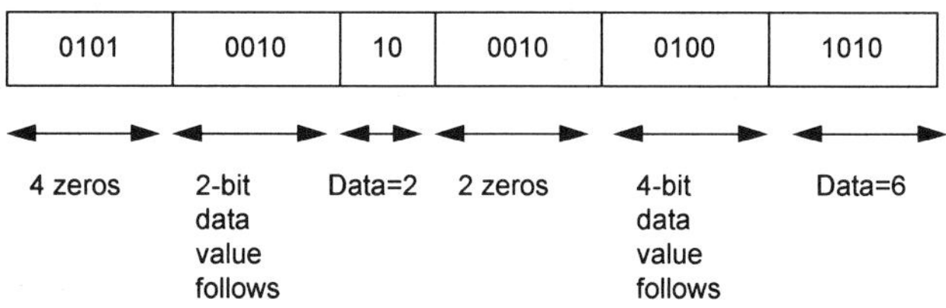

Figure 5.5 Storage of AC components.

5.3 JPEG decoding

JPEG decoding involves reversing the process:

- Uncompression.
- Unquantizing (using the stored table or a standard table of factors).
- Reverse DCT.
- Block regeneration.

The DCT is reversed with the transform:

$$f(x,y)=\frac{1}{4}\left[\sum_{u=0}^{7}\sum_{v=0}^{7}C(u)C(v)F(u,v)\cos\frac{(2x+1)u\pi}{16}\cos\frac{(2y+1)v\pi}{16}\right]$$

Program 5.3 uses this inverse transform and contains the normalized and quantized coefficients from the previous example. To recap, the input data was:

144	139	149	155	153	155	155	155
151	151	151	159	156	156	156	158
151	156	160	162	159	151	151	151
158	163	161	160	160	160	160	161
158	160	161	162	160	155	155	156
161	161	161	161	160	157	157	157
162	162	161	160	161	157	157	157
162	162	161	160	163	157	158	154

The applied normalization matrix was:

5	3	4	4	4	3	5	4
4	4	5	5	5	6	7	12
8	7	7	7	7	15	11	11
9	12	13	15	18	18	17	15
20	20	20	20	20	20	20	20
20	20	20	20	20	20	20	20
20	20	20	20	20	20	20	20
20	20	20	20	20	20	20	20

Program 5.3

```
#include <stdio.h>
#include <math.h>
#define  PI 3.1415926535897

int    main(void)
{
int    x,y,u,v;
int    in[8][8]=    {{251,0,-2,-1,0,0,0,0},
                 {-5,-3,0,0,0,0,0,0},
                 {-1,-1,0,0,0,0,0,0},
                 {0,0,0,0,0,0,0,0},
                 {0,0,0,0,0,0,0,0},
                 {0,0,0,0,0,0,0,0},
                 {0,0,0,0,0,0,0,0}};

float norm[8][8]= {{5,3,4,4,4,3,5,4},
       {4,4,5,5,5,6,7,12},
       {8,7,7,7,7,15,11,11},
       {9,12,13,15,18,18,17,15},
       {20,20,20,20,20,20,20,20},
       {20,20,20,20,20,20,20,20},
       {20,20,20,20,20,20,20,20},
       {20,20,20,20,20,20,20,20}};

float out[8][8];
float sum,Cu,Cv;

      for (x=0;x<8;x++)
      {
         for (y=0;y<8;y++)
         {
            sum=0;
            for (u=0;u<8;u++)
               for (v=0;v<8;v++)
               {
                  if (u==0) Cu=1/sqrt(2); else Cu=1;
                  if (v==0) Cv=1/sqrt(2); else Cv=1;

                  sum=sum+Cu*Cv*norm[u][v]*in[u][v]*cos(((2.0*x+1)*u*PI)/16.0)*
                     cos(((2.0*y+1)*v*PI)/16.0);
```

```
            }
            out[x][y]=1/4.0*sum;
            printf("%8.0f ",out[x][y]);
        }
        printf("\n");
    }
    printf("\n");
    return(0);
}
```

Sample run 5.3 shows that the values are similar to the input data values.

Sample run 5.3

```
146     148     151     153     154     154     155     156
148     150     153     154     155     155     155     156
153     154     156     157     157     156     156     156
157     158     159     159     158     157     156     156
159     160     161     161     159     157     156     156
160     161     162     162     160     158     157     156
159     160     162     161     160     158     157     157
158     160     161     161     160     158     157     157
```

The errors in the decoding are thus:

```
-2      -9    -2    2     -1    1     0     -1
3       1     -2    5     1     1     1     2
-2      2     4     5     2     -5    -5    -5
1       5     2     1     2     3     4     5
-1      0     0     1     1     -2    -1    0
1       0     -1    -1    0     -1    0     1
3       2     -1    -1    1     -1    0     0
4       2     0     -1    3     -1    1     -3
```

It can be seen that these errors are all less than 10, thus the decoding as not produced any significant errors.

5.4 JPEG file format

JPEG is a standard compression technique. A JPEG file normally complies with JFIF (JPEG file interchange format) which is a defined standard file format for storing a gray scale or YC_bC_r color image. Data within the JFIF contains segments separated by a 2-byte marker. This marker has a binary value of 1111 1111 (FFh) followed by a defined marker field. If a 1111 1111 (FFh) bit field occurs anywhere within the file (and it isn't a marker), the stuffed 0000 0000 (00h) byte is inserted after it so that it cannot be read as a false marker. The uncompression program must then discard the stuffed 00h byte.

Table 5.1 outlines some of the markers. For example, the code FFC0h the file is a baseline DCT frame with Huffman coding. Program 5.4 uses these codes to display the markers in a sample JPG file, and Sample run 5.1 show a sample run with a JPG file.

It can be seen that the markers in the test run are:

- Start of image (FFD8h). The segments can be organized in any order but the start-of-image marker is normally the first 2 bytes of the file.
- Application-specific type 0 (FFE0h). The JFIF header is placed after this marker.
- Define quantization table (FFDBh). Lists the quantization table(s).
- Baseline DCT, Huffman coding (FFC0h). Defines the type of coding used.
- Define Huffman table (FFC4h). Defines Huffman table(s).

5.4.1 JFIF header information

The header information of the JFIF file is contained after the application-specific type 0 marker (FFE0h). Figure 5.6 shows its format and Program 5.5 reads some of the header information from a JFIF file. It can be seen that in Sample run 5.5 the length of the segment is 4096 bytes and the file is a JFIF file, as it has the JFIF string at the correct location. The version in this case is 1.01 and the units are given in pixels per inch. Finally, it shows that the horizontal and vertical pixel density are both 11265. For comparison, another sample run for a different file is shown in Sample run 5.6.

Table 5.1 Typical standard compressed graphics formats.

Marker	*Description*	*Marker*	*Description*
C0h	Baseline DCT frame, Huffman coded	C1h	Baseline sequential DCT frame, Huffman coded
C2h	Extended sequential DCT frame, Huffman coded	C3h	Progressive DCT frame, Huffman coded
C4h	Define Huffman table	C5h	Differential sequential DCT frame, Huffman coded
C6h	Differential progressive DCT frame, arithmetic coded	C7h	Differential lossless frame, Huffman coded
C8h	Reserved	C9h	Extended sequential DCT frame, arithmetic coded
CAh	Progressive DCT frame, arithmetic coded	CBh	Lossless frame, arithmetic coded
CDh	Differential extended sequential DCT frame, arithmetic coded	CEh	Differential progressive DCT frame, arithmetic coded
CFh	Differential lossless frame, arithmetic	D8h	Start of image
D9h	End of image	E0h	Application-specific type 0

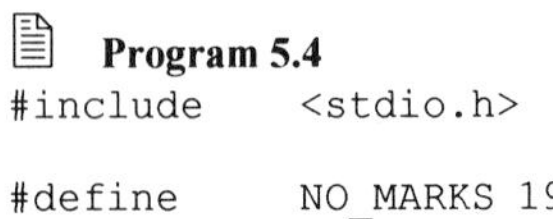

Program 5.4

```
#include    <stdio.h>

#define     NO_MARKS 19
```

```
int    main(void)
{
FILE   *in;
int    i,ch;
int    markers[NO_MARKS]={0xC0,0xC1,0xC2,0xC3,0xC4,
          0xC5,0xC6,0xC7,0xC8,0xC9,0xCA,0xCB,
          0xCD,0xCE,0xCF,0xD8,0xD9,0xDB,0xE0};
char   fname[BUFSIZ];
char   *msgs[NO_MARKS]={"Baseline DCT, Huff","Extended DCT, Huff",
          "Progress DCT, Huff","Lossless frame, Huff",
          "Define Huffman table", "Diff encoded DCT frame, Huff coded",
          "Diff progressive DCT frame, Huff", "Diff lossless frame, Huff",
          "Reserved", "Extended sequential DCT frame, arith coded",
          "Progressive DCT frame, arith coded",
          "Lossless frame, arith coded",
          "Diff extended sequential DCT frame, arith coding",
          "Diff progressive DCT frame, arith coding",
          "Diff lossless frame, arith coding",
          "Start of image", "End of image",
          "Define Quantization Tables", "Application specific type 0"};

       printf("Enter JPG file>>");
       gets(fname);

       if ((in=fopen(fname,"r"))==NULL)
       {
          printf("Can't find file %s\n",fname);
          return(1);
       }

       do
       {
          ch=getc(in);
          if (ch==0xff)
          {
             ch=getc(in);
             printf("%x",ch);
             for (i=0;i<NO_MARKS;i++)
                if (ch==markers[i])  printf("Found:%s\n",msgs[i]);
          }

       } while (!feof(in));

       fclose(in);
       return(0);
}
```

Sample run 5.4

```
Enter JPG file>> marble.jpg
d8:Found:Start of image
e0:Found:Application specific type 0
db:Found:Define Quantization Tables
db:Found:Define Quantization Tables
c0:Found:Baseline DCT, Huff
c4:Found:Define Huffman table
c4:Found:Define Huffman table
```

The reason that the segment is 4096 bytes is that a thumbnail version of the image is stored within the segment, after the basic header information defined in Figure 5.6.

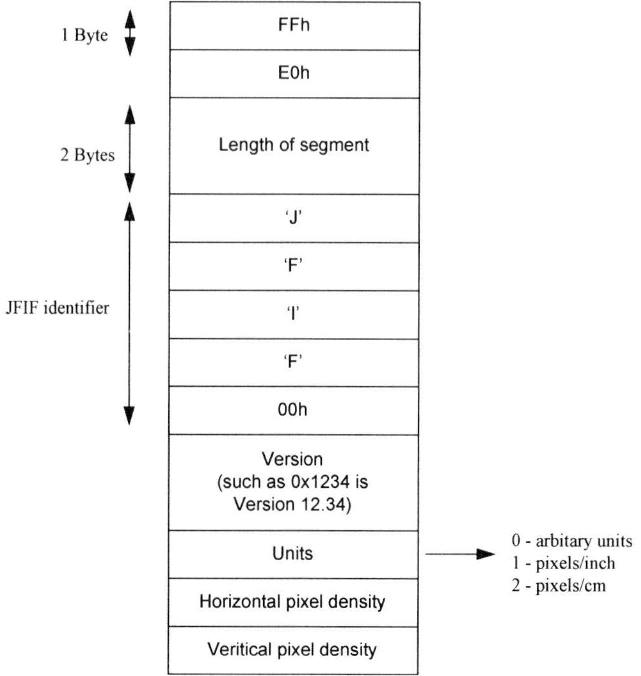

Figure 5.6 JFIF header information.

Program 5.5

```
#include    <stdio.h>

int   main(void)
{
FILE  *in;
int   i,ch,version,length,units,pixelden_X,pixelden_Y;
char  fname[BUFSIZ],str[BUFSIZ];
char  *Units[3]={"Artibrary","Pixels per inch","Pixels per cm"};

      printf("Enter JPG file>>");
      gets(fname);

      if ((in=fopen(fname,"r"))==NULL)
      {
         printf("Can't find file %s\n",fname);
         return(1);
      }

      do
      {
         ch=getc(in);

         if (ch==0xff)
         {
            ch=getc(in);
            if (ch==0xe0)
            {
               fread(&length,2,1,in); printf("Length: %d\n",length);
               fread(str,5,1,in); printf("Marker: %s\n",str);
               fread(&version,2,1,in); printf("Version: %0x\n",version);
```

```
                fread(&units,1,1,in); printf("Units: %s\n",Units[units]);
                fread(&pixelden_X,2,1,in); printf("X den: %d\n",pixelden_X);
                fread(&pixelden_Y,2,1,in); printf("Y den: %d\n",pixelden_Y);
            }
        }

    } while (!feof(in));

    fclose(in);
    return(0);
}
```

Sample run 5.5

```
Enter JPG file>> marble.jpg
Length: 4096
Marker: JFIF
Version: 101
Units: Pixels per inch
X den: 11265
Y den: 11265
```

Sample run 5.6

```
Enter JPG file>> test.jpg
Length: 4096
Marker: JFIF
Version: 101
Units: Artibrary
X den: 256
Y den: 256
```

5.4.2 Quantization table

The quantization table is defined after the quantization table marker (FFDBh). Its format, after the marker, is:

- 2 bytes for the length of the segment.
- 1 byte, of which the high 4 bits define the precision (a 0 defines a table with 8-bit entries, a 1 defines 16-bit entries), and the low 4 bits give the table's ID (such as 0 for the first, 1 for the second, and so on).
- 64 entries for the table (either 8-bit or 16-bit entries). These entries are stored in a zigzag manner (see Figure 5.3).

Program 5.6 can be used to list the quantization table and Sample run 5.7 gives a sample run from a JPG file. It can be seen that in this case the factor varies from 3 to 28.

Program 5.6

```
#include    <stdio.h>
int   main(void)
{
FILE  *in;
int      i,ch,length;
char  fname[BUFSIZ], table, entry;

      printf("Enter JPG file>>");
      gets(fname);
```

```
    if ((in=fopen(fname,"r"))==NULL)
    {
       printf("Can't find file %s\n",fname);
       return(1);
    }

    do
    {
       ch=getc(in);

       if (ch==0xff)
       {
          ch=getc(in);
          if (ch==0xdb)
          {
             fread(&length,2,1,in); printf("Length: %d\n",length);
             fread(&table,1,1,in); printf("Marker: %x\n",table);

             for (i=0;i<64;i++)
             {
                fread(&entry,1,1,in);
                printf("%d ",entry);
             }
          }
       }

    } while (!feof(in));
    fclose(in);
    return(0);
}
```

Sample run 5.7

```
Enter JPG file>> test.jpg
Length: 17152
Marker: 0
5 3 4 4 4 3 5 4 4 4 5 5 5 6 7 12 8 7 7 7 7 15 11 11 9 12 17 15 18 18 17 15 17
17 19 22 28 23 19 20 20 20 20 20 20 20 20 20 20 20 20 20 20 20 20 20 20 20 20
20 20 20 20 20
```

5.4.3 Huffman tables

Huffman tables are defined after the define Huffman table marker (FFC4h). One table defines the DC components, $F(0,0)$, and the other defines the AC components. Its format, after the marker, is:

- 2 bytes for the length of the segment.
- 1 byte, of which the high 4 bits defines the table class (0 for DC codes and 1 for AC codes), and the low 4 bits gives the table's ID (such as 0 for the first, 1 for the second, and so on).
- 16 bytes for the code lengths.
- A variable number of bytes which contain the Huffman codes (the code length defines the number of bits used for each code).

For example if the code lengths are:

3, 3, 3, 4, 4, 4

the packed Huffman code of:

0000010100001001010000

would be separated as:

000 001 010 0001 0010 1000

5.5 JPEG modes

JPEG has three main modes of compression:

- Progressive mode – this mode allows the viewing of a rough outline of an image while decoding the rest of the file. It is useful when an image is being received over a relatively slow transfer channel (such as over a modem or from the Internet). There are two main methods used: spectral-selection mode and successive-approximation mode. The successive-approximation mode first sends the high-order bits of each of the encoded values and then the lower-order lower bits. Spectral-selection mode first sends the low-frequency components of each of the 8×8 blocks then sends the high-frequency terms.
- Hierarchical mode – in this mode the image is stored in increasing resolution. For example a 1280×960 image might be stored as 160×120, 320×240, 640×480 and 1280×960. The viewing program can then show the image in increasing resolution as it reads (or receives) the file. Most systems do not implement this facility.
- Lossless mode – this mode allows data to be stored and recovered in exactly in its original state. It does not use DCT conversion or subsampling.

6 Video Signals

6.1 Introduction

This chapter and the next chapter discuss the main technologies using in TV signals and outlines the usage of digital TV. The five main classes of motion video are:

- High-definition television (HDTV).
- Studio-quality digital television (SQDV).
- Current broadcast-quality television.
- VCR-quality television.
- Low-speed video conferencing quality.

There are three main types of TV-type video signal:

- NTSC (American National Television Standards Committee).
- PAL (phase alternation line).
- SECAM (séquential couleur à mémoire).

These signals are based on composite color video which uses gaps in the black and white signal to transmit bursts of color information. In the UK this video signal is known as PAL (phase alternation line), whereas in the USA it is known as NTSC (named after the American National Television Standard Committee, who defined the original standard). Most of the countries of the world now use either PAL or NTSC. The only other standard format is SECAM, popular in French-speaking parts of the world.

6.2 Color-difference signals

A motion video is basically single images scanned at a regular rate. If these images are displayed fast enough, the human eye sees the repetitive images in a smooth way. Video signals use three primary colors: red, green and blue.

An image is normally scanned a row and one pixel at a time. When a signal row has been scanned, the scanner goes to the next row, and so on until it has scanned the last row, when it goes back to the start. The samples are then split into the three primary colors: red, green and blue, as illustrated in Figure 6.1.

The three main characteristics of color signals are:

- Its brightness (luminance).

- Its colors (hues).
- Its color saturation (the amount of color).

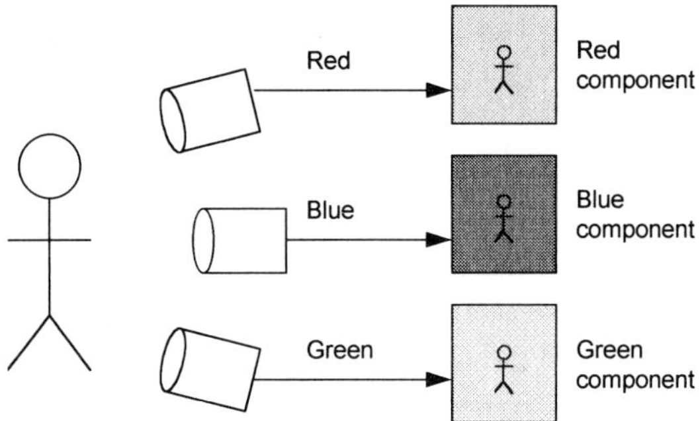

Figure 6.1 Conversion of an image into RGB components.

Any hue can vary from very pale to very deep; the amount of the color is its saturation. When transmitting a TV signal it is important to isolate the luminance signal (the black and white element) so it can be sent to monochrome receivers. This is done by carefully adding a mixture of each of the RGB signals. PAL, NSTC and SECAM use the mixture:

$$Y = 0.3R + 0.59G + 0.11B$$

where the red, green and blue signals vary between 0 and 1. Thus the maximum level for luminance will be 1 (fully saturated, pure white color) and the minimum level will be 0 (black). It can be seen from this formula that the green signal has a much greater effect than the red, which in turn has a much higher influence than the blue. In terms of luminance, the green signal has almost six times the effect of the blue signal and twice the effect of the red.

Color information is then added to the luminance signal so that color receivers can obtain the additional color information for hue and saturation (often known as chroma signal). These signals are transmitted on a subcarrier which is suppressed at the transmitter and recreated at the receiver. The subcarrier is modulated with the three color-difference signals (but only two color-difference signals need to be transmitted).

The three color-difference signals are:

Red−luminance or $(R-Y)$
Green−luminance or $(G-Y)$
Blue−luminance or $(B-Y)$

There is no need to send all of the signals because if the $R-Y$ and $B-Y$ signals are sent then the $G-Y$ difference signal can be easily recovered. This is because the Y signal is sent in the luminance signal. Thus the red and blue signal are recovered by adding the luminance signal to red − luminance and blue − luminance signals, then using the correct RGB weighting given to give:

$$G = \frac{Y - 0.3R - 0.11B}{0.59}$$

The reason why the $R-Y$ and $B-Y$ signals are sent is that as the Y signal contains a great deal of green information, thus the $G-Y$ signal is likely to be smaller than the $R-Y$ and the $B-Y$ signals. The small $G-Y$ signal is then more likely to be affected by noise in the transmission system.

The receiver must be told when the signal is at its start, thus the transmitter sends a strong sync pulse to identify the start of the image. Figure 6.2 illustrates the transmitter with the baseband video signals, often known as composite video signals.

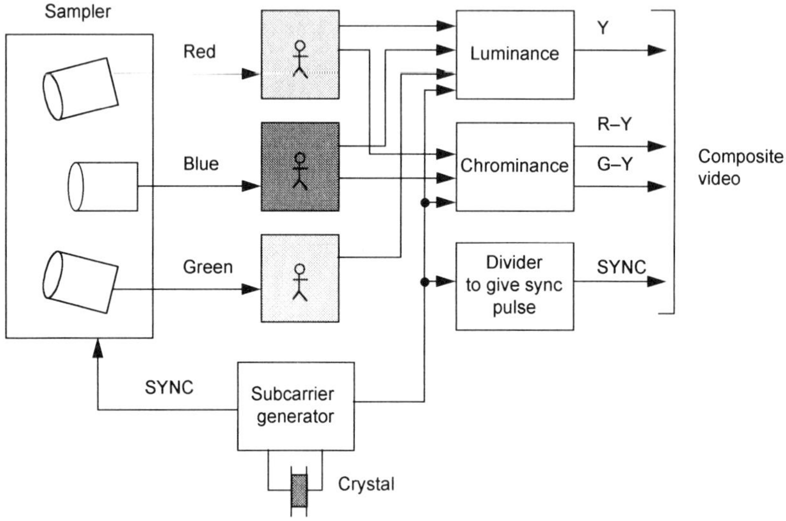

Figure 6.2 Conversion of an image into composite video components

6.3 Quadrature modulation

PAL and NSTC use an amplitude-modulated subcarrier wave. This subcarrier wave (4.433 6187 MHz for PAL and 3.579 545 MHz for NSTC) is amplitude modulated using weighted $R-Y$ and by a weighted $B-Y$ signals shifted by 90° (quadrate modulation), as illustrated in Figure 6.3.

In PAL, these difference components are called U and V, whereas in NSTC they are I and Q. SECAM uses frequency modulation, where the color-difference terms are D_R and D_B. These terms are defined by:

$$
\begin{aligned}
U &= 0.62R-0.52G-0.10B \text{ (PAL)}\\
V &= -0.15R-0.29G+0.44B \text{ (PAL)}\\
I &= 0.60R-0.28G-0.32B \text{ (NSTC)}\\
Q &= 0.21R-0.52G+0.31B \text{ (NSTC)}\\
D_R &= -1.33R+1.11G+0.22B \text{ (SECAM)}\\
D_B &= -0.45R-0.88G+1.33B \text{ (SECAM)}
\end{aligned}
$$

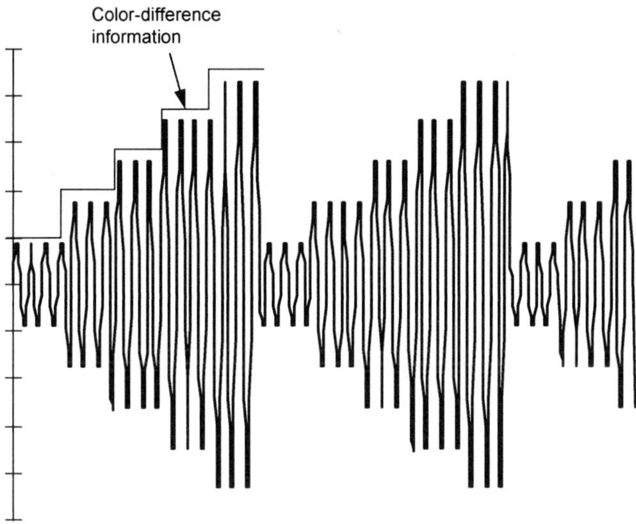

Figure 6.3 A PAL or NSTC waveform.

It can be seen that if *R*, *G* and *B* are the same value, the color-difference terms will be zero (no color difference). PAL, NSTC and SECAM use:

$$Y=0.30R+0.59G+0.11B$$

Then, in PAL, these weighting values are:

$$V=0.877(R-Y)$$
$$U=0.493(B-Y)$$

In NSTC these are:

$$I=0.74(R-Y)-0.27(B-Y)$$
$$Q=0.48(R-Y)+0.41(B-Y)$$

The fundamental feature of this mode of signal addition is that by special detection at the receiver, it becomes possible to isolate the V/U or I/Q signals again and thus extract the original $R-Y$ and $B-Y$ modulation signals.

Figure 6.4 shows examples of the color-difference signals in relation to various colors. In PAL, it can be seen that yellow has a strong Y component, a positive V component $(R-Y)$ and a negative U component $(B-Y)$. Whereas blue has a low Y component, a strong negative U component and a positive V component. Table 6.1 outlines the components for each color.

Figure 6.5 shows a phasor diagram with the V component at a phase difference of 90° from the U component. When the red signal is greater than the luminance, the V signal will be positive; and if the blue signal is greater than the luminance, the U signal will be positive. The signal vector will thus be in the first quadrant. The colors in this quadrant are likely to be lacking in green, but strong in reds and blues. Thus, the first quadrant contains purple-type colors. The example in Figure 6.5 shows how a *U* component of +0.5 and a *V* component of +0.5 give a color of magenta.

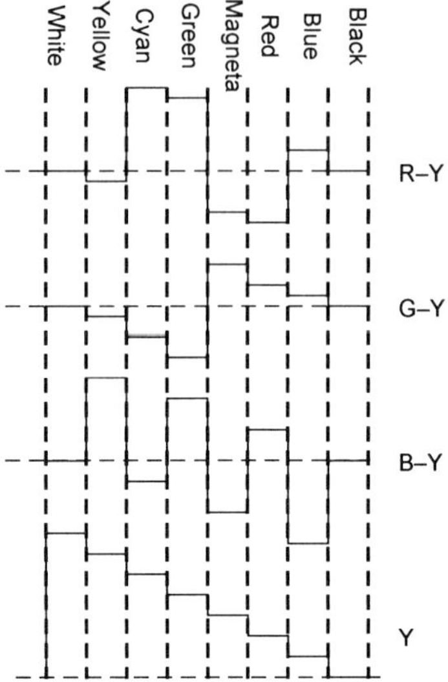

Figure 6.4 Color-difference signals for various colors.

The colors in the second quadrant have a positive *V* component and a negative *U* component. For this to occur the red component must be greater than the luminance (for *V* to be positive) and the luminance must be greater than the blue component. This gives bright colors with a strong red component. An example of the color yellow is given in Figure 6.5.

Table 6.1 Color examples with luminance and chrominance values.

Color	*Y*	*B – Y*	*R – Y*	*U*	*V*	*Amp.*	*Angle*
Black	0	0	0	0	0	0	–
Blue	0.11	+0.89	–0.11	+0.439	–0.096	0.44	347
Red	0.3	-0.3	+0.7	–0.148	–0.614	0.63	103
Magenta	0.41	+0.59	+0.59	+0.291	+0.517	0.59	61
Green	0.59	–0.59	–0.59	–0.291	–0.517	0.59	241
Cyan	0.7	+0.3	–0.7	+0.148	–0.614	0.63	283
Yellow	0.89	–0.89	+0.11	–0.449	+0.096	0.44	167
White	1.0	0	0	0	0	0	–

The third quadrant has a negative value for the *U* and *V* component. This means that the luminance is greater than both the blue and red components. Thus the colors have a strong green factor.

In the fourth quadrant the U component is positive (the blue component is greater than the luminance) and the V component is negative (the luminance is greater than the red component). The colors in this quadrant will be strong in blues with a low luminance. An example of cyan is shown in Figure 6.5.

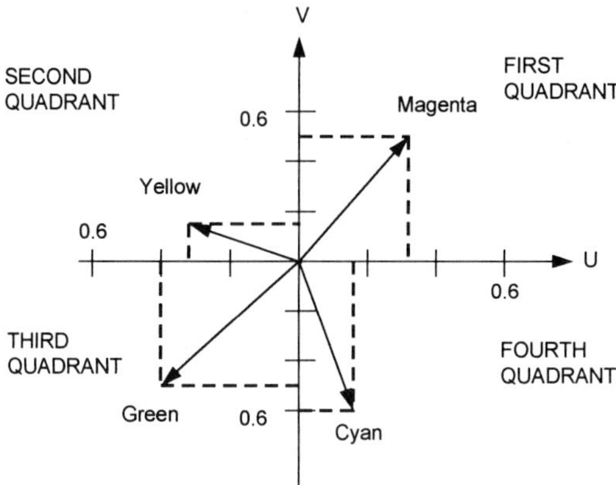

Figure 6.5 Phasors for four colors.

6.4 Baseband video signals

Baseband video signals, or composite video, are electrical signals which are used in TV systems to send video information. In TV, the video signal is traced with interlacing lines, where initially, the top left-hand corner pixel is transmitted, followed by each pixel in turn on a single line. After the last pixel on the first line, the video traces back to the start of the next displayable line. This continues until it reaches the bottom of the screen. After this the video trace returns to the top left-hand pixel and starts again, as illustrated in Figure 6.6.

With PAL, the screen refresh rate is based on the 50 Hz mains frequency, while in NTSC it is based on the 60 Hz mains frequency. A 50 Hz refresh rate causes the screen to be updated 50 times every second. On an interlaced system, each frame (or picture) is sent by sending the odd lines of the frame on the first update and then the even lines on the next screen update, and so on. Interlaced scanning is illustrated in Figure 6.7. For a 50 Hz system, the frame rate (or picture rate) is thus 25 Hz which means that one picture is drawn every 1/25 of a second (40 ms).

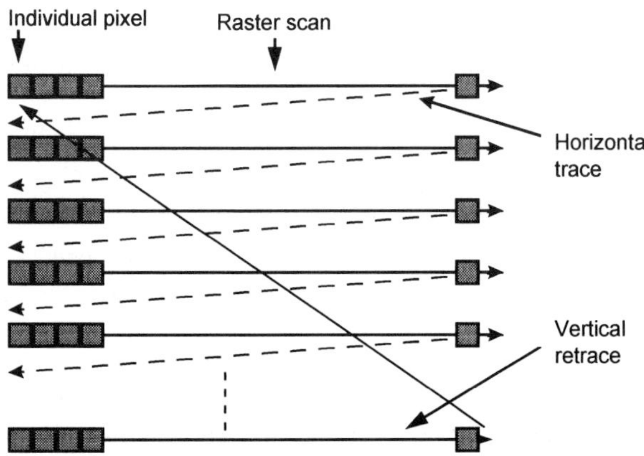

Figure 6.6 Video screen showing raster lines and retrace.

A raster line is the smallest subdivision of this horizontal line. PAL systems have 625 video raster lines at a rate of 25 frames per second and NSTC uses 525 lines at a rate of 30 frames per second.

Thus, if the screen refresh rate is 50 Hz, the screen is updated once every 20 ms. Each update contains half a picture so one complete picture is sent every 40 ms. Thus 625 lines are sent every 40 ms, the time to transmit one raster line will thus be:

$$t_{\text{raster}} = \frac{40}{625} ms = 64 \mu s$$

With a black and white signal the voltage amplitude of the waveform at any instant gives the brightness of each part of the displayed picture. A negative voltage sync pulse of 4.7 μs indicates the start of the 625 lines. A blanking level then indicates the start of each line, as illustrated in Figure 6.8. The largest voltage defines white while a zero voltage gives black.

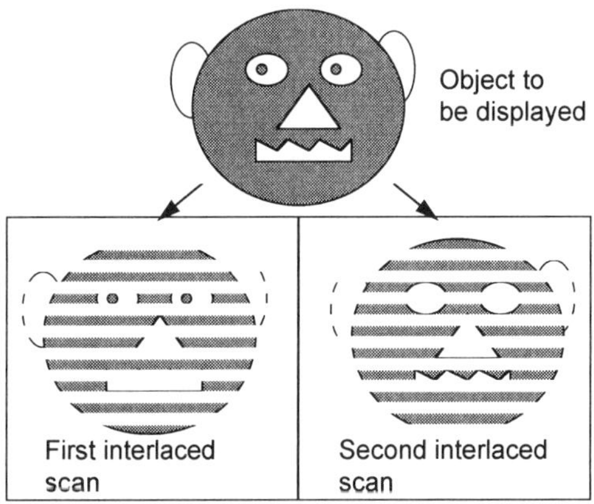

Figure 6.7 Interlaced scanning.

Composite color video signals use gaps in the black and white signal to transmit bursts of color information. A composite video signal consists of a luminance (brightness) component and two chromatic (color) components. These chromatic components are transmitted simultaneously as the amplitude modulation sidebands of a pair of suppressed subcarriers which are identical in frequency but are in phase quadrature.

Figure 6.9 shows the frequency band of a PAL composite video signal. The black and white (luminance) signal takes up the band from DC to 4.2 MHz, and a subcarrier is added in the higher-frequency portion of the band, at 4.43361875 MHz (3.58 MHz for NTSC); the modulated subcarrier can be thought of as superimposing itself on the luminance signal, as illustrated in Figure 6.10. The bandwidth for a PAL system is approximately 5.5 MHz.

6.4.1 Differences in color modulation between NSTC, PAL and SECAM

The start of the horizontal line is preceded by a sync pulse and then a color burst. This provides a reference for the phase information in the color signal. Figure 6.11 shows the horizontal blanking interval.

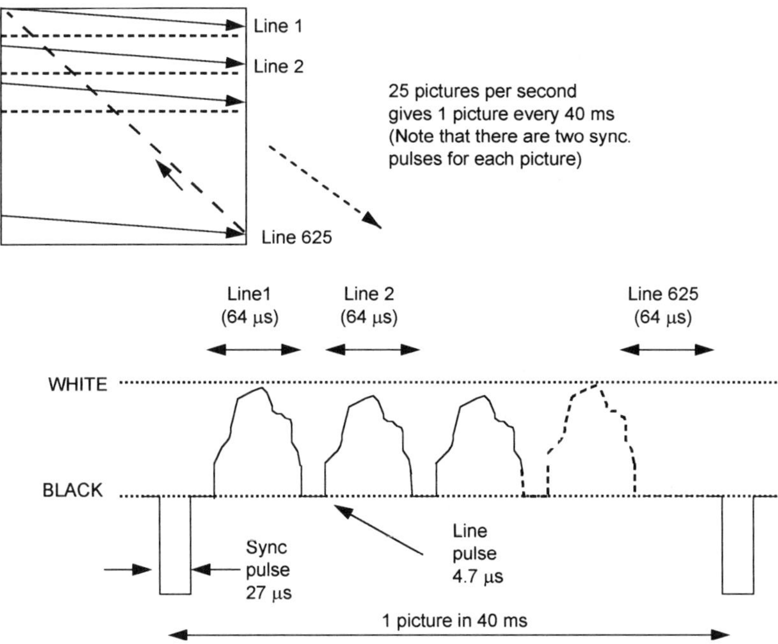

Figure 6.8 TV line waveform.

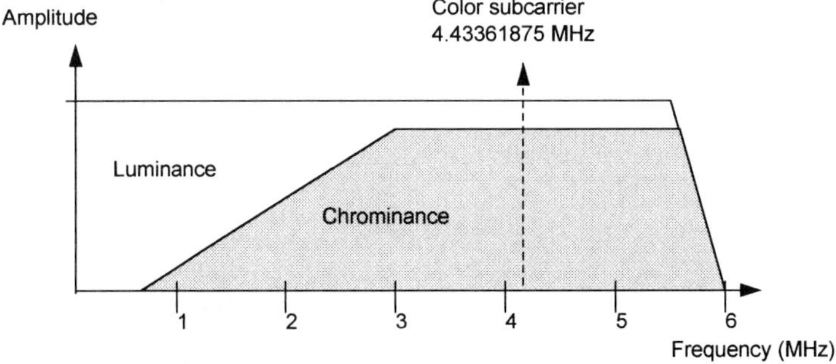

Figure 6.9 Composite video frequency spectrum.

In NSTC, the chrominance subcarrier is suppressed-carrier amplitude modulated on a 3.579 545 MHz. It is modulated by the *I* (for In-phase) and *Q* (for Quadrature) components, with the *I* component modulating the subcarrier at 0° and the *Q* component modulating at 90°. The reference burst lasts 2.67 μs and is at an angle of 57° with respect to the *I* carrier. In NSTC, the *Y* bandwidth is 4.5 MHz while the *I* bandwidth is only 1.5 MHz and the *Q* bandwidth is 0.5 MHz (although many TV receivers allow a bandwidth of 0.5 MHz for *I* and *Q* and give perfectly acceptable results).

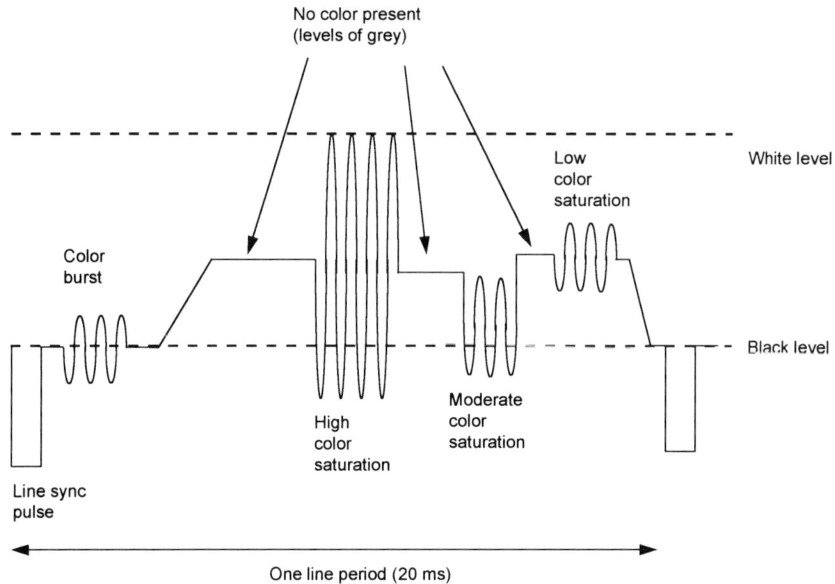

Figure 6.10 Composite video signal.

PAL is similar but has a carrier frequency of 4.433 618 7 MHz. The modulating components are referred to as *U* (0°) and *V* (90°). The V component is alternated 180° on a line-by-line basis. The reference burst last 2.25 μs and also alternates on a line-by-line basis between an angle of +135° and –135° relative to the *U* component.

SECAM uses a frequency-modulated (FM) color subcarrier for the chromatic signals, transmitting one of the color difference signals every other line, and the other color difference signal on the remaining lines.

Table 6.2 outlines the main parameters for the three technologies.

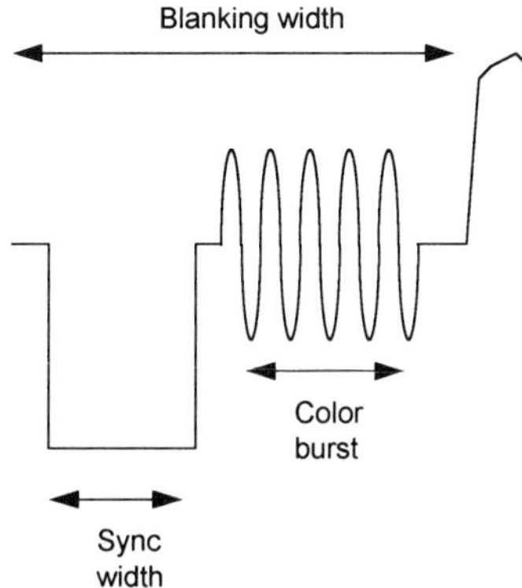

Figure 6.11 Color burst.

Table 6.2 The main differences between NSTC, PAL and SECAM.

	NSTC	PAL	SECAM
Lines/frame	525	625	625
Frame/second	30	25	25
Interlace ratio	2:1	2:1	2:1
Color subcarrier	3.579 545 MHz	4.433 619 MHz	Two FM-modulated carriers at 4.40625 MHz and 4.25 MHz
Line period	63.55 μs	64.0 μs	64.0 μs
Horizontal sync pulse width	4.7 μs	4.7 μs	4.7 μs
Color burst width	2.67 μs	2.25 μs	
Vertical sync pulse width	27.1 μs	27.3 μs	27.3 μs

6.5 Extra notes

1. In Europe, the standard adopted for whiteness is the color temperature is 6500K.
2. Multicasting is the technique that allows several program streams in a single channel. It uses digital video compression (DVC).

7 Digital TV

7.1 Introduction

Composite video signals must be sampled at twice the highest frequency of the signal. To standardize this sampling, the ITU CCIR-601 (often known as ITU-R) has been devised. It defines three signal components: Y (for luminance), C_r (for $R-Y$) and C_b (for $B-Y$).

The biggest problem with sampling is that SECAM and PAL use 625 lines and NTSC use 525 lines. The ITU-R 601 standard defines that SECAM and PAL are sampled 858 times for each line and NTSC is sampled 864 times for each line. This produces the same scanning frequency in both cases:

$$\text{Scanning frequency (PAL, SECAM)} = 864 \times 625 \times 25 = 13.5\,\text{MHz}$$

$$\text{Scanning frequency (NTSC)} = 858 \times 525 \times 30 = 13.5\,\text{MHz}$$

With this technique, the luminance (the black and white level) is digitized at a rate of 13.5 MHz and color is sampled at an equivalent rate of 6.75 MHz. Each luminance and color sample is coded as 8 bits, thus the digitized rate is:

$$\text{Digitized video signal rate} = 13.5 \times 8 + 6.75 \times 8 = 162\,\text{Mbps}$$

Thus the sample rate for all three systems will be:

$$\text{Sample rate} = \frac{1}{13.5 \times 10^{6}} = 74\text{ ns}$$

Normally the Y is sampled at 13.5 MHz and the chromatic components (such as U/V or C_b/C_r) are sampled at a quarter of this rate (that is, 3.375 MHz). The bits are then interleaved to give 12-bit YUV bundles, as illustrated in Figure 7.1. This shows that there are 12 bits transmitted every 74 ns, thus the bit rate is:

$$\text{Bit rate} = \frac{12}{74 \times 10^{-9}} = 162\,\text{Mbps}$$

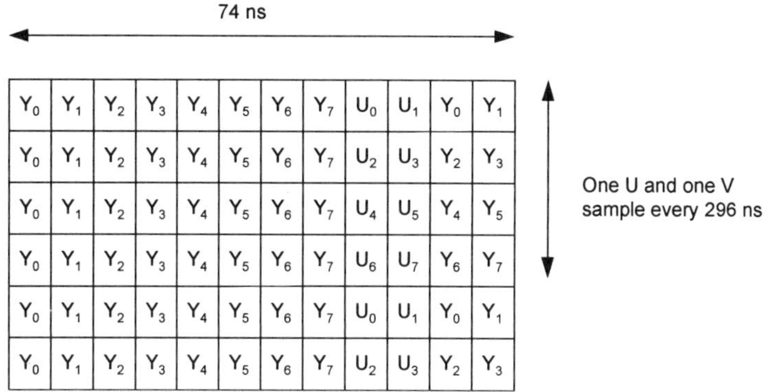

Figure 7.1 Interleaved luminance and chrominance data.

7.1.1 Subsampling

Subsampling reduces the digitized bit rate by sampling the luminance and chrominance at different rates. The standard format uses integer values as a ratio, such as:

Y sampling frequency : C_{d1} sampling frequency : C_{d2} sampling frequency

where C_d is the sampling frequency for the chominance. For example, standard studio-quality TV uses a 4:2:2 sampling ratio for the Y:C_r:C_b ratio. This means that the number of samples for the color difference is half of the luminance. For example, if the image is:

R (720×486), G (720×486) and B (720×486) (NSTC)
R (720×576), G (720×576) and B (720×576) (PAL)

then using 4:2:2 gives the resultant YC_rC_b form:

Y (720×486), C_b (360×486) and C_r (360×486) (NSTC)
Y (720×576), C_b (360×576) and C_r (360×576) (PAL)

Thus, in NSTC, the number of terms has been reduced from 1 049 760 to 699 840 (a saving of 33%).

The H.261 standard, as used in video conferencing, uses a 4:1:1 ratio. Thus the color difference components are sampled at one-quarter of the luminance rate. For example, if the image is:

R (352×288), G (352×288) and B (352×288)

then using 4:1:1 gives the resultant YC_rC_b form:

Y (352×288), C_b (176×144) and C_r (176×144)

Thus, in NSTC, the number of terms has been reduced from 304 128 to 152 064 (a saving a 50%).

7.1.2 CCIR-601 active lines

CCIR defines a constant sampling rate of 13.5 MHz. Unfortunately, a delay is required to go from the end of one line to the start of the next (called horizontal retrace). A delay is also required when the scanning reaches the end of a frame and returns to the top of the frame (called vertical retrace). The time for active pixels has been defined as 720 per line, or 53 ns (in CCIR-601 the time of scan for a line is 64 μs). This is the same for NSTC (which samples at 858 per line) and PAL (which samples at 864 per line).

7.1.3 CCIR-601 quantization

Each of the samples for luminance and chrominance has 8 bits, which gives 256 levels. Only 220 values are used, the black level is coded as 16 and the peak white levels are coded as 235. Values from 0 to 16 and 235 to 255 are reserved for special code words. The color-difference signal can take on 225 different values; the zero corresponds to coded value 128 and the peak saturation to values 16 and 240. The values from 0 to 16 and 240 to 255 are reserved for special code words.

7.2 100 Hz pictures

Digital transmission of video signals not only improves the transmission of the video signals, it can also be used to increase picture quality. Two typical problems with PAL, NSTC and SECAM systems are:

- Interline flicker – the flickering of sharp horizontal edges around the edges of objects.
- Large-area flicker – most noticeable on large screens.

Research indicates these problems disappear when the frame rate is 90 Hz eliminates all flickering. Thus as the video data is digitized it can be stored in a memory and recalled at any rate. A possible technique, for PAL/625, is to store the incoming video data in memory and then to read it out at a rate of 100 Hz. This will then be displayed to the speed at double-speed lines and field scan rates.

7.3 Analogue component hybrid systems

The IBA in the UK developed a digitally based system that permits time division multiplexing of YUV analogue signals into a form suitable for DBS/FM transmission within the 27 MHz channels (19 MHz carrier separation). In the early 1980s, it was adopted for DBS in Europe in the early 1980s. It is termed Multiplexed Analogue Components (MAC), where the separate YUV signals are digitally time compressed and then time division multiplexed to occupy the active line periods of the PAL (625 line) standard. Digital sound and data signals are then inserted into the video channel during the line sync periods, which are then frequency modulated onto a single carrier. Thus the video signal is transmitted in a digital form, but still contains digital information. At the receiver the signal is demultiplexed and thus restores the YUV signals, which are used to display as PAL signals.

A number of other MAC systems have been developed for the multiplexing of signal within a studio or for Outside Broadcast applications. These include:

1. B-MAC. Developed by Scientific-Atlanta for satellite cable distribution links or DBS and can carry up to six digital audio channels as baseband data symbols. It is designed to be simple in multiplexing. and has a fixed format, where the data packets are not addressable).
2. C-MAC/packet which transmits up to eight high-quality digital audio channels. These are directly modulated on to the carrier during the line blanking period and use binary phase shift keying. The resultant data bit rate is 20.25 Mbps.
3. D-MAC/packet has the same data rate as C-MAC, but rather that using the blanking period it is modulated on the baseband signal with a duobinary form.
4. D2-MAC/packet which is similar to D-MAC, but the data rate 10.125 Mbps. It can thus be used to transmit four high-quality digital audio channels.

The 625 D-MAC/packet system has been adopted as the UK DBS standard.

7.4 Compressed TV

TV signals and motion video have a massive amount of redundant information. This is mainly because each image has redundant information and because there are very few changes from one image to the next. A typical standard is the MPEG standard which allows compression of about 130:1.

With MPEG-2 this transmission rate can be reduced to 4 Mbps for PAL and SECAM and 3 Mbps for NTSC, giving a compression ratio of 40:1 to high quality TV. MPEG-1 typically compresses TV signals to 1.2 Mbps, giving a compression ratio of 130:1. Unfortunately, the quality is reduced to near VCR-type quality. Table 7.1 outlines these parameters.

The base bit rate for a standard Ethernet network is 10 Mbps. This allows compressed video to be transmitted over the network when there is no other traffic on the network. The 4 Mbps rate will load the network by approximately 50%. Standardized and compression techniques will be discussed in the next chapter.

Table 7.1 Motion video compression.

Type	*Bit rate*	*Compression*	*Comment*
Uncompressed TV	162 Mbps	1:1	
MPEG-1	4 Mbps	40:1	VCR quality
MPEG-2	1.2 Mbps	130:1	PAL, SECAM TV quality

7.5 HDTV quality

HDTV (high-definition TV) has been supported by many companies for many years. Standards such as the European High-Definition MAC, a mainly analogue-based system, have been promoted then abandoned.

The main parameters in a TV system are the frame rate and the picture resolution; the main improvements are:

- HDTV-quality gives a higher picture resolution with a higher frame rate (1920×1080 at 60 frames per second). This gives excellent images of 1920 pixels per lines, 1080 lines per frame and 60 frames per second.
- HDTV-quality with a high resolution and a conventional frame rate (1920×1080 at 24 frames per second). The advantage with this system is that is gives much high resolution with a frame rate which is similar to the frame rate of current system (25 frames per second for PAL and 30 frames per second for NTSC).
- Improved resolution/conventional frame rate (1280×730 at 30 frames per second). The advantage of this system is that it gives an intermediate response between conventional systems and the alternatives given above. This technology may allow the best intermediate migration between current TV and high-resolution technology. Its screen resolution is also similar to SVGA monitors.

Another important parameter in TV systems is the aspect ratio. Conventional TVs use an aspect ratio of 4:3 (which is defined as the width of the screen divided by its height). This aspect ratio does not really suit showing movies or sports events, and HDTV improves this to a 16:9 aspect ratio. HDTV will be covered in the next chapter.

8 Motion Video Compression

8.1 Motion video

Motion video contains massive amounts of redundant information. This is because each image has redundant information and also because there are very few changes from one image to the next.

Motion video image compression relies on two facts:

- Images have a great deal of redundancy (repeated images, repetitive, superfluous, duplicated, exceeding what is necessary, and so on).
- The human eye and brain have limitations on what they can perceive.

Chapter 5 discussed how JPEG encodes still images. Motion video is basically a series of still images. The basis of motion JEPG (MPEG) is to treat the video information as a series of compressed images and to allow for compression of around 130:1.

8.2 MPEG-1 overview

The Motion Picture Experts Group (MPEG) developed an international open standard for the compression of high-quality audio and video information. At the time, CD-ROM single-speed technology allowed a maximum bit rate of 1.2 Mbps and it is this rate that the standard was built around. These days, ×20 and ×24 CD-ROM bit rates are common.

MPEG's main aim was to provide good quality video and audio using hardware processors (and in some cases, on workstations with sufficient computing power, to perform the tasks using software). Figure 8.1 shows the main processing steps of encoding:

- Image conversion – normally involves converting images from RGB into YUV (or YC_rC_b) terms with optional color sub-sampling.
- Conversion into slices and macroblocks – a key part of MPEG-1's compression is the detection of movement within a frame. To detect motion a frame is subdivided into slices then each slice is divided into a number of macroblocks. Only the luminance component is then used for the motion calculations. In the subblock, luminance (Y) values use a 16×16 pixel macroblock, whereas the two chrominance components have 8×8 pixel macroblocks.
- Motion estimation – MPEG-1 uses a motion estimation algorithm to search for multiple blocks of pixels within a given search area and tries to track objects which move across the image.

- DCT conversion – as with JPEG, MPEG-1 uses the DCT method. This transform is used because it exploits the physiology of the human eye. It converts a block of pixels from the spatial domain into the frequency domain. This allows the higher-frequency terms to be reduced as the human eye is less sensitive to high-frequency changes.
- Encoding – the final stages are run-length encoding and fixed Huffman coding to produce a variable-length code.

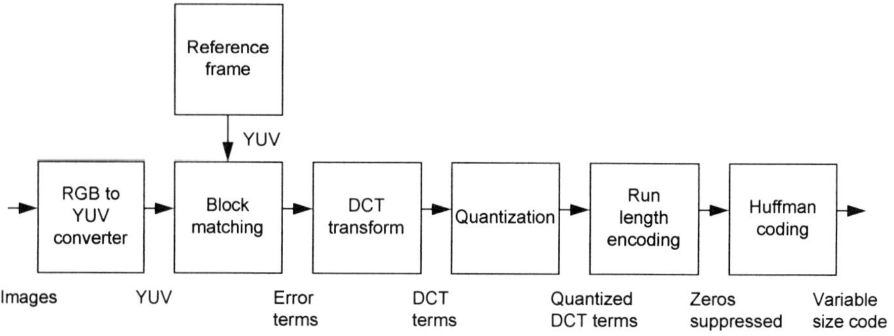

Figure 8.1 MPEG encoding with block matching.

8.3 MPEG-1 video compression

MPEG-1 typically uses the CIF format for its input, which has the following parameters:

- For NTSC, 352×240 pixels for luminance and 176×120 pixels for *U* and *V* color-difference components (that is, 4:1:1 subsampling).
- For PAL/SECAM, 352×288 pixels for luminance and 176×144 pixels for *U* and *V* color difference components (i.e. 4:1:1 subsampling).

This gives a picture quality which is similar to VCR technology. MPEG-1 differs from conventional TV in that it is non-interlaced (known as progressive scanning), but the frame rate is the same as conventional TV, i.e. 25 fps (for PAL and SECAM) and 30 fps (for NSTC). Note that MPEG-1 can also use larger pixel frames, such as CCIR-601 740×480, but the CIF format is the most frequency used.

Taking into account the interlacing effect, the CIF format is actually derived from the CCIR-601 format. The CCIR-601 digital television standard defines a picture size of 720×243 (or 240) by 60 fields per second. Note that a frame actually comprises two fields, where the odd and even information is interlaced to create the full picture. When the interlaced luminance information occupies the full 720×480 frame, the chrominance components are reduced by 4:2:2 subsampling to give 360×243 (or 240) by 60 fields per second.

MPEG-1 also reduces the chrominance components by reducing the pixel data by half in the vertical, horizontal and time directions. It also reduces the image size so that the number of pixels for it is divisible by 8 or 16. This is because the motion analysis and DCT conversion operate on 16 × 16 or 8 × 8 pixel blocks. As a result, the number of lines changes for an MPEG-1 encoded move between the NSTC standard and PAL and SECAM standards. The final figure for PAL and SECAM is 288 at 50 fps; for NSTC it is 240 at 60 fps. These require the same number of bits to encode the streams.

The MPEG encoded bitstream comprises three components: compressed video, compressed audio and system-level information. To provide easier synchronization and lip synching the audio and video streams are time stamped using a 90 kHz reference clock.

8.4 MPEG-1 compression process

8.4.1 Color space conversion

The first stage of MPEG encoding is to convert a video image into the correct color space format. In most cases, the incoming data is in 24-bit RGB color format and is converted in 4:2:2 YC_rC_b (or YUV) form. Some information will obviously be lost but it results in some compression.

8.4.2 Slices and macroblocks

MPEG-1 compression tries to detect movement within a frame. This is done by subdividing a frame into slices and then subdividing each slice into a number of macroblocks. For example, a PAL format which has:

352 × 288 pixel frame (101 376 pixels)

can, when divided into 16 × 16 blocks, give a whole number of 396 macroblocks. Dividing 288 by 16 gives a whole number of 18 slices. Dividing 352 gives 22. Thus the image is split into 22 macroblocks in the *x*-direction and 18 in the *y*-direction, as illustrated in Figure 8.2.

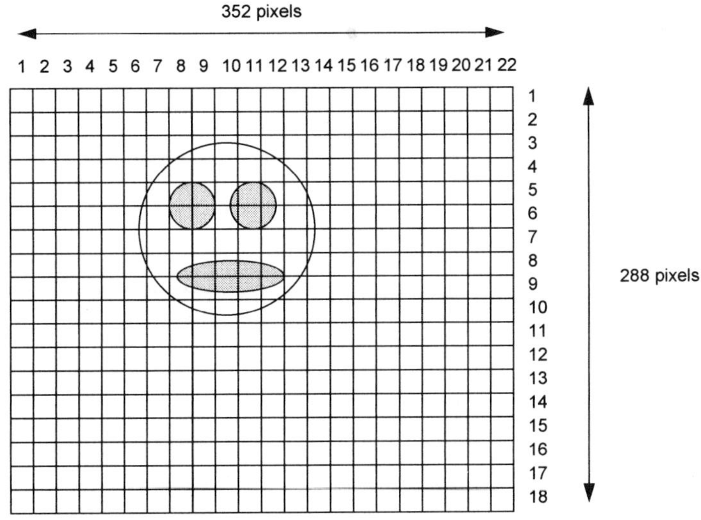

Figure 8.2 Segmentation of an image into subblocks.

Luminance (Y) values use a 16×16 pixel macroblock, whereas the two chrominance components have 8×8 pixel macroblocks. Note that only the luminance component is used for the motion calculations.

8.4.3 Motion estimation

MPEG-1 uses a motion estimation algorithm to search for multiple blocks of pixels within a given search area and tries to track objects which move across the image. Each luminance (Y) 16×16 macroblock is compared with other macroblocks within either a previous or future frame to find a close match. When a close match is found, a vector is used to describe where this block should be located, as well as any difference information from the compared block. As there tend to be very few changes from one frame to the next, it is far more efficient than using thc original data.

Figure 8.3 shows two consecutive images of 2D luminance made up into 16×5 megablocks. Each of these blocks has 16×16 pixels. It can be seen that, in this example, there are very few differences between the two blocks. If the previous image is transmitted in its entirety then the current image can be transmitted with reference to the previous image. For example, the megablocks for $(0, 1)$, $(0, 2)$ and $(0, 3)$ in the current block are the same as in the previous blocks. Thus they can be coded simply with a reference to the previous image. The $(0, 4)$ megablock is different to the previous image, but the $(0, 4)$ block is identical to the $(0, 3)$ block of the previous image, thus a reference to this block is made. This can continue with most of the block in the image being identical to the previous image. The only other differences in the current image are at $(4, 0)$ and $(4, 1)$; these blocks can be stored in their entirety or specified with their differences to a previous similar block.

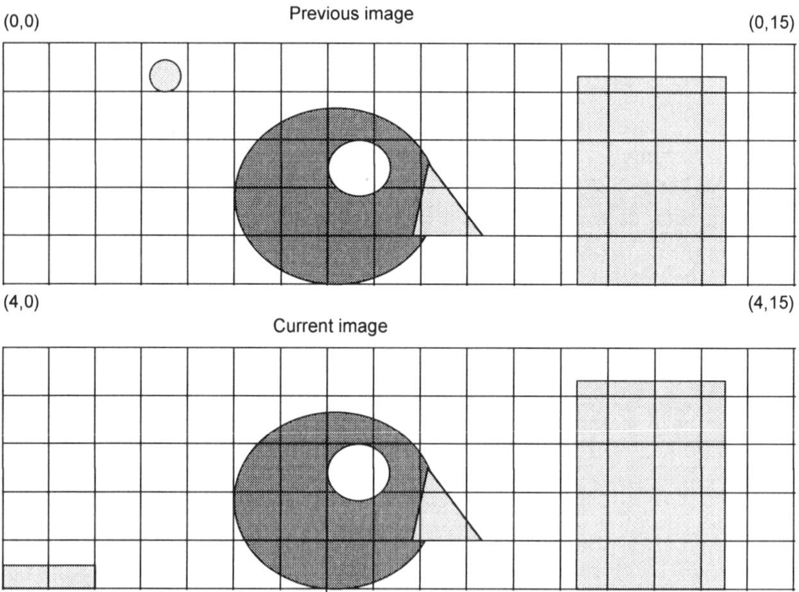

Figure 8.3 Two consecutive images.

Each macroblock is compared mathematically with another block in a previous or future frame. The offset to find another block could be over a macroblock boundary or even a pixel boundary. The comparison then repeats until a match is found or the specified search area within the frame has been exhausted. If no match is available, the search process can be repeated using a different frame or the macroblock can be stored as a complete set of data. As previously stated, if a match is found, the vector information specifying where the matching macroblock is located is specified along with any difference information.

As the technique involves very many searches over a wide search area and there are many frames to be encoded, the encoder must normally be a high-powered workstation. This has several implications:

- An asymmetrical compression process is adopted, where a relatively large amount of computing power is required for the encoder and much less for the decoder. Normally the encoding is also done in non-real time whereas the decoder reads the data in real-time. As processing power and memory capacity increase, more computers will be able to compress video information in real-time.
- Encoders influence the quality of the decoded image dramatically. Encoding shortcuts, such as limited search areas and macroblock matching, can generate poor picture quality, irrespective of the quality of the encoder.
- The decoder normally requires a large amount of electronic memory to store past and future frames, which may be needed for motion estimation.

With the motion estimation completed, the raw data describing the frame can now be converted using the DCT algorithm ready for Huffman coding.

8.4.4 I, P and B frames

MPEG video compression uses three main types of frames: I-frame, P-frame and B-frame.

Intra frame (I-frame)

An intra frame, or I-frame, is a complete image and does not require any extra information to be added to it to make it complete. Thus no motion estimation processing has been performed on the I-frame. Mainly used to provide a known starting point, it is usually the first frame to be sent.

Predictive frame (P-frame)

The predictive frame, P-frame, uses the preceding I-frame as its reference and has motion estimation processing. Each macroblock in this frame is supplied either as a vector and difference with reference to the I-frame, or if no match was found, as a completely encoded macroblock (called an intracoded macroblock). The decoder must thus retain all I-frame information to allow the P-frame to be decoded.

Bidirectional frame (B-frame)

The bidirectional frame, B-frame, is similar to the P-frame except that its reference frames are to the nearest preceding I- or P-frame and the next future I- or P-frame. When compressing the data, the motion estimation works on the future frame first, followed by the past frame. If this does not give a good match, an average of the two frames is used. If all else fails, the macroblock can be intracoded.

Needless to say, decoding B-frames requires that many I- and P-frames are retained in memory.

MPEG-1 frame sequence

MPEG-1 allows frames to be ordered in any sequence. Unfortunately a large amount of reordering requires many frame buffers that must be stored until all dependencies are cleared.

The MPEG-1 format allows random access to a video sequence, thus the file must contain regular I-frames. It also allows enhanced modes such as fast forward, which means that an I-frame is required every 0.4 seconds, or 12 frames between each I-frame (at 30 fps).

At 30 fps, a typical sequence is a starting I-frame, followed by two B-frames, a P-frame, followed by two B-frames, and so on. This is known as a group of picture (GOP).

I⇒B⇒B⇒P⇒B⇒B⇒I⇒B⇒B⇒P⇒B⇒B⇒I⇒B⇒B⇒P⇒...

When decoding, the decoder must store the I-frame, the next two B-frames are also stored until the B-frame arrives. The next two B-frames have to be stored locally until the P-frame arrives. The P-frame can be decoded using the stored I-frame and the two B-frames can be decoded using the I- and P-frames. One solution of this is to reorder the frames so that the I- and P-frames are sent together followed by the two intermediate B-frames. Another more radical solution is not to send B-frames at all, simply to use I- and P-frames.

On computers with limited memory and limited processing power, the B-frames are difficult because:

- They increase the encoding computational load and memory storage. The inclusion of the previous and future I- and P-frames as well as the arithmetic average greatly increases the processing needed. The increased frame buffers to store frames allow the encode and decode processes to proceed. This argument is again less valid with the advent of large and high-density memories.
- They do not provide a direct reference in the same way that an I- or P-frame does.

The advantage of B-frames is that they lead to an improved signal-to-noise because of the averaging out of macroblocks between I- and P-frames. This averaging effectively reduces high-frequency random noise. It is particularly useful in lower bit rate applications, but is of less benefit with higher rates, which normally have improved signal-to-noise ratios.

8.4.5 DCT conversion

As with JPEG, MPEG-1 uses the DCT. It transforms macroblocks of luminance (16×16) and chrominance (8×8) into the frequency domain. This allows the higher-frequency terms to be reduced as the human eye is less sensitive to high-frequency changes. This type of coding is the same as used in JPEG still image conversion, that was described in the previous chapter.

Frames are broken up into slices 16 pixels high, and each slice is broken up into a vector of macroblocks having 16×16 pixels. Each macroblock contains luminance and chrominance components for each of four 8×8 pixel blocks. Color decimation can be applied to a macroblock, which yields four 8×8 blocks for luminance and two 8×8 blocks (C_b and C_r) of chrominance, using one chrominance value for each of the four luminance values. This is called the 4:2:0 format; two other formats are available (4:2:2 and 4:4:4, respectively known as two luminance per chrominance and one to one), which require data rates.

For each macroblock, a spacial offset difference between a macroblock in the predicted frame and the reference frame(s) is given if one exists (a motion vector), along with a luminance value and/or chrominance difference values (an error term) if needed. Macroblocks with no differences can be skipped except in intra frames. Blocks with differences are internally compressed, using a combination of a discrete cosine transform (DCT) algorithm on pixel blocks (or error blocks) and variable quantization on the resulting frequency coefficient (rounding off values to one of a limited set of values).

The DCT algorithm accepts signed, 9-bit pixel values and produces signed 12-bit coefficient. The DCT is applied to one block at a time, and works much as it does for JPEG, converting each 8×8 block into an 8×8 matrix of frequency coefficients. The variable quantization process divides each coefficient by a corresponding factor in a matching 8×8 matrix and rounds to an integer.

8.4.6 Quantization

As with JPEG the converted data is divided, or quantized, to remove higher-frequency components and to make more of the values zero. This results in numerous zero coefficients, particularly for high-frequency terms at the high end of the matrix. Accordingly, amplitudes are recorded in run-length form following a diagonal scan pattern from low frequency to high frequency.

8.4.7 Encoding

After the DCT and quantization state, the resultant data is then compressed using Huffman coding with a set of fixed tables. The Huffman code not only specifies the number of zeros, but also the value that ended the run of zeros. This is extremely efficient in compressing the zigzag DCT encoding method.

8.5 MPEG-1 decoder

The resultant encoded bitstream contains both video and audio data. These two elements are identified using system-level coding, which specifies a multiplex data format that allows multiplexing of multiple simultaneous audio and video streams as well as privately defined data streams. This coding includes the following:

- Synchronization data for decoded audio and video frames. Each frame contains a time stamp of frames so that a decoder can synchronize the decoding and playback of audio with the correct video sequence to achieve lip synchronization. The time-stamping gives the decoder a great flexibility in the playback. It even allows variable data rates, where frames can be dropped when they cannot be processed in time, and there is no loss of synchronization. The synchronization is achieved with a 90 kHz reference clock.
- Random frame access within the stream with absolute time identification. This is important when decoding in that the time reference can be independent of the environment.
- Data buffer management to prevent overflow and underflow errors. Frames are not necessarily stored in the consecutive time sequence. Buffers must be set-up to hold data temporarily for future decoding.

8.6 MPEG-1 audio compression

This will be covered in the Chapter 11.

8.7 MPEG-2

The original MPEG-1 specification proved so successful that, as soon as it was published, the MPEG committee started work on three derivatives called MPEG-2, MPEG-3 and MPEG-4. MPEG-2 has since been published. MPEG-3 was incorporated into MPEG-2 and work continucs on the MPEG-4 standard.

The main drawback with the MPEG-1 standard are:

- It did not directly support broadcast television pictures as in the CCIR-601 specification. In particular, it did not support the interlaced mode of operation, although it could support the larger picture size of 720×480 at 30 fps.
- It was designed for a 1.5 Mbps bitstream.

Interlacing dramatically affects the motion estimation process because components could move from one field to another, and vice versa. As a result, the MPEG-1 was poor at handling interlaced images.

The main objective of the MPEG-2 standard was to make it flexible so that it supported a number of modes, called profiles, with a wide range of options. These different profiles define algorithms that may be used. Each profile has a number of associated levels which define the parameters used.

8.7.1 MPEG-2 profiles and levels

MPEG-2 defines several profiles to provide a set of known configurations for different applications. It can be used from low-level video conferencing to high-definition television. If it were a unitary standard then each encoder and decoder would have to process the signals for the entire range of applications. This would, for example, burden a video conferencing system with the capability to handle very high definition images. The cost of doing this would make MPEG-2 unworkable for video conferencing.

Table 8.1 outlines the valid profiles and modes. The four main profiles are:

- Main – supports the main area of current development.
- Simple – same as the main profile, but the B-frames are not supported (so it is mainly used in software-based applications).
- SNR – enhanced signal-to-noise ratio.
- Spacial – enhanced main profile.
- High.

There are four main levels, these are:

- Low – the low level is similar to MPEG-1 standards and supports the CIF standard of 352×240 at 30 fps (or 352×288 at 25 fps for PAL). This equates to 3.05 Mpixels per second and a bit rate of up to 4 Mbps. The low-profile applications are aimed at the consumer market and offer quality similar to a domestic VCR.

- Main – the main level is able to support a maximum frame size of 720 × 480 at 30 fps (as defined in the CCIR-601 specification). This equates to 10.4 Mpixels per second and a bit rate of up to 15 Mbps. This level is aimed at the higher-quality consumer market.
- High 1440 – the high 1440 supports a maximum frame size of 1440 × 1152 at 30 fps. This is frame size is four times the CCIR-601 specification and equates to 47 Mpixels per second, giving a bit rate of up to 60 Mbps. This level is aimed at the high-definition TV (HDT) consumer market.
- High – the high level is able to support a maximum frame size of 1920 × 1080 at 30 fps. As with high 1440, the frame size is four times the CCIR-601 specification and gives a bit rate of up to 80 Mbps. This level is also aimed at the HDTV consumer market.

Table 8.1 MPEG-2 profiles and levels.

	Simple (SP)	*Main (MP)*	*SNR*	*Spatial*	*High*
HIGH (HL)	Illegal	1920 × 1152, 60 fps	Illegal	Illegal	1920 × 1152, 60 fps 960 × 576, 30 fps
HIGH-1440 (H-14)	Illegal	1920 × 1152, 60 fps	Illegal	1440×1152, 60 fps 720 × 576, 30 fps	1440 × 1152, 60 fps 720 × 576, 30 fps
Main (ML)	720 × 576, 30 fps	720 × 576, 30 fps	720 × 576, 30 fps	Illegal	720 × 576, 30 fps 352×288, 30 fps
Low (LL)	Illegal	352×288, 30 fps	352×288, 30 fps	Illegal	Illegal

8.8 MPEG-2 system layer

The MPEG data stream consists of two layers:

- A compression layer.
- A system layer.

The system decoder splits the data stream into video and audio, each to be processed by separate decoders. Every 700 ms (or faster) a 33-bit system clock reference (SCR) is inserted into the data. For synchronization, the video and audio clocks are periodically set to the same value, every 700 ms (or faster), using 33-bit presentation time stamps (PSTs). These serve to invoke a particular picture or audio sequence.

The topmost layer of MPEG-1, the video sequence layer, can be expressed as:

```
video sequence is
{
    next start code
    repeat
    {
        sequence header
        repeat
        {
            group of pictures
        } while (next word in stream is group start code)
    } while (next word in stream is sequence header code)
    sequence end code
}
```

The video stream comprises a header, a series of frames, an end-of-sequence code. The stream contain periodic I-frames. These provide full images to be used as periodic references, and so allow reasonably random access to the data stream. Other frames are "predicted" using either preceding I-pictures (which create P-pictures) or a combination of preceding and following I-pictures (which creates B-pictures). The encoder decides how I, B and P pictures are interspersed and ordered. A typical sequence would be I-B-B-P. The order of pictures in the data stream is not the order of display; for example, the previous sequence would be sent as I-P-B-B.

8.9 Other MPEG-2 enhancements

MPEG-2 adds, among other features, an alternate scan order which further improves compression. All control data, vectors and DCT coefficients are further compressed using Huffman-like variable-length encoding.

8.10 MPEG-2 bit rate

With MPEG-2 this transmission rate can be reduced to 4 Mbps for PAL and SECAM and 3 Mbps for NTSC, thus giving a compression ratio of 40:1 to high-quality TV. MPEG-1 typically compresses TV signals to 1.2 Mbps, giving a compression ratio of 130:1. Unfortunately the quality is reduced to near VCR-type quality. Table 8.2 outlines these parameters. The base bit rate for a standard Ethernet network is 10 Mbps. This allows compressed video to be transmitted over the network when there is no other traffic on the network. The 4 Mbps rate will load the network by approximately 50%. Standardized and compression techniques will be discussed in the next chapter.

Table 8.2 Motion video compression.

Type	*Bit rate*	*Compression*	*Comment*
Uncompressed TV	162 Mbps	1:1	
MPEG-1	4 Mbps	40:1	VCR quality
MPEG-2	1.2 Mbps	130:1	PAL, SECAM TV quality

8.11 Practical MPEG compression process

Most MPEG encoder can create MPEG-1 audio streams, MPEG-1 and MPEG-2 video streams, and MPEG-1 system streams. Each can be modified by a number of parameters. These are specified next.

8.11.1 Frame rate and data rate

The frame rate and data rate are the two main parameters which effect the quality of the encoded bitstream. The frame rate is normally set by the frame rate of the input format. Standard MPEG input frame rates are 23.976, 24, 25, 29.97, 30, 50, 59.94, and 60 frames/sec. Many encoders do not support all of these rates for the output so there are two modes which can be used to reduce or increase the frame rate, these are:

- Keep original number of frames. In this mode the frames are encoded frames as they are ordered in the input file, but MPEG players will play these files at the wrong speed.
- Keep original duration. In this mode the encoder either duplicates (to increase rate) or skip some input frames (to reduce the rate) to provide correct playback frame rate.

Most encoding systems will allow the user to specify the data rate if the encoded bitstream. The encode will then try to keep to this limit when it is encoding the input bitstream. For example, a single-speed CD-ROM requires a maximum data rate of 150 KB/sec. This rate is relatively low and there may be some degradation of quality to produce this. Reasonable quality requires about 300 to 600 KBs/sec.

8.11.2 Maximum motion vectors

MPEG uses motion estimation to reduce the encoded data rate. Motion estimation involves searching for the closest block of pixels on the previously encoded frame. It is a highly computing intensive task where each square block of pixels on the currently encoded frame is marched to the most similar block of pixels on the previously encoded frame. A pair of horizontal and vertical displacement values are then used to represent the pixel block, these values are named the motion vector.

Most MPEG encoders allow the user to select the maximum values of vertical and horizontal components of motion vectors. The larger these values then the greater the probability that a good match will be found to suit the output data rate. The two disadvantages of this are:

- The processing time increases as there are many more pixel block to search through.
- The number of bits required to represent the motion vectors in the output file also increases. This increases the output data rate, but may not necessarily improve the quality of the output bitstream.

MPEG has three different types of encoded frames: I, P, and B frames. Most encoding systems allow the user to specify the sequence of I, P, and B frames. A typical sequence is to put several B frames between each of the nearest P frames, and several groups of P and B frames between the nearest I frames, such as:

I⇒B⇒B⇒P⇒B⇒B⇒P⇒B⇒B⇒P⇒B⇒B⇒I ...

Thus the rules for the maximum motion vectors will be:

- P frames. The motion vectors for these must be long enough to represent the motion of objects between the nearest P frames.
- B frames. The motion vectors must be long enough to represent the motion of objects between the current B frame and each of (or at least one of) the nearest P frames.

Thus, the motion vectors for P frames are normally greater than for B frames. When selecting the maximum motion vectors, the type of input video file should be analyzed. For example, if there are faster horizontal movements than vertical movements, the maximum values of the horizontal motion vectors can be larger than the vertical vectors. Two examples are:

- Fast motion video and large frame sizes. P frame maximum horizontal, vertical motion vector: 36, 30. B frame maximum horizontal, vertical motion vector: 24, 16.
- Slow motion video and small frame sizes. P frame maximum horizontal, vertical motion vector: 24, 16. B frame maximum horizontal, vertical motion vector: 16, 8.

8.11.3 I, P and B frames

Increasing the number of intermediate frames (such as P and B frames) can make significant savings. Each group of frames are coupled to the nearest I frames and consist of a fixed number of subgroups, each of which have one P frame and several B frames.

The user can select the following:

- The number of B frames between the nearest P frames.
- The number of P frames (subgroups comprising P and B frames) between the nearest I frames.

I frames are always larger that P or B frames and P frames are larger than B frames. Increasing the number of P frames reduces the number of P frames, but too large a value of P frames reduces the resolution of a backward/forward play system, as this requires I frames. As P frames require motion estimation, an increase in P frames causes an increase in the encoding time, but reduces the overall bitrate. B frames takes slightly more time to encode that encoding P frames as they require two types of motion estimation, one applied to the previous P or I frame and one applied to the next P or I frame. For example a typical setting is:

- I BBB P BBB P BBB I ... P frames = 2, B frames = 3.
- I BB P BB P BB P BB P BB P BB I ... P frames = 5, B frames = 2.
- I P P P P P P P P P P I... P frames = 10, B frames = 0.
- I I I I I I I I I I I... P frames = 0, B frames = 0.

8.11.4 Typical values for the encoding parameters

Typical encoding parameters for MPEG-1 are:

Frame rate	25 frames/sec
Encoded bitstream	150 KB/sec
Maximum P frame motion vectors	32, 32
Maximum B frame motion vectors	16, 16
Number of P frames	3
Number of B frames	2 (between P frames)
Audio stream bitstream	64 Kbps (mono)/128 Kbps (stereo)
MPEG-2 profile	Main profile (see Table 8.1)
MPEG-2 level	Main level (see Table 8.1)
Format	YUV 4:2:0

9 Speech and Audio Signals

9.1 Introduction

Speech and audio signals are normally converted into PCM, which can be stored or transmitted as a PCM code, or compressed to reduce the number of bits used to code the samples. Speech generally has a much smaller bandwidth than audio.

9.2 PCM parameters

Digital systems tend to be less affected by noise than analogue. The main source of noise is quantization noise, which is caused by the finite number of quantization levels converting to a digital code.

The main parameters in determining the quality of a PCM system are the dynamic range (DR) and the signal-to-noise ratio (SNR).

9.2.1 Quantization error

The maximum error between the original level and the quantized level occurs when the original level falls exactly halfway between two quantized levels. This error will be half the smallest increment or

$$\text{Max error} = \pm \frac{1}{2} \frac{\text{Full scale}}{2^N}$$

9.2.2 Dynamic range (DR)

The dynamic range is the ratio of the largest possible signal magnitude to the smallest possible signal magnitude. If the input signal uses the full range of the ADC then the maximum signal will be the full-scale voltage. The smallest signal amplitude is one which toggles between one quantization level and the level above, or below. This signal amplitude, for an n-bit ADC, is the full-scale voltage divided by the number of quantization levels (that is, 2^n). Thus, for a linearly quantized signal:

$$\text{Dynamic range} = \frac{V_{\max}}{V_{\min}}$$

$$\text{Number of levels} = 2^n - 1$$

$$\text{Dynamic range} = 20\log\frac{V_{max}}{V_{max}/2^n - 1} = 20\log(2^n - 1)\ \text{dB}$$

if 2^n is much greater that 1, then

$$\text{Dynamic range} \approx 20\log 2^n = 20n\log 2 \approx 6.02n\ \text{dB}$$

Table 9.1 outlines the DR for a given number of bits. Normally the maximum number of bits is less than 20. The voltage ratio of a given number of bits is also given in square brackets [*ratio*]. For example an 8-bit system has a DR of 48.18 dB and the largest voltage amplitude is 256 times the smallest voltage amplitude. A 16-bit system has a DR of 96.33 dB and the largest voltage amplitude is 65 536 times the smallest voltage amplitude.

Table 9.1 Dynamic range of a digital system.

Number of bits	*DR (dB) [ratio]*	*Number of bits*	*DR (dB) [ratio]*
1	6.02 [2]	11	66.23 [2048]
2	12.04 [4]	12	72.25 [4096]
3	18.06 [8]	13	78.27 [8192]
4	24.08 [16]	14	84.29 [16384]
5	30.10 [32]	15	90.31 [32768]
6	36.12 [64]	16	96.33 [65536]
7	42.14 [128]	17	102.35 [131072]
8	48.16 [256]	18	108.37 [262144]
9	54.19 [512]	19	114.39 [524288]
10	60.21 [1024]	20	120.41 [1048576]

9.2.3 Signal-to-noise ratio (SNR)

It can be shown that the SNR for a linearly quantized digital system is:

$$\text{SNR} = 1.76 + 6.02\,n\,\text{dB}$$

This proof is given in Appendix C. Table 9.2 outlines the SNR for a given number of bits. Normally the maximum number of bits is less than 20. The voltage ratio of a given number of bits is also given in square brackets [*ratio*]. For example an 8-bit system has an SNR of 49.92 dB and the largest rms voltage is 313.33 times the smallest rms voltage. A 16-bit system has an SNR of 96.33 dB and the largest rms voltage is 80 167.81 times the smallest rms voltage.

Table 9.2 Signal-to-noise ratio of a digital system.

Number of bits	*SNR (dB) [ratio]*	*Number of bits*	*SNR (dB) [ratio]*
7	43.90 [156.68]	14	86.04 [20044.72]
8	49.92 [313.33]	15	92.06 [40086.67]
9	55.94 [626.61]	16	98.08 [80167.81]
10	61.96 [1253.14]	17	104.10 [160324.5]
11	67.98 [2506.11]	18	110.12 [320626.9]
12	74.00 [5011.87]	19	116.14 [641209.6]
13	80.02 [10023.05]	20	122.16 [1282331]

9.3 Differential encoding

Differential coding is a source-coding method which is used when there is a limited change from one value to the next. It is well suited to video and audio signals, especially audio, where the sampled values can only change within a given range. It is typically used in PCM (pulse code modulation) schemes to encode audio and video signals.

9.3.1 Delta modulation PCM

PCM coverts analogue samples into a digital code. Delta PCM uses a single-bit code to represent an analogue signal. With delta modulation a '1' is transmitted (or stored) if the analogue input is higher than the previous sample or a '0' if it is lower. It must obviously work at a higher rate than the Nyquist frequency, but because it uses only 1 bit, it normally uses a lower output bit rate. Figure 9.1 shows a delta modulation transmitter.

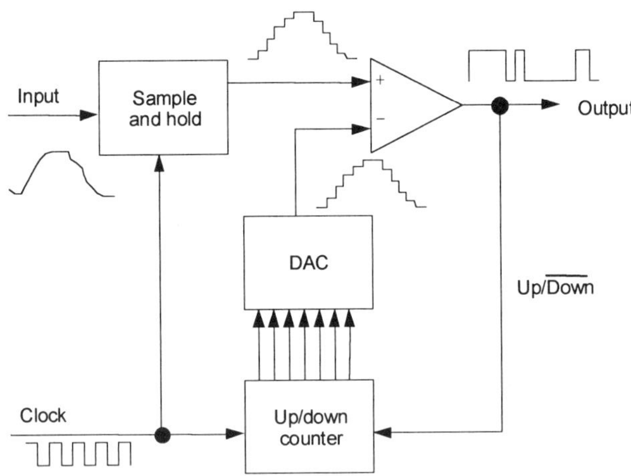

Figure 9.1 Delta modulation.

Initially the counter is set to zero. A sample is taken and if it is greater than the analogue value on the DAC output, the counter is incremented by 1, or it is decremented. This continues at a time interval given by the clock. Each time the present sample is greater than the previous sample, a '1' is transmitted; otherwise a '0' is transmitted. Figure 9.2 shows an example signal. The sampling frequency is chosen so that the tracking DAC can follow the input signal. This results in a higher sampling frequency, but because it only transmits one bit at a time, the output bit rate is normally reduced. Figure 9.3 shows that the receiver is almost identical to the transmitter except that it has no comparators.

Two problems with delta modulation are:

- Slope overload. This occurs when the signal changes too fast for the modulator to keep up; see Figure 9.4. It is possible to overcome this problem by increasing the clock frequency or increasing the step size.
- Granular noise. This occurs when the signal changes slowly in amplitude, as illustrated in Figure 9.5. The reconstructed signal contains a noise which is not present at the input. Granular noise is equivalent to quantization noise in a PCM system. It can be reduced by

decreasing the step size, though there is a compromise between smaller step size and slope overload.

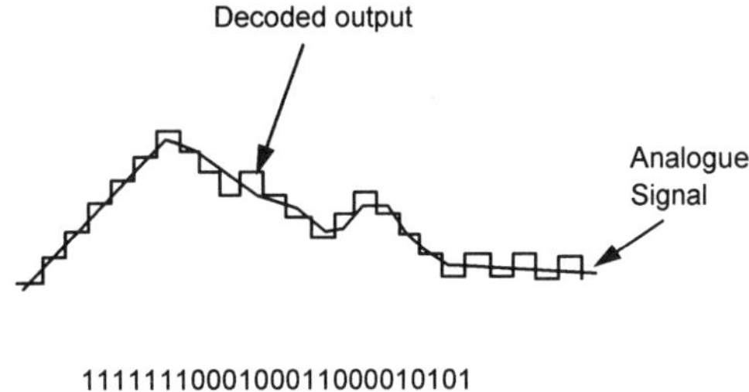

Figure 9.2 Delta modulator signal.

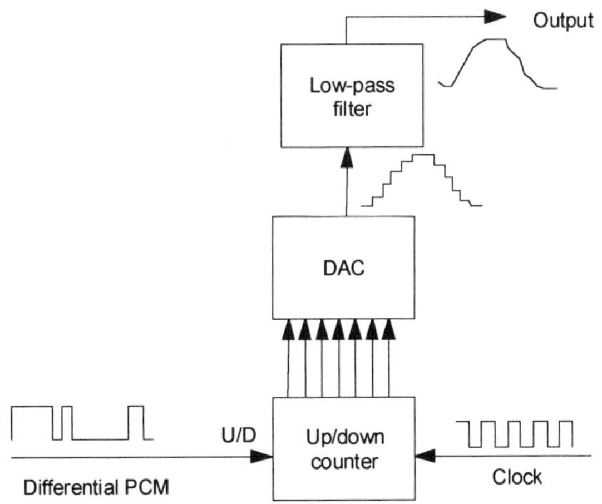

Figure 9.3 Delta modulator receiver.

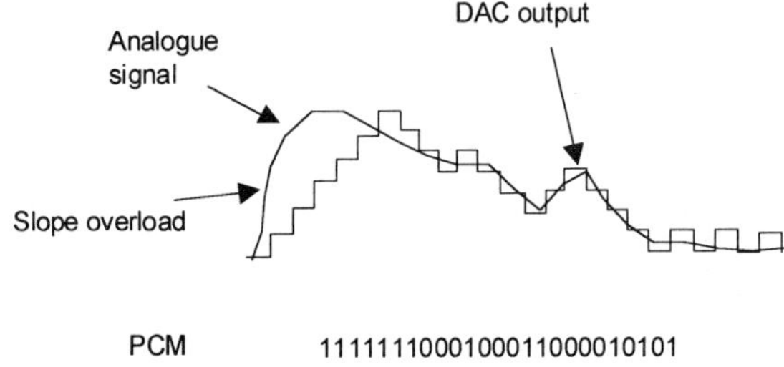

Figure 9.4 Slope overload.

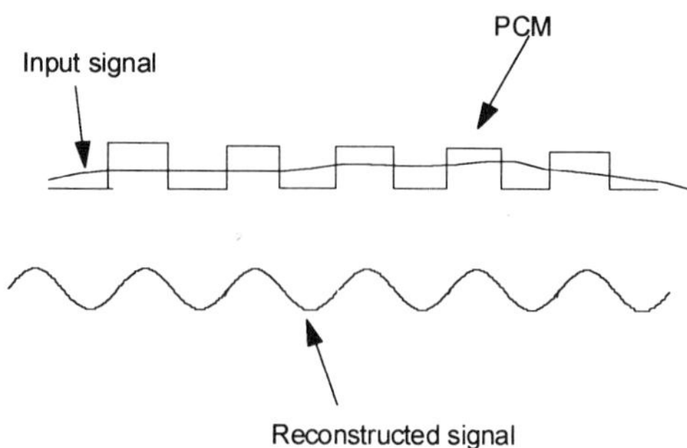

Figure 9.5 Granular noise.

9.3.2 Adaptive delta modulation PCM

Unfortunately delta modulation cannot react to very rapidly changing signals and will thus take a relatively long time to catch them up (known as slope overload). It also suffers when the signal does not change much as this ends up in a square wave signal (known as granular noise). One method of reducing granular noise and slope overload is to use adaptive delta PCM. With this method the step size is varied by the slope of the input signal. The larger the slope, the larger the step size; see Figure 9.6. The algorithms usually depend on the system and the characteristics of the signal. A typical algorithm is to start with a small step and increase it by a multiple until the required level is reached. The number of slopes will depend on the number of coded bits, such as 4 step sizes for 2 bits, 8 for 3 bits, and so on.

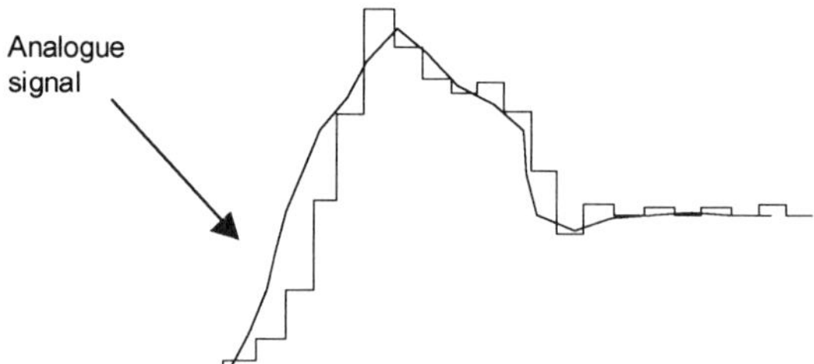

Figure 9.6 Variation of step size.

9.3.3 Differential PCM (DPCM)

Speech signals tend not to change much between two samples. Thus similar codes are sent, which leads to a degree of redundancy. For example, in a certain sample it is likely the signal will only change within a range of voltages, as illustrated in Figure 9.7.

DPCM reduces the redundancy by transmitting the difference in the amplitude of two consecutive samples. Since the range of sample differences is typically less than the range of individual samples, fewer bits are required for DPCM than for conventional PCM.

Figure 9.8 shows a simplified transmitter and receiver. The input signal is filtered to half the sampling rate. This filter signal is then compared with the previous DPCM signal. The difference between them is then coded with the ADC.

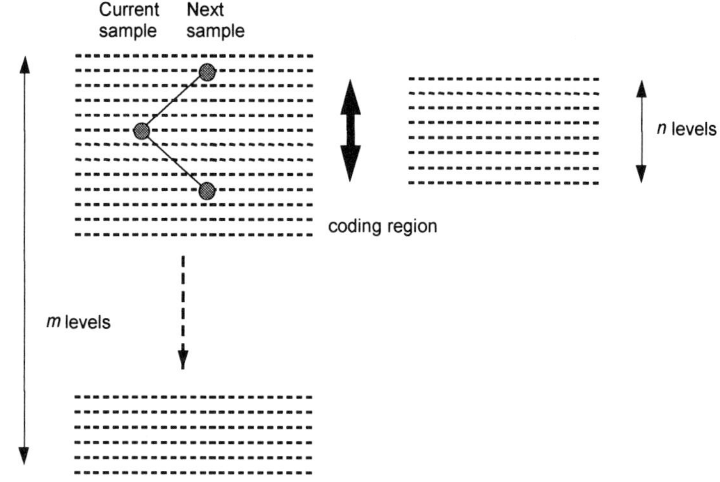

Figure 9.7 Normal and differential quantization.

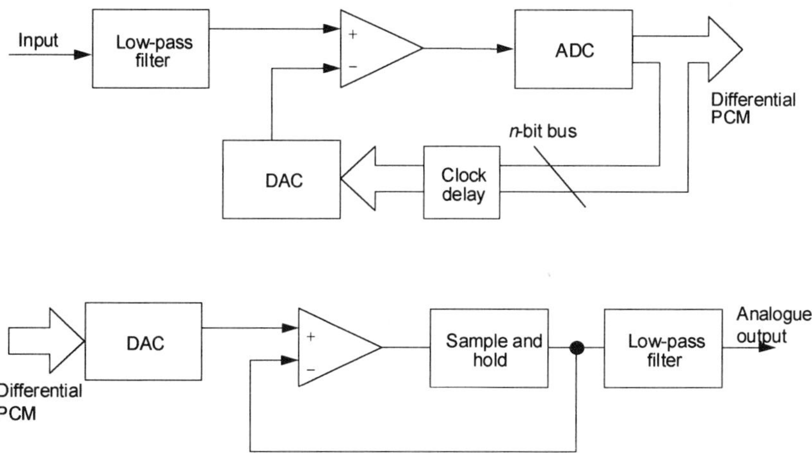

Figure 9.8 DPCM transmitter/receiver.

9.3.4 Adaptive differential PCM (ADPCM)

ADPCM allows speech to be transmitted at 32 kbps with little noticeable loss of quality. As with differential PCM the quantizer operates on the difference between the current and previous samples. The adaptive quantizer uses a uniform quantization step *M*, but when the signal moves towards the limits of the quantization range, the step size *M* is increased. If it is around the center of the ranges, the step size is decreased. Within any other regions the step size hardly changes. Figure 9.9 illustrates this operation with a signal quantized to 16 levels. This results in 4-bit code.

The change of the quantization step is done by multiplying the quantization level, M, by a number slightly greater, or less, than 1 depending on the previously quantized level.

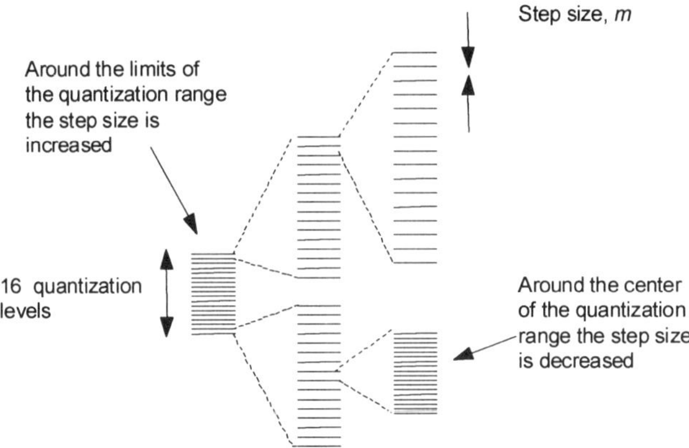

Figure 9.9 ADPCM quantization.

9.4 Speech compression

Subjective and system tests have found that 12-bit coding is required to code speech signals, which gives 4096 quantization levels. If linear quantization is applied then the quantization step is the same for quiet levels as for loud levels. Any quantization noise in the signal will be more noticeable at quiet levels than at loud levels. When the signal is loud, the signal itself swamps the quantization noise, as illustrated in Figure 9.10. Thus an improved coding mechanism is to use small quantization steps at low input levels and a higher one at high levels. This is achieved using non-linear compression.

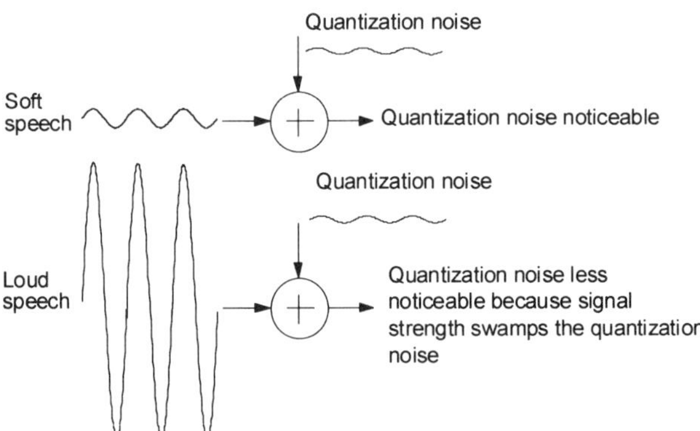

Figure 9.10 Quantization noise is more noticeable with low signal levels.

The two most popular types of compression are A-Law (in European systems) and μ-Law (in the USA). These laws are similar and compress the 12-bit quantized speech code into an 8-bit compressed code. An example compression curve is shown in Figure 9.11.

As an approximation the two laws are split into 16 line segments. Starting from the origin and moving outwards, left and right, each segment has half the slope of the previous.

Using an 8-bit compressed code at a sample rate of 8000 samples per second gives a bit rate of 64 kbps. ISDN uses this bit rate to transmit digitized speech. Figure 9.12 shows a basic transmission system.

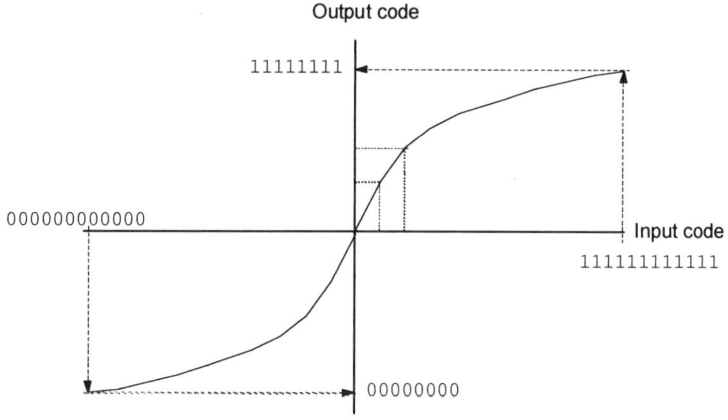

Figure 9.11 12-bit to 8-bit non-linear compression.

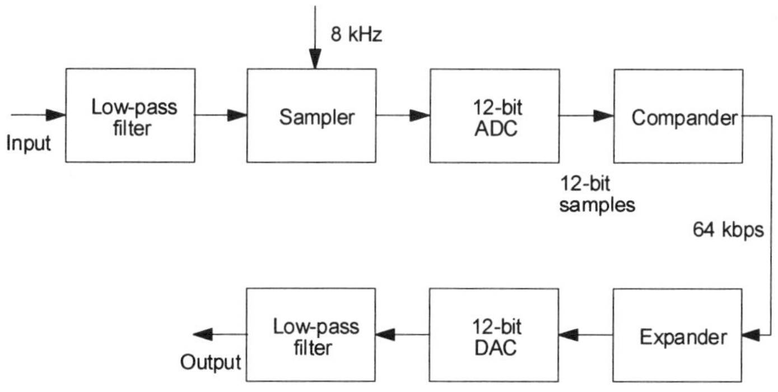

Figure 9.12 Typical PCM speech system.

9.5 A-Law and μ-Law companding

The companding and expansion encoding is normally implemented using either μ-Law or A-Law. A-Law is used in Europe and in many other countries, whereas μ-Law is used in North America and Japan. Both were defined by the CCITT in the G.711 recommendation and both use non-uniform quantization step sizes which increase logarithmically with signal level. μ-Law uses the compression characteristic of:

$$y = \frac{\log(1+\mu x)}{\log(1+x)} \quad \text{for } x \geq 0$$

where y is the output magnitude x is the input magnitude μ is a positive factor which is chosen for the required compression characteristics

Figure 9.13 shows an example of μ-Law using $\mu=1$, $\mu=50$ and $\mu=255$. Using $\mu=0$ gives uniform conversion (linear quantization). Normally speech systems use $\mu=255$ as this characteristic is well matched to human hearing. An 8-bit implementation can achieve a small SNR a and dynamic range equivalent to that of a 12-bit uniform system.

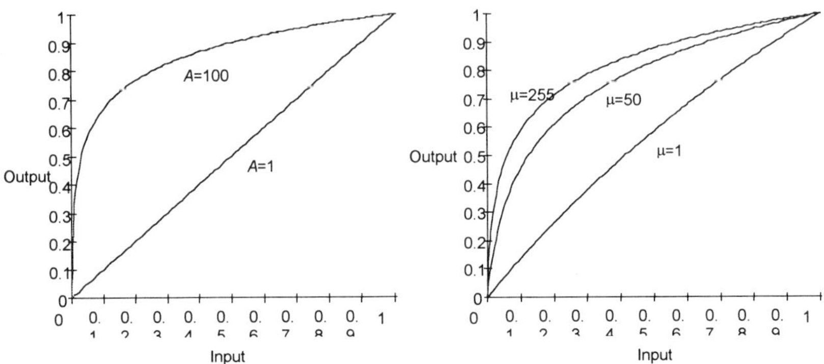

Figure 9.13 A-Law and μ-Law characteristics.

The A-law also uses quantization characteristics that vary logarithmically. Figure 9.13 shows an example of A-Law using $A=1$ and $A=100$. Most A-Law speech systems use $A=87.56$. The compression characteristic is:

$$y=\begin{cases}\dfrac{Ax}{1+\log A} & \text{for } 0\le |x|\le \dfrac{1}{A}\\[2ex] \dfrac{1+\log(Ax)}{1+\log A} & \text{for } \dfrac{1}{A}\le |x|\le 1\end{cases}$$

where A is a positive integer.

Figure 9.14 shows two input waveforms, 1 V peak to peak and 0.1 V peak to peak. It can be seen that the companding processes amplifies the lower amplitudes more than the large amplitudes. This causes low-amplitude speech signals to be boosted compared with loud speech. Also notice that the waveform has been distorted because the low amplitudes are amplified more than the large amplitudes.

9.5.1 Digitally linearizable log-companding

The mathematical formulas for A-Law and μ-Law are normally approximated to a series of linear segments. This permits more precise control of the quantization characteristics. The chosen approximation used is to make the step sizes in consecutive segments change by a factor of 2. Figure 9.15 shows the characteristic of the piecewise linear conversion. It can be seen that the slope of each segment is twice the slope of the previous segment (although in A-Law 98.56, segment 0 and segment 1 have the same slope). Each segment has 16 quantization levels and there are 16 segments (8 for positive inputs and 8 for negative inputs). Thus 1 bit identifies the sign bit, 3 bits identify the segment (in the positive or negative part) and 4 bits identifies the quantization level. The 8-bit companded values thus take the form:

```
SLLLQQQQ
```

where `S` is the sign bit, `LLL` is the segment number and `QQQQ` is the quantization level within the segment.

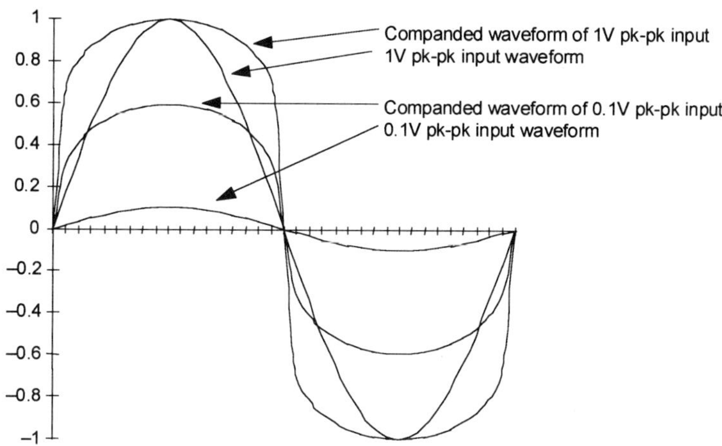

Figure 9.14 Effects of waveforms with μ-255 encoding.

Table 9.3 shows the conversion for A-Law 87.56. For example, if the input value is between 16 and 17, the companded value will be 001 0000. If this value is positive then the most significant bit will be a 1, thus the companded value will be 1001 0000.

Table 9.3 shows that the step sizes for the first two segments are the same (unity step size). Table 9.4 shows the μ-Law encoding table.

Consider A-Law with the input range between +5 V and –5 V. An input voltage of +1 V will correspond to the input level of:

$$\text{Input} = \frac{1}{5} \times 2048 = 409.6$$

Referring to Table 9.3, this is within the segment from 256 to 512. The code will thus be S101XXXX. The level within the segment will be:

$$\text{Level} = \frac{409.6 - 256}{16} = 9.6$$

which corresponds to quantization level 9. Thus the companded value is:

```
01011001
```

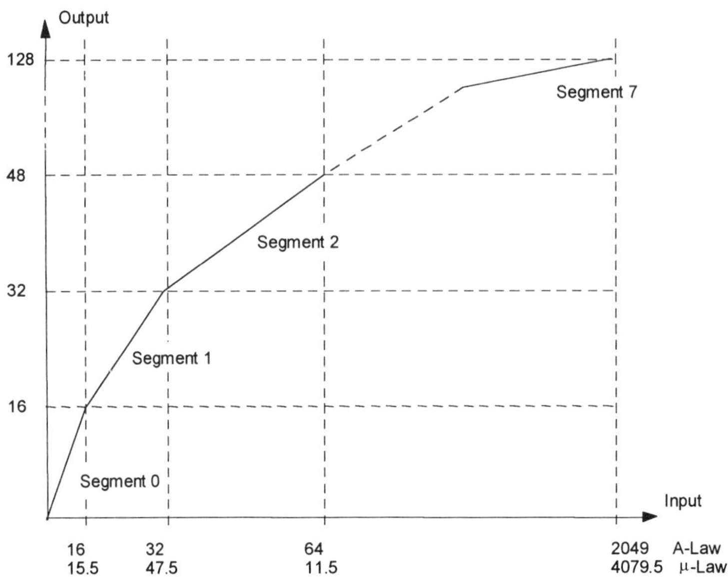

Figure 9.15 Piecewise linear compression for A-Law and μ-Law.

Table 9.3 A-Law 87.56 encoding/decoding.

Input	*Companded*	*Decoder level*	*Decoded level number*	*Step size*
0–1	000 0000	0	0.5	1
…	…	…	…	
15–16	000 1111	15	15.5	
16–17	001 0000	16	16.5	1
…	…	…	…	
31–32	001 1111	31	31.5	
32–34	010 0000	32	33	2
…	…	…	…	
62–64	010 1111	47	63	
64–68	011 0000	48	66	4
…	…	…	…	
124–128	011 1111	63	126	
128–136	100 0000	64	132	8
…	…	…	…	
248–256	100 1111	79	252	
256–272	101 0000	80	264	16
…	…	…	…	
496–512	101 1111	95	504	
512–544	110 0000	96	528	32
…	…	…	…	
992–1024	110 1111	111	1008	
1024–1088	111 0000	112	1056	64
…	…	…	…	
1984–2048	111 1111	127	2016	

Table 9.4 μ-255 encoding/decoding.

Input	*Companded*	*Decoder level*	*Decoded level number*	*Step size*
0–0.5	000 0000	0	0	1
...	...	...	...	
14.5–15.5	000 1111	15	15	
15.5–17.5	001 0000	16	16.5	2
...	...	...	...	
45.5–47.5	001 1111	31	46.5	
47.5–51.5	010 0000	32	49.5	4
...	...	...	...	
107.5–111.5	010 1111	47	109.5	
111.5–119.5	011 0000	48	115.5	8
...	...	...	...	
231.5–239.5	011 1111	63	235.5	
239.5–255.5	100 0000	64	247.5	16
...	...	...	...	
479.5–495.5	100 1111	79	487.5	
497.5–527.5	101 0000	80	511.5	32
...	...	...	...	
975.5–1007.5	101 1111	95	991.5	
1007.5–1071.5	110 0000	96	1039.5	64
...	...	...	...	
1967.5–2031.5	110 1111	111	1999.5	
2031.5–2159.5	111 0000	112	2095.5	128
...	...	...	...	
3951.5–4079.5	111 1111	127	4015.5	

9.6 Speech sampling

With telephone-quality speech the signal bandwidth is normally limited to 4 kHz, thus it is sampled at 8 kHz. If each sample is coded with 8 bits then the basic bit rate will be:

Digitized speech signal rate = 8×8 kbps = 64 kbps

Table 9.5 outlines the main compression techniques for speech. The G.722 standard allows the best-quality signal. The maximum speech frequency is 7 kHz rather than 4 kHz in normal coding systems; this is equivalent of 14 coding bits. The G.728 allows extremely low bit rates (16 kbps).

9.7 PCM-TDM systems

Multiple channels of speech can be sent over a single line using time division multiplexing (TDM). In the UK a 30-channel PCM system is used, whereas the USA uses 24.

Table 9.5 Speech compression standards.

ITU standard	*Technology*	*Bit rate*	*Description*
G.711	PCM	64 kbps	Standard PCM
G.721	ADPCM	32 kbps	Adaptive delta PCM where each value is coded with 4 bits
G.722	SB-ADPCM	48, 56 and 64 kbps	Subband ADPCM allows for higher-quality audio signals with a sampling rate of 16 kHz
G.728	LD-CELP	16 kbps	Low-delay code excited linear prediction for low bit rates

With a PCM-TDM system, several voice band channels are sampled, converted to PCM codes, these are then time division multiplexed onto a single transmission media.

Each sampled channel is given a time slot and all the time slots are built up into a frame. The complete frame usually has extra data added to it such as synchronization data, and so on. Speech channels have a maximum frequency content of 4 kHz and are sampled at 8 kHz. This gives a sample time of 125 μs. In the UK a frame is built up with 32 time slots from TS0 to TS31. TS0 and TS16 provide extra frame and synchronization data. Each of the time slots has 8 bits, therefore the overall bit rate is:

Bits per time slot	=	8
Number of time slots	=	32
Time for frame	=	125 μs

$$\text{Bit rate} = \frac{\text{No of bits}}{\text{Time}} = \frac{32 \times 8}{125 \times 10^{-6}} = 2048 \text{ kbps}$$

In the USA and Japan this bit rate is 1.544 Mbps. These bit rates are known as the primary rate multipliers. Further interleaving of several primary rate multipliers increases the rate to 6.312, 44.736 and 139.264 Mbps (for the USA) and 8.448, 34.368 and 139.264 Mbps (for the UK).

The UK multiframe format is given in Figure 9.16. In the UK format the multiframe has 16 frames. Each frame time slot 0 is used for synchronization and time slot 16 is used for signaling information. This information is sub-multiplexed over the 16 frames. During frame 0 a multiframe-alignment signal is transmitted in TS16 to identify the start of the multiframe structure. In the following frames, the eight binary digits available are shared by channels 1–15 and 16–30 for signaling purposes. TS16 is used as follows:

Frame	0	`0000XXXX`
Frames	1–15	`1234 5678`

where `1234` are the four signaling bits for channels 1, 2, 3, …, 15 in consecutive frames, and `5678` are the four signaling bits for channels 16, 17, 18, … 31 in consecutive frames.

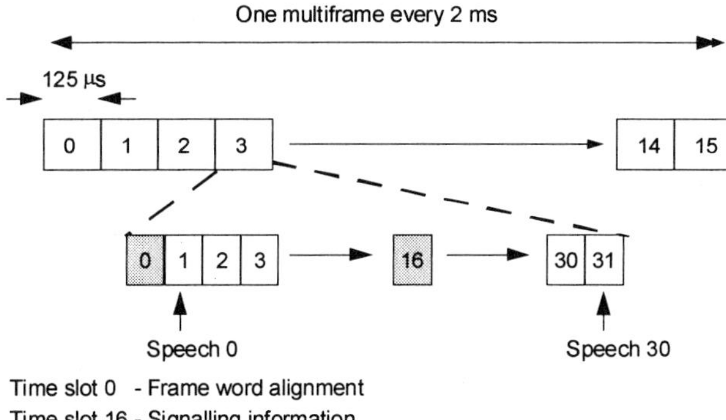

Figure 9.16 PCM-TDM multiframe format with 30 speech channels.

Thus in the first frame the 0000XXXX code word is sent, in the next frame the first channel and the 16th channel appear in TS16, the next will contain the second and the 17th, and so on. Typical 4-bit signal information is:

1111 – circuit idle/busy
1101 – disconnection

TS0 contains a frame-alignment signal which enables the receiver to synchronize with the transmitter. The frame-alignment signal (X0011011) is transmitted in alternative frames. In the intermediate frames a signal known as a not-word is transmitted (X10XXXXX). The second binary digit is the complement of the corresponding binary digit in the frame-alignment signal. This reduces the possibility of demultiplexed misalignment to imitative frame-alignment signals.

Alternative frames:

TS0: X0011011
TS0: X10XXXXX

where X stands for don’t care conditions.

10 Audio Signals

10.1 Introduction

The benefits of converting from analogue audio to digital audio are:

- The quality of the digital audio system only depends on the conversion process, whereas the quality of an analogue audio system depends on the component parts of the system.
- Digital components tend to be easier and cheaper to produce than high-specification analogue components.
- Copying digital information is relatively easy and does not lead to a degradation of the signal.
- Digital storage tends to use less physical space than equivalent analogue forms.
- It is easier to transmit digital data.
- Information can be added to digital data so that errors can be corrected.
- Improved signal-to-noise ratios and dynamic ranges are possible with a digital audio system.

Audio signals normally use PCM codes which can be compressed to reduce the number of bits used to code the samples.

For high-quality monochannel audio, the signal bandwidth is normally limited to 20 kHz, thus it is sampled at 44.1 kHz. If each sample is coded with 16 bits then the basic bit rate will be:

$$\text{Digitized audio signal rate} = 44.1 \times 16\,\text{kbps} = 705.6\,\text{kbps}$$

For stereo signals the bit rate would be 1.4112 Mbps. Many digital audio systems add extra bits from error control and framing. This increases the bit rate.

10.2 Principles

Digital audio normally involves the processes of:

- Filtering the input signal.
- Sampling the signal at twice the highest frequency.
- Converting it into a digital form with an ADC (analogue-to-digital converter).
- Converting the parallel data into a serial form.
- Storing or transmitting the serial information.

- When reading (or receiving) the data the clock information is filtered out using a PLL (phase-locked loop).
- The recovered clock is then used with a SIPO (serial-in parallel-out) converter to convert the data back into a parallel form.
- Converting the digital data back into an analogue voltage.
- Filtering the analogue voltage.

These steps are illustrated in Figure 10.1. The clock recovery part is important; there is no need to save or transmit separate clock information because it can be embedded into the data. It also has the advantage that a clock becomes jittery when it is affected by noise. Thus, if the clock information is transmitted over relatively long distances it will be jittery.

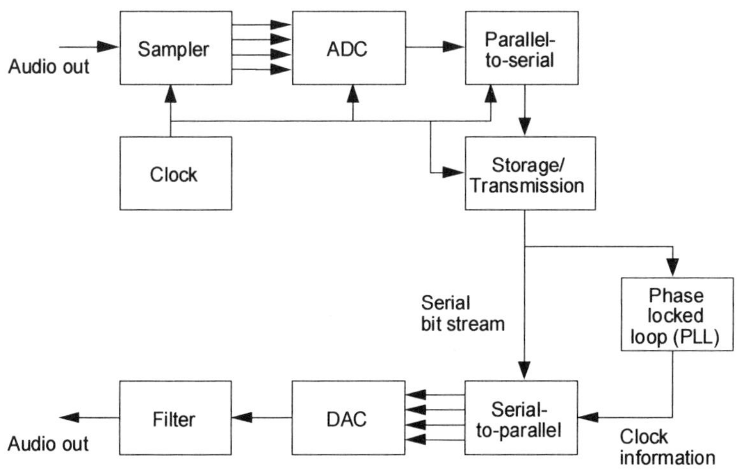

Figure 10.1 A digital audio system.

10.3 Digital audio standards

Many techniques are available for compressed audio, such as:

- Broadcast audio distribution with a sample rate of 32 kHz (to provide a 15 kHz bandwidth) and with 14-bit linear coding.
- Professional recording with a sample rate of 48 kHz with 16- or 20-bit coding.
- Compact disk recording with a sample rate of 44.1 kHz (CD/EIAJ/RDAT) with 14- or 16-bit coding.

Most methods of coding for hi-fi applications use 16 bits per sample, but professional applications use 20 bits to give an extremely wide dynamic range of 120.41 dB.

Hi-fi audio is either sampled at 32 kHz, 44.1 kHz or 48 kHz. The 48 kHz sampling rate gives the highest bandwidth. It is possible to change from one sampling rate to another, but this normally degrades the signal quality.

10.3.1 SDIF-2 interconnection

The SDIF-2 (Sony Digital Interface) protocol was developed by the Sony Corporation and allows for a standard interface between professional audio equipment. In its standard form it uses a single-ended signal with three 75 Ω coaxial cables, one for each audio channel and another for data word clock synchronization. This clock is a symmetrical square wave at the sampling frequency and is common to both channels. Thus for a sampling frequency of 44.1 kHz the word clock period will be 22.67574 μs. This is the time to transmit a single 32-bit word.

The bits are transmitted over a serial communications line using the NRZ (non-return to zero) line code, which represents a 1 as a high level and a 0 as a low level.

Figure 10.2 shows the format of the 32-bit word. The bit fields are:

- Bits 1–20 are a 20-bit PCM data sample with the most significant bit sent first. If any bits are unused, they are padded with 0's.
- Bits 21–25 are used for control purposes and are reserved for future development.
- Bits 26 and 27 are emphasis bits which are determined when the sample is converted to a digital value (00 specifies that emphasis is not used and 11 specifies that emphasis is used).
- Bit 28 is the dubbing prohibition bit (0 specifies that dubbing is possible, whereas 1 specifies that dubbing is prohibited).
- Bit 29 is the block flag bit. This is used to identify the beginning of an SDIF-2 block. If it is a 1 then it specifies the start of a 256-word block, else it is a 0.
- Bits 30–32 are used for the synchronization pattern. The time period of the field is $3T$ and it consists of one transition from low to high, if bit 29 is high, or from high to low, if bit 29 is low. This transition always occurs halfway through the bit field (that is at $1.5T$). When the receiver detects this pattern it knows to expect a following data block (which contains 256 data samples.

Data is transmitted in blocks of 256 PCM samples. The start of a block is identified with a 1 in bit 29 and a low-to-high transition in bit positions 30 to 32. Each of the following samples in the block then have a 0 in bit 29 and a high-to-low transition.

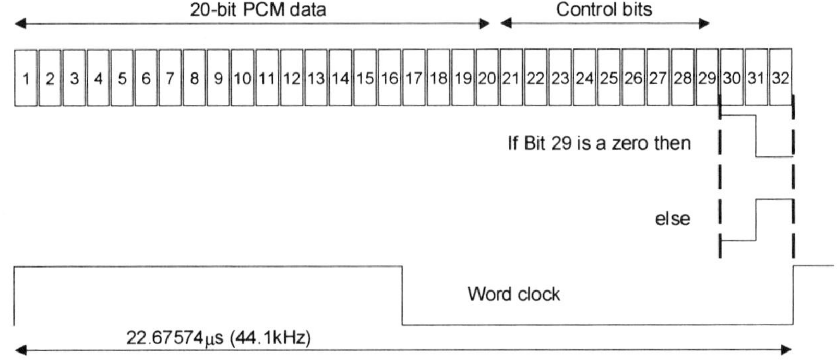

Figure 10.2 SDIF-2 word format.

10.3.2 AES/EBU/IEC professional interface format

In 1985, the Audio Engineering Society (AES) defined a standard interface known as the AES3. Then, in 1992, it was further enhanced to create the AES3-1995 standard. It uses serial transmission formation with linear PCM. It has the advantage over SDIF-2 in that a single channel carries the two audio channels and also allows for non-audio data. It is also self-clocking and self-synchronizing.

AES3 is similar to the EBU Tech. 3250-E standard developed by the European Broadcast Union (EBU) and the interface is commonly known as the AES/EBU digital interface. Other similar standards include:

- ANSI S4.40-1985 which was developed by the American National Standards Institute (ANSI).
- CCIR-Rec. 647 which was developed by the International Radio Consultative Committee (CCIR) and provides Rec. 647.
- EIAJ CP-340-type I which was developed by the Electronic Industries Association of Japan (EIAJ).
- IEC-958 which was developed by the International Electrotechnical Commission (IEC).

The EBU/AES/IEC system can use any sampling rate but is typically either 32 kHz, 44.1 kHz or 48 kHz. The line code used is bi-phase mark coding, which has the following rules:

- A 1 is coded as a high-to-low transition.
- A 0 is coded alternatively as a high or a low.

Thus a 0 has the same level for the complete bit period, whereas a 1 changes its level. This type of code is often known as binary frequency modulation. It has the advantage that it has no DC content and, in the worst case, there are two transitions for each transmitted bit (one of the transitions always occurs at the start of a bit period). These transitions allow the clock to be recovered from the signal. And all of the information is contained in bit transitions, not in levels. This makes the code less sensitive to noise, as noise tends to affect voltage levels more than it affects transitions. Figure 10.3 shows an example of biphase mark coding.

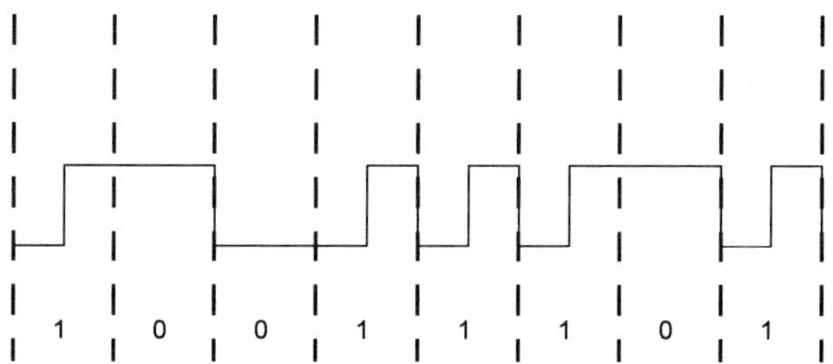

Figure 10.3 Biphase mark coding.

Each 20-bit PCM data sample is framed with a 32-bit data word, as illustrated in Figure 10.4. The data word starts with a 4-bit sync/preamble block followed by 4-bit auxiliary data. Next follows the 20-bit data sample (the least significant bit is sent first). Finally, there is a 4-

bit auxiliary field which contains a validity flag, user data flag, channel status flag and a single parity bit. The error checking is obviously very limited and it should thus only be used on low bit error rate channels.

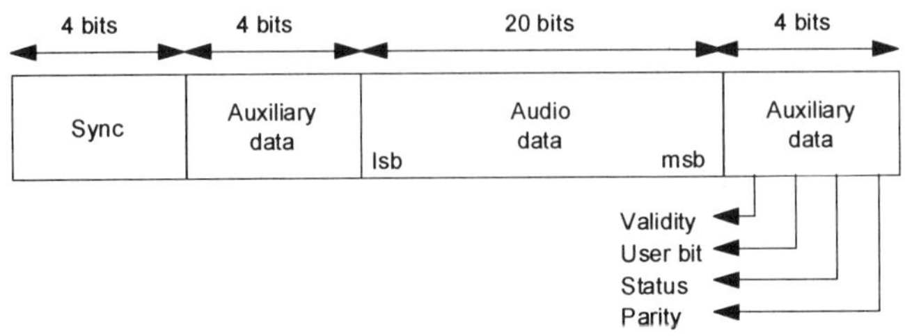

Figure 10.4 AES/EBU audio subframe format.

For a 44.1 kHz sampling rate, the time for the complete packet will be 22.67574 μs. Thus the overall bit rate will be:

$$\text{Bit rate} = \frac{2 \times 32}{22.67574} = 2.822 \text{ Mbps}$$

10.3.3 Frame format

Each frame contains 32 bits and has four status bits at the end. These status bits are:

- V – which is the validity bit. If this bit is set to a 0 then the transmitted audio value is valid if it is a 1 then the value is invalid.
- U – which is used to added extra user information.
- C – which is the channel status data. Used to add extra information, its format is described in the next section.
- P – which is the parity bit. The parity bit is added so that the number of 1's in the frame is even (even parity).

A single frame contains two subframes, one for each channel. These are then used to build up to 192 frames, as illustrated in Figure 10.5.

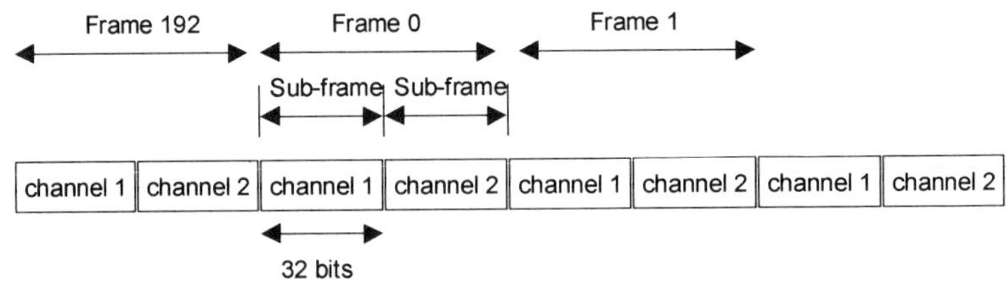

Figure 10.5 AES3 format of frames and subframes.

The 192 bits from the subframes build up into a 24-byte channel status block. The format of these is:

- Bit 0. Professional flag. If it is a consumer application it is set to a 0, else if it is a professional application it is set to a 1.
- Bit 1. Audio flag. If this is set to a 0 then the data is audio data, else if it is a 1 it is non-audio data.
- Bits 2–4. Emphasis flag. A value of 000 indicates emphasis not indicated, 100 indicates no emphasis, 110 indicates 50/15 µs emphasis and 111 indicates CCITT J.17 emphasis.
- Bit 5. Lock flag. If it is a 0 then the sampling frequency is locked, else if it is a 1 then the sampling frequency is unlocked.
- Bit 6–7. Sampling frequency. If it is 00 then the sampling frequency is not indicated, if it is 01 then the sampling frequency is 48 kHz, 10 indicates 44.1 kHz and 11 indicates 32 kHz.
- Bits 8–11. Encoded channel mode. 0000 (not indicated), 0001 (two-channel), 0010 (single-channel), 0011 (primary/secondary), 0100 (stereo).
- Bits 12–15. User bits management. 0000 (default), 0001 (192-bit block,) 0010 (AES18) and 0011 (user defined).
- Bits 16–18. Auxiliary sample bits. 000 (20 bits), 001 (24 bits), 011 (reserved).
- Bits 19–21. Sample wordlength. For a 20-bit wordlength: 000 (default), 001 (19 bits), 010 (18 bits), 011 (17 bits), 100 (16 bits), 101 (20 bits).
- Bits 22–23 and 24–31. Reserved.
- Bits 32–33. Reference.
- Bits 34–47. Reserved.
- Bits 48–79. Alphanumeric channel origin data (7-bit ASCII).
- Bits 80–111. Alphanumeric channel destination data (7-bit ASCII).
- Bits 112–143. Load sample address code (32-bit binary).
- Bits 144–176. Time-of-day code (32-bit binary).
- Bits 176–179. Reserved.
- Bits 180–183. Reliability flags.
- Bits 184–191. CRC using $x^8+x^4+x^3+x^2+1$.

10.3.4 S/PDIF consumer interconnection

The EAS3 standard is most often used in professional audio equipment. Consumer equipment normally use the S/PDIF (Sony/Philips Digital Interface) format, which is similar to the IEC-958 consumer format (known as type II). In some applications, the EIAJ CP-340 type II format is used. The S/PDIF format is similar to EAS3 and in some cases S/PDIF equipment can be connected to EAS3 equipment.

The S/PDIF status channel status block differs from the professional channel status block and only uses the first 32 bits of the 192-bit block. The format of these is:

- Bit 0. Consumer flag. If it is a consumer application it is set to a 0, else if it is a professional application it is set to a 1.
- Bit 1. Audio flag. If this is set to a 0 then the data is audio data, else if it is a 1 it is non audio data.
- Bit 2. Copy/copyright flag. If this is set to a 0 then the copyright is asserted and copying may be inhibited, else if it is a 1 then copying is permitted.
- Bits 3–4. Emphasis flag. A value of 000 indicates emphasis not indicated, 110 indicates 50/15 µs emphasis and 111 is reserved.

- Bit 6–7. Mode. If set to 00 then the mode is set to mode 0. The following bits define mode 0.
- Bits 8–14. Category code.
  - 00000000 (general).
  - 00000001 (experimental).
  - 0001xxx (solid-state memory).
  - 001xxxx (broadcast reception of audio: 0010000 is Japan, 0010011 is USA, 0011000 is Europe and 0010001 is electronic software delivery).
  - 010xxxx (digital/digital converters). 0100000 (PCM encoder/decoder), 0100010 (digital sound sampler), 0100100 (digital signal mixer), 0101100 (sample-rate converter).
  - 01100xx (ADC without copy info).
  - 01101xx (ADCs with copy info).
  - 0111xxx (broadcast reception of digital audio).
  - 100xxxx (Laser-Optical). 1000000 (CD – compatible with IEC-908), 1001000 (CD not compatible with IEC-908), 1001001 (MD – mini-disk).
  - 101xxxx (musical instruments, microphones, and so on). 1010000 (synthesizer), 1011000 (microphone).
  - 110xxxx (magnetic tape or disk). 110000 (DAT – digital audio tape), 1101000 (VCR) or 1100001 (DCC – digital compact cassette).
  - 111xxxx (reserved).
- Bit 15. L generation status. When category code is 100xxxx, 001xxxx or 0111xxx then 0 is original/commercial prerecorded) and 1 is no indication/first generation or higher. Else, 0 is no indication/first generation and 1 is original/commercially prerecorded.
- Bits 16–19. Source number. 0000 (unspecified), 0001 (source 1), 0010 (source 2) ... 1111 (source 15).
- Bits 20–23. Channel number. 0000 (unspecified), 0001 (channel A), 0011 (channel B) ... 1111 (channel O).
- Bits 24–27. Sample frequency. 0000 (44.1 kHz), 0100 (48 kHz), 1100 (32 kHz).
- Bits 28–29. Clock accuracy. 00 (±1000 ppm), 01 (variable pitch), 10 (±50 ppm)
- Bits 30–191. Reserved.

10.3.5 Serial copy management system

Many originators of hi-fi audio are extremely worried about consumers taking perfect copies of digital information. This has forced hi-fi manufacturers to install a serial copy management system (SCMS). With this system a user is allowed to make a copy from the original copyrighted material but cannot make a copy from the copy. It should be noted that SCMS allows a user to make multiple copies from the original source.

SCMS is included in S/PDIF (and IEC-958) but there are no copy protection bits for AES3. The SCMS circuit tests the following:

1. Detects the consumer bit. If it is a 1 (professional) then copying will always occur. If it is a 0 then copying depends on the following two cases.
2. Examines the second bit which is the copyright bit. If it is set to a 0 then copyright is enabled, else it is disabled. Thus copying will always occur when the copyright bit is a 1 (copyright disabled). If it is a 0 then copying will only occur when the L bit is set to indicate that it is an original.

3. Examines the L bit (bit 15) which indicates the generation of the copy. A 0 indicates that it is a copy, else it is an original. (For laser optical products and broadcasts this bit has an opposite effect: a 1 indicates a copy and a 0 indicates the original.)

Thus if the consumer bit is set (i.e. a 0) and the copyright bit is enabled (i.e. a 0) then a copy will only be made if the L bit is set to original (i.e. a 0).

It is the law in many countries that SCMS circuitry must be fitted in digital copying audio equipment. When a consumer copies from a CD source to a DAT tape, the L bit is set to indicate it is copy. The DAT source cannot then be copied to another DAT source.

10.3.6 Other AES standards

Other AES standards do exist:

AES10-1991 – which defines a multichannel interface format. This allows the interconnection of multichannel digital audio equipment. It is also equivalent to ANSI S4.43-1991.
AES11-1990 – which defines a digital audio reference signal (DARS). This allows the synchronization of digital audio equipment in studio situations.
AES18-1992 – which defines a method of adding extra information AES3 frames. The added information may include scripts, copyright, editing information, and so on.

10.4 Error control

Digital audio conversion has the advantage of being able to add extra bits to make it possible to either detect errors or correct errors. Generally, the code used is either an error detection scheme (such as parity and CRC) or an error correction scheme (such as Hamming and Reed-Solomon codes). An error correction scheme allows a number of bits to be corrected but normally requires more bits for the error correction information. Error detection and correction will be discussed in Chapters 13 and 14.

10.5 Interleaving

Digital audio allows for other techniques which reduce losses in audio information. Interleaving is a technique which is used to conceal error by interleaving the samples in time. Errors normally occur in bursts thus consecutive samples which are interleaved will cause fewer consecutive errors. For example if data is sampled as:

1, 2, 3, 4, 5, 6, 7, 8, 9, 10, 11, 12, 13, 14, 15, 16, 17, 18, 19, 20, 21, 22, 23, 24, 25

To interleave them the samples are arranged into a number of columns and rows. In this case the 25 samples can be arranged into 5 rows with 5 columns to give:

```
1  2  3  4  5
6  7  8  9  10
```

11 12 13 14 15
16 17 18 19 20
21 22 23 24 25

Next the columns are read one at a time and reordered into a series again:

1, 6, 11, 16, 21, 2, 7, 12, 17, 22, 3, 8, 13,
18, 23, 4, 9, 14, 19, 24, 5, 10, 15, 20, 25.

This sequence can then be stored (or transmitted). If a burst error occurs (typically in a CD or in a communications channel) then it will have a lesser effect than storing the values consecutively. For example, suppose the 5th, 6th and 7th samples were corrupted. Then the recovered samples would be:

1, 6, 11, 16, X, X, X, 12, 17, 22, 3, 8, 13,
18, 23, 4, 9, 14, 19, 24, 5, 10, 15, 20, 25.

These would then be reordered to give:

1, X, 3, 4, 5, 6, X, 8, 9, 10, 11, 12, 13,
14, 15, 16, 17, 18, 19, 20, X, 22, 23, 24, 25.

It can be seen that in this sequence there are no consecutive errors. The values that are in error can be concealed by taking the average of the two samples on each side of them. For example, the 2nd sample would be estimated as the average of the 1st and the 3rd sample; this is illustrated in Figure 10.6. If three consecutive samples had been in error then the resultant concealment would be much worse as there is a greater difference between the error-free samples.

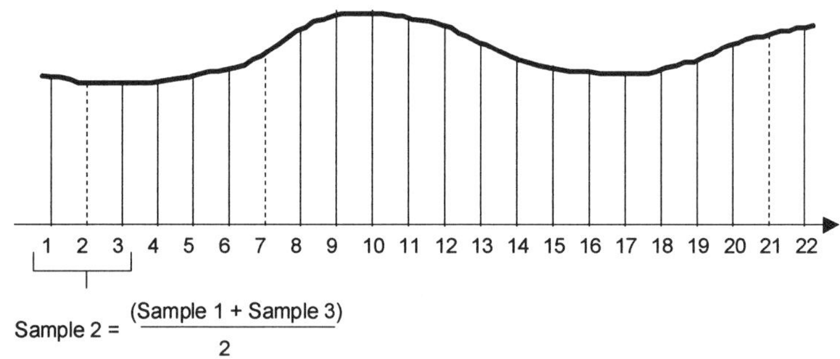

Figure 10.6 Concealment of errors.

10.6 CD audio system

The CD audio system provides a good example of the techniques used in digital audio. Figure 10.7 shows a CD. The data is recorded onto the CD in a serial manner (that is one bit at a time) and it is stored in a continuous groove that spirals around the disk from the outside to the inside. A pit (or the lack of one) differentiates between a 0 and a 1. The diagram shows that a pit is only 1.6 μm long and that there is only 0.5 μm between grooves.

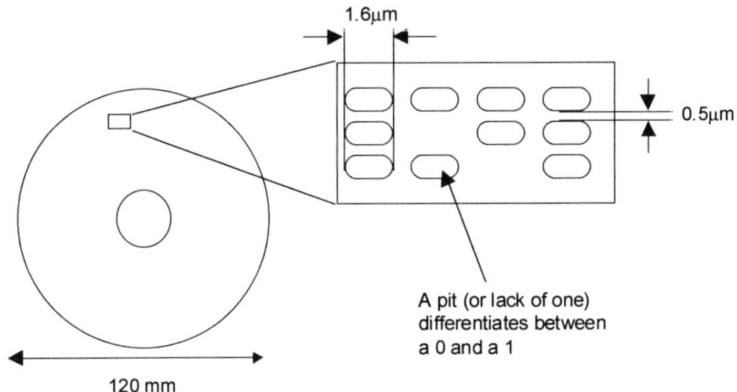

Figure 10.7 CD showing pits.

A CD uses a 44.1 kHz sampling rate with 16-bit PCM coding; it thus allows for 65 536 different levels. It can be proved that the dynamic range of a CD systems is 96.32 dB and the SNR is 98.08 dB.

As CDs are susceptible to scratches and fingerprints then errors are likely to occur in bursts. Thus, the CD uses an interlacing method to store the samples. It organizes the samples into 4 columns and 5 rows. A 20 sample input block would be arranged logically as:

1	2	3	4
5	6	7	8
9	10	11	12
13	14	15	16
17	18	19	20
21	22	23	24

This would then be stored on the disk as:

1, 5, 9, 13, 17, 21, 2, 6, 10, 14, 1, 8, 22,
3, 7, 11, 15, 19, 23, 4, 8, 12, 16, 20, 24.

Figure 10.8 shows the encoding of the data onto the disk. Initially the 16-bit word is divided into two 8-bit symbols. These symbols are then modulated by a process known as 8-to-14 bit modulation (EFM), where the 8-bit data is changed to 14-bit data through a lookup table. The EFM process reduces the CD system's sensitivity to optical tolerances in the disk player. After this, 3 bits of subcode are added to the 14-bit code. This subcode contains sync, index and time information. The resulting 17-bit code is then stored to the disc with two consecutive left-channel samples then right channels. An error correcting code is added at the end of each block.

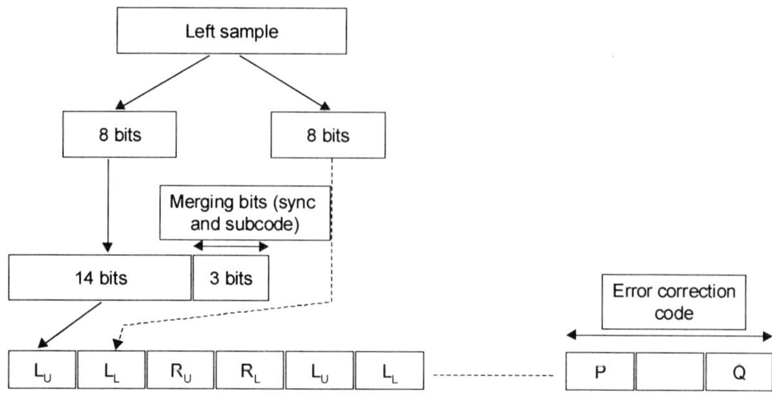

Figure 10.8 Encoding of data onto a CD.

The EFM code represents 8-bit symbols with a 14-bit code. There are 256 combinations of 8-bit symbols out of a possible 16 384 (2^{14}) different encoded values. Table 10.1 shows a portion of the codetable. The code has been chosen so that the maximum run-length of 0's will never be greater that 11.

Table 10.1 EFM encoding.

Hex	*8-bit data*	*14-bit encoded data*	*Hex*	*8-bit data*	*14-bit encoded data*
64	01100100	01000100100010	72	01110010	11000100100010
65	01100101	00000000100010	73	01110011	00100001000010
66	01100110	01000000100100	74	01110100	00000010000010
67	01100111	00100100100010	75	01110101	00000010000010
68	01101000	01001001000010	76	01110110	00010010000010
69	01101001	10000001000010	77	01110111	00100010000010
6A	01101010	10010001000010	78	01111000	01001000000010
6B	01101011	10001001000010	79	01111001	00001001001000
6C	01101100	01000001000010	7A	01111010	10010000000010
6D	01101101	00000001000010	7B	01111011	10001000000010
6E	01101110	00010001000010	7C	01111100	01000000000010
6F	01101111	00100001000010	7F	01111101	00001000000010
70	01110000	10000000100010	7E	01111110	00010000000010
71	01110001	10000010000010	7F	01111111	00100000000010

For example, the DSV for the first three codes is:

14-bit encoded data	*Waveform*	*DSV*
01000100100010	LHHHHLLLHHHHLL	+2
00000000100010	LLLLLLLLHHHHLL	−6
01000000100100	LHHHHHHHLLLHHH	+6

It can be seen that these codes give a DSV between +6 and −6. For this purpose CD discs have 3 extra packing bits added, these bits try to bring the DSV back to 0. It also inverts the code to give the inverse (for example the 01000100100010 can give either a DSV of +2 (LH HHHL LLHH HHLL) or −2 (HL LLLH HHLL LLHH).

A few example codes for the 8-to-14 conversion are:

01101010 10010001000010 01101011 10001001000010
01101100 01000000100010

The advantage of the interleaving is not only that errors are concealed by using interpolation, but also that it reduces the chances of losing sync information (which may cause the disk to jump).

10.7 Digital audio compression

There are several standard schemes for the compression of audio data; they include:

- MUSICAM – which has been adopted as one of the layers of the MPEG standard. It reduces the bit rate to 192 kbps.
- AC-1 – which is used in satellite relays of television and FM programs and gives data rates of around 512 kbps (3:1 ratio).
- AC-2 – which is applied to many applications, including sound cards, studio/transmitter links and so on. It gives data rates of around 256 kbps (6:1 ratio).
- AC-3 – which gives 6 channels of audio (left, right, surround-left, surround-right, center and sub-woofer) and gives data rates of around 512 kbps (6:1 ratio).
- MPEG – can compress to around 64 kbps for a single audio channel (monophonic channel). MPEG-3 compression is now common and is used in MiniDisk recorders.

Table 10.2 Audio compression.

Type	*Bit rate*	*Compression*	*Comment*
Uncompressed audio (mono)	705.6 kbps	1:1	
Uncompressed audio (stereo)	1.4112 Mbps	1:1	
AC-1	512 kbps	3:1	
AC-2	256 kbps	6:1	See above
AC-3	512 kbps	6:1	See above
MPEG (mono)	64 kbps	11:1	Monophonic
MUSICAM (stereo)	192 kbps	7:1	Stereophonic

10.8 The 44.1 kHz sampling rate

The 44.1 kHz sampling rate is actually derived from composite video applications. In the early days of digital audio, video recorders were adapted to store digital audio information. Thus the sampling rate was constrained to relate to the field rate and field structure of the television standard used. Thus there had to be an integer number of samples for each usable TV line in the field.

NSTC has a refresh rate of 60 Hz, 525 lines (of which 490 are active lines). As the signal is interlaced, only 245 lines are active for each transmission. Therefore, if three samples are taken for each line then the sampling rate is:

$$60 \times 245 \times 3 = 44.1\ \text{kHz}$$

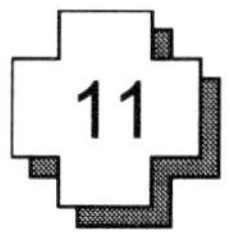

11 Audio Compression (MPEG-Audio and Dolby AC-3)

11.1 Introduction

CD-quality stereo audio requires a bit rate of 1.411200 Mbps (2×16 bits $\times$ 44.1 kHz). A single-speed CD-ROM can only transfer at a rate of 1.5 Mbps and this rate must include both audio and video. Thus, there is a great need for compression of both the video and audio data. The need to compress high-quality audio is also an increasing need as consumers expect higher-quality sound from TV systems and the increasing usage of digital audio radio.

A number of standards have been put forward for digital audio coding for TV/video systems. One of the first was MUSICAM which is now part of the MPEG-1 coding system. The FCC Advisory Committee considered several audio systems for advanced television systems. There was generally no agreement on the best technology but finally they decided to conduct a side-by-side test. The winner was Dolby AC-3 followed closely by MPEG. Many cable and satellite TV systems now use either MPEG or Dolby AC-3 coding.

11.2 Psycho-acoustic model

MPEG and Dolby AC-3 use the psycho-acoustic model to reduce the data rate, which exploits the characteristics of the human ear. This is similar to the method used in MPEG video compression which uses the fact that the human eye has a lack of sensitivity to the higher-frequency video components (that is, sharp changes of color or contrast). The psycho-acoustic model allows certain frequency components to be reduced in size without affecting the perceived audio quality as heard by the listener.

11.2.1 Masking effect

A well-known effect is the masking effect. This is where noise is only heard by a person when there are no other sounds to mask it. A typical example is high-frequency hiss from a compact cassette when there are quiet passages of music. In normal periods of music, the louder music masks out the quieter hiss and the noise is not heard. In reality, the brain is actually masking the part of the sound it wants to hear, although the noise component is still there. When there is no music to mask the sound then the noise is heard.

Noise, itself, tends to occur across a wide range of frequencies, but the masking effect also occurs with certain sounds. A loud sound at a certain frequency masks out a quieter sound at a similar frequency. Therefore the sound heard by the listener appears only to contain the loud sounds; the quieter sounds are masked out. The psycho-acoustic model tries to reduce the levels to those that would be perceived by the brain.

Figure 11.1 illustrates this psycho-acoustic process. In this case, a masking level has been applied and all the amplitudes below this level have been reduced in size. Since these frequencies have been reduced in amplitude, then any noise associated with them is also significantly reduced. This has the effect of limiting the bandwidth of the signal to the key frequency ranges and also limiting the noise bandwidth.

The psycho-acoustic model also takes into account non-linearities in the sensitivity of the ear. Its peak sensitivity is between 2 and 4 kHz (the range of the human voice) and it is least sensitive around the extremes of the frequency range (i.e. high and low frequencies). Any noise in the less sensitive frequency ranges is more easily masked, but it is important to minimize any noise in the peak range because it has a greater impact.

Masking can also be applied in the time domain, where it is applied just before and after a strong sound (such as a change of between 30 and 40 dB). Typically, premasking occurs for about 2–5 ms before the sound is perceived by the listener and the post-masking effect lasts for about 100 ms after the end of the source.

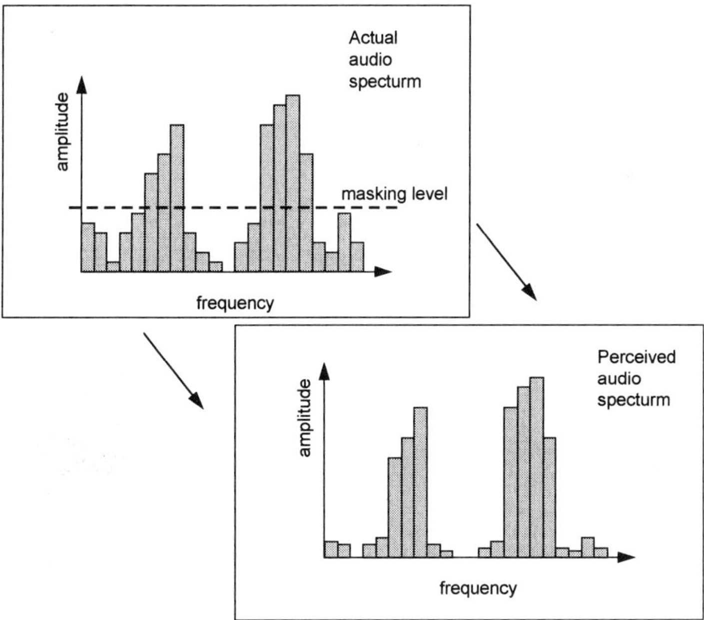

Figure 11.1 Difference between actual and perceived audio spectrum.

11.3 MPEG audio coding

MPEG basically has three different levels:

- MPEG-Audio Level I – uses a psycho-acoustic model to mask and reduce the sample size. It is basically a simplified version of MUSICAM and has a quality which is nearly equivalent to CD-quality audio. Its main advantage is that it allows the construction of simple encoders and decoders with medium performance and which operate well at 192 or 256 kbps.

- MPEG-Audio Level II – which is identical to the MUSICAM standard. It is also nearly equivalent to CD-quality audio and is optimized for a bit rate of 96 or 128 kbps per monophonic channel.
- MPEG-Audio Level III – which is a combination of the MUSICAM scheme and ASPEC, a sound compression scheme designed in Erlangen, Germany. Its main advantage is that it targets a bit rate of 64 kbps per audio channel. At that speed, the quality is very close to CD quality and produces a sound quality which is better than MPEG Level-II operating at 64 kbps.

The three levels are basically supersets of each other with Level III decoders being capable of decoding both Level I and Level II data. Level II is the most frequently used of the three standards. Level I is the simplest, while Level III gives the highest compression but is the most computational in coding.

The forward and backward compatible MPEG-2 system, following recommendations from SMPTE, EBU and others, has increased the audio capacity to five channels. Figure 11.1 shows an example of a 5-channel system; the key elements are:

- A center channel.
- Left and right surround channels.
- Left and right channels (as hi-fi stereo).

MPEG-2 also includes a low-frequency effects channel (called LFE, essentially a sub-woofer). This has a much lower bandwidth that the other channels. This type of system is often called a 5.1-channel system (5 main channels and LFE channel).

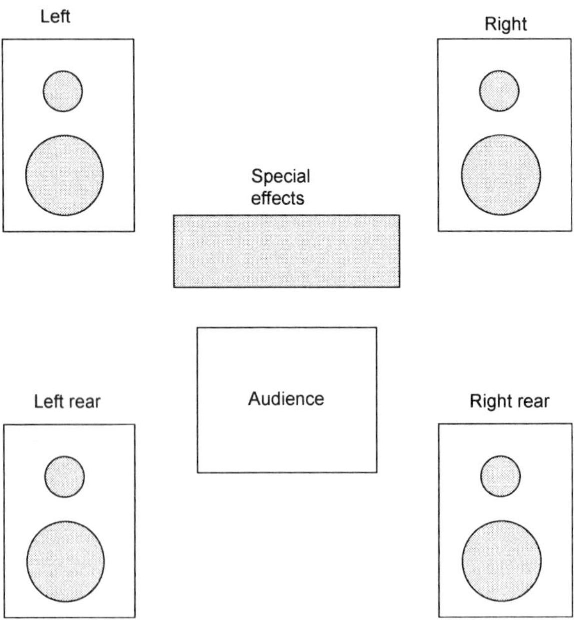

Figure 11.2 A 5.1 channel audio surround sound system.

One of the main objectives of MPEG-2 was to make it backward compatible with MPEG-1. For this reason, the MPEG-1 specification was extended to include capability for 5.1 channels so that it allowed other MPEG-1 decoders to extract the basic stereo pairs while newer decoders could recover all the channels. MPEG decoders are also forward compatible as they can recover MPEG-1 compliant audio.

MPEG audio systems break up the input signal into uniform segments with a certain number of audio samples, called frames. These frames are then filtered by digital filters into subbands of audio energy for each frame. Normally there are 32 subbands, so the subband frequency is divided by 32 ($F_s/32$).

As with MPEG and JPEG video compression, the subband sub-samples, are then quantized by means of a quantization scale factor. This compands the data and compresses the dynamic range of the audio in each subband. As with video compression, it results in a set of coefficients for each subband for each frame period.

At the same time as the calculation of the subband coefficients, a frequency-dependent masking threshold is calculated for each frame. A psycho-acoustic perceptual model is used to calculate this threshold. It is then used to alter the bit allocation process for each subband where bits in the encoder output are dynamically allocated to each subband, subject to the ratio of the signal within the subband and its masking function. This process is illustrated in Figure 11.3.

As with video compression, various methods are then applied to the quantized data, such as variable-length coding.

MPEG audio has three main levels. Level 1 gives the basic functionality while Level II adds more sophisticated processing in deriving the quantization scale factor and additional coding of the bit allocation. Level III provides for additional decomposition of the frequency components in each subband using a modified discrete cosine transform (MDCT).

The decoding process is essentially reversed, with the bits for each subband de-quantized and applied to band-limited digital filters that reconstruct a PCM audio signal for their subband.

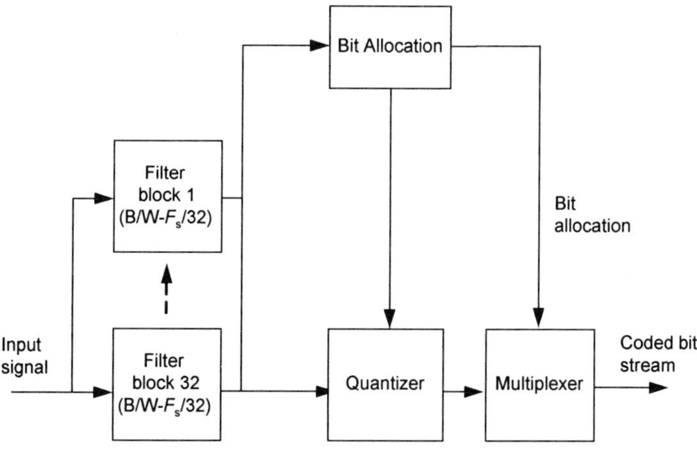

Figure 11.3 Forward bit allocation encoder.

11.3.1 MPEG levels

The three different levels of MPEG vary in the method they use to compress the audio and in their bit rates.

Level I

The MPEG Level I coder gives medium performance with bit rates of 128 kbps per channel. Philips has used it in the digital compact cassette (DCC). The Level I coder initially transforms the audio signal into frequency information using a fast Fourier transform. Next the frequency information is passed into the 32 subbands filters. A 44.1 kHz sampling rate gives a subband bandwidth of 689.0625 Hz. These filters allow the processing of each subband using a slightly different acoustic model and thus achieve a better and more accurate encoding. A polyphase filter band is normally used to create the subbands as they are relatively fast.

After the subband processing, the output information is processed using the psycho-acoustic model. There are two acoustic models: model 1 and model 2. Levels I and Level II use model 1 and Level III uses model 2. In general, model 1 is the least complex.

Model 1 processes 512 samples at a time (it uses a 512-sample window). Therefore, the audio is analyzed and processed in separate blocks. Level I encoding uses 32 samples from each of the 32 subbands. This gives a frame size of 384 samples, which are centered by calculating an offset into the middle of the 512-sample frame.

Level II

Level II is aimed at bit rates of about 128 kbps and has a slightly higher compression rate than Level I. As with Level I it also uses model 1. The sample size is increase from 384 to 1152, and the sample window is increased to 1024. This sample window is not quite large enough to fit all the samples; to cope with this, the encode performs two analyses. The first analysis takes the first 576 samples and centers them in a 1024-sample window, and the second analysis repeats the process with the second 576 samples. This results in higher signal-to-noise ratios for the final output as it effectively selects the best noise-masking values.

Level III

Level II is aimed at bit rates of about 64 kbps per channel and has a slightly higher compression rate than Level II. The sample size is the same as Level II and the same split processing techniques are used. The increase in compression is achieved by employing a more sophisticated model 2 with several important differences.

11.3.2 MPEG-1

The basic specification MPEG-1 system provides two channels of stereo at sampling rates of either:

- 32 kHz (broadcast communications equipment).
- 44.1 kHz (CD-quality audio).
- 48 kHz (professional sound equipment).

Table 11.1 gives the main parameters for MPEG-1 audio.

Table 11.1 MPEG-1 parameters.

	32 kHz	*44.1 kHz*	*48 kHz*
Audio bandwidth	15 kHz	20.6 kHz	22.5 kHz
Frame duration – Level I (384 samples)	12 ms	8.7 ms	8 ms
Frame duration – Level II (1152 samples)	36 ms	26.25 ms	24 ms
Frame duration – Level III (1152 samples)	36 ms	26.25 ms	24 ms
Number of subbands	32	32	32
Subband sampling ($F_s/32$)	1 kHz	1.378 kHz	1.5 kHz
Subband bandwidth ($F_s/64$)	500 Hz	689.06 kHz	750 Hz
Bit rate (Level 2/stereo)	192 kbps	256 kbps	256 kbps
Bit rate (Level 2/5.1-channel)	256 kbps	384 kbps	384 kbps

Levels II and III have 1152 samples. Thus for 32 kHz sampling the frame duration will be 36 ms, which gives one sample every 31.25 μs, as illustrated in Figure 11.3. Level I has only 384 samples thus the frame time is reduced to one-third of the value for Levels II and III. Each frame has 32 subbands, and the sampling frequency at the output of each filter in the sampling frequency (F_s) is divided by 32 ($F_s/32$). Since the Nyquist criterion states that the highest frequency that can be represented in a sampled system is half the sampling rate, the bandwidth of the subband is $F_s/64$. For 32 kHz sampling, the filters have output sample rates of 1 kHz and bandwidths of 500 Hz. The reason that the audio bandwidth is slightly less than half the Nyquist frequency is that it takes into account imperfect filters.

The MPEG-1 version of the MPEG audio system provides for joint stereo coding, which exploits stereophonic irrelevance. The method is called intensity stereo coding; it removes redundancy in stereo audio information by retaining only the energy envelope of the right channel and the left channel at high frequencies.

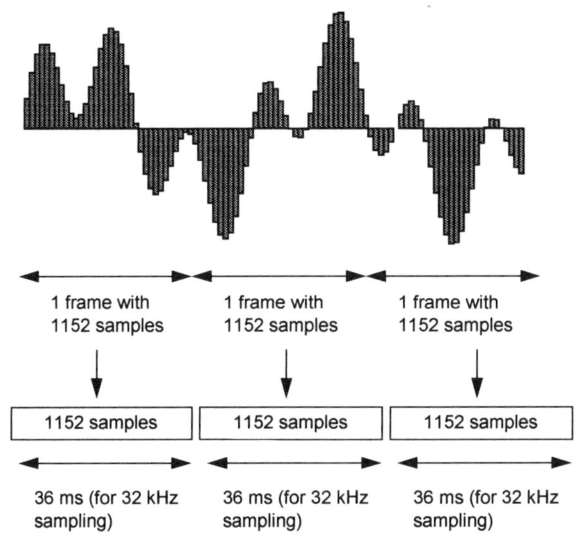

Figure 11.4 Frame for MPEG Levels 2 and 3.

The basic MPEG-1 audio frame, after encoding, includes four basic types of information:

- A header.
- A cyclic redundancy code (CRC) for error detection.
- Audio data.
- Ancillary data.

The MPEG file also contains bit allocation information, scale factor selection information, scale factors, and subband samples.

11.3.3 MPEG-2

MPEG-2 is an extension of MPEG-1 and is both forward and backward compatible with it. It adds the following:

- Three additional sampling frequencies (16 kHz, 22.05 kHz and 24 kHz).
- Up to four more channels, permitting 5-channel surround sound.
- Support for low-frequency effects.
- Support for separate audio material in different configurations. This can be used in multi-lingual transmission of a program or to include a single channel of clean dialogue for the hard of hearing, or even a commentary channel for the visually impaired.

MPEG-2 can deliver hi-fi quality stereo sound with 256 kbps and 5.1-channel surround sound with LFE in 384 kbps. A 16 kHz sampling rate, quality audio, gives a bit rate of 128 kbps (which is compatible with standard rate ISDN). This gives an audio bandwidth of 7.5 kHz, which is large enough for speech and reasonable for low-quality audio. MPEG-2 achieves compatibility with MPEG-1 because it hides the extra data with an ancillary data field. An MPEG-1 decoder ignores the ancillary data file. Table 11.2 gives the main parameters of MPEG-2. Like MPEG-1 all the sampling rates have 32 subbands.

Table 11.2 MPEG-2 coding parameters.

	16 kHz	*22.05 kHz*	*24 kHz*	*32 kHz*	*44.1 kHz*	*48 kHz*
Audio bandwidth	7.5 kHz	10.3 kHz	11.25 kHz	15 kHz	20.6 kHz	22.5 kHz
Frame duration (Level I)	24 ms	17.4 ms	16 ms	12 ms	8.7 ms	8 ms
Frame duration (Level II)	72 ms	52.5 ms	48 ms	36 ms	26.25 ms	24 ms
Frame duration (Level III)	72 ms	52.5 ms	48 ms	36 ms	26.25 ms	24 ms
Subband sampling ($F_s/32$)	500 Hz	689.0625 Hz	750 Hz	1000 Hz	1378.125 Hz	1.5 kHz
Subband bandwidth ($F_s/64$)	250 Hz	344.5313 Hz	375 Hz	500 Hz	689.0625 Hz	750 Hz
Bit rate (Level II with stereo)	96 kbps	128 kbps	128 kbps	192 kbps	256 kbps	256 kbps
Bit rate - (Level II with 5.1 channel)	128 kbps	192 kbps	192 kbps	256 kbps	384 kbps	384 kbps

11.4 Backward/forward adaptive bit allocation methods

MPEG uses a forward adaptive bit allocation method, as shown in Figure 11.5. This technique makes bit allocation decisions adaptively, based on signal content. These decisions are then sent from the encoder to the decoder so that the decoder can properly dequantize values sent to it.

Figure 11.6 shows the backward adaptive allocation method. This method represents the spectral envelope as exponents at a given sample time. The spectral envelope is encoded and sent to the decoder; it is also used for deciding which coefficients are significant to the sound at any instant and then for controlling the bit allocation to each of the subband coefficients.

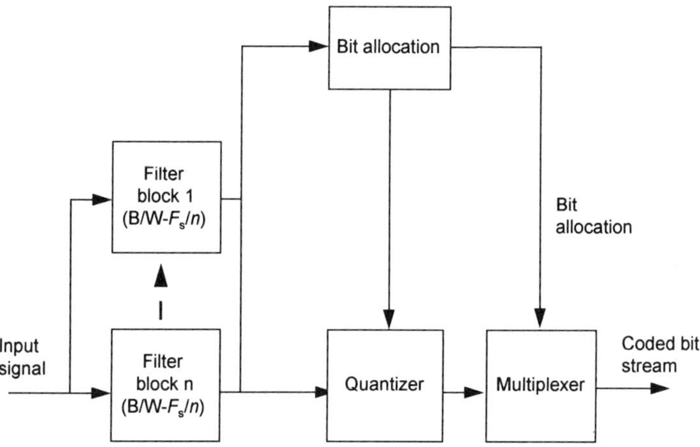

Figure 11.5 Forward bit allocation encoder.

Bit allocation quantizes the coefficients of each subband according to the calculated allocation. This quantization is similar to the method used in MPEG audio and controls the number of bits used to represent a value by changing the granularity, or the fineness of amplitude resolution, with which the value is expressed. For example, a 2-bit coded value can represent only 4 levels, whereas an 8-bit coded value can represent 256 values.

The difference between the actual value and the quantized value leads to quantization noise; the greater the differences between them, the lower the signal-to-noise ratio. The filter bank at the decoder then limits the quantization noise in any subband to have nearly the same frequency as the signal in that subband. Thus, the quantization noise is masked by the strong signal, thus yielding a higher signal-to-noise ratio.

The output samples of the subband filter are the coefficients that are processed for transmission to the decoder. Each coefficient has a combination of exponent and mantissa. The exponent contains the major level information for its subband, and the mantissa contains the detailed level of each sample. As the exponent is exponential it can be thought of as containing the amplitude "range" information of the subband. These two values are multiplexed onto the coded bitstream.

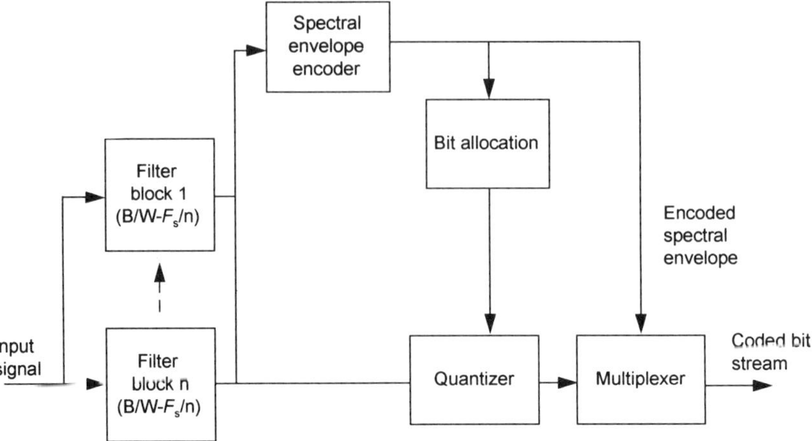

Figure 11.6 Backward adaptive bit allocation encoder.

Figure 11.7 illustrates the operation of the backward adaptive bit allocation decoder. It initially reconstructs the spectral envelope in order to calculate the bit allocation. Since both the encoder and the decoder use the same spectral envelope then the same bit allocation decisions are made at both ends. This has the advantage that the decoder can dequantize the quantized mantissas without the actual bit allocation data being sent. This results in a smaller amount of data being transmitted because all of the transmitted data is devoted to audio coding. The method is defined as a backward adaptive bit allocation because the content-controlled bit allocation is calculated identically at both ends.

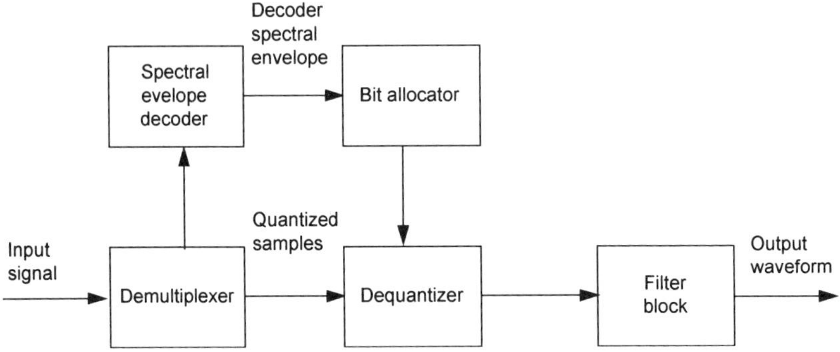

Figure 11.7 Backward adaptive bit allocation decoder.

11.5 Comparison between forward and backward adaptive methods

Table 11.3 outlines the main advantages of the forward and backward adaptive methods. As bit allocation information is encoded in the bitstream, the forward allocation method can be as accurate as required. Unfortunately, a large amount of data is devoted to transporting the bit allocation information. When transients are significant it is often desirable update the bit allocation with a finer time resolution. With the forward adaptive bit allocation method there is a direct increase in the channel information data rate. With steady-state conditions, it is desirable to provide finer frequency resolution on the spectral analysis and in the bit allocation. Again this increases the data rate.

Table 11.3 Advantages and disadvantages of adaptive bit allocation methods.

Advantages	*Disadvantages*
FORWARD	
The psycho-acoustic model resides only in the encoder.	Since the bit allocation and audio data are encoded, a substantial amount of information is devoted to non-audio data.
As the model only resides in the encoder, it can be improved over time as better models of the human auditory system are developed.	Finer resolution in the time or frequency domain will increase the data rate.
BACKWARD	
Since both the encoder and decoder use the same information (derived from the spectral envelope) to derive the bit allocation, none of the channel data capacity must be wasted in sending specific bit allocation instructions.	The decoder must calculate the bit allocation entirely from information in the coefficient bitstream. The information sent to the decoder has limited accuracy and therefore may contain small errors.
Finer resolution in the time of frequency domain need not increase the data rate.	The processing of the envelope increases the complexity of the decoder.
	The bit allocation algorithm becomes fixed as soon as the first decoders are deployed, and it then becomes impossible to update the psycho-acoustic model.

The main advantage of the backward adaptive bit allocation method is that none of the encoded data contains bit allocation information. This is because the bit allocation is derived from the spectral envelope. Bit allocation can thus have time or frequency resolution as all the information is in the envelope. The method is thus more efficient in its encoded data and can have a finer time or frequency resolution when required.

The main disadvantage with the backward adaptive bit allocation method is that the information in the envelope will not be totally accurate. As this information is used in the bit allocation it leads to inaccuracies. Another disadvantage is that the processes of bit allocation and envelope generation increases the complexity of the decoder. This obviously increases the cost of the decoder hardware. A final disadvantage is that the bit allocation algorithm is fixed, it becomes impossible to update the psycho-acoustic model.

11.6 Dolby AC-1 and AC-2

Dolby AC-1 was designed for stereo satellite relays of television and for FM programs. Using adaptive delta modulation and analogue companding, it has an audio bandwidth of 20 kHz into a 512 kbps bitstream (3:1).

It uses a perceptual coder using a low-complexity block transform. Initially it divides the wideband signal into multiple subband using a 512-point 50% overlap FFT algorithm performing the MDCT (modified discrete cosine transform). These coefficients are then grouped into subbands containing from 1 to 15 coefficients to model critical bandwidths. Each of the subbands has a number of preallocated bits. Lower frequency subbands have more preallocated bits than higher-frequency subbands. Any requirement for additional bits can be drawn upon depending on bit allocation calculations. The number of subbands is 40 (for 48 kHz sampling) and 43 (for 32 kHz sampling).

AC-2 is used in many applications, such as in PC soundcards and professional equipment. It gives high audio quality at a low data rate of 256 kbps. Typical compression rates are 6.1:1 (for 48 kHz sampling) and 5.4:1 (for 32 kHz sampling).

11.7 Dolby AC-3 coding

Dolby AC-3 uses a hybrid of forward and backward adaptive bit allocation, as illustrated in Figure 11.8. It uses backward adaptive bit allocation with core backward adaptive bit allocation. This core bit system is used in both the encoder and the decoder and can be modified in its operation by forward adaptive bit allocation side information. This allows both the encoder and decoder core bit allocations systems to run independently using the spectral envelope, as in the backward adaptive case, but they are altered by forward adaptive information, which improves their accuracy.

The core bit allocation system is based on a fixed psycho-acoustic model with two types of modifications:

- Modification of the parameters of the psycho-acoustic model.
- Differences to the bit allocations that result from the current psycho-acoustic model.

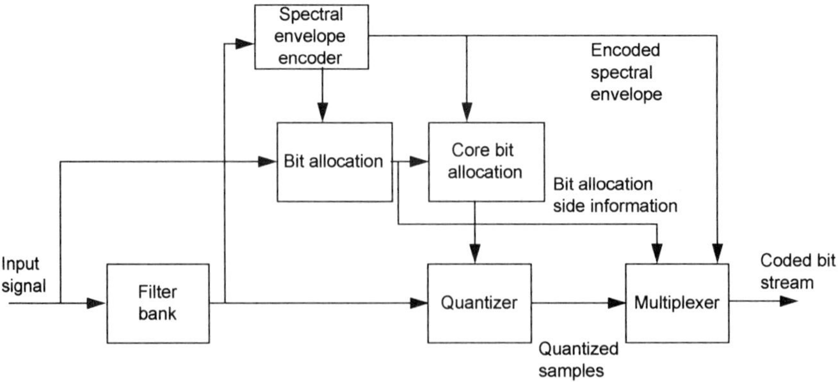

Figure 11.8 Hybrid backward/forward adaptive bit allocation encoder.

The main advantage of the hybrid approach is that the modification data sent to the core bit allocation routine is substantially less than would be required for normal forward adaptation. The psycho-acoustic model can also be updated dynamically.

11.8 AC-3 parameters

Dolby AC-3 uses a frame which consists of six blocks, as illustrate in Figure 11.9. Each block can be varied in length to accommodate different applications, typical sizes have 256 and 512 samples. Normally the data within the blocks overlaps on each side of a block. This results in the same sample being contained in two blocks. Thus for a 512-sample block, the last 256 samples will be the same as the first 256 samples of the next block. Table 11.4 outlines the main Dolby AC-3 parameters. It can be seen that a 32 kHz sampling gives a block duration of 16 ms and 44.1 kHz gives a block duration of 11.61 ms.

MPEG audio coding uses 32 subband filters, whereas Dolby AC-3 has 256 subbands. For example, 44.1 kHz sampling gives frequency domains with a 62.5 Hz bandwidth.

MPEG encoders typically use DFT or DCT techniques to generate the time-to-frequency conversion. These use an *N*-point transform that generates *N* unique non-zero transform coefficients. Dolby AC-3 uses an odd-stacked time-division aliasing cancellation (TDAC) technique which is a modified discrete cosine transform (MDCT). The main advantage with this method is that the 50% overlay is achieved without any increase in the resulting bit rate.

In TDAC, each MDCT transform of block size *N* generates only *N*/2 unique non-zero transform coefficients, so critical sampling is achieved with the 50% block overlap already described. The MDCT method is efficient in its computation and in its usage of memory.

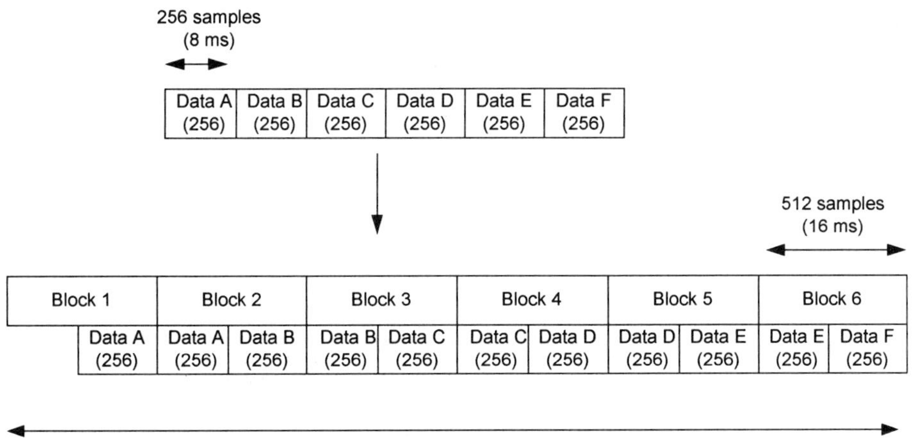

Figure 11.9 Frame and block format.

Table 11.4 Dolby AC-3 coding parameters.

	32 kHz	*44.1 kHz*	*48 kHz*
Audio bandwidth	15 kHz	20.6 kHz	22.5 kHz
Block length	512 samples	512 samples	512 samples
Block duration	16 ms	11.61 ms	10.66 ms
Subbands	256	256	256
Subband bandwidth (F_s/256)	62.5 Hz	86.133 Hz	93.75 Hz
Block repetition rate	125 Hz	173.26 Hz	187.5 Hz
Block repetition period	8 ms	5.805 ms	5.333 ms
Bit rate (stereo)	192 kbps	192 kbps	192 kbps
Bit rate (5.1-channel)	384 kbps	384 kbps	384 kbps

12 Error Coding Principles

12.1 Introduction

Error bits are added to data either to correct or to detect transmission errors. Normally, the more bits that are added, the better the correction or detection. Error detection allows the receiver to determine if there has been a transmission error. It cannot rebuild the correct data and must either request a retransmission or discard the data. With error correction the receiver detects an error and tries to correct as many error bits as possible. Again, the more error coding bits are used, the more bits can be corrected. An error correction code is normally used when the receiver cannot request a retransmission.

In a digital communication system, a single transmission symbol can actually contain many bits. If a single symbol can represent M values then it is described as an M-ary digit. An example of this is in modem communication where 2 bits are sent as four different phase shifts, e.g. 0° for 00, 90° for 01, 180° for 10 and 270° for 11. To avoid confusion it is assumed in this chapter that the digits input to the digital coder and the digital decoder are binary digits and this chapter does not deal with M-ary digits.

12.2 Modulo-2 arithmetic

Digital coding uses modulo-2 arithmetic where addition becomes the following operations:

$0+0=0$ $\quad$ $1+1=0$

$0+1=1$ $\quad$ $1+0=1$

It performs the equivalent operation to an exclusive-OR (XOR) function. For modulo-2 arithmetic, subtraction is the same operation as addition:

$0-0=0$ $\quad$ $1-1=0$

$0-1=1$ $\quad$ $1-0=1$

Multiplication is performed with the following:

$0\times 0=0$ $\quad$ $0\times 1=0$

$1\times 0=0$ $\quad$ $1\times 1=1$

which is an equivalent operation to a logical AND operation.

12.3 Binary manipulation

Binary digit representation, such as 101110, is difficult to use when multiplying and dividing. A typical representation is to manipulate the binary value as a polynomial of bit powers. This technique represents each bit as an x to the power of the bit position and then adds each of the bits. For example:

10111	x^4+x^2+x+1
1000 0001	x^7+1
1111 1111 1111 1111	$x^{11}+x^{10}+x^9+x^8+x^7+x^6+x^5+x^4+x^3+x^2+x+1$
10101010	$x^6+x^4+x^2+x$

For example:	101×110
is represented as:	$(x^2+1)\times(x^2+x)$
which equates to:	$x^4+x^3+x^2+x$
which is thus:	11110

The addition of the bits is treated as a modulo-2 addition, that is, any two values which have the same powers are equal to zero. For example:

$$x^4+x^4+x^2+1+1$$

is equal to x^2 as x^4+x^4 is equal to zero and 1+1 is equal to 0 (in modulo-2). An example which shows this is the multiplication of 10101 by 01100.

Thus:	10101×01110
is represented as:	$(x^4+x^2+1)\times(x^3+x^2+x)$
which equates to:	$x^7+x^6+x^4+x^5+x^4+x^3+x^3+x^2+x$
which equates to:	$x^7+x^6+x^5+x^4+x^4+x^3+x^3+x^2+x$
which equates to:	$x^7+x^6+x^5+0+0+x^2+x$
which equates to:	$x^7+x^6+x^5+x^2+x$
which is thus:	11100110

This type of multiplication is easy to implement as it just involves AND and XOR operations.

The division process uses exclusive-OR operation instead of subtraction and can be implemented with a shift register and a few XOR gates. For example 101101 divided by 101 is implemented as follows:

```
         1011
    100 |101101
         100
         ---
          110
          100
          ---
           101
           100
           ---
             1
```

Thus the modulo-2 division of 101101 by 100 is 1011 remainder 1. As with multiplication this modulo-2 division can also be represented with polynomial values (an example of this is given in Section 12.5.2).

Normally, pure integer or floating point multiplication and division require complex hardware and can cause a considerable delay in computing the result. Error coding multiplication and division circuits normally use a modified version of multiplication and division which uses XOR operations and shift registers.

12.4 Hamming distance

The Hamming distance, $d(C_1,C_2)$, between two code words C_1 and C_2 is defined as the number of places in which they differ. For example, the codes:

101101010 and 011101100

have a Hamming distance of 4 as they differ in 4 bit positions. Also $d(11111,00000)$ is equal to 5.

The Hamming distance can be used to determine how well the code will cope with errors. The minimum Hamming distance $\min\{d(C_1,C_2)\}$ defines by how many bits the code must change so that one code can become another code.

It can be shown that:

- A code C can detect up to N errors for any code word if $d(C)$ is greater than or equal to $N+1$ (that is, $d(C)\geq N+1$).
- A code C can correct up to M errors in any code word if $d(C)$ is greater than or equal to $2M+1$ (that is, $d(C)\geq 2M+1$).

For example the code:

{00000, 01101, 10110, 11011}

has a minimum Hamming distance of 3. Thus the number of errors which can be detected is given by:

$$d(C)\geq N+1$$

since, in this case, $d(C)$ is 3 then N must be 2. This means that one or two errors in the code word will be detected as an error. For example the following have 2 bits in error, and will thus be received as an error:

00011, 10101, 11010, 00011

The number of errors which can be corrected (M) is given by:

$$d(C)\geq 2M+1$$

thus M will be 1, which means that only one bit in error can be corrected. For example if the received code word was:

01111

then this code is measured against all the other codes and the Hamming distance calculated. Thus 00000 has a Hamming distance of 4, 01101 has a Hamming distance of 1, 10110 has a Hamming distance of 3, and 11011 has a Hamming distance of 2. Thus the received code is nearest to 01101.

12.5 General probability theory

Every digital system is susceptible to errors. These errors may happen once every few seconds or once every hundred years. The rate at which these occur is governed by the error probability.

If an event X has a probability of $P(X)$ and event Y has a probability of $P(Y)$ then the probability that either might occur is:

$$P(X \text{ or } Y) = P(A) + P(B) - P(X \text{ and } Y)$$

If one event prevents the other from happening, then they are mutually exclusive, thus $P(X \text{ and } Y)$ will be zero. This will give:

$$P(X \text{ or } Y) = P(X) + P(Y)$$

If an event X has a probability of P(X) and event Y has a probability of P(Y) then the probability that both might occur is:

$$P(X \text{ and } Y) = P(X) \times P(Y \mid X)$$
$$P(X \text{ and } Y) = P(Y) \times P(X \mid Y)$$

where $P(Y \mid X)$ is the probability that event Y will occur, assuming that event X has already occurred, and $P(X \mid Y)$ is the probability that event X will occur, assuming that event Y has already occurred. If the two events are independent then $P(X \mid Y)$ will be $P(X)$ and $P(Y \mid X)$ will be $P(Y)$. This results with:

$$P(X \text{ and } Y) = P(X) \times P(Y)$$

For example, rolls of a die are independent of each other. If a die is rolled twice then the probability of a rolling two sixes will be:

$$\begin{aligned} P(6 \text{ and } 6) &= P(6) \times P(6) \\ &= \frac{1}{6} \times \frac{1}{6} = \frac{1}{36} \end{aligned}$$

This formula is used as one roll of the die is mutually exclusive to the next throw.

The probability of throwing a three or a two in a single throw of the dice will be:

$$P(2 \text{ or } 3) = P(2) + P(3)$$
$$= \frac{1}{6} + \frac{1}{6} = \frac{1}{3}$$

12.6 Error probability

Each digital system has its own characteristics and thus will have a different probability of error. Most calculations determine the 'worst-case' situation. The transmission channel normally assumes the following:

- That each bit has the same probability of being received in error.
- That there is an equal probability of a 0 and of a 1 (that is, the probability of a 0 is 0.5 and the probability of a 1 is a 1.0).

Thus if the probability of a binary digit being received incorrectly is p, then the probability of no error is $1-p$. This is illustrated in Figure 12.1.

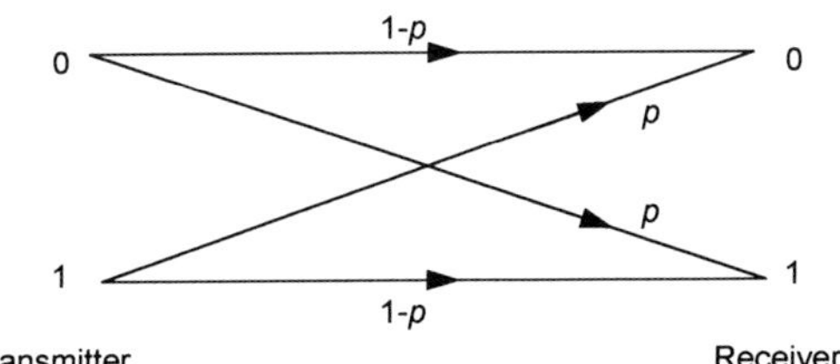

Figure 12.1 Probability or error model for a single bit.

If the probability of no errors on a signal bit is $(1-p)$, then the probability of no errors of data with n bits will thus be:

$$\text{Probability of no errors} = (1-p)^n$$

The probability of an error will thus be:

$$\text{Probability of an error} = 1-(1-p)^n$$

The probability of a single error can be determined by assuming that all the other bits are received correctly, thus this will be:

$$(1-p)^{n-1}$$

Thus the probability of a single error at a given position will be this probability multiplied by the probability of an error on a single bit, thus:

$$p(1-p)^{n-1}$$

As there are n bit positions then the probability of a single bit error will be:

$$\text{Probability of single error} = \quad n.p(1-p)^{n-1}$$

For example if the received data has 256 bits and the probability of an error in a single bit is 0.001 then:

$$\begin{aligned}\text{Probability of no error} &= (1-0.001)^{256}\\ &= 0.774\end{aligned}$$

Thus the probability of an error is 0.226, and

$$\begin{aligned}\text{Probability of single error} &= 8 \times 0.001\ (1-0.001)^{256-1}\\ &= 0.0062\end{aligned}$$

12.7 Combinations of errors

Combinational theory can be used in error calculation to determine the number of combinations of error bits that occur in some n-bit data. For example, in 8 bit data there are 8 combination of single-bit errors: (1), (2), (3), (4), (5), (6), (7) and (8). With 2 bits in error there are 28 combinations $(1,2)$, $(1,3)$, $(1,4)$, $(1,5)$, $(1,6)$, $(1,7)$, $(1,8)$, $(2,3)$, $(2,4)$, $(2,5)$, $(2,6)$, $(2,7)$, $(2,8)$, $(3,4)$, $(3,5)$...$(6,6)$, $(6,7)$, $(6,8)$, $(7,8)$. In general the formula for the number of combinations of m-bit errors for n bits is:

$$\binom{n}{m} = \frac{n!}{m!(n-m)!}$$

Thus the number of double-bit errors that can occur in 8 bits is:

$$\binom{8}{2} = \frac{8!}{2!(8-2)!} = \frac{8!}{2!6!} = \frac{8 \times 7}{2} = 28$$

Table 12.1 shows the combinations for bit error with 8-bit data. It can thus be seen that there are 255 different error conditions (8 single-bit errors, 28 double-bit errors, 56 triple-bit errors, and so on).

Table 12.1 Combinations.

No of bit errors	*Combinations*	*No of bit errors*	*Combinations*
1	8	5	56
2	28	6	28
3	56	7	8
4	70	8	1

To determine the probability with m bits at specific places, use the probability that $(n-m)$ bits will be received correctly:

$$(1-p)^{n-m}$$

Thus the probability that m bits, at specific places, will be received incorrectly is:

$$p^m(1-p)^{n-m}$$

The probability of an m-bit error in n bits is thus:

$$P_e(m) = \binom{n}{m}.p^m.(1-p)^{n-m}$$

Thus the probability of error in n n-bit data is:

$$P_e = \sum_{m=1}^{n}\binom{n}{m}.p^m.(1-p)^{n-m}$$

which is in the form of a binomial distribution.

Question

Determine the probability of error for a 4-bit data block using the formula:

$$P_e = \sum_{m=1}^{n}\binom{n}{m}.p^m.(1-p)^{n-m}$$

and prove this answer using $P_e=1-(1-p)^n$.

Answer

The probability of error will be:

$$P_e = \binom{4}{1} \cdot p \cdot (1-p)^3 + \binom{4}{2} \cdot p^2 \cdot (1-p)^2 + \binom{4}{3} \cdot p^3 \cdot (1-p)^1 + \binom{4}{4} \cdot p^4 \cdot (1-p)^0$$
$$= 4p(1-p)^3 + 6p^2(1-2p-p^2) + 4p^3 - 4p^4 + p^4$$
$$= 4p(1-p)(1-p)^2 + 6p^2 - 12p^3 + 6p^4 + 4p^3 - 4p^4 + p^4$$
$$= 4p(1-p)(1-2p+p^2) + 6p^2 - 8p^3 + 3p^4$$
$$= 4p - 8p^2 + 4p^3 - 4p^2 + 8p^3 - 4p^4 + 6p^2 - 8p^3 + 3p^4$$
$$= 4p - 6p^2 + 4p^3 - p^4$$

To prove this result, the formula $P_e = 1-(1-p)^n$ can be used to give:

$$P_e = 1-(1-p)^4$$
$$= 1-(1-p)^2(1-p)^2$$
$$= 1-(1-2p-p^2)(1-2p-p^2)$$
$$= 1-(1-2p+p^2-2p+4p^2-2p^3+p^2-2p^3+p^4)$$
$$= 1-1+2p-p^2+2p-4p^2+2p^3-p^2+2p^3-p^4$$
$$= 4p-6p^2+4p^3-p^4$$

which is the same result as the previous derivation.

Question
For 4-bit data and a probability of error equal to 0.001, determine the actual probability of errors for 1, 2, 3 and 4 bit errors. Verify that the summation of the error probabilities is given by the formula derived in the previous question.

Answer
Table 12.2 shows the results using the formula:

$$P_e(m) = \binom{n}{m} .p^m .(1-p)^{n-m}$$

with $n=4$ and $p=0.001$.

It can be seen from the table that the probability of a single error is 3.988×10^{-3}. The probability of two errors is 5.99×10^{-6}, the probability of three errors is 4×10^{-9} and the probability of four errors is 1×10^{-12}. The summation of the probabilities is thus 3.994×10^{-3}. The formula derived earlier also gives this value for the probability.

$$P_e = 4p - 6p^2 + 4p^3 - p^4$$
$$= 4 \times 0.001 - 6 \times 0.001^2 - 4 \times 0.001^3 - 0.001^4$$
$$= 0.003994$$

Table 12.2 Probability of error.

No of errors	*Probability*
1	3.988×10^{-3}
2	5.99×10^{-6}
3	4×10^{-9}
4	1×10^{-12}
Summation	3.994×10^{-3}

12.8 Linear and cyclic codes

A linear binary code is one in which the sum of any two code words is also a code word. For example:

$$\{00, 01, 10, 11\} \quad \text{and} \quad \{00000, 01101, 10110, 11011\}$$

are linear codes, because any of the code words added (using modulo-2 addition) to another gives another valid code word. For example, in the second code, $01101 + 10110$ gives 11011, which is a valid code word.

Cyclic codes often involve complex mathematics but are extremely easy to implement with XOR gates and shift registers. A code is cyclic if:

- It is a linear code.
- When any cyclic shift of a code word is also a code word, i.e. whenever $a_0a_1 \ldots a_{n-1}$ is a code word then so is $a_{n-1}a_0a_1 \ldots a_{n-2}$.

For example the code {0000, 0110, 0011, 1001, 1100, 0001, 0010, 0100, 1000} is cyclic because a shift in the bits, either left or right, produces a valid code word. Whereas the code {000, 010, 011, 100, 001} is not cyclic as a shift in the code word 011 to the left or right does not result in a valid code word. One of the most widely used codes, cyclic redundancy check (CRC) is an example of a cyclic code.

12.9 Block and convolutional coding

The two main types of error coding are block codes and convolutional codes. A block code splits the data into k data bits and forms them into n blocks. This type of code is described as an (n, k) code. For example, an (12,8) code has 12 blocks of 8 bits, as illustrated in Figure 12.2. In a block code, the coder and decoder treat each block separately from all the other blocks.

Convolutional coding, on the other hand, treats the input and output bits as a continuous stream of binary digits. If the coder has an input of k bit(s) and outputs n bit(s), then the coder is described as k/n rate code. For example if the code takes 3 bits at a time and outputs 5 bits for each 3-bit input then it is described as a $3/5$ rate code.

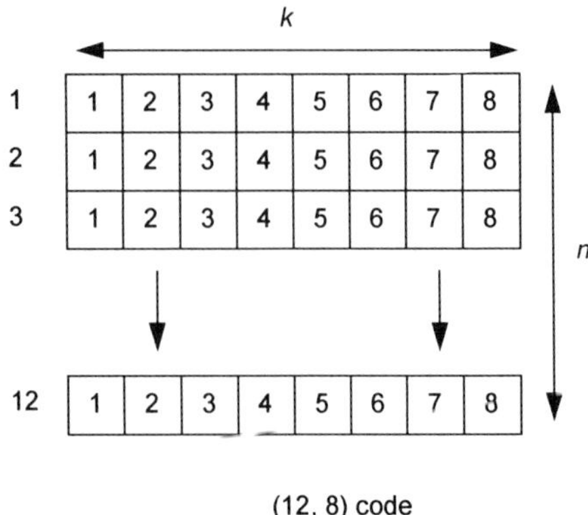

Figure 12.2 An (12,8) block code.

12.10 Systematic and unsystematic coding

Systematic code includes the input data bits in an unmodified form. Parity, or check bits, are then added to the unmodified bits. The main advantage of a systematic code is that the original data is still embedded in its original form within the coded data. Thus a receiver may ignore the extra error coding bits and simply read the uncoded data bits. Another decoder implementation could read the data and check the error bits for error detection, while a more powerful decoder might perform full error correction on the data. An unsystematic code modifies the input data bits and embeds the error coding into the coded bitstream. Thus, the decoder must normally decode the complete data stream.

12.11 Feedforward and feedback error correction

An error code can give information on error detection or correction information. Normally error correcting codes require many more bits to be added to the data and they require more processing at the decoder. The error coding method used normally depends on many factors, including:

- The locality of the transmitter (which contains the encoder) and receiver (which contains the decoder). If the transmitter is located fairly near to the receiver and can respond quickly for a retransmission then it may be possible for the receiver simply to detect an error and ask for a retransmission. If this is not the case, the receiver must either discard the data or the system must use an error correction code. For example, a CD-ROM must contain powerful error correction codes because it is not possible for a CD player to ask for the data to be retransmitted, but two computers on the same network segment can easily ask for a retransmission.

- The characteristics of the transmission system. Normally the more errors that occur in the system, the greater the need for error correcting/detecting codes and the greater the need for a powerful coding scheme.
- The types of errors that can occur; for example, do the errors occur in bursts or in single bits at a time.
- The type of data. Normally computer-type data must be transmitted error-free whereas a few errors in an image or on audio data are unlikely to cause many problems. Thus with computer-type data, a correction code can correct the data as it is received (error correction) or there may be some method for contacting the transmitter to request a retransmission (error detection).
- The power and speed of the encoder and decoder. For example, a powerful error correcting code it may not be possible to implement within a given time because of the need for low-cost decoder hardware. On the other hand, a powerful code may be inefficient for simple requirements.

With error detection the receiver must either discard the data or ask for a re-transmission. Many systems use an acknowledgment (ACK) from the receiver to acknowledge the receipt of data and a negative acknowledgment (NACK) to indicate an incorrect transmission. Typically, data is sent in data packets which contains the error code within them. Each packet also contains a value which identifies the packet number. The receiver then sends back an acknowledgment with the sequence number of the packet it is acknowledging. All previous packets before this value are automatically acknowledged. This method can be implemented in a number of modes, including:

- The transmitter transmits packets, up to a given number, and then waits for the receiver to acknowledge them. If it does not receive an acknowledgment it may resend them or reset the connection.
- The receiver sends back a NACK to inform the transmitter that it has received one or more frames in error, and would like the transmitter to retransmit them, starting with the packet number contained in the NACK.
- The receiver sends back a SREJ to selectively reject a single packet. The transmitter then retransmits this packet.

12.12 Error types

Many errors can occur in a system. Normally, in a digital system, they are caused either when the noise level overcomes the signal level or when the digital pulses become distorted by the system (normally in the transmission systems). This causes a 0 to be interrupted as a 1, or a 1 as a 0.

Noise is any unwanted signal and has many causes, such as static pickup, poor electrical connections, electronic noise in components, crosstalk, and so on. It makes the reception of a signal more difficult and can also produce unwanted distortion on the unmodulated signal.

The main sources of noise on a communication system are:

- Thermal noise – thermal noise arises from the random movement of electrons in a conductor and is independent of frequency. The noise power can be predicted from the formula $N = kTB$ where N is the noise power in watts, k is Boltzmann's constant (1.38×10^{-23} J/K), T is the temperature (in K) and B is the bandwidth of the channel (Hz). Thermal noise is predictable and is spread across the bandwidth of the system. It is unavoidable but can be reduced by reducing the temperature of the components causing the thermal noise. Many receivers which detect very small signals require to be cooled to a very low temperature in order to reduce thermal noise. An example is in astronomy where the receiving sensor is reduced to almost absolute zero. Thermal noise is a fundamental limiting factor in the performance any communications system.
- Crosstalk – electrical signals propagate with an electric field and a magnetic field. If two conductors are laid beside each other then the magnetic field from one can couple into the other. This is known as crosstalk, where one signal interferes with another. Analogue systems tend to be more affected more by crosstalk than digital systems.
- Impulse noise – impulse noise is any unpredictable electromagnetic disturbance, such as from lightning or energy radiated from an electric motor. A relatively high-energy pulse of short duration normally characterizes it. It is of little importance to an analogue transmission system as it can usually be filtered at the receiver. However, impulse noise in a digital system can cause the corruption of a significant number of bits.

A signal can be distorted in many ways, by the electrical characteristics of the transmitter and receiver, and by the characteristics of the transmission media. An electrical cable possesses inductance, capacitance and resistance. The inductance and capacitance have the effect of distorting the shape of the signal whereas resistance causes the amplitude of the signal to reduce (and to lose power).

If, in a digital system, an error has the same probability of occurring at any time then the errors are random. If errors occur in several consecutive bits then they are called burst errors. A typical cause of burst errors is interference, often from lightning or electrical discharge.

If there is the same probability of error for both 1 and 0 then the channel is called a binary symmetric channel and the probability of error in binary digits is known as the bit error rate (BER).

12.13 Coding gain

The effectiveness of an error correcting code is commonly measured with the coding gain and can therefore be used to compare codes. It can be defined as the saving in bits relative to an uncoded system delivering the same bit error rate.

13 Error Coding (Detection)

13.1 Introduction

The most important measure of error detection is the Hamming distance. This defines the number of changes in the transmitted bits that are required in order for a code word to be received as another code word. The more bits that are added, the greater the Hamming distance can be, and the objective of a good error detecting code is to be able to maximize the minimum Hamming distance between codes. For example, a code which has a minimum Hamming distance of 1 cannot be used to detect errors. This is because a single error in a specific bit in one or more code words causes the received code word to be received as a valid code word. A minimum Hamming distance of 2 will allow one error to be detected. In general, a code C can detect up to N errors for any code word if $d(C)$ is greater than or equal to $N+1$ (i.e. $d(C) \geq N+1$). For this it can be shown that:

$$\text{The number of errors detected} = d - 1$$

where d is the minimum Hamming distance.

Error detection allows the receiver to determine if there has been a transmission error. It cannot rebuild the correct data and must either request a retransmission or discard the data.

13.2 Parity

Simple parity adds a single parity bit to each block of transmitted symbols. This parity bit either makes them have an even number of 1's (even parity) or an odd number of 1's (odd parity). It is a simple method of error detection and requires only exclusive-OR (XOR) gates to generate the parity bit. This output can be easily added to the data using a shift register.

Parity bits can only detect an odd number of errors, i.e. 1, 3, 5, and so on. If an even number of bits is in error then the parity bit will be correct and no error will be detected. This type of coding is normally not used on its own or where there is the possibility of several bits being in error.

13.3 Block parity

Block parity is a block code which adds a parity symbol to the end of a block of code. For example a typical method is to transmit the one's complement (or sometimes the two's complement) of the modulo-2 sum of the transmitted values. Using this coding, and a transmitted block code after every 8 characters, the data:

1, 4, 12, –1, –6, 17, 0, –10

would be arranged as:

1	0000 0001
4	0000 0100
12	0000 1100
–1	1111 1111
–6	1111 1010
17	0001 0001
0	0000 0000
–10	1111 0110
	1110 1011

It can be seen that modulo-2 addition is 1110 1011 (which is –21 in decimal). Thus the transmitted data would be:

0000 0001 0000 0100 0000 1100 1111 1111 1111
1010 0001 0001 0000 0000 1111 0110 1110 1011 ...

In this case, a single error will cause the checksum to be wrong. Unfortunately, as with simple parity, even errors in the same column will not show-up an error, but single errors in different columns will show up as an error. Normally when errors occur they are either single-bit errors or large bursts of errors. With a single-bit error the scheme will detect an error and it is also likely to detect a burst of errors, as the burst is likely to affect several columns and also several rows.

This error scheme is used in many systems as it is simple and can be implemented easily in hardware with XOR gates or simply calculated with appropriate software.

The more symbols are used in the block, the more efficient the code will be. Unfortunately, when an error occurs the complete block must be retransmitted.

13.4 Checksum

The checksum block code is similar to the block parity method but the actual total of the values is sent. Thus it is it very unlikely that an error will go undiscovered. It is typically used when ASCII characters are sent to represent numerical values. For example, the previous data was:

1, 4, 12, –1, –6, 17, 0, –10

which gives a total of 17. This could be sent in ASCII characters as:

'1' SPACE '4' SPACE '1' '2' SPACE '–' '1' SPACE '–' '6' SPACE '1' '7' SPACE '0' SPACE '–' '1' '0' SPACE '1' '7' where the SPACE character is the delimiting character between each of the transmitted values. Typically, the transmitter and receiver will agree the amount of numbers that will be transmitted before the checksum is transmitted.

13.5 Cyclic redundancy checking (CRC)

CRC is one of the most reliable error detection schemes and can detect up to 95.5% of all errors. The most commonly used code is the CRC-16 standard code which is defined by the CCITT.

The basic idea of a CRC can be illustrated using an example. Suppose the transmitter and receiver were both to agree that the numerical value sent by the transmitter would always be divisible by 9. Then should the receiver get a value which was not divisible by 9 would know it knows that there had been an error. For example, if a value of 32 were to be transmitted it could be changed to 320 so that the transmitter would be able to add to the least significant digit, making it divisible by 9. In this case the transmitter would add 4, making 324. If this transmitted value were to be corrupted in transmission then there would only be a 10% chance that an error would not be detected.

In CRC-CCITT, the error correction code is 16 bits long and is the remainder of the data message polynomial $G(x)$ divided by the generator polynomial $P(x)$ ($x^{16}+x^{12}+x^5+1$, i.e. 10001000000100001). The quotient is discarded and the remainder is truncated to 16 bits. This is then appended to the message as the coded word.

The division does not use standard arithmetic division. Instead of the subtraction operation an exclusive-OR operation is employed. This is a great advantage as the CRC only requires a shift register and a few XOR gates to perform the division.

The receiver and the transmitter both use the same generating function $P(x)$. If there are no transmission errors then the remainder will be zero.

The method used is as follows:

1. Let $P(x)$ be the generator polynomial and $M(x)$ the message polynomial.
2. Let n be the number of bits in $P(x)$.
3. Append n zero bits onto the right-hand side of the message so that it contains $m+n$ bits.
4. Using modulo-2 division, divide the modified bit pattern by $P(x)$. Modulo-2 arithmetic involves exclusive-OR operations, i.e. 0 – 1 = 1, 1 – 1 = 0, 1 – 0 = 1 and 0 – 0 = 0.
5. The final remainder is added to the modified bit pattern.

Example: For a 7-bit data code 1001100 determine the encoded bit pattern using a CRC generating polynomial of $P(x)=x^3+x^2+x^0$. Show that the receiver will not detect an error if there are no bits in error.

Answer

$P(x)=x^3+x^2+x^0$ (1101)
$G(x)=x^6+x^3+x^2$ (1001100)

Multiply by the number of bits in the CRC polynomial.

$$x^3(x^6+x^3+x^2)$$
$$x^9+x^6+x^5 \qquad (1001100000)$$

Figure 13.1 shows the operations at the transmitter. The transmitted message is thus:

1001100001

and Figure 13.2 shows the operations at the receiver. It can be seen that the remainder is zero, so there have been no errors in the transmission.

```
         1111101
1101 | 1001100000
       1101
        1001
        1101
         1000
         1101
          1010
          1101
           1110
           1101
             1100
             1101
              001
```

Figure 13.1 CRC coding example.

```
         1111101
1101 | 1001100001
       1101
        1001
        1101
         1000
         1101
          1010
          1101
           1110
           1101
             1101
             1101
              000
```

Figure 13.2 CRC decoding example.

The CRC-CCITT is a standard polynomial for data communications systems and can detect:

- All single and double bit errors.
- All errors with an odd number of bits.

- All burst errors of length 16 or less.
- 99.997% of 17-bit error bursts.
- 99.998% of 18-bit and longer bursts.

Table 13.1 lists some typical CRC codes. CRC-32 is used in Ethernet, Token Ring and FDDI networks, whereas ATM uses CRC-8 and CRC-10.

Table 13.1 Typical schemes.

Type	*Polynomial*	*Polynomial binary equivalent*
CRC-8	$x^8+x^2+x^1+1$	100000111
CRC-10	$x^{10}+x^9+x^5+x^4+x^1+1$	11000110011
CRC-12	$x^{12}+x^{11}+x^3+x^2+1$	1100000001101
CRC-16	$x^{16}+x^{15}+x^2+1$	11000000000000101
CRC-CCITT	$x^{16}+x^{12}+x^5+1$	10001000000100001
CRC-32	$x^{32}+x^{26}+x^{23}+x^{16}+x^{12}+x^{11}$ $+x^{10}+x^8+x^7+x^5+x^4+x^2+x+1$	1000001001000000100011101101 10111

13.5.1 Mathematical representation of the CRC

The main steps to CRC implementation are:

1. Prescale the input polynomial of $M'(x)$ by the highest order of the generator polynomial $P(x)$. Thus:

$$M'(x) = x^n M(x)$$

2. Next divide $M'(x)$ by the generator polynomial to give:

$$\frac{M'(x)}{G(x)} = \frac{x^n M(x)}{G(x)} = Q(x) + \frac{R(x)}{G(x)}$$

which gives:

$$x^n M(x) = G(x)Q(x) + R(x)$$

and rearranging gives:

$$x^n M(x) + R(x) = G(x)Q(x)$$

This means that the transmitted message ($x^n M(x) + R(x)$) is now exactly divisible by $G(x)$.

13.5.2 CRC example

Question A

A CRC system uses a message of $1+x^2+x^4+x^5$. Design a FSR cyclic encoder circuit with generator polynomial $G(x)=1+x^2+x^3$ and having appropriate gating circuitry to enable/disable the shift out of the CRC remainder.

Answer - Part A

The generator polynomial is $G(x)=1+x^2+x^3$, the circuit is given in Figure 13.3.

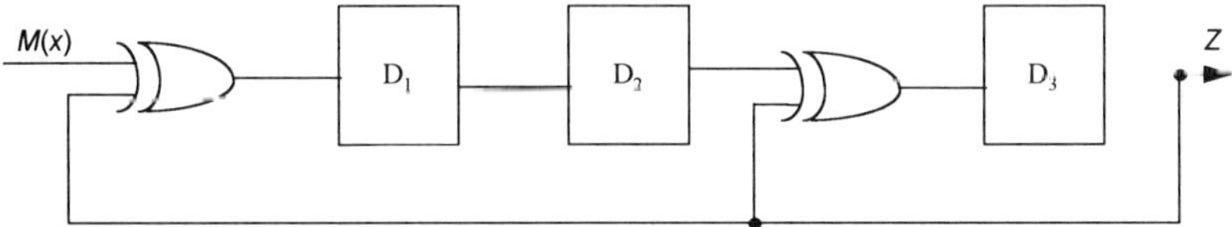

Figure 13.3 CRC coder.

Now to prove that this circuit does generate the polynomial. The output $Z(x)$ will be:

$$Z(x) = Z(x)x^{-1} + \left[M(x)x^{-2} + Z(x)x^{-2} \right] x^{-1}$$
$$= Z(x)\left(x^{-3} + x^{-1}\right) + M(x)x^{-3}$$

Thus:

$$M(x) = \frac{Z(x)\left[1 + x^{-1} + x^{-3}\right]}{x^{-3}}$$

giving:

$$P(x) = \frac{M(x)}{Z(x)} = x^3 + x^2 + 1$$

Question B

If the previous CRC system uses a message of $1+x^2+x^4+x^5$ then determine the sequence of events that occur and hence determine the encoded message as a polynomial $T(x)$. Synthesize the same code algebraically using modulo-2 division.

Answer B

First prescale the input polynomial of $M(x)$ by x^3, the highest power of $G(x)$, thus:

$$M'(x)=x^3.M(x)= x^3+x^5+x^7+x^8$$

The input is thus $x^3+x^5+x^7+x^8$ (000101011), and the generated states are:

Time	$M'(x)$	D_1	D_2	D_3	D_4	
1	000101011	0	0	0	0	
2	00010101	1	0	0	0	← MSD
3	0001010	1	1	0	0	
4	000101	0	1	1	1	
5	00010	0	0	0	0	
6	0001	0	0	0	0	
7	000	1	0	0	0	
8	00	0	1	0	0	
9	0	0	0	1	1	← LSD
10		**1**	**0**	**1**		

The remainder is thus 101, so $R(x)$ is x^2+1. The transmitted polynomial will be:

$$T(x)=x^3 M(x)+R(x)=x^8+x^7+x^5+x^3+x^2+1 \ (110101101)$$

To check this, use either modulo-2 division to give:

$$\begin{array}{r|l}
 & x^5 \qquad\qquad +1 \\
\hline
x^3+x^2+1 & x^8+x^7+x^5+x^3 \\
 & x^8+x^7+x^5 \\
\hline
 & \qquad\qquad x^3 \\
 & \qquad\qquad x^3+x^2+1 \\
\hline
\text{Remainder} \rightarrow & \qquad\qquad \boxed{x^2+1}
\end{array}$$

This gives the same answer as the state table, i.e. x^2+1.

Question C
Prove that the transmitted message does not generate a remainder when divided by $P(x)$.

Answer C
The transmitted polynomial, $T(x)$, is $x^8+x^7+x^5+x^3+x^2+1$ (110101101) and the generator polynomial, $G(x)$, is $1+x^2+x^3$. Thus:

$$\begin{array}{r|l}
 & x^5 \qquad\qquad +1 \\
\hline
x^3+x^2+1 & x^8+x^7+x^5+x^3+x^2+1 \\
 & x^8+x^7+x^5 \\
\hline
 & \qquad\qquad x^3+x^2+1 \\
 & \qquad\qquad x^3+x^2+1 \\
\hline
\text{Remainder} \rightarrow & \qquad\qquad \boxed{0}
\end{array}$$

As there is a zero remainder, there is no error.

14 Error Coding (Correction)

14.1 Introduction

Error bits are added to data either to correct or to detect transmission errors. Normally, the more bits that are added, the better the detection or correction. Error detection allows the receiver to determine if there has been a transmission error. It cannot rebuild the correct data and must either request a retransmission or discard the data. With error correction the receiver detects an error and tries to correct as many error bits as possible. Again, the more error coding bits are used, the more bits can be corrected. An error correction code is normally used when the receiver cannot request a retransmission.

14.2 Longitudinal/vertical redundancy checks (LRC/VRC)

RS-232 uses vertical redundancy checking (VRC) when it adds a parity bit to the transmitted character. Longitudinal (or horizontal) redundancy checking (LRC) adds a parity bit for all bits in the message at the same bit position. Vertical coding operates on a single character and is known as character error coding. Horizontal checks operate on groups of characters and described as message coding. LRC always uses even parity and the parity bit for the LRC character has the parity of the VRC code.

In the example given next, the character sent for LRC is thus `10101000` (28h) or a '`(`'. The message sent is '`F`', '`r`', '`e`', '`d`', '`d`', '`y`' and '`(`'.

Without VRC checking, LRC checking detects most errors but does not detect errors where an even number of characters have an error in the same bit position. In the previous example if bit 2 of the '`F`' and '`r`' were in error then LRC would be valid.

This problem is overcome if LRC and VRC are used together. With VRC/LRC the only time an error goes undetected is when an even number of bits, in an even number of characters, in the same bit positions of each character are in error. This is of course very unlikely.

On systems where only single-bit errors occur, the LRC/VRC method can be used to detect and correct the single-bit error. For systems where more than one error can occur it is not possible to locate the bits in error, so the receiver prompts the transmitter to retransmit the message.

Example

A communications channel uses ASCII character coding and LRC/VRC bits are added to each word sent. Encode the word 'Freddy' and, using odd parity for the VRC and even parity for the LRC; determine the LRC character.

Answer

	F	r	e	d	d	y	LRC
b0	0	0	1	0	0	1	**0**
b1	1	1	0	0	0	0	**0**
b2	1	0	1	1	1	0	**0**
b3	0	0	0	0	0	1	**1**
b4	0	1	0	0	0	1	**0**
b5	0	1	1	1	1	1	**1**
b6	1	1	1	1	1	1	**0**
VRC	0	1	1	0	0	0	**1**

14.3 Hamming code

Hamming code is a forward error correction (FEC) scheme which can be used to detect and correct bit errors. The error correction bits are known as Hamming bits and the number that need to be added to a data symbol is determined by the expression:

$$2^n \geq m + n + 1$$

where *m* is number of bits in the data symbol
n is number of Hamming bits

Hamming bits are inserted into the message character as desired. Typically, they are added at positions that are powers of 2, i.e. the 1st, 2nd, 4th, 8th, 16th bit positions, and so on. For example to code the character `011001` then, starting from the right-hand side, the Hamming bits would be inserted into the 1st, 2nd, 4th and 8th bit positions.

The character is `011001`
The Hamming bits are `HHHH`
The message format will be `01H100H1HH`

10	9	8	7	6	5	4	3	2	1
0	1	H	1	0	0	H	1	H	H

Next each position where there is a 1 is represented as a binary value. Then each position value is exclusive-OR'ed with the others. The result is the Hamming code. In this example:

Position	Code
9	1001
7	0111
3	0011
XOR	1101

The Hamming code error bits are thus `1101` and the message transmitted will be `0111001101`.

10	9	8	7	6	5	4	3	2	1
0	1	1	1	0	0	1	1	0	1

At the receiver all bit positions where there is a 1 are exclusive-OR'ed. The result gives either the bit position error or no error. If the answer is zero there was no single-bit errors, it gives the bit error position.

Position	*Code*
Hamming	1101
9	1001
7	0111
3	0011
XOR	0000

If an error has occurred in bit 5 then the result is 5.

Position	*Code*
Hamming	1101
9	1001
7	0111
5	0101
3	0011
XOR	0101

14.4 Representations of Hamming code

For a code with 4 data bits and 3 Hamming bits, the Hamming bits are normally inserted into the power-of-2 bit positions, thus code is known as (7,4) code. The transmitted bit are $P_1P_2D_1P_3D_2D_3D_4$. In a mathematical form the parity bits are generated by:

$$P_1 = D_1 \oplus D_2 \oplus D_4$$
$$P_2 = D_1 \oplus D_3 \oplus D_4$$
$$P_3 = D_2 \oplus D_3 \oplus D_4$$

111 110 101 **100** 011 **010** 001
D_4 D_3 D_2 P_3 D_1 P_2 P_1

At the receiver the check bits are generated by:

$$S_1 = P_1 \oplus D_1 \oplus D_2 \oplus D_4$$
$$S_2 = P_2 \oplus D_1 \oplus D_3 \oplus D_4$$
$$S_3 = P_3 \oplus D_2 \oplus D_3 \oplus D_4$$

Hamming coding can also be represented in a mathematical form. The steps are:

1. Calculate the number of Hamming bits using the formula $2^n \geq m+n+1$, where m is number of bits in the data and n is number of Hamming bits. The code is known as an $(m+n, m)$ code. For example, (7,4) code uses 4 data bits and 3 Hamming bits.

2. Determine the bit positions of the Hamming bits (typically they will be inserted in the power-of-2 bit positions, i.e. 1, 2, 4, 8, ...).

3. Generate the transmitted bit pattern with data bits and Hamming bits. For example if there are 4 data bits ($D_1D_2D_3D_4$) and 3 Hamming bits ($P_1P_2P_3$) then the transmitted bit pattern will be:

$$\mathrm{T} = \begin{bmatrix} P_1 & P_2 & D_1 & P_3 & D_2 & D_3 & D_4 \end{bmatrix}$$

4. Transpose the **T** matrix to give $\mathbf{T}^{\mathrm{T}}$; message bits $D_1D_2D_3D_4$ and Hamming bits $P_1P_2P_3$ would give:

$$\mathrm{T}^{\mathrm{T}} = \begin{bmatrix} P_1 \\ P_2 \\ D_1 \\ P_3 \\ D_2 \\ D_3 \\ D_4 \end{bmatrix}$$

5. The Hamming matrix **H** is generated by an $[n, m+n]$ matrix, where n is the number of Hamming bits and m is the number of data bits. Each row identifies the Hamming bit and a 1 is placed in the row and column if that Hamming bit checks the transmitted bit. For example, in the case of a transmitted message of $P_1P_2D_1P_3D_2D_3D_4$ then if P_1 checks the D_1, D_2 and D_4, and P_2 checks the D_1, D_3 and D_4, and P_1 checks the D_2, D_3 and D_4, then the Hamming matrix will be:

$$\mathrm{H} = \begin{bmatrix} 1 & 0 & 1 & 0 & 1 & 0 & 1 \\ 0 & 1 & 1 & 0 & 0 & 1 & 1 \\ 0 & 0 & 0 & 1 & 1 & 1 & 1 \end{bmatrix} \begin{matrix} \longleftarrow \text{Check of } P_1 \\ \longleftarrow \text{Check of } P_2 \\ \longleftarrow \text{Check of } P_3 \end{matrix}$$

The resulting matrix calculation of:

$$\mathrm{HT}^{\mathrm{T}} = \begin{bmatrix} 1 & 0 & 1 & 0 & 1 & 0 & 1 \\ 0 & 1 & 1 & 0 & 0 & 1 & 1 \\ 0 & 0 & 0 & 1 & 1 & 1 & 1 \end{bmatrix} \begin{bmatrix} P_1 \\ P_2 \\ D_1 \\ P_3 \\ D_2 \\ D_3 \\ D_4 \end{bmatrix}$$

gives the syndrome matrix **S** which is a $[1,3]$ matrix. The resulting terms for the syndrome will be:

$$S_1 = P_1 \oplus D_1 \oplus D_2 \oplus D_4$$
$$S_2 = P_2 \oplus D_1 \oplus D_3 \oplus D_4$$
$$S_3 = P_3 \oplus D_2 \oplus D_3 \oplus D_4$$

6. The parity bits are calculated to make all the terms of the syndrome zero. Using the current example:

$$\mathbf{S} = \mathbf{HT}^{\mathrm{T}} = \begin{bmatrix} 0 \\ 0 \\ 0 \end{bmatrix}$$

At the receiver the steps are:

1. The Hamming matrix is multiplied by the received bits to give the syndrome matrix. Using the current example:

$$\mathbf{S} = \mathbf{HR}^{\mathrm{T}} = \begin{bmatrix} 1 & 0 & 1 & 0 & 1 & 0 & 1 \\ 0 & 1 & 1 & 0 & 0 & 1 & 1 \\ 0 & 0 & 0 & 1 & 1 & 1 & 1 \end{bmatrix} \begin{bmatrix} P_1 \\ P_2 \\ D_1 \\ P_3 \\ D_2 \\ D_3 \\ D_4 \end{bmatrix} = \begin{bmatrix} S_1 \\ S_2 \\ S_3 \end{bmatrix}$$

2. A resulting syndrome of zero indicates no error, while any other values give an indication of the error position. If the Hamming bits are inserted into the bit positions in powers of 2 then the syndrome gives the actual position of the bit.

14.4.1 Example

Question

(a) A Hamming coded system uses 4 data bits and 3 Hamming bits. The Hamming bits are inserted in powers of 2 and they check the following bit positions:

1, 3, 5, 7   2, 3, 6, 7   4, 5, 6, 7

If the data bits are 1010 then find the coded message using matrix notation.

(b) If the received message is 1011110 determine the syndrome to indicate the position of the error.

Answer

(a) The transmitted message is:

$$\mathbf{T} = \begin{bmatrix} P_1 & P_2 & D_1 & P_3 & D_2 & D_3 & D_4 \end{bmatrix}$$

and $D_1=1, D_2=0, D_3=1, D_4=0$

for even parity:

P_1 checks the 1st, 3rd, 5th and 7th, so $P_1 \oplus D_1 \oplus D_2 \oplus D_4 = 0$; thus $P_1 = 1$
P_2 checks the 2nd, 3rd, 6th and 7th, so $P_2 \oplus D_1 \oplus D_3 \oplus D_4 = 0$; thus $P_2 = 0$
P_3 checks the 4th, 5th, 6th and 7th, so $P_3 \oplus D_2 \oplus D_3 \oplus D_4 = 0$; thus $P_3 = 1$

$$\mathbf{H} = \begin{bmatrix} 1 & 0 & 1 & 0 & 1 & 0 & 1 \\ 0 & 1 & 1 & 0 & 0 & 1 & 1 \\ 0 & 0 & 0 & 1 & 1 & 1 & 1 \end{bmatrix} \quad \text{and} \quad \mathbf{T}^{\mathrm{T}} = \begin{bmatrix} 1 \\ 0 \\ 1 \\ 1 \\ 0 \\ 1 \\ 0 \end{bmatrix} \begin{matrix} \\ \\ \longleftarrow D_1 \\ \\ \longleftarrow D_2 \\ \longleftarrow D_3 \\ \longleftarrow D_4 \end{matrix}$$

Thus $\mathbf{HT}^{\mathrm{T}}$ should equal zero. To check:

$$\mathbf{HT}^{\mathrm{T}} = \begin{bmatrix} 1 & 0 & 1 & 0 & 1 & 0 & 1 \\ 0 & 1 & 1 & 0 & 0 & 1 & 1 \\ 0 & 0 & 0 & 1 & 1 & 1 & 1 \end{bmatrix} \begin{bmatrix} 1 \\ 0 \\ 1 \\ 1 \\ 0 \\ 1 \\ 0 \end{bmatrix} = \begin{bmatrix} 1.1 \oplus 0.0 \oplus 1.1 \oplus 0.1 \oplus 1.0 \oplus 0.1 \oplus 1.0 \\ 0.1 \oplus 1.0 \oplus 1.1 \oplus 0.1 \oplus 0.0 \oplus 1.1 \oplus 1.0 \\ 0.1 \oplus 0.0 \oplus 0.1 \oplus 1.1 \oplus 1.0 \oplus 1.1 \oplus 1.0 \end{bmatrix} = \begin{bmatrix} 0 \\ 0 \\ 0 \end{bmatrix}$$

(b) The received message is:

$$\mathbf{R} = \begin{bmatrix} 1 & 0 & 1 & 1 & 1 & 1 & 0 \end{bmatrix}$$

Thus the syndrome is determine by:

$$\mathbf{HR}^{\mathrm{T}} = \begin{bmatrix} 1 & 0 & 1 & 0 & 1 & 0 & 1 \\ 0 & 1 & 1 & 0 & 0 & 1 & 1 \\ 0 & 0 & 0 & 1 & 1 & 1 & 1 \end{bmatrix} \begin{bmatrix} 1 \\ 0 \\ 1 \\ 1 \\ 1 \\ 1 \\ 0 \end{bmatrix} = \begin{bmatrix} 1.1 \oplus 0.0 \oplus 1.1 \oplus 0.1 \oplus 1.1 \oplus 0.1 \oplus 1.0 \\ 0.1 \oplus 1.0 \oplus 1.1 \oplus 0.1 \oplus 0.1 \oplus 1.1 \oplus 1.0 \\ 0.1 \oplus 0.0 \oplus 0.1 \oplus 1.1 \oplus 1.1 \oplus 1.1 \oplus 1.0 \end{bmatrix} = \begin{bmatrix} 1 \\ 0 \\ 1 \end{bmatrix}$$

Thus the resultant syndrome is not equal to zero, which means there is an error condition. Since $\mathbf{S} = 1\,0\,1$, the error must be in bit position 5, so inverting this bit gives the received message, 1011010.

14.5 Single error correction/double error detection Hamming code

The Hamming code presented can only be used to correct a single error. To correct 2 bits, another parity bit is added to give an overall parity check. Thus for 4 data bits the transmitted code would be:

$$\mathbf{T} = \begin{bmatrix} P_1 & P_2 & D_1 & P_3 & D_2 & D_3 & D_4 & P_4 \end{bmatrix}$$

where P_4 gives an overall parity check. This can be removed at the decoder and Hamming code single error detection can be carried out as before. This then leads to four conditions:

- If the syndrome is zero and the added parity is the correct parity. There is no error (as before).
- If the syndrome is zero and the added parity is the incorrect parity. There is an error in the added parity bit.
- If the syndrome is non-zero and the added parity is the incorrect parity, there is a single error. The syndrome then gives an indication of the bit position of the error.
- If the syndrome is non-zero and the added parity is the correct parity, there is a double error.

Using the example of Section 13.4:

$$\mathbf{H} = \begin{bmatrix} 1 & 0 & 1 & 0 & 1 & 0 & 1 \\ 0 & 1 & 1 & 0 & 0 & 1 & 1 \\ 0 & 0 & 0 & 1 & 1 & 1 & 1 \end{bmatrix} \quad \text{and} \quad \mathbf{T}^{\mathrm{T}} = \begin{bmatrix} 1 \\ 0 \\ 1 \\ 1 \\ 0 \\ 1 \\ 0 \end{bmatrix}$$

Then the parity bit would be a zero. Thus is if parity bit P_4 is a 1 and the syndrome is zero, it is the parity bit that is in error. If a single-bit is in error then the parity bit will be incorrect, so the syndrome will give the bit position in error. If there are two bits in error then the parity bit will be correct, thus if the syndrome is non-zero and the parity bit is correct then there are two errors (unfortunately the syndrome will be incorrect and the received message must be discarded).

14.6 Reed-Solomon coding

In most cases pure Hamming code can only correct a single-bit in error. A more powerful coding system is Reed-Solomon coding, which can correct multiple bits in error. It is a cyclic code and was devised in 1960 by Irvine Reed and Gustave Solomon at MIT. It is suitable for correcting bursts of errors.

14.7 Convolution codes

The block codes, such as VRC/LRC and CRC have the disadvantage that many bits are sent before the message is actually checked. Convolution codes, on the other hand, embed the parity checks in the data stream. They feed the data bits into the coder one bit at a time through a shift register and the output bit(s) are generated with exclusive-OR operations. An example coder is shown in Figure 14.1. The total of the bits considered in the continuous data stream is called the constraint length. Figure 14.1 shows a coder with a constraint length of 3.

A convolution code takes groups of k bit digits at a time and produces groups of n output binary digits. As k is the input data rate and n is the output data time step then the code is known as a k/n code. For example if the coder takes one input bit at a time and outputs two then it is a 1/2 coder. The coder in Figure 14.1 has $k=1$, $n=2$ and has a constraint length, L, of 3.

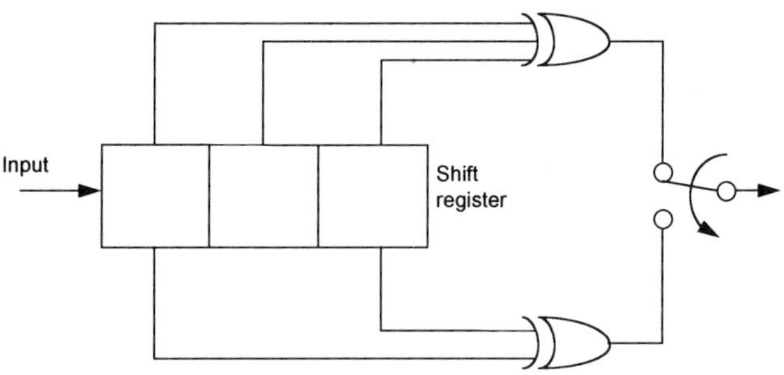

Figure 14.1 A Convolutional encoder.

At any point in time the digits in the shift register define the current state of the coder at that time step.

For example, let the input be 1101. Thus an extra two 0's are added to the input data so that the complete data can be clocked into the shift register. Table 14.1 gives the state table of the encoder. The output, from first to last, will thus be 11 01 01 00 10 11.

Table 14.1 State table of the encoder.

Input	*A*	*B*	*C*	*Output*
00101**1**	1	0	0	1,1
0010**1**	1	1	0	0,1
001**0**	0	1	1	0,1
00**1**	1	0	1	0,0
00	0	1	0	1,0
0	0	0	1	1,1

The system can also be analyzed for any input. First a coding tree is drawn up which defines the present and next state within the shift register. A 3-bit shift register will have a total of 8 states, and each of these states will have 2 next states, one for a 1 input and the other for a 0 input. Table 14.2 defines the coding tree in this case.

Table 14.2 Coding tree.

ABC	*Next state 0 input*	*Output*	*Next state 1 input*	*Output*
000	000	00	100	11
001	000	00	100	11
010	001	11	101	00
011	001	11	101	00
100	010	10	110	01
101	010	10	110	01
110	011	01	111	10
111	011	01	111	10

It can be seen that the next state when *ABC* is either 000 or 001 will always be 000 when a 0 is entered. These states can therefore be taken as the same, as can 010 and 011, 100 and 101, and 110 and 111. This also occurs with a 1 input, thus 4 of the states can be merged. This results in a coding tree with only 4 states, as given in Table 14.3.

Next a state diagram can be produced to determine the change of the output state for a given input. This diagram represents the current state within a circle and the next state is linked by an arrow with an associated value *x/yy* denoting that the input is *x* and the output is *yy*. It is initially assumed that the circuit has been reset and that the initial state is *ABC*=000 (state *a*). An input of 0 causes the circuit to stay in that state and thus to output 00. If the input is a 1 then the next state is 10X (state *b*) and 11 is output. This produces the state diagram of Figure 14.2.

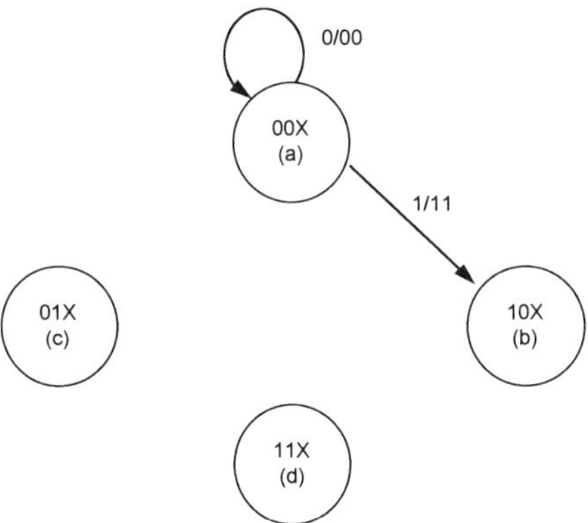

Figure 14.2 Intermediate state diagram.

Next, from state *b*, a zero input causes ABC to become 01X (state *c*) and to output 10. If a 1 is input then the next state will be 11X (state *d*) and the output a 01. This is then continues until the state diagram is complete, as shown in Figure 14.3.

Next, from the state diagram a trellis diagram can be drawn. This has the number of states down the left-hand column; each time step is mapped with its mapping to the previous state and gives an indication of the generated output. The upper of the two lines represents a 0 input, and the lower represent a 1 input. For example the circuit starts in state *a*, then a 0 outputs 00 and stays in the same state, while a 1 puts the circuit into state *b* and outputs a 11. This is shown on the trellis diagram in Figure 14.4.

Table 14.3 Coding tree.

ABC	*Next state 0 input*	*Output*	*Next state 1 input*	*Output*
00X	00X	00	10X	11
01X	00X	11	10X	00
10X	01X	10	11X	01
11X	01X	01	11X	10

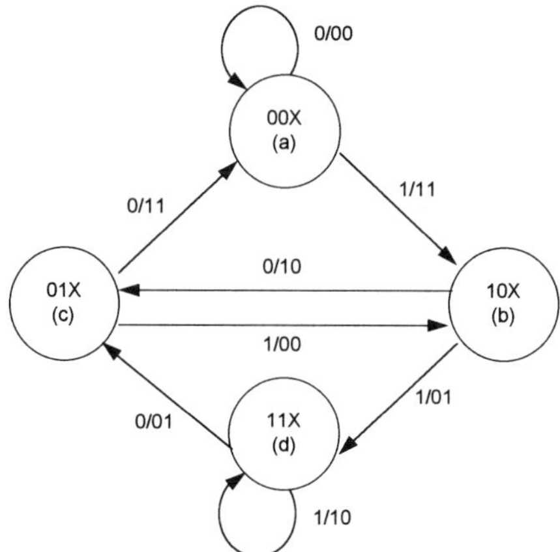

Figure 14.3 Final state diagram.

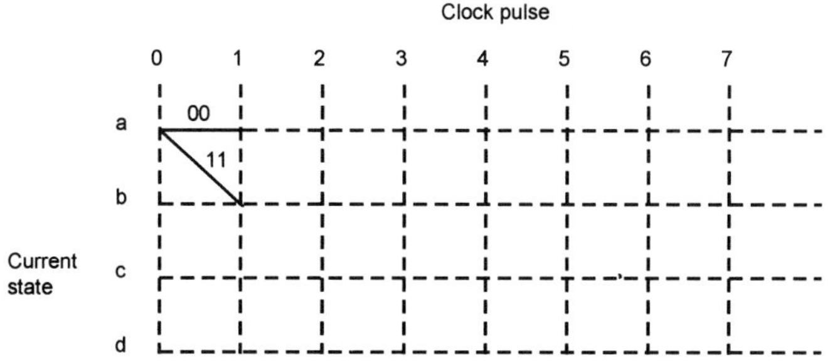

Figure 14.4 Initial trellis diagram.

If the current state is *a* then a 0 input will make it stay in the same state and a 1 will take the next state to *b*. If the current state is *b* then a 0 will make the state change to *c*, else a 1 will change it to *d*. This is illustrated in Figure 14.5.

The rest of the states on the trellis diagram can now be mapped; the next state is given in Figure 14.6 and the final trellis diagram is shown in Figure 14.7. Note that the final two input states are both 0.

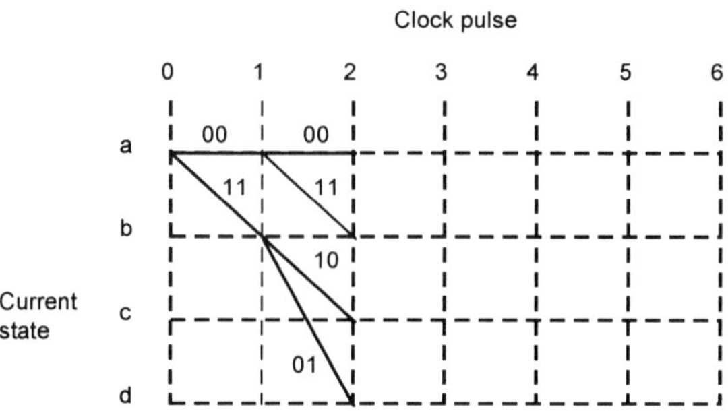

Figure 14.5 Intermediate trellis diagram.

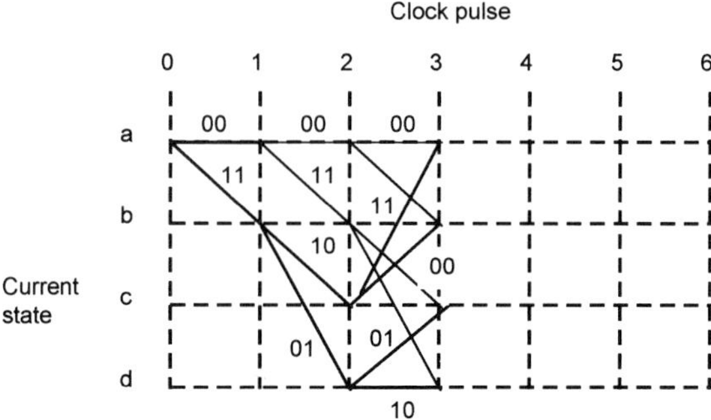

Figure 14.6 Intermediate trellis diagram.

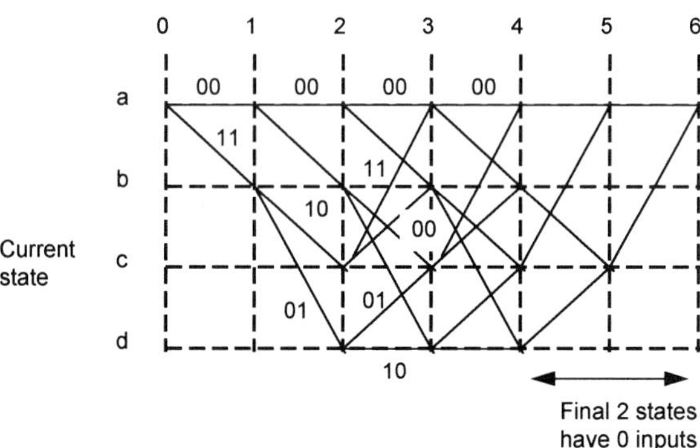

Figure 14.7 Final trellis diagram.

14.7.1 Viterbi decoder

Convolution coding allows error correcting because only some of the possible sequences are valid outputs from the coder. These sequences correspond to the possible paths through the trellis. A Viterbi decoder tries to find the best match for the received bitstream with valid paths. The best path is called the maximum likelihood and discrepancies between the received bitstream and the possible transmitted stream are called path metrics.

The decoder stores the best single path to each of the time-step nodes (in the previous example this was 4: *a*, *b*, *c*, and *d*). Any other path to that node which has a greater number of discrepancies causes that path to be ignored. The number of remembered paths will thus be equal to the number of nodes and will remain constant. For each node, the decoder must store the best path to that node and the total metric corresponding to that path. By comparing the actual *n* received digits at time step $i+1$ with those corresponding to each possible path on the trellis from step *i* to step $i+1$, the decoder calculates the additional metric for each path. From these, it selects the best path to each node at time step $i+1$, and updates the stored records of the paths and metrics.

The best way to illustrate this process is with an example. For example if the input bit sequence is:

100110

this will give an output of:

11 10 11 11 01 01

with a tail of:

11 00 00

Input:	1 0 0 1 1 0 0 0 0
Output:	11 10 11 11 01 01 11 00 00
State:	b c a b d c a a a

This gives an output of:

111011110101110000

Let's assume there are two errors in the bit stream and the receive stream is:

110011010101110000

First the difference in the number of bits is represented on the trellis diagram, as shown in Figure 14.8. Thus if the circuit stays in state *a* then the received input would require two changes to the transmitted data for it to be received as 11. There is no difference in the received bits when going from state *a* to state *b*. Thus the upper route scores a value of 2 and the lower route a value of 0.

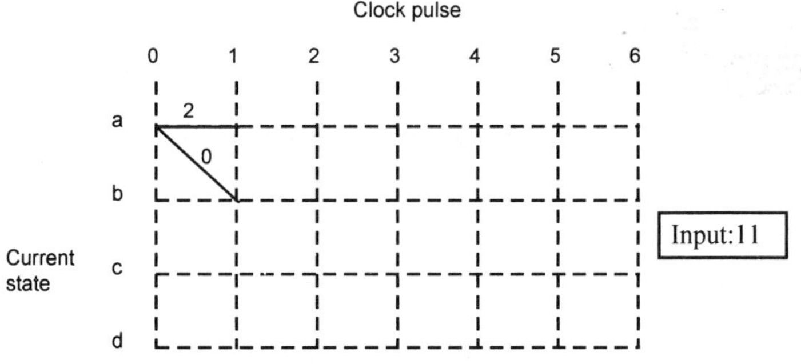

Figure 14.8 Representing the difference in the number of bits.

Next 00 is received, if the circuit had been in state *a* and stayed in state *a* then 00 would have been transmitted. So this will have a discrepancy of 0 and the resulting metric for that route will be 2. If the circuit is in state *a* and moves to state *b* then 11 would have been transmitted, thus there would be a discrepancy of 2, giving the total path metrics as 4.

If the circuit is in state *b* and a 1 is received then the circuit goes into state *d*, else a 0 causes the circuit to go into state *c*. In going into state *c* the circuit will output a 10 (a discrepancy of 1, which gives a total path metric of 1) and going into state *d* will cause an output of 01 (a discrepancy of 1, which gives a total path metric of 1). Figure 14.9 shows the resulting trellis diagram. The resulting total metrics are:

Path	Metrics	Path	Metrics
a → a → a	2	a → a → b	4
a → b → c	1	a → b → d	1

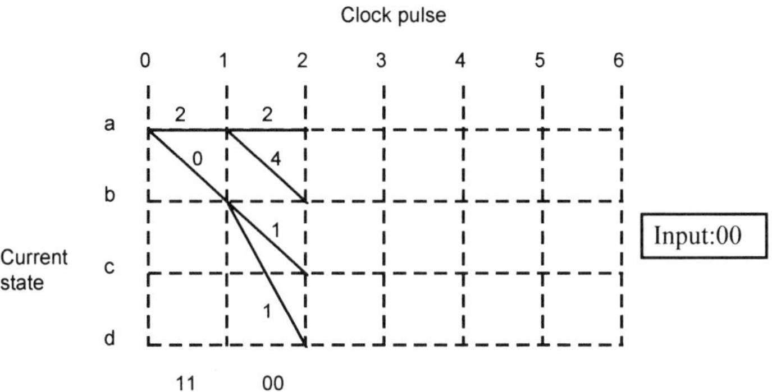

Figure 14.9 Calculating a path metric.

Figure 14.10 shows the next state. This time two routes converge on the same state. The Viterbi decoder computes the metrics for each of the two paths then rejects the route with the greater metric. It can be seen that the top route (i.e. transmitted pattern 00, 00, 00) gives a metric of 4 but the other route to that state has a metric of 1. The route with a metric of 4 will be rejected. It can be seen that the lowest metric has a value of 1. Figure 14.11 shows the resulting preferred routes.

Figure 14.12 shows the next state and Figure 14.13 shows the resultant preferred routes. It can be seen that there are now three routes which have metric of 2 and one with a metric of 3.

Figure 14.14 shows the next state and Figure 14.15 shows the resulting preferred routes. Now the lowest metric is 2.

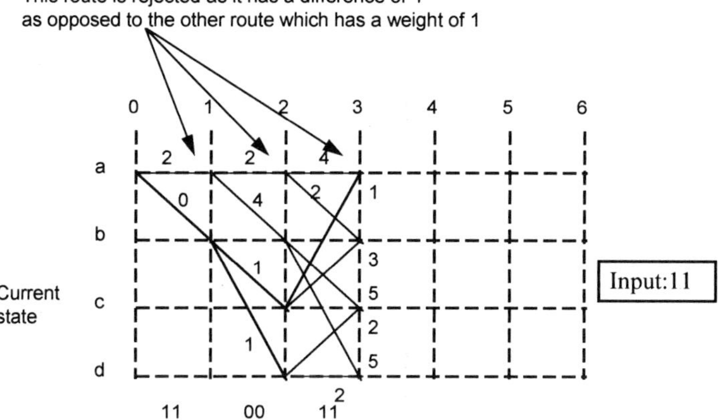

Figure 14.10 In each pair the route with the greater metric is rejected.

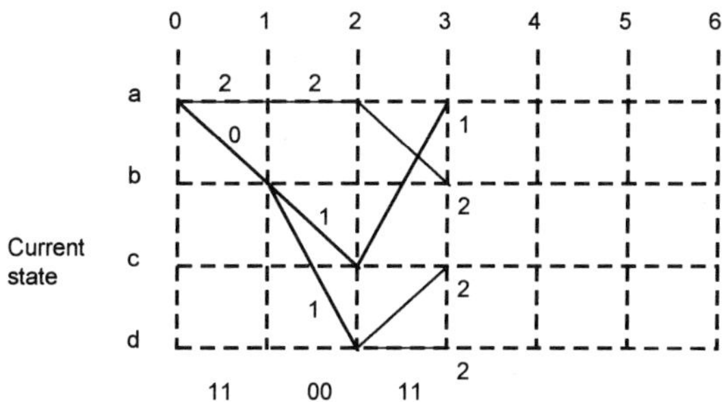

Figure 14.11 Trellis diagram up to state 3.

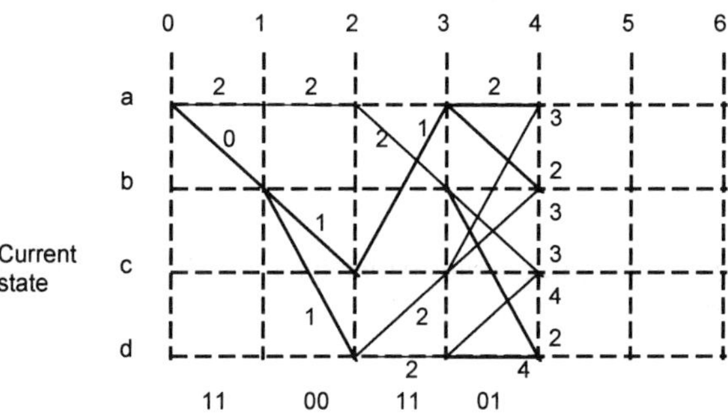

Figure 14.12 Trellis diagram up to state 4.

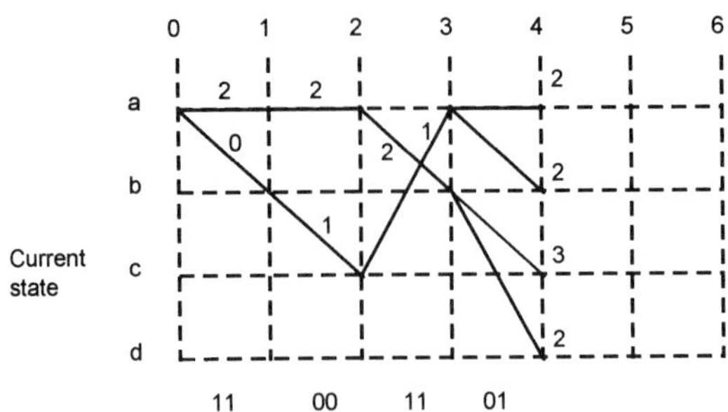

Figure 14.13 Preferred routes in Figure 14.12.

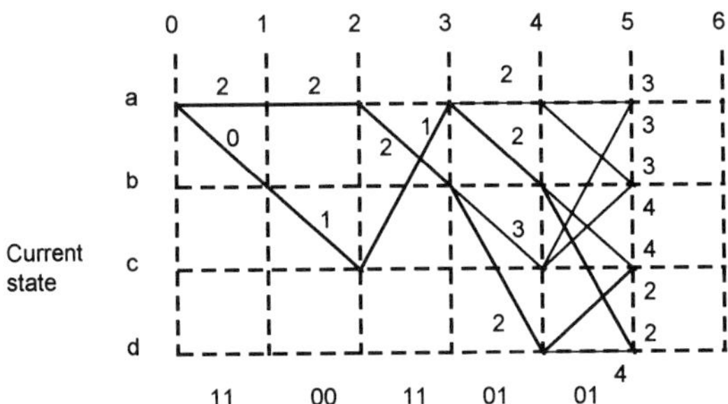

Figure 14.14 Trellis diagram up to state 5.

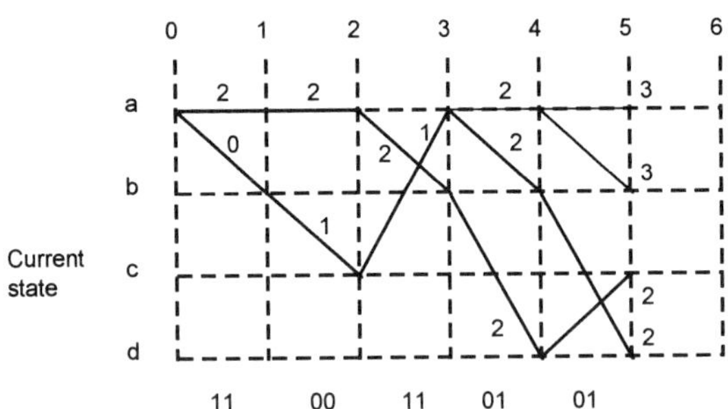

Figure 14.15 Preferred routes in Figure 14.14.

Figure 14.16 shows the next state. It can be seen that there is now one preferred route with a metric of 2 (as expected as there are two errors in the received data). The highlighted route then continues to be the most favored route as there are no more errors in the received bit pattern. Thus all the other scores will gain and the favored route will stay constant (until another error comes along). Following this route gives the decoded output of 100110, which is the data pattern transmitted. The state transition is *a–b* (input is a 1), *b–c* (input is a 0), *c–a* (input is a 0), *a–b* (input is a 1), *b–d* (input is a 1) and *d–c* (input is a 0).

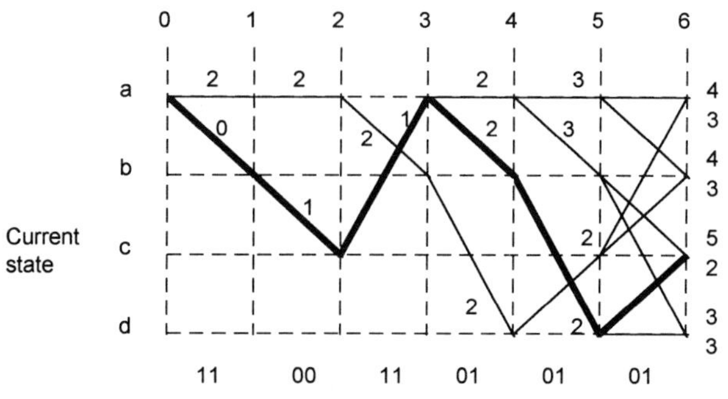

Figure 14.16 Trellis diagram up to state 6.

14.7.2 Convolution coding analysis program

Program 14.1 gives a simple C program for analyzing the convolutional coder given in Figure 14.1. The operator ^ is the exclusive-OR function and the & operator is the bitwise AND function. In the program the input bitstream is specified with the array `inseq[]`. The circuit iterations for a given number of clock cycles (defined by `NO_CLOCKS`) and calculates the output for each clock tick. Note that the AND operator is used to mask off the least significant bit of the output values. Sample run 14.1 gives a sample output.

Program 14.1

```
#include <stdio.h>

#define      NO_CLOCKS      6

int    main(void)
{
int    inseq[NO_CLOCKS]={1,0,1,1,0,0};
int    i,out1,out2,s1,s2,s3;

       puts("Bit 1 Bit 2");

       for (i=0;i<NO_CLOCKS;i++)
       {
          s1=inseq[i]; /* no shift */
          if (i>0) s2=inseq[i-1]; /* one bit shift */
          if (i>1) s3=inseq[i-2]; /* two bit shifts */
          out1=(s1^s2^s3) & 1; /* EX-OR and mask lsb */
          out2=(s1^s3) & 1; /* EX-OR and mask lsb */
          printf("  %d      %d\n",out1,out2);
       }
       return(0);
}
```

Sample run 14.1

```
Bit 1 Bit 2
  1     1
  1     0
  0     0
  0     1
  0     1
  1     1
```

15 Data Encryption Principles

15.1 Introduction

The increase in electronic mail has also increased the need for secure data transmission. An electronic mail message can be easily incepted as it transverse the world's communication networks. Thus there is a great need to encrypt the data contained in it. Traditional mail messages tend to be secure as they are normally taken by a courier or postal service and transported in a secure environment from source to destination. Over the coming years more individuals and companies will be using electronic mail systems and these must be totally secure.

Data encryption involves the science of cryptographics (note that the word *crytopgraphy* is derived from the Greek words which means hidden, or secret, writing). The basic object of cryptography is to provide a mechanism for two people to communicate without any other person being able to read the message.

Encryption is mainly applied to text transmission as binary data can be easily scrambled so it becomes almost impossible to unscramble. This is because text-based information contains certain key pointers:

- Most lines of text have the words 'the', 'and', 'of' and 'to'.
- Every line has a full stop.
- Words are separated by a space (the space character is the most probable character in a text document).
- The characters 'e', 'a' and 'i' are more probable than 'q', 'z' and 'x'.

Thus to decode a message an algorithm is applied and the decrypted text is then tested to determine whether it contains standard English (or the required language).

15.2 Government pressure

Many institutions and individuals read data which is not intended for them; they include:

- Government departments. Traditionally governments around the world have reserved the right to tap into any communications which they think may be against the national interest.
- Spies who tap into communications for industrial or governmental information.
- Individuals who like to read other people's messages.
- Individuals who 'hack' into systems and read secure information.
- Criminals who intercept information in order to use it for crime, such as intercepting PIN numbers on bank cards.

Governments around the world tend to be against the use of encryption as it reduces their chances to tap into information and determine its message. It is also the case that governments do not want other countries to use encryption because it also reduces their chances of reading their secret communications (especially military maneuvers). In order to reduce this threat they must do either of the following:

- Prevent the use of encryption.
- Break the encryption code.
- Learn everyone's cryptographic keys.

Most implementations of data encryption are in hardware. This makes it easier for governments to control their access. For example the US government has proposed to beat encryption by trying to learn everyone's cryptographic key with the Clipper chip. The US government keeps a record of all the serial numbers and encryption keys for each Clipper chip manufactured.

15.3 Cryptography

The main object of cryptography is to provide a mechanism for two (or more) people to communicate without anyone else being able to read the message. Along with this it can provide other services, such as:

- Giving a reassuring integrity check – this makes sure the message has not been tampered with by non-legitimate sources.
- Providing authentication – this verifies the sender identity.

Initially plaintext is encrypted into ciphertext, it is then decrypted back into plaintext, as illustrated in Figure 15.1. Cryptographic systems tend to use both an algorithm and a secret value, called the key. The requirement for the key is that it is difficult to keep devising new algorithms and also to tell the receiving party that the data is being encrypted with the new algorithm. Thus, using keys, there are no problems with everyone having the encryption/decryption system, because without the key it is very difficult to decrypt the message.

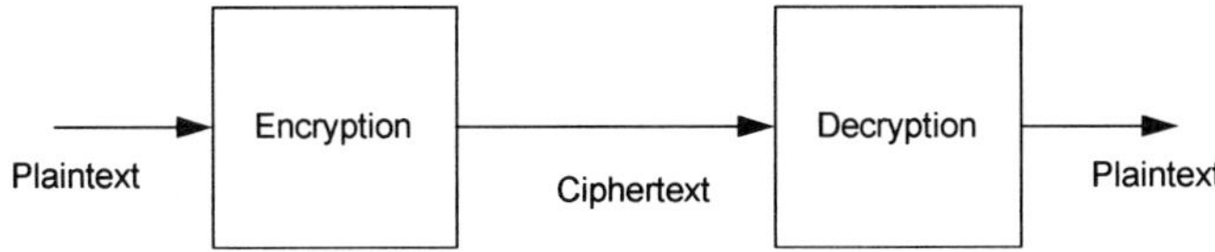

Figure 15.1 Encryption/decryption process.

15.3.1 Public key versus private key

The encrption process can either use a public key or a secret key. With a secret key the key is only known to the two communicating parties. This key can be fixed or can be passed from the two parties over a secure communications link (perhaps over the postal network or a leased line). The two most popular private key techniques are DES (Data Encryption Standard) and IDEA (International Data Encryption Algorithm).

In public-key encryption, each user has both a public and a private key. The two users can communicate because they know each other's public keys. Normally in a public-key system, each user uses a public enciphering transformation which is widely known and a private deciphering transform which is known only to that user. The private transformation is described by a private key, and the public transformation by a public key derived from the private key by a one-way transformation. The RSA (after its inventors Rivest, Shamir and Adleman) technique is one of the most popular public-key techniques and is based on the difficulty of factoring large numbers. It is discussed in more detail in Chapter 17.

15.3.2 Computational difficulty

Every code is crackable and the measure of the security of a code is the amount of time it takes persons not addressed in the code to break that code. Normally to break the code a computer tries all the possible keys until it finds a match. Thus a 1-bit code would only have 2 keys, a 2-bit code would have 4 keys, and so on. Table 15.1 shows the number of keys as a function of the number of bits in the key. It can be seen that a 64-bit code has:

18 400 000 000 000 000 000

different keys. If one key is tested every 10 μs then it would take 1.84×10^{14} seconds (5.11×10^{10} hours or 2.13×10^{8} days or 5 834 602 years). So, for example, if it takes 1 million years for a person to crack the code then it can be considered safe. Unfortunately the performance of computer systems increases by the year. For example if a computer takes 1 million years to crack a code, then assuming an increase in computing power of a factor of 2 per year, then it would only take 500 000 years the next year. Table 15.2 almost shows that after almost 20 years it would take only 1 year to decrypt the same message.

The increasing power of computers is one factor in reducing the processing time, another is the increasing usage of parallel processing. Data decryption is well suited to parallel processing as each processor or computer can be assigned a number of keys to check the encrypted message. Each of them can then work independently of the other (this differs from many applications in parallel processing which suffer from interprocess(or) communication). Table 15.3 gives typical times, assuming a doubling of processing power each year, for processor arrays of 1, 2, 4 ... 4096 elements. It can be seen that with an array of 4096 processing elements it takes only 7 years before the code is decrypted within 2 years. Thus an organization which is serious about decripering messages will have the resources to invest in large arrays of processors or networked computers. It is likely that many governments have computer systems with thousands or tens of thousands of processors operating in parallel. A prime use of these systems will be in decrypting messages.

15.4 Legal issues

Patent laws and how they are implemented vary around the world. Like many good ideas, most of the cryptographic techniques are covered by patents. The main commercial techniques are:

- DES (Data Encryption Standard) which is patented but royalty-free.
- IDEA (International Data Encryption Algorithm) which is also patented and royalty-free for the non-commercial user.

Table 15.1 Number of keys related to the number of bits in the key.

Code size	*Number of keys*	*Code size*	*Number of keys*	*Code size*	*Number of keys*
1	2	12	4 096	52	4.5×10^{15}
2	4	16	65 536	56	7.21×10^{16}
3	8	20	1 048 576	60	1.15×10^{18}
4	16	24	16 777 216	64	1.84×10^{19}
5	32	28	2.68×10^{8}	68	2.95×10^{20}
6	64	32	4.29×10^{9}	72	4.72×10^{21}
7	128	36	6.87×10^{10}	76	7.56×10^{22}
8	256	40	1.1×10^{12}	80	1.21×10^{24}
9	512	44	1.76×10^{13}	84	1.93×10^{25}
10	1 024	48	2.81×10^{14}	88	3.09×10^{26}

Table 15.2 Time to decrypt a message assuming an increase in computing power.

Year	*Time to decrypt (years)*	*Year*	*Time to decrypt (years)*
0	1 million	10	977
1	500 000	11	489
2	250 000	12	245
3	125 000	13	123
4	62 500	14	62
5	31 250	15	31
6	15 625	16	16
7	7 813	17	8
8	3 907	18	4
9	1 954	19	2

Table 15.3 Time to decrypt a message with increasing power and parallel processing.

Processors	*Year 0*	*Year 1*	*Year 2*	*Year 3*	*Year 4*	*Year 5*	*Year 6*	*Year 7*
1	1 000 000	500 000	250 000	125 000	62 500	31 250	15 625	7 813
2	500 000	250 000	125 000	62 500	31 250	15 625	7 813	3 907
4	250 000	125 000	62 500	31 250	15 625	7 813	3 907	1 954
8	125 000	62 500	31 250	15 625	7 813	3 907	1 954	977
16	62 500	31 250	15 625	7 813	3 907	1 954	977	489
32	31 250	15 625	7 813	3 907	1 954	977	489	245
64	15 625	7 813	3 907	1 954	977	489	245	123
128	7 813	3 907	1 954	977	489	245	123	62
256	3 906	1 953	977	489	245	123	62	31
512	1 953	977	489	245	123	62	31	16
1 024	977	489	245	123	62	31	16	8
2 048	488	244	122	61	31	16	8	4
4 096	244	122	61	31	16	8	4	2

Access to a global network normally requires the use of a public key. The most popular public-key algorithm is one developed at MIT and is named RSA (after its inventors Rivest, Shamir and Adleman). All public-key algorithms are patented, and most of the important patents have been acquired by Public Key Partners (PKP). As the US government funded much of the work, there are no license fees for US government use. RSA is only patented in the US, but Public Key Partners (PKP) claim that the international Hellman-Merkle patent also covers RSA. The patent on RSA runs out in the year 2000. Public keys are generated by licensing software from a company called RSA Data Security Inc. (RSADSI).

The other widely used technique is Digital Signature Standard (DSS). It is freely licensable but in many respects it is technically inferior to RSA. The free licensing means that it is not necessary to reach agreement with RSADSI or PKP. Since it was announced, PKP have claimed the Helleman-Merkle patent covers all public-key cryptography. It has also strengthened its position by acquiring rights to a patent by Schnorr which is closely related to DSS.

15.5 Basic encryption principles

Encryption codes have been used for many centuries. They have tended to be used in military situations where secret messages have to be sent between troops without the risk of them being read by the enemy.

15.5.1 Alphabet shifting

A simple encryption code is to replace the letters with a shifted equivalent alphabet. For example moving the letters two places to the right gives:

```
ABCDEFGHIJKLMNOPQRSTUVWXYZ
YZABCDEFGHIJKLMNOPQRSTUVWX
```

Thus a message:

```
THE BOY STOOD ON THE BURNING DECK
```

would become:

```
RFC ZMW QRMMB ML RFC ZSPLGLE BCAI
```

This code has the problem of being reasonably easy to decode, as there are only 26 different code combinations. The first documented use of this type of code was by Julius Caesar who used a 3-letter shift.

15.5.2 Code mappings

Code mappings have no underlying mathematical relationship; they simply use a codebook to represent the characters, often known as a monoalphabetic code. An example could be:

```
Input:      abcdefghijklmnopqrstuvwxyz
Encrypted:  mgqoafzbcdiehxjklntqrwsuvy
```

Program 15.1 shows a C program which uses this code mapping to encrypt entered text and Sample run 15.1 shows a sample run.

The number of different character maps can be determined as follows:

- Take the letter 'A' then this can be mapped to 26 different letters.
- If 'A' is mapped to a certain letter then 'B' can only map to 25 letters.
- If 'B' is mapped to a certain letter then 'C' can be mapped to 24 letters.
- Continue until the alphabet is exhausted.

Thus, in general, the number of combinations will be:

$$26 \times 25 \times 24 \times 23 \ldots 4 \times 3 \times 2 \times 1$$

Thus the code has 26! different character mappings (approximately 4.03×10^{26}). It suffers from the fact that the probabilities of the mapped characters will be similar to those in normal text. Thus if there is a large amount of text then the character having the highest probability will be either an 'e' or a 't'. The character with the lowest probability will tend to be a 'z' or a 'q' (which is also likely be followed by the character map for a 'u').

Program 15.1

```
#include <stdio.h>
#include <ctype.h>

int   main(void)
{
int   key,ch,i=0,inch;
char  text[BUFSIZ];
char input[26]="abcdefghijklmnopqrstuvwxyz";
char output[26]="mgqoafzbcdiehxjklntqrwsuvy";

      printf("Enter text >>");
      gets(text);

      ch=text[0];
      do
      {
         if (ch!=' ')   inch=output[(tolower(ch)-'a')];
         else inch='#';

         putchar(inch);
         i++;
         ch=text[i];
      } while (ch!=NULL);
      return(0);
}
```

Sample run 15.1

```
Enter text >> This is an example piece of text
qbct#ct#mx#aumhkea#kcaqa#jf#qauq
```

A code mapping encryption scheme is easy to implement but unfortunately, once it has been 'cracked', it is easy to decrypt the encrypted data. Normally this type of code is implemented with an extra parameter which changes its mapping, such as changing the code mapping over time depending on the time of day and/or date. Only parties which are allowed to decrypt the message know the mappings of the code to time and/or date. For example each day of the week may have a different code mapping.

15.5.3 Applying a key

To make it easy to decrypt, a key is normally applied to the text. This makes it easy to decrypt the message if the key is known but difficult to decrypt the message if the key is not known. An example of a key operation is to take each of the characters in a text message and then exclusive-OR (XOR) the character with a key value. For example the ASCII character 'A' has the bit pattern:

100 0001

and if the key had a value of 5 then 'A' exclusive-OR'ed with 5 would give:

'A'	100 0001
Key (5)	000 0101
Ex-OR	100 0100

The bit pattern 100 0100 would be encrypted as character 'D'. Program 15.2 is a C program which can be used to display the alphabet of encrypted characters for a given key. In this program the ^ operator represents exclusive-OR. Sample run 15.2 shows a sample run with a key of 5.

Program 15.2

```
#include <stdio.h>

int   main(void)
{
int   key,ch;

      printf("Enter key value >>");
      scanf("%d",&key);

      for (ch='A';ch<='Z';ch++)
         putchar(ch^key);

      return(0);
}
```

Sample run 15.2

```
Enter key value >> 5
DGFA@CBMLONIHKJUTWVQPSR]\_
```

Program 15.3 is an encryption program which reads some text from the keyboard, then encrypts it with a given key and saves the encryted text to a file. Program 15.4 can then be used to read the encrypted file for a given key; only the correct key will give the correct results.

Program 15.3

```
/* Encryt.c */
#include <stdio.h>
```

```
int   main(void)
{
FILE *f;
char  fname[BUFSIZ],str[BUFSIZ];
int   key,ch,i=0;

      printf("Enter output file name >>");
      gets(fname);

      if ((f=fopen(fname,"w"))==NULL)
      {
         puts("Cannot open input file");
         return(1);
      }
      printf("Enter text to be save to file>>");
      gets(str);

      printf("Enter key value >>");
      scanf("%d",&key);

      ch=str[0];

      do
      {
         ch=ch^key; /* Exclusive-OR character with itself */
         putc(ch,f);
         i++;
         ch=str[i];
      } while (ch!=NULL); /* test if end of string */
      fclose(f);
      return(0);
}
```

Sample run 15.3

```
Enter output filename >> out.dat
Enter text to be saved to file>> The boy stood on the burning deck
Enter key value >> 3
```

File listing 15.1 gives a file listing for the saved encrypted text. One obvious problem with this coding is that the SPACE character is visible in the coding. As the SPACE character is 010 0000, the key can be determined by simply XORing 010 0000 with the '#' character, thus:

SPACE	010 0000
'#'	010 0011
Key	000 0011

Thus the key is 000 0011 (decimal 3).

File listing 15.1

```
Wkf#alz#pwllg#lm#wkf#avqmjmd#gf`h
```

Program 15.4

```
/* Decryt.c */
#include <stdio.h>
#include <ctype.h>
```

```
int   main(void)
{
FILE  *f;
char  fname[BUFSIZ];
int   key,ch;

      printf("Enter encrypted filename >>");
      gets(fname);
      if ((f=fopen(fname,"r"))==NULL)
      {
         puts("Cannot open input file");
         return(1);
      }

      printf("Enter key value >>");
      scanf("%d",&key);

      do
      {
         ch=getc(f);
         ch=ch^key;
         if (isascii(ch)) putchar(ch); /* only print ASCII char */
      } while (!feof(f));
      fclose(f);
      return(0);
}
```

Program 15.5 uses the exclusive-OR operator and reads from an input file and outputs to an output file. The format of the run (assuming that the source code file is called `key.c`) is:

`key` *infile.dat outfile.enc*

where *infile.dat* is the name of the file (text or binary) and *outfile.enc* is the name of the output file.

The great advantage of this program is that the same program is used for encryption and for decryption. Thus:

`key` *outfile.enc newfile.dat*

will convert the encrypted file back into the orginal file.

Program 15.5

```
#include <stdio.h>

int main(int argc, char *argv[])
{
FILE *in,*out;
char fname[BUFSIZ],key,ch,fout[BUFSIZ],fext[BUFSIZ],*str;


      printf("Enter key >>");
      scanf("%c",&key);

      if ((in=fopen(argv[1],"rb"))==NULL)
      {
         printf("Cannot open");
         return(1);
      }
```

```
    out=fopen(argv[2],"wb");

    do
    {
        fread(&ch,1,1,in); /* read a byte from the file */
        ch=((ch & 0xff) ^ (key & 0xff)) & 0xff;
        if (!feof(in)) fwrite(&ch,1,1,out); /* write a byte */

    } while (!feof(in));

    fclose(in); fclose(out);
}
```

15.5.4 Applying a bit shift

A typical method used to encrypt text is to shift the bits within each character. For example ASCII characters only use the lower 7 bits of an 8-bit character. Thus, shifting the bit positions one place to the left will encrypt the data to a different character. For a left shift a 0 or a 1 can be shifted into the least significant bit; for a right shift the least significant bit can be shifted into the position of the most significant bit. When shifting more than one position a rotate left or rotate right can be used. Note that most of the characters produced by shifting may not be printable, thus a text editor (or viewer) cannot be viewed them. For example, in C the characters would be processed with:

```
ch=ch << 1;
```

which shifts the bits of `ch` one place to the left, and decrypted by:

```
ch=ch >> 1;
```

which shifts the bits of `ch` one place to the right.

16 Data Encryption

16.1 Introduction

Encryption techniques can use either public-keys or secret keys. Secret-key encryption techniques use a secret key which is only known by the two communicating parities. This key can be fixed or can be passed from the two parties over a secure communications link (for example over the postal network or a leased line). The two most popular private-key techniques are DES (Data Encryption Standard) and IDEA (International Data Encryption Algorithm) and a popular public-key technique is RSA (named after its inventors, Rivest, Shamir and Adleman).

16.2 Cracking the code

A cryptosystem converts plaintext into cipertext using a key. There are several methods that a hacker can use to crack a code, including:

- Known plaintext attack. Where the hacker knows part of the ciphertext and the corresponding plaintext. The known cipertext and plaintext can then be used to decrypt the rest of the cipertext.
- Chosen-cipertext. Where the hacker sends a message to the target, this is then encrypted by the target's private-key and the hacker then analyses the encrypted message. For example, a hacker may send an email to the encryption file server and the hacker spies on the delivered message.
- Exhaustive search. Where the hacker uses brute force to decrypt the cipertext and tries every possible key.
- Active attack. Where the hacker inserts or modifies messages.
- Man in the middle. Where the hacker is hidden between two parties and impersonates each of them to the other.
- The replay system. Where the hacker takes a legitimate message and sends it into the network at some future time.
- Cut and paste. Where the hacker mixes parts of two different encrypted messages and, sometimes, is able to create a new message. This message is likely to make no sense, but may trick the receiver into doing something that helps the hacker.
- Time resetting. Some encryption schemes use the time of the computer to create the key. Resetting this time or determining the time that the message was created can give some useful information to the hacker.

16.3 Random number generators

One way to crack a code is to exploit a weakness in the generation of the encryption key. The hacker can then guess which keys are more likely to occur. This is known as a statistical attack.

Many programming languages use a random number generator which is based on the current system time (such as `rand()`). This method is no good in data encryption as the hacker can simply determine the time that the message was encrypted and the algorithm used.

An improved source of randomness is the time between two keystrokes (as used in PGP – pretty good privacy). However this system has been criticized as a hacker can spy on a user over a network and determine the time between keystrokes. Other sources of true randomness have also been investigated, including noise from an electronic device and noise from an audio source.

16.4 Survey of private-key cryptosystems

16.4.1 DES

DES (Data Encryption Standard) is a block cipher scheme which operates on 64-bit block sizes. The private key has only 56 useful bit as eight of its bits are used for parity. This gives 2^{56} or 10^{17} possible keys. DES uses a complex series of permutations and substitutions, the result of these operations is XOR'ed with the input. This is then repeated 16 times using a different order of the key bits each time. DES is a very strong code and has never been broken, although several high-powered computers are now available which, using brute force, can crack the code. A possible solution is 3DES (or triple DES) which uses DES three times in a row. First to encrypt, next to decrypt and finally to encrypt. This system allows a key-length of more than 128 bits.

16.4.2 MOSS

MOSS (MIME object security service) is an Internet RFC and is typically used for sound encryption. It uses symmetric encryption and the size of the key is not specified. The only public implementation is TIS/MOSS 7.1 which is basically an implementation of 56-bit DES code with a violation.

16.4.3 IDEA

IDEA (International Data Encryption Algorithm) is discussed in Section 15.2.2 and is similar to DES. It operates on 64-bit blocks of plaintext and uses a 128-bit key.

IDEA operates over 17 rounds with a complicated mangler function. During decryption this function does not have to be reversed and can simply be applied in the same way as during encryption (this also occurs with DES). IDEA uses a different key expansion for encryption and decryption, but every other part of the process is identical. The same keys are used in DES decryption but in the reverse order.

The key is devised in eight 16-bit blocks; the first six are used in the first round of encryption the last two are used in the second run. It is free for use in non-commercial version and appears to be a strong cipher.

16.4.4 RC4/RC5

RC4 is a cipher designed by RSA Data Security, Inc and was a secret until information on it appeared on the Internet. The Netscape secure socket layer (SSL) uses RC4. RC4 uses a pseudo random number generator where the output of the generator is XOR'ed with the plaintext. It is has a fast algorithm and can use any key-length. Unfortunately the same key cannot be used twice. Recently a 40-bit key version was broken in 8 days without special computer power.

RC5 is a fast block cipher designed by Rivest for RSA Data Security. It has a parameterized algorithm with a variable block size (32, 64 or 128 bits), a variable key size (0 to 2048 bits) and a variable number of rounds (0 to 255).

It has a heavy use of data dependent rotations and the mixture of different operations. This assures that RC5 is secure. Kaliski and Yin found that RC5 with a 64-bit block size and 12 or more rounds gives good security.

16.4.5 SAFER

SAFER (Secure and Fast Encryption Routine) is a non-proprietary block-cipher developed by Massey in 1993. It operates on a 64-bit block size and has a 64-bit or 128-bit key size. SAFER has up to 10 rounds (although a minimum of 6 is recommended). Unlike most recent block ciphers, SAFER has a slightly different encryption and decryption procedure. The algorithm operates on single bytes at a time and it thus can be implemented on systems with limited processing power, such as on smart-cards applications.

A typical implementation is SAFER K-64 which uses a 40-bit key and has been shown that it is immune from most attacks when the number of rounds is greater than 6.

16.4.6 SKIPJACK

Skipjack is new block cipher which operates on 64-bit blocks. It uses an 80-bit key and has 32 rounds. The NSA have classified details of Skipjack and its algorithm is only available in hardware implementation called Clipper Chips. The name Clipper derives from an earlier implementation of the algorithm. Each transmission contains the session key encrypted in the header. The licensing of clipper chips allows US government to decrypt all SKIPJACK messages.

16.5 Private-key encryption

16.5.1 Data Encryption Standard (DES)

In 1977, the National Bureau of Standards (now the National Institute of Standards and Technology) published the DES for commercial and unclassified US government applications. DES is based on an algorithm known as the Lucifer cipher designed by IBM. It maps a 64-bit input block to a 64-bit output block and uses a 56-bit key. The key itself is actually 64 bits long but as 1 bit in each of the 8 bytes is used for odd parity on each byte, the key only contains 56 meaningful bits.

DES overview

The main steps in the encryption process are as follows:

- Initially the 64-bit input is permuted to obtain a 64-bit result (this operation does little to the security of the code).
- Next, there are 16 iterations of the 64-bit result and the 56-bit key. Only 48 bits of the key are used at a time. The 64-bit output from each iteration is used as an input to the next iteration.
- After the 16th iteration, the 64-bit output goes through another permutation, which is the inverse of the initial permutation.

Figure 16.1 shows the basic operation of DES encryption.

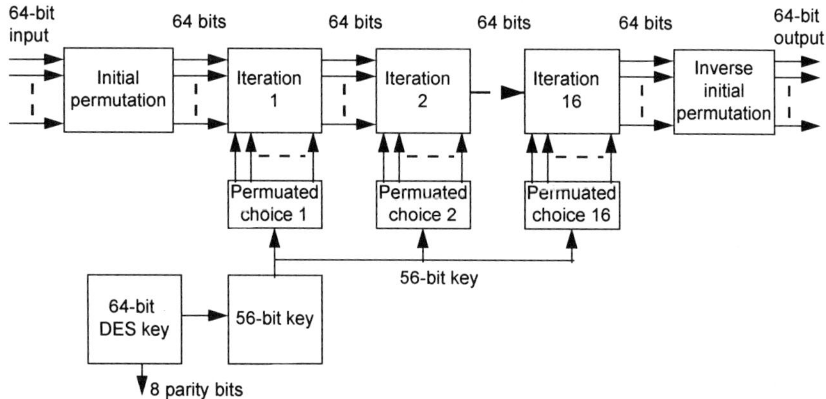

Figure 16.1 Overview of DES operation.

Permutation of the data

Before the first iteration and after the last iteration, DES performs a permutation on the data. The permutations is as follows:

Initial permutation:

```
58 50 42 34 26 18 10 2  60 52 44 36 28 20 12 4  62 54 46 38 30 22 14 6
64 56 48 40 32 24 16 8  57 49 41 33 25 17 9  1  59 51 43 35 27 19 11 3
61 53 45 37 29 21 13 5  63 55 47 39 31 23 15 7
```

Final permutation:

```
40 8  48 16 56 24 64 32 39 7  47 15 55 23 63 31 38 6  46 14 54 22 62 30
37 5  45 13 53 21 61 29 36 4  44 12 52 20 60 28 35 3  43 11 51 19 59 27
34 2  42 10 50 18 58 26 33 1  41 9  49 17 57 25
```

These numbers specify the bit numbers of the input to the permutation and the order of the numbers corresponds to the output bit position. Thus, input permutation:

- Input bit 58 moves to output bit 1 (58 is in the 1st bit position).
- Input bit 50 moves to output bit 2 (50 is in the 2nd bit position).

- Input bit 42 moves to output bit 3 (42 is in the 3rd bit position).
- Continue until all bits are exhausted.

In addition, the final permutation could be:

- Input bit 58 moves to output bit 1 (1 is in the 58th bit position).
- Input bit 50 moves to output bit 2 (2 is in the 50th bit position).
- Input bit 42 moves to output bit 3 (3 is in the 42nd bit position).
- Continue until all bits are exhausted.

Thus, the input permutation is the reverse of the output permutation. Arranged as blocks of 8 bits, it gives:

```
58 50 42 34 26 18 10 2
60 52 44 36 28 20 12 4
62 54 46 38 30 22 14 6
 :                 :
61 53 45 37 29 21 13 5
63 55 47 39 31 23 15 7
```

It can be seen that the first byte of input gets spread into the 8th bit of each of the other bytes. The second byte of input gets spread into the 7th bit of each of the other bytes, and so on.

Generating the per-round keys

The DES key operates on 64-bit data in each of the 16 iterations. The key is made of a 56-bit key used in the iterations and 8 parity bits. A 64-bit key of:

$k_1 k_2 k_3 k_4 k_5 k_6 k_7 k_8 k_9 k_{10} k_{11} k_{12} k_{13 \ldots} k_{64}$

contains the parity k_8, k_{16}, $k_{32 \ldots} k_{64}$. The iterations are numbered I_1, I_2, ... I_{16}. The initial permutation of the 56 useful bits of the key is used to generate a 56-bit output. It divides into two 28-bit values, called C_0 and D_0. C_0 is specified as:

$k_{57} k_{49} k_{41} k_{33} k_{25} k_{17} k_9 k_1 k_{58} k_{50} k_{42} k_{34} k_{26} k_{18} k_{10} k_2 k_{59} k_{51} k_{43} k_{35} k_{27} k_{19} k_{11} k_3 k_{60} k_{52} k_{44} k_{36}$

And D_0 is:

$k_{63} k_{55} k_{47} k_{39} k_{31} k_{23} k_{15} k_7 k_{62} k_{54} k_{46} k_{38} k_{30} k_{22} k_{14} k_6 k_{61} k_{53} k_{45} k_{37} k_{29} k_{21} k_{13} k_5 k_{28} k_{20} k_{12} k_4$

Thus the 28-bit C_0 key will contain the 57th bit of the DES key as the first bit, the 49th as the second bit, and so on. Notice that none of the 28-bit values contains the parity bits.

Most of the rounds have a 2-bit rotate left shift, but rounds 1, 2, 9 and 16 have a single-bit rotate left (ROL). A left rotation moves all the bits in the key to the left and the bit which is moved out of the left-hand side is shifted into the right end.

The key for each iteration (K_i) is generated from C_i (which makes the left half) and D_i (which makes the right half). The permutations of C_i that produces the left half of K_i is:

$c_{14} c_{17} c_{11} c_{24} c_1 c_5 c_3 c_{28} c_{15} c_6 c_{21} c_{10} c_{23} c_{19} c_{12} c_4 c_{26} c_8 c_{16} c_7 c_{27} c_{20} c_{13} c_2$

and the right half of K_i is:

$d_{41}\, d_{52}\, d_{31}\, d_{37}\, d_{47}\, d_{55}\, d_{30}\, d_{40}\, d_{51}\, d_{45}\, d_{33}\, d_{48}\, d_{44}\, d_{49}\, d_{39}\, d_{56}\, d_{34}\, d_{53}\, d_{46}\, d_{42}\, d_{50}\, d_{36}\, d_{29}\, d_{32}$

Thus the 56-bit key is made up of:

$c_{14}\, c_{17}\, c_{11}\, c_{24}\, c_{1}\, c_{5}\, c_{3}\, c_{28}\, c_{15}\, c_{6}\, c_{21}\, c_{10}\, c_{23}\, c_{19}\, c_{12}\, c_{4}\, c_{26}\, c_{8}\, c_{16}\, c_{7}\, c_{27}\, c_{20}\, c_{13}\, c_{2}\, d_{41}\, d_{52}\, d_{31}\, d_{37}\, d_{47}\, d_{55}\, d_{30}\, d_{40}$
$d_{51}\, d_{45}\, d_{33}\, d_{48}\, d_{44}\, d_{49}\, d_{39}\, d_{56}\, d_{34}\, d_{53}\, d_{46}\, d_{42}\, d_{50}\, d_{36}\, d_{29}\, d_{32}$

Figure 16.2 illustrates the process (note that only some of the bit positions have been shown).

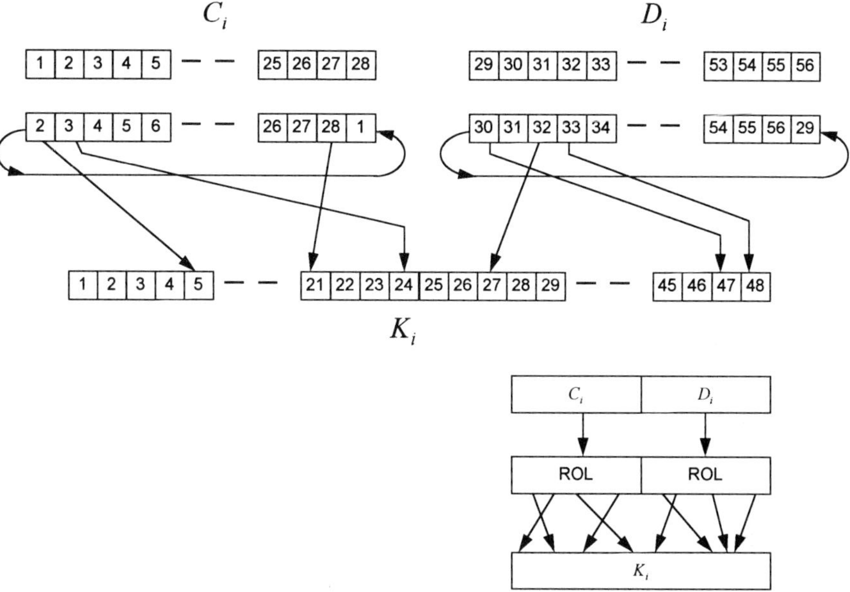

Figure 16.2 Generating the iteration key.

Iteration operations

Each iteration takes the 64-bit output from the previous iteration and operates on it with a 56-bit per iteration key. Figure 16.3 shows the operation of each iteration. The 64-bit input is split into two parts, L_i and R_i. R_i is operated on with an expansion/permutation (E-table) to give 48 bits. The output from the E-table conversion is then exclusive-OR'ed with the permuated 48-bit key. Next a substitute/ choice stage (S-box) is used to transform the 48-bit result to 32 bits. These are then XORed with L_i to give the resulting R_{i+1} (which is R_i for the next iteration). The operation of expansion/XOR/substitution is often known as the mangler function. The R_i input is also used to produce L_{i+1}.

The mangler function takes the 32-bit R_i and the 48-bit K_i and produces a 32-bit output (which when XORed with L_i produces R_i+1. It initially expands R_i from 32 bits to 48 bits. This is done by splitting R_i into eight 4-bit chunks and then expanding each of the chunks into 6 bits by taking the adjacent bits and concatenating them onto the chunk. The leftmost and rightmost bits of R are considered adjacent. For example, if R_i is:

1011 0011 1111 1010 0000 1100 1010 0110

then this is expanded into:

010110 100111 111111 110100 000001 011001 010100 001101

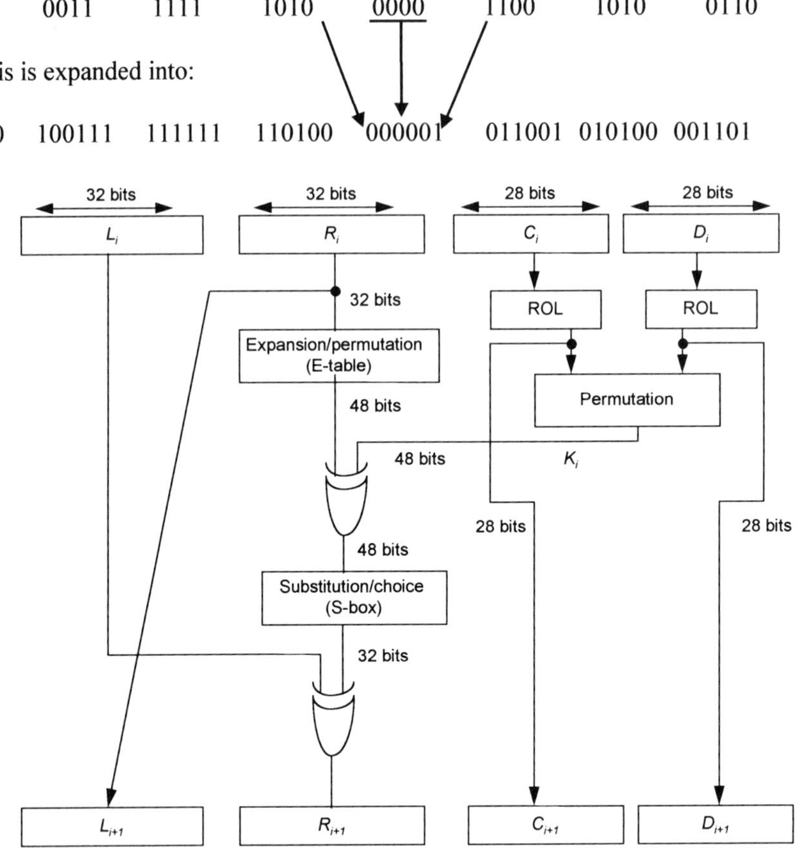

Figure 16.3 Iteration step.

The output from the expansion is then XORed with K_i and the output of this is fed into the S-box. Each 6-bit chunk of the 48-bit output from the XOR operation is then substituted with a 4-bit chunk using a lookup look-up. An S-box table for the first 6-bit chunk is given in Table 16.1. Thus, for example, the input bit sequence of:

000000 000001 *XXXXXX* ...

would be converted to:

1110 0000 *xxxx* ...

Concluding remarks

The design of the DES scheme was constructed behind closed doors so there are few pointers to the reasons for the construction of the encryption. One of the major weaknesses of the scheme is the usage of a 56-bit key, which means there are only 2^{56} or 7.2×10^{16} keys. Thus, as the cost of hardware reduces and the power of computers increases, the time taken to exhaustively search for a key becomes smaller each year.

Table 16.1 S-box conversion for first 6-bit chunk.

Input	*Output*	*Input*	*Output*	*Input*	*Output*	*Input*	*Output*
000000	1110	010000	0011	100000	0100	110000	1111
000001	0000	010001	1010	100001	1111	110001	0101
000010	0100	010010	1010	100010	0001	110010	1100
000011	1111	010011	0110	100011	1100	110011	1011
000100	1101	010100	0100	100100	1110	110100	1001
000101	0111	010101	1100	100101	1000	110101	0011
000110	0001	010110	1100	100110	1000	110110	0111
000111	0100	010111	1011	100111	0010	110111	1110
001000	0010	011000	0101	101000	1101	111000	0011
001001	1110	011001	1001	101001	0100	111001	1010
001010	1111	011010	1001	101010	0110	111010	1010
001011	0010	011011	0101	101011	1001	111011	0000
001100	1011	011100	0000	101100	0010	111100	0101
001101	1101	011101	0011	101101	0001	111101	0110
001110	1000	011110	0111	101110	1011	111110	0000
001111	0001	011111	1000	101111	0111	111111	1101

In the past, there was concern about potential weaknesses in the design of the eight S-boxes. This appears to have been misplaced as no one has found any weaknesses yet. Indeed several researchers have found that swapping the S-boxes significantly reduces the security of the code.

A new variant, called Triple DES, has been proposed by Tuchman and has been standardized in financial applications. The technique uses two keys and three executions of the DES algorithm. A key, K_1, is used in the first execution, then K_2 is used and finally K_1 is used again. These two keys give an effective key length of 112 bits, that is 2×64 key bits minus 16 parity bits. The Triple DES process is illustrated in Figure 16.4.

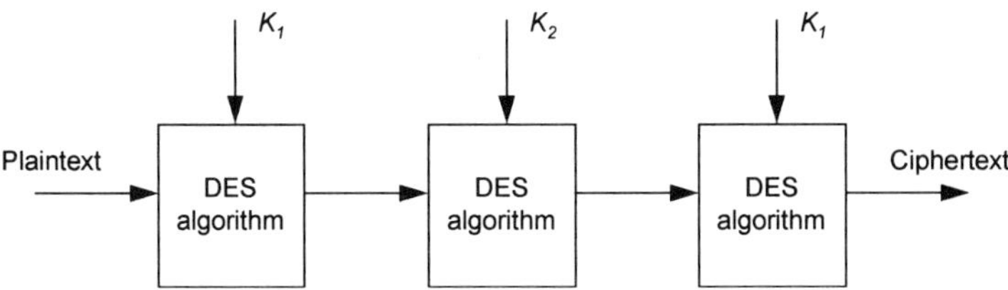

Figure 16.4 Triple DES process.

16.5.2 IDEA

IDEA (International Data Encryption Algorithm) is a private-key encryption process with is similar to DES. It was developed by Xuejia Lai and James Massey of ETH Zuria and is intended for implementation in software. IDEA operates on 64-bit blocks of plaintext; using a 128-bit key, it converts them into 64-bit blocks of ciphertext. Figure 16.5 shows the basic encryption process.

IDEA operates over 17 rounds with a complicated mangler function. During decryption this function does not have to be reversed and can simply be applied in the same way as during encryption (this also occurs with DES). IDEA uses a different key expansion for encryption and decryption, but every other part of the process is identical. The same keys are used in DES decryption but in the reverse order.

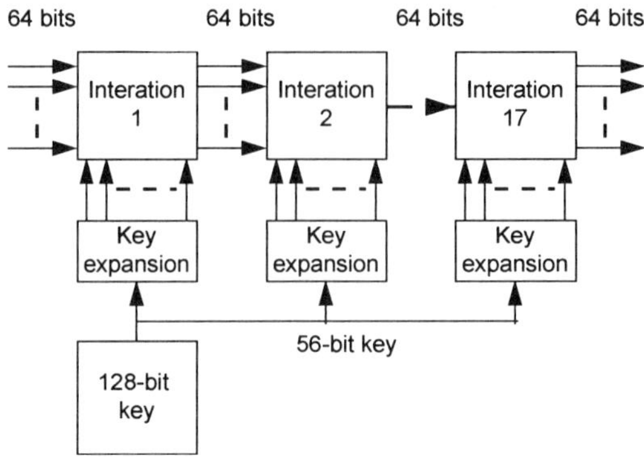

Figure 16.5 IDEA encryption.

Operation

Each primitive operation in IDEA maps two 16-bit quantities into a 16-bit quantity. IDEA uses three operations, all easy to compute in software, to create a mapping. The three basic operations are:

- Exclusive-OR (⊕).
- Slightly modified add (+), and ignore any bit carries.
- Slightly modified multiply (⊗) and ignore any bit carries. Multiplying involves first calculating the 32-bit result, then taking the remainder when divided by $2^{16}+1$ (mod $2^{16}+1$).

Key expansions

The 128-bit key is expanded into fifty-two 16-bit keys, K_1, K_2, ... K_{52}. The key is generated differently for encryption than for decryption. Once the 52 keys are generated, the encryption and decryption processes are the same.

The 52 encryption keys are generated as follows:

- Keys 1–8: write out the 128-bit key and, starting from the left, chop off 16 bits at a time. This generates eight 16-bit keys. Thus the 128-bit key of *AAAAAAAAAAAAAAAA*... *HHHHHHHHHHHHHHHH* will generate eight keys of *AAAAAAAAAAAAAAAA*, *BBBBBBBBBBBBBBBB*, and so on.
- Keys 9–16: the next eight keys are generated at bit 25, and wrapped around to the beginning when the end.
- Keys 17–24: the next eight keys are generated at bit 50, and wrapped around to the beginning when the end.
- The rest of the keys are generated by offsetting by 25 bits and wrapped around to the beginning until the end.

The 64-bit (or 32-bit) per round, keys used are made up of 4 (or 2) of the encryption keys:

Key 1:	$K_1K_2K_3K_4$	Key 2:	K_5K_6	Key 3:	$K_7K_8K_9K_{10}$
Key 4:	$K_{11}K_{12}$	Key 5:	$K_{13}K_{14}K_{15}K_{16}$	Key 6:	$K_{17}K_{18}$

Key 7:	$K_{19}K_{20}K_{21}K_{22}$	Key 8:	$K_{23}K_{24}$	Key 9:	$K_{25}K_{26}K_{27}K_{28}$
Key 10:	$K_{29}K_{30}$	Key 11:	$K_{31}K_{32}K_{33}K_{34}$	Key 12:	$K_{35}K_{36}$
Key 13:	$K_{37}K_{38}K_{39}K_{40}$	Key 14:	$K_{41}K_{42}$	Key 15:	$K_{43}K_{44}K_{45}K_{46}$
Key 16:	$K_{47}K_{48}$	Key 17:	$K_{49}K_{50}K_{51}K_{52}$		

16.5.3 Iteration

Odd rounds have a different process to even rounds. Each odd round uses a 64-bit key and even rounds use a 32-bit key.

Odd rounds are simple; the process is:

If the input is a 64-bit key of $I_1 I_2 I_3 I_4$, where the I_1 is the most significant 16 bits, I_2 is the next most significant 16 bits, and so on. The output of the iteration is also a 64-bit key of $O_1 O_2 O_3 O_4$ and the applied key is $K_a K_b K_c K_d$. The iteration for the odd iteration is then:

$$O_1 = I_1 \otimes K_a$$
$$O_2 = I_3 + K_c$$
$$O_3 = I_2 + K_b$$
$$O_4 = I_4 \otimes K_d$$

An important feature is that this operation is totally reversible: multiplying O_1 by the inverse of K_a gives I_1, and multiplying O_4 by the inverse of K_d gives I_4. Adding O_2 to the negative of K_c gives I_3, and adding O_3 to the negative of K_b gives I_2.

Even rounds are less simple, the process is as follows. Suppose the input is a 64-bit key of $I_1I_2I_3I_4$, where I_1 is the most significant 16 bits, I_2 is the next most significant 16 bits, and so on. The output of the iteration is also a 64-bit key of $O_1O_2O_3O_4$ and the applied key is 32 bits of K_aK_b. The iteration for the even round performs a mangler function of:

$$A = I_1 \otimes I_2 \qquad B = I_3 \otimes I_4$$
$$C = ((K_a \otimes A) + B)) \otimes K_b) \qquad D = (K_a \otimes A) + C)$$

$$O_1 = I_1 \oplus C \qquad O_2 = I_2 \oplus C$$
$$O_3 = I_3 \oplus D \qquad O_4 = I_4 \oplus D$$

The most amazing thing about this iteration is that the inverse of the function is simply the function itself. Thus, the same keys are used for encryption and decryption (this differs from the odd round, where the key must be either the negative or the inverse of the encryption key.

16.5.4 IDEA security

There are no known methods that can be used to crack IDEA, apart from exhaustive search. Thus, as it has a 128-bit code it is extremely difficult to break, even with modern high-performance computers.

Public-key Encryption

17.1 Introduction

Public-key algorithms use a secret element and a public element to their code. One of the main algorithms is RSA. Compared with DES it is relatively slow but it has the advantage that users can choose their own code whenever they need one. The most commonly used public-key cryptosystems are covered in the next sections.

17.1.1 RSA

RSA stands for Rivest, Shamir and Adelman, and is the most commonly used public-key cryptosystem. It is patented only in the USA and is secure for key-length of over 728 bits. The algorithm relies of the fact that it is difficult to factorize large numbers. Unfortunately, it is particularly vulnerable to chosen plaintext attacks and a new timing attack (spying on key-stroke time) was announced on the 7 December 1995. This attack would be able to break many existing implementations of RSA.

17.1.2 Elliptic curve

Elliptic curve is a new kind of public-key cryptosystem. It suffers from speed problems, but this has been overcome with modern high-speed computers.

17.1.3 DSS

DSS (digital signature standard) is related to the DSA (digital signature algorithm). This standard has been selected by the NIST and the NSA, and is part of the Capstone project. It uses 512-bit or 1024-bit key size. The design presents some lack in key-exchange capability and is slow for signature-verification.

17.1.4 Diffie-Hellman

Diffie-Hellman is commonly used for key-exchange. The security of this cipher relies on both the key-length and the discrete algorithm problem. This problem is similar to the factorizing of large numbers. Unfortunately, the code can be cracked and the prime number generator must be carefully chosen.

17.1.5 LUC

Peter Smith developed LUC which is a public-key cipher that uses Lucas functions instead of exponentiation. Four other algorithms have also been developed, these are:

- LUCDIF (a key-negotiation method).
- LUCELG PK (equivalent to EL Gamel encryption).
- LUCELG DS (equivalent to EL Gamel data signature system).
- LUCDSA (equivalent to the DSS).

17.2 RSA

RSA is a public-key encryption/decryption algorithm and is much slower than IDEA and DES. The key length is variable and the block size is also variable. A typical key length is 512 bits. RSA uses a public-key and a private key, and uses the fact that large prime numbers are extremely difficult to factorize. The following steps are taken to generate the public and private keys:

1. Select two large prime numbers, a and b (each will be roughly 256 bits long). The factors a and b remain secret and n is the result of multiplying them together. Each of the prime numbers is of the order of 10^{100}.
2. Next the public-key is chosen. To do this a number e is chosen so that e and $(a-1)\times(b-1)$ are relatively prime. Two numbers are relatively prime if they have no common factor greater than 1. The public-key is then $<e,n>$ and results in a key which is 512 bits long.
3. Next the private key for decryption, d, is computed so that:

   $$d=e^{-1} \bmod [(a-1)\times(b-1)]$$

   This then gives a private key of $<d,n>$.The values p and q can then be discarded (but should never be disclosed to anyone).

The encryption process to ciphertext, c, is then defined by:

$$c=m^e \bmod n$$

The message, m, is then decrypted with:

$$m=c^d \bmod n$$

It should be noted that the message block m must be less than n. When n is 512 bits then a message which is longer than 512 bits can be broken up into blocks of 512 bits.

17.2.1 Encryption/decryption keys

When two parties, P_1 and P_2, are communicating they encrypt data using a pair of public/private key pairs. Party P_1 encrypts their message using P_2's public-key. Then party P_2 uses their private key to decrypt this data. When party P_2 encrypts a message it sends to P_1 using P_1's public-key and P_1 decrypts this using their private key. Notice that party P_1 cannot decrypt this message that it has sent to P_2 as only P_2 has the required private key.

A great advantage of RSA is that the key has a variable number of bits. It is likely that, in the coming few years, that powerful computer systems will determine all the factors to 512-bit values. Luckily the RSA key has a variable size and can easily be changed. Some users are choosing keys with 1024 bits.

17.2.2 Simple RSA example

Initially the PARTY1 picks two prime numbers. For example:

a=11 and b=3

Next, the n value is calculated. Thus:

$$n = a \times b = 11 \times 3 = 33$$

Next PHI is calculated by:

$$PHI = (a-1)(b-1) = 20$$

The public exponent *e* is then generated so that greater common divisor of *e* and *PHI* is 1 (e is relatively prime with PHI). Thus, the smallest value for *e* is:

$$e = 3$$

The *n* (33) and the *e* (3) values are the public keys. The private key (*d*) is the inverse of *E* modulo *PHI*.

$$d = e^{-1} \text{ mod } [(p-1)\times(q-1)]$$

This can be calculated by using extended Euclidian algorithm, to give the private key, *d* of 7.

Thus n=33, e=3 and d=7.

The PARTY2 can be given the public keys of *e* and *n*. So that PARTY2 can encrypt the message with them. PARTY1, using d and n can then decrypt the encrypted message.

For example, if the message value to decrypt is 4, then:

$$c = m^{e} \text{ mod n} = 4^{3} \text{ mod } 33 = 31$$

Therefore, the encrypted message (*c*) is 31.

The encrypted message (*c*) is then decrypted by PARTY1 with:

$$m = c^{d} \text{ mod } n = 31^{7} \text{ mod } 33 = 4$$

which is equal to the message value.

17.2.3 Simple RSA program

An example program which has a limited range of prime numbers is given next.

```
#include <stdio.h>
#include <math.h>

#define  TRUE  1
#define  FALSE 0

void  get_prime( long *val);
long  getE( long PHI);
long  get_common_denom( long e, long PHI);
long  getD( long e,  long PHI);
```

```
long   decrypt(long c,long n, long d);

int    main(void)
{
long   a,b,n,e,PHI,d,m,c;


       get_prime(&a);
       get_prime(&b);
       n=a*b;
       PHI=(a-1)*(b-1);
       e=getE(PHI);

       d= getD(e,PHI);
       printf("Enter input value >> "); scanf("%ld",&m);

       c=(long)pow(m,e) % n;

       printf("a=%ld b=%ld n=%ld PHI=%ld\n",a,b,n,PHI);
       printf("e=%ld d=%ld c=%ld\n",e,d,c);

       m=decrypt(c,n,d); /* this function required as c to      */
                  /*the power of d causes and overflow   */
       printf("Message is %ld ",m);
       return(0);
}

long   decrypt(long c,long n, long d)
{
long   i,g,f;

if (d%2==0) g=1; else g=c;

for (i=1;i<=d/2;i++)
          {
          f=c*c % n;
          g=f*g % n;
       }
     return(g);
}

long getD( long e,  long PHI)
{
long u[3]={1,0,PHI};
long v[3]={0,1,e};
long q,temp1,temp2,temp3;

       while (v[2]!=0)
       {
          q=floor(u[2]/v[2]);
          temp1=u[0]-q*v[0];
          temp2=u[1]-q*v[1];
          temp3=u[2]-q*v[2];
          u[0]=v[0];
          u[1]=v[1];
          u[2]=v[2];
          v[0]=temp1;
          v[1]=temp2;
          v[2]=temp3;
       }
       if (u[1]<0) return(u[1]+PHI);
       else return(u[1]);
}

long   getE( long PHI)
{
```

```
 long great=0, e=2;

      while (great!=1)
      {
         e=e+1;
         great = get_common_denom(e,PHI);
      }
      return(e);
}

long get_common_denom(long e, long PHI)
{
long great,temp,a;

      if (e >PHI)
      {
         while (e % PHI != 0)
         {
            temp= e % PHI;
            e =PHI;
            PHI = temp;
         }
         great = PHI;
      } else
      {
         while (PHI % e != 0)
         {
            a = PHI % e;
            PHI = e;
            e = a;
         }
         great = e;
      }
      return(great);
}

void  get_prime( long *val)
{
#define NO_PRIMES 11
long  primes[NO_PRIMES]={3,5,7,11,13,17,19,23,29,31,37};
long  prime,i;

      do
      {
         prime=FALSE;
         printf("Enter a prime number >> ");
         scanf("%ld",val);
         for (i=0;i<NO_PRIMES;i++)
            if (*val==primes[i]) prime=TRUE;
      } while (prime==FALSE);
}
```

A sample run of the program is given next.

```
Enter a prime number >> 11
Enter a prime number >> 3
Enter input value >> 4
a=11 b=3 n=33 PHI=20
e=3 d=7 c=31
Message is 4
```

Transmission Control Protocol (TCP) and Internet Protocol (IP)

18.1 Introduction

Networking technologies, such as Ethernet, Token Ring and FDDI provide a data link layer function, that is, they allow a reliable connection between one node and another on the same network. They do not provide for inter-networking where data can be transferred from one network to another or one network segment to another. For data to be transmitted across networks requires an addressing scheme which is read by a bridge, gateway or router. The interconnection of networks is known as internetworking (or internet). Each part of an internet is a subnetwork (or subnet).

The Transmission Control Protocol (TCP) and Internet Protocol (IP are a pair of protocols that allow one subnet to communicate with another. A protocol is a set of rules that allow the orderly exchange of information. The IP part corresponds to the Network layer of the OSI model and the TCP part to the Transport layer. As prevously mentioned their operation should be transparent to the Physical and Data Link layers and can thus be used on Ethernet, FDDI or Token Ring networks. This is illustrated in Figure 18.1. The address of the Data Link layer corresponds to the physical address of the node, such as the MAC address (in Ethernet and Token Ring) or the telephone number (for a modem connection). The IP address is assigned to each node on the internet and is used to identify the location of the network and any subnets.

TCP/IP was originally developed by the US Defense Advanced Research Projects Agency (DARPA). Their objective was to connect a number of universities and other research establishments to DARPA. The resultant internet is now known as the Internet. It has since outgrown this application and many commercial organizations and home users now connect to the Internet. The Internet uses TCP/IP as a standard to transfer data. Each node on the Internet is assigned a unique network address, called an IP address. Note that any organization can have its own internets, but if it they want to connect these to the Internet then their addresses must conform to the Internet addressing format.

The ISO have adopted TCP/IP as the basis for the standards relating to the network and transport layers of the OSI model. This standard is known as ISO-IP. Most currently available systems conform to the IP addressing standard.

Common applications that use TCP/IP communications are remote login and file transfer. Typical programs used in file transfer and log-in over TCP communication are `ftp` for file transfer program and `telnet` that allows remote login into another computer. The `ping` program determines if a node is responding to TCP/IP communications.

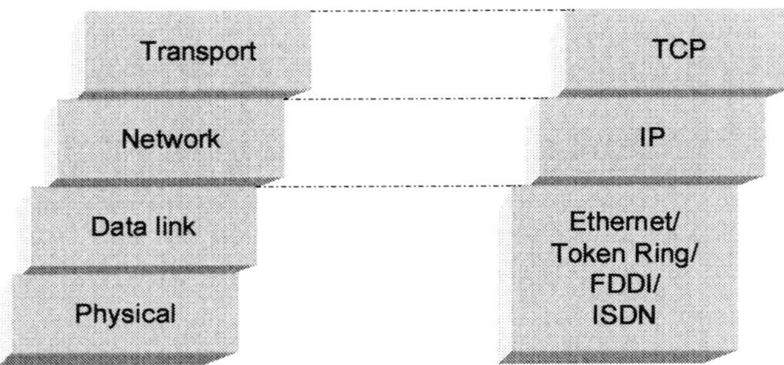

Figure 18.1 TCP/IP and the OSI model.

18.2 TCP/IP gateways and hosts

TCP/IP hosts are nodes which communicate over interconnected networks using TCP/IP communications. A TCP/IP gateway node connects one type of network to another. It contains hardware to provide the physical link between the different networks and the hardware and software to convert frames from one network to the other. Typically, it converts a Token Ring MAC layer to an equivalent Ethernet MAC layer, and vice versa.

A router connects a network to another of the same kind through a point-to-point link. The main operational difference between a gateway, a router, and a bridge, is that, for a Token Ring and Ethernet network, the bridge uses the 48-bit MAC address to route frames, whereas the gateway and router use the IP network address. As an analogy to the public telephone system, the MAC address is equivalent to a randomly assigned telephone number, whereas the IP address would contain the information on the logical located of the telephone, such as which country, area code, and so on.

Figure 18.2 shows how a gateway routes information. The gateway reads the frame from the computer on network A. It then reads the IP address contained in the frame and makes a decision as to whether it is routed out of network A to network B. If it does then it relays the frame to network B.

18.3 Function of the IP protocol

The main functions of the IP protocol are to:

- Route IP data frames – which are called internet datagrams – around an internet. The IP protocol program running on each node knows the location of the gateway on the network. The gateway must then be able to locate the interconnected network. Data then passes from node to gateway through the internet.
- Fragment the data into smaller units if it is greater than a given amount (64 KB).
- Report errors. When a datagram is being routed or is being reassembled an error can occur. If this happen then the node that detects the error reports back to the source node. Datagrams are deleted from the network if they travel through the network for more than a set time. Again, an error message is returned to the source node to inform it that the internet routing could not find a route for the datagram or that the destination node, or network, does not exist.

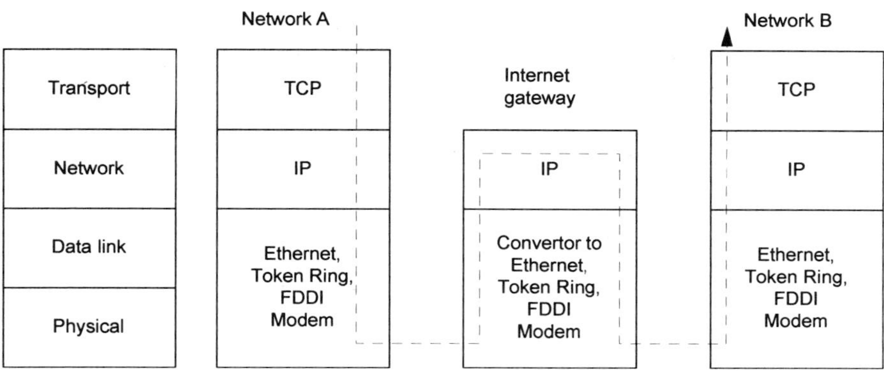

Figure 18.2 Internet gateway layers.

18.4 Internet datagram

The IP protocol is an implementation of the OSI network layer. It adds a data header onto the information passed from the transport layer, the resultant data packet is known as an internet datagram. The header contains information such as the destination and source IP addresses, the version number of the IP protocol and so on. Figure 18.3 shows its format.

The datagram contains up to 65 536 bytes (64 KB) of data. If the data to be transmitted is less than, or equal to, 64 KB, then it is sent as one datagram. If it is more than this then the sender splits the data into fragments and sends multiple datagrams. When transmitted from the source each datagram is routed separately through the internet and the received fragments are finally reassembled at the destination.

The fields in the IP datagram are:

- **Version.** The TCP/IP `version number` helps gateways and nodes correctly interpret the data unit. Differing versions may have a different format or the IP protocol interprets the header differently.
- **Type of service.** The `type of service` bit field is an 8-bit bit pattern in the form `PPPDTRXX`, where `PPP` defines the priority of the datagram (from 0 to 7), `D` sets a low delay service, `T` sets high throughput, `R` sets high reliability and `XX` are currently not used.
- **Header length.** The `header length` defines the size of the data unit in multiplies of 4 bytes (32 bits). The minimum length is 5 bytes and the maximum is 65 536 bytes. Padding bytes fill any unused spaces.
- **D and M bits.** A gateway may route a datagram and split it into smaller fragments. The `D` bit informs the gateway that it should not fragment the data and thus signifies that a receiving node should receive the data as a single unit or not at all. The `M` bit is the more fragments bit and identifies data fragments. The `fragment offset` contains the fragment number.
- **Header checksum.** The `header checksum` contains a 16-bit pattern for error detection.

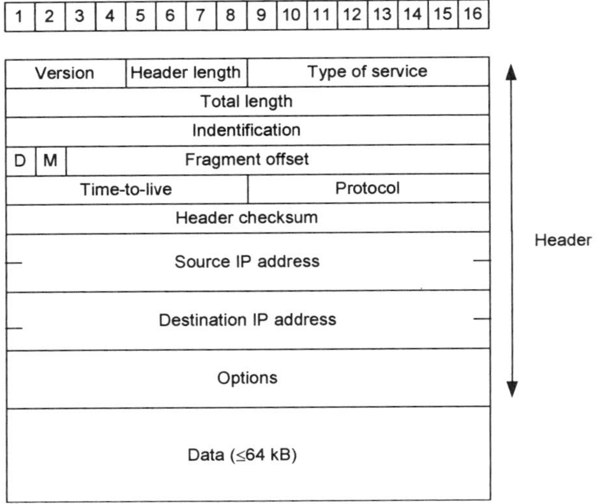

Figure 18.3 Internet datagram format and contents.

- **Time-to-live**. A datagram could propagate through the internet indefinitely. To prevent this, the 8-bit `time-to-live` value is set to the maximum transit time in seconds and is set initially by the source IP. Each gateway then decrements this value by a defined amount. When it becomes zero the datagram is discarded. It also defines the maximum amount of time that a destination IP node should wait for the next datagram fragment.
- **Protocol**. Different IP protocols can be used on the datagram. The 8-bit `protocol` field defines the type to be used.
- **Source and destination IP addresses**. The `source` and `destination IP addresses` are stored in the 32-bit source and destination IP address fields.
- **Options**. The `options` field contains information such as debugging, error control and routing information.

18.5 ICMP

Messages, such as control data, information data and error recovery data, are carried between Internet hosts using the Internet Control Message Protocol (ICMP). These messages are sent with a standard IP header. Typical messages are:

- Destination unreachable (message type 3) – which is sent by a host on the network to say that the destination host is unreachable. It can also include the reason the host cannot be reached.
- Echo request/echo reply (message type 8 or 0) – which are used to check the connectivity between two hosts. The ping command uses this message, where it sends an ICMP ‘echo request’ message to the target host and waits for the destination host to reply with an ‘echo reply’ message.
- Redirection (message type 5) – which is sent by a router to a host that is requesting its routing services. This helps to find the shortest path to a desired host.

- Source quench (message type 4) – which is used when a host cannot receive anymore IP packets at the present.

The ICMP message starts with three fields, as shown in Figure 18.4. The message type has 8 bits and identifies the type of message, these are identified in Table 18.1. The code field is also 8 bits long and a checksum field is 16 bits long. The information after this field depends on the type of message, such as:

- For echo request and reply. An 8-bit identifier follows the message header, then an 8-bit sequence number, followed by the original IP header.
- For destination unreachable, source quelch and time. The message header is followed by 32-bits and then the original IP header.
- For timestamp request. A 16-bit identifier follows the message header, then by a 16-bit sequence number, followed by a 32-bit originating timestamp.

Table 18.1 Message type field value.

Value	Message type	Value	Message type
0	Echo reply	12	Parameter problem
3	Destination unreachable	13	Timestamp request
4	Source quench	14	Timestamp reply
5	Redirect	17	Address mask request
8	Echo request	18	Address mask reply
11	Time-to-live exceeded		

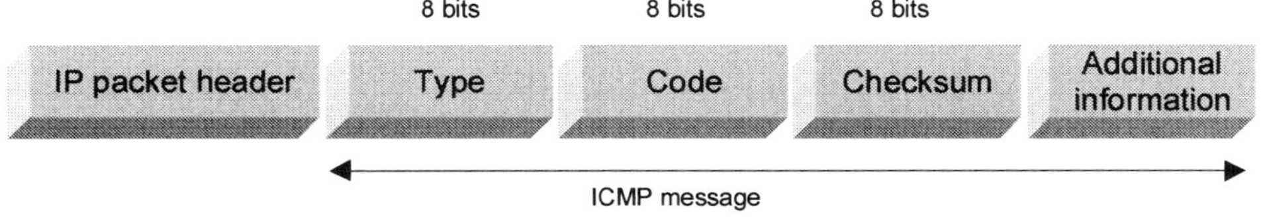

Figure 18.4 ICMP message format.

18.6 TCP/IP internets

Figure 18.5 illustrates a sample TCP/IP implementation. A gateway `MERCURY` provides a link between a token ring network (NETWORK A) and an Ethernet network (ETHER C). Another gateway `PLUTO` connects NETWORK B to ETHER C. The TCP/IP protocol thus allows a host on NETWORK A to communicate with `VAX01`.

18.6.1 Selecting internet addresses

Each node using TCP/IP communications requires an IP address which is then matched to its Token Ring or Ethernet MAC address. The MAC address allows nodes on the same segment to communicate with each other. In order for nodes on a different network to communicate, each must be configured with an IP address.

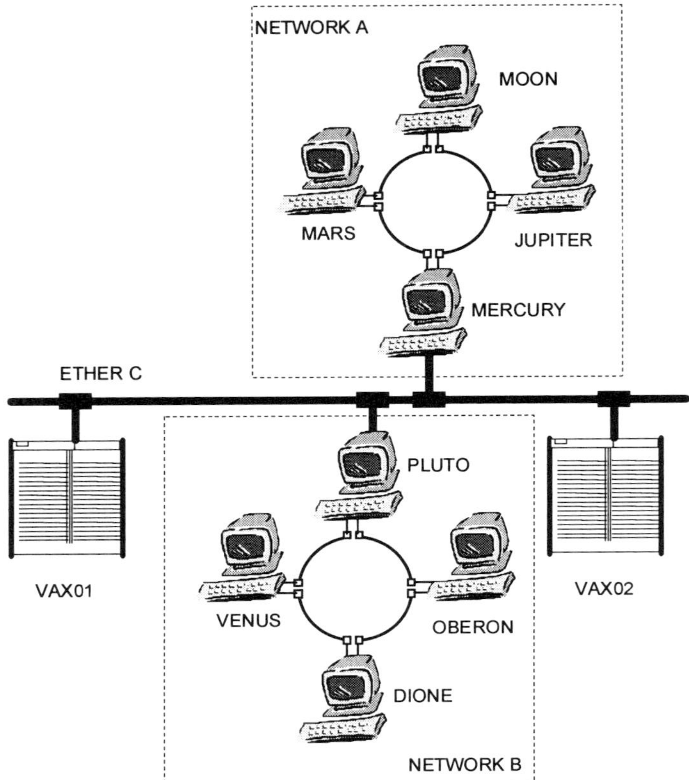

Figure 18.5 Example internet.

Nodes on a TCP/IP network are either hosts or gateways. Any nodes that run application software or are terminals are hosts. Any node that routes TCP/IP packets between networks is called a TCP/IP gateway node. This node must have the necessary network controller boards to physically interface to the other networks it connects with.

18.6.2 Format of the IP address

A typical IP address consists of two fields: the left field (or the network number) identifies the network, and the right number (or the host number) identifies the particular host within that network. Figure 18.6 illustrates this.

The IP address is 32 bits long and can address over four billion physical addresses (2^{32} or 4 294 967 296 hosts). There are three main address formats; as shown in Figure 18.7.

Each of these types is applicable to certain types of networks. Class A allows up to 128 (2^7) different networks and up to 16 777 216 (2^{24}) hosts on each network. Class B allows up to 16 384 networks and up to 65 536 hosts on each network. Class C allows up to 2 097 152 networks each with up to 256 hosts.

The Class A address is thus useful where there are a small number of networks with a large number of hosts connected to them. Class C is useful where there are many networks with a relatively small number of hosts connected to each network. Class B addressing gives a good compromise of networks and connected hosts.

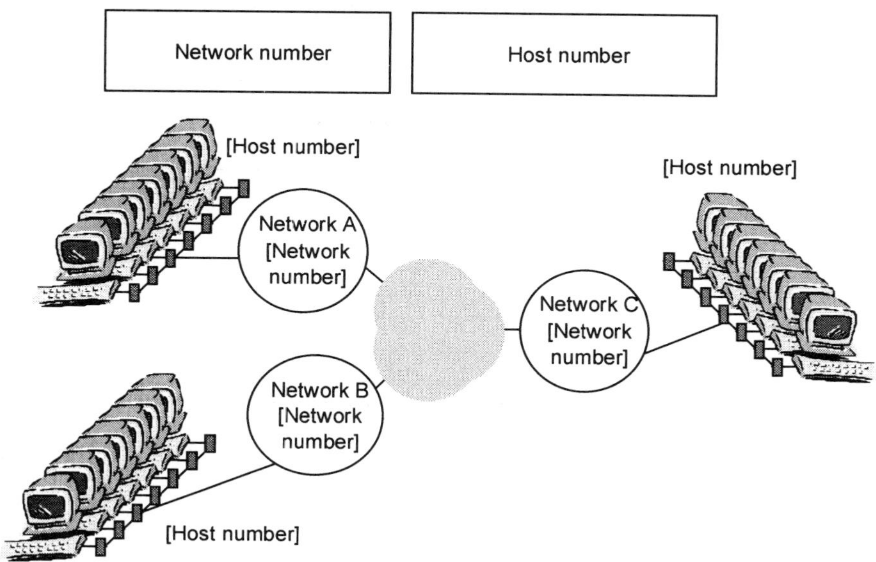

Figure 18.6 IP addressing over networks.

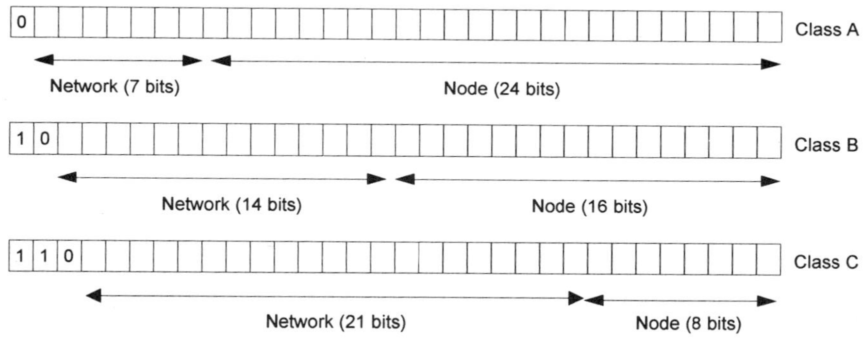

Figure 18.7 Type A, B and C IP address classes.

When selecting internet addresses for the network, the address can be specified simply with decimal numbers within a specific range. The standard DARPA IP addressing format is of the form:

```
W.X.Y.Z
```

where `W`, `X`, `Y` and `Z` represent 1 byte of the IP address. As decimal numbers they range from 0 to 255. The 4 bytes together represent both the network and host address.

Figure 18.7 gives the valid range of the different IP addresses is given and Table 18.2 defines the valid IP addresses. Thus for a class A type address there can be 127 networks and 16 711 680 (256×256×255) hosts. Class B can have 16 320 (64×255) networks and class C can have 2 088 960 (32×256×255) networks and 255 hosts.

Addresses above `223.255.254` are reserved, as are addresses with groups of zeros.

Table 18.2 Ranges of addresses for type A, B and C internet address.

Type	*Network portion*	*Host portion*
A	1 - 126	0.0.1 - 255.255.254
B	128.1 - 191.254	0.1 - 255.254
C	192.0.1 - 223.255.254	1 - 254

18.6.3 Creating IP addresses with subnet numbers

Besides selecting IP addresses of internets and host numbers, it is also possible to designate an intermediate number called a subnet number. Subnets extend the network field of the IP address beyond the limit defined by the type A, B, C scheme. They allow a hierarchy of internets within a network. For example, it is possible to have one network number for a network attached to the internet, and various subnet numbers for each subnet within the network. Figure 18.8 illustrated this.

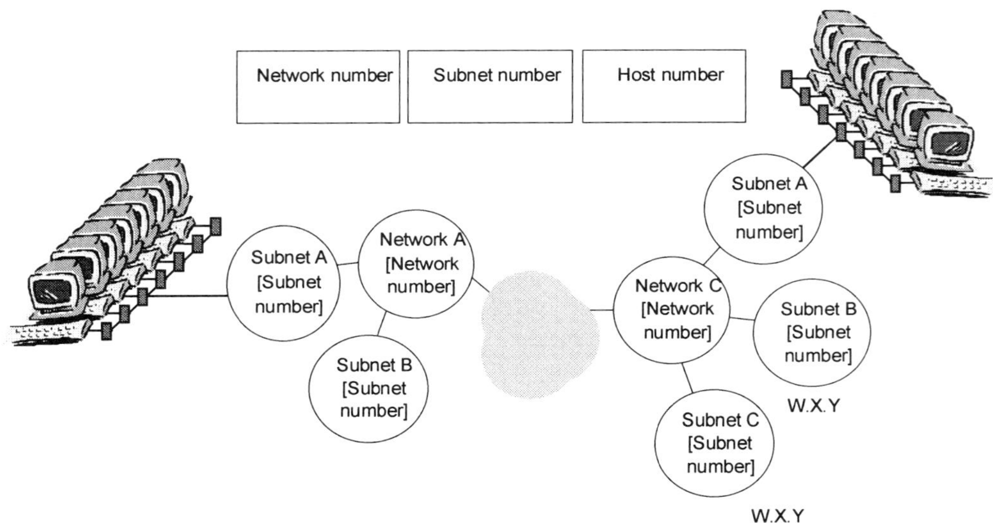

Figure 18.8 IP addresses with subnets.

For an address `W.X.Y.Z`, and for a type A, the address `W` specifies the network and `X` the subnet. For type B the `Y` field specifies the subnet, as illustrated in Figure 18.9.

To connect to a global network a number is normally assigned by a central authority. For the Internet network it is assigned by the Network Information Center (NIC). Typically, on the Internet an organization is assigned a type B network address. The first two fields of the address specify the organization network, the third specifies the subnet within the organization and the final specifies the host.

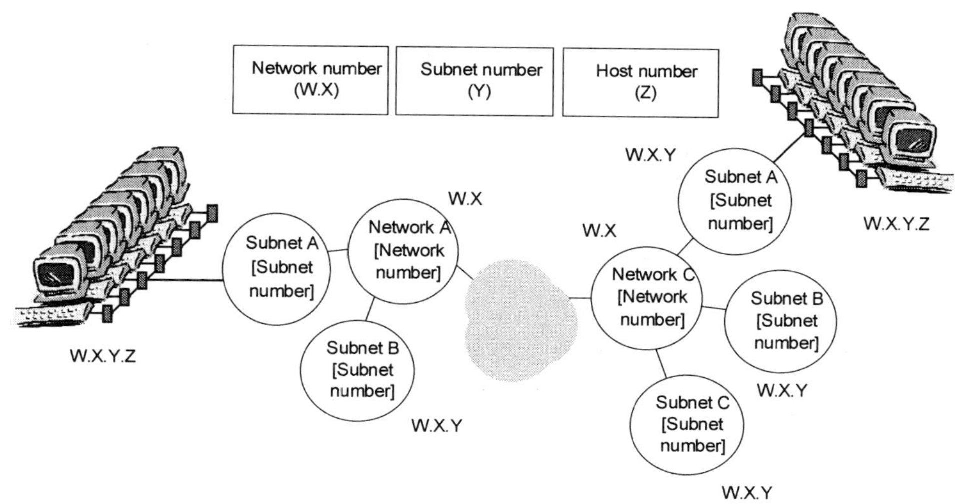

Figure 18.9 Internet addresses with subnets.

18.6.4 Specifying subnet masks

If a subnet is used then a bit mask, or subnet mask, must be specified to show which part of the address is the network part and which is the host.

The subnet mask is a 32-bit number that has 1s for bit positions specifying the network and subnet parts and 0s for the host part. A text file called `hosts` is normally used to set up the subnet mask. Table 18.3 shows example subnet masks.

Table 18.3 Default subnet mask for type A, B and C IP addresses.

Address Type	*Default mask*
Class A	255.0.0.0
Class B	255.255.0.0
Class C and Class B with a subnet	255.255.255.0

To set up the default mask the following line is added to the `hosts` file.

Hosts file

```
255.255.255.0  defaultmask
```

18.7 Domain name system

An IP address can be defined in the form WWW.XXX.YYY.ZZZ, where XXX, YYY, ZZZ and WWW are integer values in the range 0 to 255. On the Internet it is WWW.XXX.YYY that normally defines the subnet and WWW that defines the host. Such names may be difficult to remember. A better method is to use symbolic names rather than IP addresses.

Users and application programs can then use symbolic names rather than IP addresses. The directory network services on the Internet determines the IP address of the named destination user or application program. This has the advantage that users and application programs can move around the Internet and are not fixed to an IP address.

An analogy relates to the public telephone service. A telephone directory contains a list of subscribers and their associated telephone number. If someone looks for a telephone number, first the user name is looked up and their associated telephone number found. The telephone directory listing thus maps a user name (symbolic name) to an actual telephone number (the actual address).

Table 18.4 lists some Internet domain assignments for World Wide Web (WWW) servers. Note that domain assignments are not fixed and can change their corresponding IP addresses, if required. The binding between the symbolic name and its address can thus change at any time.

Table 18.4 Internet domain assignments for web servers.

Web server	*Internet domain names*	*Internet IP address*
NEC	`web.nec.com`	`143.101.112.6`
Sony	`www.sony.com`	`198.83.178.11`
Intel	`www.intel.com`	`134.134.214.1`
IEEE	`www.ieee.com`	`140.98.1.1`
University of Bath	`www.bath.ac.uk`	`136.38.32.1`
University of Edinburgh	`www.ed.ac.uk`	`129.218.128.43`
IEE	`www.iee.org.uk`	`193.130.181.10`
University of Manchester	`www.man.ac.uk`	`130.88.203.16`

18.8 Internet naming structure

The Internet naming structure uses labels separated by periods (full stops); an example is `eece.napier.ac.uk`. It uses a hierarchical structure where organizations are grouped into primary domain names. These are `com` (for commercial organizations), `edu` (for educational organizations), `gov` (for government organizations), `mil` (for military organizations), `net` (Internet network support centers) or `org` (other organizations). The primary domain name may also define the country in which the host is located, such as `uk` (United Kingdom), `fr` (France), and so on. All hosts on the Internet must be registered to one of these primary domain names.

The labels after the primary field describe the subnets within the network. For example in the address `eece.napier.ac.uk`, the `ac` label relates to an academic institution within the `uk`, `napier` to the name of the institution and `eece` the subnet with that organization. Figure 18.10 gives an example structure.

18.9 Domain name server

Each institution on the Internet has a host that runs a process called the domain name server

(DNS). The DNS maintains a database called the directory information base (DIB) which contains directory information for that institution. On adding a new host, the system manager adds its name and its IP address. After this, it can then access the Internet.

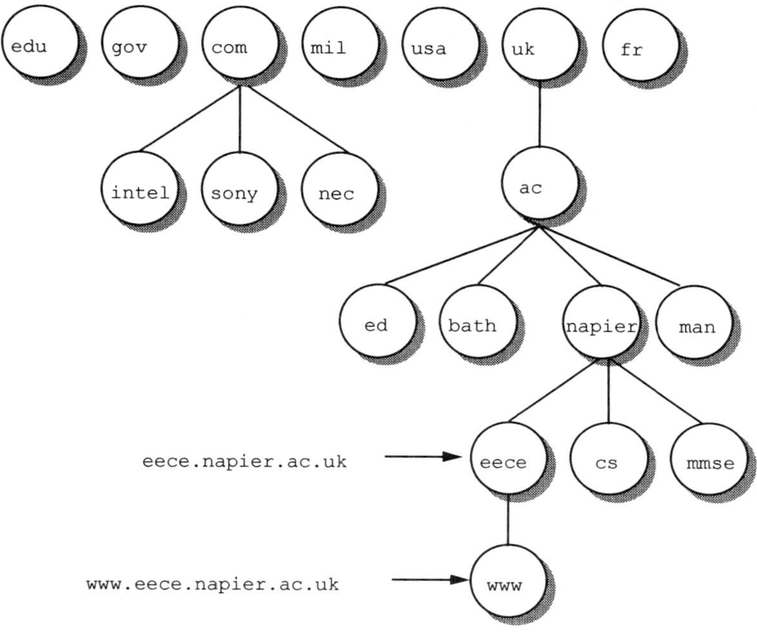

Figure 18.10 Example domain naming.

18.9.1 DNS program

The DNS program is typically run on a Lynx-based PC with a program called `named` (located in `/usr/sbin`) with an information file of `named.boot`. To run the program the following is used:

```
/usr/bin/named -b /usr/local/adm/named/named.boot
```

The following shows that the DNS program is currently running.

```
$ ps -ax
  PID TTY STAT  TIME COMMAND
  295 con S     0:00 bootpd
   35 con S     0:00 /usr/sbin/lpd
  272 con S     0:00 /usr/sbin/named -b /usr/local/adm/named/named.boot
  264 p 1 S     0:01 bash
  306 pp0 R     0:00 ps -ax
```

In this case the data file `named.boot` is located in the `/usr/local/adm/named` directory. A sample `named.boot` file is:

```
/usr/local/adm/named - scabasefile
              eece.napier.ac.uk -main record of computer names
```

```
net/net144    -reverse look-up database
net/net145       "      "
net/net146       "      "
net/net147       "      "
net/net150       "      "
net/net151       "      "
```

This file specifies that the reverse look-up information on computers on the subnets 144, 145, 146, 147, 150 and 150 is contained in the `net144`, `net145`, `net146`, `net147`, `net150` and `net151` files, respectively. These are stored in the `net` subdirectory. The main file which contains the DNS information is, in this case, `eece.napier.ac.uk`.

Whenever a new computer is added onto a network, in this case, the `eece.napier.ac.uk` file and the `net/net1`** (where ** is the relevant subnet name) are updated to reflect the changes. Finally, the serial number at the top of these data files is updated to reflect the current date, such as 19970321 (for 21st March 1997).

The DNS program can then be tested using `nslookup`. For example:

```
$ nslookup
Default Server:  ees99.eece.napier.ac.uk
Address:  146.176.151.99

> src.doc.ic.ac.uk
Server:  ees99.eece.napier.ac.uk
Address:  146.176.151.99

Non-authoritative answer:
Name:    swallow.doc.ic.ac.uk
Address:  193.63.255.4
Aliases:  src.doc.ic.ac.uk
```

18.10 Bootp protocol

The bootp protocol allocates IP addresses to computers based on a table of network card MAC addresses. When a computer is first booted, the bootp server interrogates its MAC address and then looks up the bootp table for its entry. The server then grants the corresponding IP address to the computer. The computer then uses it for connections. This is one method of limiting access to the Internet.

18.10.1 Bootp program

The bootp program is typically run on a Lynx-based PC with the `bootp` program. The following shows that the `bootp` program is currently running on a computer:

```
$ ps -ax
   PID TTY STAT  TIME COMMAND
     1 con S     0:06 init
    31 con S     0:01 /usr/sbin/inetd
 14142 con S     0:00 bootpd -d 1
    35 con S     0:00 /usr/sbin/lpd
    49 p 3 S     0:00 /sbin/agetty 38400 tty3
 14155 pp0 R     0:00 ps -ax
 10762 con S     0:18 /usr/sbin/named -b /usr/local/adm/named/named.boot
```

For the bootp system to operate then it must use a table to reconcile the MAC addresses of the card to an IP address. In the previous example this table is contained in the bootptab file which is located in the /etc directory. The following file gives an example bootptab:

Contents of bootptab file

```
# /etc/bootptab: database for bootp server
# Blank lines and lines beginning with '#' are ignored.
#
# Legend:
#     first field -- hostname
#           (may be full domain name and probably should be)
#
#     hd -- home directory
#     bf -- bootfile
#     cs -- cookie servers
#     ds -- domain name servers
#     gw -- gateways
#     ha -- hardware address
#     ht -- hardware type
#     im -- impress servers
#     ip -- host IP address
#     lg -- log servers
#     lp -- LPR servers
#     ns -- IEN-116 name servers
#     rl -- resource location protocol servers
#     sm -- subnet mask
#     tc -- template host (points to similar host entry)
#     to -- time offset (seconds)
#     ts -- time servers
#
#hostname:ht=1:ha=ether_addr_in_hex:ip=ip_addr_in_dec:tc=allhost:
.default150:\
      :hd=/tmp:bf=null:\
      :ds=146.176.151.99 146.176.150.62 146.176.1.5:\
      :sm=255.255.255.0:gw=146.176.150.253:\
      :hn:vm=auto:to=0:
.default151:\
      :hd=/tmp:bf=null:\
      :ds=146.176.151.99 146.176.150.62 146.176.1.5:\
      :sm=255.255.255.0:gw=146.176.151.254:\
      :hn:vm=auto:to=0:
pc345:   ht=ethernet:   ha=0080C8226BE2:  ip=146.176.150.2: tc=.default150:
pc307:   ht=ethernet:   ha=0080C822CD4E:  ip=146.176.150.3: tc=.default150:
pc320:   ht=ethernet:   ha=0080C823114C:  ip=146.176.150.4: tc=.default150:
pc331:   ht=ethernet:   ha=0080C823124B:  ip=146.176.150.5: tc=.default150:
:       :
pc460:   ht=ethernet:   ha=0000E8C7BB63:  ip=146.176.151.142: tc=.default151:
pc414:   ht=ethernet:   ha=0080C8246A84:  ip=146.176.151.143: tc=.default151:
pc405:   ht=ethernet:   ha=0080C82382EE:  ip=146.176.151.145: tc=.default151:
```

The format of the file is:

```
#hostname:ht=1:ha=ether_addr_in_hex:ip=ip_addr_in_dec:tc=allhost:
```

where `hostname` is the hostname, the value defined after `ha=` is the Ethernet MAC address, the value after `ip=` is the IP address and the name after the `tc=` field defines the host information script. For example:

```
pc345:   ht=ethernet:   ha=0080C8226BE2:  ip=146.176.150.2: tc=.default150:
```

defines the hostname of `pc345`, indicates it is on an Ethernet network, and shows its IP address is `146.176.150.2`. The MAC address of the computer is `00:80:C8: 22:6B:E2` and it is defined by the script `.default150`. This file defines a subnet of 255.255.255.0 and has associated DNS of

```
146.176.151.99 146.176.150.62 146.176.1.5
```

and uses the gateway at:

```
146.176.150.253
```

18.11 Example network

A university network is shown in Figure 18.11. The connection to the outside global Internet is via the Janet gateway node and its IP address is `146.176.1.3`. Three subnets, `146.176.160`, `146.176.129` and `146.176.151`, connect the gateway to departmental bridges. The Computer Studies bridge address is `146.176.160.1` and the Electrical Department bridge has an address `146.176.151.254`.

The Electrical Department bridge links, through other bridges, to the subnets `146.176.144`, `146.176.145`, `146.176.147`, `146.176.150` and 146.176.151.

The topology of the Electrical Department network is shown in Figure 18.12. The main bridge into the department connects to two Ethernet networks of PCs (subnets `146.176.150` and `146.176.151`) and to another bridge (`Bridge 1`). `Bridge 1` connects to the subnet `146.176.144`. Subnet `146.176.144` connects to workstations and X-terminals. It also connects to the gateway `Moon` that links the Token Ring subnet `146.176.145` with the Ethernet subnet `146.176.144`. The gateway `Oberon`, on the `146.176.145` subnet, connects to an Ethernet link `146.176.146`. This then connects to the gateway `Dione` that is also connected to the Token Ring subnet `146.176.147`.

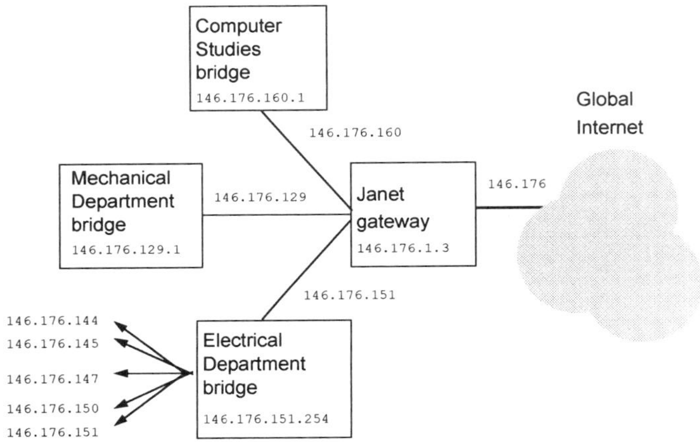

Figure 18.11 A university network.

Each node on the network is assigned an IP address. The *hosts* file for the setup in Figure 18.12 is shown next. For example, the IP address of `Mimas` is `146.176.145.21` and for `miranda` it is `146.176.144.14`. Notice that the gateway nodes: `Oberon`, `Moon` and `Dione` all have two IP addresses.

Contents of host file

```
146.176.1.3          janet
146.176.144.10       hp
146.176.145.21       mimas
146.176.144.11       mwave
146.176.144.13       vax
146.176.144.14       miranda
146.176.144.20       triton
146.176.146.23       oberon
146.176.145.23       oberon
146.176.145.24       moon
146.176.144.24       moon
146.176.147.25       uranus
146.176.146.30       dione
146.176.147.30       dione
146.176.147.31       saturn
146.176.147.32       mercury
146.176.147.33       earth
146.176.147.34       deimos
146.176.147.35       ariel
146.176.147.36       neptune
146.176.147.37       phobos
146.176.147.39       io
146.176.147.40       titan
146.176.147.41       venus
146.176.147.42       pluto
146.176.147.43       mars
146.176.147.44       rhea
146.176.147.22       jupiter
146.176.144.54       leda
146.176.144.55       castor
146.176.144.56       pollux
146.176.144.57       rigel
146.176.144.58       spica
146.176.151.254      cubridge
146.176.151.99       bridge_1
146.176.151.98       pc2
146.176.151.97       pc3
                :::::
146.176.151.71       pc29
146.176.151.70       pc30
146.176.151.99       ees99
146.176.150.61       eepc01
146.176.150.62       eepc02
255.255.255.0        defaultmask
```

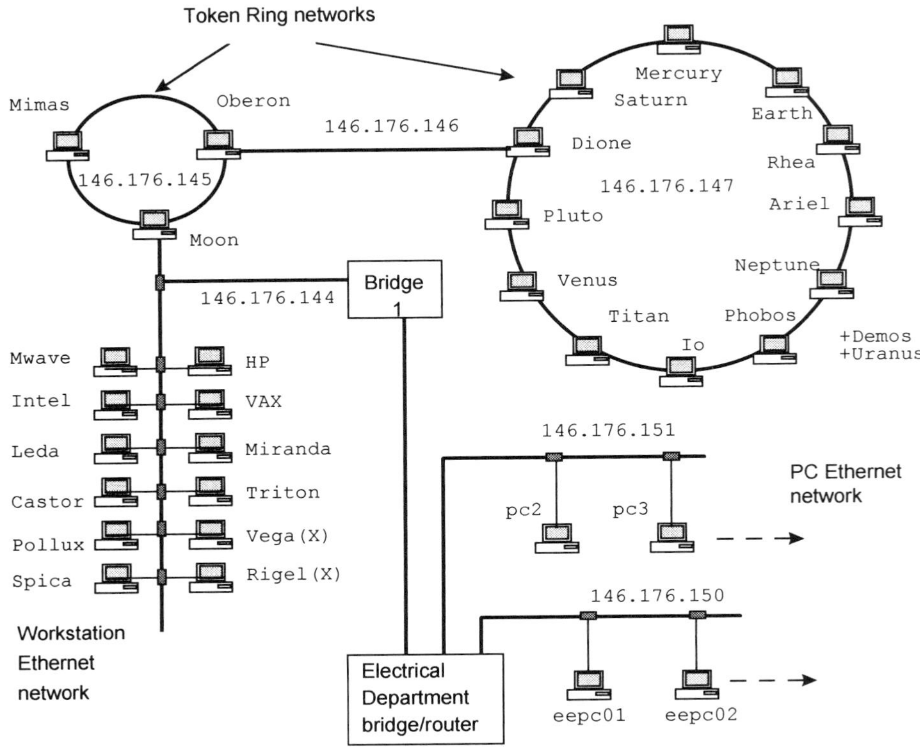

Figure 18.12 Network topology for the Department network.

TCP, IP Ver6 and TCP/IP Command

19.1 Introduction

TCP and IP are extremely important protocols as they allow hosts to communicate over the Internet in a reliable way. TCP provides a connection between two hosts and supports error handling. This chapter discusses TCP in more detail and shows how a connection is established then maintained. An important concept of TCP/IP communications is the usage of ports and sockets. A port identifies the process type (such as FTP, TELNET, and so on) and the socket identifies a unique connection number. In this way, TCP/IP can support multiple simultaneous connections of applications over a network.

The IP header is added to higher-level data. This header contains a 32-bit IP address of the destination node. Unfortunately, the standard 32-bit IP address is not large enough to support the growth in nodes connecting to the Internet. Thus a new standard, IP Version 6, has been developed to support a 128-bit address, as well as additional enhancements.

This chapter also discusses some of the TCP/IP programs which can be used to connect to other hosts and also to determine routing information.

19.2 IP Ver6

TCP and IP are extremely important protocols as they allow hosts to communicate over the Internet in a reliable way. TCP provides a connection between two hosts and supports error handling. This section discusses TCP in more detail and shows how a connection is established then maintained. An important concept of TCP/IP communications is the usage of ports and sockets. A port identifies the process type (such as FTP, TELNET, and so on) and the socket identifies a unique connection number. In this way, TCP/IP can support multiple simultaneous connections of applications over a network.

The IP header (IP Ver4) is added to higher-level data. This header contains a 32-bit IP address of the destination node. Unfortunately, the standard 32-bit IP address is not large enough to support the growth in nodes connecting to the Internet. Thus a new standard, IP Version 6 (IP Ver6), has been developed to support a 128-bit address, as well as additional enhancements, such as authentication and encryption of data.

The main techniques being investigated are:

- TUBA (TCP and UDP with bigger addresses).
- CATNIP (common architecture for the Internet).
- SIPP (simple Internet protocol plus).

It is likely that none of these will provide the complete standard and the resulting standard will be a mixture of the three.

Figure 19.1 shows the basic format of the IP Ver6 header. The main fields are:

- Version number (4 bits) – contains the version number, such as 6 for IP Ver6. It is used to differentiate between IP Ver4 and IP Ver6.
- Priority (4 bits) – indicates the priority of the datagram. For example:
  - 0 defines no priority.
  - 1 defines background traffic.
  - 2 defines unattended transfer.
  - 4 defines attended bulk transfer.
  - 6 defines interactive traffic.
  - 7 defines control traffic.
- Flow label (24 bits) – still experimental, but will be used to identify different data flow characteristics.
- Payload length (16 bits) – defines the total size of the IP datagram (and includes the IP header attached data).
- Next header – this field indicates which header follows the IP header. For example:
  - 0 defines IP information.
  - 6 defines TCP information.
  - 43 defines routing information.
  - 58 defines ICMP information.
- Hop limit – defines the maximum number of hops that the datagram takes as it traverses the network. Each router decrements the hop limit by 1; when it reaches 0 it is deleted.
- IP addresses (128 bits) – defines IP address. There will be three main groups of IP addresses: unicast, multicast and anycast. A unicast address identifies a particular host, a multicast address enables the hosts with a particular group to receive the same packet, and the anycast address will be addressed to a number of interfaces on a single multicast address.

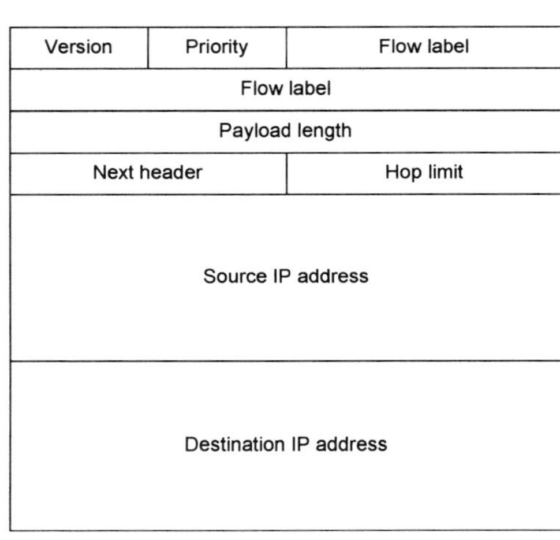

Figure 19.1 IP Ver6 header format.

19.3 Transmission control protocol

In the OSI model, TCP fits into the transport layer and IP fits into the network layer. TCP thus sits above IP, which means that the IP header is added onto the higher-level information (such as transport, session, presentation and application). The main functions of TCP are to provide a robust and reliable transport protocol. It is characterized as a reliable, connection-oriented, acknowledged and datastream-oriented server. IP, itself, does not support the connection of two nodes, whereas TCP does. With TCP, a connection is initially established and is then maintained for the length of the transmission.

The TCP information contains simple acknowledgement messages and a set of sequential numbers. It also supports multiple simultaneous connections using destination and source port numbers, and manages them for both transmission and reception. As with IP, it supports data fragmentation and reassembly, and data multiplexing/demultiplexing.

The setup and operation of TCP is as follows:

1. When a host wishes to make a connection, TCP sends out a request message to the destination machine that contains a unique number, called a socket number and a port number. The port number has a value which is associated with the application (for example a TELNET connection has the port number 23 and an FTP connection has the port number 21). The message is then passed to the IP layer, which assembles a datagram for transmission to the destination.
2. When the destination host receives the connection request, it returns a message containing its own unique socket number and a port number. The socket number and port number thus identify the virtual connection between the two hosts.
3. After the connection has been made the data can flow between the two hosts (called a data stream).

After TCP receives the stream of data, it assembles the data into packets, called TCP segments. After the segment has been constructed, TCP adds a header (called the protocol data unit) to the front of the segment. This header contains information such as a checksum, port number, destination and source socket numbers, socket number of both machines and segment sequence numbers. The TCP layer then sends the packaged segment down to the IP layer, which encapsulates it and sends it over the network as a datagram.

19.3.1 Ports and sockets

As previously mentioned, TCP adds a port number and socket number for each host. The port number identifies the required service, whereas the socket number is a unique number for that connection. Thus, a node can have several TELNET connections with the same port number but each connection will have a different socket number. A port number can be any value but there is a standard convention that most systems adopt. Table 19.1 defines some of the most common values. Standard applications normally use port values from 0 to 255, while unspecified applications can use values above 255.

19.3.2 TCP header format

The sender's TCP layer communicates with the receiver's TCP layer using the TCP protocol data unit. It defines parameters such as the source port, destination port, and so on, and is illustrated in Figure 19.2. The fields are:

Table 19.1 Typical TCP port numbers.

Port	*Process name*	*Notes*
20	FTP-DATA	File Transfer Protocol - data
21	FTP	File Transfer Protocol - control
23	TELNET	Telnet
25	SMTP	Simple Mail Transfer Protocol
49	LOGIN	Login Protocol
53	DOMAIN	Domain Name Server
79	FINGER	Finger
161	SNMP	SNMP

- Source and destination port number – which are 16-bit values that identify the local port number (source number and destination port number or destination port).
- Sequence number – which identifies the current sequence number of the data segment. This allows the receiver to keep track of the data segments received. Any segments that are missing can be easily identified.
- Data offset – which is a 32-bit value that identifies the start of the data.
- Flags – the flag field is defined as `UAPRSF`, where `U` is the urgent flag, `A` the acknowledgement flag, `P` the push function, `R` the reset flag, `S` the sequence synchronize flag and `F` the end-of-transmission flag.
- Windows – which is a 16-bit value and gives the number of data blocks that the receiving host can accept at a time.
- Checksum – which is a 16-bit checksum for the data and header.
- UrgPtr – which is the urgent pointer and is used to identify an important area of data (most systems do not support this facility).

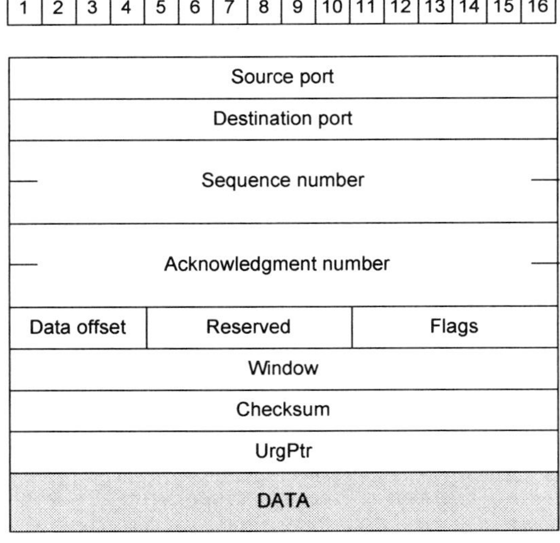

Figure 19.2 TCP header format.

19.4 TCP/IP commands

There are several standard programs available over TCP/IP connections. The example sessions is this section relate to the network outlined in Figure 18.12. These applications may include:

- FTP (File Transfer Protocol) – transfers file between computers.
- HTTP (Hypertext Transfer Protocol) – which is the protocol used in the World Wide Web (WWW) and can be used for client-server applications involving hypertext.
- MIME (Multipurpose Internet Mail Extension) – gives enhanced electronic mail facilities over TCP/IP.
- SMTP (Simple Mail Management Protocol) – gives simple electronic mail facilities.
- TELNET – allows remote login using TCP/IP.
- PING – determines if a node is responding to TCP/IP communications.

19.4.1 ping

The `ping` program (Packet Internet Gopher) determines whether a node is responding to TCP/IP communication. It is typically used to trace problems in networks and uses the Internet Control Message Protocol (ICMP) to send a response request from the target node. Sample run 19.1 shows that `miranda` is active and `ariel` isn't.

Sample run 19.1: Using PING command

```
C:\WINDOWS>ping miranda
miranda (146.176.144.14) is alive
C:\WINDOWS>ping ariel
no reply from ariel (146.176.147.35)
```

The `ping` program can also be used to determine the delay between one host and another, and also if there are any IP packet losses. In Sample run 19.2 the local host is `pc419.eece.napier.ac.uk` (which is on the `146.176.151` segment); the host `miranda` is tested (which is on the `146.176.144` segment). It can be seen that, on average, the delay is only 1 ms and there is no loss of packets.

Sample run 19.2: Using PING command

```
225 % ping miranda
PING miranda.eece.napier.ac.uk: 64 byte packets
64 bytes from 146.176.144.14: icmp_seq=0. time=1. ms
64 bytes from 146.176.144.14: icmp_seq=1. time=1. ms
64 bytes from 146.176.144.14: icmp_seq=2. time=1. ms
3 packets transmitted, 3 packets received, 0% packet loss
round-trip (ms)  min/avg/max = 1/1/1
```

In Sample run 19.3 the destination node (`www.napier.ac.uk`) is located within the same building but is on a different IP segment (`147.176.2`). It is also routed through a bridge. It can be seen that the packet delay has increased to between 9 and 10 ms. Again, there is no packet loss.

Sample run 19.3: Using PING command

```
226 % ping www.napier.ac.uk
PING central.napier.ac.uk: 64 byte packets
64 bytes from 146.176.2.3: icmp_seq=0. time=9. ms
64 bytes from 146.176.2.3: icmp_seq=1. time=9. ms
64 bytes from 146.176.2.3: icmp_seq=2. time=10. ms
3 packets transmitted, 3 packets received, 0% packet loss
round-trip (ms)  min/avg/max = 9/9/10
```

Sample run 19.4 shows a connection between Edinburgh and Bath in the UK (`www.bath.ac.uk` has an IP address of `138.38.32.5`). This is a distance of approximately 500 miles and it can be seen that the delay is now between 30 and 49 ms. This time there is 25% packet loss.

Sample run 19.4: Using PING command

```
222 % ping www.bath.ac.uk
PING jess.bath.ac.uk: 64 byte packets
64 bytes from 138.38.32.5: icmp_seq=0. time=49. ms
64 bytes from 138.38.32.5: icmp_seq=2. time=35. ms
64 bytes from 138.38.32.5: icmp_seq=3. time=30. ms
4 packets transmitted, 3 packets received, 25% packet loss
round-trip (ms)  min/avg/max = 30/38/49
```

Finally, in Sample run 19.5 the `ping` program tests a link between Edinburgh, UK, and a WWW server in the USA (`home.microsoft.com`, which has the IP address of `207.68.137.51`). It can be seen that in this case, the delay is between 447 and 468 ms, and the loss is 60%.

A similar utility program to `ping` is `spray` which uses Remote Procedure Call (RPC) to send a continuous stream of ICMP messages. It is useful when testing a network connection for its burst characteristics. This differs from `ping`, which waits for a predetermined amount of time between messages.

Sample run 19.5: Ping command with packet loss

```
224 % ping home.microsoft.com
PING home.microsoft.com: 64 byte packets
64 bytes from 207.68.137.51: icmp_seq=2. time=447. ms
64 bytes from 207.68.137.51: icmp_seq=3. time=468. ms
----home.microsoft.com PING Statistics----
5 packets transmitted, 2 packets received, 60% packet loss
```

19.4.2 ftp (file transfer protocol)

The `ftp` program uses the TCP/IP protocol to transfer files to and from remote nodes. If necessary, it reads the *`hosts`* file to determine the IP address. Once the user has logged into the remote node, the commands that can be used are similar to DOS commands such as `cd` (change directory), `dir` (list directory), `open` (open node), `close` (close node), `pwd` (present working directory). The `get` command copies a file from the remote node and the `put` command copies it to the remote node.

The type of file to be transferred must also be specified. This file can be ASCII text (the command `ascii`) or binary (the command `binary`).

19.4.3 telnet

The `telnet` program uses TCP/IP to remotely log in to a remote node.

19.4.4 nslookup

The `nslookup` program interrogates the local `hosts` file or a DNS server to determine the IP address of an Internet node. If it cannot find it in the local file then it communicates with gateways outside its own network to see if they know the address. Sample run 19.6 shows that the IP address of `www.intel.com` is `134.134.214.1`.

Sample run 19.6: Example of nslookup

```
C:\> nslookup
Default Server:  ees99.eece.napier.ac.uk
Address:  146.176.151.99
> www.intel.com
Server:  ees99.eece.napier.ac.uk
Address:  146.176.151.99
Name:    web.jf.intel.com
Address:  134.134.214.1
Aliases:  www.intel.com
230 % nslookup home.microsoft.com
Non-authoritative answer:
Name:    home.microsoft.com
Addresses:  207.68.137.69, 207.68.156.11, 207.68.156.14, 207.68.156.56
207.68.137.48, 207.68.137.51
```

19.4.5 netstat (network statistics)

On a UNIX system the command `netstat` can be used to determine the status of the network. The `-r` option shown in Sample run 19.7 shows that this node uses `moon` as a gateway to another network.

Sample run 19.7: Using Unix netstat command

```
[54:miranda :/net/castor_win/local_user/bill_b ] % netstat -r
Destination         Gateway              Flags      Refs      Use  Interface
localhost           localhost            UH            0    27306  lo0
default             moon                 UG            0  1453856  lan0
146.176.144         miranda              U             8  6080432  lan0
146.176.1           146.176.144.252      UGD           0       51  lan0
146.176.151         146.176.144.252      UGD          11     5491  lan0
```

19.4.6 traceroute

The `traceroute` program traces the route of an IP packet through the Internet. It uses the IP protocol time-to-live field and attempts to get an ICMP TIME_EXCEEDED response from each gateway along the path to a defined host. The default probe datagram length is 38 bytes (although the sample runs use 40 byte packets by default). Sample run 19.8 shows an example of `traceroute` from a PC (`pc419.eece.napier.ac.uk`). It can be seen that initially it goes through a bridge (`pcbridge.eece.napier.ac.uk`) and then to the destination (`miranda.eece.napier.ac.uk`).

Sample run 19.9 shows the route from a PC (`pc419.eece.napier.ac.uk`) to a destination node (`www.bath.ac.uk`). Initially, from the originator, the route goes through a gateway (`146.176.151.254`) and then goes through a routing switch (`146.176.1.27`) and onto EaStMAN ring via `146.176.3.1`. The route then goes round the EaStMAN to a gateway at the University of Edinburgh (smds-gw.ed.ja.net). It is then routed onto the SuperJanet network and reaches a gateway at the University of Bath (smds-gw.bath.ja.net). It then goes to another gateway (jips-gw.bath.ac.uk) and finally to its destination (jess.bath.ac.uk). Figure 19.15 shows the route the packet takes.

Sample run 19.8: Example traceroute

```
www:~/www$ traceroute miranda
traceroute to miranda.eece.napier.ac.uk (146.176.144.14), 30 hops max,
     40 byte packets
1  pcbridge.eece.napier.ac.uk (146.176.151.252)  2.684 ms  1.762 ms 1.725 ms
2  miranda.eece.napier.ac.uk (146.176.144.14)  2.451 ms  2.554 ms   2.357 ms
```

Note that gateways 4 and 8 hops away either don't send ICMP 'time exceeded' messages or send them with time-to-live values that are too small to be returned to the originator.

Sample run 19.10 shows an example route from a local host at Napier University, UK, to the USA. As before, it goes through the local gateway (146.176.151.254) and then goes through three other gateways to get onto the SMDS SuperJANET connection. The data packet then travels down this connection to University College, London (gw5.ulcc.ja.net). It then goes onto high speed connects to the USA and arrives at a US gateway (mcinet-2.sprintnap.net). Next it travels to core2-hssi2-0.WestOrange.mci.net before reaching the Microsoft Corporation gateway in Seattle (microsoft.Seattle.mci.net). It finally finds it way to the destination (207.68.145.53). The total journey time is just less than half a second.

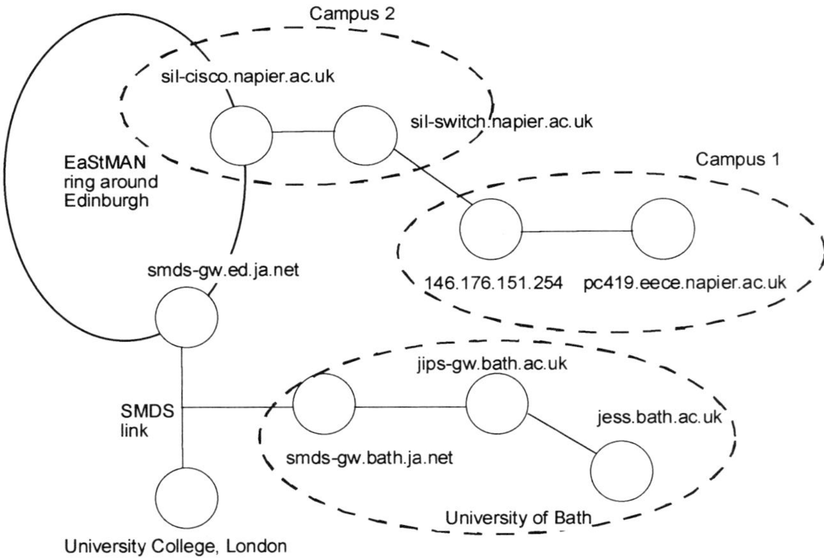

Figure 19.3 Route between local host and the University of Bath.

Sample run 19.9: Example traceroute

```
www:~/www$ traceroute www.bath.ac.uk
traceroute to jess.bath.ac.uk (138.38.32.5), 30 hops max, 40 byte packets
1  146.176.151.254 (146.176.151.254)  2.806 ms  2.76 ms  2.491 ms
2  sil-switch.napier.ac.uk (146.176.1.27)  19.315 ms  11.29 ms  6.285 ms
3  sil-cisco.napier.ac.uk (146.176.3.1)  6.427 ms  8.407 ms  8.872 ms
4  * * *
5  smds-gw.ed.ja.net (193.63.106.129)  8.98 ms  30.308 ms  398.623 ms
6  smds-gw.bath.ja.net (193.63.203.68)  39.104 ms  46.833 ms  38.036 ms
7  jips-gw.bath.ac.uk (146.97.104.2)  32.908 ms  41.336 ms  42.429 ms
8  * * *
9  jess.bath.ac.uk (138.38.32.5)  41.045 ms *  41.93 ms
```

Sample run 19.10: Example traceroute

```
> traceroute home.microsoft.com
 1  146.176.151.254 (146.176.151.254)  2.931 ms  2.68 ms  2.658 ms
2  sil-switch.napier.ac.uk (146.176.1.27)  6.216 ms  8.818 ms  5.885 ms
3  sil-cisco.napier.ac.uk (146.176.3.1)  6.502 ms  6.638 ms  10.218 ms
4  * * *
5  smds-gw.ed.ja.net (193.63.106.129)  18.367 ms  9.242 ms  15.145 ms
6  smds-gw.ulcc.ja.net (193.63.203.33)  42.644 ms  36.794 ms  34.555 ms
7  gw5.ulcc.ja.net (128.86.1.80)  31.906 ms  30.053 ms  39.151 ms
8  icm-london-1.icp.net (193.63.175.53)  29.368 ms  25.42 ms  31.347 ms
9  198.67.131.193 (198.67.131.193)  119.195 ms  120.482 ms  67.479 ms
10  icm-pen-1-H2/0-T3.icp.net (198.67.131.25)  115.314 ms  126.152 ms
      149.982 ms
11  icm-pen-10-P4/0-OC3C.icp.net (198.67.142.69)  139.27 ms  197.953 ms
      195.722 ms
12  mcinet-2.sprintnap.net (192.157.69.48)  199.267 ms  267.446 ms 287.834 ms
13  core2-hssi2-0.WestOrange.mci.net (204.70.1.49)  216.006 ms  688.139 ms
      228.968 ms
14  microsoft.Seattle.mci.net (166.48.209.250)  310.447 ms  282.882 ms
      313.619 ms
15  * microsoft.Seattle.mci.net (166.48.209.250)  324.797 ms  309.518 ms
16  * 207.68.145.53 (207.68.145.53)  435.195 ms *
```

19.4.7 arp

The arp program displays the IP to Ethernet MAC address mapping. It can also be used to delete or manually change any included address table entries. Within a network, a router forwards data packets depending on the destination IP address of the packet. Each connection must also specify a MAC address to transport the packet over the network, thus the router must maintain a list of MAC addresses. The arp protocol thus maintains this mapping. Addresses within this table are added on an as-needed basis. When a MAC address is required, an arp message is sent to the node with an arp REQUEST packet which contains the IP address of the requested node. It will then reply with an arp RESPONSE packet which contains its MAC address and its IP address.

20 WinSock Programming

20.1 Introduction

The Windows Sockets specification describes a common interface for networked Windows programs. WinSock uses TCP/IP communications and provides for binary and source code compatibility for different network types.

The Windows Sockets API (WinSock API) is a library of functions that implement the socket interface by the Berkley Software Distribution of UNIX. WinSock augments the Berkley socket implementation by adding Windows-specific extensions to support the message-driven nature of Windows system.

20.2 Windows Sockets

The main WinSock API calls are:

`socket()`.	Creates a socket.
`accept()`.	Accepts a connection on a socket.
`connect()`.	Establishes a connection to a peer.
`bind()`.	Associates a local address with a socket.
`listen()`.	Establishes a socket to listen for incoming connection.
`send()`.	Sends data on a connected socket.
`recv()`.	Receives data from a socket.
`shutdown()`.	Disables send or receive operations on a socket.
`closesocket()`.	Closes a socket.

Figure 20.1 shows the operation of a connection from a client to a server. The server is defined as the computer which waits for a connection, the client is the computer which initially makes contact with the server.

On the server the computer initially creates a socket with the `socket()` function, and this is bound to a name with the `bind()` function. After this the server listens for a connection with the `listen()` function. When the client calls the `connection()` function the server then accepts the connection with `accept()`. After this the server and client can send and receive data with the `send()` or `recv()` functions. When the data transfer is complete the `closesocket()` is used to close the socket.

20.2.1 socket()

The `socket()` function creates a socket. Its syntax is:

```
SOCKET socket ( int af, int type, int protocol)
```

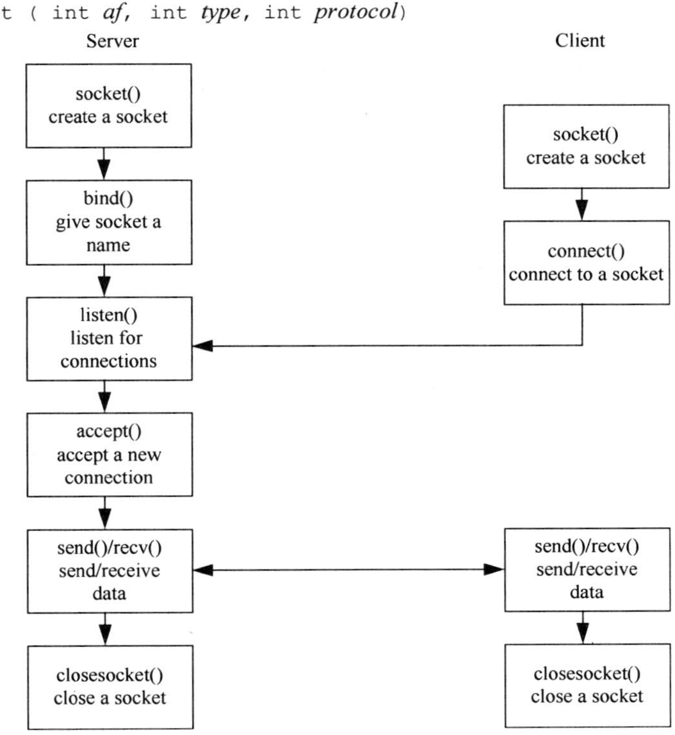

Figure 20.1 WinSock connection.

where

af A value of `PF_INET` specifies the ARPA Internet address format specification.

type Socket specification, which is either `SOCK_STREAM` or `SOCK_DGRAM`. The `SOCK_STREAM` uses TCP and provides a sequenced, reliable, two-way, connection-based stream. `SOCK_DGRAM` uses UDP and provides for connectionless datagrams. This type of connection is not recommended.

protocol Defines the protocol to be used with the socket. If it is zero then the caller does not wish to specify a protocol.

If the `socket` function succeeds then the return value is a descriptor referencing the new socket. Otherwise, it returns `SOCKET_ERROR`, and the specific error code can be tested with `WSAGetLastError`. An example creation of a socket is given next:

```
SOCKET s;

      s=socket(PF_INET,SOCK_STREAM,0);
      if (s == INVALID_SOCKET)
      {
         cout << "Socket error"
      }
```

20.2.2 bind()

The `bind()` function associates a local address with a socket. It is before calls to the `connect` or `listen` functions. When a socket is created with `socket`, it exists in a name space (address family), but it has no name assigned. The `bind` function gives the socket a local association (host address/port number) . Its syntax is:

```
int bind(SOCKET s, const struct sockaddr FAR * addr, int namelen);
```

where

s	A descriptor identifying an unbound socket.
namelen	The length of the *addr*.
addr	The address to assign to the socket. The `sockaddr` structure is defined as follows:

```
struct sockaddr
{
   u_short    sa_family;
   char       sa_data[14];
};
```

In the Internet address family, the `sockadd_in` structure is used by Windows Sockets to specify a local or remote endpoint address to which to connect a socket. This is the form of the `sockaddr` structure specific to the Internet address family and can be cast to `sockaddr`. This structure can be filled with the `sockaddr_in` structure which has the following form:

```
struct SOCKADDR_IN
{
    short                sin_family;
    unsigned short       sin_port;
    struct   in_addr sin_addr;
    char                 sin_zero[8];
}
```

where

`sin_family`	must be set to `AF_INET`.
`sin_port`	IP port.
`sin_addr`	IP address.
`sin_zero`	Padding to make structure the same size as `sockaddr`.

If an application does not care what address is assigned to it, it may specify an Internet address equal to `INADDR_ANY`, a port equal to 0, or both. An Internet address equal to `INADDR_ANY` causes any appropriate network interface be used. A port value of 0 causes the Windows Sockets implementation to assign a unique port to the application with a value between 1024 and 5000.

If no error occur then it returns a zero value. Otherwise, it returns `INVALID_SOCKET`, and the specific error code can be tested with `WSAGetLastError`.

If an application needs to bind to an arbitrary port outside of the range 1024 to 5000 then the following outline code can be used:

```
#include <windows.h>
#include <winsock.h>

int main(void)
{

SOCKADDR_IN    sin;
SOCKET         s;
      s = socket(AF_INET,SOCK_STREAM,0);

      if (s == INVALID_SOCKET)
      {
         // Socket failed
      }

      sin.sin_family = AF_INET;
      sin.sin_addr.s_addr = 0;

      sin.sin_port = htons(100); // port=100

      if (bind(s, (LPSOCKADDR)&sin, sizeof (sin)) == 0)
      {
       // Bind failed
      }
      return(0);
}
```

The Windows Sockets `htons` function converts an unsigned short (`u_short`) from host byte order to network byte order.

20.2.3 connect()

The `connect()` function establishes a connection with a peer. If the specified socket is unbound then unique values are assigned to the local association by the system and the socket is marked as bound. Its syntax is:

```
int connect (SOCKET s, const struct sockaddr FAR * name,
               int namelen)
```

where

s	Descriptor identifying an unconnected socket.
name	Name of the peer to which the socket is to be connected.
namelen	Name length.

If no error occur then it returns a zero value. Otherwise, it returns `SOCKET_ERROR`, and the specific error code can be tested with `WSAGetLastError`.

20.2.4 listen()

The `listen()` function establishes a socket which listens for an incoming connection. The sequence to create and accept a socket is:

- `socket()`. Creates a socket.

- `listen()`. This creates a queue for incoming connections and is typically used by a server that can have more than one connection at a time.
- `accept()`. These connections are then accepted with accept.

The syntax of `listen()` is:

```
int listen (SOCKET s, int backlog)
```

where

s Describes a bound, unconnected socket.
backlog Defines the queue size for the maximum number of pending connections may grow (typically a maximum of 5).

If no error occur then it returns a zero value. Otherwise, it returns `SOCKET_ERROR`, and the specific error code can be tested with `WSAGetLastError`.

```
#include <windows.h>
#include <winsock.h>

int main(void)
{

SOCKADDR_IN    sin;
SOCKET         s;

       s = socket(AF_INET,SOCK_STREAM,0);
       if (s == INVALID_SOCKET)
       {
          // Socket failed
       }

       sin.sin_family = AF_INET;
       sin.sin_addr.s_addr = 0;

       sin.sin_port = htons(100); // port=100

       if (bind(s, (struct sockaddr FAR *)&sin, sizeof (sin)) == SOCKET_ERROR)
       {
        // Bind failed
     }

       if (listen(s,4)==SOCKET_ERROR)
       {
          // Listen failed
       }
       return(0);
}
```

20.2.5 accept()

The `accept()` function accepts a connection on a socket. It extracts any pending connections from the queue and creates a new socket with the same properties as the specified socket. Finally, it returns a handle to the new socket. Its syntax is:

```
SOCKET accept(SOCKET s, struct sockaddr FAR *addr, int FAR  *addrlen );
```

where

s	Descriptor identifying a socket that is in listen mode.
addr	Pointer to a buffer that receives the address of the connecting entity, as known to the communications layer.
addrlen	Pointer to an integer which contains the length of the address *addr*.

If no error occur then it returns a zero value. Otherwise, it returns `INVALID_SOCKET`, and the specific error code can be tested with `WSAGetLastError`.

```
#include <windows.h>
#include <winsock.h>

int main(void)
{

SOCKADDR_IN    sin;
SOCKET         s;
int            sin_len;

      s = socket(AF_INET,SOCK_STREAM,0);
      if (s == INVALID_SOCKET)
      {
         // Socket failed
      }

      sin.sin_family = AF_INET;
      sin.sin_addr.s_addr = 0;
      sin.sin_port = htons(100); // port=100

      if (bind(s, (struct sockaddr FAR *)&sin, sizeof (sin)) == SOCKET_ERROR)
      {
       // Bind failed
      }

      if (listen(s,4)<0)
      {
         // Listen failed
      }
      sin_len = sizeof(sin);
      s=accept(s,(struct sockaddr FAR *) & sin,(int FAR *) &sin_len);
      if (s==INVALID_SOCKET)
      {
         // Accept failed
      }
      return(0);
}
```

20.2.6 send()

The `send()` function sends data to a connected socket. Its syntax is:

`int send (SOCKET s, const char FAR *buf, int len, int flags)`

where

s	Connected socket descriptor.
buf	Transmission data buffer.
len	Buffer length.

flags Calling flag.

The *flags* parameter influences the behavior of the function. These can be:

MSG_DONTROUTE Specifies that the data should not be subject to routing.
MSG_OOB Send out-of-band data.

If send() succeeds then the return value is the number of characters set (which can be less than the number indicated by *len*). Otherwise, it returns SOCKET_ERRO, and the specific error code can be tested with WSAGetLastError.

```
#include <windows.h>
#include <winsock.h>
#include <string.h>
#define  STRLENGTH 100

int main(void)
{

SOCKADDR_IN    sin;
SOCKET         s;
int sin_len;
char  sendbuf[STRLENGTH];

      s = socket(AF_INET,SOCK_STREAM,0);
      if (s == INVALID_SOCKET)
      {
         // Socket failed
      }
      sin.sin_family = AF_INET;
      sin.sin_addr.s_addr = 0;
      sin.sin_port = htons(100); // port=100
      if (bind(s, (struct sockaddr FAR *)&sin, sizeof (sin)) == SOCKET_ERROR)
      {
       // Bind failed
      }

      if (listen(s,4)<0)
      {
         // Listen failed
      }
      sin_len = sizeof(sin);

      s=accept(s,(struct sockaddr FAR *) & sin,(int FAR *) &sin_len);

      if (s<0)
      {
         // Accept failed
      }

      while (1)
      {
         // get message to send and put into sendbuff
         send(s,sendbuf,strlen(sendbuf),80);
      }
      return(0);
}
```

20.2.7 recv()

The recv() function receives data from a socket. It waits until data arrives and its syntax is:

`int recv(SOCKET` *s*`, char FAR *`*buf*`, int` *len*`, int` *flags*`)`

where

s	Connected socket descriptor.
buf	Incoming data buffer.
len	Buffer length.
flags	Specifies the method by which the data is received.

If `recv()` succeeds then the return value is the number of bytes received (a zero identifies that the connection has been closed). Otherwise, it returns `SOCKET_ERRO`, and the specific error code can be tested with `WSAGetLastError`.

The flags parameter may have one of the following values:

`MSG_PEEK`	Peek at the incoming data. Any received data is copied into the buffer, but not removed from the input queue.
`MSG_OOB`	Process out-of-band data.

```
#include <windows.h>
#include <winsock.h>

#define  STRLENGTH 100

int main(void)
{

SOCKADDR_IN     sin;
SOCKET          s;
int             sin_len,status;
char            recmsg[STRLENGTH];

      s = socket(AF_INET,SOCK_STREAM,0);

      if (s == INVALID_SOCKET)
      {
         // Socket failed
      }

      sin.sin_family = AF_INET;
      sin.sin_addr.s_addr = 0;

      sin.sin_port = htons(100); // port=100

      if (bind(s, (struct sockaddr FAR *)&sin, sizeof (sin)) == SOCKET_ERROR)
      {
       // Bind failed
      }

      if (listen(s,4)<0)
      {
         // Listen failed
      }
      sin_len = sizeof(sin);

      s=accept(s,(struct sockaddr FAR *) & sin,(int FAR *) &sin_len);

      if (s<0)
      {
         // Accept failed
```

```
      }
      while (1)
      {
         status=recv(s,recmsg,STRLENGTH,80);

         if (status==SOCKET_ERROR)
         {
            // no socket
            break;
         }

         recmsg[status]=NULL; // terminate string
         if (status)
         {
            // szMsg contains received string
         }
         else
         {
            break;
            // connection broken
         }
      }
      return(0);
}
```

20.2.8 shutdown()

The `shutdown()` function disables send or receive operations on a socket and does not close any opened sockets. Its syntax is:

```
int shutdown(SOCKET s, int how);
```

where

s	Socket descriptor.
how	Flag that identifies operation types that will no longer be allowed. These are: 0 – Disallows subsequent receives. 1 – Disallows subsequent sends. 2 – Disables send and receive.

If no error occur then it returns a zero value. Otherwise, it returns `INVALID_SOCKET`, and the specific error code can be tested with `WSAGetLastError`.

20.2.9 closesocket()

The `closesocket()` function closes a socket. Its syntax is:

```
int closesocket (SOCKET s);
```

where

s	Socket descriptor.

If no error occur then it returns a zero value. Otherwise, it returns `INVALID_SOCKET`, and the specific error code can be tested with `WSAGetLastError`.

21 Electronic Mail

21.1 Introduction

Electronic mail (email) is one use of the Internet which, according to most businesses, improves productivity. Traditional methods of sending mail within an office environment are inefficient, as it normally requires an individual requesting a secretary to type the letter. This must then be proofread and sent through the internal mail system, which is relatively slow and can be open to security breaches.

A faster method, and more secure method of sending information is to use electronic mail, where messages are sent almost in an instant. For example a memo with 100 words will be sent in a fraction of a second. It is also simple to send to specific groups, various individuals, company-wide, and so on. Other types of data can also be sent with the mail message such as images, sound, and so on. It may also be possible to determine if a user has read the mail. The main advantages are:

- It is normally much cheaper than using the telephone (although, as time equates to money for most companies, this relates any savings or costs to a user's typing speed).
- Many different types of data can be transmitted, such as images, documents, speech, and so on.
- It is much faster than the postal service.
- Users can filter incoming email easier than incoming telephone calls.
- It normally cuts out the need for work to be typed, edited and printed by a secretary.
- It reduces the burden on the mailroom.
- It is normally more secure than traditional methods.
- It is relatively easy to send to groups of people (traditionally, either a circulation list was required or a copy to everyone in the group was required).
- It is usually possible to determine whether the recipient has actually read the message (the electronic mail system sends back an acknowledgment).

The main disadvantages are:

- It stops people using the telephone.
- It cannot be used as a legal document.
- Electronic mail messages can be sent on the spur of the moment and may be regretted later on (sending by traditional methods normally allows for a rethink). In extreme cases messages can be sent to the wrong person (typically when replying to an email message, where a message is sent to the mailing list rather than the originator).
- It may be difficult to send to some remote sites. Many organizations have either no electronic mail or merely an intranet. Large companies are particularly wary of Internet connections and limit the amount of external traffic.

- Not everyone reads their electronic mail on a regular basis (although this is changing as more organizations adopt email as the standard communications medium).

The main standards that relate to the protocols of email transmission and reception are:

- Simple Mail Transfer Protocol (SMTP) – which is used with the TCP/IP protocol suite. It has traditionally been limited to the text-based electronic messages.
- Multipurpose Internet Mail Extension (MIME) – which allows the transmission and reception of mail that contains various types of data, such as speech, images and motion video. It is a newer standard than SMTP and uses much of its basic protocol.

21.2 Shared-file approach versus client/server approach

An email system can use either a shared-file approach or a client/server approach. In a shared-file system the source mail client sends the mail message to the local post office. This post office then transfers control to a message transfer agent, which then stores the message for a short time before sending it to the destination post office. The destination mail client periodically checks its own post office to determine if it has mail for it. This arrangement is often known as store and forward, and the process is illustrated in Figure 21.1. Most PC-based email systems use this type of mechanism.

A client/server approach involves the source client setting up a real-time remote connection with the local post office, which then sets up a real-time connection with the destination, which in turn sets up a remote connection with the destination client. The message will thus arrive at the destination when all the connections are complete.

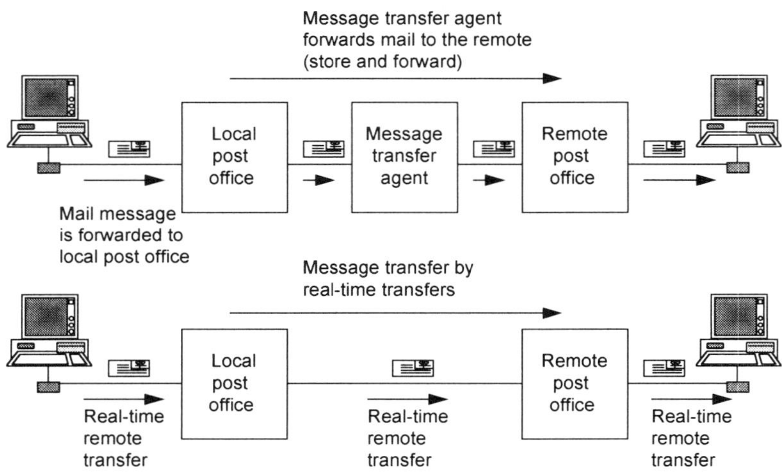

Figure 21.1 Shared-file versus client/server.

21.3 Electronic mail overview

Figure 21.2 shows a typical email architecture. It contains four main elements:

1. Post offices – where outgoing messages are temporally buffered (stored) before transmission and where incoming messages are stored. The post office runs the server software capable of routing messages (a message transfer agent) and maintaining the post office database.
2. Message transfer agents – for forwarding messages between post offices and to the destination clients. This software can either reside on the local post office or on a physically separate server.
3. Gateways – which provide part of the message transfer agent functionality. They translate between different email systems, different email addressing schemes and messaging protocols.
4. Email clients – normally the computer which connects to the post office. It contains three parts:

    - Email Application Program Interface (API), such as MAPI, VIM, MHS and CMC.
    - Messaging protocol. The main messaging protocols are SMTP or X.400. SMTP is defined in RFC 822 and RFC 821. X.400 is an OSI-defined email message delivery standard (Sections 18.5 and 18.6).
    - Network transport protocol, such as Ethernet, FDDI, and so on.

    The main APIs are:

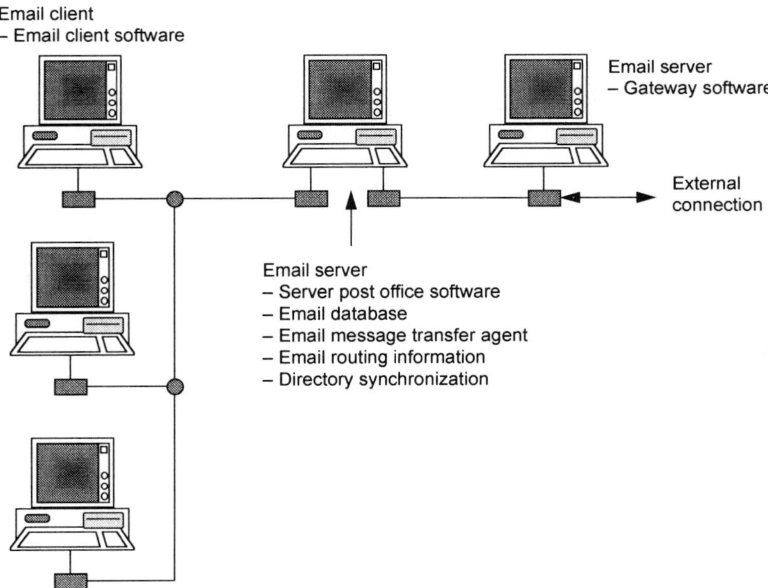

Figure 21.2 Email architecture.

* MAP (messaging API) – Microsoft part of Windows Operation Services Architecture.
* VIM (vendor-independent messaging) – Lotus, Apple, Novell and Borland derived email API.
* MHS (message handling service) – Novell network interface which is often used as an email gateway protocol.
* CMC (common mail call) – Email API associated with the X.400 native messaging protocol.

Gateways translate the email message from one system to another, such as from Lotus cc:Mail to Microsoft Mail. Typical gateway protocols are:

* MHS (used with Novell NetWare).
* SMTP.MIME (used with Internet environment).
* X.400 (used with X.400).
* MS Mail (used with Microsoft Mail).
* cc:Mail (used with Lotus cc:Mail).

A PC-based email package is Lotus cc:Mail (Figure 21.3 shows a sample screen).

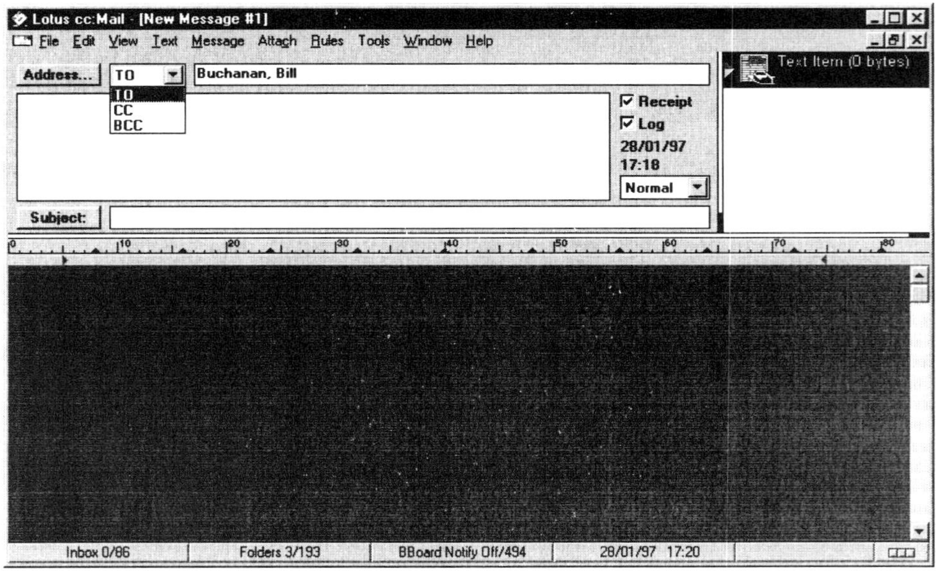

Figure 21.3 A Lotus cc:Mail screen.

21.4 Internet email address

The Internet email address is in the form of a name (such as `f.bloggs`), followed by an '`@`' and then the domain name (such as `anytown.ac.uk`). For example:

```
f.bloggs@anytown.ac.uk
```

No spaces are allowed in the address; periods are used instead. Figure 21.4 shows an example Internet address builder from Lotus cc:Mail.

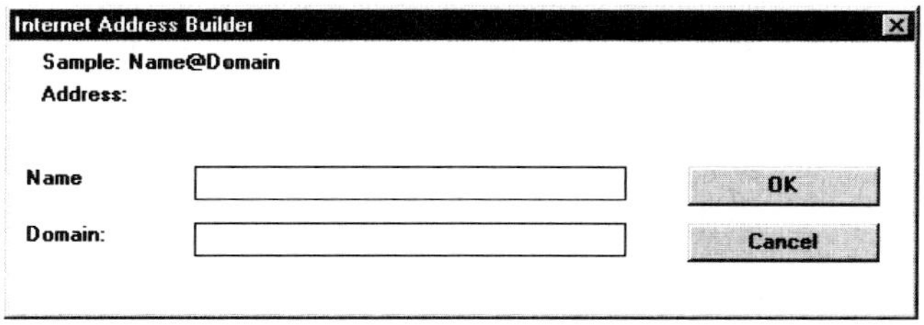

Internet Address Builder
Sample: Name@Domain
Address:
Name
Domain:
OK
Cancel

Figure 21.4 Internet address format.

21.5 SMTP

The IAB have defined the protocol SMTP in RFC 821 (Refer to Appendix E for a list of RFC standards). This section discusses the protocol for transferring mail between hosts using the TCP/IP protocol.

As SMTP is a transmission and reception protocol it does not actually define the format or contents of the transmitted message except that the data has 7-bit ASCII characters and that extra log information is added to the start of the delivered message to indicate the path the message took. The protocol itself is only concerned in reading the address header of the message.

21.5.1 SMTP operation

SMTP defines the conversation that takes place between an SMTP sender and an SMTP receiver. Its main functions are the transfer of messages and the provision of ancillary functions for mail destination verification and handling.

Initially the message is created by the user and a header is added which includes the recipient's email address and other information. This message is then queued by the mail server, and when it has time, the mail server attempts to transmit it.

Each mail may be have the following requirements:

- Each email can have a list of destinations; the email program makes copies of the messages and passes them onto the mail server.
- The user may maintain a mailing list, and the email program must remove duplicates and replace mnemonic names with actual email addresses.
- It allows for normal message provision, e.g. blind carbon copies (BCCs).

An SMTP mail server processes email messages from an outgoing mail queue and then transmits them using one or more TCP connections with the destination. If the mail message is transmitted to the required host then the SMTP sender deletes the destination from the message's destination list. After all the destinations have been sent to, the sender then deletes the message from the queue.

If there are several recipients for a message on the same host, the SMTP protocol allows a single message to be sent to the specified recipients. Also, if there are several messages to be sent to a single host, the server can simply open a single TCP connection and all the messages can be transmitted in a single transfer (there is thus no need to set up a connection for each message).

SMTP also allows for efficient transfer with error messages. Typical errors include:

- Destination host is unreachable. A likely cause is that the destination host address is incorrect. For example, `f.bloggs@toy.ac.uk` might actually be `f.bloggs@toytown.ac.uk`.
- Destination host is out of operation. A likely cause is that the destination host has developed a fault or has been shut down.
- Mail recipient is not available on the host. Perhaps the recipient does not exist on that host, the recipient name is incorrect or the recipient has moved. For example, `fred.bloggs@toytown.ac.uk` might actually be `f.bloggs@toytown.ac.uk`. To overcome the problem of user names which are similar to a user's name then some systems allow for certain aliases for recipients, such as `f.bloggs`, `fred.bloggs` and `freddy.bloggs`, but there is a limit to the number of aliases that a user can have. If a user has moved then some systems allow for a redirection of the email address. UNIX systems use the `.forward` file in the user's home directory for redirection. For example on a UNIX system, if the user has moved to `fred.bloggs@toytown.com` then this address is simply added to the `.forward` file.
- TCP connection failed on the transfer of the mail. A likely cause is that there was a timeout error on the connection (maybe due to the receiver or sender being busy or there was a fault in the connection).

SMTP senders have the responsibility for a message up to the point where the SMTP receiver indicates that the transfer is complete. This only indicates that the message has arrived at the SMTP receiver; does not indicate that:

- The message has been delivered to the recipient's mailbox.
- The recipient has read the message.

Thus, SMTP does not guarantee to recover from lost messages and gives no end-to-end acknowledgment on successful receipt (normally this is achieved by an acknowledgment message being returned). Nor are error indications guaranteed. However, TCP connections are normally fairly reliable.

If an error occurs in reception, a message will normally be sent back to the sender to explain the problem. The user can then attempt to determine the problem with the message.

SMTP receivers accept an arriving message and either place it in a user's mailbox or, if that user is located at another host, copies it to the local outgoing mail queue for forwarding.

Most transmitted messages go from the sender's machine to the host over a single TCP connection. But sometimes the connection will be made over multiple TCP connections over multiple hosts. The sender specifying a route to the destination in the form of a sequence of servers can achieve this.

21.5.2 SMTP overview

An SMTP sender initiates a TCP connection. When this is successful it sends a series of commands to the receiver, and the receiver returns a single reply for each command. All commands and responses are sent with ASCII characters and are terminated with the carriage return (CR) and line feed (LF) characters (often known as CRLF).

Each command consists of a single line of text, beginning with a four-letter command code followed by in some cases an argument field. Most replies are a single line, although multiple-line replies are possible. Table 21.1 gives some sample commands.
SMTP replies with a three-digit code and possibly other information. Some of the responses are listed in Table 21.2. The first digit gives the category of the reply, such as 2*xx* (a positive completion reply), 3*xx* (a positive intermediate reply), 4*xx* (a transient negative completion reply) and 5*xx* (a permanent negative completion reply). A positive reply indicates that the requested action has been accepted, and a negative reply indicates that the action was not accepted.

Positive completion reply indicates that the action has been successful, and a positive intermediate reply indicates that the action has been accepted but the receiver is waiting for some other action before it can give a positive completion reply. A transient negative completion reply indicates that there is a temporary error condition which can be cleared by other actions and a permanent negative completion reply indicates that the action was not accepted and no action was taken.

Table 21.1 SMTP commands.

Command	*Description*
HELO *domain*	Sends an identification of the domain
MAIL FROM: *sender-address*	Sends identification of the originator (sender-address)
RCPT FROM: *receiver-address*	Sends identification of the recipient (receiver-address)
DATA	Transfer text message
RSEY	Abort current mail transfer
QUIT	Shut down TCP connection
EXPN *mailing-list*	Send back membership of mailing list
SEND FROM: *sender-address*	Send mail message to the terminal
SOML FROM: *sender-address*	If possible, send mail message to the terminal, otherwise send to mailbox
VRFY username	Verify user name (username)

21.5.3 SMTP transfer

Figure 21.4 shows a successful email transmission. For example if:

```
f.bloggs@toytown.ac.uk
```

is sending a message to:

```
a.person@place.ac.de
```

Then a possible sequence of events is:

- Set up TCP connection with receiver host.
- If the connection is successful, the receiver replies back with a 220 code (server ready). If it is unsuccessful, it returns back with a 421 code.
- Sender sends a `HELO` command to the hostname (such as `HELO toytown.ac.uk`).
- If the sender accepts the incoming mail message then the receiver returns a `250 OK` code. If it is unsuccessful then it returns a 421, 451, 452, 500, 501 or 552 code.
- Sender sends a `MAIL FROM:` *`sender`* command (such as `MAIL FROM: f.bloggs@toytown.ac.uk`).
- If the receiver accepts the incoming mail message from the sender then it returns a `250 OK` code. If it is unsuccessful then it returns codes such as 251, 450, 451, 452, 500, 501, 503, 550, 551, 552 or 553 code.
- Sender sends an `RCPT TO:` *`receiver`* command (such as `RCPT TO: a.person@place.ac.de`).
- If the receiver accepts the incoming mail message from the sender then it returns a `250 OK` code.
- Senders sends a `DATA` command.

Table 21.2 SMTP responses.

CMD	*Description*	*CMD*	*Description*
211	System status	500	Command unrecognized due to a syntax error
214	Help message	501	Invalid parameters or arguments
220	Service ready	502	Command not currently implemented
221	Service closing transmission channel	503	Bad sequence of commands
250	Request mail action completed successfully	504	Command parameter not currently implemented
251	Addressed user does not exist on system but will forward to receiver-address	550	Mail box unavailable, request action not taken
354	Indicate to the sender that the mail message can now be sent. The end of the message is identified by two CR, LF characters	551	The addressed user is not local, please try receiver-address
421	Service is not available	552	Exceeded storage allocation, requested mail action aborted
450	Mailbox unavailable and the requested mail action was not taken	553	Mailbox name not allowed, requested action not taken
451	Local processing error, requested action aborted	554	Transaction failed
452	Insufficient storage, requested action not taken		

- If the receiver accepts the incoming mail message from the sender then it returns a `354` code (start transmission of mail message).
- The sender then transmits the email message.
- The end of the email message is sent as two LF, CR characters.
- If the reception has been successful then the receiver sends back a `250 OK` code. If it is unsuccessful then it returns a 451, 452, 552 or 554 code.
- Sender starts the connection shutdown by sending a `QUIT` command.
- Finally the sender closes the TCP connection.

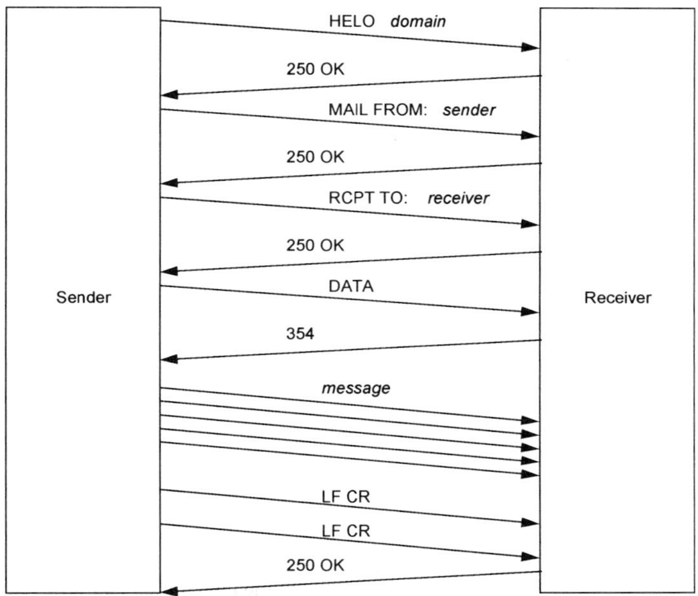

Figure 21.5 Sample SMTP email transmission.

21.5.4 RFC 822

SMTP uses RFC 822 which defines the format of the transmitted message. RFC 822 contains two main parts:

- A header – which is basically the mail header and contains information for the successful transmission and delivery of a message. This typically contains the email addresses for sender and receiver, the time the message was sent and received. Any computer involved in the transmission can added to the header.
- The contents.

Normally the email-reading program will read the header and format the information to the screen to show the sender's email address; it splits off the content of the message and displays it separately from the header.

An RFC 822 message contains a number of lines of text in the form of a memo (such as To:, From:, Bcc:, and so on). A header line usually has a keyword followed by a colon and then followed by keyword arguments. The specification also allows for a long line to be broken up into several lines.

Here is an RFC 822 message with the header shown in italics and the message body in bold. Table 21.3 explains some of the RFC 822 items in the header.

```
From FREDB@ACOMP.CO.UK Wed Jul  5 12:36:49 1995
Received: from ACOMP.CO.UK ([154.220.12.27]) by central.napier.ac.uk
(8.6.10/8.6.10) with SMTP id MAA16064 for <w.buchanan@central.napier.ac.uk>;

Wed, 5 Jul 1995 12:36:43 +0100

Received: from WPOAWUK-Message_Server by ACOMP.CO.UK
      with Novell_GroupWise; Wed, 05 Jul 1995 12:35:51 +0000

Message-Id: <sffa8725.082@ACOMP.CO.UK >

X-Mailer: Novell GroupWise 4.1

Date: Wed, 05 Jul 1995 12:35:07 +0000

From: Fred Bloggs <FREDB@ACOMP.CO.UK>

To: w.buchanan@central.napier.ac.uk
Subject:  Technical Question
Status: REO

Dear Bill
      I have a big problem. Please help.
Fred
```

Table 21.3 Header line descriptions.

Header line	*Description*
From FREDB@ACOMP.CO.UK Wed Jul 5 12:36:49 1995	Sender of the email is FREDB@ ACOM.CO.UK
Received: from ACOMP.CO.UK ([154.220.12.27]) by central.napier.ac.uk (8.6.10/8.6.10) with SMTP id MAA16064 for <w.buchanan@central.napier.ac.uk>; Wed, 5 Jul 1995 12:36:43 +0100	It was received by CENTRAL.NAPIER.AC.UK at 12:36 on 5 July 1995
Message-Id: <sffa8725.082@ACOMP.CO.UK >	Unique message ID
X-Mailer: Novell GroupWise 4.1	Gateway system
Date: Wed, 05 Jul 1995 12:35:07 +0000	Date of original message
From: Fred Bloggs <FREDB@ACOMP.CO.UK>	Sender's email address and full name
To: w.buchanan@central.napier.ac.uk	Recipient's email address
Subject: Technical Question	Mail subject

21.6 X.400

RFC 821 (the transmission protocol) and RFC 822 (the message format) have now become the de-facto standards for email systems. The CCITT, in 1984, defined new email recommendations called X.400. The RFC821/822 system is simple and works relatively well, whereas X.400 is complex and is poorly designed. These points have helped RFC 821/822 to become a de-facto standard, whereas X.400 is now almost extinct (see Figure 21.6 for the basic X.400 address builder and Figure 21.7 for the extended address builder).

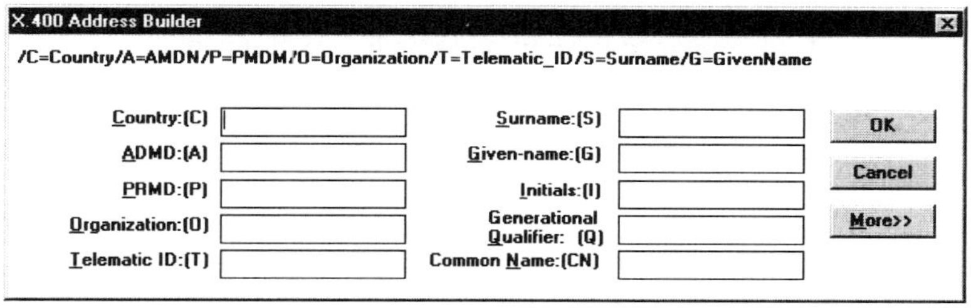

Figure 21.6 X.400 basic addressing.

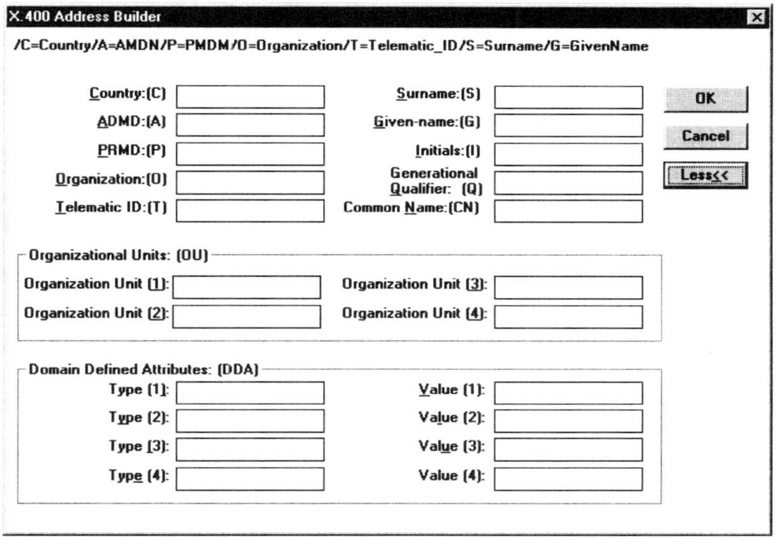

Figure 21.7 X.400 extended addressing.

21.7 MIME

SMTP suffers from several drawback, such as:

- SMTP can only transmit ASCII characters and thus cannot transmit executable files or other binary objects.

- SMTP does not allow the attachment of files, such as images and audio.
- SMTP can only transmit 7-bit ASCII character thus it does support an extended ASCII character set.

A new standard, Multipurpose Internet Mail Extension (MIME), has been defined for this purpose, which is compatible with existing RFC 822 implementations. It is defined in the specifications RFC 1521 and 1522. Its enhancements include the following:

- Five new message header fields in the RFC 822 header, which provide extra information about the body of the message.
- Use of various content formats to support multimedia electronic mail.
- Defined transfer encodings for transforming attached files.

The five new header fields defined in MIME are:

- MIME-version – a message that conforms to RFC 1521 or 1522 is MIME-version 1.0.
- Content-type – this field defines the type of data attached.
- Content-transfer-encoding – this field indicates the type of transformation necessary to represent the body in a format which can be transmitted as a message.
- Content-id – this field is used to uniquely identify MIME multiple attachments in the email message.
- Content-description – this field is a plain-text description of the object with the body. It can be used by the user to determine the data type.

These fields can appear in a normal RFC 822 header. Figure 21.8 shows an example email message. It can be seen, in the right-hand corner, that the API has split the message into two parts: the message part and the RFC 822 part. The RFC 822 part is shown in Figure 21.9. It can be seen that, in this case, the extra MIME messages are:

```
MIME-Version: 1.0
Content-Type: text/plain; charset=us-ascii
Content-Transfer-Encoding: 7bit
```

This defines it as MIME Version 1.0; the content-type is text/plain (standard ASCII) and it uses the US ASCII character set; the content-transfer-encoding is 7-bit ASCII.

RFC 822 example file listing (refer to Figure 21.9)

```
Received: from pc419.eece.napier.ac.uk by ccmailgate.napier.ac.uk (SMTPLINK
V2.11.01)
    ; Fri, 24 Jan 97 11:13:41 gmt
Return-Path: <w.buchanan@napier.ac.uk>
Message-ID: <32E90962.1574@napier.ac.uk>
Date: Fri, 24 Jan 1997 11:14:22 -0800
From: Dr William Buchanan <w.buchanan@napier.ac.uk>
Organization: Napier University
X-Mailer: Mozilla 3.01 (Win95; I; 16bit)
MIME-Version: 1.0
To: w.buchanan@napier.ac.uk
Subject: Book recommendation
Content-Type: text/plain; charset=us-ascii
Content-Transfer-Encoding: 7bit
```

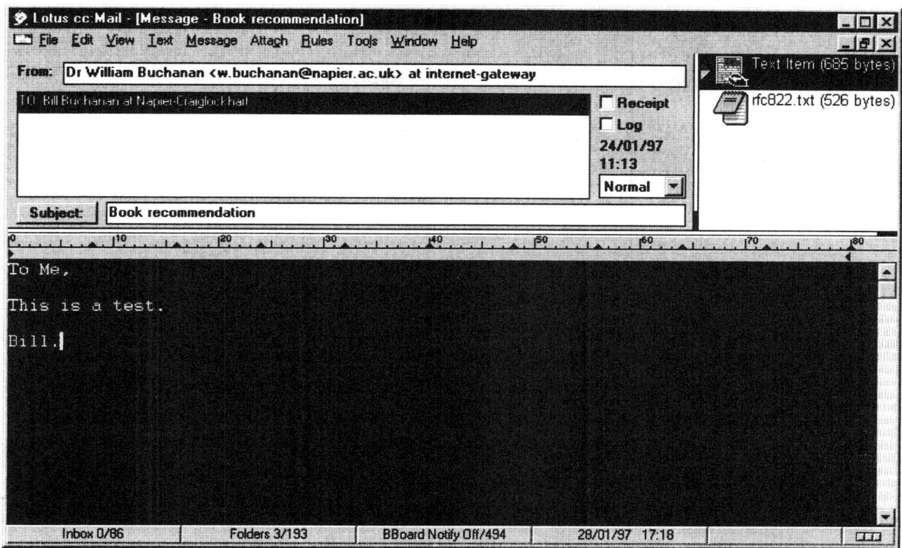

Figure 21.8 Sample email message showing message and RFC822 part.

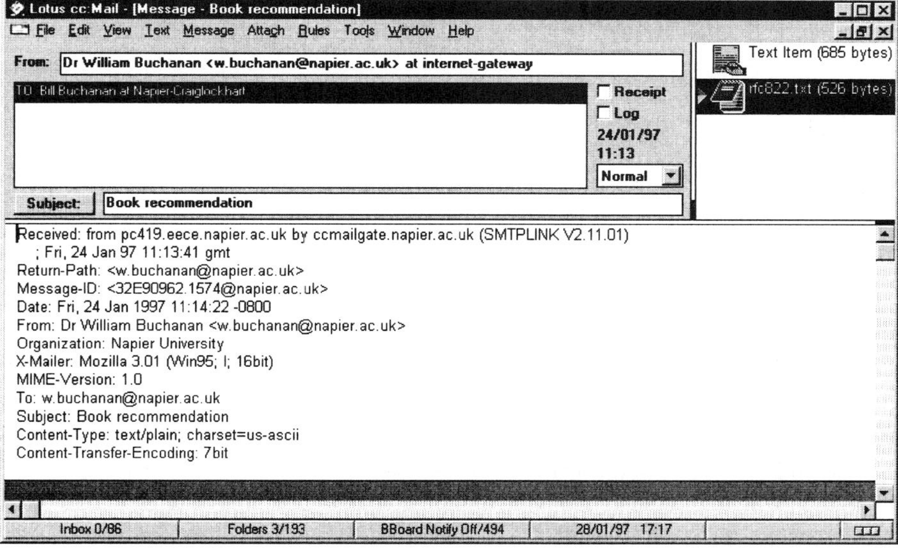

Figure 21.9 RFC 822 part.

21.7.1 MIME content types

Content types define the format of the attached files. There are a total of 16 different content types in seven major content groups. If the text body is pure text then no special transformation is required. RFC 1521 defines only one subtype, text/plain; this gives a standard ASCII character set.

A MIME-encoded email can contain multiple attachments. The content-type header field includes a boundary which defines the delimiter between multiple attachments. A boundary always starts on a new line and has the format:

```
-- boundary name
```

The final boundary is:

```
-- boundary name --
```

For example, the following message contains two parts:

Example MIME file with 2 parts

```
From: Dr William Buchanan <w.buchanan@napier.ac.uk>
MIME-Version: 1.0
To: w.buchanan@napier.ac.uk
Subject: Any subject
Content-Type: multipart/mixed; boundary="boundary name"

This part of the message will be ignored.

-- boundary name
Content-Type: multipart/mixed; boundary="boundary name"

This is the first mail message part.

-- boundary name

And this is the second mail message part.

-- boundary name --
```

Table 21.4 MIME content-types.

Content type	*Description*
text/plain	Unformatted text, such as ASCII
text/richtext	Rich text format which is similar to HTML
multipart/mixed	Each attachment is independent from the rest and all should be presented to the user in their initial ordering
multipart/parallel	Each attachment is independent from the others but the order is unimportant
multipart/alternative	Each attachment is a different version of the original data
multipart/digest	This is similar to multipart/mixed but each part is message/rfc822
message/rfc822	Contains the RFC 822 text
message/partial	Used in fragmented mail messages
message/external-body	Used to define a pointer to an external object (such as an ftp link)
image/jpeg	Defines a JPEG image using JFIF file format
image/gif	Defines GIF image
video/mpeg	Defines MPEG format
audio/basic	Defines 8-bit μ-Law encoding at 8 kHz sampling rate
application/postscript	Defines postscript format
application/octet-stream	Defines binary format which consists of 8-bit bytes

The part of the message after the initial header and before the first boundary can be used to add a comment. This is typically used to inform users that do not have a MIME-compatible program about the method used to encode the received file. A typical method for converting binary data into ASCII characters is to use the programs UUENCODE (to encode a binary file into text) or UUDECODE (to decode a uuencoded file).

The four subtypes of multipart type can be used to sequence the attachments; the main subtypes are:

- multipart/mixed subtype – which is used when attachments are independent but need to be arranged in a particular order.
- multipart/parallel subtype – which is used when the attachments should be present at the same; a typical example is to present an animated file along with an audio attachment.
- multipart/alternative subtype – which is used to represent an attachment in a number of different formats.

21.7.2 Example MIME

The following file listing shows the message part of a MIME-encoded email message (i.e. it excludes the RFC 822 header part). It can be seen that the sending email system has added the comment about the MIME encoding. In this case the MIME boundaries have been defined by:

```
-- IMA.Boundary.760275638
```

Example MIME file

```
This is a Mime message, which your current mail reader
may not understand. Parts of the message will appear as
text. To process the remainder, you will need to use a Mime
compatible mail reader. Contact your vendor for details.

--IMA.Boundary.760275638

Content-Type: text/plain; charset=US-ASCII
Content-Transfer-Encoding: 7bit
Content-Description: cc:Mail note part

This is the original message .....

--IMA.Boundary.760275638--
```

21.7.3 Mail fragments

A mail message can be fragmented using the content-type field of message/partial and then reassembled back at the source. The standard format is:

```
Content-type: message/partial;
   id="idname" ; number=x; total=y
```

where *idname* is the message identification (such as `xyz@hostname`, *x* is the number of the fragment out of a total of *y* fragments. For example, if a message had three fragments, they could be sent as:

Example MIME file with 3 fragments (first part)

```
From: Fred Bloggs <f.bloggs@toytown.ac.uk>
MIME-Version: 1.0
To: a.body@anytown.ac.uk
Subject: Any subject
Content-Type: message/partial;
      id="xyz@toytown.ac.uk"; number=1; total=3
Content=type: video/mpeg

First part of MPEG file
```

Example MIME file with 3 fragments (second part)

```
From: Fred Bloggs <f.bloggs@toytown.ac.uk>
MIME-Version: 1.0
To: a.body@anytown.ac.uk
Subject: Any subject
Content-Type: message/partial;
      id="xyz@toytown.ac.uk"; number=2; total=3
Content=type: video/mpeg

Second part of MPEG file
```

Example MIME file with 3 fragments (third part)

```
From: Fred Bloggs <f.bloggs@toytown.ac.uk>
MIME-Version: 1.0
To: a.body@anytown.ac.uk
Subject: Any subject
Content-Type: message/partial;
      id="xyz@toytown.ac.uk"; number=3; total=3
Content=type: video/mpeg

Third part of MPEG file
```

21.7.4 Transfer encodings

MIME allows for different transfer encodings within the message body:

- 7bit – no encoding, and all of the characters are 7-bit ASCII characters.
- 8bit – no encoding, and extended 8-bit ASCII characters are used.
- quoted-printable – encodes the data so that non-printing ASCII characters (such as line feeds and carriage returns) are displayed in a readable form.
- base64 – encodes by mapping 6-bit blocks of input to 8-bit blocks of output, all of which are printable ASCII characters.
- x-token – another non-standard encoding method.

When the transfer encoding is:

```
Content-transfer-encoding: quoted-printable
```

then the message has been encoded so that all non-printing characters have been converted to printable characters. A typical transform is to insert =*xx* where *xx* is the hexadecimal equivalent for the ASCII character. A form feed (FF) would be encoded with '=0C',

A transfer encoding of base64 is used to map 6-bit characters to a printable character. It is a useful method in disguising text in an encrypted form and also for converting binary data into a text format. It takes the input bitstream and reads it 6 bits at a time, then maps this to an 8-bit printable character. Table 21.5 shows the mapping.

Table 21.5 MIME base64 encoding.

Bit value	*Encoded character*	*Bit value*	*Encoded character*	*Bit value*	*Encoded character*	*Bit value*	*Encoded character*
0	A	16	Q	32	g	48	w
1	B	17	R	33	h	49	x
2	C	18	S	34	i	50	y
3	D	19	T	35	j	51	z
4	E	20	U	36	k	52	0
5	F	21	V	37	l	53	1
6	G	22	W	38	m	54	2
7	H	23	X	39	n	55	3
8	I	24	Y	40	o	56	4
9	J	25	Z	41	p	57	5
10	K	26	a	42	q	58	6
11	L	27	b	43	r	59	7
12	M	28	c	44	s	60	8
13	N	29	d	45	t	61	9
14	O	30	e	46	u	62	+
15	P	31	f	47	v	63	/

Thus if a binary file had the bit sequence:

```
101000101010100010101010
```

It would first be split into groups of 6 bits, as follows:

```
101000      101010  100010   101010  000000
```

This would be converted into the ASCII sequence:

```
YsSqA
```

which is in a transmittable form.

Thus the 7-bit ASCII sequence ‘FRED’ would use the bit pattern:

```
1000110 1010010 1000101 1000100
```

which would be split into groups of 6 bits as:

```
100011 010100 101000 101100 010000
```

which would be encoded as:

```
jUosQ
```

21.7.5 Example

The following parts of the RFC 822 messages.

(a)

```
Received: from publish.co.uk by ccmail1.publish.co.uk (SMTPLINK V2.11.01)
Return-Path: <FredB@local.exnet.com>
Received: from mailgate.exnet.com ([204.137.193.226]) by zeus.publish.co.uk with
SMTP id <17025>; Wed, 2 Jul 1997 08:33:29 +0100
Received: from exnet.com (assam.exnet.com) by mailgate.exnet.com with SMTP id
AA09732 (5.67a/IDA-1.4.4 for m.smith@publish.co.uk); Wed, 2 Jul 1997 08:34:22
+0100
Received: from maildrop.exnet.com (ceylon.exnet.com) by exnet.com with SMTP id
AA10740 (5.67a/IDA-1.4.4 for <m.smith@publish.co.uk>); Wed, 2 Jul 1997 08:34:10
+0100
Received: from local.exnet.com by maildrop.exnet.com (4.1/client-1.2DHD)
    id AA22007; Wed, 2 Jul 97 08:25:21 BST
From: FredB@local.exnet.com (Arthur Chapman)
Reply-To: FredB@local.exnet.com
To: b.smith@publish.co.uk
Subject: New proposal
Date: Wed, 2 Jul 1997 09:36:17 +0100
Message-Id: <66322430.1380704@local.exnet.com>
Organization: Local College
```

(b)

```
Received: from central.napier.ac.uk by ccmailgate.napier.ac.uk (SMTPLINK V2.11.01
Return-Path: <fred@singnetw.com.sg>
Received: from server.singnetw.com.sg (server.singnetw.com.sg [165.21.1.15]) by
central.napier.ac.uk (8.6.10/8.6.10) with ESMTP id DAA18783 for
<w.buchanan@napier.ac.uk>; Sun, 29 Jun 1997 03:15:27 GMT
Received: from si7410352.ntu.ac.sg (ts900-1908.singnet.com.sg [165.21.158.60])
    by melati.singnet.com.sg (8.8.5/8.8.5) with SMTP id KAA08773
    for <w.buchanan@napier.ac.uk.>; Sun, 29 Jun 1997 10:14:59 +0800 (SST)
Message-ID: <33B5C33B.6CCC@singnetw.com.sg>
Date: Sun, 29 Jun 1997 10:06:51 +0800
From: Fred Smith <fred@singnetw.com.sg>
X-Mailer: Mozilla 2.0 (Win95; I)
MIME-Version: 1.0
To: w.buchanan@napier.ac.uk
Subject: Chapter 15
Content-Type: text/plain; charset=us-ascii
Content-Transfer-Encoding: 7bit
```

(c)

```
Received: from central.napier.ac.uk by ccmailgate.napier.ac.uk (SMTPLINK V2.11.01
Return-Path: <bertb@scms.scotuni.ac.uk>
Received: from master.scms.scotuni.ac.uk ([193.62.32.5]) by central.napier.ac.uk
(8.6.10/8.6.10) with ESMTP id MAA20373 for <w.buchanan@napier.ac.uk>; Tue, 1 Jul
1997 12:25:38 GMT
Received: from cerberus.scms.scotuni.ac.uk (cerberus.scms.scotuni.ac.uk
[193.62.32.46]) by master.scms.scotuni.ac.uk (8.6.9/8.6.9) with ESMTP id MAA10056
for <w.buchanan@napier.ac.uk>; Tue, 1 Jul 1997 12:24:32 +0100
From: David Davidson <bertb@scms.scotuni.ac.uk>
Received: by cerberus.scms.scotuni.ac.uk (SMI-8.6/Dumb)
    id MAA03334; Tue, 1 Jul 1997 12:23:17 +0100
Date: Tue, 1 Jul 1997 12:23:17 +0100
Message-Id: <199707011123.MAA03334@cerberus.scms.scotuni.ac.uk>
To: w.buchanan@napier.ac.uk
Subject: Advert
Mime-Version: 1.0
Content-Type: text/plain; charset=us-ascii
Content-Transfer-Encoding: 7bit
Content-MD5: TzKyk+NON+vy6Cm6uqy9Cg==
```

22 The World Wide Web

22.1 Introduction

The areas of modern life that have more jargon words and associated acronyms than anything else are the World-Wide Web (WWW) and the Internet. Words, such as:

gopher, ftp, telnet, TCP/IP stack, intranets, Web servers, clients, browsers, hypertext, URLs, Internet access providers, dial-up connections, UseNet servers, firewalls

have all become common in the business vocabulary.

The WWW was initially conceived in 1989 by CERN, the European particle physics research laboratory in Geneva, Switzerland. Its main objective was:

to use the hypermedia concept to support the interlinking of various types of information through the design and development of a series of concepts, communications protocols, and systems

One of its main characteristics is that stored information tends to be distributed over a geographically wide area. The result of the project has been the worldwide acceptance of the protocols and specifications used. A major part of its success was due to the full support of the National Center for Supercomputing Applications (NCSA), which developed a family of user interface systems known collectively as Mosaic.

The WWW, or Web, is basically an infrastructure of information. This information is stored on the WWW on Web servers and it uses the Internet to transmit data around the world. These servers run special programs that allow information to be transmitted to remote computers which are running a Web browser, as illustrated in Figure 22.1. The Internet is a common connection in which computers can communicate using a common addressing mechanism (IP) with a TCP/IP connection.

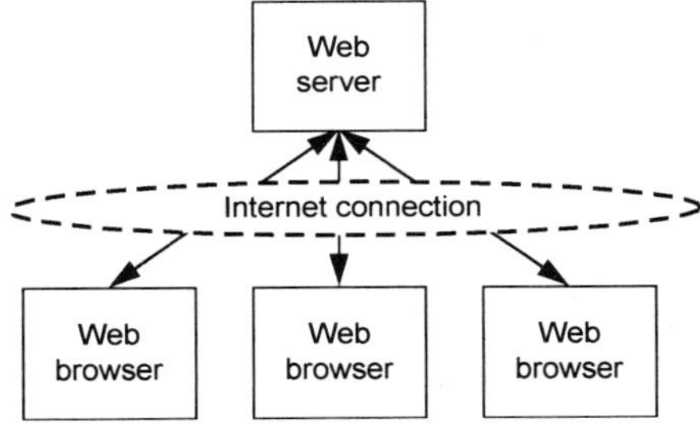

Figure 22.1 Web servers and browsers.

The information is stored on Web servers and is accessed by means of pages. These pages can contain text and other multimedia applications such as graphic images, digitized sound files and video animation. There are several standard media files (with typical file extensions):

- GIF or JPEG files for compressed images (GIF or JPG).
- QuickTime movies for video (QT or MOV).
- Postscript files (PS or EPS).
- MS video (AVI).
- Audio (AU, SND or WAV).
- MPEG files for compressed video (MPG).
- Compressed files (ZIP, Z or GZ).
- JavaScript (JAV, JS or MOCHA).
- Text files (TEX or TXT).

Each page contains text known as hypertext, which has specially reserved keywords to represent the format and the display functions. A standard language known as HTML (Hypertext Markup Language) has been developed for this purpose.

Hypertext pages, when interpreted by a browser program, display an easy-to-use interface containing formatted text, icons, pictorial hot spots, underscored words, and so on. Each page can also contain links to other related pages.

The topology and power of the Web now allows for distributed information, where information does not have to be stored locally. To find information on the Web the user can use powerful search engines to search for related links. Figure 22.2 shows an example of Web connections. The user initially accesses a page on a German Web server, this then contains a link to a Japanese server. This server contains links to UK, Swedish and French servers. This type of arrangement leads to the topology that resembles a spider's web, where information is linked from one place to another.

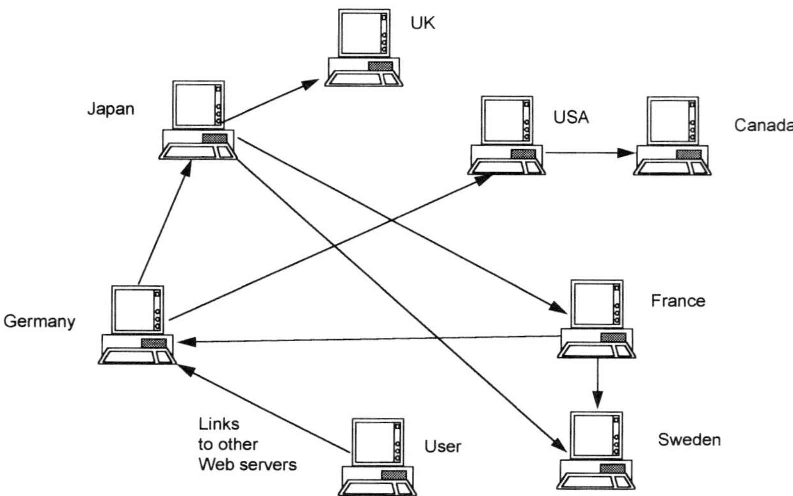

Figure 22.2 Example Web connections.

22.2 Advantages and disadvantages of the WWW

The WWW and the Internet tend to produce a polarization of views. Thus, before analyzing the WWW for its technical specification, a few words must be said on some of the subjective advantages and disadvantages of the WWW and the Internet. It should be noted that some of these disadvantages can be seen as advantages to some people, and vice versa. For example, freedom of information will be seen as an advantage to a freedom-of-speech group but often a disadvantage to security organizations. Table 22.1 outlines some of advantages and disadvantages.

Table 22.1 Advantages and disadvantages of the Internet and the WWW.

	Advantages	*Disadvantages*
Global information flow	Less control of information by the media, governments and large organizations.	Lack of control on criminal material, such as certain types of pornography and terrorist activity.
Global transmission	Communication between people and organizations in different countries which should create the Global Village.	Data can easily get lost or state secrets can be easily transmitted over the world.
Internet connections	Many different types of connections are possible, such as dial-up facilities (perhaps over a modem or with ISDN) or through frame relays. The user only has to pay for the service and the local connection.	Data once on the Internet is relatively easy to tap into and possibly easy to change.
Global information	Creation of an ever-increasing global information database.	Data is relatively easy to tap into and possibly easy to change.
Multimedia integration	Tailor-made applications with good presentation tools.	Lack of editorial control leads to inferior material, which is hacked together.
Increasing WWW usage	Helps to improve its chances of acceptance into the home.	Increased traffic swamps the global information network and slows down commercial traffic.
WWW links	Easy to set up and leads users from one place to the next in a logical manner.	WWW links often fossilize where the link information is out-of-date or doesn't even exist.
Education	Increased usage of remote teaching with full multimedia education.	Increase in surface learning and lack of deep research. It may lead to an increase in time-wasting (too much surfing and too little learning).

22.3 Client/server architecture

The WWW is structured with clients and servers, where a client accesses services from the server. These servers can either be local or available through a global network connection. A local connection normally requires the connection over a local area network but a global connection normally requires connection to an Internet provider. These providers are often known as an Internet access provider (IAPs), sometimes as an Internet connectivity provider (ICP) or Internet Service Providers (ISPs). They provide the mechanism to access the Internet and have the required hardware and software to connect from the user to the Internet. This access is typically provided through one of the following:

- Connection to a client computer though a dial-up modem connection (typically at 14.4 kbps or 28.8 kbps).
- Connection to a client computer though a dial-up ISDN connection (typically at 64 kbps or 128 kbps).
- Connection of a client computer to a server computer which connects to the Internet though a frame relay router (typically 56 kbps or 256 kbps).
- Connection of a client computer to a local area network which connects to the Internet though a T-1, 1.544 Mbps router.

These connections are illustrated in Figure 22.3. A router automatically routes all traffic to and from the Internet whereas the dial-up facility of a modem or ISDN link requires a connection to be made over a circuit-switched line (that is, through the public telephone network). Home users and small businesses typically use modem connections (although ISDN connections are becoming more common). Large corporations which require global Internet services tend to use frame routers. Note that an IAP may be a commercial organization (such as CompuServe or America On-line) or a support organization (such as giving direct connection to government departments or educational institutions). A commercial IAP organization is likely to provide added services, such as electronic mail, search engines, and so on.

An Internet Presence Provider (IPP) allows organizations to maintain a presence on the Internet without actually having to invest in the Internet hardware. The IPPs typically maintain WWW pages for a given charge (they may also provide sales and support information).

22.4 Web browsers

Web browsers interpret special hypertext pages which consist of the hypertext markup language (HTML) and JavaScript. They then display it in the given format. There are currently four main Web browsers:

Netscape Navigator – Navigator is the most widely used WWW browser and is available in many different versions on many systems. It runs on PCs (running Windows 3.1, Windows NT or Windows 95/98), UNIX workstations and Macintosh computers. Figure 22.4 shows Netscape Navigator Version 4 for Windows 95/98/NT. It has become the standard WWW browser and has many add-ons and enhancements, which have been added through continual development by Netscape. The basic package also has many compatible software plug-ins which are developed by third-party suppliers. These add extra functionality such as video players and sound support.

- NSCA Mosaic – Mosaic was originally the most popular Web browser when the Internet first started. It has now lost its dominance to Microsoft Internet Explorer and Netscape Navigator. NSCA Mosaic was developed by the National Center for Supercomputer Applications (NCSA) at the University of Illinois.
- Lynx – Lynx is typically used on UNIX-based computers with a modem dial-up connection. It is fast to download pages but does not support many of the features supported by Netscape Navigator or Mosaic.
- Microsoft Internet Explorer – Explorer now comes as a standard part of Windows 95/98/NT and as this will become the most popular computer operating system then so will this browser.

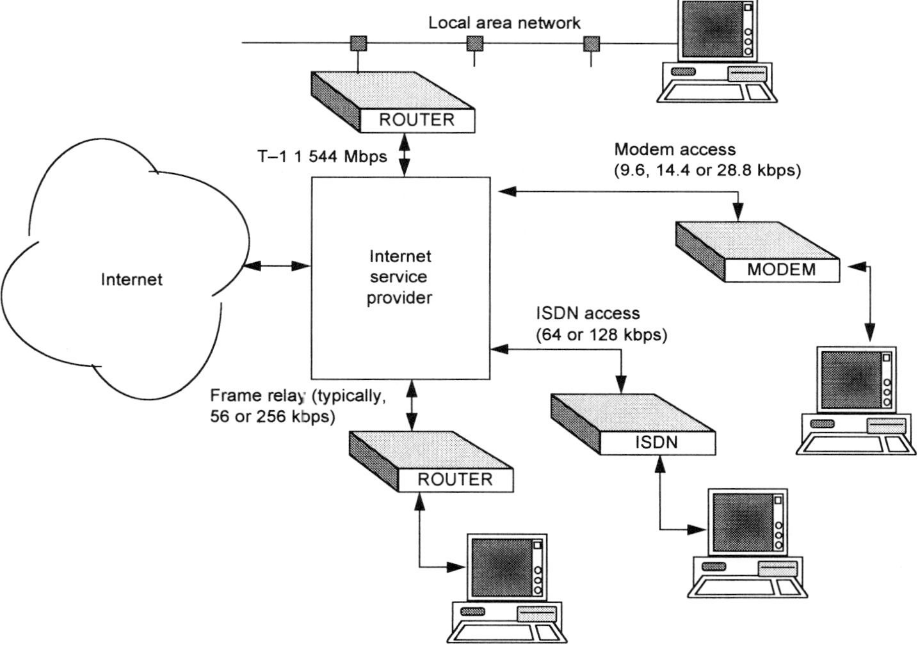

Figure 22.3 Example connection to the Internet.

22.5 Internet resources

The Internet expands by the day as the amount of servers and clients which connect to the global network increases and the amount of information contained in the network also increases. The three major services which the Internet provides are:

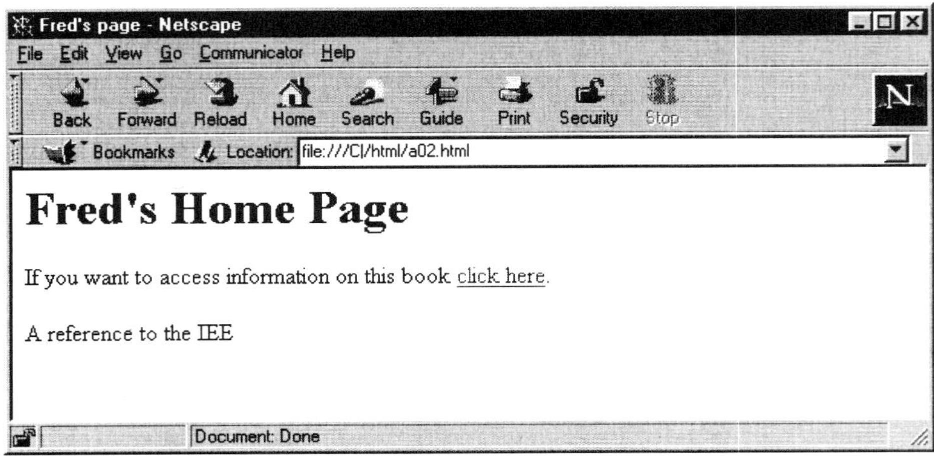

Figure 22.4 Netscape Navigator Version 4.0.

- The World Wide Web.
- Global electronic mail.
- Information sources.

The main information sources, apart from the WWW, are from FTP, Gopher, WAIS and UseNet servers. These different types of servers will be discussed in the next section.

22.6 Universal resource locators (URLs)

Universal resource locators (URLs) are used to locate a file on the WWW. They provide a pointer to any object on a server connected over the Internet. This link could give FTP access, hypertext references, and so on. URLs contains:

- The protocol of the file (the scheme).
- The server name (domain).
- The pathname of the file.
- The filename.

URL standard format is:

<scheme>:<scheme-specific-part>

and can be broken up into four parts, these are:

```
aaaa://bbb.bbb.bbb/ccc/ccc/ccc?ddd
```

where

aaaa: is the access method and specifies the mechanism to be used by the browser to communicate with the resource. The most popular mechanisms are:

- http:. HyperText Transfer Protocol. This is the most commonly used mechanism and is typically used to retrieve an HTML file, a graphic file, a sound file, an animation sequence file, a file to be executed by the server, or a word processing file.
- https:. HyperText Transfer Protocol. It is a variation on the standard access method and can be used to provide some level of transmission security.
- file:. Local File Access. This causes the browser to load the specified file from the local disk.
- ftp:. File Transport Protocol. This method allows files to be downloaded using an FTP connection.
- mailto:. E-Mail Form. This method allows access to a destination e-mail address. Normally the browser automatically generates an input form for entering the e-mail message.
- news:. USENET News. This method defines the access method for a news group.
- nntp:. Local Network News Transport Protocol.
- wais:. Wide Area Information Servers.
- gopher:. GOPHER protocol.
- telnet:. TELNET. The arguments following the access code are the login arguments to the telnet session as user[:password]@host.
- cid:. Content identifiers for MIME body part.
- mid:. Message identifiers for electronic mail.
- afs:. AFS File Access.
- prospero:. Prospero Link.
- x-exec:. Executable Program.

//bbb.bbb.bbb is the Internet node and specifies the node on the Internet where the file is located. If a node is not given then the browser defaults to the computer which is running the browser. A colon may follow the node address and the port number (most browsers default to port 80, which is also what most servers use to reply to the browser).

/ccc/ccc/ccc is the file path (including subdirectories and the filename). Typically systems restrict the access to a system by allocating the root directory as a subdirectory of the main file system.

?ddd is the arguments which depending upon the access method, and the file accessed. For example, with an HTML document a '#' identifies the fragment name internal to an HTML document which is identified by the A element with the NAMEattribute.

An example URL is:

```
http://www.toytown.anycor.co/fred/index.html
```

where http is the file protocol (Hypertext Translation Protocol), www.toytown.anycor.co is the server name, /fred is the path of the file and the file is named index.html.

22.6.1 Electronic mail address

The `mailto` scheme defines a link to an Internet email address. An example is:

```
mailto: fred.bloggs@toytown.ac.uk
```

When this URL is selected then an email message will be sent to the email address `fred.bloggs@toytown.ac.uk`. Normally, some form of text editor is called and the user can enter the required email message. Upon successful completion of the text message; it is sent to the addressee.

22.6.2 File Transfer Protocol (FTP)

The `ftp` URL scheme defines that the files and directories specified are accessed using the FTP protocol. In its simplest form it is defined as:

`ftp:`//*<hostname>*/*<directory-name>*/*<filename>*

The FTP protocol normally requests a user to log into the system. For example, many public domain FTP servers use the login of:

```
anonymous
```

and the password can be anything (but it is normally either the user's full name or their Internet email address). Another typical operation is changing directory from a starting directory or the destination file directory. To accommodate this, a more general form is:

`ftp:`//*<user>*:*<password>*@*<hostname>*:*<port>*/*<cd1>*/*<cd2>*/
.../*<cdn>*/*<filename>*

where the user is defined by *<user>* and the password by *<password>* . The host name, *<hostname>*, is defined after the @ symbol and change directory commands are defined by the *cd* commands. The node name may take the form //user[:password]@host. Without a user name, the user `anonymous` is used.

For example the reference to the standard related to HTML Version 2 can be downloaded using the URL:

```
ftp://ds.internic.net/rfc/rfc1866.txt
```

and draft Internet documents from:

```
ftp://ftp.isi.edu/internet-drafts/
```

22.6.3 Host-specific file names

The file URL defines that the file is local to the computer and is not accessed through a server application. An example, taken from a PC, for a file C:\WWW\1.HTM is accessed with the URL:

```
file:///C|/WWW/1.HTM
```

22.6.4 Hypertext Transfer Protocol (HTTP)

HTTP is the protocol which is used to retrieve information connected with hypermedia links. The client and server initially perform a negotiation procedure before the HTTP transfer takes place. This negotiation involves the client sending a list of formats it can support and the server replying with data in the required format. This will be discussed in more detail in the next chapter.

Users generally move from a link on one server to another server. Each time the user moves from one server to another, the client sends an HTTP request to the server. Thus the client does not permanently connect to the server, and the server views each transfer as independent from all previous accesses. This is known as a stateless protocol.

22.6.5 Reference to interactive sessions (TELNET)

The `telnet` URL allows users to interactively perform a telnet operation, where a user must login to the referred system.

22.6.6 UseNet news

UseNet or NewsGroup servers are part of the increasing use of general discussion news groups which share text-based news items. The news URL scheme defines a link to either a news group or individual articles with a group of UseNet news.

22.6.7 Gopher Protocol

Gopher is widely used over the Internet and is basically a distribution system for the retrieval and delivery of documents. Users retrieve documents through a series of hierarchical menus, or through keyword searches. Unlike HTML documents they are not based on the hypertext concept.

22.6.8 Wide area information servers (WAIS)

WAIS is a public domain, fully text-based, information retrieval system over the Internet which performs text-based searches. The communications protocol used is based on the ANSI standard Z39.50, which is designed for networking library catalogs.

WAIS services include index generation and search engines. An indexer generates multiple indexes for organizations or individuals who offer services over the Internet. A WAIS search engine searches for particular words or text string indexes located across multiple Internet attached information servers of various types.

22.7 Universal resource identifier

The universal resource identifier (URI) is defined as a generically designated string of characters which refers to objects on the WWW. A URL is an example of a URI, with a designated access protocol and a specific Internet address.

Specifications have still to be completed, but URIs will basically be used to define the syntax for encoding arbitrary naming or addressing schemes. This should decouple the name of a resource from its location and also from its access method. For example, the file:

```
MYPIC.HTM
```

would be automatically associated with an HTTP protocol.

23 Intranets

23.1 Introduction

An organization may experience two disadvantages in having a connection to the WWW and the Internet:

- The possible usage of the Internet for non-useful applications (by employees).
- The possible connection of non-friendly users from the global connection into the organization's local network.

For these reasons many organizations have shied away from connection to the global network and have set up intranets. These are in-house, tailor-made internets for use within the organization and provide limited access (if any) to outside services and also limit the external traffic into the intranet (if any). An intranet might have access to the Internet but there will be no access from the Internet to the organization's Intranet.

Organizations which have a requirement for sharing and distributing electronic information normally have three choices:

- Use a propriety groupware package, such as Lotus Notes.
- Set up an intranet.
- Set up a connection to the Internet.

Groupware packages normally replicate data locally on a computer whereas intranets centralize their information on central servers which are then accessed by a single browser package. The stored data is normally open and can be viewed by any compatible WWW browser. Intranet browsers have the great advantage over groupware packages in that they are available for a variety of clients, such as PCs, UNIX workstations, Macs, and so on. A client browser also provides a single GUI interface which offers easy integration with other applications, such as electronic mail, images, audio, video, animation, and so on.

The main elements of an intranet are:

- Intranet server hardware.
- Intranet server software.
- TCP/IP stack software on the clients and server.
- WWW browsers.
- A firewall.

Typically the intranet server consists of a PC running the Lynx (PC-based UNIX-like) operating system. The TCP/IP stack is software installed on each computer and allows communications between a client and a server using TCP/IP.

A firewall is the routing computer which isolates the intranet from the outside world. This will be discussed in the next section.

23.2 Firewalls

A firewall (or security gateway) protects a network against intrusion from outside sources. They tend to differ in their approach but can be characterized as follows:

- Firewalls which block traffic.
- Firewalls which permit traffic.

23.2.1 Packet filters

The packet filter is the simplest form of firewall. It basically keeps a record of allowable source and destination IP addresses, and deletes all packets which do not have them. This technique is known as address filtering. The packet filter keeps a separate source and destination table for both directions, i.e. into and out of the intranet. This type of method is useful for companies which have geographically spread sites, as the packet filter can allow incoming traffic from other friendly sites, but block other non-friendly traffic. This is illustrated by Figure 23.1.

Unfortunately this method suffers from the fact that IP addresses can be easily forged. For example, a hacker might determine the list of good source addresses and then add one of them to any packets which are addressed to the intranet.

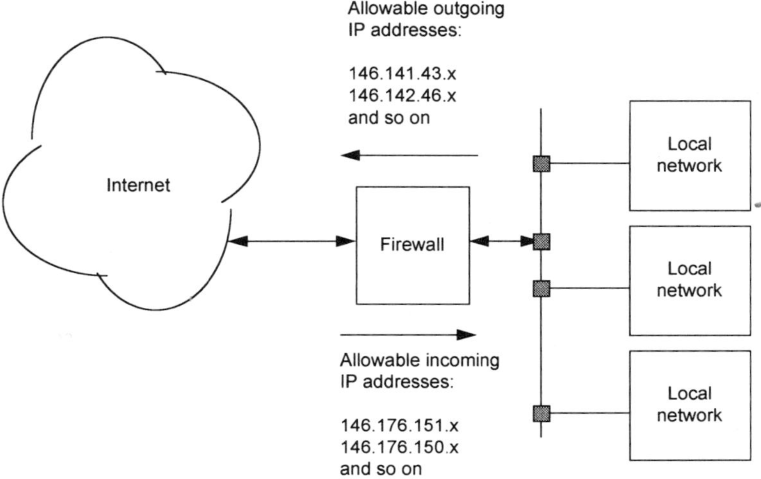

Figure 23.1 Packet filter firewalls.

23.2.2 Application level gateway

Application level gateways provide an extra layer of security when connecting an intranet to the Internet. They have three main components:

- A gateway node.
- Two firewalls which connect on either side of the gateway and only transmit packets which are destined for or to the gateway.

Figure 23.2 shows the operation of an application level gateway. In this case, Firewall A discards anything that is not addressed to the gateway node, and discards anything that is not sent by the gateway node. Firewall B, similarly discards anything from the local network that is not addressed to the gateway node, and discards anything that is not sent by the gateway node. Thus, to transfer files from the local network into the global network, the user must do the following:

- Log onto the gateway node.
- Transfer the file onto the gateway.
- Transfer the file from the gateway onto the global network.

To copy a file from the network an external user must:

- Log onto the gateway node.
- Transfer from the global network onto the gateway.
- Transfer the file from the gateway onto the local network.

A common strategy in organizations is to allow only electronic mail to pass from the Internet to the local network. This specifically disallows file transfer and remote login. Unfortunately electronic mail can be used to transfer files. To overcome this problem the firewall can be designed specifically to disallow very large electronic mail messages, so it will limit the ability to transfer files. This tends not to be a good method as large files can be split up into small parts then sent individually.

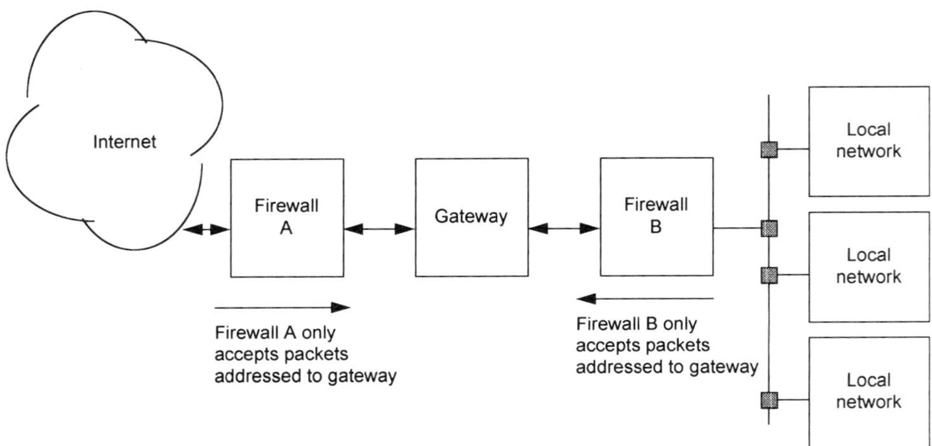

Figure 23.2 Application level gateway.

23.2.3 Encrypted tunnels

Packet filters and application level gateways suffer from insecurity, which can allow non-friendly users into the local network. Packet filters can be tricked with fake IP addresses and application level gateways can be hacked into by determining the password of certain users of the gateway then transferring the files from the network to the firewall, on to the gateway, on to the next firewall and out. The best form of protection for this type of attack is to allow only a limited number of people to transfer files onto the gateway.

The best method of protection is to encrypt the data leaving the network then to decrypt it on the remote site. Only friendly sites will have the required encryption key to receive and send data. This has the extra advantage that the information cannot be easily tapped-into.

Only the routers which connect to the Internet require to encrypt and decrypt, as illustrated in Figure 23.3.

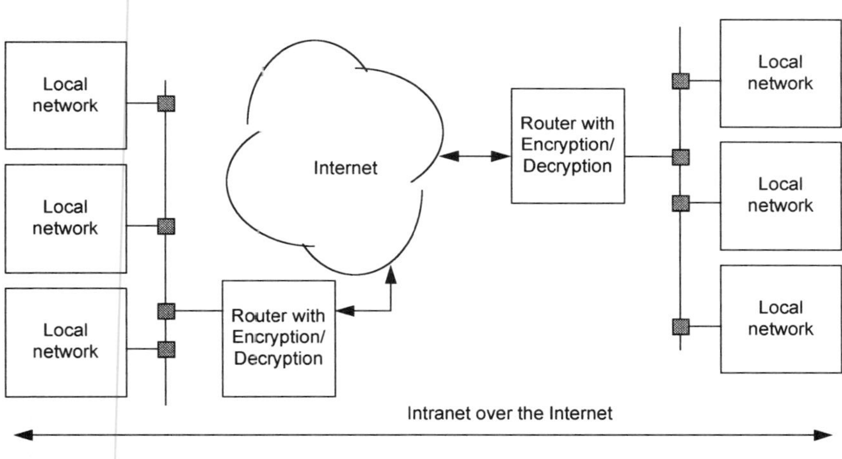

Figure 23.3 Encryption tunnels.

23.3 Extranets

Extranets (external Intranets) allow two or more companies to share parts of their Intranets related to joint projects. For example two companies may be working on a common project, an Extranet would allow them to share files related with the project.

23.4 Network security

Hackers into network come in many forms. The least determined hacker will simply be put off by simple security measures, whereas a determined hacker might attack the network in a number of ways. The main methods of security are:

- Hub security. Hub is configured so that it only recognizes authorized MAC addresses.
- Switch security. Switch is configured to filter given types, such as MAC address or protocol.
- Router security. Routers give certain filtering types, such as IP addresses, subnets, and so on.

Hub Security

Hubs with a management module can be configured with intruder prevention and eavesdrop prevention. Intruder prevention can be set in a number of ways. At the highest level of intruder prevention, a single device (identified by its hard-wired station address) is authorized to use a particular port on the hub. If another device attempts to transmit through that port, the hub disables the port and notifies the network administrator. This if an intruder connects on to the cables of a hub it will start transmitting with a different MAC address to the authorized node. The hub will detect this and disconnect the port from the network. This means that an intruder could listen to the traffic coming to the authorized node, but couldn't actively explore the network. Figure 23.4 shows an example of an intruder connected to connection from node A to the hub.

In eavesdrop prevention the hub examines the destination address in each packet. It sends the packet in its original form only to the port attached to the destination node and sends a meaningless string of 1s and 0s to the other ports. Thus, in Figure 23.4, nodes B to E can communicate with each other without the frames getting sent to node A.

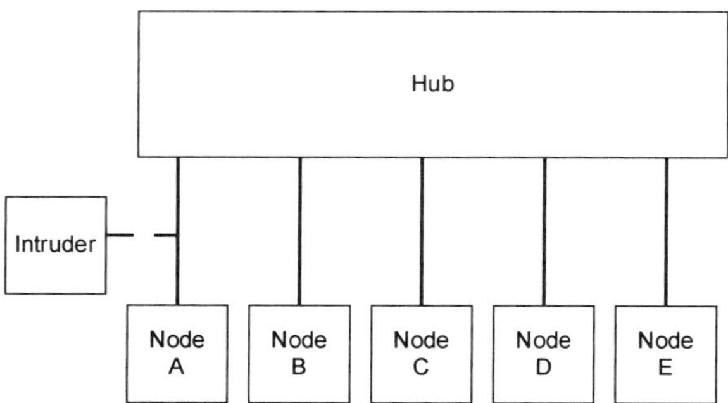

Figure 23.4 Hub with a connected intruder.

Switch Security

Switch security operates by filtering. This is achieved by the switch examining incoming packets for certain characteristics and discarding any packets that fail to meet pre-established filtering criteria. Many switches use address filtering and make use of the address table that the switch maintains. Filters can be specified of the following types:

- Multicast (the default).
- Protocol.
- Source port.
- Source MAC.

For MAC address filtering the network administrator makes a permanent entry in the bridge's address table that will cause the switch to discard any packets from the specified address.

23.4.2 Router Security

Router security is similar to switch security where it operates by filtering packets. Routers are more complex devices than switches and thus offer an increased number of filters (these depend on the routing protocol being used).

Routing protocols extend filtering beyond that provided by bridges by allowing filtering based on the origin's subnet address. This thus provides for firewall protection and allows to keep traffic local within certain parts of a network. They may also support authentication methods including SecurID, CHAP and PAP.

24 HTTP

24.1 Introduction

Chapter 22 discussed the WWW. The foundation protocol of the WWW is the Hypertext Transfer Protocol (HTTP) which can be used in any client/server application involving hypertext. It is used in the WWW for transmitting information using hypertext jumps and can support the transfer of plaintext, hypertext, audio, images, or any Internet-compatible information. The most recently defined standard is HTTP 1.1, which has been defined by the IETF standard.

24.2 HTTP operation

HTTP is a stateless protocol where each transaction is independent of any previous transactions. The advantage of being stateless is that it allows the rapid access of WWW pages over several widely distributed servers. It uses the TCP protocol to establish a connection between a client and a server for each transaction then terminates the connection once the transaction completes.

HTTP also support many different formats of data. Initially a client issues a request to a server which may include a prioritized list of formats that it can handle. This allows new formats to be added easily and also prevents the transmission of unnecessary information.

A client's WWW browser (the user agent) initially establishes a direct connection with destination server which contains the required WWW page. To make this connection the client initiates a TCP connection between the client and the server. After this is established the client then issues an HTTP request, such as the specific command (the method), the URL, and possibly extra information such as request parameters or client information. When the server receives the request, it attempts to perform the requested action. It then returns an HTTP response, which includes status information, a success/error code, and extra information itself. After the client receives this, the TCP connection is closed.

24.3 Intermediate systems

The previous section discussed the direct connection of a client to a server. Many system organizations do not wish a direct connection to an internal network. Thus HTTP supports other connections which are formed through intermediate systems, such as:

- A proxy.
- A gateway.
- A tunnel.

Each intermediate system is connected by a TCP and acts as a relay for the request to be sent out and returned to the client. Figure 24.1 shows the setup of the proxies and gateways.

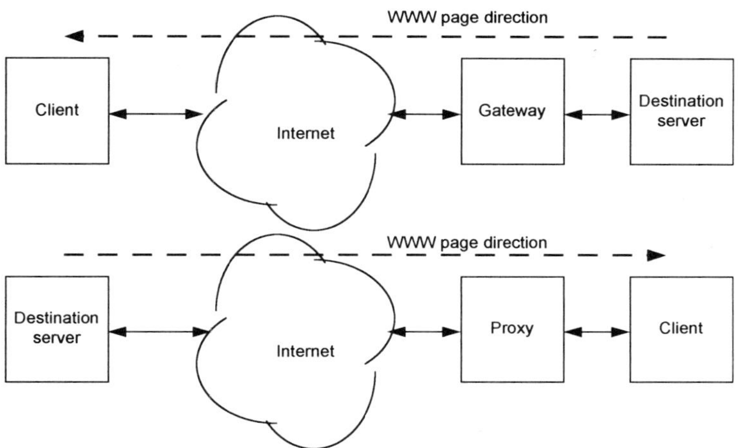

Figure 24.1 Usage of proxies and gateways.

24.3.1 Proxy

A proxy connects to a number of clients; it acts on behalf of other clients and sends requests from the clients to a server. It thus acts as a client when it communicates with a server, but as a server when communicating with a client. A proxy is typically used for security purposes where the client and server are separated by a firewall. The proxy connects to the client side of the firewall and the server to the other side of the firewall. Thus the server must authenticate itself to the firewall before a connection can be made with the proxy. Only after this has been authenticated will the proxy pass requests through the firewall. A proxy can also be used to convert between different version of HTTP.

24.3.2 Gateway

Gateways are servers that act as if they are the destination server. They are typically used when clients cannot get direct access to the server, Gateways are typically used in security applications where the gateway acts as a firewall so that the gateway communicates with the Internet and server only communicates with the Internet through the gateway. The client must then authenticate itself to the proxy, which can then pass the request on to the server.

They can also be used when the destination is a non-HTTP server. Web browsers have built into them the capability to contact servers for protocols other than HTTP, such as FTP and Gopher servers. This capability can also be provided by a gateway. The client makes an HTTP request to a gateway server. The gateway server than contacts the relevant FTP or Gopher server to obtain the desired result. This result is then converted into a form suitable for HTTP and transmitted back to the client.

24.3.3 Tunnel

A tunnel does not perform any operation on the HTTP message; it passes messages onto the client or server unchanged. This differs from a proxy or a gateway, which modify the HTTP messages. Tunnels are typically used as firewalls, where the firewall authenticates the connection but simply relays the HTTP messages.

24.4 Cache

In a computer system a cache is an area of memory that stores information likely to be accessed in a fast access memory area. For example a cache controller takes a guess on which information the process is likely to access next. When the processor wishes to access the disk then, if it has guessed right it will load, the cache controller will load from the electronic memory rather than loading it from the disk. A WWW cache stores cacheable responses so that there is a reduction in network traffic and an improvement in access times.

24.5 HTML messages

HTTP message are either requests from clients to servers or responses from servers to clients. The message is either a simple-request, a simple-response, full-request or a full-response. HTTP Version 0.9 defines the simple request/ response messages whereas HTTP Version 1.1 defines full requests/responses.

24.5.1 Simple requests/responses

The simple request is a `GET` command with the requested URI such as:

```
GET   /info/dept/courses.html
```

The simple response is a block containing the information identified in the URI (called the entity-body).

24.5.2 Full requests/responses

Very few security measures or enhanced services are built into the simple requests/responses. HTTP Version 1/1.1 improves on the simple requests/ responses by adding many extra requests and responses, as well as adding extra information about the data supported. Each message header consists of a number of fields which begin on a new line and consist of the field name followed by a colon and the field value. This follows the format of RFC 822 (as shown in Section 21.5.4) and allows for MIME encoding. It is thus similar to MIME-encoded email. A full request starts with a request line command (such as `GET`, `MOVE` or `DELETE`) and is then followed by one or more of the following:

- General-headers which contain general fields that do not apply to the entity being transferred (such as MIME version, date, and so on).
- Request-headers which contain information on the request and the client (e.g. the client's name, its authorization, and so on).

- Entity-headers which contain information about the resource identified by the request and entity-body information (such as the type of encoding, the language, the title, the time when it was last modified, the type of resource it is, when it expires, and so on).
- Entity-body which contains the body of the message (such as HTML text, an image, a sound file, and so on).

A full response starts with a response status code (such as `OK`, `Moved Temporarily`, `Accepted`, `Created`, `Bad Request`, and so on) and is then followed by one or more of the following:

- General-headers, as with requests, contain general fields which do not apply to the entity being transferred (MIME version, date, and so on).
- Response-headers which contain information on the response and the server (e.g. the server's name, its location and the time the client should retry the server).
- Entity-headers, as with request, which contain information about the resource identified by the request and entity-body information (such as the type of encoding, the language, the title, the time when it was last modified, the type of resource it is, when it expires, and so on).
- Entity-body, as with requests, which contains the body of the message (such as HTML text, an image, a sound file, and so on).

The following example shows an example request. The first line is always the request method, in this case it is `GET`. Next there are various headers. The general-header field is `Content-Type`, the request-header fields are `If-Modified-Since` and `From`. There are no entity parts to the message as the request is to get an image (if the command had been to `PUT` then there would have been an attachment with the request). Notice that the end of the message is delimited by a single blank line as this indicates the end of a request/response. Note that the headers are case sensitive, thus `Content-Type` with the correct types of letters (and `GET` is always in uppercase letters).

Example HTTP request

```
GET mypic.jpg
Content-Type: Image/jpeg
If-Modified-Since: 06 Mar 1997 12:35:00
From: Fred Bloggs <FREDB@ACOMP.CO.UK>
```

Request messages

The most basic request message is to `GET` a URI. HTTP/1.1 adds many more requests including:

```
COPY      DELETE    GET       HEAD      POST
LINK      MOVE      OPTIONS   PATCH     PUT
TRACE     UNLINK    WRAPPED
```

As before, the `GET` method requests a WWW page. The `HEAD` method tells the server that the client wants to read only the header of the WWW page. If the `If-Modified-Since` field is included then the server checks the specified date with the date of the URI and verifies whether it has not changed since then.

A PUT method requests storage of a WWW page and POST appends to a named resource (such as electronic mail). LINK connects two existing resources and UNLINK breaks the link. A DELETE method removes a WWW page.

The request-header fields are mainly used to define the acceptable type of entity that can be received by the client; they include:

Accept	Accept-Charset	Accept-Encoding
Accept-Language	Authorization	From
Host	If-Modified-Since	If-Modified-Since
Proxy-Authorization	Range	Referer
Unless	User-Agent	

The Accept field is used to list all the media types and ranges that client can accept. An Accept-Charset field defines a list of character sets acceptable to the server and Accept-Encoding is a list of acceptable content encodings (such as the compression or encryption technique). The Accept-Language field defines a set of preferred natural languages.

The Authorization field has a value which authenticates the client to the server. A From field defines the email address of the user who is using the client (e.g. From: fred.blogg@anytown.uk) and the Host field specifies the name of the host of the resource being requested.

A useful field is the If-Modified-Since field, used with the GET method. It defines a date and time parameter and specifies that the resource should not be sent if it has not been modified since the specified time. This is useful when a client has a local copy of the resource in a local cache and, rather than transmitting the unchanged resource, it can use its own local copy.

The Proxy-Authorization field is used by the client to identify itself to a proxy when the proxy requires authorization. A Range field is used with the GET message to get only a part of the resource.

The Referer field defines the URI of the resource from which the Request-URI was obtained and enables the server to generate list of back-links. An Unless field is used to make a comparison based on any entity-header field value rather than a date/time value (as with GET and If-Modified-Since).

The User-Agent field contains information about the user agent originating this request.

Response messages

In HTTP/0.9 the response from the server was either the entity or no response. HTTP/1.1 includes many other response, these include:

Accepted	Bad Gateway
Bad Request	Conflict
Continue	Created
Forbidden	Gateway Timeout
Gone	Internal Server Error
Length Required	Method Not Allowed
Moved Permanently	Moved Temporarily
Multiple Choices	No Content
Non-Authoritative Info	None Acceptable
Not Found	Not Implemented
Not Modified	OK

```
Partial Content                     Payment Required
Proxy Authorization Required        Request Timeout
Reset Content                       See Other
Service Unavailable                 Switching Protocols
Unauthorized                        Unless True
Use Proxy
```

These responses can be put into five main groupings:

- Client error – Bad Request, Conflict, Forbidden, Gone, Payment required, Not Found, Method Not Allowed, None Acceptable, Proxy Authentication Required, Request Timeout, Length Required, Unauthorized, Unless True.
- Informational – Continue, Switching Protocol.
- Redirection – Moved Permanently, Moved Temporarily, Multiple Choices, See Other, Not Modified, User Proxy.
- Server error – Bad Gateway, Internal Server Error, Not Implemented, Service Unavailable, Gateway Timeout.
- Successful – Accepted, Created, OK, Non-Authoritative Info. The OK field is used when the request succeeds and includes the appropriate response information.

The response header fields are:

```
Location          Proxy-Authenticate          Public
Retry-After       Server                      WWW-Authenticate
```

The Location field defines the location of the resource identified by the Request-URI. A Proxy-Authenticate field contains the status code of the Proxy Authorization Required response.

The Public field defines non-standard methods supported by this server. A Retry-After field contains values which define the amount of time a service will be unavailable (and is thus sent with the Service Unavailable response).

The WWW-Authenticate field contains the status code for the Unauthorized response.

General-header fields

General-header fields are used either within requests or within responses; they include:

```
Cache-Control   Connection     Date                  Forwarded
Keep-Alive      MIME-Version   Pragma        Upgrade
```

The Cache-Control field gives information on the caching mechanism and stops the cache controller from modifying the request/response. A Connection field specifies the header field names that apply to the current TCP connection.

The Date field specifies the date and time at which the message originated; this is obviously useful when examining the received message as it gives an indication of the amount of time the message took to arrive at is destination. Gateways and proxies use the Forwarded field to indicate intermediate steps between the client and the server. When a gateway or proxy reads the message, it can attach a Forwarded field with its own URI (this can help in tracing the route of a message).

The Keep-Alive field specifies that the requester wants a persistent connection. It may indicate the maximum amount of time that the sender will wait for the next request before closing the connection. It can also be used to specify the maximum number of additional requests on the current persistent connection.

The `MIME-Version` field indicates the MIME version (such as `MIME-Version: 1.0`). A `Pragma` field contains extra information for specific applications.

In a request the `Upgrade` field specifies the additional protocols that the client supports and wishes to use, whereas in a response it indicates the protocol to be used.

Entity-header fields

Depending on the type of request or response, an entity-header can be included:

```
Allow              Content-Encoding             Content-Language
Content-Length     Content-MD5                  Content-Range
Content-Type       Content-Version              Derived-From
Expires            Last-Modified                Link
Title              Transfer-encoding
URI-Header extension-header
```

The `Allow` field defines the supported methods supported by the resource identified in the `Request-URI`. A `Content-Encoding` field indicates content encodings, such as ZIP compression, that have been applied to the resource (`Content-Encoding: zip`).

The `Content-Language` field identifies natural language(s) of the intended audience for the enclosed entity (e.g. `Content-language: German`) and the `Content-Length` field defines the number of bytes in the entity.

The `Content-Range` field designates a portion of the identified resource that is included in this response, while `Content-Type` indicates the media type of the entity body (such as `Content-Type=text/html`, `Content-Type=text/plain`, `Content-Type=image/gif` or `Content-type=image/jpeg`). The version of the entity is defined in the `Content-Version` field.

The `Expires` field defines the date and time when the entity is considered stale. The `Last-Modified` field is the date and time when the resource was last modified.

The `Link` field defines other links and the `Title` field defines the title for the entity. A `Transfer-Encoding` field indicates the transformation type that is applied so the entity can be transmitted.

25 HTML (Introduction)

25.1 Introduction

HTML is a standard hypertext language for the WWW and has several different versions. Most WWW browsers support HTML 2 and most of the new versions of browser's support HTML 3. WWW pages are created and edited with a text editor, a word processor or, as is becoming more common, within the WWW browser.

HTML tags contain special formatting commands and are contained within a less than (<) and a greater than (>) symbol (which are also known as angled brackets). Most tags have an opening and closing version; for example, to highlight bold text the bold opening tag is `<B>` and the closing tag is `</B>`. Table 25.1 outlines a few examples.

HTML script 1 gives an example script and Figure 25.1 shows the output from the WWW browser. The first line is always `<HTML>` and the last line is `</HTML>`. After this line the HTML header is defined between `<HEAD>` and `</HEAD>`. The title of the window in this case is `My first HTML page`. The main HTML text is then defined between `<BODY>` and `</BODY>`.

Table 25.1 Example HTML tags.

Open tag	*Closing tag*	*Description*
<HTML>	</HTML>	Start and end of HTML
<HEAD>	</HEAD>	Defines the HTML header
<BODY>	</BODY>	Defines the main body of the HTML
<TITLE>	</TITLE>	Defines the title of the WWW page
<I>	</I>	Italic text
<B>	</B>	Bold text
<U>	</U>	Underlined text
<BLINK>	</BLINK>	Make text blink
<EM>	</EM>	Emphasize text
<FONT SIZE=+1>	</FONT>	Increase font size by one increment
<FONT SIZE=-1>	</FONT>	Reduce font size by one increment
<CENTER>	</CENTER>	Center text
<H1>	</H1>	Section header, level 1
<H2>	</H2>	Section header, level 2
<H3>	</H3>	Section header, level 3
<P>		Create a new paragraph
 		Create a line break
<!-->	-->	Comments
<SUPER>	</SUPER>	Superscript
_		Subscript

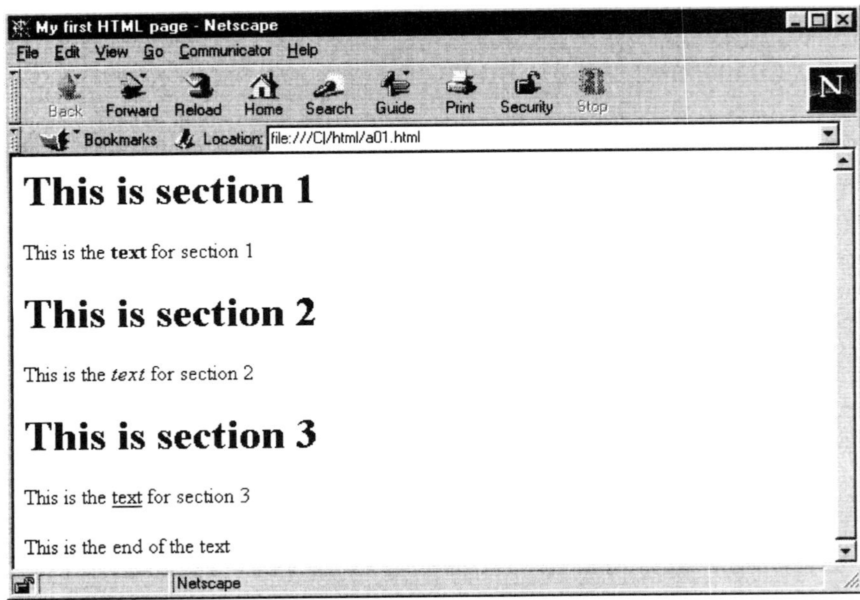

Figure 25.1 Example window from example HTML script.

HTML script 25.1

```
<HTML>
<HEAD>
<TITLE>My first HTML page</TITLE>
</HEAD>
<BODY>
<H1> This is section 1</H1>
This is the <b>text</b> for section 1
<H1> This is section 2</H1>
This is the <i>text</i> for section 2
<H1> This is section 3</H1>
This is the <u>text</u> for section 3
<p>
This is the end of the text
</BODY>
</HTML>
```

The WWW browser fits text into the window size and does not interpret line breaks in the HTML source. To force a new line the `<BR>` (line break) or a new paragraph (`<P>`) is used. The example also shows bold, italic and underlined text.

25.2 Links

The topology of the WWW is set-up using links where pages link to other related pages. A reference takes the form:

```
<A HREF="url"> Reference Name </A>
```

where *url* defines the URL for the file, *Reference Name* is the name of the reference and `</A>` defines the end of the reference name. HTML script 25.2 shows an example of the uses of references and Figure 25.2 shows a sample browser page. The background color is set using the `<BODY BGCOLOR= "#FFFFFF">` which sets the background color to white. In this case the default text color is black and the link is colored blue.

📖 HTML script 25.2

```
<HTML>

<HEAD>
<TITLE>Fred's page</TITLE>
</HEAD>

<BODY BGCOLOR="#FFFFFF">

<H1>Fred's Home Page</H1>

If you want to access information on
this book <A HREF="softbook.html">click here</A>.

<P>A reference to the <A REF="http:www.iee.com/">IEE</A>
</BODY>
</HTML>
```

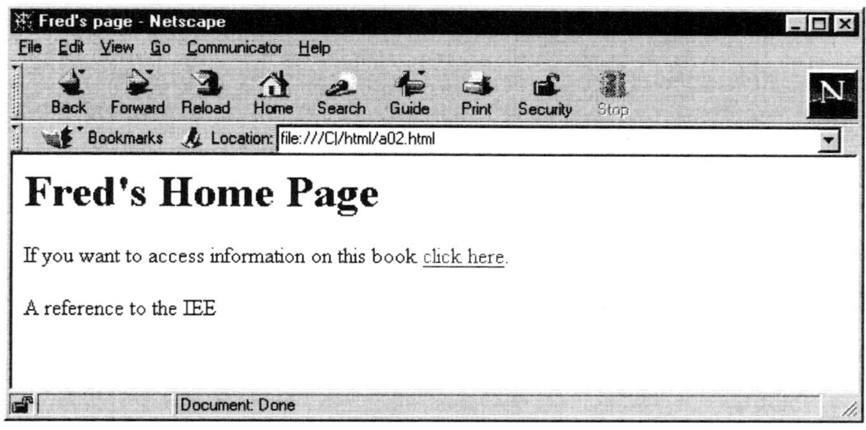

Figure 25.2 Example window from example HTML script 25.2.

25.2.1 Other links

Links can be set-up to send to e-mail addresses and newsgroups. For example:

```
<A HREF="news:sport.tennis"> Newsgroups for tennis</A>
```

to link to a tennis newsgroup and

```
<A HREF="mailto:f.bloggs@fredco.co.uk">Send a message to me</A>
```

to send a mail message to the e-mail address: `f.bloggs@ fredco.co.uk`.

25.3 Lists

HTML allows ordered and unordered lists. Lists can be declared anywhere in the body of the HTML.

25.3.1 Ordered lists

The start of an ordered list is defined with `<OL>` and the end of the list by `</OL>`. Each part of the list is defined after the `<LI>` tag. Unordered lists are defined between the `<UL>` and `</UL>` tags. HTML script 25.3 gives examples of an ordered and an unordered list. Figure 25.3 shows the output from the browser.

HTML script 25.3

```
<HTML><HEAD><TITLE>Fred's page</TITLE></HEAD>
<BODY BGCOLOR="#FFFFFF">
<H1>List 1</H1>
<OL>
<LI>Part 1
<LI>Part 2
<LI>Part 3
</OL>
<H1>List 2</H1>
<UL>
<LI>Section 1
<LI>Section 2
<LI>Section 3
</UL>
</BODY></HTML>
```

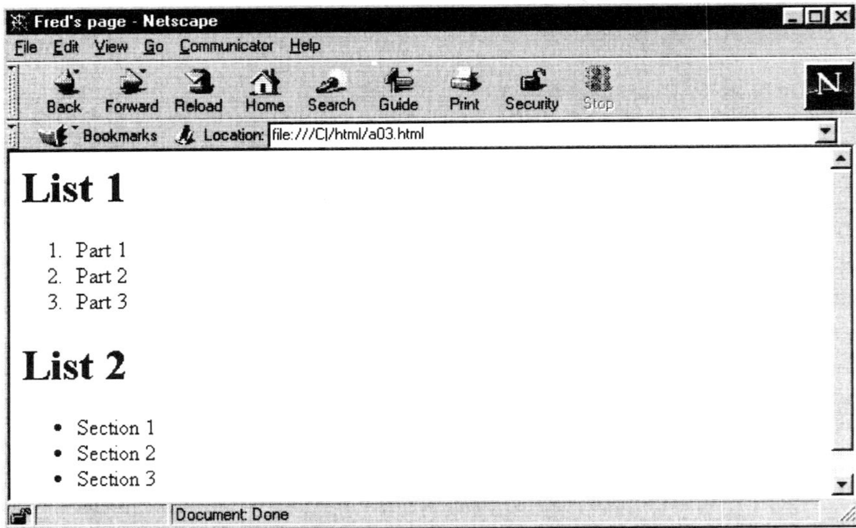

Figure 25.3 WWW browser with an ordered and unordered lists.

Some browsers allow the type of numbered list to be defined with the `<OL TYPE=x>`, where *x* can either be:

- A for capital letters (such as A, B, C, and so on).
- a for small letters (such as a, b, c, and so on).
- I for capital roman letters (such as I, II, III, and so on).
- i for small roman letters (such as i, ii, iii, and so on).
- I for numbers (which is the default).

```
<OL Type=I>
<LI> List 1
<LI> List 2
<LI> List 3
</OL>

<OL Type=A>
<LI> List 1
<LI> List 2
<LI> List 3
</OL>
```

would be displayed as:

I. List 1
II. List 2
III. List 3
A. List 1
B. List 2
C. List 3

The starting number of the list can be defined using the `<LI VALUE=n>` where *n* defines the initial value of the defined item list.

25.3.2 Unordered lists

Unordered lists are used to list a series of items in no particular order. They are defined between the `<UL>` and `</UL>` tags. Some browsers allow the type of bullet point to be defined with the `<LI TYPE=shape>`, where *shape* can either be:

- *disc* for round solid bullets (which is the default for first level lists).
- *round* for round hollow bullets (which is the default for second level lists).
- *square* for square bullets (which is the default for third).

HTML script 25.4 gives an example of an unnumbered list and Figure 25.4 shows the WWW page output for this script. It can be seen from this that the default bullets for level 1 lists are discs, for level 2 they are round and for level 3 they are square.

HTML script 25.4

```
<HTML><HEAD><TITLE>Example list</TITLE></HEAD>
<H1> Introduction </H1>
<UL>
<LI> OSI Model
<LI> Networks
      <UL>
      <LI> Ethernet
         <UL>
         <LI> MAC addresses
```

```
            </UL>
      <LI> Token Ring
      <LI> FDDI
      </UL>
<LI> Conclusion
</UL>
<H1> Wide Area Networks </H1>
<UL>
<LI> Standards
<LI> Examples
      <UL>
      <LI> EastMan
      </UL>
<LI> Conclusion
</UL>
</BODY>
</HTML>
```

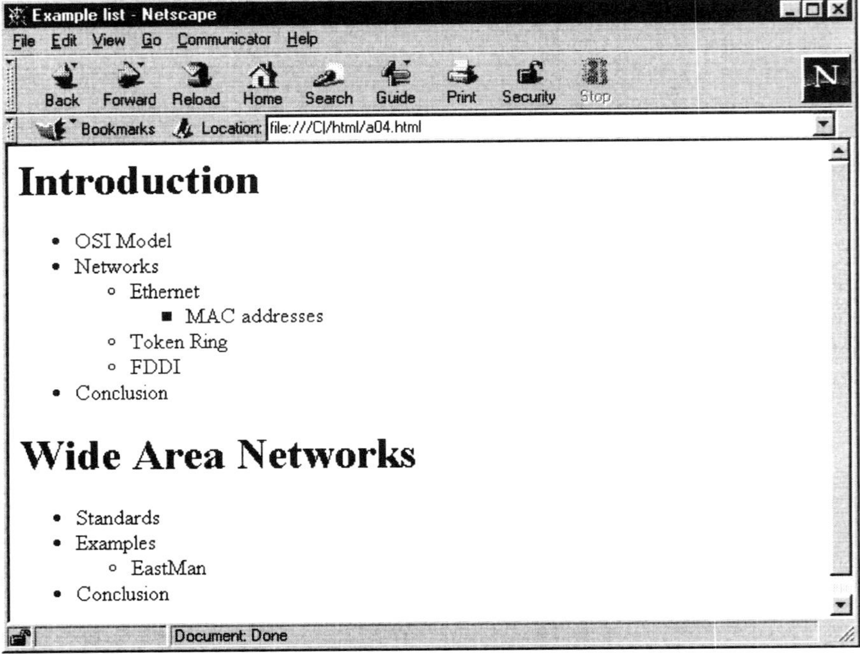

Figure 25.4 WWW page with an unnumbered list.

25.3.3 Definition lists

HTML uses the <DL> and </DL> tags for definition lists. These are normally used when building glossaries. Each entry in the definition is defined by the <DT> tag and the text associated with the item is defined after the <DD> tag. The end of the list is defined by </DL>. HTML script 25.5 shows an example with a definition list and Figure 25.5 gives a sample output. Note that it uses the <EM> tag to emphasize the definition subject.

🕮 HTML script 25.5

```
<HTML><HEAD><TITLE>Example list</TITLE></HEAD>
<H1> Glossary </H1>
<DL>
<DT> <EM> Address Resolution Protocol (ARP) </EM>
<DD> A TCP/IP process which maps an IP address to an Ethernet address.
<DT> <EM> American National Standards Institute (ANSI) </EM>
<DD> ANSI is a non-profit organization which is made up of expert committees
that publish standards for national industries.
<DT> <EM> American Standard Code for Information Interchange (ASCII) </EM>
<DD> An ANSI-defined character alphabet which has since been adopted as a
standard international alphabet for the interchange of characters.
</DL></BODY></HTML>
```

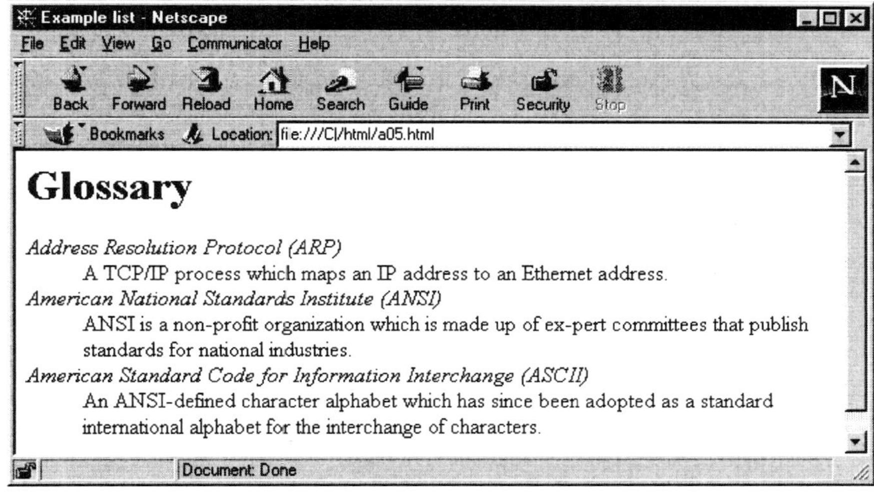

Figure 25.5 WWW page with definition list.

25.4 Colors

Colors in HTML are defined in the RGB (red/green/blue) strength. The format is `#rrggbb`, where `rr` is the hexadecimal equivalent for the red component, `gg` the hexadecimal equivalent for the green component and `bb` the hexadecimal equivalent for the blue component. Table 25.2 lists some of the codes for certain colors.

Individual hexadecimal numbers use base 16 and range from `0` to `F` (in decimal this ranges from 0 to 15). A two-digit hexadecimal number ranges from `00` to `FF` (in decimal this ranges from 0 to 255). Table 25.3 outlines hexadecimal equivalents.

HTML uses percentage strengths for the colors. For example, `FF` represents full strength (100%) and `00` represent no strength (0%). Thus, white is made from `FF` (red), `FF` (green) and `FF` (blue) and black is made from `00` (red), `00` (green) and `00` (blue). Grey is made from equal weighting of each of the colors, such as `43`, `43`, `43` for dark grey (`#434343`) and `D4`, `D4` and `D4` for light grey (`#D4D4D4`). Thus, pure red with be `#FF0000`, pure green will be `#00FF00` and pure blue with be `#0000FF`.

Table 25.2 Hexadecimal colors.

Color	*Code*	*Color*	*Code*
White	#FFFFFF	Dark red	#C91F16
Light red	#DC640D	Orange	#F1A60A
Yellow	#FCE503	Light green	#BED20F
Dark green	#088343	Light blue	#009DBE
Dark blue	#0D3981	Purple	#3A0B59
Pink	#F3D7E3	Nearly black	#434343
Dark gray	#777777	Grey	#A7A7A7
Light gray	#D4D4D4	Black	#000000

Table 25.3 Hexadecimal to decimal conversions.

Hex.	*Dec.*	*Hex.*	*Dec.*	*Hex.*	*Dec.*	*Hex.*	*Dec.*
0	0	1	1	2	2	3	3
4	4	5	5	6	6	7	7
8	8	9	9	A	10	B	11
C	12	D	13	E	14	F	15

Each color is represented by 8 bits, thus the color is defined by 24 bits. This gives a total of 16 777 216 colors (2^{24} different colors). Note that some video displays will not have enough memory to display 16.777 million colors in a certain mode so that colors may differ depending on the WWW browser and the graphics adapter.

The colors of the background, text and the link can be defined with the BODY tag. An example with a background color of white, a text color of orange and a link color of dark red is:

```
<BODY BGCOLOR="#FFFFFF" TEXT="#F1A60A"  LINK="#C91F16">
```

and for a background color of red, a text color of green and a link color of blue:

```
<BODY BGCOLOR="#FF0000" TEXT="#00FF00"  LINK="#0000FF">
```

When a link has been visited its color changes. This color itself can be changed with the VLINK. For example, to set-up a visited link color of yellow:

```
<BODY VLINK="#FCE503" "TEXT=#00FF00"  "LINK=#0000FF">
```

Note that the default link colors are:

Link:	#0000FF	(Blue)
Visited link:	#FF00FF	(Purple)

25.5 Background images

Images (such as GIF and JPEG) can be used as a background to a WWW page. For this purpose the option BACKGROUND='*src.gif*' is added to the <BODY> tag. An HTML script with a background of CLOUDS.GIF is given in HTML script 25.6. A sample output from a browser is shown in Figure 25.6.

HTML script 25.6

```
<HTML><HEAD><TITLE>Fred's page</TITLE></HEAD>
<BODY BACKGROUND="clouds.gif">
<H1>Fred's Home Page</H1>
If you want to access information on
this book <A HREF="gbook.html">click here</A>.<P>
A reference to the <A HREF="http://www.iee.com/">IEE</A>
</BODY></HTML>
```

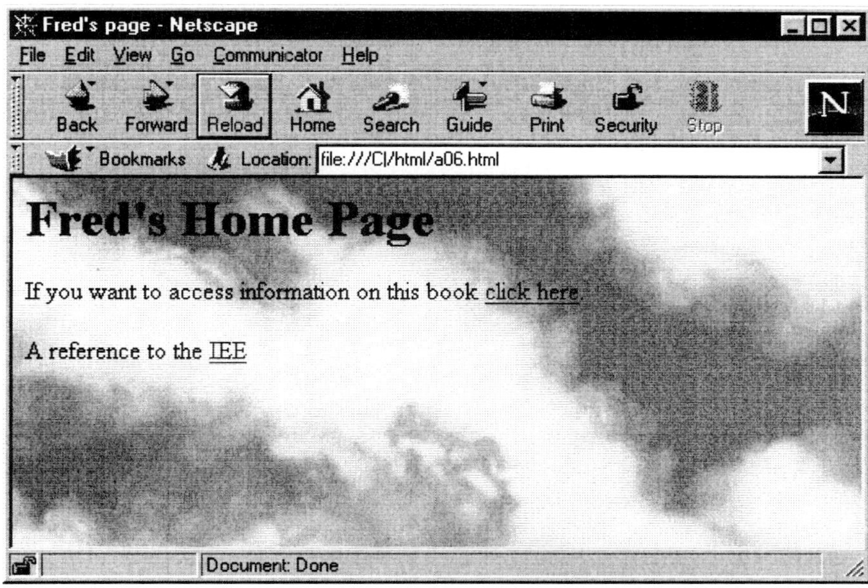

Figure 25.6 WWW page with CLOUDS.GIF as a background.

25.6 Displaying images

WWW pages can support graphics images within a page. The most common sources of images are either JPEG or GIF files, as these types of images normally have a high degree of compression. GIF images, as was previously mentioned, support only 256 colors from a pallet of 16.7 million colors, whereas JPEG supports more than 256 colors.

25.6.1 Inserting an image

Images can be displayed within a page with the `<IMG SRC=`*'src.gif'*`>` which inserts the graphic *src.gif.* HTML script 25.7 contains three images: `mypic1.jpg`, `me.gif` and `mypic2.jpg`. These are aligned either to the left or the right using the `ALIGN` option within the `<IMG SRC>` tag. The first image (`mypic1.jpg`) is aligned to the right, while the second image (`mypic2.jpg`) is aligned to the left. Figure 25.7 shows a sample output from this script. Note that images are left aligned by default.

HTML script 25.7

```
<HTML><HEAD>
<TITLE>My first home page</TITLE>
</HEAD>
<BODY BGCOLOR="#ffffff">
<IMG SRC ="mypic1.jpg" width=120 ALIGN=RIGHT>
<H1> Picture gallery </H1>
<P><P>
Here are a few pictures of me and my family. To the right
is a picture of my sons taken in the garden. Below to the
left is a picture of my two youngest sons under an
umbrella and to the right is a picture of me taken in my office.
<P>
<IMG SRC ="mypic2.jpg" ALIGN=LEFT width=200>
<IMG SRC ="me.gif" ALIGN=RIGHT width=300>
</BODY></HTML>
```

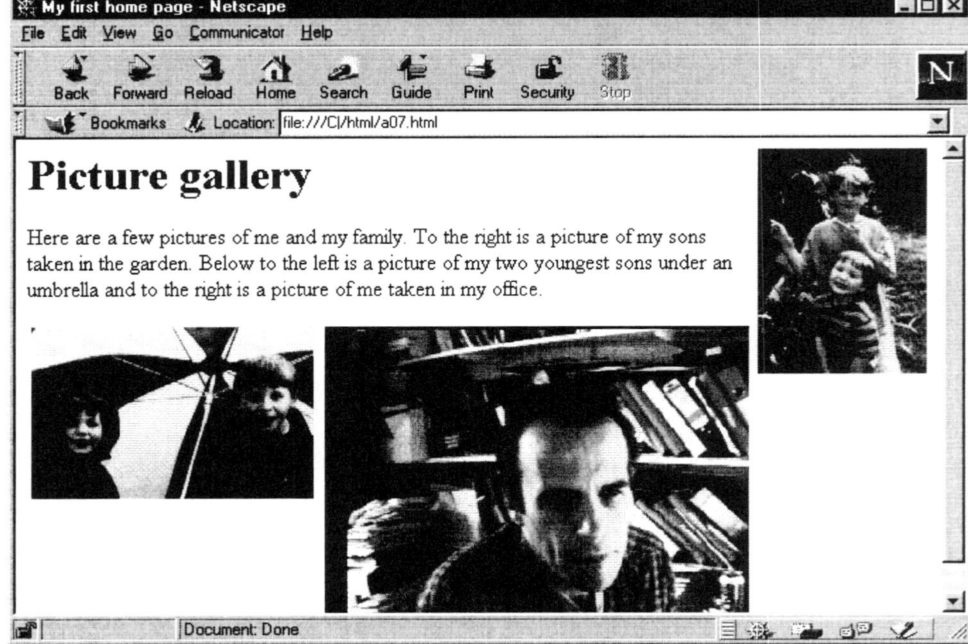

Figure 25.7 WWW page with three images.

25.6.2 Alternative text

Often users choose not to view images in a page and select an option on the viewer which stops the viewer from displaying any graphic images. If this is the case then the HTML page can contain substitute text which is shown instead of the image. For example:

```
<IMG SRC ="mypic1.jpg" ALT="In garden" ALIGN=RIGHT>
<IMG SRC ="mypic2.jpg" ALT="Under umbrella" ALIGN=LEFT>
<IMG SRC ="me.gif" ALT="Picture of me" ALIGN=RIGHT>
```

25.6.3 Other options

Other image options can be added, such as:

- `HSPACE=`*x* `VSPACE=`*y* defines the amount of space that should be left around images. The *x* value defines the number of pixels in the x-direction and the *y* value defines the number of pixels in the y-direction.
- `WIDTH=` *x* `HEIGHT=`*y* defines the scaling in the x- and y-direction, where *x* and *y* are the desired pixel width and height, respectively, of the image.
- `ALIGN=`*direction* defines the alignment of the image. This can be used to align an image with text. Valid options for aligning with text are *texttop*, *top*, *middle*, *absmiddle*, *bottom*, *baseline* or *absbottom*. HTML script 25.8 shows an example of image alignment with the image `a.gif` (which is just the letter 'A' as a graphic) and Figure 25.8 shows a sample output. It can be seen that *texttop* aligns the image with highest part of the text on the line, *top* aligns the image with the highest element in the line, *middle* aligns with the middle of the image with the baseline, *absmiddle* aligns the middle of the image with the middle of the largest item, *bottom* aligns the bottom of the image with the bottom of the text and *absbottom* aligns the bottom of the image with the bottom of the largest item.

HTML script 25.8

```
<HTML><HEAD><TITLE>My first home page</TITLE></HEAD>
<BODY BGCOLOR="#ffffff">
<IMG SRC ="a.gif" ALIGN=texttop>pple<P>
<IMG SRC ="a.gif" ALIGN=top>pple<P>
<IMG SRC ="a.gif" ALIGN=middle>pple<P>
<IMG SRC ="a.gif" ALIGN=bottom>pple<P>
<IMG SRC ="a.gif" ALIGN=baseline>pple<P>
<IMG SRC ="a.gif" ALIGN=absbottom>pple
</BODY></HTML>
```

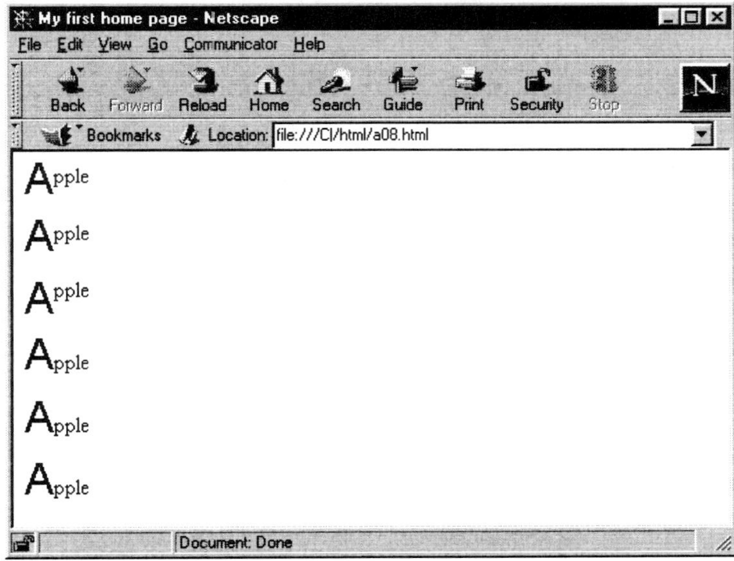

Figure 25.8 WWW page showing image alignment.

25.7 Horizontal lines

A horizontal line can be added with the `<HR>` tag. Most browsers allow extra parameters, such as:

`SIZE=` *n* – which defines that the height of the rule is *n* pixels.
`WIDTH=`*w* – which defines that the width of the rule is *w* pixels or as a percentage.
`ALIGN=`*direction* – where direction refers to the alignment of the rule. Valid options for *direction* are *left*, *right* or *center*.
`NOSHADE` – which defines that the line should be solid with no shading.

HTML script 25.9 gives some example horizontal lines and Figure 25.9 shows an example output.

HTML script 25.9

```
<HTML><HEAD><TITLE>My first home page</TITLE></HEAD>
<BODY BGCOLOR="#ffffff">
<IMG SRC ="a.gif">pple<P>
<HR>
<IMG SRC ="a.gif">pple<P>
<HR WIDTH=50% ALIGN=CENTER>
<IMG SRC ="a.gif">pple<P>
<HR SIZE=10 NOSHADE>
</BODY></HTML>
```

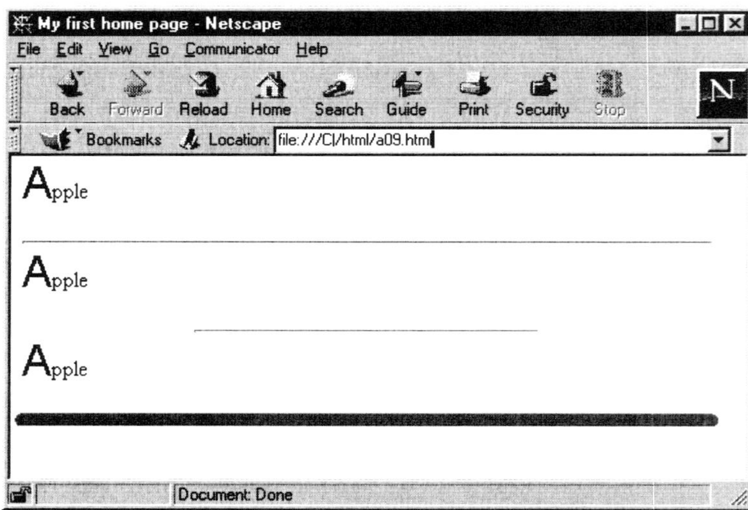

Figure 25.9 WWW page showing horizontal lines.

26 HTML (Tables, Forms and Helpers)

26.1 Introduction

Chapter 25 introduced HTML; this chapter discusses some of HTML's more advanced features. HTML differs from compiled languages, such as C and Pascal, in that the HTML text file is interpreted by an interpreter (the browser) while languages such as C and Pascal must be precompiled before they can be run. HTML thus has the advantage that it does not matter if it is the operating system, the browser type or the computer type that reads the HTML file, as the file does not contain any computer specific code. The main disadvantage of interpreted files is that the interpreter does less error checking as it must produce fast results.

The basic pages on the WWW are likely to evolve around HTML and while HTML can be produced manually with a text editor, it is likely that, in the coming years, there will be an increase in the amount of graphically-based tools that will automatically produce HTML files. Although these tools are graphics-based they still produce standard HTML text files. Thus a knowledge of HTML is important as it defines the basic specification for the presentation of WWW pages.

26.2 Anchors

An anchor allows users to jump from a reference in a WWW page to another anchor point within the page. The standard format is:

```
<A NAME="anchor name">
```

where *anchor name* is the name of the section which is referenced. The </A> tag defines the end of an anchor name. A link is specified by:

```
<A HREF="#anchor name">
```

followed by the `</A>` tag. HTML script 26.1 shows a sample script with four anchors and Figure 26.1 shows a sample output. When the user selects one of the references, the browser automatically jumps to that anchor. Figure 26.2 shows the output screen when the user selects the `#Token` reference. Anchors are typically used when an HTML page is long or when a backwards or forwards reference occurs (such as a reference within a published paper).

HTML script 26.1

```
<HTML><HEAD><TITLE>Sample page</TITLE></HEAD>
<BODY BGCOLOR="#FFFFFF">
<H2>Select which network technology you wish information:</H2>
<P><A HREF="#Ethernet">Ethernet</A>
<P><A HREF="#Token">Token Ring</A>
<P><A HREF="#FDDI">FDDI</A>
<P><A HREF="#ATM">ATM</A>
<H2><A NAME="Ethernet">Ethernet</A></H2>
Ethernet is a popular LAN which works at 10Mbps.
<H2><A NAME="Token">Token Ring</A></H2>
Token ring is a ring based network which operates at 4 or 16Mbps.
<H2><A NAME="FDDI">FDDI</A></H2>
FDDI is a popular LAN technology which uses a ring of fibre optic cable and
operates at 100Mbps.
<H2><A NAME="ATM">ATM</A></H2>
ATM is a ring based network which operates at 155Mbps.
</BODY></HTML>
```

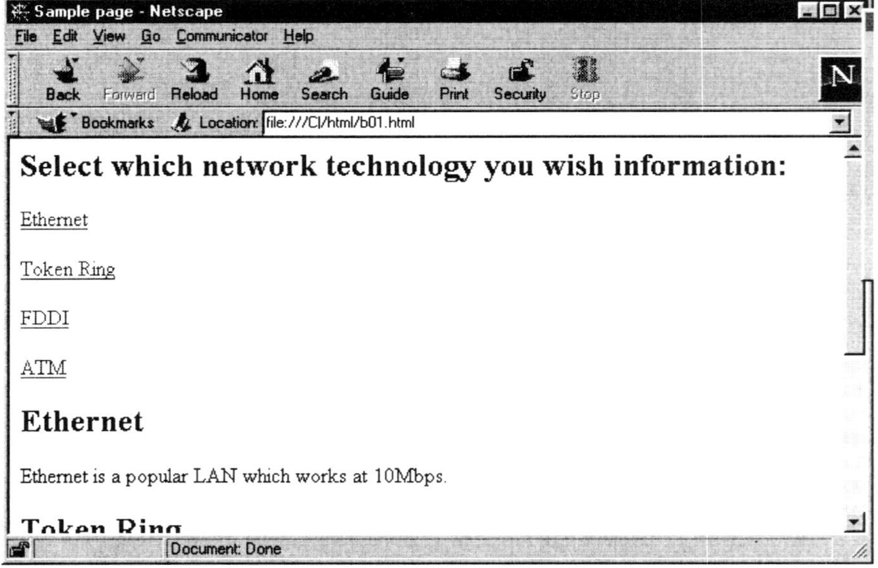

Figure 26.1 Example window with references.

26.3 Tables

Tables are one of the best methods to display complex information in a simple way. Unfortunately, in HTML they are relatively complicated to set up. The start of a table is defined with the `<TABLE>` tag and the end of a table by `</TABLE>`. A row is defined between the `<TR>` and `</TR>`, while a table header is defined between `<TH>` and `</TH>`. A regular table entry is defined between `<TD>` and `</TD>`. HTML script 26.2 shows an example of a table with links to other HTML pages. The `BORDER=`*n* option has been added to the `<TABLE>` tag to define the thickness of the table border (in pixels). In this case the border size has a thickness of 10 pixels.

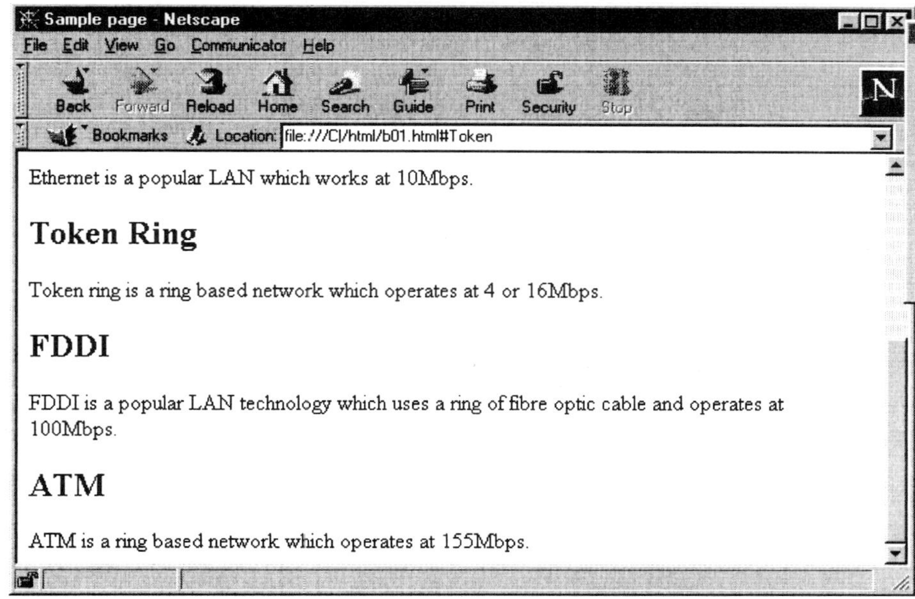

Figure 26.2 Example window with references.

HTML script 26.2

```
<HTML><HEAD>
<TITLE> Fred Bloggs</TITLE>
</HEAD>
<BODY TEXT="#000000" BGCOLOR="#FFFFFF">
<H1>Fred Bloggs Home Page</H1>
I'm Fred Bloggs. Below is a tables of links.<HR><P>
<TABLE BORDER=10>
<TR>
      <TD><B>General</B></TD>
      <TD><A HREF="res.html">Research</TD>
      <TD><A HREF="cv.html">CV</TD>
      <TD><A HREF="paper.html">Papers Published</TD>
</TR>
<TR>
      <TD><B>HTML Tutorials</B></TD>
      <TD><A HREF="intro.html">Tutorial 1</TD>
      <TD><A HREF="inter.html">Tutorial 2</TD>
      <TD><A HREF="adv.html">Tutorial 3</TD>
</TR>
<TR>
      <TD><B>Java Tutorials</B></TD>
      <TD><A HREF="java1.html">Tutorial 1</TD>
      <TD><A HREF="java2.html">Tutorial 2</TD>
      <TD><A HREF="java3.html">Tutorial 3</TD>
</TR>
</TABLE>
</BODY>
</HTML>
```

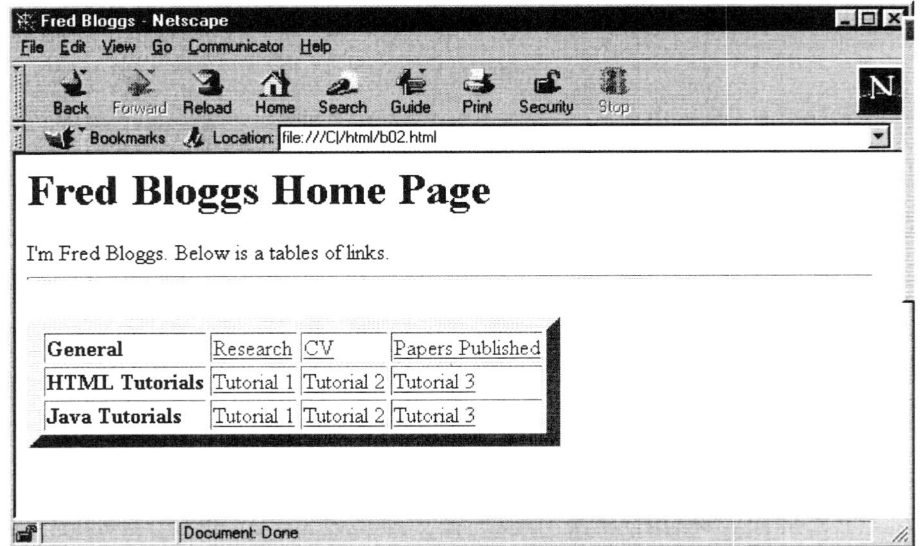

Figure 26.3 Example window from example HTML script 26.2.

Other options in the <TABLE> tag are:

- WIDTH=*x*, HEIGHT=*y* – defines the size of the table with respect to the full window size. The parameters *x* and *y* are either absolute values in pixels for the height and width of the table or are percentages of the full window size.
- CELLSPACING=*n* – defines the number of pixels desired between each cell where *n* is the number of pixels (note that the default cell spacing is 2 pixels).

An individual cell can be modified by adding options to the <TH> or <TD> tag. These include:

- WIDTH=*x*, HEIGHT=*y* – defines the size of the table with respect to the table size. The parameters *x* and *y* are either absolute values in pixels for the height and width of the table or are percentages of the table size.
- COLSPAN=*n* – defines the number of columns the cell should span.
- ROWSPAN=*n* – defines the number of rows the cell should span.
- ALIGN=*direction* – defines how the cell's contents are aligned horizontally. Valid options are *left*, *center* or *right*.
- VALIGN=*direction* – defines how the cell's contents are aligned vertically. Valid options are *top*, *middle* or *baseline*.
- NOWRAP – informs the browser to keep the text on a single line (that is, with no line breaks).

HTML script 26.3 shows an example use of some of the options in the <TABLE> and <TD> options. In this case the text within each row is center aligned. On the second row the second and third cells are merged using the COLSPAN=2 option. The first cell of the second and third rows have also been merged using the ROWSPAN=2 option. Figure 26.4 shows an example output. The table width has been increased to 90% of the full window, with a width of 50%.

HTML script 26.3

```
<HTML><HEAD><TITLE> Fred Bloggs</TITLE></HEAD>
<BODY TEXT="#000000" BGCOLOR="#FFFFFF">
<H1>Fred Bloggs Home Page</H1>
I'm Fred Bloggs. Below is a table of links.
<HR>
<P>
<TABLE BORDER=10 WIDTH=90% LENGTH=50%>
<TR>
      <TD><B>General</B></TD>
      <TD><A HREF="res.html">Research</TD>
      <TD><A HREF="cv.html">CV</TD>
      <TD><A HREF="paper.html">Papers Published</TD>
      <TD></TD>
</TR>
<TR>
      <TD ROWSPAN=2><B>HTML/Java Tutorials</B></TD>
      <TD><A HREF="intro.html">Tutorial 1</TD>
      <TD COLSPAN=2><A HREF="inter.html">Tutorial 2</TD>
</TR>
<TR>
      <TD><A HREF="java1.html">Tutorial 1</TD>
      <TD><A HREF="java2.html">Tutorial 2</TD>
      <TD><A HREF="java3.html">Tutorial 3</TD>
</TR>
</TABLE>
</BODY></HTML>
```

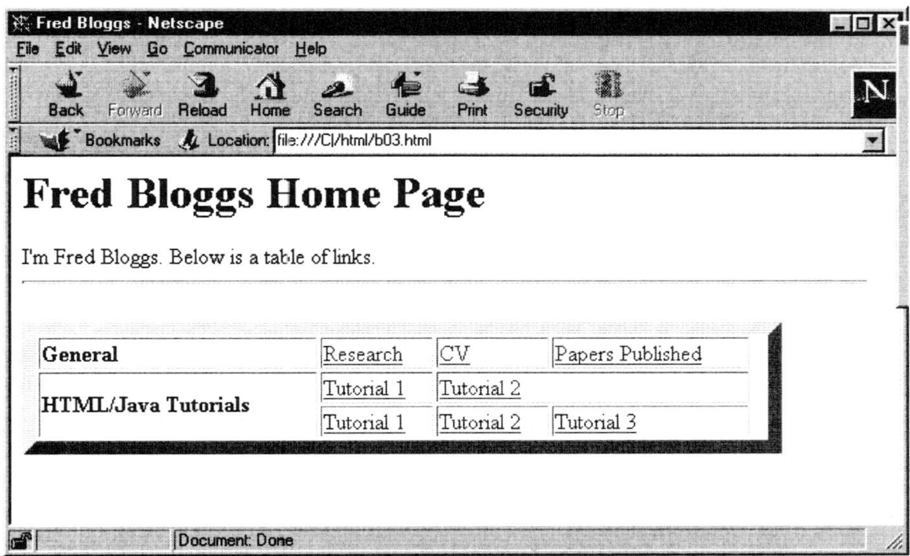

Figure 26.4 Example window from example script 26.3.

26.4 CGI scripts

CGI (Common Gateway Interface) scripts are normally written in C, Visual Basic or Perl and are compiled to produce an executable program. They can also come precompiled or in the form of a batch file. Perl has the advantage in that it is a script that can be easily run on any computer, while a precompiled C or Visual Basic program requires to be precompiled for the server computer.

CGI scripts allow the user to interact with the server and store and request data. They are often used in conjunction with forms and allow an HTML document to analyze, parse and store information received from a form. On most UNIX-type systems the default directory for CGI scripts is `cgi-bin`.

26.5 Forms

Forms are excellent methods of gathering data and can be used in conjunction with CGI scripts to collect data for future use.

A form is identified between the `<FORM>` and `</FORM>` tags. The method used to get the data from the form is defined with the `METHOD="POST"`. The `ACTION` option defines the URL script to be run when the form is submitted. Data input is specified by the `<INPUT TYPE>` tag. HTML script 26.4 form has the following parts:

- `<form action="/cgi-bin/AnyForm2" method="POST">` – which defines the start of a form and when the `"submit"` option is selected the cgi script `/cgi-bin/AnyForm2` will be automatically run.
- `<input type="submit" value="Send Feedback">` – which causes the program defined in the action option in the `<form>` tag to be run. The button on the form will contain the text `"Send Feedback"`, see Figure 26.5 for a sample output screen.
- `<input type="reset" value="Reset Form">` – which resets the data in the form. The button on the form will contain the text `"Reset Form"`, see Figure 26.5 for a sample output screen.
- `<input type="hidden" name="AnyFormTo" value= "f.bloggs @toytown. ac.uk">` – which passes a value of `f.bloggs@toytown.ac.uk` which has the parameter name of `"AnyFormTo"`. The program `AnyForm2` takes this parameter and automatically sends it to the email address defined in the value (that is, `f.bloggs@toytown.ac.uk`).
- `<input type="hidden" name="AnyFormSubject" value="Feedback form">` – which passes a value of `Feedback form` which has the parameter name of `"AnyFormSubject"`. The program `AnyForm2` takes this parameter and adds the text `"Feedback form"` in the text sent to the email recipient (in this case, `f.bloggs@toytown.ac.uk`).
- `Surname <input name="Surname">` – which defines a text input and assigns this input to the parameter name `Surname`.
- `<textarea name="Address" rows=2 cols=40> </textarea>` – which defines a text input area which has two rows and has a width of 40 characters. The thumb bars appear at the right-hand side of the form if the text area exceeds more than 2 rows, see Figure 26.5.

🕮 HTML script 26.4

```
<HTML><HEAD>
<TITLE>Example form</TITLE>
</HEAD>
<H1><CENTER>Example form</CENTER></H1><P>
<form action="/cgi-bin/AnyForm2" method="POST">
<input type="hidden" name="AnyFormTo" value="f.bloggs@toytown.ac.uk">
<input type="hidden" name="AnyFormSubject" value="Feedback form">
Surname <input name="Surname">
First Name/Names <input name="First Name"><P>
Address (including country)<P>
<textarea name="Address" rows=2 cols=40></textarea><P>
Business Phone <input name="Business Phone">
Place of study (or company) <input name="Study"><P>
E-mail    <input name="E-mail">
Fax Number <input name="Fax Number"><P>
<input type="submit" value="Send Feedback">
<input type="reset" value="Reset Form">
</Form><HTML>
```

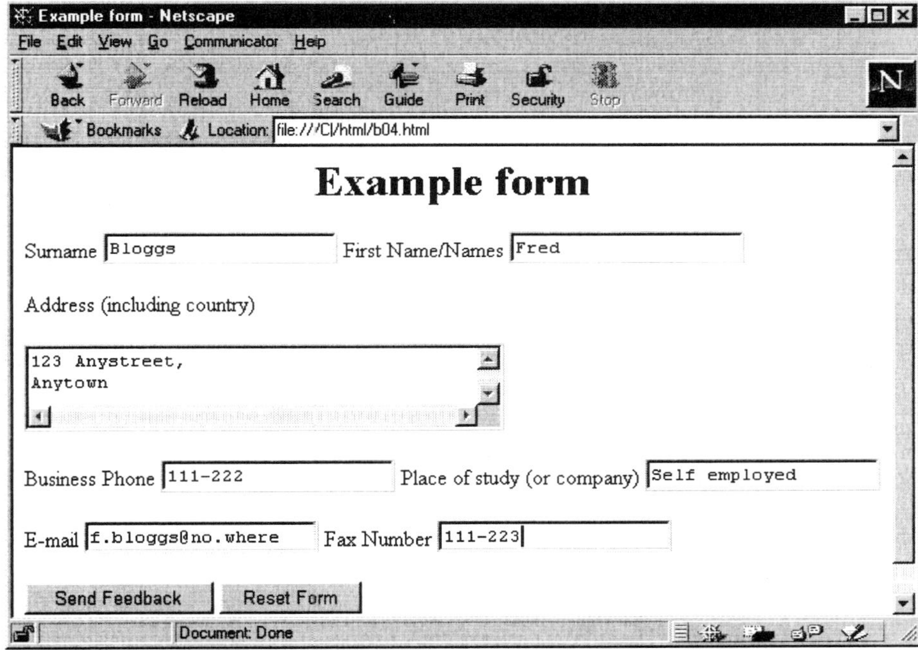

Figure 26.5 Example window showing an example form.

In this case the recipient (`f.bloggs@toytown.ac.uk`) will receive an email with the contents:

```
Anyform Subject=Example form
Surname=Bloggs
First name=Fred
Address=123 Anystreet, Anytown
Business Phone=111-222
Place of study (or company)=Self employed
Email= f.bloggs@nowhere
Fax Number=111-2223
```

The extra options to the `<input>` tag are `size="n"`, where *n* is the width of the input box in characters, and `maxlength="m"`, where *m* is the maximum number of characters that can be entered, in characters. For example:

```
<input type="text"  size="15" maxlength="10">
```

defines that the input type is text, the width of the box is 15 characters and the maximum length of input is 10 characters.

26.5.1 Input types

The type options to the `<input>` tag are defined in Table 26.1. HTML script 26.5 gives a few examples of input types and Figure 26.6 shows a sample output.

Table 26.1 Input type options.

TYPE=	*Description*	*Options*
"text"	The input is normal text.	NAME="*nm*" where *nm* is the name that will be sent to the server when the text is entered. SIZE="*n*" where *n* is the desired box width in characters. SIZE="*m*" where *m* is the maximum number of input characters.
"password"	The input is a password which will be displayed with *s. For example if the user inputs a 4-letter password then only **** will be displayed.	SIZE="*n*" where *n* is the desired box width in characters. SIZE="*m*" where *m* is the maximum number of input characters.
"radio"	The input takes the form of a radio button (such as ⊙ or O). They are used to allow the user to select a single option from a list of options.	NAME="*radname*" where *radname* defines the name of the button. VALUE="*val*" where *val* is the data that will be sent to the server when the button is selected. CHECKED is used to specify that the button is initially set.
"checkbox"	The input takes the form of a checkbox (such as ☒ or ☐). They are used to allow the user to select several options from a list of options.	NAME="*chkname*" where *chkname* defines the common name for all the checkbox options. VALUE="*defval*" where *defval* defines the name of the option. CHECKED is used to specify that the button is initially set.

🕮 HTML script 26.5

```
<HTML>
<HEAD>
<TITLE>Example form</TITLE>
</HEAD>
<FORM METHOD="Post" >
<H2>Enter type of network:</H2><P>
<INPUT TYPE="radio" NAME="network" VALUE="ethernet" CHECKED>Ethernet
```

```
<INPUT TYPE="radio" NAME="network" VALUE="token"> Token Ring
<INPUT TYPE="radio" NAME="network" VALUE="fddi" >FDDI
<INPUT TYPE="radio" NAME="network" VALUE="atm" >ATM
<H2>Enter usage:</H2><P>
<INPUT TYPE="checkbox" NAME="usage" VALUE="multi" >Multimedia
<INPUT TYPE="checkbox" NAME="usage" VALUE="word" >Word Processing
<INPUT TYPE="checkbox" NAME="usage" VALUE="spread" >Spread Sheets
<P>Enter Password<INPUT TYPE="password" NAME="passwd" SIZE="10">
</FORM></HTML>
```

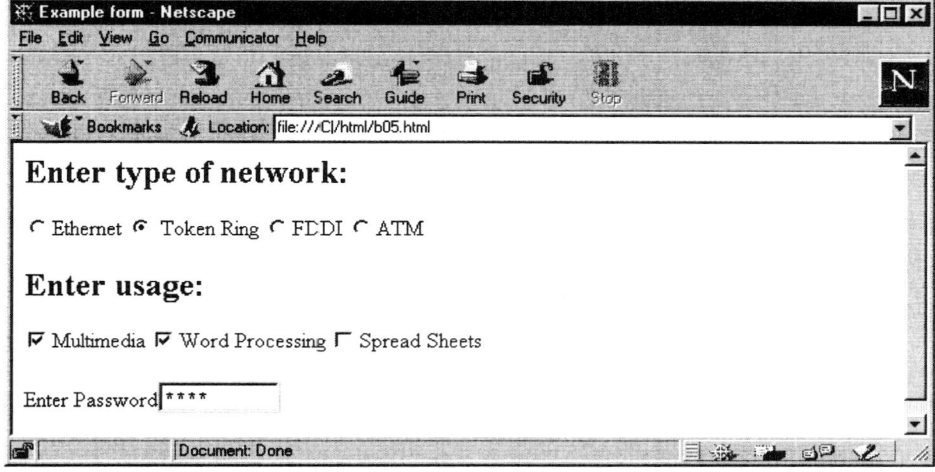

Figure 26.6 Example window with different input options.

26.5.2 Menus

Menus are a convenient method of selecting from multiple options. The `<SELECT>` tag is used to define the start of a list of menu options and the `</SELECT>` tag defines the end. Menu elements are then defined with the `<OPTION>` tag. The options defined within the `<SELECT>` are:

- `NAME="`*name*`"` – which defines that *name* is the variable name of the menu. This is used when the data is collected by the server.
- `SIZE="`*n*`"` – which defines the number of options which are displayed in the menu.

HTML script 26.6 shows an example of a menu. The additional options to the `<OPTION>` tag are:

- `SELECTED` – which defines the default selected option.
- `VALUE="`*val*`"` – where *val* defines the name of the data when it is collected by the server.

HTML script 26.6

```
<HTML>
<HEAD><TITLE>Example form</TITLE> </HEAD>
<FORM METHOD="Post" >
Enter type of network:
<select Name="network" size="1">
<option>Ethernet
```

```
<option SELECTED>Token Ring
<option>FDDI
<option>ATM
</select>
</FORM></HTML>
```

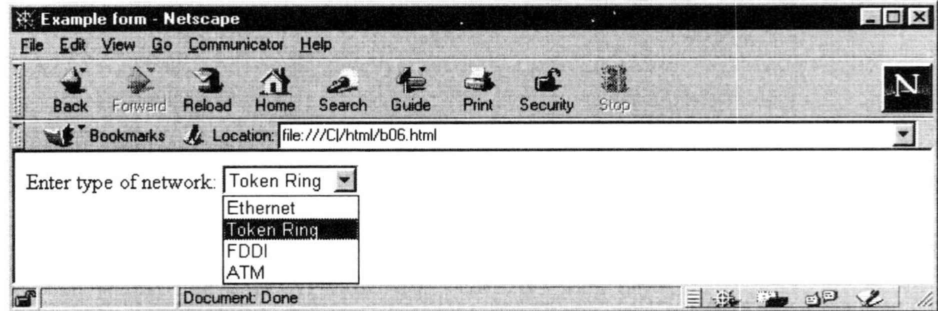

Figure 26.7 Example window showing an example form.

26.6 Multimedia

If the browser cannot handle all the file types it may call on other application helpers to process the file. This allows other 'third-party' programs to integrate into the browser. Figure 26.8 shows an example of the configuration of the helper programs. The options in this case are:

- View in browser.
- Save to disk.
- Unknown: prompt user.
- Launch an application (such as an audio playback program or MPEG viewer).

For certain applications the user can select as to whether the browser processes the file or another application program processes it. Helper programs make upgrades in helper applications relatively simple and also allow new file types to be added with an application helper. Typically when a program is installed which can be used with a browser it will prompt the user to automatically update the helper application list so that it can handle the given file type(s).

Each file type is defined by the file extension, such as `.ps` for postscript files, `.exe` for a binary executable file, and so on. These file extensions have been standardized in MIME (Multipurpose Internet Mail Extensions) specification. Table 26.2 shows some typical file extensions.

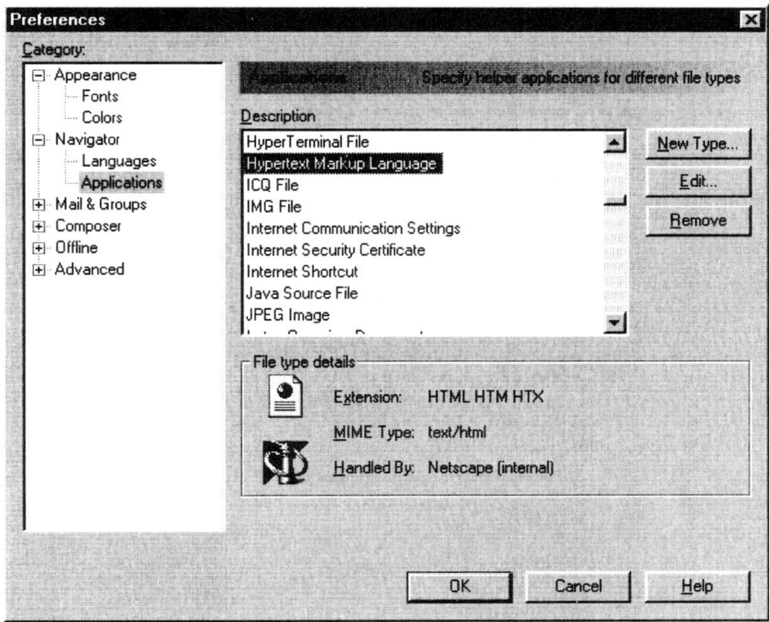

Figure 26.8 Example window showing preferences.

Table 26.2 Input type options.

Mime type	*Extension*	*Typical action*
application/octet-stream	exe, bin	Save
application/postscript	ps, ai, eps	Ask user
application/x-compress	Z	Compress program
application/x-gzip	gz	GZIP compress program
application/x-javascript	js, mocha	Ask user
application/x-msvideo	avi	Audio player
Mime type	*Extension*	*Typical action*
application/x-perl	pl	Save
application/x-tar	tar	Save
application/x-zip-compressed	zip	ZIP program
audio/basic	au, snd	Audio player
image/gif	gif	Browser
image/jpeg	jpeg, jpg, jpe	Browser
image/tiff	tif, tiff	Graphics viewer
image/x-MS-bmp	bmp	Graphics viewer
text/html	htm, html	Browser
text/plain	text, txt	Browser
video/mpeg	mpeg, mpg, mpe, mpv, vbs, mpegv	Video player
video/quicktime	qt, mov, moov	Video player

27 Java (Introduction)

27.1 Introduction

Java 1.0 was first released in 1995 and was quickly adopted as it fitted well with Internet-based programming. It was followed by Java 1.1 which gave faster interpretation of Java applets and included many new features. This book documents the basic features of Java 1.0 and the enhancements that have been made in Java 1.1.

It is a general-purpose, concurrent, class-based, object-oriented language and has been designed to be relatively simple to built complex applications. Java is developed from C and C++, but some parts of C++ have been dropped and others added.

Java has the great advantage over conventional software languages in that it produces code which is computer hardware independent. This is because the compiled code (called bytecodes) is interpreted by the WWW browser. Unfortunately this leads to slower execution, but, as much of the time in a graphical user interface program is spent updating the graphics display, then the overhead is, as far as the user is concerned, not a great one.

The other advantages that Java has over conventional software languages include:

- It is a more dynamic language than C/C++ and Pascal, and was designed to adapt to an evolving environment. It is extremely easy to add new methods and extra libraries without affecting existing applets and programs. It is also useful in Internet applications as it supports most of the standard compressed image, audio and video formats.
- It has networking facilities built into it. This provides support for TCP/IP sockets, URLs, IP addresses and datagrams.
- While Java is based on C and C++ it avoids some of the difficult areas of C/C++ code (such as pointers and parameter passing).
- It supports client/server applications where the Java applet runs on the server and the client receives the updated graphics information. In the most extreme case the client can simply be a graphics terminal which runs Java applets over a network. The small 'black-box' networked computer is one of the founding principles of Java, and it is hoped in the future that small Java-based computers could replace the complex PC/workstation for general-purpose applications, like accessing the Internet or playing network games. This 'black-box' computer concept is illustrated in Figure 27.1.

Most existing Web browsers are enabled for Java applets (such as Internet Explorer 3.0 and Netscape 2.0 and later versions). Figure 27.2 shows how Java applets are created. First the source code is produced with an editor, next a Java compiler compiles the Java source code into bytecode (normally appending the file name with `.class`). An HTML page is then constructed which has the reference to the applet. After this a Java-enabled browser or applet viewer can then be used to run the applet.

The Java Development Kit (JDK) is available, free, from Sun Microsystems from the WWW site `http://www.javasoft.com`. This can be used to compile Java applets and standalone programs. There are versions for Windows NT/95/98, Apple Mac and UNIX-based systems with many sample applets.

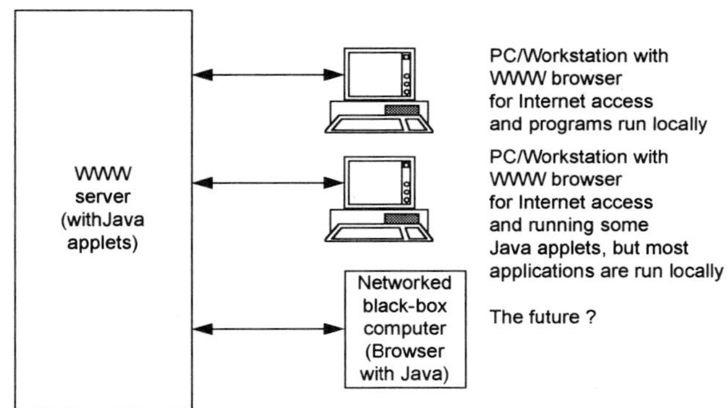

Figure 27.1 Internet accessing.

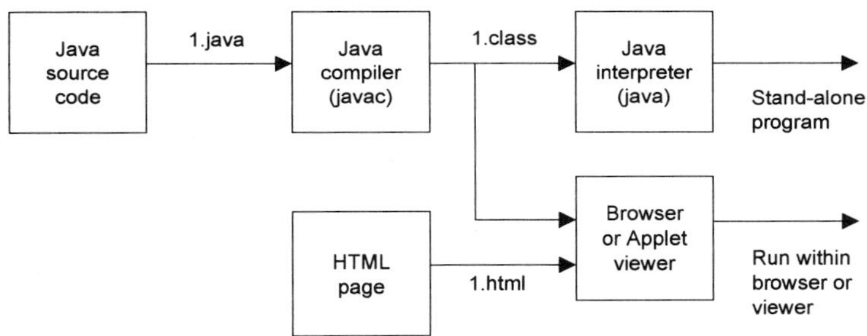

Figure 27.2 Constructing Java applets and standalone programs.

Table 27.1 shows the main files used in the PC version and Figure 27.3 shows the directory structure of the JDK tools. The Java compiler, Java interpreter and applet viewer programs are stored in the `bin` directory. On the PC, this directory is normally set up in the PATH directory, so that the Java compiler can be called while the user is in another directory. The following is a typical setup (assuming that the home directory is `C:\`*javahome*):

```
PATH=C:\WINDOWS;C:\WINDOWS\COMMAND;C:\javahome\BIN
CLASSPATH=C:\javahome\LIB;.;C:\javahome
```

The `lib` directory contains the `classes.zip` file which is a zipped-up version of the Java class files. These class files are stored in the directories below the `src/java` directory. For example, the `io` classes (such as `File.java` and `InputStream.java`) are used for input/output in Java, the `awt` classes (such as `Panel.java` and `Dialog.java`) are used to create and maintain windows. These and other classes will be discussed later.

The `include` directory contains header files for integrating C/C++ programs with Java applets and the `demo` directory contains some sample Java applets.

Table 27.1 JDK programs.

File	*Description*
`javac.exe`	Java compiler
`java.exe`	Java interpreter
`appletViewer.exe`	Applet viewer for testing and running applets
`classes.zip`	It is needed by the compiler and interpreter
`javap.exe`	Java class disassembler
`javadoc.exe`	Java document generator
`javah.exe`	C Header and Stub File Generator
`jar.exe`	Java Archive Tool which combines class files and other resources into a single jar file.
`jbd.exe`	Java debugger

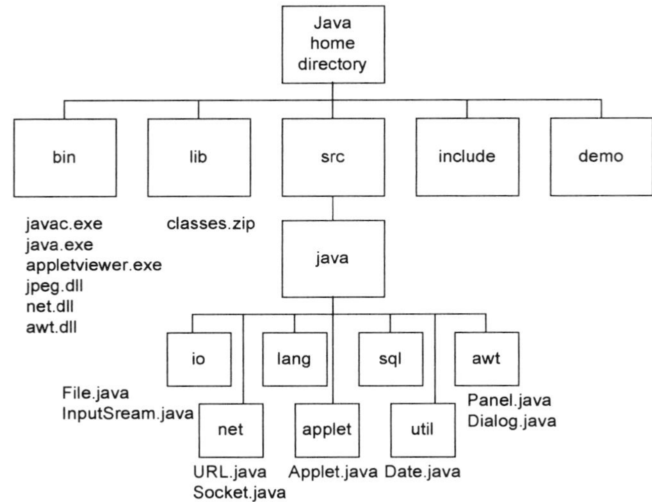

Figure 27.3 Sample directory structure of JDK for a PC-based system.

27.2 Standalone programs

A Java program can be run as a standalone program or as an applet. A standalone program allows the Java program to be run without a browser and is normally used when testing the Java applet. The method of output to the screen is:

```
System.out.println("message");
```

which prints a message (`message`) to the display. This type of debugging is messy as these statements need to be manually inserted in the program. It is likely that later versions of the JDK toolkit will contain a run-time debugger which will allow developers to view the execution of the program.

To run a standalone program the `java.exe` program is used and the user adds output statements with the `System.out.println()` method. Note that there is no output to the main graphics applet screen with this method.

Java program 27.1 gives a simple example of a standalone program. The `public static void main(Strings[] args)` defines the main method. Sample run 27.1 shows how the Java program is created (with `edit`) and then compiled (with `javac.exe`), and then finally run (with `java.exe`).

Java program 27.1 (chap27_1.java)

```
public class chap27_1
{
   public static void main(String[] args)
   {
   int i;
   i=10;
      System.out.println("This is an example of the ");
      System.out.println("output from the standalone");
      System.out.println("program");
      System.out.println("The value of i is " + i);
   }
}
```

Sample run 27.1

```
C:\DOCS\notes\java>edit chap27_01.java
C:\DOCS\notes\java>javac chap27_01.java
C:\DOCS\notes\java>java chap27_01
This is an example of the
output from the standalone
program
The value of i is 10
```

The process of developing a standalone Java program is:

- Create a Java program. This is the main Java program file and is created by a text editor (such as `edit` in a PC-based system). These files are given the `.java` file extension. In Sample run 27.1 the file created is `chap27_01.java`.
- Compile program to a class file. This is created by the Java compiler (`javac.exe`) when there is a successful compilation. By default, the file produced has the same filename as the source file and the `.class` file extension. In Sample run 27.1 the file created is `chap27_01.class`.
- Run program with the Java interpreter. The Java interpreter (`java.exe`) reads from the class file (which contains the bytecode) and gives the required output.

27.3 Data types

Variables within a program can be stored as either boolean values, numbers or as characters. For example, the resistance of a copper wire would be stored as a number (a real value), whether it exists or not (as a boolean value) and the name of a component (such as "`R1`") would be stored as characters.

An integer is any value without a decimal point. Its range depends on the number of bytes used to store it. A floating-point value is any number and can include a decimal point. This value is always in a signed format. Again, the range depends on the number of bytes used.

Integers either take up 1, 2, 4 or 8 bytes in memory for a `byte`, `short`, `int` and `long`, respectively. These are all represented in 2's complement notation and thus can store positive and negative integer values. Table 27.2 gives some typical ranges for data types.

Table 27.2 Typical ranges for data types.

Type	*Storage (bytes)*	*Range*
`boolean`	1-bit	True or False
`byte`	1	–128 to 127
`short`	2	–32 768 to 32 767
`int`	4	–2 147 483 648 to 2 147 483 647
`long`	8	2 223 372 036 854 775 808 to –2 223 372 036 854 775 809
`char`	2	16-bit unsigned integers representing Unicode characters.
`float`	4	$\pm 3.4\times 10^{-38}$ to $\pm 3.4\times 10^{38}$
`double`	8	$\pm 1.7\times 10^{-308}$ to $\pm 1.7\times 10^{308}$

27.4 Characters and strings

Typically, characters are stored using either ASCII or EBCDIC codes. ASCII is an acronym for American Standard Code for Information Interchange and EBCDIC for Extended Binary Coded Decimal Interchange Code.

ASCII characters from decimal 0 to 32 are non-printing characters that are used either to format the output or to control the hardware. Program 27.2 displays an ASCII character for an entered decimal value. The `print()` method displays the ASCII character. Sample run 27.2 shows a sample run (note that some of the displayed characters are non-printing). Characters of type `char` are stored as 2-byte Unicode characters (0x0000 to 0xFFFF). This allows for internationalization of the character set. The characters from 0 to 255 (0x0000 to 0x00FF) are standard extended ASCII character set (ISO8859-1, or Latin-1), where the characters are stored as the binary digits associated with the character. For example, the ASCII code for the character '`A`' is 65 decimal (0x41); the binary storage for this character is thus `0100 0001`.

Java program 27.2 (`chap27_2.java`)

```
public class chap27_2
{
 public static void main (String args[])
 {
 char ch;
      for (ch=0;ch<256;ch++)
         System.out.print(" " + ch); // print from 0 to 255
 }
}
```

Sample run 27.2

```
□ □ □ □ □ □ □ □ □   □
□ □ □ □ □ □ □ □ □ □ □ □ □ □ □ -   ! " # $ % & ' ( ) * + , - . / 0 1 2 3 4 5
6 7 8 9 : ; < = > ? @ A B C D E F G H I J K L M N O P Q R S T U V W X Y Z [ \
] ^ _ ` a b c d e f g h i j k l m n o p q r s t u v w x y z { | } ~ □ □ □ , ƒ
„ … † ‡ ˆ ‰ Š ‹ Œ □ □ □ □ ' ' " " • – — ˜ ™ š › œ □ □ Ÿ   ¡ ¢ £ ¤ ¥ ¦ § ¨ © ª
« ¬ - ® ¯ ° ± ² ³ ´ µ ¶ · ¸ ¹ º » ¼ ½ ¾ ¿ À Á Â Ã Ä Å Æ Ç È É Ê Ë Ì Í Î Ï Ð Ñ
Ò Ó Ô Õ Ö × Ø Ù Ú Û Ü Ý Þ ß à á â ã ä å æ ç è é ê ë ì í î ï ð ñ ò ó ô õ ö ÷ ø
ù ú û ü ý þ ÿ
```

The `println()` method sends a formatted string to the standard output (the display). This string can include special control characters, such as new lines (`'\n'`), backspaces (`'\b'`) and tabspaces (`'\t'`); these are listed in Table 27.3.

The `println()` method writes a string of text to the standard output and at the end of the text a new line is automatically appended, whereas the `print()` method does not append the output with a new line.

Special control characters use a backslash to inform the program to escape from the way they would be normally be interpreted. The carriage return (‘`\r`’) is used to return the current character pointer on the display back to the start of the line (on many displays this is the leftmost side of the screen). A form-feed control character (`'\f'`) is used to feed line printers on a single sheet and the horizontal tab (`'\t'`) feeds the current character position forward one tab space.

Quotes enclose a single character, for example ‘`a`’, whereas inverted commas enclose a string of characters, such as “`Java programming`”. Java has a special String object (`String`). Java program 27.3 shows an example of declaring two strings (`name1` and `name2`) and Sample run 27.3 shows a sample run. The ‘`\"`’ character is used to display inverted commas and the backspace character has been used to delete an extra character in the displayed string. The BELL character is displayed with the ‘`\007`’ and ‘`\u0007`’ escape sequence characters. Other escape characters used include the horizontal tab (`'\t'`) and the new line character (`'\n'`).

Strings can be easily concatenated using the ‘+’ operator. For example to build a string of two strings (with a space in-between) then following can be implemented:

```
String name1, name2, name3;
```

name3=name1+" " + name2;

Table 27.3 Special control (or escape sequence) characters.

Characters	Function
`\"`	Double quotes (")
`\'`	Single quote (')
`\\`	Backslash (\ )
\unnnn	Unicode character in hexadecimal code, e.g. `\u041` gives '!'
\0nn	Unicode character in octal code, e.g. `\041` gives '!'
`\b`	Backspace (move back one space)
`\f`	Form-feed
`\n`	New line (line-feed)
`\r`	Carriage return
`\t`	Horizontal tab spacing

Java program 27.3 (chap27_3.java)

```
public class chap27_3
{

 public static void main (String args[])
 {

 String name1="Bill", name2="Buchanan";

      System.out.println("Ring the bell 3 times \u0007\007\007");
      System.out.print("\"My name is Bill\"\n");
      System.out.println("\t\"Buchh\banan\"");

      System.out.println(name1 + " " + name2);
 }
}
```

Sample run 27.3

```
C:\java\src\chap27>java chap27_03
Ring the bell 3 times
"My name is Bill"
        "Buchanan"
Bill Buchanan
```

27.5 Java operators

Java has a rich set of operators, of which there are four main types:

- Arithmetic
- Logical
- Bitwise
- Relational

27.5.1 Arithmetic

Arithmetic operators operate on numerical values. The basic arithmetic operations are add (+), subtract (-), multiply (*), divide (/) and modulus division (%). Modulus division gives the remainder of an integer division.

The assignment operator (=) is used when a variable 'takes on the value' of an operation. Other short-handed operators are used with it, including add equals (+=), minus equals (-=), multiplied equals (*=), divide equals (/=) and modulus equals (%=).

Table 27.4 summarizes the arithmetic operators.

27.5.2 Relationship

The relationship operators determine whether the result of a comparison is TRUE or FALSE. These operators are greater than (>), greater than or equal to (>=), less than (<), less than or equal to (<=), equal to (==) and not equal to (!=). Table 27.5 lists the relationship operators.

Table 27.4 Arithmetic operators.

Operator	*Operation*	*Example*
-	subtraction or minus	5-4→1
+	addition	4+2→6
*	multiplication	4*3→12
/	division	4/2→2
%	modulus	13%3→1
+=	add equals	x += 2 is equivalent to x=x+2
-=	minus equals	x -= 2 is equivalent to x=x-2
/=	divide equals	x /= y is equivalent to x=x/y
*=	multiplied equals	x *= 32 is equivalent to x=x*32
=	assignment	x = 1
++	increment	Count++ is equivalent to Count=Count+1
--	decrement	Sec-- is equivalent to Sec=Sec-1

27.5.3 Logical (TRUE or FALSE)

A logical operation is one in which a decision is made as to whether the operation performed is TRUE or FALSE. If required, several relationship operations can be grouped together to give the required functionality. Java assumes that a numerical value of 0 (zero) is FALSE and that any other value is TRUE. Table 27.6 lists the logical operators.

Table 27.5 Relationship operators.

Operator	*Function*	*Example*	*TRUE Condition*
>	greater than	(b>a)	when b is greater than a
>=	greater than or equal	(a>=4)	when a is greater than or equal to 4
<	less than	(c<f)	when c is less than f
<=	less than or equal	(x<=4)	when x is less than or equal to 4
==	equal to	(x==2)	when x is equal to 2
!=	not equal to	(y!=x)	when y is not equal to x

Table 27.6 Logical operators.

<table>
<tr><th>Operator</th><th>Function</th><th>Example</th><th>TRUE condition</th></tr>
<tr><td>&&</td><td>AND</td><td>((x==1) && (y<2))</td><td>when x is equal to 1 and y is less than 2</td></tr>
<tr><td>||</td><td>OR</td><td>((a!=b) || (a>0))</td><td>when a is not equal to b or a is greater than 0</td></tr>
<tr><td>!</td><td>NOT</td><td>(!(a>0))</td><td>when a is not greater than 0</td></tr>
</table>

For example, if a has the value 1 and b is also 1, then the following relationship statements would apply:

Statement	*Result*
`(a==1) && (b==1)`	TRUE
`(a>1) && (b==1)`	FALSE
`(a==10) || (b==1)`	TRUE
`!(a==12)`	TRUE

Java program 27.4 shows a Java program which proves the above table and Sample run 27.4 shows a sample run.

Java program 27.4 (chap27_4.java)

```
public class chap27_4
{
 public static void main (String args[])
 {
 int a=1,b=1;
      if ((a==1) && (b==1)) System.out.println("TRUE");
      else System.out.println("FALSE");
      if ((a>1) && (b==1)) System.out.println("TRUE");
      else System.out.println("FALSE");
      if ((a==10) || (b==1)) System.out.println("TRUE");
      else System.out.println("FALSE");
      if (!(a==10)) System.out.println("TRUE");
      else System.out.println("FALSE");
 }
}
```

Sample run 27.4

```
C:\java\src\chap27>java chap27_4
TRUE
FALSE
TRUE
TRUE
```

27.5.4 Bitwise

The bitwise operators are similar to the logical operators but they should not be confused as their operation differs. Bitwise operators operate directly on the individual bits of an operand(s), whereas logical operators determine whether a condition is TRUE or FALSE.

Numerical values are stored as bit patterns in either an unsigned integer format, signed integer (2's complement) or floating-point notation (an exponent and mantissa). Characters are normally stored as ASCII characters.

The basic bitwise operations are AND (`&`), OR (`|`), 1's complement or bitwise inversion (`~`), XOR (`^`), shift left (<<), shift right with sign (>>) and right shift without sign (>>>). Table 27.12 gives the results of the AND, OR and XOR bitwise operation on two bits `Bit1` and `Bit2`.

The bitwise operators operate on each of the individual bits of the operands. For example, if two decimal integers 58 and 41 (assuming 8-bit unsigned binary values) are operated on using the AND, OR and EX-OR bitwise operators, then the following applies.

	AND	OR	EX-OR
58	00111010b	00111010b	00111010b
41	00101001b	00101001b	00101001b
Result	00101000b	00111011b	00010011b

The results of these bitwise operations are as follows:

```
58 & 41 = 40        (that is, 00101000b)
58 | 41 = 59        (that is, 00111011b)
58 ^ 41 = 19        (that is, 00010011b)
```

Java Program 27.5 shows a program which tests these operations and Sample run 27.5 shows a test run.

The 1's complement operator operates on a single operand. For example, if an operand has the value of 17 (`00010001b`) then the 1's complement of this, in binary, will be `11101110b`.

Java program 27.5 (`chap27_5.java`)

```
public class chap27_5
{
 public static void main (String args[])
 {
 int a=58,b=41,val;
        val=a&b; System.out.println("AND "+ val);
        val=a|b; System.out.println("OR "+ val);
        val=a^b; System.out.println("X-OR "+ val);
 }
}
```

Sample run 27.5

```
C:\java\src\chap1>java chap1_08
AND 40
OR 59
X-OR 19
```

To perform bit shifts, the `<<`, `>>` and `>>>` operators are used. These operators shift the bits in the operand by a given number defined by a value given on the right-hand side of the operation. The left-shift operator (<<) shifts the bits of the operand to the left and zeros fill the result on the right. The right-shift operator (>>) shifts the bits of the operand to the right and zeros fill the result if the integer is positive; otherwise it will fill with 1s. The right shift with sign (>>>) shifts the bits and ignores the sign flag; it thus treats signed integers as unsigned integers. The standard format for the three shift operators is:

```
operand >> no_of_bit_shift_positions
operand << no_of_bit_shift_positions
operand >>> no_of_bit_shift_positions
```

For example, if `y = 59` (`00111011`), then `y >> 3` will equate to 7 (`00000111`) and `y<<2` to 236 (`11101100`). Table 27.7 gives a summary of the basic bitwise operators.

Table 27.7 Bitwise operators.

Operator	*Function*	*Example*
&	AND	`c = A & B`
|	OR	`f = Z	y`
^	XOR	`h = 5 ^ f`
~	1's complement	`x = ~y`
>>	shift right	`x = y >> 1`
<<	shift left	`y = y << 2`

27.6 Selection statements

27.6.1 if...else

A decision is made with the `if` statement. It logically determines whether a conditional expression is TRUE or FALSE. For a TRUE, the program executes one block of code; a FALSE causes the execution of another (if any). The keyword `else` identifies the FALSE block. In Java, braces (`{}`) are used to define the start and end of the block.

Relationship operators, include:

- Greater than (`>`).
- Less than (<).
- Greater than or equal to (>=).
- Less than or equal to (<=).
- Equal to (==).
- Not equal to (`!=`).

These operations yield a TRUE or FALSE from their operation. Logical statements (`&&`, `||`, `!`) can then group these together to give the required functionality. These are:

- AND (`&&`)
- OR (`||`)
- NOT (`!`)

f the operation is not a relationship, such as bitwise or an arithmetic operation, then any non-zero value is TRUE and a zero is FALSE. The following is an example syntax of the `if` statement. If the statement block has only one statement then the braces (`{ }`) can be excluded.

```
if (expression)
{
   statement block
}
```

The following is an example format with an `else` extension.

```
if (expression)
{
```

```
   statement block1
}
else
{
   statement block2
}
```

It is possible to nest `if..else` statements to give a required functionality. In the next example, *statement block1* is executed if `expression1` is TRUE. If it is FALSE then the program checks the next expression. If this is TRUE the program executes *statement block2*, else it checks the next expression, and so on. If all expressions are FALSE then the program executes the final `else` statement block, in this case, *statement block4*:

```
if (expression1)
{
   statement block1
}
else if (expression2)
{
   statement block2
}
else if (expression3)
{
   statement block3
}
else
{
   statement block4
}
```

Java program 27.6 gives an example of a program which uses the if...else statement. In this case the variable `col` is tested for its value. When it matches a value from 0 to 6 the equivalent color code is displayed. If it is not between 0 and 6 then the default message is displayed (`"Not Defined Yet!"`). Sample run 27.6 shows a sample run.

Java program 27.6 (chap27_6.java)

```
public class chap27_6
{
   public static void main (String args[])
   {
   int col;
      col=4;
      if (col==0) System.out.println("BLACK");
      else if (col==1) System.out.println("BROWN");
      else if (col==2) System.out.println("RED");
      else if (col==3) System.out.println("ORANGE");
      else if (col==4) System.out.println("YELLOW");
      else if (col==5) System.out.println("GREEN");
      else System.out.println("Not Defined Yet!");
   }
}
```

Sample run 27.6

```
C:\java\src\chap2> edit chap2_01.java
C:\java\src\chap2> javac chap2_01.java
C:\java\src\chap2> java chap2_01
YELLOW
```

27.6.2 switch

The switch statement is used when there is a multiple decision to be made. It is normally used to replace the if statement when there are many routes of execution the program execution can take. The syntax of switch is as follows.

```
switch (expression)
{
   case const1:   statement(s) : break;
   case const2:   statement(s) ; break;
   :              :
   default:          statement(s) ; break;
}
```

The switch statement checks the expression against each of the constants in sequence (the constant must be an integer or character data type). When a match is found the statement(s) associated with the constant is (are) executed. The execution carries on to all other statements until a break is encountered or to the end of switch, whichever is sooner. If the break is omitted, the execution continues until the end of switch. If none of the constants matches the switch expression a set of statements associated with the default condition (default:) is executed. The data type of the switch constants can be byte, char, short, int or long.

Java program 27.7 is the equivalent of Java program 27.6 but using a switch statement. Sample run 27.7 shows a sample run.

Java program 27.7 (chap27_7.java)

```
import java.lang.Math;

public class chap27_7
{

   public static void main (String args[])
   {
   int col;
      col=4;
      switch (col)
      {
      case 0:  System.out.println("BLACK");  break;
      case 1:  System.out.println("BROWN");  break;
      case 2:  System.out.println("RED");       break;
      case 3:  System.out.println("ORANGE");    break;
      case 4:  System.out.println("YELLOW");    break;
      case 5:  System.out.println("GREEN");  break;
      default:    System.out.println("Not defined yet!");
      }

   }
}
```

Sample run 27.7

```
C:\java\src\chap2>java chap2_03
YELLOW
```

27.7 Loops

27.7.1 for()

Many tasks within a program are repetitive, such as prompting for data, counting values, and so on. The `for` loop allows the execution of a block of code for a given control function. The following is an example format; if there is only one statement in the block then the braces can be omitted.

```
for (starting condition;test condition;operation)
{
      statement block
}
```

where :

`starting condition`	–	the starting value for the loop;
`test condition`	–	if `test condition` is TRUE the loop will continue execution;
`operation`	–	the operation conducted at the end of the loop.

Program 27.8 displays ASCII characters for entered start and end decimal values. Sample run 27.8 displays the ASCII characters from decimal 40 (‘`(`’) to 50 (‘`2`’). The type conversion `(char)` is used to convert an integer to a `char`.

Java program 27.8 (`chap27_8.java`)

```
public class chap27_8
{
   public static void main (String args[])
   {
   int start,end,ch;
      start=40; end=50;
      for (ch=start;ch<=end;ch++)
         System.out.println((int)ch+" "+(char)ch);
   }

}
```

Sample run 27.8

```
C:\java\src\chap27>java chap27_8
40 (
41 )
42 *
43 +
44 ,
45 -
46 .
47 /
48 0
49 1
50 2
```

27.7.2 while( )

The `while()` statement allows a block of code to be executed while a specified condition is TRUE. It checks the condition at the start of the block; if this is TRUE the block is executed, else it will exit the loop. The syntax is:

```
while (condition)
{
   :          :
   statement block
   :          :
}
```

If the statement block contains a single statement then the braces may be omitted (although it does no harm to keep them).

27.7.3 do...while()

The `do...while()` statement is similar in its operation to `while()` except that it tests the condition at the bottom of the loop. This allows *statement block* to be executed at least once. The syntax is:

```
do
{
      statement block
} while (condition);
```

As with `for()` and `while()` loops the braces are optional. The `do...while()` loop requires a semicolon at the end of the loop, whereas the `while()` does not.

The following is an example of the usage of the `do...while()` loop. Octal numbers uses base eight. To convert a decimal value to an octal number the decimal value is divided by 8 recursively and each remainder noted. The first remainder gives the least significant digit and the final remainder the most significant digit. For example, the following shows the octal equivalent of the decimal number 55:

```
8 | 55
  |  6  r 7  <<< LSD (least significant digit)
  |  0  r 6  <<< MSD (most significant digit)
```

Thus the decimal value 55 is equivalent to 67o (where the o represents octal). Program 27.9 shows a program which determines an octal value for an entered decimal value. Unfortunately, it displays the least significant digit first and the most significant digit last, thus the displayed value must be read in reverse. Sample run 27.9 shows a sample run.

Java program 27.9 (chap27_9.java)

```
public class chap27_9
{
   public static void main (String args[])
   {
   int val,remainder;

      val=55;
```

```
        System.out.println("Conversion to octal (in reverse)");
        do
        {
           remainder=val % 8;     // find remainder with modulus
           System.out.print(remainder);
           val=val / 8;
        } while (val>0);
    }
}
```

Sample run 27.9

```
Conversion to octal (in reverse)
76
```

27.8 Classes

Classes are a general form of structures, which are common in many languages. They basically gather together data members, and in object-oriented design, they also include methods (known as functions in C and procedures in Pascal) which operate on the class. Everything within Java is contained within classes.

In C a program is normally split into modules named functions. Typically, these functions have parameters passed to them or from them. In Java these functions are named methods and operate within classes. Java program 27.10 includes a `Circle` class which contains two methods:

- `public float area(double r)`. In which the value of `r` is passed into the method and the return value is equal to πr^2 (`return(3.14159*r*r)`). The preceding `public double` defines that this method can be accessed from another class (`public`) and the `double` defines that the return type is of type `double`.
- `public float circum(double r)`. In which the value of `r` is passed into the method and the return value is equal to $2\pi r$ (`return(2*3.14159*r)`). The preceding `public double` defines that this method can be accessed from another class (`public`) and the `double` defines that the return type is of type `double`.

In defining a new class the program automatically defines a new data type (in Program 27.10 this new data type is named `Circle`). An instance of a class must first be created, thus for the `Circle` it can be achieved with:

```
Circle cir;
```

this does not create a `Circle` object, it only refers to it. Next the object can be created with the `new` keyword with:

```
cir = new Circle();
```

These two lines can be merged together into a single line with:

```
Circle cir = new Circle();
```

which creates an instance of a `Circle` and assigns a variable to it. The methods can then be used to operate on the object. For example to apply the `area()` method:

```
val=cir.area(10);
```

can be used. This passes the value of 10 into the `radius` variable in the `area()` method and the return value will be put into the `val` variable. Sample run 27.10 shows a sample run.

Java program 27.10 (chap27_10.java)

```
public class chap27_10
{
   public static void main(String[] args)
   {
      Circle cir=new Circle();
      System.out.println("Area is "+cir.area(10));
      System.out.println("Circumference is "+cir.circum(10));
   }
}
class Circle         // class is named Circle
{
   public double circum(double radius)
   {
      return(2*3.14159*radius);           // 2πr
   }
   public double area(double radius)
   {
      return(3.14159*radius*radius);      // πr²
   }
}
```

Sample run 27.10

```
C:\java\src\chap27>java chap27_01
Area is 314.159
Circumference is 62.8318
```

The data and methods within a class can either be:

- Private. These are variables (or methods) which can only be used within the class and have a preceding `private` keyword. By default variables (the members of the class) and methods are private (restricted).
- Public. These are variables (or methods) which can be accessed from other classes and have a preceding `public` keyword.

It is obvious that all classes must have a public content so that they can be accessed by external functions. In Program 27.11 the `Circle` class has three public parts:

- The methods `area()` and `circum()`, which determine the area and circumference of a circle.
- The `Circle` class variable `radius`.

Once the `Circle` class has been declared then the class variable `radius` can be accessed from outside the `Circle` class using:

```
    cir.radius=10;
```

which sets the class variable (`radius`) to a value of 10. The methods then do not need to be passed the value of radius as it is now set within the class (and will stay defined until either a new value is set or the class is deleted).

Java program 27.11 (chap27_11.java)

```
public class chap27_11
{
  public static void main(String[] args)
  {
  Circle cir=new Circle();
    cir.radius=10;
    System.out.println("Area is "+c.area());
    System.out.println("Circumference is "+c.circum());
  }
}
class Circle
{
public float radius;

  public double circum()
  {
    return(2*3.14159*radius);
  }
  public double area()
  {
    return(3.14159*radius*radius);
  }
}
```

Many instances of a class can be initiated and each will have their own settings for their class variables. For example, in Program 27.12, two instances of the `Circle` class have been declared (`cir1` and `cir2`). These are circle objects. The first circle object (`cir1`) has a radius of 15 and the second (`cir2`) has a radius of 10. Sample run 27.11 shows a sample run.

Java program 27.12 (chap27_12.java)

```
public class chap27_12
{
  public static void main(String[] args)
  {
  Circle cir1, cir2;

    cir1=new Circle();
    cir2=new Circle();

    cir1.radius=15;
    cir2.radius=10;


    System.out.println("Area1 is "+cir1.area());
    System.out.println("Area2 is "+cir2.area());

  }
}
class Circle
{
public float radius;
```

```
    public double circum()
    {
       return(2*3.14159*radius);
    }
    public double area()
    {
       return(3.14159*radius*radius);
    }
}
```

Sample run 27.11

```
C:\java\src\chap27>java chap27_03
Area1 is 706.85775
Area2 is 314.159
```

27.9 Constructors

Constructors allow for the initialization of a class. It is a special initialization method that is automatically called whenever a class is declared. The constructor always has the same name as the class name, and no data types are defined for the argument list or the return type.

Program 27.13 has a class which is named `Circle`. The constructor for this class is `Circle()`. Sample run 27.12 shows a sample run. It can be seen that initially when the program is run the message "Constructing a circle" is displayed when the object is created.

Java program 27.13 (chap27_13.java)

```
public class chap27_13
{
   public static void main(String[] args)
   {
   Circle c1,c2;
   double area1,area2;
      c1=new Circle();  c2=new Circle();

      c1.radius=15;     area1=c1.area();
      c2.radius=10;        area2=c2.area();
      System.out.println("Area1 is "+area1);
      System.out.println("Area2 is "+area2);
   }
}
class Circle
{
public float radius;

   public Circle()       // constructor called when object created
   {
      System.out.println("Constructing a circle");
   }
   public double circum()
   {
      return(2*3.14159*radius);
   }
   public double area()
   {
      return(3.14159*radius*radius);
   }
}
```

Sample run 27.12

```
C:\java\src\chap27>java chap27_13
Constructing a circle
Constructing a circle
Areal is 706.85775
Area2 is 314.159
```

C++ has also a destructor which is a member of a function and is automatically called when the class is destroyed. It has the same name as the class name but is preceded by a tilde (~). Normally a destructor is used to clean-up when the class is destroyed. Java normally has no need for destructors as it implements a technique known as garbage collection which gets rids of objects which are no longer needed. If a final clear-up is required then the `finalize()` method can be used. This is called just before the garbage collection. For example:

```
class Circle
{

   public Circle()       // constructor called when object created
   {
      System.out.println("Constructing a circle");
   }
   public finalize()     // called when object deleted
   {
      System.out.println("Goodbye. I'm out with the trash");
   }

   public double circum()
   {
      return(2*3.14159*radius);
   }

   public double area()
   {
      return(3.14159*radius*radius);
   }

}
```

27.10 Method overloading

Often the programmer requires to call a method in a number of ways but wants the same name for the different implementations. Java allows this with method overloading. With overloading the programmer defines a number of methods, each of which has the same name but which are called with a different argument list or return type. The compiler then automatically decides which one should be called. For example in Java program 27.14 the programmer has defined two square methods named `sqr()` and two for `max()`, which is a maximum method. The data type of the argument passed is of a different type for each of the methods, that is, either an `int` or a `double`. The return type is also different. The data type of the parameters passed to these methods is tested by the compiler and it then determines which of the methods it requires to use. Sample run 27.13 shows a sample run.

Java program 27.14 (chap27_14.java)

```
public class chap27_14
{
   public static void main(String[] args)
   {
   MyMath m;
   int val1=4;
   double val2=4.1;

      m=new MyMath();

      System.out.println("Sqr(4)="+m.sqr(val1));
      System.out.println("Sqr(4.1)="+m.sqr(val2));
      System.out.println("Maximum (3,4)="+m.max(3,4));
      System.out.println("Maximum (3.0,4.0)="+m.max(3.0,4.0));
   }
}
class MyMath
{
   public int sqr(int val)
   {
      return(val*val);
   }
   public double sqr(double val)
   {
      return(val*val);
   }
   public int max(int a, int b)
   {
      if (a>b) return(a);
      else return(b);
   }
   public double max(double a, double b)
   {
      if (a>b) return(a);
      else return(b);
   }
}
```

Sample run 27.13

```
C:\java\src\chap27>java chap27_05
Sqr(4)=16
Sqr(4.1)=16.81
Maximum (3,4)=4
Maximum (3.0,4.0)=4.0
```

The argument list of the overloaded function does not have to have the same number of arguments for each of the overloaded functions. Program 27.15 shows an example of an overloaded method which has a different number of arguments for each of the function calls. In this case the `max()` function can either be called with two integer values or by passing an array to it. Arrays will be covered in Section 38.16.

Java program 27.15 (chap27_15.java)

```
public class chap27_15
{
   public static void main(String[] args)
   {
   MyMath   m;
   int      val1=4, arr[]={1,5,-3,10,4};  // array has 5 elements
```

```
   double   val2=4.1;

      m=new MyMath();

      System.out.println("Sqr(4)="+m.sqr(val1));
      System.out.println("Sqr(4.1)="+m.sqr(val2));
      System.out.println("Maximum (3,4)="+m.max(3,4));
      System.out.println("Maximum (array)="+m.max(arr));
   }
}

class MyMath
{
   public int sqr(int val)
   {
      return(val*val);
   }
   public double sqr(double val)
   {
      return(val*val);
   }
   public int max(int a, int b)
   {
      if (a>b) return(a);
      else return(b);
   }
   public int max(int a[])
   {
   int i,max;
      max=a[0];                 // set max to first element
      for (i=1;i<a.length;i++)  // a.length returns array size
         if (max<a[i]) max=a[i];
      return(max);
   }
}
```

Sample run 27.14

```
C:\java\src\chap27>java chap27_15
Sqr(4)=16
Sqr(4.1)=16.81
Maximum (3,4)=4
Maximum (array)=10
```

27.11 Static methods

Declaring an object to get access to the methods in the `MyMath` class is obviously not efficient as every declaration creates a new object. If we just want access to the methods in a class then the methods within the class are declared as static methods. The methods are then accessed by preceding the method with the class name. Static methods are associated with a class and not an object, thus there is no need to create an object with them. Thus in Program 27.16 the methods are accessed by:

```
val=MyMath.sqr(val1);    val=MyMath.max(3,4);
val=MyMath.max(arr);
```

Sample run 27.15 shows a sample run.

Java program 27.16 (chap27_16.java)

```
public class chap27_16
{
   public static void main(String[] args)
   {
   int      val1=4, arr[]={1,5,-3,10,4};     // array has 5 elements
   double   val2=4.1;
      System.out.println("Sqr(val1) "+MyMath.sqr(val1));
      System.out.println("Sqr(arr) "+MyMath.sqr(val2));
      System.out.println("Max(3.0,4.0) "+MyMath.max(3,4));
      System.out.println("Max(arr) "+MyMath.max(arr));
   }
}
class MyMath
{
   public static int sqr(int val)
   {
      return(val*val);
   }
   public static double sqr(double val)
   {
      return(val*val);
   }
   public static int max(int a, int b)
   {
      if (a>b) return(a);
      else return(b);
   }
   public static int max(int a[])
   {
   int i,max;
      max=a[0];                    // set max to first element
      for (i=1;i<a.length;i++)   //a.length returns array size
         if (max<a[i]) max=a[i];
      return(max);
   }
}
```

Sample run 27.15

```
C:\java\src\chap27>java chap27_06
Sqr(val1) 16
Sqr(arr) 16.81
Max(3.0,4.0) 4
Max(arr) 10
```

27.12 Constants

Classes can contain constants which are defined as `public static` class variables. Such as:

```
class MyMath
{
   public static final double E = 2.7182818284590452354;
   public static final double PI = 3.14159265358979323846;
   public int sqr(int val)
   {
      return(val*val);
   }
   public double sqr(double val)
```

```
    {
       return(val*val);
    }
}
```

In this case the value of π is referenced by:

```
omega=2*MyMath.PI*f
```

The static class variables are declared as final so that they cannot be modified when an object is declared. Thus the following is INVALID:

```
MyMath.PI=10.1;
```

Program 27.17 shows a sample program and Sample run 27.16 shows a sample run.

Java program 27.17 (chap27_17.java)

```
public class chap27_17
{
   public static void main(String[] args)
   {
      System.out.println("PI is "+m.PI);
      System.out.println("E is ="+m.E);
   }
}
class MyMath
{
   public static final double E = 2.7182818284590452354;
   public static final double PI = 3.14159265358979323846;
   public static int sqr(int val)
   {
      return(val*val);
   }
   public static double sqr(double val)
   {
      return(val*val);
   }
   public static int max(int a, int b)
   {
      if (a>b) return(a);
      else return(b);
   }
   public static int max(int a[])
   {
   int i,max;

      max=a[0];                     // set max to first element
      for (i=1;i<a.length;i++)    //a.length returns array size
         if (max<a[i]) max=a[i];
      return(max);
   }
}
```

Sample run 27.16

```
C:\java\src\chap27>java chap_08
PI is 3.141592653589793
E is =2.718281828459045
```

27.13 Package statements

The `package` statement defines that the classes within a Java file are part of a given package. The full name of a class is:

package.classFilename

The fully qualified name for a method is:

package.classFilename.method_name ()

Each class file with the same package name is stored in the same directory. For example, the `java.applet` package contains several files, such as:

```
applet.java            appletcontent.java
appletstub.java        audioclip.java
```

Each has a first line of:

```
package java.applet;
```

and the fully classified names of the class files are:

```
java.applet.applet          java.applet.appletcontent
java.applet.appletstub      java.applet.audioclip
```

These can be interpreted as in the `java/applet` directory. An example listing from the class library given in next.

```
java/
java/lang/
java/lang/Object.class
java/lang/Exception.class
java/lang/Integer.class
```

Normally when a Java class is being developed it is not part of a package as it is contained in the current directory. The main packages are:

```
java.applet          java.awt                    java.awt.datatransfer
java.awt.event       java.awt.image              java.awt.peer
java.beans           java.io                     java.lang
java.lang            java.lang.reflect           java.math
java.net             java.rmi                    java.rmi.dgc
java.rmi.registry    java.rmi.server             java.security
java.security.acl    java.security.interfaces    java.sql
java.text            java.util                   java.utils.zip
```

27.14 Import statements

The `import` statement allows previously written code to be included in the applet. This code is stored in class libraries (or packages), which are compiled Java code. For the JDK tools, the Java source code for these libraries is stored in the `src/java` directory.

For example a Java program which uses maths methods will begin with:

```
import java.lang.Math;
```

This includes the `math` class libraries (which is in the `java.lang` package). The default Java class libraries are stored in the `classes.zip` file in the `lib` directory. This file is in a compressed form and should not be unzipped before it is used. The following is an outline of the file.

```
Searching ZIP: CLASSES.ZIP
Testing: java/
Testing: java/lang/
Testing: java/lang/Object.class
Testing: java/lang/Exception.class
Testing: java/lang/Integer.class
   ::          ::
Testing: java/lang/Win32Process.class
Testing: java/io/
Testing: java/io/FilterOutputStream.class
Testing: java/io/OutputStream.class
   ::          ::
Testing: java/io/StreamTenizer.class
Testing: java/util/
Testing: java/util/Hashtable.class
Testing: java/util/Enumeration.class
   ::          ::
Testing: java/util/Stack.class
Testing: java/awt/
Testing: java/awt/Toolkit.class
Testing: java/awt/peer/
Testing: java/awt/peer/WindowPeer.class
   ::          ::
Testing: java/awt/peer/DialogPeer.class
Testing: java/awt/Image.class
Testing: java/awt/MenuItem.class

Testing: java/awt/MenuComponent.class
Testing: java/awt/image/
   ::          ::
   ::          ::
Testing: java/awt/ImageMediaEntry.class
Testing: java/awt/AWTException.class
Testing: java/net/
Testing: java/net/URL.class
Testing: java/net/URLStreamHandlerFactory.class
   ::          ::
Testing: java/net/URLEncoder.class
Testing: java/applet/
Testing: java/applet/Applet.class
Testing: java/applet/AppletContext.class
Testing: java/applet/AudioClip.class
Testing: java/applet/AppletStub.class
```

The other form of the `import` statement is:

```
import package.*;
```

which will import all the classes within the specified package. Table 27.8 lists the main class libraries and some sample libraries.

It can be seen that upgrading the Java compiler is simple, as all that is required is to replace the class libraries with new ones. For example, if the basic language is upgraded then `java.lang.*` files is simply replaced with a new version. The user can also easily add new class libraries to the standard ones.

Table 27.8 Class libraries.

Class libraries	*Description*	*Example libraries*

`java.lang.*`	Java language	`java.lang.Class`
		`java.lang.Number`
		`java.lang.Process`
		`java.lang.String`

`java.io.*`	I/O routines	`java.io.InputStream` `java.io.OutputStream`
`java.util.*`	Utilities	`java.util.BitSet` `java.util.Dictionary`

`java.awt.*`	Windows, menus and graphics	`java.awt.Point`
		`java.awt.Polygon`
		`java.awt.MenuComponent`
		`java.awt.MenuBar`
		`java.awt.MenuItem`

`java.net.*`	Networking (such as sockets, URLs, ftp, telnet and HTTP)	`java.net.ServerSocket`
		`java.net.Socket`
		`java.net.SocketImpl`

`java.applet.*`	Code required to run an applet	`java.applet.AppletContext`
		`java.applet.AppletStub`
		`java.applet.AudioClip`

27.15 Mathematical operations

Java has a basic set of mathematics methods which are defined in the `java.lang.Math` class library. Table 27.9 outlines these methods. An example of a method in this library is `abs()` which can be used to return the absolute value of either a `double`, an `int` or a `long` value. Java automatically picks the required format and the return data type will be of the same data type of the value to be operated on.

As the functions are part of the `Math` class they are preceded with the `Math.` class method. For example:

```
val2=Math.sqrt(val1);
val3=Math.abs(val2);
z=Math.min(x,y);
```

Java program 27.18 shows a few examples of mathematical operations and Sample run 27.17 shows a sample compilation and run session.

Table 27.9 Methods defined in `java.lang.Math`.

Method	*Description*
`double` **`abs`**`(double a)`	Absolute double value of `a`.
`float` **`abs`**`(float a)`	Absolute float value of `a`.
`int` **`abs`**`(int a)`	Absolute integer value of `a`.
`long` **`abs`**`(long a)`	Absolute long value of `a`.
`double` **`acos`**`(double a)`	Inverse cosine of `a`, in the range of 0.0 to `Pi`.
`double` **`asin`**`(double a)`	Inverse sine of `a`, in the range of –`Pi`/2 to `Pi`/2.
`double` **`atan`**`(double a)`	Inverse tangent of `a`, in the range of –`Pi`/2 to `Pi`/2.
`double` **`atan2`**`(double a, double b)`	Converts rectangular co-ordinates (a, b) to polar (r, theta).
`double` **`ceil`**`(double a)`	Smallest whole number greater than or equal to `a`.
`double` **`cos`**`(double a)`	Cosine of an angle.
`double` **`exp`**`(double a)`	Exponential number e (2.718…) raised to the power of `a`.
`double` **`floor`**`( double a)`	Largest whole number less than or equal to `a`.
`double` **`IEEEremainder`**`( double f1, double f2)`	Remainder of `f1` divided by `f2` as defined by IEEE 754.
`double` **`log`**`(double a)`	Natural logarithm (base e) of `a`.
`double` **`max`**`(double a, double b)`	Greater of two `double` values, `a` and `b`.
`double` **`max`**`(float a, float b)`	Greater of two `float` values, `a` and `b`.
`int` **`max`**`(int a, int b)`	Greater of two `int` values, `a` and `b`.
`long` **`max`**`(long a,long b)`	Greater of two `long` values, `a` and `b`.
`double` **`min`**`(double a, double b)`	Smaller of two `double` values, `a` and `b`.
`float` **`min`**`(float a, float b)`	Smaller of two `float` values, `a` and `b`.
`int` **`min`**`(int a, int b)`	Smaller of two `int` values, `a` and `b`.
`long` **`min`**`(long a, long b)`	Smaller of two `long` values, `a` and `b`.
`double` **`pow`**`(double a, double b)`	Value of a raised to the power of `b`.
`double` **`random`**`()`	Random number between 0.0 and 1.0.
`double` **`rint`**`(double b)`	Double value converted into an integer value.
`long` **`round`**`(double a)`	Rounded value of a.
`int` **`round`**`(float a)`	Rounded value of a.
`double` **`sin`**`(double a)`	Sine of `a`.
`double` **`sqrt`**`(double a)`	Square root of `a`.
`double` **`tan`**`(double a)`	Tangent of a.

Java program 27.18 (chap27_18.java)

```
import java.lang.Math;
public class chap27_18
{
   public static void main(String[] args)
   {
   double x,y,z;
   int i;
      i=10;
      y=Math.log(10.0);
      x=Math.pow(3.0,4.0);
      z=Math.random(); // random number from 0 to 1
      System.out.println("Value of i is " + i);
      System.out.println("Value of log(10) is " + y);
      System.out.println("Value of 3^4 is " + x);
      System.out.println("A random number is " + z);
      System.out.println("Square root of 2 is " + Math.sqrt(2));
   }
}
```

Sample run 27.17

```
C:\java\src\chap27>javac chap27_18.java
C:\java\src\chap27>java chap27_18
Value of i is 10
Value of log(10) is 2.30259
Value of 3^4 is 81
A random number is 0.0810851
Square root of 2 is 1.41421
```

Java has also two predefined mathematical constants. These are:

- `Pi` is equivalent to 3.14159265358979323846
- `E` is equivalent to 2.71828182845904523354

27.16 Arrays

An array stores more than one value, of a common data type, under a collective name. Each value has a unique slot and is referenced using an indexing technique. For example a circuit with five resistor components could be declared within a program with five simple float declarations. If these resistor variables were required to be passed into a method then all five values would have to be passed through the parameter list. A neater way uses arrays to store all of the values under a common name (in this case `R`). Then a single array variable can then be passed into any method that uses it.

The declaration of an array specifies the data type, the array name and the number of elements in the array in brackets (`[]`). The following gives the standard format for an array declaration.

data_type array_name`[];`

The array is then created using the `new` keyword. For example, to declare an integer array named `new_arr` with 200 elements then the following is used:

```
int new_arr[];
   new_arr=new int[200];
```

or, in a single statement, with:

```
int new_arr[]=new int[200];
```

Java Program 27.19 gives an example of this type of declaration where an array (`arr`) is filled with 20 random numbers (Figure 27.18 shows a sample run).

Like C, the first element of the array is indexed `0` and the last element as `size-1`. The compiler allocates memory for the first element `array_name[0]` to the last array element `array_name[size-1]`. The number of bytes allocated in memory will be the number of elements in the array multiplied by the number of bytes used to store the data type of the array.

Java program 27.19 (chap27_19.java)

```
public class chap27_19
{
   public static void main(String[] args)
   {
      double arr[]=new double[20];
      int    i;

      for (i=0;i<20;i++)   arr[i]=Math.random();
      for (i=0;i<20;i++)   System.out.println(arr[i]);
   }
}
```

Sample run 27.18

```
C:\java\src\chap27>java chap27_19
0.6075765411193292
0.7524300612559963
0.8100796233691735
0.45045015538577704
0.32390753542869755
0.34033464565015836
0.5079716192482706
0.6426253967106341
0.7691175624480434
0.6475110502592946
0.1416366173783874
0.21181433233783153
0.21758072702009412
0.24203490620407764
0.7587570097412505
0.4470154908107362
0.19823448357551965
0.7340429664182364
0.7402367706819387
0.8975606689180567
```

Another way to create and initialize an array is to define the elements within the array within curly brackets (`{}`). A comma separates each element in the array. The size of the array is then equal to the number of elements in the array. For example:

```
int      arr1[]={-3, 4, 10, 100, 30, 22};
```

```
String   menus[]={"File", "Edit", "View", "Insert", "Help"};
```

A particular problem in most programming languages (such as C and Pascal) exists when accessing array elements which do not exist, especially by accessing an array element which is greater than the maximum size of the array. Java overcomes this by being able to determine the size of the array. This is done with the `length` field. For example, the previous example can be modified with:

```
for (i=0;i<arr.length;i++) arr[i]=Math.random();
for (i=0;i<arr.length;i++) System.out.println(arr[i]);
```

Java Program 27.20 gives an example of an array of strings. In this case the array contains the names of playing cards. When run, the program displays five random playing cards. Sample run 27.19 shows a sample run.

Java program 27.20 (chap27_20.java)

```
public class chap27_20
{
   public static void main(String[] args)
   {
   int cards,pick;
   String   Card[]={"Ace","King","Queen","Jack","10",
                 "9", "8", "7", "6", "5", "4", "3", "2"};

      for (cards=0;cards<5;cards++)
      {
         pick=(int)Math.round((Card.length)*Math.random());
         System.out.print(Card[pick] + " ");
      }
   }
}
```

Sample run 27.19

```
Ace  10  King  2  3
```

Multi-dimensional arrays are declared in a similar manner. For example an array with 3 rows and 4 columns is declared with either of the following:

```
int arr[][]=new int[3][4];
```

or if the initial values are known with:

```
int arr[][]= { {1,2,3,4}, {5,6,7,8}, {9,10,11,12} } ;
```

where `arr[0][0]` is equal to 1, `arr[1][0]` is equal to 5, `arr[2][3]` is equal to 12, and so on. This is proved with Java Program 27.21 and Sample run 27.20.

Java program 27.21 (chap27_21.java)

```
public class chap27_21
{
   public static void main(String[] args)
   {
   int   row,col;
```

```
   int    arr[][]={ {1,2,3,4},{5,6,7,8}, {9,10,11,12} };

   for (row=0;row<3;row++)
      for (col=0;col<4;col++)
         System.out.println("Arr["+row+"]["+col+"]="+arr[row][col]);
   }
}
```

Sample run 27.20

```
C:\java\src\chap27>java chap27_05
Arr[0][0]=1
Arr[0][1]=2
Arr[0][2]=3
Arr[0][3]=4
Arr[1][0]=5
Arr[1][1]=6
Arr[1][2]=7
Arr[1][3]=8
Arr[2][0]=9
Arr[2][1]=10
Arr[2][2]=11
Arr[2][3]=12
```

Java (Events and Windows)

28.1 Introduction

As has been previously discussed a Java program can either be run as an applet within a WWW browser (such as Microsoft Internet Explorer or Netscape Communicator) or can be interpreted as a standalone program. The basic code within each program is almost the same and they can be easily converted from one to the other (typically a Java program will be run through an interpreter to test its results and then converted to run as an applet).

28.2 Applet tag

An applet is called from within an HTML script with the `APPLET` tag, such as:

```
<applet code="Test.class" width=200 height=300></applet>
```

which loads an applet called `Test.class` and sets the applet size to 200 pixels wide and 300 pixels high. Table 28.1 discusses some optional parameters.

Table 28.1 Other applet HTML parameters.

Applet parameters	*Description*
`CODEBASE=`*codebaseURL*	Specifies the directory (*codebaseURL*) that contains the applet's code.
`CODE=`*appletFile*	Specifies the name of the file (*appletFile*) of the compiled applet.
`ALT=`*alternateText*	Specifies the alternative text that is displayed if the browser cannot run the Java applet.
`NAME=`*appletInstanceName*	Specifies a name for the applet instance (*appletInstanceName*). This makes it possible for applets on the same page to find each other.
`WIDTH=`*pixels* `HEIGHT=`*pixels*	Specifies the initial width and height (in *pixels*) of the applet.
`ALIGN=`*alignment*	Specifies the *alignment* of the applet. Possible values are: `left`, `right`, `top`, `texttop`, `middle`, `absmiddle`, `baseline`, `bottom` and `absbottom`.
`VSPACE=`*pixels* `HSPACE=`*pixels*	Specifies the number of *pixels* above and below the applet (`VSPACE`) and on each side of the applet (`HSPACE`).

28.2.1 Applet viewer

A useful part of the JDK tools is an applet viewer which is used to test applets before they are run within the browser. The applet viewer on the PC version is `AppletViewer.exe` and the supplied argument is the HTML file that contains the applet tag(s). It then runs all the associated applets in separate windows.

28.3 Creating an applet

Java applet 28.1 shows a simple Java applet which displays two lines of text and HTML script 28.1 shows how the applet integrates into an HTML script.

First the Java applet (`chap28_1.java`) is created. In this case the `edit` program is used. The directory listing below shows that the files created are `chap28_1.java` and `chap28_1.html` (note that Windows NT/95/98 displays the 8.3 filename format on the left-hand side of the directory listing and the long filename on the right-hand side).

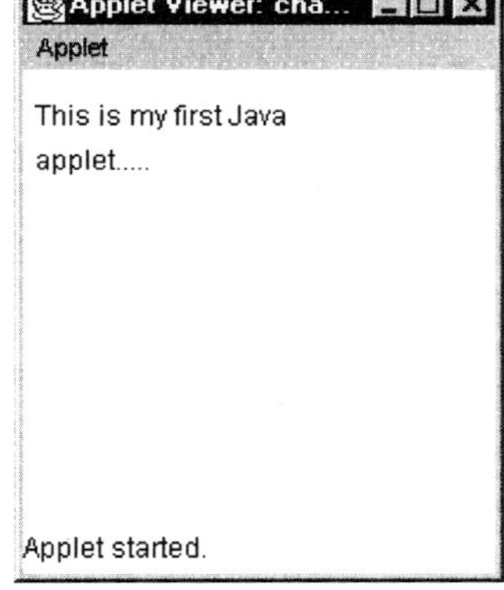

Java applet 28.1 (`chap28_1.java`)

```
import java.awt.*;
import java.applet.*;

public class chap28_1 extends Applet
{
  public void paint(Graphics g)
  {
   g.drawString("This is my first Java",5,25);
   g.drawString("applet.....",5,45);
  }
}
```

HTML script 28.1 (`chap28_1.html`)

```
<HTML><TITLE>First Applet</TITLE>
<APPLET CODE=chap28_1.class WIDTH=200
HEIGHT=200></APPLET></HTML>
```

Sample run 28.1

```
C:\java\src\chap28> edit chap28_1.java
C:\java\src\chap28> edit chap28_1.html
C:\java\src\chap28> dir
CHAP28_~1 HTM          111  14/05/98  18:40 chap28_1.html
CHAP28_~1 JAV          228  13/05/98  22:35 chap28_1.java
```

Next the Java applet is compiled using the `javac.exe` program. It can be seen from the listing that, if there are no errors, the compiled file is named `chap28_1.class`. This can then be used, with the HTML file, to run as an applet.

Sample run 28.2

```
C:\java\src\chap28> javac chap28_1.java
C:\java\src\chap28> dir
CHAP3_~1 HTM          111  14/05/98  18:40 chap28_1.html
CHAP3_~1 JAV          228  14/05/98  18:43 chap28_1.java
CHAP3_~1 CLA          460  14/05/98  18:43 chap28_1.class
C:\java\src\chap28> appletviewer chap28_1.html
```

28.4 Applet basics

Java applet 28.1 recaps the previous Java applet. This section analyses the main parts of this Java applet.

Java applet 28.1 (chap28_1.java)

```
import java.awt.*;
import java.applet.*;
public class chap28_1 extends Applet
{
  public void paint(Graphics g)
  {
   g.drawString("This is my first Java",5,25);
   g.drawString("applet.....",5,45);
  }
}
```

28.4.1 Applet class

The start of the applet code is defined in the form:

```
public class chap28_1 extends Applet
```

which informs the Java compiler to create an applet named `chap28_1` that extends the existing Applet class. The `public` keyword at the start of the statement allows the Java browser to run the applet, while if it is omitted the browser cannot access the applet.

The `class` keyword is used to creating a class object named `chap28_1` that extends the applet class. After this the applet is defined between the left and right braces (grouping symbols).

28.4.2 Applet methods

Methods allow Java applets to be split into smaller sub-tasks (just as C uses functions). These methods have the advantage that:

- They allow code to be reused.
- They allow for top-level design.
- They make applet debugging easier as each method can be tested in isolation to the rest of the applet.

A method has the `public` keyword, followed by the return value (if any) and the name of the method. After this the parameters passed to the method are defined within rounded brackets. Recapping from the previous example:

```
public void paint(Graphics g)
{
   g.drawString("This is my first Java",5,25);
   g.drawString("applet.....",5,45);
}
```

This method has the `public` keyword which allows any user to execute the method. The `void` type defines that there is nothing returned from this method and the name of the method is `paint()`. The parameter passed into the method is `g` which has the data type of `Graphics`.

Within the `paint()` method the `drawString()` method is called. This method is defined in `java.awt.Graphics` class library (this library has been included with the `import java.awt.*` statement. The definition for this method is:

```
public abstract void drawString(String str, int x, int y)
```

which draws a string of characters using the current font and colour. The `x,y` position is the starting point of the baseline of the string (`str`).

It should be noted that Java is case sensitive and the names given must be referred to in the case that they are defined as.

28.5 The paint() object

The `paint()` object is the object that is called whenever the applet is redrawn. It will thus be called whenever the applet is run and then it is called whenever the applet is redisplayed.

Java applet 28.2 shows how a `for()` loop can be used to display the square and cube of the values from 0 to 9. Notice that the final value of `i` within the `for()` loop is 9 because the end condition is `i<10` (while `i` is less than 10).

Java applet 28.2 (`chap28_2.java`)

```
import java.awt.*;
import java.applet.*;
public class chap28_2 extends Applet
{
  public void paint(Graphics g)
  {
  int     i;
   g.drawString("Value Square Cube",5,10);
   for (i=0;i<10;i++)
   {
     g.drawString(""+ i,5,20+10*i);
     g.drawString(""+ i*i ,45,20+10*i);
     g.drawString(""+ i*i*i,85,20+10*i);
   }
  }
}
```

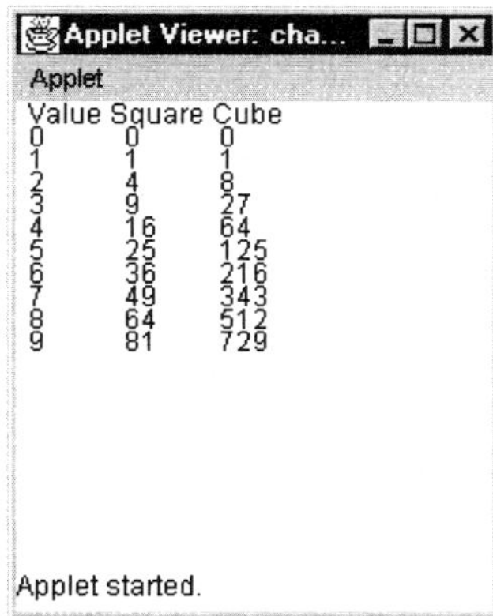

HTML script 28.2 (`chap28_2.html`)

```
<HTML> <TITLE>First Applet</TITLE>
<APPLET CODE=chap28_2.class WIDTH=200 HEIGHT=200>
</APPLET></HTML>
```

28.6 Java events

The previous chapters have discussed the Java programming language. This chapter investigates event-driven programs. Traditional methods of programming involve writing a program which flows from one part to the next in a linear manner. Most programs are designed using a top-down structured design, where the task is split into a number of sub-modules, these are then called when they are required. This means that it is relatively difficult to interrupt the operation of a certain part of a program to do another activity, such as updating the graphics display.

In general Java is event-driven where the execution of a program is not predefined and its execution is triggered by events, such as a mouse click, a keyboard press, and so on. The main events are:

- Initialization and exit methods (`init()`, `start()`, `stop()` and `destroy()`).
- Repainting and resizing (`paint()`).
- Mouse events (`mouseUp()`, `mouseDown()` and `mouseDrag()` for Java 1.0, and `mousePressed()`, `mouseReleased()` and `mouseDragged()` for Java 1.1).
- Keyboard events (`keyUp()` and `keyDown()` for Java 1.0, and `keyPressed()` and `keyReleased()` for Java 1.1).

28.7 Java 1.0 and Java 1.1

There has been a big change between Java 1.0 and Java 1.1. The main change is to greatly improve the architecture of the AWT, which helps in compatibility. Java 1.0 programs will work with most browsers, but only upgraded browsers will work with Java 1.1. The main reasons to upgrade though are:

- Java 1.1 adds new features.
- Faster architecture with more robust implementations of the AWT.
- Support for older facilities will be phased out.

28.7.1 Deprecation

Older facilitates which are contained with Java 1.0 and are still supported Java 1.1, but the Java compiler gives a deprecation warning. This warning means that the facility will eventually be phased-out. The warning is in the form of:

```
C:\jdk1.1.6\src\chap28>javac chap28_1.java
Note: chap28_1.java uses a deprecated API.  Recompile with "-deprecation" for
details.
1 warning
```

The full details on the deprecation can be found by using the –deprecation flag. For example:

```
C:\jdk1.1.6\src\chap28>javac -deprecation chap28_1.java
chap28_1.java:9: Note: The method boolean mouseUp(java.awt. Event, int, int)
in class java.awt.Component has been deprecated, and class chap28_1 (which is
not deprecated) overrides it.

  public boolean mouseUp(Event event,
                 ^
Note: chap28_1.java uses a deprecated API.  Please consult the documentation
for a better alternative.
1 warning
```

28.8 Initialization and exit methods

Java applets have various reserved methods which are called when various events occur. Table 28.2 shows typical initialization methods and their events, and Figure 28.1 illustrates how they are called.

Table 28.2 Java initialization and exit methods.

Method	*Description*
`public void init()`	This method is called each time the applet is started. It is typically used to add user interface components.
`public void stop()`	This method is called when the user moves away from the page on which the applet resides. It is thus typically used to stop processing while the user is not accessing the applet. Typically it is used to stop animation or audio files, or mathematical processing. The `start()` method normally restarts the processing.
`public void paint(Graphics g)`	This method is called when the applet is first called and whenever the user resizes or moves the windows.
`public void destroy()`	This method is called when the applet is stopped and is normally used to release associated resources, such as freeing memory, closing files, and so on.

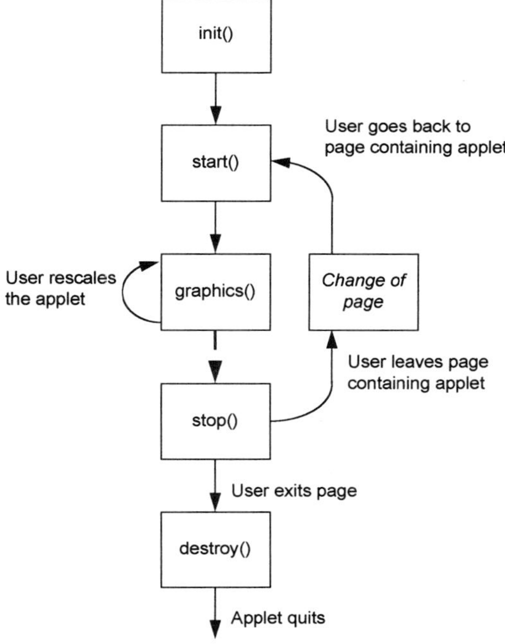

Figure 28.1 Java initialization and exit methods.

Java applet 28.3 gives an example using the `init()` and `start()` methods. The variable `i` is declared within the applet and it is set to a value of 5 in the `init()` method. The `start()` method then adds 6 onto this value. After this the `paint()` method is called so that it displays the value of `i` (which should equal 11).

Java applet 28.3 (chap28_3.java)

```
import java.awt.*;
import java.applet.*;

public class chap28_3 extends Applet
{
int     i;
  public void init()
  {
    i=5;
  }
  public void start()
  {
    i=i+6;
  }
  public void paint(Graphics g)
  {
   g.drawString("The value of i is "
                + i,5,25);
  }
}
```

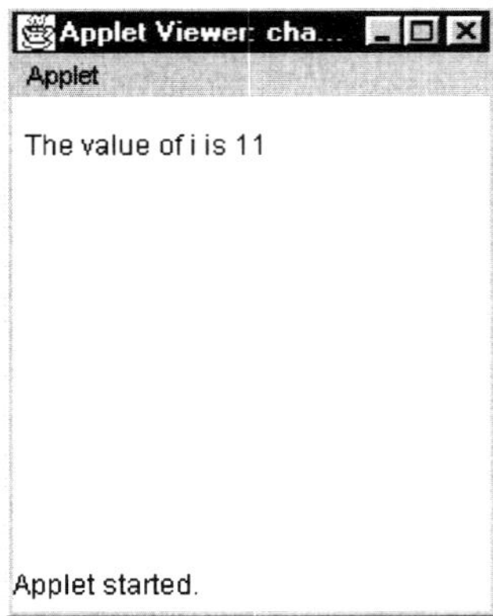

HTML script 28.3 (chap28_3.html)

```
<HTML>
<TITLE>Applet</TITLE>
<APPLET CODE=chap28_3.class
WIDTH=200 HEIGHT=200></APPLET></HTML>
```

28.9 Mouse events in Java 1.0

Most Java applets require some user interaction, normally with the mouse or from the keyboard. A mouse operation causes mouse events. The six basic mouse events which are supported in Java 1.0 are:

- `mouseUp(Event evt, int x, int y)`
- `mouseDown(Event evt, int x, int y)`
- `mouseDrag(Event evt, int x, int y)`
- `mouseEnter(Event evt, int x, int y)`
- `mouseExit(Event evt, int x, int y)`
- `mouseMove(Event evt, int x, int y)`

Java applet 28.4 uses three mouse events to display the current mouse cursor. Each of the methods must return a true value to identify that the event has been handled successfully (the return type is of data type boolean thus the return could only be a true or a false). In the example applet, on moving the mouse cursor with the left mouse key pressed down the `mouseDrag()` method is automatically called. The x and y coordinate of the cursor is stored in the x and y variable when the event occurs. This is used in the methods to build a message string (in the case of the drag event the string name is `MouseDragMsg`).

The `mouseEnter()` method is called when the mouse enters the component, `mouseExit()` is called when the mouse exits the component and `mouseMove()` when the mouse moves (the mouse button is up).

Java applet 28.4 (chap28_4.java)

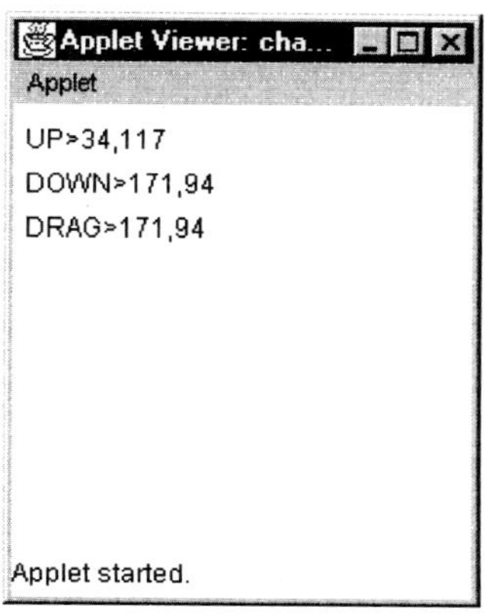


```
import java.awt.*;
import java.applet.*;
public class chap28_4 extends Applet
{
String  MouseDownMsg=null;
String  MouseUpMsg=null;
String  MouseDragMsg=null;

  public boolean mouseUp(Event event,
   int x, int y)
  {
    MouseUpMsg = "UP>" +x + "," + y;
    repaint();   // call paint()
    return(true);
  }
  public boolean mouseDown(Event event,
   int x, int y)
  {
    MouseDownMsg = "DOWN>" +x + "," + y;
    repaint();   // call paint()
    return(true);
  }

  public boolean mouseDrag(Event event,
   int x, int y)
  {
    MouseDragMsg = "DRAG>" +x + "," + y;
    repaint();   // call paint()
    return(true);
  }

  public void paint(Graphics g)
  {
    if (MouseUpMsg !=null)
      g.drawString(MouseUpMsg,5,20);
    if (MouseDownMsg !=null)
      g.drawString(MouseDownMsg,5,40);
    if (MouseDragMsg !=null)
      g.drawString(MouseDragMsg,5,60);
  }
}
```

HTML script 28.4 (chap28_4.html)

```
<HTML><TITLE>Applet</TITLE>
<APPLET CODE=chap28_4.class WIDTH=200
HEIGHT=200></APPLET></HTML>
```

28.10 Mouse event handling in Java 1.1

Java 1.1 has changed the event handling. In its place is the concept of listeners. Each listener receivers notification about the types of events that it is interested in. For mouse handling the two listeners are:

- `MouseListener`. This has the associated methods of:
  - `mousePressed()` which is equivalent to `mouseDown()` in Java 1.0
  - `mouseReleased()` which is equivalent to `mouseUp()` in Java 1.0
  - `mouseEntered()` which is equivalent to `mouseEnter()` in Java 1.0
  - `mouseExited()` which is equivalent to `mouseExit()` in Java 1.0
  - `mouseClicked()`
- `MouseMotionListener`. This has the associated methods of:
  - `mouseDragged()` which is equivalent to `mouseDrag()` in Java 1.0
  - `mouseMoved()` which is equivalent to `mouseMove()` in Java 1.0

28.10.1 Mouse methods

The arguments passed to the methods have also changed, in that there are no `x` and `y` integers passed, and there is no return from them. Their syntax is as follows:

```
public void mousePressed(MouseEvent event) {};
public void mouseReleased(MouseEvent event) {};
public void mouseClicked(MouseEvent event) {};
public void mouseExited(MouseEvent event) {};
public void mouseEntered(MouseEvent event) {};
public void mouseDragged(MouseEvent event) {};
public void mouseMoved(MouseEvent event) {};
```

The x and y coordinates of the mouse event can be found by accessing the `getX()` and `getY()` methods of the event, such as:

```
x=event.getX();    y=event.getY();
```

28.10.2 Event class

The other main change to the Java program is to add the `java.awt.event` package, with:

```
import java.awt.event.*;
```

28.10.3 Class declaration

The class declaration is changed so that the appropriate listener is defined. If both mouse listeners are required then the class declaration is as follows:

```
public class class_name extends Applet
   implements MouseListener, MouseMotionListener
```

28.10.4 Defining components that generate events

The components which generate events must be defined. In the case of a mouse event these are added as:

```
         this.addMouseListener(this);
         this.addMouseMotionListener(this);
```

28.10.5 Updated Java program

Java applet 28.5 gives the updated Java program with Java 1.1 updates.

Java applet 28.5 (chap28_5.java)

```
import java.awt.*;
import java.applet.*;
import java.awt.event.*;

public class chap28_5 extends Applet
implements MouseListener, MouseMotionListener
{
String  MouseDownMsg=null;
String  MouseUpMsg=null;
String  MouseDragMsg=null;

  public void init()
  {
        this.addMouseListener(this);
        this.addMouseMotionListener(this);
  }

  public void paint(Graphics g)
  {
    if (MouseUpMsg !=null)   g.drawString(MouseUpMsg,5,20);
    if (MouseDownMsg !=null) g.drawString(MouseDownMsg,5,40);
    if (MouseDragMsg !=null) g.drawString(MouseDragMsg,5,60);
  }

  public void mousePressed(MouseEvent event)
  {
    MouseUpMsg = "UP>" +event.getX() + "," + event.getY();
    repaint();   // call paint()
  }
  public void mouseReleased(MouseEvent event)
  {
    MouseDownMsg = "DOWN>" +event.getX() + "," + event.getY();
    repaint();   // call paint()
  }
  public void mouseClicked(MouseEvent event) {};
  public void mouseExited(MouseEvent event) {};
  public void mouseEntered(MouseEvent event) {};

   public void mouseDragged(MouseEvent event)
  {
    MouseDragMsg = "DRAG>" +event.getX() + "," + event.getY();
    repaint();   // call paint()
  }
  public void mouseMoved(MouseEvent event) {};
}
```

28.11 Mouse selection in Java 1.0

In many applets the user is prompted to select an object using the mouse. To achieve this the x and y position of the event is tested to determine if the cursor is within the defined area. Java applet 28.6 is a program which allows the user to press the mouse button on the applet screen. The applet then uses the mouse events to determine if the cursor is within a given area of the screen (in this case between 10,10 and 100,50). If the user is within this defined area then the message displayed is HIT, else it is MISS. The graphics method `g.drawRect(x1,y1,x2,y2)` draws a rectangle from (x1,y1) to (x2,y2).

Java applet 28.6 (chap28_6.java)

```
import java.awt.*;
import java.applet.*;
public class chap28_6 extends Applet
{
String  Msg=null;
int     x_start,y_start,x_end,y_end;

  public void init()
  {
    x_start=10;    y_start=10;
    x_end=100;     y_end=50;
  }

  public boolean mouseUp(Event event,
   int x, int y)
  {
    if ((x>x_start) && (x<x_end) &&
            (y>y_start) && (y<y_end))
               Msg = "HIT";
    else Msg="MISS";
    repaint();   // call paint()
    return(true);
  }

  public boolean mouseDown(Event event,
   int x, int y)
  {
    if ((x>x_start) && (x<x_end) &&
            (y>y_start) && (y<y_end))
               Msg = "HIT";
    else Msg="MISS";
    repaint();   // call paint()
    return(true);
  }

  public void paint(Graphics g)
  {

  g.drawRect(x_start,y_start,x_end,y_end);
    g.drawString("Hit",30,30);
    if (Msg !=null)
     g.drawString("HIT OR MISS: "
         + Msg,5,80);
  }
}
```


HTML script 28.5 (chap28_5.html)

```
<HTML>
<TITLE>Applet</TITLE>
<APPLET CODE=chap28_6.class WIDTH=200
HEIGHT=200></APPLET></HTML>
```

Java applet 28.7 gives the updated Java program with Java 1.1 updates.

Java applet 28.7 (chap28_7.java)

```
import java.awt.*;
import java.applet.*;
import java.awt.event.*;

public class chap28_7 extends Applet implements MouseListener
{
```

```
String  Msg=null;
int     x_start,y_start,x_end,y_end;

  public void init()
  {
    x_start=10;    y_start=10;
    x_end=100;     y_end=50;
    this.addMouseListener(this);
  }

  public void mousePressed(MouseEvent event)
  {
    int x,y;
    x=event.getX(); y=event.getY();

    if ((x>x_start) && (x<x_end) && (y>y_start) && (y<y_end))
                Msg = "HIT";
    else Msg="MISS";
    repaint();    // call paint()
  }

  public void mouseReleased(MouseEvent event)
  {
    int x,y;
    x=event.getX();  y=event.getY();
    if ((x>x_start) && (x<x_end) && (y>y_start) && (y<y_end))
                Msg = "HIT";
    else Msg="MISS";
    repaint();    // call paint()
  }
  public void mouseEntered(MouseEvent event) {};
  public void mouseExited(MouseEvent event) {};
  public void mouseClicked(MouseEvent event) {};

  public void paint(Graphics g)
  {
    g.drawRect(x_start,y_start,x_end,y_end);
    g.drawString("Hit",30,30);
    if (Msg !=null)
     g.drawString("HIT OR MISS: " + Msg,5,80);
  }
}
```

28.12 Keyboard input in Java 1.0

Java 1.0 provides for two keyboard events, these are:

- `keyUp(Event evt, int key)`. Called when a key has been released
- `keyDown(Event evt, int key)`. Called when a key has been pressed

The parameters passed into these methods are `event` (which defines the keyboard state) and an integer `Keypressed` which describes the key pressed.

The event contains an identification as to the type of event it is. When one of the function keys is pressed then the variable `event.id` is set to the macro `Event. KEY_ACTION` (as shown in Java applet 28.8). Other keys, such as the Cntrl, Alt and Shift keys, set bits in the `event.modifier` variable. The test for the Cntrl key is:

```
if ((event.modifiers & Event.CTRL_MASK)!=0)
 Msg="CONTROL KEY "+KeyPress;
```

This tests the `CTRL_MASK` bit; if it is a 1 then the Cntrl key has been pressed. Java applet 28.8 shows its uses.

Java applet 28.8 (chap28_8.java)

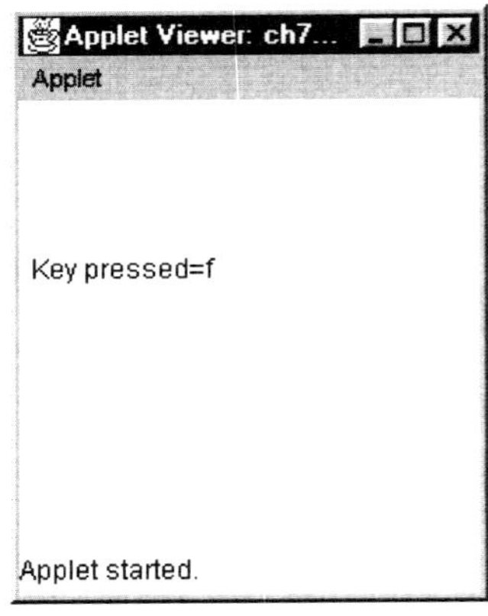


```
import java.awt.*;
import java.applet.*;

public class chap28_8 extends Applet
{
String  Msg=null;

 public boolean keyUp(Event event,
  int KeyPress)
 {
   Msg="Key pressed="+(char)KeyPress;
   repaint();   // call paint()
   return(true);
  }
  public void paint(Graphics g)
  {
   if (Msg !=null)
         g.drawString(Msg,5,80);
  }
}
```

HTML script 28.6 (chap28_6.html)

```
<HTML><TITLE>Applet</TITLE>
<APPLET CODE=chap28_8.class WIDTH=200
HEIGHT=200></APPLET></HTML>
```

Java applet 28.9 (chap28_9.java)

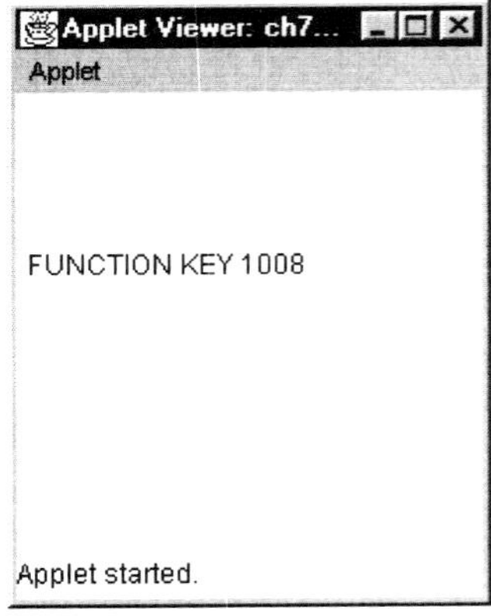


```
import java.awt.*;
import java.applet.*;

public class chap28_9 extends Applet
{
String  Msg=null;
 public boolean keyDown(Event event, int KeyPress)
 {
 if (event.id == Event.KEY_ACTION)
   Msg="FUNCTION KEY "+KeyPress;
 else if ((event.modifiers & Event.SHIFT_MASK)!=0)
   Msg="SHIFT KEY "+KeyPress;
 else if ((event.modifiers & Event.CTRL_MASK)!=0)
   Msg="CONTROL KEY "+KeyPress;
 else if ((event.modifiers & Event.ALT_MASK)!=0)
   Msg="ALT KEY "+KeyPress;
 else Msg=""+(char)KeyPress;
 repaint();   // call paint()
 return(true);
 }
 public void paint(Graphics g)
 {
  if (Msg!=null)
    g.drawString(Msg,5,80);
 }
}
```

28.13 Keyboard events in Java 1.1

Java 1.1 has changed the event handling. In its place is the concept of listeners. Each listener receivers notification about the types of events that it is interested in. For keyboard handling the two listeners are:

- `KeyListener`. This has the associated methods of:
  - `keyPressed()` which is equivalent to `keyDown()` in Java 1.0
  - `keyReleased()` which is equivalent to `keyUp()` in Java 1.0
  - `keyTyped()`

28.13.1 Key methods

The arguments passed to the methods have also changed. Their syntax is as follows:

```
public void keyPressed(KeyEvent event) {}
public void keyReleased(KeyEvent event) {}
public void keyTyped(KeyEvent event) {}
```

28.13.2 Event class

Another change to the Java program is to add the `java.awt.event` package, with:

```
import java.awt.event.*;
```

28.13.3 Class declaration

The class declaration is changed so that the appropriate listener is defined. If the key listener is required then the class declaration is as follows:

```
public class class_name extends Applet implements KeyListener
```

28.13.4 Defining components that generate events

The components which generate events must be defined. In the case of a key event these are added as:

```
compname.addKeyListener(this);
```

28.13.5 Updated Java program

Java applet 28.10 gives the updated Java program with Java 1.1 updates. In this case a `TextField` component is added to the applet (`text`). When a key is pressed on this component then the `keyPressed` event listener is called, when one is released the `keyReleased` is called. Figure 28.2 gives a sample run.

The `getKeyCode()` method is used to determine the key that has been activated. In the event method the `KeyEvent` defines a number of `VK_` constants, such as:

`VK_F1`	Function Key F1	`VK_A`	Character 'A'	`VK_ALT`	Alt key
`VK_CONTROL`	Control Key	`VK_0`	Character '0'	`VK_SHIFT`	Shift key

Java applet 28.10 (chap28_10.java)

```
import java.awt.*;
import java.applet.*;
import java.awt.event.*;

public class chap28_10 extends Applet implements KeyListener
{
String  Msg=null;
TextField text;


      public void init()
      {
         text=new TextField(20);

         add(text);
         text.addKeyListener(this);
      }

      public void keyPressed(KeyEvent event)
      {
      int KeyPress;

         KeyPress=event.getKeyCode();

         if (KeyPress == KeyEvent.VK_ALT)  Msg="ALT KEY";
         else if (KeyPress == KeyEvent.VK_CONTROL)  Msg="Cntrl KEY ";
         else if (KeyPress == KeyEvent.VK_SHIFT) Msg="SHIFT KEY ";
         else if (KeyPress == KeyEvent.VK_RIGHT) Msg="RIGHT KEY ";
         else if (KeyPress == KeyEvent.VK_LEFT)  Msg="LEFT KEY ";
         else if (KeyPress == KeyEvent.VK_F1)    Msg="Function key F1";
         else Msg="Key:"+(char)KeyPress;

         text.setText(Msg);
      }
      public void keyReleased(KeyEvent event) { }
      public void keyTyped(KeyEvent event)  { }

}
```

Figure 28.2 Sample run.

28.14 Buttons and events

Java applet 28.11 creates three `Button` objects. These are created with the `add()` function which displays the button in the applet window.

Java applet 28.11 (`chap28_11.java`)

```
import java.awt.*;
import java.applet.*;

public class chap28_11 extends Applet
{
  public void init()
  {
    add(new Button("Help"));
    add(new Button("Show"));
    add(new Button("Exit"));
  }
}
```

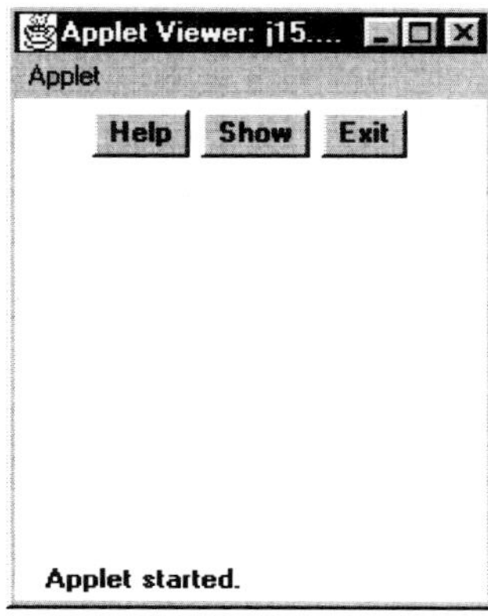

An alternative approach to creating buttons is to declare them using the Button type. For example the following applet is equivalent to Java applet 28.11. The names of the button objects, in this case, are `button1`, `button2` and `button3`.

```
import java.applet.*;
import java.awt.*;

public class chap28_11 extends Applet
{
Button button1= new Button("Help");
Button button2= new Button("Show");
Button button3= new Button("Exit");


    public void init()
    {
        add(button1);
        add(button2);
        add(button3);
    }
}
```

28.15 Action with Java 1.0

The action function is called when an event occurs, such as a keypress, button press, and so on. The information on the event is stored in the Event parameter. Its format is:

```
public boolean action(Event evt, Object obj)
```

where event is made with the specified target component, time stamp, event type, x and y coordinates, keyboard key, state of the modifier keys and argument. These are:

- `evt.target` is the target component.
- `evt.when` is the time stamp.
- `evt.id` is the event type.
- `evt.x` is the *x* coordinate.
- `evt.y` is the *y* coordinate.
- `evt.key` is the key pressed in a keyboard event.
- `evt.modifiers` is the state of the modifier keys.
- `evt.arg` is the specified argument.

Java applet 28.12 contains an example of the action method. It has two buttons (named New 1 and New 2). When any of the buttons is pressed the action method is called. Figure 28.3 shows the display when either of the buttons is pressed. In the left-hand side of Figure 28.3 the New 1 button is pressed and the right-hand side shows the display after the New 2 button is pressed. It can be seen that differences are in the `target`, `arg` parameter and the `x`, `y` coordinate parameters.

Java applet 28.12 (chap28_12.java)

```
import java.applet.*;
import java.awt.*;

public class chap28_12 extends Applet
{

String Msg1=null, Msg2, Msg3, Msg4;

    public void init()
    {
        add (new Button("New 1"));
        add (new Button("New 2"));
    }

    public boolean action(Event evt, Object obj)
    {
        Msg1= "Target= "+evt.target;
        Msg2= "When= " + evt.when + " id=" + evt.id +
                " x= "+ evt.x + " y= " + evt.y;
        Msg3= "Arg= " + evt.arg + " Key= " + evt.key;
        Msg4= "Click= " + evt.clickCount;
        repaint();
        return true;
    }

    public void paint(Graphics g)
    {
        if (Msg1!=null)
        {
                g.drawString(Msg1,30,80);
                g.drawString(Msg2,30,100);
                g.drawString(Msg3,30,120);
                g.drawString(Msg4,30,140);
        }
    }
}
```

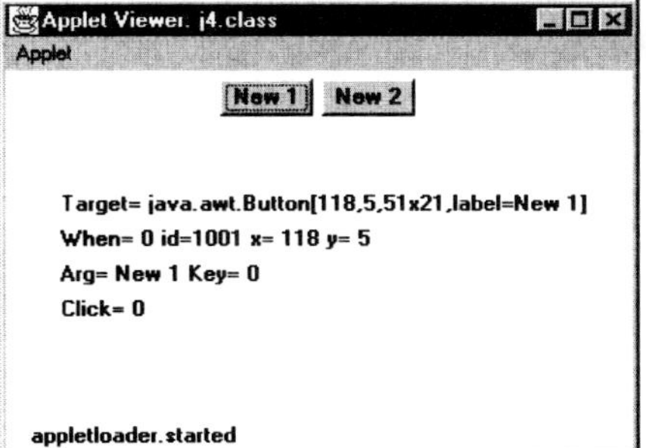

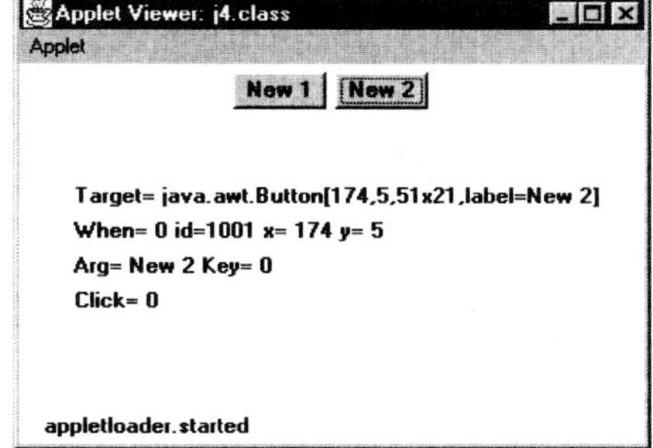

Figure 28.3 Sample runs.

Thus to determine the button that has been pressed the `evt.arg` string can be tested. Java applet 28.13 shows an example where the `evt.arg` parameter is tested for its string content.

Java applet 28.13 (chap28_13.java)

```
import java.applet.*;
import java.awt.*;

public class chap28_13 extends Applet
{

String Msg=null;

    public void init()
    {
        add (new Button("New 1"));
        add (new Button("New 2"));
    }

    public boolean action(Event evt, Object obj)
    {
        if (evt.arg=="New 1") Msg= "New 1 pressed";
        else if (evt.arg=="New 2") Msg= "New 2 pressed";
        repaint();
        return true;
    }

    public void paint(Graphics g)
    {
        if (Msg!=null)
        {
                g.drawString(Msg,30,80);
        }
    }
}
```

Java applet 28.14 uses the action function which is called when an event occurs. Within this function the event variable is tested to see if one of the buttons caused the event. This is achieved with:

```
if (event.target instanceof Button)
```

If this test is true then the `Msg` string takes on the value of the Object, which holds the name of the button that caused the event.

Java applet 28.14 (chap28_14.java)

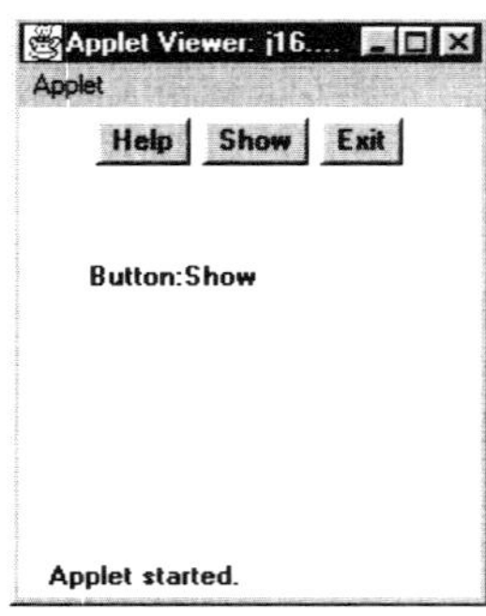


```
import java.awt.*;
import java.applet.*;
public class chap28_14 extends Applet
{
String  Msg=null;

  public void init()
  {
   add(new Button("Help"));
   add(new Button("Show"));
   add(new Button("Exit"));
  }
  public boolean action(Event event, Object object)
  {
   if (event.target instanceof Button)
   {
     Msg = (String) object;
     repaint();
   }
   return(true);
  }
  public void paint(Graphics g)
  {
   if (Msg!=null)
   g.drawString("Button:" + Msg,30,80);
  }
}
```

28.16 Action Listener in Java 1.1

As with mouse events, buttons, menus and textfields are associated with an action listener (named `ActionListener`). When an event associated with these occurs then the `actionPerformed` method is called. Its format is:

```
public void actionPerformed(ActionEvent evt)
```

where `evt` defines the event. The associated methods are:

- `getActionCommand()` is the action command
- `evt.getModifiers()` is the state of the modifier keys
- `evt.paramString()` is the parameter string

Java applet 28.15 contains an example of the `action` method. It has two buttons (named `New1` and `New2`). When any of the buttons is pressed the action method is called. Each of the buttons has an associated listener which is initiated with:

```
button1.addActionListener(this);
button2.addActionListener(this);
```

Figure 28.4 shows the display when either of the buttons are pressed. In the left-hand side of Figure 28.4 the `New1` button is pressed and the right-hand side shows the display after the `New2` button is pressed.

Java applet 28.15 (`chap28_15.java`)

```
import java.applet.*;
import java.awt.*;
import java.awt.event.*;

public class chap28_15 extends Applet implements ActionListener
{
Button   button1, button2;
String   Msg1=null, Msg2, Msg3;

      public void init()
      {

         button1 = new Button("New 1");
         button2 = new Button("New 2");
         add(button1); add(button2);
         button1.addActionListener(this);
         button2.addActionListener(this);
      }

      public void actionPerformed(ActionEvent evt)
      {
         Msg1= "Command= "+evt.getActionCommand();
         Msg2= "Modifiers= " + evt.getModifiers();
         Msg3= "String= " + evt.paramString();
         repaint();
      }

      public void paint(Graphics g)
      {
         if (Msg1!=null)
         {
            g.drawString(Msg1,30,80);
            g.drawString(Msg2,30,100);
            g.drawString(Msg3,30,120);
         }
      }
}
```

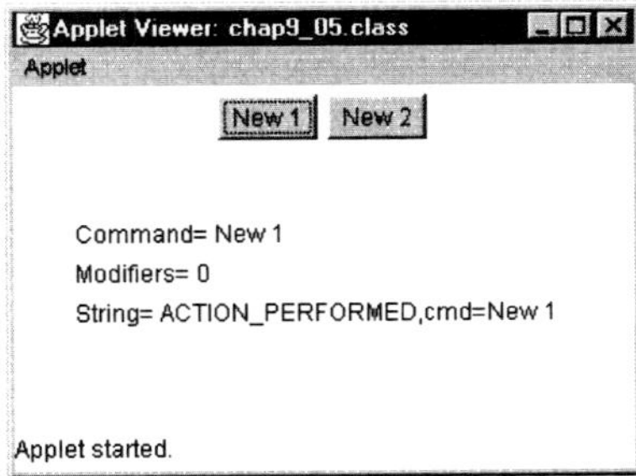

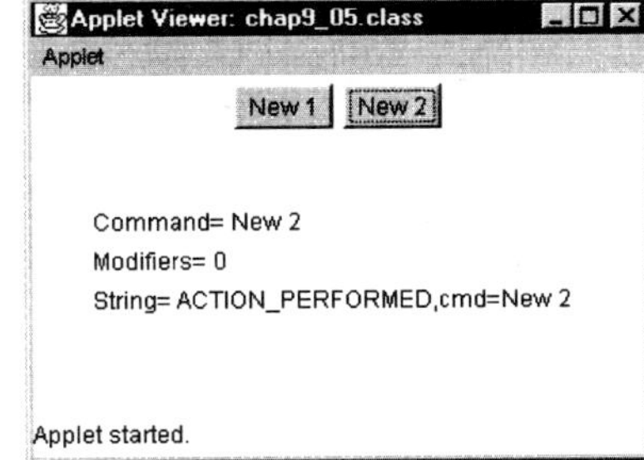

Figure 28.4 Sample run.

Thus to determine the button that has been pressed the `getActionCommand()` method is used. Java applet 28.16 shows an example where the `getActionCommand()` method is tested for its string content. Figure 28.5 shows a sample run.

📖 Java applet 28.16 (↯Java 1.1)

```
import java.applet.*;
import java.awt.*;
import java.awt.event.*;

public class chap28_16 extends Applet implements ActionListener
{

Button button1, button2;
String Msg=null;

    public void init()
    {

        button1 = new Button("New 1");
        button2 = new Button("New 2");
        add(button1); add(button2);
        button1.addActionListener(this);
        button2.addActionListener(this);
    }

    public void actionPerformed(ActionEvent evt)
    {
    String command;

        command=evt.getActionCommand();

        if (command.equals("New 1")) Msg="New 1 pressed";
        if (command.equals("New 2")) Msg="New 2 pressed";

        repaint();
    }

    public void paint(Graphics g)
    {
        if (Msg!=null)
        {
                g.drawString(Msg,30,80);
        }
    }
}
```

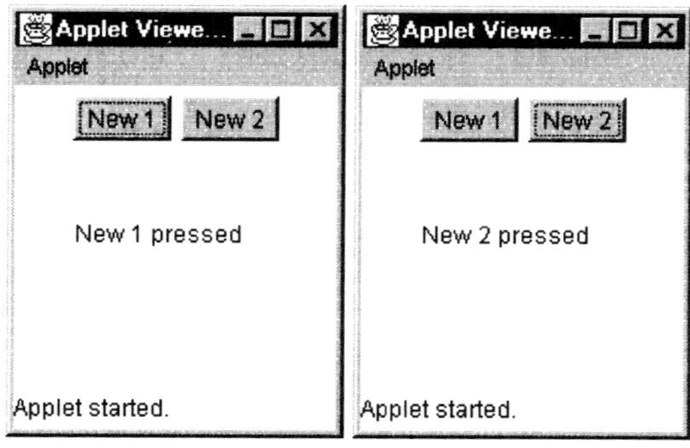

Figure 28.5 Sample run.

28.17 Checkboxes

Typically, checkboxes are used to select from a number of options. Java applet 28.17 shows how an applet can use checkboxes. As before, the `action` method is called when a checkbox changes its state and within the method the `event.target` parameter is tested for the checkbox with:

```
if (event.target instanceof Checkbox)
```

If this is true, then the method `DetermineCheckState()` is called which tests `event.target` for the checkbox value and its state (true or false).

Java applet 28.17 (Java 1.0)

```
import java.awt.*;
import java.applet.*;
public class chap28_17 extends Applet
{
String  Msg=null;
Checkbox fax, telephone, email, post;

 public void init()
 {
  fax=new Checkbox("FAX");
  telephone=new Checkbox("Telephone");
  email=new Checkbox("Email");
  post=new Checkbox("Post",null,true);
  add(fax); add(telephone);
  add(email); add(post);
 }
 public void DetermineCheckState(
     Checkbox Cbox)
 {
  Msg=Cbox.getLabel()+"  "+ Cbox.getState();
  repaint();
 }
 public boolean action(Event event,
     Object object)
 {
  if (event.target instanceof Checkbox)
    DetermineCheckState(
      (Checkbox)event.target);
  return(true);
 }
 public void paint(Graphics g)
 {
  if (Msg!=null)
   g.drawString("Check box:" + Msg,30,80);
 }
}
```

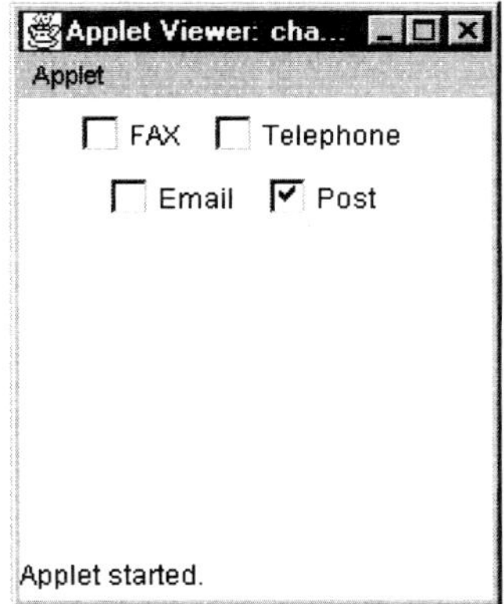

28.18 Item listener in Java 1.1

As with mouse events, checkboxes and lists are associated with an item listener (named `ItemListener`). When an event associated with these occur then the `itemStateChanged` method is called. Its format is:

```
public void itemStateChanged(ItemEvent event)
```

where `event` defines the event. The associated methods are:

- `getItem()` is the item selected
- `getStateChange()` is the state of the checkbox
- `paramString()` is the parameter string

Java applet 28.18 contains an example of checkboxes and Figure 28.6 shows a sample run. Each of the checkboxes has an associated listener which is initiated in the form:

chbox`.addItemListener(this);`

Java applet 28.18 (⭍Java 1.1)

```
import java.awt.*;
import java.applet.*;
import java.awt.event.*;

public class chap28_18 extends Applet implements ItemListener
{
String      Msg1=null,Msg2,Msg3;
Checkbox    fax, telephone, email,post;

    public void init()
    {
          fax=new Checkbox("FAX");
          telephone=new Checkbox("Telephone");
          email=new Checkbox("Email");
          post=new Checkbox("Post",null,true);
          add(fax);
          add(telephone);
          add(email);
          add(post);

          fax.addItemListener(this);
          email.addItemListener(this);
          telephone.addItemListener(this);
          post.addItemListener(this);
      }

      public void itemStateChanged(ItemEvent event)
      {
         Msg1=""+event.getItem();
         Msg2=""+event.getStateChange();
         Msg3=event.paramString();
         repaint();
      }

      public void paint(Graphics g)
      {
         if (Msg1!=null)
         {
            g.drawString(Msg1,30,80);
            g.drawString(Msg2,30,110);
            g.drawString(Msg3,30,150);
         }
      }
}
```

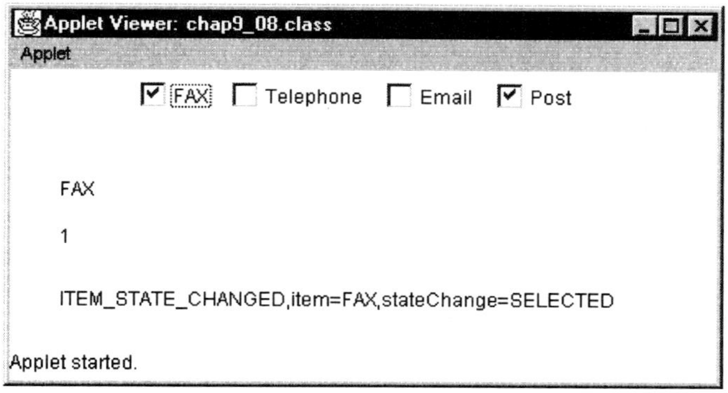

Figure 28.6 Sample run.

28.19 Radio buttons

The standard checkboxes allow any number of options to be selected. A radio button allows only one option to be selected at a time. The program is changed by:

- Adding checkbox names (such as fax, tele, email and post).
- Initializing the checkbox with `CheckboxGroup()` to a checkbox group identifier.
- Adding the identifier of the checkbox group to the `Checkbox()` method.
- Testing the target property of the event to see if it equals a checkbox name.

Java applet 28.19 shows how this is achieved with Java 1.1.

Java applet 28.19 (Java 1.1)

```
import java.awt.*;
import java.awt.event.*;
import java.applet.*;

public class chap28_19 extends Applet implements ItemListener
{
String  Msg=null;
Checkbox fax, tele, email, post;

 public void init()
  {

      CheckboxGroup RadioGroup = new CheckboxGroup();

      add(fax=new Checkbox("FAX",RadioGroup,true));
      add(tele=new Checkbox("Telephone",RadioGroup,false));
      add (email=new Checkbox("Email",RadioGroup,false));
      add (post=new Checkbox("Post",RadioGroup,false));
         fax.addItemListener(this);
         tele.addItemListener(this);
         email.addItemListener(this);
         post.addItemListener(this);
```

```
  }

  public void itemStateChanged(ItemEvent event)
  {
  Object obj;
         obj=event.getItem();

         if (obj.equals("FAX")) Msg="FAX";
         else if (obj.equals("Telephone")) Msg="Telephone";
         else if (obj.equals("Email")) Msg="Email";
         else if (obj.equals("Post")) Msg="Post";
         repaint();
     }

     public void paint(Graphics g)
     {
         if (Msg!=null)    g.drawString("Check box:" + Msg,30,80);
     }
}
```

This sets the checkbox type to `RadioGroup` and it can be seen that only one of the checkboxes is initially set (that is, 'FAX'). Figure 28.7 shows a sample run. It should be noted that grouped checkboxes use a round circle with a dot (⊙), whereas ungrouped checkboxes use a square box with a check mark (☑).

Figure 28.7 Sample run.

28.20 Pop-up menu choices

To create a pop-up menu the `Choice` object is initially created with:

```
Choice mymenu = new Choice();
```

After this the menu options are defined using the `addItem` method. Java applet 28.20 shows an example usage of a pop-up menu.

🕮 Java applet 28.20 (ϟJava 1.0)

```
import java.awt.*;
import java.applet.*;

public class chap28_20 extends Applet
{
String  Msg=null;
Choice  mymenu= new Choice();

  public void init()
  {
    mymenu.addItem("FAX");
    mymenu.addItem("Telephone");
    mymenu.addItem("Email");
    mymenu.addItem("Post");
    add(mymenu);
  }
  public void DetermineCheckState(
     Choice mymenu)
  {
    Msg=mymenu.getItem(
       mymenu.getSelectedIndex());
    repaint();
  }
  public boolean action(Event event,
      Object object)
  {
   if (event.target instanceof Choice)
    DetermineCheckState(
       (Choice)event.target);
   return(true);
  }

  public void paint(Graphics g)
  {
  if (Msg!=null)
   g.drawString("Menu select:"+Msg,30,120);
  }
}
```

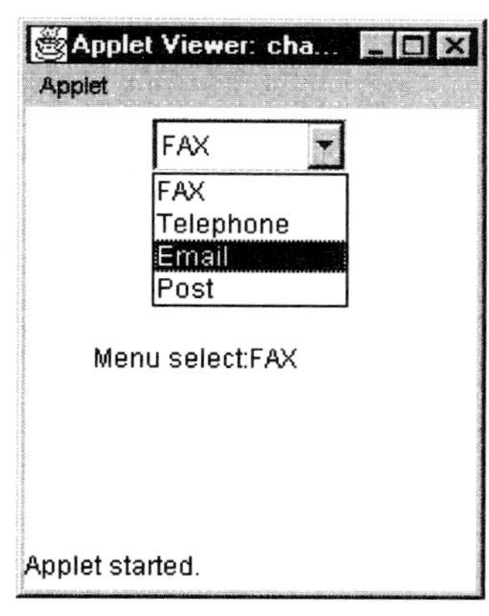

As before the `arg` property of the event can also be tested as shown in Java applet 28.21. Java applet 28.22 gives the Java 1.1 equivalent.

🕮 Java applet 28.21 (ϟJava 1.0)

```
import java.awt.*;
import java.applet.*;

public class chap28_21 extends Applet
{
String  Msg=null;
Choice  mymenu= new Choice();

  public void init()
  {
    mymenu.addItem("FAX");
    mymenu.addItem("Telephone");
    mymenu.addItem("Email");
    mymenu.addItem("Post");
    add(mymenu);
  }
  public boolean action(Event event, Object object)
```

```
  {

   if (event.arg=="FAX") Msg="FAX";
   else if (event.arg=="Telephone") Msg="Telephone";
   else if (event.arg=="Email") Msg="Email";
   else if (event.arg == "Post") Msg="Post";
   repaint();
   return(true);
  }
  public void paint(Graphics g)
  {
    if (Msg!=null)
      g.drawString("Menu select:" + Msg,30,120);
  }
}
```

Java applet 28.22 (↯Java 1.1)

```
import java.awt.*;
import java.awt.event.*;
import java.applet.*;

public class chap28_22 extends Applet implements ItemListener
{
String  Msg=null;
Choice  mymenu= new Choice();

  public void init()
  {
    mymenu.addItem("FAX");
    mymenu.addItem("Telephone");
    mymenu.addItem("Email");
    mymenu.addItem("Post");
    add(mymenu);
    mymenu.addItemListener(this);
  }
  public void itemStateChanged(ItemEvent event)
  {
  Object obj;

   obj=event.getItem();

   if (obj.equals("FAX")) Msg="FAX";
   else if (obj.equals("Telephone")) Msg="Telephone";
   else if (obj.equals("Email")) Msg="Email";
   else if (obj.equals("Post")) Msg="Post";
   repaint();
  }
  public void paint(Graphics g)
  {
    if (Msg!=null)
      g.drawString("Menu select:" + Msg,30,120);
  }
}
```

28.21 Other pop-up menu options

The `java.awt.Choice` class allows for a pop-up menu. It includes the following methods:

`public void addItem(String item);`	Adds a menu item to the end.
`public void addNotify();`	Allows the modification of a list's appearance without changing its functionality.
`public int countItems();`	Returns the number of items in the menu.
`public String getItem(int index);`	Returns the string of the menu item at that index value.
`public int getSelectedIndex();`	Returns the index value of the selected item.
`public String getSelectedItem();`	Returns the string of the selected item.
`protected String paramString();`	Returns the parameter String of the list.
`public void select(int pos);`	Selects the menu item at a given index.
`public void select(String str);`	Selects the menu item with a given string name.

The `countItems` method is used to determine the number of items in a pop-up menu, for example:

```
Msg= "Number of items is " + mymenu.countItems()
```

The `getItem(int index)` returns the string associated with the menu item, where the first item has a value of zero. For example:

```
Msg= "Menu item number 2 is " + mymenu.getItem(2);
```

Java applet 28.23 uses the `select` method to display the second menu option as the default and the `getItem` method to display the name of the option.

Java applet 28.23 (Java 1.1)

```
import java.awt.*;
import java.awt.event.*;
import java.applet.*;

public class chap28_23 extends Applet
     implements ItemListener
{
String  Msg=null;
Choice  mymenu= new Choice();

 public void init()
 {
   mymenu.addItem("FAX");
   mymenu.addItem("Telephone");
   mymenu.addItem("Email");
   mymenu.addItem("Post");
```

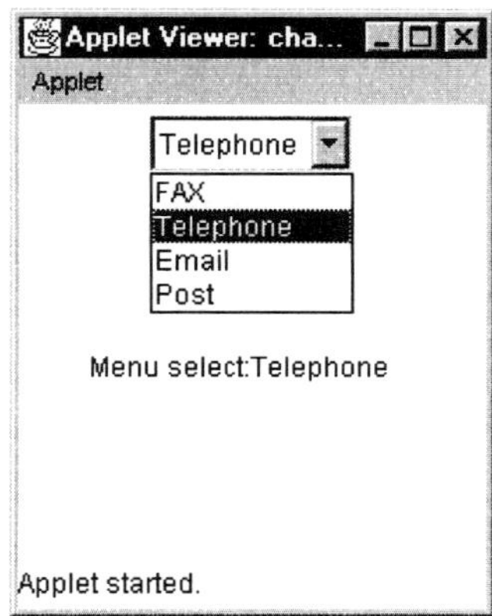

```
    add(mymenu);
    mymenu.addItemListener(this);
    mymenu.select(1);
        // Select item 1 (Telephone)
}

public void itemStateChanged(ItemEvent evt)
{
Object obj;

 obj=evt.getItem();

 if (obj.equals("FAX"))
     Msg=mymenu.getItem(0);
 else if (obj.equals("Telephone"))
     Msg=mymenu.getItem(1);
 else if (obj.equals("Email"))
     Msg=mymenu.getItem(2);
 else if (obj.equals("Post"))
     Msg=mymenu.getItem(3);
 repaint();
}
public void paint(Graphics g)
{
 if (Msg!=null)
   g.drawString("Menu select:"+Msg,30,120);
}
}
```

28.22 Multiple menus

Multiple menus can be created in a Java applet and the `action` event can be used to differentiate between the menus. Java applet 28.24 has two pull-down menus and two buttons (`age`, `gender`, `print` and `close`). The event method `getItem` is then used to determine which of the menus was selected. In this case the `print` button is used to display the options of the two pull-down menus and `close` is used to exit from the applet.

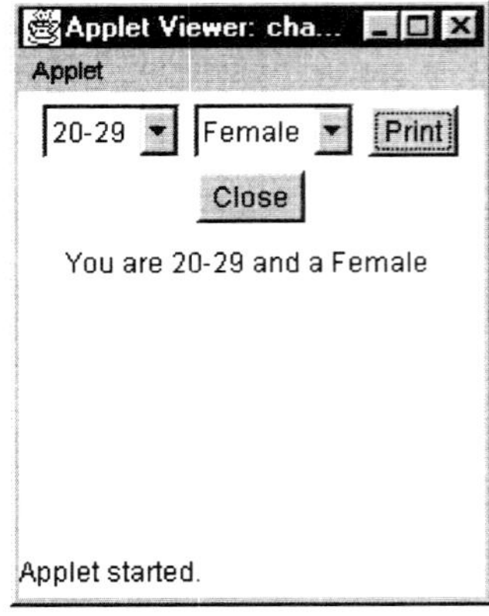

Java applet 28.24 (Java 1.1)

```
import java.applet.*;
import java.awt.*;
import java.awt.event.*;

public class chap28_24 extends Applet
  implements ItemListener, ActionListener
{
Choice age = new Choice();
Choice gender = new Choice();
Button print= new Button("Print");
Button close= new Button("Close");
String gendertype=null, agetype=null;

String Msg, Options[];

  public void init()
  {
      age.addItem("10-19");
      age.addItem("20-29");
      age.addItem("30-39");
```

```
      age.addItem("40-49");
      age.addItem("Other");
      add(age);

      gender.addItem("Male");
      gender.addItem("Female");
      add(gender);
      add(print);
      add(close);

      age.addItemListener(this);
      gender.addItemListener(this);
      print.addActionListener(this);
      close.addActionListener(this);
  }

 public void itemStateChanged(ItemEvent evt)
 {
  int i;
  Object obj;

  obj=evt.getItem();

  if (obj.equals("10-19")) agetype="10-19";
  else if (obj.equals("20-29"))
     agetype="20-29";
  else if (obj.equals("30-39"))
    agetype="30-39";
  else if (obj.equals("40-49"))
    agetype="40-49";
  else if (obj.equals("Other"))
    agetype="Other";
  else if (obj.equals("Male"))
    gendertype="Male";
  else if (obj.equals("Female"))
    gendertype="Female";
  }
 public void actionPerformed(ActionEvent evt)
 {
 String str;

  str=evt.getActionCommand();
  if (str.equals("Print"))  repaint();
  else if (str.equals("Close"))
    System.exit(0);
 }
 public void paint(Graphics g)
 {
  if ((agetype!=null) && (gendertype!=null))
   Msg="Your are " + agetype + " and a "
              + gendertype;
   else Msg="Please select age and gender";

  if (Msg!=null) g.drawString(Msg,20,80);
 }
}
```

28.23 Menu bar

Menu bars are now familiar in most GUIs (such as Microsoft Windows and Motif). They consist of a horizontal menu bar with pull-down submenus.

The `java.awt.MenuBar` class contains a constructor for a menu bar. Its format is:

```
public MenuBar();
```

and the methods which can be applied to it are:

`public Menu` **`add`**`(Menu m);`	Adds the specified menu to the menu bar.
`public void` **`addNotify`**`();`	Allows a change of appearance of the menu bar without changing any of the menu bar's functionality.
`public int` **`countMenus`**`();`	Counts the number of menus on the menu bar.
`public Menu` **`getHelpMenu`**`();`	Gets the help menu on the menu bar.
`public Menu` **`getMenu`**`(int i);`	Gets the specified menu.
`public void` **`remove`**`(int index);`	Removes the menu located at the specified index from the menu bar.
`public void` **`remove`**`(MenuComponent m);`	Removes the specified menu from the menu bar.
`public void` **`removeNotify`**`();`	Removes notify.
`public void` **`setHelpMenu`**`(Menu m);`	Sets the help menu to the specified menu on the menu bar.

Java program 28.25 gives an example of using a menu bar. Initially the menu bar is created with the `MenuBar()` constructor, and submenus with the `Menu` constructors (in this case, `mfile`, `medit` and `mhelp`). Items are added to the submenus with the `MenuItem` constructor (such as `New`, `Open`, and so on). A `handleEvent()` method has been added to catch a close window operation. The `addSeparator()` method has been added to add a line between menu items. Note that this program is not an applet so that it can be run directly with the Java interpreter (such as `java.exe`).

🕮 Java program 28.25 (↯Java 1.0)

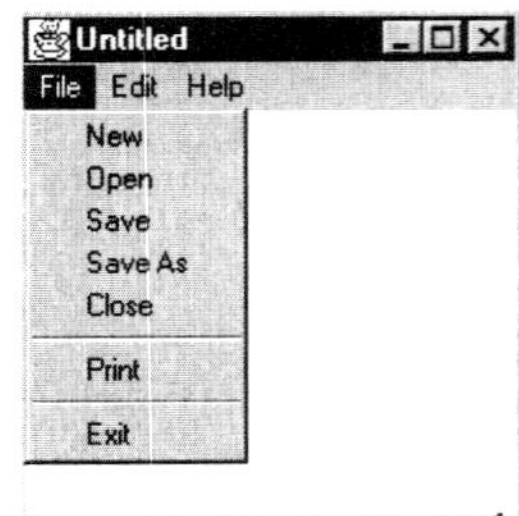


```
import java.awt.*;

public class gomenu extends Frame
{
MenuBar mainmenu = new MenuBar();
Menu mfile = new Menu("File");
Menu medit = new Menu("Edit");
Menu mhelp = new Menu("Help");

    public gomenu()
    {
        mfile.add(new MenuItem("New"));
        mfile.add(new MenuItem("Open"));
        mfile.add(new MenuItem("Save"));
        mfile.add(new MenuItem("Save As"));
        mfile.add(new MenuItem("Close"));
        mfile.addSeparator();
```

```
        mfile.add(new MenuItem("Print"));
        mfile.addSeparator();
        mfile.add(new MenuItem("Exit"));

        mainmenu.add(mfile);

        medit.add(new MenuItem("Cut"));
        medit.add(new MenuItem("Copy"));
        medit.add(new MenuItem("Paste"));
        mainmenu.add(medit);

        mhelp.add(new MenuItem("Commands"));
        mhelp.add(new MenuItem("About"));
        mainmenu.add(mhelp);

        setMenuBar(mainmenu);
    }

    public boolean action(Event evt, Object obj)
    {
        if (evt.target instanceof MenuItem)
        {
             if (evt.arg=="Exit") System.exit(0);
        }
        return true;
    }

    public boolean handleEvent(Event evt)
    {
        if (evt.id == Event.WINDOW_DESTROY)
              System.exit(0);
        return true;
    }

    public static void main(String args[])
    {
        Frame f = new gomenu();
        f.resize(400,400);
        f.show();
    }
}
```

28.24 List box

A `List` component creates a scrolling list of options (where in a pull-down menu only one option can be viewed at a time). The `java.awt.List` class contains the `List` constructor which can be used to display a list component, which is in the form:

```
public List();
public List(int rows, boolean multipleSelections);
```

where `row` defines the number of rows in a list and `multipleSelections` is true when the user can select a number of selections, else it is false.

The methods that can be applied are:

`public void` **`addItem`**`(String item);`	Adds a menu item at the end.
`public void` **`addItem`**`(String item, int index);`	Add a menu item at the end.

`public void` **`addNotify`**`();`	Allows the modification of a list's appearance without changing its functionality.
`public boolean` **`allowsMultipleSelections`**`();`	Allows the selection of multiple selections.
`public void` **`clear`**`();`	Clears the list.
`public int` **`countItems`**`();`	Returns the number of items in the list.
`public void` **`delItem`**`(int position);`	Deletes an item from the list.
`public void` **`delItems`**`(int start, int end);`	Deletes items from the list.
`Public void` **`deselect`**`(int index);`	Deselects the item at the specified index.
`Public String` **`getItem`**`(int index);`	Gets the item associated with the specified index.
`public int` **`getRows`**`();`	Returns the number of visible lines in this list.
`public int` **`getSelectedIndex`**`();`	Gets the selected item on the list.
`public int[]` **`getSelectedIndexes`**`();`	Gets selected items on the list.
`public String` **`getSelectedItem`**`();`	Returns the selected item on the list as a string.
`public String[]` **`getSelectedItems`**`();`	Returns the selected items on the list as an array of strings.
`public int` **`getVisibleIndex`**`();`	Gets the index of the item that was last made visible by the method `makeVisible`.
`public boolean` **`isSelected`**`(int index);`	Returns true if the item at the specified index has been selected.
`public void` **`makeVisible`**`(int index);`	Makes a menu item visible.
`public Dimension` **`minimumSize`**`();`	Returns the minimum dimensions needed for the list.
`public Dimension` **`minimumSize`**`(int rows);`	Returns the minimum dimensions needed for the number of rows in the list.
`protected String` **`paramString`**`();`	Returns the parameter String of the list.

`public Dimension preferredSize();`	Returns the preferred size of the list.
`public Dimension preferredSize(int rows);`	Returns the preferred size of the list.
`public void removeNotify();`	Removes notify.
`public void replaceItem(String newValue, int index);`	Replaces the item at the given index.
`Public void select(int index);`	Selects the item at the specified index.
`public void setMultipleSelections(boolean v);`	Allows multiple selections.

Java program 28.26 shows an example of a program with a list component. Initially the list is created with the `List` constructor. The `addItem` method is then used to add the four items ("Pop", "Rock", "Classical" and "Jazz"). Within `actionPerformed` the program uses the `Options` array of strings to build up a message string (`Msg`). The `Options.length` parameter is used to determine the number of items in the array.

🕮 Java applet 28.26 (⚡Java 1.1)

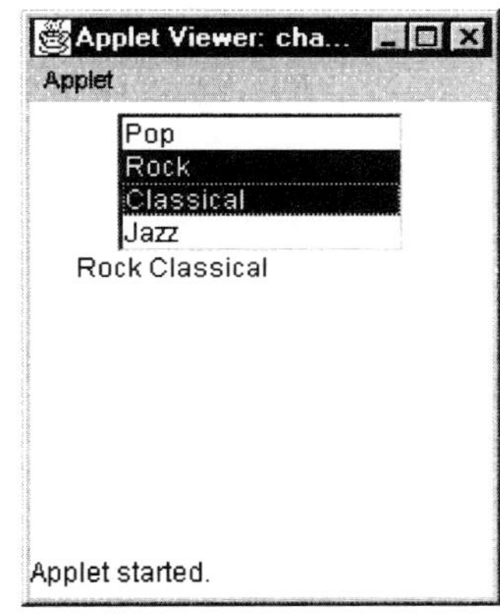


```
import java.awt.*;
import java.awt.event.*;
import java.applet.*;

public class chap28_26 extends Applet
                    implements ActionListener
{
List lmenu = new List(4,true);
String Msg, Options[];

    public void init()
    {
        lmenu.addItem("Pop");
        lmenu.addItem("Rock");
        lmenu.addItem("Classical");
        lmenu.addItem("Jazz");
        add(lmenu);
        lmenu.addActionListener(this);
  }
 public void actionPerformed(
                        ActionEvent evt)
 {
    int i;
    String str;

     str=evt.getActionCommand();
     Options=lmenu.getSelectedItems();
     Msg="";
     for (i=0;i<Options.length;i++)
            Msg=Msg+Options[i] + " ";
     repaint();
 }
 public void paint(Graphics g)
 {

   if (Msg!=null) g.drawString(Msg,20,80);
 }
}
```

28.25 File dialog

The `java.awt.Filedialog` class contains the `FileDialog` constructor which can be used to display a dialog window. To create a dialog window the following can be used:

```
public FileDialog(Frame parent, String title);
public FileDialog(Frame parent, String title, int mode);
```

where the `parent` is the owner of the dialog, `title` is the title of the dialog window and the `mode` is defined as whether the file is to be loaded or save. Two fields are defined for the mode, these are:

```
public final static int LOAD;
public final static int SAVE;
```

The methods that can be applied are:

`public void addNotify();`	Allows applications to change the look of a file dialog window without changing its functionality.
`public String getDirectory();`	Gets the initial directory.
`public String getFile();`	Gets the file that the user specified.
`public FilenameFilter getFilenameFilter();`	Sets the default file filter.
`public int getMode();`	Indicates whether the file dialog box is for file loading from or file saving.
`protected String paramString();`	Returns the parameter string representing the state of the file dialog window.
`public void setDirectory(String dir);`	Gets the initial directory.
`public void setFile(String file);`	Sets the selected file for this file dialog window to be the specified file.
`public void setFilenameFilter(FilenameFilter filter);`	Sets the filename filter for the file dialog window to the specified filter.

29 Java (Networking)

29.1 Introduction

Java is one of the fastest growing development languages and has the great advantage that it was developed after the Internet and WWW were created and thus has direct WWW/Internet support. This includes the use of HTTP and socket programming, and so on. This chapter provides an introduction to the usage of Java over a local or global area network.

29.2 Java networking functions

Java directly supports TCP/IP communications and has the following classes:

- `java.net.ContentHandler`. Class which reads data from a URLConnection and also supports MIME (Multipurpose Internet Mail Extension).
- `java.net.DatagramPacket`. Class representing a datagram packet which contains packet data, packet length, Internet addresses and the port number.
- `java.net.DatagramSocket`. Class representing a datagram socket class.
- `java.net.InetAddress`. Class representing Internet addresses.
- `java.net.ServerSocket`. Class representing Socket server class.
- `java.net.Socket`. Class representing Socket client classes.
- `java.net.SocketImpl`. Socket implementation class.
- `java.net.URL`. Class URL representing a Uniform Reference Locator (URL) which is a reference to an object on the WWW.
- `java.net.URLConnection`. Class representing an active connection to an object represented by a URL.
- `java.net.URLEncoder`. Converts strings of text into URLEncoded format.
- `java.net.URLStreamHandler`. Class for opening URL streams.

When an error occurs in the connection or in the transmission and reception of data it causes an exception. The classes which handle these are:

- `java.io.IOException`. To handle general errors.
- `java.net.MalformedURLException`. Malformed URL.
- `java.net.ProtocolException`. Protocol error.
- `java.net.SocketException`. Socket error.
- `java.net.UnknownHostException`. Unknown host error.
- `java.net.UnknownServiceException`. Unknown service error.

29.2.1 Class java.net.InetAddress

This class represents Internet addresses. The methods are:

`public static synchronized InetAddress[] getAllByName(String host)`
: This returns an array with all the corresponding `InetAddresses` for a given host name (`host`).

`public static synchronized InetAddress getByName(String host)`
: This returns the network address of an indicated host. A host name of null returns the default address for the local machine.

`public String getHostAddress()`
: This returns the IP address string (`WW.XX.YY.ZZ`) in a string format.

`public byte[] getAddress()`
: This returns the raw IP address in network byte order. The array position 0 (`addr[0]`) contains the highest order byte.

`public String getHostName()`
: Gets the hostname for this address. If the host is equal to null, then this address refers to any of the local machine's available network addresses.

`public static InetAddress getLocalHost()`
: Returns the local host.

`public String toString()` Converts the `InetAddress` to a String.

Java applet 29.1 uses the `getAllByName` method to determine all the IP addresses associated with an Internet host. In this case the host is named `www.microsoft.com`. It can be seen from the test run that there are 18 IP addresses associated with this domain name. It can be seen that the applet causes an exception error as the loop tries to display 30 such IP addresses. When the program reaches the 19th `InetAddress`, the exception error is displayed (`ArrayIndexOutOfBoundsException`).

Java applet 29.1 (`chap29_1.java`)

```
import java.net.*;
import java.awt.*;
import java.applet.*;
public class chap29_1 extends Applet
{
InetAddress[]  address;
int            i;
      public void start()
      {
         System.out.println("Started");
         try
         {
            address=InetAddress.getAllByName("www.microsoft.com");

            for (i=0;i<30;i++)
            {
               System.out.println("Address " + address[i]);
            }
         }
         catch (Exception e)
         {
            System.out.println("Error :" + e);
         }
      }
}
```

Sample run 29.1

```
C:\java\temp>appletviewer chap29_01.html
Started
Address www.microsoft.com/207.68.137.59
Address www.microsoft.com/207.68.143.192
Address www.microsoft.com/207.68.143.193
Address www.microsoft.com/207.68.143.194
Address www.microsoft.com/207.68.143.195
Address www.microsoft.com/207.68.156.49
Address www.microsoft.com/207.68.137.56
Address www.microsoft.com/207.68.156.51
Address www.microsoft.com/207.68.156.52
Address www.microsoft.com/207.68.137.62
Address www.microsoft.com/207.68.156.53
Address www.microsoft.com/207.68.156.54
Address www.microsoft.com/207.68.137.65
Address www.microsoft.com/207.68.156.73
Address www.microsoft.com/207.68.156.61
Address www.microsoft.com/207.68.156.16
Address www.microsoft.com/207.68.156.58
Address www.microsoft.com/207.68.137.53
Error :java.lang.ArrayIndexOutOfBoundsException: 18
```

Java applet 29.2 overcomes the problem of displaying the exception. In this case the exception is caught by inserting the address display within a `try {}` statement then having a `catch` statement which does nothing. Test run 29.2 shows a sample run.

Java applet 29.2 (chap29_2.java)

```
import java.net.*;
import java.awt.*;
import java.applet.*;
public class chap29_2 extends Applet
{
InetAddress[]  address;
int            i;

      public void start()
      {
         System.out.println("Started");
         try
         {
            address=InetAddress.getAllByName("www.microsoft.com");
            try
            {
               for (i=0;i<30;i++)
               {
                  System.out.println("Address " + address[i]);
               }
            }
            catch(Exception e)
            { /* Do nothing about the exception, as it is not really
                  an error */}
         }
         catch (Exception e)
         {
            System.out.println("Error :" + e);
         }
      }
}
```

Sample run 29.2

```
C:\java\temp>appletviewer chap29_02.html
Started
Address www.microsoft.com/207.68.137.59
Address www.microsoft.com/207.68.143.192
Address www.microsoft.com/207.68.143.193
Address www.microsoft.com/207.68.143.194
Address www.microsoft.com/207.68.143.195
Address www.microsoft.com/207.68.156.49
Address www.microsoft.com/207.68.137.56
Address www.microsoft.com/207.68.156.51
Address www.microsoft.com/207.68.156.52
Address www.microsoft.com/207.68.137.62
Address www.microsoft.com/207.68.156.53
Address www.microsoft.com/207.68.156.54
Address www.microsoft.com/207.68.137.65
Address www.microsoft.com/207.68.156.73
Address www.microsoft.com/207.68.156.61
Address www.microsoft.com/207.68.156.16
Address www.microsoft.com/207.68.156.58
Address www.microsoft.com/207.68.137.53
```

Java applet 29.3 shows an example of displaying the local host name (`getLocalHost`), the host name (`getHostName`) and the host's IP address (`getHostAddress`). Test run 29.3 shows a sample run.

Java applet 29.3 (chap29_3.java)

```
import java.net.*;
import java.awt.*;
import java.applet.*;

public class chap29_3 extends Applet
{
InetAddress host;
String str;
int i;
      public void start()
      {
         System.out.println("Started");

         try
         {
            host=InetAddress.getLocalHost();
            System.out.println("Local host " + host);

            str=host.getHostName();
            System.out.println("Host name: " + str);

            str=host.getHostAddress();
            System.out.println("Host address: " + str);
         }

         catch (Exception e)
         {
            System.out.println("Error :" + e);
         }
      }
}
```

Sample run 29.3

```
C:\java\temp>appletviewer chap29_03.html
Started
Local host toshiba/195.232.26.125
Host name: toshiba
Host address: 195.232.26.125
```

The previous Java applets have all displayed their output to the output terminal (with `System.out.println`). Java applet 29.4 uses the `drawString` method to display the output text to the Applet window. Figure 29.2 shows a sample run.

Java applet 29.4 (chap29_4.java)

```
import java.net.*;
import java.awt.*;
import java.applet.*;

public class chap29_4 extends Applet
{
InetAddress[] address;
int i;

	public void paint(Graphics g)
	{
	   g.drawString("Addresses for WWW.MICROSOFT.COM",5,10);
	   try
	   {
	      address=InetAddress.getAllByName("www.microsoft.com");

	      for (i=0;i<30;i++)
	      {
	         g.drawString("  "+address[i].toString(),5,20+10*i);
	      }
	   }
	   catch (Exception e)
	   {
	      System.out.println("Error :" + e);
	   }
	}
}
```

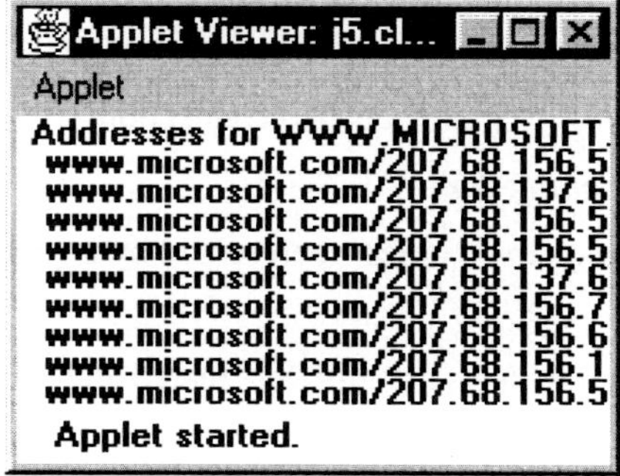

Figure 29.1 Sample run.

29.2.2 class java.net.URL

The URL (Uniform Reference Locator) class is used to reference to an object on the World Wide Web. The main constructors are:

public **URL**(String protocol, String host, int port, String file)
: Creates an absolute URL from the specified protocol (`protocol`), host (`host`), port (`port`) and file (`file`).

public **URL**(String protocol, String host, String file)
: Creates an absolute URL from the specified protocol (`protocol`), host (`host`) and file (`file`).

public **URL**(String spec)
: Creates a URL from an unparsed absolute URL (`spec`).

public **URL**(URL context, String spec)
: Creates a URL from an unparsed absolute URL (`spec`) in the specified context.

The methods are:

public int **getPort**()
: Returns a port number. A return value of –1 indicates that the port is not set.

public String **getProtocol**()
: Returns the protocol name.

public String **getHost**()
: Returns the host name.

public String **getFile**()
: Returns the file name.

public boolean **equals**(Object obj)
: Compares two URLs, where `obj` is the URL to compare against.

public String **toString**()
: Converts to a string format.

public String **toExternalForm**()
: Reverses the URL parsing.

public URLConnection **openConnection**()
: Creates a URLConnection object that contains a connection to the remote object referred to by the URL.

public final InputStream **openStream**()
: Opens an input stream.

public final Object **getContent**()
: Gets the contents from this opened connection.

29.2.3 class java.net.URLConnection

Represents an active connection to an object represented by a URL. The main methods are:

public abstract void **connect**()
: URLConnection objects are initially created and then they are connected.

public URL **getURL**()
: Returns the URL for this connection.

public int **getContentLength**()
: Returns the content length, a –1 if not known.

public String **getContentType**()
: Returns the content type, a null if not known.

`public String` **`getContentEncoding`**`()`

Returns the content encoding, a null if not known.

`public long` **`getExpiration`**`()`

Returns the expiration date of the object, a 0 if not known.

`public long` **`getDate`**`()`

Returns the sending date of the object, a 0 if not known.

`public long` **`getLastModified`**`()`

Returns the last modified date of the object, a 0 if not known.

`public String` **`getHeaderField`**`(String name)`

Returns a header field by name (`name`), a null if not known.

`public Object` **`getContent`**`()`

Returns the object referred to by this URL.

`public InputStream` **`getInputStream`**`()`

Used to read from objects.

`public OutputStream` **`getOutputStream`**`()`

Used to write to objects.

`public String` **`toString`**`()`

Returns the String URL representation.

29.2.4 class java.net.URLEncoder

This class converts text strings into x-www-form-urlencoded format.

`public static String` **`encode`**`(String s)`

Translates a string (s) into x-www-form-urlencoded format.

29.2.5 class java.net.URLStreamHandler

Abstract class for URL stream openers. Subclasses of this class know how to create streams for particular protocol types.

`protected abstract URLConnection` **`openConnection`**`(URL u)`

Opens an input stream to the object referenced by the URL (`u`).

`protected void` **`parseURL`**`(URL u, String spec, int start, int limit)`

Parses the string (`spec`) into the URL (`u`), where `start` and `limit` refer to the range of characters in `spec` that should be parsed.

`protected String` **`toExternalForm`**`(URL u)`

Reverses the parsing of the URL.

`protected void` **`setURL`**`(URL u,  String protocol,String host, int port, String file, String ref)`

Calls the (protected) set method out of the URL given.

29.2.6 java.applet.AppletContext

The AppletContext can be used by an applet to obtain information from the applet's environment, which is usually the browser or the applet viewer. Related methods are:

`public abstract void` **`showDocument`**`(URL url)`

Shows a new document.

`public abstract void` **`showDocument`**`(URL url,  String target)`

Shows a new document in a target window or frame.

29.3 Connecting to a WWW site

Java applet 29.5 shows an example of an applet that connects to a WWW site. In this case, it connects to the www.microsoft.com site.

Java applet 29.5 (chap29_5.java) (Java 1.1)

```
import java.net.*;
import java.awt.*;
import java.awt.event.*;
import java.applet.*;
public class chap29_5 extends Applet implements ActionListener
{
URL      urlWWW;
Button   btn;

    public void init()
    {
        btn = (new Button("Connect to Microsoft WWW site"));
        add(btn);
        btn.addActionListener(this);
    }
    public void start()
    {
        try //Check for valid URL
        {
            urlWWW  = new URL("http://www.microsoft.com");
        }
        catch (MalformedURLException e)
        {
            System.out.println("URL Error: " + e);
        }
    }
    public void actionPerformed(ActionEvent evt)
    {
         if (evt.getActionCommand().equals("Connect to Microsoft WWW site"))
         {
            getAppletContext().showDocument(urlWWW);
         }
    }
}
```

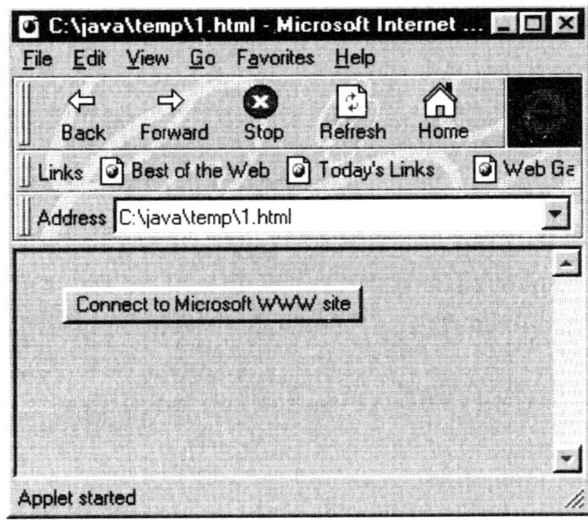

Figure 29.2 Sample run.

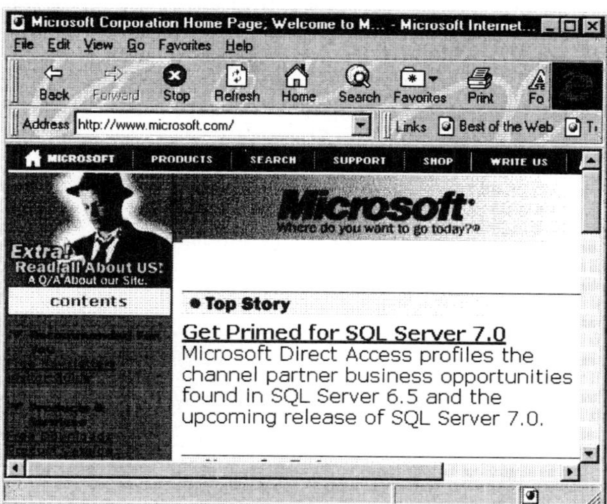

Figure 29.3 WWW connection.

Java applet 29.6 extends the previous applet by allowing the user to enter a URL and it also shows a status window (`status`). Figures 29.5, 29.6 and 29.7 show sample runs. In Figure 29.5 the user has added an incorrect URL (www.sun.com). The status windows shows that this is an error. In Figure 29.6 the user enters a correct URL (http://www.sun.com) and Figure 29.7 show the result after the Connect button is pressed.

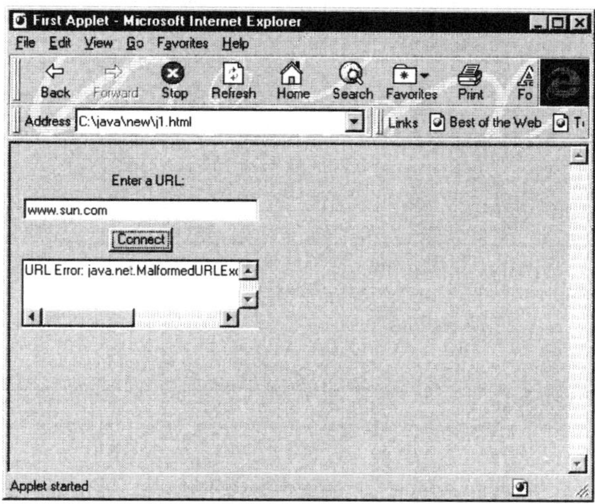

Figure 29.4 WWW connection.

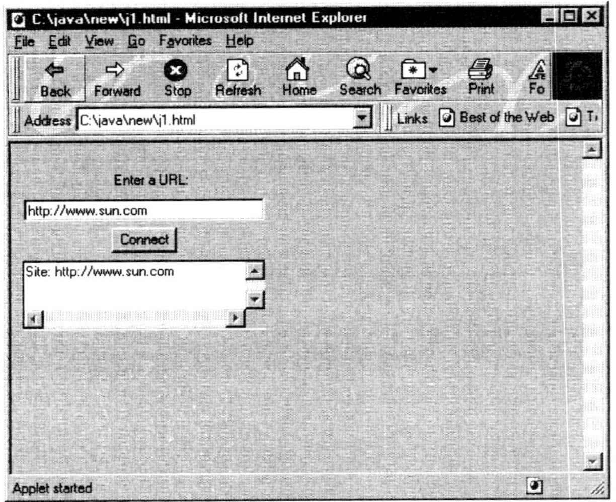

Figure 29.5 WWW connection.

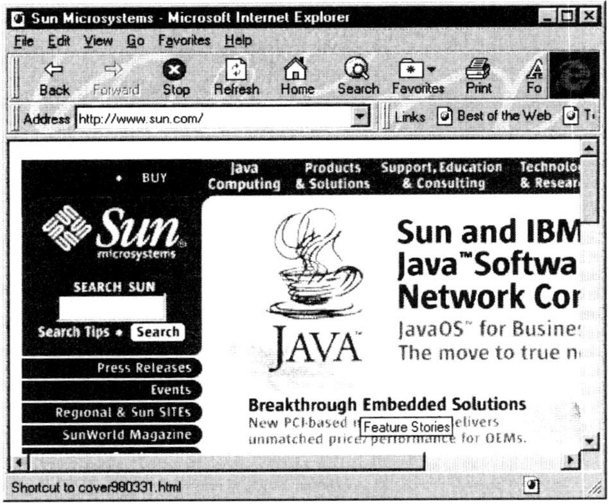

Figure 29.6 WWW connection.

Java applet 29.6 (chap29_6.java)

```
import java.net.*;
import java.awt.*;
import java.awt.event.*;
import java.applet.*;
public class chap29_06 extends Applet implements ActionListener
{
URL      urlWWW;
Button   btn;
Label    label = new Label("Enter a URL:");
TextField inURL = new TextField(30);
TextArea status = new TextArea(3,30);
```

```
        public void init()
        {
           add(label);
           add(inURL);
           btn = (new Button("Connect"));
           add(btn);
           add(status);
           btn.addActionListener(this);
        }
        public void getURL()//Check for valid URL
        {
           try
           {
           String str;
              str=inURL.getText();
              status.setText("Site: " + str);
              urlWWW  = new URL(str);
           }
           catch (MalformedURLException e)
           {
              status.setText("URL Error: " + e);
           }
        }
        public void actionPerformed(ActionEvent evt)
        {
        String str;
           str=evt.getActionCommand();
           if (str.equals("Connect"))
           {
              status.setText("Connecting...\n");
              getURL();
              getAppletContext().showDocument(urlWWW);
           }
        }
}
```

29.4 Socket programming

The main calls in standard socket programming are:

`socket()` Creates a socket.
`accept()` Accepts a connection on a socket.
`connect()` Establishes a connection to a peer.
`bind()` Associates a local address with a socket.
`listen()` Establishes a socket to listen for incoming connection.
`getInputStream()`
Gets an input data stream for a socket. This can be used to create a `receive()` method.
`getOutputStream()`
Gets an output data stream for a socket. This can be used to create a `send()` method.
`close()` Closes a socket.

Figure 29.8 shows the operation of a connection of a client to a server. The server is defined as the computer which waits for a connection, the client is the computer which initially makes contact with the server.

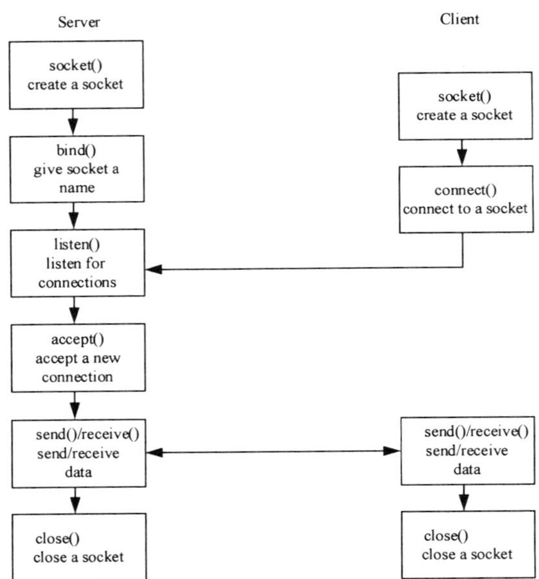

Figure 29.7 Socket connection.

On the server the computer initially creates a socket with `socket()` method, this is bound to a name with the `bind()` method. After this the server listens for a connection with the `listen()` method. When the client calls the `connect()` method the server then accepts the connection with `accept()`. After this the server and client can send and receive data with the `send()` or `receive()` functions (these are created with the `getInputStream()` and `getOutputStream()` methods). When the data transfer is complete the `close()` method is used to close the socket.

The implementation for `send()` and `receive()` is:

```
public static void send(Socket client,String str)
{
 try
 {
  DataOutputStream send=new DataOutputStream(client.getOutputStream());
  send.writeUTF(str);
 }
 catch (IOException e)
 {
   //something
 }
}
public static String receive(Socket client)
{
 String str="";
 try
 {
   DataInputStream receive=new DataInputStream(client.getInputStream());
   str=receive.readLine();
 }
 catch (IOException e)
 {
       //something
```

```
 }
 return str;
}
```

29.4.1 class java.net.Socket

The TCP protocol links two computers using sockets and ports. The constructors for `java.net.Socket` are:

`public` **`Socket`**`(InetAddress address,  int port)`
: Creates a stream socket and connects it to the specified address (`address`) on the specified port (`port`).

`public` **`Socket`**`(InetAddress address, int port, boolean stream)`
: Creates a socket and connects it to the specified address (`address`) on the specified port (`port`). The boolean value `stream` indicates whether this is a stream or datagram socket.

`public` **`Socket`**`(String host, int port)`
: Creates a stream socket and connects it to the specified port (`port`) on the specified host (`host`).

`public` **`Socket`**`(String host, int port, boolean stream)`
: Creates a socket and connects it to the specified port (`port`) on the specified host (`host`). The boolean value `stream` indicates whether this is a stream or datagram socket.

The methods are:

`public synchronized void` **`close`**`()`
: Closes the socket.

`public InetAddress` **`getInetAddress`**`()`
: Returns the address to which the socket is connected.

`public InputStream` **`getInputStream`**`()`
: Returns the `InputStream` for this socket.

`public int` **`getLocalPort`**`()` Returns the local port to which the socket is connected.

`public OutputStream` **`getOutputStream`**`()`
: Returns an `OutputStream` for this socket.

`public int` **`getPort`**`()` Returns the remote port to which the socket is connected.

`public String` **`toString`**`()` Converts the Socket to a String.

29.4.2 class java.net.SocketImpl

The `SocketImpl` class implements sockets. The methods are:

`protected abstract void` **`create`**`(boolean stream)`
: Creates a socket where `stream` indicates whether the socket is a stream or a datagram.

`protected abstract void` **`connect`**`(String host, int port)`
: Connects the socket to the specified port (`port`) on the specified host (`host`).

`protected abstract void` **`connect`**`(InetAddress address, int port)`
: Connects the socket to the specified address (`address`) on the specified port (`port`).

`protected abstract void` **`bind`**`(InetAddress host, int port)`
: Binds the socket to the specified port (`port`) on the specified host (`host`).

`protected abstract void` **`listen`**`(int backlog)`

This specifies the number of connection requests (`backlog`) the system will queue up while waiting to execute `accept()`.

`protected abstract void` **`accept`**`(SocketImpl s)`

Accepts a connection (`s`).

`protected abstract InputStream` **`getInputStream`**`()`

Returns an `InputStream` for a socket.

`protected abstract OutputStream` **`getOutputStream`**`()`

Returns an `OutputStream` for a socket.

`protected abstract int` **`available`**`()`

Returns the number of bytes that can be read without blocking.

`protected abstract void` **`close`**`()`

Closes the socket.

`protected InetAddress` **`getInetAddress`**`()`
`protected int` **`getPort`**`()`
`protected int` **`getLocalPort`**`()`
`public String` **`toString`**`()`

Returns the address and port of this Socket as a String.

29.5 Creating a socket

Java applet 29.7 sets a constructs a socket for www.sun.com using port 4001 (`Socket remote = new Socket("www.sun.com",4001)`). After this the data stream is created and assigned to `DataIn`. The `readLine()` method is then used to get the text from the stream.

Java applet 29.7 (`chap29_7.java`)

```
import java.io.*;
import java.net.*;
import java.awt.*;
import java.applet.*;

public class chap29_7 extends Applet
{
      public void init()
      {

      String          Instr;
      InputStream     Instream;

         try
         {
            Socket remote = new Socket("www.sun.com",4001);
            Instream = remote.getInputStream();
            DataInputStream DataIn = new DataInputStream(Instream);
            do
            {

               Instr = DataIn.readLine();
               if (Instr!=null)  System.out.println(str);
            } while (Instr!=null);
         }
         catch (UnknownHostException err)
         {
            System.out.println("UNKNOWN HOST: "+err);
         }
```

```
          catch (IOException err)
          {
             System.out.println("Error" + err); }
          }
}
```

Java program 29.8 contacts a server on a given port and returns the local and remote port. It uses command line arguments where the program is run in the form:

`java chap29_8` *host port*

where `java` is the Java interpreter, `chap29_8` is the name of the class file, *host* is the name of the host to contact and *port* is the port to use. The `args.length` parameter is used to determine the number of command line options, anything other than two will display the following message:

```
Usage : chap29_8 host port
```

Java program 29.8 (`chap29_8.java`)

```
import java.net.*;
import java.io.*;

public class chap29_8
{
 public static void main (String args[])
 {
  if (args.length !=2)
   System.out.println(" Usage : chap29_8 host port");

  else
  {
   String inp;
   try
   {
    Socket sock = new Socket(args[0], Integer.valueOf(args[1]).intValue());
    DataInputStream is = new DataInputStream(sock.getInputStream());

    System.out.println("address : " + sock.getInetAddress());
    System.out.println("port : " + sock.getPort());
    System.out.println("Local address : " + sock.getLocalAddress());
    System.out.println("Localport : " + sock.getLocalPort());

    while((inp = is.readLine()) != null)
    { System.out.println(inp);}
   }
   catch (UnknownHostException e)
   {
    System.out.println(" Known Host : " + e.getMessage());
   }
   catch (IOException e)
   {
    System.out.println("error I/O : " + e.getMessage());
   }
   finally
   {
    System.out.println("End of program");
   }
  }
 }
}
```

Sample run 29.4 shows a test run which connects to port 13 on `www.eece.napier.ac.uk`. It can be seen that the connection to this port causes the server to return the current date and time. Sample run 29.5 connects into the same server, in this case on port 19. It can be seen that a connection to this port returns a sequence of characters.

Sample run 29.4

```
>> java chap29_4 www.eece.napier.ac.uk 13
Host and IP address : www.eece.napier.ac.uk/146.176.151.139
port : 13
Local address :pc419.eece.napier.ac.uk
Localport : 1393
Fri May  8 13:19:59 1998
End of program
```

Sample run 29.5

```
>> java chap29_5 www.eece.napier.ac.uk 19
Host and IP address : www.eece.napier.ac.uk/146.176.151.139
port : 19
Local IP address :pc419.eece.napier.ac.uk
Localport : 1403
 !"#$%&'()*+,-./0123456789:;<=>?@ABCDEFGHIJKLMNOPQRSTUVWXYZ[\]^_`abcdefg
!"#$%&'()*+,-./0123456789:;<=>?@ABCDEFGHIJKLMNOPQRSTUVWXYZ[\]^_`abcdefgh
"#$%&'()*+,-./0123456789:;<=>?@ABCDEFGHIJKLMNOPQRSTUVWXYZ[\]^_`abcdefghi
#$%&'()*+,-./0123456789:;<=>?@ABCDEFGHIJKLMNOPQRSTUVWXYZ[\]^_`abcdefghij
$%&'()*+,-./0123456789:;<=>?@ABCDEFGHIJKLMNOPQRSTUVWXYZ[\]^_`abcdefghijk
%&'()*+,-./0123456789:;<=>?@ABCDEFGHIJKLMNOPQRSTUVWXYZ[\]^_`abcdefghijkl
&'()*+,-./0123456789:;<=>?@ABCDEFGHIJKLMNOPQRSTUVWXYZ[\]^_`abcdefghijklm
'()*+,-./0123456789:;<=>?@ABCDEFGHIJKLMNOPQRSTUVWXYZ[\]^_`abcdefghijklmn
()*+,-./0123456789:;<=>?@ABCDEFGHIJKLMNOPQRSTUVWXYZ[\]^_`abcdefghijklmno
)*+,-./0123456789:;<=>?@ABCDEFGHIJKLMNOPQRSTUVWXYZ[\]^_`abcdefghijklmnop
*+,-./0123456789:;<=>?@ABCDEFGHIJKLMNOPQRSTUVWXYZ[\]^_`abcdefghijklmnopq
+,-./0123456789:;<=>?@ABCDEFGHIJKLMNOPQRSTUVWXYZ[\]^_`abcdefghijklmnopqr
,-./0123456789:;<=>?@ABCDEFGHIJKLMNOPQRSTUVWXYZ[\]^_`abcdefghijklmnopqrs
-./0123456789:;<=>?@ABCDEFGHIJKLMNOPQRSTUVWXYZ[\]^_`abcdefghijklmnopqrst
./0123456789:;<=>?@ABCDEFGHIJKLMNOPQRSTUVWXYZ[\]^_`abcdefghijklmnopqrstu
/0123456789:;<=>?@ABCDEFGHIJKLMNOPQRSTUVWXYZ[\]^_`abcdefghijklmnopqrstuv
0123456789:;<=>?@ABCDEFGHIJKLMNOPQRSTUVWXYZ[\]^_`abcdefghijklmnopqrstuvw
123456789:;<=>?@ABCDEFGHIJKLMNOPQRSTUVWXYZ[\]^_`abcdefghijklmnopqrstuvwx
```

29.6 Client/server program

A server is a computer which runs a special program which waits for another computer (a client) to connect to it. This server normally performs some sort of special operation, such as:

- File Transfer Protocol. Transferring files
- Telnet. Remote connection
- WWW service

Java program 29.9 acts as a server program and waits for a connection on port 1111. When a connection is received on this port it sends its current date and time back to the client. This program can be run with Java program 29.8 (which is running on a remote computer) with a connection to the server's IP address (or domain name) and using port 1111. When the client connects to the server, the server responds back to the client with its current date and time.

Test run 29.6 shows a sample run from the server (NOTE THE SERVER PROGRAM MUST BE RUN BEFORE THE CLIENT IS STARTED). It can be seen that it has received connection from the client with the IP address of `146.176.150.120` (Test run 29.7 shows the client connection).

Java program 29.9 (chap29_9.java)

```
import java.net.*;
import java.io.*;
import java.util.*;

class chap29_9
{
 public static void main( String arg[])
 {
  try
  {
   ServerSocket sock = new ServerSocket(1111);
   Socket sock1 = sock.accept();
   System.out.println(sock1.toString());
   System.out.println("address : " + sock1.getInetAddress());
   System.out.println("port    : " + sock1.getPort());

   DataOutputStream out = new DataOutputStream(sock1.getOutputStream());

   out.writeBytes("Welcome "+ sock1.getInetAddress().getHostName()+
         ". We are "+ new Date()+ "\n");
   sock1.close();
   sock.close();
  }
  catch(IOException err)
  {
   System.out.println(err.getMessage());
  }
  finally
  {
   System.out.println("End of the program");
  }
 }
}
```

Sample run 29.6 (Server)

```
>> java chap29_03
address : pc419.eece.napier.ac.uk/146.176.151.130
port : 1111
End of program
```

Sample run 29.7 (Client)

```
>> java chap29_02 146.176.150.120 1111
Host and IP address : pc419.eece.napier.ac.uk/146.176.151.130
port : 1111
Local address :pc419.eece.napier.ac.uk
Localport : 1393
Fri May  8 13:19:59 1998
End of program
```

30 JavaScript

30.1 Introduction

Programming languages can either be compiled to produce an executable program or they can be interpreted while the user runs the program. Java is a program language which needs to be compiled before it is used. It thus cannot be used unless the user has the required Java compiler. JavaScript, on the other hand, is a language which is interpreted by the browser. It is similar in many ways to Java but allows the user to embed Java-like code into an HTML page. JavaScript supports a small number of data types representing numeric, Boolean and string values, and is supported by most modern WWW browsers, such as Microsoft Internet Explorer and Netscape Navigator.

HTML is useful when pages are short and do not contain expressions, loops or decisions. JavaScript allows most of the functionality of a high-level language for developing client and server Internet applications. It can be used to respond to user events such as mouse clicks, form input and page navigation.

A major advantage of JavaScript over HTML is that it supports the use of functions without any special declarative requirements. It is also simpler to use than Java because it has easier syntax, specialized built-in functionality and minimal requirements for object creation.

Important concepts in Java and JavaScript are objects. Objects are basically containers for values. The main differences between JavaScript and Java are:

- JavaScript is interpreted by the client, whereas Java is compiled on the server before it is executed.
- JavaScript is embedded into HTML pages, whereas Java applets are distinct from HTML and accessed from HTML pages.
- JavaScript has loose typing for variables (i.e. a variable's data type does not have to be declared), whereas Java has strong typing (i.e. a variable's data type must always declared before it is used).

JavaScript has dynamic binding where object references are checked at run-time. Java has static binding where object references must exist at compile-time.

Figure 30.1 shows the main functional differences between a high-level language, a Java applet and JavaScript. JavaScript is interpreted by the browser, whereas a Java applet is compiled to a virtual machine code which can be run on any computer system. The high-level language produces machine-specific code.

JavaScript 30.1 shows an HTML script which contains the JavaScript in boldface. Figure 30.2 gives the browser output.

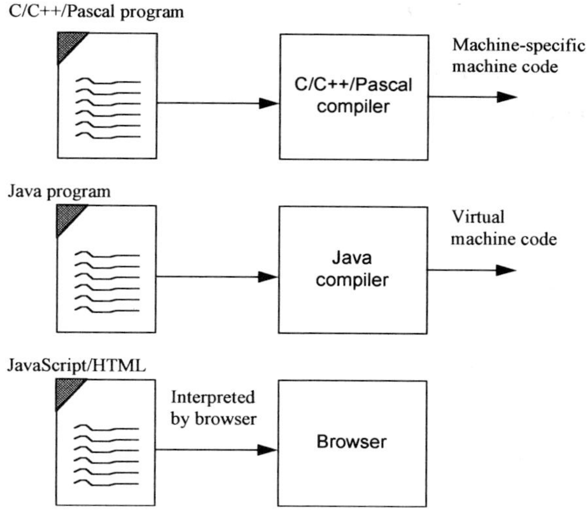

Figure 30.1 Differences between C++/Java and JavaScript.

JavaScript 30.1

```
<HTML> <HEAD><TITLE>My Java</TITLE></HEAD>
<BODY>
<SCRIPT language="javascript">
document.writeln("This is my first JavaScript");
for (i=0;i<10;i++)
 document.write("<center><font size=+1><b>Loop</b> ",i);
</SCRIPT></BODY></HTML>
```

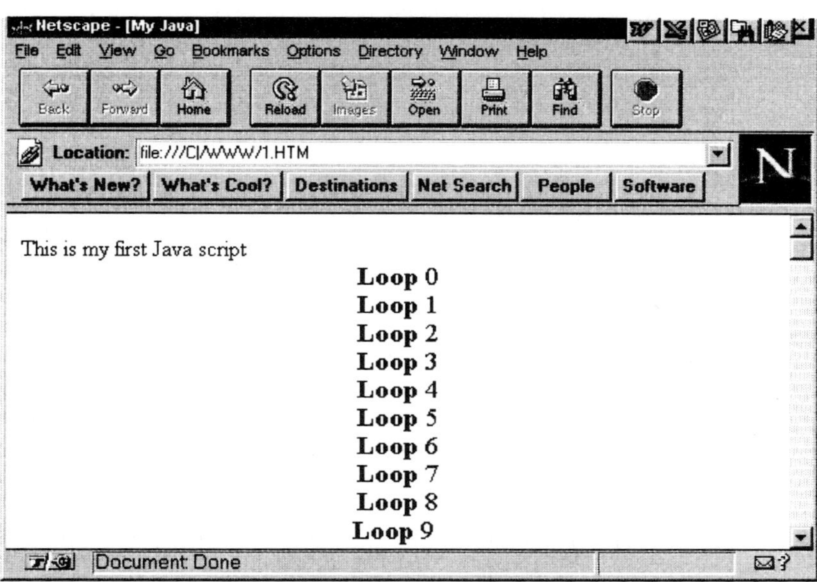

Figure 30.2 Browser output produced by JavaScript 30.1.

30.2 JavaScript values, variables and literals

JavaScript values, variable and literals are similar to the C programming language. Their syntax is discussed in this section.

30.2.1 Values

The four different types of values in JavaScript are:

- Numeric value, such as 12 or 91.5432.
- Boolean values which are either TRUE or FALSE.
- Strings types, such as "Fred Bloggs".
- A special keyword for a NULL value

Numeric values differ from most programming languages in that there is no explicit distinction between a real value (such as 91.5432) and an integer (such as, 12).

30.2.2 Data type conversion

JavaScript differs from Java in that variables do not need to have their data type defined when they are declared (loosely typed). Data types are then automatically converted during the execution of the program. Thus a variable could be declared with a numeric value as:

```
var value

      value = 19
```

and then in the same script it could be assigned a string value, such as:

```
      value = "Enter your name >>"
```

The conversion between numeric values and strings in JavaScript is easy, as numeric values are automatically converted to an equivalent string. For example:

```
<HTML><HEAD><TITLE>My Java</TITLE></HEAD>
<BODY BGCOLOR="#ffffff">
<SCRIPT language="javascript">
var x,y,str

        x=13
        y=10
        str= x + " added to " + y + " is " + x+y
        document.writeln(str)

        z=x+y
        str= x + " added to " + y + " is " + z
        document.writeln(str)
</SCRIPT>
</BODY></HTML>
```

Sample run 30.1 gives the output from this script. It can be seen that `x` and `y` have been converted to a string value (in this case, `"13"` and `"10"`) and that `x+y` in the string conversion statement has been converted to `"1310"`. If a mathematical operation is carried out (`z=x+y`) then `z` will contain `23` after the statement is executed.

Sample run 30.1

```
13 added to 10 is 1310  13 added to 10 is 23
```

JavaScript provides several special functions for manipulating string and numeric values:

- The `eval(`*string*`)` function which converts a string to a numerical value.
- The `parseInt(`*string* [*,radix*]`)` function which converts a string into an integer of the specified radix (number base). The default radix is base-10.
- The `parseFloat(`*string*`)` function which converts a string into a floating-point value.

30.2.3 Variables

Variables are symbolic names for values within the script. A JavaScript identifier must either start with a letter or an underscore ("_"), followed by letters, an underscore or any digit (0–9). Like C, JavaScript is case sensitive so that variables with the same character sequence but with different cases for one or more characters are different. The following are different variable names:

```
i=5
I=10

valueA=3.543
VALUEA=10.543
```

30.2.4 Variable scope

A variable can be declared by either simply assigning it a value or by using the `var` keyword. For example the following declares to variables `Value1` and `Value2`:

```
var   Value1;
         Value2=23;
```

A variable declared within a function is taken as a local variable and can only be used within that function. A variable declared outside a function is a global variable and can be used anywhere in the script. A variable which is declared locally which is already declared as a global variable needs to be declared with the `var` keyword, otherwise the use of the keyword is optional.

30.2.5 Literals

Literal values have fixed values with the script. Various reserved forms can be used to identify special types, such as hexadecimal values, exponent format, and so on. With an integer the following are used:

- If the value is preceded by 0x then the value is a hexadecimal value (i.e. base 16). Examples of hexadecimal values are 0x1FFF, 0xCB.
- If the value is preceded by 0 then the value is an octal value (i.e. base 8). Examples of octal values are 0777, 010.
- If it is not preceded by either 0x or 0 then it is a decimal integer.

Floating-point values

Floating-point values are typically represented as a real value (such as 1.342) or in exponent format. Some exponent format value are:

Value	*Exponent format*
`0.000001`	`1e-6`
`1342000000`	`1.342e9`

Boolean

The true and false literals are used with Boolean operations.

Strings

In C a string is represented with double quotes ("*str*") whereas JavaScript accepts a string within double (") or single (') quotation marks. Examples of strings are:

```
"A string"
'Another string'
```

C uses an escape character sequence to represent special characters within a string. This character sequence always begins with a '\' character. For example, if the escape sequence '\n' appears in the string then this sequence is interpreted as a new-line sequence. Valid escape sequences are:

Character	*Meaning*	*Character*	*Meaning*
`\b`	backspace	`\f`	form feed
`\n`	new line	`\r`	carriage return
`\t`	tab	`\\`	backslash character
`\"`	prints a " character		

30.3 Expressions and operators

The expressions and operators used in Java and JavaScript are based on C and C++. This section outlines the main expressions and operators used in JavaScript.

30.3.1 Expressions

As with C, expressions are any valid set of literals, variables, operators, and expressions that evaluate to a single value. There are basically two types of expression, one which assign a value to a variable and the other which simply gives a single value. A simple assignment is:

```
value = 21
```

which assigns the value of 21 to `value` (note that the result of the expression is 21).

The result from a JavaScript expression can be:

- A numeric value.

- A string.
- A logical value (true or false).

30.4 JavaScript operators

Java and JavaScript have a rich set of operators; there are four main types:

- Arithmetic.
- Logical.
- Bitwise.
- Relational.

30.4.1 Arithmetic

Arithmetic operators operate on numerical values. The basic arithmetic operations are add (+), subtract (–), multiply (*), divide (/) and modulus division (%). Modulus division gives the remainder of an integer division. The following gives the basic syntax of two operands with an arithmetic operator.

```
operand operator operand
```

The assignment operator (=) is used when a variable 'takes on the value' of an operation. Other shorthand operators are used with it, including add equals (+=), minus equals (–=), multiply equals (*=), divide equals (/=) and modulus equals (%=). The following examples illustrate their uses.

Statement	*Equivalent*
`x+=3.0`	`x=x+3.0`
`voltage/=sqrt(2)`	`voltage=voltage/sqrt(2)`
`bit_mask *=2`	`bit_mask=bit_mask*2`

In many applications it is necessary to increment or decrement a variable by 1. For this purpose Java has two special operators; ++ for increment and -- for decrement. They can either precede or follow the variable. If they precede the variable, then a pre-increment/decrement is conducted, whereas if they follow it, a post-increment/decrement is conducted.
Here are some:

Statement	Equivalent
`no_values++`	`no_values=no_values+1`
`i--`	`i=i-1`

Table 30.1 summarizes the arithmetic operators.

Table 30.1 Arithmetic operators.

Operator	*Operation*	*Example*
`-`	subtraction, minus	`5-4→1`
`+`	addition	`4+2→6`
`*`	multiplication	`4*3→12`
`/`	division	`4/2→2`
`%`	modulus	`13%3→1`
`+=`	add equals	`x += 2` is equivalent to `x=x+2`
`-=`	minus equals	`x -= 2` is equivalent to `x=x-2`
`/=`	divide equals	`x /= y` is equivalent to `x=x/y`
`*=`	multiplied equals	`x *= 32` is equivalent to `x=x*32`
`=`	assignment	`x = 1`
`++`	increment	`Count++` is equivalent to `Count=Count+1`
`--`	decrement	`Sec--` is equivalent to `Sec=Sec-1`

30.4.2 Relationship

The relationship operators determine whether the result of a comparison is TRUE or FALSE. These operators are greater than (>), greater than or equal to (>=), less than (<), less than or equal to (<=), equal to (==) and not equal to (`!=`). Table 30.2 lists the relationship operators.

Table 30.2 Relationship operators.

Operator	*Function*	*Example*	*TRUE Condition*
`>`	greater than	`(b>a)`	when `b` is greater than `a`
`>=`	greater than or equal	`(a>=4)`	when `a` is greater than or equal to 4
`<`	less than	`(c<f)`	when `c` is less than `f`
`<=`	less than or equal	`(x<=4)`	when `x` is less than or equal to 4
`==`	equal to	`(x==2)`	when `x` is equal to 2
`!=`	not equal to	`(y!=x)`	when `y` is not equal to `x`

30.4.3 Logical (TRUE or FALSE)

A logical operation is one in which a decision is made as to whether the operation performed is TRUE or FALSE. If required, several relationship operations can be grouped together to give the required functionality. C assumes that a numerical value of 0 (zero) is FALSE and that any other value is TRUE. Table 30.2 lists the logical operators.

The logical AND operation will yields TRUE only if all the operands are TRUE. Table 30.4 gives the result of the AND (`&&`) operator for the operation `A && B`. The logical OR operation yields a `TRUE` if any one of the operands is `TRUE`. Table 30.4 gives the logical results of the `OR` (`||`) operator for the statement `A|| B`. Table 30.4 also gives the logical result of the NOT (!) operator for the statement `!A`.

Table 30.3 Logical operators.

Operator	*Function*	*Example*	*TRUE condition*
`&&`	AND	`((x==1) && (y<2))`	when x equal 1 *and* y is less than 2
`||`	OR	`((a!=b) || (a>0))`	when a is not equal to b *or* a is greater than 0
`!`	NOT	`(!(a>0))`	when a is *not* greater than 0

Table 30.4 Logical operations.

A	B	AND (`&&`)	OR (`||`)	NOT (`!A`)
FALSE	FALSE	FALSE	FALSE	TRUE
FALSE	TRUE	FALSE	TRUE	TRUE
TRUE	FALSE	FALSE	TRUE	FALSE
TRUE	TRUE	TRUE	TRUE	FALSE

30.4.4 Bitwise

The bitwise logical operators work conceptually as follows:

- The operands are converted to 32-bit integers and expressed as a series of bits (0s and 1s).
- Each bit in the first operand is paired with the corresponding bit in the second operand: first bit to first bit, second bit to second bit, and so on.
- The operator is applied to each pair of bits, and the result is constructed bitwise.

The bitwise operators are similar to the logical operators but they should not be confused as their operation differs. Bitwise operators operate directly on the individual bits of any operands, whereas logical operators determine whether a condition is TRUE or FALSE.

Numerical values are stored as bit patterns in either an unsigned integer format, signed integer (two's complement) or floating-point notation (an exponent and mantissa). Characters are normally stored as ASCII characters.

The basic bitwise operations are AND (`&`), OR (`|`), one's complement or bitwise inversion (~), XOR (^), shift left (<<) and shift right (>>). Table 30.5 gives the results of the AND bitwise operation on two bits *A* and *B*.

The Boolean bitwise instructions operate logically on individual bits. The XOR function yields a 1 when the bits in a given bit position differ, the AND function yields a 1 only when the given bit positions are both 1s. The OR operation gives a 1 when any one of the given bit positions are a 1. For example:

```
     00110011           10101111             00011001
AND  11101110       OR  10111111        XOR  11011111
     --------           --------             --------
     00100010           10111111             11000110
     ========           ========             ========
```

Table 30.5 Bitwise operations

A	*B*	*AND*	*OR*	*EX-OR*
0	0	0	0	0
0	1	0	1	1
1	0	0	1	1
1	1	1	1	0

To perform bit shifts, the <<, >> and >>> operators are used. These operators shift the bits in the operand by a given number defined by a value given on the right-hand side of the operation. The left shift operator (<<) shifts the bits of the operand to the left and zeros fill the result on the right. The sign-propagating right shift operator (>>) shifts the bits of the operand to the right and zeros fill the result if the integer is positive; otherwise it will fill with 1s. The zero-filled right shift operator (>>>) shifts the bits of the operand to the right and fills the result with zeros. The standard format is:

```
operand >>  no_of_bit_shift_positions
operand >>> no_of_bit_shift_positions
operand <<  no_of_bit_shift_positions
```

30.4.5 Precedence

There are several rules for dealing with operators:

- Two operators, apart from the assignment operator, should never be placed side by side. For example, `x * % 3` is invalid.
- Groupings are formed with parentheses; anything within parentheses will be evaluated first. Nested parentheses can also be used to set priorities.
- A priority level or precedence exists for operators. Operators with a higher precedence are evaluated first; if two operators have the same precedence, then the operator on the left-hand side is evaluated first. The priority levels for operators are as follows:

HIGHEST PRIORITY

<table>
<tr><td>() [] .</td><td>primary</td></tr>
<tr><td>! ~ ++ -- -</td><td>unary</td></tr>
<tr><td>* / %</td><td>multiplicative</td></tr>
<tr><td>+ -</td><td>additive</td></tr>
<tr><td><< >> >>></td><td>shift</td></tr>
<tr><td>< > <= >=</td><td>relational</td></tr>
<tr><td>== !=</td><td>equality</td></tr>
<tr><td>&</td><td rowspan="3">bitwise</td></tr>
<tr><td>^</td></tr>
<tr><td>|</td></tr>
<tr><td>&&</td><td rowspan="2">logical</td></tr>
<tr><td>||</td></tr>
<tr><td>= += -=</td><td>assignment</td></tr>
</table>

LOWEST PRIORITY

The assignment operator has the lowest precedence. The following example shows how operators are prioritized in a statement (=> shows the steps in determining the result):

```
23 + 5 % 3 / 2 << 1   =>
23 + 2 / 2 << 1       =>
23 + 1 << 1           =>
23 + 2                => 25
```

30.4.6 Conditional expressions

Conditional expressions can produce one of two values depending on a condition. The syntax is:

```
(expression) ? value1 : value2
```

If the expression is true then `value1` is executed else `value2` is executed. For example:

```
(val >= 0) ? sign="postive" : sign="negative"
```

The expression will assign the string "positive" to `sign` if the value of `val` is greater than or equal to 0, else it will assign "negative".

30.4.7 String operators

The normal comparison operators, such as <, >, >=, ==, and so on can be used with strings. In addition the concatenation operator (+) can be used to concatenate two string values together. For example,

```
str="This is " + "an example"
```

will result in the string

```
"This is an example"
```

30.5 JavaScript statements

JavaScript statements are similar to C and allow a great deal of control of the execution of a script. The basic categories are:

- Conditional statements, such as `if...else`.
- Repetitive statements, such as `for`, `while`, `break` and `continue`.
- Comments, using either the C++ style for single-line comments (`//`) or standard C multiline comments (`/*...*/`).
- Object manipulation statements and operators, such as `for...in`, `new`, `this`, and `with`.

30.6 Conditional statements

Conditional statements allow a program to make decisions on the route through a program.

30.6.1 `if...else`

A decision is made with the `if` statement. It logically determines whether a conditional expression is TRUE or FALSE. For TRUE, the program executes one block of code; FALSE causes the execution of another (if any). The keyword `else` identifies the FALSE block. Braces are used to define the start and end of the block.

Relationship operators (>,<,>=,<=,==,!=) yield TRUE or FALSE from their operation. Logical statements (&&, ||, !) can then group them together to give the required functionality. If the operation is not a relationship, such as a bitwise or arithmetic operation, then any non-zero value is TRUE and a zero is FALSE.

The following is an example of the `if` statement syntax. If the statement block has only one statement the braces ({ }) can be excluded.

```
if (expression)
{
        statement block
}
```

The following is an example of an `else` extension.

```
if (expression)
{
        statement block1
}
else
{
        statement block2
}
```

It is possible to nest `if...else` statements to give a required functionality. In the next example, *statement block1* is executed if `expression1` is TRUE. If it is FALSE then the program checks the next expression. If this is TRUE the program executes *statement block2*, else it checks the next expression, and so on. If all expressions are FALSE then the program executes the final `else` statement block, in this case *statement block 3*:

```
if (expression1)
{
        statement block1
}
else if (expression2)
{
        statement block2
}
else
{
        statement block3
}
```

30.7 Loops

30.7.1 `for()`

Many tasks within a program are repetitive, such as prompting for data, counting values, and so on. The `for` loop allows the execution of a block of code for a given control function. The following is an example format; if there is only one statement in the block then the braces can be omitted.

```
for (starting condition;test condition;operation)
```

```
{
        statement block
}
```

where

`starting condition`	means the starting value for the loop
`test condition`	means if `test condition` is TRUE the loop will continue execution
`operation`	means the operation conducted at the end of the loop.

30.7.2 `while()`

The `while()` statement allows a block of code to be executed while a specified condition is TRUE. It checks the condition at the start of the block; if this is TRUE the block is executed, else it will exit the loop. The syntax is:

```
while (condition)
{
      :          :  statement block
      :          :
}
```

If the statement block contains a single statement then the braces may be omitted (although it does no harm to keep them).

30.8 Comments

Comments are author annotations that explain what a script does. Comments are ignored by the interpreter. JavaScript supports Java-style comments:

- Comments on a single line are preceded by a double-slash (`//`).
- Multiline comments can be preceded by `/*` and followed by `*/`.

The following example shows two comments:

```
// This is a single-line comment.
/* This is a multiple-line comment. It can be of any length, and you can put
whatever you want here. */
```

30.9 Functions

JavaScript supports modular design using functions. A function is defined in JavaScript with the `function` reserved word and the code within the function is defined within curly brackets. The standard format is:

```
function myfunct(param1, param2 ...)
{
    statements
    return(val)
}
```

where the parameters (`param1`, `param2`, and so on) are the values passed into the function. Note that the return value (`val`) from the function is only required when a value is returned from the function.

JavaScript 30.2 gives an example with two functions (`add()` and `mult()`). In this case the values, `value1` and `value2`, are passed into the variables, `a` and `b`, within the `add()` function; the result is then sent back from the function into `value3`.

JavaScript 30.2

```
<HTML><TITLE>Example</TITLE>
<BODY BGCOLOR="#FFFFFF">

<SCRIPT>
var value1,value2,value3,value4;

value1=15;
value2=10;
value3=add(value1,value2)
value4=mult(value1,value2)
document.write("Added is ",value3)
document.write("<P>Multiplied is ",value4)

function add(a,b)
{
var   c
      c=a+b
      return(c)
}
function mult(a,b)
{
var   c
      c=a*b
      return(c)
}
</SCRIPT></FORM></HTML>
```

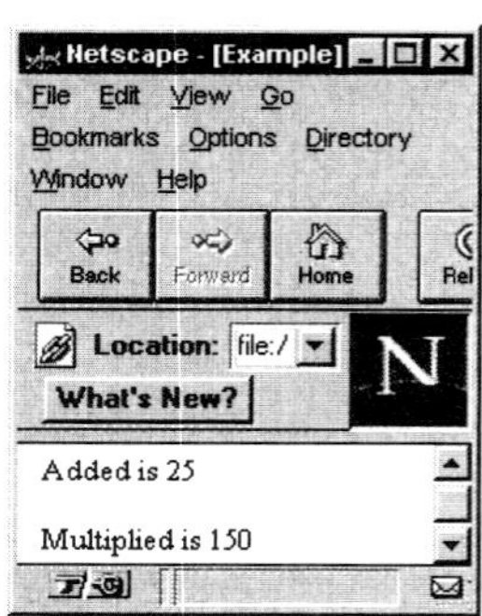

30.10 Objects and properties

JavaScript is based on a simple object-oriented paradigm, where objects are a construct with properties that are JavaScript variables. Each object has properties associated with it and can be accessed with the dot notation, such as:

objectName.propertyName

31 Windows 95/98/NT

31.1 Introduction

DOS has long been the Achilles heel of the PC and has limited its development. It has also been its strength in that it provides a common platform for all packages. DOS and Windows 3.*x* operated in a 16-bit mode and had limited memory accessing. Windows 3.0 provided a great leap in PC systems as it provided an excellent graphical user interface to DOS. It suffered from the fact that it still used DOS as the core operating system. Windows 95/98 and Windows NT have finally moved away from DOS and operate as full 32-bit protected-mode operating systems. Their main features are:

- Run both 16-bit and 32-bit application programs.
- Allow access to a large virtual memory (up to 4 GB).
- Support for pre-emptive multitasking and multithreading of Windows-based and MS-DOS-based applications.
- Support for multiple file systems, including 32-bit installable file systems such as VFAT, CDFS (CD-ROM) and network redirectors. These allow better performance, use of long file names, and are an open architecture to support future growth.
- Support for 32-bit device drivers which give improved performance and intelligent memory usage.
- A 32-bit kernel which includes memory management, process scheduling and process management.
- Enhanced robustness and cleanup when an application ends or crashes.
- Enhanced dynamic environment configuration.

The three most widely used operating systems are: MS-DOS, Microsoft Windows and UNIX. Microsoft Windows comes in many flavors; the main versions are outlined below and Table 31.1 lists some of their attributes.

- Microsoft Windows 3.*x* – 16-bit PC-based operating system with limited multitasking. It runs from MS-DOS and thus still uses MS-DOS functionality and file system structure.
- Microsoft Windows 95/98 – robust 32-bit multitasking operating system (although there are some 16-bit parts in it) which can run MS-DOS applications, Microsoft Windows 3.*x* applications and 32-bit applications.
- Microsoft Windows NT – robust 32-bit multitasking operating system with integrated networking. Networks are around NT servers and clients. As with Microsoft Windows 95/98 it can run MS-DOS, Microsoft Windows 3.*x* applications and 32-bit applications.

Windows NT and 95/98 provide excellent network support as they can communicate directly with many different types of networks, protocols and computer architectures. They can networks to make peer-to-peer connections and also connections to servers for access to file systems and print servers.

Windows NT has more security in running programs than Windows 95/98 as programs and data are insulated from the operation of other programs. The operating system parts of Windows NT and Windows 95/98 run at the most trusted level of privilege of the Intel processor, which is ring zero. Application programs run at least trusted level of privilege, which is ring three. These programs can use either a 32-bit flat mode or any of the memory models, such as large, medium, compact or small.

There was a great leap in performance between the 16-bit Windows 3.*x* operating system (which was built on DOS) to Windows 95/98 and Window NT. Apart from running in a dual 16-bit and 32-bit mode they also allow for application robustness. Figure 31.1 outlines the internal architecture of Windows 95/98.

Table 31.1 Windows comparisons.

	Windows 3.1	*Windows 95/98*	*Windows NT*
Pre-emptive multitasking		✓	✓
32-bit operating system		✓	✓
Long file names		✓	✓
TCP/IP	✓	✓	✓
32-bit applications		✓	✓
Flat memory model		✓	✓
32-bit disk access	✓	✓	✓
32-bit file access	✓	✓	✓
Centralized configuration storage		✓	✓
OpenGL 3D graphics			✓

31.2 Windows registry

On DOS-based systems, the main configuration files were AUTOEXEC.BAT, CONFIG. SYS and INI files. INI files were a major problem in that each application program and device driver configuration required one or more of these file to store default settings (such as IRQ, I/O addresses, default directories, and so on). Several important INI files are:

- WIN.INI. Information about the appearance of the Windows environment.
- SYSTEM.INI. System-specific information on the hardware and device driver configuration of the system.

Windows 95/98/NT use a central database called the Registry, which stores user-specific and configuration-specific information at a single location. This location could be on the local computer or stored on a networked computer. It thus allows network managers to standardize the configuration of networked PCs.

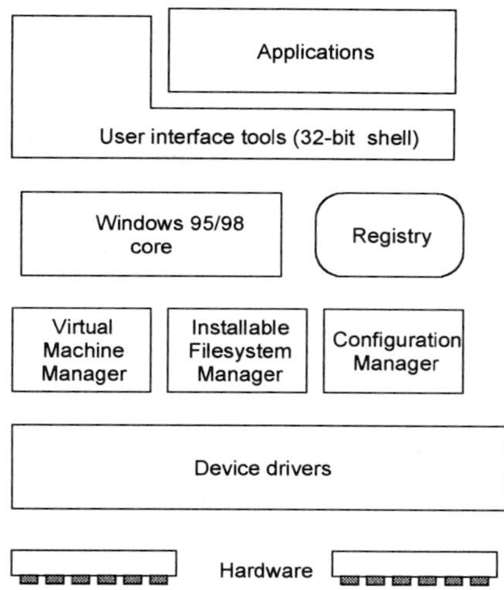

Figure 31.1 Windows 95/98 architecture.

When a computer is initially upgraded from Windows 3.*x* to Windows 95/98 the upgrade program reads the SYSTEM.INI file and system-specific information which it then puts into the Registry. Many INI files are still retained on the system as many Win16-based applications use them. For example, Microsoft Word Version 6 uses the WINWORD6.INI to store package information, such as: location of filters, location of spell checker, location of grammar checker, and so on. An example is:

```
[Microsoft Word]
WPHelp=0
Hyphenate 1033,0=C:\MSOFFICE\WINWORD\HYPH.DLL,C:\MSOFFICE\WINWORD\HY_EN.
NoLongNetNames=Yes
USER-DOT-PATH=C:\MSOFFICE\WINWORD\TEMPLATE
PICTURE-PATH=C:\MSOFFICE\WINWORD
PROGRAMDIR=C:\MSOFFICE\WINWORD
TOOLS-PATH=C:\MSOFFICE\WINWORD
STARTUP-PATH=C:\DOCS\NOTES\
INI-PATH=C:\MSOFFICE\WINWORD
DOC-PATH=C:\DOCS\NOTES\
Hyphenate 2057,0=C:\MSOFFICE\WINWORD\HYPH.DLL,C:\MSOFFICE\WINWORD\HY_EN.
```

An important role for the Registry is to store hardware-specific information, which can be used by hardware detection and Plug-and-Play programs. The Configuration Manager determines the configuration of installed hardware (such as, IRQs, I/O addresses, and so on) and it uses this information to update the Registry. This allows new devices to be installed and checked to see if they conflict with existing devices. If they are Plug-and-Play devices then the system assigns hardware parameters which do not conflict with existing devices.

The advantages of the Registry over INI files include:

- **No limit to size and data type**. The Registry has no size restriction and can include binary and text values (INI files are text based and are limited to 64 KB in size).
- **Hierarchical information**. The Registry is hierarchically arranged, whereas INI files are non-hierarchical and support only two levels of information.
- **Standardized setup**. The Registry provides a standardized method of setting up pro-

grams, whereas many INI files contain a whole host of switches and entries, and are complicated to configure.

- **Support for user-specific information**. The Registry allows the storage of user-specific information, using the `Hkey_Users` key. This allows each user of a specific computer (or a networked computer) to have their own user-specific information. INI files do not support this.
- **Remote administration and system policies**. The Registry can be used to remotely administer and set system policies (which are stored as Registry values). These can be downloaded from a central server each time a new user logs on.

The Registry will be discussed in more detail in the next chapter. Figure 31.2 shows an example of the Registry in Windows 95/98.

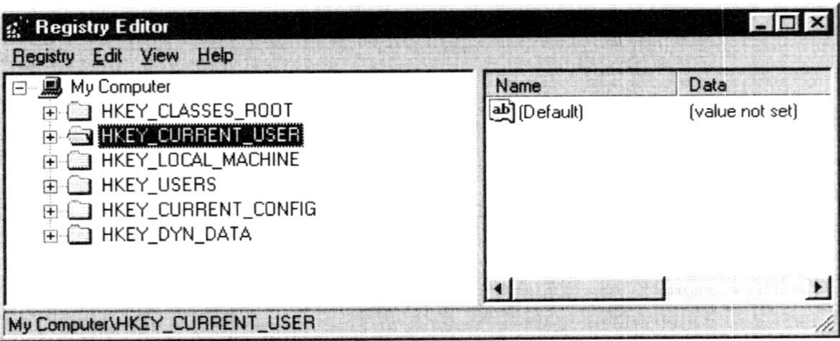

Figure 31.2 Example registry.

31.3 Device drivers

In Windows 3.*x*, device drivers were complex entities and were, in part, static and unchanging. Windows 95/98/NT now provide enhanced support for hardware devices and peripherals including disk devices. Windows NT will be discussed in Section 31.11. Windows 95/98 uses a universal driver/mini-driver architecture that makes writing device-specific code much easier.

The universal driver provides for most of the code for a specific class of device (such as for printers or mice) and the mini-driver is a relatively small and simple driver that provides for the additional information for the hardware.

The actual system interface to the hardware (or some software parts) is through a virtual device driver (VxD), which is a 32-bit, protected-mode driver. These keep track of the state of the device for each application and ensure that the device is in the correct state whenever an application continues. This allows for multitasking programming and also for multi-access for a single device. VxD files also support hardware emulation, such as in the case of the MS-DOS device driver, where any calls to the PC hardware can be handled by the device driver and not by the physical hardware. Typical VxD drivers are:

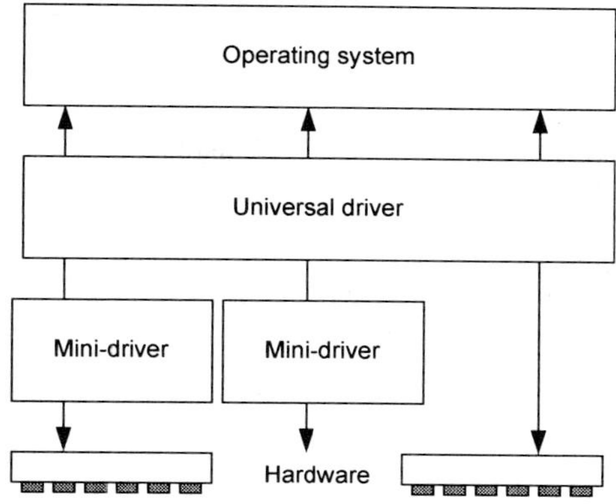

Figure 31.3 Device drivers.

EISA.VXD	EISA bus driver	ISAPNP.VXD	ISA Plug and Play
SERIAL.VXD	Serial port	LPTENUM.VXD	Parallel port
MSMOUSE.VXD	MS Mouse	PARALINK.VXD	Parallel port
PCI.VXD	PCI	QC117.VXD	Tape backup
IRCOMM.VXD	Infra-red comms.	UNIMODEM.VXD	Modem
WSOCK.VXD	WinSock	LPT.VXD	LPT
VMM32.VXD	Memory management	JAVASUP.VXD	JavaScript
PPPMAC.VXD	PPP connection	NDIS.VXD	NDIS
NDIS2SUP.VXD	NDIS 2.0	NETBEUI.VXD	Net BEUI
NWREDIR.VXD	NetWare Redirect	VNETBIOS.VXD	Net BIOS
WSIPX.VXD	IPX	WSHTCP.VXD	TCP

In Windows 95/98, VxD files are loaded dynamically and are thus only loaded when they are required, whereas in Windows 3.*x* they were loaded statically (and thus took up a lot of memory). In Window 3.*x* these virtual device drivers have a 386 file extension.

31.4 Configuration Manager

A major drawback with Windows 3.*x* and DOS is that they did not automate PC configuration. For this purpose, Windows 95/98 has a Configuration Manager. The left-hand side of Figure 31.4 shows how it integrates into the system and the right side of Figure 31.4 shows an example device connection of a PC. Its aim is to:

- Determine, with the aid of several subcomponents, each bus and each device on the system, and their configuration settings. This is used to ensure that each device has unique IRQs and I/O port addresses and that there are no conflicts with other devices. With Plug-and-Play, devices can be configured so that they do not conflict with other devices.
- Monitor the PC for any changes to the number of devices connected and also the device types. If it detects any changes then it manages the reconfiguration of the devices.

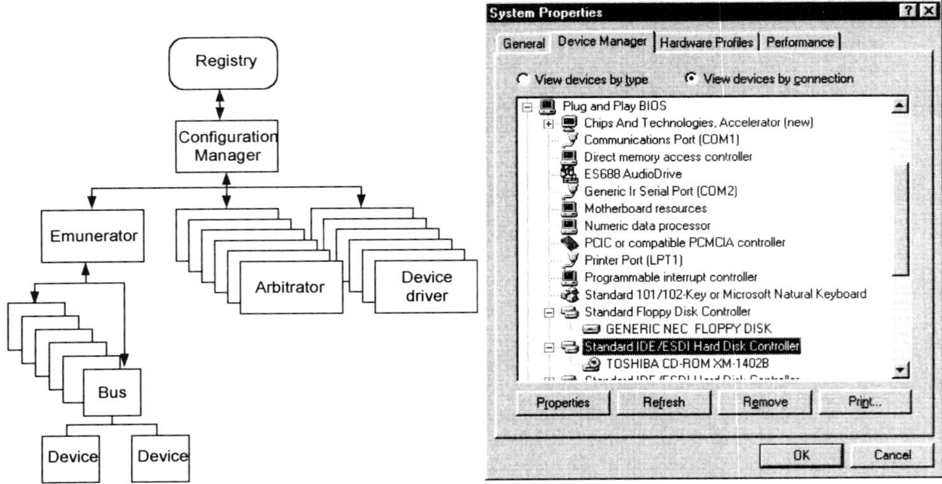

Figure 31.4 Configuration Manager and example connection of devices.

The operation is as follows:

1. The Configuration Manager communicates with each of the bus enumerators and asks them to identify all the devices on the buses and their respective resource requirements. A bus enumerator is a driver that is responsible for creating a hardware tree, which is a hierarchical representation of all the buses and devices on a computer. Figure 31.5 shows an example tree.
2. The bus enumerator locates and gathers information from either the device drivers or the BIOS services for that particular device type. For example, the CD-ROM bus enumerator calls the CD-ROM drivers to gather information.

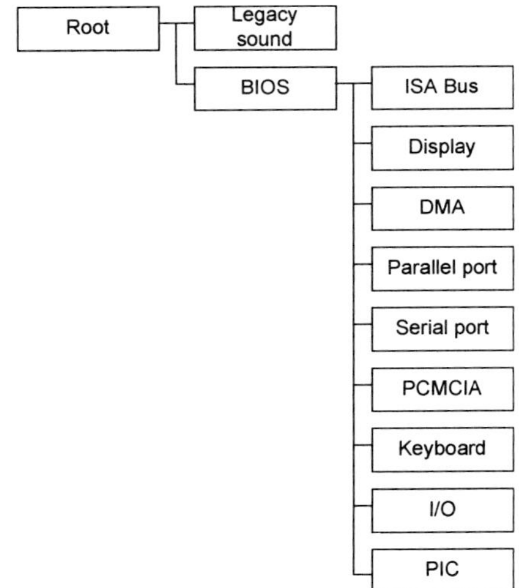

Figure 31.5 Hierarchical representation of the system.

3. Each of the drivers is then loaded and they wait for the Configuration Manager to assign their specific resources (such as IRQs, I/O addresses, and so on).
4. Configuration Manager calls on resource arbitrators to allocate resources for each device.
5. Resource arbitrators identify any devices which are conflicting and tries to resolve them.
6. The Configuration Manager informs all device drivers of their device configuration. This process is repeated when the BIOS or one of the other bus enumerators informs Configuration Manager about a system configuration change.

31.5 Virtual Machine Manager (VMM)

The perfect environment for a program is to run on a stand-alone, dedicated computer, which does not have any interference from any other programs and can have access to any device when it wants. This is the concept of the Virtual Machine. In Windows 95/98 the Virtual Machine Manager (VMM) provides each application with the system resources when it needs them. It creates and maintains the virtual machine environments in which applications and system processes run (in Windows 3.*x* the VMM was called WIN386.EXE).

The VMM is responsible for three areas:

- **Process scheduling**. This is responsible for scheduling processes. It allows for multiple applications to run concurrently and also for providing system resources to the applications and other processes that run. This allows multiple applications and other processes to run concurrently, using either cooperative multitasking or pre-emptive multitasking.
- **Memory paging**. Windows 95/98/NT uses a demand-paged virtual memory system, which is based on a flat, linear address space accessed using 32-bit addresses. The system allocates each process a unique virtual address space of 4 GB. The upper 2 GB is shared, while the lower 2 GB is private to the application. This virtual address space is divided into equal blocks (or pages).
- **MS-DOS Mode support**. Provides support for MS-DOS-based applications which must have exclusive access to the hardware. When an MS-DOS-based application runs in this mode then no other applications or processes are allowed to compete for system resources. The application thus has sole access to the resources.

Windows 95/98 has a single VMM (named System VMM) in which all system processes run. Win32-based and Win16-based applications run within this VMM. Each MS-DOS-based application runs in its own VM.

31.5.1 Process scheduling and multitasking

This allows multiple applications and other processes to run concurrently, using either cooperative multitasking and pre-emptive multitasking. In Windows 3.*x*, applications ran using cooperative multitasking. This method requires that applications check the message queue periodically and to give-up control of the system to other applications. Unfortunately, applications that do not check the message queue at frequent intervals can effectively 'hog' the processor and prevent other applications from running. As this does not provide effective multi-processing, Windows 95/98/NT uses pre-emptive multitasking for Win32-based applications (but also supports cooperative multitasking for computability reasons). Thus, the operating system takes direct control away from the application tasks.

Win16 programs need to yield to other tasks in order to multitask properly, whereas Win32-based programs do not need to yield to share resources. This is because Win32-based applications (called processes) use multithreading, which provides for multi-processing. A thread in a program is a unit of code that can get a time slice from the operating system to run concurrently with other code units. Each process consists of one or more execution threads that identify the code path flow as it is run on the operating system. A Win32-based application can have multiple threads for a given process. This enhances the running of an application by improving throughput and responsiveness. It allows processes for smooth background processing.

31.5.2 Memory paging

Windows 95/98/NT use a demand-paged virtual memory system, which is based on a flat, linear address space using 32-bit addresses. The system allocates each process a unique virtual address space of 4 GB (which should be enough for most applications). The upper 2 GB is shared, while the lower 2 GB is private to the application. This virtual address space divides into equal blocks (or pages), as illustrated in Figure 31.6.

Demand paging is a method by which code and data are moved in pages from physical memory to a temporary paging file on disk. When required, information is then paged back into physical memory.

The functions of the Memory Pager are:

- To map virtual addresses from the process's address space to physical pages in memory. This then hides the physical organization of memory from the process's threads and ensures that the thread can access the required memory when required. It also stops other processes from writing to another memory location.

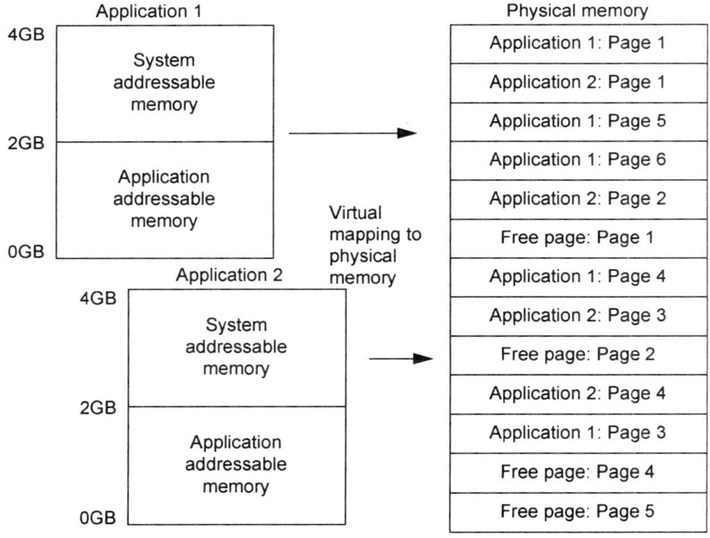

Figure 31.6 Memory paging.

- To support a 16-bit segmented memory model for Windows 3.*x* and MS-DOS applications. In this addressing scheme the addresses are made from a 16-bit segment address and a 16-bit offset address (see Section 1.3).

Windows 95/98/NT use the full addressing capabilities of the 80x86/Pentium processors by supporting a flat, linear memory model for 32-bit operating system functionality and Win32-based applications. This linear addressing model simplifies the development process for application vendors, and removes the performance penalties of a segmented memory architecture.

31.6 Multiple file systems

Windows 95/98/NT supports a layered file system architecture that directly supports multiple file systems (such as, FAT and CDFS). Windows 95/98/NT have great performance improvements over Windows 3.*x*, for example:

- Support for 32-bit protected-mode code when reading and writing information to and from a file system.
- Support for 32-bit dynamically allocated cache size.
- Support for an open file system architecture to enhance future system support.

Figure 31.7 shows the file system architecture used by Windows 95/98. It has the following components:

- IFS (Installable File System) Manager. This is the arbiter for the access to different file system components. On MS-DOS and Windows 3.*x* it was provided by interrupt 21h. Unfortunately, some add-on components did not run correctly and interfered with other installed drivers. It also did not directly support multiple network redirections (the IFS Manager can have an unlimited number of 32-bit redirectors).
- File system drivers. These provide support file systems, such as FAT-based disk devices, CD-ROM file systems and redirected network devices. They are ring 0 components, whereas with Windows 3.*x* supported them through MS-DOS. The two enhanced file systems are:
  - 32-bit VFAT. The 'legacy' 16-bit FAT file system suffers from many problems, such as the 8.3 file format. The 32-bit VFAT format is an enhanced form which works directly in the protected mode, and thus provides smooth multitasking as it is reentrant and multithreaded (a non-reentrant system does not allow an interrupt within an interrupt). It uses the VFAT.VXD driver and uses 32-bit code for all file accesses. Another advantage is that it provides for real-mode disk caching (VCACHE), where cache memory is automatically allocated or deallocated when it is required (In Windows 3.*x* this was provided by the SMARTDRV.EXE program).
  - 32-bit CDFS. The 32-bit, protected-mode CDFS format (as defined in the ISO 9660 standard) gives improved CD-ROM access and support for a dynamic cache (in Windows 3.*x* the MSCDEX driver provided to access CD-ROMs).

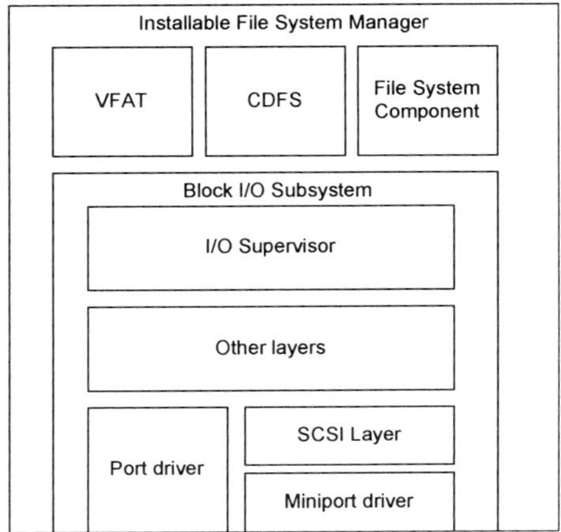

Figure 31.7 File system architecture.

- Block I/O subsystem. This is responsible for the actual physical access to the disk drive. Its components are:
  - **Input/Output Supervisor** (IOS). This component provides for an interface between the file systems and drivers. It is responsible for the queuing of file service requests and for routing the requests to the appropriate file system driver.
  - **Port driver**. This component is a 32-bit, protected-mode driver that communicates with a specific IDE disk device. It implements the functionality of the SCSI manager and miniport driver.
  - **SCSI layer**. This component is a 32-bit, protected-mode, universal driver model architecture for communicating with SCSI devices. It provides all the high-level SCSI functionality, and then uses a miniport driver to handle device-specific I/O calls.
  - **Miniport driver**. In Windows 95/98 these miniport driver models are used to write device-specific code. The Windows 95/98 miniport driver is a 32-bit protected-mode code, and is binary-compatible with Windows NT miniport drivers.

In Windows 95/98, the I/O Supervisor (IOS) is a VxD that controls and manages all protected-mode file system and block device drivers. It loads and initializes protected-mode device drivers and provides services needed for I/O operations. In Windows 3.*x* the I/O Supervisor was *BLOCKDEV. Other responsibilities of the IOS include:

- Registering drivers.
- Routing and queuing I/O requests, and sending asynchronous notifications to drivers as needed.
- Providing services that drivers can use to allocate memory and complete I/O requests.

On Windows 95/98, the IOS stores port drivers, miniport and VxD drivers in the SYSTEM\IOSUBSYS directory. The PDR file extension identifies the port drivers, MPD identifies miniport drivers and VxD (or 386) identifies the VxD drivers. Other clients or virtual device drivers should be stored in other directories and explicitly loaded using device=

entries in SYSTEM.INI. A sample listing of the IOSUBSYS directory is given next:

```
Directory of C:\WINDOWS\SYSTEM\IOSUBSYS
AIC78XX.MPD     AMSINT.MPD      APIX.VXD          ATAPCHNG.VXD
BIGMEM.DRV      CDFS.VXD        CDTSD.VXD         CDVSD.VXD
DISKTSD.VXD     DISKVSD.VXD     DRVSPACX.VXD      ESDI_506.PDR
HSFLOP.PDR      NCRC710.MPD     NCRC810.MPD       NECATAPI.VXD
RMM.PDR         SCSI1HLP.VXD    SCSIPORT.PDR      TORISAN3.VXD
VOLTRACK.VXD
```

31.7 Core system components

The core of Windows 95/98 has three components: User, Kernel, and GDI (graphical device interface), each of which has a pair of DLLs (one for 32-bit accesses the other for 16-bit accesses). The 16-bit DLLs allow for Win16 and MS-DOS computability.

Figure 31.8 shows that the lowest-level services provided by the Windows 95/98 Kernel are implemented as 32-bit code. In Windows 95/98 the names of the files are GDI32.DLL, KERNAL32.DLL and USER32.DLL; these are contained in the \WINDOWS\SYSTEM directory.

31.7.1 User

The User component provides input and output to and from the user interface. Input is form the keyboard, mouse, and any other input device and the output is to the user interface. It also manages interaction with the sound driver, timer, and communications ports.

Win32 applications and Windows 95/98 use an asynchronous input model for system input. With this devices have an associated interrupt handler (for example, the keyboard interrupts with IRQ1) which converts the interrupt into a message. This message is then sent to a raw input thread area, which then passes the message to the appropriate message queue. Each Win32 application can have its own message queue, whereas all Win16 applications share a common message queue.

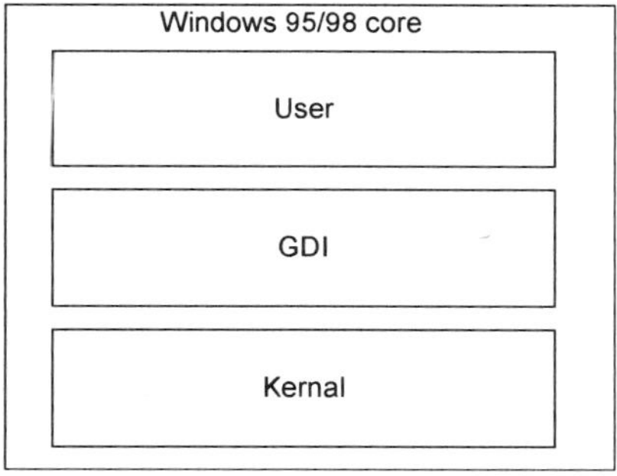

Figure 31.8 Core components.

31.7.2 Kernel

The Kernel provides for core operating system components including file I/O services, virtual memory management, task scheduling and exception handling, such as:

- File I/O services.
- Exceptions. These are events that occur as a program runs and call additional software which is outside of the normal flow of control. For example, if an application generates an exception, the Kernel is able to communicate that exception to the application to perform the necessary functions to resolve the problem. A typical exception is caused by a divide-by-zero error in a mathematical calculation, an exception routine can be designed so that it handles the error and does not crash the program.
- Virtual memory management. This resolves import references and supports demand paging for the application.
- Task scheduling. The Kernel schedules and runs threads of each process associated with an application.
- Provides services to both 16-bit and 32-bit applications by using a thunking process which is the translation process between 16-bit and 32-bit formats. It is typically used by a Win16 program to communicate with the 32-bit operating system core.

Virtual memory allows processes to allocate more memory than can be physically allocated. The operating system allocates each process a unique virtual address space, which is a set of addresses available for the process's threads. This virtual address space appears to be 4 GB in size, where 2 GB are reserved for program storage and 2 GB for system storage.

Figure 31.9 illustrates where the system components and applications reside in virtual memory. Its contents are:

- 3 GB–4 GB. All Ring 0 components.
- 2 GB–3 GB. Operating system core components and shared DLLs. These are available to all applications.
- 4 MB–2 GB. Win32-based applications, where each has its own address space. This memory is protected so that other programs cannot corrupt or otherwise hinder the application.
- 0–640 KB. Real-mode device drivers and TSRs.

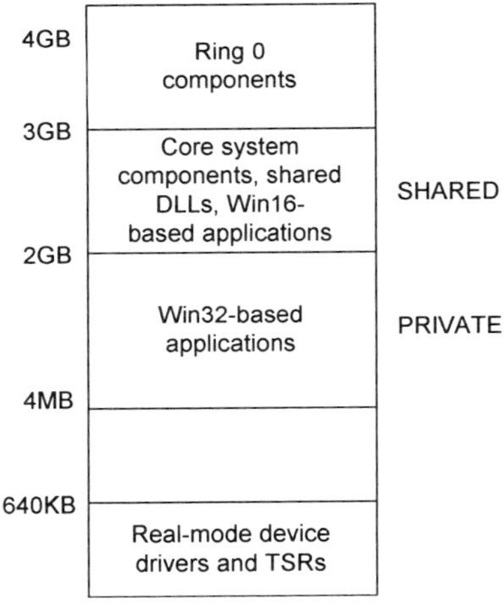

Figure 31.9 System memory usage.

31.7.3 GDI

The Graphical Device Interface (GDI) is the graphical system that:

- Manages information that appears on the screen.
- Draws graphic primitives and manipulates bitmaps.
- Interacts with device-independent graphics drivers, such as display and printer drivers.

The graphics subsystem provides input and output graphics support. Windows uses a 32-bit graphics engine (known as DIB, Device-independent Bitmaps) which:

- Directly controls the graphics output on the screen.
- Provides a set of optimized generic drawing functions for monochrome, 16-color, 16-bit high color, 256-color, and 24-bit true color graphic devices. It also supports Bézier curves and paths.
- Support for Image Color Matching for better color matching between display and color output devices.

The Windows graphics subsystem is included as a universal driver with a 32-bit mini-driver. The mini-driver provides only for the hardware-specific instructions.

The 32-bit Windows 95/98 printing subsystem has several enhancements over Windows 3.*x*; these include:

- They use a background thread processing to allow for smooth background printing.
- Smooth printing where the operating system only passes data to the printer when it is ready to receive more information.
- They send enhanced metafile (EMF) format files, rather than raw printer data. This EMF information is interpreted in the background and the results are then sent to the printer.
- Support for deferred printing, where a print job can be sent to a printer and then stored until the printer becomes available.
- Support for bi-directional communication protocols for printers using the Extended Communication Port (ECP) printer communication standard. ECP mode allows printers to send messages to the user or to application programs. Typical messages are: 'Paper Jam', 'Out-of-paper', 'Out-of-Memory', 'Toner Low', and so on.
- Plug-and-play.

31.8 Multitasking and threading

Multitasking involves running several tasks at the same time. It normally involves running a process for a given amount of time, before it is released and allowing another process a given amount of time. There are two forms of multitasking; these are:

- Pre-emptive multitasking. This type of multitasking involves the operating system controlling how long a process stays on the processor. This allows for smooth multitasking and is used in Windows NT/95/98 32-bit programs.

- Co-operative multitasking. This type of multitasking relies on a process giving up the processor. It is used with Windows 3.*x* programs and suffers from processor hogging, where a process can stay on a processor and the operating system cannot kick it off.

The logical extension to multitasking programs is to split a program into a number a parts (threads) and run each of these on the multitasking system (multithreading). A program which is running more than one thread at a time is known as a multithreaded program. Multi-threaded programs have many advantages over non-multithreaded programs, including:

- They make better use of the processor, where different threads can be run when one or more threads are waiting for data. For example, a thread could be waiting for keyboard input, while another thread could be reading data from the disk.
- They are easier to test, where each thread can be tested independently of other threads.
- They can use standard threads, which are optimized for a given hardware.

They also have disadvantages, including:

- The program has to be planned properly so that threads must know on which threads they depend.
- A thread may wait indefinitely for another thread which has crashed or terminated.

The main difference between multiple processes and multiple threads is that each process has independent variables and data, while multiple threads share data from the main program.

31.8.1 Scheduling

Scheduling involves determining which thread should be run on the process at a given time. This element is time is named a time slice, and its actual value depends on the system configuration.

Each thread currently running has a base priority. The programmer who created the program sets this base priority level of the thread. This value defines how the thread is executed in relation to other system threads. The thread with the highest priority gets use of the processor.

NT and 95/98 have 32 priority levels. The lowest priority is 0 and the highest is 31. A scheduler can change a threads base priority by increasing or decreasing it by two levels. This changes the threads priority.

The scheduler is made up from two main parts:

- **Primary scheduler**. This scheduler determines the priority numbers of the threads which are currently running. It then compares their priority and assigns resources to them depending on their priority. Threads with the highest priority are executed for the current time slice. When two or more threads have the same priority then the threads are put on a stack. One thread is run and then put to the bottom of the stack, then the next is run and it is put to the bottom, and so on. This continues until all threads with the same priority have been run for a given time slice.
- **Secondary scheduler**. The primary scheduler runs threads with the highest priority, whereas the secondary scheduler is responsible for increasing the priority of non-executing threads (which are all other threads apart from the currently executed thread).

It is thus important for giving low priority threads a chance to run on the operating system. Threads which are given a higher or lower priority are:

- A thread which is waiting for user input has its priority increased.
- A thread that has completed a voluntary wait also has its priority increased.
- Threads with a computation-bound thread get their priorities reduced. This prevents the blocking of I/O operations.

Apart from these, all threads get a periodic increase. This prevents lower-priority threads hogging shared resources that are required by higher-priority threads.

31.8.2 Priority inheritance boosting

One problem that can occur is when a low priority thread access resources which are required by a higher priority thread. For example, an RS-232 program could be loading data into memory while another program requires to access the memory. One method which can be used to overcome this is Priority Inheritance Boosting. In this case, low priority threads gets a boost so that they can quickly release resources. For example, if a system has three threads: Thread A, Thread B and Thread C. If Thread A has the highest priority and it requires a resource from Thread C then Thread C gets a boost in its priority. Thread A remains blocked until Thread C releases the required resource. When it does release it then Thread C goes back to its normal priority and Thread A then gets access to the resource.

31.9 Plug-and-play process

Plug-and-play allows the operating system to configure hardware as required. On system startup, the configuration manager scans the system hardware. When it finds a new plug-and-play device it does the following:

- **Sets the device into configuration mode**. This is achieved by using 3 I/O ports. Some data (the initiation key) is written to one of the ports and enables the Plug-and-Play logic.
- **Isolate and identify each device**. Each device is isolated, one at a time. The method used is to assign each device a unique number, which is a unique handle for the device. This number is made from a device ID and a serial number.
- **Determine device specifications**. Each device sends its functionality to the operating system, such as how many joysticks it supports, its audio functions, its networking modes, and so on.
- **Allocate resources**. The operating system then allocates resources to the device depending on its functionality and the Plug-and-Play device is informed of the allocated resources (such as IRQs, I/O addresses, DMA channels, and so on). It also checks for conflicts on these resources.
- **Activate device**. When the above have been completed the device is enabled. Only the initiation key can re-initialize the device.

31.10 Windows NT architecture

Windows NT uses two modes:

- User mode. This is a lower privileged mode than kernal mode. It has no direct access to the hardware or to memory. It interfaces to the operating system through well-defined API (Application Program Interface) calls.
- Kernal mode. This is a privileged mode of operation and allows all code direct access to the hardware and memory, including memory allocated to user mode processes. Kernal mode processes also have a higher priority over user mode processes.

Figure 31.10 shows an outline of the architecture of NT. It can be seen that only the kernal mode has access to the hardware. This kernal includes an executive services which include managers (for I/O, interprocess communications, and so on) and device drivers (which control the hardware). Its parts include:

- Microkernel. Controls basic operating system services, such as interrupt handling and scheduling.
- HAL. This a library of hardware-specific programs which give a standard interface between the hardware and software. This can either be Microsoft written or manufacturer provided. They have the advantage of allowing for transportability of programs across different hardware platforms.
- Win32 Window Manager. Supports Win32, MS-DOS and Windows 3.*x* applications.

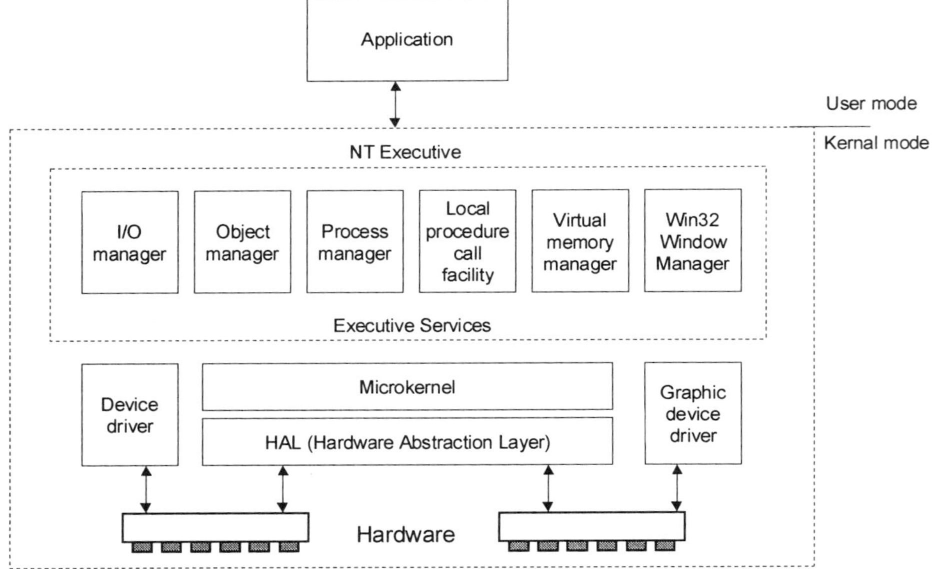

Figure 31.10 NT architecture.

31.10.1 MS-DOS support

Windows NT supports MS-DOS-based applications with an NT Virtual DOS Machine (NTVDM), where each MS-DOS application has its own NTVDM. The NTVDM is started by the application Ntvdm.exe and when this has started the application communicates with two system files Ntio.sys (equivalent to IO.SYS) and Ntdos.sys (equivalent to MSDOS.SYS). Note that the AUTOEXEC.BAT and CONFIG.SYS files have also been replace by Autoexec.nt and Config.nt (which are normally located in `\WINNT\System32`).

Multiple NTVDMs have the advantage of being reliable because if one NTVDM fails then it does not effect any others. It also allows MS-DOS-based applications to be multitasked. Unfortunately, each NTVDM needs at least 1 MB of physical memory.

Some MS-DOS applications require direct access to the hardware. NT supports this by providing virtual device drivers (VDDs). These detect a call to hardware and communicate with the NT 32-bit device driver.

Windows NT communicates with hardware through device drivers. These drivers have a `.sys` file extension. An example listing of these is given next:

```
Directory of C:\WINNT\system32\drivers
afd.sys         atapi.sys       atdisk.sys        beep.sys
cdaudio.sys     cdfs.sys        cdrom.sys         changer.sys
cirrus.sys      disk.sys        diskdump.sys      diskperf.sys
fastfat.sys     floppy.sys      ftdisk.sys        hpscan16.sys
i8042prt.sys    kbdclass.sys    ksecdd.sys        modem.sys
mouclass.sys    msfs.sys        mup.sys           ndis.sys
netdtect.sys    npfs.sys        ntfs.sys          null.sys
parallel.sys    parport.sys     parvdm.sys        pcmcia.sys
scsiport.sys    scsiprnt.sys    scsiscan.sys      serial.sys
sfloppy.sys     streams.sys     tape.sys          tdi.sys
vga.sys         videoprt.sys
```

With this, virtual memory applications can have access to the full available memory but NT then maps this to a private memory range (called a virtual memory space). It maps physical memory to virtual memory in 4 KB blocks (called pages). This was previously illustrated in Figure 31.6. The driver used to perform the page file access is Pagefile.sys (which is normally found in the top-level directory).

Windows NT has 32 levels of priority (0 to 31). Levels 0 to 15 are used for dynamic applications (such as non-critical operations) and 16 to 31 are used for real-time applications (such as Kernal operations). NT provides a virtual memory by paging file(s) onto the hard disk. Priority levels 0 to 15 can be paged, but levels 16 to 31 cannot.

A summary of the `system32` directory is shown below. The `wowdeb.exe` and `wowexec.exe` files allow Windows 3.*x* programs to run in a 32-bit environment.

```
Directory of C:\winnt\system32
ansi.sys        append.exe      at.exe          atsvc.exe
attrib.exe      autoexec.nt     backup.exe      bootok.exe
bootvrfy.exe    cacls.exe       chcp.com        chkdsk.exe
clipsrv.exe     comm.drv        command.com     comp.exe
compact.exe     config.nt       control.exe     convert.exe
country.sys     csrss.exe       dcomcnfg.exe    ddeshare.exe
ddhelp.exe      ebug.exe        diskcomp.com    diskcopy.com
diskperf.exe    doskey.exe      dosx.exe        DRIVERS
edit.com        exe2bin.exe     expand.exe      fastopen.exe
fc.exe          find.exe        findstr.exe     finger.exe
fontview.exe    forcedos.exe    format.com      ftp.exe
gdi.exe         graftabl.com    graphics.com    grpconv.exe
```

```
help.exe       himem.sys      inetins.exe    internat.exe
kb16.com       keyb.com       keyboard.drv   keyboard.sys
krnl386.exe    label.exe      lights.exe     lodctr.exe
mem.exe        mode.com       more.com       mpnotify.exe
mscdexnt.exe   nddeagnt.exe   nddeapir.exe   net.exe
nlsfunc.exe    notepad.exe    ntdos.sys      ntio.sys
ntvdm.exe      os2ss.exe      pax.exe        pentnt.exe
ping.exe       portuas.exe    posix.exe      print.exe
psxss.exe      rdisk.exe      recover.exe    redir.exe
replace.exe    restore.exe    rpcss.exe      rundll32.exe
runonce.exe    savedump.exe   setup.exe      setver.exe
share.exe      shmgrate.exe   skeys.exe      smss.exe
sort.exe       SPOOL          sprestrt.exe   subst.exe
syncapp.exe    sysedit.exe    systray.exe    taskman.exe
taskmgr.exe    telnet.exe     tree.com       unlodctr.exe
ups.exe        user.exe       userinit.exe   VIEWERS
win.com        winhlp32.exe   winspool.exe   winver.exe
wowdeb.exe     wowexec.exe
```

31.11 Novell NetWare networking

Novell NetWare is one of the most popular systems for PC LANs and provides file and print server facilities. The protocol used is SPX/IPX. This is also used by Windows NT to communicate with other Windows NT nodes and with NetWare networks. The Internet Packet Exchange (IPX) protocol is a network layer protocol for transportation of data between computers on a Novell network. IPX is very fast and has a small connectionless datagram protocol. Sequenced Packet Interchange (SPX) provides a communications protocol which supervises the transmission of the packet and ensures its successful delivery.

Novell uses the Open Data-Link Interface (ODI) standard to simplify network driver development and to provide support for multiple protocols on a single network adapter. It allows Novell NetWare drivers to be written to without concern for the protocol that will be used on top of them (similar to NDIS in Windows NT). The link support layer (LSL or LSL.COM) provides a foundation for the MAC layer to communicate with multiple protocols (similar to NDIS in Windows NT). The IPX.COM (or IPXODI.COM) program normally communicates with the LSL and the applications. The MAC driver is a device driver or NIC driver. It provides low-level access to the network adapter by supporting data transmission and some basic adapter management functions. These drivers also pass data from the physical layer to the transport protocols at the network and transport layers. NetWare and IPX/SPX are covered in more detail in the next chapter.

31.12 Servers, workstations and clients

Microsoft Windows NT is a 32-bit, preemptive, multitasking operating system. One of the major advantages it has over UNIX is that it can run PC-based software. A Windows NT network normally consists of a server and a number of clients. The server provides file and print servers as well as powerful networking applications, such as electronic mail applications, access to local and remote peripherals, and so on.

The Windows NT client can either:

- Operate as a standalone operating system.
- Connect with a peer-to-peer connection.
- Connect to a Windows NT server.

A peer-to-peer connection is when one computer logs into another computer. Windows NT provides unlimited outbound peer-to-peer connections and typically up to 10 simultaneous inbound connections.

31.13 Workgroups and domains

Windows NT assigns users to workgroups which are collection of users who are grouped together with a common purpose. This purpose might be to share resources such as file systems or printers, and each workgroup has its own unique name. With workgroups each Windows NT workstation interacts with a common group of computers on a peer-to-peer level. Each workstation then manages its own resources and user accounts. Workgroups are useful for small groups where a small number of users require to access resources on other computers

A domain in Windows NT is a logical collection of computers sharing a common user accounts database and security policy. Thus each domain must have at least one Windows NT server.

Windows NT is designed to operate with either workgroups or domains. Figure 31.11 illustrates the difference between domains and workgroups.

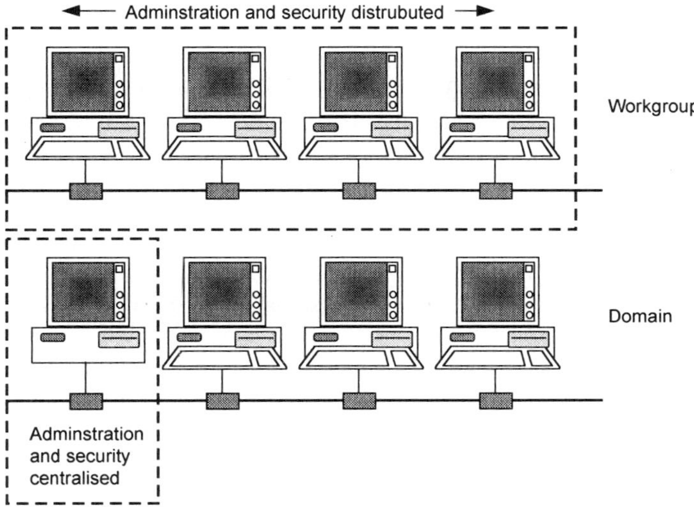

Figure 31.11 Workgroups and domains.

Domains have the advantages that:

- Each domain forms a single administrative unit with shared security and user account information. This domain has one database containing user and group information and security policy settings.

- They segment the resources of the network so that users, by default, can view all networks for a particular domain.
- User accounts are automatically validated by the domain controller. This stops invalid users from gaining access to network resources.

31.14 User and group accounts

Each user within a domain has a user account and is assigned to one or more groups. Each group is granted permissions for the file system, accessing printers, and so on. Group accounts are useful because they simplify an organization into a single administrative unit. They also provide a convenient method of controlling access for several users who will be using Windows NT to perform similar tasks. By placing multiple users in a group, the administrator can assign rights and/or permissions to the group.

Each user on a Windows NT system has the following:

- A user name (such as `fred_bloggs`).
- A password (assigned by the administrator then changed by the user).
- The groups in which the user account is a member (e.g. `staff`).
- Any user rights for using the assigned computer.

Each time a user attempts to perform a particular action on a computer, Windows NT checks the user account to determine whether the user has the authority to perform that action (such as read the file, write to the file, delete the file, and so on).

Normally there are three main default user accounts: Administrator, Guest and an 'Initial User' account. The system manager uses the Administrator account to perform such tasks as installing software, adding/deleting user accounts, setting up network peripherals, installing hardware, and so on.

Guest accounts allow occasional users to log on and be granted limited rights on the local computer. The system manager must be sure that the access rights are limited so that hackers or inexperienced users cannot do damage to the local system.

The 'Initial User' account is created during installation of the Windows NT workstation. This account, assigned a name during installation, is a member of the Administrator's group and therefore has all the Administrator's rights and privileges.

After the system has been installed the Administrator can allocate new user accounts, either by creating new user accounts, or by copying existing accounts.

31.15 New user accounts

Typically the system manager creates new accounts by copying existing users accounts. The items copied directly from an existing user account to a new user account are as follows:

- The description of the user (such as `Fred Bloggs, Ext 4444`).

- Group account membership (such as `Production`).
- Profile settings (such as home directory).
- If set, the attribute to stop the user from changing their password (sometimes the manager does not want the user to change the default password).
- If set, the attribute that causes the password to remain unexpired (sometimes the system manager forces users to change their passwords from time to time).

The items which are cleared and completed by the system manager are:

- The username and full name (see Figure 31.12).
- The attribute that prompts users to change their passwords when they next login (normally a default password is initially set up and is changed when a user initially logs in).
- The attribute which disables the account (the manager must reset this before a user can log in).

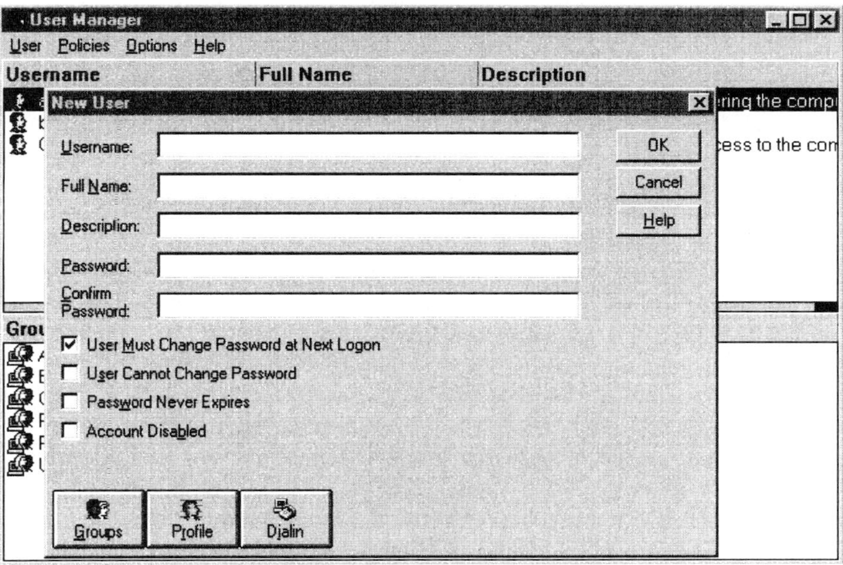

Figure 31.12 Adding a user.

31.16 File systems

Windows NT supports three different types of file system:

- FAT (file allocation table) – as used by MS-DOS, OS/2 and Windows NT. A single volume can be up to 2 GB (now increased to 4GB). The maximum file size is 4GB. It has no built-in security but can be access through Windows 95/98, MS-DOS and Windows NT.
- HPFS (high performance file system) – a UNIX-style file system which is used by OS/2

and Windows NT. A single volume can be up to 8 GB. MS-DOS applications cannot access files.
- NTFS (NT file system) – as used by Windows NT. A single volume can be up to 64 TB (based on current hardware, but, theoretically, 16 exabytes). It has built-in security and also supports file compression/decompression. MS-DOS applications, themselves, cannot access the file system but they can when run with Windows NT, nor can Windows 95/98.

The FAT file system is widely used and supported by a variety of operating systems, such as MS-DOS, Windows NT and OS/2. If a system is to use MS-DOS it must be installed with a FAT file system.

31.16.1 FAT

The standard MS-DOS FAT file and directory-naming structure allows an 8-character file name and a 3-character file extension with a dot separator (.) between them (the 8.3 file name). It is not case sensitive and the file name and extension cannot contain spaces and other reserved characters, such as:

```
" / \ : ; | = , ^ * ? .
```

With Windows NT and Windows 95/98 the FAT file system supports long file names which can be up to 255 characters. The name can also contain multiple spaces and dot separators. File names are not case sensitive, but the case of file names is preserved (a file named `FredDocument.XYz` will be displayed as `FredDocument.XYz` but can be accessed with any of the characters in upper or lower case.

Each file in the FAT table has four attributes (or properties): read-only, archive, system and hidden (as shown in Figure 31.5). The FAT uses a linked list where the file's directory entry contains its beginning FAT entry number. This FAT entry in turn contains the location of the next cluster if the file is larger than one cluster, or a marker that designates this is included in the last cluster. A file which occupies 12 clusters will have 11 FAT entries and 10 FAT links.

The main disadvantage with FAT is that the disk is segmented into allocated units (or clusters). On large-capacity disks these sectors can be relatively large (typically 512 bytes/sector). Disks with a capacity of between 256 MB and 512 MB use 16 sectors per cluster (8 KB) and disks from 512 MB to 1 GB use 32 sectors per cluster (16 KB). Drives up to 2 GB use 64 sectors per cluster (32 KB). Thus if the disk has a capacity of 512 MB then each cluster will be 8 KB. A file which is only 1 KB will thus take up 8 KB of disk space (a wastage of 7 KB), and a 9 kB file will take up 16 KB (a wastage of 7 KB). Thus a file system which has many small files will be inefficient on a cluster-based system. A floppy disk normally uses one cluster per sector (512 bytes).

Windows 95/98 and Windows NT support up to 255 characters in file names; unfortunately, MS-DOS and Windows 3.*x* applications cannot read them. To accommodate this, every long file name has an autogenerated short file name (in the form `XXXXXXXX.YYY`). Table 13.2 shows three examples. The conversion takes the first six characters of the long name then adds a *~number* to the name to give it a unique name. File names with the same initial six characters are identified with different *numbers*. For example, `Program Files` and `Program Directory` would be stored as `PROGRA~1` and `PROGRA~2`, respectively. Sample listing 31.1 shows a listing from Windows NT. The left-hand column shows the short file name and the far right-hand column shows the long file name.

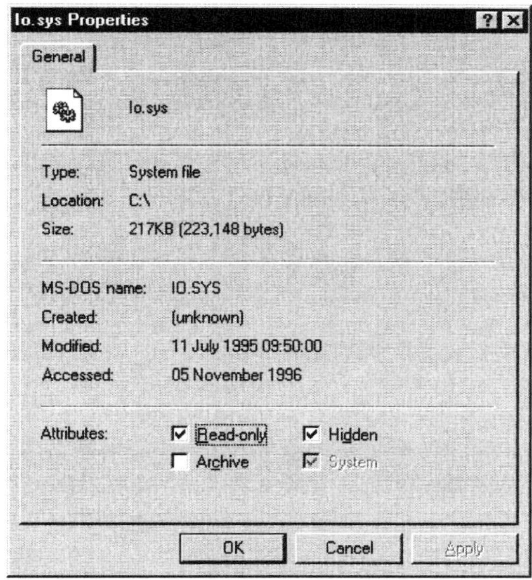

Figure 31.13 File attributes.

Table 31.2 File name conversions.

Long file name	*Short file name*
`Program Files`	`PROGRA~1`
`Triangular.bmp`	`TRIANG~1.BMP`
`Fredte~1.1`	`FRED.TEXT.1`

Sample run 31.1

```
EXAMPL~1 DOC  4,608  05/11/96  23:36 Example Document 1.doc
EXAMPL~2 DOC  4,608  05/11/96  23:36 Example Document 2.doc
EXAMPL~3 DOC  4,608  05/11/96  23:36 Example Document 3.doc
EXAMPL~4 DOC  4,608  05/11/96  23:36 Example Document 4.doc
EXAMPL~5 DOC  4,608  05/11/96  23:36 Example Document 5.doc
EXAMPL~6 DOC  4,608  05/11/96  23:36 Example Document 6.doc
EXAMPL~7 DOC  4,608  05/11/96  23:36 Example Document 7.doc
EXAMPL~8 DOC  4,608  05/11/96  23:39 Example assignment A.doc
EXAMPL~9 DOC  4,608  05/11/96  23:40 Example assignment B.doc
EXAMP~10 DOC  4,608  05/11/96  23:40 Example assignment C.doc
```

31.16.2 HPFS (high-performance file system)

HPFS is supported by OS/2 and is typically used to migrate from OS/2 to Windows NT. It allows long file names of up to 254 characters with multiple extensions. As with the Windows 95/98 and Windows NT FAT system the file names are not case sensitive but preserve the case. HPFS uses B-tree format to store the file system directory structure. The B-tree format stores directory entries in an alphabetic tree, and binary searches are used to search for the target file in the directory list. The reserved characters for file names are:

```
" / \ : < > | * ?
```

31.16.3 NTFS (NT file system)

NTFS is the preferred file system for Windows NT as it makes more efficient usage of the disk and it offers increased security. It allows for file systems up to 16 EB (16 exabytes, or 1 billion gigabytes, or 2^{64} bytes). As with HPFS it uses B-tree format for storing the file systems directory structure. Its main objectives are:

- To increase reliability. NTFS automatically logs all directory and file updates which can be used to redo or undo failed operations resulting from system failures such as power losses, hardware faults, and so on.
- To provide sector sparing (or hot fixing). When NTFS finds errors in a bad sector, it causes the data in that sector to be moved to a different section and the bad sector to be marked as bad. No other data is then written to that sector. Thus, the disk fixes itself as it is working and there is no need for disk repair programs (FAT only marks bad areas when formatting the disk).
- Increases file system size (up to 16 EB).
- To enhance security permissions.
- To support POSIX requirements, such as case-sensitive naming, addition of a time stamp to show the time the file was last accessed and hard links from one file (or directory) to another.

The reserved characters for file names are:

```
" / \ : < > | * ?
```

31.17 Windows NT networking

Networks must use a protocol to transmit data. Typical protocols are:

- IPX/SPX – used with Novell NetWare, it accesses file and printer services.
- TCP/IP – used for Internet access and client/server applications.
- SNA DLC – used mainly by IBM mainframes and minicomputers.
- AppleTalk – used by Macintosh computers.
- NetBEUI – used in some small LANs (stands for NetBIOS Extended User Interface).

Novell NetWare is installed in many organization to create local area networks of PCs. It uses IPX/SPX for transmitting data and allows access to file servers and network printing services. TCP/IP is the standard protocol used when accessing the Internet and also for client/server applications (such as remote file transfer and remote login).

A major advantage of Windows NT is that networking is built into the operating system. Figure 31.14 shows how it is organized in relation to the OSI model. Windows NT has the great advantage of being protocol-independent and will work with most standard protocols, such as TCP/IP, IPX/SPX, NetBEUI, DLC and AppleTalk. The default protocol is NetBEUI.

Figure 31.15 shows a sample configuration for the Windows NT Client; the display shows the default network configurations.

There are two main boundaries in Windows NT and NDIS and TDI. The Network Device Interface Standard (NDIS) boundary layer interfaces to several network interface adapters (such as Ethernet, Token Ring, RS-232, modems, and so on) with different protocols. It allows for an unlimited number of network interface cards (NICs) and protocols to be connected to be used with the operating system. In Windows NT, a single software module, `NDIS.SYS`, (the NDIS wrapper) interfaces with the manufacturer-supplied NDIS NIC device driver. The wrapper provides a uniform interface between the protocol drivers (such as TCP/IP or IPX/SPX) and the NDIS device driver. Figure 31.16 shows the setup of the network adapter.

31.17.1 TCP/IP

See Section 31.18.

31.17.2 IPX/SPX

NetWare networks use SPX/IPX and is supported through Windows NT using the NWLink protocol stack. This protocol is covered in more detail in the next chapter.

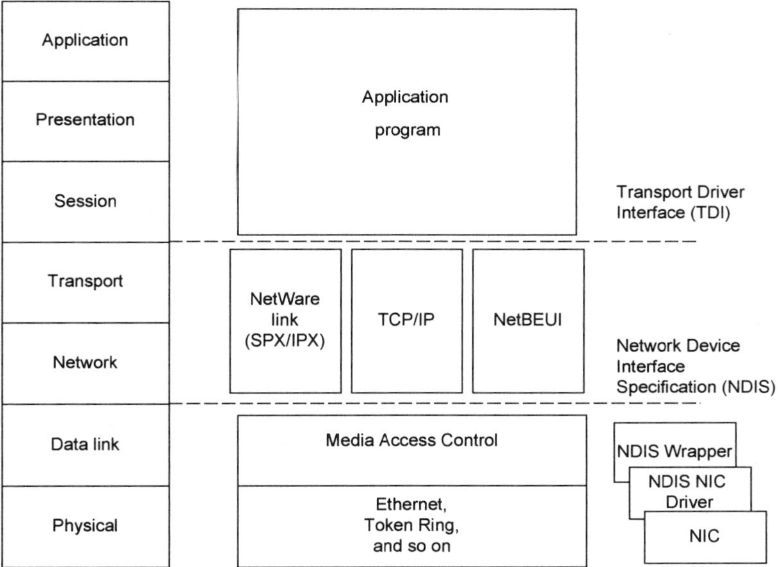

Figure 31.14 Windows NT network interfaces.

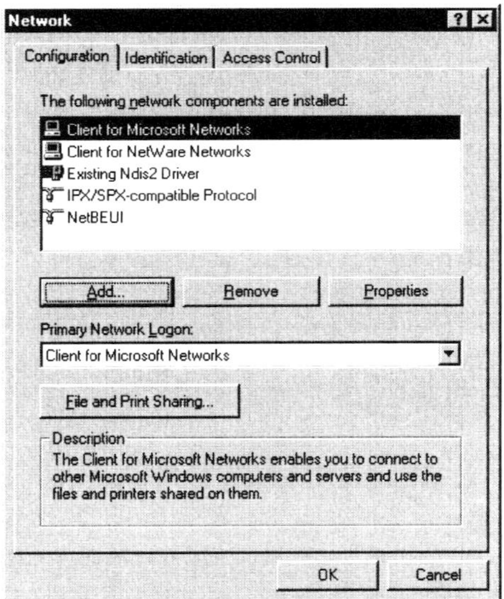

Figure 31.15 Windows NT network interfaces.

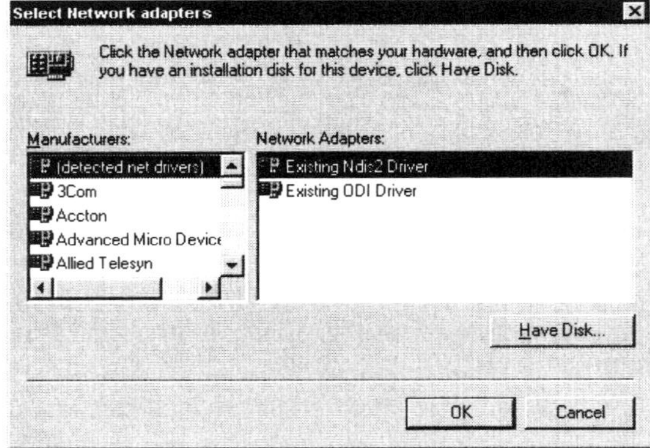

Figure 31.16 Windows NT network adapter drivers.

31.17.3 NetBEUI

NetBEUI (NetBIOS Extended User Interface) has been used with network operating systems, such as Microsoft LAN manager and OS/2 LAN server. In Windows NT, the NetBEUI frame (NBF) protocol stack gives backward compatibility with existing NetBEUI implementations and also provides for enhanced implementations. NetBEUI is the standard technique that NT clients and servers use to intercommunicate.

NBF is similar to TCP/IP and SPX/IPX, it is used to establish a session between a client and a server, and also to provide the reliable transport of the data across the connection-oriented session. Thus NetBEUI tries to provide reliable data transfer through error checking and acknowledgment of each successfully received data packet. In the standard form of

NetBEUI each packet must be acknowledged after its delivery. This is wasteful in time. Windows NT uses NBF which improves NetBEUI as it allows several packets to be sent before requiring an acknowledgment (called an adaptive sliding window protocol).

Each NetBEUI is assigned a 1-byte session number and thus allows a maximum of 254 simultaneously active session (as two of the connection numbers are reserved). NBF enhances this by allowing 254 connections to computers with 254 sessions for each connection (thus there is a maximum of 254×254 sessions).

31.17.4 AppleTalk

The AppleTalk protocol allows Windows NT to share a network with Macintosh clients. It can also act as an AppleShare server.

31.17.5 DLC

Data link control (DLC) is a communications protocol which is used with IBM mainframes. Windows NT interfaces to a DLC network.

31.18 Setting up TCP/IP networking on Windows NT

The default internetworking protocol for Windows NT/95 is NetBEUI. To use any Internet applications (such as FTP, TELNET, WWW browsers, and so on) the TCP/IP protocol must be installed. To achieve this the network icon in the control panel is selected, as shown in Figure 31.17. Next the network configuration screen is shown.

Windows NT has a DHCP (Dynamic Host Configuration Program) which assigns IP addresses from a pool of addresses. It relieves the system manager from assigning IP addresses to individual workstations and maintaining those addresses. Windows NT also has a name resolution service called WINS (Windows Internet Name Service). This program maps a computer name to an IP address; for example, `www.napier.ac.uk` is mapped to the IP address `146.176.131.10`. This facility is similar to the DNS server which is used on many TCP/IP networks. Note that a Windows NT server can support both WINS and DNS.

The settings for TCP/IP, in Windows 95/98, are set up by selecting Control Panel → Network and then, if they are not already set up, select Add → Protocol → Microsoft → TCP/IP (as shown in Figure 31.18). After the network adapter is selected (such as NDIS) then select Properties from the TCP/IP option. This then gives the settings for:

- TCP/IP properties – which is used to set the IP address of the host node. In the example in Figure 31.19 the node has an IP address of 146.176.151.130 and the subnet mask is 255.255.255.0.
- DNS configuration – which sets the IP address of the DNS server. In the case in Figure 31.20 there are two DNS servers, 146.176.150.62 and 146.176.151.99. The host is named `pc419` in the domain of `eece.napier.ac.uk`.
- Gateway – which is used to define the gateway node. In Figure 31.21 the gateway node is defined as 146.176.151.254.
- WINS server – which is used to define a WINS node (this functions as a DNS server). In Figure 31.22 the WINS node is defined as 146.176.151.50.

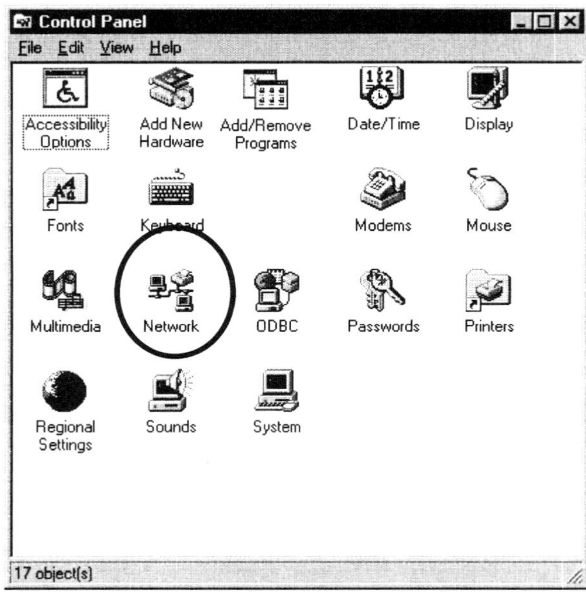

Figure 31.17 Windows 95/98 selection of networking configuration.

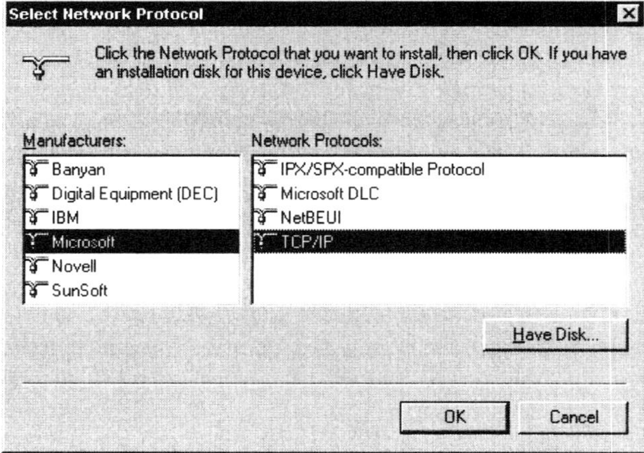

Figure 31.18 Setting-up for TCP/IP.

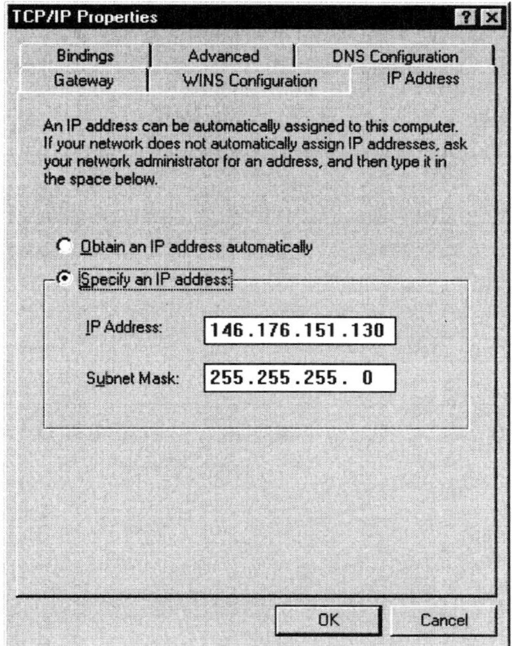

Figure 31.19 Setting-up an IP address.

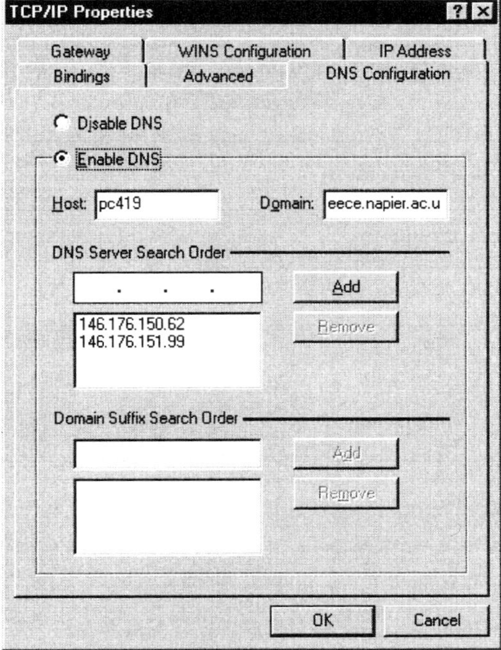

Figure 31.20 Windows 95/98 DNS set-up.

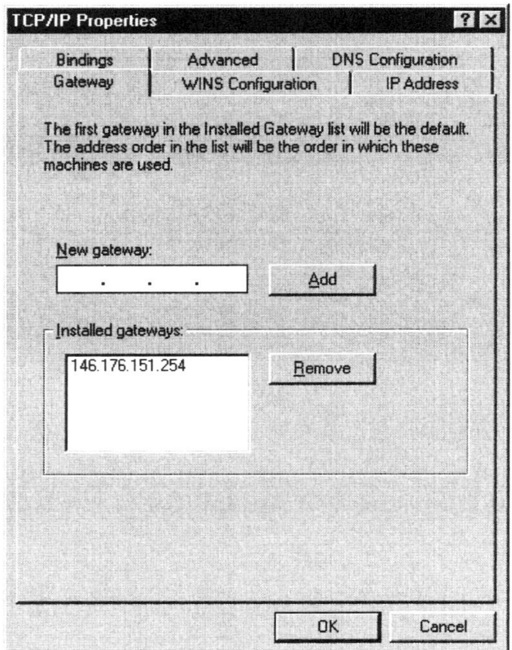

Figure 31.21 Setting up a gateway.

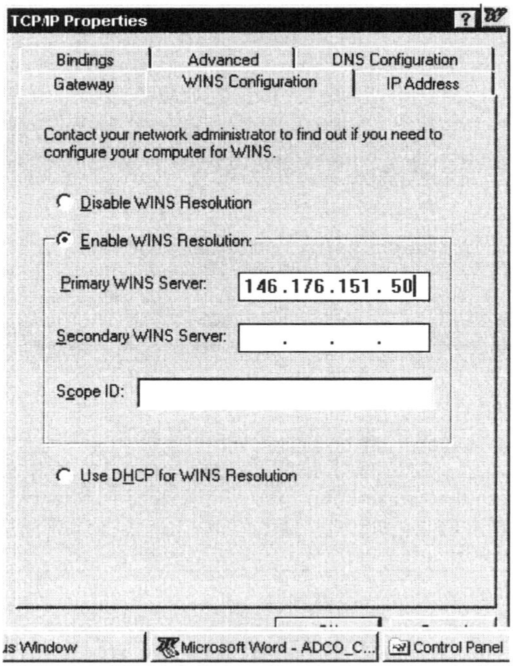

TCP/IP Properties
Bindings
Advanced
DNS Configuration
Gateway
WINS Configuration
IP Address
Contact your network administrator to find out if you need to configure your computer for WINS.
Disable WINS Resolution
Enable WINS Resolution:
Primary WINS Server: 146.176.151. 50
Secondary WINS Server:
Scope ID:
Use DHCP for WINS Resolution
Microsoft Word - ADCO_C...
Control Panel

Figure 31.22 Windows 95/98 WINS server IP address set-up.

31.19 Windows sockets

A Windows socket (WinSock) is a standard method that allows nodes over a network to communicate with each other using a standard interface. It supports internetworking protocols such as TCP/IP, IPX/SPX, AppleTalk and NetBEUI. WinSock communicates through the TDI interface and uses the file `WINSOCK.DLL` or `WINSOCK32.DLL`. These DLLs (dynamic link libraries) contain a number of networking functions which are called in order to communicate with the transport and network layers (such as TCP/IP or SPX/IPX). As it communicates with these layers it is independent of the networking interface (such as Ethernet or FDDI).

31.20 Robust networking

Windows NT provides fault tolerance in a number of ways. These are outlined in the following sections.

31.20.1 Disk mirroring

Network servers normally support disk mirroring which protects against hard disk failure. It uses two partitions on different disk drives which are connected to the same controller. Data written to the first (primary) partition is mirrored automatically to the secondary partition. If the primary disk fails then the system uses the partition on the secondary disk. Mirroring also allows unallocated space on the primary drive to be allocated to the secondary drive. On a disk mirroring system the primary and secondary partitions have the same drive letter (such as C: or D:) and users are unaware that disks are being mirrored.

31.20.2 Disk duplexing

Disk duplexing means that mirrored pairs are controlled by different controllers. This provides for fault tolerance on both disk and controller. Unfortunately, it does not support multiple controllers connected to a single disk drive.

31.20.3 Striping with parity

Network servers normally support disk striping with parity. This technique is based on RAID 5 (Redundant Array of Inexpensive Disks), where a number of partitions on different disks are combined to make one large logical drive. Data is written in stripes across all of the disk drives and additional parity bits. For example, if a system has four disk drives then data is written to the first three disks and the parity is written to the fourth drive. Typically the stripe is 64 KB, thus 64 KB will be written to Drive 1, the same to Drive 2 and Drive 3, then the parity of the other three to the fourth. The following example illustrates the concept of RAID where a system writes the data `110`, `000`, `111`, `100` to the first three drives, this gives parity bits of `1`, `1`, `0` and `0`.

If one of the disk drives fails then the addition of the parity bit allows the bits on the failed disk to be recovered. For example, if disk 3 fails then the bits from the other disk are simply XOR-ed together to generate the bits from the failed drive. If the data on the other disk drives is `111` then the recovered data gives `0`, `001` gives `0`, and so on.

Disk 1	*Disk 2*	*Disk 3*	*Disk 4 (Odd parity)*
1	1	0	1
0	0	0	1
1	1	1	0
1	0	0	0

The 64 KB stripes of data are also interleaved across the disks. The parity block is written to the first disk drive, then in the next block to the second, and so on. A system with four disk drives would store the following data:

Disk 1	*Disk 2*	*Disk 3*	*Disk 4*
Parity block 1	Data block A	Data block B	Data block C
Data block D	Parity block 2	Data block E	Data block F
Data block G	Data block H	Parity block 3	Data block I

Each of the data blocks will be 64 KB, which is also equal to the parity block. The interlacing of the data ensures that the parity stripes are not all on the same disk. Thus there is no single point of failure for the set.

Striping of data improves reading performance when each of the disk drives has a separate controller, because the data is simultaneously read by each of the controllers and simultaneously passed to the systems. It thus provides fast reading of data but only moderate writing performance (because the system must calculate the parity block).

The main advantages of RAID 5 can be summarized as:

- It recovers data when a single disk drive or controller fails (RAID level 0 does not use a parity block thus it cannot regenerate lost data).
- It allows a number of small partitions to be built into a large partition.
- Several disks can be mounted as a single drive.
- Performance can be improved with multiple disk controllers.

The main disadvantages of RAID 5 are:

- It requires increased memory because of the parity block.
- Performance is reduced when one of the disks fails because of the need to regenerate the failed data.
- It increases the amount of disk space as it has an overhead due to the parity block (although the overhead is normally less than disk mirroring, which has a 50% overhead).
- It requires at least three disk drives.

31.20.4 Tape backup

Windows NT provides for automatic tape based on the Maynard tape system. Tape backup allows data to be recovered when faults occur. Normally the system manager backs up the network at the start of the week and then does an incremental backup each day.

31.20.5 UPS services

Windows NT provides services to uninterruptable power supplies (UPSs). UPS systems provide power, from batteries, to a computer system when there is a glitch in the supply, power sags or power failure. Windows NT detects signals from a UPS unit and performs an orderly shutdown of applications, services and file systems as the stored energy in the UPS is depleted.

31.21 Security model

Windows NT treats all its resources as objects that can only be accessed by authorized users and services. Examples of objects are directories, printers, processes, ports, devices and files. On an NTFS partition the access to an object is controlled by the security descriptor (DS) structure which contains an access control list (ACL) and security identifier (SI). The SD contains the user (and group) accounts that have access and permissions to the object. The system always checks the ACL of an object to determine whether the user is allowed to access it.

The main parts of the SI are:

OWNER	Indicates the user account for the object.
GROUP	Indicates the group the object belongs to.
User ACL	The user-controller ACL.
System ACL	System manager controlled ACL.

The ACL file access rights are:

```
Full control          (All)      (All)
Read                  (RX)       (RX)
Change                (RWXD)     (RWXD)
Add                   (WX)
Change                (RWXD)
List                  (RX)
Change Permissions    (P)
```

31.22 TCP/IP applications

After the TCP/IP protocol has been installed (and all other TCP/IP drivers have been removed) then the system will be ready to run any TCP/IP applications. For example Windows NT and 95/98 are installed with Telnet and ftp. In Windows NT they are run by entering either Telnet or ftp from the run command, as illustrated in Figures 31.23 and 31.24.

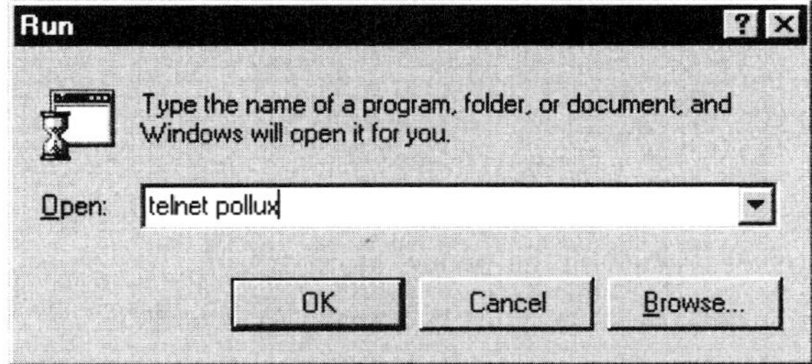

Figure 31.23 Running telnet.

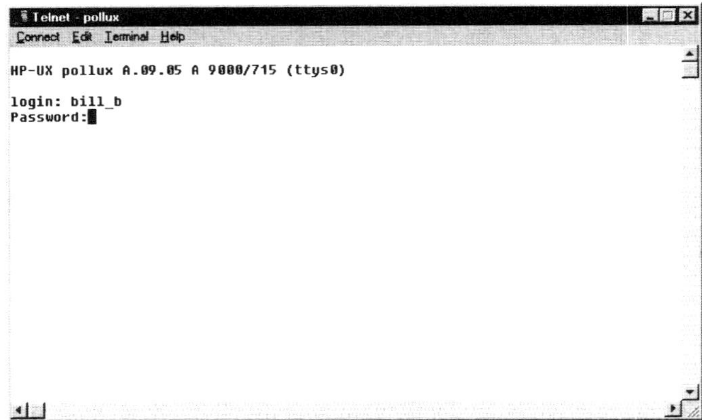

Figure 31.24 Example Telnet program.

31.23 Windows NT network drives

Windows NT and 95/98 displays the currently mounted network drives within the group My Computer. Figure 31.25 shows drives which are either local (C: D: and E:) or mounted using NetWare (F: G: and so on). Windows NT and 95/98 also automatically scan the neighing networks to find network servers. An example is shown in Figures 31.26 and 31.27. Figure 31.26 shows the currently mounted servers (e.g. EECE_1) and by selecting the Global Network icon all the other connected local servers can be shown. Figure 31.27 shows an example network.

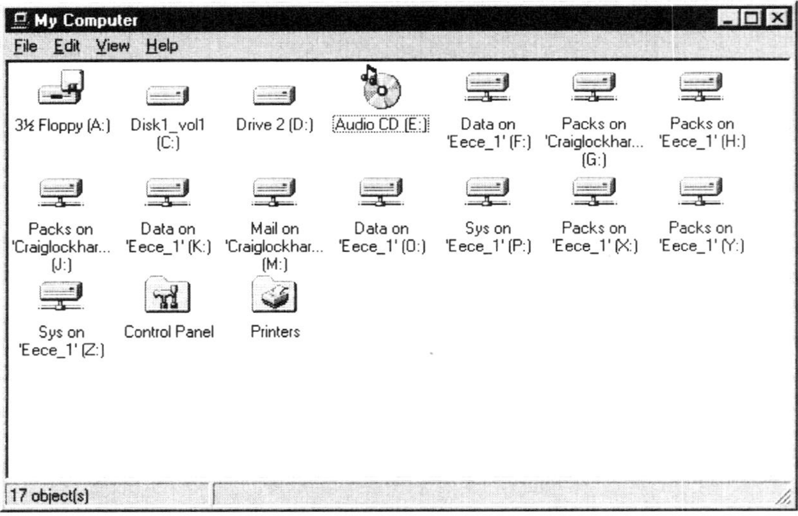

Figure 31.25 Mounted network and local drives.

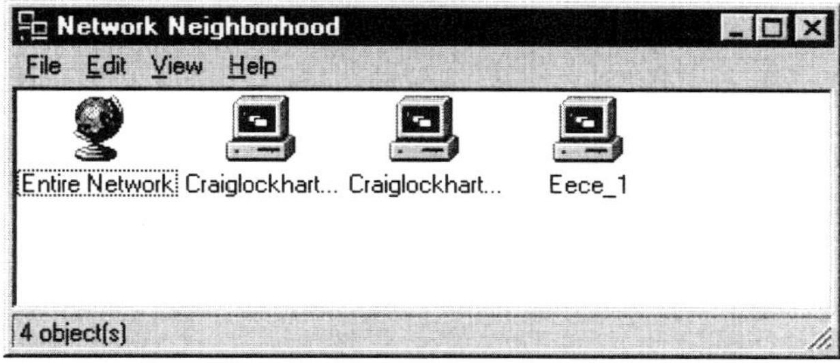

Figure 31.26 Network neighborhood.

Figure 31.27 Local neighborhood servers.

32 NetWare

32.1 Novell NetWare networking

Novell NetWare is one of the most popular network operating systems for PC LANs and provides file and print server facilities. Its default network protocol is normally SPX/IPX. This can also be used with Windows NT to communicate with other Windows NT nodes and with NetWare networks. The Internet Packet Exchange (IPX) protocol is a network layer protocol for transportation of data between computers on a NetWare network. IPX is very fast and has a small connectionless datagram protocol. The Sequenced Packet Interchange (SPX) provides a communications protocol which supervises the transmission of the packet and ensures its successful delivery.

32.2 NetWare and TCP/IP integration

NetWare is typically used in organizations and works well on a local network. Network traffic which travels out on the Internet or that communicates with UNIX networks must be in TCP/IP form. This section outlines possible methods used to integrate NetWare with TCP/IP traffic.

32.2.1 IP tunneling

IP tunneling encapsulates the IPX packet within the IP packet. This can then be transmitted into the Internet network. When the IP packet is received by the destination NetWare gateway; the IP encapsulation is stripped off. IP tunneling thus relies on a gateway into each IPX-based network that also runs IP. The NetWare gateway is often called an IP tunnel peer.

32.3 NetWare architecture

NetWare provides many services, such as file sharing, printer sharing, security, user administration and network management. The interface between the network interface card (NIC) and the SPX/IPX stack is ODI (Open Data-link Interface). NetWare clients run software which connects them to the server, the supported client operating systems are DOS, Windows, Windows NT, UNIX, OS/2 and Macintosh.

With NetWare Version 3, DOS and Windows 3 clients use a NetWare shell called NETx.COM. This shell is executed when the user wants to log into the network and stay resident. It acts as a command redirector and processes requests which are either generated by application programs or from the keyboard. It then decides whether they should be handled by the NetWare network operating system or passed to the client's local DOS operating system. NETx builds its own tables to keep track of the location of network-attached resources rather than using DOS tables. Figure 32.1 illustrates the relationship between the NetWare shell and DOS, in a DOS-based client. Note that Windows 3 uses the DOS operating system, but Windows NT and 95 have their own operating systems and only emulate DOS. Thus, Windows NT and 95 do not need to use the NETx program.

The ODI allows NICs to support multiple transport protocols, such as TCP/IP and IPX/SPX, simultaneously. Also, in an Ethernet interface card, the ODI allows simultaneous support of multiple Ethernet frame types such as Ethernet 802.3, Ethernet 802.2, Ethernet II, and Ethernet SNAP. Figure 32.2 shows a configuration of the frame type for IPX/SPX protocol.

To install NetWare, the server must have a native operating system, such as DOS or Windows NT, and it must be installed on its own disk partition. NetWare then adds a partition in which the NetWare partition is added. This partition is the only area of the disk the NetWare kernel can access.

32.3.1 NetWare loadable modules (NLMs)

NetWare allows enhancements from third-party suppliers using NLMs. The two main categories are:

- Operating systems enhancements – these allow extra operating system functions, such as a virus checker and also client hardware specific modules, such as a network interface drivers.
- Application programs – these programs actually run on the NetWare server rather than on the client machine.

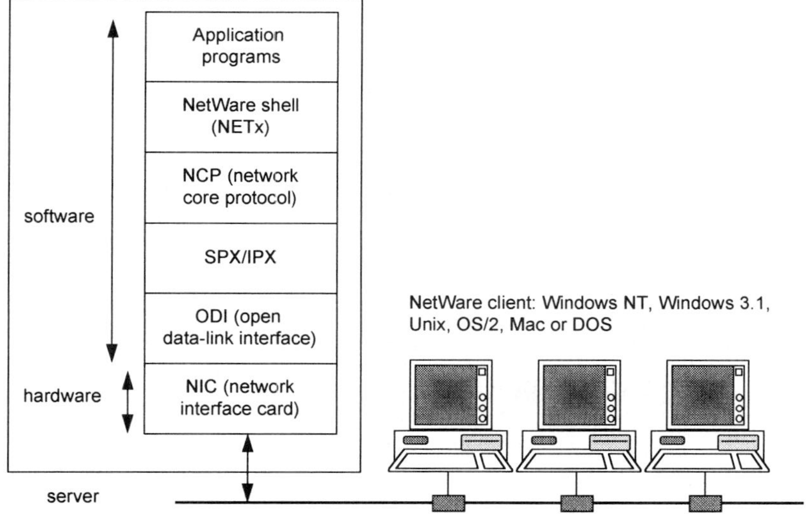

Figure 32.1 NetWare architecture.

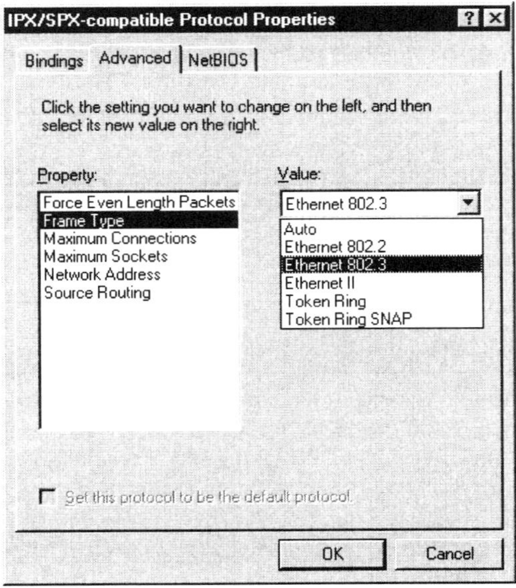

Figure 32.2 IPX/SPX frame types.

32.3.2 Bindery services

NetWare must keep track of users and their details. Typically, NetWare must keep track of:

- User names and passwords.
- Groups and group rights.
- File and directory rights.
- Print queues and printers.
- User restrictions (such as allowable login times, the number of times a user can simultaneously log in to the network).
- User/group administration and charging (such as charging for user login).
- Connection to networked peripherals.

This information is kept in the bindery files. Whenever a user logs in to the network their login details are verified against the information in the bindery files.

The bindery is organized with objects, properties and values. Objects are entities that are controlled or managed, such as users, groups, printers (servers and queues), disk drives, and so on. Each object has a set of properties, such as file rights, login restrictions, restrictions to printers, and so on. Each property has a value associated with it. Here are some examples:

Object	*Property*	*Value*
User	Login restriction	Wednesday 9 am till 5pm
User	Simultaneous login	2
Group	Access to printer	No

Objects, properties and values are stored in three separate files which are linked by pointers on every NetWare server:

1. `NET$OBJ.SYS` (contains object information).
2. `NET$PROP.SYS` (contains property information).
3. `NET$VAL.SYS` (contains value information).

If multiple NetWare servers exist on a network then bindery information must be exchanged manually between the servers so that the information is the same on each server. In a multiserver NetWare 3.x environment, the servers send SAP (service advertising protocol) information between themselves to advertise available services. Then the bindery services on a particular server update their bindery files with the latest information regarding available services on other reachable servers. This synchronization is difficult when just a few servers exist but is extremely difficult when there are many servers. Luckily, NetWare 4.1 has addressed this problem with NetWare directory services; this will be discussed later.

32.4 NetWare protocols

NetWare uses IPX (Internet Packet Exchange) for the network layer and either SPX (Sequenced Packet Exchange) or NCP (NetWare Core Protocols) for the transport layer. The routing information protocol (RIP) is also used to transmit information between NetWare gateways. These protocols are illustrated in Figure 32.3.

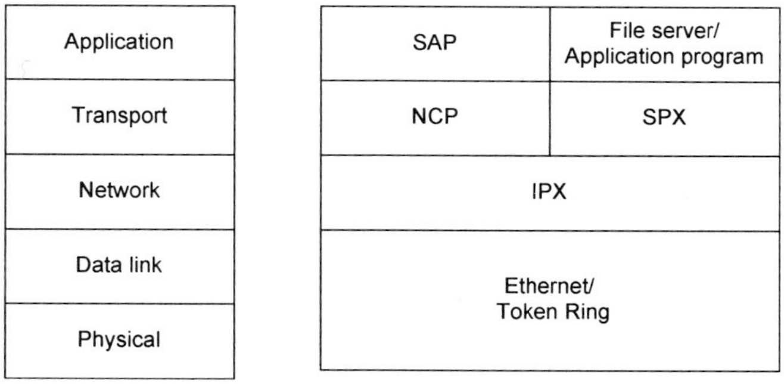

Application	SAP	File server/ Application program
Transport	NCP	SPX
Network	IPX	
Data link	Ethernet/ Token Ring	
Physical		

Figure 32.3 NetWare reference model.

32.5 IPX

IPX performs a network function that is similar to IP. The higher information is passed to the IPX layer which then encapsulates it into IPX envelopes. It is characterized by:

- A Connectionless connection – each packet is sent into the network and must find its own way through the network to the final destination (connections are established with SPX).

- It is unreliable – as there is only basic error checking and no acknowledgment (acknowledgments are achieved with SPX).

IPX uses a 12-byte station address (whereas IP uses a 4-byte address). The IPX fields are:

- Checksum (2 bytes) – this field is rarely used in IPX, as error checking is achieved in the SPX layer. The lower-level data link layer also provides an error detection scheme (both Ethernet and Token Ring support a frame check sequence).
- Length (2 bytes) – this gives the total length of the packet in bytes (i.e. header+DATA). The maximum number of bytes in the DATA field is 546, thus the maximum length will be 576 bytes (2+2+1+1+12+12+546).
- Transport control (1 byte) –this field is incremented every time the frame is processed by a router. When it reaches a value of 16 it is deleted. This stops packets from traversing the network for an infinite time. It is also typically known as the time-to-live field or hop counter.
- Packet type (1 byte) –this field identifies the upper layer protocol so that the DATA field can be properly processed.
- Addressing (12 bytes) –this field identifies the address of the source and destination station. It is made up of three fields: a network address (4 bytes), a host address (6 bytes) and a socket address (2 bytes). The 48-bit host address is the 802 MAC LAN address. NetWare supports a hierarchical addressing structure where the network and host addresses identify the host station and the socket address identifies a process or application and thus supports multiple connections (up to 50 per node).

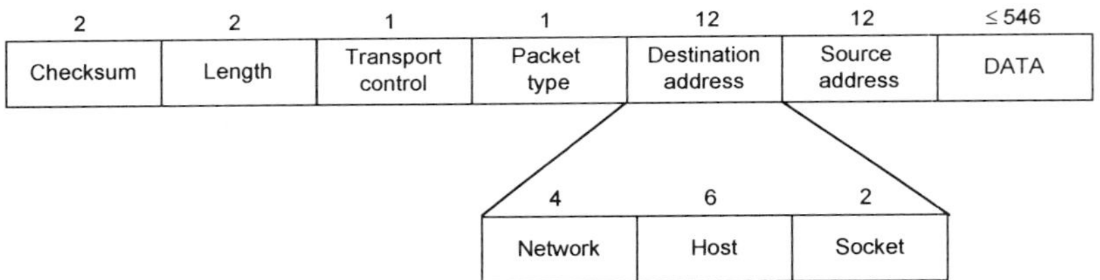

Figure 32.4 IPX packet format.

32.5.1 SPX

On a NetWare network the level above IPX is either NCP or SPX. The SPX protocol sets up a virtual circuit between the source and the destination (just like TCP). Then all SPX packets follow the same path and will thus always arrive in the correct order. This type of connection is described as connection-oriented.

SPX also allows for error checking and an acknowledgment to ensure that packets are received correctly. Each SPX packet has flow control and also sequence numbers. Figure 32.5 illustrates the SPX packet.

The fields in the SPX header are:

- Connection control (1 byte) – this is a set of flags which assist the flow of data. These flags include an acknowledgment flag and an end-of-message flag.
- Datastream type (1 byte) – this byte contains information which can be used to determine the protocol or information contained within the SPX data field.

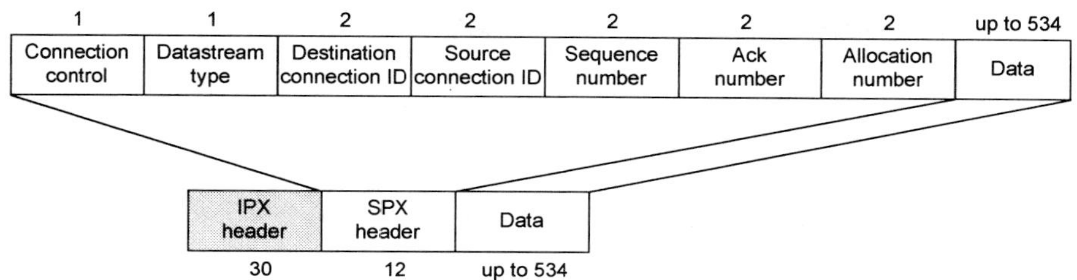

Figure 32.5 SPX packet format.

- Destination connection ID (2 bytes) – the destination connection ID allows the routing of the packet through the virtual circuit.
- Source connection ID (2 bytes) – the source connection ID identifies the source station when it is transmitted through the virtual circuit.
- Sequence number (2 bytes) – this field contains the sequence number of the packet sent. When the receiver receives the packet, the destination error checks the packet and sends back an acknowledgment with the previously received packet number in it.
- Acknowledgment number (2 bytes) – this acknowledgment number is incremented by the destination when it receives a packet. It is in this field that the destination station puts the last correctly received packet sequence number.
- Allocation number (2 bytes) – this field informs the source station of the number of buffers the destination station can allocate to SPX connections.
- DATA (up to 534 bytes).

32.5.2 RIP

The NetWare Routing Information Protocol (RIP) is used to keep routers updated on the best routes through the network. RIP information is delivered to routers via IPX packets. Figure 32.6 illustrates the information fields in an RIP packet. The RIP packet is contained in the field which would normally be occupied by the SPX packet.

Routers are used within networks to pass packets from one network to another in an optimal way (and error-free with a minimal time delay). A router reads IPX packets and examines the destination address of the node. If the node is on another network then it routes the packet in the required direction. This routing tends not to be fixed as the best route will depend on network traffic at given times. Thus the router needs to keep the routing tables up to date; RIP allows routers to exchange their current routing tables with other routers.

The RIP packet allows routers to request or report on multiple reachable networks within a single RIP packet. These routes are listed one after another (Figure 32.6 shows two routing entries). Thus each RIP packet has only one operation field, but has multiple entries of the network number, the number of router hops, and the number of tick fields, up to the length limit of the IPX packet.

The fields are:

- Operation (2 bytes) – this field indicates that the RIP packet is either a request or a response.

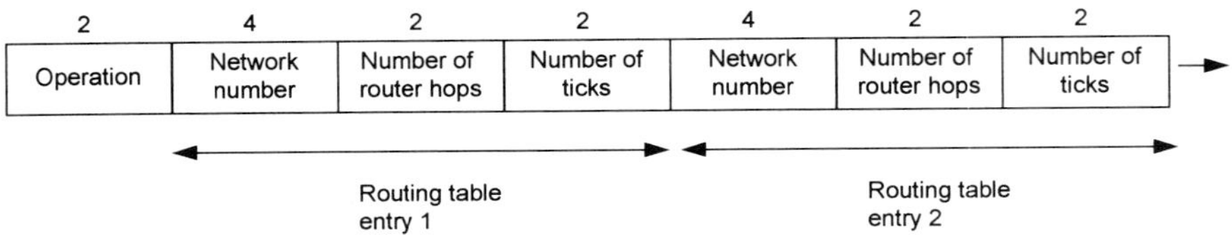

Figure 32.6 RIP packet format.

- Network number (4 bytes) – this field defines the assigned network address number to which the routing information applies.
- Number of router hops (2 bytes) – this field indicates the number of routes that a packet must go through in order to reach the required destination. Each router adds a single hop.
- Number of ticks (2 bytes) – this field indicates the amount of time (in 1/18 second) that it takes a packet to reach the given destination. Note that a route which has the fewest hops may not necessarily be the fastest.

RIP packets add to the general network traffic as each router broadcasts its entire routing table every 60 seconds. This shortcoming has been addressed by NetWare 4.1.

32.5.3 SAP

Every 60 seconds each server transmits a SAP (Service Advertising Protocol) packet which gives its address and tells other servers which services it offers. These packets are read by special agent processes running on the routers which then construct a database that defines which servers are operational and where they are located.

When the client node is first booted it transmits a request in the network asking for the location of the nearest server. The agent on the router then reads this request and matches it up to the best server. This choice is then sent back to the client. The client then establishes an NCP (NetWare Core Protocol) connection with the server, from which the client and server negotiate the maximum packet size. After this, the client can access the networked file system and other NetWare services.

Figure 32.7 illustrates the contents of a SAP packet. It can be seen that each SAP packet contains a single operation field and data on up to seven servers. The fields are:

- Operation type (2 bytes) – defines whether the SAP packet is server information request or a broadcast of server information.
- Server type (2 bytes) – defines the type of service offered by a server. These services are identified by a binary pattern, such as:

File server	0000 1000	Job server	0000 1001
Gateway	0000 1010	Print server	0000 0111
Archive server	0000 1001	SNA gateway	0010 0001
Remote bridge server	0010 0100	TCP/IP gateway	0010 0111
NetWare access server	1001 1000		

- Server name (48 bytes) – which identifies the actual name of the server or host offering the service defined in the service type field.
- Network address (4 bytes) – which defines the address of the network to which the server is attached.

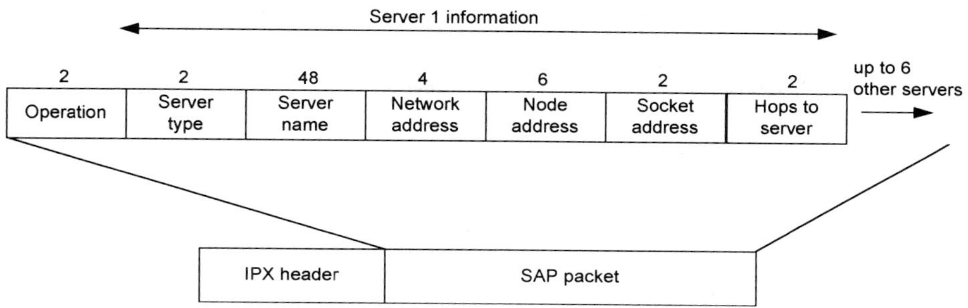

Figure 32.7 SAP packet format.

- Node Address (6 bytes) – which defines the actual MAC address of the server.
- Socket address (6 bytes) – which defines the socket address on the server assigned to this particular type of service.
- Hops to server (2 bytes) – which indicates the number of hops to reach the particular service.

32.5.4 NCP

The clients and servers communicate using the NetWare Core Protocols (NCPs). They have the following operation:

- The NETx shell reads the application program request and decides whether it should direct it to the server.
- If it does redirect, then it sends a message within an NCP packet, which is then encapsulated within an IPX packet and transmitted to the server.

Figure 32.8 illustrates the packet layout and encapsulation of an NCP packet.

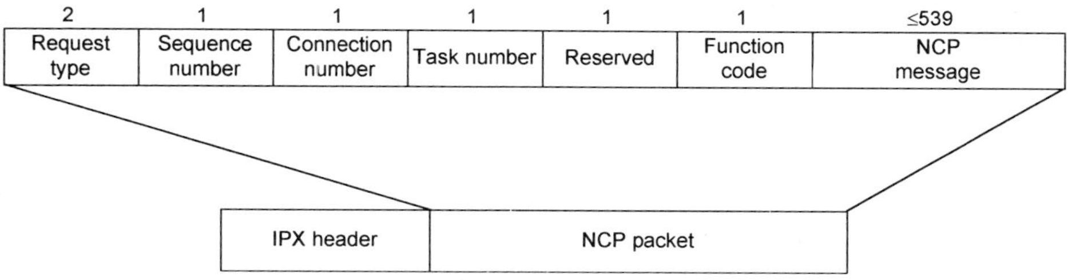

Figure 32.8 NCP packet format.

The fields are:

- Request type (2 bytes) – which gives the category of NCP communications. Among the possible types are:

Busy message	1001 1001 1001 1001
Create a service	0001 0001 0001 0001

Service request from workstation	0010 0010 0010 0010
Service response from server	0011 0011 0011 0011
Terminate a service connection	0101 0101 0101 0101

For example the create-a-service request is initiated at login time and a terminate-a-connection request is sent at logout.

- Sequence number (1 byte) – which contains a request sequence number. The client reads the sequence number so that it knows the request to which the server is responding to.
- Connection number (1 byte) –a unique number which is assigned when the user logs into the server.
- Task number (1 byte) – which identifies the application program on the client issued the by service request.
- Function code (1 byte) – which defines the NCP message or commands. Example codes are:

Close a file	0100 0010	Create a file	0100 1101
Delete a file	0100 0100	Get a directory entry	0001 1111
Get file size	0100 0000	Open a file	0100 1100
Rename a file	0100 0101	Extended functions	0001 0110

  Extended functions can be defined after the 0001 0110 field.

- NCP message (up to 539 bytes) – the NCP message field contains additional information which is passed between the clients and servers. If the function code contains 0001 0110 then this field will contain subfunction codes.

32.6 Novel NetWare setup

NetWare 3.x and 4.1 use the Open Data-Link Interface (ODI) to interface NetWare to the NIC. Figure 32.9 shows how the NetWare 3.x fits into the OSI model. ODI is similar to NDIS in Windows NT and was developed jointly between Apple and Novell. It provides a standard vendor-independent method to interface the software and the hardware (Figure 32.10 shows that Windows NT can choose between NDIS2 and ODI).

A typical login procedure for a NetWare 3.x network is:

```
LSL.COM
NE2000
IPXODI
NETx /PS=EECE_1
F:
LOGIN
```

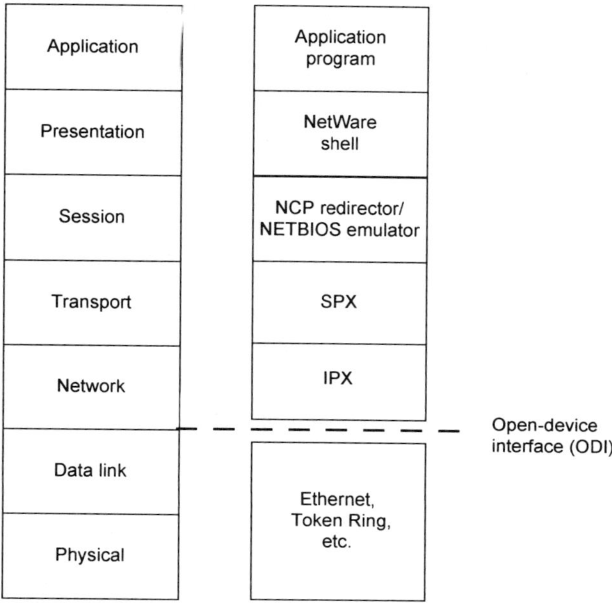

Figure 32.9 OSI model and NetWare 3.x.

The program LSL (link support layer) provides a foundation for the MAC layer to communicate with multiple protocols. An interface adapter driver (in this case NE2000) provides a MAC layer driver and is used to communicate with the interface card. This driver is known as a multilink interface driver (MLID). After this driver is installed, the program IPXODI is then installed. This program normally communicates with LSL and applications.

The NETx program communicates with the server and sets up a connection with the server EECE_1. This then sets up a local disk partition of F: (onto which the user's network directory will be mounted). Next the user logs in to the network with the command LOGIN.

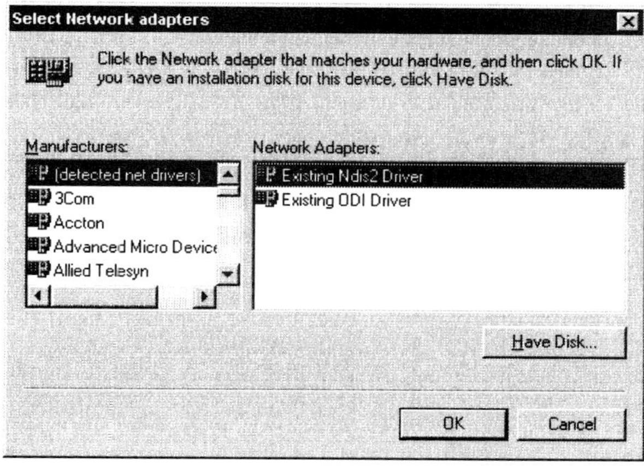

Figure 32.10 Network adapter driver.

32.6.1 ODI

ODI allows users to load several protocol stacks (such as TCP/IP and SPX/IPX) simultaneously for operation with a single NIC. It also allows support to link protocol drivers to adapter drivers. Figure 32.11 shows the architecture of the ODI interface. The LSL layer supports multiple protocols and it reads from a file `NET.CFG`, which contains information on the network adapter and the protocol driver, such as the interface adapter, frame type and protocol.

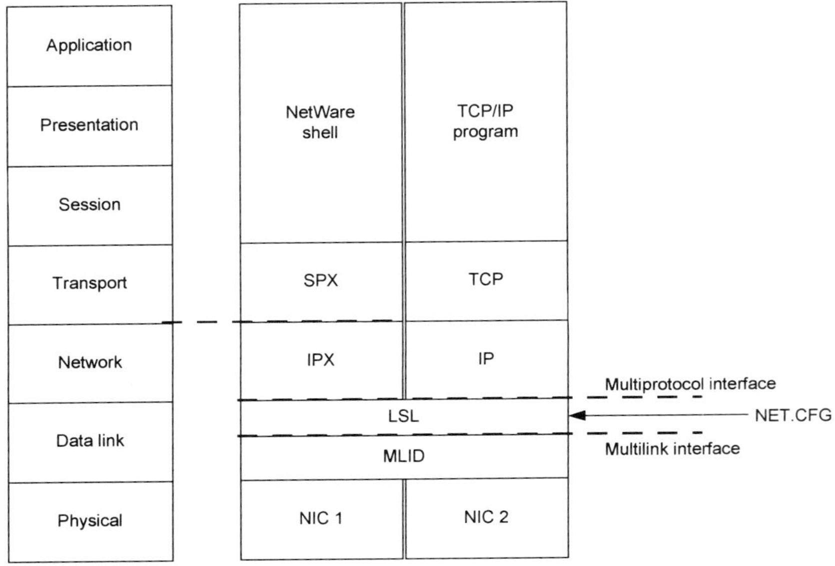

Figure 32.11 ODI architecture.

Here is a sample `NET.CFG` file is:

```
Link Driver NE2000
      Int #1 11
      Port #1 320
      Frame Ethernet_II
      Frame Ethernet_802.3
      Protocol IPX 0 Ethernet_802.3
        Protocol Ethdev 0 Ethernet_II
```

This configuration file defines the interface adapter as using interrupt line 11, having a base address of 320h, operating IPX, Ethernet 802.3 frame type and following the Ethernet II protocol. Network interface card drivers (such as NE2000, from the previous setup) are referred to as a multilink interface driver (MLID).

32.7 NetWare 4.1

The main disadvantages of NetWare 3.x are:

- It uses SPX/IPX which is incompatible with TCP/IP traffic.
- It is difficult to synchronize servers with user information.
- The file structure is local to individual servers.
- Its server architecture is flat; it cannot be organized into a hierarchy structure.

These disadvantages have been addressed with NetWare 4.1 which has:

- A hierarchical server structure.
- Network-wide users and groups.
- Global objects.
- System-wide login with a single password.
- Support for a distributed file system.

32.7.1 NetWare directory services (NDS)

One of the major changes between NetWare 3.x and NetWare 4.1 is NDS. A major drawback of the NetWare 3.x bindery files is that they were independently maintained on each server. NDS addresses this by setting up a single logical database which contains information on all network-attached resources. It is logically a single database but may be physically located on different servers over the network. As the database is global to the network, a user can log in to all authorized network-attached resources, rather than requiring to login into each separate server. Thus administration is focused on the single database.

As with NetWare 3.x bindery services, NDS organizes network resources by objects, properties, and values. NDS differs from the bindery services in that it defines two types of object:

- Leaf objects – which are network resources such as disk volumes, printers, printer queues, and so on.
- Container objects – which are cascadable organization units that contain leaf objects. A typical organizational unit might be company, department or group.

NDS organizes networked resources in a hierarchical or tree structure (as most organizations are structured). The top of the tree is the root object, to which there is only a single root for an entire global NDS database. Servers then use container objects to connect to branches coming off the root object. This structure is similar to the organization of a directory file structure and can be used to represent the hierarchical structure of an organization. Figure 32.12 illustrates a sample NDS database with root, container and leaf objects. In this case the organization splits into four main containers: electrical, mechanical, production and administration. Each of the containers has associated leaf objects, such as disk drivers, printer queues, and so on. This is a similar approach to Workgroups in Microsoft Windows.

To improve fault tolerance, NDS allows branches of the tree (or partitions) to be stored on multiple file servers. These copies are then synchronized to keep them up-to-date. Another advantage of replicating partitions is that local copies of files can be stored so that network traffic is reduced.

32.7.2 Virtual loadable modules (VLMs)

The NETx redirector shell has been replaced with DOS client software known as the requester. Its main advantage is that it allows NetWare client to easily add or update their functionality by using VLMs. This is controlled through the DOS-based VLM management program (VLM.EXE). It differs from NETx in that the requester uses DOS tables of network-attached resources rather than creating and maintaining its own. The main difference between NETx and the requester is that it is the DOS system which controls whether the NetWare DOS request is called to handle network requests.

Various VLM modules can be added onto the client, such as:

- Bindery-based services.
- File management.
- IPX and NCP protocol stacks.
- NDS services.
- NetWare support for multiple protocol stacks (e.g. TCP/IP, SPX/IPX).
- NETx shell emulation.
- Printer redirector to network print queues.
- TCP/IP and NCP protocol stacks.

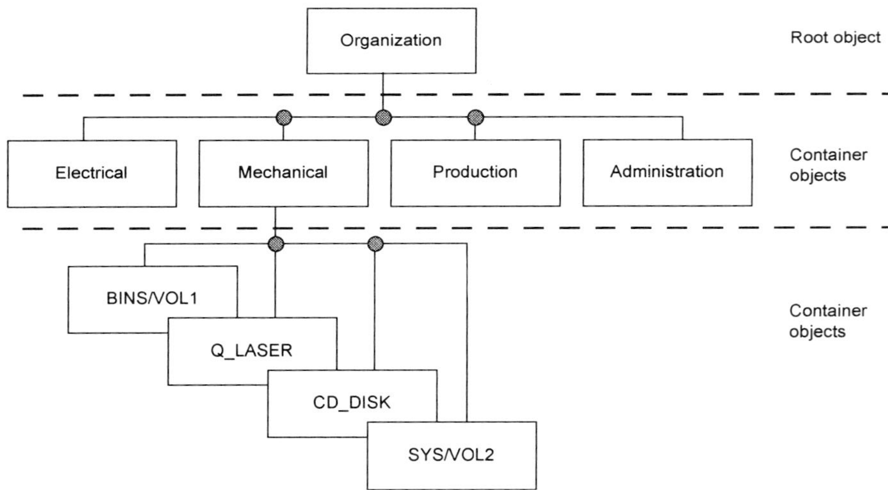

Figure 32.12 NDS structure.

32.7.3 Fault tolerance

The previous chapter discussed how Windows NT servers use RAID (Redundant Array of Inexpensive Disks) with parity to allow the recovery of data when a disk drive fails. NetWare 4.1 allows disk mirroring of partitions when a disk drive fails.

Another major fault occurs when a server becomes inoperative. NetWare 4.1 uses a novel technique, known as SFT III, which allows server duplexing. In this technique, the contents of the disk, memory and CPU are synchronized between primary and duplexed servers. When the primary server fails then the duplexed server takes over transparently. These servers are synchronized using the mirror server link (MSL), a dedicated link between the two servers, as illustrated in Figure 32.13. The MSL is a dedicated link because it prevents general network traffic from swamping the data.

It may seem expensive to have a backup server doing nothing apart from receiving data, but if it is costed with the loss of business or data when the primary server goes down then it is extremely cheap.

32.7.4 Communications protocols

NetWare 4.x has improved existing protocols and added the support for other standard network protocols, especially TCP/IP.

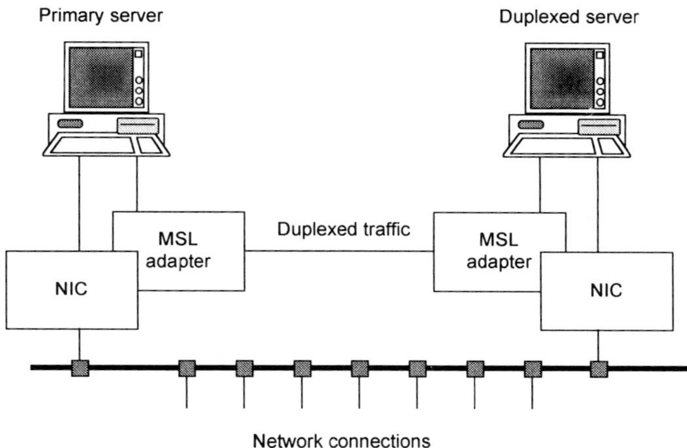

Figure 32.13 Mirror server link.

TCP/IP

TCP/IP is supported with NetWare/IP which is included with NetWare 4.1; NetWare/IP servers can support IP, IPX or IP and IPX traffic.

Large IPX packets (LIPs)

Most networks have become less prone to error. Thus larger data packets can be transmitted with a low risk of errors occurring. LIP allows NetWare clients to increase the size of their DATA field by negotiating with routers as to the size of the IPX frame (normally its has a maximum of 576 bytes). Unfortunately, an error in the packet causes the complete packet to be retransmitted (thus causing inefficiencies). In addition, the router must support the use of LIP. The software-based Novell router has a multiprotocol router which supports LIP. Unfortunately, other vendors may not support the LIP protocol.

NetWare Link-State Routing Protocol (NLSP)

RIP has several disadvantages, these include:

- File servers transmit their routing table every 60 seconds. This can have a great effect on the network loading, especially for interconnected networks.
- RIP only supports 16 hops before an RIP packet is discarded, thus limiting the physical size of the internetwork linking NetWare LAN

NLSP, which is included in NetWare 4.1, overcomes these problems. With the routing table, NLSP only broadcasts when a change occurs, with a minimum update of once every 2 hours. This can significantly reduce the router-to-router traffic. As with LIP, Novell routers support NLSP, but other vendors may not necessarily support it.

NLSP supports an increased hop size. A great advantage with NLSP is that it can coexist with RIP and is thus backward compatible. This allows a gradual migration of network segments to NLSP.

32.7.5 NetWare 4.1 SMP

One of the great improvements in computer processing and power will be achievable through the use of parallel processing. This processing can either be realized using multiple local processors or network processors, called symmetrical multiprocessing (SMP). To maintain compatibility with a previous release, NetWare 4.1 SMP loads the SMP kernel which works cooperatively with the operating system kernel. The main processor runs the main operating system while the SMP kernel runs the 2nd, 3rd and 4th processors.

32.7.6 Other enhancements

Other enhancements have been added, such as:

- File compression, which is controllable on a file-by-file basis.
- Increased supervisor security.
- Increased support for printers (up to 255 can be connected).

33 UNIX

33.1 Introduction

UNIX is an extremely popular operating system and dominates in the high-powered, multi-tasking workstation market. It is relatively simple to use and to administer, and also has a high degree of security. UNIX computers use TCP/IP communications, which they use to mount, disk resources from one machine onto another. Its main characteristics are:

- Multi-user.
- Preemptive multitasking.
- Multiprocessing.
- Multithreaded applications.
- Memory management with paging (organizing programs so that the program is loading into pages of memory) and swapping (which involves swapping the contents of memory to disk storage).

The two main families of UNIX are UNIX System V and BSD (Berkeley Software Distribution) Version 4.4. System V is the operating system most often used and has descended from a system developed by the Bell Laboratories and was recently sold to SCO (Santa Cruz Operation). Popular UNIX systems are:

- AIX (on IBM workstations and mainframes).
- HP-UX (on HP workstations).
- Linux (on PC-based systems).
- OSF/1 (on DEC workstations).
- Solaris (on Sun workstations).

An initiative by several software vendors has resulted in a common standard for the user interface and the operation of UNIX. The user interface standard is defined by the common desktop environment (CDE). This allows software vendors to write calls to a standard CDE API (application-specific interface). The common UNIX standard has been defined as Spec 1170 APIs. Compliance with the CDE and Spec 1170 API are certified by X/Open, which is a UNIX standard organization.

33.2 Network setup

Modern UNIX-based systems tend to be based around four main components:

- UNIX operating system.
- TCP/IP communications.
- Network file system (NFS).
- X-Windows interface.

X-Windows presents a machine-independent user interface for client/server applications. This will be discussed in more detail in the next chapter. TCP/IP allows for network communications and NFS allows disk drives to be linked together to make a global file system.

33.3 TCP/IP protocols

UNIX uses the normal range of TCP/IP protocols, grouped into transport, routing, network addresses, users services, gateway and other protocols.

Routing

Routing protocols manage the addressing of the packets and provide a route from the source to the destination. Packets may also be split up into smaller fragments and reassembled at the destination. The main routing protocols are:

- ICMP (Internet Control Message Protocol) which support status messages for the IP protocol these may be errors or network changes that can affect routing.
- IP (Internet Protocol) which defines the actual format of the IP packet.
- RIP (Routing Information Protocol) which is a route determining protocol

Transport

The transport protocols are used by the transport layer to transport a packet around a network. The protocols used are:

- TCP (Transport Control Protocol) which is a connection-based protocol where the source and the destination make a connection and maintain the connection for the length of the communications.
- UDP (User Datagram Protocol) which is a connectionless server where there is no communication between the source and the destination.

Network addresses

The network address protocols resolve IP addresses with their symbolic names, and vice versa. These are:

- ARP (Address Resolution Protocol) determines the IP address of nodes on a network.
- DNS (Domain Name System) which determines IP addresses from symbolic names (such as `anytown.ac.uk` might be resolved to 112.123.33.22).

User services

These are applications to which users have direct access.

- BOOTP (Boot Protocol) which is typically used to start up a diskless networked node. Thus rather than reading boot information from its local disk it reads the data from a server. Typically X-Windows terminals use it to get their IP address.
- FTP (File Transfer Protocol) which is used to transfer files from one node to another.
- Telnet which is used to remotely log in to another node.

Gateway protocols

The gateway protocols provide help for the routing process, these protocols include:

- EGP (Exterior Gateway Protocol) which transfers routing information for external network.
- GGP (Gateway-to-Gateway Protocol) which transfers routing information between Internet gateways.
- IGP (Interior Gateway Protocol) which transfers routing information for internal networks.

Others services

Other important services provide support for networked files systems, electronic mail and time synchronization as well as helping maintain a global network database. The main services are:

- NFS (Network File System) which allows disk drives on remote nodes to be mounted on a local node and thus create a global file system. This will be discussed in more detail in Section 26.4.
- NIS (Network Information Systems) which maintain a network-wide database for user accounts and thus allows users to log in to any computer on the network. Any changes to a user's accounts are made over the whole network.
- NTP (Network Time Protocol) which is used synchronize clocks of nodes on the network.
- RPC (Remote Procedure Call) which enables programs running on different nodes on a network to communicate with each other using standard function calls.
- SMTP (Simple Mail Transfer Protocol) which is a standard protocol for transferring electronic mail messages.
- SNMP (Simple Network Management Protocol) which maintains a log of status messages about the network.

33.4 NFS

The Network File System (NFS) allows computers to share the same files over a network and was originally developed by Sun Microsystems. It has the great advantage that it is independent of the host operating system and can provide data sharing among different types of systems (heterogeneous systems).

NFS uses a client/server architecture where a computer can act as an NFS client, an NFS server or both. An NFS client makes requests to access data and files on servers; the server then makes that specific resource available to the client.

NFS servers are passive and stateless. They wait for requests from clients and they maintain no information on the client. One advantage of servers being stateless is that it is possible to reboot servers without adverse consequences to the client.

The components of NFS are as follows:

- NFS remote file access may be accompanied by network information service (NIS).
- External data representation (XDR) is a universal data representation used by all nodes. It provides a common data representation if applications are to run transparently on a heterogeneous network or if data is to be shared among heterogeneous systems. Each node translates machine-dependent data formats to XDR format when sending and translating data. It is XDR that enables heterogeneous nodes and operating systems to communicate with each other over the network.
- Remote Procedure Call (RPC) provides the ability for clients to transparently execute procedures on remote systems of the network. NFS services run on top of the RPC, which corresponds to the session layer of the OSI model.
- Network lock manager (`rpc.lockd`) allows users to coordinate and control access to information on the network. It supports file locking and synchronizes access to shared files.

Figure 33.1 shows how the protocols fit into the OSI model.

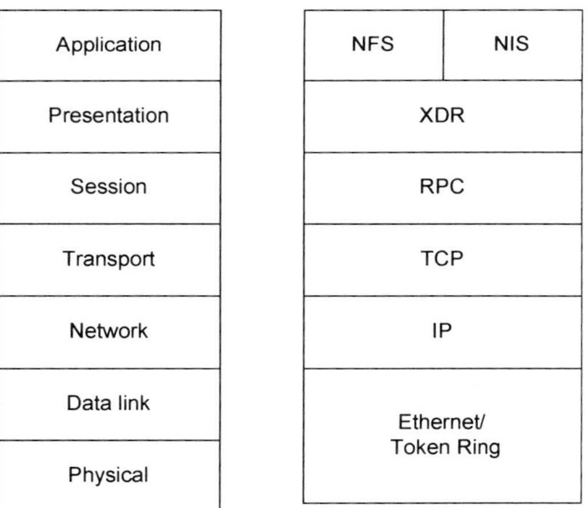

Figure 33.1 NFS services protocol stack.

33.4.1 Network Information Service (NIS)

NIS is an optional network control program which maintains the network configuration files over a network. Previously it was named *Yellow Pages* (YP), but has had a name change as this is a registered trademark of the company British Telecommunications. It normally administers the network configuration files such as `/etc/group` (which defines the user groups), /etc/hosts (which defines the IP address and symbolic names of nodes on a network), /etc/passwd (which contains information, such as user names, encrypted passwords, home directories, and so on). Here is an except from a passwd file:

```
root:FDEc6.32:1:0:Super unser:/user:/bin/csh
fred:jt.06hLdiSDaA:2:4:Fred Blogs:/user/fred:/bin/csh
fred2:jtY067SdiSFaA:3:4:Fred Smith:/user/fred2:/bin/csh
```

This passwd file has three defined users, these are root, fred and fred2. The encrypted password is given in the second field (between the first and second colon). The third field is a unique number that defines the user (in this case fred is 2 and fred2 is 3). The fourth field in this case defines the group number (which ties up with the /etc/groups file. An example of a groups file is given next. It can be seen from this file that group 4 is defined as freds_group, and contains three users: fred, fred2 and fred3. The fifth field is simply a comment field and in this case it contains the user's names. In the next field each user's home directory is defined and the final field contains the initial UNIX shell (in this case it is the C-shell).

```
root::0:root
other::1:root,hpdb
bin::2:root,bin
sys::3:root,uucp
freds_grp::4:fred,fred2,fred3
```

A sample listing of a directory shows that a file owned by fred has the group name freds_grp.

```
> ls -l
-r-sr-xr-x    1 fred      freds_grp   24576     Apr 22   1997 file1
-r-xr-xr-x   13 fred      freds_grp   40        Apr 22   1997 file2
dr-xr-xr-x    2 fred      freds_grp   1024      Aug  5    14:01 myfile
-r-xr-xr-x    1 fred      freds_grp   32768     Apr 22   1997 text1.ps
-r-xr-sr-x    1 fred      freds_grp   24576     Apr 22   1997 text2.ps
-r-xr-xr-x    2 fred      freds_grp   16384     Apr 22   1997 temp1.txt
-r-xr-xr-x    1 fred      freds_grp   16384     Apr 22   1997 test.doc
```

An excerpt from the /etc/hosts file is shown next.

```
138.38.32.45       bath
198.4.6.3          compuserve
193.63.76.2        niss
148.88.8.84        hensa
146.176.2.3        janet
146.176.151.51     sun
```

The /etc/protocols file contains information with known protocols used on the Internet.

```
# The form for each entry is:
# <official protocol name> <protocol number> <aliases>
#
# Internet (IP) protocols
#

ip       0   IP       # internet protocol, pseudo protocol number
icmp     1   ICMP     # internet control message protocol
ggp      3   GGP      # gateway-gateway protocol
```

```
tcp      6  TCP      # transmission control protocol
egp      8  EGP      # exterior gateway protocol
pup      12 PUP      # PARC universal packet protocol
udp      17 UDP      # user datagram protocol
hmp      20 HMP      # host monitoring protocol
xns-idp  22 XNS-IDP  # Xerox NS IDP
rdp      27 RDP      # "reliable datagram" protocol
```

The `/etc/netgroup` file defines network-wide groups used for permission checking when doing remote mounts, remote logins, and remote shells. Here is a sample file:

```
# The format for each entry is: groupname  member1  member2 ...
#     (hostname, username, domainname)
engineering hardware software (host3, mikey, hp)
hardware (hardwhost1, chm, hp)   (hardwhost2, dae, hp)
software (softwhost1, jad, hp)   (softwhost2, dds, hp)
```

NIS master server and slave server

With NIS, a single node on a network acts as the NIS master server, and there may be a number of NIS slave servers. The slave servers receive their NIS information from the master server. When a client first starts up it sends out a broadcast to all NIS servers (master or slaves) on the network and waits for the first to respond. The client then binds to the first that responds and addresses all NIS requests to that server. If this server becomes inoperative then an NIS client will automatically rebind to the first NIS server which responds to another broadcast.

Table 33.1 outlines the records which are used in the NIS database (or NIS map). This file consists of logical records with a search key and a related value for each record. For example, in the `passwd.byname` map, the users' login names are the keys and the matching lines from `/etc/passwd` are the values.

NIS domain

An NIS domain is a logical grouping of the set of maps contained on NIS servers. The rules for NIS domains are:

- All nodes in an NIS domain have the same domain name.
- Only one master server exists on an NIS domain.
- Each NIS domain can have zero or more slave servers.

An NIS domain is a subdirectory of `/usr/etc/yp` on each NIS server, where the name of the subdirectory is the name of the NIS domain. All directories that appear under `/usr/etc/yp` are assumed to be domains that are served by an NIS server. Thus to remove a domain being served, the user deletes the domain's subdirectory name from `/etc/etc/yp` on all of its servers.

The start up file on most UNIX systems is the `/etc/rc` file. This automatically calls the `/etc/netnfsrc` file which contains the default NIS domain name, which uses the program `domainname`. Appendix E gives an part of an example `netnfsrc` file.

Table 33.1 NIS database components.

NIS map	File maintained	Description
group.bygid group.byname	/etc/group	Maintains user groups.
hosts.byaddr hosts.byname	/etc/hosts	Maintains a list of IP addresses and symbolic names.
netgroup.byhost netgroup.byuser	/etc/netgroup	Contains a mapping of network group names to a set of node, user and NIS domain names.
networks.byaddr networks.byname	/etc/network	Defines network-wide groups used for permission checking when doing remote mounts, remote logins, and remote shells.
passwd.byname passwd.byuid	/etc/passwd	Contains details, such as user names and encrypted passwords.
protocols.byname protocols.bynumber	/etc/protocols	Contains information with known protocols used on the Internet.
rpc.bynumber rpc.byname	/etc/rpc	Maps the RPC program names to the RPC program numbers and vice versa. This file is static; it is already correctly configured.
services.byname servi.bynp	/etc/services	
mail.byaddr mail.aliases	/etc/aliases	

33.4.2 NFS remote file access

To initially mount a remote directory (or file system) onto a local the superuser must do the following:

- On the server, exports the directory to the client.
- On the client, mounts (or imports) the directory.

For example if the remote directory /user is to be mounted on to be host miranda as the directory `/win`. To achieve these operations the following are setup:

1. The superuser logs on to the remote server and edits the file `/etc/exports` adding the `/user` directory.
2. The superuser then runs the program `exportfs` to make the `/user` directory available to the client.

```
% exportfs -a
```

3. The superuser then logs into the client and creates a mount point `/win` (empty directory).

```
% mkdir /mnt
```

4. The remote directory can then be mounted with:

```
% mount miranda:/user /win
```

NFS maintains the file `/etc/mnttab` which contains a record of the mounted file systems. The general format is:

```
special_file_name  dir  type  opts  freq  passno  mount_time  cnode_id
```

where *mount_time* contains the time the file system was mounted using mount. Sample contents of `/etc/mnttab` could be:

```
/dev/dsk/c201d6s0 / hfs defaults 0 1 850144122 1
/dev/dsk/c201d5s0 /win hfs defaults 1 2 850144127 1
castor:/win /net/castor_win nfs rw,suid 0 0 850144231 0
miranda:/win /net/miranda_win nfs rw,suid 0 0 850144291 0
spica:/usr/opt /opt nfs rw,suid 0 0 850305936 0
triton:/win /net/triton_win nfs rw,suid 0 0 850305936 0
```

In this case there are two local drivers (`/dev/dsk/c201d6s0` is mounted as the root directory and `/dev/dsk/c201d5s0` is mounted locally as `/win`). There are also four remote directories which are mounted from remote servers (`castor`, `miranda`, `spica` and `triton`). The directory mounted from castor is the `/win` directory and it is mounted locally as `/net/castor_win`. `hfs` defines a UNIX format disk and `nfs` defines that the disk is mounted over NFS.

A disk can be unmounted from a system using the umount command, e.g.

```
% umount miranda:/win
```

33.4.3 NIS commands

NIS commands allow the maintenance of network information. The main commands are as follows:

- `domainname` which displays or changes the current NIS domain name.
- `ypcat` which lists the specified NIS map contents.
- `ypinit` which, on a master server, builds a map using the networking files in `/etc`. On a slave server the map is built using the master server.
- `ypmake` which is a script that builds standard NIS maps from files such as `/etc/passwd`, `/etc/groups`, and so on.
- `ypmatch` which prints the specified NIS map data (values) associated with one or more keys.
- `yppasswd` which can be used to change (or install) a user's password in the NIS `passwd` map.
- `ypwhich` which is used to print the host name of the NIS server supplying NIS services to an NIS client.
- `ypxfr` which transfers the NIS map from one slave server to another.

For example the command:

```
ypcat group.byname
```

lists the group name, the group ID and the members of the group. Here is an example of changing a user's password for the NIS domain. In this case, the user `bill_b` changes the network-wide password on the master server `pollux`.

```
% yppasswd
Changing NIS password for bill_b...
Old NIS password: ********
New password: *******
Retype new password: *******
The NIS passwd has been changed on pollux, the master NIS passwd server.
```

The next example uses the `ypcat` program.

```
% ypcat group
students:*:200:msc01,msc02,msc03,msc04
nogroup:*:-2:
daemon::5:root,daemon
users::20:root,msc08
other::1:root,hpdb
root::0:root
```

The next example shows the `ypmake` command file which rebuild the NIS database.

```
# ypmake
For NIS domain eece:
The passwd map(s) are up-to-date.
Building the group map(s)... group build complete.
   Pushing the group map(s):  group.bygid  group.byname
The hosts map(s) are up-to-date.
The networks map(s) are up-to-date.
The rpc map(s) are up-to-date.
The services map(s) are up-to-date.
The protocols map(s) are up-to-date.
The netgroup map(s) are up-to-date.
The vhe_list map(s) are up-to-date.
ypmake complete:
```

33.5 Network configuration files

The main files used to set up networking are as follows:

`/etc/checklist` is a list of directories or files are automatically mounted at boot time.

`/etc/exports` contains a list of directories or files that clients may import.

`/etc/inded.conf` contains information about servers started by `inetd`.

`/etc/netgroup` contains a mapping of network group names to a set of node, user, and NIS domain names.

`/etc/netnfsrc` is automatically started at run time and initiates the required daemons and servers, and defines the node as a client or server.

`/etc/rpc` maps the RPC program names to the RPC program numbers and vice versa.

`/usr/adm/inetd.sec` checks the Internet address of the host requesting a service against the list of hosts allowed to use the service.

33.5.1 Daemons

Networking programs normally initiate networking daemons which are background processes and are always running. Their main function is to wait for a request to perform a task. Typical daemons are:

`biod` which is asynchronous block I/O daemons for NFS clients.

`inetd` which is an Internet daemon that listens to service ports. It listens for service requests and calls the appropriate server. The server it calls depends on the contents of the `/etc/inetd.conf` file.

`nfsd` which is the NFS server daemon. It is used by the client for reading and writing to a remote directory and it sends a request to the remote server `nfsd` process.

`pcnfsd` which is a PC user authentication daemon.

`portmap` which is an RPC program to port number conversion daemon. When a client makes an RPC call to a given program number, it first contacts `portmap` on the server node to determine the port number where RPC requests should be sent.

Here is an extract from the processes that run a networked UNIX workstation:

```
  UID    PID  PPID  C     STIME TTY      TIME COMMAND
 root    100     1  0  Dec   9  ?        0:00 /etc/portmap
 root    138     1  0  Dec   9  ?        0:00 /etc/inetd
 root     93     1  0  Dec   9  ?        0:00 /etc/rlbdaemon
 root    104     1  0  Dec   9  ?        9:20 /usr/etc/ypserv
 root    106     1  0  Dec   9  ?        0:00 /etc/ypbind
 root    122   120  0  Dec   9  ?        0:00 /etc/nfsd 4
 root    120     1  0  Dec   9  ?        0:00 /etc/nfsd 4
 root    116     1  0  Dec   9  ?        0:00 /usr/etc/rpc.yppasswdd
 root    123   120  0  Dec   9  ?        0:00 /etc/nfsd 4
 root    124   120  0  Dec   9  ?        0:00 /etc/nfsd 4
 root    125     1  0  Dec   9  ?        0:02 /etc/biod 4
 root    126     1  0  Dec   9  ?        0:02 /etc/biod 4
 root    127     1  0  Dec   9  ?        0:02 /etc/biod 4
 root    128     1  0  Dec   9  ?        0:02 /etc/biod 4
 root    131     1  0  Dec   9  ?        0:00 /etc/pcnfsd
 root    133     1  0  Dec   9  ?        0:00 /usr/etc/rpc.statd
 root    135     1  0  Dec   9  ?        0:00 /usr/etc/rpc.lockd
 root   4652     1  0 14:33:15 ?        0:00 /etc/pcnfsd
 root   4649     1  0 14:33:15 ?        0:00 /usr/etc/rpc.mountd
```

33.6 XDR format

As previous mentioned XDR is a standard technique which is used to describe and encode data. This standard form allows for transferring data between different computer architectures. It fits into the presentation layer and uses a language, which is similar to C, to describe data formats. This language allows data formats to be described in a concise manner. XDR assumes that bytes are defined as 8 bits of data.

33.6.1 Basic block size

The basic definition of the blocks are:

- Items are defined in multiples of four bytes (32 bits) of data.
- These bytes are numbered from 0 to n–1.
- Bytes are read (or written) to a byte stream so that byte m always precedes byte m+1.
- If the number of bytes (n) in the data is not divisible by 4 then the bytes are followed by enough (0 to 3) residual zero bytes (r) to make the count a multiple of 4.

33.6.2 Data types

The basic data types are defined in this section.

Unsigned Integer and Signed Integer

A signed integer has 32 bits and thus has a range from –2 147 483 648 to +2 147 483 647. It uses a 2's complement notation with the first byte the most significant and byte 3 the least significant. Integers are declared as follows:

```
int identifier;
```

and can be represented as:

```
(MSB)                       (LSB)
+-------+-------+-------+-------+
|byte 0 |byte 1 |byte 2 |byte 3 |
+-------+-------+-------+-------+
<------------32 bits------------>
```

An unsigned integer has 32 bits and thus has a range from 0 to +4 294 967 295. The most and least significant bytes are 0 and 3, respectively. Unsigned integers are declared as follows:

```
unsigned int identifier;
```

Enumeration

Enumerations have the same representation as signed integers and are useful in defining subsets of the integers. They are declared as follows:

```
enum { name-identifier = constant, ... } identifier;
```

For example, three menu options (FILE, EDIT and VIEW) could be described by an enumerated type:

```
enum { FILE=1, EDIT=2, VIEW = 3 } menu_options
```

Boolean

Booleans are declared as follows:

```
bool val;
```

which is equivalent to:

```
enum { FALSE = 0, TRUE = 1 } val;
```

Hyper Integer and Unsigned Hyper Integer

A hyper integer is a 64-bit value and allows greater ranges for integer values. The signed integer format uses 2's completed. In a hyper integer the most significant byte is 0 and the least significant is 7. They are declared as:

```
hyper identifier; unsigned hyper identifier;
```

and can be represented by:

```
  (MSB)                                                  (LSB)
+-------+-------+-------+-------+-------+-------+-------+-------+
|byte 0 |byte 1 |byte 2 |byte 3 |byte 4 |byte 5 |byte 6 |byte 7 |
+-------+-------+-------+-------+-------+-------+-------+-------+
<---------------------------64 bits---------------------------->
```

Floating-point

A float data type has 32 bits and uses the standard IEEE standard for normalized single-precision floating-point numbers. It has three fields:

- S (sign). A 1-bit value which represents a positive number as a 0 and a negative number as a 1.
- E (exponent). An 8-bit value which represents the exponent of the number in base 2, minus 127.
- F (fractional part). A 23-bit value which represents the base-2 fractional part of the number's mantissa.

The floating-point value is thus represented by:

$$\text{Value}= -1^{S}\times 2^{(E-127)} \times 1.F$$

It is declared as follows:

```
float identifier;
```

and can be represented by:

```
+-------+-------+-------+-------+
|byte 0 |byte 1 |byte 2 |byte 3 |
S|   E   |          F           |
+-------+-------+-------+-------+
1|<- 8 ->|<-------23 bits------>|
<------------32 bits------------>
```

Double-precision Floating-point

A double data type has 64 bits and uses the standard IEEE standard for normalized double-precision floating-point numbers. It has three fields:

- S (sign). A 1-bit value which represents a positive number as a 0 and a negative number as a 1.
- E (exponent). An 11-bit value which represents the exponent of the number in base 2, minus 1023.
- F (fractional part). A 52-bit value which represents the base-2 fractional part of the number's mantissa.

The floating-point value is thus represented by:

$$\text{Value} = -1^{S} \times 2^{(E-1023)} \times 1.F$$

It is declared as follows:

```
double identifier;
```

and can be represented by:

```
+------+------+------+------+------+------+------+------+
|byte 0|byte 1|byte 2|byte 3|byte 4|byte 5|byte 6|byte 7|
S|    E   |                   F                          |
+------+------+------+------+------+------+------+------+
1|<--11-->|<-----------------52 bits------------------->|
<-----------------------64 bits------------------------->
```

Fixed-length and Variable-length Opaque Data

Opaque data is uninterpreted data and consists of a number of bytes (either fixed or variable). It is declared as following

```
opaque identifier[n];
```

where the constant n is the (static) number of bytes necessary to contain the opaque data. If n is not divisible by 4 then a number of residual bytes are added. This can be represented as follows:

```
    0        1       ...
+--------+--------+...+--------+--------+...+--------+
| byte 0 | byte 1 |...|byte n-1|    0   |...|    0   |
+--------+--------+...+--------+--------+...+--------+
|<-----------n bytes---------->|<------r bytes------>|
|<-----------n+r (where (n+r) mod 4 = 0)------------>|
```

Variable-length opaque data is defined as a sequence of n (numbered 0 through n-1) arbitrary bytes. The first 4 bytes define the number (as an unsigned integer) of encoded bytes in the sequence. If this value is not divisible by 4 then a number of residual bytes are added. It is declared as following:

```
opaque identifier<m>;
```

where the constant m denotes an upper bound of the number of bytes that the sequence may contain. It can be represented by:

```
   0     1     2     3     4     5   ...
+-----+-----+-----+-----+-----+-----+...+-----+-----+...+-----+
|        length n       |byte0|byte1|...| n-1 |  0  |...|  0  |
+-----+-----+-----+-----+-----+-----+...+-----+-----+...+-----+
|<-------4 bytes------->|<------n bytes------>|<---r bytes--->|
                        |<----n+r (where (n+r) mod 4 = 0)---->|
```

String

A string contains a number of ASCII bytes numbered 0 through n-1). The first value is an unsigned 4-byte integer which is the number of bytes in the string. If this value is not divisible by 4 then a number of residual bytes are added. It is declared as following:

```
string object<m>;
```

It can be represented by:

```
   0     1     2     3     4     5   ...
+-----+-----+-----+-----+-----+-----+...+-----+-----+...+-----+
|        length n       |byte0|byte1|...| n-1 |  0  |...|  0  |
+-----+-----+-----+-----+-----+-----+...+-----+-----+...+-----+
|<-------4 bytes------->|<------n bytes------>|<---r bytes--->|
                        |<----n+r (where (n+r) mod 4 = 0)---->|
```

Fixed-length Array

Fixed-length arrays of homogeneous elements are declared as follows:

```
type-name identifier[n];
```

where the elements are numbered from 0 to n–1 and each element contains 4 bytes. It can be represented by:

```
+---+---+---+---+---+---+---+---+...+---+---+---+---+
|   element 0   |   element 1   |...|  element n-1  |
+---+---+---+---+---+---+---+---+...+---+---+---+---+
|<--------------------n elements------------------->|
```

Variable-length Array

A variable-length array is represented by:

```
type-name identifier<m>;
```

where m specifies the maximum acceptable element count of an array. The first 4 bytes of the array contains the number of elements in the array. It can be represented as:

```
  0  1  2  3
+--+--+--+--+--+--+--+--+--+--+--+--+...+--+--+--+--+
|     n     | element 0 | element 1 |...|element n-1|
+--+--+--+--+--+--+--+--+--+--+--+--+...+--+--+--+--+
|<-4 bytes->|<--------------n elements------------->|
```

Structure

Structures are declared as follows:

```
struct {
   first-declaration;
   second-declaration;
   ...
} identifier;
```

Each component has 4 bytes and can be represented as:

```
+-----------------+-----------------+...
| 1st declaration | 2nd declaration |...
+-----------------+-----------------+...
```

Others

XDR declares several other data types, these include:

- Void. This is a 0-byte quantity which can be used for describing operations that take no data as input or no data as output.
- Constant. This allows the definition of a constant value. Its syntax is:

```
const name-identifier = n;
```

- Typedef. This is used to declared a different data type. Its syntax is:

```
typedef declaration;
```

34 X-Windows

34.1 Introduction

An operating system allows application programs to access the hardware of computer. A user interface allows a user to access application programs in an easy-to-use way. In a text-based system the user enters text commands which are then interpreted by the operating system. Typical commands are to run application programs, copy files and so on. With a graphical interface, graphical objects (icons) represent applications programs and options are selected using menus. Rather than entering commands from the keyboard a mouse pointer is normally used to select objects. This usage of Windows, Icons, Menus and Pointers (WIMPs) is typically known as a Graphical User Interface (GUI).

The two main graphical user interfaces are Microsoft Windows and X-Windows. Newer versions of these implement layers 5, 6 and 7 of the OSI model and use standard networking technologies, such as Ethernet and networking protocols like TCP/IP. Both are becoming de facto standards for computer systems.

X-Windows runs on most computer systems but is typically used on Unix workstations. It is a portable user interface and can be used to run programs remotely over a network. Massachusetts Institute of Technology (MIT) developed it and it has become a de facto standard because of its manufacturer independence, its portability, its versatility and its ability to operate transparently across most network technologies.

The main features of X-Windows are:

- that it is network transparent. The output from a program can either be sent to the local graphics screen or to a remote node on the network. Application programs can output simultaneously to displays on the network, as illustrated in Figure 34.1. The communication mechanism used is machine-independent and operating system independent.
- that many different styles of user interface can be supported. The management of the user interface, such as the placing, sizing and stacking of windows is not embedded in the system, but is controlled by an application program which can easily be changed.
- that since X isn't embedded into an operating system it can be easily ported to a wide range of computer systems.
- that calls are made from application programs to the X-windows libraries which control WIMPs. The application program thus does not have to create any of these functions.

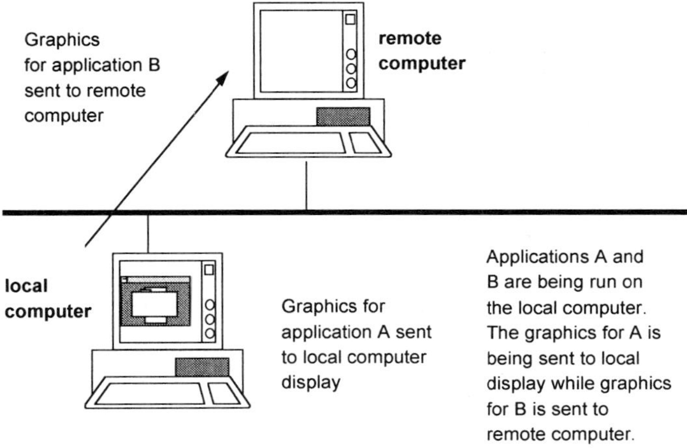

Figure 34.1 X programs can display to remote and local machines.

34.2 Fundamentals of X

There are three main parts of X software:

- A 'server' to control the physical display and input devices;
- 'Client' programs which request the server to perform particular operations on specified windows;
- A 'communications channel' through which the client programs and the server talk to each other.

This relationship is shown in Figure 34.2.

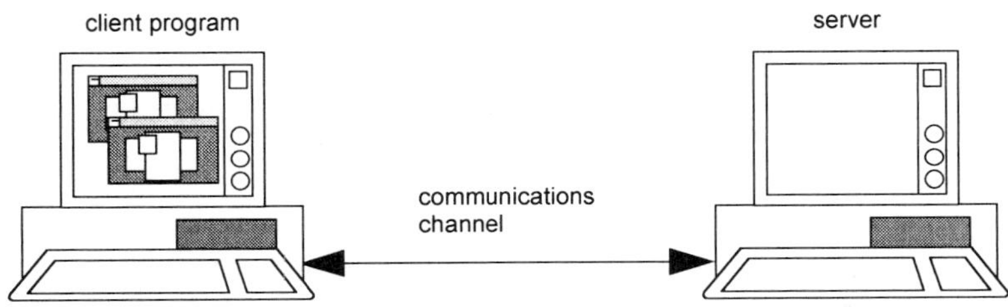

Figure 34.2 Fundamentals components of X.

34.2.1 Server

The server is the software which creates windows and draw images and text within them. This is done in response to the client programs.

34.2.2 Client

An application program makes use of the system's window facilities. Application programs in X are called 'clients' as they are customers of the server and ask the server to perform task on their behalf. For example a client may request the server to 'display the text Input a value' in window USER1 or to 'draw a rectangle in window TEMP'. This obviously reduces the burden to processing graphics on the client but increasing the processing on the server.

34.2.3 Communications channel

Clients send requests to the server via the communications channel and vice versa. X-Windows is transparent to the network technology and it supports different types of communications channel in the basic X-Windows library. All dependence of the types of communication is isolated to this library, and all communication between client and server is via this library (known as Xlib in the standard X implementation) as illustrated in Figure 34.3.

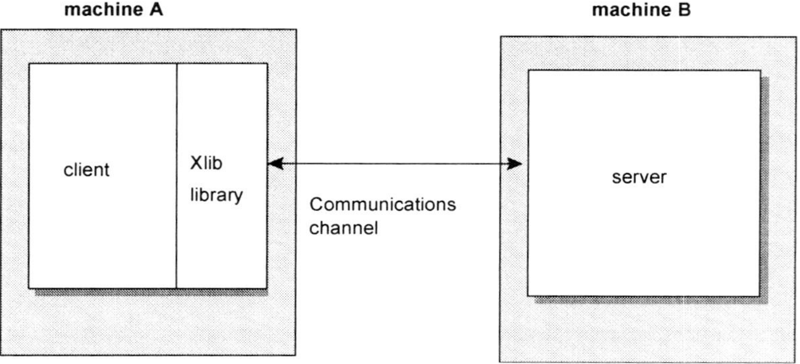

Figure 34.3 Function of the Xlib library.

There are two main modes of communications between the client and server, these are:

1. when the server and client are running on the same computer. Here they can communicate using any method of inter-process communications (ICP) available on the machine. When running in this mode, X is effectively operating like many conventional window systems.
2. when the client is running on one machine, but the display (and its server) is on another. The client and server communicate across the network using a mutually agreed protocol, such as TCP/IP.

X allows programs to communicate user information which is network transparent. This feature is useful in building multi-purpose networks of co-operating machines.

As the server and client are completely separate then a computer can run the application programs and that the client is required to do is to provide the keyboard, pointing device and graphics screen. This has led to a new type of display called an X terminal. An X terminal is simply a stripped-down computer which is dedicated to running the X server and nothing else. It has a keyboard, mouse and screen and some way of communicating across the network. It does not have its own file-system and cannot support general-purpose programs. Consequently, these programs have to be run elsewhere on the network.

34.3 Network aspects of X

The main advantages of the client/server relationship are:

- That a client can use a powerful remote server. This server could be a supercomputer, have a special processor, enhanced floating-point accelerator, and so on. An example of several X terminals and a computer using powerful workstations is illustrated in Figure 34.4.
- That if the server is a file server providing most of the disk resources to the local network then the disk network traffic is reduced as applications are run remotely on the server particular applications which are highly disk-intensive such as a large application programs. In this way only the results from a program need to be reported over the network and intermediate data is not required to be sent over the network.
- The remote server may have special software facilities available on it alone. With the growth of workstations, it is increasingly common to have some software licensed to only a few machines on the network and several clients run the software from these licensed machines. This may reduce the license fee, as only the servers may need to be paid for.
- It allows an enhanced user interface for remote logging into networked machines.
- When a new computer is added which can run X clients, it can be immediately used by any device running X.
- if a new display is added it can immediately make use of all the existing X client applications on any machine.
- There is a requirement to output to several displays.

The main disadvantage of X-Windows is that application programs become dependent on servers. If a server becomes overburdened or develops a fault then it can seriously affect the performance of application programs. This problem is similar to network problems in star network where the network performance is dependent upon the central server.

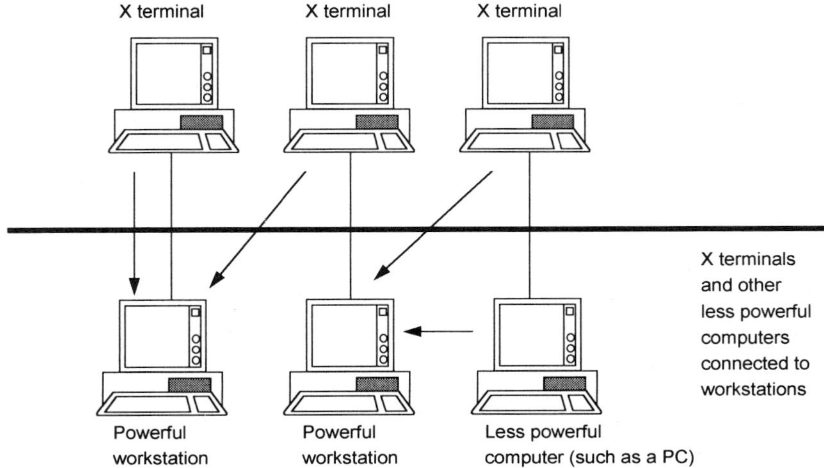

Figure 34.4 Client and server on separate machines.

The more programs that run on the server then the more main memory it needs to cope with the loading. If it does not have enough main memory then it can use a local disk to compensate. This will slow the performance of application programs.

An X-based network must thus be planned so as not to burden the server nodes. Typically, only two X terminals connect to a server. A well-planned network can normally cope with server faults by having several computers which can act as servers to X terminals, or remote computers.

34.4 History of X

In 1984, MIT started the development of X. Their main objective was to create a good windows system for Unix machines. Many versions evolved from this and by 1985 it was decided that X would be available to anyone who wanted for a nominal cost.

In 1985, version 10 was distributed to organizations outside MIT and by 1986 DEC produced the first commercial X product. During that year it was clear that version 10 could not evolve to satisfy all the requirements that were being asked for it. For this purpose MIT and DEC undertook a complete redesign of the protocol. The result was X, version 11, this has since become known as X.11.

34.5 X system programs

Any program can use the X-Windows libraries. Several standard programs exist which comprise the basic system:

- `X` is the display server. This software controls the keyboard, mouse and screen. It is the heart of X and creates and destroys windows. It writes text and draws objects within windows at the request of other 'client' programs. The command `xstart` or `xinit` normally starts X.
- `xterm` is the X terminal emulator. Many programs were not written specifically for X-Windows and must be run within a text window. The `xterm` program converts the text output to the graphics display.
- `xhost` controls which other hosts on the network are allowed access to display screen.
- `xkill` kills unwanted applications.
- `xwd` captures an image from the screen.
- `xpr` prints a previously captured screen to the printer.
- `xmag` magnifies a selected portion of the screen.
- `xclock` to display an analogue or digital clock.
- `xcalc` to be used a calculator.
- `xload` to displays the system loading of the machine.

35 Networking Elements

35.1 Introduction

The interconnection of PCs over a network is becoming more important especially as more hardware is accessed remotely and PCs intercommunicate with each other. This chapter gives an introduction to networking technology.

Computers communicate with other digital equipment over a local area network (LAN). A LAN is defined as a collection of computers within a single office or building that connect to a common electronic connection – commonly known as a network backbone. A LAN can be connected to other networks either directly or through a wide area network (WAN), as illustrated in Figure 35.1.

A WAN normally connects networks over a large physical area, such as in different buildings, towns or even countries. Figure 35.1 shows three local area networks, LAN A, LAN B and LAN C, some of which are connected by the WAN. A modem connects a LAN to a WAN when the WAN connection is an analogue line. For a digital connection a gateway connects one type of LAN to another LAN, or WAN, and a bridge connects a LAN to similar types of LAN.

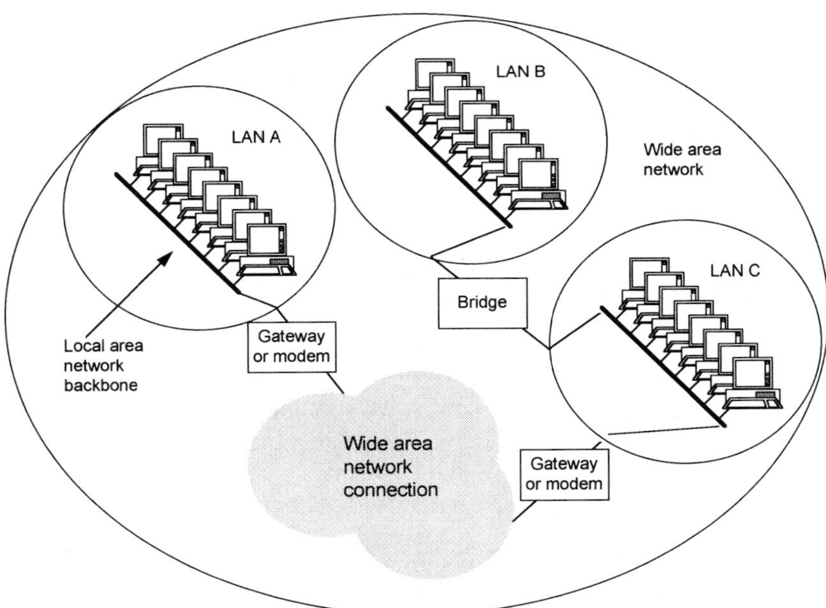

Figure 35.1 Interconnection of LANs to make a WAN.

The public switched telecommunications network (PSTN) provides long-distance analogue lines. These public telephone lines can connect one network line to another using circuit switching. Unfortunately, they have a limited bandwidth and can normally only transmit frequencies from 400 to 3 400 Hz. For a telephone line connection, a modem is used to convert the digital data into a transmittable form. Figure 35.2 illustrates the connection of computers to a PSTN. These computers can connect to the WAN through a service provider (such as CompuServe) or through another network which is connected by modem. The service provider has the required hardware to connect to the WAN.

A public switched data network (PSDN) allows the direct connection of digital equipment to a digital network. This has the advantage of not requiring the conversion of digital data into an analogue form. The integrated services digital network (ISDN) allows the transmission of many types of digital data into a truly global digital network. Transmittable data types include digitized video, digitized speech and computer data. Since the switching and transmission are digital, fast access times and relatively high bit rates are possible. Typical base bit rates may be 64 kbps. All connections to the ISDN require network termination equipment (NTE).

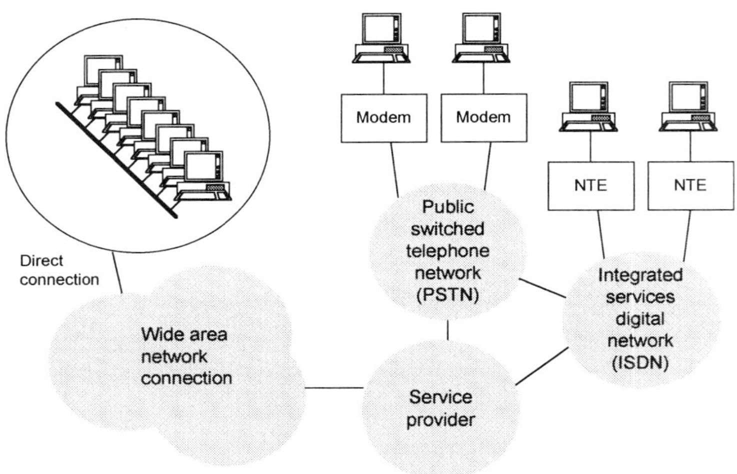

Figure 35.2 Connection of nodes to a PSTN.

35.2 OSI model

An important concept in understanding data communications is the OSI (open systems interconnection) model. It allows manufacturers of different systems to interconnect their equipment through standard interfaces. It also allows software and hardware to integrate well and be portable on differing systems. The International Standards Organization (ISO) developed the model and it is shown in Figure 35.3.

Data is passed from the top layer of the transmitter to the bottom then up from the bottom layer to the top on the recipient. However, each layer on the transmitter communicates directly with the recipient's corresponding layer. This creates a virtual data flow between layers.

The top layer (the application layer) initially gets data from an application and appends it with data that the recipient's application layer will read. This appended data passes to the next layer (the presentation layer). Again, it appends its own data, and so on, down to the physical layer. The physical layer is then responsible for transmitting the data to the recipient. The data sent can be termed a data packet or data frame.

Figure 35.4 shows the basic function of each of the layers. The physical link layer defines the electrical characteristics of the communications channel and the transmitted signals. This includes voltage levels, connector types, cabling, and so on.

The data link layer ensures that the transmitted bits are received in a reliable way. This includes adding bits to define the start and end of a data frame, adding extra error detection/correction bits and ensuring that multiple nodes do not try to access a common communications channel at the same time.

The network layer routes data frames through a network. If data packets require to go out of a network then the transport layer routes them through interconnected networks. Its task may involve splitting up data for transmission and reassembling it upon reception.

The session layer provides an open communications path to the other system. It involves setting up, maintaining and closing down a session. The communications channel and the internetworking of the data should be transparent to the session layer.

The presentation layer uses a set of translations that allow the data to be interpreted properly. It may have to carry out translations between two systems if they use different presentation standards such as different character sets or different character codes. For example, on a UNIX system a text file uses a single ASCII character for new line (the carriage return), whereas on a DOS-based system there are two, the line feed and the carriage return. The presentation layer would convert from one computer system to another so that the data could be displayed correctly, in this case by either adding or taking away a character. The presentation layer can also add data encryption for security purposes.

The application layer provides network services to application programs such as file transfer and electronic mail.

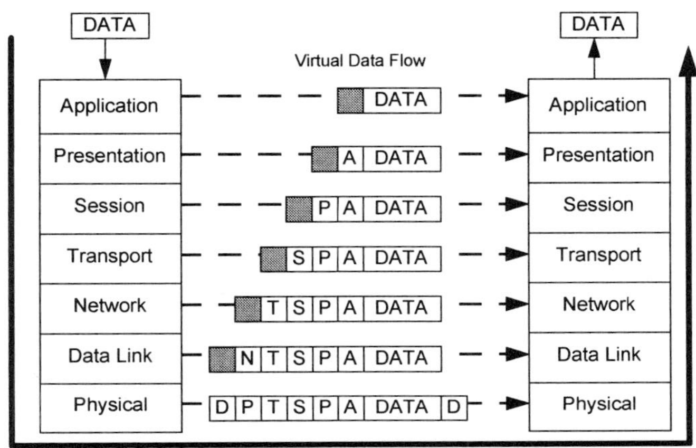

Figure 35.3 Seven-layer OSI model.

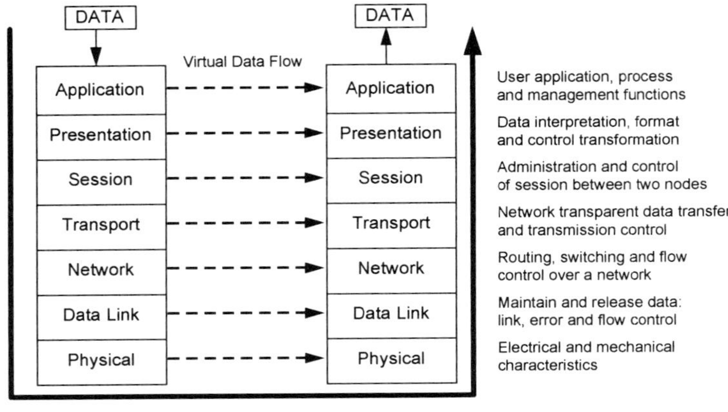

Figure 35.4 ISO open systems interconnection (OSI) model.

Figure 35.5 shows an example with two interconnected networks, Network A and Network B. Network A has four nodes N1, N2, N3 and N4, and Network B has nodes N5, N6, N7 and N8. If node N1 were to communicate with node N7 then a possible path would be via N2, N5 and N6. The data link layer ensures that the bits transmitted between nodes N1 and N2, nodes N2 and N5, and so on, are transmitted in a reliable way.

The network layer would then be responsible for routing the data packets through Network A and through Network B. The transport layer routes the data through interconnections between the networks. In this case, it would route data packets from N2 to N5. If other routes existed between N1 and N7 it might use another route.

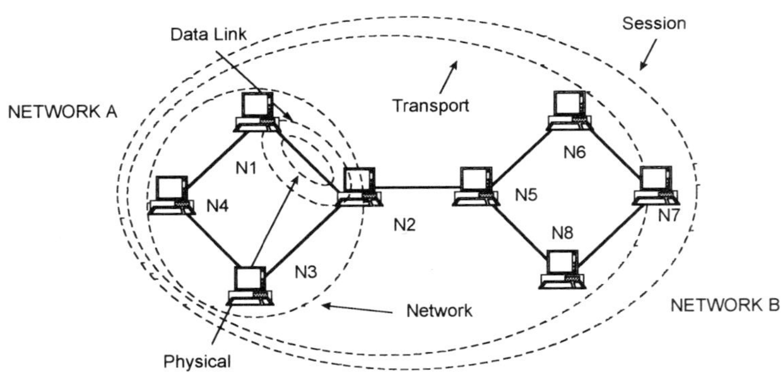

Figure 35.5 Scope of concern of OSI layers.

35.3 Communications standards and the OSI model

The following sections look at practical examples of data communications and networks, and how they fit into the layers of the OSI model. Unfortunately, many currently available technologies do not precisely align with the layers of this model. For example, RS-232 provides a standard for a physical layer but it also includes some data link layer functions, such as adding error detection and framing bits for the start and end of a packet.

Figure 35.6 shows the main technologies. These are split into three basic sections: asynchronous data communication, local area networks (LANs) and wide area networks (WANs).

The most popular types of LAN are Ethernet and Token Ring. Standards for Ethernet include Ethernet 2.0 and IEEE 802.3 (with IEEE 802.2). For Token Ring the standards are IBM Token Ring and IEEE 802.5 (with IEEE 802.2). Ethernet uses carrier sense multiple access/collision detect (CMSA/CD) technology which is why the IEEE standard includes the name CSMA/CD.

One of the main standards for the interconnection of networks is the Transport Control Protocol/Internet Protocol (TCP/IP). IP routes data packets through a network and TCP routes data packets between interconnected networks. An equivalent to the TCP/IP standard used in some PC networks; is called SPX/IPX.

For digital connections to WANs the main standards are CCITT X.21, HDLC and CCITT X.25.

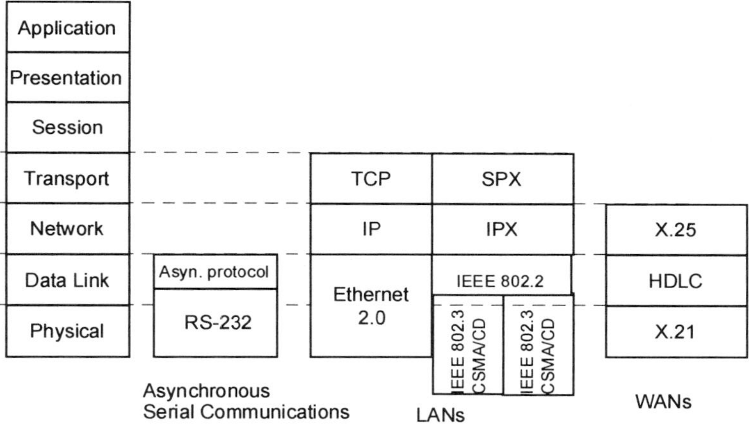

Figure 35.6 ISO open systems interconnection (OSI) model.

35.4 Standards agencies

There are six main international standards agencies that define standards for data communications systems. They are the International Standards Organization (ISO), the Comité Consultatif International Télégraphique et Telecommunications (CCITT), the Electrical Industries Association (EIA), the International Telecommunications Union (ITU), the American National Standards Institute (ANSI) and the Institute of Electrical and Electronic Engineers (IEEE).

The ISO and the IEEE have defined standards for the connection of computers to local area networks and the CCITT (now know as the ITU) has defined standards for the interconnection of national and international networks. The CCITT standards covered in this book split into three main sections: these are asynchronous communications (V.xx standards), PSDN connections (X.xxx standards) and ISDN (I.4xx standards). The main standards are given in Table 35.1. The EIA has defined standards for the interconnection of computers using serial communications. The original standard was RS-232-C; this gives a maximum bit rate of 20 kbps over 20 m. It has since defined several other standards, including RS-422 and RS-423, which provide a data rate of 10 Mbps.

Table 35.1 Typical standards.

Standard	*Equivalent ISO/CCITT*	*Description*
EIA RS-232C	CCITT V.28	Serial transmission up to 20 kps/20 m
EIA RS-422	CCITT V.11	Serial transmission up to 10 Mbps/1200 m
EIA RS-423	CCITT V.10	Serial transmission up to 300 Kbps/1200 m
ANSI X3T9.5		LAN: Fiber-optic FDDI standard
IEEE 802.2	ISO 8802.2	LAN: IEEE standard for logical link control
IEEE 802.3	ISO 8802.3	LAN: IEEE standard for CSMA/CD
IEEE 802.4	ISO 8802.4	LAN: Token passing in a Token Ring network
IEEE 802.5	ISO 8802.5	LAN: Token Ring topology
	CCITT X.21	WAN: Physical layer interface to a PSDN
HDLC	CCITT X.212/ 222	WAN: Data layer interfacing to a PSDN
	CCITT X.25	WAN: Network layer interfacing to a PSDN
	CCITT I430/1	ISDN: Physical layer interface to an ISDN
	CCITT I440/1	ISDN: Data layer interface to an ISDN
	CCITT I450/1	ISDN: Network layer interface to an ISDN

35.5 Network cable types

The cable type used on a network depends on several parameters, including:

- The data bit rate.
- The reliability of the cable.
- The maximum length between nodes.
- The possibility of electrical hazards.
- Power loss in the cables.
- Tolerance to harsh conditions.
- Expense and general availability of the cable.
- Ease of connection and maintenance.
- Ease of running cables, and so on.

The main types of cables used in networks are twisted-pair, coaxial and fiber-optic, they are illustrated in Figure 35.7. Twisted-pair and coaxial cables transmit electric signals, whereas fiber-optic cables transmit light pulses. Twisted-pair cables are not shielded and thus interfere with nearby cables. Public telephone lines generally use twisted-pair cables. In LANs they are generally used up to bit rates of 10 Mbps and with maximum lengths of 100 m.

Coaxial cable has a grounded metal sheath around the signal conductor. This limits the amount of interference between cables and thus allows higher data rates. Typically they are used at bit rates of 100 Mbps for maximum lengths of 1 km.

The highest specification of the three cables is fiber-optic. This type of cable allows extremely high bit rates over long distances. Fiber-optic cables do not interfere with nearby cables and give greater security, more protection from electrical damage by external equipment and greater resistance to harsh environments; they are also safer in hazardous environments.

A typical bit rate for a LAN using fiber-optic cables is 100 Mbps; in other applications this reaches several gigabits per second. The maximum length of the fiber-optic cable depends on the electronics in the transmitter and receiver, but a single length of 20 km is possible.

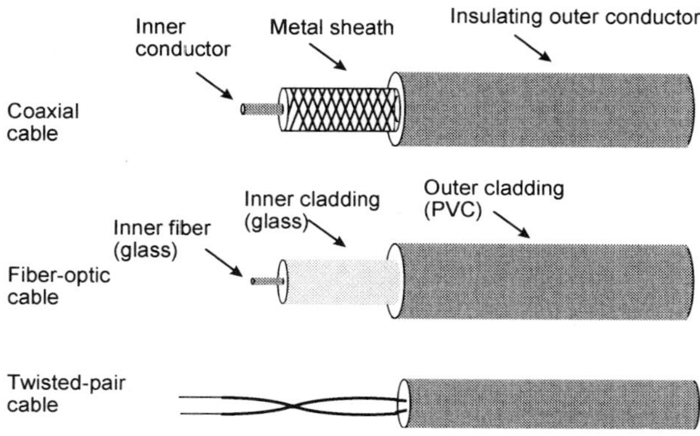

Figure 35.7 Types of network cable.

35.6 LAN topology

Computer networks are ever expanding, and a badly planned network can be inefficient and error prone. Unfortunately networks tend to undergo evolutionary change instead of revolutionary change and they can become difficult to manage if not planned properly. Most modern networks have a backbone, which is a common link to all the networks within an organization. This backbone allows users on different network segments to communicate and also allows data into and out of the local network. Figure 35.8 shows that a local area network contains various segments: LAN A, LAN B, LAN C, LAN D, LAN E and LAN F. These are connected to the local network via the BACKBONE1. Thus if LAN A talks to LAN E then the data must travel out of LAN A, onto BACKBONE1, then into LAN C and through onto LAN E.

Networks are partitioned from other networks using a bridge, a gateway or a router. A bridge links two networks of the same type, such as Ethernet to Ethernet, or Token Ring to Token Ring. A gateway connects two networks of dissimilar type. Routers operate rather like gateways and can either connect two similar networks or two dissimilar networks. The key operation of a gateway, bridge or router is that it only allows data traffic through itself when the data is intended for another network which is outside the connected network. This filters traffic and stops traffic not intended for the network from clogging up the backbone. Modern bridges, gateways and routers are intelligent and can determine the network topology.

A spanning-tree bridge allows multiple network segments to be interconnected. If more than one path exists between individual segments then the bridge finds alternative routes. This is useful in routing frames away from heavy traffic routes or around a faulty route. Conventional bridges can cause frames to loop around forever. Spanning-tree bridges have built-in intelligence and can communicate with other bridges. This allows them to build up a picture of the complete network and thus to make decisions on where frames are routed.

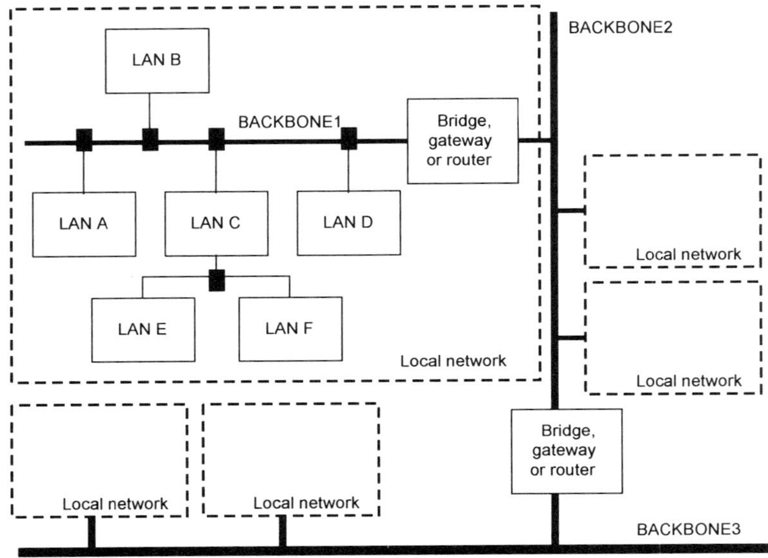

Figure 35.8 Interconnection of local networks.

35.7 Internetworking connections

Networks connect to other networks through repeaters, bridges or routers. A repeater corresponds to the physical layer and always routes signals from one network segment to another. Bridges route using the data link layer and routers route using the network layer. Figure 35.9 illustrates the three interconnection types.

35.7.1 Repeaters

All types of network connections suffer from attenuation and pulse distortion; for a given cable specification and bit rate, each has a maximum length of cable. Repeaters can be used to increase the maximum interconnection length and will do the following:

- Clean signal pulses.
- Pass all signals between attached segments.
- Boost signal power.
- Possibly translate between two different media types (e.g. fiber-optic to twisted-pair cable).

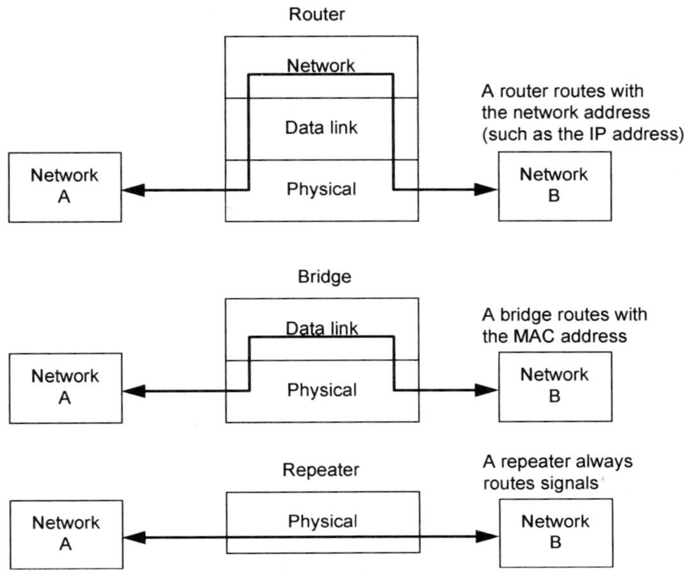

Figure 35.9 Repeaters, bridges and routers.

35.7.2 Bridges

Bridges filter input and output traffic so that only packets intended for a network are actually routed into the network and only packets intended for the outside are allowed out of the network.

The performance of a bridge is governed by two main factors (which are measured in packets per second or frames per second):

- The filtering rate – the bridge reads the MAC address of the Ethernet/Token Ring/FDDI node and then decides if it should forward the packet into the internetwork. When a bridge reads the destination address on an Ethernet frame or Token Ring packet and decides whether that packet should be allowed access to the internetwork. Filter rates for bridges range from around 5000 to 70 000 pps (packets per second). A typical bridge has a filtering rate of 17 500 pps.
- The forward rate – once the bridge has decided to route the packet into the internetwork then the bridge must forward the packet onto the internetwork media. This is a forwarding operation and typical rates range from 500 to 140 000 pps. A typical forwarding rate is 90 000 pps.

A typical Ethernet bridge has the following specifications:

Bit rate:	10 Mbps
Filtering rate:	17 500 pps
Forwarding rate:	11 000 pps
Connectors:	2 DB15 AUI (female), 1 DB9 male console port, 2 BNC (for 10BASE2) or 2 RJ-45 (for 10BASE-T)

Algorithm: Spanning-tree protocol. It automatically learns the addresses of all devices on both interconnected networks and builds a separate table for each network

And for a Token Ring bridge:

Bit rate: 4/16 Mbps
Filtering rate: 120 000 pps
Forwarding rate: 3400 pps
Connectors: 1 DB9 male console port, 2 DB9 connectors
Algorithm: Source routing transparent

Spanning-tree architecture (STA) bridges

The spanning-tree algorithm has been defined by the standard IEEE 802.1. It is normally implemented as software on STA-compliant bridges. On power-up they automatically learn the addresses of all nodes on both interconnected networks and build a separate table for each network.

They can also have two connections between two LANs so that when the primary path becomes disabled, the spanning-tree algorithm can re-enable the previously disabled redundant link. The path management is achieved by each bridge communicating using configuration bridge protocol data units (configuration BPDU).

Source route bridging

With source route bridging a source device, not the bridge, is used to send special explorer packets which are then used to determine the best path to the destination. Explorer packets are sent out from the source routing bridges until they reach their destination workstation. Then each source routing bridge along the route enters its address in the routing information field of the explorer packet. The destination node then sends back the completed RIF field to the source node. When the source device (normally a PC) has determined the best path to the destination, it sends the data message along with the path instructions to the local bridge. It then forwards the data message according to the received path instructions.

Although the source routing bridge receives the data, there is a 7-hop limit on the number of internetwork connections. This is because of the limited space in the router information field (RIF) of the explorer packet.

35.7.3 Routers

Routers examine the network address field (such as IP or IPX) and determine the best route for the packet. They have the great advantage that they normally support several different types of network layer protocol.

Normally routers which only read one type of protocol have high filtering and forwarding rates. If they support multiple protocols then there is normally an overhead in that the router must detect the protocol and look in the correct place for the destination address.

Typical network layer protocols and their associated network operating systems or upper layer protocols are:

AFP	AppleTalk	IP	TCP/IP	IPX	NetWare
OSI	Open Systems	VIP	Vines	XNS	3Com

Routers can also be used to connect to data link layer protocols without network layer addressing schemes. Typically, the data link frames are encapsulated within a routable network layer protocol (such as IP).

Routing protocols

Routers need to communicate with other routers so they can exchange routing information. Most network operating systems have associated routing protocols which support the transfer of routing information. Typical routing protocols and their associated network operating systems are:

- BGP (Border Gateway Protocol) – TCP/IP.
- EGP (Exterior Gateway Protocol) – TCP/IP.
- IS-IS (Immediate System to Intermediate Systems) – DECnet, OSI.
- NLSP (NetWare Link State Protocol) – NetWare 4.1.
- OSPF (Open Shortest Path First) – TCP/IP.
- RIP (Routing Information Protocol) – XNS, NetWare, TCP/IP.
- RTMP (Routing Table Maintenance Protocol) – AppleTalk.

Most routers support IP, IPX/SPX and AppleTalk network protocols using RIP, and EGP. The main Internet-based protocols are discussed in the next section. In the past RIP was the most popular router protocol standard. Its widespread use is due in no small part to the fact that it was distributed along with the Berkeley Software Distribution (BSD) of UNIX (from which most commercial versions of UNIX are derived). It suffers from several disadvantages and has been largely replaced by OSFP and EGP. These protocols have an advantage over RIP in that they can handle large internetworks as well as reducing routing table update traffic.

RIP uses a distance vector algorithm, which measures the number of hops, up to a maximum of 16, to the destination router. The OSPF and EGP protocols use a link state algorithm, which can decide between multiple paths to the destination router. They are based not only hops, but on other parameters such as delay capacity, reliability and throughput.

With distance vector routing each router maintains its table by communicating with neighboring routers. The number of hops in its own table is then computed, as it knows the number of hops to local routers. Unfortunately, the routing table can take some time to be updated when changes occur, because it takes time for all the routers to communicate with each other (known as slow convergence).

OSPF, EGP, BGP and NLSP use link state protocols and differ from distance vector routing in that they use network information received from all routers on a given network, rather than just from neighboring routes. This then overcomes slow convergence.

35.7.4 Example bridge/router

Here is an example of a typical bridge/router, which has a RISC-based processor, and can operate as a bridge and/or a router:

WAN port:	DB25 connector
LAN port:	2 BNC, 2 AUI (for Ethernet connections)
WAN line speed:	256 kbps
Forwarding rate:	59 kpps

Filtering rate:	14.88 kpps
Routable protocol:	IP, IPX/SPX, AppleTalk
Routing protocol:	RIP, HELLO, EGP
WAN interface:	RS-232/V.24, V.35, RS-530

35.8 Network topologies

There are three basic topologies for LANs, which are shown in Figure 35.10, these are:

- A star network.
- A ring network.
- A bus network.

There are other topologies which are either a combination of two or more topologies or are derivatives of the main types. A typical topology is a tree topology that is essentially a star and a bus network combined, as illustrated in Figure 35.11. A concentrator (or hub) is used to connect the nodes onto the network.

35.8.1 Star network

In a star topology, a central server switches data around the network. Data traffic between nodes and the server will thus be relatively low. Its main advantages are:

- Since the data rate is relatively low between central server and the node, a low-specification twisted-pair cable can be used connect the nodes to the server.
- A fault on one of the nodes will not affect the rest of the network. Typically, mainframe computers use a central server with terminals connected to it.

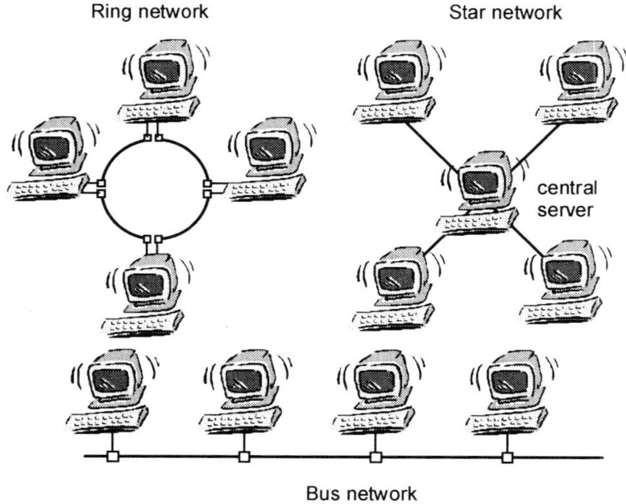

Figure 35.10 Network topologies.

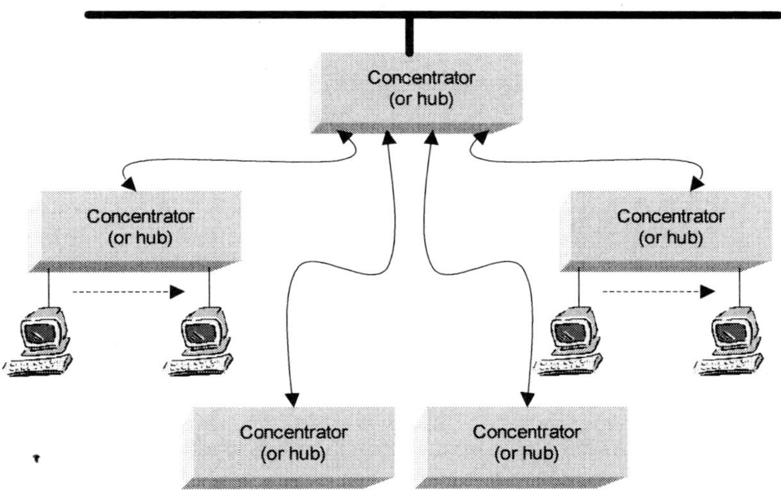

Figure 35.11 Tree topology.

The main disadvantage of this type of topology is that the network is highly dependent upon the operation of the central server. If it were to slow down significantly then the network becomes slow. In addition, if it were to become un-operational then the complete network would shut down.

35.8.2 Ring network

In a ring network the computers link together to form a ring. To allow an orderly access to the ring a single electronic token is passed from one computer to the next around the ring, as illustrated in Figure 35.17. A computer can only transmit data when it captures a token. In a manner similar to the star network each link between nodes is a point-to-point link and allows almost any transmission medium to be used. Typically, twisted-pair cables allow a bit rate of up to 16 Mbps, but coaxial and fiber-optic cables are normally used for extra reliability and higher data rates.

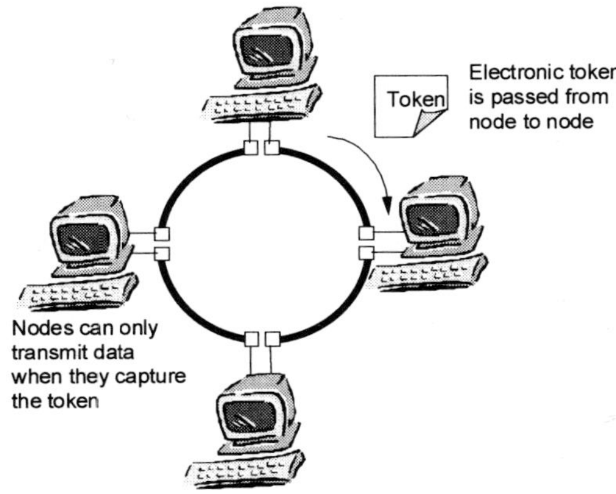

Figure 35.12 Token passing ring network.

A typical ring network is IBM Token Ring. The main advantage of token ring networks is that all nodes on the network have an equal chance of transmitting data. Unfortunately it suffers from several problems; the most severe is that if one of the nodes goes down then the whole network may go down.

35.8.3 Bus network

A bus network uses a multi-drop transmission medium, as shown in Figure 35.13. All nodes on the network share a common bus and all share communications. This allows only one device to communicate at a time. A distributed medium access protocol determines which station is to transmit. As with the ring network, data packets contain source and destination addresses. Each station monitors the bus and copies frames addressed to itself.

Twisted-pair cables give data rates up to 100 Mbps. Coaxial and fiber-optic cables give higher bit rates and longer transmission distances. A bus network is a good compromise over the other two topologies as it allows relatively high data rates. Also, if a node goes down then it does not affect the rest of the network. The main disadvantage of this topology is that it requires a network protocol to detect when two nodes are transmitting at the same time. A typical bus network is Ethernet 2.0.

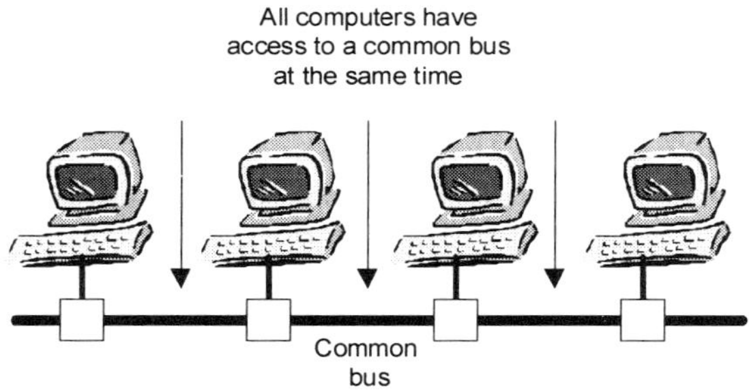

Figure 35.13 Bus topology.

35.9 Network loading

The data traffic on a network will not be constant over time. It will vary each minute, each hour, each day, each month and so on. Loading will tend to be similar from day to day. An example loading is given in Figure 35.14. It shows that, in this case, the peak network traffic occurs at around 11 am and also between 3 pm and 5 pm. Network managers often have to perform network maintenance and file backups. In this case the network manager has tried to even out network traffic by performing network backups at times when the network loading was low (i.e. at night).

Much of the network traffic is due to disk transfers. By analyzing network statistics it is possible to determine when hot spots occur. From these statistics the network manager may ask some users to change the way they operate. For example, by staggering lunch breaks the network loading traffic could be evened out. The system manager may also even out the network traffic by allowing only certain applications (or users) to use the network at certain times of the day. It is usual for heavy processing or network-intensive tasks to run during periods when there is a light network loading, typically at night.

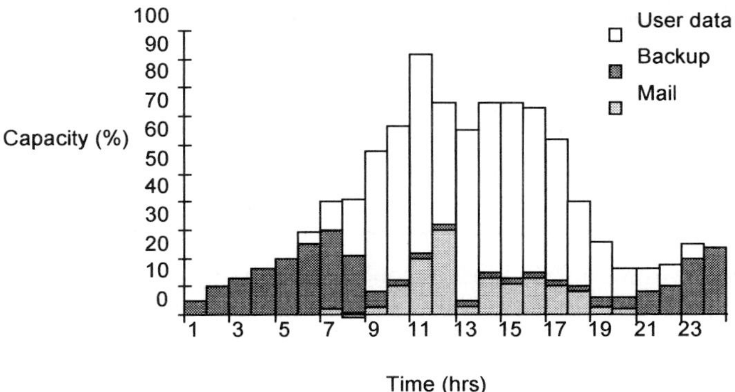

Figure 35.14 Example of network traffic over 24 hours.

36 Internet Routing Protocols

36.1 Introduction

Routers filter the network traffic so that the only traffic that flows into and out of a network is required to go into a network. In many cases, there are several possible routes that can be taken between to nodes on different networks. Consider the network in Figure 36.1. In this case, the upper network shows the connection between two nodes A and B through routers 1 to 6. In can be seen from the lower diagram that there are four routes that the data can take. To stop traffic taking a long route or even one that does not exist, each router must maintain a routing table so that it knows where the data must be sent when it receives data destined for a remote node.

For routers to find the best way, they must communicate with their neighbors to find the best way through the network. This measure can be defined in a number of ways, such as the number of router hops to the remote node. Unfortunately the number of router hops is not a good measure on the delay of data transmission as a route with a fewer number of hops may be congested.

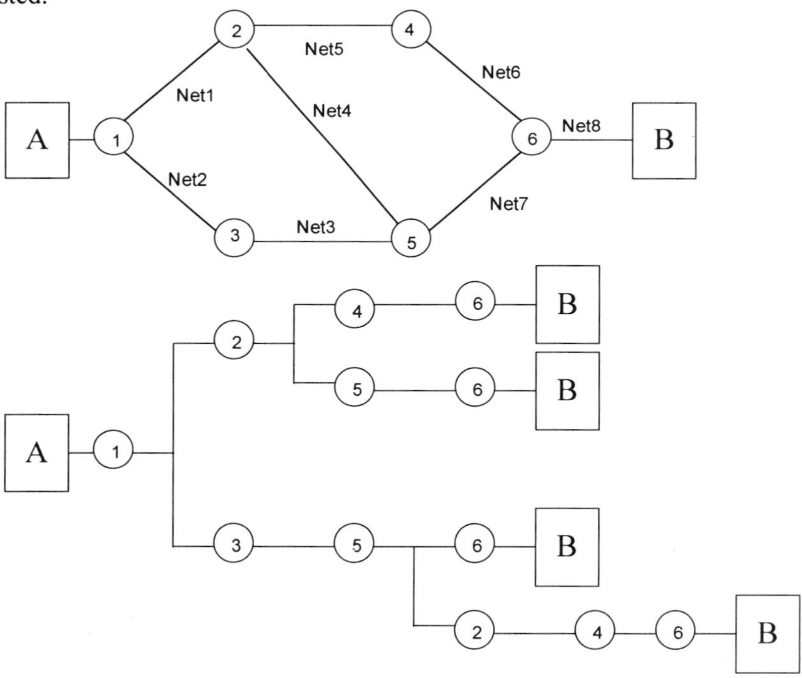

Figure 36.1 Example routing.

Each router communicates with its neighbors to build-up a routing table. For example, in Figure 36.1 the routing table for router 1 could be:

Destination	*Distance (hops)*	*Next router*	*Output port*
Net5	1	2	(Net1)
Net7	2	3	(Net2)
Net8	3	2	(Net1)
Net3	1	3	(Net2)
		And so on.	

It can be seen that the best route (measured by hops) from node A to Net8 to go via Net1. This is the only information that the router needs to store. When the data gets to router 2 it has the choice of whether to send it to Net4 or Net5, as both routes get to Net8 in two hops.

A better method of determining the best route is to have some measure of the delay. For this routers pass delay information about their neighbors. For example, if the relative delay in Net5 was 1.5 and in Net6 it was 1.25, and the relative delay in Net4 was 1.1 and in Net7 was 1.6, then the relative delay between router2 and router6 can be calculated as:

Route(2,5,6) = 1.5 + 1.26 = 2.76 Route(2,4,6) = 1.1 + 1.3 = 2.4

Thus the best route is via router 4.

Another technique used to determine the best route is error probability. In this case the probability of an error is multiplied to give the total probability. The route with the lowest error probability will be the most reliable route. For example:

$P_e(2-5)=0.01$ $P_e(5-6)=0.15$
$P_e(2-4)=0.05$ $P_e(4-6)=0.1$

Thus,

$P_{noerror}(2,5,6)=(1-0.01)\times(1-0.15)=0.8415$
$P_{noerror}(2,5,6)=(1-0.05)\times(1-0.1)=0.855$

Thus, the route via router 5 is the most reliable.

36.2 RIP

Figure 36.2 recalls from Chapter 32 the RIP packet format. The fields are:

- Operation (2 bytes) – this field gives an indication that the RIP packet is either a request or a response. The first 8 bits of the field give the command/request name and the next 8 bits give the version number.
- Network number (4 bytes of IP addresses) – this field defines the assigned network address number to which the routing information applies (note that, although 4 bytes are

shown, there are in fact 14 bytes reserved for the address. In RIP version 1 (RIPv1), with IP traffic, 10 of the bytes were unused; RIPv2 uses the 14-byte address field for other purposes, such as subnet masks.

- Number of router hops (2 bytes) – this field indicates the number of routes that a packet must go through in order to reach the required destination. Each router adds a single hop, the minimum number is 1 and the maximum is 16.
- Number of ticks (2 bytes) – this field indicates the amount of time (in 1/18 second) it will take for a packet to reach a given destination. Note that a route which has the fewest hops may not necessarily be the fastest route.

RIP packets add to the general network traffic as each router broadcasts its entire routing table every 30–60 seconds.

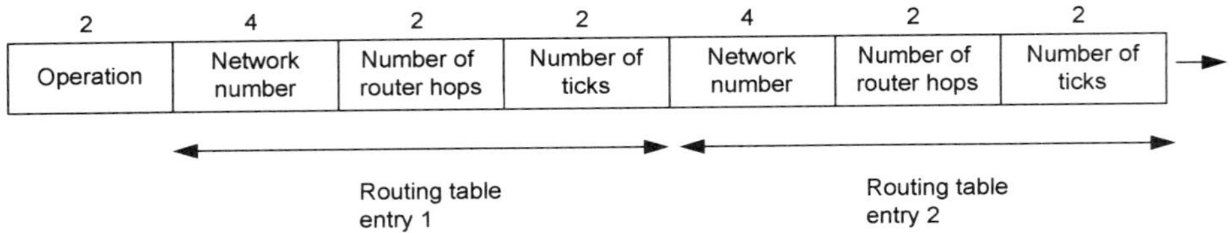

Figure 36.2 RIP packet format.

36.3 OSPF

The OSPF is an open, non-proprietary standard which was created by the IEFF (Internet Engineering Task Force), a task force of the IAB. It is a link state routing protocol and is thus able to maintain a complete and more current view of the total internetwork than distance vector routing protocols. Link state routing protocols have the features:

- They use link state packets (LSPs) which are special datagrams that determine the names of and the cost or distance to any neighboring routers and associated networks.
- Any information learned about the network is then passed to all known routers, and not just neighboring routers, using LSPs. Thus all routers have a fuller knowledge of the entire internetwork than the view of only the immediate neighbors (as with distance vector routing).

OSPF adds to these features with:

- Additional hierarchy. OSPF allows the global network to be split into areas. Thus a router in a domain does not necessarily have to know how to reach all the networks with a domain, it simply has to send to the right area.
- Authentication of routing messages using an 8-byte password. This length is not long enough to stop unauthorized users from causing damage. Its main purpose is to reduce the traffic from misconfigured routers. Typically a misconfigured router will inform the network that it can reach all nodes with no overhead.

- Load balancing. OSPF allows multiple routes to the same place to be assigned the same cost and will cause traffic to be distributed evenly over those routes.

Figure 36.3 shows the OSPF header. The fields in the header are:

- A version number (1 byte) which, in current implementations, has the version number of 2.
- The type field (1 byte) which can range from 1 to 5. Type 1 is the HELLO message and the others are to request, send and acknowledge the receipt of link state messages. HELLO messages are used by nodes to convince their neighbors that they are alive and reachable. If a router fails to receive these messages from one of its neighbors for some period of time, it assumes that the node is no longer directly reachable and updates its link state information accordingly.
- SourceAddr (4 bytes) identifies the sender of the message.
- Areald (4 bytes) is an identifier to the area in which the node is located.
- An authentication field can either be set to 0 (none) or 1. If it is set to 1 then the authentication contains an 8-byte password.

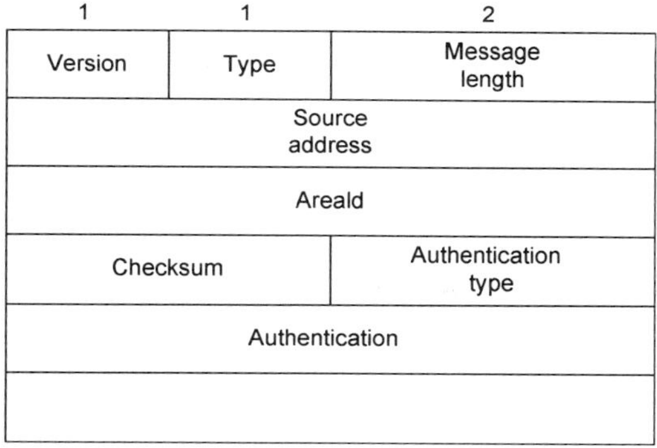

Figure 36.3 OSPF header format.

36.4 EGP/BGP

The two main interdomain routing protocols in recent history are EGP and BGP. EGP suffers from several limitations. Its principle limitation is that it treats the Internet as a tree like structure. A treelike structure, as illustrated in Figure 36.4, is normally made up of parents and children, with a single backbone. A more typical topology for the Internet is illustrated in Figure 36.5.

BGP is an improvement on EGP (the fourth version of BGP is known as BGP-4). Unfortunately it is more complex than EGP, but not as complex as OSPF.

BGP assumes that the Internet is made up of an arbitrarily interconnected set of nodes. It then assumes the Internet connects to a number of AANs (autonomously attached networks), as illustrated in Figure 36.6. These may create boundaries around an organization, an Internet service provider, and so on. It then assumes that, once they are in the AAN, the packets the packets will be properly routed.

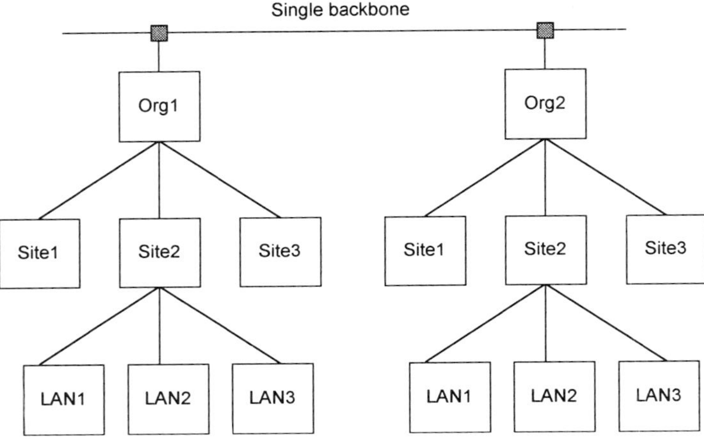

Figure 36.4 Tree-like topology.

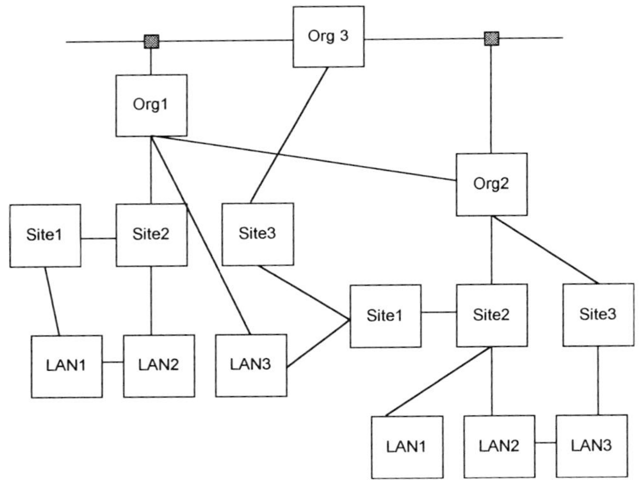

Figure 36.5 Network with multiple backbones.

BGP differs in that it tries to find any paths through the network. Thus the main goal is reachability instead of the number of hops to the destination. So finding a path which is nearly optimal is a good achievement. The AAN administrator selects at least one node to be a BGP speaker and also one or more border gateways. These gateways simply route the packet into and out of the AAN. The border gateways are the routers through which packets reach the AAN. Most routing algorithms try to find the quickest way through the network.

The speaker on the AAN broadcasts its reachability information to all the networks within its AAN. This information states only whether a destination AAN can be reached; it does not describe any other metrics.

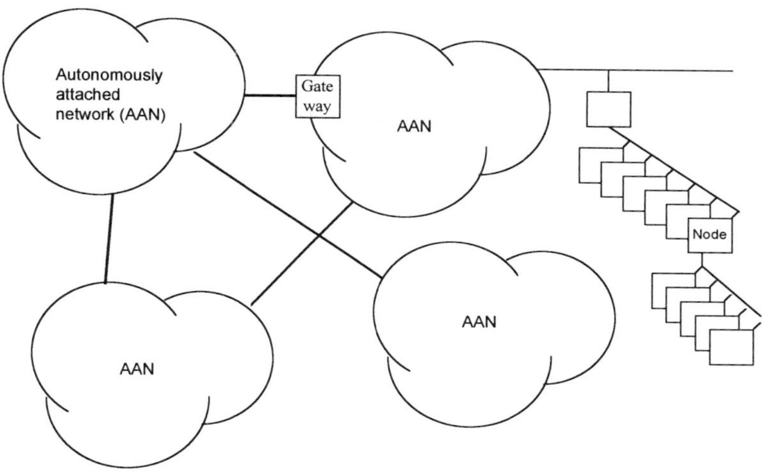

Figure 36.6 Autonomously attached networks.

The BGP update packet also contains information on routes which cannot be reached (withdrawn routes). The content of the BGP-4 update packet is:

- Unfeasible routes length (2 bytes).
- Withdrawn routes (variable length).
- Total path attribute length (2 bytes).
- Path attributes (variable length).
- Network layer reachability information (variable length).

The network layer reachability information can contain extra information, such as 'use AAN 1 in preference to AAN 2'.

An important point is that BGP is not a distance vector or link state protocol because it transmits complete routing information instead of partial information.

36.5 BGP specification

Border Gateway Protocol (BGP) is an inter-Autonomous System routing protocol, which builds on EGP. The main function of a BGP-based system to communicate network reachability information with other BGP systems. Initially two systems exchange messages to open and confirm the connection parameters, and then transmit the entire BGP routing table. After this incremental updates are sent as the routing tables change.

Each message has a fixed-size header and may or may not be proceeded by a data portion. The fields are:

- Marker. Contains a value that the receiver of the message can predict. It can be used to detect a loss of synchronization between a pair of BGP peers, and to authenticate incoming BGP messages. 16 bytes.

- Length. Indicates the total length, in bytes, of the message, including the header. It must always be greater than 18 and no greater than 4096. 2 bytes.
- Type. Indicates the type of message, such as 1 – OPEN, 2 – UPDATE, 3 – NOTIFICATION and 4 – KEEPALIVE.

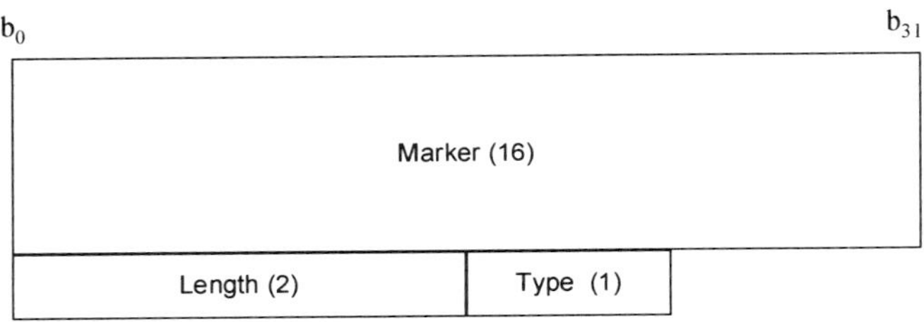

Figure 36.7 BGP message header.

36.5.1 OPEN message

The OPEN message is the first message sent after a connection has been made. A KEEPALIVE message is sent back confirming the OPEN message. After this the UPDATE, KEEPALIVE, and NOTIFICATION messages can be exchanged.

Figure 36.7 shows the extra information added to the fixed-size BGP header. It has the following fields:

- Version. Indicates the protocol version number of the message. Typical values are 2, 3 or 4. 1 byte.
- My Autonomous System. Identifies the senders Autonomous System number. 2 bytes.
- Hold Time. Indicates the maximum number of seconds that can elapse between the receipt of successive KEEPALIVE and/or UPDATE and/or NOTIFICATION messages. 2 bytes.
- Authentication Code. Indicates the authentication mechanism being used. This should define the form and meaning of the Authentication Data and the algorithm for computing values of Marker fields.
- Authentication Data. The form and meaning of this field is a variable-length field depends on the Authentication Code.

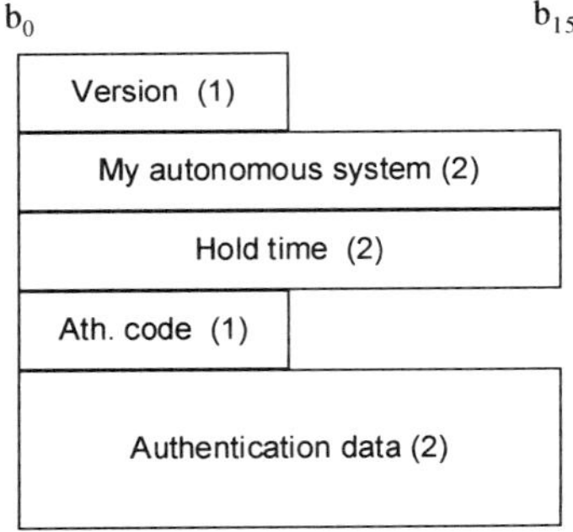

Figure 36.8 BGP OPEN message data.

36.5.2 UPDATE message format

The UPDATE message is used to transfer routing information between BGP peers. This information is used to construct a graph describing the relationships of the various Autonomous Systems. The extra information added to the fixed-size BGP header are as follows:

- Total Path Attribute Length. Indicates the length of the Path Attributes field in bytes. Its value must allow the number of Network fields to be determined as specified below. 2 bytes.
- Path Attributes. A variable length sequence of path attributes. Each path contains the attributes for attribute type, attribute length and attribute value. These are:

    - Attribute Type. Consists of the Attribute Flags byte (b_0-b_7) followed by the Attribute Type Code byte (b_8-b_{15}). 2 bytes. The format of the Attribute flags are:

    b_0. Optional bit. Defines whether the attribute is optional (if set to 1) or well-known (if set to 0).

    b_1 Transitive bit. Defines whether an optional attribute is transitive (if set to 1) or non-transitive (if set to 0). See Section 34.4.6 for a discussion of transitive attributes.

    b_2 Partial bit. Defines whether the information contained in the optional transitive attribute is partial (if set to 1) or complete (if set to 0). For well-known attributes and for optional non-transitive attributes the Partial bit must be set to 0.

    b_3 Extended Length bit. Defines whether the Attribute Length is one byte (if set to 0) or two bytes (if set to 1).

    b_4-b_7 Unused and set to zero.

    - Attribute Type Code contains the Attribute Type Code (See Section 34.4.6)
- Network. Each Internet network number indicates one network whose Inter-Autonomous System routing is described by the Path Attributes. 4-byte.

The total number of Network fields in the UPDATE message can be determined by the formula:

$$\text{Message\_Length} = 19 + \text{Total\_Path\_Attribute\_Length} + 4 \times \text{No\_of\_networks}$$

36.5.3 KEEPALIVE message format

KEEPALIVE message consists of only message header (and are thus only 19 bytes long) and are used to determine if peers are reachable. Unlike other routing protocols, such as RIP, BGP does not continually poll its peers to determine if they are still reachable, instead peers exchange KEEPALIVE messages (which must be less than the hold time of the OPEN message). A typical maximum time between KEEPALIVE messages is one-third of the Hold Time period.

36.5.4 NOTIFICATION message format

The NOTIFICATION message is sent when an error condition occurs and the BGP connection is immediately closed after sending it. The extra information added to the fixed-size BGP header are as follows:

- Error Code. Indicates the type of NOTIFICATION. 1 byte. Defined error codes are:

1	Message Header Error	2	OPEN Message Error
3	UPDATE Message Error	4	Hold Timer Expired
5	Finite State Machine Error	6	Cease

- Error subcode. Provides more specific information about the error. 1 byte. Each Error Code can have one or more Error subcodes associated with it. If no appropriate. Error subcodes are:

  - Message Header Error subcodes:
    1 – Connection Not Synchronized. 2 – Bad Message Length.
    3 – Bad Message Type.

  - OPEN Message Error subcodes:
    1 – Unsupported Version Number. 2 – Bad Peer AS.
    3 – Unsupported Authentication Code. 4 – Authentication Failure.

  - UPDATE Message Error subcodes:
    1 – Malformed Attribute List. 2 – Unrecognized Well–known Attribute.
    3 – Missing Well–known Attribute. 4 – Attribute Flags Error.
    5 – Attribute Length Error. 6 – Invalid ORIGIN Attribute
    7 – AS Routing Loop. 8 – Invalid NEXT_HOP Attribute.
    9 – Optional Attribute Error. 10 – Invalid Network Field.

- Data. Diagnostic data for the reason for the NOTIFICATION. Its contents depends on the Error Code and Error Subcode. Variable-length.

The message length can be determined by:

Message_Length = 21 + Data_Length

36.5.5 Path attributes

Path attributes in the UPDATE message fall into four separate categories:

- Well-known mandatory.
- Well-known discretionary.
- Optional transitive.
- Optional non-transitive.

Attributes which are well known must be recognized by all BGP implementations. If they are mandatory they must be included in every UPDATE message. Discretionary attributes may or may not be sent in an UPDATE message. Table 36.1 defines the well-known attributes.

Table 36.1 Well-known attributes.

Attribute Name	*Type code*	*Length*	*Attribute category*	*Description*
ORIGIN	1	1	Well-known, mandatory	Defines the origin of the path information. Values are: 0 IGP - network(s) are interior to the originating AS. 1 EGP - network(s) learned via EGP. 2 INCOMPLETE - network(s) learned by some other means.
AS_PATH	2	variable	Well-known, mandatory	AS_PATH attribute enumerates the ASs that must be traversed to reach the networks listed in the UPDATE message.
NEXT_HOP	3	4	Well-known, mandatory	Defines the IP address of the border router that should be used as the next hop to the networks listed in the UPDATE message.
UNREACHABLE	4	0	Well-known, discretionary	Used to notify a BGP peer that some of the previously advertised routes have become unreachable.
INTER-AS METRIC	5	2	Optional, non-transitive	May be used on external (inter-AS) links to discriminate between multiple exit or entry points to the same neighboring AS.

36.5.6 BGP state transitions and actions

BGP states are:

1 – Idle 2 – Connect
3 – Active 4 – OpenSent
5 – OpenConfirm 6 – Established

and the events are:

1 – BGP Start 2 – BGP Stop
3 – BGP Transport connection open 4 – BGP Transport connection closed
5 – BGP Transport connection open failed 6 – BGP Transport fatal error
7 – ConnectRetry timer expired 8 – Holdtime timer expired
9 – KeepAlive timer expired 10 – Receive OPEN message
11 – Receive KEEPALIVE message 12 – Receive UPDATE messages
13 – Receive NOTIFICATION message

The following defines the state transitions of the BGP FSM and the actions that occur.

EVENT	ACTIONS	MESSAGE_SENT	NEXT_STATE
Idle (1)			
1	Initialize resources Start ConnectRetry timer Initiate a transport connection	none	2
Connect(2)			
1	none	none	2
3	Complete initialization Clear ConnectRetry timer	OPEN	4
5	Restart ConnectRetry timer	none	3
7	Restart ConnectRetry timer Initiate a transport connection	none	2
Active (3)			
1	none	none	3
3	Complete initialization Clear ConnectRetry timer	OPEN	4
5	Close connection Restart ConnectRetry timer		3
7	Restart ConnectRetry timer Initiate a transport connection	none	2
OpenSent(4)			
1	none	none	4
4	Close transport connection Restart ConnectRetry timer	none	3
6	Release resources	none	1
10	Process OPEN is OK	KEEPALIVE	5
	Process OPEN failed	NOTIFICATION	1
OpenConfirm (5)			
1	none	none	5
4	Release resources	none	1
6	Release resources	none	1
9	Restart KeepAlive timer	KEEPALIVE	5
11	Complete initialization Restart Holdtime timer	none	6
13	Close transport connection Release resources		1
Established (6)			
1	none	none	6
4	Release resources	none	1
6	Release resources	none	1
9	Restart KeepAlive timer	KEEPALIVE	6
11	Restart Holdtime timer	KEEPALIVE	6
12	Process UPDATE is OK	UPDATE	6
	Process UPDATE failed	NOTIFICATION	1
13	Close transport connection Release resources		1

37 SNMP, Wins and DHCP

37.1 Introduction

SNMP (Simple Network Management Protocol) is a well-supported standard which can be used to monitor and control devices. It typically runs of hubs, switches and bridges. Many SNMP devices provides both general network management and device management through a serial cable, modem, or over the network from a remote computer. It involves a primary management station communicating with different management processes. Figure 37.1 shows an outline of an SNMP-based system. A SNMP agent runs SNMP management software. An SNMP server sends commands to the agent which responses back with the results. In this figure the server asks the agent for its routing information and the agent responds with its routing table. These responses can either be polled (the server sends a request for information) or interrupt-driven (where the agent sends its information at given events). A polled system tends to increase network traffic as the agent may not have any updated information (and the server must re-poll for the information).

The SNMP (Simple Network Management Protocol) protocol is initially based in the RFC1157 document. It defines a simple protocol which gives network element management information base (MIB). There are two types of MIB: MIB-1 and MIB-2. MIB-1 was defined in 1988 and has 114 table entries, divided into two groups. MIB-2 is a 1990 enhancement which has 171 entries organized into 10 groups. Most devices are MIB-1 compliant and newer one with both MIB-1 and MIB-2.

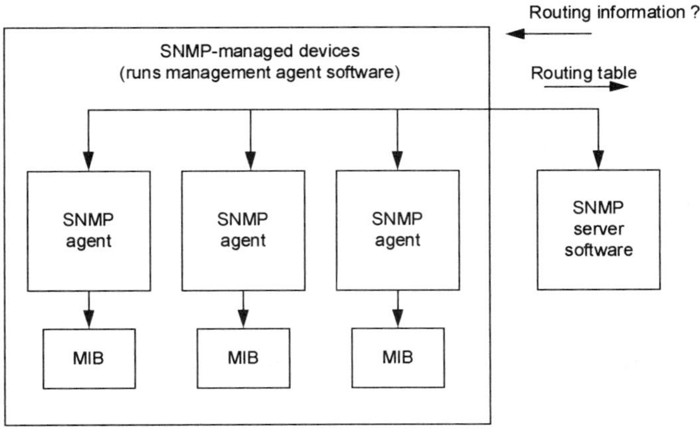

Figure 37.1 SNMP architecture.

The database contains entries with four fields:

- Object type. Defines the name of the entry.
- Syntax. Gives the actual value (as string or an integer).
- Access field. Defines whether the value is read-only, read/write, write-only and not accessible.
- Status field. Contains an indication on whether the entry in the MIB is mandatory (the managed device must implement the entry), optional (the managed device may implement the entry) or obsolete (the entry is not used).

SNMP is a very simple protocol but suffers from the fact that it is based on connectionless, unreliable, UDP. The IAB have recommended that the Common Management Information Services (CMIS) and Common Management Information Protocol (CMIP) be accepted as standard for future TCP/IP systems. The two main version of SNMP are SNMP Ver1 and SNMP Ver2. SNMP has added security to stop intruders determining network loading or the state of the network.

37.2 SNMP

The SNMP architecture is based on a collection of:

- Network management stations. These execute management applications which monitor and control network elements.
- Network elements. These are devices such as hosts, gateways, terminal servers, and so on and have management agents which perform network management functions replying to requests from network management stations.

The SNMP management information represented by a subset of the ASN.1 language. SNMP models all management agent functions as alterations or inspections of variables. Thus, a protocol entity on a logically remote host interacts with the management agent resident on the network element in order to retrieve (get) or alter (set) variables.

SNMP determines the current state of the network by polling for appropriate information from monitoring center(s).

37.2.1 Protocol specification

The network management protocol operates by inspecting or altering variables on an agent's MIB (management information base). They communicate by exchanging messages within UDP datagrams. These messages are defined using ASN.1 and are specified in listing 37.1. They consist of:

- A Version identifier (`version`). An integer value defining the version number.
- SNMP community name (`community`). An eight character string defining the community name.
- A protocol data unit (`data`). All SNMP implementations five PDUs: `GetRequest-PDU`, `GetNextRequest-PDU`, `GetResponse-PDU`, `SetRequest-PDU`, and `Trap-PDU`.

The protocol receives messages from:

- UDP port 161. For all messages apart from report traps (Trap-PDU).
- UDP port 162. Report trap Messages

Listing 37.1

```
RFC1157-SNMP DEFINITIONS ::= BEGIN
IMPORTS
  ObjectName, ObjectSyntax, NetworkAddress, IpAddress, TimeTicks
      FROM RFC1155-SMI;

-- top-level message
    Message ::=
        SEQUENCE {
          version     -- version-1 for this RFC
            INTEGER {
              version-1(0)
            },
          community   -- community name
            OCTET STRING,
          data       -- e.g., PDUs if trivial
            ANY    -- authentication is being used
        }

-- protocol data units
    PDUs ::=
        CHOICE {
          get-request           GetRequest-PDU,
          get-next-request      GetNextRequest-PDU,
          get-response          GetResponse-PDU,
          set-request           SetRequest-PDU,
          trap                  Trap-PDU
          }
-- the individual PDUs and commonly used
-- data types will be defined later
END
```

Common constructs

Listing 37.2 defines the ANS.1 constructs used in the PDUs. `RequestID` distinguishes between outstanding requests. `ErrorStatus` is an integer (0-5) which defines the error level, a zero indicates no error, a 1 indicates `tooBig`, and so on. When an error occurs, the `ErrorIndex` parameter is typically used to indicate which variable in a list caused the error.

The `VarBind` parameter (variable binding) refers to the link between the name of a variable and the variable's value, and `VarBindList` is a list of variable names and their corresponding values.

Listing 37.2

```
-- request/response information
  RequestID ::=
      INTEGER
  ErrorStatus ::=
      INTEGER {
        noError(0), tooBig(1), noSuchName(2),
        badValue(3), readOnly(4), genErr(5)
      }
  ErrorIndex ::=
      INTEGER
```

```
-- variable bindings
VarBind ::=
    SEQUENCE {
      name   ObjectName,
      value  ObjectSyntax
    }
VarBindList ::=
    SEQUENCE OF
      VarBind
```

GetRequest-PDU

GetRequest-PDU gets objects the MIB of the receiptiant. These objects are defined in the variable binding field. On no errors, the receiptiant sends back `GetResponse-PDU`, with:

- Each object named in the variable-bindings field of the received message.
- Corresponding component of the `GetResponse-PDU` representing by the name.
- Value of each of the variables.

If all the objects are found, the value of the error-status field and the error-index field of the `GetResponse-PDU` must be `noError` and zero, respectively. The request-id field will be the same as the received message.

If any of the objects cannot be found in the DIB then the receiptiant sends to the originator a `GetResponse-PDU` with an error-status field value of is `noSuchName`. The value of error-index is set to the value of the error-index field of the object name component in the received message.

Listing 37.3 defines the general form of the `GetRequest-PDU`.

Listing 37.3

```
GetRequest-PDU ::= [0]
     IMPLICIT SEQUENCE {
       request-id          RequestID,
       error-status        ErrorStatus,
       error-index         ErrorIndex,
       variable-bindings   VarBindList
     }
```

GetNextRequest-PDU

GetNextRequest-PDU is sent when the protocol entity makes a request of its SNMP application entity. If no error occurs then the receiving protocol entity sends the GetResponse-PDU containing noError in the error-status field and a zero value in the errorindex field. The value of the request-id field of the GetResponse-PDU is that of the received message.

Listing 37.4 defines the format in ASN.1 language. The only difference between this PDU and the GetRequest-PDU is defined with type value of 0 (rather than 1).

Listing 37.4

```
GetNextRequest-PDU ::=
    IMPLICIT SEQUENCE {
        request-id          RequestID,
          error-status       ErrorStatus,
          error-index        ErrorIndex,
          variable-bindings VarBindList
    }
```

GetResponse-PDU

GetResponse-PDU occurs after a protocol entity receives a GetRequest-PDU, GetNextRequest-PDU, or a SetRequest-PDU. After it is received the receiving protocol entity sends its contents to its SNMP application entity. Listing 37.5 defines the format in ASN.1 language.

Listing 37.5

```
GetResponse-PDU ::=
    IMPLICIT SEQUENCE {
         request-id             RequestID,
         error-status           ErrorStatus,
         error-index            ErrorIndex,
         variable-bindings      VarBindList
    }
```

The SetRequest-PDU

The protocol entity sends a SetRequest-PDU at the request of its SNMP application entity. When received the receiving entity sends the corresponding value is assigned to the variable for each object named in the variable-bindings field of the received message. Listing 37.6 defines the format in ASN.1 language.

Listing 37.6

```
SetRequest-PDU ::=
   IMPLICIT SEQUENCE {
      request-id             RequestID,
      error-status           ErrorStatus,
      error-index            ErrorIndex,
      variable-bindings      VarBindList
    }
```

Trap-PDU

Trap-PDU occurs when the protocol entity receives a request from a SNMP application entity. When received, the receiving protocol entity sends its contents to its SNMP application entity. The variable-bindings component of the Trap-PDU is implementation-specific, these are:

- coldStart (0). Sending protocol entity is reinitializing itself, thus the agent's configuration may be altered.
- warmStart (1). Sending protocol entity is reinitializing itself, so that neither the agent configuration nor the protocol entity implementation will be altered.
- linkDown (2). Sending protocol entity recognizes a failure in one of the communication links represented in the agent's configuration.
- linkUp (3). Sending protocol entity recognizes that one of the communication links represented in the agent's configuration has come up.
- authenticationFailure (4). Sending protocol entity is the addressee of a protocol message that is not properly authenticated.
- egpNeighborLoss (5). EGP neighbor for whom the sending protocol entity was an EGP peer no longer exists.
- enterpriseSpecific (6). Sending protocol entity recognizes an enterprise-specific event.

The form of the Trap-PDU is:

```
 Trap-PDU ::=     [4]
IMPLICIT SEQUENCE {
  enterprise     -- type of object generating type

    OBJECT IDENTIFIER,
  agent-addr           -- address of object generating
    NetworkAddress,    -- trap
  generic-trap         -- generic trap type
    INTEGER {
      coldStart(0), warmStart(1),
      linkDown(2),  linkUp(3),
      authenticationFailure(4),  egpNeighborLoss(5),
      enterpriseSpecific(6)
    },
  specific-trap  -- specific code, present even
    INTEGER,     -- if generic-trap is not
                 -- enterpriseSpecific
   time-stamp    -- time elapsed between the last
   TimeTicks,    -- (re)initialization of the network
          -- entity and the generation of the
           trap
  variable-bindings  -- "interesting" information
    VarBindList
}
```

37.3 RMON (Remote Monitoring)

In addition to SNMP, many systems support two additional types of network management: EASE (Embedded Advanced Sampling Environment) and RMON. RMOS is SNMP MIB that specifies the types of information listed in a number of special MIB groups that are commonly used for traffic management. Typical groups are Statistics, History, Alarms, Hosts, Hosts Top N, Matrix, Filters, Events, and Packet Capture. The information can be gathered by dedicated hardware devices or by software built in to data communications equipment, such as routers, bridges, or switches (often known as an embedded agent).

37.4 EASE (Embedded Advanced Sampling Environment)

EASE is a Hewlett-Packard developed system which samples LAN data to build traffic matrices and monitor users of the network. It can be used standard network types (such as 10BASE-2 and Token Ring) and to high speed networks (such as 100VG-AnyLAN, 100Base-T and FDDI)

37.5 DHCP

Dynamic Host Configuration Protocol (DHCP) allows for the transmission of configuration information over a TCP/IP network. Microsoft implemented Dynamic Host Configuration Protocol (DHCP) on their Windows NT operating system and many other vendors are incorporating it into the systems. It is based on the Bootstrap Protocol (BOOTP) and adds additional services, such as:

- Automatic allocation of reusable IP network addresses.
- Additional TCP/IP configuration options.

It has two components:

- A protocol for delivering host-specific configuration parameters from a DHCP server to a host
- A mechanism for allocation of network addresses to hosts.

DHCP has been fully defined in the following RFCs:

RFC1533. DCHP options and Bootp vendor extensions.
RFC1534. Interoperation between DHCP and BOOTP.
RFC1541. DHCP.
RFC1542. Clarifications and Extensions for Bootstrap Protocol.
RFC2131. DHCP.
RFC2240. DHCP for Novell.

DHCP uses a client-server architecture, where the designated DHCP server hosts (servers) allocate network addresses and deliver configuration parameters to dynamically configured hosts (clients).

The three techniques that DHCP uses to assign IP addresses are:

- Automatic allocation. DHCP assigns a permanent IP address to a client.
- Dynamic allocation. DHCP assigns an IP address to a client for a limited period of time or when the client releases the address. It allows for automatic reuse of IP addresses that are no longer used by clients. It is typically used when there is a limited pool of IP addresses (which is less than the number of hosts) so that a host can only connect when it can get one of the IP addresses from the pool.
- Manual allocation. DHCP is used to convey an IP address which has been assigned by the network administrator. This allows DHCP to be used to eliminate assigning an IP address to a host through its operating system.

Networks can use several of these techniques.

DHCP messages are based on BOOTP messages, this allows DHCP to listen to a BOOTP relay agent and to allow integration of BOOTP clients and DHCP servers. A BOOTP relay agent or relay agent is an Internet host or router that passes DHCP messages between DHCP clients and DHCP servers. DHCP uses the same relay agent behavior as the BOOTP protocol specification. BOOTP relay agents are useful because it eliminates the need of having a DHCP server on each physical network segment.

Some of the objectives of DHCP are:

- DHCP should be a mechanism rather than a policy. DHCP must allow local system administrators control over configuration parameters where desired.
- No requirements for manual configuration of clients.
- DHCP does not require a server on each subnets and should communicate with routers and BOOTP relay agents and clients.
- Ensure that the same IP address cannot be used use by more than one DHCP client at a time.
- Restore DHCP client configuration when the client is rebooted.
- Provide automatic configuration for new clients.
- Support fixed or permanent allocation of configuration parameters.

37.5.1 Host configuration parameters

The host configuration parameters are:

Be a router	on/off
Non-local source routing	on/off
Policy filters for non-local source routing	(list)
Maximum reassembly size	integer
Default TTL	integer
PMTU aging timeout	integer
MTU plateau table	(list)

IP-layer parameters are:

IP address	(address)
Subnet mask	(address mask)
MTU	integer
All-subnets-MTU	on/off
Broadcast address flavor	00000000h/FFFFFFFFh
Perform mask discovery	on/off
Be a mask supplier	on/off
Perform router discovery	on/off
Router solicitation address	(address)
Default routers, list of:	
router address	(address)
preference level	integer
Static routes, list of:	
destination	(host/subnet/net)
destination mask	(address mask)
type-of-service	integer
first-hop router	(address)
ignore redirects	on/off
PMTU	integer
perform PMTU discovery	on/off

Link-layer parameters, per interface:

Trailers	on/off
ARP cache timeout	integer
Ethernet encapsulation	(RFC 894/RFC 1042)

TCP parameters, per host:

TTL	integer
Keep-alive interval	integer
Keep-alive data size	0/1

Where MTU is Path MTU Discovery and RD is Router Discovery.

37.5.2 Protocol outline

Figure 37.2 defines the format of a DHCP message. The numbers in parentheses indicate the size of each field in octets.

- Op. Defines message code/ op code (1 for BOOTREQUEST and 2 for BOOTREPLY). 1 byte.
- Htype. Defines hardware address type, such as, 1 for 10 Mbps Ethernet. 1 byte.
- Hlen. Hardware address length such as 6 for Ethernet. 1 byte.
- Hops. Client sets to 0, optionally used by relay agents when booting through a relay agent. 1 byte.
- Xid. Transaction ID which is a random number chosen by the client and used by the client and the server to associate messages and responses. 4 bytes.
- Secs. Sets by client for the number of seconds elapsed since client began address acquisition. 2 bytes.
- Flags. Flags, the format is BRRR…R where R bits are reserved for future use (and must always be zero) and B defines the broadcast bit which overcomes the problem where some clients that cannot accept IP unicast datagrams before the TCP/IP software is configured. 2 bytes.
- Ciaddr. Defines the client's IP address. It is only addressed when the client is in the BOUND, RENEW or REBINDING state and can respond to ARP requests. 4 bytes.
- Yiaddr. Clients IP address. 4 bytes.
- Siaddr. IP address of next server to use in bootstrap; returned in DHCPOFFER, DHCPACK by server. 4 bytes.
- Giaddr. Relay agent IP address and is used in booting through a relay agent. 4 bytes.
- Chaddr. Clients MAC address. 16 bytes.
- Sname. Optional null terminated server host name string. 64 bytes.
- File. Boot null-terminated file name string. 128 bytes.
- Options. Optional parameters field. Variable number of bytes.

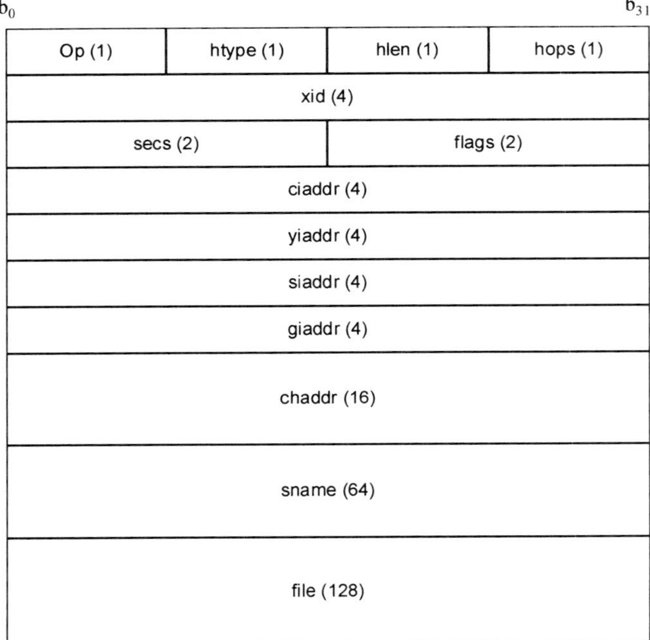

Figure 37.2 DHCP message format.

37.5.3 Allocating a network address

The protocol between clients and server is defined by DHCP messages, these are:

- DHCPDISCOVER. The client broadcasts this to locate available servers. When a client initially is started it binds with an address of 0.0.0.0. It then sends out the DHCPDISCOVER message in a UDP packet to port 67 (which is the DHCP/BOOTP server port).
- DHCPOFFER. After a DHCPDISCOVER message, a server may response to a client with a DHCPOFFER message that includes an available network address in the 'yiaddr' field. When allocating a new address, servers normally check that the offered network address is not already in use. Sending out an ICMP Echo Request to the new address can do this. If the client receives one or more DHCPOFFER messages from one or more servers then the client may choose to wait for multiple responses. It chooses the server based on the configuration parameters offered in the DHCPOFFER messages. The server sends out the DHCPOFFER message in a UDP packet to port 68 (which is the DHCP/BOOTP client port). This is send as a broadcast as the client does not currently have an IP address.
- DHCPREQUEST. Sent by clients to servers when either requesting offered parameters from one server and implicitly declining offers from all others, confirming parameter allocation (such as after a system boot), or extending the time on a particular network address. The server selected in the DHCPREQUEST message responds with a DHCPACK message containing the configuration parameters for the requesting client. The 'client

identifier' ('chaddr') and assigned network address define a unique identifier for the client's and are used by both the client and server to identify the lease. Servers not selected by the DHCPREQUEST message use the message as notification that the client has declined that server's offer.

- DHCPACK. Sent by the server to a client with configuration parameters, including committed network address. The client receives the DHCPACK message with configuration parameters, after which the client is setup. If the client detects that the address is already in use then the client sends back a DHCPDECLINE message to the server and restarts the configuration process.
- DHCPNAK. Sent by the server to the client indicating that the clients network address is wrong (for example, client has moved and does not have the correct subnet) or the time allocation for its network address has expired. On receiving a DHCPNAK message, the client restarts the configuration process.
- DHCPDECLINE. Sent by a client to the server indicating network address is already in use.
- DHCPRELEASE. Sent by a client to the server relinquishing network address. It identifies the lease to be released with its 'client identifier (chaddr) and network address in the DHCPRELEASE message.
- DHCPINFORM. Sent by a client to the server, asking only for local configuration parameters.

37.5.4 Time allocations

Client acquires the lease of a network time for specified time (either finite or infinite time). The units of time are unsigned integer value in seconds, although the value FFFFFFFFh is used to represent infinity. This gives a range of between 0 and approximately 100 years.

When a client cannot contact the local DHCP server and has knowledge of a previous network address then it may continue to use the previous assigned network address until the lease expires. If the lease expires before the client can contact a DHCP server then the client immediately stops using the previous network address and informs local users of the problem.

37.6 WINS

The Windows Internet Naming Service (WINS) is an excellent companion to DHCP. WINS provides a name registration and resolution on TCP/IP. It extends the function of DNS which will only map static IP addresses to TCP/IP host names. WINS is designed to resolved NetBIOS names on TCP/IP to dynamic network addresses assigned by DHCP. As it resolves NetBIOS names it is obviously aimed at Microsoft Windows-based (and DOS) networks.

37.6.1 Name registration

The WINS server stores a WINS database which maps IP addresses to NetBIOS names. The operation is as follows:

1. Startup. WINS client send a Name Registration Request in a UDP packet to the WINS server to registers its NetBIOS name and IP address. When the WINS server receives it then it checks its stored database to make sure that the requested name is not already in

use on the network.

2. Unsuccessful registration. When a client tries to register a name that is currently in use then the server sends the client a Denial message. The user of the client will then be informed that the computer's name is already in use on that network.
3. Successful registration. On successful name registration, the server sends a Name Registration Acknowledgement to the client. It fills the Time-to-Live (TTL) field to define the amount of time that the name registration will be active, after this time the server will cancel it
4. Initially re-registering name. The WINS client must send a Name Refresh Request a given time interval so that its name will not expire. The first request is made after one-eight of the TTL time, and then if unsuccessful after periods of one-eight of the TTL time. It has not been able to contact the primary WINS server after half the time defined in the TTL field, then the client tries to contact the secondary WINS server (if there is one).
5. Re-registering a name. After the first registration, the following registration will be made a 50 percent of the TTL time (instead of one-eight of the time).
6. Client shutdown. When a WINS client shuts down, it sends a Name Release Request to the WINS server, releasing its name from the WINS database.

37.6.2 Name resolution

A client that requires the resolution of a NetBIOS name to an IP will go through the following:

1. Check the name is actually the local computer then looks in its own name resolution cache for a match.
2. Sends a required for a directed name lookup from the WINS server. If it finds one the WINS server sends its IP address to the client.
3. If the WINS server does not find a match then the client broadcasts to the network for help.
4. If there is no response from a broadcast then the client looks into its local LMHOSTS file, else it will look in its local HOSTS file.

37.6.3 WINS proxy agents

Many older Windows systems do not support WINS clients. To allow them to communicate with a WINS server a WINS proxy agent can be used. This agent listens to the network for clients broadcasting for NetBIOS names resolution. The WINS proxy agent then redirects then to the WINS server, which will then pass the IP address resolution to the proxy agent, which in turn will pass it onto the client.

38 Ethernet

38.1 Introduction

Ethernet is the most widely used networking technology used in LANs (local area networks). On its own, Ethernet cannot make a network; it needs some other protocol such as TCP/IP to allow nodes to communicate. Unfortunately, Ethernet in its standard form does not cope well with heavy traffic, but has many advantages, including:

- Its networks are easy to plan and cheap to install.
- Its network components are cheap and well supported.
- It is well-proven technology which is fairly robust and reliable.
- It is simple to add and delete computers on the network.
- It is supported by most software and hardware systems.

A major problem with Ethernet is that because computers compete for access to the network there is no guarantee that a particular computer will get access within a given time. And contention causes problems when two computers try to communicate at the same time; they must both back off and no data can be transmitted. In its standard form Ethernet allows a bit rate of 10 Mbps. New standards for fast Ethernet systems minimize the problems of contention and also increase the bit rate to 100 Mbps. Ethernet uses coaxial or twisted-pair cable.

DEC, Intel and the Xerox Corporation initially developed Ethernet, and the IEEE 802 committee has since defined standards for it. The most common standards for Ethernet are Ethernet 2.0 and IEEE 802.3. It uses a shared-media, bus-type network topology where all nodes share a common bus. It is a contention-type network where only one node communicates at a time. Data is transmitted in frames which contain the MAC (media access control) source and destination addresses of the sending and receiving node, respectively. The local shared-media is known as a segment. Each node on the network monitors the segment and copies any frames addressed to itself.

Ethernet uses carrier sense multiple access with collision detection (CSMA/CD). On a CSMA/CD network, nodes monitor the bus (or Ether) to determine if it is busy. A node wishing to send data waits for an idle condition then transmits its message. Unfortunately collision can occur when two nodes transmit at the same time, thus nodes must monitor the cable when they transmit. When this happens both nodes stop transmitting frames and transmit a jamming signal. This informs all nodes on the network that a collision has occurred. Each of the nodes then waits a random period before attempting a retransmission. As each node has a random delay time, there can be a prioritization of the nodes on the network. Nodes thus contend for the network and are not guaranteed access to it. Collisions generally slow down the network. Each node on the network must be able to detect collisions and must be capable of transmitting and receiving simultaneously.

38.2 IEEE standards

The IEEE is the main standards organization for LANs; it calls the standard for Ethernet CSMA/CD (carrier sense multiple access/collision detect). Figure 38.1 shows how the IEEE standards for Token Ring and CSMA/CD fit into the OSI model. The two layers of the IEEE standards correspond to the physical and data link layers of the OSI model. A Token Ring network uses IEEE 802.5 (ISO 8802.5) and a CSMA/CD network uses IEEE 802.3 (ISO 8802.3). On Ethernet networks, most hardware will comply with the IEEE 802.3 standard. The object of the MAC layer is to allow many nodes to share a single communication channel. It also adds start and end frame delimiters, error detection bits, access control information and source and destination addresses.

The IEEE 802.2 (ISO 8802.2) logical link control (LLC) layer conforms to the same specification for both types of network.

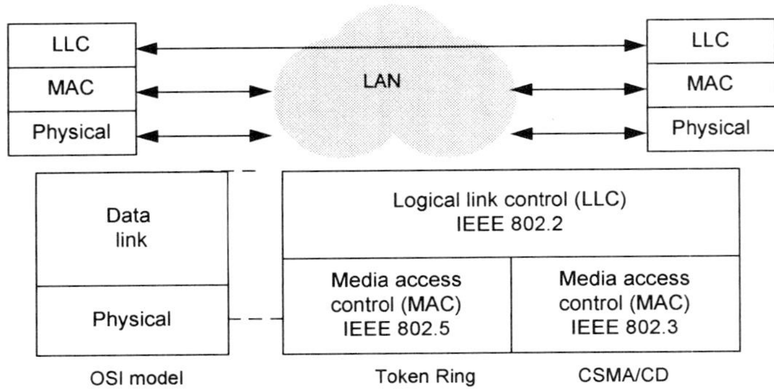

Figure 38.1 Standards for IEEE 802 LANs.

38.3 Ethernet – media access control (MAC) layer

When sending data the MAC layer takes the information from the LLC link layer. Figure 38.2 shows the IEEE 802.3 frame format. It contains 2 or 6 bytes for the source and destination addresses (16 or 48 bits each), 4 bytes for the CRC (32 bits), and 2 bytes for the LLC length (16 bits). The LLC part may be up to 1500 bytes long. The preamble and delay components define the start and end of the frame. The initial preamble and start delimiter are, in total, 8 bytes long and the delay component is a minimum of 96 bytes long.

A 7-byte preamble precedes the Ethernet 802.3 frame. Each byte has a fixed binary pattern of `10101010` and each node on the network uses it to synchronize their clocks and transmission timings. It also informs nodes that a frame is to be sent and for them to check the destination address in the frame.

The end of the frame is a 96-byte delay period which provides the minimum delay between two frames. This slot time delay allows for the worst-case network propagation delay.

The start delimiter field (SDF) is a single byte (or octet) of `10101011`. It follows the preamble and identifies that there is a valid frame being transmitted. Most Ethernet systems use a 48-bit MAC address for the sending and receiving nodes. Each Ethernet node has a unique MAC address, which is normally defined using hexadecimal digits, such as:

```
4C - 31 - 22 - 10  - F1 - 32
or      4C31 : 2210: F132
```

A 48-bit address field allows 2^{48} different addresses (or approximately 281 474 976 710 000 different addresses).

The LLC length field defines whether the frame contains information or whether it can be used to define the number of bytes in the logical link field. The logical link field can contain up to 1500 bytes of information and has a minimum of 46 bytes; its format is given in Figure 38.2. If the information is greater than the upper limit then multiple frames are sent. Also, if the field is less than the lower limit then it is padded with extra redundant bits.

The 32-bit frame check sequence (FCS) is an error detection scheme. It is used to determine transmission errors and is often called a cyclic redundancy check (CRC) or simply a checksum.

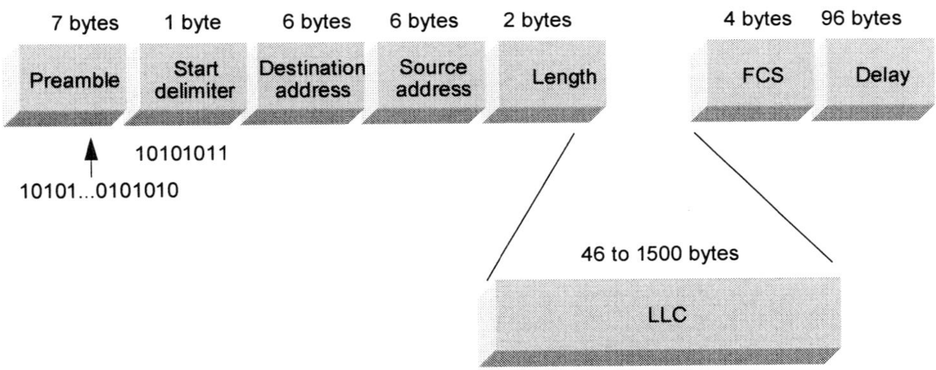

Figure 38.2 IEEE 802.3 frame format.

38.3.1 Ethernet II

The first standard for Ethernet was Ethernet I. Most currently available systems implement either Ethernet II or IEEE 802.3 (although most networks are now defined as being IEEE 802.3 compliant). An Ethernet II frame is similar to the IEEE 802.3 frame; it consists of 8 bytes of preamble, 6 bytes of destination address, 6 bytes of source address, 2 bytes of frame type, between 46 and 1500 bytes of data, and 4 bytes of the frame check sequence field.

When the protocol is IPX/SPX the type field contains the bit pattern 1000 0001 0011 0111, but when the protocol is TCP/IP the type field contains 0000 1000 0000 0000.

38.4 IEEE 802.2 and Ethernet SNAP

The LLC is embedded in the Ethernet frame and is defined by the IEEE 802.2 standard. Figure 38.3 illustrates how the LLC fields are inserted into the IEEE 802.3 frame. The DSAP and SSAP fields define the types of network protocol used. A SAP code of 1110 0000 identifies the network operating system layer as NetWare, whereas 0000 0110 identifies the

TCP/IP protocol. These SAP numbers are issued by the IEEE. The control field is, among other things, for the sequencing of frames.

In some cases it was difficult to modify networks to be IEEE 802-compliant. Thus an alternative method was to identify the network protocol, known as Ethernet SNAP (Subnetwork Access Protocol). This was defined to ease the transition to the IEEE 802.2 standard and is illustrated in Figure 38.4. It simply adds an extra two fields to the LLC field to define an organization ID and a network layer identifier. NetWare allows for either Ethernet SNAP or Ethernet 802.2 (as Novell used Ethernet SNAP to translate to Ethernet 802.2).

Non-compliant protocols are identified with the DSAP and SSAP code of 1010 1010, and a control code of 0000 0011. After these fields:

- Organization ID which indicates where the company that developed the embedded protocol belongs. If this field contains all zeros it indicates a non-company-specific generic Ethernet frame.
- EtherType field which defines the networking protocol. A TCP/IP protocol uses 0000 1000 0000 0000 for TCP/IP, while NetWare uses 1000 0001 0011 0111. NetWare frames adhering to this specification are known as NetWare 802.2 SNAP.

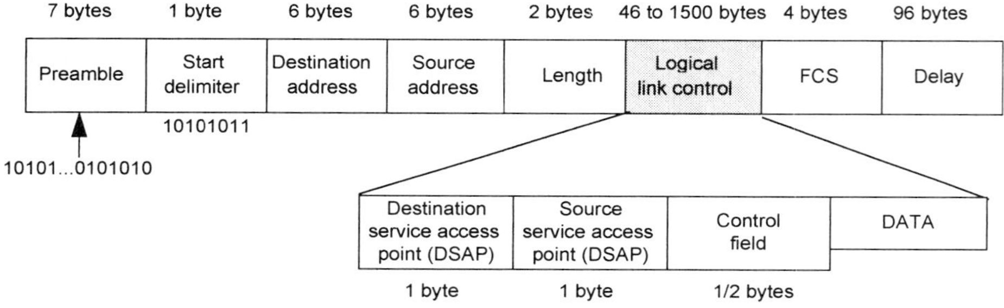

Figure 38.3 Ethernet IEEE 802.3 frame with LLC.

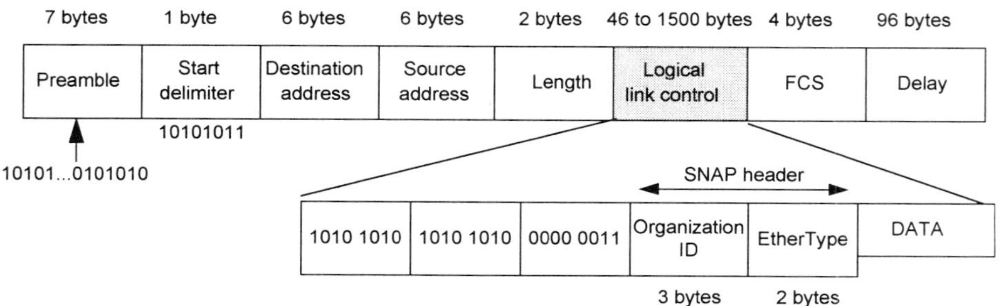

Figure 38.4 Ethernet IEEE 802.3 frame with LLC containing SNAP header.

38.4.1 LLC protocol

The 802.3 frame provides some of the data link layer functions, such as node addressing (source and destination MAC addresses), the addition of framing bits (the preamble) and error control (the FCS). The rest of the functions of the data link layer are performed with the control field of the LLC field; these functions are:

- Flow and error control. Each data frame sent has a frame number. A control frame is sent from the destination to a source node informing that it has or has not received the frames correctly.
- Sequencing of data. Large amounts of data are sliced and sent with frame numbers. The spliced data is then reassembled at the destination node.

Figure 38.5 shows the basic format of the LLC frame. There are three principal types of frame: information, supervisory and unnumbered. An information frame contains data, a supervisory frame is used for acknowledgment and flow control, and an unnumbered frame is used for control purposes. The first two bits of the control field determine which type of frame it is. If they are 0X (where X is a don't care) then it is an information frame, 10 specifies a supervisory frame and 11 specifies an unnumbered frame.

An information frame contains a send sequence number in the control field which ranges from 0 to 127. Each information frame has a consecutive number, N(S) (note that there is a roll-over from frame 127 to frame 0). The destination node acknowledges that it has received the frames by sending a supervisory frame. The function of the supervisory frame is specified by the 2-bit S-bit field. This can either be set to Receiver Ready (RR), Receiver Not Ready (RNR) or Reject (REJ). If an RNR function is set then the destination node acknowledges that all frames up to the number stored in the receive sequence number N(R) field were received correctly. An RNR function also acknowledges the frames up to the number N(R), but informs the source node that the destination node wishes to stop communicating. The REJ function specifies that frame N(R) has been rejected and all other frames up to N(R) are acknowledged.

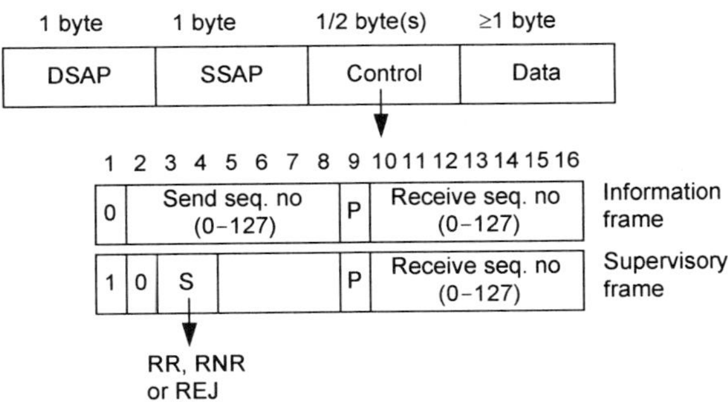

Figure 38.5 LLC frame format.

38.5 OSI and the IEEE 802.3 standard

Ethernet fits into the data link and the physical layer of the OSI model. These two layers only deal with the hardware of the network. The data link layer splits into two parts: the LLC and the MAC layer.

The IEEE 802.3 standard splits into three sublayers:

- MAC (media access control).
- Physical signaling (PLS).
- Physical media attachment (PMA).

The interface between PLS and PMA is called the attachment unit interface (AUI) and the interface between PMA and the transmission media is called the media dependent interface (MDI). This grouping into modules allows Ethernet to be very flexible and to support a number of bit rates, signaling methods and media types. Figure 38.6 illustrates how the layers interconnect.

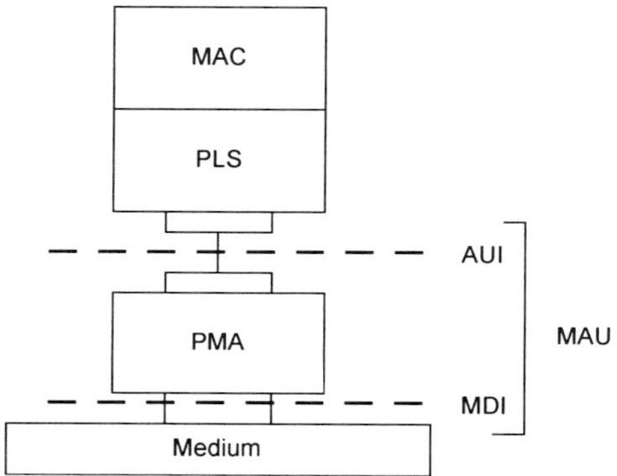

Figure 38.6 Organization of the IEEE 802.3 standard.

38.5.1 Media access control (MAC)

CSMA/CD is implemented in the MAC layer. The functions of the MAC layers are:

- When sending frames: receive frames from LLC; control whether the data fills the LLC data field, if not add redundant bits; make the number of bytes an integer, and calculate the FCS; add the preamble, SFD and address fields to the frame; send the frame to the PLS in a serial bit stream.
- When receiving frames: receive one frame at a time from the PLS in a serial bit stream; check whether the destination address is the same as the local node; ensures the frame contains an integer number of bytes and the FCS is correct; remove the preamble, SFD, address fields, FCS and remove redundant bits from the LLC data field; send the data to the LLC.
- Avoid collisions when transmitting frames and keep the right distance between frames by not sending when another node is sending; when the medium gets free, wait a specified period of time before starting to transmit.
- Handle any collision that appears by sending a jam signal; generate a random number and back off from sending during that random time.

38.5.2 Physical signaling (PLS) and physical medium attachment (PMA)

PLS defines transmission rates, types of encoding/decoding and signaling methods. In PMA, a further definition of the transmission media is accomplished, such as coaxial, fiber or twisted-pair. PMA and MDI together form the media attachment unit (MAU), often known as the transceiver.

38.6 Ethernet transceivers

Ethernet requires a minimal amount of hardware. The cables used to connect it are either unshielded twisted-pair cable (UTP) or coaxial cables. These cables must be terminated with their characteristic impedance, which is 50 Ω for coaxial cables and 100 Ω for UTP cables.

Each node has transmission and reception hardware to control access to the cable and also to monitor network traffic. The transmission/reception hardware is called a transceiver (short for *trans*mitter/re*ceiver*) and a controller builds up and strips down the frame. The transceiver builds transmit bits at a rate of 10 Mbps and thus the time for one bit is $1/10 \times 10^6$ which is 0.1 μs.

The Ethernet transceiver transmits onto a single Ether. When none of the nodes are transmitting then the voltage on the line is +0.7 V. This provides a carrier sense signal for all nodes on the network; it is also known as the heartbeat. If a node detects this voltage then it knows that the network is active and that no nodes are currently transmitting.

Thus when a node wishes to transmit a message it listens for a quiet period. Then if two or more transmitters transmit at the same time, a collision results. When they detect the signal, each node transmits a 'jam' signal. The nodes involved in the collision then wait for a random period of time (ranging from 10 to 90 ms) before attempting to transmit again. Each node on a network also awaits a retransmission. Thus, collisions are inefficient in a network as they stop nodes from transmitting. Transceivers normally detect a collision by monitoring the dc (or average) voltage on the line.

When transmitting, a transceiver unit transmits the preamble of consecutive 1s and 0s. The coding used is a Manchester code which represents a 0 as a high-to-low voltage transition and a 1 as a low-to-high voltage transition. A low voltage is –0.7 V and a high is +0.7 V. Thus when the preamble is transmitted the voltage will change between +0.7 and –0.7 V; this is illustrated in Figure 38.7. If after the transmission of the preamble no collisions are detected then the rest of the frame is sent.

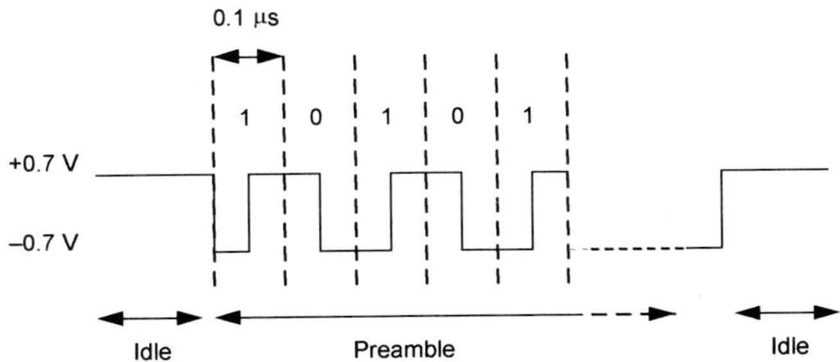

Figure 38.7 Ethernet digital signal.

38.7 NIC

When receiving data, the function of the NIC is to copy all data transmitted on the network, decode it and transfer it to the computer. An Ethernet NIC contains three parts:

- Physical medium interface. The physical medium interface corresponds to the PLS and PMA in the standard and is responsible for the electrical transmission and reception of data. It consists of two parts: the transceiver, which receivers and transmits data from or onto the transmission media; and a code converter that encodes/decodes the data. It also recognizes a collision of the media.
- Data link controller. The controller corresponds to the MAC layer.
- Computer interface.

It can be split into four main functional blocks:

- Network interface.
- Manchester decoder.
- Memory buffer.
- Computer interface.

38.7.1 Network interface

The network interface function is to listen, recreate the waveform transmitted on the cable into a digital signal and transfer the digital signal to the Manchester decoder. The network interface consists of three parts:

- BNC/RJ-45 connector.
- Reception hardware. The reception hardware translates the waveforms transmitted on the cable to digital signals then copies them to the Manchester decoder.
- Isolator. The isolator is connected directly between the reception hardware and the rest of the Manchester decoder; it guarantees that no noise from the network affects the computer, and vice-versa.

The reception hardware is called a receiver and is the main component in the network interface. Basically it has the function of an earphone, listening and copying the traffic on the cable. Unfortunately, the Ether and transceiver electronics are not perfect. The transmission line contains resistance and capacitance which distort the shape of the bit stream transmitted onto the Ether. Distortion in the system causes pulse spreading, which leads to intersymbol interference. There is also a possibility of noise affecting the digital pulse as it propagates through the cable. Therefore, the receiver also needs to recreate the digital signal and filter noise.

Figure 38.8 shows a block diagram of an Ethernet receiver. The received signal goes through a buffer with high input impedance and low capacitance to reduce the effects of loading on the coaxial cable. An equalizer passes high frequencies and attenuates low frequencies from the network, flattening the network passband. A 4-pole Bessel low-pass filter provides the average dc level from the received signal. The squelch circuit activates the line driver only when it detects a true signal. This prevents noise activating the receiver.

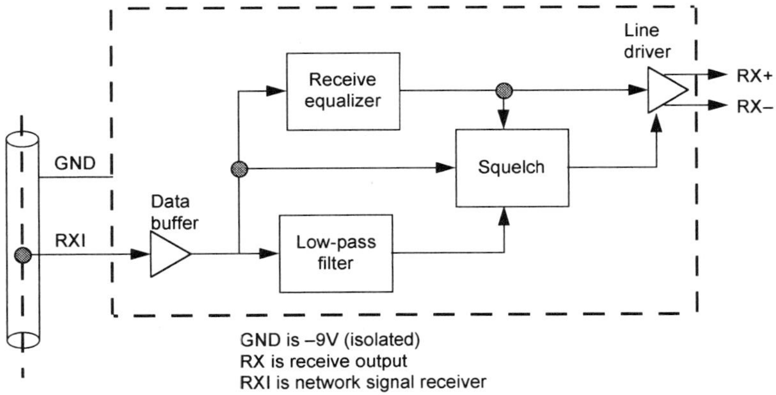

Figure 38.8 Ethernet receiver block diagram.

38.7.2 Manchester decoder

Manchester coding has the advantage of embedding timing (clock) information within the transmitted bits. A positively edged pulse (low → high) represents a 1 and a negatively edged pulse (high → low) a 0, as shown in Figure 38.9. Another advantage of this coding method is that the average voltage is always zero when used with equal positive and negative voltage levels.

Figure 38.10 is an example of transmitted bits using Manchester encoding. The receiver passes the received Manchester-encoded bits through a low-pass filter. This extracts the lowest frequency in the received bit stream, i.e., the clock frequency. With this clock the receiver can then determine the transmitted bit pattern.

For Manchester decoding, the Manchester-encoded signal is first synchronized to the receiver (called bit synchronization). A transition in the middle of each bit cell is used by a clock recovery circuit to produce a clock pulse in the center of the second half of the bit cell. In Ethernet the bit synchronization is achieved by deriving the clock from the preamble field of the frame using a clock and data recovery circuit. Many Ethernet decoders used the SEEQ 8020 Manchester code converter, which uses a phase-locked loop (PLL) to recover the clock. The PLL is designed to lock onto the preamble of the incoming signal within 12-bit cells. Figure 38.11 shows a circuit schematic of bit synchronization using Manchester decoding and a PLL.

The PLL is a feedback circuit which is commonly used for the synchronization of digital signals. It consists of a phase detector (such as an EX-OR gate) and a voltage-controlled oscillator (VCO) which uses a crystal oscillator as a clock source.

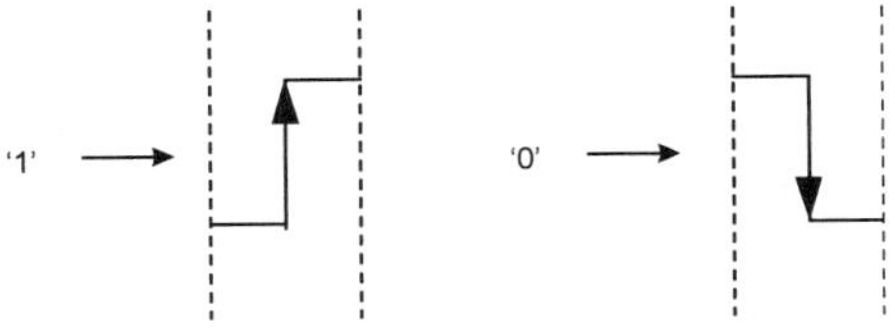

Figure 38.9 Manchester encoding.

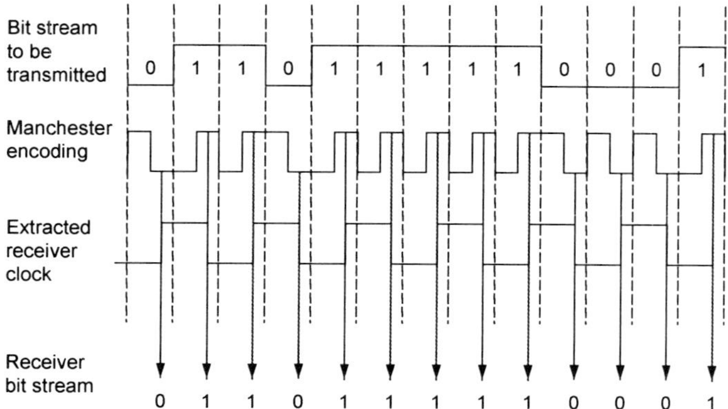

Figure 38.10 Example of Manchester coding.

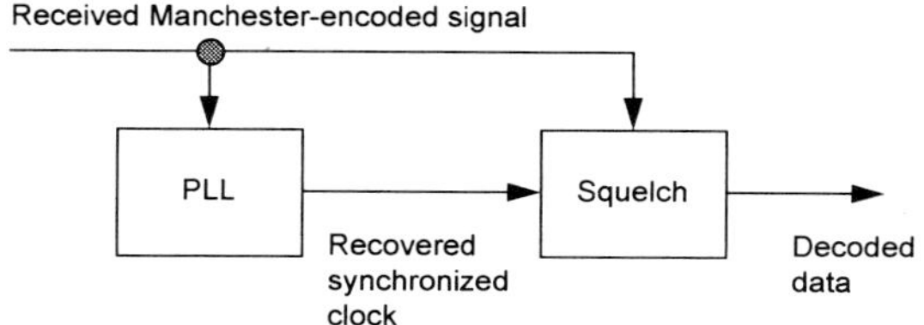

Figure 38.11 Manchester decoding with bit synchronization.

The frequency of the crystal is twice the frequency of the received signal. It is so constant that it only needs irregular and small adjustments to be synchronized to the received signal. The function of the phase detector is to find irregularities between the two signals and adjusts the VCO to minimize the error. This is accomplished by comparing the received signals and the output from the VCO. When the signals have the same frequency and phase the PLL is locked. Figure 38.12 shows the PLL components and the function of the EX-OR.

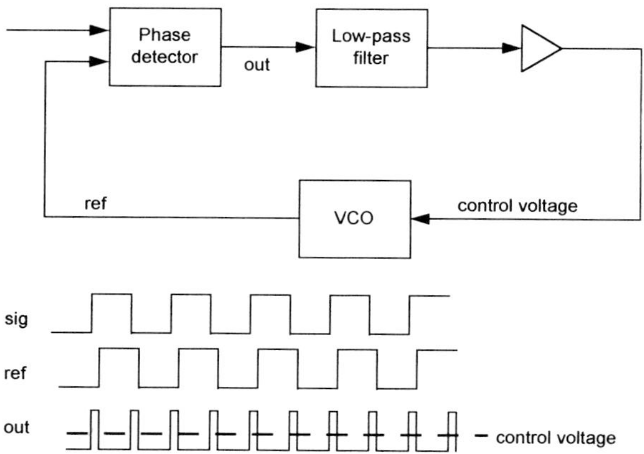

Figure 38.12 PLL and example waveform for the phase detector.

38.7.3 Memory buffer

The rate at which data is transmitted on the cable differs from the data rate used by the receiving computer, and the data appears in bursts. To compensate for the difference between the data rate, a first-in first-out (FIFO) memory buffer is used to produce a constant data rate. An important condition is that the average data input rate should not exceed the frequency of the output clock; if this is not the case the buffer will be filled up regardless of its size.

A FIFO is a RAM that uses a queuing technique where the output data appears in the same order that it went in. The input and output are controlled by separate clocks, and the FIFO keeps track of the data that has been written and the data that has been read and can thus be overwritten. This is achieved with a pointer. Figure 38.13 shows a block diagram of the FIFO configuration. The FIFO status is indicated by flags, the empty flag (EF) and the full flag (FF), which show whether the FIFO is either empty or full.

38.7.4 Ethernet implementation

The completed circuit for the Ethernet receiver is given in on the WWW page given in the Preface and is outlined in Figure 38.14. It uses the SEEQ Technologies 82C93A Ethernet transceiver as the receiver and the SEEQ 8020 Manchester code converter which decodes the Manchester code. A transformer and dc-to-dc converter isolates the SEEQ 82C92A and the network cable from the rest of the circuit (and the computer). The isolated dc-to-dc converter converts a 5 V supply to the −9 V needed by the transceiver.

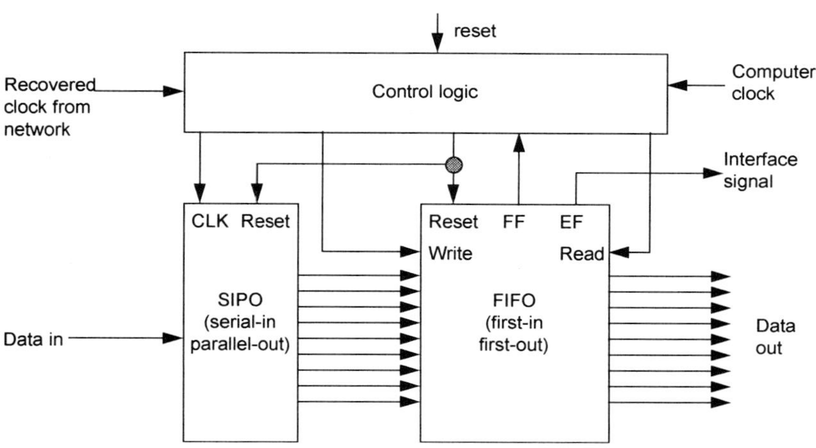

Figure 38.13 Memory buffering.

The memory buffer used is the AMD AM7204 FIFO which has 4096 data words with 9-bit words (but only 8 bits are actually used). The output of the circuit is 8 data lines, the control lines $\overline{FF}$, $\overline{EF}$, $\overline{RS}$, $\overline{R}$ and $\overline{W}$, and the +5 V and GND supply rails.

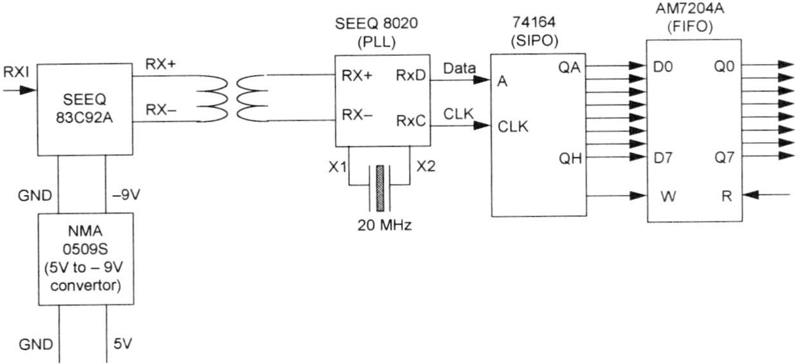

Figure 38.14 Ethernet receiver.

38.8 Standard Ethernet limitations

The standard Ethernet CSMA/CD specification places various limitations on maximum cable lengths. This limitation is due to maximum signal propagation times and the clock period.

38.8.1 Length of segments

Twisted-pair and coaxial cables have a characteristic impedance and a cable must be terminated with the correct characteristic impedance so that there is no loss of power and no reflections at terminations. For twisted-pair cables the characteristic impedance is normally $100\,\Omega$ and for coaxial cables it is $50\,\Omega$. The Ethernet connection may consist of many spliced coaxial sections. One or many sections constitute a cable segment, which is a stand-alone network. A segment must not exceed 500 m. This is shown in Figure 38.15.

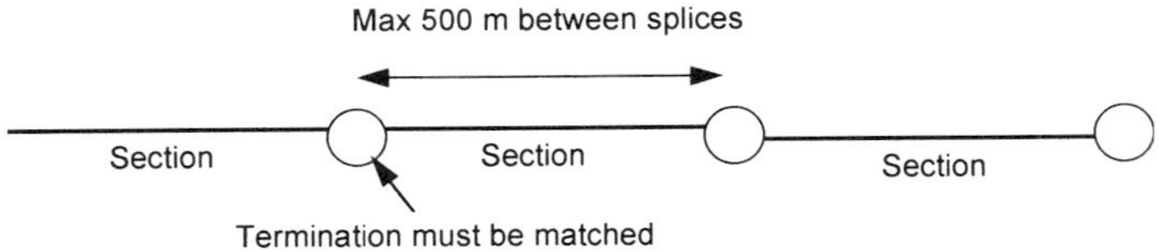

Figure 38.15 Connection of sections.

38.8.2 Repeater lengths

A repeater is added between segments to boost the signal. A maximum of two repeaters can be inserted into the path between two nodes. The maximum distance between two nodes connected via repeaters is 1500 m; this is illustrated in Figure 38.16.

38.8.3 Maximum links

The maximum length of a point-to-point coaxial link is 1500 m. A long run such as this is typically used as a link between two remote sites within a single building.

38.8.4 Distance between transceivers

Transceivers should not be placed closer than 2.5 m. Additionally, each segment should not have more than 100 transceiver units, as illustrated in Figure 38.17. Transceivers which are placed too close to each other can cause transmission interference and also an increased risk of collision.

Each node transceiver lowers network resistance and dissipates the transmission signal. A sufficient number of transceivers reduce the electrical characteristic of the network below the specified operation threshold.

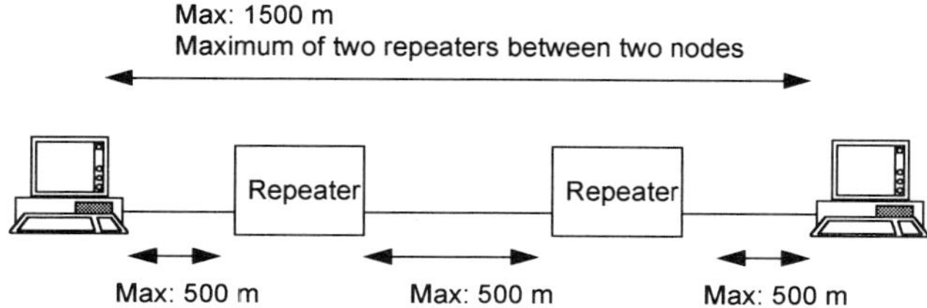

Figure 38.16 Maximum number of repeaters between two nodes.

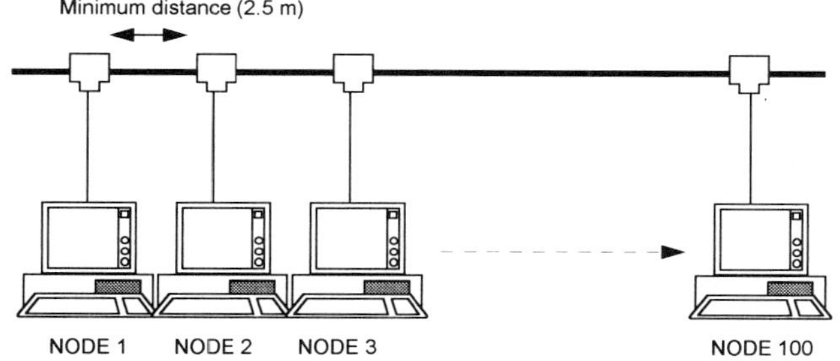

Figure 38.17 Connection of sections.

38.9 Ethernet types

The five main types of standard Ethernet are:

- Standard, or thick-wire, Ethernet (10BASE5).
- Thinnet, or thin-wire Ethernet, or Cheapernet (10BASE2).
- Twisted-pair Ethernet (10BASE-T).
- Optical fiber Ethernet (10BASE-FL).
- Fast Ethernet (100BASE-TX or 100VG-Any LAN).

The thin- and thick-wire types connect directly to an Ethernet segment, these are shown in Figure 38.18 and Figure 38.19. Standard Ethernet, 10BASE5, uses a high specification cable

(RG-50) and N-type plugs to connect the transceiver to the Ethernet segment. A node connects to the transceiver using a 9-pin D-type connector. A vampire (or bee-sting) connector can be used to clamp the transceiver to the backbone cable.

Thin-wire, or Cheapernet, uses a lower specification cable (it has a smaller inner conductor diameter). The cable connector required is also of a lower specification, that is, BNC rather than N-type connectors. In standard Ethernet the transceiver unit is connected directly onto the backbone tap. On a Cheapernet network the transceiver is integrated into the node.

Many modern Ethernet connections are to a 10BASE-T hub, which connects UTP cables to the Ethernet segment. An RJ-45 connector is used for 10BASE-T. The fiber-optic type, 10BASE-FL, allows long lengths of interconnected lines, typically up to 2 km. They use either SMA connectors or ST connectors. SMA connectors are screw-on types while ST connectors are push-on. Table 38.1 shows the basic specifications for the different types.

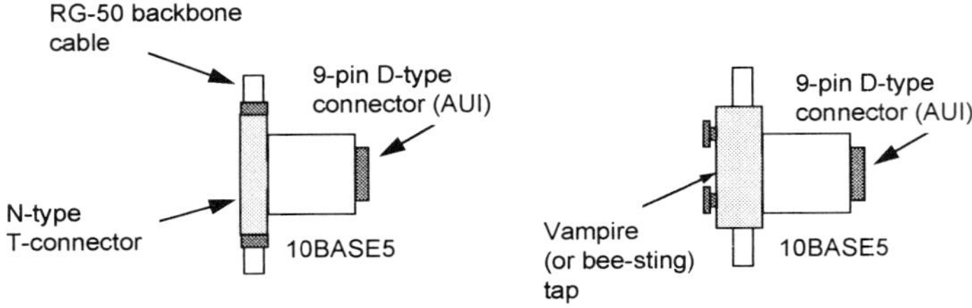

Figure 38.18 Ethernet connections for Thick Ethernet.

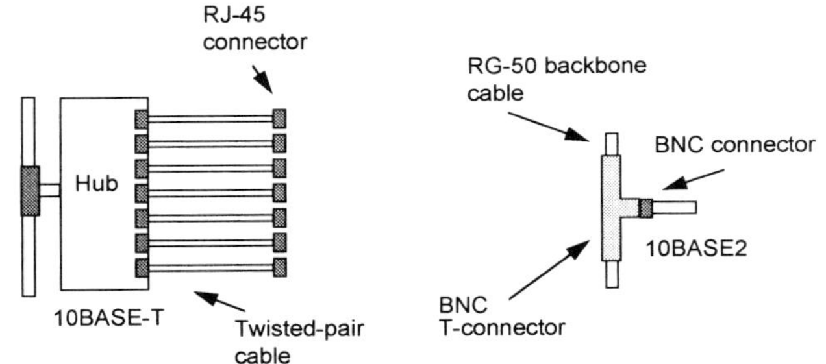

Figure 38.19 Ethernet connections for Thin Ethernet and 10BASE-T.

38.10 Twisted-pair hubs

Twisted-pair Ethernet (10BASE-T) nodes normally connect to the backbone using a hub, as illustrated in Figure 38.20. Connection to the twisted-pair cable is via an RJ-45 connector. The connection to the backbone can either be to thin- or thick-Ethernet. Hubs can also be stackable where one hub connects to another. This leads to concentrated area networks (CANs) and limits the amount of traffic on the backbone. Twisted-pair hubs normally improve network performance.

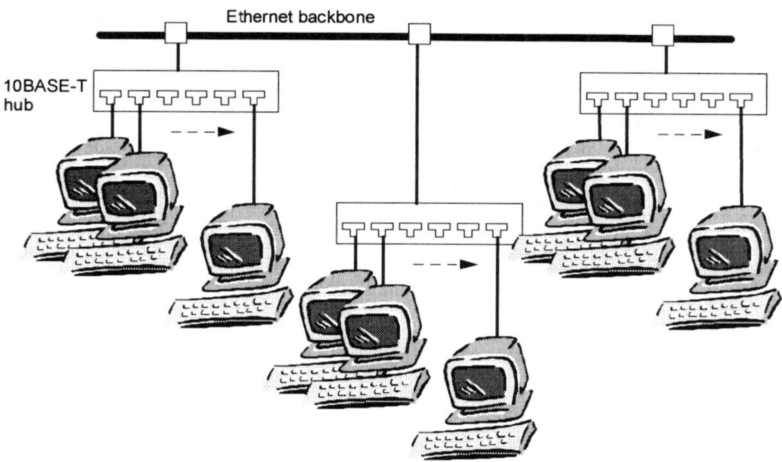

Figure 38.20 10BASE-T connection.

10BASE-T uses 2 twisted-pair cables, one for transmit and one for receive. A collision occurs when the node (or hub) detects that it is receiving data when it is currently transmitting data.

Table 38.1 Ethernet network parameters.

Parameter	*10BASE5*	*10BASE2*	*10BASE-T*
Common name	Standard or thick-wire Ethernet	Thinnet or thin-wire Ethernet	Twisted-pair Ethernet
Data rate	10 Mbps	10 Mbps	10 Mbps
Maximum segment length	500 m	200 m	100 m
Maximum nodes on a segment	100	30	3
Maximum number of repeaters	2	4	4
Maximum nodes per network	1024	1024	
Minimum node spacing	2.5 m	0.5 m	No limit
Location of transceiver electronics	Cable connection	integrated into node	in a hub
Typical cable type	RG-50	RG-6	UTP cables
Connectors	N-type	BNC	RJ-45/Telco
Cable impedance	50 Ω	50 Ω	100 Ω

38.11 Ethernet security

Ethernet itself provides no security to be it is designed as a simple and open physical medium for data transmission. Furthermore, it is not immune to snooping and spying. The vulnerabilities of the Ethernet are:

- It is an open architecture where any node can transmit and/or receive.
- It uses broadcast communications.
- It is easy to tap into.
- It has no hardware security.
- It is easy to jam a network.

39 Fast Ethernet and Switches

39.1 Introduction

Standard 10 Mbps Ethernet does not perform well when many users are running multimedia applications. Two improvements to the standard are Fast Ethernet and 100VG-AnyLAN. The IEEE has defined standards for both of them, IEEE 802.3u for Fast Ethernet and 802.12 for 100VG-AnyLAN. They are supported by many manufacturers and use bit rates of 100 Mbps. This gives at least 10 times the performance of standard Ethernet.

New standards relating to 100 Mbps Ethernet are now becoming popular:

- 100BASE-TX (twisted-pair) – which uses 100 Mbps over two pairs of Cat-5 UTP cable or two pairs of Type 1 STP cable.
- 100BASE-T4 (twisted-pair) – which is the physical layer standard for 100 Mbps bit rate over Cat-3, Cat-4 or Cat-5 UTP.
- 100VG-AnyLAN (twisted-pair) – which uses 100 Mbps over two pairs of Cat-5 UTP cable or two pairs of Type 1 STP cable.
- 100BASE-FX (fiber-optic cable) – which is the physical layer standard for 100 Mbps bit rate over fiber-optic cables.

Fast Ethernet, or 100BASE-T, is simply 10BASE-T running at 10 times the bit rate. It is a natural progression from standard Ethernet and thus allows existing Ethernet networks to be easily upgraded. Unfortunately, as with standard Ethernet, nodes contend for the network, reducing the network efficiency when there are high traffic rates. Also, as it uses collision detect, the maximum segment length is limited by the amount of time for the farthest nodes on a network to properly detect collisions. On a Fast Ethernet network with twisted-pair copper cables this distance is 100 m and for a fiber-optic link it is 400 m. Table 39.1 outlines the main network parameters for Fast Ethernet.

Since 100BASE-TX standards are compatible with 10BASE-TX networks then the network allows both 10 Mbps and 100 Mbps bit rates on the line. This makes upgrading simple, as the only additions to the network are dual-speed interface adapters. Nodes with the 100 Mbps capabilities can communicate at 100 Mbps, but they can also communicate with slower nodes, at 10 Mbps.

The basic rules of a 100BASE-TX network are:

- The network topology is a star network and must be no loops.
- All four pairs are required in a 4-pair UTP network. In other words, do not use the remaining two pairs in 100Base-TX for anything else.
- Cat-5 cable is used.
- Up to two hubs can be cascaded in a network.
- Each hub is equivalent of 5 meters in latency.

- Segment length is limited to 100 meters.
- Network diameter must not exceed 205 meters.

Table 39.1 Fast Ethernet network parameters.

	100BASE-TX	*100VG-AnyLAN*
Standard	IEEE 802.3u	IEEE 802.12
Bit rate	100 Mbps	100 Mbps
Actual throughput	Up to 50 Mbps	Up to 96 Mbps
Maximum distance (hub to node)	100 m (twisted-pair, Cat-5) 400 m (fiber)	100 m (twisted-pair, Cat-3) 200 m (twisted-pair, Cat-5) 2 km (fiber)
Scaleability	None	Up to 400 Mbps
Advantages	Easy migration from 10BASE-T	Greater throughput, greater distance

39.2 100BASE-4T

100BASE-4T allows the use of standard Cat-3 cables. These contain eight wires made up of four twisted pairs. 100BASE-4T uses all of the pairs to transmit at 100 Mbps. This differs from 10BASE-T in that 10BASE-T uses only two pairs, one for transmit and one for receive. 100BASE-T allows compatibility with 10BASE-T in that the first two pairs (Pair 1 and Pair 2) are used in the same way as 10BASE-T connections. 100BASE-T then uses the other two pairs (Pair 3 and Pair 4) with half-duplex links between the hub and the node. The connections are illustrated in Figure 39.1.

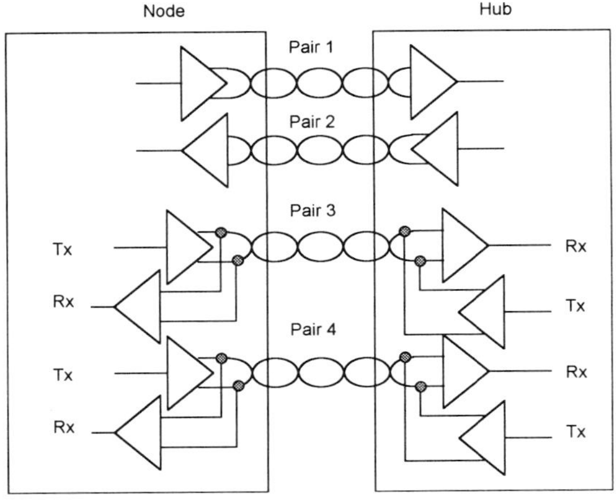

Figure 39.1 100BASE-4T connections.

39.2.1 8B6T

100BASE-4T uses four separate Cat-3 twisted-pair wires. The maximum clock rate that can be applied to Cat-3 cable is 30 Mbps. Thus, some mechanism must be devised which can reduce the line bit rate to under 30 Mbps but give a symbol rate of 100 Mbps. This is achieved with a 3-level code (+, – and 0) and is known as 8B6T. The code converts eight binary digits into six ternary symbols. Table 39.3 gives the part of the codetable. Thus, the bit sequence 00000000 will be coded as a positive voltage, a negative voltage, a zero voltage, a zero voltage, a positive voltage and a negative voltage.

Apart from reducing the frequencies with the digital signal, the 8B6T code has the advantage of reducing the dc content of the signal. Most of the codes contain the same number of positive and negative voltages. This is because only 256 of the possible 729 (3^6) codes are actually used. The codes have also chosen to have at least two transitions in every code word, so that clock information is embedded into signal.

Unfortunately, it is not possible to have all codes with the same number of negative voltages as positive voltages. Thus there are some codes which have a different number of negatives and positives, these include:

```
0100 0001      +0–00++
0111 1001      +++–0–
```

The technique used to overcome this is to invert consecutive codes which have a weighting of +1. For example, suppose the line code were:

```
+0++––    ++0+––    +++––0 +++––0
```

it would actually be coded as:

```
+0++––    ––0–++    +++––0 –––++0
```

The receiver detects the –1 weighted codes as an inverted pattern.

Table 39.2 8B6T code.

8-bit data	*Encoded data*	*8-bit data*	*Encoded data*
00000000	+-00+-	00010000	+0+--0
00000001	0+-+-+	00010001	++0-0
00000010	+-0+-0	00010010	+0+-0-
00000011	-0++-0	00010011	0++-0-
00000100	-0+0+-	00010100	0++--0
00000101	0+--0+	00010101	++00--
00000110	+-0-0+	00010110	+0+0--
00000111	-0+-0+	00010111	0++0--
00001000	-+00+-	00011000	0+-0+-
00001001	0-++-0	00011001	0+-0-+
00001010	-+0+-0	00011010	0+-++-
00001011	+0-+-0	00011011	0+-00+
00001100	+0-0+-	00011100	0-+00+
00001101	0-+-0+	00011101	0-+++-
00001110	-+0-0+	00011110	0-+0-+
00001111	+0--0+	00011111	0-+0+-

39.3 100VG-AnyLAN

The 100VG-AnyLAN standard (IEEE 802.12) was developed mainly by Hewlett Packard and overcomes the contention problem by using a priority-based round-robin arbitration method, known as the demand priority access method (DPAM). Unlike Fast Ethernet, nodes always connect to a hub which regularly scans its input ports to determine whether any nodes have requests pending.

100VG-AnyLAN has the great advantage that it supports both IEEE 802.3 (Ethernet) and IEEE 802.5 (Token Ring) frames and can thus integrate well with existing 10BaseT and Token Ring networks.

100VG-AnyLAN has an in-built priority mechanism with two priority levels: a high-priority request and a normal-priority request. A normal-priority request is used for non-real-time data, such as data files, and so on. High-priority requests are used for real-time data, such as speech or video data. At present, there is limited usage of this feature and there is no support mechanism for this facility after the data has left the hub.

100VG-AnyLAN allows up to seven levels of hubs (i.e. one root and six cascaded hubs) with a maximum distance of 150 m between nodes. Unlike other forms of Ethernet, it allows any number of nodes to be connected to a segment.

39.3.1 5B6B

The 100VG-AnyLAN standard uses 5B6B to transmit an Ethernet frame between the hub and the node. This code is used so that there is an increase the number of transitions in the transmitted waveform.

In 100VG-AnyLAN, a 100 Mbps bit stream is multiplexed onto four 25 Mbps streams and transmitted over the four twisted-pair cables. The encoding process thus increases the bit rate on each twisted pair cable to 30 Mbps (as six encoded bits are sent for every five bit stream bits). Figure 39.2 illustrated this and Table 39.3 gives the 5B/6B encoding.

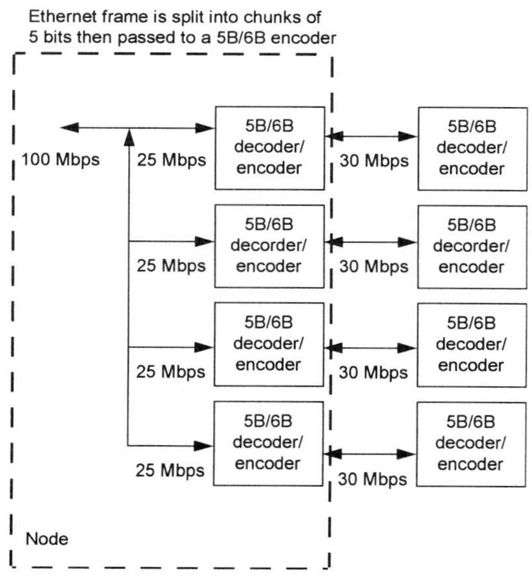

Figure 39.2 Encoding of the bit stream in 100VG-AnyLAN.

Table 39.3 5B/6B encoding.

5-bit data	*Mode 2 encoding*	*Mode 4 encoding*	*5-bit data*	*Mode 2 encoding*	*Mode 4 encoding*
00000	001100	110011	10000	000101	111010
00001	**101100**	101100	10001	**100101**	100101
00010	100010	101110	10010	001001	110110
00011	**001101**	001101	10011	**010110**	010110
00100	001010	110101	10100	**111000**	111000
00101	**010101**	010101	10101	011000	100111
00110	**001110**	001110	10110	**011001**	011001
00111	**001011**	001011	10111	100001	011110
01000	**000111**	000111	11000	**110001**	110001
01001	**100011**	100011	11001	**101010**	101010
01010	**100110**	100110	11010	010100	101011
01011	000110	111001	11011	**110100**	110100
01100	101000	010111	11100	**011100**	011100
01101	**011010**	011010	11101	**010011**	010011
01110	100100	100100	11110	010010	101101
01111	**101001**	101001	11111	**110010**	110010

Unfortunately, it is not possible to code each one of the 6-bit encoded values with an equal number of 0s and 1s as there are only 20 encoded values which have an equal number of 0s and 1s, these are highlighted in Table 39.3. Thus, two modes are used with the other 12 values having either two 0s and four 1s or four 0s and two 1s. The data is then transmitted in two modes:

- Mode 2. Where the encoded data has either an equal number of 0s and 1s or has four 1s and two 0s.
- Mode 4. Where the encoded data has either an equal number of 1s and 0s or has four 0s and two 1s.

These modes alternate and this gives, on average, digital sum value (DSV) of zero.

39.3.2 Connections

100BASE-TX, 100BASE-T4 and 100VG-AnyLAN use the RJ-45 connector, which has eight connections. 100BASE-TX uses pairs 2 and 3, whereas 100BASE-T4 and 100VG-AnyLAN use pairs 1, 2, 3 and 4. The connections for the cables are defined in Table 39.4. The white/orange color identifies the cable which is white with an orange stripe, whereas orange/white identifies an orange cable with a white stripe.

Table 39.4 Cable connections for 100BASE-TX

Pin	*Cable color*	*Cable color*	*Pair*
1	white/orange	white/orange	Pair 4
2	orange/white	orange/white	Pair 4
3	white/green	white/green	Pair 3
4	blue/white	blue/white	Pair 3
5	white/blue	white/blue	Pair 1
6	green/white	green/white	Pair 1
7	white/brown	white/brown	Pair 2
8	brown/white	brown/white	Pair 2

39.3.3 Migration to Fast Ethernet

If an existing network is based on standard Ethernet then, in most cases, the best network upgrade is either to Fast Ethernet or 100VG-AnyLAN. Since the protocols and access methods are the same there is no need to change any of the network management software or application programs. The upgrade path for Fast Ethernet is simple and could be:

- Upgrade high data rate nodes, such as servers or high-powered workstations to Fast Ethernet.
- Gradually upgrade NICs (network interface cards) on Ethernet segments to cards which support both 10BASE-T and 100BASE-T. These cards automatically detect the transmission rate to give either 10 or 100 Mbps.

The upgrade path to 100VG-AnyLAN is less easy as it relies on hubs and, unlike Fast Ethernet, most NICs have different network connectors, one for 10BASE-T and the other for 100VG-AnyLAN (although it is likely that more NICs will have automatic detection). A possible path could be:

- Upgrade high data rate nodes, such as servers or high-powered workstations to 100VG-AnyLAN.
- Install 100VG-AnyLAN hubs.
- Connect nodes to 100VG-AnyLAN hubs and change over connectors.

It is difficult to assess the performance differences between Fast Ethernet and 100VG-AnyLAN. Fast Ethernet uses a well-proven technology but suffers from network contention. 100VG-AnyLAN is a relatively new technology and the handshaking with the hub increases delay time. The maximum data throughput of a 100BASE-TX network is limited to around 50 Mbps, whereas 100VG-AnyLAN allows rates up to 96 Mbps.

The 100BASE-TX standard does not allow future upgrading of the bit rate, whereas 100VG-AnyLAN allows possible upgrades to 400 Mbps.

39.4 Switches and switching hubs

A switch is a very fast, low-latency, multiport bridge that is used to segment LANs. They are typically also used to increase communication rates between segments with multiple parallel conversations and also between technologies (such as between FDDI and 100Base-TX).

A switching hub is a repeater that contains a number of network segments (typically 4, 8 or 16). Through software, any of the ports on the hub can directly connect to any of the four segments, at any time. Thus, for a 4-port switching hub connected to 10 Mbps segments, gives a maximum capacity of 40 Mbps in a single hub.

Ethernet switches overcome the contention problem of normal CSMA/CD networks. They segment traffic by giving each connection a guaranteed bandwidth allocation. Figure 39.3 and Figure 39.4 show the two types of switches, their main features are:

- Desktop switch (or workgroup switch). These connect directly to nodes. They are economical with fixed configuration for end-node connections and are designed for standalone networks or distributed workgroups in a larger network.

- Segment switch. These connect both 10Mbps workgroup switches and 100Mbps interconnect (backbone) switches that are used to interconnect hubs and desktop switches. They are modular, high-performance switches for interconnecting workgroups in mid- to large-size networks.

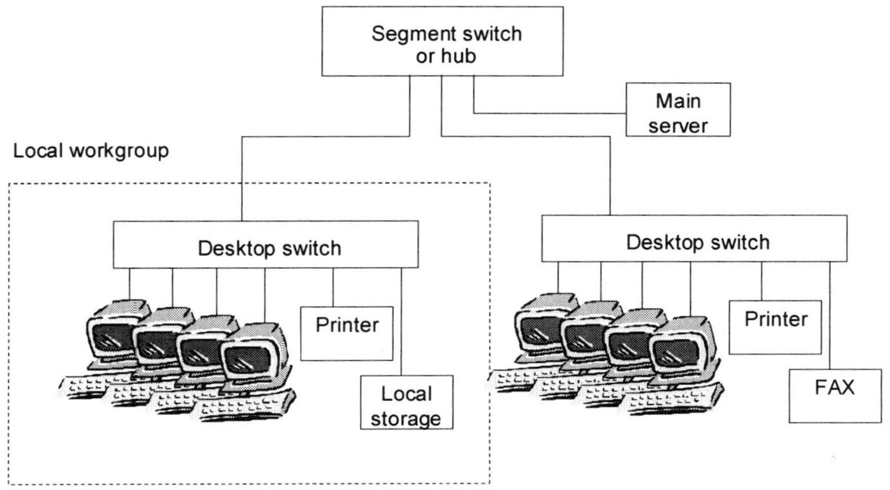

Figure 39.3 Desktop switch.

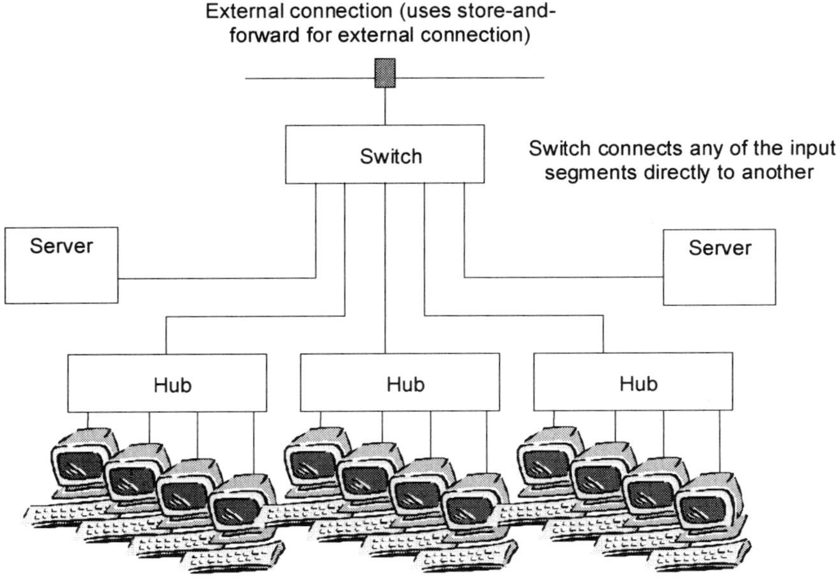

Figure 39.4 Segment switch.

39.4.1 Segment switch

A segment switch allows simultaneous communication between any client and any server. These switches can simply replace existing Ethernet hubs. Figure 39.4 shows a switch with five ports, each transmitting at 10 Mbps. This allows up to five simultaneous connections giving a maximum aggregated bandwidth of 50 Mbps. If the nodes support 100 Mbps communication then the maximum aggregated bandwidth will be 500 Mbps. To optimize the network nodes should be connect to the switch that connects to the server that they most often communicate with. This allows for a direct connection with that server.

39.4.2 Desktop switch

A desktop switch can simply replace an existing 10BASE-T/100BASE-T hubs. It has the advantage that any of the ports can connect directly to another. In the network in Figure 39.3, any of the computers in the local workgroup can connect directly to any others, or to the printer or to the local disk drive. This type of switch works well if there is a lot of local traffic, typically between a local server and local peripherals.

39.4.3 Store-and-forward switching

Store-and-forwarding techniques have been used extensively in bridges and routers, and they are now being used with switches. It involves reading the entire Ethernet frame, before forwarding it, with the required protocol and at the correct speed, to the destination port. This has the advantages of:

- Improved error checking. Bad frames are blocked from entering a network segment.
- Protocol filtering. Store-and-forwarding allows the switch to convert from one protocol to another.
- Speed matching. Typically, for Ethernet, reading at 10 Mbps or 100 Mbps and transmitting at 100 Mbps or 10 Mbps. Also a matching between ATM (155 Mbps), FDDI (100 Mbps), Token Ring (4/16 Mbps) and Ethernet (10/100 Mbps).

The main disadvantage is:

- System delay. As the frame must be totally read before it is transmitted there is a delay in the transmission. The improvement in error checking normally overcomes this disadvantage.

39.4.4 Switching technology

A switch uses store-and-forward packets to switch between ports. The main technologies used are:

- Shared bus. This method uses a high-speed backplane to interconnect the switched ports. It is frequently used to build modular switches that give a large number of ports and to interconnect multiple LAN technologies, such as FDDI, 100VG-AnyLAN, 100Base-T, and ATM.
- Shared memory. These use a common memory area (several MBs) in which data is passed between the ports. It is very common in low-cost, small-scale switches and has the advantage that it can cope with different types of network, which may be operating at different speeds. The main types of memory allocation are:

- Pooled memory. Memory is allocated as it is need by the ports from a common memory pool.
- Dedicated shared memory. Memory is fixed for each shared port pair.
- Distributed memory. Memory is fixed and dedicated to each port.

39.5 Comparison of Fast Ethernet

Table 39.5 compares Fast Ethernet with other types of networking technologies.

Table 39.5 Comparison of Fast Ethernet.

	100VG-AnyLAN (Cat 3, 4, or 5)	*100Base-T (TX/FX/T4)*	*Gigabit Ethernet (802.3z)*
Maximum segment length	100 m	100m (Cat-5) 412m (Fiber)	100 m (Cat 5) 1000m (Fiber)
Maximum network diameter with repeater(s)	6000m	320m	To be determined by the standard
Bitrate	100 Mbps	100 Mbps	1 Gbps
Media access method	Demand Priority	CSMA/CD	CSMA/CD
Maximum nodes on each domain	1024	Limited by hub	To be determined
Frame type	Ethernet and Token Ring	Ethernet	Ethernet
Multimedia support	✓	✗	YES (with 802.1p)
Integration with 10BASE2	YES with bridges, switches and routers	YES with switches	YES with 10/100 Mbps switching
Relative cost	Low	Low	Medium
Relative complexity	Low	Low	Low

40 Token Ring

40.1 Introduction

Token Ring networks were developed by several manufacturers, the most prevalent being the IBM Token Ring. Token Ring networks cope well with high network traffic loadings. They were at one time extremely popular but their popularity has since been overtaken by Ethernet. Token Ring networks have, in the past, suffered from network management problems and poor network fault tolerance.

Token Ring networks are well suited to situations which have large amounts of traffic and also work well with most traffic loadings. They are not suited to large networks or networks with physically remote stations. The main advantage of Token Ring is that it copes better with high traffic rates than Ethernet, but requires a great deal of maintenance, especially when faults occur or when new equipment is added to or removed from the network. Many of these problems have now been overcome by MAUs (multistation access units), which are similar to the hubs using in Ethernet.

The IEEE 802.5 standard specifies the MAC layer for a Token Ring network with a bit rate of either 4 Mbps or 16 Mbps. There are two main types of Token Ring networks. Type 1 Token Ring uses Type 1 Token Ring cable (shielded twisted-pair) with IBM style universal connectors. Type 3 Token Ring use either Cat-3, Cat-4 or Cat-5 unshielded twisted-pair cables with modular connectors. Cat-3 has the advantage of being cheap to install and is typically used in telephone connections. Unfortunately the interconnection distance is much less than for Cat-4 and Cat-5 cables.

40.2 Operation

A Token Ring network circulates an electronic token (named a control token) around a closed electronic loop. Each node on the network reads the token and repeats it to the next node. The control token circulates around the ring even when there is no data being transmitted.

Nodes on a Token Ring network wishing to transmit must await a token. When they get it, they fill a frame with data and add the source and destination addresses then send it to the next node. The data frame then circulates around the ring until it reaches the destination node. It then reads the data into its local memory area (or buffer) and marks an acknowledgment on the data frame. This then circulates back to the source (or originating) node. When it receives the frame it tests it to determine whether it contains an acknowledgment. If it does then the source nodes knows that the data frame was received correctly, else the node is not responding. If the source node has finished transmitting data then it transmits a new token, which can be used by other nodes on the ring.

Figure 40.1(a)–(d) shows a typical interchange between node B and node A. Initially, in (a), the control token circulates between all the nodes. This token does not contain any data and is only 3 bytes long. When node B finally receives the control token it then transmits a data frame, as illustrated in (b). This data frame is passed to node C, then to node D and finally onto A. Node A will then read the data in the data frame and return an acknowledgment to node B, as illustrated in (c). After node B receives the acknowledgment, it passes a control token onto node C and this then circulates until a node wishes to transmit a data frame. Nodes are not allowed to transmit data unless they have received a valid control token. A distributed control protocol determines the sequence in which nodes transmit. This gives each node equal access to the ring.

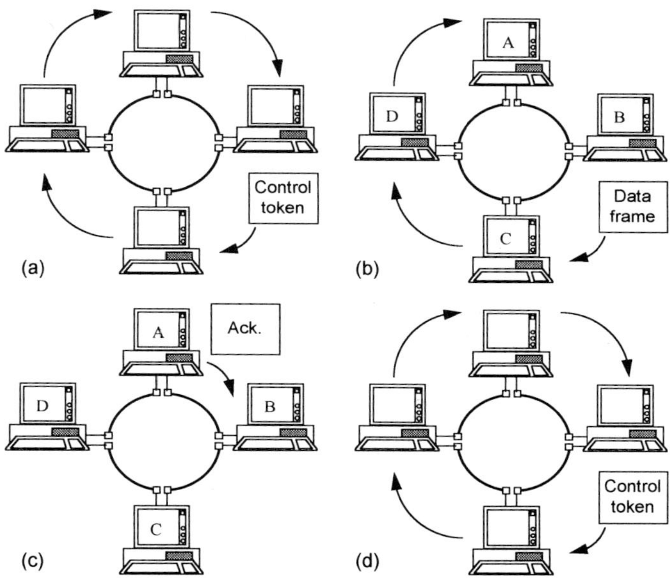

Figure 40.1 Example data exchange.

40.3 Token Ring - media access control (MAC)

Token passing allows nodes controlled access to the ring. Figure 40.2 shows the token format for the IEEE 802.5 specification. There are two main types of frame: a control token and a data frame. A control token contains only a start and end delimiter, and an access control (AC) field. A data frame has start and end delimiters (SD/ED), an access control field, a frame control field (CF), a destination address (DA), a source address (SA), frame check sequence (FCS), data and a frame status field (FS).

The access control and frame control fields contain information necessary for managing access to the ring. This includes priority reservation, priority information and information on whether the data is user data or control information. It also contains an express indicator which informs networked nodes that an individual node requires immediate action from the network management node.

The destination and source addresses are 6 bytes in length. Logical link control information has variable length and is shown in Figure 40.2. It can either contain user data or network control information.

The frame check sequence (FCS) is a 32-bit cyclic redundancy check (CRC) and the frame control field is used to indicate whether a destination node has read the data in the token.

The start and end delimiters are special bit sequences which define the start and end of the frame and thus cannot occur anywhere within the frame. As with Ethernet the bits are sent using Manchester coding. The start and end delimiters violate the standard coding scheme. The standard Manchester coding codes a 1 as a low-to-high transition and a 0 as a high-to-low transition. In the start and end delimiters, two of the bits within the delimiters are set to either a high level (`H`) or a low level (`L`). These bits disobey the standard coding as there is no change in level, i.e. from a high to a low or a low to a high. When the receiver detects this violation and the other standard coded bits in the received bit pattern, it knows that the accompanying bits are a valid frame. The coding is as follows:

- If the preceding bit is a 1 then the start delimiter is `HL0HL000`, else
- If the preceding bit is a 0 then the start delimiter is `LH0LH000`.

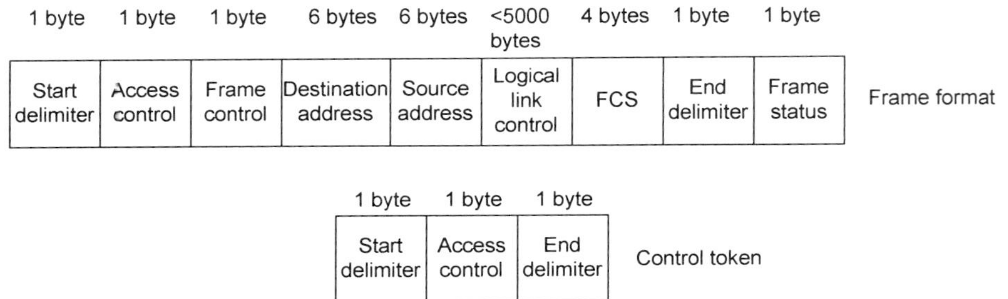

Figure 40.2 IEEE 802.5 frame format.

They are shown in Figure 40.3. The end delimiter is similar to the start delimiter, but 0's are replaced by 1's. An error detection bit (`E`) and a last packet indicator bit (`I`) are added.

If the bit preceding the end delimiter is a 1 then the end delimiter is `HL1HL1IE`. If it is a 0 then it is `LH1LH1IE`. The `E` bit is used for error detection and is initially set by the originator to a 0. If any of the nodes on the ring detects an error the `E` bit is set to a 1. This indicates to the originator that the frame has developed an error as it was sent. The `I` bit determines whether the data being sent in a frame is the last in a series of data frames. If the `I` bit is a 0 then it is the last, else it is an intermediate frame.

The access control field controls the access of nodes on the ring. It takes the form of `PPPTMRRR`, where:

`PPP` – indicates the priority of the token; this indicates which type of token the destination node can transmit.

`T` – is the token bit and is used to discriminate between a control token and a data token.

`M` – is the monitor bit and is used by an active ring monitor node to stop tokens from circulating around a network continuously.

`RRR` – are the reservation bits and allow nodes with a high priority to request the next token.

The frame control field contains control information for the MAC layer. It takes the form `FFDDDDDD`, where:

`FF` – indicates whether the frame is a data frame; if it is not then the `DDDDDD` bits control the operation of the Token Ring MAC protocol.

`DDDDDD` – controls the operation of the Token Ring MAC protocol.

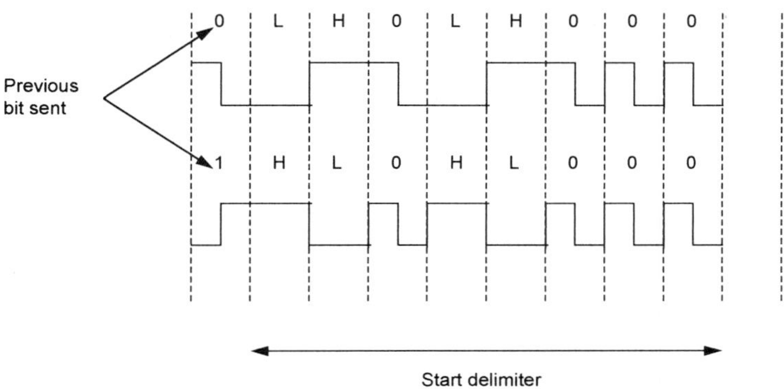

Figure 40.3 Start delimiter.

The source and destination addresses can either be 2 or 6 bytes (that is, 16 or 48 bits) in length. This size must be the same for all nodes on a ring. The first bit specifies the type of address. If it is a 0 then the address is an individual node address, else it is a group address. An individual node address is used to transmit to a single node, whereas a group address transmits to all nodes with the same group address. The source address will always by an individual address as it indicates the node which originated the token. A special destination address of all 1's is used to transmit to all nodes on a ring.

The frame status field contains information on how a frame has been operated upon as it circulates round the ring. It takes the form `ACXXACXX`, where:

`A` – indicates if the destination address has been recognized. It is initially set to a 0 by the source node and is set to a 1 when the destination reads the data. If the source node detects that this bit has not been set then it knows the destination is either not present on the network or is not responding.

`C` – indicates that a destination node has copied a frame into its memory. This bit is also initially set to a 0 by the source node. When the destination node reads the data from the frame it is set to a 1. By testing this bit and the `A` bit, the source node can determine whether the destination node is active but not reading data from the frame.

40.4 Token Ring maintenance

A Token Ring system requires considerable maintenance; it must perform the following function:

- Ring initialization – when the network is started, or after the ring has been broken, it must be reinitialized. A cooperative decentralized algorithm sorts out which nodes start a new token, which goes next, and so on.
- Adding to the ring – if a new node is to be physically connected to the ring then the network must be shut down and reinitialized.
- Deletion from the ring – a node can disconnect itself from the ring by joining together its predecessor and its successor. Again, the network may have to be shut down and reinitialized.
- Fault management – typical Token Ring errors occur when two nodes think it is their turn to transmit or when the ring is broken as no node thinks that it is their turn.

40.5 Token Ring multistation access units (MAUs)

The problems of connecting and deleting nodes to or from a ring network are significantly reduced with a multistation access unit (MAU). Figure 40.4 shows two 3-way MAUs connected to produce a 6-node network. Normally, an MAU allows nodes to be switched in and out of a network using a changeover switch or by automatic electronic switching (known as auto-loopback). This has the advantage of not shutting down the network when nodes are added and deleted or when they develop faults.

If the changeover switches in Figure 40.4 are in the down position then the node is bypassed; if they are in the up position then the node connects to the ring.

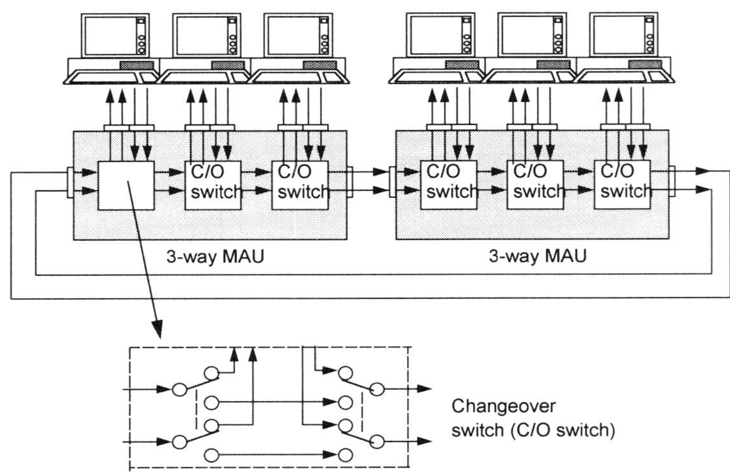

Figure 40.4 Six-node Token Ring network with two MAUs.

A single coaxial (or twisted-pair) cable connects one MAU to another and two coaxial (or twisted-pair) cables connect a node to the MAU (for the in ports and the out ports). Most modern application use STP cables.

The IBM 8228 is a typical passive MAU. It can operate at 4 Mbps or 16 Mbps and has 10 connection ports, i.e. 8 passive node ports along with ring in (RI) and ring out (RO) connections. The maximum distance between MAUs is typically 650 m (at 4 Mbps) and 325 m (at 16 Mbps).

Most MAUs either have 2, 4 or 8 ports and can automatically detect the speed of the node (i.e. either 4 or 16 Mbps). Figure 40.4 shows a 32-node Token Ring network using four 8-port MAUs. Typical connectors are RJ-45 and IBM Type A connectors. The ring cable is normally either twisted-pair (Type 3), fiber-optic or coaxial cable (Type 1). MAU are intelligent devices and can detect faults on the cables supplying nodes then isolate them from the rest of the ring. Most MAUs are passive devices in that they do not require a power supply. If there are large distances between nodes then an active unit is normally used.

Modern Token Ring networks normally use twisted-pair cables instead of coaxial cables. These twisted-pair cables can either be high-specification shielded twisted-pair (STP) or lower-specification unshielded twisted-pair (UTP). Cabling is discussed in the next section.

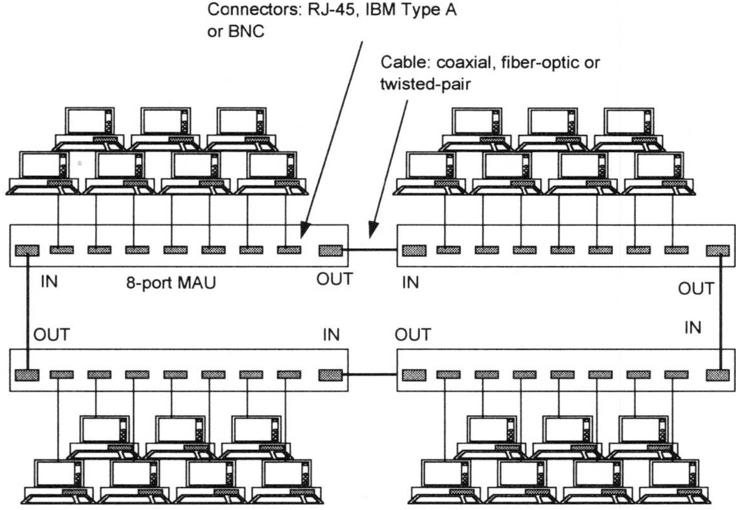

Figure 40.5 A 32-node Token Ring network with 4 MAUs.

40.6 Cabling and connectors

There are two main types of cabling used in Token Ring networks: Type 1 and Type 3. Type 1 uses STP (shielded twisted-pair) cables with IBM style male-female connectors. Type 3 networks uses Cat-3 or Cat-5 UTP (unshielded twisted-pair) cables with RJ-45 connectors. Unfortunately, Cat-3 cables are unshielded which reduces the maximum length of the connection. Type 1 networks can connect up to 260 nodes, whereas Type 3 networks can only connect up to 72 nodes.

A further source of confusion comes from the two different types of modern STP cables used in Token Ring networks. IBM type 1 cable has four cores with a screen tinned copper braid around them. Each twisted-pair is screened from each other with aluminized polyester tape. The characteristic impedance of the twisted-pairs is 150 Ω. The IBM type 6 cable is a lightweight cable which is preferred in office environments. It has a similar construction but, because it has a thinner core, signal loss is higher.

40.7 Repeaters

A repeater is used to increase either main-ring or lobe lengths in a Token Ring LAN. The main-ring length is the distance between MAUs. The lobe length is the distance from an MAU to a node. Table 40.1 shows some typical maximum cable lengths for different bit rates and cable types. Fiber-optic cables provide the longest distances with a range of 1 km. The next best are STP cables, followed by Cat-5 and finally the lowest specification Cat-3 cables. Figure 40.6 shows the connection of two MAUs with repeaters. In this case, four repeaters are required as each repeater has only two ports (IN and OUT). The token will circulate clockwise around the network.

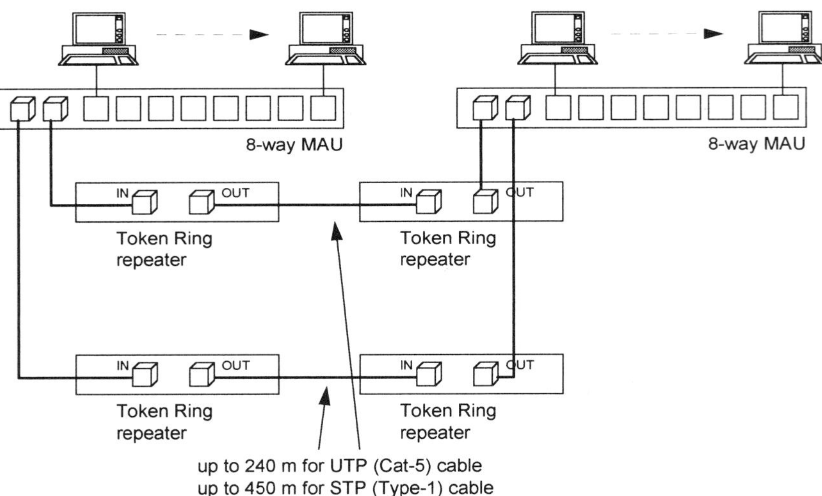

Figure 40.6 16-node Token Ring network with repeaters.

Table 40.1 Typical maximum cable lengths for different cables and bit rates.

Type	*Bit rate*	*Cable type*	*Maximum distance*
Type 1	4 Mbps	STP	730 m
Type 3	4 Mbps	UTP (Cat-3 cable)	275 m
Type 3	16 Mbps	UTP (Cat-5 cable)	240 m
Type 1	16 Mbps	STP	450 m
Type 1	16 Mbps	Fiber	1000 m

40.8 Jitter suppression

Jitter can be a major problem with Token Ring networks. It is caused when the nodes on the network operate with different clock rates. It can lead to network slowdown, data corruption and station loss. Jitter is the reason that the number of nodes on a Token Ring is limited to 72 at 16 Mbps. With a jitter suppressor the number of nodes can be increased to 256 nodes. It also allows Cat-3 cable to be used at 16 Mbps. Normally a Token Ring Jitter Suppresser is connected to a group of MAUs. Thus, the network in Figure 40.6 could have one jitter suppresser unit connected to two of the MAUs (this would obviously limit to 7 the number of nodes connected to these MAUs).

41 FDDI

41.1 Introduction

A token-passing mechanism allows orderly access to a network. Apart from Token Ring the most commonly used token-passing network is the Fiber Distributed Data Interchange (FDDI) standard. This operates at 100 Mbps and, to overcome the problems of line breaks, has two concentric Token Rings, as illustrated in Figure 41.1. Fiber optic cables have a much high-specification than copper cables and allow extremely long connections. The maximum circumference of the ring is 100 km (62 miles), with a maximum 2 km between stations (FDDI nodes are also known as stations). It is thus an excellent mechanism for connecting networks across a city or over a campus. Up to 500 stations can connect to each ring with a maximum of 1000 stations for the complete network. Each station connected to the FDDI highway can be a normal station or a bridge to a conventional local area network, such as Ethernet or Token Ring.

The two rings are useful for fault conditions but are also used for separate data streams. This effectively doubles the data-carrying capacity of FDDI (to 200 Mbps). However, if the normal traffic is more than the stated carrying capacity, or if one ring fails, then its performance degrades.

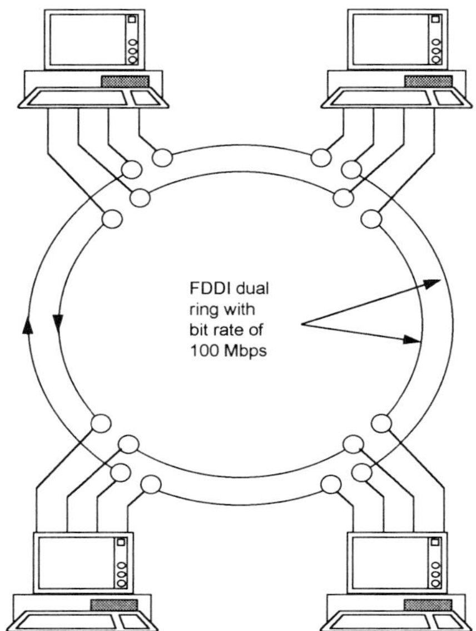

Figure 41.1 FDDI network.

The main features of FDDI are:

- Point-to-point Token Ring topology.
- A secondary ring for redundancy.
- Dual counter rotating ring topology.
- Distributed clock for the support of large numbers of stations on the ring.
- Distributed FDDI management – equal rights and duties for all stations.
- Data integrity ensured through sophisticated encoding techniques.

41.2 Operation

As with Token Ring, FDDI uses a token passing medium access method. Unlike Token Ring there are two types of (control) token:

- A restricted token – which is the normal token. The restricted token circulates around the network. A station wishing to transmit data captures the unrestricted token. It then transmits frames for a period of time made up of a fixed part (T_f) and a variable part (T_v). The variable time depends on the traffic on the ring. When the traffic is light a station may keep the token much longer than when it is heavy.
- An unrestricted token – which is used for extended data interchange between two stations. To enter into an extended data interchange a station must capture the unrestricted token and change it to a restricted token. This token circulates round the network until the exchange is complete, or the extended time is over. Other stations on the network may use the ring only for the fixed period of time T_f. Once complete the token is changed back to an unrestricted type.

FDDI uses a timed token-passing protocol to transmit data because a station can hold the token no longer than a specified amount of time. Therefore, there is a limit to the amount of data that a station can transmit on any given opportunity.

The sending station must always generate a new token once it has transmitted its data frames. The station directly downstream from a sending station has the next opportunity to capture the token. This feature and the timed token ensures that the ring's capacity is divided almost equally among the stations on the ring.

41.3 FDDI layers

The ANSI-defined FDDI standard defines four key layers:

- Media access control (MAC) layer.
- Physical layer (PHY).
- Physical media dependent (PMD).
- Station management (SMT) protocol.

FDDI covers the first two layers of the OSI model; Figure 41.2 shows how these layers fit into the model.

The MAC layer defines addressing, scheduling and data routing. Data is formed into data packets with a PHY layer. It encodes and decodes the packets into symbol streams for transmission. Each symbol has 4 bits and FDDI then uses the 4B/5B encoding to encode them into a group of 5 bits. The 5 bits are chosen to contain, at most, two successive zeros. Table 41.1 shows the coding for the bits. This type of coding ensures that there will never be more than four consecutive zeros. FDDI uses NRZI (non-return to zero with inversion) to transmit the bits. With NRZI a 1 is coded with an alternative light (or voltage) level transition for each 1, and a zero does not change the light (or voltage) level. Figure 41.3 shows an example.

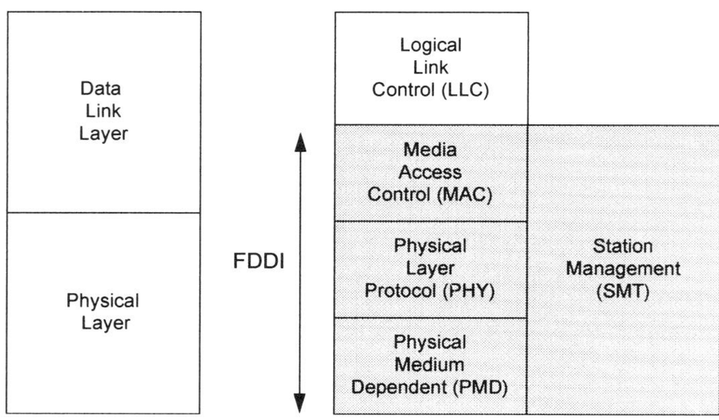

Figure 41.2 FDDI network.

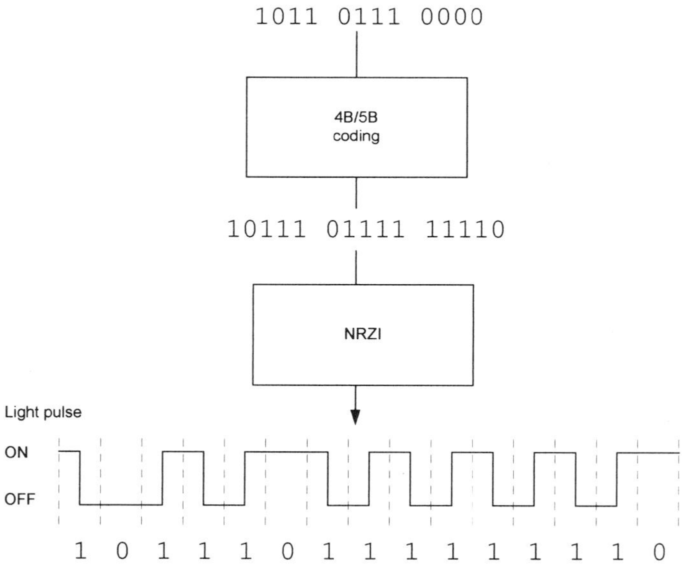

Figure 41.3 Example of bit encoding.

In this case the input bitsteam is 1011, 0111 and 0000. This is encoded, as in Table 41.1, as 10111, 01111 and 11110. This is then transmitted in NRZI. The first bit sent is a 1 and is represented as a high-to-low transition. The next bit is a zero and thus has no transition. After this a 1 is encoded, this will be transmitted as a low-to-high transition, and so on. The main advantage of NRZI coding is that the timing information is inherent within the transmitted signal and can be easily filtered out.

Apart from the 16 encoded bit patterns given Table 41.1 there are also eight control and eight violation patterns. Table 41.2 shows the eight other control symbols (QUIET, HALT, IDLE, J, K, T, R and S) and eight other violation symbols (the encoded bitstream binary values are 00001, 00010, 00011, 00101, 00110, 01000, 01100 and 10000).

The coding of the data symbols is chosen so that there are no more than three consecutive zeros in a row. This is necessary to ensure that all the stations on the ring have their clocks synchronized with all the others. Each station on an FDDI has its own independent clock. The control and violation symbols allow the reception of four or more zero bits in a row.

Table 41.1 4B/5B coding.

Symbol	*Binary*	*Bitstream*	*Symbol*	*Binary*	*Bitstream*
0	0000	11110	8	1000	10010
1	0001	01001	9	1001	10011
2	0010	10100	A	1010	10110
3	0011	10101	B	1011	10111
4	0100	01010	C	1100	11010
5	0101	01011	D	1101	11011
6	0110	01110	E	1110	11100
7	0111	01111	F	1111	11101

Table 41.2 4B/5B coding.

Symbol	*Bitstream*	*Symbol*	*Bitstream*
QUIET	00000	K	10001
HALT	00100	T	01101
IDLE	11111	R	00111
J	11000	S	11001

41.4 SMT protocol

The SMT protocol handles the management of the FDDI ring, which includes:

- Adjacent neighbor indication.
- Fault detection and reconfiguration.
- Insertion and deletion from the ring.
- Traffic statistics monitoring.

41.5 Physical connection management

Within each FDDI station there are SMT entities called PCM (physical connection management). The number of PCM entities with a station is exactly equal to the number of ports the station has. This is because each PCM is responsible for one port. The PCM entities are the parts of the SMT, which control the ports. In order to make a connection, two ports must be physically connected to each other by means of a fiber or copper cable.

41.6 Fault tolerance method

When a station on a ring malfunctions or there is a break in one of the rings then the rest of the stations can still use the other ring. When a station on the network malfunctions then both of the rings may become inoperative. FDDI allows other stations on the network to detect this and to implement a single rotating ring. Figure 41.4 shows an FDDI network with four connected stations. In this case, the link between the upper stations have developed a fault. These stations will quickly determine that there is a fault in both cables and will inform the other stations on the network to implement a single rotating ring with the outer ring transmitting in the clockwise direction and the inner ring in the counterclockwise direction. This fault tolerance method also makes it easier to insert and delete stations from the ring.

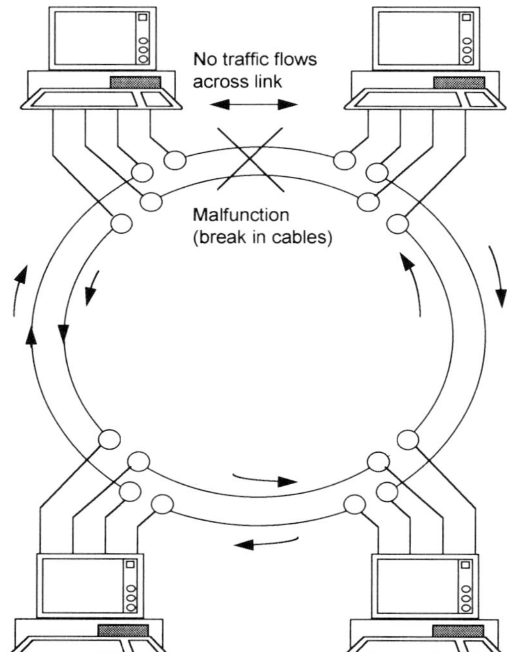

Figure 41.4 Fault tolerant network.

41.7 FDDI token format

A token circulates around the ring until a station captures it and then transmits its data in a data frame (see next section). Figure 41.5 shows the basic token format. The preamble (PA) field has four or more symbols of idle (bitstream of 11111). This is followed by the start delimiter (SD) which has a fixed pattern of 'J' (bitstream of 11000) and 'K' (bitstream of 10001). The end delimiter (ED) is two 'T' symbols (01101). The start and end delimiter bit patterns cannot occur anywhere else in the frame as they violate the standard 4B/5B coding (i.e. they may contain three or more consecutive zeros).

41.8 FDDI Frame format

Figure 41.6 shows the FDDI data frame format, which is similar to the IEEE 802.6 frame. The PA, SD and ED fields are identical to the token fields. The frame control field (FC) contains information on the kind of frame that is to follow in the INFO field.

The fields are:

- Preamble. The preamble field contains 16 idle symbols which allow stations to synchronize their clocks.
- Start delimiter. This contains a fixed field of the 'J' and 'K' symbols.
- Control field. The format of the control field is SAFFxxxx where S indicates whether it is synchronous or asynchronous. A indicates whether it is a 16-bit or 48-bit address; FF indicates whether this is an LLC (01), MAC control (00) or reserved frame. For a control frame, the remaining 4 bits (xxxx) are reserved for control types. Typical (decoded) codes are:

0100 0000 – void frame 0101 0101 – station management frame
1100 0010 – MAC frame 0101 0000 – LLC frame

When the frame is a token the control field contains either 10000000 or 11000000.

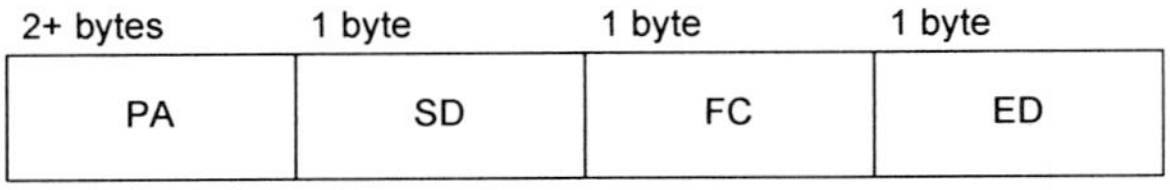

2+ bytes	1 byte	1 byte	1 byte
PA	SD	FC	ED

PA – preamble (4 or more symbols of idle)
SD – start delimiter ('J' and 'K')
FC – frame control (2 symbols)
ED – end delimiter (two 'T' symbols)

Figure 41.5 FDDI token format.

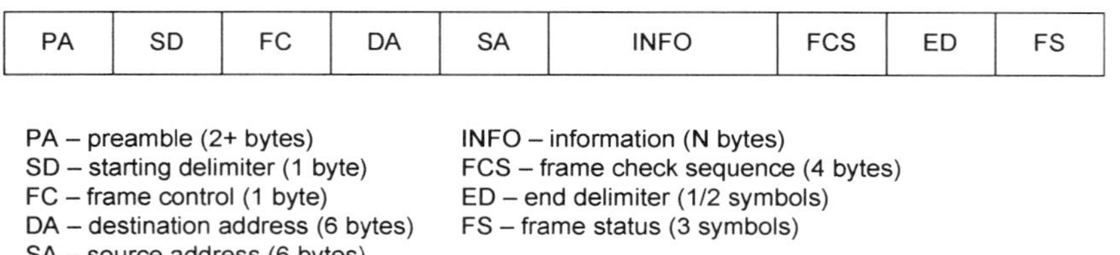

Figure 41.6 FDDI data frame format.

- The destination address (DA) and source address (SA) are 12-symbol (6-byte) codes which identify the address of the station from where the frame has come or to where the frame is heading. Each station has a unique address and each station on the ring compares it with its own address. If the frame is destined for a particular station, that station copies the frame's contents into its buffer.

  A frame may also be destined for a group of stations. Group addresses are identified by the start bit 1. If the start bit is 0 then the frame is destined for an individual station. A broadcast address of all 1's is used to send information to all the stations on the ring.

  Station addresses can either be locally or globally administered. For global addresses the first six symbols are the manufacturer's OUI; each manufacturer has a unique OUI for all its products. The last six symbols of the address differentiate between stations of the same manufacturer.

  The second bit in the address field identifies whether the address is local or global. If it is set (a 1) then it is a locally administered address, if it is unset then it is a globally administered address. In a locally administered network the system manager sets the addresses of all network stations.
- The information field (INFO) can contain from 0 to 4 478 bytes of data. Thus the maximum frame size will be as follows:

Field	*Number of bytes*
Start delimiter	1
Frame control	1
Destination address	6
Source address	6
DATA (maximum)	4 478
Frame check sequence	4
End delimiter	2
End of frame sequence	2
TOTAL	4 500

- The frame check sequence contains a 32-bit CRC which is calculated from the FC, DA, SA and information fields.
- The ending delimiter contains two 'T' symbols.
- The frame status contains extra bits which identify the current status, such as frame copied indicators (F), errors detected (E) and address recognized (A).

FDDI supports either synchronous or asynchronous traffic, but the terms are actually confusing. Frames that are transmitted during their capacity allocation are known as synchronous.

41.9 MAC protocol

The MAC protocol of FDDI MAC is similar to IEEE 802.5. It can be thought of as train, filled with passengers, traveling around a track. The train travels around the ring continuously. When a passenger wishes to get on the train they get in front of the train, which then pushes them round the ring. The passenger then travels around the ring, and delivers their message to the destination station. The passenger stays on the train until the train reaches the source again, where they will get off. Other passengers can get on the train and deliver their own messages while there are others on the train. These passengers go in between the train and the existing passengers.

The actual operational parts work like this:

- A node cannot transmit until it captures a token (the train).
- When a node captures the token, it transmits its frame then the token.
- The frame travels around the ring from source back to source station.
- If another station wishes to transmit a frame it waits for the end of the frames currently on the ring, adds its frame after these frames then appends the token onto the end.
- The station which initiates a frame is responsible for taking it off the ring.
- Each station reads the frame as it circulates around the ring.
- Each station can modify the status bits as the frame passes. If a station detects an error then it sets the E bit; if it has copied the frame into its buffer then it sets the C bit; if it detects its own address it sets the A bit.
- Thus a correctly received frame will be received back at the source node with the C and A bits set.
- A frame which is received back at the source with the E bit is not automatically retransmitted. A message is sent to higher layer in the protocol (such as the LLC layer).

41.10 Applications of FDDI networks

As was seen in Chapter 38, Ethernet is an excellent method of attaching stations to a network cheaply but is not a good transport mechanism for a backbone network or with high traffic levels. It also suffers, in its standard form, from a lack of speed. FDDI networks overcome these problems as they offer a much higher bit rate, higher reliability and longer interconnections. Thus typical applications of FDDI networks are:

- As a backbone network in an internetwork connection.
- Any applications which requires high security and/or a high degree of fault tolerance. Fiber-optic cables are generally more reliable and are difficult to tap into without it being detected.
- As a subnetwork connecting high-speed computers and their peripheral devices (such as storage units).
- As a network connecting stations where an application program requires high-speed transfers of large amounts of data (such as computer-aided design – CAD). Maximum data traffic for an FDDI network is at least 10 times greater than for standard Ethernet and Token Ring networks. As it is a token-passing network it is less susceptible to heavy traffic loads than Ethernet.

41.11 FDDI backbone network

The performance of the network backbone is extremely important as many users on the network depend on it. If the traffic is too heavy, or if it develops a fault, then it affects the performance of the whole network. An FDDI backbone helps with these problems because it has a high bit rate and normally increases the reliability of the backbone.

Figure 41.7 shows an FDDI backbone between four campuses. In this case, the FDDI backbone only carries traffic which is transmitted between campuses. This is because the router only routes traffic out of the campus network when it is intended for another campus. As tokens circulate round both rings, two data frames can be transmitted round the rings at the same time.

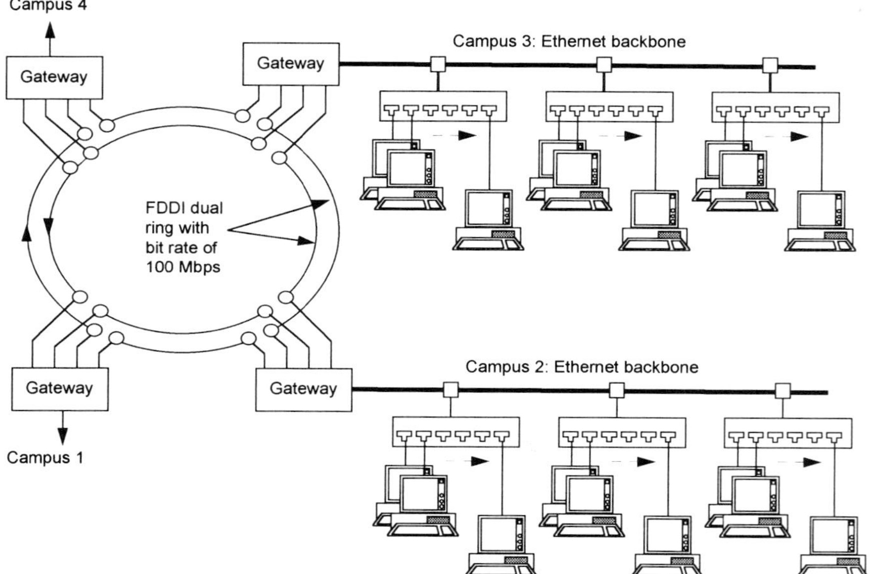

Figure 41.7 FDDI backbone network.

41.12 FDDI media

FDDI networks can use two types of fiber-optic cable, either single-mode or multimode. The mode refers to the angle at which light rays are reflected and propagated through the fiber core. Single-mode fibers have a narrow core, such as 10 μm for the core and 125 μm for the cladding (known as 10/125 micron cable). This type allows light to enter only at a single angle. Multimode fiber has a relatively thick core, such as 62.5 μm for the core and 125 μm for the cladding (known as 62.5/125 micron cable). Multi-mode cable reflects light rays at many angles. The disadvantage of these multiple propagation paths is that it can cause the light pulses to spread out and thus limit the rate at which data is accurately received. Thus, single-mode fibers have a higher bandwidth than multimode fibers and allow longer interconnection distances. The fibers most commonly used in FDDI are 62.5/125, and this type of cable is defined in the ANSI X3T9.5 standard.

41.13 FDDI attachments

There are four types of station which can attach to an FDDI network:

- Dual attachment stations (DAS).
- Single attachment stations (SAS).
- Dual attachment concentrators (DAC).
- Single attachment concentrators (SAC).

Figure 41.8 shows an FDDI network configuration that includes all these types of station. An SAS connects to the FDDI rings through a concentrator, so it is easy to add, delete or change its location. The concentrator automatically bypasses disconnected stations.

Each DAS and DAC requires four fibers to connect it to the network: Primary In, Primary Out, Secondary In and Secondary Out. The connection of an SAS only requires two fibers. Normally Slave In and Slave Out on the SAS are connected to the Master Out and Master In on the concentrator unit, as shown in Figure 41.9.

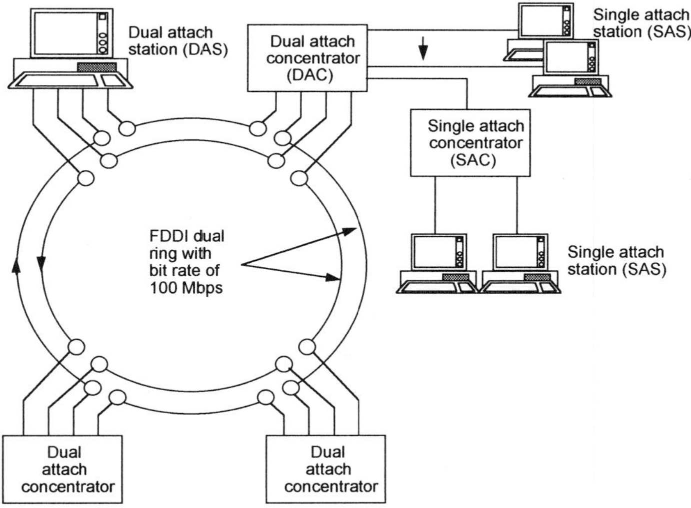

Figure 41.8 FDDI network configuration.

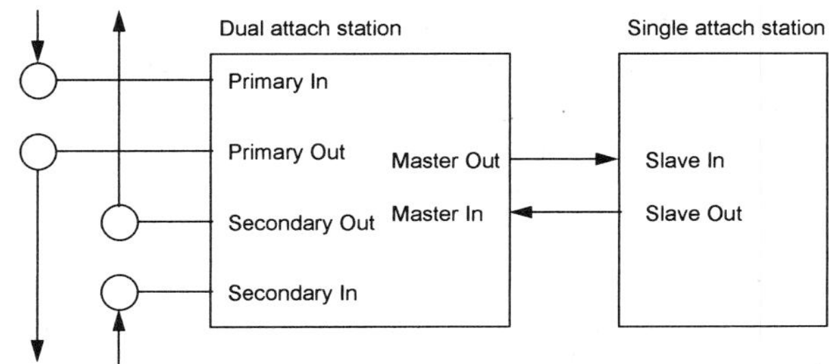

Figure 41.9 Connection of a DAS and a SAS.

FDDI stations attach to the ring using a media interface connector (MIC). An MIC receptacle connects to the stations and an MIC plug on the network end of the connection. A dual attachment station has two MIC receptacles. One provides Primary Ring In and Secondary Ring Out, the other has Primary Ring Out and Secondary Ring In, as illustrated in Figure 41.10.

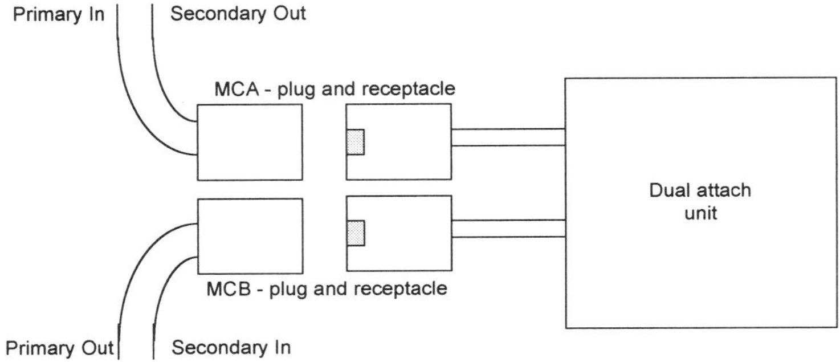

Figure 41.10 Connection of dual attach units.

41.14 FDDI specification

Table 41.3 describes the basic FDDI specification. Notice that the maximum interconnection distance for multimode cable is 2 km as compared with 20 km for single-mode cable.

Table 41.3 Basic FDDI specification.

Parameter	*Description*
Topology	Token Ring
Access method	Time token passing
Transport media	Optical fiber, shield twisted pair, unshielded twisted pair
Maximum number of stations	500 each ring (1000 total)
Data rate	100 Mbps
Maximum data packet size	4 500 bytes
Maximum total ring length	100 km
Maximum distance between stations	2 km (for multimode fiber cable), 20 km (for single-mode fiber cable)
Attenuation budget	11 dB (between stations), 1.5 dB/km at 1300 nm for 62.5/125 fiber
Link budget	< 11 dB

41.15 FDDI-II

FDDI-II is an upward-compatible extension to FDDI that adds the ability to support circuit-switched traffic in addition to the data frames supported by the original FDDI. With FDDI-II, it is possible to set up and maintain a constant data rate connection between two stations.

The circuit-switched connection consists of regularly repeating time slots in the frame, often called an isochronous frame. This type of data is common when real-time signals, such as speech and video, are sampled. For example, speech is sampled 8000 times per second, whereas high-quality audio is sampled at 44 000 times per second.

Figure 41.11 shows a layer diagram of an FDDI-II station. The physical layer and the station management are the same as the original FDDI. Two new layers have been added to the MAC layer; known as hybrid ring control, they consist of:

- Hybrid multiplexer (HMUX).
- Isochronous MAC (IMAC).

The IMAC module provides an interface between FDDI and the isochronous service, represented by the circuit-switched multiplexer (CS-MUX). The HMUX multiplexes the packet data from the MAC and the isochronous data from the IMAC.

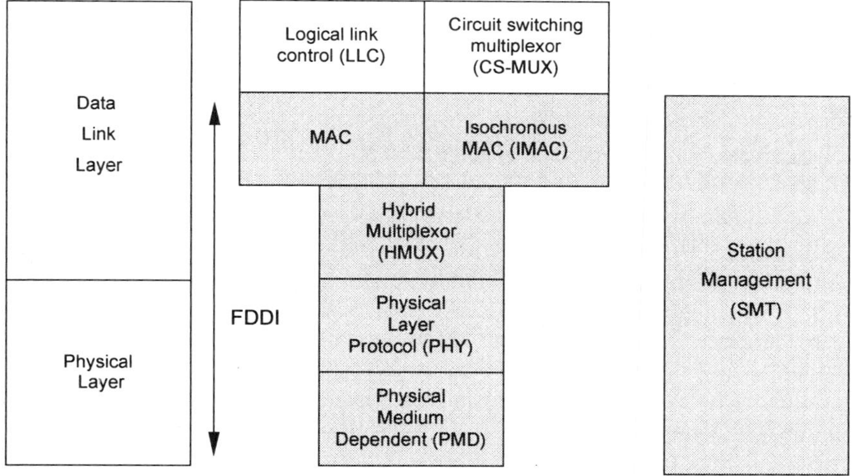

Figure 41.11 FDDI-II layered model.

An FDDI-II network operates in a basic mode or a hybrid mode. In the basic mode the FDDI ring operates as the original FDDI specification where tokens rotate around the network. In the hybrid mode a connection can either be circuit-switched or packet-switched. It uses a continuously repeating protocol data unit known as a cycle. A cycle is a data frame that is similar to synchronous transmission systems. The content of the cycle is visible to all stations as it circulates around the ring. A station called the cycle master generates a new cycle 8000 times a second. At 100 Mbps, this gives a cycle size of 12 500 bits. As each cycle completes its circuit of the ring, its is stripped by the cycle master. Figure 41.12 shows the two different types of transmission.

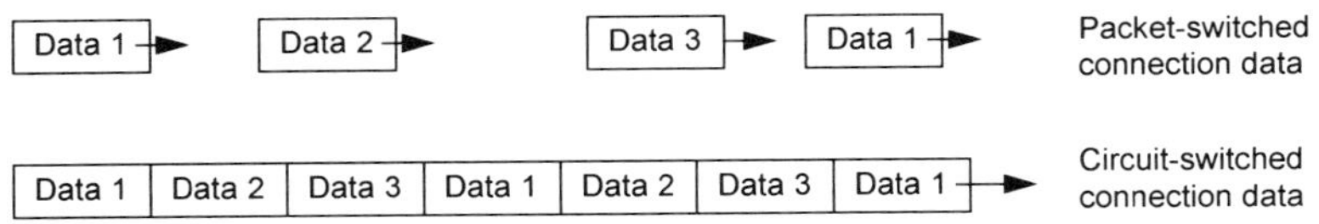

Figure 41.12 Circuit-switched and packet-switched data.

41.16 Standards

The FDDI standard was defined by the ANSI committee X3T9.5. This has since been adopted by the ISO as ISO 9314 which defines FDDI using five main layers:

- ISO 9314-1: Physical Layer Protocol (PHY).
- ISO 9314-2: Media Access Control (MAC).
- ISO 9314-3: Physical Media Dependent (PMD).
- ISO 9314-4: Station Management (SMT).
- ISO 9314-5: Hybrid Ring Control (HRC), FDDI-II.

41.17 Practical FDDI network – EaStMAN

The EaStMAN (Edinburgh and Stirling Metropolitan Area Network) consortium comprises seven institutions of Higher Education in the Edinburgh and Stirling area of Scotland. The main institutions are: the University of Edinburgh, the University of Stirling, Napier University, Heriot-Watt University, Edinburgh College of Art, Moray House Institute of Education and Queen Margaret College. FDDI and ATM networks have been installed around Edinburgh with an optical link to Stirling. It was funded jointly by the Scottish Higher Education Funding Council (SHEFC), the Joint Information Systems Committee (JISC) and the individual institutions.

Figure 41.13 shows the connections of Phase 1 of the project and Figure 41.14 shows the rings. The total circumference of the rings is 58 km (which is less than the maximum limit of 100 km).

The FDDI ring provides intercampus communications and also a link to the SuperJANET (Joint Academic NETwork) and the ATM ring is for future development. The ATM ring will be discussed in more detail in Chapter 42.

Future plans for the network are to link it to the FDDI networks of FaTMAN (Fife and Tayside), ClydeNet (Glasgow network), and AbMAN (Aberdeen network).

41.17.1 Fiber optic cables

The new optical fiber network is leased from Scottish Telecom (a subsidiary of Scottish Power). The FDDI ring is 58 km and, for the purpose of FDDI standardization across the MAN, the outer FDDI ring is driven anticlockwise and the inner ring clockwise. Figure 41.15

shows the attenuation rates and distance between each site (in dBs). Note that an extra 0.4 dB should be added onto each fiber connection to take into account the attenuation at the fiber termination. Thus, the total attenuation between the New College and Moray House will be approximately 2.4 dB.

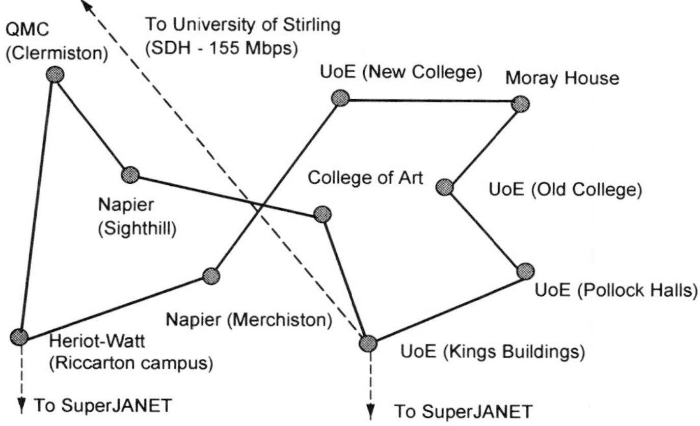

Figure 41.13 EaStMAN Phase 1 connections.

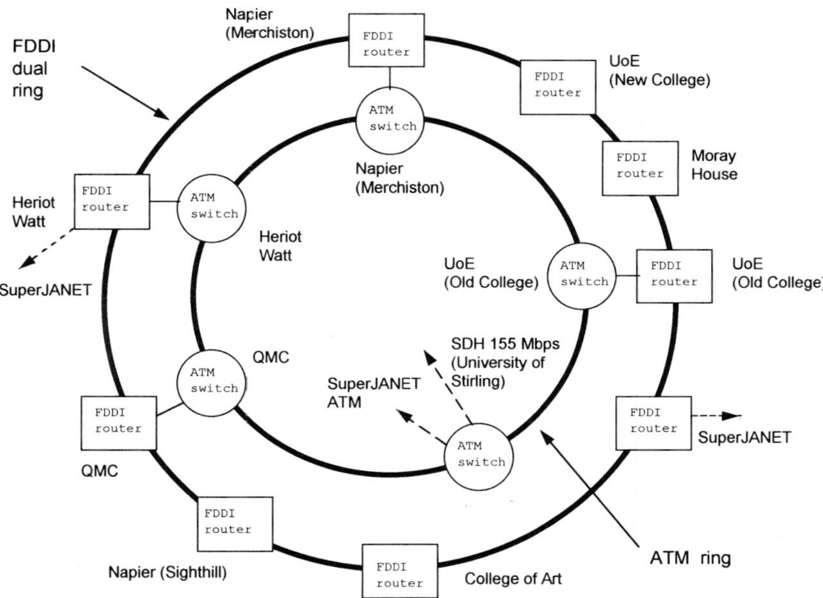

Figure 41.14 EaStMAN ring connections.

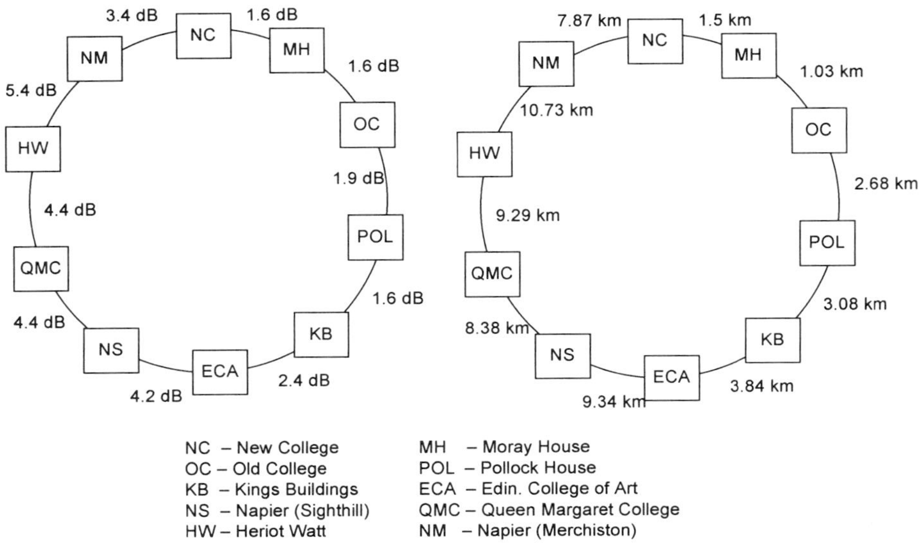

Figure 41.15 Attenuation and distances of EaStMAN Phase I

42 ATM

42.1 Introduction

Most of the networking technologies discussed so far are good at carrying computer-type data and they provide a reliable connection between two nodes. Unfortunately they are not as good at carrying real-time sampled data, such as digitized video or speech. Real-time data from speech and video requires constant sampling and these digitized samples must propagate through the network with the minimum of delay. Any significant delay in transmission can cause the recovered signal to be severely distorted or for the connection to be lost. Ethernet, Token Ring and FDDI simply send the data into the network without first determining whether there is a communication channel for the data to be transported.

Figure 42.1 shows some traffic profiles for sampled speech and computer-type data (a loading of 1 is the maximum loading). It can be seen that computer-type data tends to burst in periods of time. These bursts have a relatively heavy loading on the network. On the other hand, sampled speech has a relatively low loading on the network but requires a constant traffic throughput. It can be seen that if these traffic profiles were to be mixed onto the same network then the computer-type data would swamp the sampled speech data at various times.

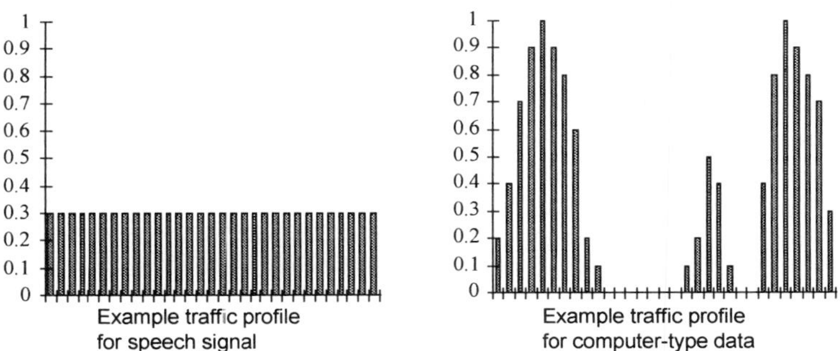

Figure 42.1 Traffic profiles for sampled speech and computer-type data.

Asynchronous transfer mode (ATM) overcomes the problems of transporting computer-type data and sampled real-time data by:

- Analyzing the type of connection to be made. The type of data dictates the type of connection; for example, computer data requires a reliable connection, whereas real-time sampled data requires a connection with a low propagation time.
- Analyzing the type of data to be transmitted and knowing its traffic profile. Computer data tends to create bursts of traffic whereas real-time data will be constant traffic.

- Reserving a virtual path for the data to allow the data profile to be transmitted within the required quality of service.
- Splitting the data into small packets which have the minimum overhead in the number of extra bits. These 'fast-packets' traverse the network using channels which have been reserved for them.

ATM has been developed mainly by the telecommunications companies. Unfortunately two standards currently exist. In the USA the ANSI T1S1 subcommittee have supported and investigated ATM and in Europe it has been investigated by ETSI. There are small differences between the two proposed standards, but they may converge into one common standard. The CCITT has also dedicated study group XVIII to ATM-type systems with the objective of merging differences and creating one global standard for high-speed networks throughout the world.

42.2 Real-time sampling

Before introducing the theory of ATM, first consider the concept of how analogue signals (such as speech, audio or video) are converted into a digital form. The basic principle involves sampling theory and pulse code modulation.

42.2.1 Sampling theory

Recall from Section 1.6 that for a signal to be reconstructed as the original signal it must be sampled at a rate defined by the Nyquist criterion (Figure 42.2). This states:

the sampling rate must be twice the highest frequency of the signal

For telephone speech channels the maximum signal frequency is limited to 4 kHz and must therefore be sampled at least 8000 times per second (8 kHz). This gives one sample every 125 μs. Hi-fi quality audio has a maximum signal frequency of 20 kHz and must be sampled at least 40 000 times per second (many professional hi-fi sampling systems sample at 44.1 kHz). Video signals have a maximum frequency of 6 MHz, so a video signal must be sampled at 12 MHz (or once every 83.3 ns).

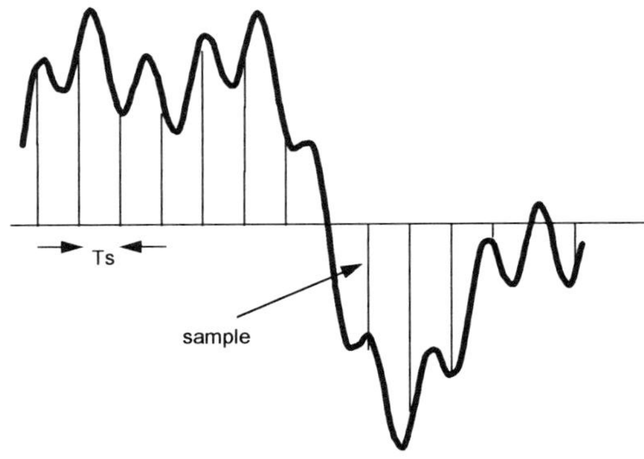

Figure 42.2 The sampling process.

42.2.2 Pulse code modulation

Once analogue signals have been sampled for their amplitude they can be converted into a digital format using pulse code modulation (PCM). The digital form is then transmitted over the transmission media. At the receiver the digital code is converted back into an analogue form.

The accuracy of the PCM depends on the number of bits used for each analogue sample. This gives a PCM-based system a dependable response over an equivalent analogue system because an analogue system's accuracy depends on component tolerance, producing a differing response for different systems.

42.3 PCM-TDM systems and ISDN

ATM tries to integrate real-time data and computer-type data. The main technology currently used in transmitting digitized speech over the public switched telephone network (PSTN) is PCM-TDM (PCM time division multiplexing). PCM-TDM involves multiplexing the digitized speech samples in time. Each sample is assigned a time slot which is reserved for the total time of the connection, as illustrated in Figure 42.3. This example shows the connection of Telephone 1–4 to Telephone A–D, respectively. The digitizer, in this case, consists of a sampler (at 8 kHz) and an analogue-to-digital converter (ADC). Four input channels are time-division multiplexed onto a signal line, the time between one sample from a certain channel and the next must be 125 μs.

The integrated services digital network (ISDN) is to be covered in Chapter 32. It uses a base bit rate of 64 kbps and can be used to transmit 8-bit samples at a rate of 8 kHz. ISDN is similar to a telephone connection but allows the direct connection of digital equipment. As with the PSTN the connection between the transmitter and the receiver is set up by means of a switched connection. On an ISDN network the type of data carried is transparent to the network. Higher bit rates are achieved by splitting the data into several channels and transmitting each channel at 64 kbps.

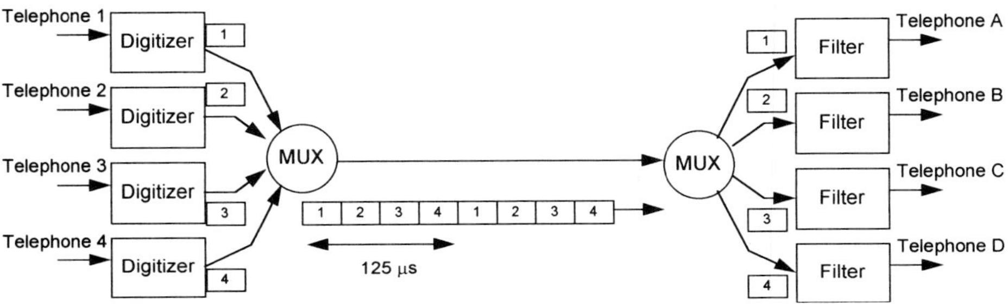

Figure 42.3 PCM-TDM with four connections.

42.4 Objectives of ATM

The major objective of ATM is to integrate real-time data (such as voice and video signals) and non-real-time data (such as computer data and file transfer). Computer-type data can typically be transferred in non-real-time but it is important that the connection is free of errors. In many application programs a single bit error can cause serious damage. On the other hand, voice and video data require a constant sampling rate and low propagation delays, but are more tolerant to errors and any losses of small parts of the data.

An ATM network relies on user-supplied information to profile traffic flows so that a connection has the desired service quality. Table 42.1 gives four basic data types. These are further complicated by differing data types either sending data in a continually repeating fashion (such as telephone data) or with a variable frequency (such as interactive video). For example a high-resolution video image may need to be sent as several megabytes of data in a short time burst, but then nothing for a few seconds. For speech the signal must be sampled 8000 times per second.

Computer data will typically be sent in bursts. Sometimes a high transfer rate is required (perhaps when running a computer package remotely over a network) or a relatively slow transfer (such as when reading text information). Conventional circuit-switched technology (such as ISDN and PCM-TDM) are thus wasteful in their connection because they either allocate a switched circuit (ISDN) or reserve a fixed time slot (PCM-TDM), no matter whether there data is being transmitted at that time. And it may not be possible to service high burst rates by allocating either time slots or switched circuits when all of the other time slots are full, or because switched circuits are being used.

Table 42.1 Four basic categories of data.

Data type	*Error or loss sensitive*	*Delay sensitive*
Real-time control system	yes	yes
Telephone/hi-fi music	no	yes
File transfer, application programs	yes	no
Teletex information	no	no

42.5 ATM versus ISDN and PCM-TDM

ISDN and PCM-TDM use a synchronous transfer mode (STM) technique where a connection is made between two devices by circuit switching. The transmitting device is assigned a given time slot to transmit the data. This time slot is fixed for the period of the transmission. The main problems with this type of transmission are:

- Not all the time slots are filled by data when there is light data traffic; this is wasteful in data transfer.
- When a connection is made between two endpoints a fixed time slot is assigned and data from that connection is always carried in that time slot. This is also wasteful because there may be no data being transmitted in certain time periods.

ATM overcomes these problems by splitting the data up into small fixed-length packets, known as cells. Each data cell is sent with its connection address and follows a fixed route through the network. The packets are small enough that, if they are lost, possibly due to congestion, they can either be requested (for high reliability) or cause little degradation of the signal (typically in voice and video traffic).

The address of devices on an ATM network are identifier by a virtual circuit identifier (VCI), instead of by a time slot as in an STM network. The VCI is carried in the header portion of the fast packet.

42.6 Statistical multiplexing

Fast packet switching attempts to solve the problem of unused time slots of STM. This is achieved by statistically multiplexing several connections on the same link based on their traffic characteristics. Applications, such as voice traffic, which require a constant data transfer are allowed safe routes through the network. Whereas several applications, which have bursts of traffic, may be assigned to the same link in the hope that statistically they will not all generate bursts of data at the same time. Even if some of them were to burst simultaneously, then their data could be buffered and sent at a later time. This technique is called statistical multiplexing and allows the average traffic on the network to be evened out over a relatively short time period. This is impossible on an STM network.

42.7 ATM user network interfaces (UNIs)

A user network interface (UNI) allows users to gain access to an ATM network. The UNI transmits data into the network with a set of agreed specifications and the network must then try to ensure the connection stays within those requirements. These requirements define the required quality of service for the entire duration of the connection.

It is likely that there will be several different types of ATM service provision. One type will provide an interface to one or more of the LAN standards (such as Ethernet or Token Ring) or FDDI. The conversion of the LAN frames to ATM cells will be done inside the UNI at the source and destination endpoints respectively. Typically it will be used as a bridge for two widely separated LANs. This provides a short-term solution to justifying the current investment in LAN technology and allows a gradual transition to a complete ISDN/ATM network.

The best long-term solution is to connect data communication equipment directly onto an ATM network. This allows computer equipment, telephones, video, and so on, to connect directly to a global network. The output from an ATM multiplexer interfaces with the UNI of a larger ATM backbone network.

A third type of ATM interface connects existing STM networks to ATM networks. This allows a slow migration of existing STM technology to ATM.

42.8 ATM cells

The ATM cell, as specified by the ANSI T1S1 subcommittee, has 53 bytes, as shown in Figure 42.4. The first five bytes are the header and the remaining bytes are the information field which can hold 48 bytes of data. Optionally the data can contain a 4-byte ATM adaptation layer and 44 bytes of actual data. A bit in the control field of the header sets the data to either 44 or 48 bytes. The ATM adaptation layer field allows for fragmentation and reassembly of cells into larger packets at the source and destination respectively. The control field also contains bits which specify whether this is a flow control cell or an ordinary data cell, a bit to indicate whether this packet can be deleted in a congested network, and so on.

The ETSI definition of an ATM cell also contains 53 bytes with a 5-byte header and 48 bytes of data. The main differences are the number of bits in the VCI field, the number of bits in the header checksum, and the definitions and position of the control bits.

The IEEE 802.6 standard for the MAC layer of the metropolitan area network (MAN) DQDB (distributed queue dual bus) protocol is similar to the ATM cell.

42.9 Routing cell within an ATM network

In STM networks, data can change its position in each time slot in the interchanges over the global network. This can occur in ATM where the VCI label changes between intermediate nodes in the route.

When a transmitting node wishes to communicate through the network it makes contact with the UNI and negotiates parameters such as destination, traffic type, peak and traffic requirements, delay and cell loss requirement, and so on. The UNI forwards this request to the network. From this data the network computes a route based on the specified parameters and determines which links on each leg of the route can best support the requested quality of service and data traffic. It sends a connection setup request to all the nodes in the path en route to the destination node.

Figure 42.5 shows an example of ATM routing. In this case User 1 connects to Users 2 and 3. The virtual path set up between User 1 and User 2 is through the ATM switches 2, 3 and 4, whereas User 1 and User 3 connect through ATM switches 1, 5 and 6. A VCI number of 12 is assigned to the path between ATM switches 1 to 2, in the connection between User 1 and User 2. When ATM switch 2 receives a cell with a VCI number of 12 then it sends the cell to ATM switch 3 and gives it a new VCI number of 6. When it gets to ATM switch 3 it is routed to ATM switch 4 and given the VCI number of 22. The virtual circuit for User 1 to User 2 is through ATM switches 1, 5 and 6, and the VCI numbers used are 10 and 15. Once a connection is terminated the VCI labels assigned to the communications are used for other connections.

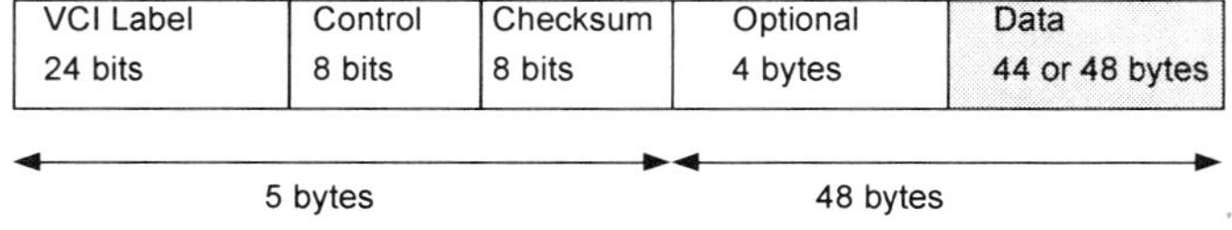

Figure 42.4 ATM cell.

Certain users, or applications, can be assigned reserved VCI labels for special services that may be provided by the network. However, as the address field only has 24 bits it is unlikely that many of these requests would be granted. ATM does not provide for acknowledgments when the cells arrive at the destination.

Note that as there is a virtual circuit setup between the transmitting and receiving node then cells are always delivered in the same order as they are transmitted. This is because cells cannot take alternative routes to the destination. Even if the cells are buffered at a node, they will still be transmitted in the correct sequence.

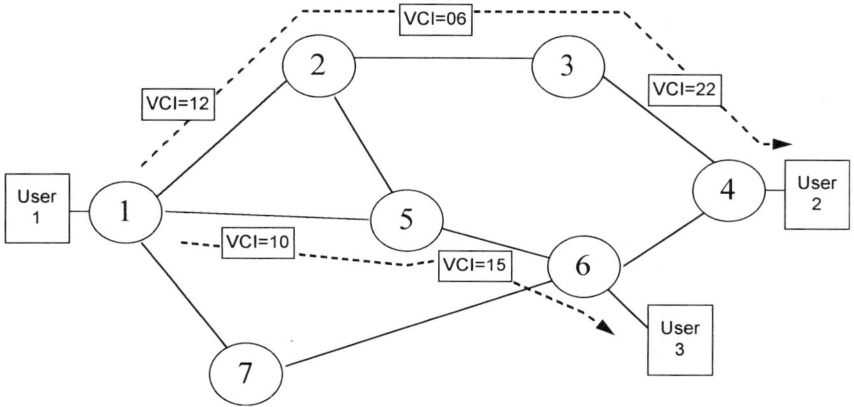

Figure 42.5 A virtual ATM virtual connection.

42.9.1 VCI header

The 5-byte ATM user network cell header is split into six main fields:

- GFC (4 bits) which is the generic control bit. It is used only for local significance and is not transmitted from the sender to the receiver.
- VPI (8 bits) which is the path connection identifier (VPI). See next section for an explanation of virtual path.
- VCI (16 bits) which is the virtual path/channel identifier (VCI). Its usage was described in the previous section. Each part of the route is allocated VCI number.
- PT (3 bits) which is the payload type field. This is used to identify the higher-layer application or data type.
- CLP (1 bit) which is the cell loss priority bit and indicates if a cell is expendable. When the network is busy an expendable cell may be deleted.
- HEC (8 bits) which is the header error control field. This is an 8-bit checksum for the header.

Note that the user to network cell differs from the network to network cell. A network to network cell uses a 12-bit VPI field and does not have a GFC field. Otherwise it is identical.

42.10 Virtual channels and virtual paths

Virtual circuits are set up between two users when a connection is made. Cells then travel over this fixed path through a reserved path. Often several virtual circuits take the same path. These circuits can be grouped together to form a virtual path.

A virtual path is defined as a collection of virtual channels which have the same start and end points. These channels will take the same route. This makes the network administration easier and allows new virtual circuit, with the same route, to be set up easily.

Some of the advantages of virtual paths are:

- Network user groups or interconnected networks can be mapped to virtual paths and are thus easily administered.
- Simpler network architecture which consists of groups (virtual paths) with individual connections (virtual circuits).
- Less network administration and shorter connection times arise from fewer setup connections.

Virtual circuits and virtual paths allows two levels of cell routing through the network. A VC switch routes virtual circuits and a VP switch routes virtual paths. Figure 42.7 shows a VP switch and a VC switch. In this case the VP switch contains the routing table which maps VP1 to VP2, VP2 to VP3 and VP3 to VP1. This switch does not change the VCI number of the incoming virtual circuits (for example VC1 goes in as VC1 and exits as VC1).

The diagram shows the concepts between both types of switches. The VP switch of the left will redirect the contents of a virtual path to a different virtual path. The virtual connections it contains are unchanged. This is similar to switching an input cable to a different physical cable. In a VC switch the virtual circuits are switched. In the case of Figure 42.7 the routing table will contain VC1 mapped to VC5, VC2 to VC6, and so on. The VC switch thus ignores the VP number and only routes the VC number. Thus the input and output VP number can change.

A connection is made by initially sending routing information cells through the network. When the connection is made, each switch in the route adds a link address for either a virtual path and or a virtual connection.

The combination of VP and VC addressing allows for the support of any addressing scheme, including subscriber telephone numbering or IP addresses. Each of these address can be broken down in a chain of VPI/VCI addresses.

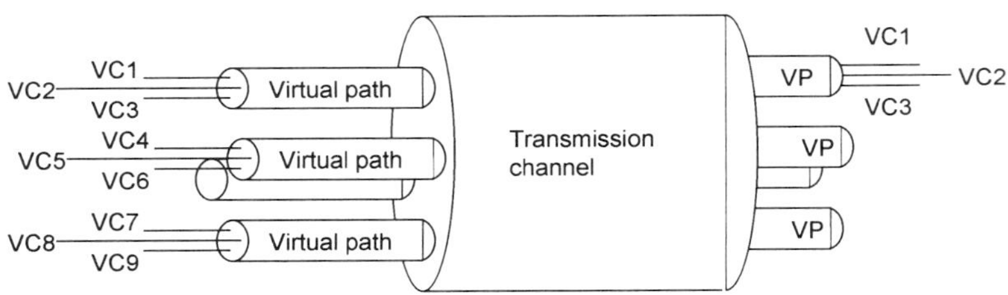

Figure 42.6 Virtual circuits and virtual paths.

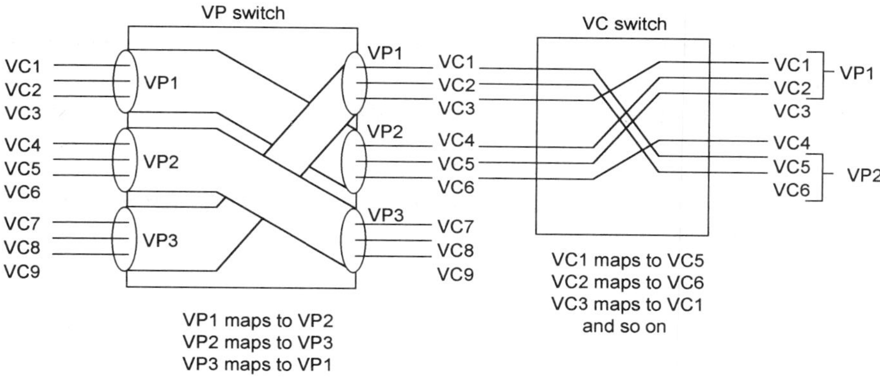

Figure 42.7 Virtual circuits and virtual paths.

42.11 ATM and the OSI model

The basic ATM cell fits roughly into the data link layer of the OSI model, but contains some network functions, such as end-to-end connection, flow control, and routing. It thus fits into layers 2 and 3 of the model, as shown in Figure 42.8. The layer 4 software layer, such as TCP/IP (as covered in Chapter 18), can communicate directly with ATM.

The ATM network provides a virtual connection between two gateways and the IP protocol fragments IP packets into ATM cells at the transmitting UNI which are then reassembled into the IP packet at the destination UNI.

With TCP/IP each host is assigned an IP address as is the ATM gateway. Once the connection has been made then the cells are fragmented into the ATM network and follow a predetermined route through the network. At the receiver the cells are reassembled using the ATM adaptation layer. This reforms the original IP packet which is then passed to the next layer.

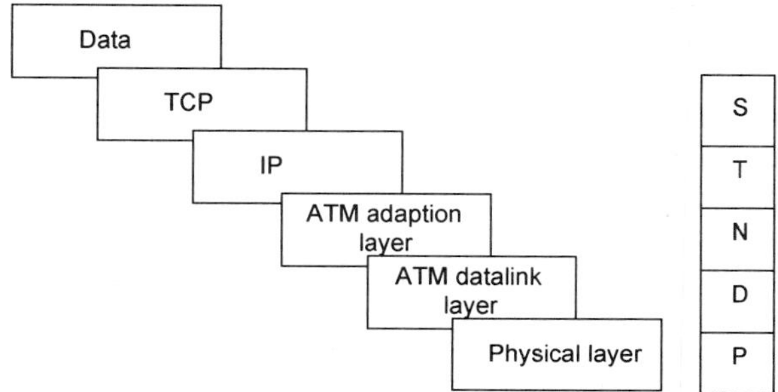

Figure 42.8 ATM and the OSI model.

The functions of the three ATM layers are:

- ATM adaptation layer (AAL) – segmentation and reassembly of data into cells and vice-versa, such as convergence (CS) and segmentation (SAR). It is also involved with quality of service (QOS).
- ATM data link (ADL) – maintenance of cells and their routing through the network, such as generic flow control, cell VPI/VCI translation, cell multiplex and demultiplex.
- ATM physical layer (PHY) – transmission and physical characteristics, such as cell rate decoupling, HEC header sequence generation/verification, cell delineation, transmission frame adaptation, transmission frame generation/recovery.

42.12 ATM physical layer

The physical layer is not an explicit part of the ATM definition, but is currently being considered by the standards organizations. T1S1 has standardized on SONET (Synchronous Optical NETwork) as the preferred physical layer, with STS-3c at 155.5 Mbps, STS-12 at 622 Mbps and STS-48 at 2.4 Gbps.

The SONET physical layer specification provides a standard worldwide digital telecommunications network hierarchy, known internationally as the Synchronous Digital Hierarchy (SDH). The base transmission rate, STS-1, is 51.84 Mbps. This is then multiplexed to make up higher bit rate streams, such as STS-3 which is 3 times STS-1, STS-12 which is 12 times STS-1, and so on. The 155 Mbps stream is the lowest bit rate for ATM traffic and is also called STM-1 (synchronous transport module - level 1).

The SDH specifies a standard method on how data is framed and transported synchronously across fiber-optic transmission links without requiring that all links and nodes have the same synchronized clock for data transmission and recovery.

42.13 AAL service levels

The AAL layer uses cells to process data into a cell-based format and to provide information to configure the level of service required.

42.13.1 Processing data

The AAL performs two essential functions for processing data as shown: the higher-level protocols present a data unit with a specific format. This data frame is then converted using a convergence sublayer with the addition of a header and trailer that give information on how the data unit should be segmented into cells and reassembled at the destination. The data is then segmented into cells, together with the convergence subsystem information and other management data, and sent through the network.

42.13.2 AAL functionality

The AAL layer, as part of the process, also defines the level of service that the user wants from the connection. The following shows the four classes supported. For each class there is an associated AAL.

Timing information between source and destination
CLASS A: Required CLASS B: Required
CLASS C: Not required CLASS D: Not required

Bit rate characteristics
CLASS A: Constant CLASS B: Variable
CLASS C: Variable CLASS D: Variable

Connection mode
CLASS A: Connection-oriented CLASS B: Connection-oriented
CLASS C: Connection-oriented CLASS D: Connectionless

42.13.3 AAL services

Thus class A supports a constant bit rate with a connection and preserves timing information. This is typically used for voice transmission. Class B is similar to A but has a variable bit rate. Typically it is used for video/audio data. Class C also has a variable bit rate and is connection-oriented, although there is no timing information. This is typically used for non-real-time data, such as computer data. Class D is the same as class C but is connectionless. This means there is no connection between the sender and the receiver before the data is transmitted.

There are four AAL services: AAL1 (for class A), AAL2 (for class B), AAL3/4 or AAL5 (for class C) and AAL3/4 (for class D).

AAL1

The AAL1 supports class A and is intended for real-time voice traffic; it provides a constant bit rate and preserves timing information over a connection. The format of the 48 bytes of the data cell consists of 47 bytes of data, such as PCM or ADPCM code and a 1-byte header. Figure 42.9 shows the format of the cell, including the cell header. The 47-byte data field is described as the SAR-PDU (segmentation and reassembly protocol data units). The header consists of:

- SN (4 bits) which is a sequence number.
- SNP (4 bits) which is the sequence number protection.

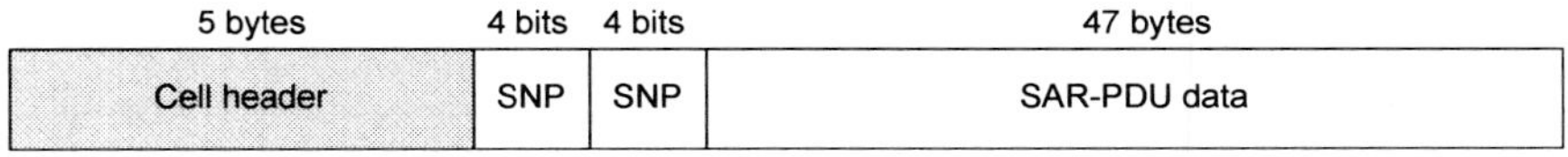

Figure 42.9 ATM and the OSI model.

AAL type 2

AAL type 2 is under further study.

AAL type 3/4

Type 3/4 is connection-oriented where the bit rate is variable and there is no need for timing information (Figure 42.10). It uses two main formats:

- SAR (segment and reassemble) which is segments of CPCS PDU with a SAR header and trailer. The extra SAR fields allow the data to be reassembled at the receiver. When the CPCS PDU data has been reassembled the header and trailer are discarded. The fields in the SAR are:
  - Segment type (ST) identifies has the SAR has been segmented. Figure 42.10 show how the CPCS PDU data has been segmented into five segments: one beginning segment, three continuation segments and one end segment. The ST field has 2 bits and can therefore contain one of four possible types:
    - SSM (single sequence message) identifies that the SAR contains the complete data.
    - BOM (beginning of message) identifies that it is the first SAR PDU in a sequence.
    - COM (continuation of message) identifies that it is an intermediate SAR PDU.
    - EOM (end of message) identifies that it is the last SAR PDU.
  - Sequence number (SN) which is used to reassemble a SAR SDU and thus verify that all of the SAR PDUs have been received.
  - Message identifier (MI) which is a unique identifier associated with the set of SAR PDUs that carry a single SAR SDU.
  - Length indication (LI) which defines the number of bytes in the SAR PDU. It can have a value between 4 and 44. The COM and BOM types will always have a value of 44. If the EOM field contains fewer than 44 bytes, it is padded to fill the remaining bytes. The LI then indicates the number of value bytes. For example if the LI is 20; there are only 20 value bytes in the SAR PDU the other 24 are padding bytes.
  - CRC which is a 10-bit CRC for the entire SAR PDU.
- CPCS (convergence protocol sublayer) takes data from the PDU. As this can be any length, the data is padded so it can be divided by 4. A header and trailer are then added and the completed data stream is converted into one or more SAR PDU format cells. Figure 42.11 shows the format of the CPCS-PDU for AAL type 3/4.

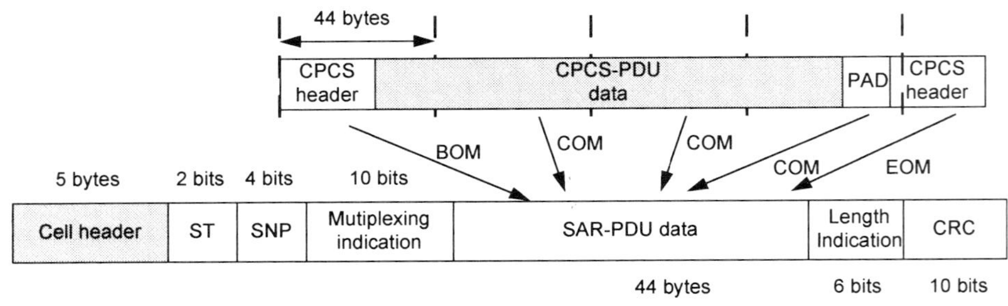

Figure 42.10 AAL type 3/4 cell format.

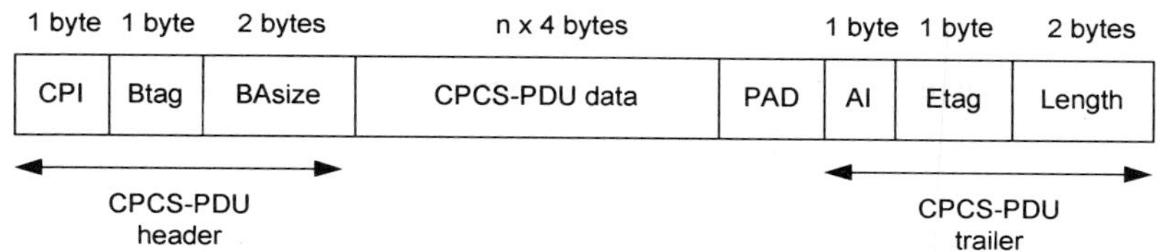

Figure 42.11 CPCS-PDU type 3/4 frame format.

The fields in the CPCS-PDU are:

- CPI (common part indicator) which indicates how the remaining fields are interpreted (currently one version exists).
- Btag (beginning tag) which is a value associated with the CPCS-PDU data. The Etag has the same value as the Btag.
- BASize (buffer allocation size) which indicates the size of the buffer that must be reserved so that the completed message can be stored.
- AI (alignment) a single byte which is added to make the trailer equal to 32 bits.
- Etag (end tag) which is the same as the Btag value.
- Length which gives the length of the CPCS PDU data field.

AAL type 5

AAL type 5 is a connectionless service; it has no timing information and can have a variable bit rate. It assumes that one of the levels above the AAL can establish and maintain a connection. Type 5 provides stronger error checking with a 32-bit CRC for the entire CPCS PDU, whereas type 3/4 only allows a 10-bit CRC which is error checking for each SAR PDU. The type 5 format is given in Figure 42.12. The fields are:

- CPCS-UU (CPCS user-to-user) indication.
- CPI (common part indicator) which indicates how the remaining fields are interpreted (currently one version exists).
- Length which gives the length of the CPCS-PDU data.
- CRC which is a 32-bit CRC field.

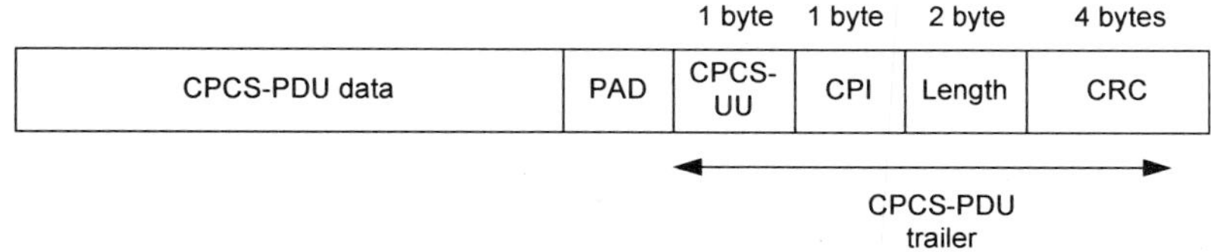

Figure 42.12 CPCS-PDU type 5 frame format.

The type 5 CPCS-PDU is then segmented into groups of 44 bytes and the ATM cell header is added. Thus type 5 does not have the overhead of the SAR-PDU header and trailer (i.e. it does not have ST, SN, MID, LI or CRC). This means it does not contain any sequence numbers. It is thus assumed that the cells will always be received in the correct order and none of the cells will be lost. Types 3/4 and 5 can be summarized as follows:

Type 3/4: SAR-PDU overhead is 4 bytes, CPCS-PDU overhead is 8 bytes.
Type 5: SAR-PDU overhead is 0 bytes, CPCS-PDU overhead is 8 bytes.

Type 5 can be characterized as:

- Strong error checking.
- Lack of sequence numbers.
- Reduced overhead of the SAR-PDU header and trailer.

42.14 ATM flow control

ATM cannot provide for a reactive end-to-end flow control because by the time a message is returned from the destination to the source, large amounts of data could have been sent along the ATM pipe, possibly making the congestion worse. The opposite can occur when the congestion on the network has cleared by the time the flow control message reaches the transmitter. The transmitter will thus reduce the data flow when there is little need. ATM tries to react to network congestion quickly, and it slowly reduces the input data flow to reduce congestion.

This rate-based scheme of flow control involves controlling the amount of data to a specified rate, agreed when the connection is made. It then automatically changes the rate based on the past history of the connection as well as the present congestion state of the network.

Data input is thus controlled by early detection of traffic congestion through closely monitoring the internal queues inside the ATM switches, as shown in Figure 42.13. The network then reacts gradually as the queues lengthen and reduces the traffic into the network from the transmitting UNI. This is an improvement over imposing a complete restriction on the data input when the route is totally congested. In summary, anticipation is better than desperation.

A major objective of the flow control scheme is to try to affect only the streams which are causing the congestion, not the well-behaved streams.

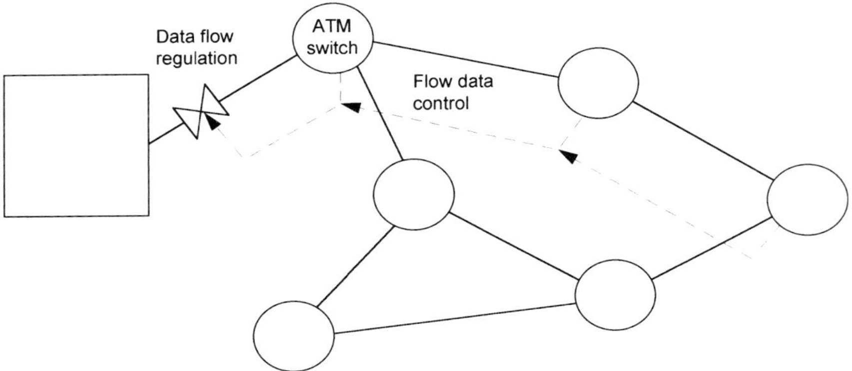

Figure 42.13 Flow control feedback from ATM switches.

42.15 Practical ATM networks

As mentioned in Chapter 41 a metropolitan area network has been set-up around Edinburgh, UK. It consists of two rings on ATM and FDDI. The two rings of FDDI and ATM have been run around the Edinburgh sites. This also connects to the University of Stirling through a 155 Mbps SDH connection. Two connections on the ring are made to the SuperJANET network, connections at Heriot-Watt University and the University of Edinburgh.

The 100 Mbps FDDI dual rings link 10 Edinburgh city sites. This ring provides for IP traffic on SuperJANET and also for high-speed metropolitan connections. Initially a 155 Mbps SDH/STM-1 ATM network connects five Edinburgh sites and the University of Stirling. This also connects to the SuperJANET ATM pilot network. Figure 42.14 shows the FDDI and ATM connections.

The two different network technologies allow the universities to operate a two-speed network. For computer-type data the well-established FDDI technology provides good reliable communications and the ATM network allows for future exploitation of mixed voice, data and video transmissions.

The JANET and SuperJANET networks provide connections to all UK universities. A gateway out of the network to the rest of the world is located at University College London (UCL).

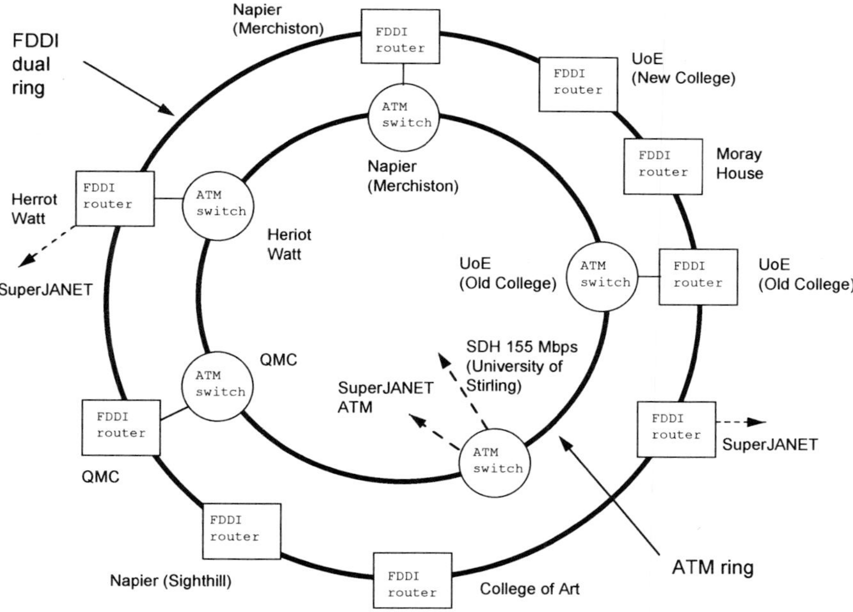

Figure 42.14 EaStMAN ring connections.

43 ISDN

43.1 Introduction

A major problem in data communications and networks is the integration of real-time sampled data with non-real-time (normal) computer. Sampled data tends to create a constant traffic flow whereas computer-type data has bursts of traffic. And sampled data normally needs to be delivered at a given time but computer-type data needs a reliable path where delays are relatively unimportant.

The basic rate for real-time data is speech. It is normally sampled at a rate of 8 kHz and each sample is coded with 8 bits. This leads to a transmission bit rate of 64 kbps. ISDN uses this transmission rate for its base transmission rate. Computer-type data can then be transmitted using this rate or can be split to transmit over several 64 kbps channels. The basic rate ISDN service uses two 64 kbps data lines and a 16 kbps control line, as illustrated in Figure 43.1. Table 43.1 summarizes the I series CCITT standards.

Typically modems are used in the home for the transmission of computer-type data. Unfortunately modems have a maximum bit rate of 28.8 kbps. This is automatically increased, on a single channel, to 64 kbps. The connections made by a modem and by ISDN are circuit-switched.

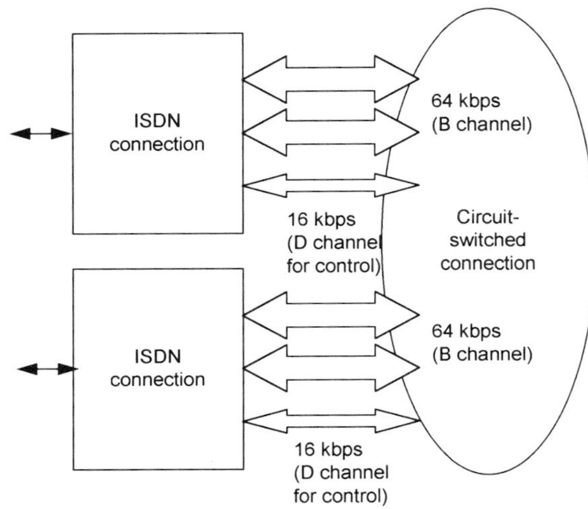

Figure 43.1 Basic rate ISDN services.

Table 43.1 CCITT standards on ISDN.

CCITT standard number	*Description*
I.1*XX*	ISDN terms and technology
I.2*XX*	ISDN services
I.3*XX*	ISDN addressing
I.430 and I.431	ISDN physical layer interface
I.440 and I.441	ISDN data layer interface
I.450 and I.451	ISDN network layer interface
I.5*XX*	ISDN internetworking
I.6*XX*	ISDN maintenance

The great advantage of an ISDN connection is that the type of data transmitted is irrelevant to the transmission and switching circuitry. Thus it can carry other types of digital data, such as facsimile, teletex, videotex and computer data. This reduces the need for modems, which convert digital data into an analogue form, only for the public telephone network to convert the analogue signal back into a digital form for transmission over a digital link. It is also possible to multiplex the basic rate of 64 kbps to give even higher data rates. This multiplexing is known as N × 64 kbps or Broadband ISDN (B-ISDN).

Another advantage of ISDN is that it is a circuit-switched connection where a permanent connection is established between two nodes. This connection is guaranteed for the length of the connection. It also has a dependable delay time and is thus suited to real-time data.

43.2 ISDN channels

ISDN uses channels to identify the data rate, each based on the 64 kbps provision. Typical channels are B, D, H0, H11 and H12. The B-channel has a data rate of 64 kbps and provides a circuit-switched connection between endpoints. A D-channel operates at 16 kbps and it controls the data transfers over the B channels. The other channels provide B-ISDN for much higher data rates. Table 43.2 outlines the basic data rates for these channels.

The two main types of interface are the basic rate access and the primary rate access. Both are based around groupings of B- and D-channels. The basic rate access allows two B-channels and one 16 kbps D-channel.

Primary rate provides B-ISDN, such as H12, which gives 30 B-channels and a 64 kbps D-channel. For basic and primary rates, all channels multiplex onto a single line by combining channels into frames and adding extra synchronization bits. Figure 43.3 gives examples of the basic rate and primary rate.

Table 43.2 ISDN channels.

Channel	*Description*
B	64 kbps
D	16 kbps signaling for channel B (ISDN)
	64 kbps signaling for channel B (B-ISDN)
H0	384 kbps (6 × 64 kbps) for B-ISDN
H11	1.536 Mbps (24 × 64 kbps) for B-ISDN
H12	1.920 Mbps (30 × 64 kbps) for B-ISDN

The basic rate ISDN gives two B-channels at 64 kbps and a signaling channel at 16 kbps. These multiplex into a frame and, after adding extra framing bits, the total output data rate is 192 kbps. The total data rate for the basic rate service is thus 128 kbps. One or many devices may multiplex their data, such as two devices transmitting at 64 kbps, a single device multiplexing its 128 kbps data over two channels (giving 128 kbps), or by several devices transmitting a sub-64 kbps data rate over the two channels. For example, four 32 kbps devices could simultaneously transmit their data, eight 16 kbps devices, and so on.

For H12, 30 × 64 kbps channels multiplex with a 64 kbps-signaling channel, and with extra framing bits, the resulting data rate is 2.048 Mbps (compatible with European PCM-TDM systems). This means the actual data rate is 1.920 Mbps. As with the basic service this could contain a number of devices with a data rate of less than or greater than a multiple of 64 kbps.

For H11, 24 × 64 kbps channels multiplex with a 64 kbps-signaling channel, and with extra framing bits, it produces a data rate of 1.544 Mbps (compatible with USA PCM-TDM systems). The actual data rate is 1.536 Mbps.

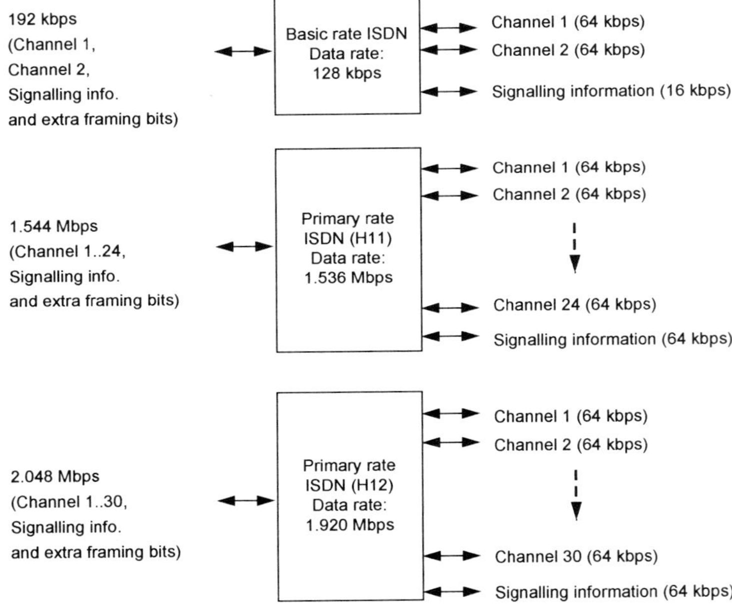

Figure 43.2 Basic rate, H11 and H12 ISDN services.

43.3 ISDN physical layer interfacing

The physical layer corresponds to layer 1 of the OSI 7-layer model and is defined in CCITT specifications I.430 and I.431. Pulses on the line are not coded as pure binary, they use a technique called alternate mark inversion (AMI).

43.3.1 Alternative mark inversion (AMI) line code

AMI line codes use three voltage levels. In pure AMI, 0 V represents a '0', and the voltage amplitude for each '1' is the inverse of the previous '1' bit. ISDN uses the inverse of this, i.e. 0 V for a '1' and an inverse in voltage for a '0', as shown in Figure 43.3. Normally the pulse amplitude is 0.75 V.

Inversion of the AMI signal (i.e. inverting a '0' rather than a '1') allows for timing information to be recovered when there are long runs of zeros, which is typical in the idle state. AMI line code also automatically balances the signal voltage, and the average voltage will be approximately zero even when there are long runs of 0's.

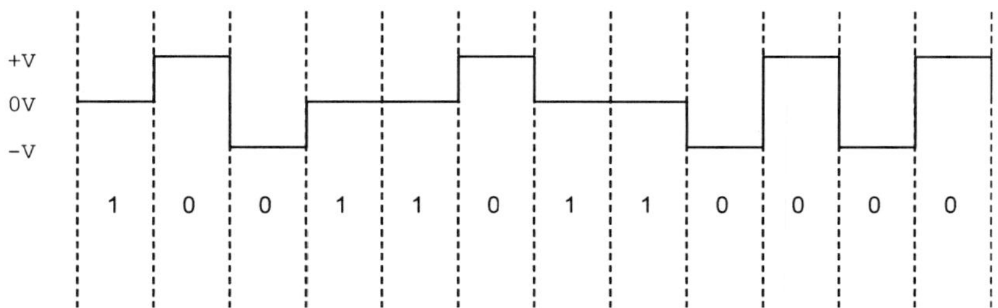

Figure 43.3 AMI used in ISDN.

43.3.2 System connections

In basic rate connections up to eight devices, or items of termination equipment (TE), can connect to the network termination (NT). They connect over a common four-wire bus using two sets of twisted-pair cables. The transmit output (T_X) on each TE connects to the transmit output on the other TEs, and the receive input (R_X) on each TE connects to all other TEs. On the NT the receive input connects to the transmit of the TEs, and the transmit output of the NT connects to the receive input of the TEs. A contention protocol allows only one TE to communicate at a time.

An 8-pin ISO 8877 connector connects a TE to the NT; this is similar to the RJ-45 connector but has two extra pin connections. Figure 43.4 shows the pin connections. Pins 3 and 6 carry the T_X signal from the TE, pins 4 and 5 provide the R_X to the TEs. Pins 7 and 8 are the secondary power supply from the NT and pins 1 and 2 the power supply from the TE (if used). The T_X/R_X lines connect via transformers thus only the AC part of the bitstream transfers into the PCM circuitry of the TE and the NT. This produces a need for a balanced DC line code such as AMI, as the DC component in the bitstream will not pass through the transformers.

43.3.3 Frame format

Figures 33.5 and 33.6 show the ISDN frame formats. Each frame is 250 μs long and contains 48 bits; this give a total bit rate of 192 kbps ($48/250 \times 10^{-6}$) made up of two 64 kbps B channels, one 16 kbps D-channel and extra framing, DC balancing and synchronization bits.

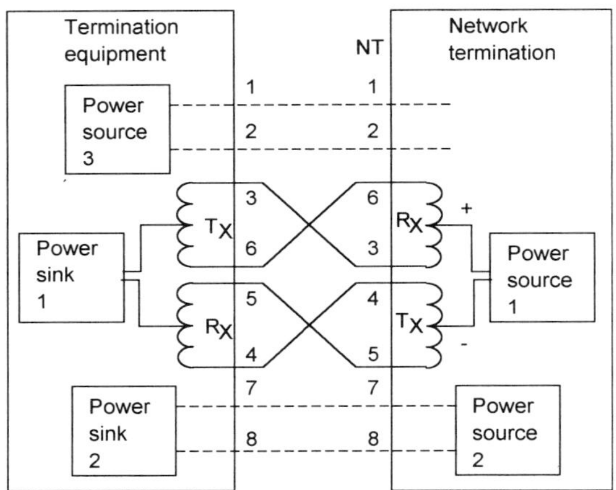

Figure 43.4 Power supplies between NT and TE.

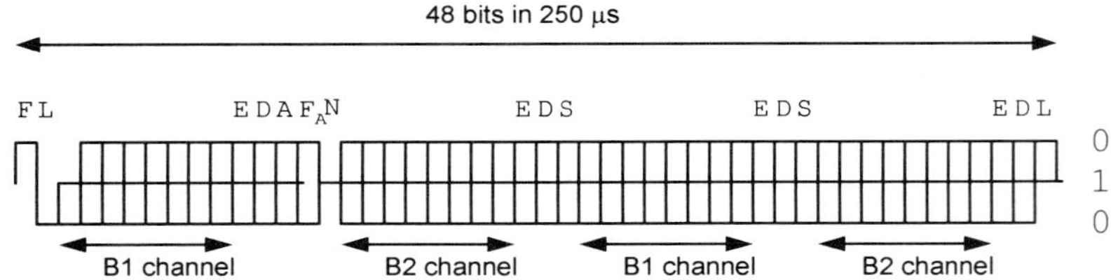

Figure 43.5 ISDN frame format for NT to TE.

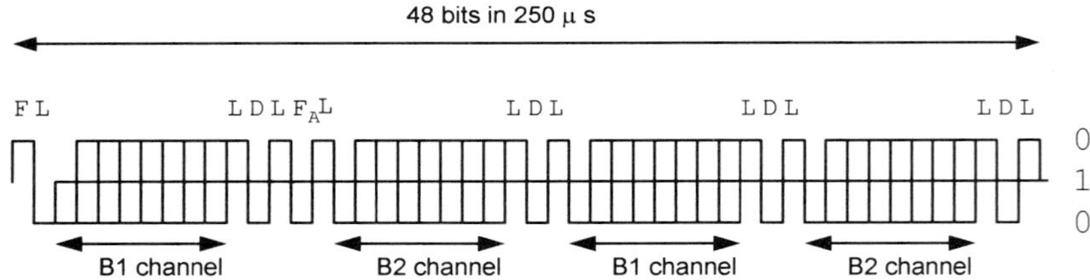

Figure 43.6 ISDN frame format for TE to NT.

where

F –	framing bit	N –	set to a 1
L –	DC balancing bit	D –	D-channel bit
E –	D-echo channel bit	F_A –	auxiliary framing bit (= 0)
S –	reserved for future use	A –	activation bit
M –	multiframing bit	B1 –	bits for channel 1
B2 –	bits for channel 2		

The F/L pair of bits identify the start of each transmitted frame. When transmitting from a TE to an NT there is a 10-bit offset in the return of the frame back to the TE. The E bits echo the D-channel bits back to the TE.

When transmitting from the NT to the TE, the bits after the F/L bits, in the B-channel, have a volition in the first 0. If any of these bits is a 0 then a volition will occur, but if they are 1's then no volition can occur. To overcome this the F_A bit forces a volition. Since it is followed by 0 (the N bit) it will not be confused with the F/L pair. The start of the frame can thus be traced backwards to find the F/L pair.

There are 16 bits for each B-channel, giving a basic data rate of 64 kbps ($16/250 \times 10^{-6}$) and there are 4 bits in the frame for the D-channel, giving a bit rate of 16 kbps ($4/250 \times 10^{-6}$).

The L bit balances the DC level on the line. If the number of 0's following the last balancing bit is odd then the balancing bit is a 0, else it is a 1. When synchronized the NT informs the TEs by setting the A bit.

43.4 ISDN data link layer

The data link layer uses a form known as the Link Access Procedure for the D-channel (LAPD). Figure 43.7 shows the frame format. The unique bit sequence 01111110 identifies the start and end of the frame. This bit pattern cannot occur in the rest of the frame due to zero bit-stuffing.

The address field contains information on the type of data contained in the frame (the service access point identifier) and the physical address of the ISDN device (the terminal endpoint identifier). The control field contains a supervisory, an unnumbered or an information frame. The frame check sequence provides error detection information.

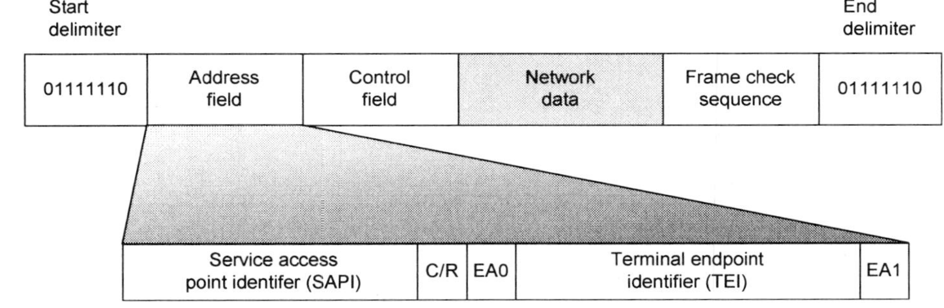

Figure 43.7 D-channel frame structure.

43.4.1 Address field

The data link address only contains addressing information to connect the TE to the NT and does not have network addresses. Figure 43.7 shows the address field format. The SAPI identifies the type of ISDN services. For example a frame from a telephone would be identified as such, and only telephones would read the frame.

All TEs connect to a single multiplexed bus, thus each has a unique data link address, known as a terminal endpoint identifier (TEI). The user or the network sets this; the ranges of available addresses are:

0–63	non-automatic assignment TEIs
64–126	automatic assignment TEIs
127	global TEI

The non-automatic assignment involves the user setting the address of each of the devices connected to the network. When a device transmits data it inserts its own TEI address and only receives data which has its TEI address. In most cases devices should not have the same TEI address, as this would cause all devices with the same TEI address, and the SAPI, to receive the same data (although, in some cases, this may be a requirement).

The network allocates addresses to devices requiring automatic assignment before they can communicate with any other devices. The global TEI address is used to broadcast messages to all connected devices. A typical example is when a telephone call is incoming to a group on a shared line where all the telephones would ring until one was answered.

The `C/R` bit is the command/response bit and `EA0/EA1` are extended address field bits.

43.4.2 Bit stuffing

With zero bit-stuffing the transmitter inserts a zero into the bitstream when transmitting five consecutive 1's. When the receiver receives five consecutive 1's it deletes the next bit if it is a zero. This stops the unique `01111110` sequence occurring within the frame. For example if the bits to be transmitted are:

`101000101011111100001010001010000111110101010`

then the with the start and end delimiter this would be:

01111110101000101*01111110*0001010001010000111110101010**01111110**

It can be seen from this bitstream that the stream to be transmitted contains the delimiter within the frame. This zero bit-insertion is applied to give:

01111110101000101*011111***0***10*00010100010100001111**0**0101010**01111110**

Notice that the transmitter has inserted a zero when five consecutive 1's occur. Thus the bit pattern 01111110 cannot occur anywhere in the bitstream. When the receiver receives five consecutive 1's it deletes the next bit if it is a zero. If it is a 1 then it is a valid delimiter. In the example the received stream will be:

01111110101000101011111100001010001010000111110101010**01111110**

43.4.3 Control field

ISDN uses a 16-bit control field for information and supervisory frames and an 8-bit field for unnumbered frames, as illustrated in Figure 43.8. Information frames contain sequenced data. The format is `0SSSSSSSXRRRRRRR`, where `SSSSSSS` is the send sequence number and `RRRRRRR` is the frame sequence number that the sender expects to receive next (`X` is the poll/final bit). Since the extended mode uses a 7-bit sequence field then information frames are numbered from 0 to 127.

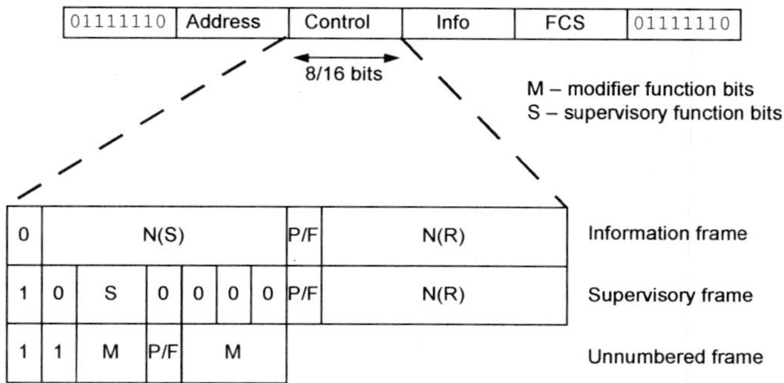

Figure 43.8 ISDN control field.

Supervisory frames contain flow control data. Table 43.3 lists the supervisory frame types and the control field bit settings. The `RRRRRRR` value represent the 7-bit receive sequence number.

Table 43.3 Supervisory frame types and control field settings.

Type	*Control field setting*
Receiver ready (RR)	`10000000PRRRRRRR`
Receiver not ready (RNR)	`10100000PRRRRRRR`
Reject (REJ)	`10010000PRRRRRRR`

Unnumbered frames set up and clear connections between a node and the network. Table 43.4 lists the unnumbered frame commands and Table 43.5 lists the unnumbered frame responses.

Table 43.4 Unnumbered frame commands and control field settings.

Type	*Control field setting*
Set asynchronous balance mode extended (SABME)	`1111P110`
Unnumbered information (UI)	`1100F000`
Disconnect mode (DISC)	`1100P010`

Table 43.5 Unnumbered frame responses and control field settings.

Type	*Control field setting*
Disconnect mode (DM)	`1111P110`
Unnumbered acknowledgment (UA)	`1100F000`
Frame reject (FRMR)	`1110P001`

In ISDN all connected nodes and the network connection can send commands and receive responses. Figure 43.9 shows a sample connection of an incoming call to an ISDN node (address `TEI_1`). The SABME mode is set up initially using the SABME command (`U[SABME,TEI_1,P=1]`), followed by an acknowledgment from the ISDN node (`U[UA,TEI_1,F=1]`). At any time, either the network or the node can disconnect the connection. In this case the ISDN node disconnects the connection with the command U[DISC,TEI_1,P=1]. The network connection acknowledges this with an unnumbered acknowledgment (U[UA,TEI_1,F=1]).

43.4.4 D-channel contention

The D-channel contention protocol ensures that only one terminal can transmit its data at a time. This happens because the start and the end of the D-channel bits have the bitstream 01111110, as shown below:

1111**01111110**XXXXXXXXX...XXXXXXXX**01111110**1111

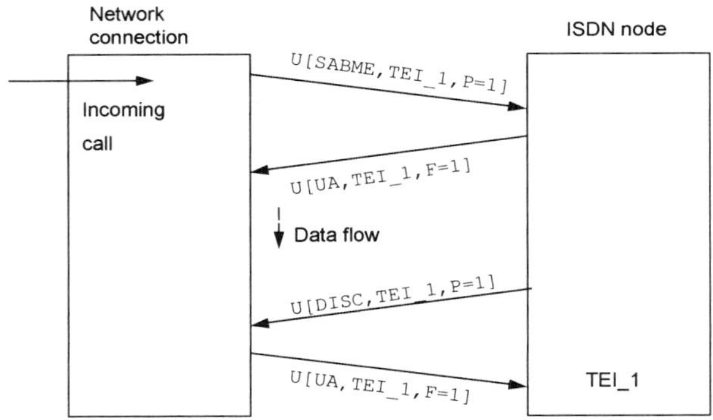

Figure 43.9 Example connection between a primary/secondary

When idle, each TE floats to a high-impedance state, which is taken as a binary 1. To transmit, a TE counts the number of 1's in the D-channel. A 0 resets this count. After a predetermined number, greater than a predetermined number of consecutive 1's, the TE transmits its data and monitors the return from the NT. If it does not receive the correct D-channel bitstream returned through the E bits then a collision has occurred. When a TE detects a collision it immediately stops transmitting and monitors the line.

When a TE has finished transmitting data it increases its count value for the number of consecutive 1's by 1. This gives other TEs an opportunity to transmit their data.

43.4.5 Frame check sequence

The frame check sequence (FCS) field contains an error detection code based on cyclic redundancy check (CRC) polynomials. It uses the CCITT V.41 polynomial, which is $G(x) = x^{16} + x^{12} + x^5 + x^1$.

43.5 ISDN network layer

The D-channel carriers network layer information within the LAPD frame. This information establishes and controls a connection. The LAPD frames contain no true data as this is carried in the B-channel. Its function is to set up and manage calls and to provide flow control between connections over the network.

Figure 43.10 shows the format of the layer 3 signaling message frame. The first byte is the protocol discriminator. In the future this byte will define different communications protocols. At present it is normally set to `0001000`. After the second byte the call reference value is defined. This is used to identify particular calls with a reference number. The length of the call reference value is defined within the second byte. As it contains a 4-bit value, up to 16 bytes can be contained in the call reference value field. The next byte gives the message type and this type defines the information contained in the proceeding field.

There are four main types of message: call establish, call information, call clearing and miscellaneous messages. Table 43.6 outlines the main messages. Figure 43.11 shows an example connection procedure. The initial message sent is SETUP. This may contain some of the following:

- Channel identification – identifies a channel with an ISDN interface.
- Calling party number.
- Calling party subaddress.
- Called party number.
- Called party subnumber.
- Extra data (2–131 bytes).

After the calling TE has sent the SETUP message, the network then returns the SETUP ACK message. If there is insufficient information in the SETUP message then other information needs to flow between the called TE and the network. After this the network sends back a CALL PROCEEDING message and it also sends a SETUP message to the called TE. When the called TE detects its TEI address and SAPI, it sends back an ALERTING message. This informs the network that the node is alerting the user to answer the call. When it is answered, the called TE sends a CONNECT to the network. The network then acknowledges this with a CONNECT ACK message, at the same time it sends a CONNECT message to the calling TE. The calling TE then acknowledges this with a CONNECT ACK. The connection is then established between the two nodes and data can be transferred.

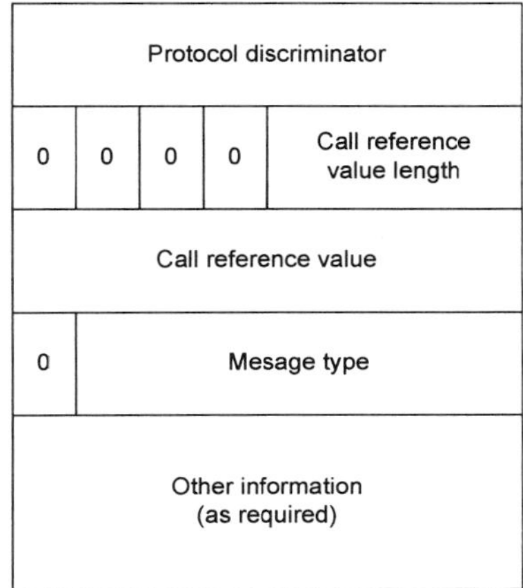

Figure 43.10 Signaling message structure.

To disconnect the connection the DISCONNECT, RELEASE and RELEASE COMPLETE messages are used.

Table 43.6 ISDN network messages

Call establish	*Information messages*	*Call clearing*
ALERTING	RESUME	DISCONNECT
CALL PROCEEDING	RESUME ACKNOWLEDGE	RELEASE
CONNECT	RESUME REJECT	RELEASE COMPLETE
CONNECT ACKNOWLEDGE	SUSPEND	RESTART
PROGRESS	SUSPEND ACKNOWLEDGE	RESTART ACKNOWLEDGE
SETUP	SUSPEND REJECT	
SETUP ACKNOWLEDGE	USER INFORMATION	

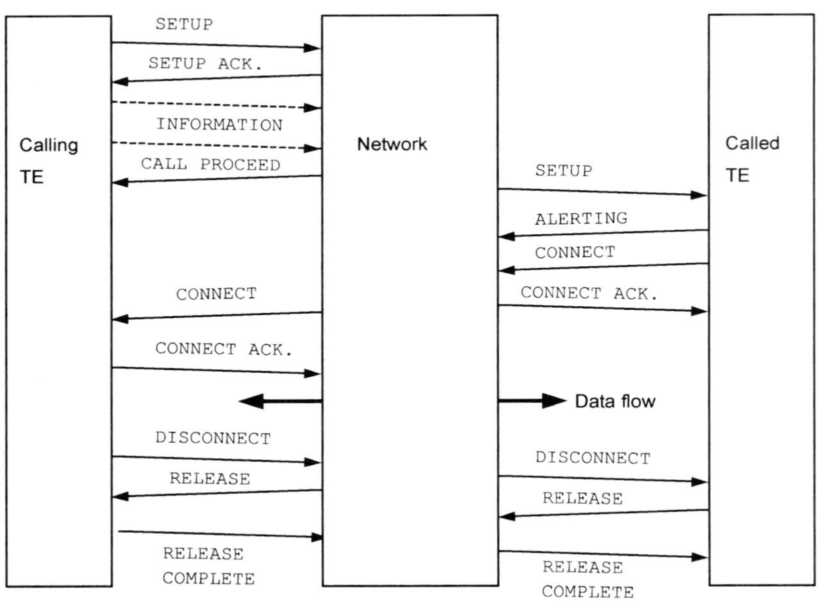

Figure 43.11 Call establishment and clearing.

44 HDLC

44.1 Introduction

The data link layer is the second layer in the OSI seven-layer model and its protocols define rules for the orderly exchange of data information between 2 adjacent nodes connected by a data link. Final framing, flow control between nodes, and error detection and correction are added at this layer. In previous chapters the data link layer was discussed in a practical manner. In this chapter its functions will be discussed with reference to HDLC.

The two types of protocol are:

- Asynchronous protocol.
- Synchronous protocol.

Asynchronous communications uses start-stop method of communication where characters are sent between nodes, as illustrated in Figure 44.1. Special characters are used to control the data flow. Typical flow control characters are End of Transmission (EOT), Acknowledgement (ACK), Start of Transmission (STX) and Negative Acknowledgement (NACK).

Synchronous communications involves the transmission of frames of bits with start and end bit characters to delimit the frame. The most popular are IBM's synchronous data link communication (SDLC) and high-level data link control (HDLC). Many network data link layers are based upon these standards, examples include the LLC layer in IEE 802.*x* LAN standards and LAPB in the X.25 packet switching standard.

Synchronous communications normally uses a bit-oriented protocol (BOP), where data is sent one bit at a time. The data link control information is interpreted on a bit-by-bit basis rather than with unique data link control characters.

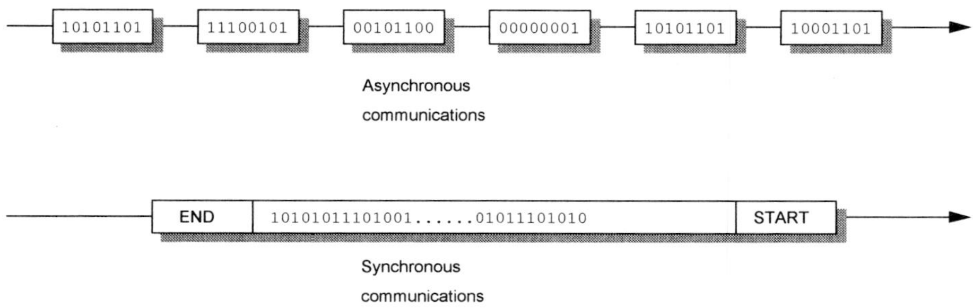

Figure 44.1 Asynchronous and synchronous communications.

HDLC is a standard developed by the ISO to provide a basis for the data link layer for point-to-point and multi-drop connections. It can transfer data either in a simplex, half-duplex, or full-duplex mode. Frames are generally limited to 256 bytes in length and a single control field performs most data link control functions.

44.2 HDLC protocol

In HDLC, a node is either defined as a primary station or a secondary station. A primary station controls the flow of information and issues commands to secondary stations. The secondary station then sends back responses to the primary. A primary station with one or more secondary stations is known as unbalanced configuration.

HDLC allows for point-to-point and multi-drop. In point-to-point communications a primary station communicates with a single secondary station. For multi-drop, one primary station communications with many secondary stations.

In point-to-point communications it is possible for a station be operate as a primary and a secondary station. At any time one of the stations can be a primary and the other the secondary. Thus commands and responses flow back and forth over the transmission link. This is known as a balanced configuration, or combined stations.

44.2.1 HDLC modes of operation

HDLC has three modes of operation. Unbalanced configurations can use the normal response mode (NRM). Secondary stations can only transmit when specifically instructed by the primary station. When used as a point-to-point or multi-drop configuration only one primary station is used. Figure 44.2 shows a multi-drop NRM configuration.

Unbalanced configurations can also use the asynchronous response mode (ARM). It differs from NRM in that the secondary is allowed to communicate with the primary without receiving permission from the primary.

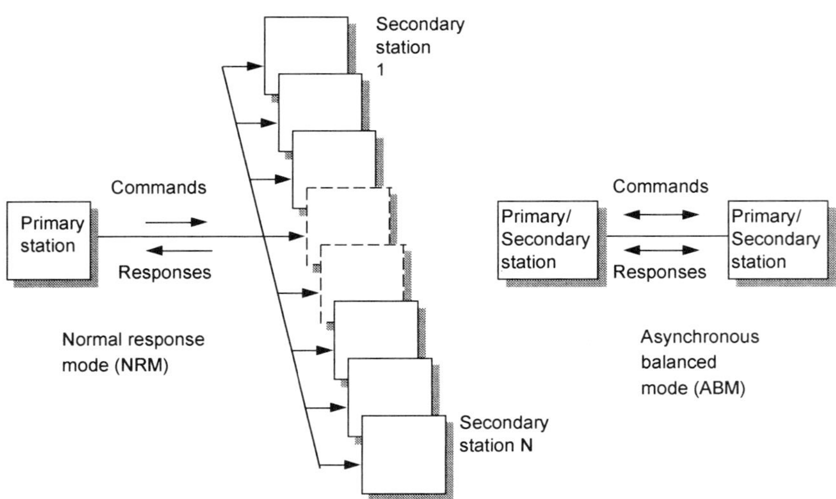

Figure 44.2 NRM and ABM mode.

In asynchronous balanced mode (ABM) all stations have the same priority and can perform the functions of a primary and secondary station.

44.2.2 HDLC frame format

HDLC frames are delimited by the bit sequence `01111110`. Figure 44.3 shows the standard format of the HDLC frame, the 5 fields are the:

- Flag field.
- Address field.
- Control field.
- Information field.
- Frame check sequence (FCS) field.

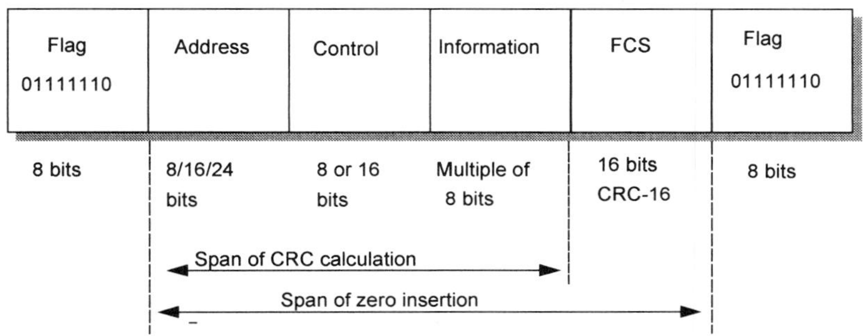

Figure 44.3 HDLC frame structure.

44.2.3 Information field

The information fields contain data, such as OSI level 3, and above, information. It contains an integer number of bytes and thus the number of bits contained is always a multiple of eight. The receiver determines the number of bytes in the data because it can detect the start and end flag. By this method it also finds the FCS field. Note that the number of characters in the information can be zero as not all frames contain data.

44.2.4 Flag field

A unique flag sequence, `01111110` (or `7Eh`), delimits the start and end of the frame. As this sequence could occur anywhere within the frame a technique called bit-insertion is used to stop this happening except at the start and end of the frame.

44.2.5 Address field

The address field is used to address connected stations an, in basic addressing, it contains an 8-bit address. It can also be extended, using extended addressing, to give any multiple of 8 bits.

When it is 8 bits wide it can address up to 254 different nodes, as illustrated in Figure 44.4. Two special addresses are `00000000` and `11111111`. The `00000000` address defines the null or void address and the `11111111` broadcasts a message to all secondaries. The other 254 addresses are used to address secondary nodes individually.

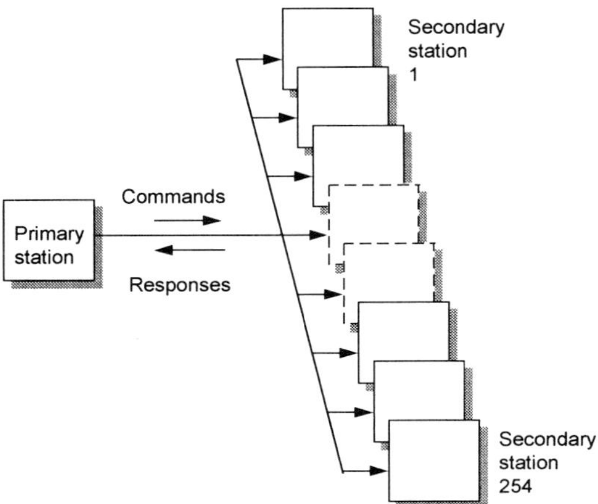

Figure 44.4 HDLC addressing range.

If there are a large number of secondary stations then extended address can be used to extend the address field indefinitely. A 0 in the first bit of the address field allows a continuation of the address, or a 1 ends it. For example:

```
XXXXXXX1 XXXXXXX0 XXXXXXX0 XXXXXXX0
```

44.2.6 Control field

The control field can either be 8 or 16 bits wide. It is used to identify the frame type and can also contain flow control information. The first two bits of the control field define the frame type, as shown in Figure 44.5. There are three types of frames, these are:

- Information frames.
- Supervisory frames.
- Unnumbered frames.

When sent from the primary the `P/F` bit indicates that it is polling the secondary station. In an unbalanced mode a secondary station cannot transmit frames unless the primary sets the poll bit.

When sending frames from the secondary, the `P/F` bit indicates whether the frame is the last of the message, or not. Thus if the `P/F` bit is set by the primary it is a poll bit (`P`), if it is set by the secondary it is a final bit (`F`).

The following sections describe 8-bit control fields. Sixteen-bit control fields are similar but reserve a 7-bit field for the frame counter variables `N(R)` and `N(S)`.

Information frame

An information frame contains sequenced data and is identified by a 0 in the first bit position of the control field. The 3-bit variable `N(R)` is used to confirm the number of transmitted frames received correctly and `N(S)` is used to number an information frame. The first frame transmitted is numbered 0 as (`000`), the next as 1 (`001`), until the eighth which is numbered `111`. The sequence then starts back at 0 again and this gives a sliding window of eight frames.

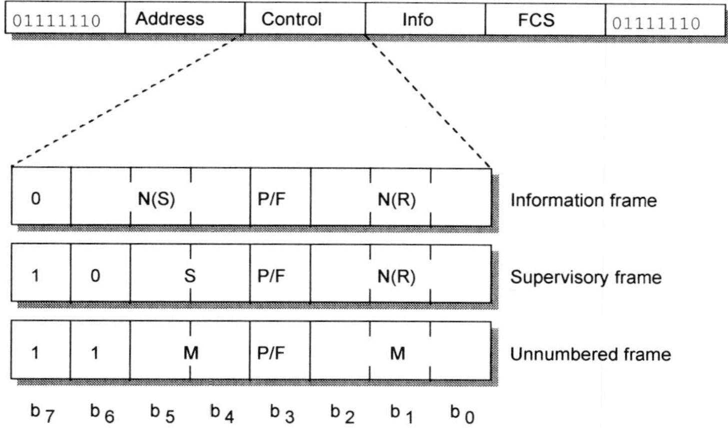

Figure 44.5 Format of an 8-bit control field.

Supervisory frame

Supervisory frames contain flow control data. They confirm or reject previously received information frames and also can indicate whether a station is ready to receive frames.

The `N(S)` field is used with the S bits to acknowledge, or reject, previously transmitted frames. Responses from the receiver are set in the S field, these are receiver ready (RR), ready not to receive (RNR), reject (REJ) and selectively reject (SREJ). Table 44.1 gives the format of these bits.

RR informs the receiver that it acknowledges the frames sent up to `N(R)`. RNR tells the transmitter that the receiver cannot receive any more frames at the present time (RR will cancel this). It also acknowledges frames up to `N(R)`. The REJ control rejects all frames after `N(R)`. The transmitter must then send frames starting at `N(R)`.

Table 44.1 Supervisory bits

b_5	b_4	*Receiver status*
0	0	Receiver ready (RR)
1	0	Receiver not ready (RNR)
0	1	Reject (REJ)
1	1	Selectively reject (SREJ)

Unnumbered frame

If the first two bits of the control field are 1's then it is an unnumbered frame. Apart from the `P/F` flag the other bits are used to send unnumbered commands. When sending commands, the `P/F` flag is a poll bit (asking for a response), and for responses it is a flag bit (end of response).

The available commands are SARM (set asynchronous response mode), SNRM (set normal response mode), SABM (set asynchronous balance mode), RSET (reset), FRMR (frame reject) and Disconnect (DISC). The available responses are UA (unnumbered acknowledge), CMDR (command reject), FRMR (frame reject) and DM (disconnect mode). Bit definitions for some of these are:

SABM	`1111P110`	DM	`1111F000`	DISC	`1100P010`
UA	`1100F110`	FRMR	`1110F001`		

44.2.7 Frame check sequence field

The frame check sequence (FCS) field contains an error detection code based on cyclic redundancy check (CRC) polynomials. It is used to check the address, control and information fields, as previously illustrated in Figure 44.2. HDLC uses a polynomial specified by CCITT V.41, which is $G(x) = x^{16} + x^{12} + x^5 + x^1$. This is also known as CRC-16 or CRC-CCITT. Refer to Chapter 13 for more information on CRCs.

44.3 Transparency

The flag sequence `01111110` can occur anywhere in the frame. To prevent this a transparency mechanism called zero-bit insertion or zero stuffing is used. There are two main rules that are applied, these are:

- In the transmitter, a 0 is automatically inserted after five consecutive 1's, except when the flag occurs.
- At the receiver, when five consecutive 1's are received and the next bit is a 0 then the 0 is deleted and removed. If it is a 1 then it must be a valid flag.

In the following example a flag sequence appears in the data stream where it is not supposed to (spaces have been inserted around it). Notice that the transmitter detects five 1's in a row and inserts a 0 to break them up.

```
Message:  00111000101000 01111110  01011111 1111010101
Sent:     00111000101000 011111010 0101111101111010101
```

44.4 Flow control

Supervisory frames (`S[]`) send flow control information to acknowledge the reception of data frames or to reject frames. Unnumbered frames (`U[]`) set up the link between a primary and a secondary, by the primary sending commands and the secondary replying with responses. Information frames (`I[]`) contain data.

44.4.1 Link connection

Figure 44.6 shows how a primary station (node A) sets up a connection with a secondary station (node B) in NRM (normal response mode). In this mode one or many secondary stations can exist. First the primary station requests a link by sending an unnumbered frame with: node B's address (`ADDR_B`), the set normal response mode (`SNRM`) command and with poll flag set (`P=1`), that is, `U[SNRM,ABBR_B,P=1]`. If the addressed secondary wishes to make a connection then it replies back with an unnumbered frame containing: its own address (`ADDR_B`), the unnumbered acknowledge (`UA`) response and the final bit set (`F=1`), i.e.

U[UA,ABBR_B,F=1]. The secondary sends back its own address because many secondaries can exist and it thus identifies which station has responded. There is no need to send the primary station address as only one primary exists.

Once the link is set up data can flow between the nodes. To disconnect the link, the primary station sends an unnumbered frame with: node B's address (ADDR_B), the disconnect (DISC) command and the poll flag set (P=1), that is, U[DISC,ABBR_B,P=1]. If the addressed secondary accepts the disconnection then it replies back with an unnumbered frame containing: its own address (ADDR_B), the unnumbered acknowledge (UA) response and the final bit set (F=1), i.e. U[UA,ABBR_B,F=1].

When two stations act as both primaries and secondaries then they use the asynchronous balanced mode (ABM). Each station has the same priority and can perform the functions of a primary and secondary station. Figure 44.7 shows a typical connection. The ABM mode is set up initially using the SABM command (U[SABM,ABBR_B,P=1]). The connection between node A and node B is then similar to the NRM but, as node B operates as a primary station, it can send a disconnect command to node A (U[DISC,ABBR_B,P=1]).

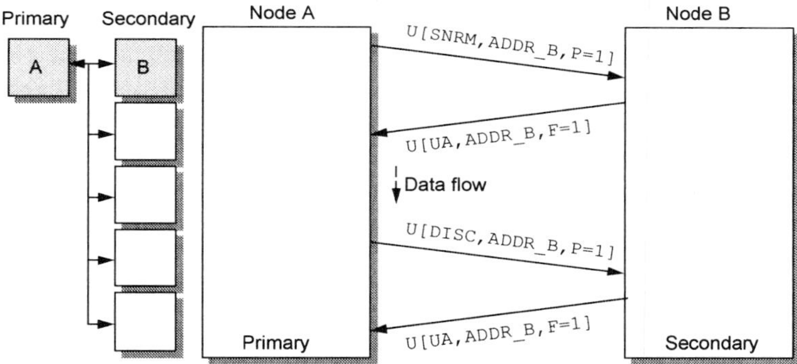

Figure 44.6 Connection between a primary and secondary in NRM.

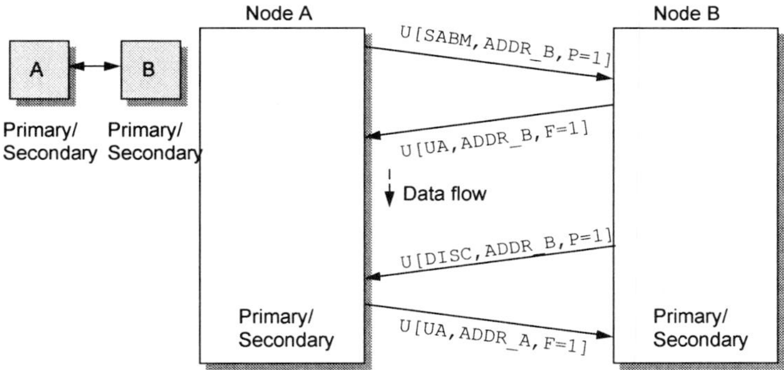

Figure 44.7 Connection between a primary/secondary in SABM.

The SABM, SARM and SNRM modes set up communications using an 8-bit control field. Three other commands exist which set up a 16-bit control field, these are SABME (set asynchronous balanced mode extended), SARME and SNRME. The format of the 16-bit control field is given in Figure 44.8.

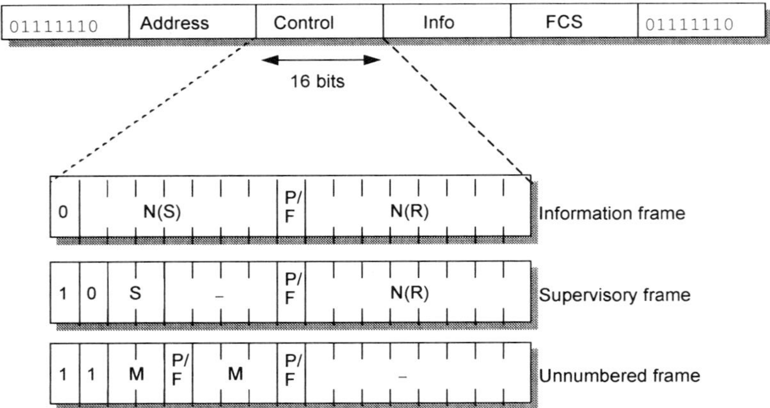

Figure 44.8 Extended control field.

Figure 44.9 shows an example conversation between a sending station (node A) and a receiving station (node B). Initially three information frames are sent numbered 2, 3 and 4 (`I[N(S)=2]`, `I[N(S)=3]` and `I[N(S)=4, P=1]`). The last of these frames has the poll bit set, which indicates to node B that node A wishes it to respond, either to acknowledge or reject previously unacknowledged frames. Node B does this by sending back a supervisory frame (`S[RR, N(R)=5]`) with the receiver ready (RR) acknowledgement. This informs node A that node B expects to receive frame number 5 next. Thus it has acknowledged all frames up to and including frame 4.

In the example in Figure 44.9 an error has occurred in the reception of frame 5. The recipient informs the sender by sending a supervisory frame with a reject flow command (`S[REJ, N(R)=5]`). After the sender receives this it resends each frame after and including frame 5.

If the receiver does not want to communicate, at the present, it sends a receiver not ready flow command. For example `S[RNR, N(R)=5]` tells the transmitter to stop sending data, at the present. It also informs the sender that all frames up to frame 5 have been accepted. The sender will transmit frames once it has received a receiver ready frame from the receiver.

Figure 44.9 shows an example of data flow in only the one direction. With ABM both stations can transmit and receive data. Thus each frame sent contains receive and send counter values. When stations send information frames the previously received frames can be acknowledged, or rejected, by piggy-backing the receive counter value. In Figure 44.10, node A sends three information frames with `I[N(S)=0,N(R)=0]`, `I[N(S)=1, N(R)=0]`, and `I[N(S)=2,N(R)=0]`. The last frame informs node B that node A expects to receive frame 0 next. Node B then sends frame 0 and acknowledges the reception of all frames up to, and including frame 2 with `I[N(S)=0,N(R)=3]`, and so on.

44.5 Derivatives of HDLC

There are many derivatives of HDLC, including:

- LAPB (link access procedure balanced) is used in X.25 packet switched networks;
- LAPM(link access procedure for modems) is used in error correction modems;
- LLC (logical link control) is used in Ethernet and Token Ring networks;
- LAPD (link access procedure D-channel) is used in Integrated Services Digital Networks (ISDNs).

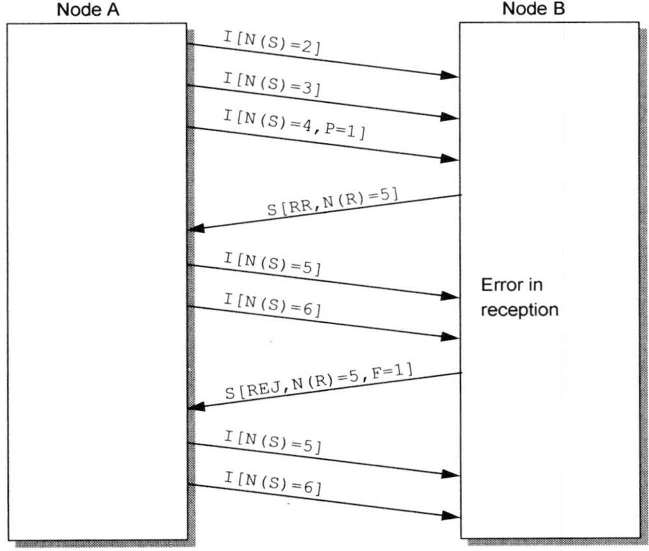

Figure 44.9 Example flow.

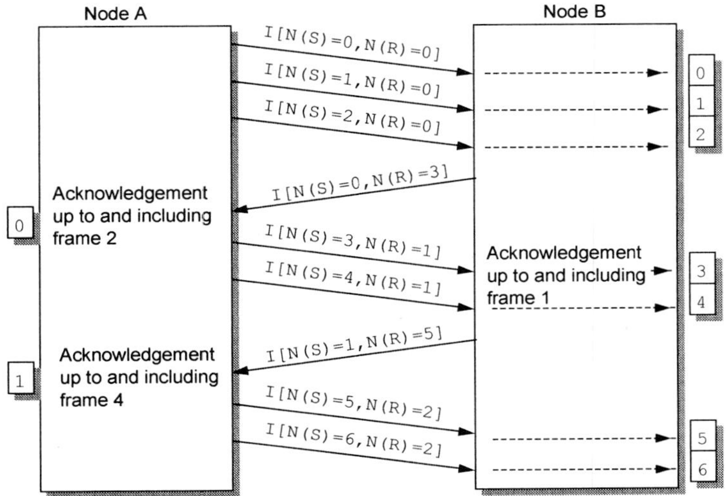

Figure 44.10 Example flow with piggy-backed acknowledgement.

X.25

45.1 Introduction

A wide area network (WAN) connects one node to another over relatively large distances via an arbitrary graph of switching nodes. For the transmission of digital data, then data is either sent through a public data network (PDN) or through dedicated company connections.

As shown in Figure 45.1, there are two main types of connection over the public telephone network, circuit-switching and packet-switching.

With circuit switched, a physical, or a reserved multiplexed, connection exists between two nodes, a typical example is the public-switched telephone network (PSTN). The connection must be made before transferring any data. In the past this connection took a relatively long time to set-up (typically over 10 seconds), but with the increase in digital exchanges it has reduced to less than a second. The usage of digital exchanges has also allowed the transmission of digital data, over PSTNs, at rates of 64 kbps and greater. This type of network is known as a circuit-switched digital network (CSDN). Its main disadvantage is that a permanent connection is set-up between the nodes. This is wasteful in time and can be costly. Another disadvantage is that the transmitting and receiving nodes must be operating at the same speed. A CSDN, also, does not perform any error detection or flow control.

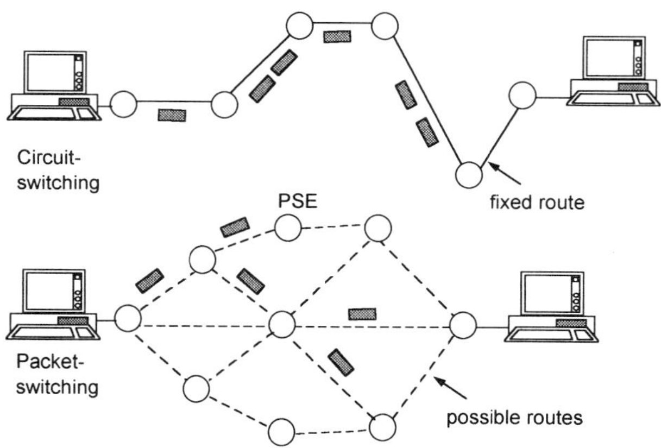

Figure 45.1 Circuit- and packet-switching.

Packet-switching involves segmenting data into packets that propagate within a digital network. They either follow a pre-determined route or are routed individually to the receiving node via packet-switched exchanges (PSE) or routers. These examine the destination addresses and based on an internal routing directory pass it to the next PSE on the route. As with circuit-switching, data can propagate over a fixed route. This differs from circuit-switching in that the path is not an actual physical circuit (or a reserved multiplexed channel). As it is not a physical circuit it is normally defined as a virtual circuit. This virtual circuit is less wasteful on channel resources as other data can be sent when there are gaps in the data flow.

Table 45.1 gives a comparison of the two types.

Table 45.1 Comparison of switching techniques

	Circuit-switching	*Packet-switching*
Investment in equipment	Minimal as it uses existing connections	Expensive for initial investment
Error and flow control	None, this must be supplied by the end users	Yes, using the FCS in the data link layer
Simultaneous transmissions and connections	No	Yes, nodes can communicate with many nodes at the same time and over many different routes
Allows for data to be sent without first setting up a connection	No	Yes, using datagrams
Response time	Once the link is set-up it provides a good reliable connection with little propagation delay	Response time depends on the size of the data packets and the traffic within the network

45.2 Packet-switching and the OSI model

The CCITT developed the X.25 standard for packet switching and it fits-in well with the OSI model. In a packet-switched network the physical layer is normally defined by the X.21 standard and the data link layer by a derivative HDLC, known as LAPB. The network, or packet, level is defined by X.25.

45.2.1 The physical layer (X.21)

The CCITT recommendation X.21 defines the physical interface between a node (the DTE) and the network connection (the DCE). Figure 45.2 shows the connections between the node and the network connection.

A second standard, known as X.21 (bis), has also been defined and is similar to the RS-232/V.24 standards. This allows RS-232 equipment to directly connect to the network.

The Transmit (`T`) line sends data from the DTE to the DCE and the Receive (`R`) sends data from the DCE to the DTE. A DCE controls the Indicate (`I`) line to indicate that it is ready to receive data. The DTE controls the Control (`C`) line to request to the DCE that it is ready to send data.

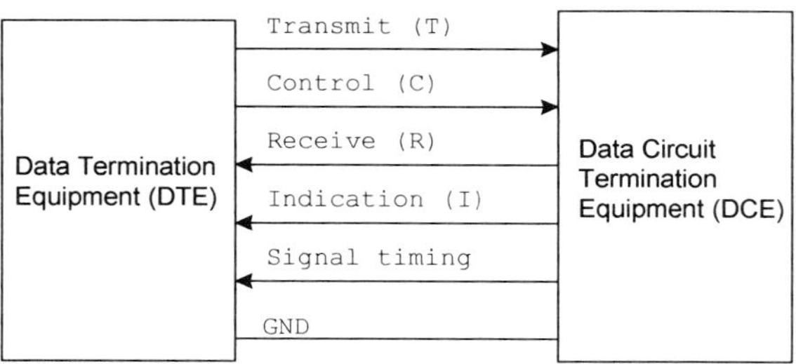

Figure 45.2 X.21 connections.

Figure 45.3 shows a simplified flow control between a sending DTE and a receiving DCE. With reference to the state numbers in the diagram the sequence of operations is as follows:

1 Initially, the Transmit (T) and Receive (R) lines are high to indicate that the DTE and the DCE are active and ready to communicate, respectively.
2 When the DTE wishes to transmit data it first sets the Control (C) line low (ON). At the same time it sets the Transmit (T) line low.
3 When the DCE accepts the data transfer it sets the Indicate (I) line low.
4-12 Data is transmitted on the Transmit (T) line and, in some modes, it is echoed back on the Receive (R) line.
12 When the DTE finishes transmitting data it sets the Control (C) line high (OFF).
13 The DCE responds to the Control (C) line going high by setting the Indication (I) line high.
14 The DCE sets the Receive (R) line high to indicate that it is active and ready to communicate.
15 The DTE sets the Transmit (T) line high to indicate that it is active and ready to communicate.

45.2.2 Data link layer (LAPB)

The data link layer provides a reliable method of transferring packets between the DTE and the local PSE. Frames sent contain no information on the addressing of the remote node, this information is contained within the packet. The standard, known as the Link Access Procedure Version B is based on HDLC. It uses ABM (asynchronous balance mode) where both the DTE and the PSE can initiate commands and responses at any time.

45.2.3 Network (packet) layer

The packet layer is equivalent to the network layer in the OSI model. Its main purpose is to route data over a network.

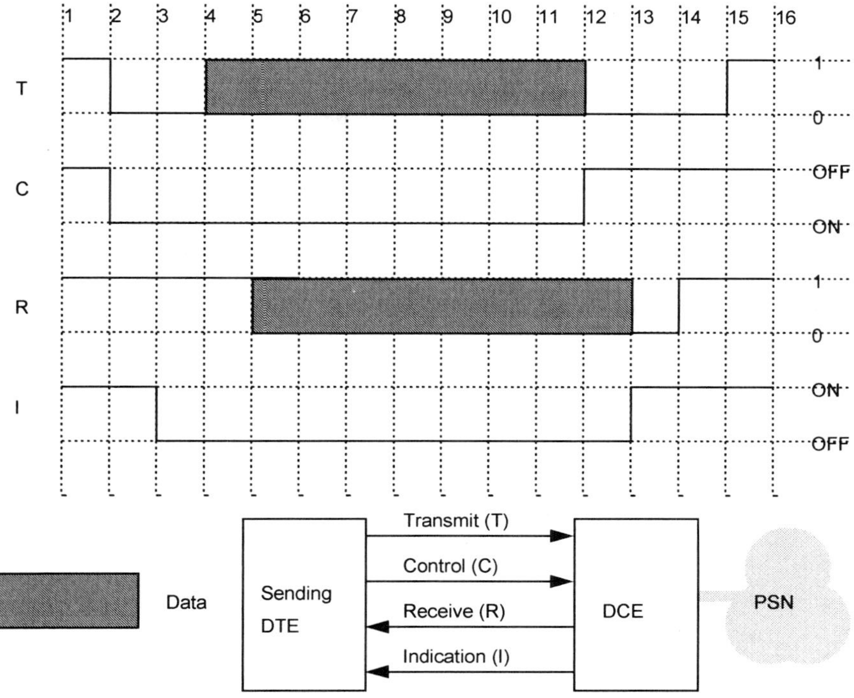

Figure 45.3 Example X.21 signals.

45.3 X.25 packets

X.25 packets contain a header and either control information and/or data. The LAPB frame envelops the packet and physical layer transmits it. Figure 45.4 shows the format of the transmitted frame.

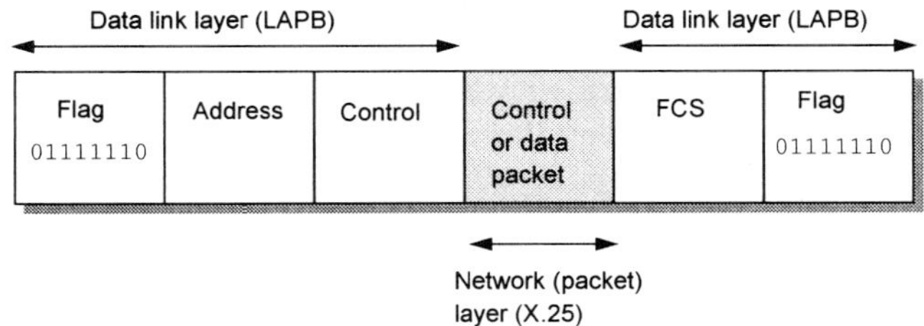

Figure 45.4 Transmitted frame.

45.3.1 Packet headers

Packet headers are 3 bytes long. Figure 45.5 shows a packet represented both as a bit stream and arranged in groups of 8 bits. The first two bytes of the header contain the group format identifier (GFI), the logical group number (LGN) and the logical channel number (LCN). The third byte identifies the packet type.

The GFI number is a 4-bit binary value of `QdYY`, where the `Q` bit is the qualifier bit and the `D` bit is the delivery confirmation bit. The `d` bit requests an acknowledgement from the remote node, this is discussed in more detail in section 7.4.3. The `YY` bits indicate the range of packet sequence numbers. If they are `01` then packets are numbered from 0 to 7 (modulo-8), or if they are `10` then packets are numbered from 0 to 127 (modulo-128). This packet sequencing is similar to the method that HDLC uses to provide confirmation of received frames. As LAPB (the HDLC-derivative) provides reliable data link error control, the sequencing of packets is mainly used as flow control rather than for error control.

The LGN and LCN together define a 12-bit virtual circuit identifier (VCI). This allows packets to find logical routes though the packet switched network. For example all packets could take the same route or each group of packets could find different routes.

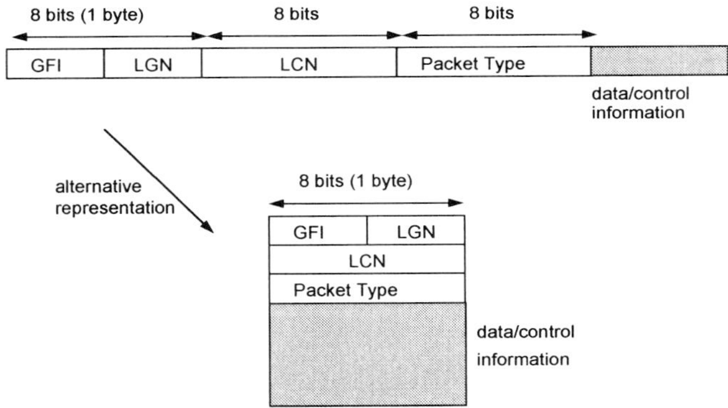

Figure 45.5 Packet header.

45.3.2 Packet types

The third byte of the packet header defines the packet type, Table 45.2 lists some of these. A 0 in the eighth bit position identifies that it is a data packet while a 1 marks it as a flow control or a call set-up packet.

The call set-up and clearing packets have differing definitions depending on whether they are sent or received. For example if the calling node sends a packet type of `00001011` then it is a Call Request packet, if it is received it is interpreted as an Incoming Call packet.

Table 45.2 The main packet types.

Packet type	*Identifier*	*Description*
Data Packet	`RRRMSSS0`	The Data packet is sent with a send sequence number (`SSS`) and a receive sequence number (`RRR`).

Call Request/ Incoming Call	`00001011`	If a node sends this packet to the network it is a Call Request packet, if the node receives it, it is an Incoming Call packet.
Call Accepted/ Call Confirmation	`00001111`	If a node sends this packet to the network it is a Call Accepted packet, if a node receives it, it is a Call Confirmation.
Receive Ready	`RRR00001`	The Receive Ready packet is sent from a node to inform the other node that it is ready to receive data. It also informs the other node that the next data packet it expects to receive should have sequence number `RRR`.
Receive Not Ready	`RRR00101`	The Receive Not Ready packet is sent from a node to inform the other node that it is not ready to receive data. It also informs the other node that the next data packet it expects to receive should have sequence number `RRR`.
Reject	`RRR01001`	The Reject packet is sent from a node to inform the other node that it rejects packet number `RRR`. All other packets before this are acknowledged.
Clear Request/ Clear Indication	`00010011`	If the calling node sends this packet to the network it is a Clear Request packet, if the called node receives it is a Clear Indication packet.
Clear Confirm/ Clear Confirm	`00010111`	If the called node sends this packet to the network it is a Clear Confirm packet, if the calling node receives it is a Clear Confirm.

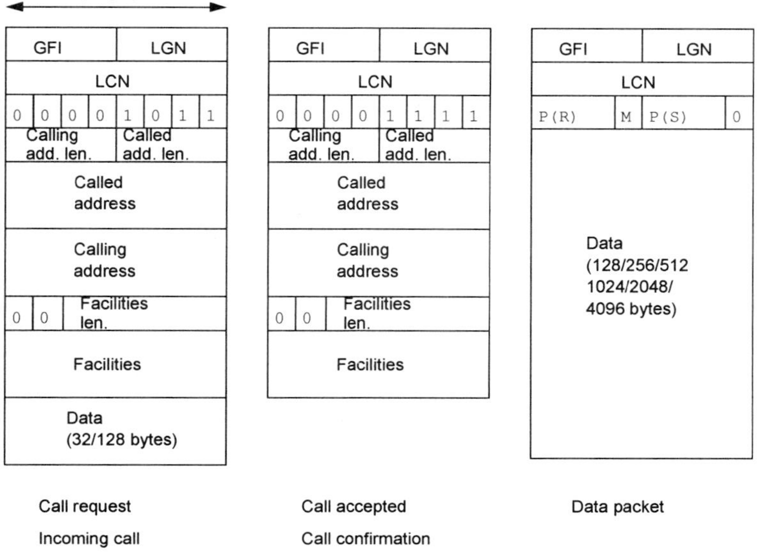

Figure 45.6 Call request, call accepted and data packets.

Figure 45.6 shows the format of the Call Request/ Incoming Call, Call Accepted/ Call Confirmation and the Data packets. With the Call Request/ Incoming Call and the Call Accepted/ Call Confirmation packets the fourth byte of the packet contains two 4-bit numbers which define the number of bytes in the calling and called address. After this byte, the called and the calling addresses are sent. Following this the next byte defines the number of bytes in the facilities field. The facilities field enables selected operational parameters to be negotiated when a call is being set up, these include:

- The data packet size (typically, 128 bytes).
- Number of packets to be received before an acknowledgement is required (typically, two).
- Data throughput, in bytes per second.
- Reverse charging.
- Usage of extended sequence numbers.

A data packet contains the standard packet header followed by a byte that contains the send and receive sequence number. The `M` bit identifies that there is more data to be sent to complete the message. Notice that the data packet does not contain either the calling or the called addresses. This is because once the connection is made then the VCI label identifies the path between the called and the calling node.

The `P(R)` variable is the sequence number of the packet that the sending node expects to receive next, and `P(S)` is the sequence number of the current packet. With modulo-8 sequencing, the packets are numbered from 0 to 7. The first packet sent is 0, the next is 1 and so on until the eighth packet that is numbered 7. The next is then numbered as 0 and so on. With modulo-128 sequencing, the packets are numbered 0 to 127.

The data size can be 128, 256, 512, 1024, 2048 or 4096 bytes, although its size is normally limited, by the public-carrier packet-switched network, to 128. This achieves a reasonable response time.

Figure 45.7 shows the format of the Receive Ready (RR), Receive Not Ready (RNR), Reject (REJ), Clear Confirmation and Clear Request packets. The RR, RNR and REJ packets contain a receive sequence number. This is the sequence number of the packet that the receiving node expects to receive next.

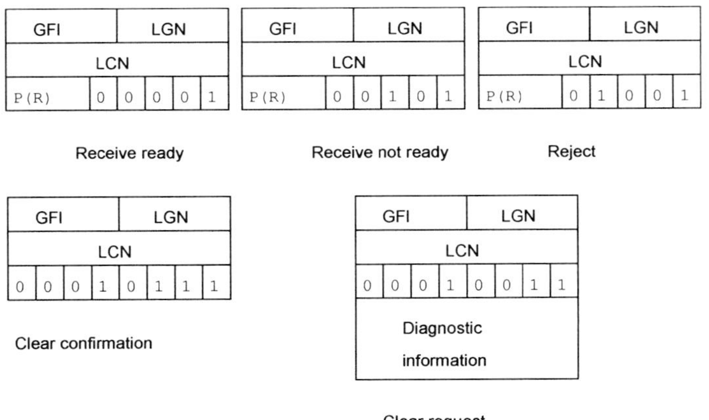

Figure 45.7 RR, RNR, REJ and Clear Confirm and Request packets.

45.4 X.25 packet flow

The three types of packets are:

- Call set-up and clearing - Call Request, Incoming Call, Clear Request, Clear Indication and Clear Confirmation.
- Data packets.
- Flow control - Receive Ready (RR), Receive Not Ready (RNR) and Reject (REJ).

45.4.1 Call set-up and clearing

Figure 45.8 shows a typical data transfer. Initially the calling node (Node A) sends a Call Request packet (`P[Call_request]`) to the network. When this propagates through the packet-switched network the receiving node (Node B) receives it as an Incoming Call packet (`P[Incoming_call]`). When Node B accepts the call it sends a Call Accepted packet (`P[Incoming_call]`), which propagates through the network and Node A receives it as a Call Confirmation (`P[Call_confirmation ]`).

The call initialization sets up a virtual circuit between the nodes and sequenced data packets and flow control information can now flow between the nodes.

To clear the connection Node A sends a Clear Request packet (`P[Clear_request]`) to the network. When this propagates through the network, Node B receives it as a Clear Indication packet (`P[Clear_indication)`. When Node B accepts that the call is to be cleared then it sends a Clear Confirmation packet (`P[Clear_confirm ]`). This propagates through the network and Node A receives it as a Clear Confirm (`P[Call_confirmation]`).

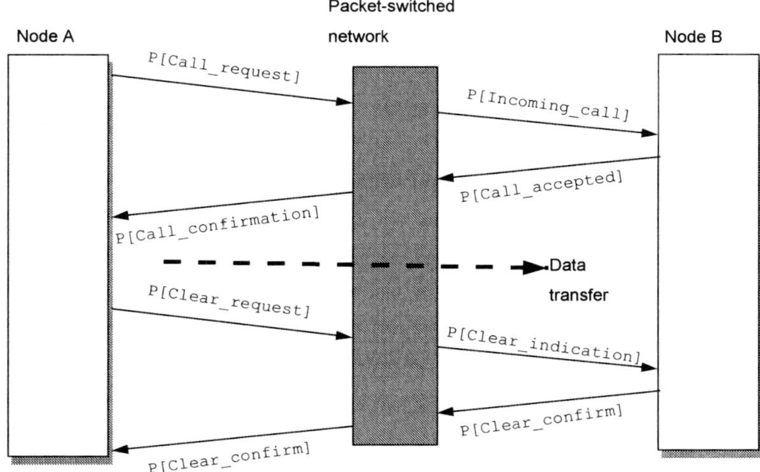

Figure 45.8 Call set-up and clearing.

45.4.2 Data transmission and flow control

After a virtual circuit has been set-up then sequenced data packets and flow control information can flow between the nodes.

A data packet contains sends two sequence numbers `P[D, N(R), N(S)]`. For 3-bit sequence numbers, N(S) and N(R) range from 0 to 7. N(S) is the sequence number of the data packet and N(R) is the sequence number of the packet that the sending node expects to receive next.

Figure 45.9 shows an example conversation between a sending node (Node A) and a receiving node (Node B). The flow control window has been set at 3. This window defines the number of packets that can be sent before the receiver must send an acknowledgement. Initially, in the example, three information frames are sent, numbered 2, 3 and 4 (P[D, N(R)=0, N(S)=2], P[D, N(R)=0, N(S)=3] and P[D, N(R)=0, N(S)=4]). The window is set to 3 thus Node B must send an acknowledgement for the packets it has received. It does this by sending a Receive Ready packet (P[RR, N(R)=5]). This informs node A that Node B expects to receive packet number 5 next. This acknowledges all frames before, and including, frame 4.

In the example in Figure 45.9 an error has occurred in the reception of frame 5. The recipient informs the sender by sending a Reject packet (P[REJ, N(R)=5]). After the sender receives this it re-sends each frame after, and including, frame 5.

If a node does not wish to communicate, at the present, it sends a Receive Not Ready packet. For example P[RNR, N(R)=5] tells the transmitter to stop sending data, at the present. It also informs the sender that all frames up to frame 5 have been accepted. The sender will transmitting frames only once it has received a Receive Ready packet from the receiver.

Figure 45.8 shows an example of data flow in only the one direction. With X.25 both stations can transmit and receive data. Thus each packet sent contains receive and send counter values. When nodes send data packets the previously received frames can be acknowledged, or rejected, by piggy-backing the receive counter value. In Figure 45.9, node A sends 3 data packets with P[D, N(R)=0, N(S)=0], P[D, N(R)=0, N(S)=1], and P[D, N(R)=0, N(S)=2]. The last data packet informs Node B that Node A expects to receive data packet 0 next. Node B then sends data packet 0 and acknowledges the reception of all frames up to, and including frame 2 with P[D, N(R)=3, N(S)=0], and so on.

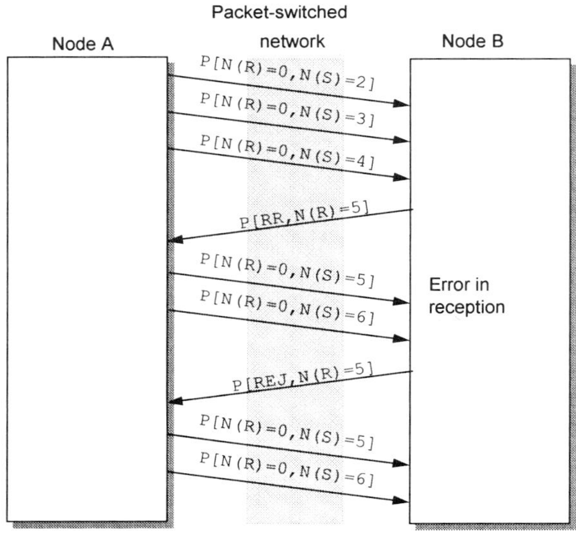

Figure 45.9 Example flow.

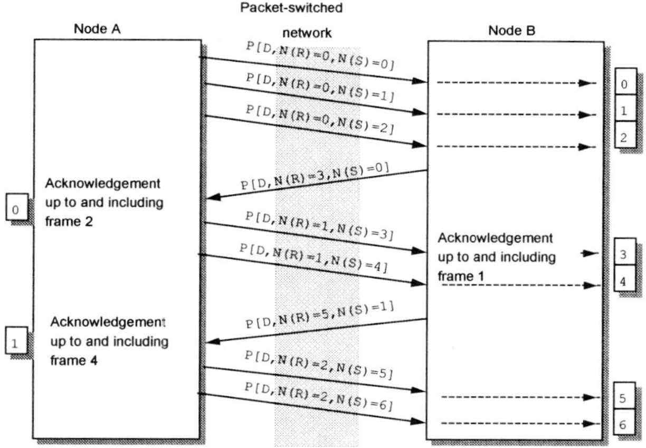

Figure 45.10 Example flow with piggy-backed acknowledgement.

45.4.3 The delivery bit

The delivery bit (d) identifies which connection should respond with an acknowledgement. If it is not set then the local network connection sends an acknowledgement for data packets. If it is set then the remote nodes sends the acknowledgement. The latter was the case in the example given in Figure 45.9. An example is given in Figure 45.11. The number of packets before an acknowledgement is set by the window. When the d-bit is not set then the window does not have any significance as the network connection returns back all packets sent.

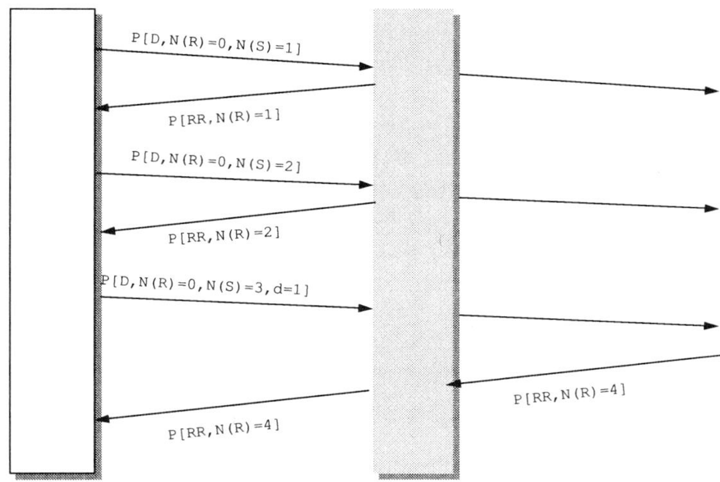

Figure 45.11 Usage of the D-bit.

45.5 Packet switching routing

There are three main types of routing used in X.25, these are:

- Permanent virtual call.
- Virtual call.
- Datagram.

A virtual call sets up a route for the two-way flow of packets between two specific nodes, whereas a datagram is sent into the network without first establishing a route. A datagram is analogous to a letter sent by the post, where a letter is addressed and sent without first finding out if the letter will be received. The virtual call is analogous to a telephone call where a direct connection is made before the call is initiated.

A datagram is normally only used where there is a small amount of data in a few packets, whereas a virtual call is set-up where there are relatively large amounts of data to be sent in many packets. With a datagram there is no need to initiate the call set-up procedures, as previously shown in Figure 45.8. The call set-up and clearing packets (Call Request, Call Indication, and so on) are only used when a virtual circuit is used. An example of a datagram might be to transmit an electronic mail message, as there is no need to establish a virtual circuit before the message is sent.

Once a virtual circuit is set-up then it is used until all the data has been transferred. A new conversation establishes a new virtual circuit. In some applications, though, a reliable permanent circuit is required, this is described as a permanent virtual call where two nodes have a permanent virtual connection. There is no need, in this case, to set-up a connection as the virtual circuit between the two nodes is dedicated to them.

When a calling node wishes to communicate through the network it makes contact with the called node and negotiates such things as the packet size, reverse charging and maximum data throughput. The window size is the number of packets that are sent before an acknowledgement must be sent back. Maximum data throughput is the maximum number of bytes that can be sent per second. These flow control parameters are contained in the facilities field of the Call Request packet. If the called node accepts them then a connection is made.

The network computes a route based on the specified parameters and determines which links on each part of the route best supports the requested flow control parameters. It sets up request to all the packet routing nodes on the path en-route to the destination node. Figure 45.12 shows an example route between packet-switched routers (PSR) from a calling node to a destination node. The route selected is `PSR1` → `PSR2` → `PSR3` → `PSR4`. Each of the router selects an unused VCI label on their respective links and reserves it for the virtual circuit in their connection lookup tables.

For example `PSR1` could use `VC_2` (for example the VCI could have a value of 17), this can be sent to `PSR2`. `PSR2` in turn picks `VC_3` and associates it with `VC_2` in its connection table. It then forwards `VC_3` to `PSR3`. `PSR3` selects `VC_4` and associates it with `VC_3`. It then forwards `VC_4` to `PSR4`. `PSR4` selects `VC_5` and associates it with `VC_4`. If the called node accepts the call then it sends back a Call Accepted packet back over the virtual circuit. Each of the nodes on way back to the calling node assigns a new VCI number. Thus the acknowledgement passes back from `PSR4` to `PSR3`, `PSR3` to `PSR2` and so on to the calling node to confirm that the connection has been established. Data packets can then be transmitted between the two nodes. As has been previously discussed, there is no need to transmit the calling or called addresses with the packet as the source and destination is identified using the VCI label. When the connection is terminated the VCI labels assigned to the communications can be used for other connections.

If a single virtual circuit is set-up then packets are always delivered in the order these were transmitted. This is because packets cannot take alternative routes to the destination. Even if the packets are buffered within a node they will still be transmitted in the correct sequence.

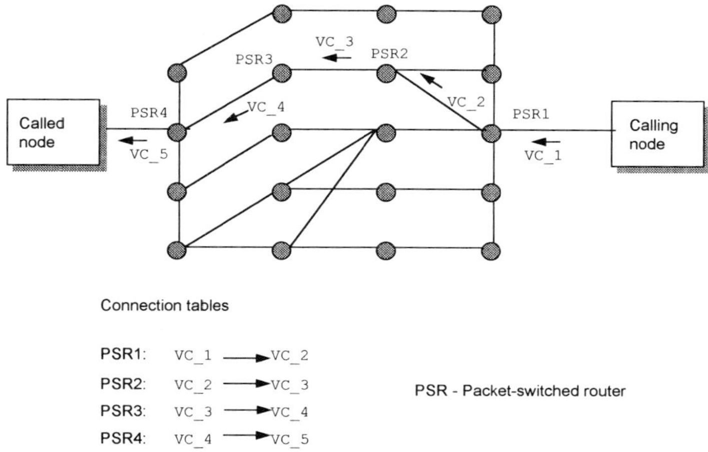

Figure 45.12 Virtual call set-up.

45.6 Logical channels

The VCI label contains a 4-bit logical group number (LGN) and a 16-bit logical channel number (LCN) to define the VCI. There can be 16 groups and within each group there can be 256 different channels. This allows for a node to communicate with several nodes simultaneously. Figure 45.13 shows an example of a node communicating with four nodes over four channels. Node A is communicating with Node B, C, D and E. The route for Node A to Node E is through routers A, B, C, G, and I. For Node A to Node D it is through routers A, F, G and H.

45.7 X.25 node addressing

Nodes on an X.25 network have an individual NSAP (network service access point) address. Since these nodes operate globally over international networks the addresses must be assigned a globally unique network address. The network on which the node is connected is usually a country-wide network. Each of these packet-switched public digital networks are known as a subnet.

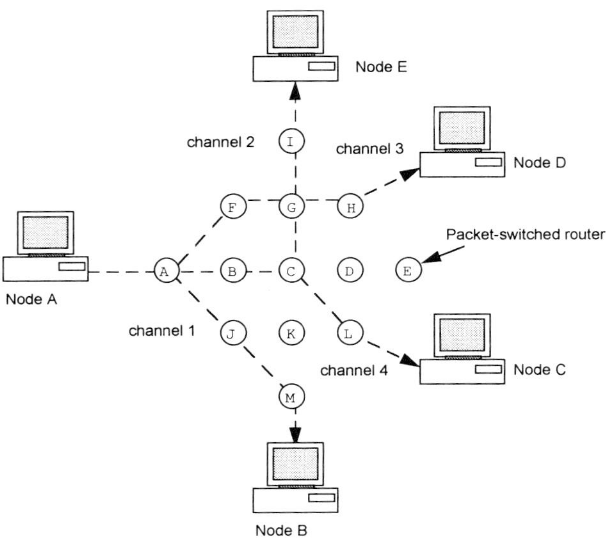

Figure 45.13 Multiple channels.

The definition of the addresses is defined either in pure binary or a binary-coded decimal (BCD) digits. For example, if the network address is defined in BCD then the binary address `0011 0110 0001 0111 1001 0011` corresponds to `361792`. The NSAP address is made up of up-to 40 decimal digits or 20 bytes (as one BCD digit is represented by 4 bits). The calling and called address length are defined within the X.25 packet.

The NSAP address is made up of parts, the initial domain part (IDP) and the domain specific part (DSP), as illustrated in Figure 45.14. An IDP is made up of two sub-parts, the authority and format identifier (AFI), and the initial domain identifier (IDI). As several authorities can grant NSAP addresses, the AFI field contains 2 BCD digits which identify the granting authority and the format of the rest of the address field.

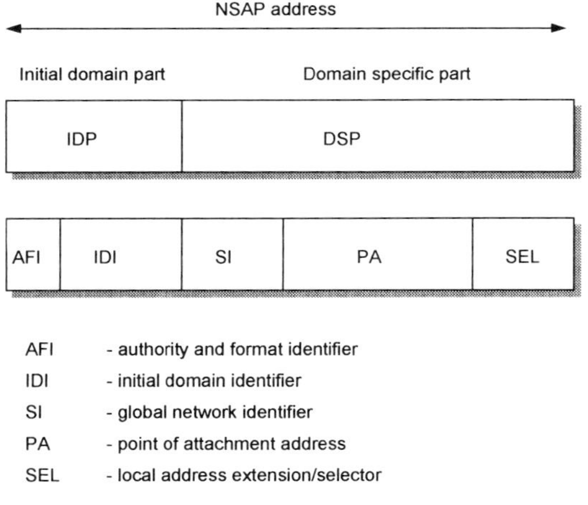

Figure 45.14 NSAP address.

For example, if the AFI value is `36` then the granting authority is the CCITT and the format is defined in the X.121 recommendation. The resulting address is:

```
36XXXXXXXXXXXXXXYYYYYYYYYYYYYYYYYYYYYYYY
```

where `XX..XXX` is the IDI part (14 digits) and `YY...YYY` is the DSP part (24 digits). This gives a total of 40 digits.

If the AFI value is `38` then the granting authority is the ISO and the format is defined by the ISO-assigned country codes, or ISO DCC. The resulting address is:

```
38XXXYYYYYYYYYYYYYYYYYYYYYYYYYYYYYYYYYY
```

where `XX..XXX` is the IDI part (3 digits) and `YY...YYY` is the DSP part (35 digits). With ISO address the IDI portion is assigned by the country the network is resident.

After the initial domain is defined then the DSP part defines a smaller and smaller subnetwork within the domain. Figure 45.15 shows an example addressing structure. The SEL part defines the local node with at the point of attachment of the packet switched network.

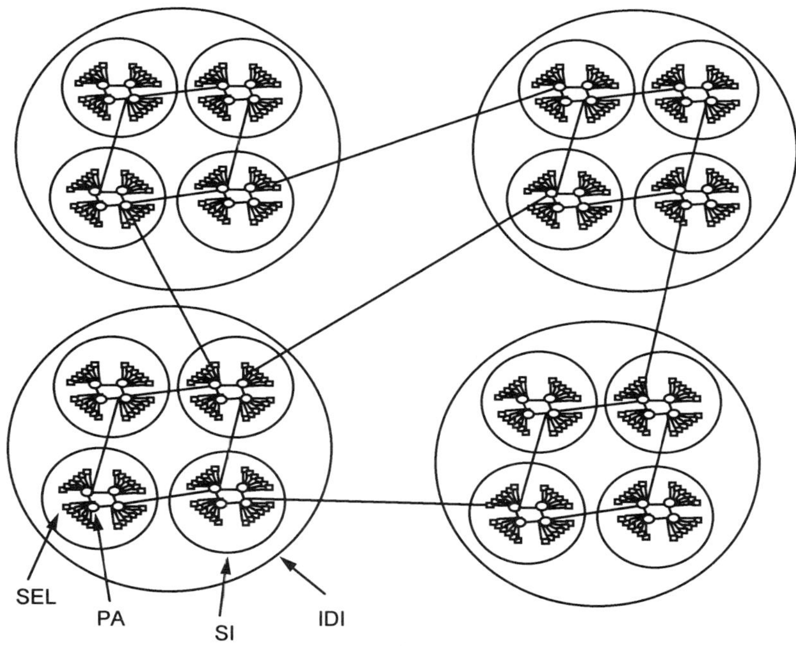

Figure 45.15 NSAP addressing structure.

Modems

46.1 Introduction

Modems (MOdulator/DEModulator) connect digital equipment to a telephone line. It connects digital equipment to a speech bandwidth-limited communications channel. Typically, modems are used on telephone lines, which have a bandwidth of between 400 Hz and 3.4 kHz. If digital pulses were applied directly to these lines, they would end up severely distorted.

Modem speeds range from 300 bps to 56 kbps. A modem normally transmits about 10 bits per character (each character has 8 bits). Thus the maximum rate of characters for a high-speed modem is 2880 characters per second. This chapter contains approximately 15 000 characters and thus to transmit the text in this chapter would take approximately 5 seconds. Text, itself, is relatively fast transfer, unfortunately even compressed graphics can take some time to be transmitted. A compressed image of 20 KB (equivalent to 20 000 characters) will take nearly 6 seconds to load on the fastest modem.

The document that was used to store this chapter occupies, in an uncompressed form, 360 KB. Thus to download this document over a modem, on the fastest modem, would take:

$$\text{Time taken} = \frac{\text{Total file size}}{\text{Characters per second}} = \frac{360\,000}{2\,800} = 125\text{ s}$$

A 14.4 kbps modem would take 250 seconds. Typically, home users connect to the Internet and WWW through a modem (although increasingly ISDN is being used). The example above shows the need to compress files when transferring them over a modem. On the WWW, documents and large files are normally compressed into a ZIP file and images and video compressed in GIF and JPG.

Most modems are able to do the following:

- Automatically dial (known as Auto-dial) another modem using either touch-tone or pulse dialing.
- Automatically answer (known as Auto-answer) calls and make a connection with another modem.
- Disconnect a telephone connection when data transfer has completed or if an error occurs.
- Automatic speed negotiation between the two modems.
- Convert bits into a form suitable for the line (modulator).
- Convert received signals back into bits (demodulator).
- Transfer data reliably with the correct type of handshaking.

Figure 46.1 shows how two computers connect to each other using RS-232 converters and modems. The RS-232 converter is normally an integral part of the computer, while the modem can either be external or internal to the computer. If it is externally connected then it is normally connected by a cable with a 25-pin male D-type connector on either end.

Modems are either synchronous or asynchronous. A synchronous modem recovers the clock at the receiver. There is no need for start and stop bits in a synchronous modem. Asynchronous modems are, by far, the most popular types. Synchronous modems have a typical speed of 56 Kbps whereas for asynchronous modems it is 33 Kbps. A measure of the speed of the modem is the baud rate or bps (bits per second).

There are two types of circuits available from the public telephone network: either direct dial or a permanent connection. The direct dial type is a dial-up network where the link is established in the same manner as normal voice calls with a standard telephone or some kind of an automatic dial/answer machine. They can use either touch-tones or pulses to make the connection. With private line circuits, the subscriber has a permanent dedicated communication link.

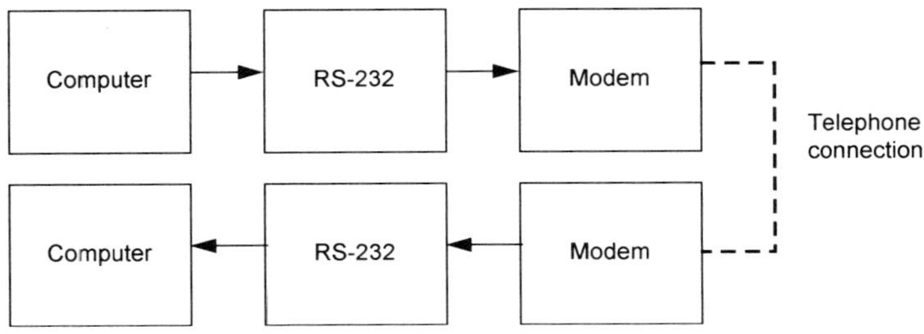

Figure 46.1 Data transfer using modems.

46.2 RS-232 communications

The communication between the modem and the computer is via RS-232. RS-232 uses asynchronous communication which has a start-stop data format. Each character is transmitted one at a time with a delay between characters. This delay is called the inactive time and is set at a logic level high as shown in Figure 46.2. The transmitter sends a start bit to inform the receiver that a character is to be sent in the following bit transmission. This start bit is always a '0'. Next 5, 6 or 7 data bits are sent as a 7-bit ASCII character, followed by a parity bit and finally either 1, 1.5 or 2 stop bits. The rate of transmission is set by the timing of a single bit. Both the transmitter and receiver need to be set to the same bit-time interval. An internal clock on both of them sets this interval. They only have to be roughly synchronized and approximately at the same rate as data is transmitted in relatively short bursts.

46.2.1 Bit rate and the baud rate

One of the main parameters for specifying RS-232 communications is the rate at which data is transmitted and received. It is important that the transmitter and receiver operate at roughly the same speed.

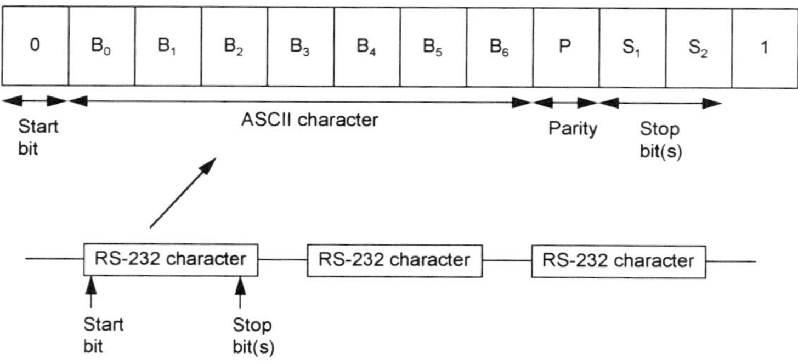

Figure 46.2 RS-232 frame format.

For asynchronous transmission the start and stop bits are added in addition to the seven ASCII character bits and the parity. Thus a total of 10 bits are required to transmit a single character. With 2 stop bits, a total of 11 bits are required. If 10 characters are sent every second and if 11 bits are used for each character, then the transmission rate is 110 bits per second (bps). The fastest modem thus has a character transmission rate of 2880 characters per second.

In addition to the bit rate, another term used to describe the transmission speed is the baud rate. The bit rate refers to the actual rate at which bits are transmitted, whereas the baud rate is to the rate at which signaling elements, used to represent bits, are transmitted. Since one signaling element encodes 1 bit, the two rates are then identical. Only in modems does the bit rate differ from the baud rate.

46.3 Modem standards

The CCITT (now known as the ITU) has defined standards which relate to RS-232 and modem communications. Each uses a V. number to define their type. Modems tend to state all the standards they comply with. An example FAX/modem has the following compatibility:

- V.32bis (14.4 Kbps). V.32 (9.6 Kbps).
- V.22bis (2.4 Kbps). V.22 (1.2 Kbps).
- Bell 212A (1.2 Kbps). Bell 103 (300 bps).
- V.17 (14.4 bps FAX). V.29 (9.6 Kbps FAX).
- V.27ter (4.8 Kbps FAX). V.21 (300 bps FAX - secondary channel).
- V.42bis (data compression). V.42 (error correction).
- MNP5 (data compression). MNP2–4 (error correction).

A 28.8 Kbps modem also supports the V.34 standard.

46.4 Modem commands

Most modems are Hayes compatible. Hayes was the company that pioneered modems and defined the standard method of programming the mode of the modem, which is the AT command language. A computer gets the attention of the modem by sending an 'AT' command. For example, 'ATDT' is the touch-tone dial command. Initially, a modem is in the command mode and accepts commands from the computer. These commands are sent at either 300 bps or 1200 bps (the modem automatically detects which of the speeds is being used).

Most commands are sent with the AT prefix. Each command is followed by a carriage return character (ASCII character 13 decimal); a command without a carriage return character is ignored (after a given time delay). More than one command can be placed on a single line and, if necessary, spaces can be entered to improve readability. Commands can be sent in either upper or lower case. Table 46.1 lists some AT commands. The complete set is defined in Appendix H.

Table 46.1 Example AT modem commands.

Command	Description
ATDT54321	Automatically phone number 54321 using touch-tone dialing. Within the number definition, a comma (,) represents a pause and a W waits for a second dial tone and an @ waits for a 5 second silence.
ATPT12345	Automatically phone number 12345 using pulse dialing.
AT S0=2	Automatically answer a call. The S0 register contains the number of rings the modem uses before it answers the call. In this case there will be two rings before it is answered. If S0 is zero then the modem will not answer a call.
ATH	Hang up telephone line connection.
+++	Disconnect line and return to on-line command mode.
AT A	Manually answer call.
AT E0	Commands are not echoed (AT E1 causes commands to be echoed). See Table 46.2.
AT L0	Low speaker volume (AT L1 gives medium volume and AT L2 gives high speaker volume).
AT MO	Internal speaker off (ATM1 gives internal speaker on until carrier detected, ATM2 gives the speaker always on, AT M3 gives speaker on until carrier detect and while dialing).
AT QO	Modem sends responses (AT Q1 does not send responses). See Table 46.2.
AT V0	Modem sends numeric responses (AT V1 sends word responses). See Table 46.2.

The modem can enter one of two states: the normal state and the command state. In the normal state the modem transmits and/or receives characters from the computer. In the command state, characters sent to the modem are interpreted as commands. Once a command is interpreted, the modem goes into the normal mode. Any characters sent to the modem are then sent along the line. To interrupt the modem so that it goes back into command mode, three consecutive '`+`' characters are sent, i.e. '`+++`'.

After the modem has received an AT command it responds with a return code. Some return codes are given in Table 46.2 (a complete set is defined in Appendix D). For example, if a modem calls another which is busy then the return code is 7. A modem dialing another modem returns the codes for OK (when the ATDT command is received), CONNECT (when it connects to the remote modem) and CONNECT 1200 (when it detects the speed of the remote modem). Note that the return code from the modem can be suppressed by sending the AT command '`ATQ1`'. The AT code for it to return the code is '`ATQ0`'; normally this is the default condition

Table 46.2 Example return codes.

Message	*Digit*	*Description*
OK	0	Command executed without errors
CONNECT	1	A connection has been made
RING	2	An incoming call has been detected
NO CARRIER	3	No carrier detected
ERROR	4	Invalid command
CONNECT 1200	5	Connected to a 1200 bps modem
NO DIALTONE	6	Dial-tone not detected
BUSY	7	Remote line is busy
NO ANSWER	8	No answer from remote line
CONNECT 600	9	Connected to a 600 bps modem
CONNECT 2400	10	Connected to a 2400 bps modem
CONNECT 4800	11	Connected to a 4800 bps modem
CONNECT 9600	13	Connected to a 9600 bps modem
CONNECT 14400	15	Connected to a 14 400 bps modem
CONNECT 19200	61	Connected to a 19 200 bps modem
CONNECT 28800	65	Connected to a 28 800 bps modem
CONNECT 1200/75	48	Connected to a 1200/75 bps modem

Figure 46.3 shows an example session when connecting one modem to another. Initially the modem is set up to receive commands from the computer. When the computer is ready to make a connection it sends the command '`ATDH 54321`' which makes a connection with telephone number 54321 using tone dialing. The modem then replies with an OK response (a 0 value) and the modem tries to make a connection with the remote modem. If it cannot make the connection it returns back a response of NO CARRIER (3), BUSY (7), NO DIALTONE (6) or NO ANSWER (8). If it does connect to the remote modem then it returns a connect response, such as CONNECT 9600 (13). The data can then be transmitted between the modem at the assigned rate (in this case 9600 bps). When the modem wants to end the connection it gets the modem's attention by sending it three '+' characters ('+++'). The modem will then wait for a command from the host computer. In this case the command is hang-up the connection (ATH). The modem will then return an OK response when it has successfully cleared the connection.

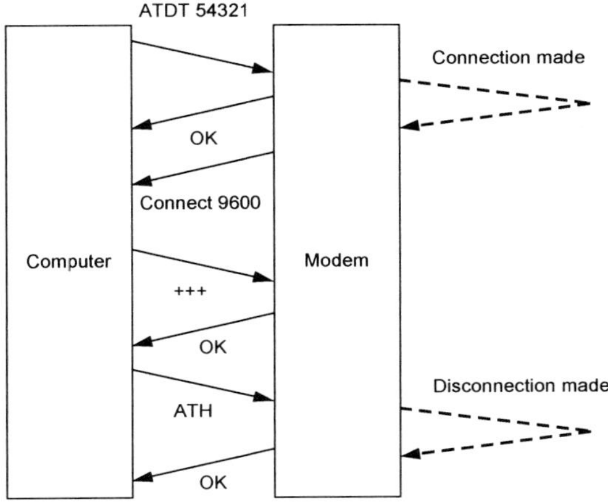

Figure 46.3 Commands and responses when making a connection.

The modem contains various status registers called the S-registers which store modem settings. Table 46.3 lists some of these registers (Appendix D gives a complete listing). The S0 register sets the number of rings that must occur before the modem answers an incoming call. If it is set to zero (0) then the modem will not answer incoming calls. The S1 register stores the number of incoming rings when the modem is rung. S2 stores the escape character, normally this is set to the '+' character and the S3 register stores the character which defines the end of a command, normally the CR character (13 decimal).

Table 46.3 Modem registers.

Register	*Function*	*Range [typical default]*
S0	Rings to Auto-answer	0–255 rings [0 rings]
S1	Ring counter	0–255 rings [0 rings]
S2	Escape character	[43]
S3	Carriage return character	[13]
S6	Wait time for dial-tone	2–255 s [2 s]
S7	Wait time for carrier	1–255 s [50 s]
S8	Pause time for automatic dialing	0–255 [2 s]

46.5 Modem setups

Figure 46.4 shows a sample window from the Microsoft Windows Terminal program (in both Microsoft Windows 3.*x* and Windows 95/98). It shows the Modem commands window. In this case, it can be seen that when the modem dials a number the prefix to the number dialed is '`ATDT`'. The hang-up command sequence is '`+++ ATH`'. A sample dialing window is shown in Figure 46.6. In this case the number dialed is `9,123456789`. A ',' character represent a delay. The actual delay is determined by the value in the S8 register (see Table 46.3). Typically, this value is about 2 seconds.

On many private switched telephone exchanges a `9` must prefix the number if an outside line is required. A delay is normally required after the `9` prefix before dialing the actual number. To modify the delay to 5 seconds, dial the number `9 0112432` and wait 30 seconds for the carrier, then the following command line can be used:

```
ATDT 9,0112432 S8=5 S7=30
```

It can be seen in Figure 46.4 that a prefix and a suffix are sent to the modem. This is to ensure there is a time delay between the transmission prefix and the suffix string. For example, when the modem is to hang-up the connection, the '`+++`' is sent followed by a delay then the '`ATH`'.

In Figure 46.6 there is an option called Originate. This string is sent initially to the modem to set it up. In this case the string is '`ATQ0V1E1S0=0`'. The Q0 part informs the modem to return a send status code. The V1 part informs the modem that the return code message is to be displayed rather than just the value of the return code; for example, it displays CONNECT 1200 rather than the code 5 (V0 displays the status code). The E1 part enables the command message echo (E0 disables it).

Figure 46.6 shows the modem setup windows for CompuServe access. The string in this case is:

```
ATS0=0 Q0 V1 &C1&D2^M
```

as previously seen, `S0` stops the modem from Auto-answering. `V1` causes the modem to respond with word responses. `&C1` and `&D2` set up the hardware signals for the modem. Finally `^M` represent Cntrl-M which defines the carriage return character.

The modem reset command in this case is `AT &F`. This resets the modem and restores the factor default settings.

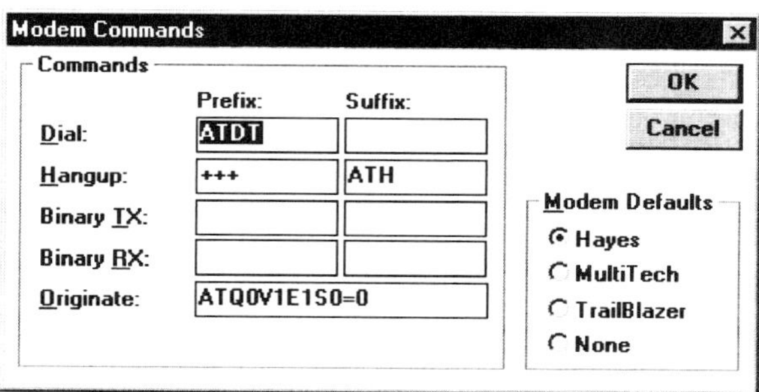

Figure 46.4 Modem commands.

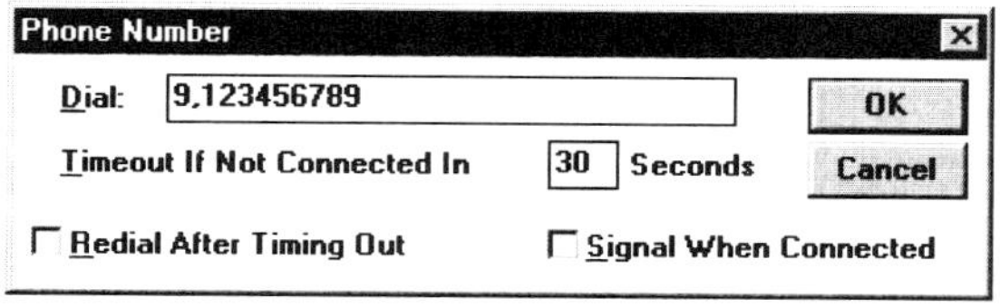

Figure 46.5 Dialing a remote modem.

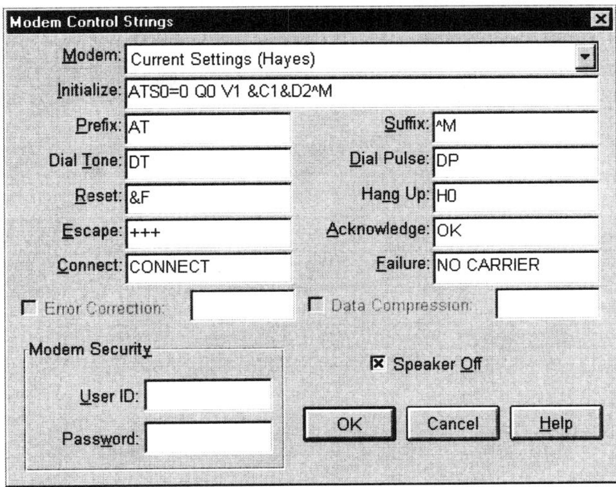

Figure 46.6 Example modem settings.

46.6 Modem indicators

Most external modems have status indicators to inform the user of the current status of a connection. Typically, the indicator lights are:

- AA – is ON when the modem is ready to receive calls automatically. It flashes when a call is incoming. If it is OFF then it will not receive incoming calls. Note that if the S0 register is loaded with any other value than 0 then the modem goes into Auto-answer mode. The value stored in the S0 register determines the number of rings before the modem answers.
- CD – is ON when the modem detects the remote modem's carrier, else it is OFF.
- OH – is ON when the modem is on-hook, else it is OFF.
- RD – flashes when the modem is receiving data or is getting a command from the computer.
- SD – flashes when the modem is sending data.
- TR – shows that the DTR line is active (i.e. the computer is ready to transmit or receive data).
- MR – shows that the modem is powered-up.

46.7 Profile viewing

The settings of the modem can be determined by using the AT command with &V. An example is shown next (which uses the program in Chapter 47). In this it can be seen that the settings include: B0 (CCITT 300 or 1200 bps for call establishment), E1 (enable command echo), L2 (medium volume), M1 (speaker is off when receiving), Q1 (prohibits modem from sending result codes to the DTE) T (set tone dial) and V1 (display result codes in a verbose form). It can be seen that the S0 register is set to 3 which means that the modem waits for three rings before it will automatically answer the call.

```
+++
AT &V
ACTIVE PROFILE:
B0 E1 L2 M1 Q1 T V1 X4 Y0 &C1 &D0 &E0 &G2 &L0 &M0 &O0 &P1 &R0 &S0 &X0 &Y1
%A000 %C1 %D1 %E1 %P0 %S0 \A3 \C0 \E0 \G0 \J0 \K5 \N6 \Q0 \T000 \V1 \X0
S00:003 S01:000 S06:004 S07:045 S08:002 S09:006 S10:014 S11:085 S12:050
S16:1FH S18:000 S21:20H S22:F6H S23:B2H S25:005 S26:001 S27:60H S28:00H
STORED PROFILE 0:
B0 E1 L2 M1 Q0 T V1 X4 Y0 &C1 &D2 &E0 &G2 &L0 &M0 &O0 &P1 &R0 &S0 &X0
%A000 %C1 %D1 %E1 %P0 %S0 \A3 \C0 \E0 \G0 \J0 \K5 \N6 \Q3 \T000 \V1 \X0
S00:000 S16:1FH S21:30H S22:F6H S23:89H S25:005 S26:001 S27:000 S28:000
STORED PROFILE 1:
B0 E0 L2 M1 Q1 T V1 X4 Y0 &C1 &D0 &E0 &G2 &L0 &M0 &O0 &P1 &R0 &S0 &X0
%A000 %C1 %D1 %E1 %P0 %S0 \A3 \C0 \E0 \G0 \J0 \K5 \N6 \Q0 \T000 \V1 \X0
S00:003 S16:1FH S21:20H S22:F6H S23:95H S25:005 S26:001 S27:096 S28:000
TELEPHONE NUMBERS:
&Z0=
&Z1=
&Z2=
&Z3=
```

46.8 Test modes

There are several modes associated with the modems.

46.8.1 Local analog loopback (&T1)

In the analog loopback test the modem connects the transmit and receive lines on its output, as illustrated in Figure 46.7. This causes all transmitted characters to be received. It is initiated with the `&T1` mode. For example:

```
AT &Q0          <Enter>
AT S18=0 &T1 <Enter>
CONNECT 9600
Help the bridge is on fire <Enter>
+++
OK
AT &T0
OK
```

The initial command `AT &Q0` sets the modem into an asynchronous mode (stop-start). Next the `AT S18=0 &T1` command sets the timer test time to zero (which disables any limit to the time of the test) and `&T1` sets an analog test. The modem responds with the message `CONNECT 9600`. Then the user enters the text `Help .... on fire` followed by an <Enter>. Next the user enters three + characters which puts the modem back into command mode. Finally the user enters `AT &T0` which disables the current test.

If a time-limited test is required then the S18 register is loaded with the number of seconds that the test should last. For example, a test that last 2 minutes will be set-up with:

```
AT S18=120 &T1
```

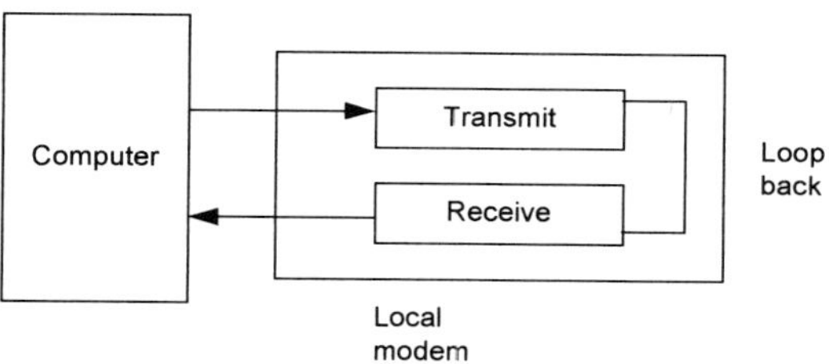

Figure 46.7 Analogue loopback with self-test.

46.8.2 Local analog loopback with self-test (&T8)

In the analog loopback test with self-test the modem connects the transmit and receive lines on its output and then automatically sends a test message which is then automatically received, as illustrated in Figure 46.8. The local error checker then counts the number of errors and displays a value when the test is complete. For example, the following test has found two errors:

```
AT &Q0        <Enter>
AT S18=0 &T8 <Enter>
+++
AT &T0
002
OK
```

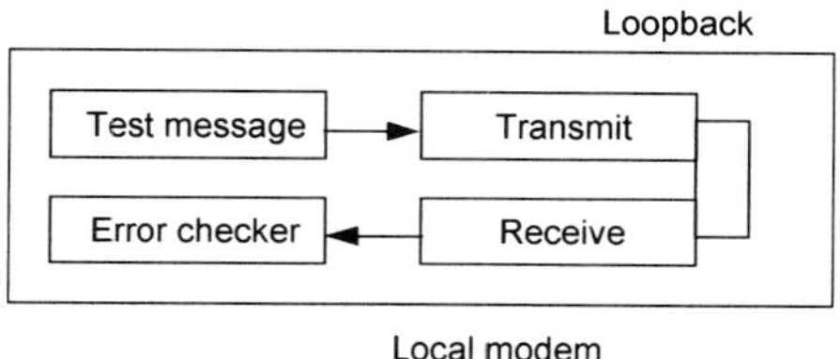

Figure 46.8 Analog loopback with self-test.

46.8.3 Remote digital loopback (&T6)

The remote digital loopback checks the local computer to modem connection, the local modem, the telephone line and the remote modem. The remote modem performs a loopback at the connection from the remote modem to its attached computer. Figure 46.9 illustrates the test setup. An example session is:

```
AT &Q0        <Enter>
AT S18=0 &T6 <Enter>
CONNECT 9600
Help the bridge is on fire <Enter>
+++
OK
AT &T0
OK
```

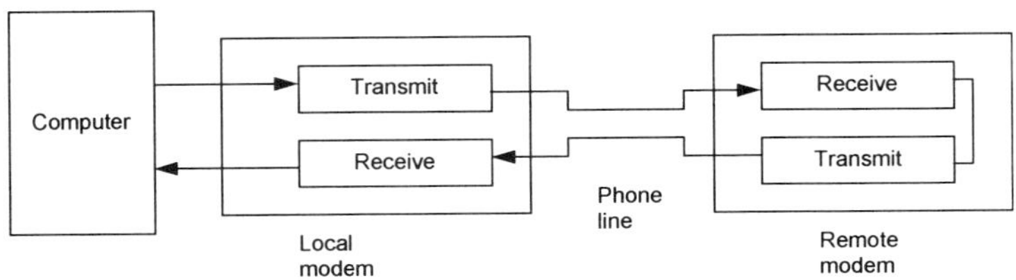

Figure 46.9 Remote digital loopback test.

46.8.4 Remote digital loopback with self-test (&T7)

The remote digital loopback with self-test checks the local computer to modem connection, the local modem, the telephone line and the remote modem. The remote modem performs a loopback at the connection from the remote modem to its attached computer. The local modem sends a test message and checks the received messages for errors. On completion of the test, the local modem transmits the number of errors. Figure 46.10 illustrates the test setup. An example session is:

```
AT &Q0         <Enter>
AT S18=0 &T7 <Enter>
+++
AT &T0
004
OK
```

or with a test of 60 seconds then the user does not have to send the break sequence:

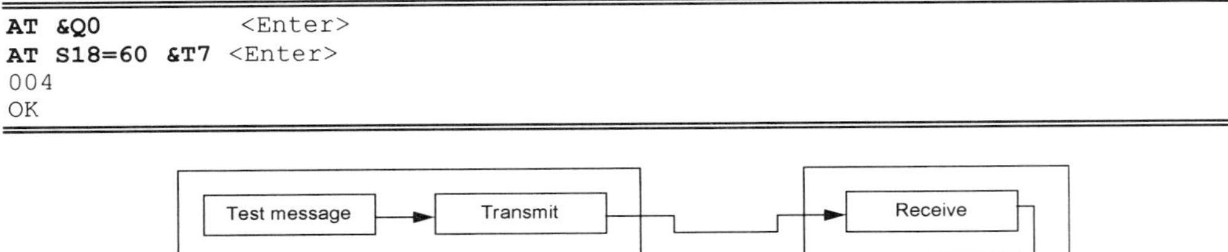

```
AT &Q0           <Enter>
AT S18=60 &T7 <Enter>
004
OK
```

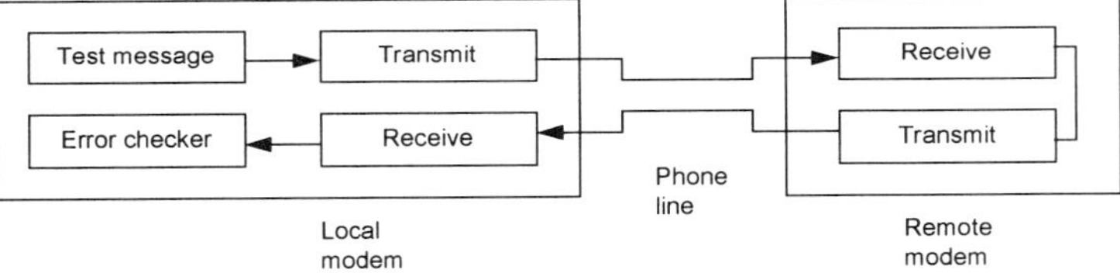

Figure 46.10 Remote digital loopback test with self-test.

46.9 Digital modulation

Digital modulation changes the characteristic of a carrier according to binary information. With a sine wave carrier the amplitude, frequency or phase can be varied. Figure 46.11 illustrates the three basic types: amplitude-shift keying (ASK), frequency-shift keying (FSK) and phase-shift keying (PSK).

46.9.1 Frequency-shift keying (FSK)

FSK, in the most basic case, represents a 1 (a mark) by one frequency and a 0 (a space) by another. These frequencies lie within the bandwidth of the transmission channel.

On a V.21, 300 bps, full-duplex modem the originator modem uses the frequency 980 Hz to represent a mark and 1180 Hz a space. The answering modem transmits with 1650 Hz for a mark and 1850 Hz for a space. The four frequencies allow the caller originator and the answering modem to communicate at the same time; that is full-duplex communication.

FSK modems are inefficient in their use of bandwidth, with the result that the maximum data rate over normal telephone lines is 1800 bps. Typically, for rates over 1200 bps, other modulation schemes are used.

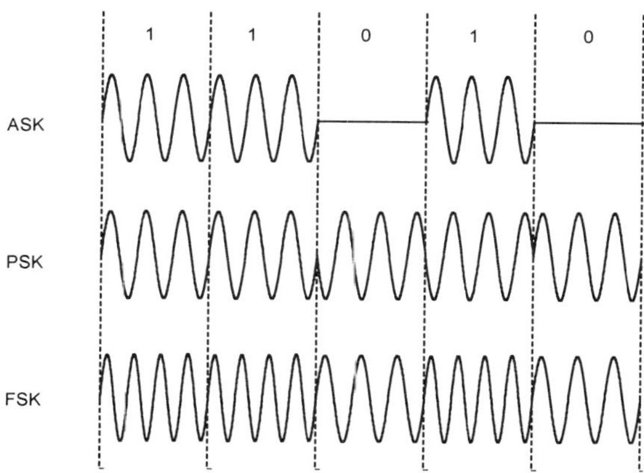

Figure 46.11 Waveforms for ASK, PSK and FSK.

46.9.2 Phase-shift keying (PSK)

In coherent PSK a carrier gets no phase shift for a 0 and a 180° phase shift for a 1, as given next:

$$0 \quad \Rightarrow \quad 0° \qquad\qquad 1 \quad \Rightarrow \quad 180°$$

Its main advantage over FSK is that since it uses a single frequency it uses much less bandwidth. It is thus less affected by noise. It has an advantage over ASK because its information is not contained in the amplitude of the carrier, thus again it is less affected by noise.

46.9.3 M-ary modulation

With *M*-ary modulation a change in amplitude, phase or frequency represents one of *M* possible signals. It is possible to have *M*-ary FSK, *M*-ary PSK and *M*-ary ASK modulation schemes. This is where the baud rate differs from the bit rate. The bit rate is the true measure of the rate of the line, whereas the baud rate only indicates the signaling element rate, which might be a half or a quarter of the bit rate.

For four-phase differential phase-shift keying (DPSK) the bits are grouped into two and each group is assigned a certain phase shift. For 2 bits there are four combinations: a 00 is coded as 0°, 01 coded as 90°, and so on:

00 ⇒	0°	01 ⇒	90°
11 ⇒	180°	10 ⇒	270°

It is also possible to change a mixture of amplitude, phase or frequency. *M*-ary amplitude-phase keying (APK) varies both the amplitude and phase of a carrier to represent *M* possible bit patterns.

M-ary quadrature amplitude modulation (QAM) changes the amplitude and phase of the carrier. 16-QAM uses four amplitudes and four phase shifts, allowing it to code 4 bits at a time. In this case, the baud rate will be a quarter of the bit rate.

Typical technologies for modems are:

FSK – used up to 1200 bps
Four-phase DPSK – used at 2400 bps
Eight-phase DPSK – used at 4800 bps
16-QAM – used at 9600 bps

46.10 Typical modems

Most modern modems operate with V.22bis (2400 bps), V.32 (9600 bps), V.32bis (14 400 bps); some standards are outlined in Table 46.4. The V.32 and V.32bis modems can be enhanced with echo cancellation. They also typically have built-in compression using either the V.42bis standard or MNP level 5.

Table 46.4 Typical modems.

ITU recommendation	*Bit rate(bps)*	*Modulation*
V.21	300	FSK
V.22	1 200	PSK
V.22bis	2 400	ASK/PSK
V.27ter	4 800	PSK
V.29	9 600	PSK
V.32	9 600	ASK/PSK
V.32bis	14 400	ASK/PSK
V.34	28 800	ASK/PSK

46.10.1 V.42bis and MNP compression

There are two main standards used in modems for compression. The V.42bis standard is defined by the ITU and the MNP (Microcom Networking Protocol) has been developed by a company named Microcom. Most modems will try to compress using V.42bis but if this fails they try MNP level 5. V.42bis uses the Lempel-Ziv algorithm which builds dictionaries of code words for recurring characters in the data stream. These code words normally take up fewer bits than the uncoded bits. V.42bis is associated with the V.42 standard which covers error correction.

46.10.2 V.22bis modems

V.22bis modems allow transmission at up to 2400 bps. It uses four amplitudes and four phases. Figure 46.12 shows the 16 combinations of phase and amplitude for a V.22bis modem. It can be seen that there are 12 different phase shifts and four different amplitudes. Each transmission is known as a symbol, thus each transmitted symbol contains 4 bits. The transmission rate for a symbol is 600 symbols per second (or 600 Baud), thus the bit rate will be 2 400 bps.

Trellis coding tries to ensure that consecutive symbols differ as much as possible.

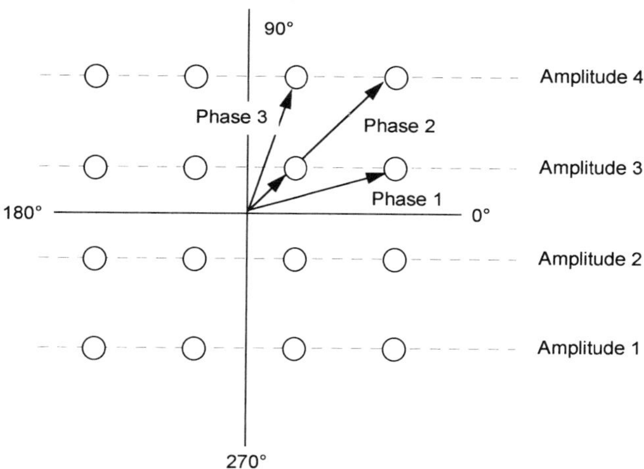

Figure 46.12 Phase and amplitude coding for V.32.

46.10.3 V.32 modems

V.32 modems include echo cancellation which allows signals to be transmitted in both directions at the same time. Previous modems used different frequencies to transmit on different channels. Echo cancellation uses DSP (digital signal processing) to subtract the sending signal from the received signal.

V.32 modems use trellis encoding to enhance error detection and correction. They encode 32 signaling combinations of amplitude and phase. Each of the symbols contains 4 data bits and a single trellis bit. The basic symbol rate is 2400 bps, thus the actual data rate will be 9600 bps. A V.32bis modem uses 7 bits per symbol, thus the data rate will be 14 400 bps (2400×6).

46.11 Fax transmission

Facsimile (fax) transmission involves the transmission of images over a telephone line using a modem. A standalone fax consists of:

- An image scanner.
- A graphics printer (normally a thermal printer).
- A transmission/reception modem.

The fax scans an A4 image with 1142 scan lines (3.85 lines per millimeter) and 1728 pixels per line. The EIA and ITU originally produced the RS-328 standard for the transmission of analogue voltage levels to represent different brightness. The ITU recommendations are known as Group I and Group II standards. The Group III standard defines the transmission of faxes using digital transmission with 1142 × 1728 pixels of black or white. Group IV is an extension to Group III but allows different gray scales and also color (unfortunately it requires a high bit rate).

An A4 scan would consist of 1 976 832 (1142 × 1728) scanned elements. If each element is scanned for black and white, then, at 9600 bps, it would take over 205 s to transmit. Using RLE coding can drastically reduced this is transmission time.

46.11.1 Modified Huffman coding

Group III compression uses modified Huffman code to compress the transmitted bit stream. It uses a table of codes in which the most frequent run lengths are coded with a short code. Typically, documents contain long runs of white or black. A compression ratio of over 10:1 is easily achievable (thus a single-page document can be sent in under 20 s, for a 9600 bps transmission rate). Table 46.5 shows some code runs of white and Table 46.6 shows some codes for runs of black. The transmitted code always starts on white code. The codes range from 0 to 63. Values from 64 to 2560 use two codes. The first gives the multiple of 64 followed by the normally coded remainder.

For example, if the data to be encoded is:

16 white, 4 black, 16 white, 2 black, 63 white, 10 black, 63 white

it would be coded as:

```
101010  011 101010  11 00110100  0000100  00110100
```

This would take 40 bits to transmit the coding, whereas it would take 304 bits (i.e. 16 + 4 + 16 + 2 + 128 + 10 + 128). This results in a compression ratio of 7.6:1.

Table 46.5 White run length coding.

Run length	*Coding*	*Run length*	*Coding*	*Run length*	*Coding*
0	00110101	1	000111	2	0111
3	1000	4	1011	5	1100
6	1110	7	1111	8	10011
9	10100	10	00111	11	01000
12	001000	13	000011	14	110100
15	110101	16	101010	17	101011
18	0100111	19	0001100	61	00110010
62	00110011	63	00110100	EOL	00000000001

Table 46.6 Black run-length coding.

Run length	*Coding*	*Run length*	*Coding*	*Run length*	*Coding*
0	0000110111	1	010	2	11
3	10	4	011	5	0011
6	0010	7	00011	8	000101
9	000100	10	0000100	11	0000101
12	0000111	13	00000100	14	00000111
15	000011000	16	0000010111	17	0000011000
18	0000001000	19	00001100111	61	000001011010
62	0000001100110	63	000001100111	EOL	00000000001

47 RS-232

47.1 Introduction

RS-232 is one of the most widely used techniques used to interface external equipment to computers. It uses serial communications where one bit is sent along a line, at a time. This differs from parallel communications which send one or more bytes, at a time. The main advantage of serial communications over parallel communications is that a single wire is need to transmit and another to receive. RS-232 is a *de facto* standard that most computer and instrumentation companies comply with. The Electronics Industries Association (EIA) standardized it in 1962. Unfortunately, this standard only allows short cable runs with low bit rates, such as a bit rate of 19 600 bps for a maximum distance of 20 meters. New serial communications standards, such as RS-422 and RS-449, allow very long cable runs and high bit rates. For example, RS-422 allows a bit rate of up to 10 Mbps over distances up to one mile, using twisted-pair, coaxial cable or optical fibers. The new standards can also be used to create computer networks. This chapter introduces the RS-232 standard and gives simple programs which can be used to transmit and receive using RS-232. Chapter 48 uses an interrupt-driven technique to further enhance the transmission and reception.

47.2 Electrical characteristics

47.2.1 Line voltages

The electrical characteristics of RS-232 define the minimum and maximum voltages of a logic '1' and '0'. A logic '1' ranges from −3 V to −25 V, but will typically be around −12 V. A logical '0' ranges from 3 V to 25 V, but will typically be around +12 V. Any voltage between −3 V and +3 V has an indeterminate logical state. If no pulses are present on the line then the voltage level is equivalent to a high level, that is −12 V. A voltage level of 0 V at the receiver is interpreted as a line break or a short circuit. Figure 47.1 shows an example transmission.

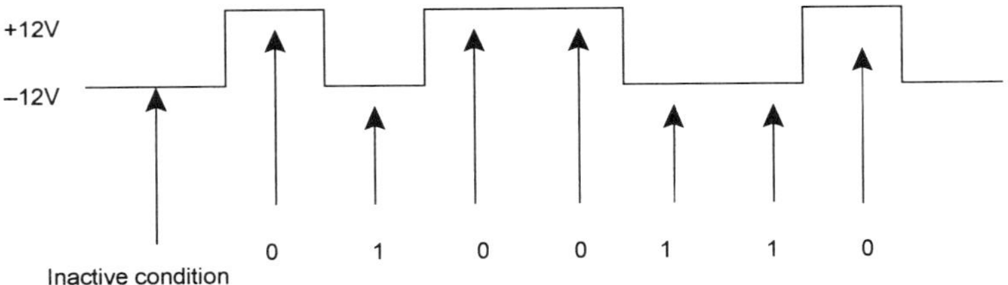

Figure 47.1 RS-232 voltage levels.

47.2.2 DB25S connector

The DB25S connector is a 25-pin D-type connector and gives full RS-232 functionality. Figure 47.2 shows the pin number assignment. A DCE (the terminating cable) connector has a male outer casing with female connection pins. The DTE (the computer) has a female outer casing with male connecting pins. There are three main signal types: control, data and ground. Table 47.1 lists the main connections. Control lines are active high, that is they are high when the signal is active and low when inactive.

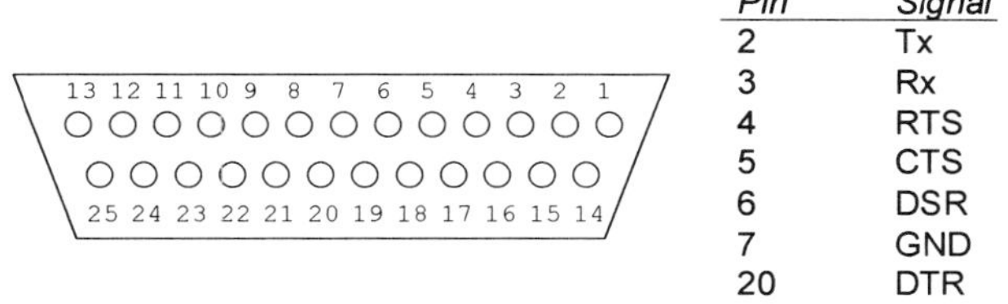

Figure 47.2 RS-232 DB25S connector.

47.2.3 DB9S connector

The 25-pin connector is the standard for RS-232 connections but as electronic equipment becomes smaller, there is a need for smaller connectors. For this purpose most PCs now use a reduced function 9-pin D-type connector rather than the full function 25-way D-type. As with the 25-pin connector the DCE (the terminating cable) connector has a male outer casing with female connection pins. The DTE (the computer) has a female outer casing with male connecting pins. Figure 47.3 shows the main connections.

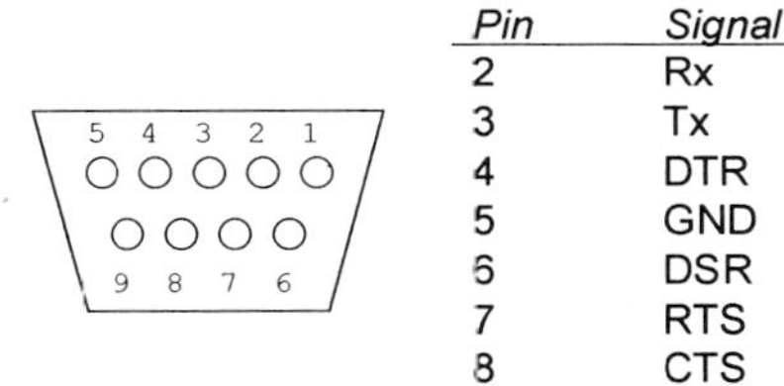

Figure 47.3 RS-232 DB9S Interface.

47.2.4 PC connectors

All PCs have at least one serial communications port. The primary port is named `COM1:` and the secondary is `COM2:`. There are two types of connectors used in RS-232 communications; these are the 25- and 9-way D-type. Most modern PCs use either a 9-pin connector for the primary (`COM1:`) serial port and a 25-pin for a secondary serial port (`COM2:`), or they use two 9-pin connectors for serial ports. The serial port can be differentiated from the parallel port in that the 25-pin parallel port (`LPT1:`) is a 25-pin female connector on the PC and a male connector on the cable. The 25-pin serial connector is a male on the PC and a female on the cable. The different connector types can cause problems in connecting devices. Thus, a 25-to-9 pin adapter is a useful attachment, especially to connect a serial mouse to a 25-pin connector.

Table 47.1 Main pin connections used in 25-pin connector.

Pin	*Name*	*Abbreviation*	*Functionality*
1	Frame Ground	FG	This ground normally connects the outer sheath of the cable and to earth ground.
2	Transmit Data	TD	Data is sent from the DTE (computer or terminal) to a DCE via TD.
3	Receive Data	RD	Data is sent from the DCE to a DTE (computer or terminal) via RD.
4	Request to Send	RTS	DTE sets this active when it is ready to transmit data.
5	Clear to Send	CTS	DCE sets this active to inform the DTE that it is ready to receive data.
6	Data Set Ready	DSR	Similar functionality to CTS but activated by the DTE when it is ready to receive data.
7	Signal Ground	SG	All signals are referenced to the signal ground (GND).
20	Data Terminal Ready	DTR	Similar functionality to RTS but activated by the DCE when it wishes to transmit data.

47.3 Frame format

RS-232 uses asynchronous communications which has a start-stop data format, as shown in Figure 47.4. Each character is transmitted one at a time with a delay between them. This delay is called the inactive time and is set at a logic level high (–12 V) as shown in Figure 47.5. The transmitter sends a start bit to inform the receiver that a character is to be sent in the following bit transmission. This start bit is always a '0'. Next, 5, 6 or 7 data bits are sent as a 7-bit ASCII character, followed by a parity bit and finally either 1, 1.5 or 2 stop bits. Figure 47.3 shows a frame format and an example transmission of the character 'A', using odd parity. The timing of a single bit sets the rate of transmission. Both the transmitter and receiver need to be set to the same bit-time interval. An internal clock on both sets this interval. These only have to be roughly synchronized and approximately at the same rate as data is transmitted in relatively short bursts.

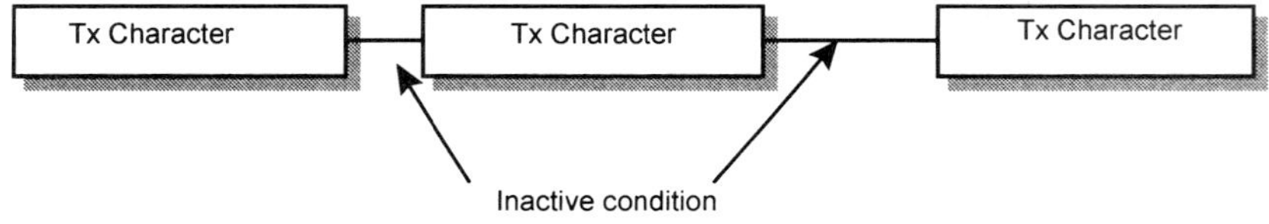

Figure 47.4 Asynchronous communications.

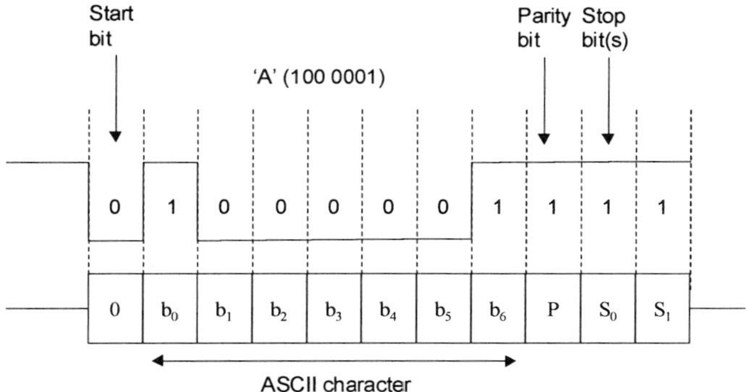

Figure 47.5 RS-232 frame format.

Example

An RS-232 serial data link uses 1 start bit, 7 data bits, 1 parity bit, 2 stop bits, ASCII coding and even parity. Determine the message sent from the following bit stream.

First bit sent
⇓
`11111010000010110000011111111111111100000111111110001100111101010011111111111`

ANSWER

The format of the data string sent is given next:

{idle} `11111` {start bit} `0` {'A'} `1000001` {parity bit} `0` {stop bits } `11` {start bit} `0` {'p'} `0000111` {parity bit} `1` {stop bits} `11` {idle} `11111111` {start bit} `0` {'p'} `0000111` {parity bit} `1` {stop bits} `11` {idle} `11` {start bit} `0` {'L'} `0011001` {parity bit} `1` {stop bits} `11`

The message sent was thus 'AppL'.

Parity

Error control is data added to transmitted data in order to detect or correct an error in transmission. RS-232 uses a simple technique known as parity to provide a degree of error detection.

A parity bit is added to transmitted data to make the number of 1s sent either even (even parity) or odd (odd parity). It is a simple method of error coding and only requires exclusive-OR (XOR) gates to generate the parity bit. The parity bit is added to the transmitted data by inserting it into the shift register at the correct bit position.

A single parity bit can only detect an odd number of errors, that is, 1, 3, 5, and so on. If there is an even number of bits in error then the parity bit will be correct and no error will be detected. This type of error coding is not normally used on its own where there is the possibility of several bits being in error.

Baud rate

One of the main parameters which specify RS-232 communications is the rate of transmission at which data is transmitted and received. It is important that the transmitter and receiver operate at, roughly, the same speed.

For asynchronous transmission the start and stop bits are added in addition to the 7 ASCII character bits and the parity. Thus a total of 10 bits are required to transmit a single character. With 2 stop bits, a total of 11 bits are required. If 10 characters are sent every second and if 11 bits are used for each character, then the transmission rate is 110 bits per second (bps).

	Bits
ASCII character	7
Start bit	1
Stop bit	2
Total	10

Table 47.2 lists how the bit rate relates to the characters sent per second (assuming 10 transmitted bits per character). The bit rate is measured in bits per second (bps).

In addition to the bit rate, another term used to describe the transmission speed is the Baud rate. The bit rate refers to the actual rate at which bits are transmitted, whereas the Baud rate relates to the rate at which signaling elements, used to represent bits, are transmitted. Since one signaling element encodes one bit, the two rates are then identical. Only in modems does the bit rate differ from the Baud rate.

Table 47.2 Bits per second related to characters sent per second.

Speed (bps)	*Characters / second*
300	30
1200	120
2400	240

Bit stream timings

Asynchronous communications is a stop-start mode of communication and both the transmitter and receiver must be set up with the same bit timings. A start bit identifies the start of transmission and is always a low logic level. Next, the least significant bit is sent followed by the rest of the 7-bit ASCII character bits. After this, the parity bit is sent followed by the stop bit(s). The actual timing of each bit relates to the Baud rate and can be determined using the following formula:

$$\text{Time period of each bit} = \frac{1}{\text{Baud rate}}\ \text{s}$$

For example, if the Baud rate is 9600 Baud (or bps) then the time period for each bit sent is 1/9600 s, or 104 μs. Table 47.2 shows some bit timings as related to the Baud rate. An example of the voltage levels and timings for the ASCII character 'V' is given in Figure 47.4.

Table 47.3 Bit timings related to Baud rate.

Baud rate	Time for each bit (μs)
1200	833
2400	417
9600	104
19200	52

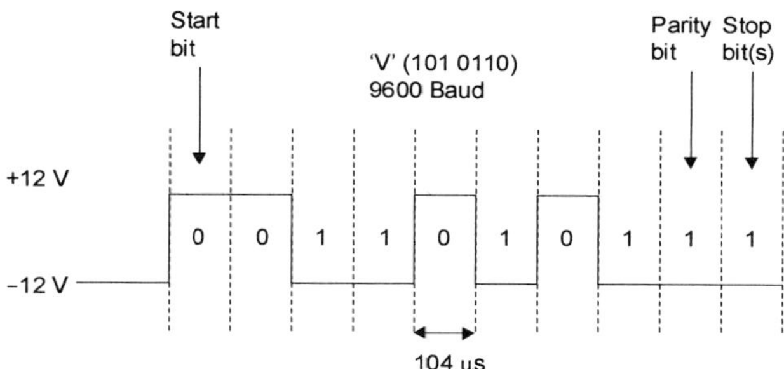

Figure 47.6 ASCII 'V' at RS-232 voltage levels.

47.4 Communications between two nodes

RS-232 is intended to be a standard but not all manufacturers abide by it. Some implement the full specification while others implement just a partial specification. This is mainly because not every device requires the full functionality of RS-232; for example, a modem requires many more control lines than a serial mouse.

The rate at which data is transmitted and the speed at which the transmitter and receiver can transmit/receive the data dictates whether data handshaking is required.

47.4.1 Handshaking

In the transmission of data there can be either no handshaking, hardware handshaking or software handshaking. If no handshaking is used then the receiver must be able to read the received characters before the transmitter sends another. The receiver may buffer the received character and store it in a special memory location before it is read. This memory location is named the receiver buffer. Typically, it may only hold a single character. If it is not emptied before another character is received then any character previously in the buffer will be overwritten. An example of this is illustrated in Figure 47.7. In this case the receiver has read the first two characters successfully from the receiver buffer, but it did not read the third character as the fourth transmitted character has overwritten it in the receiver buffer. If this condition occurs then some form of handshaking must be used to stop the transmitter sending characters before the receiver has had time to service the received characters.

Hardware handshaking involves the transmitter asking the receiver if it is ready to receive data. If the receiver buffer is empty it will inform the transmitter that it is ready to receive data. Once the data is transmitted and loaded into the receiver buffer, the transmitter is informed not to transmit any more characters until the character in the receiver buffer has been read. The main hardware handshaking lines used for this purpose are:

- `CTS` – Clear to Send.
- `RTS` – Ready to Send.
- `DTR` – Data Terminal Ready.
- `DSR` – Data Set Ready.

Software handshaking involves sending special control characters. These include the `DC1`–`DC4` control characters.

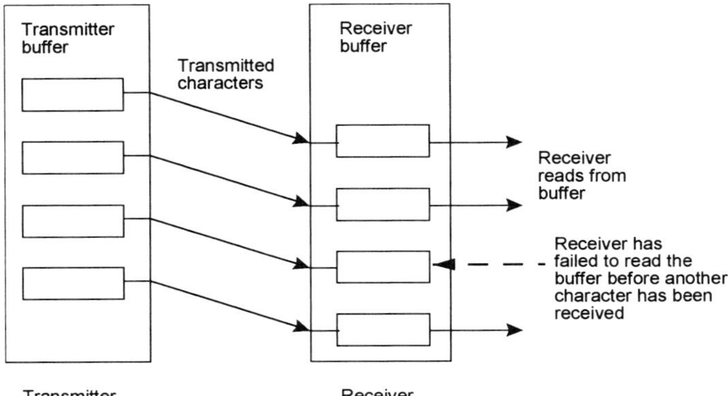

Figure 47.7 Transmission and reception of characters.

47.4.2 RS-232 set-up

Windows 95/98 allow the serial port setting to be set by selecting Control Panel → System → Device Manager → Ports (COM and LPT) → Port Settings. The settings of the communications port (the IRQ and the port address) can be changed by selecting Control Panel → System → Device Manager → Ports (COM and LPT) → Resources for IRQ and Addresses. Figure 47.8 shows example parameters and settings. The selectable baud rates are typically 110, 300, 600, 1200, 2400, 4800, 9600 and 19 200 Baud for an 8250-based device. A 16650 compatible UART speed also gives enhanced speeds of 38400, 57 600, 115 200, 230 400, 460 800 and 921 600 Baud. Notice that the flow control can either be set to software handshaking (X-ON/X-OFF), hardware handshaking or none.

The parity bit can either be set to none, odd, even, mark or space. A mark in the parity option sets the parity bit to a '1' and a space sets it to a '0'. In this case COM1: is set at 9600 Baud, 8 data bits, no parity, 1 stop bit and no parity checking.

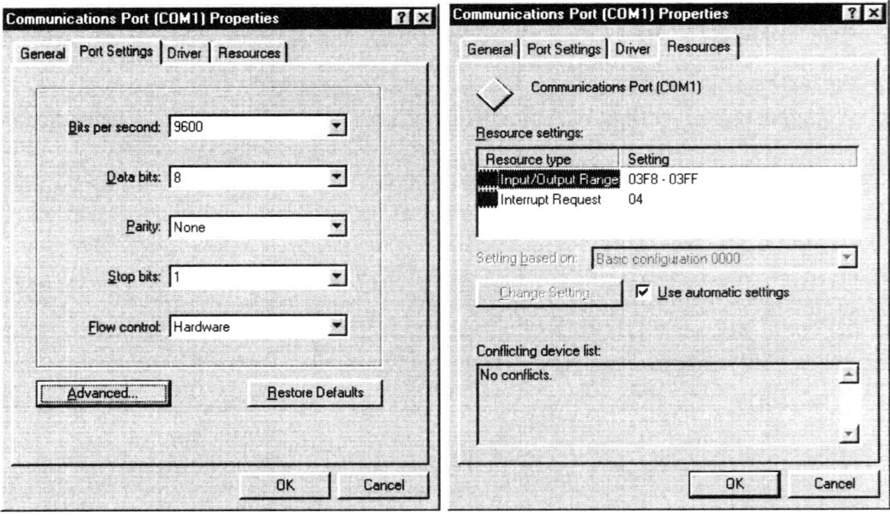

Figure 47.8 Changing port setting and parameters.

47.4.3 Simple no-handshaking communications

In this form of communication it is assumed that the receiver can read the received data from the receive buffer before another character is received. Data is sent from a TD pin connection of the transmitter and is received in the RD pin connection at the receiver. When a DTE (such as a computer) connects to another DTE, then the transmit line (TD) on one is connected to the receive (RD) of the other and vice versa. Figure 47.9 shows the connections between the nodes.

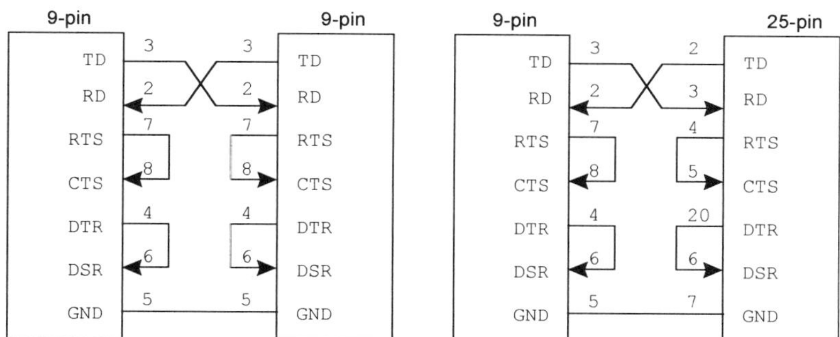

Figure 47.9 RS-232 connections with no hardware handshaking.

47.4.4 Software handshaking

Two ASCII characters start and stop communications. These are X-ON (^S, Cntrl-S or ASCII 11) and X-OFF (^Q, Cntrl-Q or ASCII 13). When the transmitter receives an X-OFF character it ceases communications until an X-ON character is sent. This type of handshaking is normally used when the transmitter and receiver can process data relatively quickly. Normally, the receiver will also have a large buffer for the incoming characters. When this buffer is full it transmits an X-OFF. After it has read from the buffer the X-ON is transmitted; see Figure 47.10.

47.4.5 Hardware handshaking

Hardware handshaking stops characters in the receiver buffer from being overwritten. The control lines used are all active high. Figure 47.11 shows how the nodes communicate. When a node wishes to transmit data it asserts the RTS line active (that is, high). It then monitors the CTS line until it goes active (that is, high). If the CTS line at the transmitter stays inactive then the receiver is busy and cannot receive data, at the present. When the receiver reads from its buffer the RTS line will automatically go active indicating to the transmitter that it is now ready to receive a character.

Receiving data is similar to the transmission of data, but the lines DSR and DTR are used instead of RTS and CTS. When the DCE wishes to transmit to the DTE the DSR input to the receiver will become active. If the receiver cannot receive the character, it sets the DTR line inactive. When it is clear to receive it sets the DTR line active and the remote node then transmits the character. The DTR line will be set inactive until the character has been processed.

47.4.6 Two-way communications with handshaking

For full handshaking of the data between two nodes the RTS and CTS lines are crossed over (as are the DTR and DSR lines). This allows for full remote node feedback (see Figure 47.12).

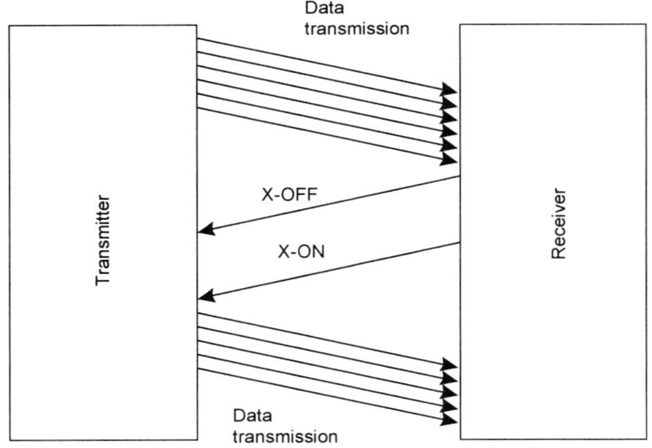

Figure 47.10 Software handshaking using X-ON and X-OFF.

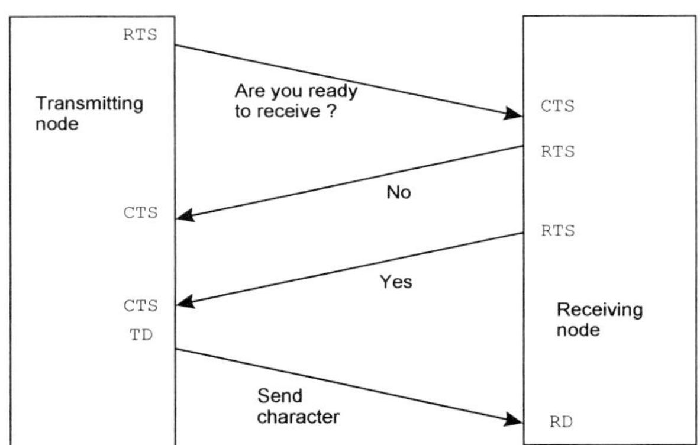

Figure 47.11 Handshaking lines used in transmitting data.

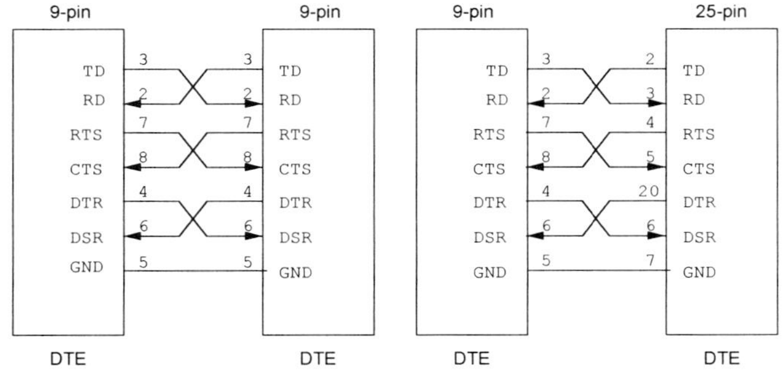

Figure 47.12 RS-232 communications with handshaking.

47.4.7 DTE-DCE connections (PC to modem)

A further problem occurs in connecting two nodes. A DTE–DTE connection requires cross-overs on their signal lines, whereas DTE–DCE connections require straight-through lines. Figure 47.13 shows an example connection for a computer to modem connection.

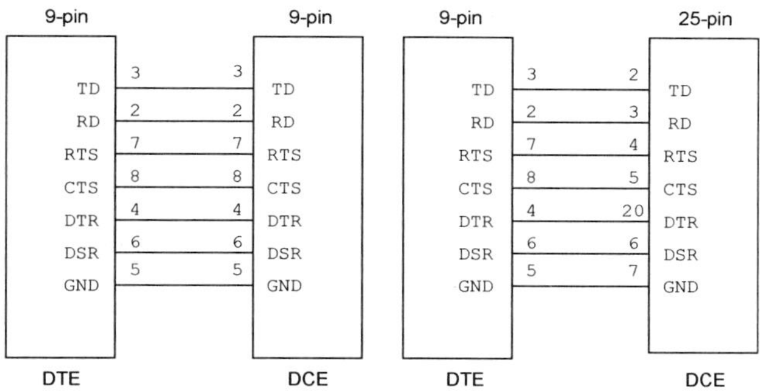

Figure 47.13 DTE to DCE connections.

47.5 Programming RS-232

Normally, serial transmission is achieved via the RS-232 standard. Although 25 lines are defined usually only a few are used. Data is sent along the TD line and received by the RD line with a common ground return. The other lines, used for handshaking, are RTS (Ready to Send) which is an output signal to indicate that data is ready to be transmitted and CTS (Clear to Send), which is an input indicating that the remote equipment is ready to receive data.

The 8250 IC is commonly used in serial communications. It can either be mounted onto the motherboard of the PC or fitted to an I/O card. This section discusses how it is programmed.

47.5.1 Programming the serial device

Normally, serial transmission is achieved via the RS-232 standard. Although 25 lines are defined usually only a few are used. Data is sent along the TD line and received by the RD line with a common ground return. The other lines, used for handshaking, are RTS (Ready to Send) which is an output signal to indicate that data is ready to be transmitted and CTS (Clear to Send), which is an input indicating that the remote equipment is ready to receive data.

The 8250 IC is commonly used in serial communications. It can either be mounted onto the motherboard of the PC or fitted to an I/O card. This section discusses how it is programmed.

Programming the serial device

The main registers used in RS-232 communications are the Line Control Register (LCR), the Line Status Register (LSR) and the Transmit and Receive buffers (see Figure 47.14). The Transmit and Receive buffers share the same addresses.

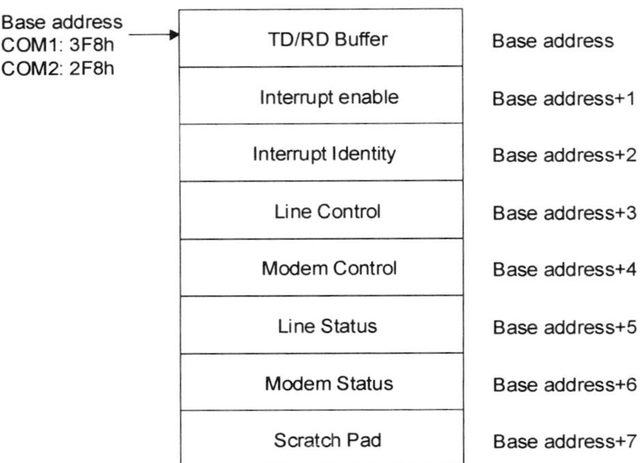

Figure 47.14 Serial communication registers.

The base address of the primary port (`COM1:`) is normally set at 3F8h and the secondary port (`COM2:`) at 2F8h. A standard PC can support up to four COM ports. These addresses are set in the BIOS memory and the address of each of the ports is stored at address locations `0040:0000` (`COM1:`), `0040:0002` (`COM2:`), `0040:0004` (`COM3:`) and `0040:0008` (`COM4:`). Program 47.1 can be used to identify these addresses. The statement:

```
ptr=(int far *)0x0400000;
```

initializes a far pointer to the start of the BIOS communications port addresses. Each address is 16 bits, thus the pointer points to an integer value. A far pointer is used as this can access the full 1 MB of memory, a non-far pointer can only access a maximum of 64 KB.

Program 47.1

```
#include <stdio.h>
#include <conio.h>
int   main(void)
{
int   far *ptr; /* 20-bit pointer */
      ptr=(int far *)0x0400000; /* 0040:0000 */ clrscr();
      printf("COM1: %04x\n",*ptr);
      printf("COM2: %04x\n",*(ptr+1));
      printf("COM3: %04x\n",*(ptr+2));
      printf("COM4: %04x\n",*(ptr+3));
      return(0);
}
```

Test run 47.1 shows a sample run. In this case there are four COM ports installed on the PC. If any of the addresses is zero then that COM port is not installed on the system.

Sample run 47.1

```
COM1: 03f8
COM2: 02f8
COM3: 03e8
COM4: 02e8
```

Line Status Register (LSR)

The LSR determines the status of the transmitter and receiver buffers. It can only be read from, and all the bits are automatically set by hardware. The bit definitions are given in Figure 47.15. When an error occurs in the transmission of a character one (or several) of the error bits is (are) set to a '1'.

One danger when transmitting data is that a new character can be written to the transmitter buffer before the previous character has been sent. This overwrites the contents of the character being transmitted. To avoid this the status bit S_6 is tested to determine if there is still a character still in the buffer. If there is then it is set to a '1', else the transmitter buffer is empty.

To send a character:

Test Bit 6 until set;
Send character;

A typical Pascal routine is:

```
repeat
   status := port[LSR] and $40;
until (status=$40);
```

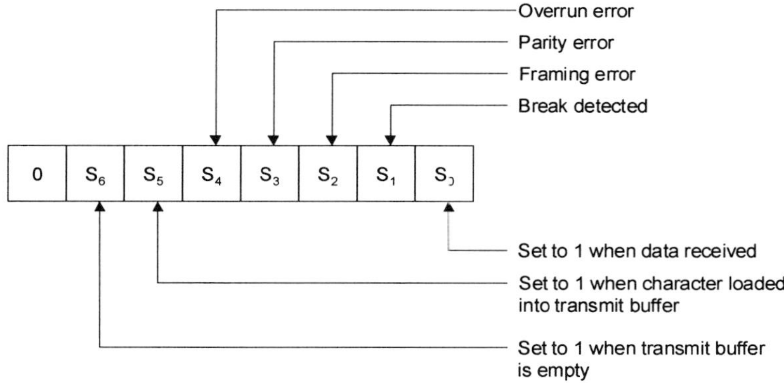

Figure 47.15 Line Status Register.

When receiving data the S_0 bit is tested to determine if there is a bit in the receiver buffer. To receive a character:

Test Bit 0 until set;
Read character;

A typical Pascal routine is:

```
repeat
      status := port[LSR] and $01;
until (status=$01);
```

Figure 47.16 shows how the LSR is tested for the transmission and reception of characters.

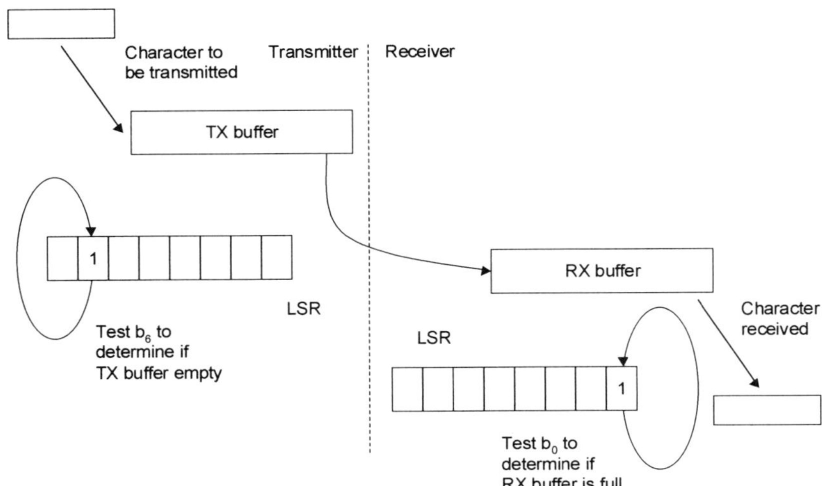

Figure 47.16 Testing of the LSR for the transmission and reception of characters.

Line Control Register (LCR)

The LCR sets up the communications parameters. These include the number of bits per character, the parity and the number of stop bits. It can be written to or read from and has a similar function to that of the control registers used in the PPI and PTC. The bit definitions are given in Figure 47.17.

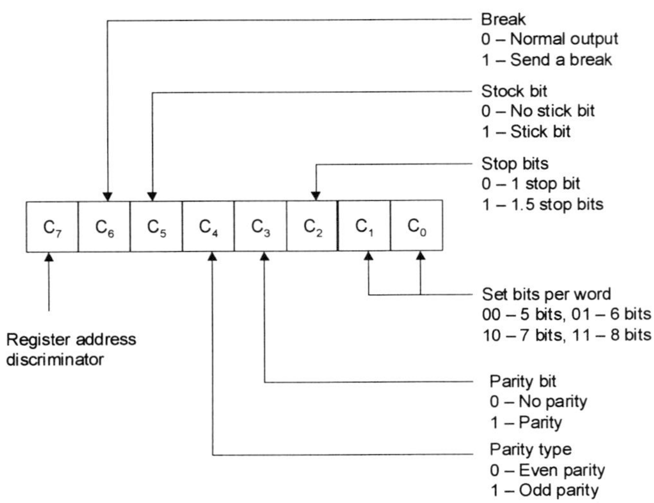

Figure 47.17 Line Control Register.

The msb, C_7, must to be set to a '0' in order to access the transmit and receive buffers, (TX/RX buffer) else if it is set to a '1', the Baud rate divider is accessed. The Baud rate is set by loading an appropriate 16-bit divisor, with the lower 8 bits of the divisor put into the TX/RX buffer address and the upper 8 bits put into the next address after the TX/RX buffer. The value loaded depends on the crystal frequency connected to the IC. Table 47.4 shows divisors for a crystal frequency is 1.8432 MHz. In general the divisor, N, is related to the Baud rate by:

$$Baud\ rate = \frac{Clock\ frequency}{16 \times N}$$

For example, for 1.8432 MHz and 9600 Baud $N = 1.8432 \times 10^6/(9600 \times 16) = 12$ (000Ch).

Table 47.4 Baud rate divisors.

Baud rate	Divisor	Baud rate	Divisor
110	0417h	2400	0030h
300	0180h	4800	0018h
600	00C0h	9600	000Ch
1200	0060h	19200	0006h
1800	0040h		

Register addresses

The addresses of the main registers are given in Table 47.5. To load the Baud rate divisor, first the LCR bit 7 is set to a '1', then the LSB is loaded into divisor LSB and the MSB into the divisor MSB register. Finally, bit 7 is set back to a '0'. For example, for 9600 Baud, `COM1` and 1.8432 MHz clock then `0Ch` is loaded in `3F8h` and `00h` into `3F9h`.

When bit 7 is set at a '0' then a read from base address reads from the RD buffer and a write operation writes to the TD buffer. An example of this is shown in Figure 47.18.

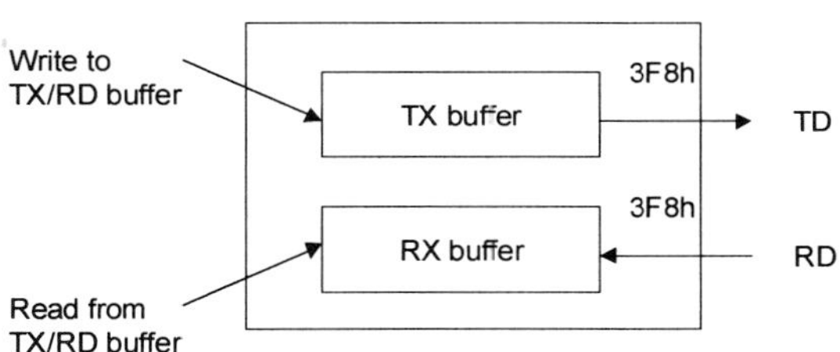

Figure 47.18 Read and write from TD/RD buffer.

Table 47.5 Serial communications addresses.

Primary	Secondary	Register	Bit 7 of LCR
3F8h	2F8h	TD buffer	'0'
3F8h	2F8h	RD buffer	'0'
3F8h	2F8h	Divisor LSB	'1'
3F9h	2F9h	Divisor MSB	'1'
3FBh	2FBh	Line Control Register	
3FDh	2FDh	Line Status Register	

47.6 RS-232 programs

Figure 47.13 shows the main RS-232 connection for 9- and 25-pin connections without hardware handshaking. The loopback connections are used to test the RS-232 hardware and the software, while the null modem connections are used to transmit characters between two computers. Program 47.2 uses a loopback on the TD/RD lines so that a character sent by the computer will automatically be received into the receiver buffer. This set-up is useful in testing the transmit and receive routines. The character to be sent is entered via the keyboard. A Cntrl-D (^D) keystroke exits the program.

Program 47.3 can be used as a sender program (send.c) and Program 47.4 can be used as a receiver program (receive.c). With these programs the null modem connections shown in Figure 47.19 are used.

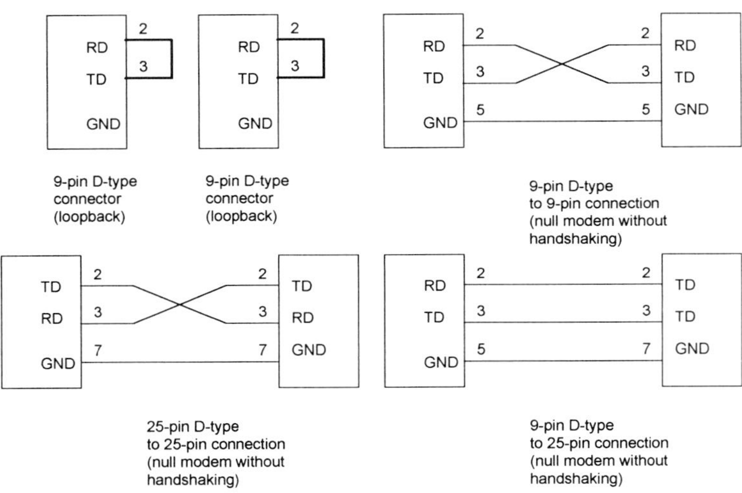

Figure 47.19 System connections.

Note that Programs 47.2 to 47.4 are written for Microsoft Visual C. For a Borland C/C++ program change `_inp` for `inportb` and `_outp` for `outportb`.

Program 47.2

```
/*    This program transmits a character from COM1: and receives */
/*    it via this port. The TD is connected to RD.               */

#define  COM1BASE    0x3F8
#define  COM2BASE    0x2F8
#define  TXDATA      COM1BASE
#define  LCR         (COM1BASE+3) /*   0x3FB line control        */
#define  LSR         (COM1BASE+5) /*   0x3FD line status         */

#include <conio.h>   /* required for getch()                     */
#include <stdio.h>

/* Some ANSI C prototype definitions   */
void  setup_serial(void);
void  send_character(int ch);
int   get_character(void);
```

```
int       main(void)
{
int       inchar,outchar;

      setup_serial();
      do
      {
         puts("Enter char to be transmitted (Cntrl-D to end)");
         outchar=getch();
         send_character(outchar);
         inchar=get_character();
         printf("Character received was %c\n",inchar);
      } while (outchar!=4);
      return(0);
}

void      setup_serial(void)
{
      _outp( LCR, 0x80);
      /* set up bit 7 to a 1  to set Register address bit   */

      _outp(TXDATA,0x0C);
      _outp(TXDATA+1,0x00);
      /* load TxRegister with 12, crystal frequency is 1.8432MHz */

      _outp(LCR, 0x0A);
      /* Bit pattern loaded is 00001010b, from msb to lsb these are:      */
      /* 0 - access TD/RD buffer ,  0 - normal output            */
      /* 0 - no stick bit  , 0 - even parity                     */
      /* 1 - parity on,  0 - 1 stop bit                          */
      /* 10 - 7 data bits                                        */
}

void  send_character(int ch)
{
char  status;
      do
      {
         status = _inp(LSR) & 0x40;
      } while (status!=0x40);
      /*repeat until Tx buffer empty ie bit 6 set*/

      _outp(TXDATA,(char) ch);
}

int   get_character(void)
{
int   status;
      do
      {
         status = _inp(LSR) & 0x01;
      } while (status!=0x01);
      /* Repeat until bit 1 in LSR is set */
      return( (int)_inp(TXDATA));
}
```

Program 47.3

```
/*       send.c                                              */
#define  TXDATA   0x3F8
#define  LSR      0x3FD
#define  LCR      0x3FB

#include <stdio.h>
```

```
#include    <conio.h>   /* included for getch           */
#include    <dos.h>
void     setup_serial(void);
void     send_character(int ch);
int      main(void)
{
int      ch;
      puts("Transmitter program. Please enter text (Cntl-D to end)");
      setup_serial();
      do
      {
         ch=getche();
         send_character(ch);
      }  while (ch!=4);
      return(0);
}

void  setup_serial(void)
{
      _outp( LCR, 0x80);
      /* set up bit 7 to a 1  to set Register address bit             */
      _outp(TXDATA,0x0C);
      _outp(TXDATA+1,0x00);
      /* load TxRegister with 12, crystal frequency is 1.8432MHz      */
      _outp(LCR, 0x0A);
      /* Bit pattern loaded is 00001010b, from msb to lsb these are:  */
      /* Access TD/RD buffer, normal output, no stick bit             */
      /* even parity, parity on, 1 stop bit, 7 data bits              */

}
void  send_character(int ch)
{
char  status;
      do
      {
         status = _inp(LSR) & 0x40;
      } while (status!=0x40);
      /*repeat until Tx buffer empty ie bit 6 set*/
      _outp(TXDATA,(char) ch);
}
```

Program 47.4

```
/*       receive.c                                     */
#define  TXDATA   0x3F8
#define  LSR      0x3FD
#define  LCR      0x3FB
#include    <stdio.h>
#include    <conio.h>   /* included for getch          */
#include    <dos.h>

void     setup_serial(void);
int      get_character(void);
int      main(void)
{
int      inchar;

      setup_serial();
      do
      {
         inchar=get_character();
         putchar(inchar);
      }  while (inchar!=4);
      return(0);
}
```

```
void setup_serial(void)
{
	_outp( LCR, 0x80);
	/* set up bit 7 to a 1  to set Register address bit                    */
	_outp(TXDATA,0x0C);
	_outp(TXDATA+1,0x00);
	/* load TxRegister with 12, crystal frequency is 1.8432MHz             */
	_outp(LCR, 0x0A);
}
int	get_character(void)
{
int	status;
	do
	{
	   status = _inp(LSR) & 0x01;
	} while (status!=0x01);
	/* Repeat until bit 1 in LSR is set */
	return( (int)_inp(TXDATA));
}
```

47.7 Standard Windows serial communications programs

Often an RS-232 communication program requires to be tested for the transmitted characters. In electrical circuits a digital meter is used to test points. The equivalent in RS-232 communications is the Terminal (in Windows 3.*x*) and HyperTerminal (in Windows 95/98). In Windows 95/98 it must be installed on the computer; this is either done automatically or can be added with Control Panel → Add/Remove Program → Windows Set Up → Communications (as shown in Figure 47.20). It can be seen that, in this case, the HyperTerminal program has been installed (as well as Direct Cable Connection and Dial-up Networking). The Direct Cable Connection allows two computers to be connected together using RS-232. Once connected the connection allows the computers to share resources such as their disk drives (hard disk and floppy disk), files, CD-ROMs, and so on. The Dial-up Networking program allows a computer to connect to a remote network using a modem.

47.7.1 HyperTerminal and Terminal

The HyperTerminal (and Terminal, in Windows 3.*x*) allow the transmission and reception of RS-232 characters. In Windows 95/98 it is selected from Start → Program → Accessories → HyperTerminal, in Windows 3.*x* it is selected from the Accessories group. The left-hand side of Figure 47.21 shows an example HyperTerminal folder. This contains previously set-up connections (such as CompuServe and MCI Mail). Activating Hypertrm accesses the HyperTerminal program. After this, the user is prompted for the connection name and the associated icon (as shown in the right-hand side of Figure 47.21).

After this the user enters the telephone number (if it connects via a modem) or the communication port. Figure 47.22 shows that, in this case, the connection is set for a direct connect to COM1. After this the program prompts the user for the communications port settings (this will be the same as the system settings). A sample window is shown on the right-hand side of Figure 47.22. Finally, the main screen is shown (as shown in Figure 47.23). The program can then be used to either send or receive characters (typically it is used to test other programs and/or communications hardware).

Figure 47.20 Install communications programs.

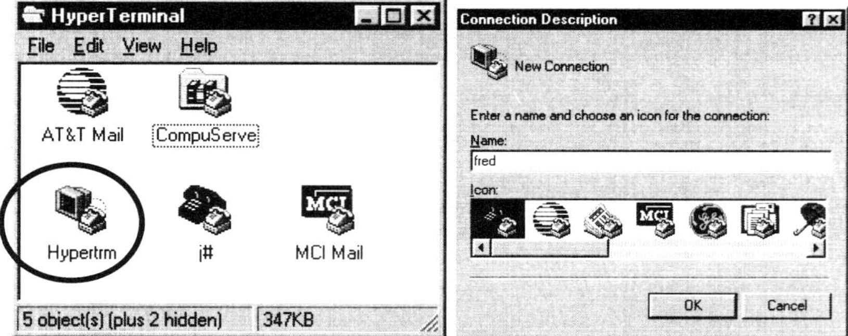

Figure 47.21 HyperTerminal folder and connection name window.

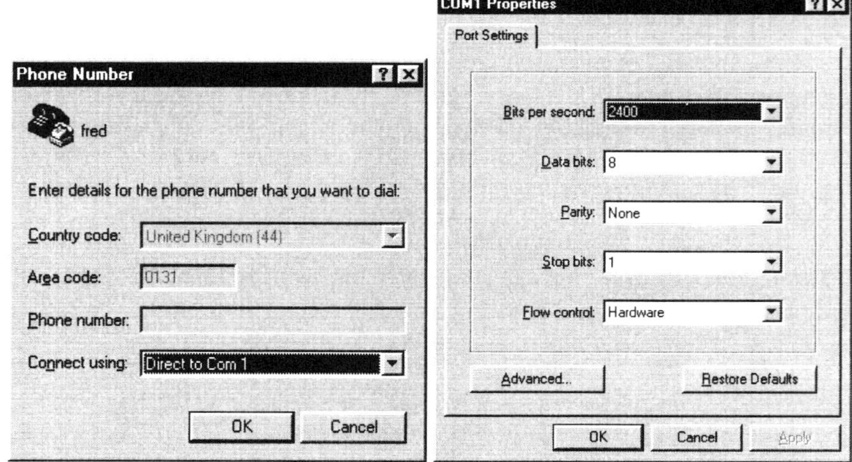

Figure 47.22 HyperTerminal set-up windows.

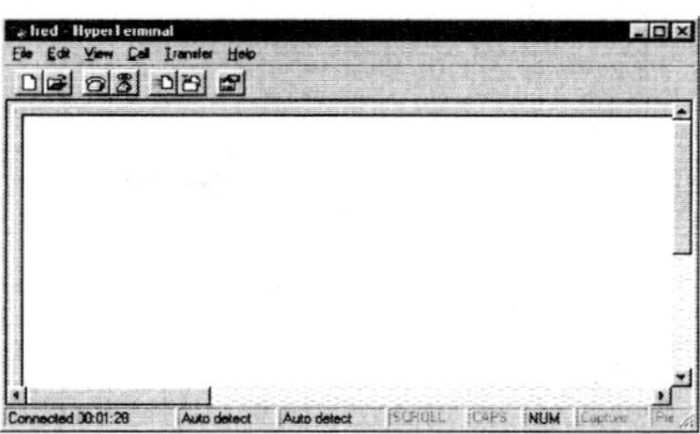

Figure 47.23 HyperTerminal.

47.7.2 Dial-up networking

The point-to-point protocol (PPP) is a set of industry standard protocols which allow a remote computer to connect to a remote network using a dial-up connection. Windows 95/98 can connect to a remote network using this protocol and the Dial-up Networking facility. It is typically used to connect to an Internet provider or to an organizational network.

To set it up in Windows 95/98 the user first selects the `Dial-Up Network` program from within `My Computer`. Next, the user selects the name of the connection, as shown in Figure 47.24.

After this the user enters the telephone number of the remote modem with the local code and country code. Figure 47.25 shows an example set-up (note that the telephone number has been set at 0131-xxx xxxx).

After the Next option is selected a connection icon is made, as shown in Figure 47.26.

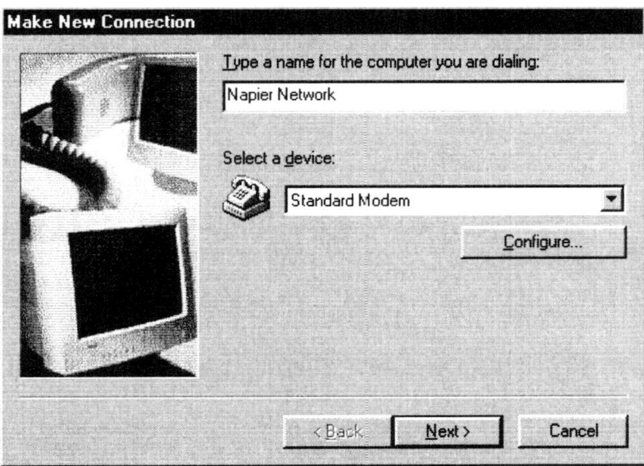

Figure 47.24 Specifying connection name.

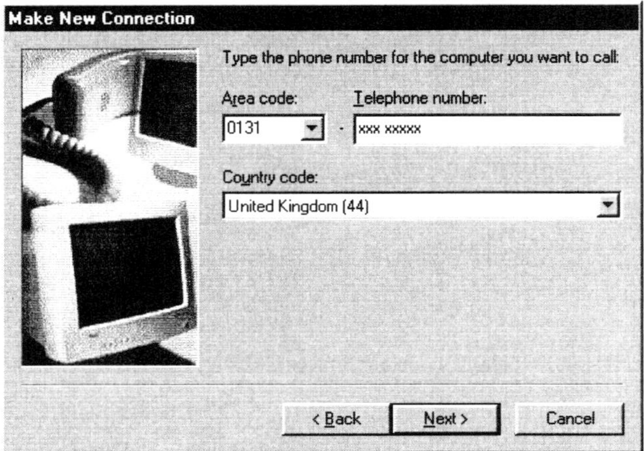

Figure 47.25 Specifying telephone number.

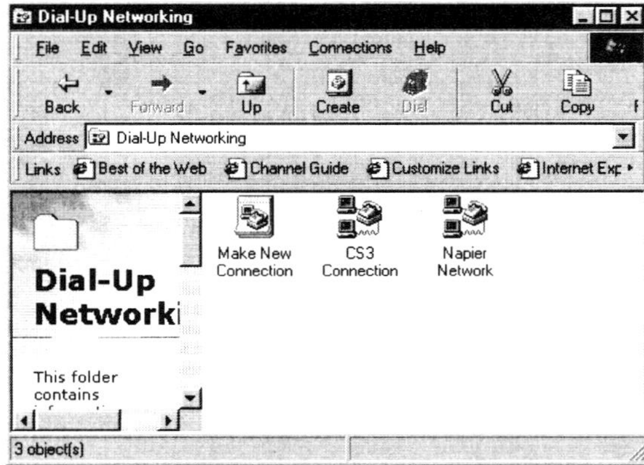

Figure 47.26 Connection icon.

Next the server type is set by selecting the File → Properties option to give the window shown in right-hand side Figure 47.27. From this window the Server Type option is selected. This will then show the window shown in Figure 47.20. The type of server should be set to:

```
PPP; Internet; Windows NT Server; Windows 98;
```

and the protocol to:

```
TCP/IP
```

After these have been set then the user can log into the remote network, as shown in Figures 47.28 and 47.29.

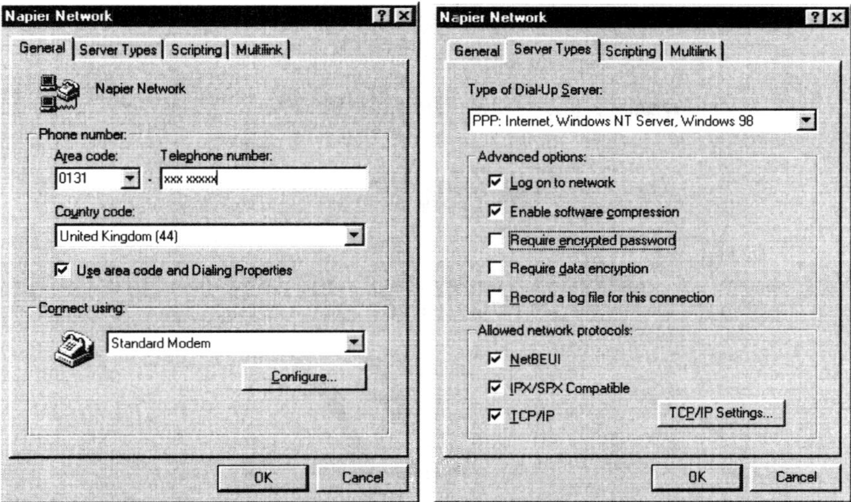

Figure 47.27 Connection settings and server types

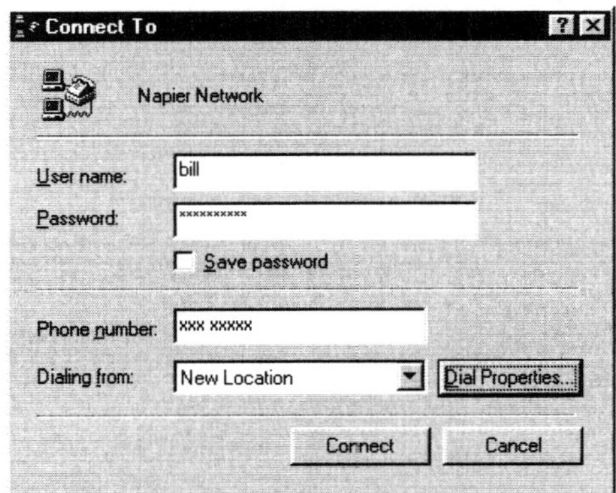

Figure 47.28 Connection setting.

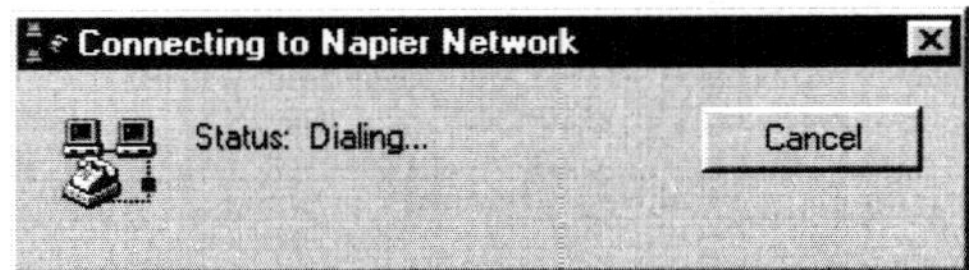

Figure 47.29 Dialing message.

47.8 Direct Connect

The Direct Cable Connection program allows two computers to connect over a serial or a parallel port connection. This allows computers to share resources, such as CD-ROM drives and hard disks.

The initial screen is given in Figure 47.30 shows the initial screen where the user is asked if the computer is to be a host or a guest. A guest setting allows the current computer to connect to a host computer and a host setting allows a guest to connect into it. Figure 47.31 shows the setup of the ports.

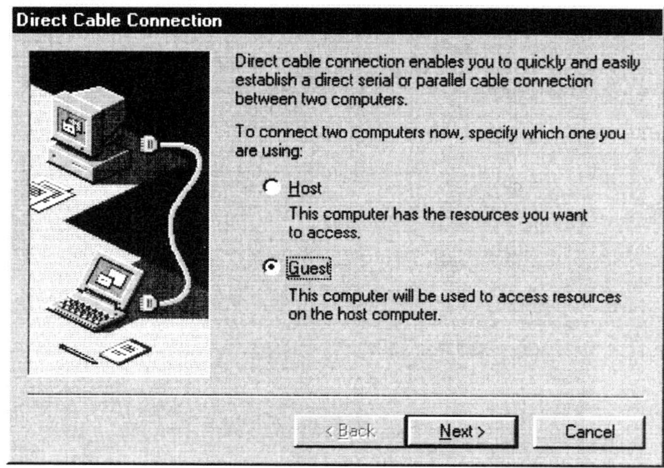

Figure 47.30 Initial message screen.

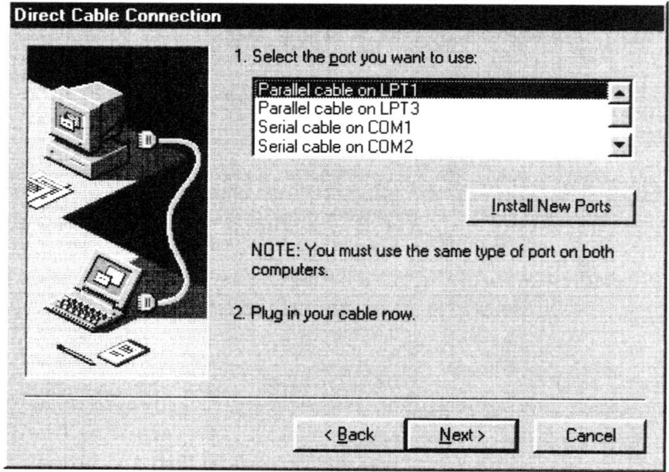

Figure 47.31 Connection type.

Interrupt-driven RS-232

48.1 Interrupt-driven RS-232

Interrupt-driven devices are efficient on processor time as they allow the processor to run a program without having to poll the devices. This allows fast devices almost instant access to the processor and stops slow devices from 'hogging' the processor. For example, a line printer tends to be slow in printing characters. If the printer only interrupted the processor when it was ready for data then the processor can do other things while the printer is printing the character. Another example can be found in serial communications. Characters sent over an RS-232 link are transmitted and received relatively slowly. In a non-interrupt-driven system the computer must poll the status register to determine if a character has been received, which is inefficient in processor time. But, if the amount of time spent polling the status register is reduced, there is a possibility of the computer missing the received character as another could be sent before the first is read from the receiver buffer. If the serial communications port was set up to interrupt the processor when a new character arrived then it is guaranteed that the processor will always process the receiver buffer.

A major disadvantage with non-interrupt-driven software is when the processor is involved in a 'heavy processing' task such as graphics or mathematical calculations. This can have the effect of reducing the amount of time that can be spent in polling and/or reading data.

48.2 Win32 programs

The problem with many C/C++ programs is that some of the application functions, such as graphics and communications, are not standardized. Thus, the graphics function to draw a circle in one development system might be different from another. Win32 overcomes this by providing a set of standard functions and is common across all development systems. In RS-232, the functions used include: OpenFile, ReadFile, SetCommState, SetCommMask, WaitCommEvent, WriteFile, and CloseFile. The program writen in this chapter illustrates the operation of interrupt-driven RS-232; it is likely in non-DOS-based operating systems (such as Windows NT and Windows 95/98) that the Win32 function will be used. These functions isolate the operation of the hardware from the program. They also automatically buffer the transmitted and received characters. The program in this chapter helps to illustrate the operation of the hardware.

48.3 DOS-based RS-232 program

Program 48.1 is a simple interrupt-driven DOS-based RS-232 program which is written for Turbo/Borland C/C++. If possible, connect two PCs together with a cable which swaps the TX and RX lines, as shown in Figure 47.19. Each of the computers should be able to transmit and receive concurrently. A description of this program is given in the next section. The header file associated with this program is **serial.h**.

Program 48.1

```
#include <dos.h>
#include <conio.h>
#include <stdio.h>
#include <bios.h>
#include "serial.h"

void  interrupt rs_interrupt(void);
void  setup_serial(void);
void  send_character(int ch);
int   get_character(void);
int   get_buffer(void);
void  set_vectors(void);
void  reset_vectors(void);
void  enable_interrupts(void);
void  disable_interrupts(void);

void  interrupt(*oldvect)();

char  buffer[RSBUFSIZE];

unsigned int   startbuf=0,endbuf = 0;

int   main(void)
{
int   ch, done  = FALSE;
      setup_serial();
      set_vectors(); /* set new interrupt vectors and store old ones      */
      enable_interrupts();
      printf("Terminal emulator, press [ESC] to quit\n");
      do
      {
         if (kbhit())
         {
            ch=getche();
            if (ch==ESC) break;
            send_character(ch);
         }
         /* empty RS232 buffer   */
         do
         {
            if ((ch=get_buffer()) != -1) putch(ch);
         } while (ch!=-1);
      } while (!done);
      disable_interrupts();
      reset_vectors();
      return(0);
}

void  interrupt rs_interrupt(void)
{
      disable();
      if ((inportb(IIR) & RX_MASK) == RX_ID)
```

```
       {
          buffer[endbuf] = inportb(RXR);
          endbuf++;
          if (endbuf == RSBUFSIZE) endbuf=0;
       }
       /* Set end of interrupt flag */
       outportb(ICR, EOI);
       enable();
}

void      setup_serial(void)
{
int       RS232_setting;
       RS232_setting=BAUD1200 | STOPBIT1 | NOPARITY | DATABITS7;
       bioscom(0,RS232_setting,COM1);
}

void   send_character(int ch)
{
char   status;
       do
       {
          status = inportb(LSR) & 0x40;
       } while (status!=0x40);
       /*repeat until Tx buffer empty ie bit 6 set*/
       outportb(TXDATA,(char) ch);
}

int    get_character(void)
{
int    status;
       do
       {
          status = inportb(LSR) & 0x01;
       } while (status!=0x01);
       /* Repeat until bit 1 in LSR is set */
       return( (int)inportb(TXDATA));
}

int    get_buffer(void)
{
int    ch;
       if (startbuf == endbuf) return (-1);
       ch = (int) buffer[startbuf];
       startbuf++;
       if (startbuf == RSBUFSIZE) startbuf = 0;
       return (ch);
}

void    set_vectors(void)
{
    oldvect = getvect(0x0C);
    setvect(0x0C, rs_interrupt);
}

/* Uninstall interrupt vectors before exiting the program */
void    reset_vectors(void)
{
    setvect(0x0C, oldvect);
}
void      disable_interrupts(void)
{
int       ch;
       disable();
       ch = inportb(IMR) | ~IRQ4; /* disable IRQ4 interrupt */
       outportb(IMR, ch);
```

```
      outportb(IER, 0);
      enable();
}
void     enable_interrupts(void)
{
int   ch;
      disable();
      /* initialize rs232 port   */
      ch = inportb(MCR) | MC_INT;
      outportb(MCR, ch);
      /* enable interrupts for IRQ4 */
      outportb(IER, 0x01);
      ch = inportb(IMR) & IRQ4;
      outportb(IMR, ch);
      enable();
}
```

Header file 48.1: serial.h

```
#define  FALSE       0
/* RS232 set up parameters */
#define COM1         0
#define COM2         1

#define DATABITS7    0x02
#define DATABITS8    0x03

#define STOPBIT1     0x00
#define STOPBIT2     0x04

#define NOPARITY     0x00
#define ODDPARITY    0x08
#define EVENPARITY   0x18

#define BAUD110      0x00
#define BAUD150      0x20
#define BAUD300      0x40
#define BAUD600      0x60
#define BAUD1200     0x80
#define BAUD2400     0xA0
#define BAUD4800     0xC0
#define BAUD9600     0xE0

#define ESC          0x1B       /* ASCII Escape character     */
#define RSBUFSIZE    10000      /* RS232 buffer size          */

#define COM1BASE     0x3F8      /* Base port address for COM1 */

#define TXDATA       COM1BASE         /* Transmit register    */
#define RXR          COM1BASE         /* Receive register     */
#define IER          (COM1BASE+1)     /* Interrupt Enable     */
#define IIR          (COM1BASE+2)     /* Interrupt ID         */
#define LCR          (COM1BASE+3)     /* Line control         */
#define MCR          (COM1BASE+4)     /* Line control         */
#define LSR          (COM1BASE+5)     /* Line Status          */

#define RX_ID        0x04
#define RX_MASK      0x07
#define MC_INT       0x08

/*    Addresses of the 8259 Programmable Interrupt Controller (PIC).*/
#define IMR          0x21  /* Interrupt Mask Register port        */
#define ICR          0x20  /* Interrupt Control Port              */
/* An end of interrupt needs to be sent to the Control Port of    */
/* the 8259 when a hardware interrupt ends.                       */
```

```
#define EOI          0x20  /* End Of Interrupt                     */

#define IRQ4         0xEF  /* COM1                                 */
```

48.3.1 Description of program

The initial part of the program sets up the required RS-232 parameters. It uses `bioscom()` to set `COM1:` with the parameters of 1200 bps, 1 stop bit, no parity and 7 data bits.

```
void     setup_serial(void)
{
int      RS232_setting;
      RS232_setting=BAUD1200 | STOPBIT1 | NOPARITY | DATABITS7;
      bioscom(0,RS232_setting,COM1);
}
```

After the serial port has been initialized the interrupt service routine for the `IRQ4` line is set to point to a new 'user-defined' service routine. The primary serial port `COM1:` sets the `IRQ4` line active when it receives a character. The interrupt associated with `IRQ4` is 0Ch (12). The `getvect()` function gets the ISR address for this interrupt, which is then stored in the variable `oldvect` so that at the end of the program it can be restored. Finally, in the `set_vectors()` function, the interrupt assigns a new 'user-defined' ISR (in this case it is the function `rs_interrupt()`).

```
void     set_vectors(void)
{
      oldvect = getvect(0x0C);       /* store IRQ4 interrupt vector       */
      setvect(0x0C, rs_interrupt);  /* set ISR to rs_interrupt()         */
}
```

At the end of the program the ISR is restored with the following code.

```
void    reset_vectors(void)
{
      setvect(0x0C, oldvect);        /* reset IRQ4 interrupt vector       */
}
```

The `COM1:` port is initialized for interrupts with the code given next. The statement

```
      ch = inportb ( MCR ) | 0x08;
```

resets the RS-232 port by setting bit 3 for the modem control register (MCR) to a 1. Some RS-232 ports require this bit to be set. The interrupt enable register (IER) enables interrupts on a port. Its address is offset by 1 from the base address of the port (that is, 0x3F9 for `COM1:`). If the least significant bit of this register is set to a 1 then interrupts are enabled, else they are disabled.

To enable the `IRQ4` line on the PIC, bit 5 of the IMR (interrupt mask register) is to be set to a 0 (zero). The statement:

```
      ch = inportb(IMR) & 0xEF;
```

achieves this as it bitwise ANDs all the bits, except for bit 4, with a 1. This is because any bit which is ANDed with a 0 results in a 0. The bit mask `0xEF` has been defined with the macro `IRQ4`.

```
void     enable_interrupts(void)
{
int    ch;
       disable();
       ch = inportb(MCR) | MC_INT; /* initialize rs232 port */
       outportb(MCR, ch);
       outportb(IER, 0x01);
       ch = inportb(IMR) & IRQ4;
       outportb(IMR, ch);   /* enable interrupts for IRQ4 */

       enable();
}
```

At the end of the program the function `disable_interrupts()` sets the IER register to all 0s. This disables interrupts on the `COM1:` port. Bit 4 of the IMR is also set to a 1 which disables `IRQ4` interrupts.

```
void     disable_interrupts(void)
{
int      ch;

       disable();
       ch = inportb(IMR) | ~IRQ4; /* disable IRQ4 interrupt */
       outportb(IMR, ch);
       outportb(IER, 0);
       enable();
}
```

The ISR for the `IRQ4` function is set to `rs_interrupt()`. When it is called, the Interrupt Status Register (this is named IIR to avoid confusion with the interrupt service routine) is tested to determine if a character has been received. Its address is offset by 2 from the base address of the port (that is, 0x3FA for `COM1:`). The first 3 bits give the status of the interrupt. A `000b` indicates that there are no interrupts pending, a `100b` that data has been received, or a `111b` that an error or break has occurred. The statement `if ((inportb(IIR) & 0x7) == 0x4)` tests if data has been received. If this statement is true then data has been received and the character is then read from the receiver buffer array with the statement `buffer[endbuf] = inportb(RXR);`. The end of the buffer variable (`endbuf`) is then incremented by 1.

At the end of this ISR the end of interrupt flag is set in the interrupt control register with the statement `outportb(ICR, 0x20);`. The `startbuf` and `endbuf` variables are global, thus all parts of the program have access to them.

Turbo/Borland functions `enable()` and `disable()` in `rs_interrupt()` are used to enable and disable interrupts, respectively.

```
void  interrupt rs_interrupt(void)
{
      disable();
      if ((inportb(IIR) & RX_MASK) == RX_ID)
      {
         buffer[endbuf] = inportb(RXR);
         endbuf++;
         if (endbuf == RSBUFSIZE) endbuf=0;
      }
      /* Set end of interrupt flag  */
      outportb(ICR, EOI);
      enable();
}
```

The `get_buffer()` function is given next. It is called from the main program and it tests the variables `startbuf` and `endbuf`. If they are equal then it returns –1 to the `main()`. This indicates that there are no characters in the buffer. If there are characters in the buffer then the function returns, the character pointed to by the `startbuf` variable. This variable is then incremented. The difference between `startbuf` and `endbuf` gives the number of characters in the buffer. Note that when `startbuf` or `endbuf` reach the end of the buffer (`RSBUFSIZE`) they are set back to the first character, that is, element 0.

```
int     get_buffer(void)
{
int     ch;

        if (startbuf == endbuf) return (-1);
        ch = (int) buffer[startbuf];
        startbuf++;
        if (startbuf == RSBUFSIZE) startbuf = 0;
        return (ch);
}
```

The `get_character()` and `send_character()` functions are similar to those developed in Chapter 47. For completeness, these are listed next.

```
void    send_character(int ch)
{
char    status;
        do
        {
           status = inportb(LSR) & 0x40;
        } while (status!=0x40);

        /*repeat until Tx buffer empty ie bit 6 set*/
        outportb(TXDATA,(char) ch);
}

int        get_character(void)
{
int        status;

        do
        {
           status = inportb(LSR) & 0x01;
        } while (status!=0x01);
        /* Repeat until bit 1 in LSR is set */
        return( (int)inportb(TXDATA));
}
```

The `main()` function calls the initialization and the de-initialization functions. It also contains a loop, which continues until the Esc key is pressed. Within this loop, the keyboard is tested to determine if a key has been pressed. If it has then the `getche()` function is called. This function returns a key from the keyboard and displays it to the screen. Once read into the variable `ch` it is tested to determine if it is the Esc key. If it is then the program exits the loop, else it transmits the entered character using the `send_character()` function. Next the `get_buffer()` function is called. If there are no characters in the buffer then a –1 value is returned, else the character at the start of the buffer is returned and displayed to the screen using `putch()`.

```
int      main(void)
{
int      ch, done  = FALSE;

      setup_serial();
      /* set new interrupt vectors and store old ones*/
      set_vectors();
      enable_interrupts();
      printf("Terminal emulator, press [ESC] to quit\n");
      do
      {
         if (kbhit())
         {
            ch=getche();
            if (ch==ESC) break;
            send_character(ch);
         }
         /* empty RS232 buffer   */
         do
         {
            if ((ch=get_buffer()) != -1) putch(ch);
         } while (ch!=-1);
      } while (!done);
      disable_interrupts();
      reset_vectors();
      return(0);
}
```

49 Visual Basic RS-232

49.1 Introduction

This chapter discusses how Visual Basic can be used to access serial communication functions. Windows hides much of the complexity of serial communications and automatically puts any received characters in a receive buffer and characters sent into a transmission buffer. The receive buffer can be read by the program whenever it has time and the transmit buffer is emptied when it is free to send characters.

49.2 Communications control

Visual Basic allows many additional components to be added to the toolbox. The Microsoft Comm component is used to add a serial communication facility.

This is added to the toolbox with: `Project → Components` (Ctrl-T)

or in Visual Basic 4 with: `Tools → Custom Controls` (Ctrl-T)

Notice that both are selected by using the Ctrl-T keystroke. Figure 49.1 shows how a component is added in Visual Basic 4 and shows how it is added in Visual Basic 5. This then adds a Comms Component into the toolbox, as shown in Figure 49.2.

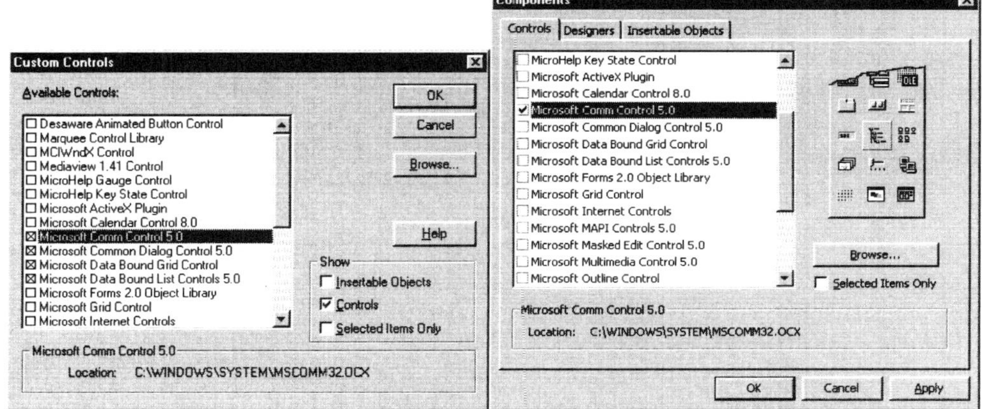

Figure 49.1 Adding Microsoft Comm component with Visual Basic 4/5.

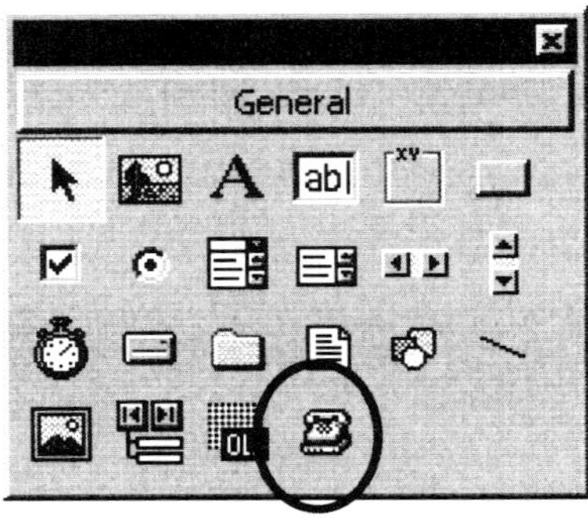

Figure 49.2 Toolbox showing Comms components.

In order to use the Comms component the files MSCOMM16.OCX (for a 16-bit module) or MSCOMM32.OCX (for a 32-bit module) must be present in the \WINDOWS\SYSTEM directory. The class name is MSComm.

The communications control provides the following two ways for handling communications:

- **Event-driven**. Event-driven communications is the best method of handling serial communication as it frees the computer to do other things. The event can be defined as the reception of a character, a change in CD (carrier detect) or a change in RTS (request to send). The OnComm event can be used to capture these events, and also to detect communications errors.
- **Polling**. CommEvent properties can be tested to determine if an event or an error has occurred. For example, the program can loop waiting for a character to be received. Once it is the character is read from the receive buffer. This method is normally used when the program has time to poll the communications receiver or that a known response is imminent.

Visual Basic uses the standard Windows drivers for the serial communication ports (such as serialui.dll and serial.vxd). The communication control is added to the application for each port. The parameters (such as the bit rate, parity, and so on) can be changed by selecting Control Panel → System → Device Manager → Ports (COM and LPT) → Port Settings. The settings of the communications port (the IRQ and the port address) can be changed by selecting Control Panel → System → Device Manager → Ports (COM and LPT) → Resources for IRQ and Addresses. Figure 49.3 shows example parameters and settings.

49.3 Properties

The Comm component is added to a form whenever serial communications are required (as shown in left-hand side of Figure 49.4). The right-hand side of Figure 49.5 shows its properties. By default, the first created object is named MSComm1 (the second is named MSComm2, and so on). It can be seen that the main properties of the object are: CommPort, DTREnable, EOFEnable, Handshaking, InBufferSize, Index, InputLen, InputMode, Left, Name, NullDiscard, OutBufferSize, ParityReplace, RThreshold, RTSEnable, Settings, SThreshold, Tag and Top. The main properties are defined in Table 49.1.

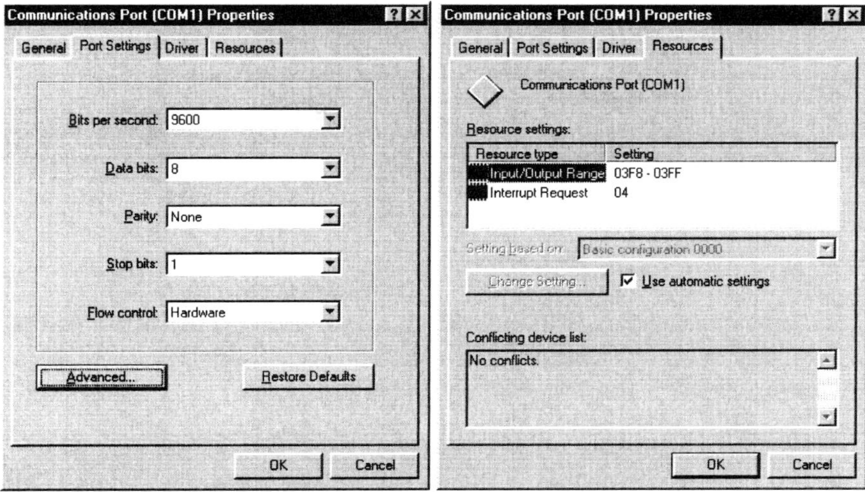

Figure 49.3 Changing port setting and parameters.

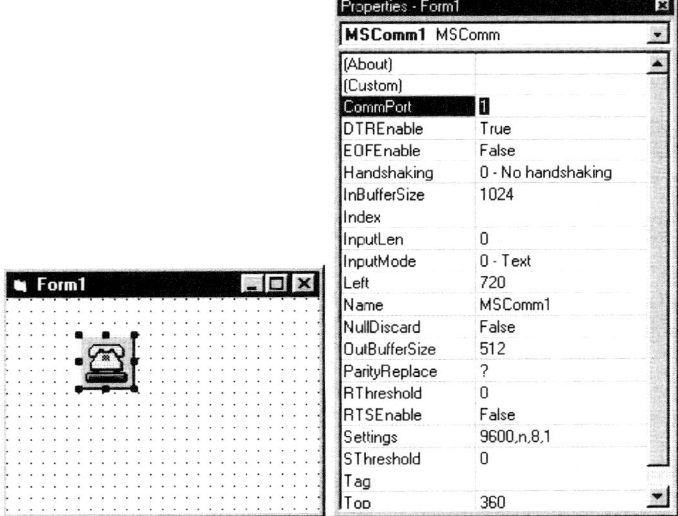

Figure 49.4 Communications control and MS Comm Properties.

Table 49.1 The main communications control properties.

Properties	*Description*
CommPort	Sets and returns the communications port number.
Input	Returns and removes characters from the receive buffer.
Output	Writes a string of characters to the transmit buffer.
PortOpen	Opens and closes a port, and gets port settings
Settings	Sets and returns port parameters, such as bit rate, parity, number of data bits and so on.

49.3.1 Settings

The Settings property sets and returns the RS-232 parameters, such as baud rate, parity, the number of data bit, and the number of stop bits. Its syntax is:

[form.]MSComm.`Settings`[= *setStr$*]

where the strStr is a string which contains the RS-232 settings. This string takes the form:

```
"BBBB,P,D,S"
```

where `BBBB` defines the baud rate, `P` the parity, `D` the number of data bits, and `S` the number of stop bits.

The following lists the valid baud rates (default is 9600 Baud):

110, 300, 600, 1200, 2400, 9600, 14 400, 19 200, 38 400, 56 000, 128 000, 256 000.
The valid parity values are (default is N): E (Even), M (Mark), N (None), O (Odd), S (Space).

The valid data bit values are (default is 8): 4, 5, 6, 7 or 8.

The valid stop bit values are (default is 1). 1, 1.5 or 2.

An example of setting a control port to 4800 Baud, even parity, 7 data bits and 1 stop bit is:

```
Com1.Settings = "4800,E,7,1"
```

49.3.2 CommPort

The CommPort property sets and returns the communication port number. Its syntax is:

[form.]MSComm.`CommPort`[= *portNumber%*]

which defines the portNumber from a value between 1 and 99. A value of 68 is returned if the port does not exist.

49.3.3 PortOpen

The PortOpen property sets and returns the state of the communications port. Its syntax is:

[form.]MSComm.`PortOpen`[= {*True* | *False*}]

A True setting opens the port, while a False closes the port and clears the receive and transmit buffers (this automatically happens when an application is closed).

The following example opens communications port number 1 (COM1:) at 4800 Baud with even parity, 7 data bits and 1 stop bit:

```
Com1.Settings = "4800,E,7,1"
Com1.CommPort = 1
Com1.PortOpen = True
```

49.3.4 Inputting data

The three main properties used to read data from the receive buffer are Input, InBufferCount and InBufferSize.

Input

The Input property returns and removes a string of characters from the receive buffer. Its syntax is:

[*form.*]*MSComm.*`Input`

To determine the number of characters in the buffer the InBufferCount property is tested (to be covered in the next section). Setting InputLen to 0 causes the Input property to read the entire contents of the receive buffer.

Program 49.1 shows an example of how to read data from the receiver buffer.

Program 49.1

```
' Check for characters in the buffer
If Com1.InBufferCount Then
      ' Read data in the buffer
      InStr$ = Com1.Input
End If
```

InBufferSize

The InBufferSize property sets and returns the maximum number of characters that can be received in the receive buffer (by default it is 1024 bytes). Its syntax is:

[*form.*]*MSComm.*`InBufferSize`[= *numBytes*%]

The size of the buffer should be set so that it can store the maximum number of characters that will be received before the application program can read them from the buffer.

InBufferCount

The InBufferCount property returns the number of characters in the receive buffer. It can also be used to clear the buffer by setting the number of characters to 0. Its syntax is:

[*form.*]*MSComm.*`InBufferCount`[= *count*%]

49.3.5 Outputting data

The three main properties used to write data to the transmit buffer are Output, OutBuffer-Count and OutBufferSize.

Output

The Output property writes a string of characters to the transmit buffer. Its syntax is:

[*form.*]*MSComm.*`Output`[= *outString*$]

Program 49.2 uses the KeyPress event on a form to send the character to the serial port.

Program 49.2

```
Private Sub Form_KeyPress (KeyAscii As Integer)
      if (Com1.OutBufferCount < Com1.OutBufferSize)
         Com1.Output = Chr$(KeyAscii)
End Sub
```

OutBufferSize

The OutBufferSize property sets and returns the number of characters in the transmit buffer (default size is 512 characters). Its syntax is:

[*form.*]*MSComm*.`OutBufferSize`[= *NumBytes%*]

OutBufferCount

The OutBufferCount property returns the number of characters in the transmit buffer. The transmit buffer can also be cleared by setting it to 0. Its syntax is:

[*form.*]*MSComm*.`OutBufferCount`[= *0*]

49.3.6 Other properties

Other properties are:

- **Break**. Sets or clears the break signal. A True sets the break signal, while a False clears the break signal. When True character transmission is suspended and a break level is set on the line. This continues until Break is set to False. Its syntax is:

  [*form.*]*MSComm*.`Break`[= {*True* | *False*}]

- **CDTimeout**. Sets and returns the maximum amount of time that the control waits for a carried detect (CD) signal, in milliseconds, before a timeout. Its syntax is:

  [*form.*]*MSComm*.`CDTimeout`[= *milliseconds&*]

- **CTSHolding**. Determines whether the CTS line should be detected. CTS is typically used for hardware handshaking. Its syntax is:

  [*form.*]*MSComm*.`CTSHolding`[= {*True* | *False*}]

- **DSRHolding**. Determines the DSR line state. DSR is typically used to indicate the presence of a modem. If is a True then the DSR line is high, else it is low. Its syntax is:

  [*form.*]*MSComm*.`DSRHolding`[= *setting*]

- **DSRTimeout**. Sets and returns the number of milliseconds to wait for the DSR signal before an OnComm event occurs. Its syntax is:

  [*form.*]*MSComm*.`DSRTimeout`[= *milliseconds&*]

- **DTEEnable**. Determines whether the DTR signal is enabled. It is typically send from the computer to the modem to indicate that it is ready to receive data. A True setting enables the DTR line (output level high). It syntax is:

  [*form.*]*MSComm.*`DTREnable`[= {*True* | *False*}]

- **RTSEnable.** Determines whether the RTS signal is enabled. Normally used to handshake incoming data and is controlled by the computer. Its syntax is:

  [*form.*]*MSComm.*`RTSEnable`[= {*True* | *False*}]

- **NullDiscard**. Determines whether null characters are read into the receive buffer. A True setting does not transfer the characters. Its syntax is:

  [*form.*]*MSComm.*`NullDiscard`[= {*True* | *False*}]

- **SThreshold.** Sets and returns the minimum number of characters allowable in the transmit buffer before the OnComm event. A 0 value disables generating the OnComm event for all transmission events, while a value of 1 causes the OnComm event to be called when the transmit buffer is empty. Its syntax is:

  [*form.*]*MSComm.*`SThreshold`[= *numChars%*]

- **Handshaking**. Sets and returns the handshaking protocol. It can be set to no handshaking, hardware handshaking (using RTS/CTS) or software handshaking (XON/XOFF). Valid settings are given in Table 49.2. Its syntax is:

  [*form.*]*MSComm.*`Handshaking`[= *protocol%*]

- **CommEvent**. Returns the most recent error message. Its syntax is:

  [*form.*]*MSComm.*`CommEvent`

Table 49.2 Settings for handshaking.

Setting	*Value*	*Description*
`comNone`	0	No handshaking (Default).
`comXOnXOff`	1	XON/XOFF handshaking.
`comRTS`	2	RTS/CTS handshaking.
`comRTSXOnXOff`	3	RTS/CTS and XON/XOFF handshaking.

When a serial communication event (OnComm) occurs then the event (error or change) can be determined by testing the CommEvent property. Table 49.3 lists the error values and Table 49.4 lists the communications events.

Table 49.3 CommEvent property.

Setting	*Value*	*Description*
`comBreak`	1001	Break signal received.
`comCTSTO`	1002	CTSTimeout. Occurs when transmitting a character and CTS was low for CTSTimeout milliseconds.
`comDSRTO`	1003	DSRTimeout. Occurs when transmitting a character and DTR was low for DTRTimeout milliseconds.
`comFrame`	1004	Framing Error.
`comOverrun`	1006	Port Overrun. The receive buffer is full and another character was written into the buffer, overwriting the previously received character.
`comCDTO`	1007	CD Timeout. Occurs CD was low for CDTimeout milliseconds, when transmitting a character.
`comRxOver`	1008	Receive buffer overflow.
`comRxParity`	1009	Parity error.
`comTxFull`	1010	Transmit buffer full.

Table 49.4 Communications events.

Setting	*Value*	*Description*
`comEvSend`	1	Character has been sent.
`comEvReceive`	2	Character has been received.
`comEvCTS`	3	Change in CTS line.
`comEvDSR`	4	Change in DSR line from a high to a low.
`comEvCD`	5	Change in CD line.
`comEvRing`	6	Ring detected.
`comEvEOF`	7	EOF character received.

49.4 Events

The Communication control generates an event (OnComm) when the value CommEvent property changes its value. Figure 49.5 shows the event subroutine and Program 49.3 shows an example event routine which tests the CommEvent property. It also shows the property window which is shown with a right click on the comms component.

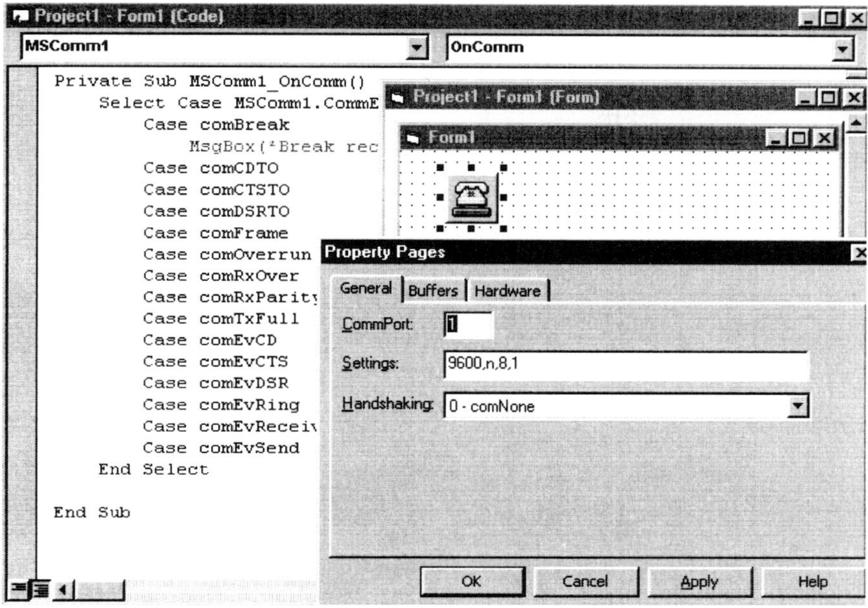

Figure 49.5 OnComm event.

Program 49.3

```
Private Sub MSComm_OnComm ()
      Select Case MSComm1.CommEvent
         Case comBreak             ' A Break was received.
         MsgBox("Break received")
         Case comCDTO              ' CD (RLSD) Timeout.
         Case comCTSTO             ' CTS Timeout.
         Case comDSRTO             ' DSR Timeout.
         Case comFrame             ' Framing Error
         Case comOverrun           ' Data Lost.
         Case comRxOver            ' Receive buffer overflow.
         Case comRxParity          ' Parity Error.
         Case comTxFull            ' Transmit buffer full.
         Case comEvCD              ' Change in the CD.
         Case comEvCTS             ' Change in the CTS.
         Case comEvDSR             ' Change in the DSR.
         Case comEvRing            ' Change in the RI.
         Case comEvReceive
         Case comEvSend
      End Select
End Sub
```

49.5 Example program

Program 49.4 shows a simple transmit/receive program which uses COM1: to transmit and receive. A loopback connection which connects the transmit line to the receive line can be used to test the communications port. All the characters that are transmitted should be automatically received. A sample form is given in Figure 49.6.

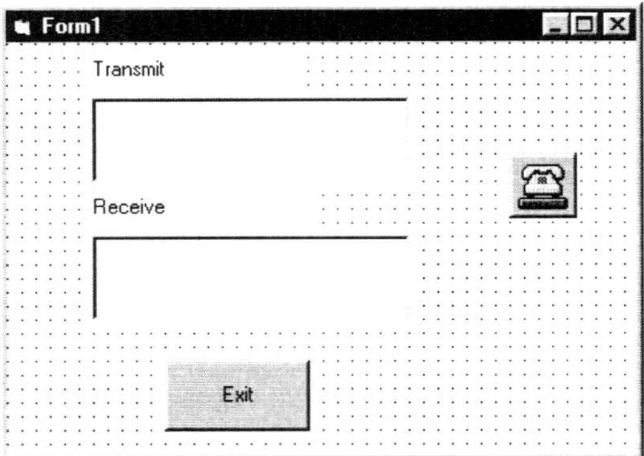

Figure 49.6 Simple serial communications transmit/receive form.

The loading of the form (Form_Load) is called when the program is initially run. This is used to set-up the communication parameters (in this case to 9600 Baud, no parity, 8 data bits and 1 stop bit). When the user presses a key on the form the Form_Keypress event is called. This is then used to transmit the entered character and display it to the Transmit text window (Text1). When a character is received the OnComm event is called and the MSComm1.CommEvent is set to 2 (comEvReceive) which identifies that a character has been received. This character is then displayed to the Receive text window (Text2). Figure 49.7 shows a sample run.

Program 49.4

```
Private Sub Form_Load()

  MSComm1.CommPort = 1                ' Use COM1.
  MSComm1.Settings = "9600,N,8,1"     ' 9600 baud, no parity, 8 data,
                                      '  and 1 stop bit.
  MSComm1.InputLen = 0                ' Read entire buffer when Input
                                      ' is used
  MSComm1.PortOpen = True             ' Open port
End Sub

Private Sub Form_KeyPress(KeyAscii As Integer)
    MSComm1.Output = KeyAscii
    Text1.Text = KeyAscii
End Sub

Private Sub MSComm1_OnComm()
    If (MSComm1.CommEvent = comEvReceive) Then
        Text2.Text = MSComm1.Input
    End If
End Sub

Private Sub Command1_Click()
    End
End Sub
```

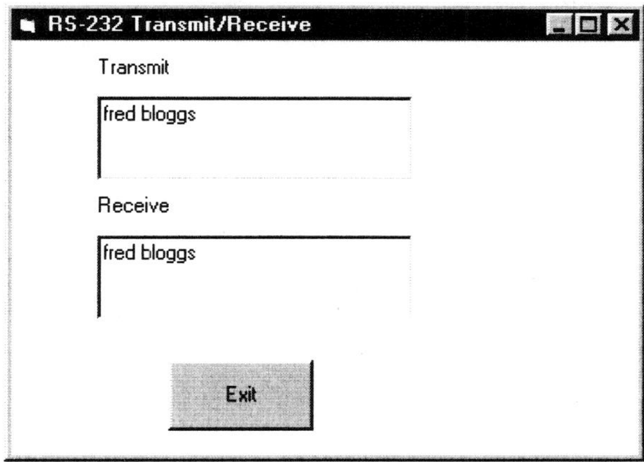

Figure 49.7 Sample run.

49.6 Error messages

Table 49.5 identifies the run-time errors that can occur with the Communications control.

Table 49.5 Error messages.

Error number	*Message explanation*	*Error number*	*Message explanation*
8000	Invalid operation on an opened port	8010	Hardware is not available
8001	Timeout value must be greater than zero	8011	Cannot allocate the queues
8002	Invalid port number	8012	Device is not open
8003	Property available only at run-time	8013	Device is already open
8004	Property is read-only at run-time	8014	Could not enable Comm notification
8005	Port already open	8015	Could not set Comm state
8006	Device identifier is invalid	8016	Could not set Comm event mask
8006	Device identifier is invalid	8018	Operation valid only when the port is open
8007	Unsupported Baud rate	8019	Device busy
8008	Invalid Byte size is invalid	8020	Error reading Comm device
8009	Error in default parameters		

49.7 RS-232 polling

The previous program used interrupt-driven RS-232. It is also possible to use polling to communicate over RS-232. Program 49.5 uses COM2 to send the message 'Hello' and then waits for a received string. It determines that there has been a response by continually testing the number of received characters in the receive buffer (InBufferCount). When there is more than one character in the input buffer it is read.

Program 49.5

```
Private Sub Form_Load()
  Dim Str As String                    ' String to hold input

  MSComm1.CommPort = 2                 ' Use COM2

  MSComm1.Settings = "9600,N,8,1"      ' 9600 baud, no parity, 8 data,
                                       ' and 1 stop bit
  MSComm1.InputLen = 0                 ' Read entire buffer when Input is used

  MSComm1.PortOpen = True              ' Open port

  Text1.Text = "Sending: Hello"

  MSComm1.Output = "Hello"             ' Send message

  ' Wait for response from port
  Do
       DoEvents
  Loop Until MSComm1.InBufferCount >= 2

  Str = MSComm1.Input                  ' Read input buffer

  Text1.Text = "Received: " + Str

  MSComm1.PortOpen = False  ' Close serial port.

End Sub
```

50 Parallel Port

50.1 Introduction

This chapter discusses parallel communications. The Centronics printer interface transmits 8 bits of data at a time to an external device, normally a printer. Normally it uses a 25-pin D-type connector to connect to a 36-pin Centronics printer interface. In the past, it has been one of the most under used parts of a PC and was not normally used to interface to other equipment. This was because, as a standard, it could only transmit data in one direction (from the PC to the external device). Some interface devices overcame this by using four of the input handshaking lines to input data and then multiplexing using an output handshaking line to multiplex them to produce 8 output bits.

As technology has improved there has been a great need for an inexpensive, external bi-directional port to connect to devices such as tape backup drives, CD-ROMs, and so on. The Centronics interface unfortunately lacks speed (150 Kbps), has limited length of lines (2 m) and very few computer manufacturers have complied with an electrical standard.

Thus, in 1991, several manufacturers (including IBM and Texas Instruments) formed a group called NPA (National Printing Alliance). Their original objective was to develop a standard for control printers over a network. To achieve this a bi-directional standard was developed which was compatible with existing software. This standard was submitted to the IEEE and was published as the IEEE 1284-1994 Standard (as it was released in 1994 and was developed by the IEEE 1284 committee).

With this standard all parallel ports use a bi-directional link in either a compatible, nibble or byte mode. These modes are relatively slow, as the software must monitor the handshaking lines (and gives rates up to 100 Kbps). To allow high speed the EPP (Enhanced Parallel Port) and ECP (Extended Capabilities Port Protocol) modes have been developed to allow high-speed data transfer using automatic hardware handshaking. In addition to the previous three modes, EPP and ECP are being implemented on the latest I/O controllers by most of the Super I/O chip manufacturers. These modes use hardware to assist in the data transfer. For example, in EPP mode, a byte of data can be transferred to the peripheral by a simple OUT instruction and the I/O controller handles all the handshaking and data transfer to the peripheral.

50.2 Data handshaking

Figure 50.1 shows the pin connections on the PC connector. The data lines (D0–D7) output data from the PC and each of the data lines has an associated ground line (GND).

The main handshaking lines are $\overline{\text{ACK}}$, BUSY and $\overline{\text{STROBE}}$. Initially the computer places the data on the data bus, then it sets the $\overline{\text{STROBE}}$ line low to inform the external device that the data on the data bus is valid. When the external device has read the data it sets the $\overline{\text{ACK}}$ lines low to acknowledge that it has read the data. The PC then waits for the printer to set the BUSY line inactive, that is, low. Figure 50.2 shows a typical handshaking operation and Table 50.1 outlines the definitions of the pins.

The parallel interface can be accessed either by direct reads to and writes from the I/O memory addresses or from a program which uses the BIOS printer interrupt. This interrupt allows a program either to get the status of the printer or to write a character to it. Table 50.2 outlines the interrupt calls.

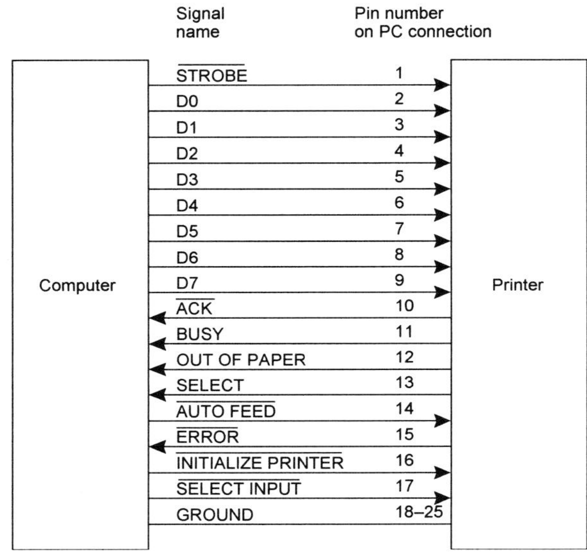

Figure 50.1 Centronics parallel interface showing pin numbers on PC connector.

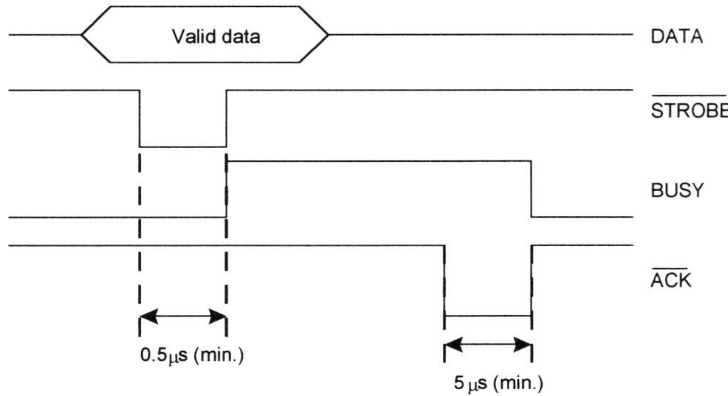

Figure 50.2 Data handshaking with the Centronics parallel printer interface.

50.2.1 BIOS printer

Program 50.1 uses the BIOS printer interrupt to test the status of the printer and outputs characters to the printer.

Table 50.1 Signal definitions.

Signal	*In/out*	*Description*
$\overline{\text{STROBE}}$	Out	Indicates that valid data is on the data lines (active low)
$\overline{\text{AUTOFEED}}$	Out	Instructs the printer to insert a line feed for every carriage return (active low)
$\overline{\text{SELECT INPUT}}$	Out	Indicates to the printer that it is selected (active low)
INIT	Out	Resets the printer
$\overline{\text{ACK}}$	In	Indicate that the last character was received (active low)
BUSY	In	Indicates that the printer is busy and thus cannot accept data
OUT OF PAPER	In	Out of paper
SELECT	In	Indicates that the printer is on-line and connected
$\overline{\text{ERROR}}$	In	Indicates that an error exists (active low)

Table 50.2 BIOS printer interrupt.

Description	*Input registers*	*Output registers*
Initialize printer port	AH = 01h DX = printer number (00h–02h)	AH = printer status bit 7: not busy bit 6: acknowledge bit 5: out of paper bit 4: selected bit 3: I/O error bit 2: unused bit 1: unused bit 0: timeout
Write character to printer	AH = 00h AL = character to write DX = printer number (00h–02h)	AH = printer status
Get printer status	AH = 02h DX = printer number (00h–02h)	AH = printer status

Program 50.1

```
#include <dos.h>
#include <stdio.h>
#include <conio.h>

#define  PRINTERR -1

void  print_character(int ch);
int   init_printer(void);

int   main(void)
```

```
{
int    status,ch;

       status=init_printer();
       if (status==PRINTERR) return(1);

       do
       {
          printf("Enter character to output to printer");
          ch=getch();
          print_character(ch);
       } while (ch!=4);
       return(0);
}

int    init_printer(void)
{
union REGS inregs,outregs;

  inregs.h.ah=0x01;  /* initialize printer */
  inregs.x.dx=0; /* LPT1: */
  int86(0x17,&inregs,&outregs);
  if (inregs.h.ah & 0x20)  { puts("Out of paper"); return(PRINTERR); }
  else if (inregs.h.ah & 0x08)   { puts("I/O error"); return(PRINTERR); }
  else if (inregs.h.ah & 0x01)   { puts("Printer timeout"); return(PRINTERR); }
  return(0);
}

void   print_character(int ch)
{
union REGS inregs,outregs;

       inregs.h.ah=0x00; /* print character */
       inregs.x.dx=0; /* LPT1: */
       inregs.h.al=ch;

       int86(0x17,&inregs,&outregs):
}
```

50.3 I/O addressing

50.3.1 Addresses

The printer port has three I/O addresses assigned for the data, status and control ports. These addresses are normally assigned to:

Printer	Data register	Status register	Control register
LPT1	378h	379h	37ah
LPT2	278h	279h	27ah

The DOS debug program is used to display the base addresses for the serial and parallel ports by displaying the 32 memory location starting at `0040:0008`. For example:

```
-d 40:00
0040:0000  F8 03 F8 02 00 00 00 00-78 03 00 00 00 00 29 02
```

The first four 16-bit addresses give the serial communications ports. In this case there are two COM ports at address 03F8h (COM1) and 02F8h (for COM2). The next four 16-bit addresses give the parallel port addressees. In this case, there are two parallel ports; one at 0378h (LPT1) and one at 0229h (LPT4).

50.3.2 Output lines

Figure 50.3 shows the bit definitions of the registers. The Data port register links to the output lines. Writing a 1 to the bit position in the port sets the output high, while a 0 sets the corresponding output line to a low. Thus to output the binary value 1010 1010b (AAh) to the parallel port data then using Borland C:

```
outportb(0x378,0xAA);    /* in Visual C this is _outp(0x378,0xAA); */
```

The output data lines are each capable of sourcing 2.6 mA and sinking 24 mA; it is thus essential that the external device does not try to pull these lines to ground.

The Control port also contains five output lines, of which the lower four bits are $\overline{\text{STROBE}}$, $\overline{\text{AUTO FEED}}$, INIT and $\overline{\text{SELECT INPUT}}$, as illustrated in Figure 50.3. These lines can be used as either control lines or as data outputs. With the data line, a 1 in the register gives an output high, while the lines in the Control port have inverted logic. Thus, a 1 to a bit in the register causes an output low.

Program 50.2 outputs the binary pattern 0101 0101b (55h) to the data lines and sets $\overline{\text{SELECT INPUT}}$=0, INIT=1, $\overline{\text{AUTO FEED}}$=1, and $\overline{\text{STROBE}}$=0, the value of the Data port will be 55h and the value written to the Control port will be XXXX 1101 (where X represents don't care). The value for the control output lines must be inverted, so that the $\overline{\text{STROBE}}$ line will be set to a 1 so that it will be output as a LOW.

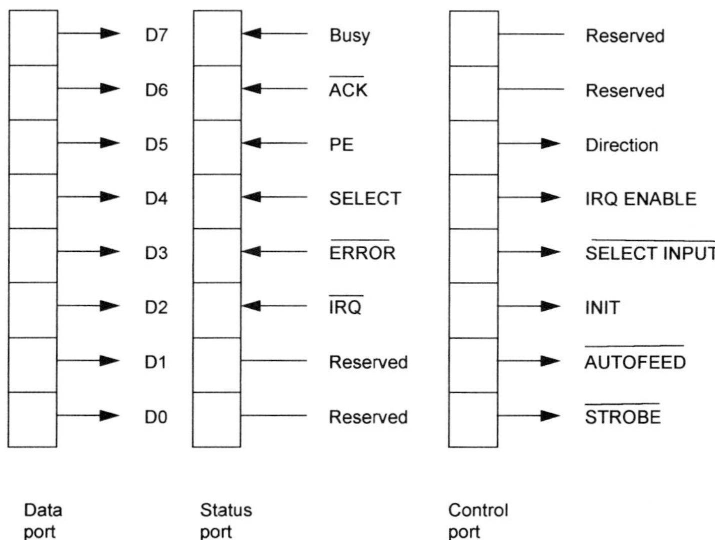

Figure 50.3 Port assignments.

Program 50.2

```
#define DATA        0x378
#define STATUS      DATA+1
#define CONTROL     DATA+2

int    main(void)
{
int out1,out2;
       out1 = 0x55;                    /* 0101 0101 */
       outportb(DATA, out1);
       out2 = 0x0D;                    /* 0000 1101 */
       outportb(CONTROL, out2);        /* STROBE=LOW, AUTOFEED=HIGH, etc */
       return(0);
}
```

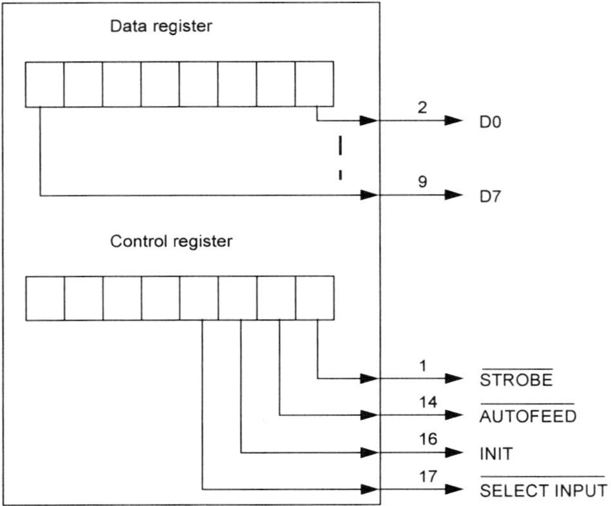

Figure 50.4 Output lines.

The setting of the output value (in this case, `out2`) looks slightly confusing, as the output is the inverse of the logical setting (that is, a 1 sets the output low). An alternative method is to exclusive-OR (EX-OR) the output value with $B which inverts the 1st, 2nd and 4th least significant bits ($\overline{\text{SELECT INPUT}}$=0, $\overline{\text{AUTOFEED}}$=1, and $\overline{\text{STROBE}}$=0), while leaving the 3rd least significant bit (INIT) untouched. Thus, the following will achieve the same as the previous program:

```
out2 = 0x06;                      /* 0000 0110 */
outportb(CONTROL, out2 ^ 0xb);    /* STROBE=LOW, AUTOFEED=HIGH, etc */
```

If the 5th bit on the control register (IRQ Enable) is written as 1 then the output on this line will go from a high to a low which will cause the processor to be interrupted.

The control lines are driven by open collector drivers pulled to +5 Vdc through 4.7 kΩ resistors. Each can sink approximately 7 mA and maintain 0.8 V down-level.

50.3.3 Inputs

There are five inputs from the parallel port (BUSY, $\overline{\text{ACK}}$, PE, SELECT and $\overline{\text{ERROR}}$). The status of these lines can be found by simply reading the upper 5 bits of the Status register, as illustrated in Figure 50.5.

Unfortunately, the BUSY line has an inverted status. Thus when a LOW is present on BUSY, the bit will actually be read as a 1. For example, Program 50.3 reads the bits from the Status register, inverts the BUSY bit and then shifts the bits three places to the right so that the 5 inputs bits are in the 5 least significant bits.

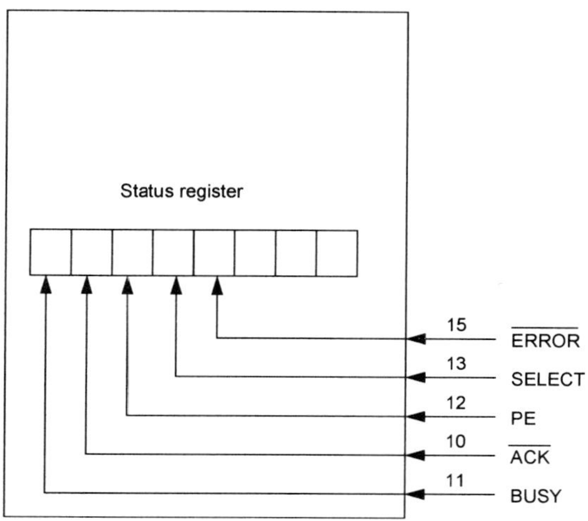

Figure 50.5 Input lines.

Program 50.3

```
#include         <stdio.h>
#define DATA     0x378
#define STATUS   DATA+1
int   main(void)
{
unsigned int in1;

      in1 = inportb(STATUS); /* read from status register */

      in1 = in1 ^ 0x80        /* invert  BUSY bit  */
      in1 = in1 >> 3;         /* move bits so that the inputs are the least
                                 significant bits */
      printf("Status bits are %d\n",in1);
      return(0);
}
```

50.3.4 Electrical interfacing

The output lines can be used to drive LEDs. Figure 50.6 shows an example circuit where a LOW output will cause the LED to be ON while a HIGH causes the output to be OFF. For an input, an open push button causes a HIGH input on the input.

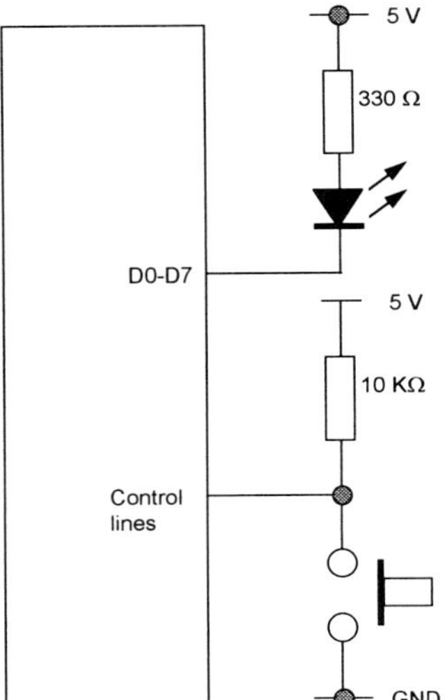

Figure 50.6 Interfacing to inputs and outputs.

50.3.5 Simple example

Program 50.4 uses a push button connected to pin 11 (BUSY). When the button is open then the input to BUSY will be a HIGH and the most significant bit in the status register will thus be a 0 (as the BUSY signal is inverted). When the button is closed then this bit will be a 1. This is tested with:

```
if (inl&0x80)==1)
```

When this condition is TRUE (that is, when the button is closed) then the output data lines (D0–D7) will flash on and off with a delay of 1 second between flashes. An output of all 1s to the data lines causes the LEDs to be off and all 0s cause the LEDs to be on.

Program 50.4

```
/*    Flash LEDs on and off when the push button connected to BUSY is closed */
#include <stdio.h>
#include <dos.h>
#define DATA       0x378
#define STATUS     DATA+1
#define CONTROL    DATA+2
int main(void)
{
int inl;
   do
   {
      inl = inportb(STATUS);   /* replace with _inp() and _outp() if required */
      if (inl&0x80)==1) {     /* if switch closed this is TRUE                */
         outportb(DATA,0x00);       /* LEDs on          */
         delay(1000);
         outportb(DATA, 0xff);      /* LEDs off         */
         delay(1000);
      }
      else    outportb(DATA,0x01); /* switch open     */
    } while (!kbhit());
    return(0);
}
```

51 Interrupt-driven Parallel Port

51.1 Introduction

The previous chapter discussed how the parallel port is used to output data. This chapter discusses how an external device can interrupt the processor. It does this by hooking onto the interrupt server routine for the interrupt that the port is attached to. Normally this interrupt routine serves as a printer interrupt (such as lack of paper, paper jam, and so on). Thus, an external device can use the interrupt service routine to transmit data to or from the PC.

51.2 Interrupts

Each parallel port is hooked to an interrupt. Normally the primary parallel port is connected to IRQ7. It is assumed in this chapter that this is the case. As with the serial port this interrupt line must be enabled by setting the appropriate bit in the interrupt mask register (IMR), which is based at address 21h. The bit for IRQ7 is the most significant bit, and it must be set to a 0 to enable the interrupt. As with the serial port, the end of interrupt signal must be acknowledged by setting the EOI signal bit of the interrupt control register (ICR) to a 1.

The interrupt on the parallel port is caused by the $\overline{\text{ACK}}$ line (pin 10) going from a high to a low (just as a printer would acknowledge the reception of a character). For this interrupt to be passed to the PIC then bit 4 of the control port (IRQ Enable) must be set to a 1.

51.3 Example program

Program 51.1 is a simple interrupt-driven parallel port Borland C program (Many other C++ compilers change `_inp` for `inportb` and `_outp` for `outportb`). The program interrupts each time the $\overline{\text{ACK}}$ line is pulled LOW. When this happens the output value should change corresponding to a binary count (0000 0000 to 1111 1111, and then back again). The user can stop the program by pressing any key on the keyboard. Figure 51.1 shows a sample setup with a push button connected to the $\overline{\text{ACK}}$ line and LEDs connected to the output data lines.

Program 51.1

```
/* Program to sample data from the parallel port   */
/* when the ACK line goes low                      */
#include <stdio.h>
#include <bios.h>
```

```
#include <conio.h>
#include <dos.h>
#define  TRUE     1
#define  FALSE    0
#define  DATA     0x378
#define  STATUS   DATA+1
#define  CONTROL  DATA+2
#define  IRQ7     0x7F  /* LPT1 interrupt                 */
#define  EOI      0x20  /* End of Interrupt               */
#define  ICR      0x20  /* Interrupt Control Register     */
#define  IMR      0x21  /* Interrupt Mask Register        */

void  interrupt far pl_interrupt(void);
void  setup_parallel (void);
void  set_vectors(void);
void  enable_interrupts(void);
void  disable_interrupts(void);
void  reset_vectors(void);
void  interrupt far (*oldvect)();
int   int_flag = TRUE;
int   outval=0;

int main(void)
{
      set_vectors();
      setup_parallel();
      do
      {
         if (int_flag)
         {
            printf("New value sent\n");
            int_flag=FALSE;
         }
      } while (!kbhit());
      reset_vectors();
      return(0);
}

void  setup_parallel(void)
{

   outportb(CONTROL, inportb(CONTROL) | 0x10);
                  /* Set Bit 4 on control port to a 1 */
}

void interrupt far pl_interrupt(void)
{
      disable();
      outportb(DATA,outval);
      if (outval!=255) outval++; else outval=0;
      int_flag=TRUE;
      outportb(ICR,EOI);
      enable();
}

void set_vectors(void)
{
int int_mask;
      disable();                    /* disable all ints        */
      oldvect=getvect(0x0f);        /* save any old vector     */
      setvect (0x0f,pl_interrupt);  /* set up for new int serv   */
}

void  enable_interrupts(void)
{
int ch;
```

```
        disable();
        ch=inportb(IMR);
        outportb(IMR, ch & IRQ7);
        enable();
}

void   disable_interrupts(void)
{
int ch;
        disable();
        outportb(IMR, ch & ~IRQ7);
        enable();
}

void   reset_vectors(void)
{
        setvect(0x0f,oldvect);
}
```

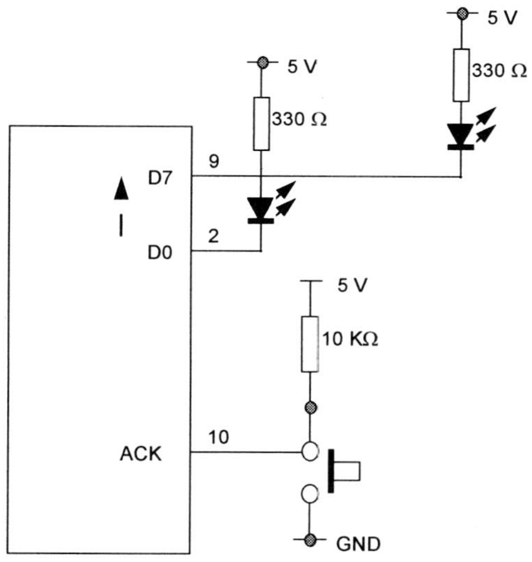

Figure 51.1 Example set-up for interrupt-driven parallel port.

51.4 Program explanation

The initial part of the program enables the interrupt on the parallel port by setting bit 4 of the control register to 1.

```
void   setup_parallel(void)
{

   outportb(CONTROL, inportb(CONTROL) | 0x10); /* Set Bit 4 on control port*/
}
```

After the serial port has been initialized the interrupt service routine for the `IRQ7` line is set to point to a new 'user-defined' service routine. The primary parallel port `LPT1:` normally sets the `IRQ7` line active when the $\overline{\text{ACK}}$ line goes from a high to a low. The interrupt associated with `IRQ7` is 0Fh (15). The `getvect()` function gets the ISR address for this interrupt, which is then stored in the variable `oldvect` so that at the end of the program it can be restored. Finally, in the `set_vectors()` function, the interrupt assigns a new 'user-defined' ISR (in this case it is the function `pl_interrupt()`).

```
void set_vectors(void)
{
int int_mask;
      disable();   /* disable all ints */
      oldvect=getvect(0x0f);   /* save any old vector */
      setvect (0x0f,pl_interrupt);   /* set up for new int serv */
}
```

At the end of the program the ISR is restored with the following code.

```
void  reset_vectors(void)
{
      setvect(0x0f,oldvect);
}
```

To enable the `IRQ7` line on the PIC, bit 5 of the IMR (interrupt mask register) is to be set to a 0 (zero). The statement:

```
      ch = inportb(IMR) & 0x7F;
```

achieves this as it bitwise ANDs all the bits, except for bit 7, with a 1. This is because any bit which is ANDed with a 0 results in a 0. The bit mask `0x7F` has been defined with the macro `IRQ7`.

```
void  enable_interrupts(void)
{
int ch;
      disable();
      ch=inportb(IMR);
      outportb(IMR, ch & IRQ7);
      enable();
}
```

At the end of the program the interrupt on the parallel port is disabled by setting bit 7 of the IMR to a 1; this disables `IRQ7` interrupts.

```
void  disable_interrupts(void)
{
int ch;
      disable();
      outportb(IMR, ch & ~IRQ7);
      enable();
}
```

The ISR for the `IRQ7` function is set to `pl_interrupt()`. It outputs the value of `outval`, which is incremented each time the interrupt is called (note that there is a roll-over statement which resets the value of `outval` back to zero when its value is 255). At the end of the ISR the end of interrupt flag is set in the interrupt control register with the statement `out-`

```
portb(ICR, EOI);.

void interrupt far pl_interrupt(void)
{
      disable();
      outportb(DATA,outval);
      if (outval!=255) outval++; else outval=0;
      int_flag=TRUE;
      outportb(ICR,EOI);
      enable();
}
```

The `main()` function calls the initialization and the de-initialization functions. It also contains a loop which continues until any key is pressed. Within this loop the keyboard is tested to determine if a key has been pressed. The interrupt service routine sets `int_flag`. If the main routine detects that it is set it displays the message 'New value sent' and resets the flag.

```
int main(void)
{
   set_vectors();
   outportb(CONTROL, inportb(CONTROL) | 0x10);
                  /* set bit 4 on control port to logic one */
      do
      {
         if (int_flag)
         {
            printf("New value sent\n");
            int_flag=FALSE;
         }
   } while (!kbhit());
   reset_vectors();
   return(0);
}
```

52 Enhanced Parallel Port

52.1 Introduction

The Centronics parallel port only allows data to be sent from the host to a peripheral. To overcome this the IEEE published the 1284 standard which is entitled 'Standard Signaling Method for a Bi-directional Parallel Peripheral Interface for Personal Computers'. It allows for bi-directional communication and high communication speeds, while it is backwardly compatible with existing parallel ports.

The IEEE 1284 standard defines the following modes:

- Compatibility mode (forward direction only). This mode defines the transfer of data between the PC and the printer (Centronics mode, as covered in Chapters 50 and 51.
- Nibble mode (reverse direction). This mode defines how 4 bits are transferred, at a time, using status lines for the input data (sometimes known as Hewlett Packard Bi-tronics). The Nibble mode can thus be used for bi-directional communication, with the data lines being used as outputs. Inputting a byte thus requires two nibble cycles.
- Byte mode (reverse direction). This mode defines how 8 bits are transferred at a time.
- Enhanced Parallel Port (EPP). This mode defines standard bi-directional communications and is used by many peripherals, such as CD-ROMs, tape drives, external hard disks, and so on.

52.2 IEEE 1284 Data Transfer Modes

In the IEEE 1284 standard, the control and status signals for nibble, byte and EPP modes have been renamed. It also classifies the modes as forward (data goes from the PC), reverse (data is sent to the PC) and bi-directional. Both the compatibility and nibble modes can be implemented with all parallel ports (as the nibble mode uses the status lines and the compatibility mode only outputs data). Some parallel ports support input and output on the data lines and thus support the byte mode. This is usually implemented by the addition of a direction bit on the control register.

52.3 Compatibility mode

The compatibility mode was discussed in Chapters 50 and 51. In this mode the program sends data to the data lines and then sets the $\overline{\text{STROBE}}$ low and then high (see Figure 52.1). This then latches the data to the printer. The operations that the program does are:

1. Data is written to the data register.
2. The program reads from the status register to test to see if the BUSY signal is low (that is, the printer is not busy).
3. If the printer is not busy then the program sets the $\overline{\text{STROBE}}$ line active low.
4. The program then makes the $\overline{\text{STROBE}}$ line high by de-asserting it.

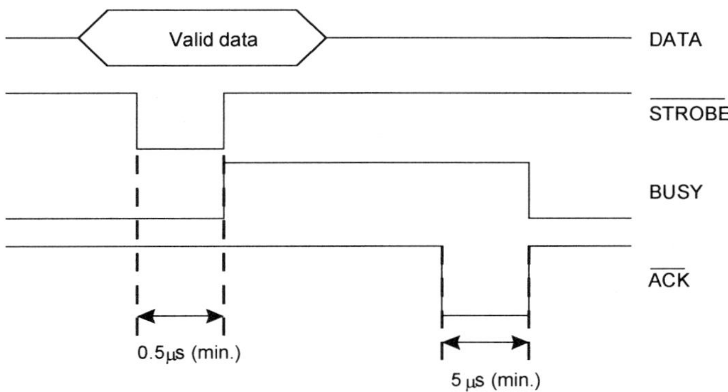

Figure 52.1 Compatibility mode transfer.

52.4 Nibble mode

This mode defines how 4 bits are transferred, at a time, using status lines for the input data (sometimes known as Hewlett Packard Bi-tronics). The Nibble mode can thus be used for bi-directional communication, with the data lines being used as outputs. To input a byte thus requires two nibble cycles.

As seen in Chapter 3 there are five inputs from the parallel port (BUSY, $\overline{\text{ACK}}$, PE, SELECT and $\overline{\text{ERROR}}$). The status of these lines can be found by simply reading the upper 5 bits of the status register. The BUSY, PE, SELECT and $\overline{\text{ERROR}}$ are normally used with $\overline{\text{ACK}}$ to interrupt the processor.

Table 52.1 defines the names of the signal in the nibble mode and Figure 52.2 shows the handshaking for this mode.

The nibble mode has the following sequence:

1. Host (PC) indicates that it is ready to receive data by setting HostBusy low.
2. The peripheral then places the first nibble on the status lines.
3. The peripheral indicates that the data is valid on the status line by setting PtrClk low.
4. The host then reads from the status lines and sets HostBusy high to indicate that it has received the nibble, but it is not yet ready for another nibble.
5. The peripheral sets PtrClk high as an acknowledgement to the host.
6. Repeat steps 1–5 for second nibble.

Table 52.1 Nibble mode signals.

Compatibility signal name	*Nibble mode name*	*In/out*	*Description*
$\overline{\text{STROBE}}$	$\overline{\text{STROBE}}$	O	Not used.
$\overline{\text{AUTO FEED}}$	HostBusy	O	Host nibble mode handshake signal. It is set low to indicate that the host is ready for nibble and set high when the nibble has been received.
$\overline{\text{SELECT INPUT}}$	1284Active	O	Set high when the host is transferring data.
$\overline{\text{INIT}}$	$\overline{\text{INIT}}$	O	Not used.
$\overline{\text{ACK}}$	PtrClk	I	Indicates valid data on the status lines. It is set low to indicate that there is valid data on the control lines and then set high when the HostBusy going high.
BUSY	PtrBusy	I	Data bit 3 for one cycle then data bit 7.
PE	AckDataReq	I	Data bit 2 for one cycle then data bit 6.
SELECT	Xflag	I	Data bit 1 for one cycle then data bit 5.
$\overline{\text{ERROR}}$	$\overline{\text{DataAvail}}$	I	Data bit 0 for one cycle then data bit 4.
D0–D7	D0–D7		Not used.

These operations are software intensive as the driver requires to set and read the handshaking lines. This limits transfer to about 50 KBytes/s. Its main advantage is that it works with all printer ports because it uses the standard Centronics setup and is normally used in low-speed bi-directional operations, such as ADC adapters, reading data from switches, and so on.

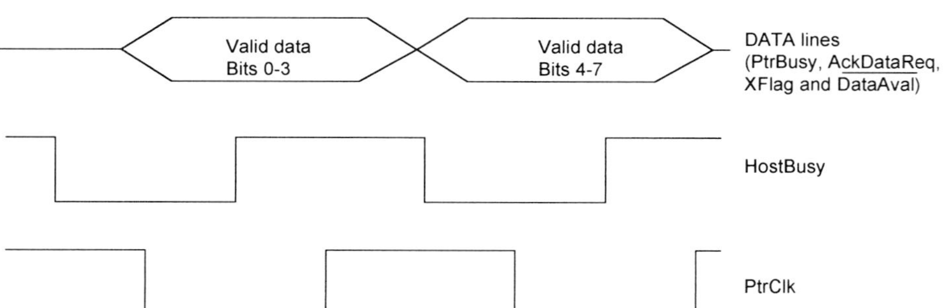

Figure 52.2 Nibble mode data transfer cycle.

52.5 Byte mode

The byte mode is often known as a bi-directional port and it uses bi-directional data lines. It has the advantage over nibble mode in that it only takes a single cycle to transfer a byte. Unfortunately, it is only compatible with newer ports. Table 52.2 defines the names of the signal in the nibble mode and Figure 52.3 shows the handshaking for this mode.

The byte mode has the following sequence:

1. The host (PC) indicates that it is ready to receive data by setting HostBusy low.
2. The peripheral then places the byte on the status lines.
3. The peripheral indicates that the data is valid on the status line by setting PtrClk low.
4. The host then reads from the data lines and sets HostBusy high to indicate that it has received the nibble, but it is not yet ready for another nibble.
5. The peripheral sets PtrClk high as an acknowledge to the host.
6. The host then acknowledges the transfer by pulsing HostClk.

Table 52.2 Byte mode signals.

Compatibility signal name	*Byte mode name*	*In/out*	*Description*
$\overline{\text{STROBE}}$	HostClk	O	Used as an acknowledgment signal. It is pulsed low after each transferred byte.
$\overline{\text{AUTO FEED}}$	HostBusy	O	It is set low to indicate that the host is ready for nibble and set high when the nibble has been received.
$\overline{\text{SELECT INPUT}}$	1284Active	O	Set high when the host is transferring data.
$\overline{\text{INIT}}$	$\overline{\text{INIT}}$	O	Not used.
$\overline{\text{ACK}}$	PtrClk	I	Indicates valid data byte. It is set low to indicate that there is valid data on the data lines and then set high when the HostBusy going high.
BUSY	PtrBusy	I	Busy status (for forward direction).
PE	AckDataReq	I	Same as $\overline{\text{DataAvail}}$.
SELECT	Xflag	I	Not used.
$\overline{\text{ERROR}}$	$\overline{\text{DataAvail}}$	I	Indicates that there is reverse data available.
D0–D7	D0–D7	I/O	Input/output data lines.

52.6 EPP

The Enhanced Parallel Port (EPP) mode defines standard bi-directional communications and is used by many peripherals, such as CD-ROMs, tape drives, external hard disks, and so on.

The EPP protocol provides four types of data transfer cycles:

1. Data read and write cycles. These involve transfers between the host and the peripheral.
2. Address read and write cycle. These pass address, channel, or command and control information.

Table 52.3 defines the names of the signals in the nibble mode and Figure 52.4 shows the handshaking for this mode. The $\overline{\text{WRITE}}$ signal occurs automatically when the host writes data to the output lines.

The data write cycle has the following sequence:

1. Program executes an I/O write cycle to the base address port + 4 (EPP Data Port); see Table 52.4. Then the following occurs with hardware:
2. The $\overline{\text{WRITE}}$ line is set low which puts the data on the data bus.
3. The $\overline{\text{DATASTB}}$ is then set low.
4. The host waits for peripheral to set the $\overline{\text{WAIT}}$ line high.
5. The $\overline{\text{DATASTB}}$ and $\overline{\text{WRITE}}$ are then set high and the cycle ends.

The important parameter is that it takes just one memory mapped I/O operation to transfer data. This gives transfer rates of up to 2 million bytes per second. While it is not as fast as a peripheral transferring over the ISA, it has the advantage that the peripheral can transfer data at a rate that is determined by the peripheral (ISA has a fixed transfer rate).

52.6.1 EPP registers

Several extra ports are defined; these are the EPP address register and the EPP data register. The EPP address register has an offset of 3 bytes from the base address and the EPP data register is offset by 4 bytes. Table 52.4 defines the registers.

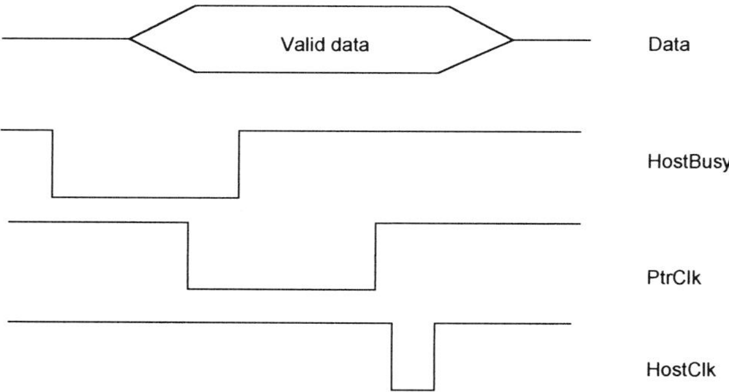

Figure 52.3 Byte mode data transfer cycle.

Table 52.3 EPP mode signals.

Compatibility signal name	*EPP mode name*	*In/out*	*Description*
$\overline{\text{STROBE}}$	$\overline{\text{WRITE}}$	O	A low for a write operation while a high indicates a read operation.
$\overline{\text{AUTO FEED}}$	$\overline{\text{DATASTB}}$	O	Indicates a data read or write operation.
$\overline{\text{SELECT INPUT}}$	$\overline{\text{ADDRSTROBE}}$	O	Indicates an address read or write operation.
$\overline{\text{INIT}}$	$\overline{\text{RESET}}$	O	Peripheral reset when low.
$\overline{\text{ACK}}$	$\overline{\text{INTR}}$	I	Peripheral sets this line low when it wishes to interrupt to the host.
BUSY	$\overline{\text{WAIT}}$	I	When it is set low it indicates that it is valid to start a cycle, else if it is high then it is valid to end the cycle.
PE	User defined	I	Can be set by each peripheral.
SELECT	User defined	I	Can be set by each peripheral.
$\overline{\text{ERROR}}$	User defined	I	Can be set by each peripheral.
D0–D7	AD0–AD7	I/O	Bi-directional address and data lines.

Table 52.4 EPP register definitions.

Port Name	*I/O address*	*Read/ write*	*Description*
Data register	BASE_AD	W	
Status register	BASE_AD +1	R	
Control register	BASE_AD +2	W	
EPP address port	BASE_AD+3	R/W	Generates EPP address read or write cycle.
EPP data port	BASE_AD+4	R/W	Generates EPP data read or write cycle.

52.7 ECP

The extended capability port (ECP) protocol was proposed by Hewlett Packard and Microsoft as an advanced mode for communication with printer and scanner type peripherals. It provides a high performance bi-directional data transfer between a host and a peripheral.

ECP provides the following cycle types in both the forward and reverse directions:

- Data cycles.
- Command cycles. The command cycles are divided into 2 types: Run Length Count and Channel address.

It supports several enhancements, such as:

- Run Length Encoding (RLE). This allows for real-time compression with compression ratios of up to 64:1. RLE allows multiple occurrences of a sequence to be sent as a short code. Typically graphics images and video information have long sequences of the same data.
- Forward and reverse channel FIFOs.
- DMA.
- Programmed I/O with a standard addressing structure.
- Channel addressing. This supports many logical devices connected to a single parallel port connection. Each of the devices can have its own connection. Typically a FAX, modem, printer and CD-ROM drive could be connected to a single parallel port connection.

In the ECP protocol the signal lines have been renamed to be consistent with an ECP handshake. Table 52.5 describes these signals.

Table 52.5 ECP mode signals.

Compatibility signal name	*ECP mode name*	*In/out*	*Description*
$\overline{\text{STROBE}}$	HostClk	O	Along with PeriphAck it is used to transfer data or address information in the forward direction.
$\overline{\text{AUTO FEED}}$	HostAck	O	Gives Command/Data status in the forward direction.
$\overline{\text{SELECT INPUT}}$	1284Active	O	Set to a high when host is a transfer mode.
$\overline{\text{INIT}}$	$\overline{\text{ReverseRequest}}$	O	Active low puts the channel into the reverse direction.
$\overline{\text{ACK}}$	PeriphClk	I	Along with HostAck it is used to transfer data in the reverse direction.
BUSY	PeriphAck	I	Along with HostClk it is used to transfer data or address information in the forward direction.
PE	nAckReverse	I	Active low to acknowledge nReverseRequest.
SELECT	Xflag	I	Extensibility flag.
$\overline{\text{ERROR}}$	nPeriphRequest	I	Active low to indicate the availability of reverse data.
D0–D7	Data[8:1]	I/O	Data lines.

Figure 52.4 shows two forward transfer cycles, a data cycle followed by a command cycle. An active HostAck signal indicates that it is a data cycle, else it is a command cycle. In the command cycle the data byte represents either:

- RLE count. If the most significant bit of the data byte is a 0 then the rest of the bytes represent the Run Length Count (0–127).
- Channel address. If the most significant bit of the data byte is a 1 then the rest of the bytes represent a channel address.

Forward Transfer phase is as follows:

1. Host sets the HostAck signal to identify if the transfer is a data or a command cycle Host puts its data on the data bus. (A).
2. Host sets HostClk low to indicate valid data. (B).
3. The peripheral sets PeriphAck high to acknowledge the transfer. (C).
4. Host sets HostClk high which clocks data into the peripheral. (D).
5. Peripheral sets PeriphAck low to indicate that it is ready for more data. (E).

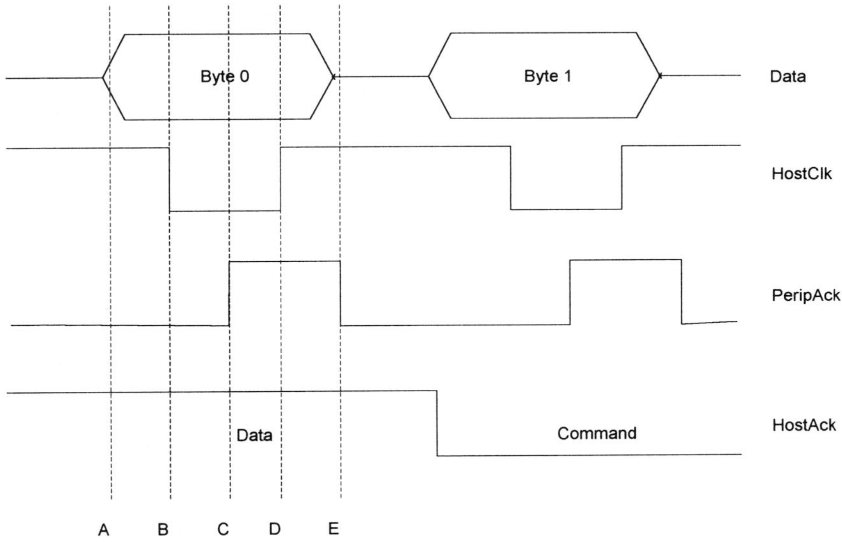

Figure 52.4 ECP forward data and Command cycle.

The reverse transfer is as follows:

1. Host identifies a reverse channel transfer by setting nReverseRequest low.
2. Peripheral acknowledges this by setting nAckReverse low.
3. Peripheral puts its data on the data bus and sets PeriphAck high.
4. Peripheral sets PeriphClk low to indicate valid data.
5. Host acknowledges this by setting HostAck high.
6. Peripheral sets PeriphClk high which clocks data into the host.
7. Host sets HostAck low to indicate that it is ready for the next transfer.

ECP registers

ECP mode has a standard set of I/O registers using a number of modes (as defined in Table 52.6). The additional registers have been added at an offset of 400h from the base port, as given in Table 52.7. Note that only extra three I/O addresses have been added. It can be seen from Figure 52.5 that the ECP driver uses the address 378h to 37Ah and 778h to 77Ah (offset from the base register by 400h). The configuration of the ECP is setup using the ECR register.

Table 52.6 ECR Register Modes

Mode	*Description*
000	SPP mode
001	Bi-directional mode (Byte mode)
010	Fast Centronics
011	ECP Parallel Port mode
100	EPP Parallel Port mode (note 1)
101	(reserved)
110	Test mode
111	Configuration mode

Table 52.7 ECP register description.

Offset	*Read/Write*	*ECP Mode*	*Function*
000	R/W	000–001	Data register
000	R/W	011	ECP address FIFO
001	R/W	All	Status register
002	R/W	All	Control register
400	R/W	010	Parallel port data FIFO
400	R/W	011	ECP data FIFO
400	R/W	110	Test FIFO
400	R	111	Configuration register A
401	R/W	111	Configuration register B
402	R/W	All	Extended control register

52.8 1284 Negotiation

The negotiation mode allows the host to determine the attached peripherals and the method used to control them. It has been designed so that is does not affect older devices, which do not respond to the negotiation phase.

In the negotiation phase, the host places a request on the data lines, such as:

- Setup a mode.
- Request a device ID.

It then goes into a negotiation sequence and uses an extensibility byte, as defined in Table 52.8.

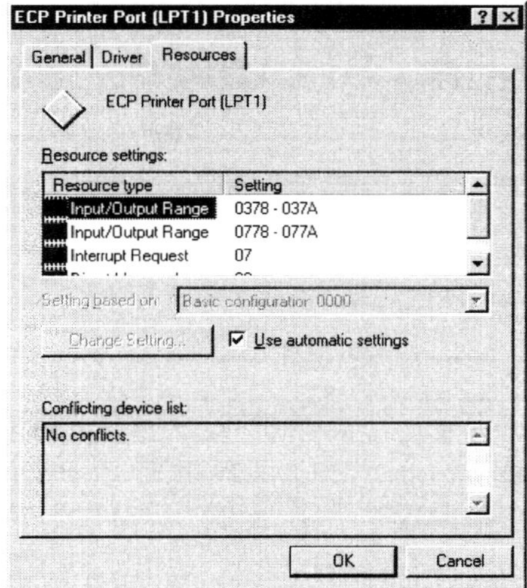

Figure 52.5 ECP register settings.

The negotiation is as follows:

1. Host puts the required extensibility byte on the data bus.
2. Host indicates the negotiation phase by setting nSelectIn high and nAutoFeed.
3. Compliant peripherals respond by setting nAck low, nError, PE high and Select high.
4. Host sets nStrobe low which clocks the extensibility byte into the peripheral.
5. Host sets nStrobe high and nAutoFeed high to acknowledge the transfer.
6. Peripheral then sets PE low, nError low and Select high.
7. Peripheral sets nAck high to signal that the negotiation sequence is over.

Table 52.8 Extensibility byte bit values.

Bit	*Description*	*Valid bit values*
8	Request Extensibility Link	1000 0000
7	Request EPP Mode	0100 0000
6	Request ECP Mode with RLE	0011 0000
5	Request ECP Mode without RLE	0001 0000
4	Reserved	0000 1000
3	Request Device ID	Return data using mode: Nibble Mode 0000 0100 Byte Mode 0000 0101 ECP Mode without RLE 0001 0100 ECP Mode with RLE0011 0100
2	Reserved	0000 0010
1	Byte Mode	0000 0001
none	Nibble Mode	0000 0000

53 Interfacing Standards

53.1 Introduction

The type of interface card used greatly affects the performance of a PC system. Early models of PCs relied on expansion options to improve their specification. These expansion options were cards that plugged into an expansion bus. Eight slots were usually available and these added memory, video, fixed and floppy disk controllers, printer output, modem ports, serial communications and so on.

There are eight main types of interface busses available for the PC. The number of data bits they handle at a time determines their classification. They are:

- PC (8-bit) ISA (16-bit)
- EISA (32-bit) MCA (32-bit)
- VL-Local Bus (32-bit) PCI bus (32/64-bit)
- SCSI (16/32-bit) PCMCIA (16-bit)

53.2 PC bus

The PC bus uses the architecture of the Intel 8088 processor which has an external 8-bit data bus and 20-bit address bus. A PC bus connector has a 62-pin printed circuit card edge connector and a long narrow or half-length plug-in card. Since it uses a 20-bit address bus it can address a maximum of 1 MB of memory. The transfer rate is fixed at 4.772 727 MHz, thus a maximum 4 772 727 bytes can be transferred every second. Dividing a crystal oscillator frequency of 14.31818 MHz by three derives this clock speed. Figure 53.1 shows a PC card. Table 53.1 defines the bus signals for the PC bus and Figure 53.2 defines the signal connections. The direction of the signal is taken as input if a signal comes from the ISA bus controller and an output if it comes from the slave device. An input/output identifies that the signal can originate from either the ISA controller or the slave device.

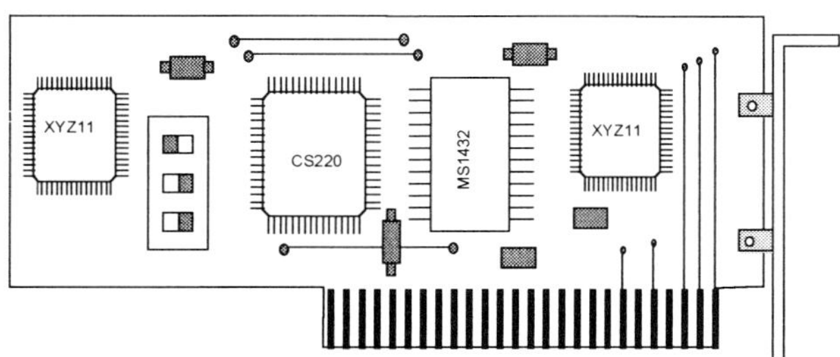

Figure 53.1 PC card.

Table 53.1 8-bit PC bus connections.

Signal	*Name*	*Description*
SA0-SA19	Address bus (input/output)	The lower 20 bits of the system address bus.
AEN	Address enable (output)	The address enable allows for an expansion bus board to disable its local I/O address decode logic. It is active high. When active, address enable indicates that either DMA or refresh are in control of the busses.
D0-D7	Data bus (input/output)	The 8 data bits that allow a transfer between the busmaster and the slave.
CLK	Clock (output)	The bus CLK is set to 4.772 727 MHz (for PC bus and 8.33 MHz for ISA bus) and provides synchronization of the data transmission (it is derived from the OSC clock).
ALE	Address latch (output)	The bus address latch indicates to the expansion bus that the address bus and bus cycle control signals are valid. It thus indicates the beginning of a bus cycle on the expansion bus.
$\overline{\text{IOR}}$	I/O read (input/output)	I/O read command signal indicates that an I/O read cycle is in progress.
$\overline{\text{IOW}}$	I/O write (input/output)	I/O write command signal indicates that an I/O write bus cycle is in progress.
$\overline{\text{SMEMR}}$	System memory read (output)	System memory read signal indicates a memory read bus cycle for the 20-bit address bus range (0h to FFFFFh).
$\overline{\text{SMEMW}}$	System memory write (output)	System memory write signal indicates a memory read bus cycle from the 20-bit address bus range (0h to FFFFFh).
IO CH RDY	Bus ready (input)	The bus ready signal allows a slave to lengthen the amount of time required for a bus cycle.
$\overline{\text{0WS}}$	Zero wait states (input)	The zero wait states (or no wait state) allows a slave to shorten the amount of time required for a bus cycle.
DRQ1-DRQ3	DMA request (input)	The DMA request indicates that a slave device is requesting a DMA transfer.
$\overline{\text{DACK1}}$ - $\overline{\text{DACK3}}$	DMA acknowledge (output)	The DMA acknowledge indicates to the requesting slave that the DMA is handling its request.
$\overline{\text{REF}}$	Refresh (output)	The refresh signal is used to inform a memory board that it should perform a refresh cycle.

T/C	Terminal count (input)	The Terminal count indicates that the DMA transfer has been successful and all the bytes have been transferred.
IRQ2-IRQ7	Interrupt request	The interrupt request signals indicate that the slave device is requesting service by the processor.
OSC	Crystal oscillator (output)	The crystal oscillator signal is 14.31818 MHz signal provided for use by expansion boards. This clock speed is three times the CLK speed.
RESET DRV	Reset drive (output)	Resets plug-in boards connected to the ISA bus.
$\overline{\text{IO CH CHK}}$	I/O check (input)	The I/O check signal indicates that a memory slave has detected a parity error.
±5V, ±12V and GND	Power (output)	

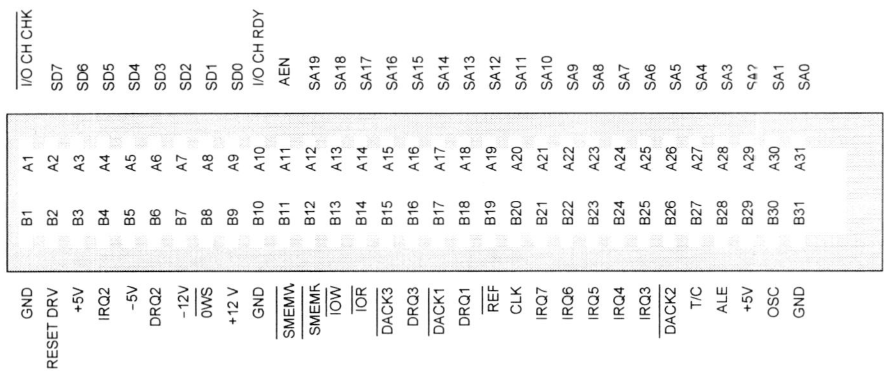

Figure 53.2 PC card connections.

53.3 ISA bus

IBM developed the ISA (Industry Standard Architecture) for their 80286-based AT (Advanced Technology) computer. It had the advantage of being able to deal with 16 bits of data at a time. An extra edge connector gives compatibility with the PC bus. This gives an extra 8 data bits and 4 address lines. Thus, the ISA bus has a 16-bit data and a 24-bit address bus, which gives a maximum of 16 MB of addressable memory and like the PC bus it uses a fixed clock rate of 8 MHz. The maximum data rate is thus two bytes (16 bits) per clock cycle, giving a maximum throughput of 16 MB/sec. In machines that run faster than 8 MHz the ISA bus runs slower than the rest of the computer.

A great advantage of PC bus cards is that they can be plugged into an ISA bus connector. ISA cards are very popular as they give good performance for most interface applications.

The components used are extremely cheap and it is a well-proven and reliable technology. Typical applications include serial and parallel communications, networking cards and sound cards. Figure 53.3 illustrates an ISA card and Table 53.2 gives the pin connections for the bus. It can be seen that there are four main sets of connections, the A, B, C and D sections. The standard PC bus connection contains the A and B sections. The A section includes the address lines A0-A19 and 8 data lines, D0–D7. The B section contains interrupt lines, IRQ0-IRQ7, power supplies and various other control signals. The extra ISA lines are added with the C and D section; these include the address lines, A17-A23, data lines D8–D15 and interrupt lines IRQ10-IRQ14.

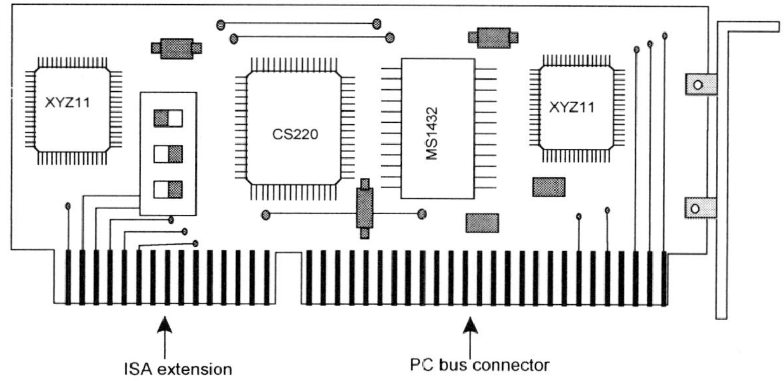

Figure 53.3 ISA card.

The Industry Standard Architecture (ISA) bus uses a 16-bit data bus (D0–D15) a 24-bit address bus (A0-A24) and the CLK signal is set to 8.33 MHz. Figure 53.4 illustrates the ISA bus pin connections and Table 53.2 lists the extra pin connections.

The $\overline{\text{SMEMR}}$ and $\overline{\text{SMEMW}}$ lines are used to transfer data for the lowest 1 MB (0h to FFFFFh) of memory (where the S prefix can be interpreted as small memory model) and the signals $\overline{\text{MEMR}}$ and $\overline{\text{MEMW}}$ are used to transfer data between 1 MB (FFFFFh) and 16 MB (FFFFFFh). For example, if reading from address 001000h then the $\overline{\text{SMEMR}}$ line is made active low, while if the address is 1F0000h then the $\overline{\text{MEMR}}$ line is made active. For a 16-bit transfer the $\overline{\text{M16}}$ and $\overline{\text{IO16}}$ lines are made active.

Table 53.2 Extra 16-bit ISA bus connections.

Signal	*Name*	*Description*
A17-A23	Address bus (input/output)	The upper 7 bits of the address of the system address bus.
$\overline{\text{SBHE}}$	System byte high enable (output)	The system byte high enable indicates that data is expected on the upper 8 bits of the data bus (D8–D15).
D8-D15	Data bus (input/output)	The upper 8 bits of the data bus provide for the second half of the 16-bit data bus.

$\overline{\text{MEMR}}$	Memory read (input/output)	The memory read command indicates a memory read when the memory address is in the range `100000h` - `FFFFFFh` (16 MB of memory).
$\overline{\text{MEMW}}$	Memory write (input/output)	The memory write command indicates a memory write when the memory address is in the range `100000h` – `FFFFFFh` (16 MB of memory).
$\overline{\text{M16}}$	16-bit memory slave	Indicates that the addressed slave is a 16-bit memory slave.
$\overline{\text{IO16}}$	16-bit I/O slave (input/output)	Indicates that the addressed slave is a 16-bit I/O slave.
DRQ0, DRQ5-DRQ7	DMA request lines (input)	Extra DMA request lines that indicate that a slave device is requesting a DMA transfer.
$\overline{\text{DACK0}}$, $\overline{\text{DACK5}}$- $\overline{\text{DACK7}}$	DMA acknowledge lines (output)	Extra DMA acknowledge lines that indicate to the requesting slave that the DMA is handling its request.
$\overline{\text{MASTER}}$	Bus ready (input)	This allows another processor to take control of the system address, data and control lines.
IRQ9-IRQ12, IRQ14, IRQ15	Interrupt requests (input)	Additional interrupt request signals that indicate that the slave device is requesting service by the processor. Note that the `IRQ13` line is normally used by the hard disk and included in the IDE bus.

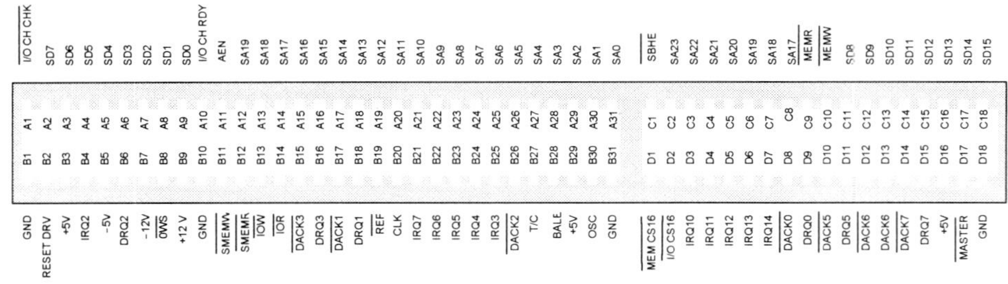

Figure 53.4 ISA bus connections.

53.3.1 Handshaking lines

Figure 53.5 shows typical connections to the ISA bus. The ALE (sometimes known as BALE) controls the address latch and, when active low, it latches the address lines A2–A19 to the ISA bus. The address is thus latched when ALE goes from a high to a low.

The Pentium's data bus is 64 bits wide, whereas the ISA expansion bus is 16 bits wide. It is the bus controller's function to steer data between the processor and the slave device for either 8-bit or 16-bit communications. For this purpose the bus controller monitors $\overline{\text{BE0}}$ - $\overline{\text{BE3}}$, W/$\overline{\text{R}}$, $\overline{\text{M16}}$, and $\overline{\text{IO16}}$ to determine the movement of data.

When the processor outputs a valid address it sets address lines (AD2–AD31), the byte enables ($\overline{\text{BE0}}$ - $\overline{\text{BE3}}$) and sets ADS active. The bus controller then picks up this address and uses it to generate the system address lines, SA0-SA19 (which are just a copy of the lines A2–A19). The bus controller then uses the byte enable lines to generate the address bits SA0 and SA1.

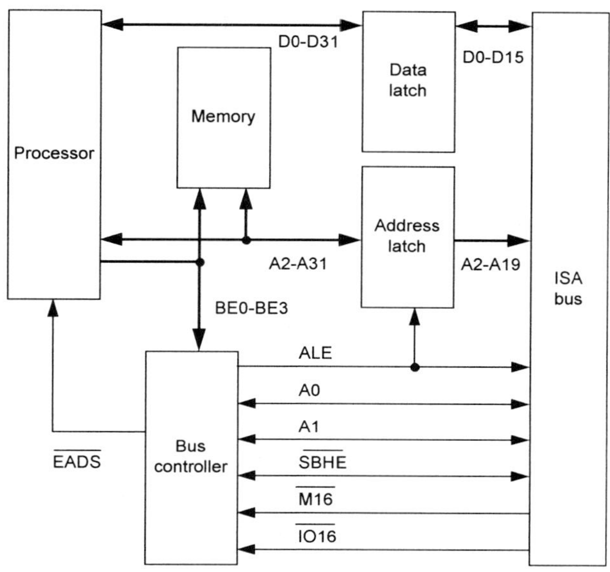

Figure 53.5 ISA bus connections.

The $\overline{\text{EADS}}$ signal returns an active low signal to the processor if the external bus controller has sent a valid address on address pins A2–A21.

It can be seen from Figure 53.6 that the $\overline{\text{BE0}}$ line accesses the addresses ending with 0h, 4h, 8h and Ch, the $\overline{\text{BE1}}$ line accesses addresses ending with 1h, 5h, 9h and Dh, the $\overline{\text{BE2}}$ line accesses addresses ending with 2h, 6h, Ah and Eh, and so on.

Thus if the $\overline{\text{BE0}}$ line is low and $\overline{\text{SBHE}}$ is high then a single byte is accessed through D0–D7. If $\overline{\text{SBHE}}$ is low then a word is accessed and D0–D15 contains the data.

Table 53.3 shows three examples of handshaking lines. The first is an example of a byte transfer with an 8-bit slave at an even address. The second example gives a byte transfer for an 8-bit slave at an odd address. Finally, the table shows a 2-byte transfer with a 16-bit slave at an even address.

If a 32-bit data is to be accessed then $\overline{\text{BE0}}-\overline{\text{BE3}}$ will be 0000 which makes 4 bytes active. The bus controller will then through cycle from SA0, SA1 = 00 to SA0, SA1 = 11. Each time 8 data bits are placed into a copy buffer, which are then passed to the processor as 32 bits.

Table 53.3 Example handshaking lines.

$\overline{\text{BE0}}$	$\overline{\text{BE1}}$	$\overline{\text{BE2}}$	$\overline{\text{BE3}}$	$\overline{\text{IO16}}$	$\overline{\text{M16}}$	$\overline{\text{SBHE}}$	SA0	SA1	*Data*
0	1	1	1	1	1	1	0	0	SD0–SD7
1	0	1	1	1	1	0	1	0	SD8–SD15
0	0	1	1	0	1	0	0	0	SD0–SD15

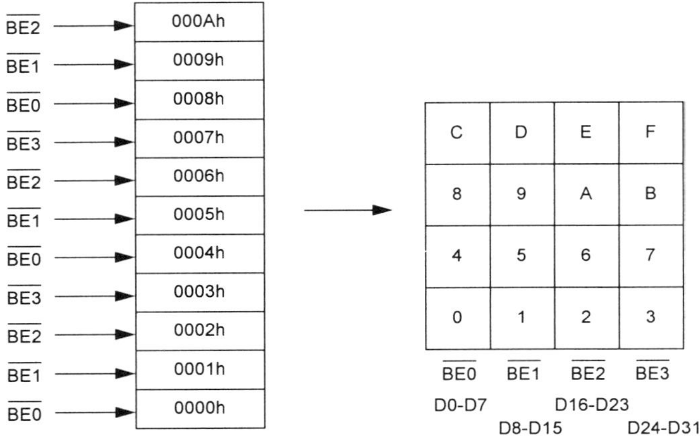

Figure 53.6 Address decoding.

53.4 MCA bus

IBM developed the Microchannel Interface Architecture (MCA) bus for their PS/2 computers. This bus is completely incompatible with the ISA bus and can operate as a 16-bit or a 32-bit data bus. The main technical difference between the MCA and PC/ISA (and EISA) is that MCA has a synchronous bus whereas PC/ISA/EISA use an asynchronous bus. An asynchronous bus works at a fixed clock rate whereas a synchronous bus data transfer is not dependent on a fixed clock. Synchronous busses take their timings from the devices involved in the data transfer (that is, the processor or system clock). The original MCA specification resulted in a maximum transfer rate of 160 MB/sec. Very few manufacturers adopted MCA technology and it is mainly found in IBM PS/2 computers.

53.5 EISA bus

Several manufacturers developed the EISA (Extended Industry Standard Architecture) bus in direct competition to the MCA bus. It provides compatibility with PC/ISA but not with MCA. The EISA connector looks like an ISA connector. It is possible to plug an ISA card into an EISA connector, but a special key allows the EISA card to insert deeper into the EISA bus connector. It then makes connections with a 32-bit data and address bus. An EISA card has twice the number of connections over an ISA card and there are extra slots that allow it to be inserted deeper into the connector. The ISA card only connects with the upper connectors because it has only a single key slot.

EISA uses an asynchronous transfer at a clock speed of 8 MHz. It has a full 32-bit data and address bus and can address up to 4 GB of memory. In theory, the maximum transfer rate is four bytes for every clock cycle. Since the clock runs at 8 MHz, the maximum data rate is 32 MB/sec.

53.6 Comparison of different types

Data throughput depends on the number of bytes being communicated for each transfer and the speed of the transfer. With the PC, ISA and EISA busses this transfer rate is fixed at 8 MHz, whereas the PCI and VL local busses use the system clock (typically, 33 MHz or 50 MHz). For many applications the ISA bus offers the best technology as it has been around for a long time, it gives a good data throughput and it is relatively cheap and reliable. It has a 16-bit data bus and can thus transfer data at a maximum rate of 16 MB/sec. The EISA bus can transfer 4 bytes for each clock cycle, thus if 4 bytes are transferred for each clock cycle, it is twice as fast as ISA. Table 53.4 shows the maximum data rates for the different interface cards.

The type of interface technology used depends on the data throughput. Table 53.5 shows some typical transfer data rates. The heaviest usage on the system are microprocessor to memory and graphics adapter transfers. These data rates depend on the application and the operating system used. Graphical user interface (GUI) programs have much greater data throughput than programs running in text mode. Notice that a high-specification sound card with recording standard quality (16-bit samples at 44.1kHz sampling rate) only requires a transfer rate of 88 KB/sec.

A standard Ethernet local area network card transfers at data rates of up 10 Mbps (approx. 1 MB/sec), although new fast Ethernet cards can transfer at data rates of up to 100 Mbps (approx. 10 MB/sec). These transfers thus require local bus type interfaces.

The PCI local bus has become a standard on most new PC systems and has replaced the VL-local bus for graphics adapters. It has the advantage over the VL-local bus in that it can transfer at much higher rates. Unfortunately, most available software packages cannot use the full power of the PCI bus because they do not use the full 64-bit data bus. PCI and VL-local bus are discussed in the next chapter.

Table 53.4 Maximum data rates for different I/O cards.

I/O card	*Maximum data rate*
PC	8 MB/sec
ISA	16 MB/sec
EISA	32 MB/sec
VL-Local bus	132 MB/sec (33 MHz system clock using 32-bit transfers)
PCI	264 MB/sec (33 MHz system clock using 64-bit transfers)
MCA	20 MB/sec (160 MB/sec burst)

Table 53.5 Typical transfer rates.

Device	*Transfer rate*	*Application*
Hard disk	4 MB/sec	Typical transfer
Sound card	88 KB/sec	16-bit, 44.1 KHz sampling
LAN	1 MB/sec	10 Mbit/sec Ethernet
RAM	66 MB/sec	Microprocessor to RAM
Serial Communications	1 KB/sec	9600 bps
Super VGA	15 MB/sec	1024×768 pixels with 256 colors

54 Local Bus and PC Motherboard

54.1 Introduction

The main problem with the PC, ISA and EISA busses is that the transfer rate is normally much slower than the system clock. This is wasteful in processor time and generally reduces system performance. For example, if the system clock is running at 50 MHz and the EISA interface operates at 8 MHz then for 84% of the data transfer time the processor is doing nothing. This chapter discusses the main local busses and the next discusses motherboard design.

54.2 VESA VL-local bus

An improvement to the ISA interface is to transfer data at the speed of the system clock. For this reason the Video Electronics Standards Association (VESA) created the VL-local bus to create fast processor-to-video card transfers. It uses a standard ISA connector with an extra connection to tap into the system bus (Figure 54.1).

Memory, graphics and disk transfers are the heaviest for data transfer rates, whereas applications such as modems, Ethernet and sound cards do not require fast transfer rates. The VL-local bus addresses this by allowing the processor, memory, graphics and disk controller access to a 33 MHz/32-bit local bus. Other applications still use the normal ISA bus, as shown in Figure 54.2. The graphics adapter and disk controller connect to the local bus whereas other slower peripherals connect to the slower 8 MHz/16-bit ISA bus. A maximum of three devices can connect to the local bus (normally graphics and disk controllers). Note that the speed of the data transfer is dependent on the clock rate of the system and that the maximum clock speed for the VL-local bus is 33 MHz.

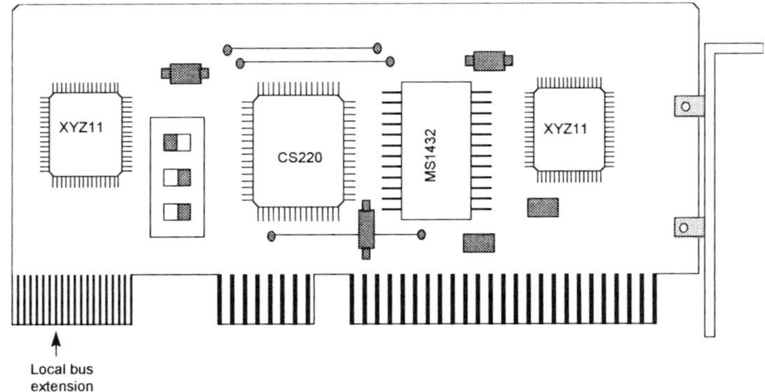

Figure 54.1 VL-local bus interface card.

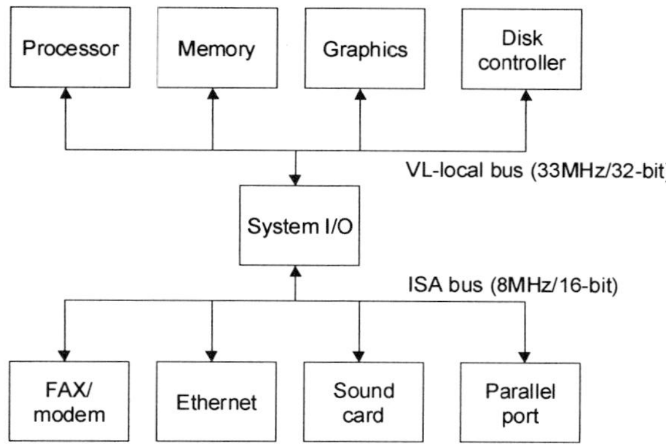

Figure 54.2 VESA VL-local bus architecture.

Table 54.1 lists the pin connections for the 32-bit VL-local bus and it shows that, in addition to the standard ISA connector, there are two sides of connections, the A and the B side. Each side has 58 connections giving 116 connections. It has a full 32-bit data and address bus. The 32 data lines are labeled `DAT00–DAT31` and 32 address lines are labeled from `ADR00` to `ADR31`. Note that while the data and address lines are contained within the extra VL-local bus extension, some of the standard ISA lines are used, such as the IRQ lines.

54.3 PCI bus

Intel have developed a new standard interface, named the PCI (Peripheral Component Interconnection) local bus, for the Pentium processor. This technology allows fast memory, disk and video access. A standard set of interface ICs known as the 82430 PCI chipset is available to interface to the bus.

As with the VL-local bus, the PCI bus transfers data using the system clock, but has the advantage over the VL-local Bus in that it can operate over a 32-bit or 64-bit data path. The high transfer rates used in PCI architecture machines limit the number of PCI bus interfaces to two or three (normally the graphics adapter and hard disk controller). If data is transferred at 64 bits (8 bytes) at a rate of 33 MHz then the maximum transfer rate is 264 MB/sec. Figure 54.3 shows PCI architecture. Notice that an I/O bridge gives access to ISA, EISA or MCA cards. Unfortunately, to accommodate for the high data rates and for a reduction in the size of the interface card, the PCI connector is not compatible with PC, ISA or EISA.

The maximum data rate of the PCI bus is 264 MB/sec, which can only be achievable using 64-bit software on a Pentium-based system. On a system based on the 80486 processor this maximum data will only be 132 MB/sec (that is, using a 32-bit data bus).

The PCI local bus is a radical re-design of the PC bus technology and is logically different from the ISA and VL-local bus. Table 54.2 lists the pin connections for the 32-bit PCI local bus and it shows that there are two lines of connections, the A and the B side. Each side has 64 connections giving 128 connections. A 64-bit, 2×94-pin connector version is also available. The PCI bus runs at the speed of the motherboard which for the Pentium processor is typically 33 MHz or 50 MHz (as compared to the VL-local bus which gives a maximum transfer rate of 33 MHz).

Table 54.1 32-bit VESA VL-local bus connections.

Pin	*Side A*	*Side B*	*Pin*	*Side A*	*Side B*
1	D0	D1	30	A17	A16
2	D2	D3	31	A15	A14
3	D4	GND	32	VCC	A12
4	D6	D5	33	A13	A10
5	D8	D7	34	A11	A8
6	GND	D9	35	A9	GND
7	D10	D11	36	A7	A6
8	D12	D13	37	A5	A4
9	VCC	D15	38	GND	$\overline{WBACK}$
10	D14	GND	39	A3	$\overline{BE0}$
11	D16	D17	40	A2	VCC
12	D18	VCC	41	NC	$\overline{BE1}$
13	D20	D19	42	$\overline{RESET}$	$\overline{BE2}$
14	GND	D21	43	$D/\overline{C}$	GND
15	D22	D23	44	$M/\overline{IO}$	$\overline{BE3}$
16	D24	D25	45	$W/\overline{R}$	$\overline{ADS}$
17	D26	GND	46	KEY	KEY
18	D28	D27	47	KEY	KEY
19	D30	D29	48	$\overline{RDYRTN}$	$\overline{LRDY}$
20	VCC	D31	49	GND	$\overline{LDEV}$
21	A31	A30	50	IRQ9	$\overline{LREQ}$
22	GND	A28	51	$\overline{BRDY}$	GND
23	A29	A26	52	$\overline{BLAST}$	$\overline{LGNT}$
24	A27	GND	53	ID0	VCC
25	A25	A24	54	ID1	ID2
26	A23	A22	55	GND	ID3
27	A21	VCC	56	LCLK	ID4
28	A19	A20	57	VCC	NC
29	GND	A18	58	$\overline{LBS16}$	$\overline{LEADS}$

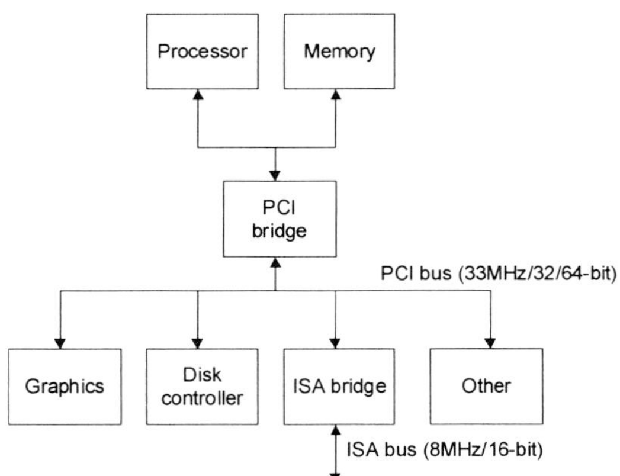

Figure 54.3 PCI bus architecture.

Table 54.2 32-bit PCI local bus connections.

Pin	*Side A*	*Side B*	*Pin*	*Side A*	*Side B*
1	–12V	$\overline{\text{TRST}}$	32	AD17	AD16
2	TCK	+12V	33	$\overline{\text{C / BE2}}$	+3.3V
3	GND	TMS	34	GND	$\overline{\text{FRAME}}$
4	TDO	TDI	35	$\overline{\text{IRDY}}$	GND
5	+5V	+5V	36	+3.3V	$\overline{\text{TRDY}}$
6	+5V	$\overline{\text{INTA}}$	37	$\overline{\text{DEVSEL}}$	GND
7	$\overline{\text{INTB}}$	$\overline{\text{INTC}}$	38	GND	$\overline{\text{STOP}}$
8	$\overline{\text{INTD}}$	+5V	39	$\overline{\text{LOCK}}$	+3.3V
9	$\overline{\text{PRSNT1}}$	Reserved	40	$\overline{\text{PERR}}$	SDONE
10	Reserved	+5V(I/O)	41	+3.3V	$\overline{\text{SBO}}$
11	$\overline{\text{PRSNT2}}$	Reserved	42	$\overline{\text{SERR}}$	GND
12	GND	GND	43	+3.3V	PAR
13	GND	GND	44	$\overline{\text{C / BE1}}$	AD15
14	Reserved	Reserved	45	AD14	+3.3V
15	GND	$\overline{\text{RST}}$	46	GND	AD13
16	CLK	+5V(I/O)	47	AD12	AD11
17	GND	$\overline{\text{GNT}}$	48	AD10	GND
18	$\overline{\text{REQ}}$	GND	49	GND	AD09
19	+5V(I/O)	Reserved	50	KEY	KEY
20	AD31	AD30	51	KEY	KEY
21	AD29	+3.3V	52	AD08	$\overline{\text{C / BE0}}$
22	GND	AD28	53	AD07	+3.3V
23	AD27	AD26	54	+3.3V	AD06
24	AD25	GND	55	AD05	AD04
25	+3.3V	AD24	56	AD03	GND
26	$\overline{\text{C / BE3}}$	IDSEL	57	GND	AD02
27	AD23	+3.3V	58	AD01	AD00
28	GND	$\overline{\text{FRAME}}$	59	+5V(I/O)	+5V(I/O)
29	AD21	AD20	60	$\overline{\text{ACK64}}$	$\overline{\text{REQ64}}$
30	AD19	GND	61	+5V	+5V
31	+3.3V	$\overline{\text{TRDY}}$	62	+5V	+5V

54.3.1 PCI operation

The PCI bus cleverly saves lines by multiplexing the address and data lines. It two modes (Figure 54.4):

- Multiplexed mode. The address and data lines are used alternately. First, the address is sent, followed by a data read or write. Unfortunately, this requires two or three clock cycles for a single transfer (either an address followed by a read or write cycle, or an address followed by read and write cycle). This causes a maximum data write transfer rate of 66 MB/s (address then write) and a read transfer rate of 44 MB/s (address, write then read), for a 32-bit data bus width.
- Burst mode. The multiplexed mode obviously slows the maximum transfer rate. Additionally, it can be operated in burst mode, where a single address can be initially sent, followed by implicitly addressed data. Thus, if a large amount of sequentially addressed memory is transferred then the data rate approach the maximum transfer of 133 MB/s for a 32-bit data bus and 266 MB/s for a 64-bit data bus.

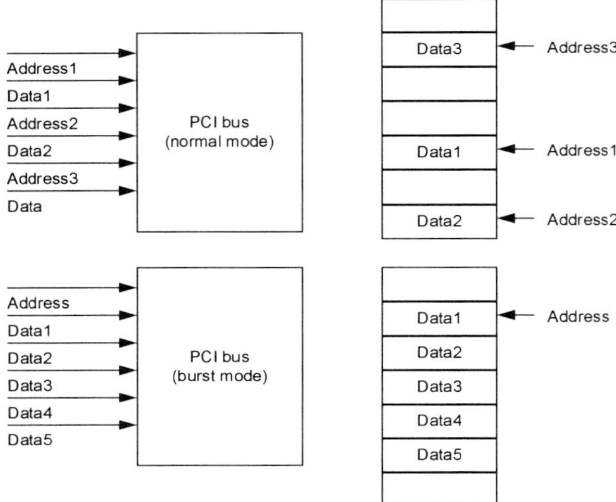

Figure 54.4 PCI bus transfer modes.

If the data from the processor is sequentially address data then PCI bridge buffers the incoming data and then releases it to the PCI bus in burst mode. The PCI bridge may also use burst mode when there are gaps in the addressed data and use a handshaking line to identify that no data is transferred for the implied address. For example in Figure 54.4 the burst mode could involve Address+1, Address+2 and Address+3 and Address+5, then the byte enable signal can be made inactive for the fourth data transfer cycle.

To accommodate the burst mode, the PCI bridge has a prefetch and posting buffer on both the host bus and the PCI bus sides. This allows the bridge to build the data access up into burst accesses. For example, the processor typically transfers data to the graphics card with sequential accessing. The bridge can detect this and buffer the transfer. It will then transfer the data in burst mode when it has enough data. Figure 54.5 shows an example where the PCI bridge buffers the incoming data and transfers it using burst mode. The transfers between the processor and the PCI bridge, and between the PCI bridge and the PCI bus can be independent where the processor can be transferring to its local memory while the PCI bus is transferring data. This helps to decouple the PCI bus from the processor.

The primary bus in the PCI bridge connects to the processor bus and the secondary bus connects to the PCI bus. The prefetch buffer stores incoming data from the connected bus and the posting buffer holds the data ready to be sent to the connected bus.

The PCI bus also provides for a configuration memory address (along with direct memory access and isolated I/O memory access). This memory is used to access the configuration register and 256-byte configuration memory of each PCI unit.

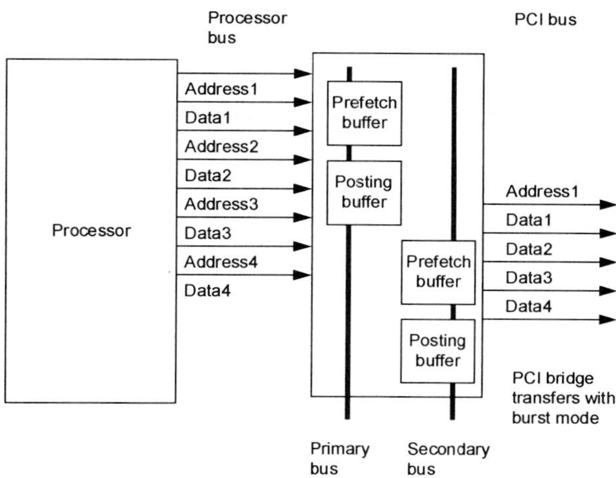

Figure 54.5 PCI bridge using buffering for burst transfer.

54.3.2 PCI bus cycles

The PCI has built-in intelligence where the command/byte enable signals ($\overline{C/BE3}-\overline{C/BE0}$) are used to identify the command. They are given by:

$\overline{C/BE3}$	$\overline{C/BE2}$	$\overline{C/BE1}$	$\overline{C/BE0}$	*Description*
0	0	0	0	INTA sequence
0	0	0	1	Special cycle
0	0	1	0	I/O read access
0	0	1	1	I/O write access
0	1	1	0	Memory read access
0	1	1	1	Memory write access
1	0	1	0	Configuration read access
1	0	1	1	Configuration write access
1	1	0	0	Memory multiple read access
1	1	0	1	Dual addressing cycle
1	1	1	0	Line memory read access
1	1	1	1	Memory write access with invalidations

The PCI bus allows any device to talk to any other device, thus one device can talk to another without the processor being involved. The device that starts the conversion is known as the initiator and the addressed PCI device is known as the target. The sequence of operation for write cycles, in burst mode, is:

- Address phase. The data transfer is started by the initiator activating the $\overline{FRAME}$ signal. The command is set on the command lines ($\overline{C/BE3}-\overline{C/BE0}$) and the address/data pins (AD31–AD0) are used to transfer the address. The bus then uses the byte enable lines ($\overline{C/BE3}-\overline{C/BE0}$) to transfer a number of bytes.
- The target sets the $\overline{TRDY}$ signal (target ready) active to indicate that the data has on the AD31–AD0 (or AD62–AD0 for a 64-bit transfer) lines is valid. In addition, the initiator indicates its readiness to the PCI bridge by setting the $\overline{IRDY}$ signal (indicator ready) active.

Figure 54.6 illustrates this.

- The transfer continues using the byte enable lines. The initiator can block transfers if it sets $\overline{\text{IRDY}}$ and the target with $\overline{\text{TRDY}}$.
- Transfer is ended by deactivating the $\overline{\text{FRAME}}$ signal.

The read cycle is similar but the $\overline{\text{TRDY}}$ line is used by the target to indicate that the data on the bus is valid.

54.3.3 PCI commands

The first phase of the bus access is the command/addressing phase. Its main commands are:

- INTA sequence. Addresses an interrupt controller where interrupt vectors are transferred after the command phase.
- Special cycle. Used to transfer information to the PCI device about the processor's status. The lower 16 bits contain the information codes, such as 0000h for a processor shutdown, 0001h for a processor halt, 0002h for *x*86 specific code and 0003h to FFFFh for reserved codes. The upper 16 bits (AD31–AD16) indicate *x*86 specific codes when the information code is set to 0002h.
- I/O read access. Indicates a read operation for I/O address memory, where the AD lines indicate the I/O address. The address lines AD0 and AD1 are decoded to define whether an 8-bit or 16-bit access is being conducted.
- I/O write access. Indicates a write operation to an I/O address memory, where the AD lines indicate the I/O address.
- Memory read access. Indicates a direct memory read operation. The byte-enable lines ($\overline{\text{C/BE3}} - \overline{\text{C/BE0}}$) identify the size of the data access.
- Memory write access. Indicates a direct memory write operation. The byte-enable lines ($\overline{\text{C/BE3}} - \overline{\text{C/BE0}}$) identify the size of the data access.

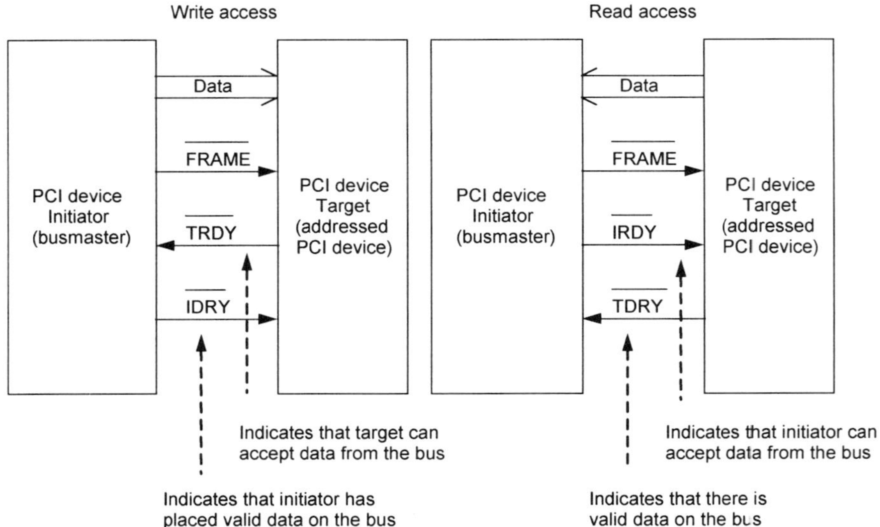

Figure 54.6 PCI handshaking.

- Configuration read access. Used when accessing the configuration address area of a PCI unit. The initiator sets the IDSEL line activated to select it. It then uses address bits AD7–AD2 to indicate the addresses of the double words to be read (AD1 and AD0 are set to 0). The address lines AD10–AD18 can be used for selecting the addressed unit in a multi-function unit.
- Configuration write access. As the configuration read access, but data is written from the initiator to the target.
- Memory multiple read access. Used to perform multiple data read transfers (after the initial addressing phase). Data is transferred until the initiator sets the $\overline{\text{FRAME}}$ signal inactive.
- Dual addressing cycle. Used to transfer a 64-bit address to the PCI device (normally only 32-bit addresses are used) in either a single or a double clock cycle. In a single clock cycle the address lines AD63–AD0 contain the 64-bit address (note that the Pentium processor only has a 32-bit address bus, but this mode has been included to support other systems). With a 32-bit address transfer the lower 32 bits are placed on the AD31–AD0 lines, followed by the upper 32 bits on the AD31–AD0 lines.
- Line memory read access. Used to perform multiple data read transfers (after the initial addressing phase). Data is transferred until the initiator sets the $\overline{\text{FRAME}}$ signal inactive.
- Memory write access with invalidations. Used to perform multiple data write transfers (after the initial addressing phase).

54.3.4 PCI interrupts

The PCI bus support four interrupts ($\overline{\text{INTA}}-\overline{\text{INTD}}$). The $\overline{\text{INTA}}$ signal can be used by any of the PCI units, but only a multi-function unit can use the other three interrupt lines ($\overline{\text{INTB}}-\overline{\text{INTD}}$). These interrupts can be steered, using system BIOS, to one of the IRQ*x* interrupts by the PCI bridge. For example, a 100 Mbps Ethernet PCI card can be set to interrupt with $\overline{\text{INTA}}$ and this could be steered to IRQ10.

54.3.5 Bus arbitration

Busmasters are devices on a bus which are allowed to take control of the bus. For this purpose, PCI uses the $\overline{\text{REQ}}$ (request) and $\overline{\text{GNT}}$ (grant) signals. There is no real standard for this arbitration, but normally the PCI busmaster activates the $\overline{\text{REQ}}$ signal to indicate a request to the PCI bus, and the arbitration logic must then activate the $\overline{\text{GNT}}$ signal so that the requesting master gains control of the bus. To prevent a bus lock-up, the busmaster is given 16 CLK cycles before a time-overrun error occurs.

54.3.6 Other PCI pins

The other PCI pins are:

- $\overline{\text{RST}}$ (Pin A15). Resets all PCI devices.
- $\overline{\text{PRSNT1}}$ and $\overline{\text{PRSNT2}}$ (Pins B9 and B11). These, individually, or jointly, show that there is an installed device and what the power consumption is. A setting of 11 (that is, $\overline{\text{PRSNT1}}$ is a 1 and $\overline{\text{PRSNT2}}$ is a 1) indicates no adapter installed, 01 indicates maximum power dissipation of 25 W, 10 indicates a maximum dissipation of 15 W and 00 indicate a maximum power dissipation of 7.5 W.
- $\overline{\text{DEVSEL}}$ (Pin B37). Indicates that addressed device is the target for a bus operation.

- TMS (test mode select), TDI (test data input), TDO (test data output), $\overline{\text{TRST}}$ (test reset), TCK (test clock). Used to interface to the JTAG boundary scan test.
- IDSEL (Pin A26). Used for device initialization select signal during the accessing of the configuration area.
- $\overline{\text{LOCK}}$ (Pin A15). Indicates that an addressed device is to be locked-out of bus transfers. All other unlocked device can still communicate.
- PAR, $\overline{\text{PERR}}$ (Pins A43 and B40). The parity pin (PAR) is used for even parity for AD31–AD0 and C/BE3–C/BE0, and $\overline{\text{PERR}}$ indicates that a parity error has occurred.
- SDONE, $\overline{\text{SBO}}$ (Pins A40 and A41). Used in snoop cycles. SDONE (snoop done) and $\overline{\text{SBO}}$ (snoop back off signal).
- $\overline{\text{SERR}}$ (Pin B42). Used to indicate a system error.
- $\overline{\text{STOP}}$ (Pin A38). Used by a device to stop the current operation.
- $\overline{\text{ACK64}}$, $\overline{\text{REQ64}}$ (Pins B60 and A60). The $\overline{\text{REQ64}}$ signal is an active request for a 64-bit transfer and $\overline{\text{ACK64}}$ is the acknowledge for a 64-bit transfer.

54.3.7 Configuration address space

Each PCI device has 256 bytes of configuration data, which is arranged as 64 registers of 32 bits. It contains a 64-byte predefined header followed by an extra 192 bytes which contain extra configuration data. Figure 54.7 shows the arrangement of the header. The definitions of the fields are:

- Unit ID and Man. ID. A Unit ID of FFFFh defines that there is no unit installed, while any other address defines its ID. The PCI SIG, which is the governing body for the PCI specification, allocates a Man. ID. This ID is normally shown at BIOS start-up. Section 54.5 gives some example Man. IDs (and Plug-and-play IDs).Ff
- Status and Command.
- Class code and Revision. The class code defines PCI device type. It splits into two 8-bit values with a further 8-bit value that defines the programming interface for the unit. The first defines the unit classification (00h for no class code, 01h for mass storage, 02h for network controllers, 03h for video controllers, 04h for multimedia units, 05h for memory controller and 06h for a bridge), followed by a subcode which defines the actual type. Typical codes are:

  - 0100h. SCSI controller.
  - 0101h. IDE controller.
  - 0102h. Floppy controller.
  - 0200h. Ethernet network adapter.
  - 0201h. Token ring network adapter.
  - 0202h. FDDI network adapter.
  - 0280h. Other network adapter.
  - 0300h. VGA video adapter.
  - 0301h. XGA video adapter.
  - 0380h. Other video adapter.
  - 0400h. Video multimedia device.
  - 0401h. Audio multimedia device.
  - 0480h. Other multimedia device.
  - 0500h. RAM memory controller.
  - 0501h. Flash memory controller.
  - 0580h. Other memory controller.
  - 0600h. Host.
  - 0601h. ISA Bridge.
  - 0602h. EISA Bridge.
  - 0603h. MAC Bridge.
  - 0604h. PCI-PCI Bridge.
  - 0680h. Other Bridge.

- Expansion ROM Base Address. Allows a ROM expansion to be placed at any position in the 32-bit memory address area.

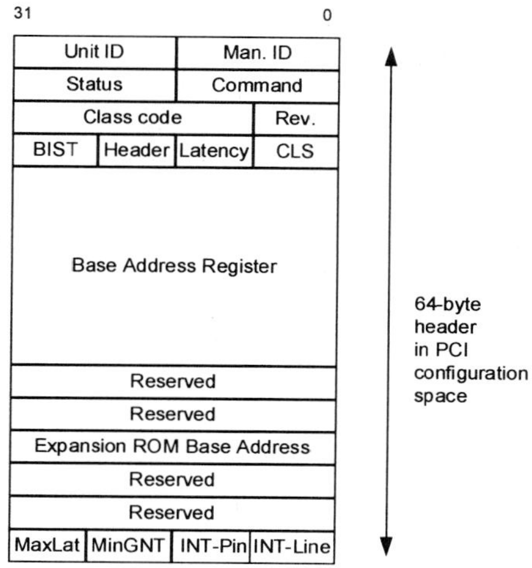

Figure 54.7 PCI configuration space.

- BIST, Header, Latency, CLS. The BIST (Built-in Self Test) is an 8-bit field, where the most significant bit defines if the device can carry out a BIST, the next bit defines if a BIST is to be performed (a 1 in this position indicates that it should be performed) and bits 3–0 define the status code after the BIST has been performed (a value of zero indicates no error). The Header field defines the layout of the 48 bytes after the standard 16-byte header. The most significant bit of the Header field defines whether the device is a multi-function device or not. A 1 defines a multi-function unit. The CLS (Cache Line Size) field defines the size of the cache in units of 32 bytes. Latency indicates the length of time for a PCI bus operation, where the amount of time is the latency+8 PCI clock cycles.
- Base Address Register. This area of memory allows the device to be programmed with an I/O or memory address area. It can contain a number of 32- or 64-bit addresses. The format of a memory address is:

Bit 64–4	Base address.
Bit 3	PRF. Prefetching, 0 identifies not possible, 1 identifies possible.
Bit 2, 1	Type. 00 – any 32-bit address, 01 – less than 1MB, 10 – any 64-bit address and 11 – reserved.
Bit 0	0. Always set to a 0 for a memory address.

For an I/O address space it is defined as:

Bit 31–2	Base address.
Bit 1, 0	01. Always set to a 01 for an I/O address.

- MaxLat, MinGNT, INT-Pin, INT-Line. The MinGNT and MaxLat registers are read-only registers that define the minimum and maximum latency values. The INT-Line field is a 4-bit field that defines the interrupt line used (IRQ0–IRQ15). A value of 0 corresponds to IRQ0 and a value of 15 corresponds to IRQ15. The PCI bridge can then redirect this interrupt to the correct IRQ line. The 4-bit INT-pin defines the interrupt line that the device is using. A value of 0 defines no interrupt line, 1 defines $\overline{\text{INTA}}$, 2 defines $\overline{\text{INTB}}$, and so on.

54.3.8 I/O addressing

The standard PC I/O addressing ranges from 0000h to FFFFh, which gives an addressable space of 64 KB, whereas the PCI bus can support a 32-bit or 64-bit addressable memory. The PCI device can be configured using one of two mechanisms.

Configuration mechanism 1

Passing two 32-bit values to two standard addresses configures the PCI bus:

Address	*Name*	*Description*
0CF8h	Configuration Address	Used to access the configuration address area.
0CFCh	Configuration Data	Used to read or write a 32-bit (double word) value to the configuration memory of the PCI device.

The format of the Configuration Address register is:

Bit 31 ECD (`Enable CONFIG_DATA`) bit. A 1 activates the `CONFIG_DATA` register, while a 0 disables it.

Bit 30–24 Reserved.

Bit 23–16 PCI bus number. Defines the number of the number of the PCI bus (to a maximum of 256).

Bit 15–11 PCI unit. Selects a PCI device (to a maximum of 32). PCI thus supports a maximum of 256 attached buses with a maximum of 32 devices on each bus.

Bit 10–8 PCI function. Selects a function within a PCI multi-function device (one of eight functions).

Bit 7–2 Register. Selects a Dword entry in a specified configuration address area (one of 64 Dwords).

Bit 1, 0 Type. 00 – decode unit, 01 – `CONFIG_ADDRESS` value copy to `AD`*x*.

Configuration mechanism 2

In this mode, each PCI device is mapped to a 4 KB I/O address range between C000h and CFFFh. This is achieved by used in the activation register CSE (Configuration Space Enable) for the configuration area at the port address 0CF8h. The format of the CSE register is located at 0CF8h and is defined as:

Bit 7–4 Key. 0000 – normal mode, 0001…1111 – configuration area activated. A value other than zero for the key activates the configuration area mapping, that is, all I/O addresses to the 4 KB range between C000h and CFFFh would be performed as normal I/O cycles.

Bit 3–1 Function. Defines the function number within the PCI device (if it represents a multi-function device).

Bit 0 SCE. 0 defines a configuration cycle, 1 defines a special cycle.

The forward register is stored at address 0CFAh and contains:

Bit 7–0 PCI bus.

The I/O address is defined by:

Bit 31–12 Contains the bit value of 0000Ch.
Bit 11–8 PCI unit.
Bit 7–2 Register index.
Bit 1, 0 Contains the bit value of 00b.

54.4 HX motherboard

54.4.1 Introduction

This chapter analyzes a Pentium-based motherboard. An example board is the Intel 430HX motherboard that supports most Pentium processors and has the following component parts:

- PCIset components. 82438 System Controller (TXC) and 82371SB PCI ISA Xcelerator (PIIX3).
- 82091AA (AIP) for serial and parallel ports, and floppy disk controller.
- DRAM main memory. These are arranged either as SIMMs or DIMMs.
- L-2 cache SRAM. Support for up to 256kB level-2 cache.
- Universal Serial Bus (USB).
- Interface slots (typically 4 PCI and 3 ISA).
- 1 Mbit flash RAM.

Figure 54.8 illustrates the main connections of the PCIset (which are the TXC and PIIX3 devices). The TXC allows for a host-to-PCI bridge, whereas the PIIX3 device supports:

- PCI-to-ISA bridge.
- Fast IDE.
- APIC (Advanced Programmable Interrupt Controller) support.
- USB host/hub controller. Connection to the Universal Serial Bus.
- Power management.

The 430HX board has 3 V and 5 V busses. PCI bus connections are 5 V and the Pentium bus is 3 V. An upgraded TX board includes the upgraded 82439 System Controller (MTXC) and the 82371AB PCI ISA Xcelerator (PIIX4).

54.4.2 Pentium processor

Figure 54.9 illustrates the main connections to the Pentium II processor (note that a # symbol after the signal name identifies an active low signal). It can be seen that it has:

- 64-bit data bus (`D0–D63`) which connects to the TXC (`HD0–HD63`).
- 32-bit address bus (`A0–A31`) which connects to the TXC (`HA0–HA31`).

- 8-byte address lines ($\overline{BE0}-\overline{BE7}$) to allow the processor to access from 1 to 8 bytes (64 bits) at a time, which connects to TXC ($\overline{HBE0}-\overline{HBE7}$).
- Read/write line ($W/\overline{R}$) which connects to TXC ($HW/\overline{R}$).
- Memory/IO ($HM/\overline{IO}$) which connects to TXC ($HM/\overline{IO}$).
- Data/control ($HD/\overline{C}$) which connects to TXC ($HD/\overline{C}$).

The host bus (the connections between the processor and the TXC) typically runs at 60/66 MHz and the PCI bus typically run at 30/33 MHz.

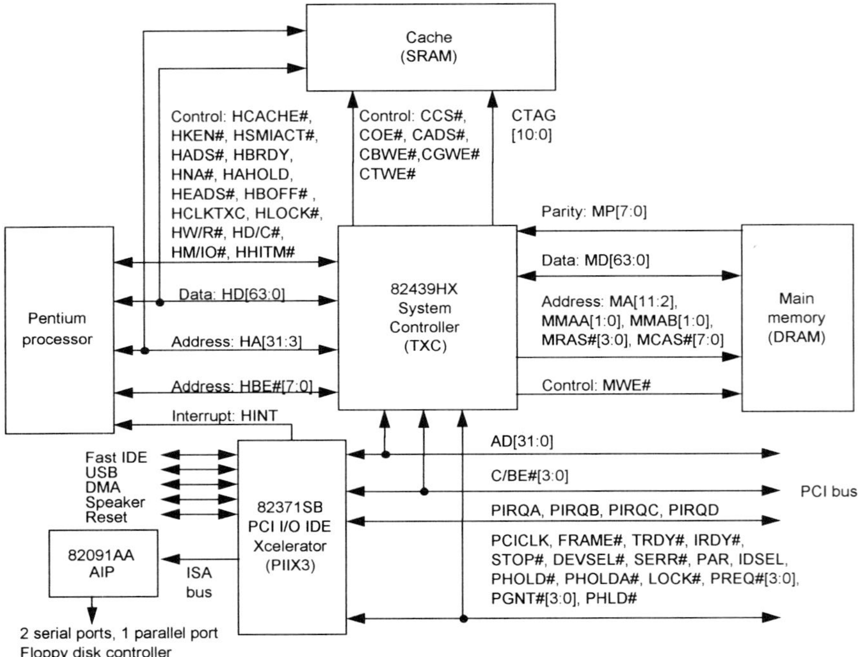

Figure 54.8 PCIset system architecture.

It can be seen that the interface between the processor and the TXC device has an address bus from `A3` to `A31`. These provide access to addresses in banks of 8 bytes. The byte enable lines ($\overline{HBE0}-\overline{HBE7}$) provide the lower address lines. Thus, 8 bits, 16 bits, 32 bits or 64 bits can be accessed at a time. For example, to address the 8 bytes from the following binary address:

0110 1110 0000 0110 0110 0110 0110 *XXXX*b

then the address:

6E00666*x*h

would be put on the address bus. Then to access the 16 bits from 6E06661h to 6E06662h then the $\overline{HBE1}$ and $\overline{HBE2}$ lines would be made active, all the other byte lines will be inactive.

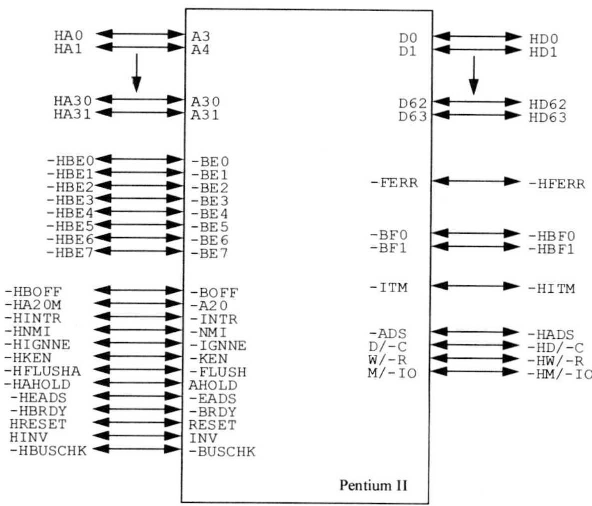

Figure 54.9 Pentium II connections.

It can be seen that the TXC device handles most of the processor signals. The PIIX3 device handles all the interrupts within the computer (IRQ0–IRQ15 and PCIRQA–PCIRQD) and connects directly to the interrupt line of the processor. This device also allows interrupts to be steered to unassigned interrupt and allow supports plug-and-play. The other direct connection from the processor to the PIIX3 device is for the math co-processor error interrupt.

54.4.3 82371SB PCI ISA Xcelerator (PIIX3)

The PIIX3 is a 208-pin QFP (Quad flat pack) IC which integrates much of the functionality of the ISA bus interface onto a single device. Table 54.3 outlines the main connections to the PIIX3 IC. The PIIX4 is a 324-pin device.

Table 54.3 PIIX3 connections.

PCI Address/ Data lines		**IRQ lines**		**ISA lines**		**ISA lines**	
Signal	*Pin*	*Signal*	*Pin*	*Signal*	*Pin*	*Signal*	*Pin*
AD0	206	IRQ1	4	BALE	64	SA8/DD0	55
AD1	205	IRQ3	58	AEN	20	SA9/DD1	50
AD2	204	IRQ4	56	LA17	86	SA10/DD2	49
AD3	203	IRQ5	34	LA18	84	SA11/DD3	48
AD4	202	IRQ6	33	LA19	82	SA12/DD4	47
AD5	201	IRQ7	32	LA20	80	SA13/DD5	46
AD6	200	-IRQ8	5	LA21	76	SA14/DD6	45
AD7	199	IRQ9	10	LA22	74	SA15/DD7	44
AD8	197	IRQ10	73	LA23	72	SA16/DD8	43
AD9	194	IRQ11	75	SA0	69	SA17/DD9	41
AD10	193	IRQ12/M	77	SA1	68	SA18/DD10	40
AD11	192	IRQ14	83	SA2	67	SA19/DD11	39
AD12	191	IRQ15	81	SA3	66	SA20/DD12	38
AD13	190			SA4	63	SA21/DD13	37
AD14	189			SA5	61	SA22/DD14	36
AD15	188			SA6	59	SA23/DD15	35
AD16	177			SA7	57	-OWS	15
AD17	176			DRQ0	87	-SMEMW	22

AD18	175	DRQ1	30	-SMEMR	19
AD19	174	DRQ2	12	-IOW	24
AD20	173	DRQ3	25	-IOR	23
AD21	172	DRQ5	91	-REFRESH	31
AD22	171	DRQ6	95	T/C	62
AD23	168	DRQ7	99	OSC	
AD24	166	-DACK0	85	-MEMCS16	70
AD25	165	-DACK1	29	-IOCS16	71
AD26	164	-DACK2	60	-MASTER	
AD27	163	-DACK3	21	IOCHK	6
AD28	162	-DACK5	89	IOCHRDY	18
AD29	161	-DACK6	93	-SBHE (DD12)	
AD30	160	-DACK7	97	-MEMR	88
AD31	159	RSTISA		-MEMW	90

USB

Signal	*Pin*	*Signal*	*Pin*	*Signal*	*Pin*
USBP1-	143	USBP1+	142	USBCLK	146
USBP0-	145	USBP0+	144		

PCI control lines

C/BE0#	198	FRAME#	179	PIRQA	149
C/BE1#	187	DEVSEL#	184	PIRQB	150
C/BE2#	178	IRDY#	180	PIRQC	151
C/BE3#	167	STOP#	185	PIRQD	152
		PHLDA#	110		
		SERR#	3		
		TRDY#	181		

PIIX3's functionality includes:

- Enhanced 7-channel DMA with two 8237 controllers. This is supported with the handshaking lines DRQ0- DRQ7 and $\overline{\text{DRQ0}}-\overline{\text{DRQ7}}$.
- ISA-PCI bridge.
- Fast IDE support for up to four disk drives (two masters and two slaves). It supports mode 4 timings which gives transfer rates of up to 22 MB/s.
- I/O APIC (Advanced Programmable Interrupt Controller) support.
- Implementation of PCI 2.1 which allows for PCI auto-configuration.
- Incorporates 82C54 timer for system timer, refresh request and speaker output tone.
- Non-maskable interrupts (NMI).
- PCI clock speed of 25/33 MHz. Motherboard configurable clock speed (normally 33 MHz).
- Plug-and-play support with one steerable interrupt line and one programmable chip select. The motherboard interrupt MIRQ0 can be steered to any one of 11 interrupts (IRQ3–IRQ7, IRQ9–IRQ12, IRQ14 and IRQ15).
- Steerable PCI interrupts for PCI device plug-and-play. The PCI interrupt lines (PIRQA–PIRQD) can be steered to one of 11 interrupts (IRQ3–IRQ7, IRQ9–IRQ12, IRQ14 and IRQ15).
- Support for PS/2-type mouse and serial port mouse. IRQ12/M can be enabled for the PS/2-type mouse or disabled for a serial port mouse.
- Support for 5 ISA slots. Typical applications for ISA include 10 Mbps Ethernet adapter cards, serial/parallel port cards, sound cards, and so on.
- System Power Management. Allows the system to operate in a low-power state without being powered-down. This can be triggered either by a software, hardware or external event. It uses the programmable $\overline{\text{SMI}}$ (system management interrupt) line.
- Math Co-processor error function. The $\overline{\text{FERR}}$ line goes active (LOW) when a math co-

processor error occurs. The PIIX3 device automatically generates an IRQ13 interrupt and sets the INTR line (HINT) to the processor. The PXII3 device then sets the $\overline{\text{IGNNE}}$ active and INTR inactive when there is a write to address F0h.

- Two 82C59 controllers with 14 interrupts. The interrupt lines IRQ1, IRQ3–IRQ15 are available (IRQ0 is used by the system time and IRQ2 by the cascaded interrupt line). When an interrupt occurs the PIIX3 uses the HINT line to interrupt the processor.
- Universal Serial Bus with root hub and two USB ports. With the USB the host controller transfers data between the system memory and USB devices. This is achieved by processing data structures set up to by the Host Controller Driver (HCD) software and generating the transaction on USB.

The PCI bus address lines (AD0–AD22) connect to the TXC IC and the available interrupt lines at IRQ1, IRQ2–IRQ12, IRQ14 and IRQ15 (IRQ0 is generated by the system timer and IRQ2 is the cascaded interrupt line). The PS/2-type mouse uses the IRQ12/M line.

54.4.4 82438 System Controller (TXC)

The 324-pin TXC BGA (ball grid array) provides an interface between the processor, DRAM and the external busses (such as the PCI, ISA, and so on). Table 54.4 outlines its main pin connections. The TXC's functionality includes:

- Supports 50 MHz, 60 MHz and 66 MHz host system bus.
- Integrated DRAM controller. Supports four CAS lines and eight RAS lines. The memory supports symmetrical and asymmetrical addressing for 1 MB, 2 MB and 4 MB-deep SIMMs and symmetrical addressing for 16 MB-deep SIMMs.
- Integrated second-level cache controller. Supports up to 512 KB of second-level cache with synchronous pipelined burst SRAM.
- Dual processor support.
- Optional parity with 1 parity bit for every 8 bits stored in the DRAM.
- Optional error checking and correction on DRAM. The ECC mode is software configurable and allows for single-bit error correction and multi-bit error detection on single nibbles in DRAM.
- Swappable memory bank support. This allows memory banks to be swapped-out.
- PCI 2.1 compliant bus.
- Supports USB.

The TXC controls the processor cycles for:

- Second-level cache transfer. The processor directly sends data to the second-level cache and the TXC controls its operation.
- All other processor cycles. The TXC directs all other processor cycles to their destination (DRAM, PCI or internal TXC configuration space).

54.4.5 Error detection and correction

Parity or error correction can be configured by software (parity is the default). The ECC mode provides single-error correction, double-error detection and detection of all errors in a single nibble for the DRAM memory.

Table 54.4 TXC connections.

PCI Address/ Data bus		**Processor Addresses bus**		**Processor Data bus**			
Signal	*Pin*	*Signal*	*Pin*	*Signal*	*Pin*	*Signal*	*Pin*
AD0	15			HD0	305	HD32	179
AD1	14			HD1	307	HD33	178
AD2	33			HD2	306	HD34	149
AD3	13	HA3	275	HD3	308	HD35	180
AD4	52	HA4	315	HD4	285	HD36	136
AD5	32	HA5	252	HD5	286	HD37	135
AD6	12	HA6	316	HD6	265	HD38	138
AD7	51	HA7	312	HD7	212	HD39	125
AD8	11	HA8	272	HD8	245	HD40	126
AD9	50	HA9	271	HD9	287	HD41	115
AD10	30	HA10	311	HD10	267	HD42	137
AD11	10	HA11	291	HD11	288	HD43	117
AD12	49	HA12	251	HD12	225	HD44	128
AD13	29	HA13	310	HD13	268	HD45	114
AD14	9	HA14	270	HD14	247	HD46	127
AD15	48	HA15	290	HD15	266	HD47	102
AD16	47	HA16	250	HD16	248	HD48	101
AD17	27	HA17	309	HD17	247	HD49	116
AD18	7	HA18	289	HD18	246	HD50	104
AD19	46	HA19	269	HD19	214	HD51	103
AD20	26	HA20	249	HD20	228	HD52	81
AD21	6	HA21	273	HD21	213	HD53	84
AD22	45	HA22	254	HD22	226	HD54	82
AD23	25	HA23	253	HD23	201	HD55	61
AD24	66	HA24	294	HD24	215	HD56	83
AD25	44	HA25	293	HD25	203	HD57	63
AD26	24	HA26	274	HD26	202	HD58	62
AD27	4	HA27	313	HD27	191	HD59	41
AD28	23	HA28	314	HD28	204	HD60	42
AD29	3	HA29	255	HD29	193	HD61	43
AD30	22	HA30	295	HD30	192	HD62	21
AD31	2	HA31	292	HD31	194	HD63	1
PCI control lines							
C/BE0#	21	FRAME#	86	PREQ0#	67	PGNT0#	68
C/BE1#	31	DEVSEL#	89	PREQ1#	69	PGNT1#	70
C/BE2#	8	IRDY#	88	PREQ2#	71	PGNT2#	72
C/BE3#	5	STOP#	91	PREQ3#	73	PGNT3#	74
		LOCK#	85				
		PHOLD#	64				
		PHLDA#	65				
		PAR	92				
		SERR#	93				

Cache Memory Tag		**DRAM Parity**		**DRAM Address lines**		**DRAM Data lines**			
Signal	*Pin*	*Signal*	*Pin*	*Signal*	*Pin*	*Signal*	*Pin*		
CTAG0	207	MP0	133			MD0	304	MD32	283
CTAG1	260	MP1	123			MD1	241	MD33	263
CTAG2	261	MP2	146	MA2	317	MD2	243	MD34	244
CTAG3	281	MP3	113	MA3	297	MD3	224	MD35	221
CTAG4	238	MP4	132	MA4	277	MD4	210	MD36	209
CTAG5	282	MP5	124	MA5	257	MD5	198	MD37	190
CTAG6	302	MP6	134	MA6	237	MD6	176	MD38	175
CTAG7	322	MP7	122	MA7	298	MD7	161	MD39	160
CTAG8	303			MA8	258	MD8	111	MD40	112
CTAG9	323			MA9	319	MD9	90	MD41	98
CTAG10	324			MA10	318	MD10	59	MD42	60
				MA11	278	MD11	58	MD43	20
				MAA0	276	MD12	38	MD44	77
				MAA1	236	MD13	36	MD45	56

MAB0	296	MD14	35	MD46	75
MAB1	256	MD15	53	MD47	54
MWE#	235	MD16	262	MD48	284
		MD17	264	MD49	242
		MD18	222	MD50	223
		MD19	208	MD51	211
		MD20	200	MD52	199
		MD21	189	MD53	188
		MD22	174	MD54	162
		MD23	148	MD55	147
		MD24	99	MD56	100
		MD25	97	MD57	79
		MD26	78	MD58	40
		MD27	19	MD59	39
		MD28	57	MD60	18
		MD29	17	MD61	37
		MD30	76	MD62	16
		MD31	34	MD63	55

Cache address lines

MRASR0#	121	MCASR0#	145	MCASR4#	130
MRASR1#	110	MCASR1#	159	MCASR5#	144
MRASR2#	109	MCASR2#	131	MCASR6#	120
MRASR3#	96	MCASR3#	173	MCASR7#	172

Cache control lines

CBWE#	321	COE#	259	CCS#	300	CADS#	299
CGWE#	320	CADV#	279	TWE#	280	GWE#	320
BWE	321						

54.4.6 PCI interface

The TXC supports up to four PCI busmasters and provides the interface between PCI and main memory. It can operate the PCI interface at 25 MHz, 30 MHz or 33 MHz. When used as a PCI master the PIIX3 runs cycles on behalf of DMA, ISA masters or a bus master IDE.

54.4.7 82091AA (AIP)

The AIP device integrates the serial ports, parallel ports and floppy disk interfaces. Figure 54.10 shows its connections and Figure 54.11 shows the interconnection between the AIP and the PIIX3 device. The OSC frequency is set to 14.21818 MHz. It can be seen that the range of interrupts for the serial, parallel and floppy disk drive is IRQ3, IRQ4, IRQ5, IRQ6 and IRQ7. Normally the settings are:

- IRQ3. Secondary serial port (COM2/COM4).
- IRQ4. Primary serial port (COM1/COM3).
- IRQ6. Floppy disk controller.
- IRQ7. Parallel port (LPT1).

Figure 54.11 shows the main connections between the TXC, PIIX3 and the AIP. It can be seen that the AIP uses many of the ISA connections (such as $\overline{\text{0WS}}$, IOCHRDY, and so on). The interface between the TXC and the PIIX3 defines the PCI bus and the interface between the PIIX3 and AIP defines some of the ISA signals. It can be seen that the AIP can support up to 8 DMA transfers (DRQ0–DRQ7).

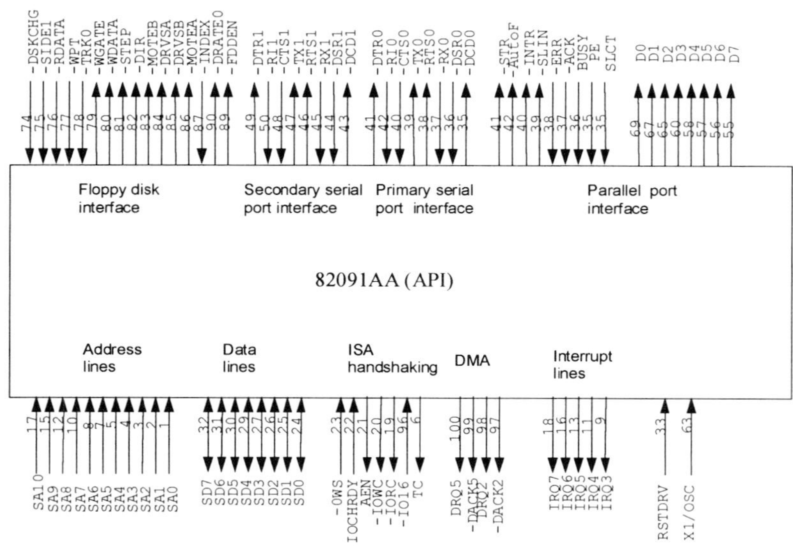

Figure 54.10 API IC.

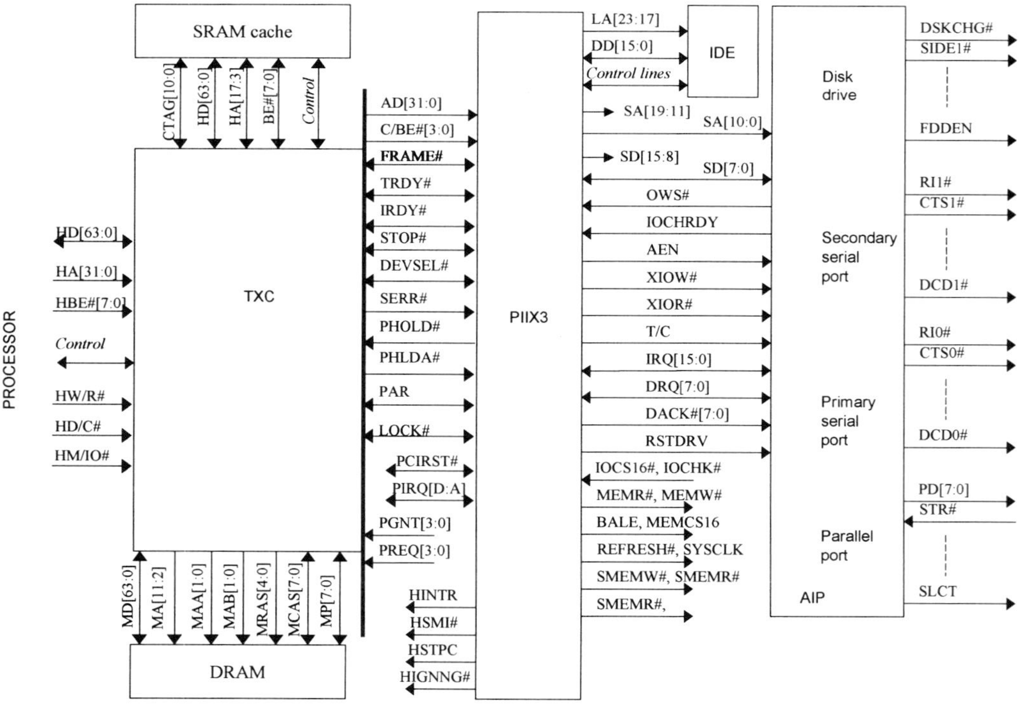

Figure 54.11 Connections between TXC, PIIX3 and AIP.

54.4.8 DRAM interface

The DRAM interface supports from 4 MB to 512 MB with eight row address lines ($\overline{\text{MRAS0}}$ – $\overline{\text{MRAS7}}$), eight column address lines ($\overline{\text{MCAS0}}$ – $\overline{\text{MCAS7}}$) and a 64-bit data path with 8 parity bits. It can use either a 3.3 V or 5 V power supply and both standard page mode and extended data out (EDO) memory are supported with a mixture of memory sizes for 1 MB, 2 MB and 4 MB-deep SIMMs and symmetrical addressing for 16 MB-deep SIMMs.

Each SIMM (single in-line memory module) has 12 input address lines and has a 32-bit data output. They are normally available with 72 pins (named tabs) on each side. These pins can read the same signal because they are shorted together on the board. For example tab 1 (pin 1) on side A is shorted to tab 1 on side B. Thus the 144 tabs only give 72 useable signal connections.

Figure 54.12 shows how the DRAM memory is organized. It shows banks 1 and 2 (and does not show banks 3 and 4). Each bank has two modules, such that modules 0 and 1 are in bank 1, modules 2 and 3 are in bank 2, and so on. The bank is selected with the $\overline{\text{MRAS}}$ lines; for example, bank 1 is selected with $\overline{\text{MRAS0}}$ and $\overline{\text{MRAS1}}$, bank 1 by $\overline{\text{MRAS2}}$ and $\overline{\text{MRAS3}}$, and so on. An even-numbered module gives the lower 32 bits (MD0–MD31) and the odd-numbered modules give the upper 32 bits (MD32–MD63). Each module also provides 4 parity bits (MP0–MP3 and MP4–MP7). Note that the MAA0 and MAA1, and MAB0 and MAB1 signals are the same.

DIMMs (dual in-line memory modules) have independent signal lines on each side of the module and are available with 72 (36 tabs on each side), 88 (44 tabs on each side), 144 (72 tabs on each side), 168 (84 tabs on each side) or 200 tabs (100 tabs on each side). They give greater reliability and density and are used in modern high performance PC servers.

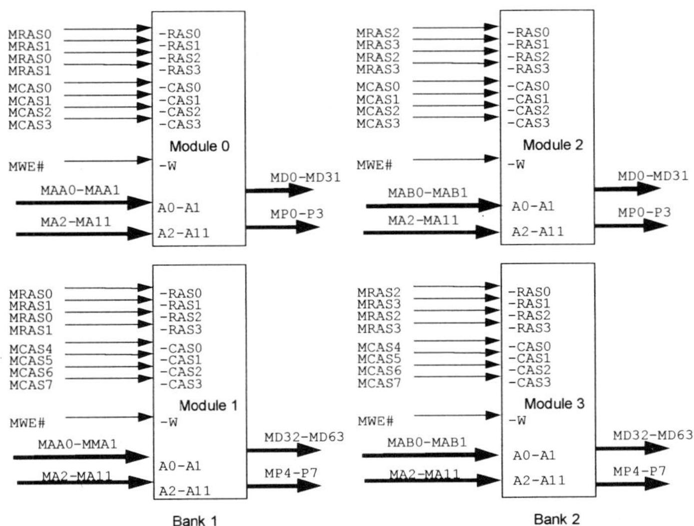

Figure 54.12 DRAM memory interface.

54.4.9 Clock rates

The system board runs at several clock frequencies. These are:

- Processor speed, such as 66 MHz.
- PCI bus speed.
- 24 or 48 MHz. USB.
- 12 MHz. Keyboard.
- 24 MHz. Floppy clock.
- 14 MHz. ISA bus OSC.
- 8 MHz. ISA bus clock

The ICS9159-02S IC provides for each of these clock speeds. The input to the device is a 14.31818 MHz crystal clock, as illustrated in Figure 54.13. Two jumpers (Jumper 1 and Jumper 2) set the system speed. These set the system clock speed to either 50 MHz, 60 MHz or 66 MHz.

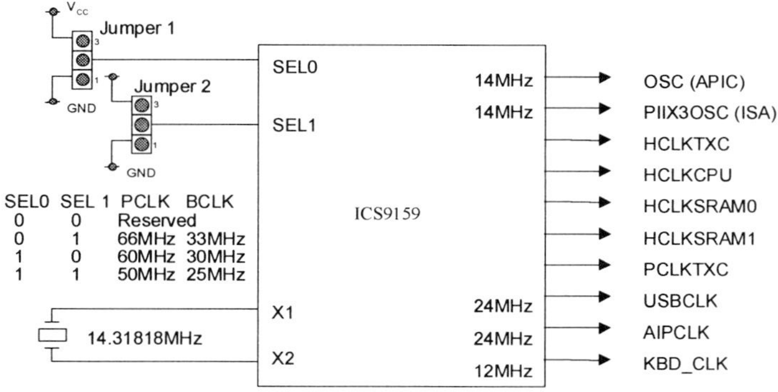

Figure 54.13 Clock generator device.

54.4.10 ISA/IDE interface

The IDE and ISA busses share several data, address and control lines. Figure 54.14 shows the connections to the busses. A multiplexor (MUX) is used to select either the ISA or IDE interface lines. The IDE interface uses the DD[12:0] and LA[23:17] lines, and the ISA uses these lines as $\overline{\text{SBHE}}$, SA[19:8], CS1S, CS3S, CS1P, CS3P and DA[2:0].

The IDE adapter is a 40-pin header connector. It is thus very easy to insert in the wrong way or to the wrong pins. For this reason all the input and output pins are short-circuit protected. The data lines connect to the IDE through 22 Ω resistors and are pulled-up with 4.7 KΩ resistors, and the address lines connect to the IDE through 33 Ω resistors.

54.4.11 DMA interface

The PIIX3 device incorporates the functionality of two 8237 DMA controllers to give seven independently programmable channel (Channels 0–3 and Channels 5–7). DMA channel 4 is used to cascade the two controllers and defaults to cascade mode in the DMA channel mode (DCM) register. Figure 54.15 shows the interface connections and that DMA channel 4 is used for the cascaded controller.

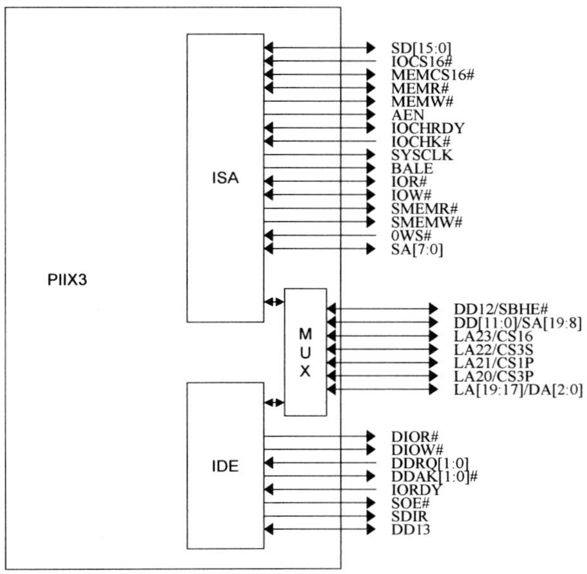

Figure 54.14 IDE/ISA interface with PIIX3.

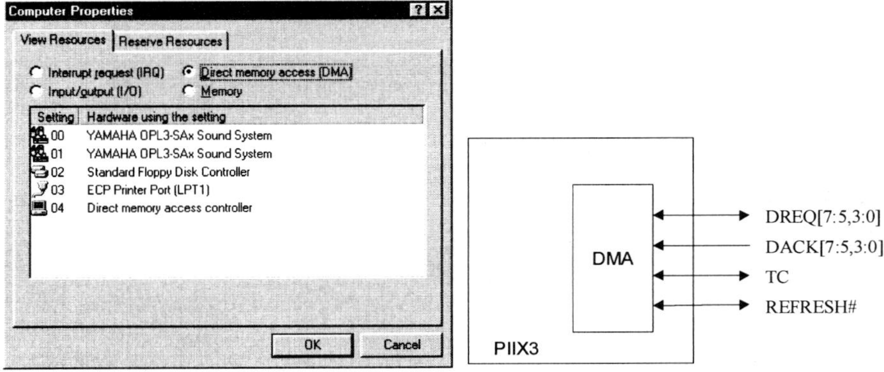

Figure 54.15 DMA interface.

54.4.12 Interval timer

The PIIX3 contains three 8251-compatible counters which are contained in one PIIX3 timer unit, referred to as Timer-1. The 14.21818 MHz counters normally use OSC as a clock source. Each counter provides an essential system function, such as:

- Counter 0, which connects to the IRQ0 line and provides a system timer interrupt for a time-of-day, diskette time-out, and so on.
- Counter 1, which generates a refresh request signal.
- Counter 2, which generates the speaker tone.

54.4.13 Interrupt controller

The PXII3 incorporates two 8259-compatible interrupt controllers and provides an ISA-compatible interrupt controller. These are cascaded to give 13 external and 3 internal interrupts. The primary interrupt controller connects to IRQ0–IRQ7 and the secondary connects to IRQ8–IRQ15. The three internal interrupts are:

- IRQ0. Used by the system timer and is connected to Timer 1, Counter 0.
- IRQ2. Used by the primary and secondary controller.
- IRQ13. Used by the math co-processor, which is connected to the $\overline{\text{FERR}}$ pin on the processor.

The interrupt unit also supports interrupt steering. PIIX3 also supports interrupt steering where the four PCI active low interrupts ($\overline{\text{PIRQA}}-\overline{\text{PIRQD}}$) can be internally routed to one of 11 interrupts (IRQ15, IRQ14, IRQ12–IRQ9, IRQ7–IRQ3).

54.4.14 Mouse function

The mouse normally either connects to one of the serial ports (COM1: or COM2:) or a PS/2-type connector. If they connect to the PS/2-type connector then IRQ12 is used (see Figure 54.16), else a serial port connected mouse uses the serial interrupts (such as IRQ4 for COM1 and IRQ3 for COM2). Thus, a system with a serial connected mouse must have the IRQ12/M interrupt disabled. This is typically done with a motherboard jumper (to enable or disable the mouse interrupt) or by BIOS steering.

54.4.15 Power management

PIIX3 has extensive power management capabilities and permits the system to operate in a low-power state without being powered-down. In a typical desktop PC, there are two states – Power On and Power Off. Leaving a system powered on, when not in use, wastes power. PIIX3 provides a fast on/off feature that creates a third state called Fast Off. When in the Fast Off state, the system consumes less power than the Power-On state.
The PIIX3's power management function is based on two modes:

- System Management Mode (SMM). Software (called SMM code) controls the transitions between the Power On state and the Fast Off state. PIIX3 invokes this software by generating an SMI to the CPU (asserting the $\overline{\text{SMI}}$ signal).
- Advanced Power Management (APM).

54.4.16 Universal serial bus

PIIX3 contains a USB. The host controller includes the root hub and two USB ports. This allows up to two USB peripheral devices to be directly connected to the PIIX3 without an external hub. If more devices are required, an external bus can be connected to either of the built-in ports. The USB's PCI configuration registers are located in function 2, PCI configuration space.

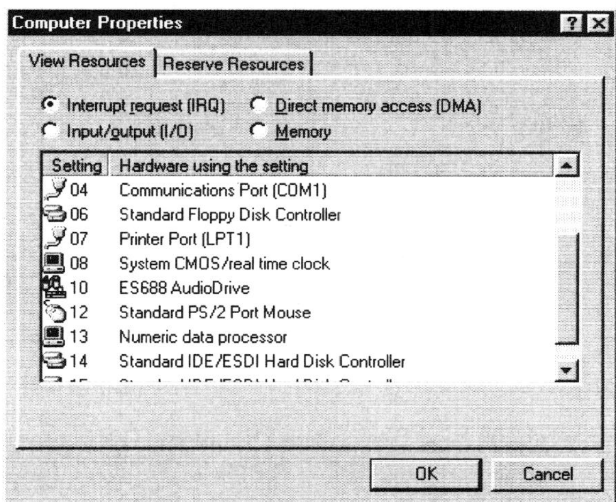

Figure 54.16 Interrupts usage showing PS/2 port mouse.

The PIIX3 host controller completely supports the standard Universal Host Controller Interface (UHCI) and thus, takes advantage of the standard software drivers written to be compatible with UHCI. Its advantages are:

- Automatic mapping of function to driver and configuration.
- Supports synchronous and asynchronous transfer types.
- Self-identifying peripherals that can be hot-plugged.
- Supports up to 127 devices.
- Supports error-handling and fault-recovery.
- Guaranteed bit rate with low delay times.

54.4.17 Mouse and keyboard interface

The mouse and keyboard interface use the 8242 device, as illustrated in Figure 54.17 It can be seen that the two interrupts which are available are `IRQ1` (the keyboard interrupt) and `IRQ12` (PS/2 style mouse). If the mouse connects to the serial port then the `IRQ12` line does not cause an interrupt. All clock frequencies are derived from the keyboard clock frequency (see Figure 54.17).

54.4.18 Example ATX motherboard

Figure 54.18 shows an example ATX motherboard. It supports the Pentium II though a Slot 1 SEC socket and is based on Intel 440LX chipset. It is similar to the HX motherboard (which has a Socket 7 processor connector) but has the following:

- DIMM connection for up to 384 MB for memory. Support for synchronous 100 MHz DRAM (SDRAM) for a 64/72-bit data path with autodetection for any combination of 4/16/64 MB DRAM modules.

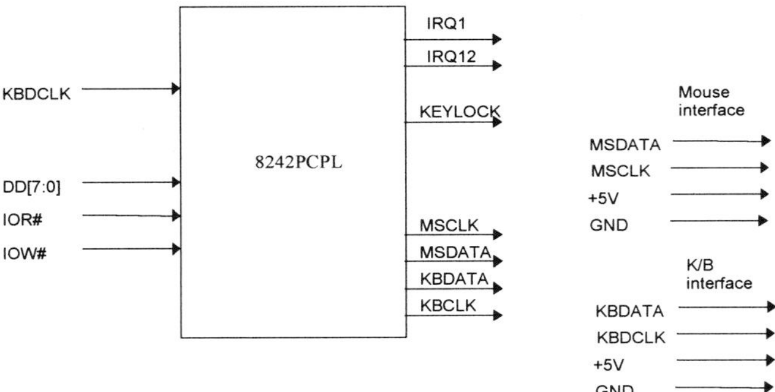

Figure 54.17 Mouse and keyboard interface.

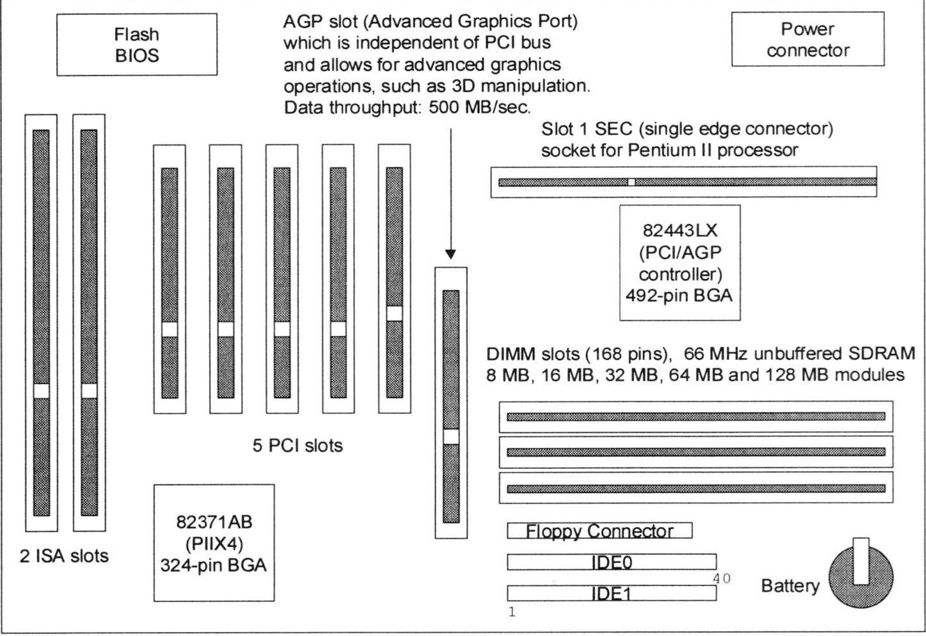

Figure 54.18 Typical LX motherboard.

- Support for Slot 1 SEC to inserting the processor into the motherboard. This allows for easy upgrade of the processor. Normally the processor is inserted into the motherboard so that the fan is near to its location. This overcomes the problem of mounting a fan on top of the Socket 7 based processor. The Pentium II also has an integrated heat sink and integrated Level-2 cache which is on the substrate for the SEC cartridge.
- Intel 82443LX PCI/AGP controller (PAC). This supports the PCI bus and the Advanced Graphics Port (AGP). The AGP transfers at 133 MHz and can achieve a maximum data rate of 500 MB/s. It is enhanced for graphics transfers and gives pipelined-memory read

and write operations that hides any memory access delays.

- 82371AB PCI/ISA IDE Xcellerator (PIIX4). Support for PCI bridge, USB controller, IDE controller, DMA controller, Interrupt controller, Power management and real-time clocks. The IDE interface supports Mode 3 and Mode 4 transfers at up to 16 MB/s and also Ultra DMA/33 which gives synchronous DMA transfers at up to 33 MB/s. It also supports ATAPI devices (such as CD-ROMs).
- National Semiconductor PC97307 Super I/O Controller. Supports 2 serial ports, a multi-mode parallel port (standard mode, EPP and ECP), floppy disk controller (DP8473 and N82077 compatible), keyboard controller, mouse controller and Infra-red communications controller. The serial ports contain two 16450/166550A-compatiable UARTs which have an integrated 16-byte FIFO buffer for storing incoming and outgoing characters. The floppy disk controller also has a 16-byte FIFO buffer.
- Support for LS-120MB floppy disk drives and Desktop Management Interface (DMI). DMI is a management system for networked PC, where a system manager can control the settings remotely.

54.5 Example manufacturer and plug-and-play IDs

Manufacturer	*Man. ID*	*PNP ID*	*Manufacturer*	*Man. ID*	*PnP ID*
NCR	1000	4096	Toshiba	102f	4143
Motorola	1057	4183	Compaq	1032	4146
Mitsubishi	1067	4199	HP	103c	4156
EPSON	1008	4104	Intel	8086	32902
Yamaha	1073	4211	Adaptec	9004	36868
Cyrix	1078	4216	Matsushita	10f7	4343
Tseng Labs	100c	4108	Creative	10f6	4342

55 IDE and Mass Storage

55.1 Introduction

This chapter and the next chapter discuss IDE and SCSI interfaces which are used to interface to disk drives and mass storage devices. Disks are used to store data reliably in the long term. Typical disk drives either store binary information as magnetic fields on a fixed disk (as in a hard disk drive), a plastic disk (as in a floppy disk or tape drive), or as optical representation (on optical disks).

The main sources of permanent read/writeable storage are:

- Magnetic tape – where the digital bits are stored with varying magnetic fields. Typical devices are tape cartridges, DAT and 8 mm video tape.
- Magnetic disk – as with the magnetic tape the bits are stored as varying magnetic fields on a magnetic disk. This disk can be either permanent (such as a Winchester hard disk) or flexible (such as a floppy disk). Large capacity hard disks allow storage of several GBs of data. Normally fixed disks are designed to a much higher specification than floppy disks and can thus store much more information.
- Optical disk – where the digital bits are stored as pits on an optical disk. A laser then reads these bits. This information can either be read-only (CD-ROM), write-once read many (WORM) or can be reprogrammable. A standard CD-ROM stores up to 650 MB of data. Their main disadvantage in the past has been their relative slowness as compared with Winchester hard disks; this is now much less of a problem as speeds have steadily increased over the years.

55.2 Tracks and sectors

A disk must be formatted before it is used, which allows data to be stored in a logical manner. The format of the disk is defined by a series of tracks and sectors on either one or two sides. A track is a concentric circle around the disk where the outermost track is track 0. The next track is track 1 and so on, as shown in Figure 55.1. Each of these tracks is divided into a number of sectors. The first sector is named sector 1, the second is sector 2, and so on. Most disks also have two sides: the first side of the disk is called side 0 and the other is side 1.

Figure 55.1 also shows how each track is split into a number of sectors, in this case there are eight sectors per track. Typically, each sector stores 512 bytes. The total disk space, in bytes, will thus be given by:

$$\text{Disk space} = \text{No. of sides} \times \text{tracks} \times \text{sectors per track} \times \text{bytes per sector}$$

For example, a typical floppy disk has two sides, 80 tracks per side, 18 sectors per track and 512 bytes per sector, so:

Disk capacity	$= 2 \times 80 \times 18 \times 512$	= 1 474 560 B
	= 1 474 560/1 024 KB	= 1 440 KB
	= 1440/1024 MB	= 1.4 MB

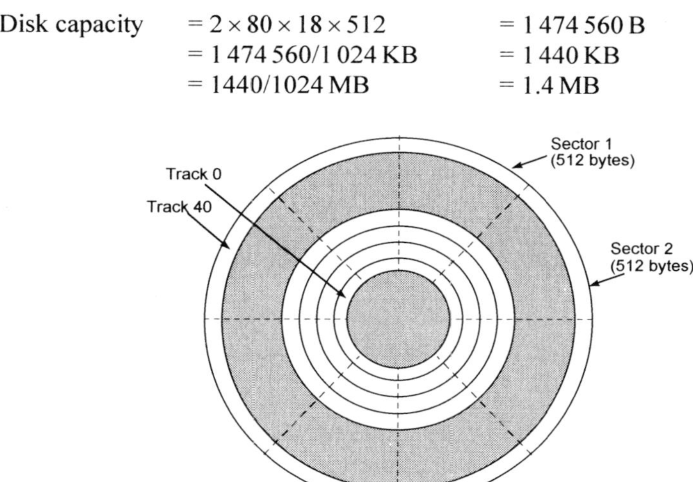

Figure 55.1 Tracks and sectors on a disk.

55.3 Floppy disks

A 3½-inch DD (double density) disk can be formatted with 2 sides, 9 sectors per track and 40 tracks per side. This gives a total capacity of 720 KB. A 3½-inch HD (high density) disk has a maximum capacity when formatted with 80 tracks per side.

A 5¼-inch DD disk can be formatted with two sides, nine sectors per disk with either 40 or 80 tracks per side. The maximum capacity of these formats is 360 KB (40 tracks) or 720 KB (80 tracks). A 5¼-inch HD disk can be formatted with 15 sectors per track which gives a total capacity of 1.2 MB. When reading data the disks rotate at 300 rpm. Table 55.1 outlines the differing formats.

Table 55.1 Capacity of different disk types

Size	*Tracks per side*	*Sectors per track*	*Capacity*
5¼"	40	9	360 KB
5¼"	80	15	1.2 MB
3½"	40	9	720 KB
3½"	80	18	1.44 MB

55.4 Fixed disks

Fixed disks store large amounts of data and vary in their capacity, from several MB to several

GB. A fixed disk (or hard disk) consists of one or more platters which spin at around 3 000 rpm (10 times faster than a floppy disk). A hard disk with four platters is shown in Figure 55.2. Data is read from the disk by a flying head, which sits just above the surface of the platter. This head does not actually touch the surface as the disk is spinning so fast. The distance between the platter and the head is only about 10 μin (which is no larger than the thickness of a human hair or a smoke particle). It must thus be protected from any outer particles by sealing it in an airtight container. A floppy disk is prone to wear as the head touches the disk as it reads but a fixed disk has no wear as its heads never touch the disk.

One problem with a fixed disk is head crashes, typically caused when the power is abruptly interrupted or if the disk drive is jolted. This can cause the head to crash into the disk surface. In most modern disk drives the head is automatically parked when the power is taken away. Older disk drives that do not have automatic head parking require a program to park the heads before the drive is powered down.

There are two sides to each platter and, like floppy disks, each side divides into a number of tracks which are subdivided into sectors. A number of tracks on fixed disks are usually named cylinders. For example, a 40 MB hard disk has two platters with 306 cylinders, four tracks per cylinder, 17 sectors per track and 512 bytes per sector, thus each side of a platter stores:

$$
\begin{aligned}
306 \times 4 \times 17 \times 512\,\text{B} &= 10\,653\,696\,\text{B} \\
&= 10\,653\,696/\,1\,048\,576\,\text{MB} \\
&= 10.2\,\text{MB}
\end{aligned}
$$

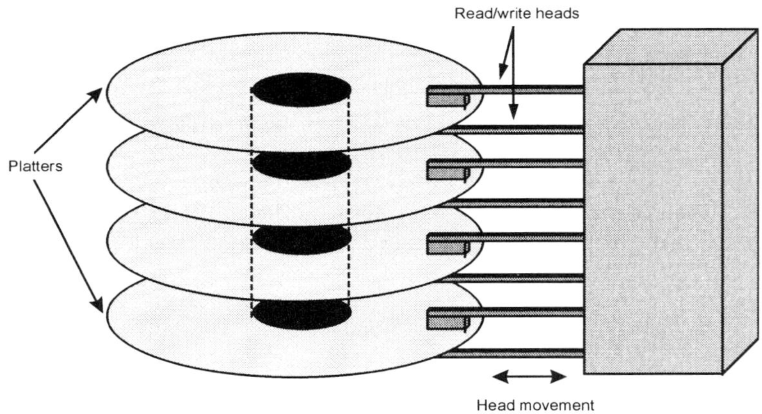

Figure 55.2 Hard disk with four platters.

55.5 Drive specifications

Access time is the time taken for a disk to locate data. Typical access times for modern disk drives range from 10 to 30 ms. The average access time is the time for the head to travel half way across the platters. Once the head has located the correct sector then there may be another wait until it locates the start of the data within the sector. If it is positioned at a point after the start of the data, it requires another rotation of the disk to locate the data. This average wait, or latency time, is usually taken as half of a revolution of the disk. If the disk spins at 3600 rpm then the latency is 8.33 ms.

The main parameters that affect the drive specification are the data transfer rate and the average access time. The transfer rate is dependent upon the interface for the controller/disk drive and system/controller and the access time is dependent upon the disk design.

55.6 Hard disk/CD-ROM interfaces

There are two main interfaces involved with hard disks (and CD-ROMs). One connects the disk controller to the system (system/controller interface) and the other connects the disk controller to the disk drive (disk/controller interface).

The controller can be interfaced by standards such as ISA, EISA, MCA, VL-local bus or PCI bus. For the interface between the disk drive and the controller then standards such as ST-506, ESDI, SCSI or IDE can be used. Seagate Technologies developed ST-506 and is used in many older machines with hard disks of a capacity less than 40 MB. The enhanced small disk interface (ESDI) is capable of transferring data between itself and the processor at rates approaching 10 MB/s.

The small computer system interface (SCSI) allows up to seven different disk drives or other interfaces to be connected to the system through the same interface controller. SCSI is a common interface for large capacity disk drives and is illustrated in Figure 55.3.

The most popular type of PC disk interface is the integrated drive electronics (IDE) standard. It has the advantage of incorporating the disk controller in the disk drive, and attaches directly to the motherboard through an interface cable. This cable allows many disk drives to be connected to a system without worrying about bus or controller conflicts. The IDE interface is also capable of driving other I/O devices besides a hard disk. It also normally contains at least 32 KB of disk cache memory. Common access times for an IDE are often less than 16 ms, where access times for a floppy disk is about 200 ms. With a good disk cache system the access time can reduce to less than 1 ms. A comparison of the maximum data rates is given in Table 55.2.

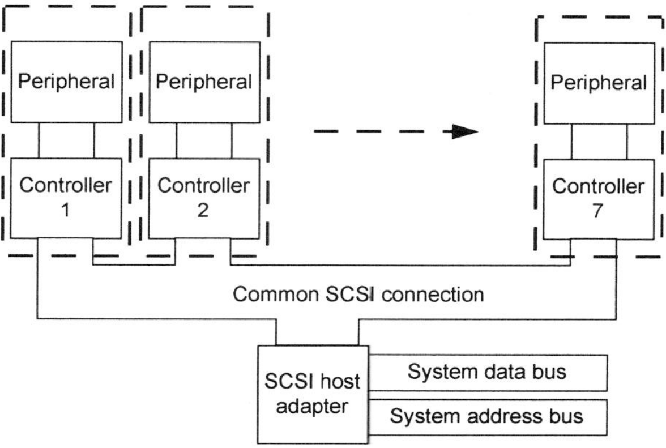

Figure 55.3 SCSI interface.

Table 55.2 Capacity of different disk types.

Interface	*Maximum data rate*
ST-506	0.6 MB/s
ESDI	1.25 MB/s
IDE	8.3 MB/s
E-IDE	16.6 MB/s
SCSI	4 MB/s
SCSI-II	10 MB/s

A typical modern PC contains two IDE connections on the motherboard, named IDE0 and IDE1. The IDE0 connection connects to the master drive (C:) and IDE1 to the slave drive (D:). These could connect either to two hard disks or, possibility, to one hard disk and a CD-ROM drive (or even a tape backup system). Unfortunately, the IDE standard only allows disk access up to 528 MB. A new standard called Enhanced-IDE (E-IDE) allows for disk capacities of over this limit. The connector used is the same as IDE but the computers' BIOS must be able to recognize the new standard. Most computers manufactured since 1993 are able to fully access E-IDE disk drives.

The specification for the IDE and E-IDE are:

- IDE.
  - Maximum of two devices (hard disks).
  - Maximum capacity for each disk of 528 MB.
  - Maximum cable length of 18 inches.
  - Data transfer rates of 3.3, 5.2 and 8.3 MB/s.
- E-IDE.
  - Maximum of four devices (hard disks, CD-ROM and tape).
  - Uses two ports (for master and slave).
  - Maximum capacity for each disk is 8.4 GB.
  - Maximum cable length of 18 inches.
  - Data transfer rates of 3.3, 5.2, 8.3, 11.1 and 16.6 MB/s.

55.7 IDE interface

The most popular interface for hard disk drives is the Integrated Drive Electronics (IDE) interface. Its main advantage is that the hard disk controller is built into the disk drive and the interface to the motherboard simply consists of a stripped-down version of the ISA bus. The most common standard is the ANSI-defined ATA-IDE standard. It uses a 40-way ribbon cable to connect to 40-pin header connectors. Table 55.3 lists the pin connections. It has a 16-bit data bus (D0–D15) and the only available interrupt line used is IRQ14 (the hard disk uses IRQ14).

The standard allows for the connection of two disk drives in a daisy chain configuration. This can cause problems because both drives have controllers within their drives. The primary drive (Drive 0) is assigned as the master and the secondary driver (Drive 1) as the slave. Setting jumpers on the disk drive sets a drive as a master or a slave. They can also be set by software using the Cable Select (CSEL) pin on the interface.

E-IDE has various modes (ANSI modes) of operation, these are:

- Mode 0. 600 ns read/write cycle time. 3.3 MB/s burst data transfer rate.
- Mode 1. 383 ns read/write cycle time. 5.2 MB/s burst data transfer rate.
- Mode 2. 240 ns read/write cycle time. 8.3 MB/s burst data transfer rate.
- Mode 3. 180 ns read/write cycle time. 11.1 MB/s burst data transfer rate.
- Mode 4. 120 ns read/write cycle time. 16.6 MB/s burst data transfer rate.

Table 55.3 IDE connections.

Pin	*IDE signal*	*AT signal*	*Pin*	*IDE signal*	*AT signal*
1	RESET	RESET DRV	2	GND	–
3	D7	SD7	4	D8	SD8
5	D6	SD6	6	D9	SD9
7	D5	SD5	8	D10	SD10
9	D4	SD4	10	D11	SD11
11	D3	SD3	12	D12	SD12
13	D2	SD2	14	D13	SD13
15	D1	SD1	16	D14	SD14
17	D0	SD0	18	D15	SD15
19	GND	–	20	KEY	–
21	DRQ3	DRQ3	22	GND	–
23	$\overline{\text{IOW}}$	$\overline{\text{IOW}}$	24	GND	–
25	$\overline{\text{IOR}}$	$\overline{\text{IOR}}$	26	GND	–
27	IOCHRDY	IOCHRDY	28	CSEL	–
29	$\overline{\text{DACK3}}$	$\overline{\text{DACK3}}$	30	GND	–
31	IRQ14	IRQ14	32	$\overline{\text{IOCS16}}$	$\overline{\text{IOCS16}}$
33	Address bit 1	SA1	34	$\overline{\text{PDIAG}}$	–
35	Address bit 0	SA0	36	Address bit 2	SA2
37	$\overline{\text{CS1FX}}$	–	38	$\overline{\text{CS3FX}}$	–
39	SP / $\overline{\text{DA}}$	–	40	GND	–

55.8 IDE communication

The IDE (or AT bus) is the *de facto* standard for most hard disks in PCs. It has the advantage over older type interfaces that the controller is integrated into the disk drive. Thus the computer only has to pass high-level commands to the unit and the actual control can be achieved with the integrated controller. Several companies developed a standard command set for an ATA (AT attachment). Commands included:

- Read sector buffer. Reads contents of the controller's sector buffer.
- Write sector buffer. Writes data to the controller's sector buffer.
- Check for active.
- Read multiple sectors.
- Write multiple sectors.
- Lock drive door.

The control of the disk is achieved by passing a number of high-level commands through a number of I/O port registers. Table 55.3 outlines the pin connections for the IDE connector. Typically pin 20 is missing on the connector cable so that it cannot be inserted in the wrong way, although most systems buffer the signals so that the bus will not be damaged if the cable is inserted in the wrong way. The five control signals, which are unique to the IDE interface (and not the AT bus), are:

- $\overline{\text{CS3FX}}$, $\overline{\text{CS1FX}}$. These are used to identify either the master or the slave.
- $\overline{\text{PDIAG}}$ (Passed diagnostic). Used by the slave drive to indicate that it has passed its diagnostic test.
- SP / $\overline{\text{DA}}$ (Slave present/drive active). Used by the slave drive to indicate that it is present and active.

The other signals are:

- IOCHRDY. This signal is optional and is used by the drive to tell the processor that it requires extra clock cycles for the current I/O transfer. A high level informs the processor that it is ready, while a low informs it that it need more time.
- DRQ3, $\overline{\text{DACK3}}$. These are used for DMA transfers.

55.8.1 AT task file

The processor communicates with the IDE controller through data and control registers (typically known as the AT task file). The base registers used are between 1F0h and 1F7h for the primary disk (170h and 177h for secondary), and 3F6h (376h for secondary), as shown in Figure 55.4. Their function is:

Port	*Function*	*Bits*	*Direction*
1F0h	Data register	16	R/W
1F1h	Error register	8	R
	Precompensation	8	W
1F2h	Sector count	8	R/W
1F3h	Sector number	8	R/W
1F4h	Cylinder LSB	8	R/W
1F5h	Cylinder MSB	8	R/W
1F6h	Drive/head	8	R/W
1F7h	Status register	8	R
	Command register	8	W
3F6h	Alternative status reg.	8	R
	Digital output reg.	8	W
3F7h	Drive address	8	R

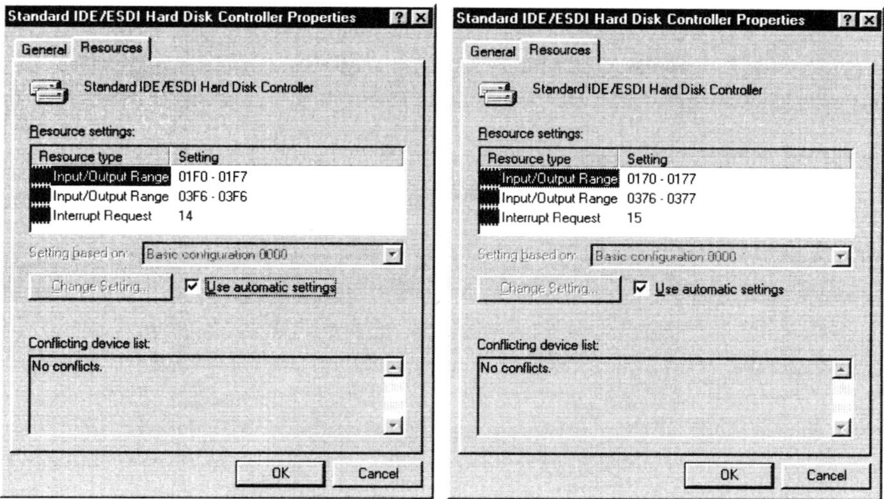

Figure 55.4 Typical hard-disk controller settings for the primary and secondary drive.

Data register (1F0h)

The data register is a 16-bit register that is used to read/write data from/to the disk.

Error register (1F1h)

The error register is read-only and contains error information relating to the last command. Its definitions are:

b_7	b_6	b_5	b_4	b_3	b_2	b_1	b_0
BBK	UNC	MC	NID	MCR	ABT	NT0	NDM

where:

- BBK. Set to 1 if the sector is bad.
- UNC. Set to 1 if there is an unrecoverable error.
- NID. Set to 1 if mark not found.
- ABT. Set to 1 if command aborted.
- NT0. Set to 1 if track 0 not found.
- MC. Set to 1 identifies that the medium has changed (E-IDE only). The E-IDE standard support disks which can be changed while the system is running (such as CD-ROMs, tape drives, and so on).
- MCR. Set to 1 identifies that the medium requires to be changed (E-IDE only).

Sector count register (1F2h)

This is a read/write 8-bit register that defines the number of sectors to be read, written or verified. Each transfer to/from the disk causes the register value to be decremented by one.

Sector number register (1F3h)

This is a read/write 8-bit register that defines the start sector to be read, written or verified. After each transfer to/from the disk, the register contains the last processed sector.

Cylinder register (1F4h/1F5h)

These are read/write 8-bit registers which define the LSB (1F4h) and MSB (1F5h) of the cylinder number. The two registers are capable of containing a 16-bit value. In standard IDE the cylinder number is 10-bit and can only vary from 0 to 1023 (0 to 2^{10}–1). For E-IDE the value can be a 16-bit value and can thus vary from 0 to 65 535 (0 to 2^{16}–1). This is one of the main reasons that E-IDE can address much more data than IDE. For example:

Drive/head register (1F6h)

This is a read/write 8-bit register that defines the currently used head. Its definitions are:

b_7	b_6	b_5	b_4	b_3	b_2	b_1	b_0
1	L	1	DRV	HD_3	HD_2	HD_1	HD_0

where:

- L. Set to a 1 if LBA (logical block addressing) mode else set to a 0 if CHS (E-IDE only).
- DRV. Set to 1 for the slave, else it is master.
- HD_3–HD_0. Identifies head number, where 0000 identifies head 0, 0001 identifies head 1, and so on.

Status register (1F7h)

The 1F7h register has two modes. If it is written-to then it is a command register (see next section), else if it is read-from then it is a status register. The status register is a read-only 8-bit register that contains status information from the previously issued command. Its definitions are:

b_7	b_6	b_5	b_4	b_3	b_2	b_1	b_0
BUSY	RDY	WFT	SKT	DRQ	COR	IDX	ERR

where:

- BUSY. Set to 1 if the drive is busy.
- RDY. Set to 1 if the drive is ready.
- WFT. Set to 1 if there is a write fault.
- SKT. Set to 1 if head seek positioning is complete.
- DRQ. Set to 1 if data can be transferred.
- COR. Set to 1 if there is a correctable data error.
- IDX. Set to 1 identifies that the disk index has just passed.
- ERR. Set to 1 identifies that the error register contains error information.

Command register (1F7h)

If the 1F7h register is written-to then it is a command register. The command register is an 8-bit register that can contain commands, such as:

Command	b_7	b_6	b_5	b_4	b_3	b_2	b_1	b_0	*Related registers*
Calibrate drive	0	0	0	1	–	–	–	–	1F6h
Read sector	0	0	1	0	–	–	L	R	1F2h–1F6h
Write sector	0	0	1	1	–	–	L	R	1F2h–1F6h
Verify sector	0	1	0	0	–	–	–	R	1F2h–1F6h
Format track	0	1	0	1	–	–	–	–	1F3h–1F6h
Seek	0	1	1	1	–	–	–	–	1F4h–1F6h
Diagnostics	1	0	0	1	–	–	–	–	1F2h,1F6h
Read sector buffer	1	1	1	0	0	1	0	0	1F6h
Write sector buffer	1	1	1	0	1	0	0	0	1F6h
Identify drive	1	1	1	0	1	1	–	–	1F6h

where R is the set to a 0 if the command is automatically retried and L identifies the long-bit.

Digital output register (3F6h)

This is a write-only 8-bit register that allows drives to be reset and also IRQ14 to be masked. Its definitions are:

b_7	b_6	b_5	b_4	b_3	b_2	b_1	b_0
–	–	–	–	–	SRST	$\overline{\text{IEN}}$	–

where:

- SRST. Set to a 1 to reset all connected drives, else accept the command.
- $\overline{\text{IEN}}$. Controls the interrupt enable. If set to 1 then IRQ14 is always masked, else interrupt after each command.

Drive address register (3F7h)

The drive address register is a read-only register that contains information on the active drive and drive head. Its definitions are:

b_7	b_6	b_5	b_4	b_3	b_2	b_1	b_0
–	$\overline{\text{WTGT}}$	$\overline{\text{HS3}}$	$\overline{\text{HS2}}$	$\overline{\text{HS1}}$	$\overline{\text{HS0}}$	$\overline{\text{DS1}}$	$\overline{\text{DS0}}$

where:

- $\overline{\text{WTGT}}$. Set to a 1 if the write gate is closed, else the write gate is open.
- $\overline{\text{HS3}} - \overline{\text{HS0}}$. 1s complement value of currently active head.
- $\overline{\text{DS1}} - \overline{\text{DS0}}$. Identifies the selected drive.

55.8.2 Command phase

The IRQ14 line is used by the disk to when it want to interrupt the processor, either when it wants to read or write data to/from memory. For example, using Microsoft C++ (for Borland replace `_outp()` and `_inp()` with `outport()` and `inportb()`) to write to a disk at cylinder 150, head 0 and sector 7:

```
#include <conio.h>

int main(void)
{
int          sectors=4, sector_no=7, cylinder=150, drive=0, command=0x33, i;
unsigned     int buff[1024], *buff_pointer;

      do
      {
         /* wait until BSY signal is set to a 1 */

      } while (( _inp(0x1f7) & 0x80) != 0x80);

      _outp(0x1f2,sectors);              /* set number of sectors          */
      _outp(0x1f3,sector_no);            /* set sector number              */
      _outp(0x1f4,cylinder & 0x0ff);     /* set cylinder number LSB        */
      _outp(0x1f5,cylinder & 0xf00);     /* set cylinder number MSB        */
      _outp(0x1f6,drive);                /* set DRV=0 and head=0           */
      _outp(0x1f7,command);              /* 0011 0011 (write sector)       */

      do
      {
         /* wait until BSY signal is set to a 1 and DRQ is set to a 1 */

      } while ( ((_inp(0x1f7) & 0x80) != 0x80) && ((_inp(0x1f7) & 0x08) !=0x08) );

      buff_pointer= buff;

      for (i=0;i<512;i++,buff_pointer++)
      {
         _outp(0x1f0,*buff_pointer); /* output 16-bits at a time */
      }
      return(0);
}
```

Note that if the L bit is set then an extra 4 ECC (error correcting code) bytes must be written to the sector (thus a total of 516 bytes are written to each sector). The code used is cyclic redundancy check which, while it cannot correct errors, is very powerful at detecting them.

55.8.3 E-IDE

The main differences between IDE and E-IDE are:

- E-IDE supports removable media.
- E-IDE supports a 16-bit cylinder value, which gives a maximum of 65 636 cylinders.
- Higher transfer rates. In mode 4, E-IDE has a 120 ns read/write cycle time, which gives a 16.6 MB/s burst data transfer rate.
- E-IDE supports LBA (logical block addressing) which differs from CHS (cylinder head sector) in that the disk drive appears to be a continuous stream of sequential blocks. The addressing of these blocks is achieved from within the controller and the system does not have to bother about which cylinder, header and sector is being used.

IDE is limited to 1024 cylinders, 16 heads (the drive/head register has only 4 bits for the number of heads) and 63 sectors, which gives:

$$\text{Disk capacity} = 1024 \times 16 \times 63 \times 512 \quad = 504\,\text{MB}$$

With enhanced BIOS this is increased to 1024 cylinders, 256 heads (8-bit definition for the number of heads) and 63 sectors, to give:

$$\text{Disk capacity} = 1024 \times 256 \times 63 \times 512 \ = 7.88\,\text{GB}$$

With E-IDE the maximum possible is 65 536 cylinders, 256 heads and 63 sectors, to give:

$$\text{Disk capacity} = 65536 \times 256 \times 63 \times 512 \ = 128\,\text{GB}$$

Normally a 3½-inch hard disk would be limited around two platters, with four heads. Thus, the capacity is around 8.1 GB.

55.9 Optical storage

Optical storage devices can store extremely large amounts of digital data. They use a laser beam which reflects from an optical disk. If a pit exists in the disk then the laser beam does not reflect back. Figure 55.5 shows the basic mechanism for reading from optical disks. A focusing lens directs the laser light to an objective lens that focuses the light onto a small area on the disk. If a pit exists then the light does not reflect back from the disk. If the pit does not exist then it is reflected and directed through the objective lens and a quarter-wave plate to the polarized prism. The quarter-wave polarizes the light by 45° and thus the reflected light will have a polarization by 90°, with respect to the original incident light in the prism. The polarized prism then directs this polarized light to the sensor.

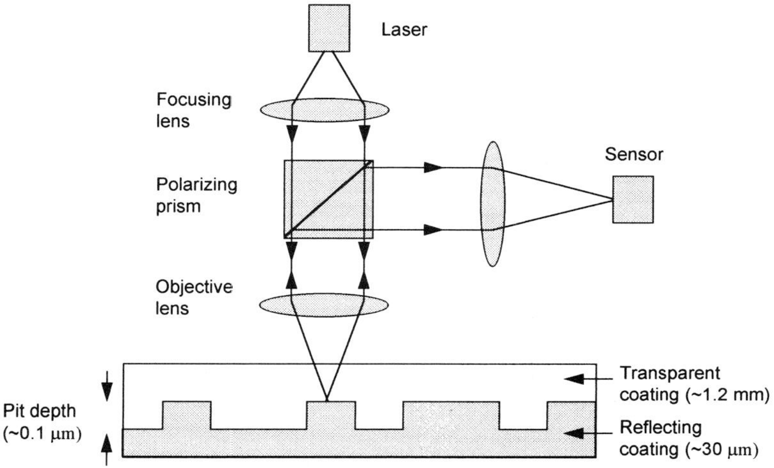

Figure 55.5 Reading from an optical disk.

55.9.1 CD-ROM

In a permanent disk (also known as compact disk, or CD) the pits are set up by pressing then onto the disk at production. The data on this type of disk is permanent and cannot be reprogrammed to store different data, and is known as CD-ROM (compact disk read-only memory). This type of disk is normally only cost effective in large quantities.

Standard CD-ROM disks have a diameter of 120 mm (4.7 inch) and a thickness of 1.2 mm. They can store up to 650 MB of data which gives around 72 minutes of compressed video (MPEG format with near VCR quality) or uncompressed hi-fi audio. The reflective coating (normally aluminum) on the disk is approximately 30 μm and the pits are approximately 0.1 μm long and deep. A protective transparent coating is applied on top of the reflective coating with a depth of 1.2 mm (the approximate thickness of the disk). The protective coating also help to focus the light beam from about 0.7 mm on the surface of the coating to the 0.1 μm pit.

Data is stored on the disk as a spiral starting from the outside and ending at the inside. The thickness of the track is 1.6 μm, which gives a total spiral length of 5.7 km.

55.9.2 WORM drives

WORM (write-once read many) disks allow data to be written to the optical disk once. The data is then permanent and thus cannot be altered. They are typically used in data logging applications and in making small volumes of CD-ROMs. A 350 mm (14 inch) WORM disk can store up to 10 GB of data (5 GB per side). This gives around 15 hours of compressed video (MPEG format with near VCR quality).

WORM disks consist of two pieces of transparent material (normally glass) with a layer of metal (typically tellurium) sandwiched in-between. Initially the metal recording surface is clear. A high-intensity laser beam then writes information to the disk by burning small pits into the surface.

55.9.3 CD-R/ CD-RW disks

CD-R (CD-Recordable) disks are write-once disks that can store up to 650 MB of data or 74 minutes of audio. For a disk to be read by any CD-ROM drive they must comply with ISO 9660 format. A CD-R disk can also be made multi-session where a new file system is written each time the disk is written to. Unfortunately this takes up around 14 MB of header data for each session. Typical parameters for sessions are:

No. of sessions	*Header information*	*Data for each session*
1	approx. 14 MB	636 MB one session
5	approx. 70 MB	116 MB each session
10	approx. 140 MB	51 MB each session
30	approx. 420 MB	7.7 MB each session

Typically CD recorders write at two (or even four) times the standard writing/playback speed of 150 KB (75 sectors) per second.

A CD-RW (CD-ReWriteable) disk allows a disk to be written-to many times, but the file format is incompatible with standard CD-ROM systems (IS0 9660). The formatting of the CD-RW disk (which can take several hours) takes up about 157 MB of disk space, which only leaves about 493 MB for data.

New CD-R/CD-RW writing systems incorporate a smart laser system that eradicates the problem of dirt on the disk. It does this by adjusts the write power of the laser using Automatic Power Control. This allows the unit to continue to write when it encounters minor media errors such as dirt, smudges, small scratches, and so on.

55.9.4 CD-ROM disk format

The two main standard for writing a CD-ROM are ISO 9660 and UDF (Universal Disk Format). The ISO 9660 disk unfortunately uses 14 MB for each session write.

In 1980, Philips NV and Sony Corporation announced the CD-DA (digital audio) format and in 1983 they released the standard for CD-ROM. Then, in 1988, they released the Red Book standard for recordable CD audio disks (CD-DA)

This served as a blueprint for the Yellow Book specification for CD-ROMs (CD-ROM and CD-ROM-XA Data Format) and the Orange Book Parts 1 and 2 specifications for CD-Recordable (CD-R/CD-E (CD-Recordable/CD-Erasable)). In the Red Book standard a disk is organized into a number of segments:

- Lead In. Contains the disk's table of contents that specifies the physical location of each track.
- Program Area. Contains the actual disk data or audio data and is divided up into 99 tracks, with a two-second gap between each track.
- Lead Out. Contains a string of zeros which is a legacy of the old Red Book standard. These zeros enabled old CD players to identify the end of a CD.

The CD is laid out in a number of sectors. Each of these sectors contains 2352 bytes, made up of 2048 bytes of data and other information such as headers, sub-headers, error detection code, and so on. The data is organized into logical blocks. After each session a logical block has a logical address, which is used by the drive to find a particular logical block number (LBN).

Within the tracks the CD can contain either audio or computer data. The most common formats for computer data are ISO 9660, hierarchical file system (HFS) and the Joliet file system.

The ISO 9660 standard was developed at a time when disks required to be mass-replicated. It thus wrote the complete file system at the time of creation, as there was no need for incremental creation. Now, with CD-R technology, it is possible to incrementally write to a disk. This is described as multi-session. Unfortunately, after each session a new Lead In and Lead Out must be written (requiring a minimum of 13 MB of disk space). This consists of:

- 13.2 MB for the Lead Out for the first session and 4.4 MB for each subsequent session.
- 8.8 MB for Lead In for each session.

Thus multi-session is useful for writing large amounts of data for each session, but is not efficient when writing many small updates. Most new CD-R systems now use a track-at-once technique which stores the data one track at a time and only writes the Lead In and Lead Out data when the session is actually finished. In this technique, the CD can be build up with data over a long time period. Unfortunately, the disk cannot be read by standard CD-ROM drives until the session is closed (and written with the ISO 9660 format). Another disadvantage is that the Red Book only specifies up to 99 tracks for each CD.

Unfortunately, ISO 9660 is not well suited for packet writing and is likely to be phased out over the coming years.

55.9.5 Magneto-optical (MO) disks

As with CD-R disks, magneto-optical (MO) disks allow data to be rewritten many times. These disks use magnetic and optical fields to store the data. Unfortunately, the disk must first be totally erased before writing data (although new developments are overcoming this limitation).

55.9.6 Transfer rates

Optical disks spin at variable speeds, they spin at a lower rate on the outside of the disk than on the inside. Thus, the disk increases its speed progressively as the data is read from the disk. The actual rate at which the drive reads the data is constant for the disk. The basic transfer rate for a typical CD-ROM is 150 KB/s. This has recently been increased to 300 KB/s (×2 CD drives), 600 KB/s (×4), 900 KB/s (×6), 1.5 MB/s (×10) and even 3.6 MB/s (×24).

55.9.7 Standards

Data disks are described in the following standards books, each of them specific to an area or application type. They can be obtained by becoming a licensed CD developer with Philips and they apply to media, hardware, operating systems, file systems and software.

Red Book	World standard for all audio Compact Disks (CD-DA).
Yellow Book	CD-ROM and CD-ROM-XA data formats.
Green Book	CD-I data formats and operating systems (photo).
White Book	CD-I (video)
Orange Book	CD-R/CD-E (CD-Recordable/CD-Eraseable).
Blue Book	CD-Enhanced (CD Extra, CD Plus).

55.9.8 Silver, green, blue or gold

CD-ROMs are available in a number of colors, these are:

- Silver. These are read-only disk which are a stamped as an original disk.
- Gold. These are recordable disks which use a basic phthalocyanine formulation which was patented by Mitsui Toatsu Chemicals (MTC) of Japan, and is licensed to other phthalocyanine media manufacturers. They generally work better with 2× writing speeds as some models of disk can not be written to at 1× writing speed.
- Green. These are recordable disks which are based on cyanine-based formulations. They are not covered by a governing patent, and are more or less unique to the individual manufacturers. An early problem was encountered with cyanine-based disk as the dye became chemically unstable in the presence of sunlight. Other problems included a wide variation in electrical performance depending on write speed and location (inner or outer portion of the disk). Eventually, in 1995, some stabilizing compounds were added. The best attempt produced a metal-stabilized cyanine dye formulation that gave excellent overall performance. Gradually the performance of these disks is approaching gold disk performance.
- Blue. These are recordable disk which are based on an Azo media. This was designed and manufactured by Mitsubishi Chemical Corporation (MCC) and marketed through its U.S. subsidiary, Verbatim Corporation.

55.10 Magnetic tape

Magnetic tapes use a thin plastic tape with a magnetic coating (normally of ferric oxide). Most modern tapes are either reel-to-reel or cartridge type. A reel-to-reel tape normally has two interconnected reels of tape with tension arms (similar to standard compact audio cassettes). The cartridge type has a drive belt to spin the reels; this mechanism reduces the strain on the tape and allows faster access speeds.

Magnetic tapes have an extremely high capacity and are relatively cheap. Data is saved in a serial manner with one bit (or one record) at a time. This has the disadvantage that they are relatively slow when moving back and forward within the tape to find the required data. Typically it may take many seconds (or even minutes) to search from the start to the end of a tape. In most applications, magnetic tapes are used to backup a system. This type of application requires large amount of data to be stored reliably over time but the recall speed is not important.

The most common types of tape are:

- Reel-to-reel tapes – the tapes have two interconnected reels with an interconnecting tape which is tensioned by tension arms. They were used extensively in the past to store computer-type data but have been replaced by the following three types (8 mm, QIC and DAT tapes).
- 8 mm video cartridge tapes – this type of tape was developed to be used in video cameras and is extremely compact. As with video tapes the tape wraps round the read/write head in a helix.
- Quarter inch cartridge (QIC) tapes – a QIC is available in two main sizes: 5¼-inch and 3½-inch. They give capacities of 40 MB to tens of GB.
- Digital audio tapes (DAT) – this type of tape was developed to be used in hi-fi applications and is extremely compact. As with the 8 mm tape, the tape wraps round the read/write head in a helix. The tape itself is 4 mm wide and can store several GBs of data with a transfer rate of several hundred Kbps.

55.10.1 QIC tapes

QIC tapes are available in two sizes: 5¼ inch and 3½ inch. The tape length ranges from 200 to 1000 feet, with a tape width of ¼ inch. Typical capacities range from 40 MB to tens of GB. A single capstan drive is driven by the tape drive. Figure 55.6 illustrates a QIC tape.

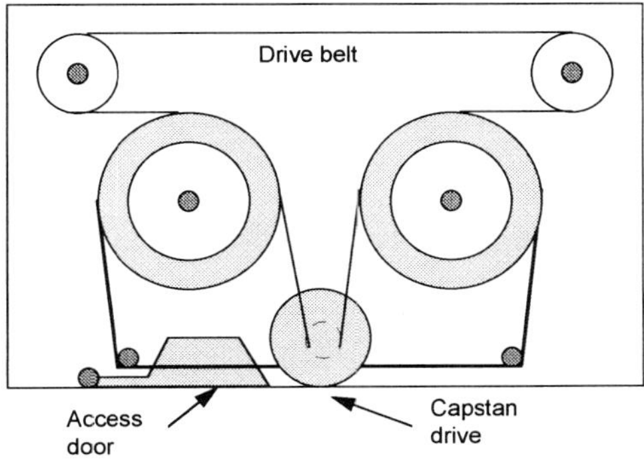

Figure 55.6 QIC tape.

55.10.2 8 mm video tape

The 8 mm video tape is a high specification tape and was originally used in video cameras. These types are also known as Exabyte after the company that originally developed a backup system using 8 mm video tapes. They can be used to store several GB of data with a transfer rate of 500 Kbps. In order to achieve this high transfer rate the read/write head spins at 2000 rpm and the tape passes it at a relatively slow speed.

55.10.3 Digital audio tape (DAT)

The DAT tape is a high specification tape and was originally used in the music industry.

56 SCSI

56.1 Introduction

SCSI is often the best choice of bus for high-specification systems. It has many advantages over IDE, these include:

- A single bus system for up to seven connected devices.
- It supports many different peripherals, such as hard disks, tape drives, CD-ROMs, and so on.
- It supports device priority where a higher SCSI-ID has priority over a lower SCSI-ID.
- It supports both high-quality connectors and cables, and low-quality connectors and ribbon cable.
- It supports differential signals, which gives longer cable lengths.
- Extended support for commands and messaging.
- Devices do not need individual IRQ lines (as they do in IDE) as the controller communicates with the devices. Thus it requires only a single IRQ line.
- It has great potential for faster transfer and enhanced peripheral support.

56.2 SCSI types

SCSI has an intelligent bus subsystem and can support multiple devices cooperating currently, where each device is assigned a priority. The main types of SCSI are:

- SCSI-I. Transfer rate of 5 MB/sec with an 8-bit data bus and seven devices per controller.
- SCSI-II. Support for SCSI-I and with one or more of the following:
  - Fast SCSI, which uses a synchronous transfer to give 10 MB/s transfer rate. The initiator and target initially negotiate to see if they can both support synchronous transfer. If they can they then go into a synchronous transfer mode.
  - Fast/Wide-SCSI-II, which doubles the data, bus width to 16 bits and gives a 20 MB/s transfer rate.
  - 15 devices per master device.
  - Tagged command queuing (TCQ) which greatly improves performance and is supported by Windows NT, NetWare and OS/2.
  - Multiple commands sent to each device.
  - Commands executed in whatever sequence will maximize device performance.
- Ultra-SCSI (SCSI-III) which operates either as 8-bit or 16-bit giving a 20 MB/sec or 40 MB/s transfer rate.

56.2.1 SCSI-II

SCSI-II supports Fast SCSI, which is basically SCSI-I operating at a rate of 10 MB/s (using Synchronous vs. Asynchronous), and Wide-SCSI which uses a 64-pin connector and has a 16-bit data bus. The SCSI-II controller is also more efficient and processes commands up to seven times faster than SCSI-I.

The SCSI-II drive latency is much less than SCSI-I because it uses tagged command queuing (TCQ) which allows multiple commands to be sent to each device. These then hold their own commands and execute them in sequence that maximizes system performance (such as by minimizing disk rotation latency). Table 56.1 shows examples of Fast SCSI-II and Fast/Wide-SCSI-II. It can be seen that both disks have predictive failure analysis (PFA) and automatic defect reallocation (ADR).

The normal 50-core cable is typical known as A-cable, while the 68-core cable is known as B-cable.

Table 56.1 Comparison of SCSI-II disks.

	Seek time (ms)	*Latency (ms)*	*Rotational Speed (rpm)*	*Sustained data read (MB/s)*	*PFA*	*ADR*
1GB SCSI-II Fast	10.5	5.56	5400	4	✓	✓
4.5GB SCSI-II Fast/Wide	8.2	4.17	7200	12	✓	✓

56.2.2 Ultra-SCSI

Ultra-SCSI (or SCSI-III) allows for 20 MB/s burst transfers on an 8-bit data path and 40 MB/s burst transfer on a 16-bit data path. It uses the same cables as SCSI-II and the maximum cable length is 1.5 m. Ultra-SCSI disks are compatible with SCSI-II controllers; however, the transfer will be at the slower speed of the SCSI controller. SCSI disks are compatible with Ultra-SCSI controllers; however, the transfer will be at the slower speed of the SCSI disk.

SCSI-I and Fast SCSI-II use a 50-pin 8-bit connector, while Fast/Wide-SCSI-II and Ultra-SCSI use a 68-pin 16-bit connector. The 16-bit connector is physically smaller than the 8-bit connector and the 16-bit connector cannot connect directly to the 8-bit connector. The cable used is called P-cable and replaces the A/B-cable.

Note that SCSI-II and Ultra-SCSI require an active terminator on the last external device. Table 56.2 compares the main types of SCSI.

Table 56.2 SCSI types.

	Data bus (bits)	*Transfer rate (MB/s)*	*Tagged command queuing*	*Parity checking*	*Maximum devices*	*Pins on cable and connector*
SCSI-I	8	5	×	×/✓ (opt.)	7	50
SCSI-II Fast	8	10	✓	✓	7	50
SCSI-II Fast/Wide	16	20	✓	✓	15	68
Ultra-SCSI	32	40	✓	✓	15	68

56.3 SCSI Interface

In its standard form the Small Computer Systems Interface (SCSI) standard uses a 50-pin header connector and a ribbon cable to connect to up to eight devices. It overcomes the problems of the IDE, where devices are assigned to be either a master or a slave. SCSI and Fast SCSI transfer data one byte at a time with a parity check on each byte. SCSI-II, Wide-SCSI and Ultra-SCSI use a 16-bit data transfer and a 68-pin connector. Table 56.3 lists the pin connections for SCSI-I (single-ended cable) and Fast SCSI (differential cable) and Table 56.4 lists the pin connections for SCSI-II, Wide-SCSI and Ultra-SCSI. With Wide-SCSI and Ultra-SCSI there are 24 data bits ($\overline{\text{D8}}-\overline{\text{D31}}$) and three associated parity bits ($\overline{\text{D(PARITY1)}}-\overline{\text{D(PARITY3)}}$).

Table 56.3 SCSI-I and Fast SCSI connections.

Single-ended cable				**Differential cable**			
Pin	*Signal*	*Pin*	*Signal*	*Pin*	*Signal*	*Pin*	*Signal*
1	GND	2	$\overline{\text{D0}}$	1	GND	2	GND
3	GND	4	$\overline{\text{D1}}$	3	$+\overline{\text{D0}}$	4	$-\overline{\text{D0}}$
5	GND	6	$\overline{\text{D2}}$	5	$+\overline{\text{D1}}$	6	$-\overline{\text{D1}}$
7	GND	8	$\overline{\text{D3}}$	6	$+\overline{\text{D2}}$	8	$-\overline{\text{D2}}$
9	GND	10	$\overline{\text{D4}}$	8	$+\overline{\text{D3}}$	10	$-\overline{\text{D3}}$
11	GND	12	$\overline{\text{D5}}$	11	$+\overline{\text{D4}}$	12	$-\overline{\text{D4}}$
13	GND	14	$\overline{\text{D6}}$	13	$+\overline{\text{D5}}$	14	$-\overline{\text{D5}}$
15	GND	16	$\overline{\text{D7}}$	15	$+\overline{\text{D6}}$	16	$-\overline{\text{D6}}$
17	GND	18	$\overline{\text{D(PARITY)}}$	17	$+\overline{\text{D7}}$	18	$-\overline{\text{D7}}$
19	GND	20	GND	19	D(PARITY)	20	$-\overline{\text{D(PARITY)}}$
21	GND	22	GND	21	DIFFSEN	22	GND
23	RESERVED	24	RESERVED	23	RESERVED	24	RESERVED
25	Open	26	TERMPWR	25	TERMPWR	26	TEMPWR
27	RESERVED	28	RESERVED	27	RESERVED	28	RESERVED
29	GND	30	GND	29	$+\overline{\text{ATN}}$	30	$-\overline{\text{ATN}}$
31	GND	32	$\overline{\text{ATN}}$	31	GND	32	GND
33	GND	34	GND	33	$+\overline{\text{RST}}$	34	$-\overline{\text{RST}}$
35	GND	36	$\overline{\text{BSY}}$	35	$+\overline{\text{ACK}}$	36	$-\overline{\text{ACK}}$
37	GND	38	$\overline{\text{ACK}}$	37	$+\overline{\text{RST}}$	38	$-\overline{\text{RST}}$
39	GND	40	$\overline{\text{RST}}$	39	$+\overline{\text{MSG}}$	40	$-\overline{\text{MSG}}$
41	GND	42	$\overline{\text{MSG}}$	41	$+\overline{\text{SEL}}$	42	$-\overline{\text{SEL}}$
43	GND	44	$\overline{\text{SEL}}$	43	$+\overline{\text{C}}$ / D	44	$-\overline{\text{C}}$ / D
45	GND	46	$\overline{\text{C}}$ / D	45	$+\overline{\text{REQ}}$	46	$-\overline{\text{REQ}}$
47	GND	48	$\overline{\text{REQ}}$	47	$+\overline{\text{I}}$ / O	48	$-\overline{\text{I}}$ / O
49	GND	50	$\overline{\text{I}}$ / O	49	GND	50	GND

56.3.1 Signals

A SCSI bus is made up of a SCSI host adapter connected to a number of SCSI units via a SCSI bus. As all units connect to a common bus, only two units can transfer data at a time, either from one SCSI unit to another or from one SCSI unit to the SCSI host. The great advantage of this transfer is that is does not involve the processor.

Table 56.4 SCSI-II, Wide-SCSI and Ultra-SCSI.

Pin	*Signal*	*Pin*	*Signal*	*Pin*	*Signal*	*Pin*	*Signal*
1	GND	18	TERMPWR	35	GND	52	$\overline{D19}$
2	GND	19	GND	36	$\overline{D8}$	53	$\overline{D20}$
3	GND	20	GND	37	$\overline{D9}$	54	$\overline{D21}$
4	GND	21	GND	38	$\overline{D10}$	55	$\overline{D22}$
5	GND	22	GND	39	$\overline{D11}$	56	$\overline{D23}$
6	GND	23	GND	40	$\overline{D12}$	57	$\overline{D(PARITY2)}$
7	GND	24	GND	41	$\overline{D13}$	58	$\overline{D24}$
8	GND	25	GND	42	$\overline{D14}$	59	$\overline{D25}$
9	GND	26	GND	43	$\overline{D15}$	60	$\overline{D26}$
10	GND	27	GND	44	$\overline{D(PARITY1)}$	61	$\overline{D27}$
11	GND	28	GND	45	$\overline{ACKB}$	62	$\overline{D28}$
12	GND	29	GND	46	GND	63	$\overline{D29}$
13	GND	30	GND	47	$\overline{REQB}$	64	$\overline{D30}$
14	GND	31	GND	48	$\overline{D16}$	65	$\overline{D31}$
15	GND	32	GND	49	$\overline{D17}$	66	ATN
16	GND	33	GND	50	TERMPWR	67	$\overline{D(PARITY3)}$
17	TERMPWR	34	GND	51	TERMPWR	68	GND

Each unit on a SCSI is assigned a SCSI-ID address. As SCSI-I has an 8-bit data bus this address can range from 0 to 7 (where 7 is normally reserved for a tape drive). The host adapter takes one of the addresses and thus a maximum of seven units can connect to the bus. Most systems allow the units to take on any SCSI-ID address, but older system required boot drives to be connected to a specific SCSI address. On boot the host adapter sends out a 'Start Unit' command to each SCSI unit. This allows each unit to start in an orderly manner (and not overload the local power supply). The host starts with the highest priority address (ID=7) and finishes with the lowest address (ID=0). Typically, the ID is set with a rotating switch selector or by three jumpers.

SCSI defines an initiator control and a target control. The initiator requests functions from a target, which then executes the function, as illustrated in Figure 56.1. The initiator effectively takes over the bus for the time to send a command, the target then executes the command and contacts the initiator to transfer data. The bus will then be free for other transfers.

The main signals are:

- $\overline{BSY}$. Indicates that the bus is busy, or not (an OR-tied signal).
- $\overline{ACK}$. Activated by the initiator to indicate an acknowledgement for a $\overline{REQ}$ information transfer handshake.
- $\overline{RST}$. When active (low) resets all the SCSI devices (an OR-tied signal).
- $\overline{ATN}$. Activated by the initiator to indicate the attention state.
- $\overline{MSG}$. Activated by the target to indicate the message phase.
- $\overline{SEL}$. Activated by the initiator and is used to select a particular target device (an OR-tied signal).
- $\overline{C}/D$ (control/data). Activated by the target to identify if there is data or control on the SCSI bus.
- $\overline{REQ}$. Activated by the target to acknowledge a request for an $\overline{ACK}$ information transfer

handshake.

- $\overline{I}/O$ (input/output). Activated by the target to show the direction of the data on the data bus. Input defines that data is an input to the initiator, else, it is an output.

Each of the control signals can be true or false. They can be:

- OR-tied driven, where the driver does not drive the signal to the false state. In this case any SCSI device can pull the signal false whenever it is released by another device. If any driver is asserted, then the signal is true. The $\overline{BSY}$, $\overline{SEL}$, and $\overline{RST}$ signals are OR-tied. In the ordinary operation of the bus, the $\overline{BSY}$ and $\overline{RST}$ signals may be simultaneously driven true by several drivers.
- Non-OR-tied driven where the signal may be actively driven false. No signals other than $\overline{BSY}$, $\overline{RST}$, and $\overline{D(PARITY)}$ are simultaneously driven by two or more drivers.

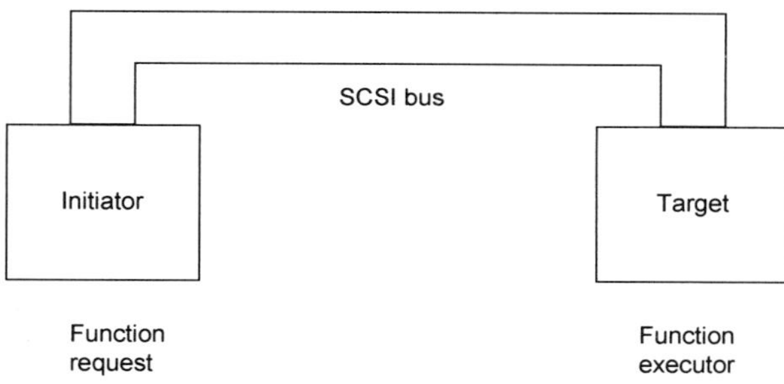

Figure 56.1 Initiator and target in SCSI.

56.4 SCSI operation

The SCSI bus allows any unit to talk to any other unit, or the host to talk to any unit. Thus there must be some way of arbitration where units capture the bus. The main phases that the bus goes through are:

- **Free-bus**. In this state no units are either transfering data or have control of the bus. It is identified by disactive $\overline{SEL}$ and $\overline{BSY}$ lines (both will be high). In this state any unit can capture the bus.
- **Arbitration**. In this state a unit can take control of the bus and become an initiator. To do this it activates the $\overline{BSY}$ signal and puts its own ID address on the data bus. Next, after a delay, it tests the data bus to determine if a higher-priority unit has put its own address on the bus. If it has then it allows the other unit(s) access to the bus. If its address is still on the bus then it asserts the $\overline{SEL}$ line. After a delay, it then has control of the bus.
- **Selection**. In this state the initiator selects a target unit and gets the target to carry out a given function, such as reading or writing data. The initiator outputs the OR-value of its SCSI-ID and the target's SCSI-ID onto the data bus (for example, if the initiator is 2 (0000 0100) and the target is 5 (0010 0000) then the OR-ed ID on the bus will be 0010

0100.). The target then determines that its ID is on the data bus and sets the $\overline{\mathrm{BSY}}$ line active. If this does not happen within a given time then the initiator deactivates the $\overline{\mathrm{SEL}}$ signal, and the bus becomes free. The target determines that it is selected when the $\overline{\mathrm{SEL}}$ signal and its SCSI-ID bit are active and the $\overline{\mathrm{BSY}}$ and $\overline{\mathrm{I}}/\mathrm{O}$ signals are false. It then asserts the $\overline{\mathrm{BSY}}$ signal within a selection abort time.

- **Reselection**. When the arbitration phase is complete, the winning SCSI device asserts the $\overline{\mathrm{BSY}}$ and $\overline{\mathrm{SEL}}$ signals and has delayed at least a bus clear delay plus a bus settle delay. The winning SCSI device sets the data bus to a value that is the logical OR of its SCSI-ID bit and the initiator's SCSI-ID bit. Sometimes the target takes a while reply to the initiator's request. The initiator determines that it is reselected when the $\overline{\mathrm{SEL}}$ and $\overline{\mathrm{I}}/\mathrm{O}$ signals and its SCSI-ID bit are true and the $\overline{\mathrm{BSY}}$ signal is false. The reselected initiator then asserts the $\overline{\mathrm{BSY}}$ signal within a selection abort time of its most recent detection of being reselected. An initiator does not respond to a reselection phase if other than two SCSI-ID bits are on the data bus. After the target detects that the $\overline{\mathrm{BSY}}$ signal is true, it also asserts the $\overline{\mathrm{BSY}}$ signal and waits a given time delay to release the $\overline{\mathrm{SEL}}$ signal. The target may then change the $\overline{\mathrm{I}}/\mathrm{O}$ signal and the data bus. After the reselected initiator detects the $\overline{\mathrm{SEL}}$ signal is false, it releases the $\overline{\mathrm{BSY}}$ signal. The target continues to assert the $\overline{\mathrm{BSY}}$ signal until it gives-up the SCSI bus.
- **Command**. The command phase is used by the target to request command information from the initiator. The target asserts the $\overline{\mathrm{C}}/\mathrm{D}$ signal and negates the $\overline{\mathrm{I}}/\mathrm{O}$ and $\overline{\mathrm{MSG}}$ signals during the $\overline{\mathrm{REQ}}/\overline{\mathrm{ACK}}$ handshake(s) of this phase.
- **Data**. The data phase covers both the data in and data out phase. In the data in phase the target requests that data is to be sent to the initiator. It (the target) asserts the $\overline{\mathrm{I}}/\mathrm{O}$ signal and negates the $\overline{\mathrm{C}}/\mathrm{D}$ and $\overline{\mathrm{MSG}}$ signals during the $\overline{\mathrm{REQ}}/\overline{\mathrm{ACK}}$ handshake(s) of this phase. In the data out phase the target requests that data be sent from the initiator to the target. For this the target negates the $\overline{\mathrm{C}}/\mathrm{D}$, $\overline{\mathrm{I}}/\mathrm{O}$, and $\overline{\mathrm{MSG}}$ signals during the $\overline{\mathrm{REQ}}/\overline{\mathrm{ACK}}$ handshake(s) of this phase.
- **Message**. The message phase covers both the message out and message in phase. The first byte transferred can be either a single-byte message or the first byte of a multiple-byte message. Multiple-byte messages are completely contained within a single message phase.
- **Status**. The status phase allows the target to request status information from the initiator. For this the target asserts the $\overline{\mathrm{C}}/\mathrm{D}$ and $\overline{\mathrm{I}}/\mathrm{O}$ signals and negate the $\overline{\mathrm{MSG}}$ signal during the $\overline{\mathrm{REQ}}/\overline{\mathrm{ACK}}$ handshake of this phase.

Typical times are:

- Arbitration delay, 2–4 μs. This is the minimum time that the SCSI device waits from asserting $\overline{\mathrm{BSY}}$ for arbitration until the data bus can be examined to see if arbitration has been won.
- Power-on to selection time, 10 s. This is the maximum time from power start-up until a SCSI target is able to respond with appropriate status and sense data.
- Selection abort time, 200 μs. This is the maximum time that a target (or initiator) takes from its most recent detection of being selected (or reselected) until asserting a $\overline{\mathrm{BSY}}$ response. This is required to ensure that a target (or initiator) does not assert $\overline{\mathrm{BSY}}$ after an aborted select (or reselection) phase.

- Selection time out delay, 250 ms. The minimum time that a SCSI device waits for a $\overline{BSY}$ response during the selection or reselection phase before starting the time out procedure.
- Disconnection delay, 200 µs. The minimum time that a target waits after releasing $\overline{BSY}$ before entering an arbitration phase when implementing a disconnect message from the initiator.
- Reset hold time, 23 µs. The minimum time for which $\overline{RST}$ is asserted.

The signals $\overline{C}/D$, $\overline{I}/O$, and $\overline{MSG}$ differentiate the different information transfer phases, as summarized in Table 56.5. The target drives these signals and thus controls the operation of the bus. The initiator requests a message out phase with the $\overline{ATN}$ signal, while the target initiates a bus free phase by releasing the $\overline{MSG}$, $\overline{C}/D$, $\overline{I}/O$, and $\overline{BSY}$ signals.

Information transfer phases use one or more $\overline{REQ}/\overline{ACK}$ handshakes to control the transfer. Each $\overline{REQ}/\overline{ACK}$ handshake allows the transfer of one byte of information. During this phase the $\overline{BSY}$ signal remain active (low) and the $\overline{SEL}$ signal false (high). Additionally, during this phase, the target continuously uses the $\overline{REQ}/\overline{ACK}$ handshake(s) with the $\overline{C}/D$, $\overline{I}/O$, and $\overline{MSG}$ signals so that these control signals are valid for a bus settle delay before assertioning the $\overline{REQ}$ signal of the first handshake and remain valid until after the negation of the $\overline{ACK}$ signal at the end of the handshake of the last transfer of the phase.

Table 56.5 Information transfer phases.

$\overline{MSG}$	$\overline{C}/D$	$\overline{I}/O$	Phase	Direction
0	1	1	–	–
0	1	0	–	–
0	0	1	Message out	Initiator→target
0	0	0	Message in	Initiator←target
1	1	1	Data out	Initiator→target
1	1	0	Data in	Initiator←target
1	0	1	Command	Initiator→target
1	0	0	Status	Initiator←target

The $\overline{I}/O$ signal controls the information direction. When low, information is transferred from the target to the initiator and when high, the transfer is from the initiator to the target.

The handshaking operation for a transfer to the initiator is as follows:

- $\overline{I}/O$ signal is set low.
- Target sets the data bus lines and asserts the $\overline{REQ}$ signal.
- Initiator reads the data bus and then indicates its acceptance of the data by asserting the $\overline{ACK}$ signal.
- Target may change or release the data bus.
- Target negates the $\overline{REQ}$ signal.
- Initiator negates the $\overline{ACK}$ signal.
- Target then transfer data using the data bus and the $\overline{REQ}$ signal, and so on.

The handshaking operation for a transfer from the initiator is as follows:

- $\overline{I}/O$ signal is set high.

- Target asserts the $\overline{\text{REQ}}$ signal (requesting information).
- Initiator put data on bus and asserts the $\overline{\text{ACK}}$ signal.
- Target reads the data bus and negates the $\overline{\text{REQ}}$ signal (acknowledging transfer).
- Initiator continues to transfer data, and so on.

56.5 SCSI pointers

SCSI provides for three pointers for each I/O process (called saved pointers), for command, data, and status. When an I/O process becomes active, its three saved pointers are copied into the initiator's set of three current pointers. These current pointers point to the next command, data, or status byte to be transferred between the initiator's memory and the target.

56.6 Message system description

The message system allows the initiator and the target to communicate over the interface connection. Each message can be one or more bytes in length. In a single message phase, one or more messages can be transmitted (but a message cannot be split between multiple message phases). Table 56.6 lists the message format, where the first byte of the message determines the format. The initiator ends the message out phase (by negating $\overline{\text{ATN}}$) when it sends certain messages identified in Table 56.7. Single-byte messages consist of a single byte transferred during a message phase.

Table 56.6 Message format.

Value	*Message format*
00h	One-byte message (command complete)
01h	Extended messages
02h–1Fh	One-byte messages
20h–2Fh	Two-byte messages
30h–7Fh	Reserved
80h–FFh	One-byte message (identify)

Table 56.7 Message codes.

Code	*Message*	*Direction*	*Description*
00h	Command complete	In	Sent from a target to an initiator to indicate a successful completion of a process. After sending this message, the target by releasing the $\overline{\text{BSY}}$ signal and the bus becomes free. The target detects a success when it detects the negation of $\overline{\text{ACK}}$ for the command complete message with the $\overline{\text{ATN}}$ signal false.
03h	Restore pointers	In	

04h	Disconnect	In/Out	Sent from target to inform an initiator that the present connection is going to be broken. After sending this message, the target releases the $\overline{\text{BSY}}$ signal goes into the bus free phase. It then considers the message transmission to be successful when it detects the negation of the $\overline{\text{ACK}}$ signal.
05h	Initiator detected error	Out	
06h	Abort	Out	Sent from initiator to reset the connection. The target then goes to the bus free phase following its receipt.
07h	Message reject	Out	Sent to indicate that the last message or message byte was invalid (or not implemented).
08h	No operation	Out	Sent when the initiator does not currently have any other valid message to send.
09h	Message parity error	Out	
0Ah	Linked command complete	In	
0Bh	Linked command complete (with flag)	In	
0Ch	Bus device reset	Out	Forces a hard reset on the selected SCSI device.
0Dh	Abort tag	Out	
0Eh	Clear queue	Out	
0Fh	Initiate recovery	In/Out	
10h	Release recovery	Out	
11h	Terminate I/O process	Out	
12h–1Fh	Reserved		
23h	Ignore wide residue (2 bytes)		
24h–2Fh	Reserved for 2-byte messages		
30h–7Fh	Reserved		
80h–FFh	Identify	In/Out	

56.7 SCSI commands

Commands are sent from the initiator to the target. The first byte of all SCSI commands contains an operation code, followed by a command descriptor block and finally the control byte.

The formats of the command descriptor block for 6-byte commands are:

Byte 0. Operation code.
Byte 1. Logical unit number (MSB, if required).
Byte 2. Logical block address.
Byte 3. Logical block address (LSB, if required).
Byte 4. Transfer length (if required) / Parameter list length (if required) / Allocation length (if required).
Byte 5. Control.

56.7.1 Operation code

Figure 56.2 shows the operation code of the command descriptor block. It has a group code field and a command code field. The 3-bit group code field provides for eight groups of command codes and the 5-bit command code field provides for 32 command codes in each group.

The group code specifies one of the following groups:

- Group 0 – 6-byte commands.
- Group 1/2 – 10-byte commands.
- Group 3/4 – reserved.
- Group 5 – 12-byte commands.
- Group 6/7 – vendor-specific.

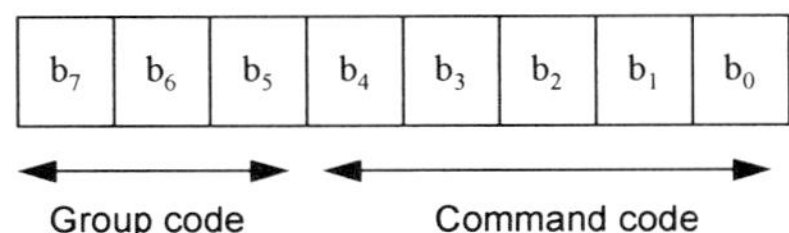

Figure 56.2 Operation code.

56.7.2 Logical unit number

The logical unit number (LUN) is defined in the identify message. The target ignores the LUN specified within the command descriptor block if an identify message was received (normally the logical unit number in the command descriptor block is set to zero).

56.7.3 Logical block address

The logical block address (LBA) on a disk drive starts a zero after a drive partition and is contiguous up to the last logical block on that device.

A 10-byte and a 12-byte command descriptor block contains a 32-bit logical block addresses, whereas a 6-byte command descriptor block contains a 21-bit logical block address.

56.7.4 Transfer length

The transfer length field specifies the amount of data to be transferred (normally the number of blocks). For several commands, the transfer length indicates the requested number of bytes to be sent as defined in the command description. Commands that uses 1 byte for the transfer length thus allow up to 256 blocks of data for one command (a value of 0 identifies a transfer bock of 256 blocks).

56.7.5 Parameter list length

The parameter list length specifies the number of bytes to be sent during the data out phase. It is typically used in command descriptor blocks for parameters that are sent to a target (such as, mode parameters, diagnostic parameters, log parameters, and so on).

56.7.6 Allocation length

The allocation length field specifies the maximum number of bytes that an initiator has allocated for returned data. The target ends the data in phase when allocation length bytes have been transferred or when all available data have been transferred to the initiator, whichever is less. The allocation length is used to limit the maximum amount of data returned to an initiator.

56.7.7 Control field

The control field is the last byte of every command descriptor block. Its format is shown in Figure 56.3. The flag bit specifies which message the target should return to the initiator. If the link bit is a 1 and the command completes without error. If the link bit is 0 and the flag bit is 1 the target returns check condition status.

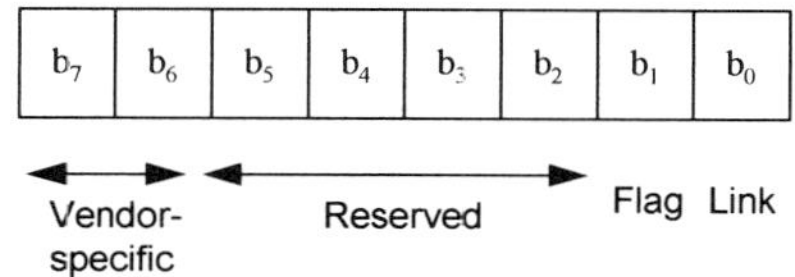

Figure 56.3 Control field.

56.7.8 Command code

Commands for all device types include (bold type identifies the mandatory commands and the operation code is given in parentheses):

- Change definition (40h). Modifies the operating definition of the selected logical unit or target.
- Compare (39h). Compares data on one logical unit with another.
- Copy (18h). Copies data from one logical unit to another or the same logical unit. The logical unit that receives and performs the copy command is the copy manager. and is responsible for copying data from the source device to the destination device.
- Copy and compare (3Ah). Performs a copy and compare command.
- **Inquiry** (12h). Requests that information regarding parameters of the target and its attached peripheral device(s) be sent to the initiator.

- Log select (4Ch). Allows an initiator to manage statistical information on by the device about the device or its logical units (used with Log sense).
- Log sense (4Dh). Allows the initiator to retrieve statistical information from the device about the device or its logical units (used with Log select).
- Mode select (15h). Allows the initiator to specify medium, logical unit, or peripheral device parameters to the target (used with Mode sense).
- Mode sense (1Ah). Allows the target to report parameters to the initiator (used with Mode select).
- Read buffer (3Ch). Used with the write buffer command as a diagnostic function for testing target memory and the SCSI bus integrity.
- Receive diagnostic results (1Ch). Requests analysis data be sent to the initiator after a send diagnostic.
- **Request sense** (03h). Requests that the target transfer sense data to the initiator.
- **Send diagnostic** (1Dh). Requests the target to perform diagnostic operations on itself, on the logical unit, or on both.
- **Test unit ready** (00h). Provides a means to check if the logical unit is ready.
- Write buffer (3Bh). Used with the read buffer command as a diagnostic for testing target memory and the SCSI bus integrity.

56.8 Status

The status phase normally occurs at the end of a command (although in some cases it may occur before transferring the command descriptor block). Figure 56.4 shows the format of the status byte and Table 56.8 defines some status byte codes. The status byte is sent from the target to the initiator during the status phase at the completion of each command unless the command is terminated by one of the following events:

- Abort message.
- Abort tag message.
- Bus device reset message.
- Clear queue message.
- Hard reset condition.
- Unexpected disconnect.

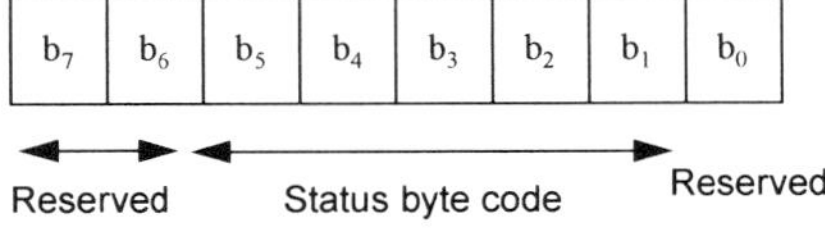

Figure 56.4 Status field.

57 PCMCIA

57.1 Introduction

The Personal Computer Memory Card International Association (PCMCIA) interface allows small thin cards to be plugged into laptop, notebook or palmtop computers. It was originally designed for memory cards (Version 1.0) but has since been adopted for many other types of adapters (Version 2.0), such as fax/modems, sound-cards, local area network cards, CD-ROM controllers, digital I/O cards, and so on. Most PCMCIA cards comply with either PCMCIA Type II or Type III. Type I cards are 3.3 mm thick, Type II cards take cards up to 5 mm thick, Type III allows cards up to 10.5 mm thick. A new standard, Type IV, takes cards which are greater than 10.5 mm. Type II interfaces can accept Type I cards, Type III accept Types I and II and Type IV interfaces accepts Types I, II and III.

The PCMCIA standard uses a 16-bit data bus (D0–D15) and a 26-bit address bus (A0–A25), which gives an addressable memory of 2^{26} bytes (64MB). The memory is arranged as:

- Common memory and attribute memory, which gives a total addressable memory of 128MB.
- I/O addressable space of 64k 8-pin ports.

The PCMCIA interface allows the PCMCIA device to map into the main memory or into the I/O address space. For example, a modem PCMCIA device would map its registers into the standard COM port addresses (such as 3F8h–3FFh for COM1 or 2F8h–2FF for COM2). Any accesses to the mapped memory area will be redirected to the PCMCIA rather that the main memory or I/O address space. These mapped areas are called windows. A window is defined with a START address and a LAST address. The PCMCIA control register contains these addresses. Table 57.1 shows the pin connections.

57.2 PCMCIA signals

The main PCMCIA signals are:

- A25–A0, D15–D0. Data bus (D15–D0) and a 26-bit memory address (A25–A0) or 16-bit I/O memory address (A15–A0).
- $\overline{\text{CARD DETECT 1}}$, $\overline{\text{CARD DETECT 2}}$. Used to detect if a card is present in a socket. When a card is inserted one of these lines is pulled to a low level.
- $\overline{\text{CARD ENABLE 1}}$, $\overline{\text{CARD ENABLE 2}}$. Used to enable the upper 8-bits of the data bus ($\overline{\text{CARD ENABLE 1}}$) and/or the lower eight bits of the data bus ($\overline{\text{CARD ENABLE 2}}$).

Table 57.1 PCMCIA connections.

Pin	*Signal*	*Pin*	*Signal*	*Pin*	*Signal*	*Pin*	*Signal*
1	GND	18	Vpp1	35	GND	52	Vpp2
2	D3	19	A16	36	*See below*	53	A22
3	D4	20	A15	37	D11	54	A23
4	D5	21	A12	38	D12	55	A24
5	D6	22	A7	39	D13	56	A25
6	D7	23	A6	40	D14	57	RFU
7	$\overline{\text{CARD ENABLE 1}}$	24	A5	41	D15	58	RESET
8	A10	25	A4	42	*See below*	59	$\overline{\text{WAIT}}$
9	$\overline{\text{OUTPUT ENABLE}}$	26	A3	43	REFRESH	60	$\overline{\text{INPACK}}$
10	A11	27	A2	44	$\overline{\text{IOR}}$	61	$\overline{\text{REGISTER SELECT}}$
11	A9	28	A1	45	$\overline{\text{IOW}}$	62	$\overline{\text{SPKR}}$
12	A8	29	A0	46	A17	63	$\overline{\text{STSCHG}}$
13	A13	30	D0	47	A18	64	D8
14	A14	31	D1	48	A19	65	D9
15	*See below*	32	D2	49	A20	66	D10
16	READY / $\overline{\text{BUSY}}$	33	$\overline{\text{IOIS16}}$	50	A21	67	$\overline{\text{CARD DETECT 2}}$
17	+5V	34	GND	51	+5V	68	GND

Pin 15 $\overline{\text{WRITE ENABLE / PROGRAM}}$
Pin 33 $\overline{\text{IOIS16}}$ (Write Protect)
Pin 36 $\overline{\text{CARD DETECT 1}}$
Pin 42 $\overline{\text{CARD ENABLE 2}}$

- $\overline{\text{OUTPUT ENABLE}}$. Set low by the computer when reading data from the PCMCIA unit.
- $\overline{\text{REGISTER SELECT}}$. Set high when accessing common memory or a low when accessing attribute memory.
- RESET. Used to reset the PCMCIA card.
- REFRESH. Used to refresh PCMCIA memory.
- $\overline{\text{WAIT}}$. Used by the PCMCIA device when it cannot transfer data fast enough and requests a wait cycle.
- $\overline{\text{WRITE ENABLE / PROGRAM}}$. Used to program the PCMCIA device.
- Vpp1, Vpp2. Programming voltages for flash memories.
- READY / $\overline{\text{BUSY}}$. Used by the PCMCIA card when it is ready to process more data (when a high) or is still occupied by a previous access (when it is a low).
- $\overline{\text{IOIS16}}$. Used to indicate the state of the write-protect switch on the PCMCIA card. A high level indicates that the write-protect switch has been set.
- $\overline{\text{INPACK}}$. Used by the PCMCIA card to acknowledge the transfer of a signal.
- $\overline{\text{IOR}}$. Used to issue an I/O read access from the PCMCIA card (must be used with an active $\overline{\text{REGISTER SELECT}}$ signal).
- $\overline{\text{IOW}}$. Used to issue an I/O write access to the PCMCIA card (must be used with an active $\overline{\text{REGISTER SELECT}}$ signal).
- $\overline{\text{SPKR}}$. Used by PCMCIA card to send audio data to the system speaker.
- $\overline{\text{STSCHG}}$. Used to identify that the card has changed its status.

57.3 PCMCIA registers

A typical PCMCIA Interface Controller (PCIC) is the 82365SL. Figure 57.1 shows the main registers for the first socket. The second socket index values are simply offset by 40h. Figure 57.2 shows that the base address of the PCIC is, in Windows, set to 3E0h, by default. Figure 57.3 shows an example of a FIRST and LAST memory address. The PCIC is accessed using two addresses: 3E0h and 3E1h. The I/O windows 0/1 are accessed through:

- 08h/0Ch for the low byte of the FIRST I/O address.
- 09h/0Dh for the high byte of the FIRST I/O address.
- 0Ah/0Eh for the high byte of the LAST I/O address.
- 0Bh/0Fh for the high byte of the LAST I/O address.

The registers are accessed by loading the register index into 3E0h and then the indexed register is accessed through the 3E1h. The memory windows 0/1/2/3/4 are accessed through:

- 10h/18h/20h/28h/30h for the low byte of the FIRST memory address.
- 11h/19h/21h/29h/31h for the high byte of the FIRST memory address.
- 12h/1Ah/22h/2Ah/32h for the low byte of the LAST memory address.
- 13h/1Bh/23h/2Bh/33h for the high byte of the LAST memory address.
- 14h/1Ch/24h/2Ch/34h for the low byte of the card offset.
- 15h/1Dh/25h/2Dh/35h for the high byte of the card offset.

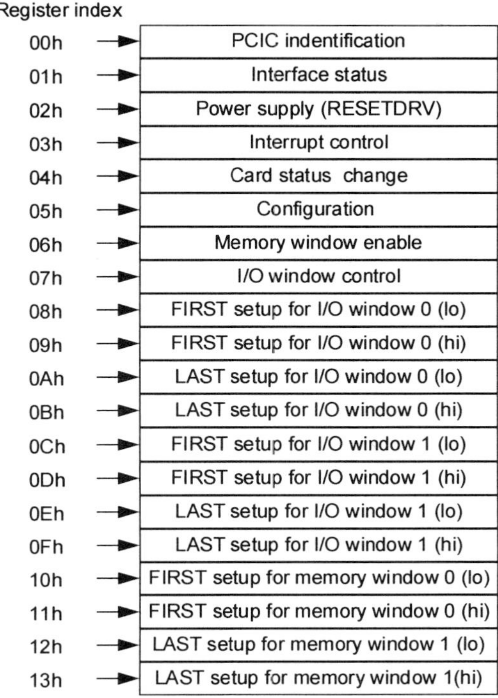

Figure 57.1 PCMCIA controller status and control registers.

For example, to load a value of 22h into the Card status change register, the following would be used:

```
_outp(0x3E0,5h);     /* point to Card status change register          */
_outp(0x3E1,22h);    /* load 22h into Card status change register     */
```

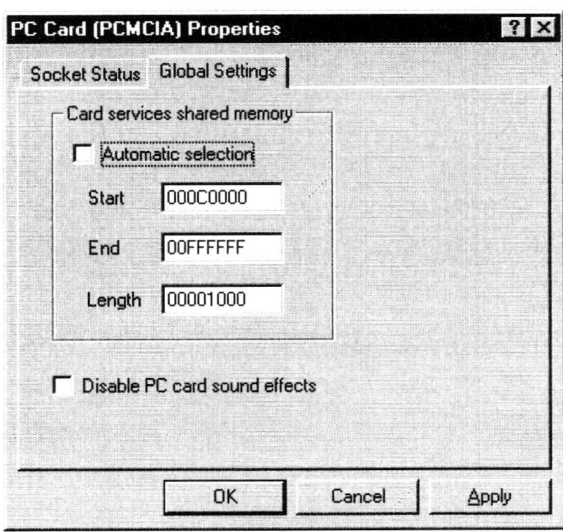

Figure 57.2 Start and end of shared memory.

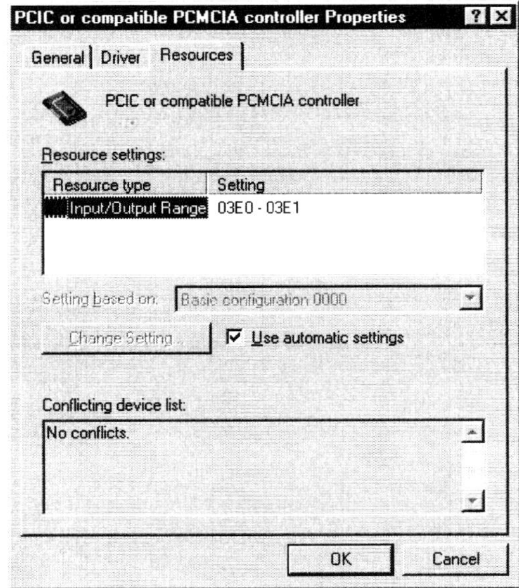

Figure 57.3 Base address of the PCIC.

57.3.1 Window enable register

The Window enable register has a register index of 06h (and 46h for the second socket). The definition of the register is:

Bit 7 IOW1. I/O window 1 enable (1)/ disable (0).
Bit 6 IOW0. I/O window 0 enable (1)/ disable (0).
Bit 5 DEC. If active (1) $\overline{\text{MEMCS16}}$ generated from A23–A12, else from A23–A17.
Bit 4 MW4. Memory window 4 enable (1)/ disable (0).
Bit 3 MW3. Memory window 3 enable (1)/ disable (0).
Bit 2 MW2. Memory window 2 enable (1)/ disable (0).
Bit 1 MW1. Memory window 1 enable (1)/ disable (0).
Bit 0 MW0. Memory window 0 enable (1)/ disable (0).

57.3.2 FIRST setup for memory window

The FIRST window memory address is made up of a low byte and a high byte. The format of the high-byte register is:

Bit 7 DS. Data bus size: 16-bit (1)/ 8-bit (0).
Bit 6 0WS. Zero wait states: no wait states (1)/ additional wait states (0).
Bit 5 SCR1. Scratch bit (not used).
Bit 4 SCR0. Scratch bit (not used).
Bit 3–0 Window start address A23–A20.

The format of the low-byte register is:

Bit 7–0 A19–A12. Window start address A19–A12.

57.3.3 LAST setup for memory window

The LAST window memory address is made up of a low byte and a high byte. The format of the high-byte register is:

Bit 7, 6 WS1, WS0. Wait state.
Bit 5, 4 Reserved.
Bit 3–0 A23–A20. Window start address A23–A20.

The format of the low-byte register is:

Bit 7–0 Window start address A19–A12.

57.3.4 Card offset setup for memory window

The card offset memory address is made up of a low byte and a high byte. The format of the high-byte register is:

Bit 7 WP Write protection: protected (1)/ unprotected (0).
Bit 6 REG $\overline{\text{REGISTER SELECT}}$ enabled. If set to a 1 then access to attribute memory, else common memory.
Bit 5–0 Window start address A25–A20.

The format of the low-byte register is:

Bit 7–0 Window start address A19–A12.

57.3.5 FIRST setup for I/O window

The FIRST window I/O address is made up of a low byte and a high byte. The format of the high-byte register is:

Bit 7–0 A15–A8.

The format of the low-byte register is:

Bit 7–0 A7–A8.

57.3.6 LAST setup for I/O window

The LAST window I/O address is made up of a low byte and a high byte. The format of the high-byte register is:

Bit 7–0 A15–A8.

The format of the low-byte register is:

Bit 7–0 A7–A8.

57.3.7 Control register for I/O address window

The Control register for the I/O address window is made up from a single byte. Its format is:

Bit 7, 3	WS1, WS0.	Wait states for Window 1 and 0.
Bit 6, 2	0WS1, 0WS0.	Zero wait states for Window 1 and 0.
Bit 5, 1	CS1, CS0.	$\overline{\text{IOIS16}}$ source. Select $\overline{\text{IOIS16}}$ from PC (1) or select data size from DS1 and DS0 (0).
Bit 4, 0	DS1, DS0.	Data size: 16-bit (1)/ 8-bit (0).

57.3.8 Examples

A typical application of the PCMCIA socket is to use it for a modem. This is an example of a setting to set-up a modem on the COM2 port. For this purpose, the socket must be set-up to map into the I/O registers from 02F8h to 02FFh. The following code will achieve this:

```
/* load 02f8 into FIRST and 02FFh into LAST registers           */
_outp(0x3E0,08h);     /* point to FIRST low byte                */
_outp(0x3E1,f8h);     /* load f8h into FIRST low byte           */

_outp(0x3E0,09h);     /* point to FIRST high byte               */
_outp(0x3E1,02h);     /* load 02h into FIRST high byte          */

_outp(0x3E0,0Ah);     /* point to LAST low byte                 */
_outp(0x3E1,ffh);     /* load ffh into LAST low byte            */

_outp(0x3E0,0Bh);     /* point to LAST high byte                */
_outp(0x3E1,02h);     /* load 02h into LAST high byte           */
```

```
/*setup control register: no wait states, 8-bit data access        */
_outp(0x3E0,07h);    /* point to I/O Control register               */
_outp(0x3E1,00h);    /* load 00h into register                      */

/* enable window 0 */
_outp(0x3E0,06h);    /* point to memory enable window               */
_outp(0x3E1,04h);    /* load 0100 0000b to enable I/O window 0      */
```

58 Cable Specifications

58.1 Introduction

The cable type used to transmit a signal depends on several parameters, including:

- The signal bandwidth.
- The reliability of the cable.
- The maximum length between nodes.
- The possibility of electrical hazards.
- Power loss in the cables.
- Tolerance to harsh conditions.
- Expense and general availability of the cable.
- Ease of connection and maintenance.
- Ease of running cables, and so on.

The main types of networking cables are twisted-pair, coaxial and fiber-optic. Twisted-pair and coaxial cables transmit electric signals, whereas fiber-optic cables transmit light pulses. Twisted-pair cables are not shielded and thus interfere with nearby cables. Public telephone lines generally use twisted-pair cables. In LANs they are generally used up to bit rates of 10 Mbps and with maximum lengths of 100 m.

Coaxial cable has a grounded metal sheath around the signal conductor. This limits the amount of interference between cables and thus allows higher data rates. Typically, they are used at bit rates of 100 Mbps for maximum lengths of 1 km.

The highest specification of the three cables is fiber-optic. This type of cable allows extremely high bit rates over long distances. Fiber-optic cables do not interfere with nearby cables and give greater security, give more protection from electrical damage by external equipment and greater resistance to harsh environments; they are also safer in hazardous environments.

58.1.1 Cable characteristics

The main characteristics of cables used in video communication are attenuation, crosstalk and characteristic impedance. Attenuation defines the reduction in the signal strength at a given frequency for a defined distance. It is normally defined in dB/100 m, which is the attenuation (in dB) for 100 m. An attenuation of 3 dB/100 m gives a signal voltage reduction of 0.5 for every 100 m. Table 58.1 lists some attenuation rates and equivalent voltage ratios; they are illustrated in Figure 58.1.

The characteristic impedance of a cable and its connectors are important, as all parts of the transmission system need to be matched to the same impedance. This impedance is normally classified as the characteristic impedance of the cable. Any differences in the matching result in a reduction of signal power and also produce signal reflections (or ghosting).

Table 58.1 Attenuation rates as a ratio.

dB	*Ratio*	*dB*	*Ratio*	*dB*	*Ratio*
0	1.000	10	0.316	60	0.001
1	0.891	15	0.178	65	0.000 6
2	0.794	20	0.100	70	0.000 3
3	0.708	25	0.056	75	0.000 2
4	0.631	30	0.032	80	0.000 1
5	0.562	35	0.018	85	0.000 06
6	0.501	40	0.010	90	0.000 03
7	0.447	45	0.005 6	95	0.000 02
8	0.398	50	0.003 2	100	0.000 01
9	0.355	55	0.001 8		

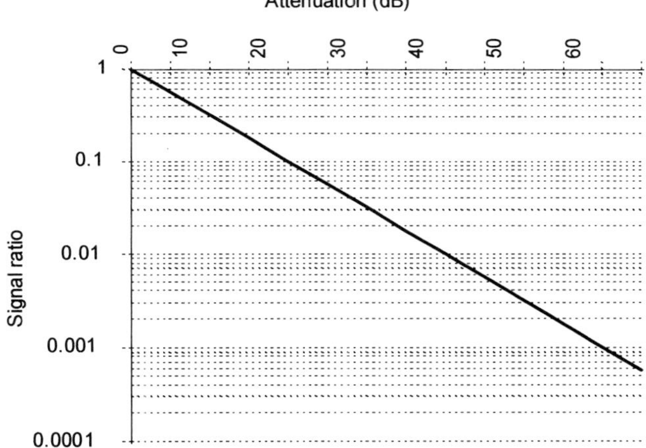

Figure 58.1 Signal ratio related to attenuation.

Crosstalk is important as it defines the amount of signal that crosses from one signal path to another. This causes some of the transmitted signal to be received back where it was transmitted.

Capacitance (pF/100 m) defines the amount of distortion in the signal caused by each signal pair. The lower the capacitance value, the lower the distortion.

The main types of cable used in networking and data communications are:

- Coaxial cable – cables with an inner core and a conducting shield having a characteristic impedance of either 75 Ω for TV signal or 50 Ω for other types.
- Cat-3 UTP cable – level 3 cables have non-twisted-pair cores with a characteristic impedance of 100 Ω (±15 Ω) and a capacitance of 59 pF/m. Conductor resistance is around 9.2 Ω/100 m.
- Cat-4 UTP cable – level 4 cables have twisted-pair cores with a characteristic impedance of 100 Ω (±15 Ω) and a capacitance of 49.2 pF/m. Conductor resistance is around 9 Ω/100 m.

- Cat-5 UTP cable – level 5 cables have twisted-pair cores with a characteristic impedance of 100 Ω (±15 Ω) and a capacitance of 45.9 pF/m. Conductor resistance is around 9 Ω/100 m.

The Electrical Industries Association (EIA) has defined five main types of cables. Levels 1 and 2 are used for voice and low-speed communications (up to 4 Mbps). Level 3 is designed for LAN data transmission up to 16 Mbps and level 4 is designed for speeds up to 20 Mbps. Level 5 cables, have the highest specification of the UTP cables and allow data speeds of up to 100 Mbps. The main EIA specification on these types of cables is EIA/TIA568 and the ISO standard is ISO/IEC11801.

Coaxial cables have an inner core separated from an outer shield by a dielectric. They have an accurate characteristic impedance (which reduces reflections), and because they are shielded they have very low crosstalk levels.

UTPs (unshielded twisted-pair cables) have either solid cores (for long cable runs) or are stranded patch cables (for shorts run, such as connecting to workstations, patch panels, and so on). Solid cables should not be flexed, bent or twisted repeatedly, whereas stranded cable can be flexed without damaging the cable. Coaxial cables use BNC connectors while UTP cables use either the RJ-11 (small connector which is used to connect the handset to the telephone) or the RJ-45 (larger connector which is used to connect LAN networks to a hub).

Table 58.2 and Figure 58.2 show typical attenuation rates (dB/100 m) for the Cat-3, Cat-4 and Cat-5 cables. Notice that the attenuation rates for Cat-4 and Cat-5 are approximately the same. These two types of cable have lower attenuation rates than equivalent Cat-3 cables. Notice that the attenuation of the cable increases as the frequency increases. This is due to several factors, such as the skin effect, where the electrical current in the conductors becomes concentrated around the outside of the conductor, and the fact that the insulation (or dielectric) between the conductors actual starts to conduct as the frequency increases.

The Cat-3 cable produces considerable attenuation over a distance of 100 m. The table shows that the signal ratio of the output to the input at 1 MHz, will be 0.76 (2.39 dB), then, at 4 MHz it is 0.55 (5.24 dB), until at 16 MHz it is 0.26. This differing attenuation at different frequencies produces not just a reduction in the signal strength but also distorts the signal (because each frequency is affected differently by the cable. Cat-4 and Cat-5 cables also produce distortion but their effects will be lessened because attenuation characteristics have flatter shapes.

Coaxial cables tend to have very low attenuation, such as 1.2 dB at 4 MHz. They also have a relatively flat response and virtually no crosstalk (due to the physical structure of the cables and the presence of a grounded outer sheath).

Table 58.2 Attenuation rates (dB/100 m) for Cat-3, Cat-4 and Cat-5 cable.

	Attenuation rate (dB/100m)		
Frequency (MHz)	*Cat-3*	*Cat-4*	*Cat-5*
1	2.39	1.96	2.63
4	5.24	3.93	4.26
10	8.85	6.56	6.56
16	11.8	8.2	8.2

Table 58.3 and Figure 58.3 show typical near end crosstalk rates (dB/100 m) for Cat-3, Cat-4 and Cat-5 cables. The higher the figure, the smaller the crosstalk. Notice that Cat-3 cables have the most crosstalk and Cat-5 have the least, for any given frequency. Notice also that the crosstalk increases as the frequency of the signal increases. Thus, high-frequency signals have more crosstalk than lower-frequency signals.

Table 58.3 Near-end crosstalk (dB/100 m) for Cat-3, Cat-4 and Cat-5 cable.

Frequency (MHz)	*Near end crosstalk (dB/100m)*		
	Cat-3	*Cat-4*	*Cat-5*
1	13.45	18.36	21.65
4	10.49	15.41	18.04
10	8.52	13.45	15.41
16	7.54	12.46	14.17

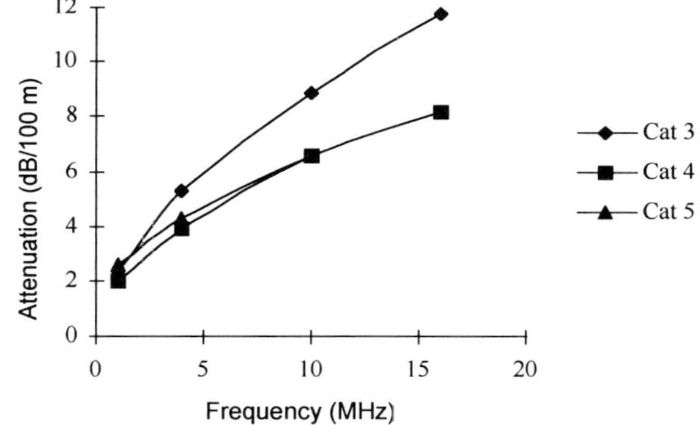

Figure 58.2 Attenuation characteristics for Cat-3, Cat-4 and Cat-5 cables.

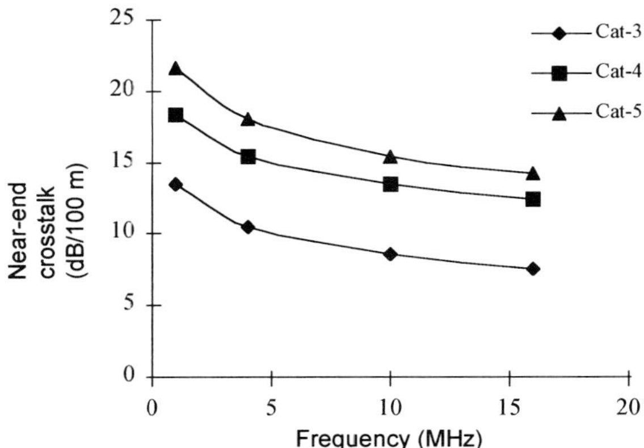

Figure 58.3 Near-end crosstalk characteristics for Cat-3, Cat-4 and Cat-5 cables.

59 Fiber Optic Systems

59.1 Introduction

One the greatest revolutions in data communications is the usage of light waves to transmit digital pulses through fiber optic cables. A light carrying system has an almost unlimited information capacity. Theoretically, it has more than 200 000 times the capacity of a satellite TV system.

Optoelectronics is the branch of electronics which deals with light. Electronic devices that use light operate within the optical part of the electromagnetic frequency spectrum, as shown in Figure 59.1. There are three main bands in the optical frequency spectrum, these are:

- Infra-red – the band of light wavelengths that are too long to been seen by the human eye;
- Visible – the band of light wavelengths that the human eye responds to;
- Ultra-violet – band of light wavelengths that are too short for the human eye to see.

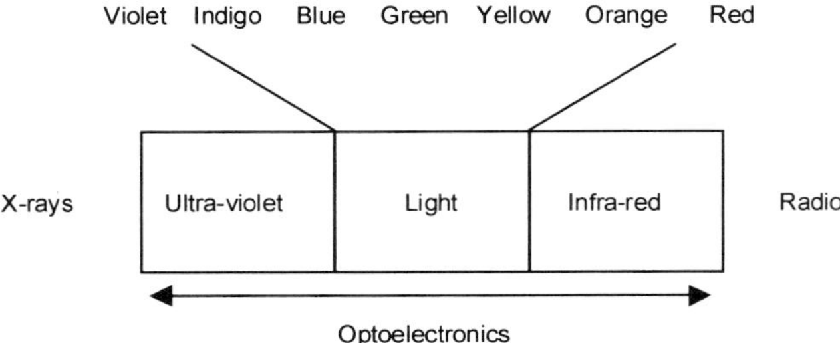

Figure 59.1 EM optoelectronics spectrum.

59.2 Light parameters

59.2.1 Wavelength

Wavelength is defined as the physical distance between two successive points of the same electrical phase. Figure 59.2 shows a wave and its wavelength. The wavelength is dependent upon the frequency of the wave *f*, and the velocity of light, *c* (3×10^8 m/s) and is given by:

$$\lambda = \frac{c}{f}$$

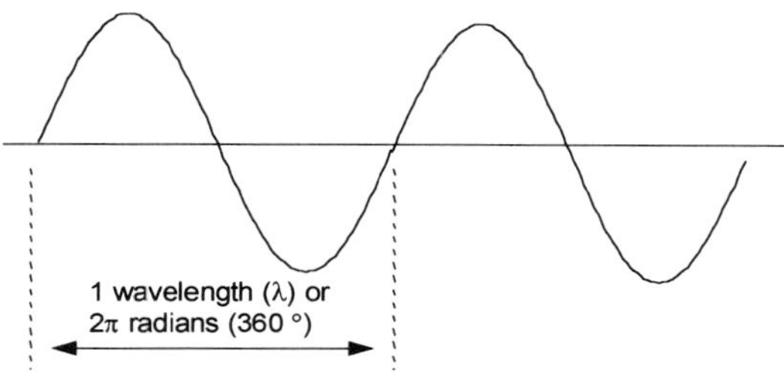

Figure 59.2 Wavelength of wave.

The optical spectrum ranges from wavelengths of 0.005 mm to 4000 mm. In frequency terms these are extremely large values from 6×10^{16} Hz to 7.5×10^{10} Hz. It is thus much simpler to talk in terms of wavelengths rather than frequencies.

59.2.2 Color

The human eye sees violet at one end of the color spectrum and red on the other. In-between, the eye sees blue, indigo, green, yellow and orange. Two beams of light that have the same wavelength are seen as the same color and the same colors usually have the same wavelength. Figure 59.3 shows the color spectrum.

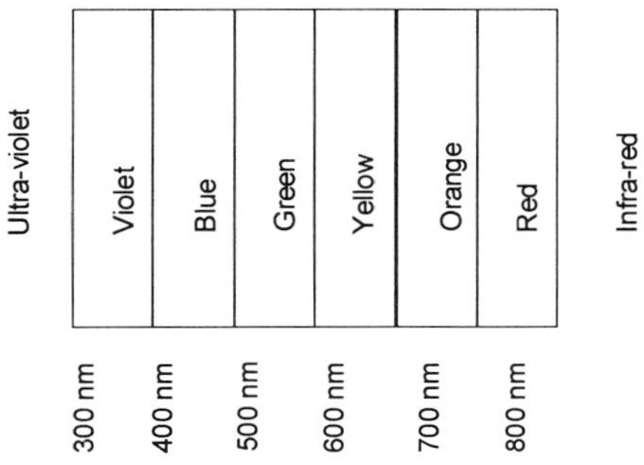

Figure 59.3 Color spectrum.

59.2.3 Velocity of propagation

In free space electromagnetic waves travel at approximately 300 000 000 m/sec (186 000 miles/sec). However, their velocity is lower when they travel through denser materials. When travelling from a material to an another which is less dense then the light ray to be refracted (or bent) away from the normal. This is shown in Figure 59.4.

59.2.4 Refractive index

The amount of bending or refraction at the interface between two materials of different densities depends on the refractive index of the two materials. This index is the ratio of the velocity of propagation of a light ray in free space to the velocity of propagation of a light ray the material, as given by:

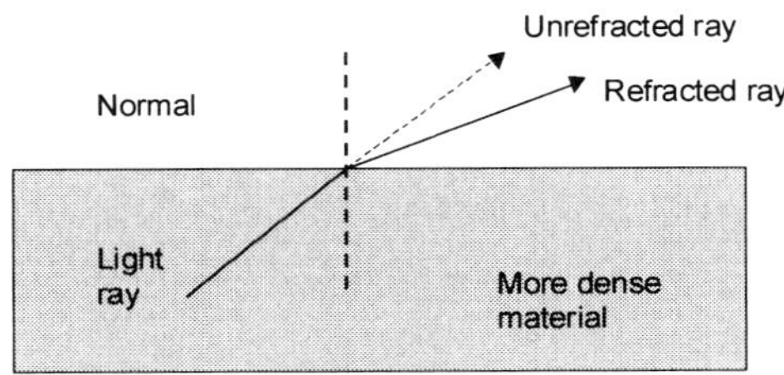

Figure 59.4 Refracted ray.

$$n = \frac{c}{v}$$

where c is speed of light in free space and v is the speed of light in a given medium. Typical refractive indexes are given in Table 59.1.

Table 59.1 Refractive index of sample materials.

Medium	*Refractive index*
Air	1.0003
Water	1.33
Glass Fiber	1.5-1.9
Diamond	2.0-2.42
Gallium Arsenide	3.6
Silicon	3.4

59.3 Light emitting diode (LED)/ Injection Laser Diode (ILD)

An LED converts electrical energy into light energy. Figure 59.5 shows a simple bias circuit with the voltage applied to the LED by a DC source. It also contains a load resistor to limit the current in the LED. This resistor is determined by subtracting the diode ON voltage from the supply voltage and dividing by the required diode current.

Normally a current of around 10 mA is necessary to produce a good intensity of light. For a GaAs diode the ON voltage is around 2 V. Thus for a GaAs diode, with an ON bias current of 10 mA and a 5 V source, the limiting resistors value would be:

$$R = \frac{5-2}{10\times10^{-3}} = 300\,\Omega$$

The wavelength of the light emitted depends upon the type of semiconductor used. Gallium Arsenide (GaAs) emits a wavelength in the infra-red range and is typically used as a source of infrared light.

Although the basic GaAs emits infra-red, it can be doped with other materials to provide a wider range of wavelengths. Gallium phosphide (GaP) emits green light and it can radiate a red light depending on the doping.

Gallium arsenide phosphide (GaAsP) emits light over an orange-red range depending on the amount of GaP in the material. With the correct amount of GaP in the material a yellow light is emitted.

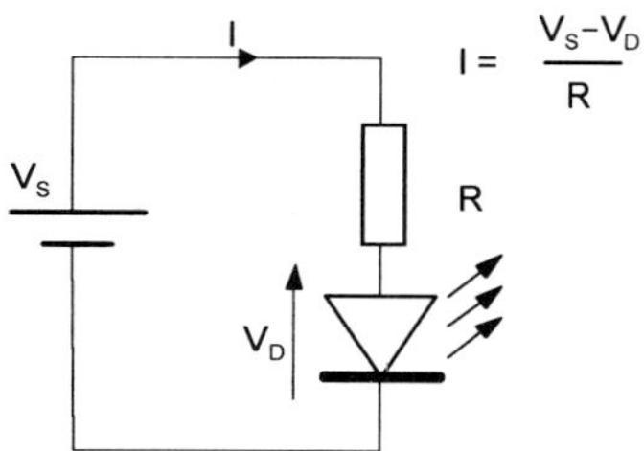

Figure 59.5 Driving an LED.

Another source of light used in optoelectronics is the laser (Light Amplification by Stimulation Emission of Radiation). An injection laser diode (ILD) is an electronic laser which emits light of a single wavelength, known as monochromatic light. ILDs have advantage over LEDs because:

- They produce a more focused light.
- Their output radiation power is greater than for an LED, typically 5 mW for ILD and 0.5 mW for an LED.
- They offer higher bit rates as they can be turned ON and OFF faster.
- They produce monochromatic light.

The main disadvantages of ILDs are that they are more expensive, have a shorter lifetime and are more temperature dependent.

59.4 Photodiodes

Photodiodes and phototransistors convert light energy (photons) into electrical energy. Their operation is based on the fact that the number of free electrons generated in a semiconductor material is proportional to the intensity of the incident light.

A photodiode must be reverse biased, the reverse biased current varies as the amount of light on the diode junctions. A basic biasing circuit for a photodiode is shown in Figure 59.6. The amount of current is normally extremely small, possibly just a few hundred μA.

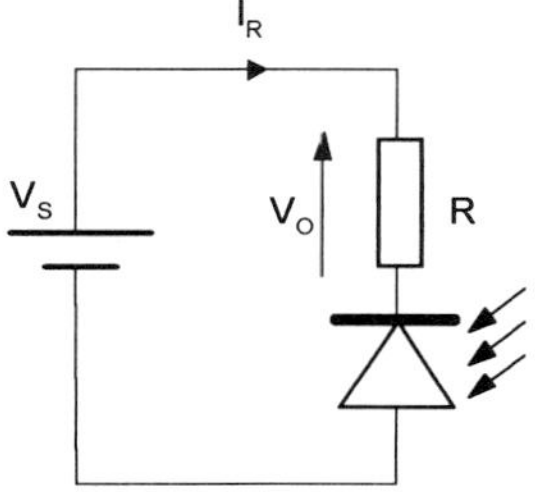

Figure 59.6 Basic biasing arrangement.

59.5 Fiber optics

59.5.1 Introduction

Optical fibers are transparent, dielectric cylinders surrounded by a second transparent dielectric cylinder. Light is transported by a series of reflections from wall to wall from the interface between a core (inner cylinder) and its cladding (outer cylinder). A cross-section of a fiber is given in Figure 59.7.

Reflections occur because the core has a higher reflective index than the cladding (it thus has a higher density). Abrupt differences in the refractive index causes the light wave to bounce from the core/cladding interface back through the core to its opposite wall. Thus the light is transported from a light source to a light detector at the other end of the fiber.

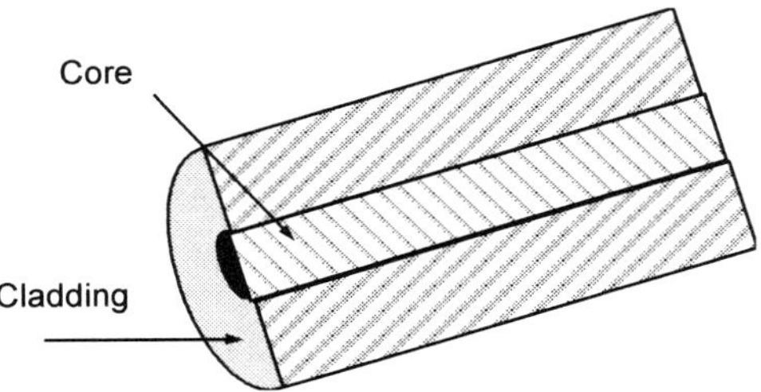

Figure 59.7 Cross-section of an optical fiber.

59.5.2 Theory

Optical fibers transmit light by total internal reflection (TIR). Light rays passing between the boundaries of two optically transparent media of different densities experience refraction, as shown in Figure 59.8. This changed direction can be determined according to Snells Law:

$$n_1 \sin\theta_1 = n_2 \sin\theta_2$$

Thus

$$\theta_2 = \sin^{-1}\left[\frac{n_1}{n_2}\sin\theta_1\right]$$

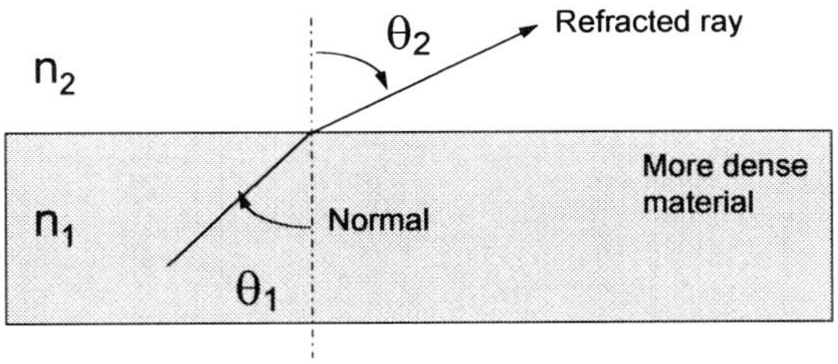

Figure 59.8 Refraction.

The angle at which the ray travels along the interface between the two materials is called the critical angle (θ_c) and is given by:

$$\theta_2 = 90°$$

$$\theta_c = \sin^{-1}\left[\frac{n_2}{n_1}\right]$$

It can be shown that when the angle of incident (θ_i) is

$$\theta_i < \sin^{-1}\left[\sqrt{n_1^2 - n_2^2}\right]$$

then the ray is totally reflected from the outer cladding. It then propagates along the fiber being reflected by the cladding on the way, as shown in Figure 59.9. This angle is called the acceptance angle.

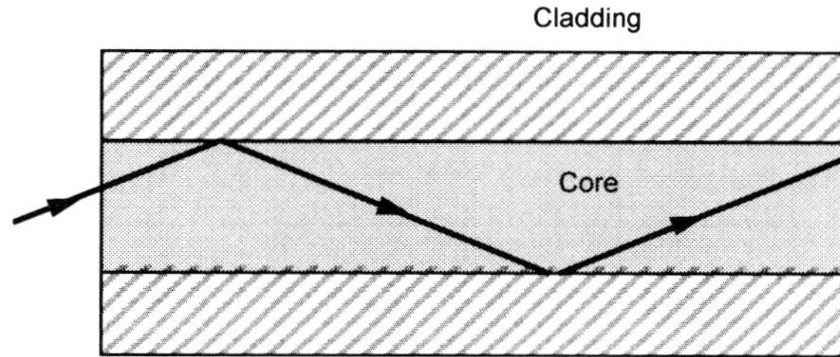

Figure 59.9 Light propagating in an optical fiber.

59.5.3 Losses in fiber optic cables

Fiber optic losses result in a lower transmitted light power. This reduces the system bandwidth, information transmission rate, efficiency and overall system capacity. The main losses are:

- Absorption losses - impurities in the glass fiber cause the transmitted wave to be absorbed and converted into heat.
- Material scattering - extremely small irregularities in the structure of the cable causes light to be diffracted. This causes the light to disperse or spread out in many directions. A greater loss occurs a visible wavelengths than at infra-red.
- Chromatic distortion - this is caused by each wavelength of light travelling at differing speeds. They thus arrive at the receiver at different times causing a distorted pulse shape. Monochromatic light reduces this type of distortion.
- Radiation losses - this is caused by small bends and kinks in the fiber which scatters the wave.
- Modal dispersion - is caused by light taking different paths through the fiber. This will each have a different propagation time to travel along the fiber. These different paths are described as modes. Figure 59.10 shows two rays taking different paths. Ray 2 will take a long time to get to the receiver than ray 1.
- Coupling losses - these losses are due to light being lost at mismatches at terminations between fiber/fiber, LED/fiber, etc.

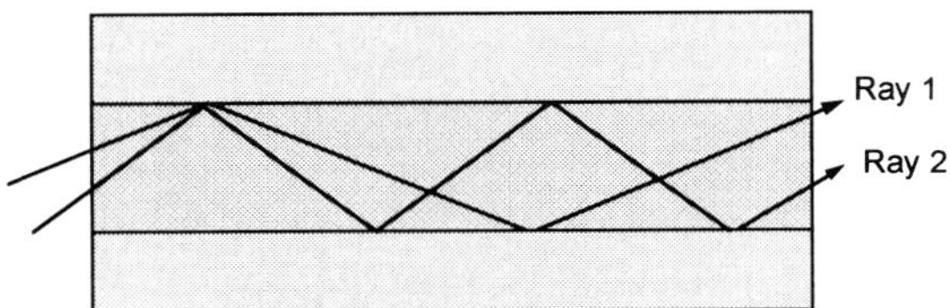

Figure 59.10 Light propagating in different modes.

59.5.4 Fiber optic link

There are three basic parts to a fiber optic system, the transmitter, the receiver and the fiber guide. The transmitter consists of an analogue-to-digital converter, a voltage-to-current converter, a light source, and a source-to-fiber light coupler. A fiber guide is either ultra-pure glass or plastic cable. The receiver has a fiber-to-light detector coupling device, a photodetector, a current-to-voltage converter, an amplifier, and a digital-to-analogue converter, as shown in Figure 59.11.

The light source is either an light emitting diode (LED) or a injection laser diode (ILD) and the amount of light emitted by either an LED or ILD is proportional to the current applied.

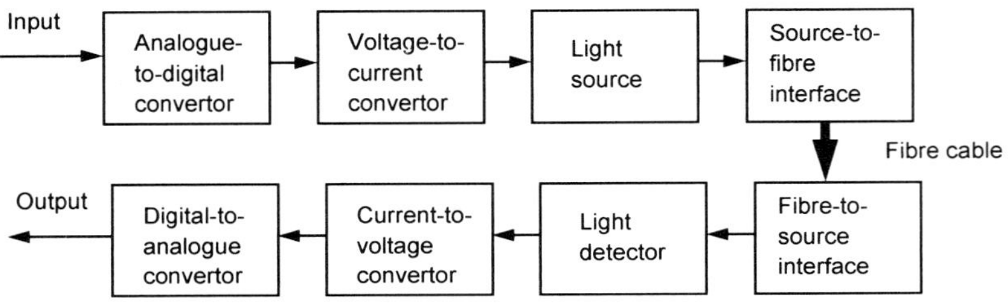

Figure 59.11 Fiber optic communications link.

59.6 Typical optical fiber characteristics

Table 59.2 shows two typical fiber optic cable characteristics. It can be seen that the inside core and the cladding diameters are relatively small i.e. fractions of a millimeter. Normally the cladding is covered in a coating which is then covered in a jacket. These give the cable mechanical strength and also makes it easier to work with. In the case of the 50/125 μm glass cable in Table 59.2 the outer diameter of the cable is 3.2 mm but the inner core diameter is 50 μm.

Normally glass fiber cables have better electrical characteristic over plastic equivalents, but are more prone to breakage and damage. It can be seen that the glass cable has improved bandwidth and lower attenuation over the plastic equivalent.

Table 59.2 Typical fiber optic cables characteristics.

	50/125 μm glass	*200 μm PCS*
Construction	Glass	Plastic coated silica (PCS)
Core diameter	50 μm	200 μm
Cladding diameter	125 μm	389 μm
Coating diameter	250 μm	600 μm
Jacket material	Polyethylene	PVC
Overall diameter	3.2 mm	4.8 mm
Connector	9 mm SMA	9 mm SMA
Bandwidth	400 MHz/km	25 MHz/km
Minimum bend radius	30 mm	50 mm
Temperature range	–15 °C to +60 °C	–10 °C to +50 °C
Attenuation @820 nm	3 dB.km^{-1}	7 dB/km

59.7 Advantages of fiber optics over copper conductors

There are many advantages in using fiber optics cable and very few disadvantages. A summary of the advantages are:

- Fiber systems have a greater capacity due to the inherently larger bandwidths available with optical frequencies. Metallic cables contain capacitance and inductance along their conductors, which cause them to act like low-pass filters and limit a signals bandwidths and also its speed of propagation;
- Fiber systems are immune from cross-talk between cables caused by magnetic induction. Glass fibers are non-conductors of electricity and therefore do not have a magnetic field associated with them. In metallic cables, the primary cause of cross-talk is magnetic induction between conductors located near each other;
- Fiber cables do not suffer from static interference caused by lightning, electric motors, fluorescent lights, and other electrical noise sources. This immunity is because fibers are non-conductors of electricity;
- Fiber systems have greater electrical isolation thus allow equipment greater protection from damage due to external sources. For example if the receiver is hit by lightning pulse them it may damage the opto-receiver but a high voltage pulse cannot travel along the optical cable and damage sensitive equipment as the source. They also prevent electrical noise travel back from the receiver to the transmitter, as illustrated in Figure 59.12
- Fiber cables do not radiate energy and therefore cannot cause interference with other communications systems. This characteristic makes fiber systems ideal for military applications, where the effect of nuclear weapons (EMP-electromagnetic pulse interference) has a devastating effect on conventional communications systems.

- Fiber cables are more resistive to environmental extremes. They operate over a larger temperature variation than copper cables and are affected less by corrosive liquids and gases.
- Fiber cables are safer and easier to install and maintain as glass and plastic as they have no electrical currents or voltages associated with them. Optical fibers can be used around volatile liquids and gases without worrying about the risk of explosions or fires. They are also smaller and more lightweight than copper cables.
- Fiber cables are more electrically secure than their copper cables and are virtually impossible to tap into without users knowing about it.

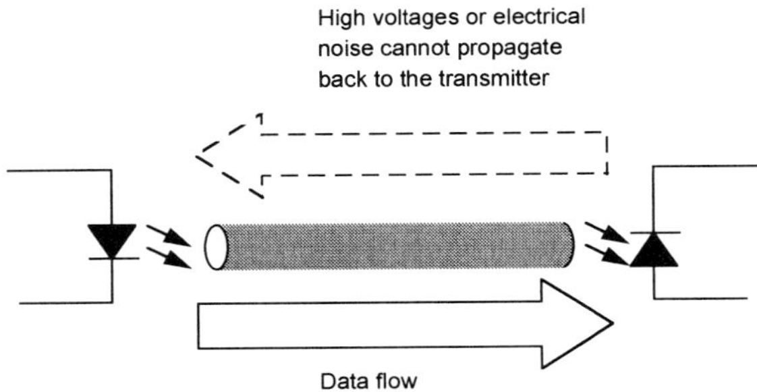

Figure 59.12 Fiber optic isolation.

60 Line Codes

60.1 Introduction

Data bits often have to be encoded before they can be stored or transmitted. This is typically done when:

- The bitstream needs to contain timing information, as a long run of the same digital value does not give much timing information.
- The Hamming distance between encoded values needs to be increased. For example, using a code which uses all the encoded bit values will mean that a single bit in error will lead to another valid code.
- To reduce DC wander, where the average value of the encoded bitstream is almost zero.
- To reduce power dissipation. Most of the information in a digital waveform is contained in the transitions, thus the DC component of the waveform does not carry much information. Therefore, the DC component of the signal leads to a waste of power. Typically, also, the power to remote equipment can be sent along the transmission line as the DC component. The signal line is then coupled to the transmission line through an isolated transformer.

60.2 NRZI

Non-return to zero (NRZI) is typically used in the transmission and storage of bits as it helps to reduces the DC content of the bitstream and also aids clock recovery. A 1 is coded with an alternative level, i.e. either a high (H) or a low (L). For example, the bitstream `110100` would be coded as:

Bitstream:	1	1	0	1	0	0
Waveform:	H	L	L	H	H	H

which has 2 lows and 4 highs, and the bitstream `0100010010` would be encoded as:

Bitstream:	0	1	0	0	0	1	0	0	1	0
Waveform:	L	H	H	H	H	L	L	L	H	H

which has 4 lows and 6 highs. This is illustrated in Figure 60.1.

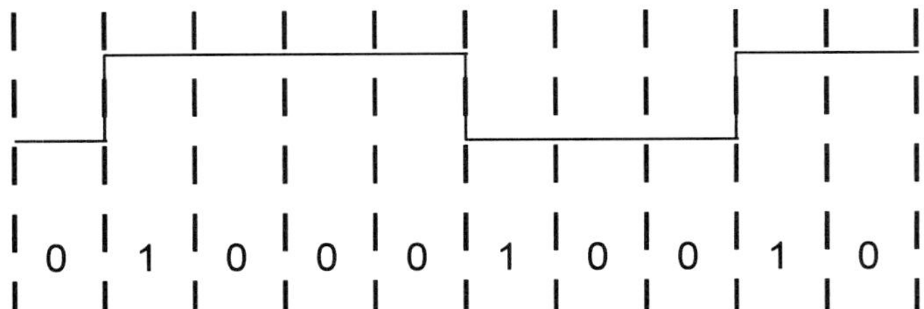

Figure 60.1 NRZI waveform for 0100010010.

Another way of viewing the NRZI encoding is that 1's are coded with alternative transitions, from low-to-high or a high-to-low. In this case the viewpoint of the waveform is at transitional points. For example, the bitstream 110100 would be viewed as:

Data:	1	1	0	1	0	0
Waveform:	LH	HL	LL	LH	HH	HH

which has 5 lows and 7 highs, and the bitstream `0100010010` would be encoded as:

Data:	0	1	0	0	0	1	0	0	1	0
Waveform:	LL	LH	HH	HH	HH	HL	LL	LL	LH	HH

9 lows and 11 highs. This is illustrated in Figure 60.2.

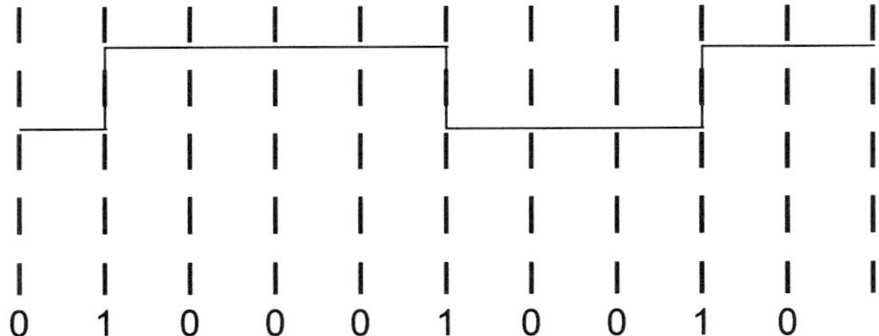

Figure 60.2 Viewpoint of NRZI waveform for 0100010010 around transitions.

60.3 DSV

An important measure of the DC content of a digital waveform is the DSV (digital sum value). This parameter keeps a running total of the deviation of the encoded waveform away from the zero level. Every high level adds one onto the value and a low takes one away from it. Thus taking the example of:

Bitstream:	1	1	0	1	0	0
Waveform:	L	H	L	H	H	H
	–	+	–	+	+	+

gives a DSV of +2. The bitstream `0100010010` would be encoded as follows:

Bitstream:	0	1	0	0	0	1	0	0	1	0
Waveform:	L	H	H	H	H	L	L	L	H	H
	–	+	+	+	+	–	–	–	+	+

which gives a DSV of +2. Note that if the first bit had a low level then the DSV would be −2, as shown next:

Bitstream:	0	1	0	0	0	1	0	0	1	0
Waveform:	H	L	L	L	L	H	H	H	L	L
	+	–	–	–	–	+	+	+	–	–

Thus it can be seen that this code will always have a DSV magnitude of 2. The technique of alternating inverting the levels can be used to either produce a positive or a negative DSV, which can be used to help the current DSV value.

60.4 4B/5B code

The 4B/5B is used in many applications, including FDDI and MADI (multichannel audio digital interface). It encodes four bits into five bit and the encoding table has been chosen so that they have the least DC component combined with a high clock content. The maximum run of zero's is limited to 3.

It can be seen that the DSV for the encoded bitstream for `11110` is:

Bitstream:	1	1	1	1	0
Waveform:	L	H	L	H	H
	–	+	–	+	+

is +1 (or −1 if the previous 1 bit is a Low).

Table 60.1 4B/5B encoding.

4-bit data	5-bit encoded data	4-bit data	5-bit encoded data
0000	11110	0001	01001
0001	10100	0011	10101
0100	01010	0101	01011
0110	01110	0111	01111
1000	10010	1001	10011
1010	10110	1011	10111
1100	11010	1101	11011
1110	11100	1111	11101

60.5 EFM (eight-to-fourteen modulation)

The EFM code is used in Compact Discs where 8-bit symbols are represented by a 14-bit code. There are 256 combinations of 8-bit symbols out of a possible 16 384 (2^{14}) different encoded values. Table 60.2 shows a portion of the codetable. The code has been chosen so that the maximum run-length of 0's will never be greater that 11.

Table 60.2 EFM encoding.

Hex	*8-bit data*	*14-bit encoded data*	*Hex*	*8-bit data*	*14-bit encoded data*
64	01100100	01000100100010	72	01110010	11000100100010
65	01100101	00000000100010	73	01110011	00100001000010
66	01100110	01000000100100	74	01110100	00000010000010
67	01100111	00100100100010	75	01110101	00000010000010
68	01101000	01001001000010	76	01110110	00010010000010
69	01101001	10000001000010	77	01110111	00100010000010
6A	01101010	10010001000010	78	01111000	01001000000010
6B	01101011	10001001000010	79	01111001	00001001001000
6C	01101100	01000001000010	7A	01111010	10010000000010
6D	01101101	00000001000010	7B	01111011	10001000000010
6E	01101110	00010001000010	7C	01111100	01000000000010
6F	01101111	00100001000010	7F	01111101	00001000000010
70	01110000	10000000100010	7E	01111110	00010000000010
71	01110001	10000010000010	7F	01111111	00100000000010

For example, the DSV for the first three codes is:

14-bit encoded data	*Waveform*	*DSV*
01000100100010	LHHHHLLLHHHHLL	+2
00000000100010	LLLLLLLLHHHHLL	−6
01000000100100	LHHHHHHHLLLHHH	+6

It can be seen that these codes give a DSV between +6 and −6. For this purpose CD discs have 3 extra packing bits added, these bits try to bring the DSV back to 0. It also inverts the code to give the inverse (for example the 01000100100010 can give either a DSV of +2 (LH HHHL LLHH HHLL) or −2 (HL LLLH HHLL LLHH).

60.6 5B6B

The 100VG-AnyLAN standard uses 5B6B to transmit an Ethernet frame between the hub and the node. This code is used so that there is an increase the number of transitions in the transmitted waveform.

In 100VG-AnyLAN, a 100 Mbps bitstream is multiplexed onto four 25 Mbps streams and transmitted over the 4 twisted-pair cables. The encoding process thus increases the bit rate on each twisted pair cable to 30Mbps (as six encoded bits are sent for every 5 bitstream bits). This is illustrated in Figure 39.2 and Table 60.3 gives the 5B/6B encoding.

Unfortunately, it is not possible to code each one of the 6-bit encoded values with an equal number of 0's and 1's as there are only 20 encoded values which have an equal number of 0's and 1's, these are highlighted in Table 60.3. Thus two modes are used with the other 12 values having either two 0's and four 1's or four 0's and two 1's. The data is then transmitted in two modes:

- Mode 2. Where the encoded data has either an equal number of 0's and 1's or has four 1's and two 0's.
- Mode 4. Where the encoded data has either an equal number of 1's and 0's or has four 0's and two 1's.

These modes alternate and this gives, on average, DSV of zero.

Table 60.3 5B/6B encoding

5-bit data	*Mode 2 encoding*	*Mode 4 encoding*	*5-bit data*	*Mode 2 encoding*	*Mode 4 encoding*
00000	001100	110011	10000	000101	111010
00001	**101100**	101100	10001	**100101**	100101
00010	100010	101110	10010	001001	110110
00011	**001101**	001101	10011	**010110**	010110
00100	001010	110101	10100	**111000**	111000
00101	**010101**	010101	10101	011000	100111
00110	**001110**	001110	10110	**011001**	011001
00111	**001011**	001011	10111	100001	011110
01000	**000111**	000111	11000	**110001**	110001
01001	**100011**	100011	11001	**101010**	101010
01010	**100110**	100110	11010	010100	101011
01011	000110	111001	11011	**110100**	110100
01100	101000	010111	11100	**011100**	011100
01101	**011010**	011010	11101	**010011**	010011
01110	100100	100100	11110	010010	101101
01111	**101001**	101001	11111	**110010**	110010

60.7 8B6T

100BASE-4T uses four separate Cat-3 twisted-pair wires. The maximum clock rate that can be applied to Cat-3 cable is 30 Mbps. Thus, some mechanism must be devised which can reduce the line bit rate to under 30 Mbps but give a symbol rate of 100 Mbps. This is achieved with a 3-level code (+, – and 0) and is known as 8B6T. The code converts eight binary digits into six ternary symbols. Table 60.4 gives the part of the codetable. Thus, the bit sequence 00000000 will be coded as a positive voltage, a negative voltage, a zero voltage, a zero voltage, a positive voltage and a negative voltage.

Apart from reducing the frequencies with the digital signal, the 8B6T code has the advantage of reducing the dc content of the signal. Most of the codes contain the same number of positive and negative voltages. This is because only 256 of the possible 729 (3^6) codes are actually used. The codes have also chosen to have at least two transitions in every code word, so that clock information is embedded into signal.

Unfortunately, it is not possible to have all codes with the same number of negative voltages as positive voltages. Thus there are some codes which have a different number of negatives and positives, these include:

```
0100 0001    +0-00++
0111 1001    +++-0-
```

The technique used to overcome this is to invert consecutive codes which have a weighting of +1. For example, suppose the line code were:

Table 60.4 8B6T code.

8-bit data	*Encoded data*	*8-bit data*	*Encoded data*
00000000	+-00+-	00010000	+0+--0
00000001	0+-+-+	00010001	++0-0
00000010	+-0+-0	00010010	+0+-0-
00000011	-0++-0	00010011	0++-0-
00000100	-0+0+-	00010100	0++--0
00000101	0+--0+	00010101	++00--
00000110	+-0-0+	00010110	+0+0--
00000111	-0+-0+	00010111	0++0--
00001000	-+00+-	00011000	0+-0+-
00001001	0-++-0	00011001	0+-0-+
00001010	-+0+-0	00011010	0+-++-
00001011	+0-+-0	00011011	0+-00+
00001100	+0-0+-	00011100	0-+00+
00001101	0-+-0+	00011101	0-+++-
00001110	-+0-0+	00011110	0-+0-+
00001111	+0--0+	00011111	0-+0+-

100BASE-4T uses four separate Cat-3 twisted-pair wires. The maximum clock rate that can be applied to Cat-3 cable is 30 Mbps. Thus, some mechanism must be devised which can reduce the line bit rate to under 30 Mbps but give a symbol rate of 100 Mbps. This is achieved with a 3-level code (+, – and 0) and is known as 8B6T. The code converts eight binary digits into six ternary symbols. Table 60.4 gives the part of the codetable. Thus, the bit sequence 00000000 will be coded as a positive voltage, a negative voltage, a zero voltage, a zero voltage, a positive voltage and a negative voltage.

Apart from reducing the frequencies with the digital signal, the 8B6T code has the advantage of reducing the dc content of the signal. Most of the codes contain the same number of positive and negative voltages. This is because only 256 of the possible 729 (3^6) codes are actually used. The codes have also chosen to have at least two transitions in every code word, so that clock information is embedded into signal.

Unfortunately, it is not possible to have all codes with the same number of negative voltages as positive voltages. Thus there are some codes which have a different number of negatives and positives, these include:

```
0100 0001    +0-00++
0111 1001    +++-0-
```

The technique used to overcome this is to invert consecutive codes which have a weighting of +1. For example, suppose the line code were:

+0++−−   ++0+−−   +++−−0 +++−−0

it would actually be coded as:

+0++−−   −−0−++   +++−−0 −−−++0

The receiver detects the −1 weighted codes as an inverted pattern.

60.8 8/10 code

The 8/10 code is used in DAT (digital audio tape) recordings where 8-bit symbols are represented by a 10-bit code. There are 256 combinations of 8-bit symbols out of a possible 1024 (2^{10}) different encoded values. Table 60.5 shows a portion of the code-table. It is used to zero DC and suppress low frequencies.

Table 60.5 Portion of the 8/10 encoding table.

8-bit data	*10-bit encoded*	*Alternative 10-bit encoded*
`00010000`	`1101010010`	
`00010001`	`0100010010`	`01110011`
`00010010`	`0101010010`	
`00010011`	`0101110010`	
`00010100`	`1101110001`	`01110111`
`00010101`	`1101110011`	`01111000`
`00010110`	`1101110110`	`01111001`

Some of the codes have a maximum DSV of +2. For example, `0100010010` and a previous 1 of a Low will give:

Bitstream:	1	1	0	1	0	1	0	0	1	0
Waveform:	H	L	L	H	H	L	L	L	H	H
	+	−	−	+	+	−	−	−	+	+

which gives a DSV of 0. If the previous 1 was a High then:

Bitstream:	1	1	0	1	0	1	0	0	1	0
Waveform:	L	H	H	L	L	H	H	H	L	L
	−	+	+	−	−	+	+	+	−	−

which also gives a DSV of 0, thus there is no DC offset. Taking the `0100010010` as an example and if the previous 1 was a Low then:

Bitstream:	0	1	0	0	0	1	0	0	1	0
Waveform:	L	H	H	H	H	L	L	L	H	H
	–	+	+	+	+	–	–	–	+	+

which gives a DSV is +2. It can be shown that if the previous 1 was a High then the DSV will be –2. The alternative code to this is `1100010010` which is encoded as:

Bitstream:	1	1	0	0	0	1	0	0	1	0
Waveform:	H	L	L	L	L	H	H	H	L	L
	+	–	–	–	–	+	+	+	–	–

which has a DSV of –2 and can be used to offset a +2 DSV. It can be seen that the only difference between the two codes is the first bit which is either a 0 or a 1. Thus:

- If the current DSV is +2 and the last 1 bit was a Low then the `1100010010` can be used to bring the DSV back to 0.
- If the current DSV is –2 and the last 1 bit was a Low then the `0100010010` can be used to bring the DSV back to 0.
- If the current DSV is +2 and the last 1 bit was a High then the `0100010010` can be used to bring the DSV back to 0.
- If the current DSV is –2 and the last 1 bit was a High then the `1100010010` can be used to bring the DSV back to 0.

61 Transmission Lines

61.1 Introduction

Digital pulses are affected by transmission systems in the following ways:

- They are attenuated along the line.
- The transmission line acts as a low-pass filter, blocking high frequencies.
- Different frequencies within pulses travel at different rates causing phase distortion of the pulse.
- Spreading of the pulses, causing them it to interfere with other pulses.
- Mismatches on the line cause reflections (and thus 'ghost pulses').

61.2 Equivalent circuit

A transmission line transmits electrical signals from a source to a receiver. It can be a coaxial cable, a twisted-pair cable, a waveguide, and so on. From a circuit point of view, the conductors of a transmission line contain series resistance and inductance and the insulation between conductors has a shunt conductance and capacitance.

If a given length of this transmission line were divided into more and more sections, the ultimate case would be an infinitesimal section of the basic elements, resistance R, conductance G, inductance L and capacitance C. The circuit that results is given in Figure 61.1. Parameters R, L, G and C are known as the primary line constants and are defined as:

- Series resistance R $\Omega.\text{meter}^{-1}$
- Series inductance L H.meter^{-1}
- Shunt conductance G S.meter^{-1} ($\text{Siemens.meter}^{-1}$)
- Shunt capacitance C F.meter^{-1}

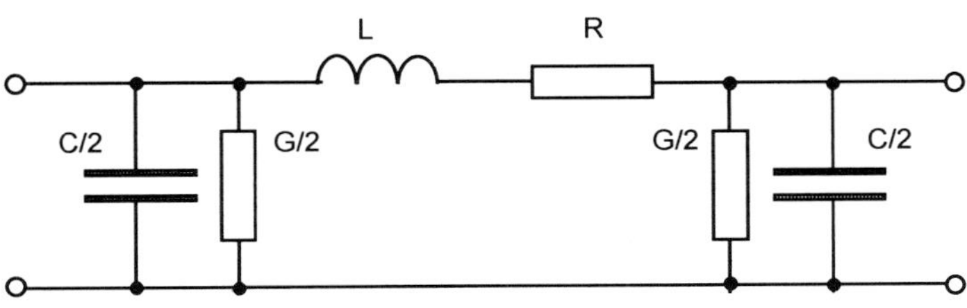

Figure 61.1 Equivalent circuit of a transmission line.

The characteristic impedance (Z_0) is the ratio of the voltage to the current for each wave propagated along a transmission line, and is given by:

$$Z_0 = \frac{V}{I}\ \Omega$$

From this it can be shown that:

$$Z_0 = \sqrt{\frac{R + j\omega L}{G + j\omega C}}\ \Omega$$

The characteristic impedance can either be measured by using an infinite length of transmission line and measuring the input impedance or by suitably terminating the line, as shown in Figure 61.2.

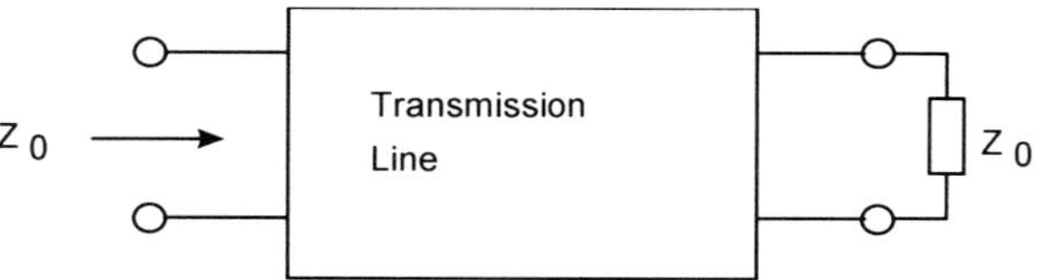

Figure 61.2 Transmission line terminated with Z_0.

The magnitude of the characteristic impedance will thus be given by:

$$Z_0 = \sqrt{\frac{\sqrt{R^2 + (2\pi f L)^2}}{\sqrt{G^2 + (2\pi f C)^2}}}\ \Omega$$

The characteristic impedance can be approximated for certain conditions:

(i) when the frequency is large, then

$\omega L >> R$ $\qquad$ $\omega C >> G$

thus $\quad Z_o = \sqrt{\frac{L}{C}}\ \Omega$

(ii) on a lossless transmission line then R = 0 and G = 0,

thus: $\quad Z_o = \sqrt{\frac{L}{C}}\ \Omega$

61.2.1 Program to determine characteristic impedance

Program 61.1 determines the magnitude of the characteristic impedance for a transmission line with entered primary line constants.

Program 61.1

```
/*    Program to determine impedance of a transmission   */
/*    line impedance of TL                                */
#include <stdio.h>
#include <math.h>

#define MILLI            1e-3
#define MICRO            1e-6
#define INFINITYFLAG     -1
#define PI               3.14159

float    calc_Zo(float r,float l,float g,float c,float f);
float    calc_mag(float x,float y);
float    calc_imp(float f, float val);

int      main(void)
{
float    Zmag,R,L,G,C,f;

      puts("Program to determine impedance of a transmission line");
      printf("Enter R,L(mH),G(mS),C(uF) and freq.>>");
      scanf("%f %f %f %f %f",&R,&L,&G,&C,&f);
      Zmag=calc_Zo(R,L*MILLI,G*MILLI,C*MICRO,f);
      if (Zmag==INFINITYFLAG)
            printf("Magnitude is INFINITY ohms\n");
      else
            printf("Magnitude is %.2f ohms\n",Zmag);
      return(0);
}

float    calc_Zo(float r,float l,float g,float c,float f)
{
float    value1,value2;

      value1=calc_mag(r,calc_imp(f,l));
      value2=calc_mag(g,calc_imp(f,c));
      /* Beware if dividing by zero */
      if (value2==0)      return(INFINITYFLAG);
      else                return(sqrt(value1/value2));
}
float    calc_mag(float x,float y)
{
      return(sqrt((x*x)+(y*y)));
}
float    calc_imp(float f, float val)
{
      return(2*PI*f*val);
}
```

Test run 61.1 shows a sample test run.

Test run 61.1

```
Program to determine impedance of a transmission line
Enter R, L (mH), G (mS), C (uF) and freq.>> 0 40 0 7 1000
Magnitude is 75.59 ohms
```

61.2.2 Propagation coefficient, γ

The propagation coefficient γ determines the variation of current, or voltage, with respect to distance x along a transmission line, as shown in Figure 61.3. The current (and voltage) distribution along a matched line is found to vary exponentially with distance, as given next:

$$I_x = I_s e^{-\gamma x}$$

$$V_x = V_s e^{-\gamma x}$$

where I_s is the magnitude of current at x=0 and V_s the magnitude of voltage at x=0.

Like the characteristic impedance, the propagation coefficient also dependent on the primary constants and the frequency of the signal, and is given by:

$$Z_o = \sqrt{(R + j\omega L)(G + j\omega C)}$$

This is a complex quantity and can be written as:

$$\gamma = \alpha + j\beta$$

The attenuation coefficient, α (nepers.meter^{-1}), determines how the voltage or current amplitude varies with distance along the line. The phase shift coefficient, β (radians.meter^{-1}), determines the phase angle of the voltage (or current) variation with distance. Since a phase shift of 2π radians (or 360°) occurs over a distance of one wavelength, λ , then

$$\beta = \frac{2\pi}{\lambda}$$

where λ is the physical wavelength.

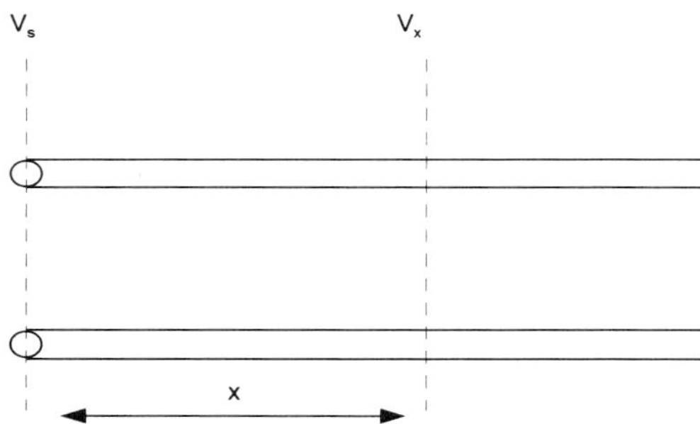

Figure 61.3 Voltage at a distance x.

61.3 Speed of propagation

The velocity of propagation u is given by:

$$u = f\lambda = \frac{2\pi}{\beta} \text{ m.s}^{-1}$$

On a lossless line, or at high frequencies then:

$$\beta = \omega\sqrt{LC}$$

$$u = \frac{f2\pi}{\omega\sqrt{LC}}$$

$$= \frac{1}{\sqrt{LC}} \text{ m.s}^{-1}$$

It can be found using field theory and the geometry of normal lines to calculate the inductance and capacitance that:

$$u = \frac{1}{\sqrt{\mu\varepsilon}} = \frac{1}{\sqrt{\mu_o\varepsilon_o\varepsilon_r}} = \frac{c}{\sqrt{\varepsilon_r}} \text{ m.s}^{-1}$$

since $\mu_r = 1$ and $c = \dfrac{1}{\sqrt{\mu_o\varepsilon_o}}$. Where c is the velocity of light (m.s^{-1}) and ε_r is the dielectric constant of transmission line.

61.4 Transmission line reflections

When a pulse meets a mismatch in an electrical circuit a reflected pulses is bounced back of the mismatch. This 'ghost' pulse is then back along the transmission line. They also cause a loss of signal power.

The characteristic impedance of transmission lines and terminations is the important factor in minimizing reflections and maximizing power transfer. Typically transmission lines, such as coaxial cables, have characteristic impedances of 50 Ω, for TV and video it 75 Ω.

61.4.1 Reflections from resistive termination's

A matched termination is when a transmission line is terminated with a resistance equal to Z_0. The line then behaves as if it has infinite length and there is thus no reflected energy.

If the line is terminated in any resistance other than Z_0, then energy is reflected from the termination. The amplitude of a reflected pulse can be determined by the impedance of the load and the characteristic impedance of the line.

The transmission line in Figure 61.4 is terminated with a resistance which is greater that Z_0. The lattice and surge diagrams show the amplitude of the pulse against time. On the surge diagram pulses are plotted against a horizontal time axis. In the lattice diagram time is plotted on the vertical axis. Both diagrams show the pulse approaching the load. In Figure 61.5 the pulse has reached the load and part of the pulse is reflected. The rest of the pulses energy is transmitted to the load.

V_i is the incident pulse voltage, V_r the reflected pulse voltage and V_t the transmitted pulse voltage. It can be shown that:

$$V_t = V_i + V_r$$

using this, the reflection coefficient (ρ) can be derived:

$$\rho = \frac{V_r}{V_i} = \frac{Z_L - Z_o}{Z_L + Z_o}$$

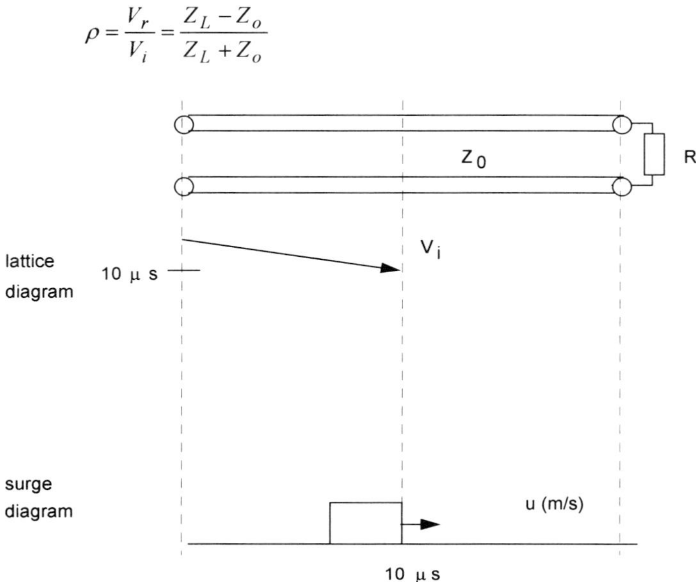

Figure 61.4 Pulse as it approaches the load.

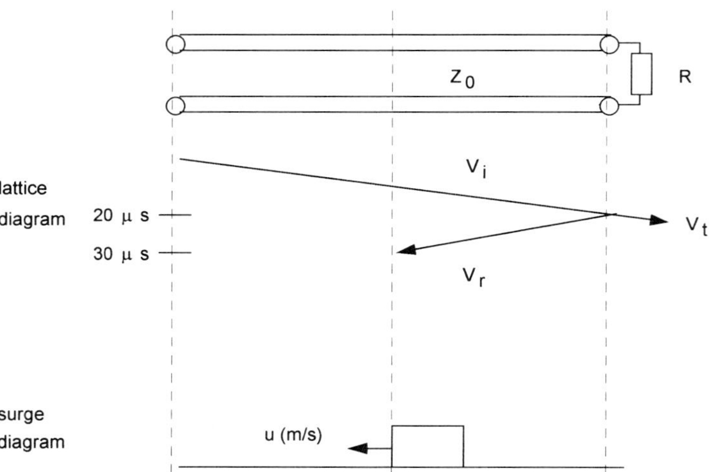

Figure 61.5 Reflected pulse from load.

If the load impedance is equal to the characteristic impedance (Z_0) then no pulse is reflected from the load (that is, ρ=0). The transmitted pulse voltage will have the same voltage as the incident pulse.

When the load impedance is larger than the transmission line characteristic impedance then the reflected pulse is positive (that is, ρ>0). This pulse travels back along the line towards the source. The resulting transmitted pulse voltage will be larger than the incident pulse as it is the addition of the incident and the reflected pulse voltage.

When the load impedance is less than the transmission line characteristic impedance then the pulse reflected is negative (that is, $\rho<0$) and travels back along the line toward the source. In this case, the transmitted pulse will be less than the incident pulse.

Note that although the transmitted pulse voltage is increased or decreased the electric current also changes. As the voltage increases and the current decreases (or vice versa) it can be proved that there is no increase in electric power. In fact, the reflected pulse gives the loss in transmitted power. It can be shown that the reflection coefficient for current is equal to the negative of the voltage reflection coefficient.

61.4.2 Reflections at junctions between two transmission lines

If two transmission lines are joined, one with a characteristic impedance of Z_{01} and the other of Z_{02} then the reflection coefficient is given by:

$$\rho = \frac{Z_{02} - Z_{01}}{Z_{02} + Z_{01}}$$

For example, if a pulse of 3 V traveling along a cable with $Z_{01} = 50\,\Omega$ meets a cable with characteristic impedance $Z_{02} = 100\,\Omega$, as shown in Figure 61.6, then the reflected and transmitted pulses can be found by:

$$\rho = \frac{100-50}{100+50} = \frac{1}{3} = \frac{V_r}{V_i}$$

$$V_r = \rho V_i = \frac{1}{3}.3 = 1\,V$$

$$\begin{aligned} V_t &= V_i + V_r \\ &= 3+1 = 4\,V \end{aligned}$$

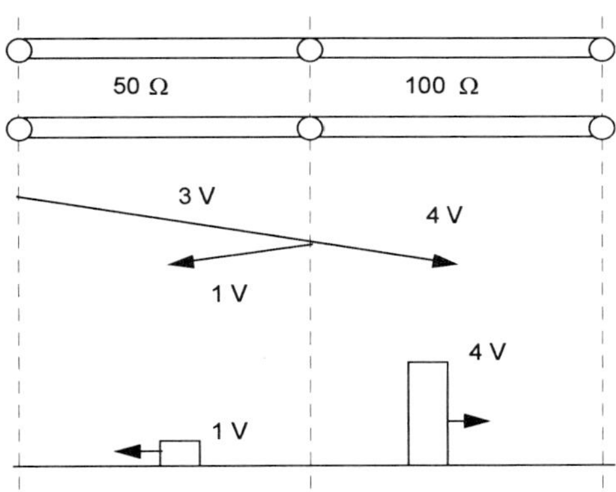

Figure 61.6 Reflection from a junction.

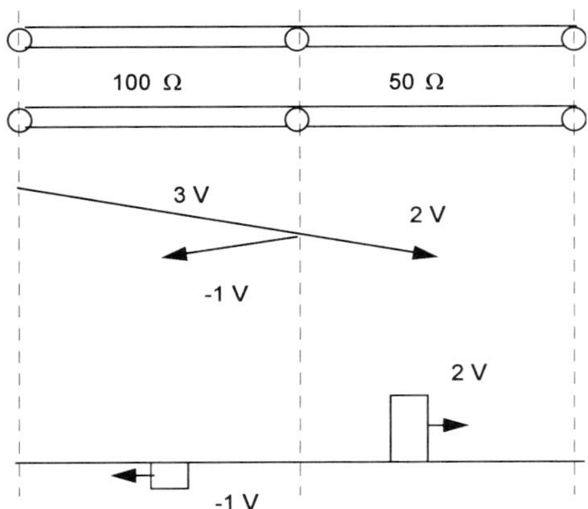

Figure 61.7 Reflection from a junction.

If the cables were changed so that Z_{01} is 100 Ω and Z_{02} is 50 Ω, as shown in Figure 61.7, then the reflected and transmitted pulses would be determined by:

$$\rho = \frac{50-100}{50+100} = -\frac{1}{3}$$

$$V_r = \rho V_i = -\frac{1}{3}.3 = -1V$$

$$V_t = V_i + V_r$$

$$= 3 - 1 = 2V$$

61.4.3 Reflections at junctions with two transmission line in parallel

In many situations two or more transmission lines are connected in parallel to a source transmission line. Figure 61.8 shows the two parallel lines connected to a single source line. If the source line has a characteristic impedance of Z_{01} and the two parallel line have characteristic impedances of Z_{02} and Z_{03}, then the equivalent input impedance at the termination is Z_{02} in parallel with Z_{03}. The equivalent load impedance at the junction between the source line and the parallel lines will be:

$$Z_p = \frac{Z_{02}Z_{03}}{Z_{02}+Z_{03}} \Omega$$

and the reflection coefficient by:

$$\rho = \frac{Z_p - Z_{01}}{Z_p + Z_{01}}$$

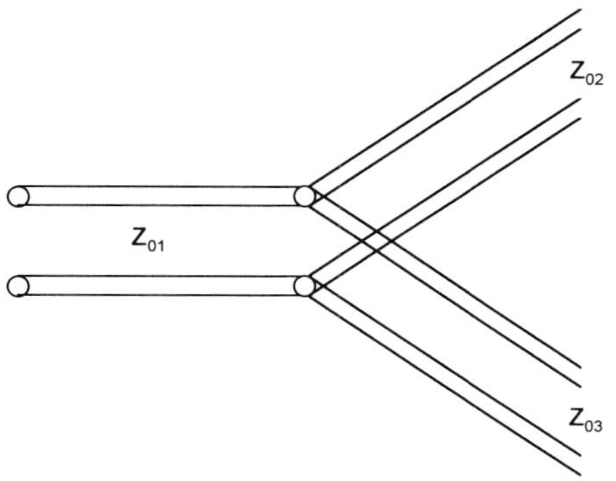

Figure 61.8 Connection to parallel transmission lines.

61.5 Matching terminations

Terminations can be matched by either inserting a series or a parallel resistance. If the characteristic impedance of the source line is higher than the connecting characteristic impedance then a resistance equal to the difference in the impedance is added in series with the connecting line. There will then be no reflections from the load transmission line (although there is a loss of power).

If the characteristic impedance of the source line is lower than the load characteristic impedance then a resistor is inserted in parallel with the junction. To match the junction the equivalent input impedance of the termination should be equal to that of the connecting line as shown in Figure 61.9.

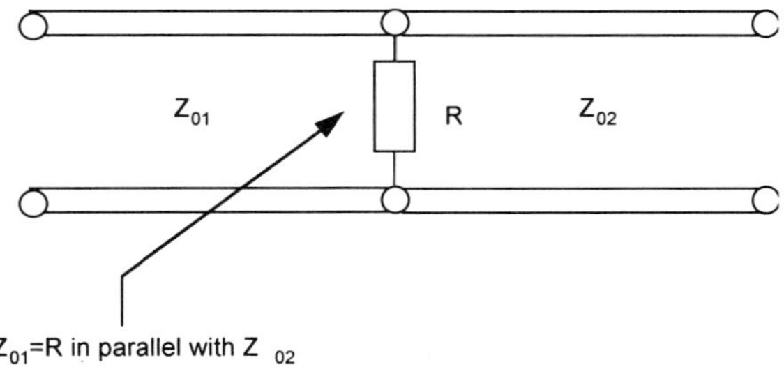

Figure 61.9 Matching a termination.

If a transmission line with characteristic impedance Z_{02} is the connected to a line with characteristic impedance Z_{01} and the pulse originates from the line with characteristic impedance Z_{01}, then:

if $Z_{01} > Z_{02}$ then series resistor required is $Z_{01} - Z_{02}\ \Omega$;

if $Z_{01} < Z_{02}$ then parallel resistor required is $\frac{Z_{02}Z_{01}}{Z_{02}+Z_{01}}\ \Omega$.

Example

(a) A uniform transmission line is 100 km long and has the following primary constants per km:

R	=	0	$\Omega.km^{-1}$
L	=	40	$mH.km^{-1}$
C	=	7	$\mu F.km^{-1}$
G	=	0	$S.km^{-1}$

If a digital pulse of 10 V is applied at the input of the line and the receiving end has a 125 Ω equivalent load, calculate:

(i) speed of propagation of the pulse;
(ii) time for pulse to reach load;
(iii) the characteristic impedance of the line;
(iv) the reflected pulse amplitude;
(v) the transmitted pulse;
(vi) the value of resistor required to match load.

ANSWER

(i) speed of propagation (v):

$$v = \frac{1}{\sqrt{LC}}$$
$$= \frac{1}{\sqrt{40\times10^{-3}\times7\times10^{-6}}}$$
$$= 1889.8\ \text{km.s}^{-1}$$

(ii) time taken (t):

$$\text{speed} = \frac{\text{distance}}{\text{time}}$$
$$\text{time} = \frac{\text{distance}}{\text{speed}}$$
$$\text{time taken} = \frac{d}{v} = \frac{100\times10^{-3}}{1889.8\times10^{3}}$$
$$= 52.9\ ms$$

(iii) the characteristic impedance (Z_0):

$$Z_0 = \sqrt{\frac{R + j\omega L}{G + j\omega C}}$$

$$Z_0 = \sqrt{\frac{L}{C}}$$

$$Z_0 = \sqrt{\frac{40\times10^{-3}}{7\times10^{-6}}}$$
$$= 75.6\ \Omega$$

(iv) the reflected pulse amplitude (V_r):

$$\rho = \frac{Z_L - Z_0}{Z_L + Z_0}$$
$$= \frac{125 - 75.6}{125 + 75.6}$$
$$= 0.25$$
$$V_r = \rho V_i$$
$$= 0.25\times10$$
$$= 2.5\ V$$

(v) the transmitted pulse amplitude (V_t):

$$V_t = V_i + V_r$$
$$= 10 + 2.5$$
$$= 12.5\ V$$

(vi) resistor to match load (R):

$$\frac{125.R}{125 + R} = 76.5$$
$$125R = 75.6R + 9450$$
$$49.4R = 9450$$
$$R = 191.3\ \Omega$$

61.6 Open and short circuit terminations

A short circuit has a zero impedance. If this were used as a load then the reflection coefficient is –1. The impedance of an open-circuit is infinity, and gives a reflection coefficient of +1. Thus with an open-circuit load, the reflected pulse is equal to the incident pulse. For a short-circuited load the reflected pulse is the same magnitude, but will be negative. The proofs for the reflection coefficients for an open and short circuit load are given next:

$$\rho_{S/C} = \frac{Z_L - Z_0}{Z_L + Z_0} = \frac{0 - Z_0}{0 + Z_0} = -1$$

$$\rho_{O/C} = \frac{\infty - Z_0}{\infty + Z_0} = \frac{\infty}{\infty} = +1$$

A typical technique used to find faults on underground cables is to send a pulse along the line and measure the time taken for a reflection pulse to return. Since the speed of propagation of the pulse is known then the distance to the fault can be found by dividing the speed of propagation by half the time taken (or twice the speed of propagation divided by the time taken). For a short circuit on the line an inverted voltage pulse is returned, else, if there is an open circuit a positive voltage pulse is returned, as illustrated in Figure 61.10.

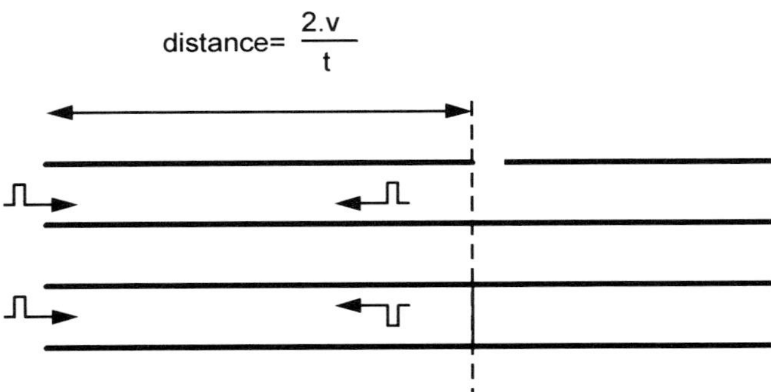

Figure 61.10 Testing for faults on a line.

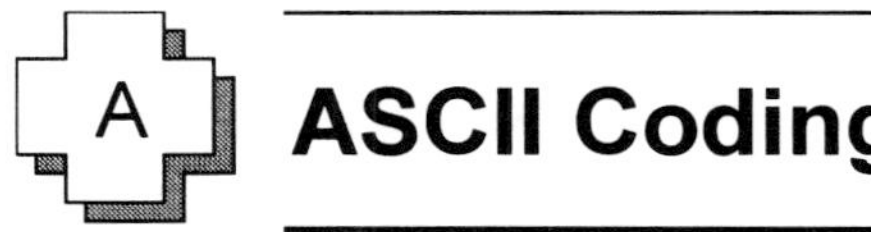

ASCII Coding

A.1 International alphabet No. 5

ANSI defined a standard alphabet known as ASCII. This has since been adopted by the CCITT as a standard, known as IA5 (International Alphabet No. 5). The following tables define this alphabet in binary, as a decimal value, as a hexadecimal value and as a character.

Binary	*Decimal*	*Hex*	*Character*	*Binary*	*Decimal*	*Hex*	*Character*
00000000	0	00	NUL	00010000	16	10	DLE
00000001	1	01	SOH	00010001	17	11	DC1
00000010	2	02	STX	00010010	18	12	DC2
00000011	3	03	ETX	00010011	19	13	DC3
00000100	4	04	EOT	00010100	20	14	DC4
00000101	5	05	ENQ	00010101	21	15	NAK
00000110	6	06	ACK	00010110	22	16	SYN
00000111	7	07	BEL	00010111	23	17	ETB
00001000	8	08	BS	00011000	24	18	CAN
00001001	9	09	HT	00011001	25	19	EM
00001010	10	0A	LF	00011010	26	1A	SUB
00001011	11	0B	VT	00011011	27	1B	ESC
00001100	12	0C	FF	00011100	28	1C	FS
00001101	13	0D	CR	00011101	29	1D	GS
00001110	14	0E	SO	00011110	30	1E	RS
00001111	15	0F	SI	00011111	31	1F	US

Binary	*Decimal*	*Hex*	*Character*	*Binary*	*Decimal*	*Hex*	*Character*
00100000	32	20	SPACE	00110000	48	30	0
00100001	33	21	!	00110001	49	31	1
00100010	34	22	"	00110010	50	32	2
00100011	35	23	#	00110011	51	33	3
00100100	36	24	$	00110100	52	34	4
00100101	37	25	%	00110101	53	35	5
00100110	38	26	&	00110110	54	36	6
00100111	39	27	/	00110111	55	37	7
00101000	40	28	(	00111000	56	38	8
00101001	41	29	)	00111001	57	39	9
00101010	42	2A	*	00111010	58	3A	:
00101011	43	2B	+	00111011	59	3B	;
00101100	44	2C	,	00111100	60	3C	<
00101101	45	2D	-	00111101	61	3D	=
00101110	46	2E	.	00111110	62	3E	>
00101111	47	2F	/	00111111	63	3F	?

Binary	*Decimal*	*Hex*	*Character*	*Binary*	*Decimal*	*Hex*	*Character*
01000000	64	40	@	01010000	80	50	P
01000001	65	41	A	01010001	81	51	Q
01000010	66	42	B	01010010	82	52	R
01000011	67	43	C	01010011	83	53	S
01000100	68	44	D	01010100	84	54	T
01000101	69	45	E	01010101	85	55	U
01000110	70	46	F	01010110	86	56	V
01000111	71	47	G	01010111	87	57	W
01001000	72	48	H	01011000	88	58	X
01001001	73	49	I	01011001	89	59	Y
01001010	74	4A	J	01011010	90	5A	Z
01001011	75	4B	K	01011011	91	5B	[
01001100	76	4C	L	01011100	92	5C	\
01001101	77	4D	M	01011101	93	5D	]
01001110	78	4E	N	01011110	94	5E	`
01001111	79	4F	O	01011111	95	5F	_

Binary	*Decimal*	*Hex*	*Character*	*Binary*	*Decimal*	*Hex*	*Character*
01100000	96	60		01110000	112	70	p
01100001	97	61	a	01110001	113	71	q
01100010	98	62	b	01110010	114	72	r
01100011	99	63	c	01110011	115	73	s
01100100	100	64	d	01110100	116	74	t
01100101	101	65	e	01110101	117	75	u
01100110	102	66	f	01110110	118	76	v
01100111	103	67	g	01110111	119	77	w
01101000	104	68	h	01111000	120	78	x
01101001	105	69	i	01111001	121	79	y
01101010	106	6A	j	01111010	122	7A	z
01101011	107	6B	k	01111011	123	7B	{
01101100	108	6C	l	01111100	124	7C	:
01101101	109	6D	m	01111101	125	7D	}
01101110	110	6E	n	01111110	126	7E	~
01101111	111	6F	o	01111111	127	7F	DEL

A.2 Extended ASCII code

The standard ASCII character has 7 bits and the basic set ranges from 0 to 127. This code is rather limited as it does not contains symbols such as Greek letters, lines, and so on. For this purpose the extended ASCII code has been defined. This fits into character numbers 128 to 255. The following four tables define a typical extended ASCII character set.

Binary	*Decimal*	*Hex*	*Character*	*Binary*	*Decimal*	*Hex*	*Character*
10000000	128	80	Ç	10010000	144	90	É
10000001	129	81	ü	10010001	145	91	æ
10000010	130	82	é	10010010	146	92	Æ
10000011	131	83	â	10010011	147	93	ô
10000100	132	84	ä	10010100	148	94	ö
10000101	133	85	à	10010101	149	95	ò
10000110	134	86	å	10010110	150	96	û
10000111	135	87	ç	10010111	151	97	ù
10001000	136	88	ê	10011000	152	98	ÿ
10001001	137	89	ë	10011001	153	99	Ö
10001010	138	8A	è	10011010	154	9A	Ü
10001011	139	8B	ï	10011011	155	9B	¢
10001100	140	8C	î	10011100	156	9C	£
10001101	141	8D	ì	10011101	157	9D	¥
10001110	142	8E	Ä	10011110	158	9E	•
10001111	143	8F	Å	10011111	159	9F	ƒ

Binary	*Decimal*	*Hex*	*Character*	*Binary*	*Decimal*	*Hex*	*Character*
10100000	160	A0	á	10110000	176	B0	
10100001	161	A1	í	10110001	177	B1	
10100010	162	A2	ó	10110010	178	B2	
10100011	163	A3	ú	10110011	179	B3	
10100100	164	A4	ñ	10110100	180	B4	
10100101	165	A5	Ñ	10110101	181	B5	
10100110	166	A6	ª	10110110	182	B6	
10100111	167	A7	º	10110111	183	B7	
10101000	168	A8	¿	10111000	184	B8	
10101001	169	A9	•	10111001	185	B9	
10101010	170	AA	¬	10111010	186	BA	
10101011	171	AB	½	10111011	187	BB	
10101100	172	AC	¼	10111100	188	BC	
10101101	173	AD	¡	10111101	189	BD	
10101110	174	AE	«	10111110	190	BE	
10101111	175	AF	»	10111111	191	BF	

Binary	Decimal	Hex	Character	Binary	Decimal	Hex	Character
11000000	192	C0		11010000	208	D0	
11000001	193	C1		11010001	209	D1	
11000010	194	C2		11010010	210	D2	
11000011	195	C3		11010011	211	D3	
11000100	196	C4		11010100	212	D4	
11000101	197	C5		11010101	213	D5	
11000110	198	C6		11010110	214	D6	
11000111	199	C7		11010111	215	D7	
11001000	200	C8		11011000	216	D8	
11001001	201	C9		11011001	217	D9	
11001010	202	CA		11011010	218	DA	
11001011	203	CB		11011011	219	DB	
11001100	204	CC		11011100	220	DC	
11001101	205	CD		11011101	221	DD	
11001110	206	CE		11011110	222	DE	
11001111	207	CF		11011111	223	DF	

Binary	Decimal	Hex	Character	Binary	Decimal	Hex	Character
11100000	224	E0		11110000	240	F0	
11100001	225	E1		11110001	241	F1	
11100010	226	E2		11110010	242	F2	
11100011	227	E3		11110011	243	F3	
11100100	228	E4		11110100	244	F4	
11100101	229	E5		11110101	245	F5	
11100110	230	E6		11110110	246	F6	
11100111	231	E7		11110111	247	F7	
11101000	232	E8		11111000	248	F8	
11101001	233	E9		11111001	249	F9	
11101010	234	EA		11111010	250	FA	
11101011	235	EB		11111011	251	FB	
11101100	236	EC		11111100	252	FC	
11101101	237	ED		11111101	253	FD	
11101110	238	EE		11111110	254	FE	
11101111	239	EF		11111111	255	FF	

RLE Program

B.1 RLE program

Program B.1 is a very simple program which scans a file `IN.DAT` and, using RLE, stores to a file `OUT.DAT`. The special character sequence is:

ZZ*cxx*

where `ZZ` is the flag sequence, *c* is the repeditive character and *xx* the number of times the character occurs. The `ZZ` flag sequence is choosen because, in a text file, it is unlikely to occur within the file. File listing B.1 shows a sample `IN.DAT` and File listing B.2 shows the RLE encoded file (`OUT.DAT`).

Program B.1

```
/*    ENCODE.C    */
#include <stdio.h>
int   main(void)
{
FILE  *in,*out;
char  previous,current;
int   count;

      if ((in=fopen("in.dat","r"))==NULL)
      {
         printf("Cannot open <in.dat>");
         return(1);
      }
      if ((out=fopen("out.dat","w"))==NULL)
      {
         printf("Cannot open <out.dat>");
         return(1);
      }
      do {
         count=1;
         previous=current;
         current=fgetc(in);
         do    {
            previous=current;
            current=fgetc(in);
            if (previous!=current) ungetc(current,in);
            else count++;
         } while (previous==current);
         if (count>1) printf(out,"ZZ%c%02d",previous,count);
         else fprintf(out,"%c",previous);
      }  while (!feof(in));
      fclose(in); fclose(out);
      return(0);
}
```

File list B.1

```
The     bbbbbbboy stood onnnnn the burning
deck         and still did.
1.000000000
3.000000010
5.000000000
```

File list B.2

```
TheZZ 05ZZb07oy stZZo02d oZZn05 the burning
deckZZ 09and stiZZl02 did.
1.ZZ009
3.ZZ00710
5.ZZ009
```

Program B.2 gives a simple C program which unencodes the RLE file produced by the previous program.

Program B.2

```
/*    UNENCODE.C       */
#include <stdio.h>

int   main(void)
{
FILE  *in,*out;
char  ch;
int   count,i;

      if ((in=fopen("out.dat","r"))==NULL)
      {
         printf("Cannot open <out.dat>");
         return(1);
      }
      if ((out=fopen("in1.dat","w"))==NULL)
      {
         printf("Cannot open <in1.dat>");
         return(1);
      }

      do
      {
         ch=fgetc(in);

         if (ch=='Z')
         {
            ch=fgetc(in);
            if (ch=='Z')
            {
               fscanf(in,"%c%02d",&ch,&count);
               for (i=0;i<count;i++)
                  fprintf(out,"%c",ch);
            }
            else ungetc(ch,in);
         }
         else fprintf(out,"%c",ch);

      } while (!feof(in));
      fclose(in);
      fclose(out);
      return(0);
}
```

The ZZ flag sequence is inefficient as it uses two characters to store the flag; a better flag could be an 8-bit character that cannot occur, such as 11111111b or ffh. Program B.3 shows an example of this and Program B.4 shows the decoder.

Program B.3

```
#include <stdic.h>
#define  FLAG  0xff  /* 1111 1111b  */

int   main(void)
{
FILE  *in,*out;
char  previous,current;
int   count;

 ;;; ;;;;;
        if (ccunt>1) fprintf(out,"%c%c%02d",FLAG,previous,count);
        else fprintf(out,"%c",previous);
      }  while (!feof(in));
      fclose(in); fclose(out);   return(0);
}
```

Program B.4

```
/*    UNENCODE.C      */
#include <stdio.h>
#define  FLAG  0xff  /* 1111 1111b  */

int   main(void)
{
FILE  *in,*out;
char  ch;
int   count,i;

      ;;; ;;;;
      do
      {
         ch=fgetc(in);
         if (ch==FLAG)  {
            ch=fgetc(in);
            fscanf(in,"%c%02d",&ch,&count);
            for (i=0;i<count;i++) fprintf(out,"%c",ch);
         }
         else fprintf(out,"%c",ch);

      }  while (!feof(in));
      fclose(in);
      fclose(out);
      return(0);
}
```

In a binary file any bit sequence can occur. To overcome this, a flag sequence, such as 10101010 can be used to identify the flag. If this sequence occurs within the data, it will be coded with two flags two consecutive flags in the data are coded with three flags and so on.

For example: 00000000 10101010 10101010 00011100 01001100

would be encoded as: 00000000 10101010 10101010 10101010 00011100 01001100

thus when the three flags are detected, one of them is deleted.

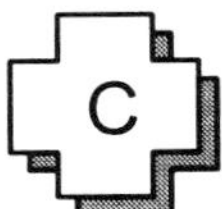

C SNR for PCM

C.1 SNR

If a waveform has a maximum signal amplitude of V, then the relative signal power will be:

$$\text{Signal power} = v_{rms}^2 = \left(\frac{V}{\sqrt{2}}\right)^2 = \frac{V^2}{2}$$

If n-bit PCM coding is used then there will be 2^n different levels, as illustrated in Figure C.1.

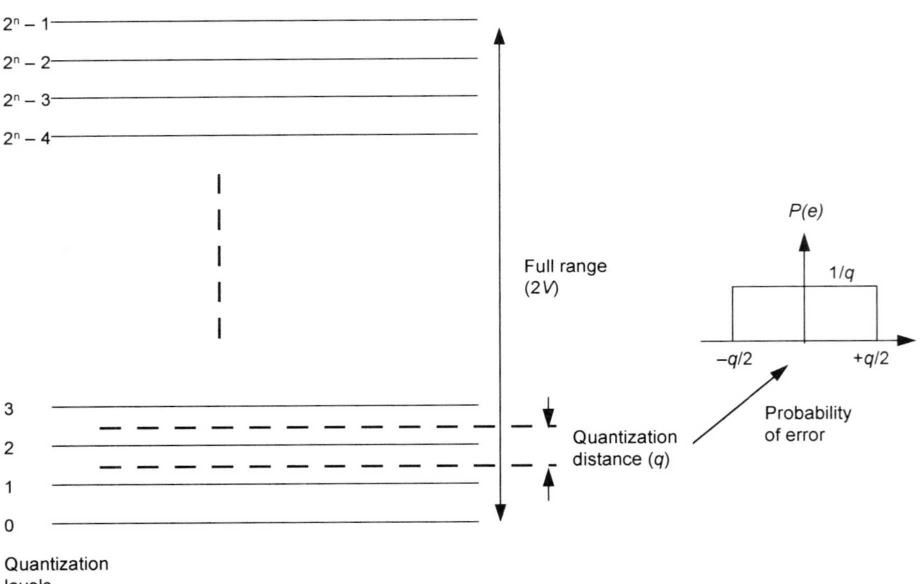

Figure C.1 Quantization.

Thus, if the input signal ranges between $+V$ and $-V$, the error in the quantization signal will range from:

$$+\frac{q}{2} \text{ to } -\frac{q}{2}$$

where q is the quantization distance and is given by:

$$q = \frac{2V}{2^n}$$

Figure C.1 shows that the probability of error will be constant from between $-q/2$ to $+q/2$. The area over the interval the *P(e)* should equal unity, hence the *y*-axis value for *P(e)* will be $1/q$. Thus the noise power will be:

$$\text{Noise power} = \int_{-\infty}^{+\infty} v^2 P(v) dv = 2 \int_{0}^{+q/2} \frac{v^2}{q} dv$$

$$= \frac{2}{q} \int_{0}^{+q/2} v^2 dv = \frac{2}{q} \left[\frac{v^3}{3} \right]_0^{+q/2} = \frac{2}{3q} \left(\frac{q^3}{2^3} \right)$$

$$= \frac{q^2}{12}$$

The signal-to-noise ratio will thus be:

$$\text{SNR(dB)} = 10 \log_{10} \left(\frac{\text{Signal power}}{\text{Noise power}} \right) = 10 \log_{10} \left(\frac{V^2/2}{q^2/12} \right) = 10 \log_{10} \left(\frac{6V^2}{q^2} \right)$$

$$= 10 \log_{10} \left(\frac{6V^2}{(2V)^2 / (2^n)^2} \right) = 10 \log_{10} \left(\frac{3}{2} (2^n)^2 \right)$$

$$= 10 \log_{10} (1.5) + 20 \log_{10} (2^\text{n}) = 10 \log_{10} (1.5) + 20 \log_{10} (2^n)$$

$$= 1.76 + 20 \text{ n} \log_{10} (2)$$

$$= 6.02 \text{ n} + 1.76$$

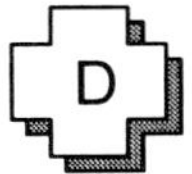

RFC Standards

The IAB (Internet Advisor Board) has published many documents on the TCP/IP protocol family. They are known as RFC (request for comment) and can be obtained using FTP from the following:

- Internet Network Information Center (NIC) at `nic.ddn.mil`, or one of several other FTP sites, such as from the InterNIC Directory and Database Services server at `ds.internic.net`.
- Through electronic mail from the automated InterNIC Directory and Database Services mail server at `mailserv@ds.internic.net`. The main body of the message should contain the command:

  `document-by-name rfc`*NNNN*

  where NNNN is the number of the RFC. Multiple requests can be made by sending a single message with each specified document separated by comma-separated list.

The main RFC documents are:

RFC768 User Datagram Protocol.
RFC775 Directory-Oriented FTP Commands.
RFC781 Specification of the Internet Protocol Timestamp Option.
RFC783 TFTP Protocol.
RFC786 User Datagram Protocol (UDP).
RFC791 Internet Protocol (IP).
RFC792 Internet Control Message Protocol (ICMP).
RFC793 Transmission Control Protocol (TCP).
RFC799 Internet Name Domains.
RFC813 Window and Acknowledgment in TCP.
RFC815 IP Datagram Reassembly Algorithms.
RFC821 Simple Mail-Transfer Protocol (SMTP).
RFC822 Standard for the Format of ARPA Internet Text Messages.
RFC823 DARPA Internet Gateway.
RFC827 Exterior Gateway Protocol (EGP).
RFC877 Standard for the Transmission of IP Datagrams over Public Data Networks.
RFC879 TCP Maximum Segment Size and Related Topics.
RFC886 Proposed Standard for Message Header Munging.
RFC893 Trailer Encapsulations.
RFC894 Standard for the Transmission of IP Datagrams over Ethernet Networks.
RFC895 Standard for the Transmission of IP Datagrams over Experimental Ethernet Networks.
RFC896 Congestion Control in TCP/IP Internetworks.
RFC903 Reverse Address Resolution Protocol.
RFC904 Exterior Gateway Protocol Formal Specifications.
RFC906 Bootstrap Loading Using TFTP.
RFC919 Broadcast Internet Datagram.

RFC920	Domain Requirements.
RFC932	Subnetwork Addressing Schema.
RFC949	FTP Unique-Named Store Command.
RFC950	Internet Standard Subnetting Procedure.
RFC951	Bootstrap Protocol.
RFC959	File Transfer Protocol.
RFC974	Mail Routing and the Domain System.
RFC980	Protocol Document Order Information.
RFC1009	Requirements for Internet Gateways.
RFC1011	Official Internet Protocol.
RFC1013	X Windows System Protocol.
RFC1014	XDR: External Data Representation Standard.
RFC1027	Using ARP to Implement Transparent Subnet Gateways.
RFC1032	Domain Administrators Guide.
RFC1033	Domain Administrators Operation Guide.
RFC1034	Domain Names - Concepts and Facilities.
RFC1035	Domain Names - Implementation and Specifications.
RFC1041	Telnet 3270 Regime Option.
RFC1042	Standard for the Transmission of IP Datagrams over IEEE 802 Networks.
RFC1043	Telnet Data Entry Terminal Option.
RFC1044	Internet Protocol on Network System's HYPERchannel.
RFC1053	Telnet X.3 PAD Option.
RFC1055	Nonstandard for Transmission of IP Datagrams over Serial Lines.
RFC1056	PCMAIL: A Distributed Mail System for Personal Computers.
RFC1058	Routing Information Protocol.
RFC1068	Background File Transfer Program (BFTP).
RFC1072	TCP Extensions of Long-Delay Paths.
RFC1073	Telnet Window Size Option.
RFC1074	NSFNET Backbone SPF-based Interior Gateway Protocol.
RFC1079	Telnet Terminal Speed Option.
RFC1080	Telnet Remote Flow Control Option.
RFC1084	BOOTP Vendor Information Extensions.
RFC1088	Standard for the Transmission of IP Datagrams over NetBIOS Network.
RFC1089	SNMP over Ethernet.
RFC1091	Telnet Terminal-Type Option.
RFC1094	NFS: Network File System Protocol Specification.
RFC1101	DNS Encoding of Network Names and Other Types.
RFC1102	Policy Routing in Internet Protocols.
RFC1104	Models of Policy-Based Routing.
RFC1112	Host Extension for IP Multicasting.
RFC1122	Requirement for Internet Hosts - Communication Layers.
RFC1123	Requirement for Internet Hosts - Application and Support.
RFC1124	Policy Issues in Interconnecting Networks.
RFC1125	Policy Requirements for Inter-Administrative Domain Routing.
RFC1127	Perspective on the Host Requirements RFC.
RFC1129	Internet Time Protocol.
RFC1143	Q Method of Implementing Telnet Option Negotiation.
RFC1147	FYI on a Network Management Tool Catalog.
RFC1149	Standard for the Transmission of IP Datagrams over Avian Carriers.
RFC1155	Structure and Identification of Management Information for TCP/IP-Based Internets.
RFC1156	Management Information Base for Network Management of TCP/IP-Based Internets.
RFC1157	Simple Network Management Protocol (SNMP).
RFC1163	Border Gateway Protocol (BGP).

RFC1164 Application of the Border Gateway Protocol in the Internet.
RFC1166 Internet Numbers.
RFC1171 Point-to-Point Protocol for the Transmission of Multi-Protocol Datagrams.
RFC1172 Point-to-Point Protocol Initial Configuration Options.
RFC1173 Responsibilities of Host and Network Managers.
RFC1175 FYI on Where to Start: A Bibliography of Internetworking Information.
RFC1178 Choosing a Name For Your Computer.
RFC1179 Line Printer Daemon Protocol.
RFC1184 Telnet Linemode Option.
RFC1187 Bulk Table Retrieval with the SNMP.
RFC1188 Proposed Standard for the Transmission of TP Datagrams over FDDI Networks.
RFC1195 Use of OSI IS-IS for Routing in TCP/IP and Dual Environments.
RFC1196 Finger User Information Protocol.
RFC1198 FYI on the X Windows System.
RFC1201 Transmitting IP Traffic over ARCNET Networks.
RFC1205 520 Telnet Interface.
RFC1208 Glossary of Networking Terms.
RFC1209 Transmission of IP Datagrams over the SMDS Service.
RFC1212 Concise MIB Definitions.
RFC1213 MIB for Network Management of TCP/IP-Based Internets.
RFC1214 OSI Internet Management: Management Information Base.
RFC1215 Convention for Defining Traps for Use with the SNMP.
RFC1219 On the Assignment of Subnet Numbers.
RFC1220 Point-to-Point Protocol Extensions for Bridges.
RFC1224 Techniques for Managing Asynchronous Generated Alerts.
RFC1227 SNMP MUX Protocol and MIB.
RFC1228 SNMP-DPI: Simple Network Management Protocol Distributed Program Interface.
RFC1229 Extensions to the Generic-interface MIB.
RFC1230 IEEE 802.4 Token Bus MIB.
RFC1231 IEEE 802.5 Token Ring MIB.
RFC1232 Definitions of Managed Objects for the DS1 Interface Type.
RFC1233 Definitions of Managed Objects for the DS3 Interface Type.
RFC1236 IP to X.121 Address Mapping for DDN IP.
RFC1238 CLNS MIB for Use with Connectionless Network Protocol.
RFC1239 Reassignment of Experiment MIBs to Standard MIBs.
RFC1243 Appletalk Management Information Base.
RFC1245 OSPF Protocol Analysis.
RFC1246 Experience with the OSPF Protocol.
RFC1247 OSPF Version 2.
RFC1253 OSPF Version 2: Management Information Base.
RFC1254 Gateway Congestion Control Survey.
RFC1267 A Border Gateway Protocol (BGP-3).
RFC1271 Remote Network Monitoring Management Information Base.
RFC1321 The MD5 Message-Digest Algorithm.
RFC1340 Assigned Numbers.
RFC1341 MIME Mechanism for Specifying and Describing the Format of Internet Message Bodies.
RFC1360 IAB Official Protocol Standards.
RFC1522 MIME (Multipurpose Internet Mail Extensions)Part Two : Message Header Extensions for Non-ASCII Text.
RFC1521 MIME (Multipurpose Internet Mail Extensions) Part One : Mechanisms for Specifying and Describing the Format of Internet Mail Message Bodies).
RFC1583 OSPF Version 2.
RFC1630 Universal Resource Identifiers in WWW.

RFC1738 Uniform Resource Identifiers (URL).
RFC1752 The Recommendation for the IP Next-Generation Protocol.
RFC1771 A Border Gateway Protocol 4 (BGP-4).
RFC1808 Relative Uniform Resource Identifiers.
RFC1809 Using the Flow Label in IPv6.
RFC1825 Security Architecture for the Internet Protocol.
RFC1826 IP Authentication Header.
RFC1827 IP Encapsulating Security Payload (ESP).
RFC1828 IP Authentication Using Keyed MD5.
RFC1829 The ESP DES-CBC Transform.
RFC1883 Internet Protocol, Version 6 Specification.
RFC1884 IP Version 6 Addressing Architecture.
RFC1885 Internet Control Message Protocol (ICMPv6) for the Internet Protocol Version-6 (IPv6) Specification.
RFC1886 DNS Extensions to Support IP Version 6.
RFC1887 An Architecture for IPv6 Unicast Address Allocation.
RFC1901 Introduction to Community-Based SNMPv2.
RFC1902 Structure of Management Information for SNMPv2.
RFC1903 Textual Conventions for SNMPv2.
RFC1904 Conformance Statements for SNMPv2.
RFC1905 Protocol Operations for SNMPv2.
RFC1906 Transport Mappings for SNMPv2.
RFC1907 Management Information Base for SNMPv2.
RFC1908 Coexistence Between Version 1 and Version 2 of the Internet-Standard Network Management Framework.
RFC1909 An Administrative Infrastructure for SNMPv2.
RFC1910 User-based Security Model for SNMPv2.
RFC1911 Voice Profile for Internet Mail.
RFC1912 Common DNS Operational and Configuration Errors.
RFC1913 Architecture of the Whois++ Index Service.
RFC1914 How to Interact with a Whois++ Mesh.
RFC1915 Variance for The PPP Connection Control Protocol and The PPP Encryption Control Protocol
RFC1916 Enterprise Renumbering: Experience and Information Solicitation.
RFC1917 An Appeal to the Internet Community to Return Unused IP Networks (Prefixes) to the IANA
RFC1918 Address Allocation for Private Internets.
RFC1919 Classical versus Transparent IP Proxies.
RFC1920 INTERNET OFFICIAL PROTOCOL STANDARDS.
RFC1922 Chinese Character Encoding for Internet Messages.
RFC1923 RIPv1 Applicability Statement for Historic Status.
RFC1924 A Compact Representation of IPv6 Addresses.
RFC1925 The Twelve Networking Truths.
RFC1926 An Experimental Encapsulation of IP Datagrams on Top of ATM.
RFC1927 Suggested Additional MIME Types for Associating Documents.
RFC1928 SOCKS Protocol Version 5.
RFC1929 Username/Password Authentication for SOCKS V5.
RFC1930 Guidelines for creation, selection, and registration of an Autonomous System (AS).
RFC1931 Dynamic RARP Extensions for Automatic Network Address Acquisition.
RFC1932 IP over ATM: A Framework Document.
RFC1933 Transition Mechanisms for IPv6 Hosts and Routers.
RFC1934 Ascend's Multilink Protocol Plus (MP+).
RFC1935 What is the Internet, Anyway?.

RFC1936 Implementing the Internet Checksum in Hardware.
RFC1937 "Local/Remote" Forwarding Decision in Switched Data Link Subnetworks.
RFC 1938 A One-Time Password System.
RFC 1939 Post Office Protocol - Version 3.
RFC 1940 Source Demand Routing: Packet Format and Forwarding Specification (Version 1).
RFC 1941 Frequently Asked Questions for Schools.
RFC1942 HTML Tables.
RFC1943 Building an X.500 Directory Service in the US.
RFC1944 Benchmarking Methodology for Network Interconnect Devices.
RFC1945 Hypertext Transfer Protocol -- HTTP/1.0.
RFC1946 Native ATM Support for ST2+.
RFC1947 Greek Character Encoding for Electronic Mail Messages.
RFC1948 Defending Against Sequence Number Attacks.
RFC1949 Scalable Multicast Key Distribution.
RFC1950 ZLIB Compressed Data Format Specification version 3.3.
RFC1951 DEFLATE Compressed Data Format Specification version 1.3.
RFC1952 GZIP file format specification version 4.3.
RFC1953 Ipsilon Flow Management Protocol Specification for IPv4 Version 1.0.
RFC1954 Transmission of Flow Labelled IPv4 on ATM Data Links Ipsilon Version 1.0.
RFC1955 New Scheme for Internet Routing and Addressing (ENCAPS) for IPNG.
RFC1956 Registration in the MIL Domain.
RFC1957 Some Observations on Implementations of the Post Office Protocol (POP3).
RFC1958 Architectural Principles of the Internet.
RFC1959 An LDAP URL Format.
RFC1960 A String Representation of LDAP Search Filters.
RFC1961 GSS-API Authentication Method for SOCKS Version 5.
RFC1962 The PPP Compression Control Protocol (CCP).
RFC1963 PPP Serial Data Transport Protocol (SDTP).
RFC1964 The Kerberos Version 5 GSS-API Mechanism.
RFC1965 Autonomous System Confederations for BGP.
RFC1966 BGP Route Reflection An alternative to full mesh IBGP.
RFC1967 PPP LZS-DCP Compression Protocol (LZS-DCP).
RFC1968 The PPP Encryption Control Protocol (ECP).
RFC1969 The PPP DES Encryption Protocol (DESE).
RFC1970 Neighbor Discovery for IP Version 6 (IPv6).
RFC1971 IPv6 Stateless Address Autoconfiguration.
RFC1972 A Method for the Transmission of IPv6 Packets over Ethernet Networks.
RFC1973 PPP in Frame Relay.
RFC1974 PPP Stac LZS Compression Protocol.
RFC1975 PPP Magnalink Variable Resource Compression.
RFC1976 PPP for Data Compression in Data Circuit-Terminating Equipment (DCE).
RFC1977 PPP BSD Compression Protocol.
RFC1978 PPP Predictor Compression Protocol.
RFC1979 PPP Deflate Protocol.
RFC1980 A Proposed Extension to HTML : Client-Side Image Maps.
RFC1981 Path MTU Discovery for IP version 6.
RFC1982 Serial Number Arithmetic.
RFC1983 Internet Users' Glossary.
RFC1984 IAB and IESG Statement on Cryptographic Technology and the Internet.
RFC1985 SMTP Service Extension for Remote Message Queue Starting.
RFC1986 Experiments with a Simple File Transfer Protocol for Radio Links using Enhanced Trivial File Transfer Protocol (ETFTP).
RFC1987 Ipsilon's General Switch Management Protocol Specification Version 1.1.

RFC1988 Conditional Grant of Rights to Specific Hewlett-Packard Patents In Conjunction With the Internet Engineering Task Force's Internet-Standard Network Management Framework.
RFC1989 PPP Link Quality Monitoring.
RFC1990 The PPP Multilink Protocol.
RFC1991 PGP Message Exchange Formats.
RFC1992 The Nimrod Routing Architecture.
RFC1993 PPP Gandalf FZA Compression Protocol.
RFC1994 PPP Challenge Handshake Authentication Protocol (CHAP).
RFC1995 Incremental Zone Transfer in DNS.
RFC1996 A Mechanism for Prompt Notification of Zone Changes.
RFC1997 BGP Communities Attribute.
RFC1998 An Application of the BGP Community Attribute in Multi-home Routing.
RFC1999 Request for Comments Summary RFC Numbers 1900-1999.
RFC2000 INTERNET OFFICIAL PROTOCOL STANDARDS.
RFC2001 TCP Slow Start, Congestion Avoidance, Fast Retransmit, and Fast
RFC2002 IP Mobility Support.
RFC2003 IP Encapsulation within IP.
RFC2004 Minimal Encapsulation within IP.
RFC2005 Applicability Statement for IP Mobility Support.
RFC2006 The Definitions of Managed Objects for IP Mobility Support using SMIv2.
RFC2007 Catalogue of Network Training Materials.
RFC2008 Implications of Various Address Allocation Policies for Internet Routing.
RFC2009 GPS-Based Addressing and Routing.
RFC2010 Operational Criteria for Root Name Servers.
RFC2011 SNMPv2 Management Information Base for the Internet Protocol using SMIv2.
RFC2012 SNMPv2 Management Information Base for the Transmission Control Protocol using SMIv2.
RFC2013 SNMPv2 Management Information Base for the User Datagram Protocol using SMIv2.
RFC2014 IRTF Research Group Guidelines and Procedures.
RFC2015 MIME Security with Pretty Good Privacy (PGP).
RFC2016 Uniform Resource Agents (URAs).
RFC2017 Definition of the URL MIME External-Body Access-Type.
RFC2018 TCP Selective Acknowledgement Options.
RFC2019 Transmission of IPv6 Packets Over FDDI.
RFC2020 IEEE 802.12 Interface MIB.
RFC2021 Remote Network Monitoring Management Information Base Version 2 using SMIv2.
RFC2022 Support for Multicast over UNI 3.0/3.1 based ATM Networks.
RFC2023 IP Version 6 over PPP.
RFC2024 Definitions of Managed Objects for Data Link Switching using SMIv2.
RFC2025 The Simple Public-Key GSS-API Mechanism (SPKM).
RFC2026 The Internet Standards Process -- Revision 3.
RFC2027 IAB and IESG Selection, Confirmation, and Recall Process: Operation of the Nominating and Recall Committees.
RFC2028 The Organizations Involved in the IETF Standards Process.
RFC2029 RTP Payload Format of Sun's CellB Video Encoding.
RFC2030 Simple Network Time Protocol (SNTP).
RFC2031 IETF-ISOC relationship.
RFC2032 RTP Payload Format for H.261 Video Streams.
RFC2033 Local Mail Transfer Protocol.
RFC2034 SMTP Service Extension for Returning Enhanced Error Codes.
RFC2035 RTP Payload Format for JPEG-compressed Video.
RFC2036 Observations on the use of Components of the Class A Address Space within the Internet.
RFC2037 Entity MIB using SMIv2.

RFC2038 RTP Payload Format for MPEG1/MPEG2 Video.
RFC2039 Applicability of Standards Track MIBs to Management of World Wide Web Servers.
RFC2040 The RC5, RC5-CBC, RC5-CBC-Pad, and RC5-CTS Algorithms.
RFC2041 Mobile Network Tracing.
RFC2042 Registering New BGP Attribute Types.
RFC2043 The PPP SNA Control Protocol (SNACP).
RFC2044 UTF-8, a transformation format of Unicode and ISO 10646.
RFC2045 Multipurpose Internet Mail Extensions (MIME) Part One: Format of Internet Message Bodies.
RFC2046 Multipurpose Internet Mail Extensions (MIME) Part Two: Media Types.
RFC2047 MIME (Multipurpose Internet Mail Extensions) Part Three: Message Header Extensions for Non-ASCII Text.
RFC2048 Multipurpose Internet Mail Extension (MIME) Part Four: Registration Procedures.
RFC2049 Multipurpose Internet Mail Extensions (MIME) Part Five: Conformance Criteria and Examples.
RFC2050 INTERNET REGISTRY IP ALLOCATION GUIDELINES.
RFC2051 Definitions of Managed Objects for APPC using SMIv2.
RFC2052 A DNS RR for specifying the location of services (DNS SRV).
RFC2053 The AM (Armenia) Domain.
RFC2054 WebNFS Client Specification.
RFC2055 WebNFS Server Specification.
RFC2056 Uniform Resource Locators for Z39.50.
RFC2057 Source Directed Access Control on the Internet.
RFC2058 Remote Authentication Dial In User Service (RADIUS).
RFC2059 RADIUS Accounting.
RFC2060 INTERNET MESSAGE ACCESS PROTOCOL - VERSION 4rev1.
RFC2061 IMAP4 COMPATIBILITY WITH IMAP2BIS.
RFC2062 Internet Message Access Protocol - Obsolete Syntax.
RFC2063 Traffic Flow Measurement: Architecture.
RFC2064 Traffic Flow Measurement: Meter MIB.
RFC2065 Domain Name System Security Extensions.
RFC2066 TELNET CHARSET Option.
RFC2067 IP over HIPPI.
RFC2068 Hypertext Transfer Protocol -- HTTP/1.1.
RFC2069 An Extension to HTTP: Digest Access Authentication.
RFC2070 Internationalization of the Hypertext Markup Language.
RFC2071 Network Renumbering Overview: Why would I want it and what is it anyway?.
RFC2072 Router Renumbering Guide.
RFC2073 An IPv6 Provider-Based Unicast Address Format.
RFC2074 Remote Network Monitoring MIB Protocol Identifiers.
RFC2075 IP Echo Host Service.
RFC2076 Common Internet Message Headers.
RFC2077 The Model Primary Content Type for Multipurpose Internet Mail Extensions.
RFC2078 Generic Security Service Application Program Interface, Version 2.
RFC2079 Definition of an X.500 Attribute Type and an Object Class to Hold Uniform Resource Identifiers (URIs).
RFC2080 RIPng for IPv6.
RFC2081 RIPng Protocol Applicability Statement.
RFC2082 RIP-2 MD5 Authentication.
RFC2083 PNG (Portable Network Graphics) Specification.
RFC2084 Considerations for Web Transaction Security.
RFC2085 HMAC-MD5 IP Authentication with Replay Prevention.
RFC2086 IMAP4 ACL extension.

RFC2087 IMAP4 QUOTA extension.
RFC2088 IMAP4 non-synchronizing literals.
RFC2089 V2ToV1 Mapping SNMPv2 onto SNMPv1 within a bi-lingual SNMP agent.
RFC2090 TFTP Multicast Option.
RFC2091 Triggered Extensions to RIP to Support Demand Circuits.
RFC2092 Protocol Analysis for Triggered RIP.
RFC2093 Group Key Management Protocol (GKMP) Specification.
RFC2094 Group Key Management Protocol (GKMP) Architecture.
RFC2095 IMAP/POP AUTHorize Extension for Simple Challenge/Response.
RFC2096 IP Forwarding Table MIB.
RFC2097 The PPP NetBIOS Frames Control Protocol (NBFCP).
RFC2098 Toshiba's Router Architecture Extensions for ATM : Overview.
RFC2099 Request for Comments Summary RFC Numbers 2000-2099.
RFC2100 The Naming of Hosts.
RFC2101 IPv4 Address Behavior Today.
RFC2102 Multicast Support for Nimrod : Requirements and Solution Approaches.
RFC2103 Mobility Support for Nimrod : Challenges and Solution Approaches.
RFC2104 HMAC: Keyed-Hashing for Message Authentication.
RFC2105 Cisco Systems' Tag Switching Architecture Overview.
RFC2106 Data Link Switching Remote Access Protocol.
RFC2107 Ascend Tunnel Management Protocol - ATMP.
RFC2108 Definitions of Managed Objects for IEEE 802.3 Repeater Devices using SMIv2.
RFC2109 HTTP State Management Mechanism.
RFC2110 MIME E-mail Encapsulation of Aggregate Documents, such as HTML (MHTML).
RFC2111 Content-ID and Message-ID Uniform Resource Locators.
RFC2112 The MIME Multipart/Related Content-type.
RFC2113 IP Router Alert Option.
RFC2114 Data Link Switching Client Access Protocol.
RFC2115 Management Information Base for Frame Relay DTEs Using SMIv2.
RFC2116 X.500 Implementations Catalog-96.
RFC2117 Protocol Independent Multicast-Sparse Mode (PIM-SM): Protocol
RFC2118 Microsoft Point-To-Point Compression (MPPC) Protocol.
RFC2119 Key words for use in RFCs to Indicate Requirement Level.
RFC2120 Managing the X.500 Root Naming Context.
RFC2121 Issues affecting MARS Cluster Size.
RFC2122 VEMMI URL Specification.
RFC2123 Traffic Flow Measurement: Experiences with NeTraMet.
RFC2124 Cabletron's Light-weight Flow Admission Protocol Specification.
RFC2125 The PPP Bandwidth Allocation Protocol (BAP) / The PPP Bandwidth Allocation Control Protocol (BACP).
RFC2126 ISO Transport Service on top of TCP (ITOT).
RFC2127 ISDN Management Information Base using SMIv2.
RFC2128 Dial Control Management Information Base using SMIv2.
RFC2129 Toshiba's Flow Attribute Notification Protocol (FANP).
RFC2130 The Report of the IAB Character Set Workshop held 29 February - 1 March, 1996.
RFC2131 Dynamic Host Configuration Protocol.
RFC2132 DHCP Options and BOOTP Vendor Extensions.
RFC2133 Basic Socket Interface Extensions for Ipv6.
RFC2134 Articles of Incorporation of Internet Society.
RFC2135 Internet Society By-Laws. ISOC Board of Trustees.
RFC2136 Dynamic Updates in the Domain Name System (DNS UPDATE).
RFC2137 Secure Domain Name System Dynamic Update.
RFC2138 Remote Authentication Dial In User Service (RADIUS).

RFC2139 RADIUS Accounting.
RFC2140 TCP Control Block Interdependence.
RFC2141 URN Syntax.
RFC2142 Mailbox Names for Common Services, Roles and Functions.
RFC2143 Encapsulating IP with the Small Computer System Interface.
RFC2145 Use and Interpretation of HTTP Version Numbers.
RFC2146 U.S. Government Internet Domain Names. Federal Networking
RFC2147 TCP and UDP over IPv6 Jumbograms.
RFC2148 Deployment of the Internet White Pages Service.
RFC2149 Multicast Server Architectures for MARS-based ATM multicasting.
RFC2150 Humanities and Arts: Sharing Center Stage on the Internet.
RFC2151 A Primer On Internet and TCP/IP Tools and Utilities.
RFC2152 UTF-7 A Mail-Safe Transformation Format of Unicode.
RFC2153 PPP Vendor Extensions.
RFC2154 OSPF with Digital Signatures.
RFC2155 Definitions of Managed Objects for APPN using SMIv2.
RFC2165 Service Location Protocol.
RFC2166 APPN Implementer's Workshop Closed Pages Document DLSw v2.0 Enhancements.
RFC2167 Referral Whois (RWhois) Protocol V1.5.
RFC2168 Resolution of Uniform Resource Identifiers using the Domain Name System.
RFC2169 A Trivial Convention for using HTTP in URN Resolution.
RFC2170 Application REQuested IP over ATM (AREQUIPA).
RFC2171 MAPOS - Multiple Access Protocol over SONET/SDH Version 1.
RFC2172 MAPOS Version 1 Assigned Numbers.
RFC2173 A MAPOS version 1 Extension - Node Switch Protocol.
RFC2174 A MAPOS version 1 Extension - Switch-Switch Protocol.
RFC2175 MAPOS 16 - Multiple Access Protocol over SONET/SDH with 16 Bit Addressing.
RFC2176 IPv4 over MAPOS Version 1
RFC2177 IMAP4 IDLE command.
RFC2178 OSPF Version 2.
RFC2179 Network Security For Trade Shows.
RFC2180 IMAP4 Multi-Accessed Mailbox Practice.
RFC2181 Clarifications to the DNS Specification.
RFC2182 Selection and Operation of Secondary DNS Servers.
RFC2183 Communicating Presentation Information in Internet Messages: The Content-Disposition Header Field.
RFC2184 MIME Parameter Value and Encoded Word Extensions: Character Sets, Languages, and Continuations.
RFC2185 Routing Aspects of IPv6 Transition.
RFC2186 Internet Cache Protocol (ICP), version 2.
RFC2187 Application of Internet Cache Protocol (ICP), version 2.
RFC2188 AT&T/Neda's Efficient Short Remote Operations (ESRO) Protocol Specification Version 1.2.
RFC2189 Core Based Trees (CBT version 2) Multicast Routing.
RFC2190 RTP Payload Format for H.263 Video Streams.
RFC2191 VENUS - Very Extensive Non-Unicast Service.
RFC2192 IMAP URL Scheme.
RFC2193 IMAP4 Mailbox Referrals.
RFC2194 Review of Roaming Implementations.
RFC2195 IMAP/POP AUTHorize Extension for Simple Challenge/Response.
RFC2196 Site Security Handbook.
RFC2197 SMTP Service Extension for Command Pipelining.
RFC2198 RTP Payload for Redundant Audio Data.

RFC2200 INTERNET OFFICIAL PROTOCOL STANDARDS.
RFC2201 Core Based Trees (CBT) Multicast Routing Architecture.
RFC2202 Test Cases for HMAC-MD5 and HMAC-SHA-1.
RFC2203 RPCSEC_GSS Protocol Specification.
RFC2204 ODETTE File Transfer Protocol.
RFC2205 Resource ReSerVation Protocol (RSVP) -- Version 1 Functional Specification.
RFC2206 RSVP Management Information Base using SMIv2.
RFC2207 RSVP Extensions for IPSEC Data Flows.
RFC2208 Resource ReSerVation Protocol (RSVP) -- Version 1 Applicability Statement Some Guidelines on Deployment.
RFC2209 Resource ReSerVation Protocol (RSVP) -- Version 1 Message Processing Rules.
RFC2210 The Use of RSVP with IETF Integrated Services.
RFC2211 Specification of the Controlled-Load Network Element Service.
RFC2212 Specification of Guaranteed Quality of Service.
RFC2213 Integrated Services Management Information Base using SMIv2.
RFC2214 Integrated Services Management Information Base Guaranteed Service Extensions using SMIv2.
RFC2215 General Characterization Parameters for Integrated Service Network Elements.
RFC2216 Network Element Service Specification Template.
RFC2217 Telnet Com Port Control Option.
RFC2218 A Common Schema for the Internet White Pages Service.
RFC2219 Use of DNS Aliases for Network Services.
RFC2220 The Application/MARC Content-type.
RFC2221 IMAP4 Login Referrals.
RFC2222 Simple Authentication and Security Layer (SASL).
RFC2223 Instructions to RFC Authors.
RFC2224 NFS URL Scheme.
RFC2226 IP Broadcast over ATM Networks.
RFC2227 Simple Hit-Metering and Usage-Limiting for HTTP.
RFC2228 FTP Security Extensions.
RFC2229 A Dictionary Server Protocol.
RFC2230 Key Exchange Delegation Record for the DNS.
RFC2231 MIME Parameter Value and Encoded Word Extensions: Character Sets, Languages, and Continuations.
RFC2232 Definitions of Managed Objects for DLUR using SMIv2.
RFC2233 The Interfaces Group MIB using SMIv2.
RFC2234 Augmented BNF for Syntax Specifications: ABNF.
RFC2235 Hobbes' Internet Timeline.
RFC2236 Internet Group Management Protocol, Version 2.
RFC2237 Japanese Character Encoding for Internet Messages.
RFC2238 Definitions of Managed Objects for HPR using SMIv2.
RFC2239 Definitions of Managed Objects for IEEE 802.3 Medium Attachment Units (MAUs) using SMIv2.
RFC2240 A Legal Basis for Domain Name Allocation.
RFC2241 DHCP Options for Novell Directory Services.
RFC2242 NetWare/IP Domain Name and Information.

UNIX Network Startup Files

E.1 netnfsrc file

This appendix documents a typical netnfsrc (NFS startup file) file. In the script portion given below the NFS_CLIENT is set to a 1 if the host is set to a client (else it will be 0) and the NFS_SERVER parameter is set to a 1 if the host is set to a server (else it will be 0). Initially the NFS clients and servers are started. Note that a host can be a client, a server, both or neither.

Next the mountd daemon is started, after which the NFS daemons (nfsd) are started (only on servers). After this the biod daemon is run.

```
NFS_CLIENT=1
NFS_SERVER=1
START_MOUNTD=0
#       Read in /etc/exports
##
if [ $LFS -eq 0 -a $NFS_SERVER -ne 0 -a -f /etc/exports ] ; then
    > /etc/xtab
    /usr/etc/exportfs -a  && echo "    Reading in /etc/exports"
    set_return
fi

if [ $NFS_SERVER -ne 0 -a $START_MOUNTD -ne 0 -a -f /usr/etc/rpc.mountd ] ;
then
    /usr/etc/rpc.mountd && echo "starting up the mountd" && echo
                             "\t/usr/etc/rpc.mountd"
    set_return
fi
##
if [ $LFS -eq 0 -a $NFS_SERVER -ne 0 -a -f /etc/nfsd ] ; then
    /etc/nfsd 4 && echo "starting up the NFS daemons" && echo "\t/etc/nfsd 4"
    set_return
fi
##
if [ $NFS_CLIENT -ne 0 ] ; then
   if [ -f /etc/biod ] ; then
        /etc/biod 4 && echo
             "starting up the BIO daemons" && echo "\t/etc/biod 4"
          set_return
fi
   /bin/cat /dev/null > /etc/nfs.up
fi
```

The next part of the netnfsrc file deals with the NIS services. There are three states: NIS_MASTER_SERVER, NIS_SLAVE_SERVER and NIS_CLIENT. A host can either be a master server or a slave server, but cannot be both. All NIS servers must also be NIS clients, so the NIS_MASTER_SERVER or NIS_SLAVE_SERVER parameters shoud be set to 1. Initally the domain name is set using the command domainname (in this case it is eece).

```
NIS_MASTER_SERVER=1
NIS_SLAVE_SERVER=0
NIS_CLIENT=1
NISDOMAIN=eece
NISDOMAIN_ERR=""

if [ "$NISDOMAIN" -a -f /bin/domainname ] ; then
    echo "\t/bin/domainname $NISDOMAIN"
    /bin/domainname $NISDOMAIN
    if [ $? -ne 0 ] ; then
   echo "Error:  NIS domain name not set" >&2
   NISDOMAIN_ERR=TRUE
    fi
else
    echo "\tNIS domain name not set"
    NISDOMAIN_ERR=TRUE
fi
```

Next portmap is started for ARPA clients.

```
if [ -f /etc/portmap ] ; then
    echo "\t/etc/portmap"
    /etc/portmap
    if [ $? -ne 0 ] ; then
   echo  "Error:  NFS portmapper NOT powered up"  >&2
   exit 1
    fi
fi
```

Next the NIS is started.

```
if [ "$NISDOMAIN_ERR" -o \( $NIS_MASTER_SERVER -eq 0 -a $NIS_SLAVE_SERVER -eq
0\
   -a $NIS_CLIENT -eq 0 \) ] ; then
    echo "    Network Information Service not started."
else
    echo "    starting up the Network Information Service"

    HOSTNAME=`hostname`

    if [ $NIS_MASTER_SERVER -ne 0 -o $NIS_SLAVE_SERVER -ne 0 ]; then
   NIS_SERVER=TRUE
    fi

    if [ $NIS_MASTER_SERVER -ne 0 -a $NIS_SLAVE_SERVER -ne 0 ]; then
     echo "NOTICE:both NIS_MASTER_SERVER and NIS_SLAVE_SERVER variables set;"
     echo "\t$HOSTNAME will be only a NIS slave server."
     NIS_MASTER_SERVER=0
    fi

 if [ $NIS_CLIENT -eq 0 ]; then
 echo "NOTICE:$HOSTNAME will be a NIS server, but the NIS_CLIENT variable is"
 echo "\tnot set; $HOSTNAME will also be a NIS client."
 NIS_CLIENT=1
 fi
```

Next the yp services are started.

```
#  The verify_ypserv function determines if it is OK to start ypserv(1M)
#  (and yppasswdd(1M) for the master NIS server).  It returns its result
#  in the variable NISSERV_OK - if non-null, it is OK to start ypserv(1M);
#  if it is null, ypserv(1M) will not be started.
#
#  First, the filesystem containing /usr/etc/yp is examined to see if it
#  supports long or short filenames.  Once this is known, the proper list
#  of standard NIS map filenames is examined to verify that each map exists
#  in the NIS domain subdirectory.  If any map is missing, verify_ypserv
#  sets NISSERV_OK to null and returns.
##

verify_ypserv() {
   ##
   #  LONGNAMES are the names of the NIS maps on a filesystem that
   #  supports long filenames.
   ##

   LONGNAMES="group.bygid.dir group.bygid.pag group.byname.dir \
         group.byname.pag hosts.byaddr.dir hosts.byaddr.pag \
         hosts.byname.dir hosts.byname.pag networks.byaddr.dir \
         networks.byaddr.pag networks.byname.dir networks.byname.pag \
         passwd.byname.dir passwd.byname.pag passwd.byuid.dir \
         passwd.byuid.pag protocols.byname.dir protocols.byname.pag \
         protocols.bynumber.dir protocols.bynumber.pag \
         rpc.bynumber.dir rpc.bynumber.pag services.byname.dir \
         services.byname.pag ypservers.dir ypservers.pag"


   ##
   #  SHORTNAMES are the names of the NIS maps on a filesystem that
   #  supports only short filenames (14 characters or less).
   ##


   SHORTNAMES="group.bygi.dir group.bygi.pag group.byna.dir \
          group.byna.pag hosts.byad.dir hosts.byad.pag \
          hosts.byna.dir hosts.byna.pag netwk.byad.dir \
          netwk.byad.pag netwk.byna.dir netwk.byna.pag \
          passw.byna.dir passw.byna.pag passw.byui.dir \
          passw.byui.pag proto.byna.dir proto.byna.pag \
          proto.bynu.dir proto.bynu.pag rpc.bynu.dir \
          rpc.bynu.pag servi.byna.dir servi.byna.pag \
          ypservers.dir ypservers.pag"

   NISSERV_OK=TRUE

   if `/usr/etc/yp/longfiles`; then
      NAMES=$LONGNAMES
   else
      NAMES=$SHORTNAMES
   fi

   for NAME in $NAMES ; do
      if [ ! -f /usr/etc/yp/$NISDOMAIN/$NAME ] ; then
         NISSERV_OK=
         return
      fi
   done
}
```

Next ypserv and ypbind are started.

```
    if [ "$NIS_SERVER" -a -f /usr/etc/ypserv ] ; then
   verify_ypserv
   if [ "$NISSERV_OK" ] ; then
      /usr/etc/ypserv && echo "\t/usr/etc/ypserv"
                set_return
   else
      echo "\tWARNING:  /usr/etc/ypserv not started:  either"
      echo "\t          -  the directory /usr/etc/yp/$NISDOMAIN does not ex-
ist or"
      echo "\t          -  some or all of the $NISDOMAIN NIS domain's"
      echo "\t             maps are missing."
      echo "\tTo initialize $HOSTNAME as a NIS server, see ypinit(1M)."
                returnstatus=1
   fi
    fi
    if [ $NIS_CLIENT -ne 0 -a -f /etc/ypbind ] ; then
   /etc/ypbind  && echo "\t/etc/ypbind "
        set_return

        ##
        #   check if the NIS domain is bound. If not disable NIS
        ##
        CNT=0;
        MAX_NISCHECKS=2
        NIS_CHECK=YES
        echo " Checking NIS binding."
        while [ ${CNT} -le ${MAX_NISCHECKS} -a "${NIS_CHECK}" = "YES" ]; do
       /usr/bin/ypwhich 2>&1 | /bin/fgrep 'not bound ypwhich' > /dev/null

       if [ $? -eq 0 ]; then
           CNT=`expr $CNT + 1`
           if [ ${CNT} -le 2 ]; then
          sleep 5
           else
          echo "  Unable to bind to NIS server using domain ${NISDOMAIN}."
          echo "  Disabling NIS"
          /bin/domainname ""
          /bin/ps -e | /bin/grep ypbind | \
            kill -15 `/usr/bin/awk '{ print $1 }'`
          NIS_CHECK=NO
          returnstatus=1
          break;
           fi
       else
           echo " Bound to NIS server using domain ${NISDOMAIN}."
           NIS_CHECK=NO
       fi
        done
    fi

    ##
    if [ $NIS_MASTER_SERVER -ne 0 -a -f /usr/etc/rpc.yppasswdd ] ; then
   if [ "$NISSERV_OK" ] ; then
         echo "\t/usr/etc/rpc.yppasswdd"
         /usr/etc/rpc.yppasswdd /etc/passwd -m passwd PWFILE=/etc/passwd
      set_return
   else
      echo "\tWARNING:  /usr/etc/rpc.yppasswdd not started:  refer to the"
      echo "\t          reasons listed in the WARNING above."
           returnstatus=1
   fi
    fi
fi
```

Finally the PC-NFS daemons (`pcnfsd`) and the lock manager daemon (`rpc.lockd`) status monitor daemon (`rpc.statd`) are started.

```
PCNFS_SERVER=1
if [ $LFS -eq 0 -a $PCNFS_SERVER -ne 0 -a -f /etc/pcnfsd ] ; then
    /etc/pcnfsd && echo "starting up the PC-NFS daemon" && echo
"\t/etc/pcnfsd"
    set_return
fi

if [ $NFS_CLIENT -ne 0 -o $NFS_SERVER -ne 0 ] ; then
    if [ -f /usr/etc/rpc.statd ] ; then
   /usr/etc/rpc.statd && echo "starting up the Status Monitor daemon" && echo
"\t/usr/etc/rpc.statd"
        set_return
    fi
    if [ -f /usr/etc/rpc.lockd ] ; then
   /usr/etc/rpc.lockd && echo "starting up the Lock Manager daemon" && echo
"\t/usr/etc/rpc.lockd"
        set_return
    fi
fi
exit $returnstatus
```

E.2 rc file

The `rc` file is executed when the UNIX node starts. It contains a number of functions (such as `localrc()`, `hfsmount()`, and so on) which are called from a main section. The example script given next contains some of the functions defined in Table E.1.

Table E.1 Sample rc functions.

Function	*Description*
`localrc()`	Add local configuration to the node. In the example script the Bones-Licensing 2.4 is started locally on the node. This part of the script will probably be the only function which is different on different nodes.
`hfsmount()`	Mounts local disk drives
`map_keyboard()`	Loads appropriate keymap
`syncer_start()`	The syncer helps to minimize file damage when this is a power failure or a system crash
`lp_start()`	Starts the lp (line printer) scheduler
`net_start()`	Starts networking through `netlinkrc`
`swap_start()`	Starts swapping on alternate swap devices

```
initialize()
{
   if [ "$SYSTEM_NAME" = "" ]
   then
   SYSTEM_NAME=pollux
      export SYSTEM_NAME
   fi
}

localrc()
{

#%%CSIBeginFeature: Bones-Licensing 2.4
   DESIGNERHOME=/win/designer-2.0
   export DESIGNERHOME

   echo -n "Starting Bones-Licensing 2.4 ..."
   if [ -f ${DESIGNERHOME}/bin/start-lmgrd ]; then
      ${DESIGNERHOME}/bin/start-lmgrd
      echo " lmgrd."
   else
      echo " failed."
   fi
}

set_date()
{
  if [ $SET_PARMS_RUN -eq 0 ] ; then
   if [ $TIMEOUT -ne 0 ] ; then
      echo "\007Is the date `date` correct? (y or n, default: y) \c"
      reply=`line -t $TIMEOUT`
      echo ""

      if [ "$reply" = y -o "$reply" = "" -o "$reply" = Y ]
      then
         return
      else
         if [ -x /etc/set_parms ]; then
            /etc/set_parms time_only
         fi
      fi
   fi

   fi # if SET_PARMS_RUN
}

hfsmount()
{
   # create /etc/mnttab with valid root entry
   /etc/mount -u >/dev/null

   # enable quotas on the root file system
   # (others are enabled by mount)
   [ -f /quotas -a -x /etc/quotaon ] && /etc/quotaon -v /

   # Mount the HFS volumes listed in /etc/checklist:
   /etc/mount -a -t hfs -v
   # (NFS volumes are mounted via net_start() function)

   # Uncomment the following mount command to mount CDFS's
   /etc/mount -a -t cdfs -v

   # Preen quota statistics
   [ -x /etc/quotacheck ] && echo checking quotas && /etc/quotacheck -aP
}
```

```
map_keyboard()
{
#
itemap_option=""
if [ -f /etc/kbdlang ]
then
   read MAP_NAME filler < /etc/kbdlang
   if [ $MAP_NAME ]
   then
      itemap_option="-l $MAP_NAME"
   fi
fi

if [ -x /etc/itemap ]
then
   itemap -i -L $itemap_option -w /etc/kbdlang
fi
}

syncer_start()
{
   if /usr/bin/rtprio 127 /etc/syncer
   then
      echo syncer started
   fi
}

lp_start()
{
   if [ -s /usr/spool/lp/pstatus ]
   then
      lpshut > /dev/null 2>&1
      rm -f /usr/spool/lp/SCHEDLOCK
      lpsched
      echo line printer scheduler started
   fi
}

clean_ex()
{
   if [ -x /usr/bin/ex ]
   then
      echo "preserving editor files (if any)"
      ( cd /tmp; expreserve -a )
   fi
}

clean_uucp()
{
   if [ -x /usr/lib/uucp/uuclean ]
   then
      echo "cleaning up uucp"
      /usr/lib/uucp/uuclean -pSTST -pLCK -n0
   fi
}

net_start()
{
   if [ -x /etc/netlinkrc ] && /etc/netlinkrc
   then
      echo NETWORKING started.
   fi
}

swap_start()
```

```
{
   if /etc/swapon -a
   then
      echo 'swap device(s) active'
   fi
}

cron_start()
{
   if [ -x /etc/cron ]
   then
      if [ -f /usr/lib/cron/log ]
      then
         mv /usr/lib/cron/log /usr/lib/cron/OLDlog
      fi
      /etc/cron && echo cron started
   fi
}

audio_start ()
{
   # Start up the audio server
   if [ -x /etc/audiorc ] && /etc/audiorc
   then
      echo "Audio server started"
   fi
}

#
# The main section of the rc script
#

# Where to find commands:
PATH=/bin:/usr/bin:/usr/lib:/etc

# Set termio configuration for output device.
stty clocal icanon echo opost onlcr ixon icrnl ignpar

if [ ! -f /etc/rcflag ]    # Boot time invocation only
then
   # /etc/rcflag is removed by /etc/brc at boot and by shutdown
   touch /etc/rcflag

   hfsmount
   map_keyboard
   setparms
   initialize
   switch_over
   uname -S $SYSTEM_NAME
   hostname $SYSTEM_NAME

   swap_start
   syncer_start
   lp_start
   clean_ex
   clean_uucp
   net_start
   audio_start
   localrc
fi
```

Ethernet Monitoring System

F.1 Ethernet receiver

The next page gives a schematic for an Ethernet monitoring system. Its component values are:

R1 = 1 kΩ
R2 = 500 Ω
R3 = 10 MΩ
R4 = 39 Ω
R5 = 1.5 kΩ
R6 = 10 Ω

L1 = 1:1, 200 μH

C1 = 100 mF
C2 = 0.1 μF
C3 = 1.5 pF

XTAL = 20 MHz

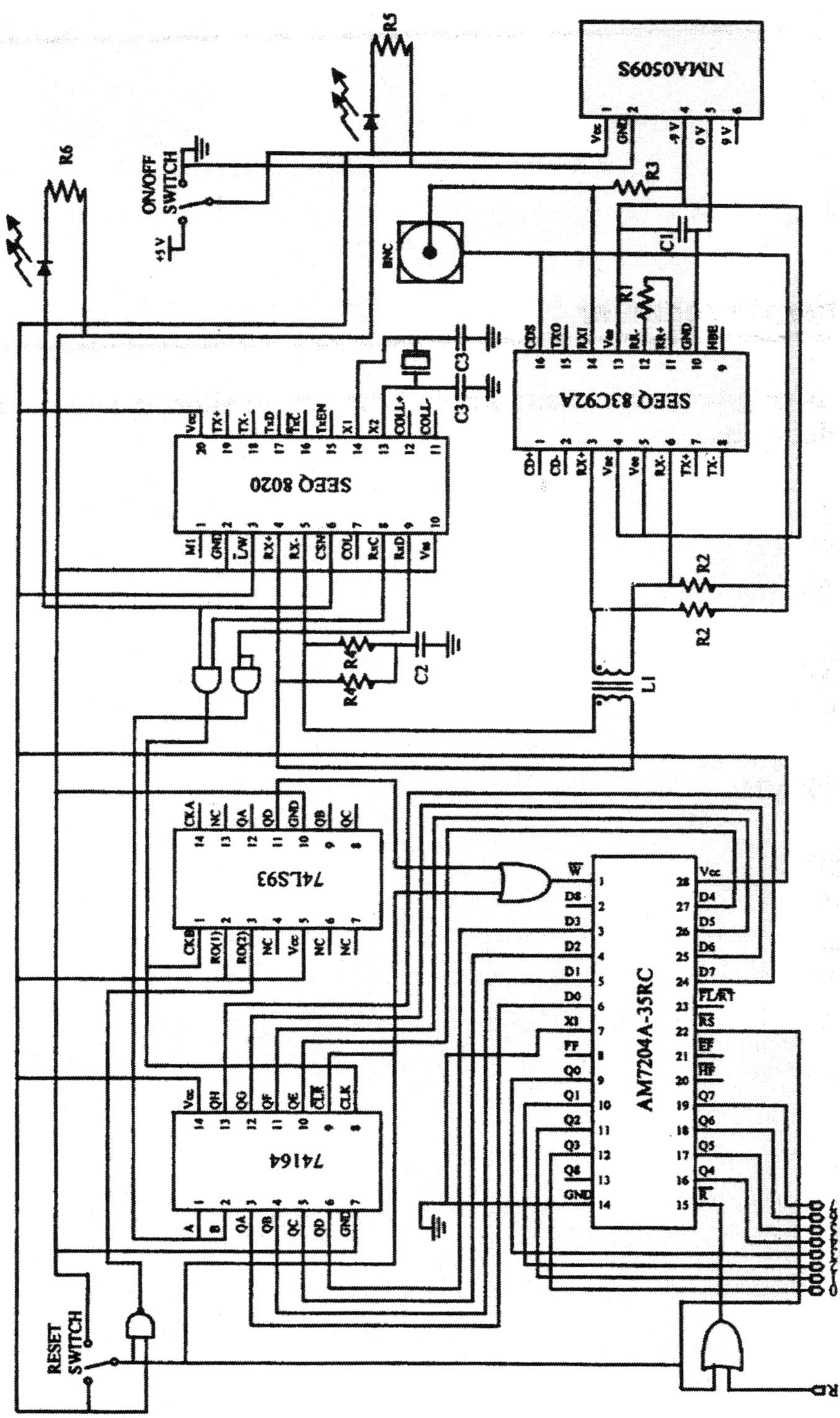
NMA0509S
SEEQ 8020
SEEQ 83C92A
74LS93
74164
AM7204A-35RC
BNC
ON/OFF SWITCH
RESET SWITCH
+5 V
R1
R2
R3
R4
R5
R6
C1
C2
C3
L1

RS-232 and Parallel Port

G.1 RS-232

G.1.1 The UART

The main device used to construct an RS-232 interface is the UART (universal asynchronous receiver and transmitter) which is a fully programmable device. It can be set-up to have a defined Baud rate, number of data bits, number of start bits, number of stop bits, type of parity, manipulation of the output and input handshaking lines. Figure G.1 shows a block diagram. The UART is fully compatible with TTL devices and has a +5 V and 0 V power supply. Typical UARTs are Intel's 8250, Motorola's MC8650, Motorola's MC68681 and National's INS 8250. The 8250/1 device is compatible with Intel's 80×86 microprocessors and is thus used in many PC applications. A 16-bit UART device is available and is named the 16450.

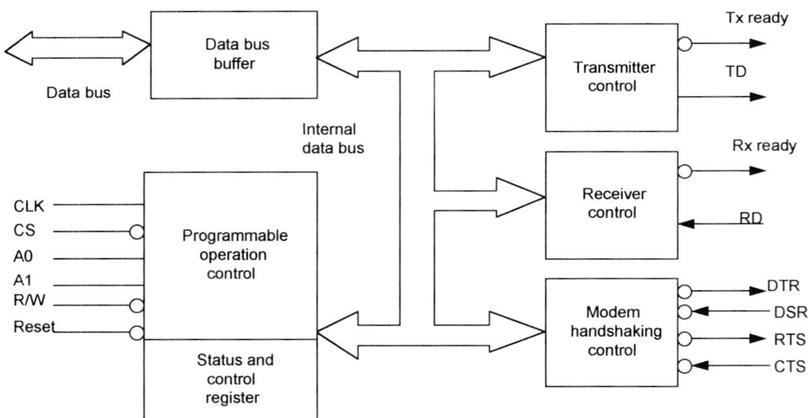

Figure G.1 UART.

G.1.2 Drivers and receivers

Driver ICs convert the TTL output/input of the UART to RS-232 voltage levels. The MC1488 quad line driver contains four NAND logic buffer gates and has supply lines of +/− 12V and GND, as illustrated in Figure G.2. If the two inputs to the drivers are tied together to create an invertor function, a table of the conversion is shown next:

- 0 V input gives +12 V output.
- 5 V input gives -12 V output.

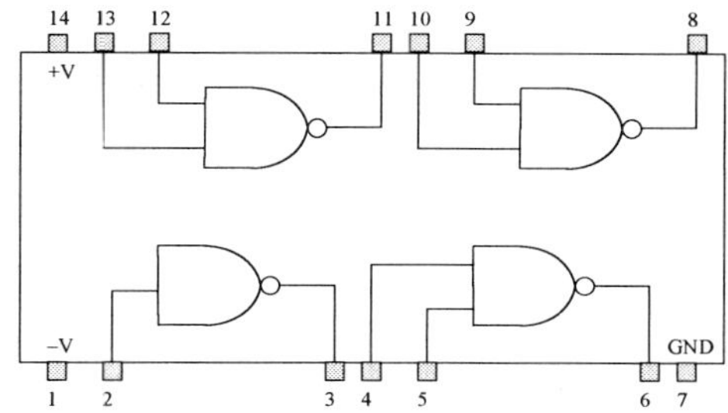

Figure G.2 MC1488 pin out.

The second input of the NAND gate can also be used to generate a Break signal on the transmit line, where a low input will cause a high output (+12 V).

The MC1489 is a line receiver that converts RS-232 line voltage back to TTL levels. This device is specially designed to receive pulses over a long and highly capacitive transmission line. Figure G.3 shows its pin connections.

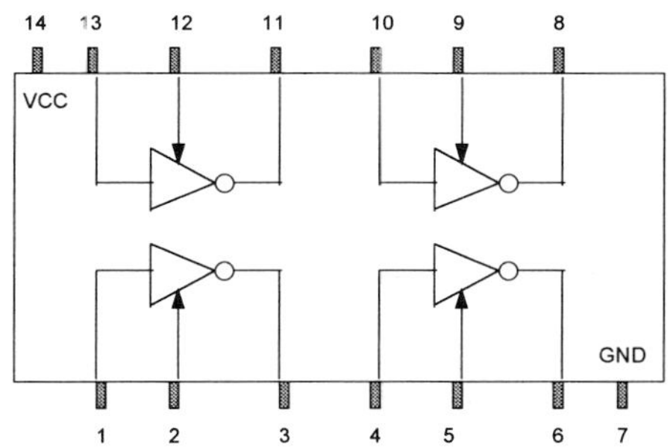

Figure G.3 MC1489 pin out.

G.1.3 Transmitter/receiver interface

The output lines from the UART are TD, DTR, and RTS and the input lines are RD, DSR, DCD and CTS. Figure G.4 shows how the these lines are buffered via the driver and receiver gates.

G.2 Parallel port

G.2.1 IEEE 1284 cables assemblies

The IEEE 1284 standard defined standards for computability between different platforms and peripherals. The most typical connector is a DB25 male on one end and a 36-pin Champ plug connector on the other. The cable has from 18 to 25 conductors and from one to eight ground wires. This cable is acceptable for low speed (10kB/sec) at a maximum distance of 2 m but will not operate properly for high bit rates (such as 2MB/sec) over relatively long runs (such as 10 m).

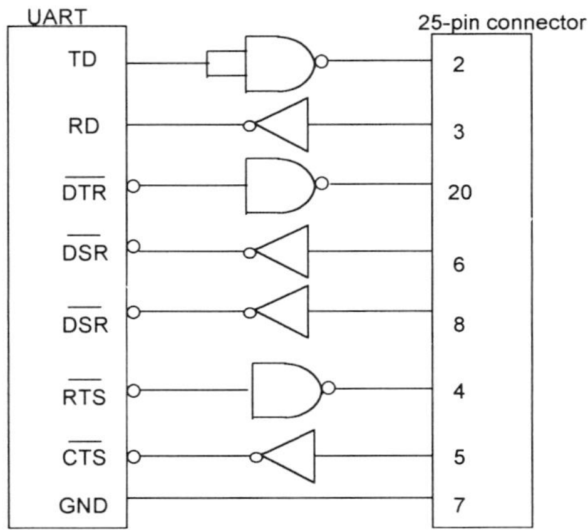

Figure G.4 Transmitter/receiver interface.

A 1284 connection complies with the following:

1. Signals connect to a twisted pair with a signal line and ground return.
2. Each signal has a characteristic unbalanced impedance of approx. 62 Ω.
3. Crosstalk is less than 10%.
4. 85% minimum coverage of braid over foil.
5. Cable shield connects to the connector backshell using a 360° concentric method (and no pigtail connections).
6. Cables are marked with the label: IEEE Std 1284-1994.

G.2.2 IEEE 1284 connectors

The 1284 standard defines the standard connectors to be used with the parallel port, for its both its mechanical and electrical specification. This ensures compatibility with existing and future applications.

The three connectors defined are:

- 1284 Type A. 25-pin DB25 connector.
- 1284 Type B. 36-pin, 0.085 centerline Champ connector with bale locks
- 1284 Type C. 36-pin, 0.050 centerline mini-connector with clip latches.

Figure G.5 illustrates these connectors. The best specification is the type C connector as:

- It has a smaller footprint than the others.
- It has a simple-to-use clip latch for cable retention.
- It has the easiest cable assembly with the optimal electrical properties.
- Its cable assembly has two extra signals, which are Peripheral Logic High and Host Logic High. These can be used to determine if the device at the other end of the cable is powered on. This allows for some degree of intelligent power management for 1284 interfaces.

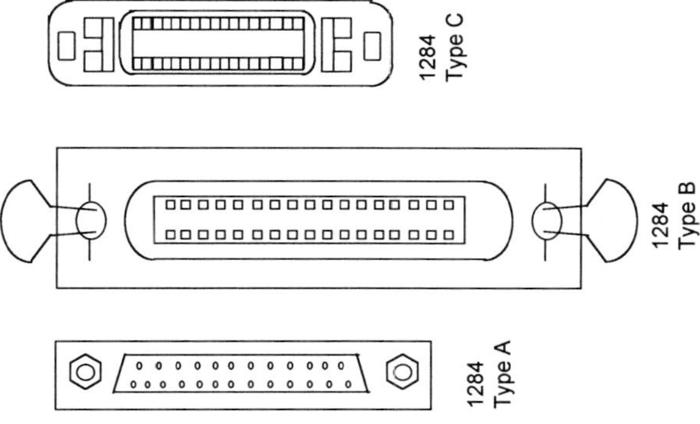

Figure G.5 1284 interface I/O connectors.

Typical assembly types are:

AMAM	Type A Male to Type A Male	AMAF	Type A Male to Type A Female
AB	Type A Male to Type B Plug	AC	Type A Male to Type C Plug
BC	Type B Plug to Type C Plug	CC	Type C Plug to Type C Plug

G.2.3 IEEE 1284 Electrical Interface

The 1284 standard defines two levels of interface compatibility:

- Level I. Designed for low-speed applications, but which need the reverse channel capabilities.
- Level II. Design for advanced modes, high bit rates and long cables.

The electrical specifications for Level II drivers and receivers are defined at the connector interface. For drivers these are:

1. Open-circuit high-level output voltage: less than +5.5 V.
2. Open-circuit low-level output voltage: not less than –0.5 V.
3. DC steady-state, high-level output voltage: greater than +2.4 V with a source current of 14 mA.
4. DC steady-state, low-level output voltage: less than +0.4 V with a sink current of 14 mA.
5. Output impedance: 50 +/– 5 Ω.
6. Output slew rate: between 0.05 and 0.40 V.ns^{-1}.

For drivers these are:

1. Peak input voltage transients without damage: between –2.0 V and +7.0 V.
2. High-level input threshold: less than 2.0 V
3. Low-level input threshold: greater than 0.8 V.
4. Input hysteresis: at least 0.2 V, but not more than 1.2 V.
5. High-level sink current: less than 20 μA at +2.0 V.
6. Low-level input source current: less than 20 μA at +0.8 V.
7. Circuit and stray capacitance: less than 50 pF.

Figure A.6 shows the recommended termination for a driver/receiver pair, where R_0 is the output impedance at the connector. This should be matched to the impedance of the cable (Z_0) so that noise and reflections can be minimized. A series resistance (R_s) can thus be added to obtain a match.

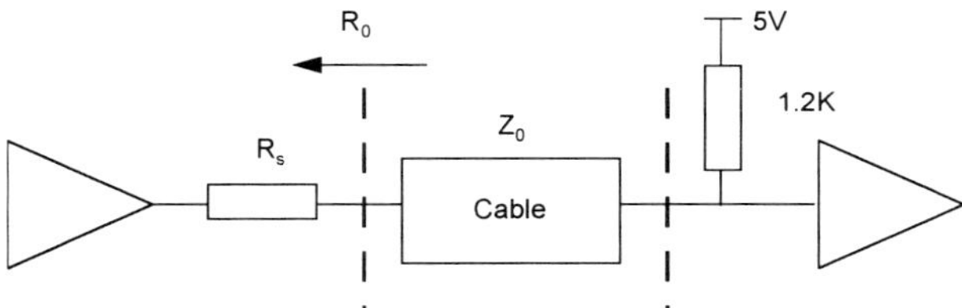

Figure G.6 Level II driver/receiver pair termination example.

G.3 PC connections

The 430HX motherboard uses the 82091AA (API) device. Thus incorporates two 8250 UARTs, a parallel port connection and a floppy disk interface. Refer to Figure 11.3 to pin out. It shows the connection between the AIP and the serial port interface. It shows the signal lines for COM1; these are fed into the GD75232SOP device which converts between 0/+5 V and +/–12 V. It then connects to a 10-pin header which is wired to the serial port.

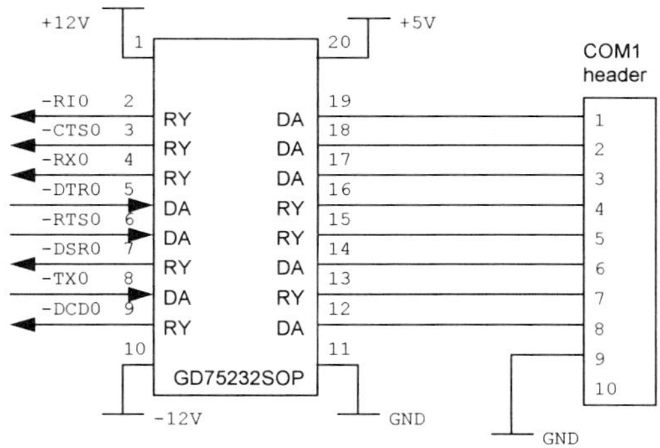

Figure G.7 COM1 connection.

G.4 Parallel port connection

Figure G.9 shows the connections from the AIP to the 26-pin parallel port header. The 33 Ω resistors on the output lines are the to protect against a short circuit on the output pins. This limits the short circuit current to less than 150 mA (5 V/33 Ω). The 1 KΩ pull-up resistor causes the input line to be a high input, if there is nothing connected to the port. Thus, when nothing is connected, the BUSY, SLCT and PE lines will be active, while ERR and ACK will be inactive. Note that the physical layout of the 26-pin header is:

1 (-STB)	2 (-AFD)
3 (PD0)	4 (-ERR)
5 (PD1)	6 (-INIT)
7 (PD2)	8 (-SLIN)
9 (PD3)	10 (GND)
11 (PD4)	12 (GND)
13 (PD5)	14 (GND)
15 (PD6)	16 (GND)
17 (PD7)	18 (GND)
19 (-ACK)	20 (GND)
21 (BUSY)	22 (GND)
23 (PE)	24 (GND)
25 (SLCT)	26 (N/C)

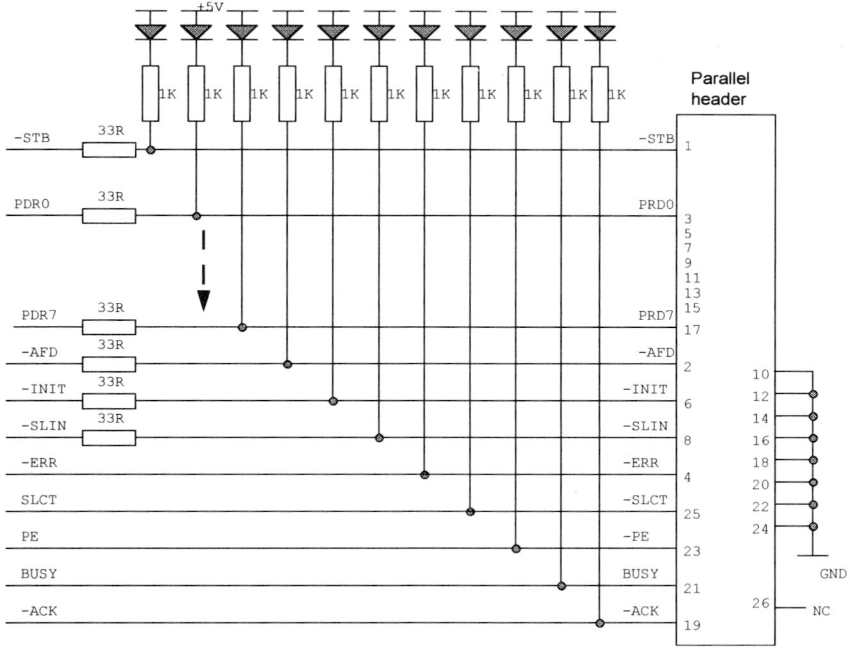

Figure G.8 Parallel port connection.

G.5 RS-232C interface

Table G.1 RS-232C connections.

9-pin D-type	*25-pin D-type*	*Name*	*RS-232 name*	*Description*	*Signal direction on DCE*
	1		AA	Protective GND	
3	2	TXD	BA	Transmit Data	IN
2	3	RXD	BB	Receive Data	OUT
7	4	RTS	CA	Request to Send	IN
8	5	CTS	CB	Clear to Send	OUT
6	6	DSR	CC	Data Set Ready	OUT
5	7	GND	AB	Signal GND	
1	8	DCD	CF	Received Line Signal detect	OUT
	9		-	RESERVED	-
	10		-	RESERVED	-
	11			UNASSIGNED	-
	12		SCF	Secondary Received Line Signal Detector	OUT
	13		SCB	Secondary Clear to Send	OUT
	14		SBA	Secondary Transmitted Data	IN
	15		DB	Transmission Signal Element Detector	OUT
	16		SBB	Secondary Received Data	OUT
	17		DD	Receiver Signal Element Time	OUT
	18			UNASSIGNED	-
	19		SCA	Secondary Request to Send	IN
4	20	DTR	CD	Data Terminal Ready	IN
	21		CG	Signal Quality Detector	OUT
9	22	RI	CE	Ring Indicator	OUT
	23		CH/CI	Data Signal Rate Selector	IN/OUT
	24		DA	Transmit Signal Element Timing	IN
	25			UNASSIGNED	-

G.6 RS-449 interface

RS-449 defines a standard for the functional/mechanical interface of DTEs/DCEs for serial communications and is usually used with synchronous transmissions. Table G.2 lists the main connections.

Table G.2 RS-449 connections.

Pin number	*Mnemonic*	*Description*
1		Shield
2	SI	Signaling Rate Indicator
3,21		Spare
4,22	SD	Sending Time
5,23	ST	Receive Data
6,24	RD	Receive Data
7,25	RS	Request to Send
8,26	RT	Receive Timing
9,27	CS	Clear to Send
10	LL	Local Loopback
11,29	DM	Data Mode
12,30	TR	Terminal Ready
13,31	RR	Receiver Ready
14	RL	Remote Loopback
15	IC	Incoming Call
16	SF/SR	Select Frequency/ Signaling Rate Select
17,37	TT	Terminal Timing
18	TM	Test Mode
19	SG	Signal Ground
20	RC	Receive Common
28	IS	Terminal in Service
32	SS	Select Standby
33	SQ	Signal Quality
34	NS	New Signal
36	SB	Standby Indicator
37	SC	Send Common

Modem Codes

H.1 AT commands

The AT commands are preceded by the attention code AT. They are:

A — **Go on-line in answer mode**
Instructs the modem to go off-hook immediately and then make a connection with a remote modem

Bn — **Select protocol to 300 bps to 1200 bps**

B0	Selects CCITT operation at 300 bps or 1200 bps
B1	Selects BELL operation at 300 bps or 1200 bps

D — **Go on-line in originate mode**
Instructs the modem to go off-hook and automatically dials the number contained in the dial string which follows the D command

En — **Command echo**

E0	Disable command echo
E1	Enables command echo (default)

Fn — **Select line modulation**

F0	Select auto-detect mode
F1	Select V.21 or Bell 103
F4	Select V.22 or Bell 212A 1200 bps
F5	Select V.22bis line modulation.
F6	Select V.32bis or V.32 4800 bps line modulation
F7	Select V.32bis or V.32 7200 bps line modulation
F8	Select V.32bis or V.32 9600 bps line modulation
F9	Select V.32bis 12000 line modulation
F10	Select V.32bis 14400 line modulation

Hn — **Hang-up**

H0	Go on-hook (hang-up connection)
H1	Goes off-hook

In — **Request product code or ROM checksum**

I0	Reports the product code
I1/I2	Reports the hardware ROM checksum
I3	Reports the product revision code
I4	Reports response programmed by an OEM
I5	Reports the country code number

Ln — **Control speaker volume**

L0	Low volume
L1	Low volume
L2	Medium volume (default)
L3	High volume

Mn — **Monitor speaker on/off**

M0/M	Speaker is always off
M1	Speaker is off while receiving carrier (default)
M2	Speaker is always on
M3	Speaker is on when dialing but is off at any other time

Nn — **Automode enable**

N0	Automode detection is disabled

N1 Automode detection is enabled

On **Return to the on-line state**

O0 Enters on-line data mode with a retrain

O1 Enters on-line data mode without a retrain

P **Set pulse dial as default**

Q **Result code display**

Q0 Send result codes to the computer

Q1 No return codes

Sn **Reading and writing to S registers**

Sn? Reads the Sn register

Sn=val Writes the value of val to the Sn register

T **Set tone dial as default**

Vn **Select word or digit result code**

V0 Display result codes in a numeric form

V1 Display result code in a long form (default)

Wn **Error correction message control**

W0 When connected report computer connection speed

W1 When connected report computer connection speed, error correcting protocol and line speed

W2 When connected report modem connection speed

Xn **Select result code**

X0 Partial connect message, dial-tone monitor off, busy tone monitor off

X1 Full connect message, dial-tone monitor off, busy tone monitor off

X2 Full connect message, dial-tone monitor on, busy tone monitor off

X3 Full connect message, dial-tone monitor off, busy tone monitor on

X4 Full connect message, dial-tone monitor on, busy tone monitor on

Yn **Enables or disables long space disconnection**

Y0 Disables long space disconnect (default)

Y1 Enables long space disconnect

Zn **Reset**

Z0 Resets modem and load stored profile 0

Z1 Resets modem and load stored profile 1

&Cn **Select DCD options**

&C0 Sets DCD permanently on

&C1 Use state of carrier to set DCD (default)

&Dn **DTR option**

This is used with the &Qn setting to determine the operation of the DTR signal

	&D0	&D1	&D2	&D3
&Q0	a	c	d	e
&Q1	b	c	d	e
&Q2	d	d	d	d
&Q3	d	d	d	d
&Q4	b	c	d	e
&Q5	a	c	d	e
&Q6	a	c	d	e

where

a – modem ignore DTR signal

b – modem disconnects and sends OK result code

c – modem goes into command mode and sends OK result code

d – modem disconnects and sends OK result code.

&F **Restore factory configuration**

&Gn **Set guard tone**

&G0 Disables guard tone (default)

&G1 Disables guard tone

&G2 Selects 1800 Hz guard tone

&Kn **DTE/modem flow control**

&K0 Disables DTE/DCE flow control

&K3 Enables RTS/CTS handshaking flow control (Default)

	&K4	Enables XON/XOFF flow control		
	&K5	Enables transparent XON/XOFF flow control		
	&K6	Enables RTS/CTS and XON/XOFF flow control		
&L	**Line selection**			
	&L0	Selects dial-up line operation (Default)		
	&L1	Selects leased line operation		
&Mn	**Communications mode**			
&Pn	**Select pulse dialing make/break ratio**			
	&P0	Sets a 39/61 make-break ratio at 10 pps (Default)		
	&P1	Sets a 33/67 make-break ratio at 10 pps (Default)		
	&P2	Sets a 39/61 make-break ratio at 20 pps (Default)		
	&P3	Sets a 33/67 make-break ratio at 20 pps (Default)		
&Qn	**Asynchronous/synchronous mode selection**			
	&Q0	Set direct asynchronous operation		
	&Q1	Set synchronous operation with asynchronous off-line		
	&Q2	Set synchronous connect mode with asynchronous off-line		
	&Q3	Set synchronous connect mode		
	&Q5	Modem negotiation for error-corrected link		
	&Q6	Set asynchronous operation in normal mode		
&Rn	**RTS/CTS option**			
	&R0	In synchronous mode, CTS changes with RTS (the delay is defined by the S26 register)		
	&R1	In synchronous mode, CTS is always ON		
&Sn	**DSR option**			
	&S0	DSR is always ON (Default)		
	&S1	DSR is active after the answer tone has been detected		
&Tn	**Testing and diagnostics**			
	&T0	Terminates any current test	&T1	Local analogue loopback test
	&T2	Local digital loopback test		
&V	**View configuration profiles**			
&Wn	**Store the current configuration in non-volatile RAM**			
	&W0	Writes current settings to profile 0 in nonvolatile RAM		
	&W1	Writes current settings to profile 1 in nonvolatile RAM		
&Xn	**Clock source selection**			
	&X0	Selects internal timing, where the modem uses its own clock for transmitted data		
	&X1	Selects external timing, where the modem gets its timing from the DTE (computer)		
	&X2	Selects slave receive timing, where the modem gets its timing from the received signal		
&Yn	**Select default profile**			
	&Y0	Use profile 0 on power-up (Default)		
	&Y1	Use profile 1 on power-up		
&Zn	**Store telephone numbers**			
	&Z0	Store telephone number 1	&Z1	Store telephone number 2
	&Z2	Store telephone number 3	&Z3	Store telephone number 4
\An	**Maximum MNP block size**			
	\A0	64 characters		
	\A1	128 characters		
	\A2	192 characters		
	\A3	256 characters		
\Bn	**Transmit break**			
	\B1	Break length 100 ms	\B2	Break length 200 ms
	\B3	Break length 300 ms (Default)	*and so on.*	
\Gn	**Modem/modem flow control**			
	\G0	Disable (Default)		
	\G1	Enable		
\Jn	**Enable/disable DTE auto rate adjustment**			
	\J0	Disable	\J1	Enable
\Kn	**Break control**			
	\K0	Enter on-line command mode with no break signal		

	\K1	Clear data buffers and send a break to the remote modem
	\K3	Send a break to the remote modem immediately
	\K5	Send a break to the remote modem with transmitted data
\Ln	**MNP block transfer control**	
	\L0	Use stream mode for MNP connection (Default)
	\L1	Use interactive MNP block mode.

H.2 Result codes

After the modem has received an AT command it responds with a return code. A complete set of return codes are given in Table H.1.

Table H.1 Modem return codes.

Message	*Digit*	*Description*
OK	0	Command executed without errors
CONNECT	1	A connection has been made
RING	2	An incoming call has been detected
NO CARRIER	3	No carrier detected
ERROR	4	Invalid command
CONNECT 1200	5	Connected to a 1200 bps modem
NO DIAL-TONE	6	Dial-tone not detected
BUSY	7	Remote line is busy
NO ANSWER	8	No answer from remote line
CONNECT 600	9	Connected to a 600 bps modem
CONNECT 2400	10	Connected to a 2400 bps modem
CONNECT 4800	11	Connected to a 4800 bps modem
CONNECT 9600	13	Connected to a 9600 bps modem
CONNECT 14400	15	Connected to a 14 400 bps modem
CONNECT 19200	16	Connected to a 19200 bps modem
CONNECT 28400	17	Connected to a 28400 bps modem
CONNECT 38400	18	Connected to a 38400 bps modem
CONNECT 115200	19	Connected to a 115200 bps modem
FAX	33	Connected to a FAX modem in FAX mode
DATA	35	Connected to a data modem in FAX mode
CARRIER 300	40	Connected to V.21 or Bell 103 modem
CARRIER 1200/75	44	Connected to V.23 backward channel carrier modem
CARRIER 75/1200	45	Connected to V.23 forwards channel carrier modem
CARRIER 1200	46	Connected to V.22 or Bell 212 modem
CARRIER 2400	47	Connected to V.22 modem
CARRIER 4800	48	Connected to V.32bis 4800 bps modem
CONNECT 7200	49	Connected to V.32bis 7200 bps modem
CONNECT 9600	50	Connected to V.32bis 9600 bps modem
CONNECT 12000	51	Connected to V.32bis 12000 bps modem
CONNECT 14400	52	Connected to V.32bis 14400 bps modem
CONNECT 19200	61	Connected to a 19 200 bps modem
CONNECT 28800	65	Connected to a 28 800 bps modem
COMPRESSION: CLASS 5	66	Connected to modem with MNP Class 5 compression
COMPRESSION: V.42bis	67	Connected to a V.42bis modem with compression
COMPRESSION: NONE	69	Connection to a modem with no data compression
PROTOCOL: NONE	70	
PROTOCOL: LAPM	77	
PROTOCOL: ALT	80	

H.3 S-registers

The modem contains various status registers called the S-registers which store modem settings. Table H.2 lists these registers.

Table H.2 Modem registers.

Register	*Function*	*Range [typical default]*
S0	Rings to Auto-answer	0–255 rings [0 rings]
S1	Ring counter	0–255 rings [0 rings]
S2	Escape character	[43]
S3	Carriage return character	[13]
S6	Wait time for dial-tone	2–255 s [2 s]
S7	Wait time for carrier	1–255 s [50 s]
S8	Pause time for automatic dialing	0–255 s [2 s]
S9	Carrier detect response time	1–255 in 0.1 s units [6]
S10	Carrier loss disconnection time	1–255 in 0.1 s units [14]
S11	DTMF tone duration	50–255 in 0.001 s units [95]
S12	Escape code guard time	0–255 in 0.02 s units [50]
S13	Reserved	
S14	General bitmapped options	[8Ah (1000 1010b)]
S15	Reserved	
S16	Test mode bitmapped options (&T)	[0]
S17	Reserved	
S18	Test timer	0–255 s [0]
S19–S20	Reserved	
S21	V.24/General bitmapped options	[04h (0000 0100b)]
S22	Speak/results bitmapped options	[75h (0111 0101b)]
S23	General bitmapped options	[37h (0011 0111b)]
S24	Sleep activity timer	0–255 s [0]
S25	Delay to DSR off	0–255 s [5]
S26	RTS–CTS delay	0–255 in 0.01 s [1]
S27	General bitmapped options	[49h (0100 1001b)]
S28	General bitmapped options	[00h]
S29	Flash dial modifier time	0–255 in 10 ms [0]
S30	Disconnect inactivity timer	0–255 in 10 s [0]
S31	General bitmapped options	[02h (0000 0010b)]
S32	XON character	[Cntrl–Q, 11h (0001 0001b)]
S33	XOFF character	[Cntrl–S, 13h (0001 0011b)]
S34–S35	Reserved	
S36	LAMP failure control	[7]
S37	Line connection speed	[0]
S38	Delay before forced hang-up	0–255 s [20]
S39	Flow control	[3]
S40	General bitmapped options	[69h (0110 1001b)]
S41	General bitmapped options	[3]
S42–S45	Reserved	
S46	Data compression control	[8Ah (1000 1010b)]
S48	V.42 negotiation control	[07h (0000 0111b)]
S80	Soft-switch functions	[0]
S82	LAPM break control	[40h (0100 0000b)]
S86	Call failure reason code	0–255
S91	PSTN transmit attenuation level	0–15 dBm [10]
S92	Fax transmit attenuation level	0–15 dBm [10]
S95	Result code message control	[0]
S99	Leased line transmit level	0–15 dBm [10]

S14 **Bitmapped options**

	0	1
Bit 0		
Bit 1	E0	**E1**
Bit 2	**Q0**	Q1
Bit 3	V0	**V1**
Bit 4	Reserved	
Bit 5	**T** (tone dial)	P (pulse dial)
Bit 6	Reserved	
Bit 7	Answer mode	**Originate mode**

S16 **Modem test mode register**

	0	1
Bit 0	Local analogue loopback terminated	Local analogue loopback test in progress
Bit 1	Unused	
Bit 2	Local digital loopback terminated	Local digital loopback test in progress
Bit 3	Remote modem analogue loopback test terminated	Remote modem analogue loopback test in progress
Bit 4	Remote modem digital loopback test terminated	Remote modem digital loopback test in progress
Bit 5	Remote modem digital self- test terminated	Remote modem digital self-test in progress
Bit 6	Remote modem analogue self- test terminated	Remote modem analogue self-test in progress
Bit 7	Unused	

S21 **Bitmapped options**

	0	1
Bit 0	**&J0**	&J1
Bit 1		
Bit 2	&R0	**&R1**
Bit 5	&C0	**&C1**
Bit 6	**&S0**	&S1
Bit 7	**Y0**	Y1

Bit 4, 3 = 00 &D0
Bit 4, 3 = 01 &D1
Bit 4, 3 = 10 **&D2**
Bit 4, 3 = 11 &D3

S22 **Speaker/results bitmapped options**

Bit 1, 0 = 00 L0
Bit 1, 0 = 01 **L1**
Bit 1, 0 = 10 L2
Bit 1, 0 = 11 L3
Bit 3, 2 = 00 M0
Bit 3, 2 = 01 **M1**
Bit 3, 2 = 10 M2
Bit 3, 2 = 11 M3
Bit 6, 5, 4 = 000 X0
Bit 6, 5, 4 = 001 Reserved
Bit 6, 5, 4 = 010 Reserved
Bit 6, 5, 4 = 011 Reserved
Bit 6, 5, 4 = 100 X1
Bit 6, 5, 4 = 101 X2
Bit 6, 5, 4 = 110 X3
Bit 6, 5, 4 = 111 **X4**
Bit 7 Reserved

S23 **Bitmapped options**

	0	1
Bit 0	&T5	**&T4**

Bit 3, 2, 1 = 000 300 bps communications rate
Bit 3, 2, 1 = 001 600 bps communications rate
Bit 3, 2, 1 = 010 1200 bps communications rate
Bit 3, 2, 1 = 011 **2400 bps communications rate**
Bit 3, 2, 1 = 100 4800 bps communications rate
Bit 3, 2, 1 = 101 960 bps communications rate
Bit 3, 2, 1 = 110 19200 bps communications rate
Bit 3, 2, 1 = 111 Reserved
Bit 5, 4 = 00 Even parity
Bit 5, 4 = 01 **Not used**
Bit 5, 4 = 10 Odd parity
Bit 5, 4 = 11 No parity
Bit 7, 6 = 00 **G0**
Bit 7, 6 = 01 G1
Bit 7, 6 = 10 G2
Bit 7, 6 = 11 G3

S23 **Bitmapped options**
Bit 3, 1, 0 = 000 &M0 or &Q0
Bit 3, 1, 0 = 001 &M1 or &Q1
Bit 3, 1, 0 = 010 &M2 or &Q2
Bit 3, 1, 0 = 011 &M3 or &Q3
Bit 3, 1, 0 = 100 &Q3
Bit 3, 1, 0 = 101 &Q4
Bit 3, 1, 0 = 110 **&Q5**
Bit 3, 1, 0 = 111 &Q6

	0	1
Bit 2	**&L0**	&L1
Bit 6	B0	**B1**

Bit 5, 4 = 00 **X0**
Bit 5, 4 = 01 X1
Bit 5, 4 = 10 X2

S28 **Bitmapped options**
Bits 0, 1, 2 Reserved
Bit 4, 3 = 00 **&P0**
Bit 4, 3 = 01 &P1
Bit 4, 3 = 10 &P2
Bit 4, 3 = 11 &P3

S31 **Bitmapped options**

	0	1
Bit 1	**N0**	N1

Bit 3, 2 = 00 **W0**
Bit 3, 2 = 01 W1
Bit 3, 2 = 10 W2

S36 **LAPM failure control**
Bit 2, 1, 0 = 000 Modem disconnect
Bit 2, 1, 0 = 001 Modem stays on-line and a direct mode connection
Bit 2, 1, 0 = 010 Reserved
Bit 2, 1, 0 = 011 Modem stays on-line and normal mode connection is established
Bit 2, 1, 0 = 100 An MNP connection is made, if it fails then the modem disconnects
Bit 2, 1, 0 = 101 An MNP connection is made, if it fails then the modem makes a direct connection
Bit 2, 1, 0 = 110 Reserved
Bit 2, 1, 0 = 111 An MNP connection is made, if it fails then the modem makes a normal mode connection

S37 **Desired line connection speed**
Bit 3, 2, 1, 0 = 0000 **Auto mode connection (F0)**
Bit 3, 2, 1, 0 = 0001 Modem connects at 300 bps (F1)
Bit 3, 2, 1, 0 = 0010 Modem connects at 300 bps (F1)
Bit 3, 2, 1, 0 = 0011 Modem connects at 300 bps (F1)

	Bit 3, 2, 1, 0 = 0100	Reserved
	Bit 3, 2, 1, 0 = 0101	Modem connects at 1200 bps (F4)
	Bit 3, 2, 1, 0 = 0110	Modem connects at 2400 bps (F5)
	Bit 3, 2, 1, 0 = 0111	Modem connects at V.23 (F3)
	Bit 3, 2, 1, 0 = 1000	Modem connects at 4800 bps (F6)
	Bit 3, 2, 1, 0 = 1001	Modem connects at 9600 bps (F8)
	Bit 3, 2, 1, 0 = 1010	Modem connects at 12000 bps (F9)
	Bit 3, 2, 1, 0 = 1011	Modem connects at 144000 bps (F10)
	Bit 3, 2, 1, 0 = 1100	Modem connects at 7200 bps (F7)
S39	**Flow control**	
	Bit 2, 1, 0 = 000	No flow control
	Bit 2, 1, 0 = 011	**RTS/CTS (&K3)**
	Bit 2, 1, 0 = 100	XON/XOFF (&K4)
	Bit 2, 1, 0 = 101	Transparent XON (&K5)
	Bit 2, 1, 0 = 110	RTS/CTS and XON/XOFF (&K6)
S39	**General bitmapped options**	
	Bit 5, 4, 3 = 000	\K0
	Bit 5, 4, 3 = 001	\K1
	Bit 5, 4, 3 = 010	\K2
	Bit 5, 4, 3 = 011	\K3
	Bit 5, 4, 3 = 100	\K4
	Bit 5, 4, 3 = 101	**\K5**
	Bit 7, 6 = 00 MNP 64 character block size (\A0)	
	Bit 7, 6 = 01 **MNP 128 character block size (\A1)**	
	Bit 7, 6 = 10 MNP 192 character block size (\A2)	
	Bit 7, 6 = 11 MNP 256 character block size (\A3)	

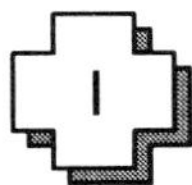

PC Interfacing and Interrupts

I.1 Software Interrupts

I.1.1 Introduction

An interrupt allows a program or an external device to interrupt the execution of a program. The generation of an interrupt can occur by hardware (hardware interrupt) or software (software interrupt). When an interrupt occurs an interrupt service routine (ISR) is called. For a hardware interrupt the ISR then communicates with the device and processes any data. When it has finished the program execution returns to the original program. A software interrupt causes the program to interrupt its execution and goes to an interrupt service routine. Typical software interrupts include reading a key from the keyboard, outputting text to the screen and reading the current date and time.

I.1.2 BIOS and the operating system

The Basic Input/Output System (BIOS) communicates directly with the hardware of the computer. It consists of a set of programs which interface with devices such as keyboards, displays, printers, serial ports and disk drives. These programs allow the user to write application programs that contain calls to these functions, without having to worry about controlling them or which type of equipment is being used. Without BIOS the computer system would simply consist of a bundle of wires and electronic devices.

There are two main parts to BIOS. The first is the part permanently stored in a ROM (the ROM BIOS). It is this part that starts the computer (or boots it) and contains programs which communicate with resident devices. The second stage is loaded when the operating system is started. This part is non-permanent.

An operating system allows the user to access the hardware in an easy-to-use manner. It accepts commands from the keyboard and displays them to the monitor. The Disk Operating System, or DOS, gained its name from its original purpose of providing a controller for the computer to access its disk drives. The language of DOS consists of a set of commands which are entered directly by the user and are interpreted to perform file management tasks, program execution and system configuration. It makes calls to BIOS to execute these. The main functions of DOS are to run programs, copy and remove files, create directories, move within a directory structure and to list files. Microsoft Windows 95/98NT call BIOS programs directly.

I.1.3 Interrupt vectors

Interrupt vectors are addresses which inform the interrupt handler as to where to find the ISR. All interrupts are assigned a number from 0 to 255. The interrupt vectors associated with each interrupt number are stored in the lower 1024 bytes of PC memory. For example, interrupt 0 is stored from `0000:0000` to `0000:0003`, interrupt 1 from `0000:0004` to

0000:0007, and so on. The first two bytes store the offset and the next two store the segment address. Each interrupt number is assigned a predetermined task, as outlined in Table I.1. An interrupt can be generated either by external hardware, software, or by the processor. Interrupts 0, 1, 3, 4, 6 and 7 are generated by the processor. Interrupts from 8 to 15 and interrupt 2 are generated by external hardware. These get the attention of the processor by activating a interrupt request (IRQ) line. The IRQ0 line connects to the system timer, the keyboard to IRQ1, and so on. Most other interrupts are generated by software.

Table I.1 Interrupt handling.

Interrupt	*Name*	*Generated by*
00 (00h)	Divide error	processor
01 (00h)	Single step	processor
02 (02h)	Non-maskable interrupt	external equipment
03 (03h)	Breakpoint	processor
04 (04h)	Overflow	processor
05 (05h)	Print screen	Shift-Print screen key stroke
06 (06h)	Reserved	processor
07 (07h)	Reserved	processor
08 (08h)	System timer	hardware via IRQ0
09 (09h)	Keyboard	hardware via IRQ1
10 (0Ah)	Reserved	hardware via IRQ2
11 (0Bh)	Serial communications (COM2)	hardware via IRQ3
12 (0Ch)	Serial communications (COM1)	hardware via IRQ4
13 (0Dh)	Reserved	hardware via IRQ5
14 (0Eh)	Floppy disk controller	hardware via IRQ6
15 (0Fh)	Parallel printer	hardware via IRQ7
16 (10h)	BIOS – Video access	software
17 (11h)	BIOS – Equipment check	software
18 (12h)	BIOS – Memory size	software
19 (13h)	BIOS – Disk operations	software
20 (14h)	BIOS – Serial communications	software
22 (16h)	BIOS – Keyboard	software
23 (17h)	BIOS – Printer	software
25 (19h)	BIOS – Reboot	software
26 (1Ah)	BIOS – Time of day	software
28 (1Ch)	BIOS – Ticker timer	software
33 (21h)	DOS – DOS services	software
39 (27h)	DOS – Terminate and stay resident	software

I.1.4 Processor interrupts

The processor-generated interrupts normally occur either when a program causes a certain type of error or if it is being used in a debug mode. In the debug mode the program can be made to break from its execution when a break-point occurs. This allows the user to test the current status of the computer. It can also be forced to step through a program one operation at a time (single step mode).

I.2 Hardware Interrupts

I.2.1 Introduction

Computer systems either use polling or interrupt-driven software to service external equipment. With polling the computer continually monitors a status line and waits for it to become active, while an interrupt-driven device sends an interrupt request to the computer, which is then serviced by an interrupt service routine (ISR). Interrupt-driven devices are normally better in that the computer is thus free to do other things, while polling slows the system down as it must continually monitor the external device. Polling can also cause problems because a device may be ready to send data and the computer is not watching the status line at that point. Figure I.1 illustrates polling and interrupt-driven devices.

The generation of an interrupt can occur by hardware or software, as illustrated in Figure I.2. If a device wishes to interrupt the processor, it informs the programmable interrupt controller (PIC). The PIC then decides whether it should interrupt the processor. If there is a processor interrupt then the processor reads the PIC to determine which device caused the interrupt. Then, depending on the device that caused the interrupt, a call to an ISR is made. The ISR then communicates with the device and processes any data. When it has finished the program execution returns to the original program.

A software interrupt causes the program to interrupt its execution and go to an interrupt service routine. Typical software interrupts include reading a key from the keyboard, outputting text to the screen and reading the current date and time.

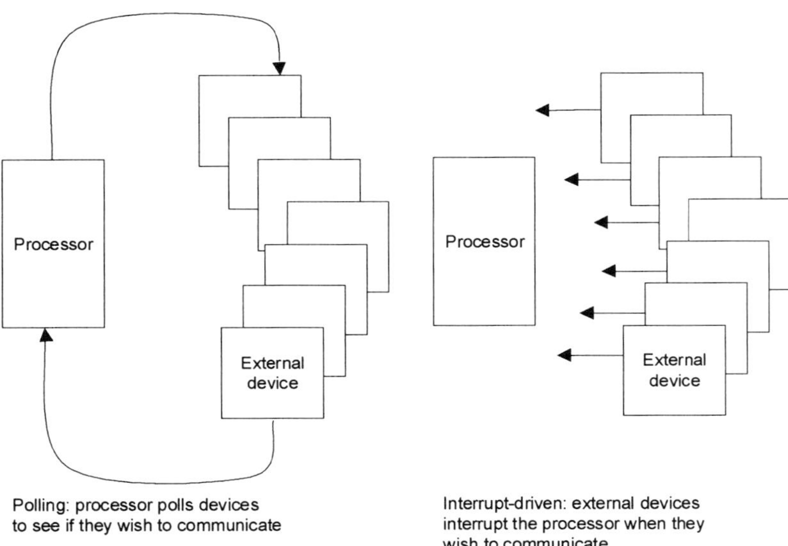
Processor
External device
Processor
External device
Polling: processor polls devices to see if they wish to communicate
Interrupt-driven: external devices interrupt the processor when they wish to communicate

Figure I.1 Polling and interrupt-driven communications.

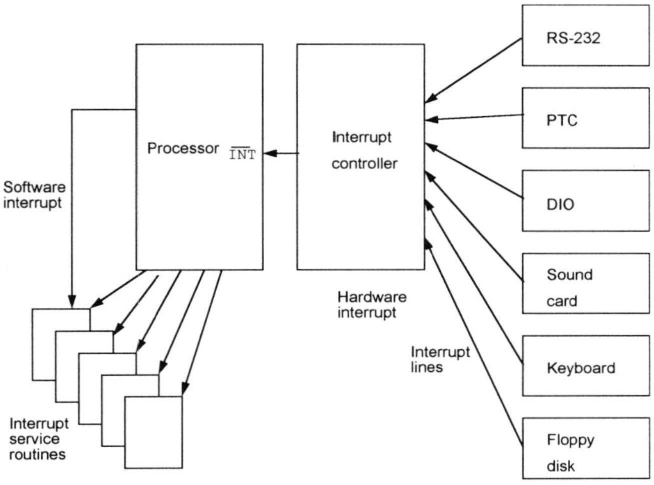

Figure I.2 Interrupt handling.

I.2.2 Hardware interrupts

Hardware interrupts allow external devices to gain the attention of the processor. Depending on the type of interrupt the processor leaves the current program and goes to a special program called an interrupt service routine (ISR). This program communicates with the device and processes any data. After it has completed its task then program execution returns to the program that was running before the interrupt occurred. Examples of interrupts include the processing of keys from a keyboard and data from a sound card.

As previously mentioned, a device informs the processor that it wants to interrupt it by setting an interrupt line on the PC. Then, depending on the device that caused the interrupt, a call to an ISR is made. Each PIC allows access to eight interrupt request lines. Most PCs use two PICs which gives access to 16 interrupt lines.

I.2.3 Interrupt vectors

Each device that requires to be 'interrupt-driven' is assigned an IRQ (interrupt request) line. Each IRQ is active high. The first eight (`IRQ0`–`IRQ7`) map into interrupts 8 to 15 (08h–0Fh) and the next eight (`IRQ8`–`IRQ15`) into interrupts 112 to 119 (70h–77h). Table I.1 outlines the usage of each of these interrupts. When `IRQ0` is made active the ISR corresponds to interrupt vector 8. `IRQ0` normally connects to the system timer, the keyboard to `IRQ1`, and so on. The standard set up of these interrupts is illustrated in Figure I.3. The system timer interrupts the processor 18.2 times per second and is used to update the system time. When the keyboard has data it interrupts the processor with the `IRQ1` line.

Data received from serial ports interrupts the processor with `IRQ3` and `IRQ4` and the parallel ports use `IRQ5` and `IRQ7`. If one of the parallel, or serial, ports does not exist then the IRQ line normally assigned to it can be used by another device. It is typical for interrupt-driven I/O cards, such as a sound card, to have a programmable IRQ line which is mapped to an IRQ line that is not being used.

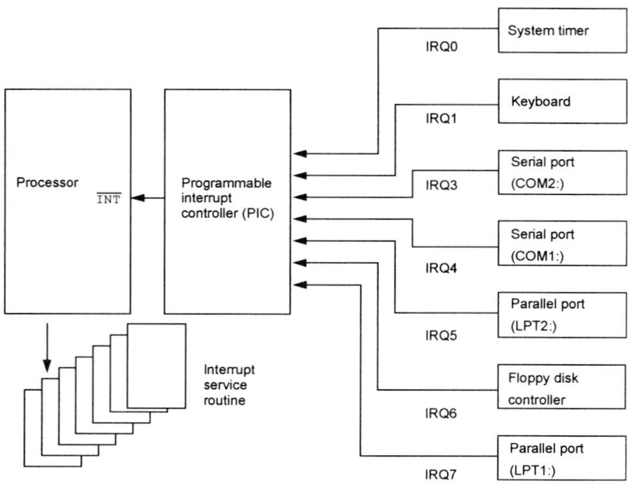

Figure I.3 Standard usage of IRQ lines.

Note that several devices can use the same interrupt line. A typical example is `COM1:` and `COM3:` sharing `IRQ4` and `COM2:` and `COM4:` sharing `IRQ3`. If they do share then the ISR must be able to poll the shared devices to determine which of them caused the interrupt. If two different types of device (such as a sound card and a serial port) use the same IRQ line then there may be a contention problem as the ISR may not be able to communicate with different types of interfaces.

Table I.2 Interrupt handling.

Interrupt	*Name*	*Generated by*
08 (08h)	System timer	IRQ0
09 (09h)	Keyboard	IRQ1
10 (0Ah)	Reserved	IRQ2
11 (0Bh)	Serial communications (COM2:)	IRQ3
12 (0Ch)	Serial communications (COM1:)	IRQ4
13 (0Dh)	Parallel port (LPT2:)	IRQ5
14 (0Eh)	Floppy disk controller	IRQ6
15 (0Fh)	Parallel printer (LPT1:)	IRQ7
112 (70h)	Real-time clock	IRQ8
113 (71h)	Redirection of IRQ2	IRQ9
114 (72h)	Reserved	IRQ10
115 (73h)	Reserved	IRQ11
116 (74h)	Reserved	IRQ12
117 (75h)	Math co-processor	IRQ13
118 (76h)	Hard disk controller	IRQ14
119 (77h)	Reserved	IRQ15

Microsoft Windows 95/98 contains a useful program which determines the usage of the system interrupts. It is selected from Control Panel by selecting System→ Device Manager→ Properties. Figure I.4 shows a sample window. In this case, it can be seen that the system timer uses `IRQ0`, the keyboard uses `IRQ1`, the PIC uses `IRQ2`, and so on. Notice that a Sound

Blaster is using `IRQ5`. This interrupt is normally reserved for the secondary printer port. If there is no printer connected then `IRQ5` can be used by another device. Some devices can have their I/O address and interrupt line changed. An example is given in Figure I.5. In this case the IRQ line is set to IRQ7 and the base address is 378h.

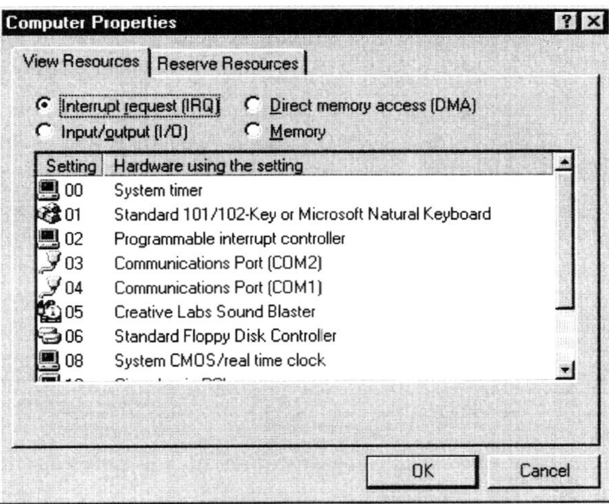

Figure I.4 Standard usage of IRQ lines.

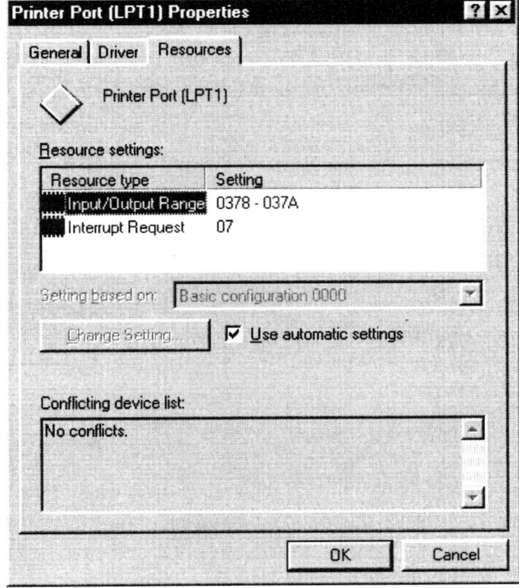

Figure I.5 Standard set up of IRQ lines.

IRQ0: System timer

The system timer uses `IRQ0` to interrupt the processor 18.2 times per second and is used to keep the time-of-day clock updated.

IRQ1: Keyboard data ready

The keyboard uses `IRQ1` to signal to the processor that data is ready to be received from the keyboard. This data is normally a scan code, but the interrupt handler performs differently for the following special keystrokes:

- *Ctrl-Break* invokes interrupt 1Bh.
- *SysRq* invokes interrupt 15h/AH=85h.
- *Ctrl-Alt-Del* performs hard or soft reboot.
- *Shift-PrtSc* invokes interrupt 05h.

IRQ2: Redirection of IRQ9

The BIOS redirects the interrupt for `IRQ9` back here.

IRQ3: Secondary serial port (COM2:)

The secondary serial port (`COM2:`) uses `IRQ3` to interrupt the processor. Typically, `COM3:` to `COM8:` also use it, although `COM3:` may use `IRQ4`.

IRQ4: Primary serial port (COM1:)

The primary serial port (`COM1:`) uses `IRQ4` to interrupt the processor. Typically, `COM3:` also uses it.

IRQ5: Secondary parallel port (LPT2:)

On older PCs the `IRQ5` line was used by the fixed disk. On new systems the secondary parallel port uses it. Typically, it is used by a sound card on PCs which have no secondary parallel port connected.

IRQ6: Floppy disk controller

The floppy disk controller activates the `IRQ6` line on completion of a disk operation.

IRQ7: Primary parallel port (LPT1:)

Printers (or other parallel devices) activate the `IRQ7` line when they become active. As with `IRQ5` it may be used by another device, if there are no other devices connected to this line.

IRQ9

Redirected to `IRQ2` service routine.

I.2.4 Programmable interrupt controller (PIC)

The PC uses the 8259 IC to control hardware-generated interrupts. It is known as a programmable interrupt controller and has eight input interrupt request lines and an output line to interrupt the processor. Originally, PCs only had one PIC and eight IRQ lines (`IRQ0`-`IRQ7`). Modern PCs can use up to 15 IRQ lines which are set up by connecting a secondary PIC interrupt request output line to the `IRQ2` line of the primary PIC. The interrupt lines on the secondary PIC are then assigned IRQ lines of `IRQ8` to `IRQ15`. This set up is shown in Figure I.6.

When an interrupt occurs on any of these lines it is sensed by the processor on the `IRQ2` line. The processor then interrogates the primary and secondary PIC for the interrupt line which caused the interrupt.

The primary and secondary PICs are programmed via port addresses 20h and 21h, as given in Table I.3.

The operation of the PIC is programmed using registers. The IRQ input lines are either configured as level-sensitive or edge-triggered interrupt. With edge-triggered interrupts, a change from a low to a high on the IRQ line causes the interrupt. A level-sensitive interrupt occurs when the IRQ line is high. Most devices generate edge-triggered interrupts.

Table I.3 Interrupt port addresses.

Port address	*Name*	*Description*
20h	Interrupt control port (ICR)	Controls interrupts and signifies the end of an interrupt
21h	Interrupt mask register (IMR)	Used to enable and disable interrupt lines

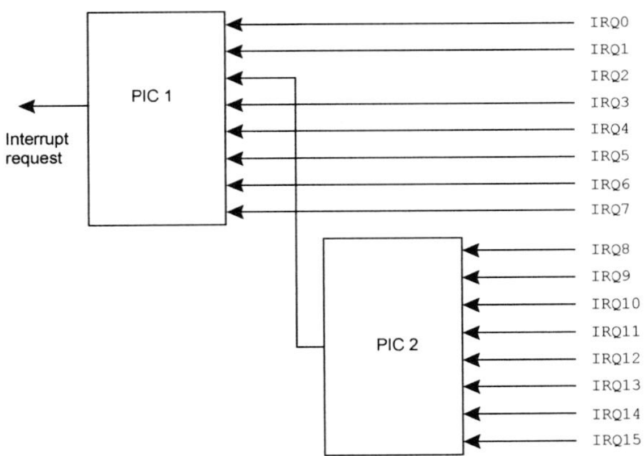

Figure I.6 PC PIC connections.

In the IMR an interrupt line is enabled by setting the assigned bit to a 0 (zero). This allows the interrupt line to interrupt the processor. Figure I.7 shows the bit definitions of the IMR. For example, if bit 0 is set to a 0 then the system timer on `IRQ0` is enabled.

In the example code given next the lines `IRQ0`, `IRQ1` and `IRQ6` are allowed to interrupt the processor, whereas, `IRQ2`, `IRQ3`, `IRQ4` and `IRQ7` are disabled.

```
outportb(0x21)=0xBC;  /*  1011 1100 enable disk  (bit 6), keyboard (1)
                          and timer (0) interrupts  */
```

When an interrupt occurs all other interrupts are disabled and no other device can interrupt the processor. Interrupts are enabled again by setting the EOI bit on the interrupt control port, as shown in Figure I.8.

The following code enables interrupts:

```
outportb(0x20,0x20); /* EOI command */
```

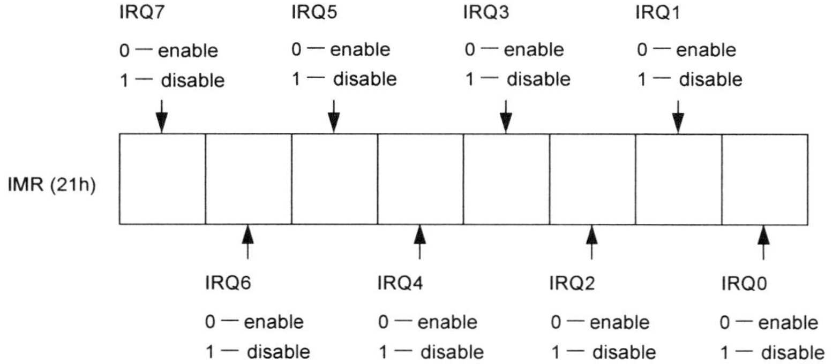

Figure I.7 Interrupt mask register bit definitions.

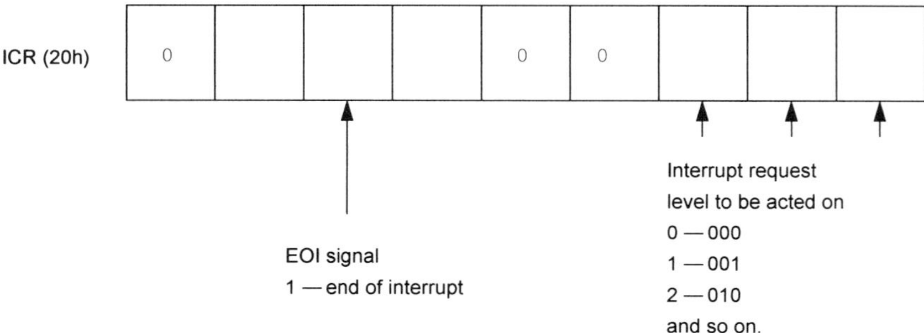

Figure I.8 Interrupt control register bit definitions.

I.3 Interfacing

I.3.1 Introduction

There are two main methods of communicating external equipment, either they are mapped into the physical memory and given a real address on the address bus (memory mapped I/O) or they are mapped into a special area of input/output memory (isolated I/O). Figure I.9 shows the two methods. Devices mapped into memory are accessed by reading or writing to the physical address. Isolated I/O provides ports which are gateways between the interface device and the processor. They are isolated from the system using a buffering system and are accessed by four machine code instructions. The `IN` instruction inputs a byte, or a word, and the `OUT` instruction outputs a byte, or a word. A high-level compiler interprets the equivalent high-level functions and produces machine code which uses these instructions.

I.3.2 Interfacing with memory

The 80x86 processor interfaces with memory through a bus controller, as shown in Figure I.10. This device interprets the microprocessor signals and generates the required memory signals. Two main output lines differentiate between a read or a write operation ($R/\overline{W}$) and between direct and isolated memory access ($M/\overline{IO}$). The $R/\overline{W}$ line is low when data is being written to memory and high when data is being read. When $M/\overline{IO}$ is high, direct memory access is selected and when low, the isolated memory is selected.

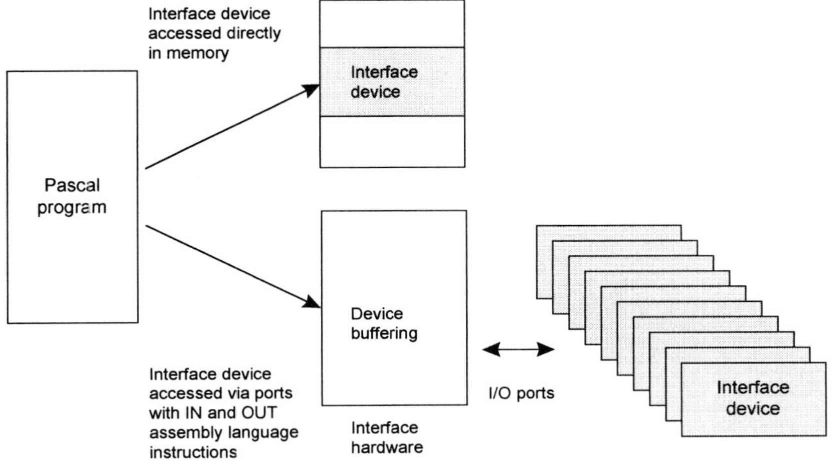

Figure I.9 Memory mapping or isolated interfacing.

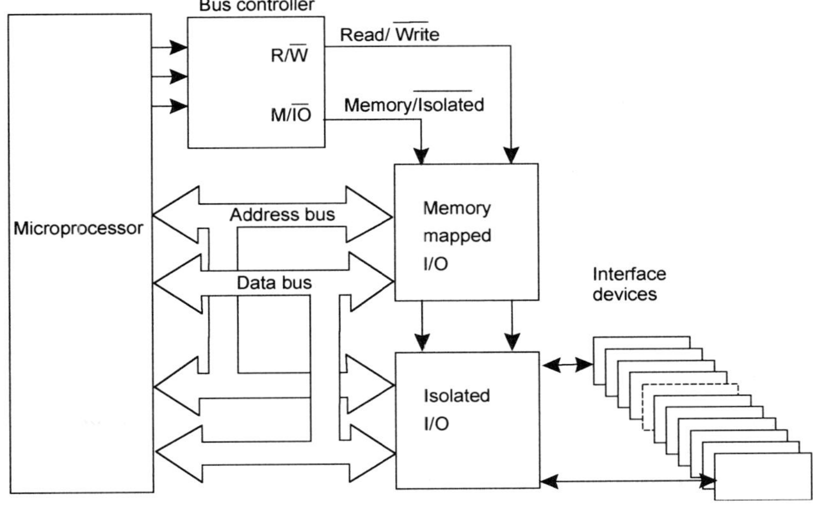

Figure I.10 Access memory mapped and isolated I/O.

I.3.3 Memory mapped I/O

Interface devices can map directly onto the system address and data bus. In a PC-compatible system the address bus is 20 bits wide, from address `00000h` to `FFFFFh` (1 MB). If the PC is being used in an enhanced mode (such as with Microsoft Windows) it can access the area of memory above the 1 MB. If it uses 16-bit software (such as Microsoft Windows 3.1) then it can address up to 16 MB of physical memory, from 000000h to FFFFFFh. If it uses 32-bit software (such as Microsoft Windows 95/98) then the software can address up to 4 GB of physical memory, from 00000000h to FFFFFFFFh. Table I.4 and Figure I.11 gives a typical memory allocation.

Table I.4 Memory allocation for a PC

Address	*Device*
00000h-00FFFh	Interrupt vectors
00400h-0047Fh	ROM BIOS RAM
00600h-9FFFFh	Program memory
A0000h-AFFFFh	EGA/VGA graphics
B0000h-BFFFFh	EGA/VGA graphics
C0000h-C7FFFh	EGA/VGA graphics

I.3.4 Isolated I/O

Devices are not normally connected directly onto the address and data bus of the computer because they may use part of the memory that a program uses or they could cause a hardware fault. On modern PCs only the graphics adaptor is mapped directly into memory, the rest communicate through a specially reserved area of memory, known as isolated I/O memory.

Isolated I/O uses 16-bit addressing from 0000h to FFFFh, thus up to 64 KB of memory can be mapped. Microsoft Windows 95/98 can display the isolated I/O memory map by selecting Control Panel → System → Device Manager, then selecting Properties. From the computer properties window the Input/output (I/O) option is selected. Figure I.11 shows an example for a computer in the range from 0000h to 0064h and Figure I.13 shows from 0378h to 03FFh.

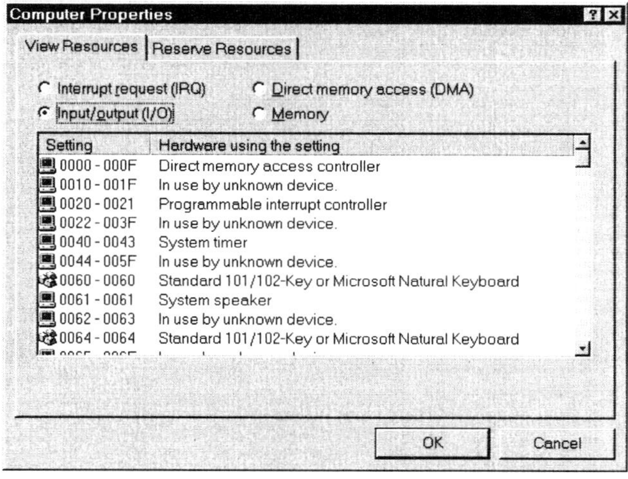

Figure I.11 Example I/O memory map from 0000h to 0064h.

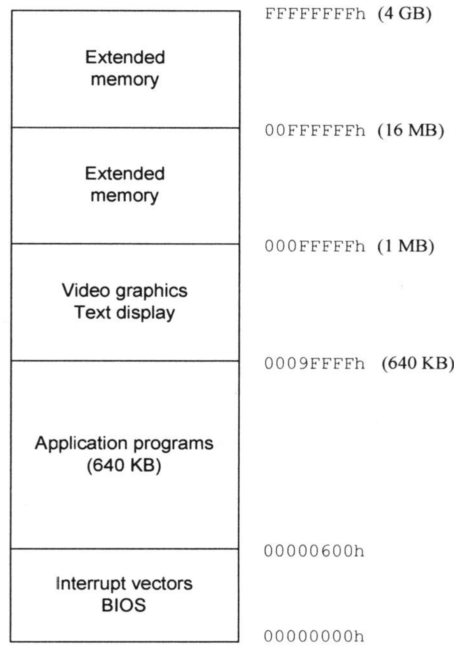

Figure I.12 Typical PC memory map.

It can be seen from Figure I.5 that the keyboard maps into address `0060h` and `0064h`, the speaker maps to address `0061h` and the system timer between `0040h` and `0043h`. Table I.5 shows the typical uses of the isolated memory area.

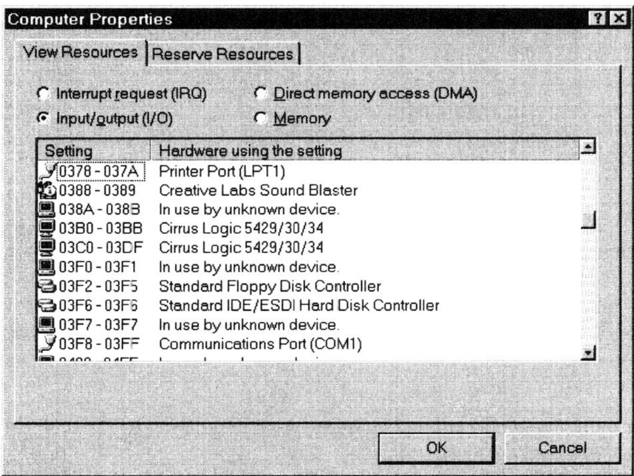

Figure I.13 Example I/O memory map from `0378h` to `03FFh`.

Table I.5 Typical isolated I/O memory map.

Address	*Device*
000h-01Fh	DMA controller
020h-021h	Programmable interrupt controller
040h-05Fh	Counter/Timer
060h-07Fh	Digital I/O
080h-09Fh	DMA controller
0A0h-0BFh	NMI reset
0C0h-0DFh	DMA controller
0E0h-0FFh	Math coprocessor
170h-178h	Hard disk (Secondary IDE drive or CD-ROM drive)
1F0h-1F8h	Hard disk (Primary IDE drive)
200h-20Fh	Game I/O adapter
210h-217h	Expansion unit
278h-27Fh	Second parallel port (LPT2:)
2F8h-2FFh	Second serial port (COM2:)
300h-31Fh	Prototype card
378h-37Fh	Primary parallel port (LPT1:)
380h-38Ch	SDLC interface
3A0h-3AFh	Primary binary synchronous port
3B0h-3BFh	Graphics adapter
3C0h-3DFh	Graphics adapter
3F0h-3F7h	Floppy disk controller
3F8h-3FFh	Primary serial port (COM1:)

Inputting a byte from an I/O port

The assembly language command to input a byte is:

```
IN AL,DX
```

where `DX` is the Data Register which contains the address of the input port. The 8-bit value loaded from this address is put into the register `AL`.

For Turbo/Borland C the equivalent function is `inportb()`. Its general syntax is as follows:

```
value=inportb(PORTADDRESS);
```

where `PORTADDRESS` is the address of the input port and `value` is loaded with the 8-bit value from this address. This function is prototyped in the header file `dos.h`.

For Turbo Pascal the equivalent is accessed via the `port[]` array. Its general syntax is as follows:

```
value:=port[PORTADDRESS];
```

where `PORTADDRESS` is the address of the input port and `value` the 8-bit value at this address. To gain access to this function the statement `uses dos` requires to be placed near the top of the program.

Microsoft C++ uses the equivalent `_inp()` function (which is prototyped in `conio.h`).

Inputting a word from a port

The assembly language command to input a word is:

```
IN AX,DX
```

where `DX` is the Data Register which contains the address of the input port. The 16-value loaded from this address is put into the register `AX`.

For Turbo/Borland C the equivalent function is `inport()`. Its general syntax is as follows:

```
value=inport(PORTADDRESS);
```

where `PORTADDRESS` is the address of the input port and `value` is loaded with the 16-bit value at this address. This function is prototyped in the header file `dos.h`.

For Turbo Pascal the equivalent is accessed via the `portw[]` array. Its general syntax is as follows:

```
value:=portw[PORTADDRESS];
```

where `PORTADDRESS` is the address of the input port and `value` is the 16-bit value at this address. To gain access to this function the statement `uses dos` requires to be placed near the top of the program.

Microsoft C++ uses the equivalent `_inpw()` function (which is prototyped in `conio.h`).

Outputting a byte to an I/O port

The assembly language command to output a byte is:

```
OUT DX,AL
```

where `DX` is the Data Register which contains the address of the output port. The 8-bit value sent to this address is stored in register `AL`.

For Turbo/Borland C the equivalent function is `outportb()`. Its general syntax is as follows:

```
outportb(PORTADDRESS,value);
```

where `PORTADDRESS` is the address of the output port and `value` is the 8-bit value to be sent to this address. This function is prototyped in the header file `dos.h`.

For Turbo Pascal the equivalent is accessed via the `port[]` array. Its general syntax is as follows:

```
port[PORTADDRESS]:=value;
```

where PORTADDRESS is the address of the output port and value is the 8-bit value to be sent to that address. To gain access to this function the statement uses dos requires to be placed near the top of the program.

Microsoft C++ uses the equivalent _outp() function (which is prototyped in conio.h).

Outputting a word

The assembly language command to input a byte is:

```
OUT DX,AX
```

where DX is the Data Register which contains the address of the output port. The 16-bit value sent to this address is stored in register AX.

For Turbo/Borland C the equivalent function is outport(). Its general syntax is as follows:

```
outport(PORTADDRESS,value);
```

where PORTADDRESS is the address of the output port and value is the 16-bit value to be sent to that address. This function is prototyped in the header file dos.h.

For Turbo Pascal the equivalent is accessed via the port[] array. Its general syntax is as follows:

```
portw[PORTADDRESS]:=value;
```

where PORTADDRESS is the address of the output port and value is the 16-bit value to be sent to that address. To gain access to this function the statement uses dos requires to be placed near the top of the program.

Microsoft C++ uses the equivalent _outp() function (which is prototyped in conio.h).

PC Processors

J.1 Introduction

Intel marketed the first microprocessor, named the 4004. This device caused a revolution in the electronics industry because previous electronic systems had a fixed functionality. With this processor the functionality could be programmed by software. Amazingly, by today's standards, it could only handle four bits of data at a time (a nibble), contained 2000 transistors, had 46 instructions and allowed 4 KB of program code and 1 KB of data. From this humble start the PC has since evolved using Intel microprocessors (Intel is a contraction of *Int*e-grated *El*ectronics).

The second generation of Intel microprocessors began in 1974. These could handle 8 bits (a byte) of data at a time and were named the 8008, 8080 and the 8085. They were much more powerful than the previous 4-bit devices and were used in many early microcomputers and in applications such as electronic instruments and printers. The 8008 has a 14-bit address bus and can thus address up to 16 KB of memory (the 8080 has a 16-bit address bus giving it a 64 KB limit).

The third generation of microprocessors began with the launch of the 16-bit processors. Intel released the 8086 microprocessor which was mainly an extension to the original 8080 processor and thus retained a degree of software compatibility. IBM's designers realized the power of the 8086 and used it in the original IBM PC and IBM XT (eXtended Technology). It has a 16-bit data bus and a 20-bit address bus, and thus has a maximum addressable capacity of 1 MB. The 8086 could handle either 8 or 16 bits of data at a time (although in a messy way).

A stripped-down, 8-bit external data bus, version called the 8088 is also available. This stripped-down processor allowed designers to produce less complex (and cheaper) computer systems. An improved architecture version, called the 80286, was launched in 1982, and was used in the IBM AT (Advanced Technology).

In 1985, Intel introduced its first 32-bit microprocessor, the 80386DX. This device was compatible with the previous 8088/8086/80286 (80×86) processors and gave excellent performance, handling 8, 16 or 32 bits at a time. It has a full 32-bit data and address buses and can thus address up to 4 GB of physical memory. A stripped-down 16-bit external data bus and 24-bit address bus version called the 80386SX was released in 1988. This stripped-down processor can thus only access up to 16 MB of physical memory.

In 1989, Intel introduced the 80486DX which is basically an improved 80386DX with a memory cache and math co-processor integrated onto the chip. It had an improved internal structure making it around 50% faster with a comparable 80386. The 80486SX was also introduced, which is merely an 80486DX with the link to the math co-processor broken. Clock doubler/trebler 80486 processors were also released. In these the processor runs at a higher speed than the system clock. Typically, systems with clock doubler processors are around

75% faster than the comparable non-doubled processors. Typical clock doubler processors are DX2-66 and DX2-50 which run from 33 MHz and 25 MHz clocks, respectively. Intel have also produced a range of 80486 microprocessors which run at three or four times the system clock speed and are referred to as DX4 processors. These include the Intel DX4-100 (25 MHz clock) and Intel DX4-75 (25 MHz clock).

The Pentium (or P-5) is a 64-bit 'superscalar' processor. It can execute more than one instruction at a time and has a full 64-bit (8-byte) data bus and a 32-bit address bus. In terms of performance, it operates almost twice as fast as the equivalent 80486. It also has improved floating-point operations (roughly three times faster) and is fully compatible with previous 80×86 processors.

The Pentium II (or P-6) is an enhancement of the P-5 and has a bus that supports up to four processors on the same bus without extra supporting logic, with clock multiplying speeds of over 300 MHz. It also has major savings of electrical power and the minimization of electromagnetic interference (EMI). A great enhancement of the P-6 bus is that it detects and corrects all single-bit data bus errors and also detects multiple-bit errors on the data bus.

J.2 8088 microprocessor

The great revolution in processing power arrived with the 16-bit 8086 processor. This has a 20-bit address bus and a 16-bit address bus, while the 8088 has an 8-bit external data bus. Figure J.1 shows the pin connections of the 8088 and also the main connections to the processor. Many of the 40 pins of the 8086 have dual functions. For example, the lines AD0–AD7 act either as the lower 8 bits of the address bus (A0–A7) or as the lower 8 bits of the data bus (D0–D7). The lines A16/S3–A19/S6 also have a dual function, S3–S6 are normally not used by the PC and thus they are used as the 4 upper bits of the address bus. The latching of the address is achieved when the ALE (address latch enable) goes from a high to a low.

The bus controller (8288) generates the required control signals from the 8088 status lines $\overline{\text{S0}}-\overline{\text{S1}}$. For example, if $\overline{\text{S0}}$ is high, $\overline{\text{S1}}$ is low and $\overline{\text{S2}}$ is low then the $\overline{\text{MEMR}}$ line goes low. The main control signals are:

- $\overline{\text{IOR}}$ (I/O read) which means that the processor is reading from the contents of the address which is on the I/O bus.
- $\overline{\text{IOW}}$ (I/O write) which means that the processor is writing the contents of the data bus to the address which is on the I/O bus.
- $\overline{\text{MEMR}}$ (memory read) which means that the processor is reading from the contents of the address which is on the address bus.
- $\overline{\text{MEMW}}$ (memory write) which means that the processor is writing the contents of the data bus to the address which is on the address bus.
- $\overline{\text{INTA}}$ (interrupt acknowledgement) which is used by the processor to acknowledge an interrupt ($\overline{\text{S0}}$, $\overline{\text{S1}}$ and $\overline{\text{S2}}$ all go low). When a peripheral wants the attention of the processor it sends an interrupt request to the 8259 which, if it is allowed, sets INTR high.

The processor either communicates directly with memory (with $\overline{\text{MEMW}}$ and $\overline{\text{MEMR}}$) or communicates with peripherals through isolated I/O ports (with $\overline{\text{IOR}}$ and $\overline{\text{IOW}}$).

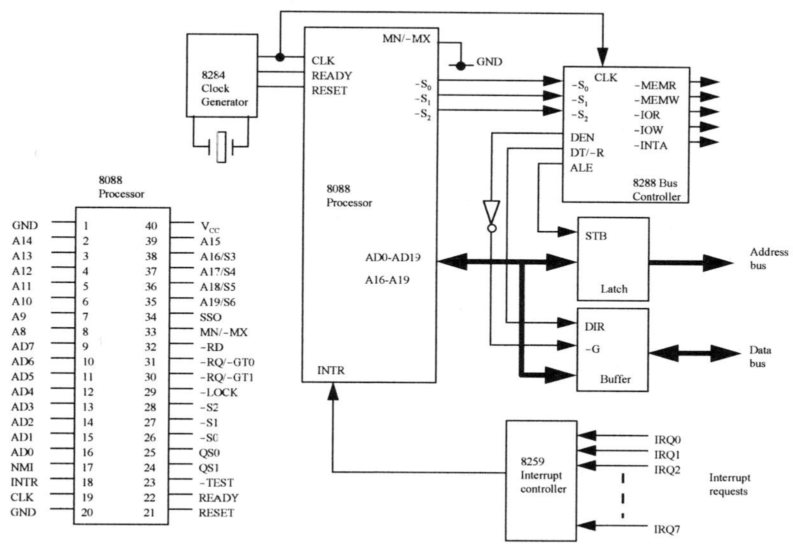

Figure J.1 8088 connections.

J.2.1 Registers

Each of the PC-based Intel microprocessors is compatible with the original 8086 processor and is normally backwardly compatible. Thus, for example, a Pentium can run 8086, 80386 and 80486 code. Microprocessors use registers to perform their operations. These registers are basically special memory locations within the processor that have special names. The 8086/88 has 14 registers which are grouped into four categories, as illustrated in Figure J.2.

General-purpose registers

There are four general-purpose registers that are AX, BX, CX and DX. Each can be used to manipulate a whole 16-bit word or with two separate 8-bit bytes. These bytes are called the lower and upper order bytes. Each of these registers can be used as two 8-bit registers, for example, AL represents an 8-bit register that is the lower half of AX and AH represents the upper half of AX.

The AX register is the most general purpose of the four registers and is normally used for all types of operations. Each of the other registers has one or more implied extra functions. These are:

- AX is the accumulator. It is used for all input/output operations and some arithmetic operations. For example, multiply, divide and translate instructions assume the use of AX.
- BX is the base register. It can be used as an address register.
- CX is the count register. It is used by instructions which require to count. Typically is it is used for controlling the number of times a loop is repeated and in bit-shift operations.
- DX is the data register. It is used for some input/output and also when multiplying and dividing.

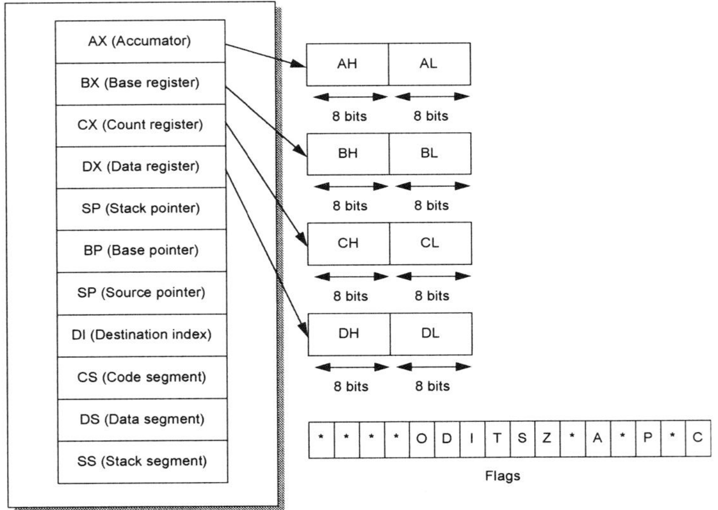

Figure J.2 8086/88 registers.

Addressing registers

The addressing registers are used in memory addressing operations, such as holding the source address of the memory and the destination address. These address registers are named BP, SP, SI and DI, which are:

- SI is the source index and is used with extended addressing commands.
- DI is the destination index and is used in some addressing modes.
- BP is the base pointer.
- SP is the stack pointer.

Status registers

Status registers are used to test for various conditions in an operations, such as 'is the result negative', 'is the result zero', and so on. The two status registers have 16 bits and are called the instruction pointer (IP) and the flag register (F):

- IP is the instruction pointer and contains the address of the next instruction of the program.
- Flag register holds a collection of 16 different conditions. Table J.1 outlines the most used flags.

Segments registers

There are four areas of memory called segments, each of which are 16 bits and can thus address up to 64 KB (from `0000h` to `FFFFh`). These segments are:

- Code segment (cs register). This defines the memory location where the program code (or instructions) is stored.
- Data segment (ds register). This defines where data from the program will be stored (ds

stands for data segment register).
- Stack segment (ss register). This defined where the stack is stored.
- Extra segment (es).

All addresses are with reference to the segment registers.

The 8086 has a segmented memory, the segment registers are used to manipulate memory within these segments. Each segment provides 64 KB of memory, this area of memory is known as the current segment. Segmented memory will be discussed in more detail in Section J.3.

Table J.1 Processor flags.

Bit	*Flag position*	*Name*	*Description*
C	0	Set on carry	Contains the carry from the most significant bit (left-hand bit) following a shift, rotate or arithmetic operation.
A	4	Set on 1/2 carry	
S	7	Set on negative result	Contains the sign of an arithmetic operation (0 for positive, 1 for negative).
Z	6	Set on zero result	Contains results of last arithmetic or compare result (0 for nonzero, 1 for zero).
O	11	Set on overflow	Indicates that an overflow has occurred in the most significant bit from an arithmetic operation.
P	2	Set on even parity	
D	10	Direction	
I	9	Interrupt enable	Indicates whether the interrupt has been disabled.
T	8	Trap	

Memory addressing

There are several methods of accessing memory locations, these are:

- Implied addressing which uses an instruction in which it is known which registers are used.
- Immediate (or literal) addressing uses a simple constant number to define the address location.
- Register addressing which uses the address registers for the addressing (such as AX, BX, and so on).
- Memory addressing which is used to read or write to a specified memory location.

J.3 Memory segmentation

The 80386, 80486 and Pentium processors run in one of two modes, either virtual or real. In virtual mode they act as a pseudo-8086 16-bit processor, known as the protected mode. In real-mode they can use the full capabilities of their address and data bus. This mode normally depends on the addressing capabilities of the operating system. All DOS-based programs use the virtual mode.

The 8086 has a 20-bit address bus so that when the PC is running 8086-compatible code it can only address up to 1 MB of physical memory. It also has a segmented memory architecture and can only directly address 64 KB of data at a time. A chunk of memory is known as a segment and hence the phrase 'segmented memory architecture'.

Memory addresses are normally defined by their hexadecimal address. A 4-bit address bus can address 16 locations from `0000b` to `1111b`. This can be represented in hexadecimal as `0h` to `Fh`. An 8-bit bus can address up to 256 locations from `00h` to `FFh`. Section J.10 outlines the addressing capabilities for a given address bus size.

Two important addressing capabilities for the PC relate to a 16- and a 20-bit address bus. A 16-bit address bus addresses up to 64 KB of memory from `0000h` to `FFFFh` and a 20-bit address bus addresses up to 1 MB from `00000h` to `FFFFFh`. The 80386/80486/ Pentium processors have a 32-bit address bus and can address from `00000000h` to `FFFFFFFFh`.

A segmented memory address location is identified with a segment and an offset address. The standard notation is `segment:offset`. A `segment` address is a 4-digit hexadecimal address which points to the start of a 64 KB chunk of data. The `offset` is also a 4-digit hexadecimal address which defines the address offset from the segment base pointer. This is illustrated in Figure J.3.

The `segment:offset` address is defined as the logical address, the actual physical address is calculated by shifting the segment address 4 bits to the left and adding the offset. The example given next shows that the actual address of `2F84:0532` is `2FD72h`.

Segment (`2F84`):	0010	1111	1000	0100	0000
Offset (`0532`):		0000	0101	0011	0010
Actual address:	0010	1111	1101	0111	0010

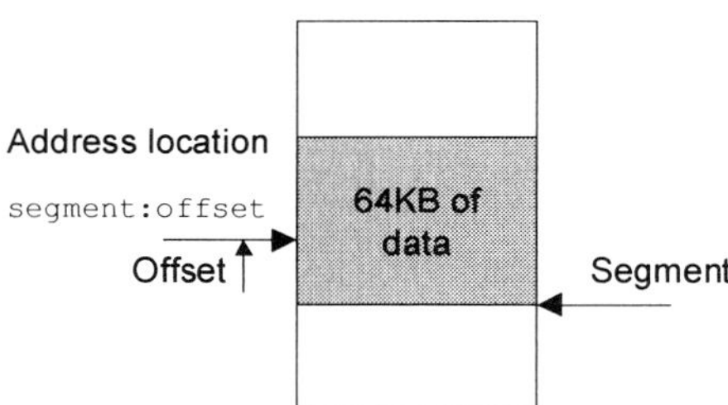

Figure J.3 Memory addressing.

J.4 80386/80486

J.4.1 Introduction

The PC had grown from the 8086 processor, which could run 8-bit or 16-bit software. This processor was fine with text-based applications, but struggled with graphical programs, especially with GUIs (Graphical User Interfaces). The original version of Microsoft Windows (Windows Version 1.0 and Version 2.0) ran on these limited processes. The great leap in computing power came with the development of the Intel 80386 processor and with Microsoft Windows 3.0. A key to the success of the 80386 was that it was fully compatible with the previous 8088/8086/80286 processors. This allowed it run all existing DOS-based program and new 32-bit applications. The DX version has full 32-bit data and address bus and can thus address up to 4 GB of physical memory. An SX version with a stripped-down 16-bit external data bus and 24-bit address bus version can access only up to 16 MB of physical memory (at its time of release this was a large amount of memory). Most of the time, with Microsoft Windows 3.0, the processor was using only 16 bits, and thus not using the full power of the processor.

The 80486DX basically consists of an improved 80386 with a memory cache and a math co-processor integrated onto the chip. An SX version had the link to the math co-processor broken. At the time, a limiting factor was the speed of the system clock (which was limited to around 25 MHz or 33 MHz). Thus clock doubler, treblers or quadrupers allow the processor to multiply the system clock frequency to a high speed. Thus, internal operations of the processor are carried out at much higher speeds, but accesses outside the processor must slow down the system clock. As most of the operations within the computer involve operations within the processor then the overall speed of the computer is improved (roughly by about 75% for a clock doubler).

J.4.2 80486 pin out

To allow for easy upgrades and to save space the 80486 and Pentium processors are available in a pin-grid array (PGA) form. A 168-pin PGA 80486 processor is illustrated in Figure J.4.

It can be seen that the 486 processor has a 32-bit address bus (A0–A31) and a 32-bit data bus (D0–D31). The pin definitions are given in Table J.2.

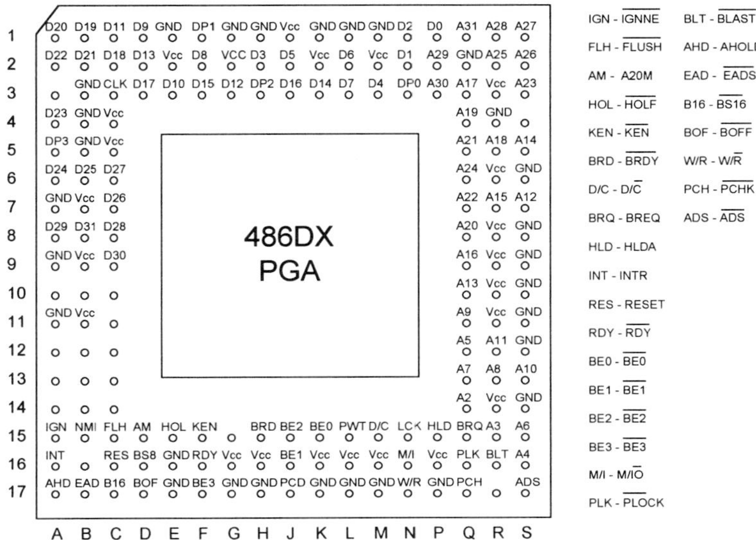

Figure J.4 i486DX processor.

Table J.3 defines the how the control signals are interpreted. For the STOP/special bus cycle, the byte enable signals ($\overline{BE0}-\overline{BE3}$) further define the cycle. These are:

- Write back cycle $\overline{BE0}=1$, $\overline{BE1}=1$, $\overline{BE2}=1$, $\overline{BE3}=0$.
- Halt cycle $\overline{BE0}=1$, $\overline{BE1}=1$, $\overline{BE2}=0$, $\overline{BE3}=1$.
- Flush cycle $\overline{BE0}=1$, $\overline{BE1}=0$, $\overline{BE2}=1$, $\overline{BE3}=1$.
- Shutdown cycle $\overline{BE0}=0$, $\overline{BE1}=1$, $\overline{BE2}=1$, $\overline{BE3}=1$.

Table J.2 80486 signal lines.

Signals	*I/O*	*Description*
A2–A31	I/O	The 30 most significant bits of the address bus.
$\overline{A20M}$	I	When active low, the processor internally masks the address bit A20 before every memory access.
$\overline{ADS}$	O	Indicates that the processor has valid control signals and valid address signals.
AHOLD	I	When active a different bus controller can have access to the address bus. This is typically used in a multi-processor system.
$\overline{BE0}-\overline{BE3}$	O	The byte enable lines indicate which of the bytes of the 32-bit data bus are active.
$\overline{BLAST}$	O	Indicates that the current burst cycle will end after the next $\overline{BRDY}$ signal.
$\overline{BOFF}$	I	The backoff signal informs the processor to deactivate the bus on the next clock cycle.
$\overline{BRDY}$	I	The burst ready signal is used by an addressed system that has sent data on the data bus or read data from the bus.
BREQ	O	Indicates that the processor has internally requested the bus.
$\overline{BS16}$, $\overline{BS8}$	I	The $\overline{BS16}$ signal indicates that a 16-bit data bus is used, the $\overline{BS8}$ signal indicates that an 8-bit data bus is used. If both are high then a 32-bit data bus is used.
DP0–DP3	I/O	The data parity bits gives a parity check for each byte of the 32-bit data bus. The parity bits are always even parity.
$\overline{EADS}$	I	Indicates that an external bus controller has put a valid address on the address bus.
$\overline{FERR}$	O	Indicates that the processor has detected an error in the internal floating-point unit.

$\overline{\text{FLUSH}}$	I	When active the processor writes the complete contents of the cache to memory.
HOLD, HLDA	I/O	The bus hold (HOLD) and acknowledge (HLDA) are used for bus arbitration and allow other bus controllers to take control of the busses.
$\overline{\text{IGNNE}}$	I	When active the processor ignores any numeric errors.
INTR	I	External devices to interrupt the processor use the interrupt request line.
$\overline{\text{KEN}}$	I	This signal stops caching of a specific address.
$\overline{\text{LOCK}}$	O	If active the processor will not pass control to an external bus controller, when it receives a HOLD signal.
$M/\overline{IO}$, $D/\overline{C}$, $W/\overline{R}$	O	See Table J.3.
NMI	I	The non-maskable interrupt signal causes an interrupt 2.
$\overline{\text{PCHK}}$	O	If it is set active then a data parity error has occurred.
$\overline{\text{PLOCK}}$	O	The active pseudo lock signal identifies that the current data transfer requires more than one bus cycle.
PWT, PCD	O	The page write-through (PWT) and page cache disable (PCD) are used with cache control.
$\overline{\text{RDY}}$	I	When active the addressed system has sent data on the data bus or read data from the bus.
RESET	I	If the reset signal is high for more than 15 clock cycles then the processor will reset itself.

The 486 integrates a processor, cache and a math co-processor onto a single IC. Figure J.5 shows the main 80386/80486 processor connections. The Pentium processor connections are similar but it has a 64-bit data bus. There are three main interface connections: the memory/IO interface, interrupt interface and DMA interface.

Table J.3 Control signals.

$M/\overline{IO}$	$D/\overline{C}$	$W/\overline{R}$	*Description*
0	0	0	Interrupt acknowledge sequence
0	0	1	STOP/special bus cycle
0	1	0	Reading from an I/O port
0	1	1	Writing to an I/O port
1	0	0	Reading an instruction from memory
1	0	1	Reserved
1	1	0	Reading data from memory
1	1	1	Writing data to memory

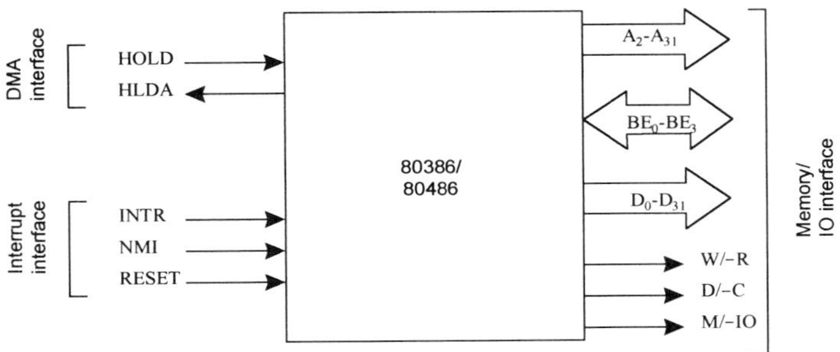

Figure J.5 Some of the 80386/80486 signal connections.

The write/read ($W/\overline{R}$) line determines whether data is written to (W) or read from ($\overline{R}$) memory. PCs can interface directly with memory or can interface to isolated memory. The signal line $M/\overline{IO}$ differentiates between the two types. If it is high then the processor accesses direct memory; else if it is low then it accesses isolated memory.

The 80386DX and 80486 have an external 32-bit data bus (D_0–D_{31}) and a 32-bit address bus ranging from A_2 to A_{31}. The two lower address lines, A_0 and A_1, are decoded to produce the byte enable signals $\overline{BE0}$, $\overline{BE1}$, $\overline{BE2}$ and $\overline{BE3}$. The $\overline{BE0}$ line activates when A_1A_0 is 00, $\overline{BE1}$ activates when A_1A_0 is 01, $\overline{BE2}$ activates when A_1A_0 and $\overline{BE3}$ actives when A_1A_0 is 11. Figure J.6 illustrates this addressing.

The byte enable lines are also used to access either 8, 16, 24 or 32 bits of data at a time. When addressing a single byte, only the $\overline{BE0}$ line is active (D_0–D_7), if 16 bits of data are to be accessed then $\overline{BE0}$ and $\overline{BE1}$ are active (D_0–D_{15}), if 32 bits are to be accessed then $\overline{BE0}$, $\overline{BE1}$, $\overline{BE2}$ and $\overline{BE3}$ are active (D_0–D_{31}).

The $D/\overline{C}$ line differentiates between data and control signals. When it is high then data is read from or written to memory, else if it is low then a control operation is indicated, such as a shutdown command.

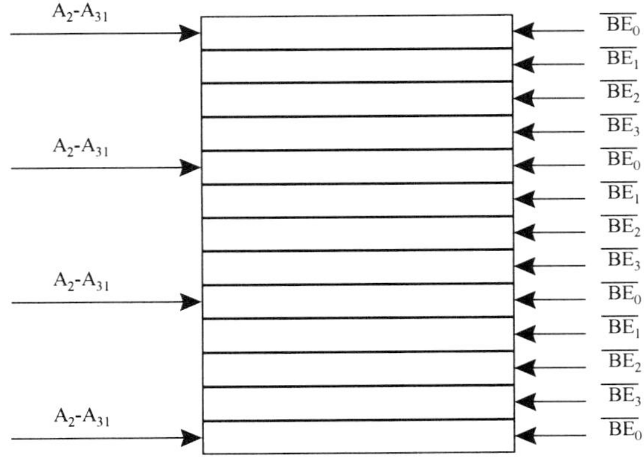

Figure J.6 Memory addressing.

The interrupt lines are interrupt request (INTR), non-maskable interrupt request (NMI) and system reset (RESET), all of which are active high signals. The INTR line is activated when an external device, such as a hard disk or a serial port, wishes to communicate with the processor. This interrupt is maskable and the processor can ignore the interrupt if it wants. NMI is a non-maskable interrupt and is always acted on. When it becomes active the processor calls the non-maskable interrupt service routine. The RESET signal causes a hardware reset and is normally made active when the processor is powered-up.

J.4.3 80386/80486 registers

The 80386 and 80486 are 32-bit processors and can thus operate on 32-bits at a time. They have expanded 32-bit registers, which can also be used as either 16-bit or 8-bit registers (mainly to keep compatibility with other processors and software). The general purpose registers, such as AX, BX, CX, DX, SI, DI and BP are expanded from the 8086 processor and are named EAX, EBX, ECX, EDX, ESI, EDI and EBP, respectively, as illustrated in Figure J.7. The CS, SS and DS registers are still 16 bits, but the flag register has been expanded to 32 bits and is named EFLAG.

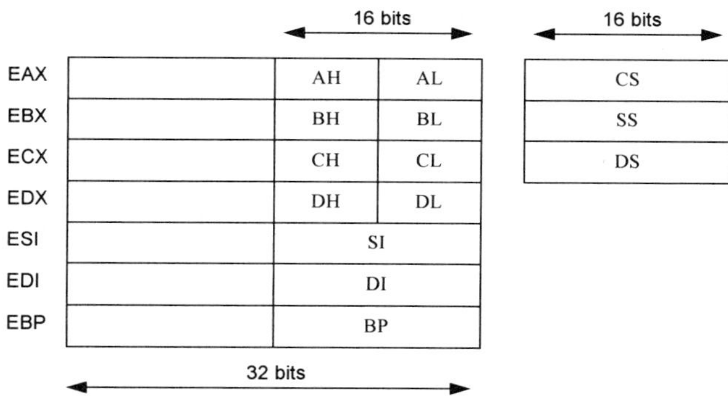

Figure J.7 80386/80486 registers.

J.4.4 Memory cache

DRAM is based on the charging and discharge of tiny capacitors. It is thus a relatively slow type of memory compared with SRAM. For example, typical access time for DRAM is 80 ns and a typical motherboard clock speed is 50 MHz. This gives a clock period of 20 ns. Thus, the processor would require five wait states before the data becomes available. A cache memory can be used to overcome this problem. This is a bank of fast memory (SRAM) that uses a cache controller to load data from main memory (typically DRAM) into it. The cache controller guesses the data the processor requires and loads this into the cache memory. Figure J.8 shows that if the controller guesses correctly then it is a cache hit, else if it is wrong it is a cache miss. A miss causes the processor to access the memory in the normal way (that is, there may be wait states as the DRAM memory need time to get the data). Typical cache memory sizes are 16 KB, 32 KB and 64 KB for 80486 processors and 256 KB and 512 KB for Pentium processors. This should be compared with the size of the RAM on a typical PC which is typically at least 32 MB.

The 80486 and Pentium have built-in cache controllers and, at least, 64 KB (or 256 KB for the Pentium) of local SRAM cache memory. This is a first-level cache and the total cache size can be increase with an off-chip (or near-chip) memory (second-level cache).

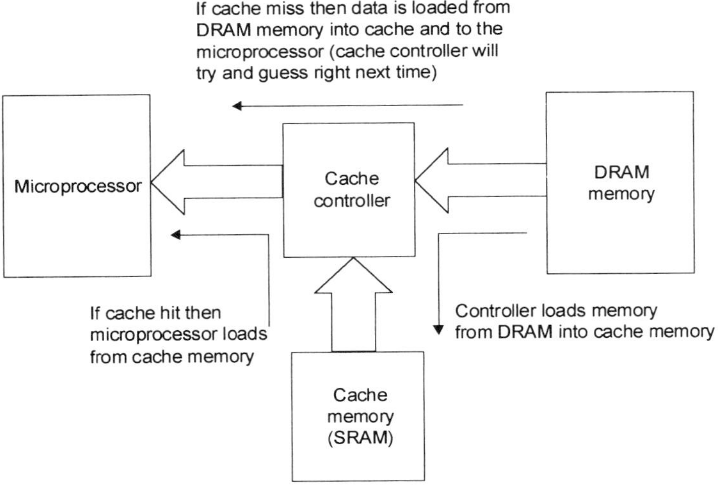

Figure J.8 Cache operation.

Cache architecture

The main cache architectures are:

- **Look-through cache**. In a look-through cache the system memory is isolated from the processor address and control busses. In this case the processor directly sends a memory request to the cache controller which then determines whether it should forward it to its own memory or the system memory. Figure J.9 illustrates this type of cache. It can be seen that the cache controls whether the processor address contents are latched through to the DRAM memory and it also controls whether the contents of the DRAM's memory is loaded onto the processor data bus (through the data transceiver). The operation is described as bus cycle forwarding.
- **Look-aside cache**. A look-aside cache is where the cache and system memory connect to a common bus. System memory and the cache controller see the beginning of the processor bus cycle at the same time. If the cache controller detects a cache hit then it must inform the system memory before it tries to find the data. If a cache miss is found then the memory access is allowed to continue.
- **Write-through cache**. With a write-through cache all memory address accesses are seen by the system memory when the processor performs a bus cycle.
- **Write-back cache**. With a write-back cache the cache controller controls all system writes. It thus does not write the system memory unless it has to.

Second-level caches

An L1-cache (first-level cache) provides a relatively small on-chip cache, where an L2-cache (second-level cache) provides an external, on-board, cache, which provides a cache memory of between 128 and 512 KB. The processor looks in its own L1-cache for a cache hit; if none is found then it searches in the on-board L2-cache. A cache hit in the L1-cache will obviously be faster than the off-chip cache.

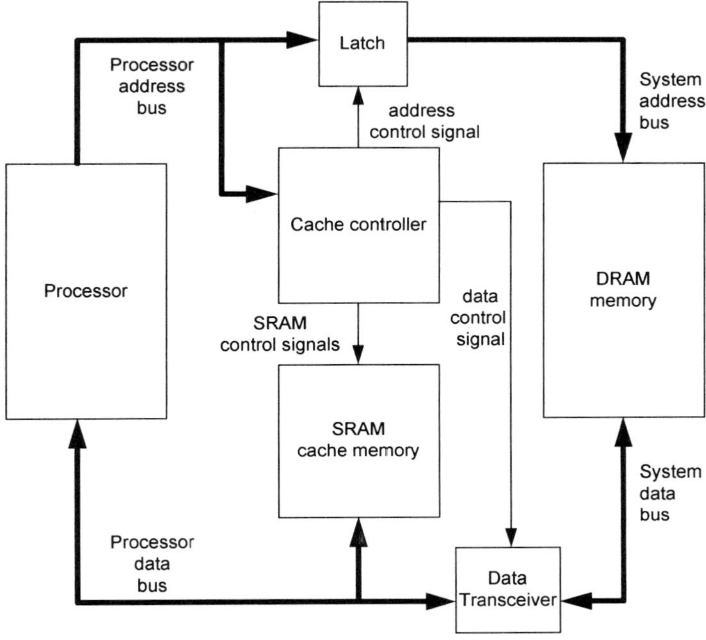

Figure J.9 Look-through cache.

J.5 Pentium/Pentium Pro

J.5.1 Introduction

Intel have gradually developed their range of processors from the original 16-bit 8086 processor to the 32-bit processors, such as the Pentium II. Table J.1 4contrasts the Intel processor range. It can also be seen from the table that the Pentium II processor is nearly more than a thousand times more powerful than an 8086 processor. The original 8086 had just 29 000 transistors and operated at a clock speed of 8 MHz. It had an external 20-bit bus and could thus only access up to 1 MB of physical memory. Compare this with the Pentium II which can operate at over 300 MHz, contains over 6 000 000 transistors and can access up to 64 GB of physical memory.

Table J.4 Processor comparison.

Processor	*Clock (when released)*	*Register size*	*External data bus*	*Max. external memory*	*Cache*	*Power (MIPs)*
8086	8 MHz	16	16	1 MB		0.8
286	12.5 MHz	16	16	16 MB		2.7
386DX	20 MHz	32	32	4 GB		6.0
486DX	25 MHz	32	32	4 GB	8 KB L-1	20
Pentium	60 MHz	32	64	4 GB	16 KB L-1	100
Pentium Pro	200 MHz	32	64	64 GB	16 KB L-1 256 K L-2	440
Pentium II	200 MHz	32	64	64 GB	16 KB L-1 512 KB L-2	700

J.6 Intel processor development

The 80386 processor was a great leap in processing power over the 8086 and 80286, but it required an on-board math co-processor to enhance its mathematical operations and it could also only execute one instruction at a time. The 80486 brought many enhancements, such as:

- The addition of parallel execution with the expansion of the Instruction decode and execution units into five pipelined stages. Each of these stages operate in parallel, with the others, on up to five instructions in different stages of execution. This allows up to five instructions to be completed at a time.
- The addition of an 8 KB on-chip cache to greatly reduce the data and code access times.
- The addition of an integrated floating-point unit.
- Support for more complex and powerful systems, such as off-board L-2 cache support and multiprocessor operation.

With the increase in notebook and palmtop computers, the 80486 was also enhanced to support many energy and system management capabilities. These processors were named the 80486SL processors. The new enhancements included:

- System Management Mode. This mode is triggered by the processor's own interrupt pin and allows complex system management features to be added to a system transparently to the operating system and application programs.
- Stop Clock and Auto Halt Powerdown. These allow the processor to either shut itself down (and preserve its current state) or run at a reduced clock rate.

The Intel Pentium processor added many enhancements to the previous processors, including:

- The addition of a second execution pipeline. These two pipelines, named u and v, can execute two instructions per clock cycle. This is known as superscalar operation.
- Increased on-chip L-1 cache, 8 KB for code and another 8 KB for data. It uses the MESI

protocol to support write-back mode, as well as the write-through mode (which is used by the 80486 processor).
- Branch prediction with an on-chip branch table that improves looping characteristics.
- Enhancement to the virtual-8086 mode to allow for 4 MB as well as 4 KB pages.
- 128-bit and 256-bit data paths are possible (although the main registers are still 32 bits).
- Burstable 64-bit external data bus.
- Addition of Advanced Programmable Interrupt Controller (APIC) to support multiple Pentium processors.
- New dual processing mode to support dual processor systems.

The Pentium processor has been extremely successful and has helped support enhanced multitasking operating systems such as Windows NT and Windows 95/98. The Intel Pentium Pro enhanced the Pentium processor with the following:

- Addition of a 3-way superscalar architecture, as opposed to a 2-way for the Pentium. This allows three instructions to be executed for every clock cycle.
- Uses enhanced prediction of parallel code (called dynamic execution micro-architecture) for the superscalar operation. This includes methods such as micro-data flow analysis, out-of-order execution, enhanced branch prediction and speculative execution. The three instruction decode units work in parallel to decode object code into smaller operations called micro-ops. These micro-ops then go into an instruction pool, and, when there are no interdependencies they can be executed out-of-order by the five parallel execution units (two integer units, two for floating-point operations and one for memory). A retirement unit retires completed micro-ops in their original program order and takes account of any branches. This recovers the original program flow.
- Addition of register renaming. Multiple instructions not dependent on each other, using the same registers, allow the source and destination registers to be temporarily renamed. The original register names are used when instructions are retired and program flow is maintained.
- Addition of a closely coupled, on-package, 256 KB L-2 cache that has a dedicated 64-bit full clock speed bus. The L-2 cache also supports up to four concurrent accesses through a 64-bit external data bus. Each of these accesses is transaction-oriented where each access is handled as a separate request and response. This allows for numerous requests while awaiting a response.
- Expanded 36-bit address bus to give a physical address size of 64 GB.

The Pentium II processor is a further enhancement to the processor range. Apart from increasing the clock speed it has several enhancements over the Pentium Pro, including:

- Integration of MMX technology. MMX instructions support high-speed multimedia operations and include the addition of eight new registers (MM0 to MM7), four MMX data types and an MMX instruction set.
- Single edge contact (SEC) cartridge packaging. This gives improved handling performance and socketability. It uses surface mount component and has a thermal plate (which accepts a standard heat sink), a cover and a substrate with an edge finger connection.
- Integrated on-chip L-1 cache 16 KB for code and another 16 KB for data.
- Increased size, on-package, 512 KB L-2 cache.

- Enhanced low-power states, such as AutoHALT, Stop-Grant, Sleep and Deep Sleep.

J.7 Terms

Before introducing the Pentium Pro (P-6) various terms have to be defined. These are:

Transaction	Used to define a bus cycle. It consists of a set of phases, which relate to a single bus request.
Bus agent	Devices that reside on the processor bus, that is, the processor, PCI bridge and memory controller.
Priority agent	The device handling reset, configuration, initialization, error detection and handling; generally the processor-to-PCI bridge.
Requesting agent	The device driving the transaction, that is, busmaster.
Addressed agent	The slave device addressed by the transaction, that is, target agent.
Responding agent	The device that provides the transaction response on $\overline{RS2} - \overline{RS0}$ signals.
Snooping agent	A caching device that snoops on the transactions to maintain cache coherency.
Implicit write-back	When a hit to a modified line is detected during the snoop phase, an implicit write-back occurs. This is the mechanism used to write-back the cache line.

J.8 Pentium II and Pentium Pro

A major objective of electronic systems design is the saving of electrical power and the minimization of electromagnetic interference (EMI). Thus gunning transceiver logic (GTL) has been used to reduce both power consumption and EMI as it has a low voltage swing. GTL requires a 1 V reference signal and signals which use GTL logic are terminated to 1.5 V. If a signal is 0.2 V above the reference voltage, that is, 1.2 V, then it is considered high. If a signal is 0.2 V below the reference voltage, that is, 0.8 V, then it is considered low.

The Pentium Pro and II support up to four processors on the same bus without extra supporting logic. Integrated into the bus structure are cache coherency signals, advanced programmable interrupt control signals and bus arbitration.

A great enhancement of the Pentium Pro bus is data error detection and correction. The Pentium Pro bus detects and corrects all single-bit data bus errors and also detects multiple-bit errors on the data bus. Address and control bus signals also have basic parity protection.

The Pentium Pro bus has a modified line write-back performed without backing off the current bus owner, where the processor must perform a write-back to memory when it detects a hit to a modified line. The following mechanism eliminates the need to back-off the current busmaster. If a memory write is being performed by the current bus owner then two writes will be seen on the bus, that is, the original one followed by the write-back. The memory controller latches, and merges the data from the two cycles, and performs one write to DRAM. If the current bus owner is performing a memory read then it accepts the data when it is being written to memory.

Other enhanced features are:

- Deferred reply transactions stop the processor from having to wait for slow devices; transactions that require a long time can be completed later, that is, deferred.
- Deeply pipelined bus transactions where the bus supports up to eight outstanding pipelined transactions.

J.9 System overview

Figure J.01 outlines the main components of a Pentium. A major upgrade is the support for up to four processors. The memory control and data path control logic provides the memory control signals, that is, memory address, $\overline{\text{RAS}}$ and $\overline{\text{CAS}}$ signals. The data path logic moves the data between the processor bus and the memory data bus. The memory interface component interfaces the memory data bus with the DRAM memory. Both interleaved and non-interleaved methods are generally supported. The memory consists of dual-in-line memory modules, that is, DIMMs. A DIMM module supports 64 data bits, and 8 parity or ECC bits.

The PCI bridge provides the interface between the processor bus and the PCI bus. The standard bridge provides an interface between the PCI bus and the EISA / ISA bus. EISA / ISA Support Component provides the EISA / ISA bus support functions, for example, timers, interrupt control, flash ROM, keyboard interface, LA/SA translation and XD bus control.

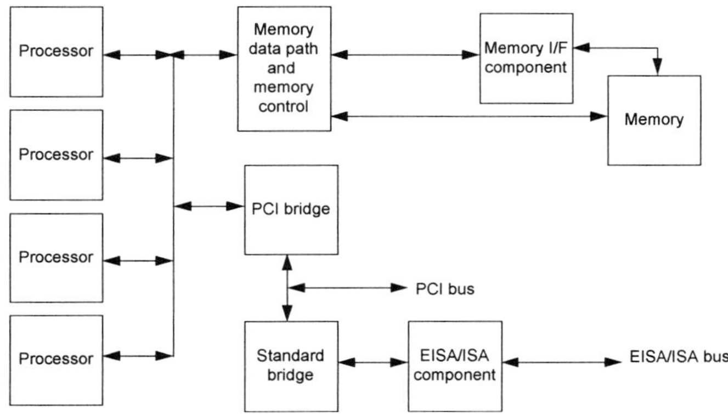

Figure J.10 P-6 architecture.

J.10 Memory address reference

Address bus size	*Addressable memory (bytes)*	*Address bus size*	*Addressable memory (bytes)*
1	2	15	32K
2	4	16	64K
3	8	17	128K
4	16	18	256K
5	32	19	512K
6	64	20	1M†
7	128	21	2M
8	256	22	4M
9	512	23	8M
10	1K*	24	16M
11	2K	25	32M
12	4K	26	64M
13	8K	32	4G‡
14	16K	64	16GG

* 1K represents 1024

† 1M represents 1 048 576 (1024 K)

‡ 1G represents 1 073 741 824 (1024 M)

HTML Reference

K.1 Introduction

K.1.1 Data Characters

Characters which are not markup text are mapped directly to strings of data characters. An ampersand followed by a character reference or a number value can be used to define a character. Table K.1 defines these characters (the equivalent ampersand character reference is given in brackets). For example:

```
Fred&#174&ampBert&iquest
```

will be displayed as:

```
Fred®&Bert¿
```

An ampersand is only recognized as markup when it is followed by a letter or a `#' and a digit:

```
Fred & Bert
```

will be displayed as:

```
Fred & Bert
```

In the HTML document character set only three control characters are allowed: Horizontal Tab, Carriage Return, and Line Feed (code positions 9, 13, and 10).

Table K.1 Character mappings.

�-	Unused		Horizontal tab
Line feed	,	Unused
Carriage Return	-	Unused
	Space	!	Exclamation mark
"	Quotation mark (")	#	Number sign
$	Dollar sign	%	Percent sign
&	Ampersand (&)	'	Apostrophe
(	Left parenthesis	)	Right parenthesis
*	Asterisk	+	Plus sign
,	Comma	-	Hyphen
.	Period (fullstop)	/	Solidus

0-9	Digits 0-9	:	Colon
;	Semi-colon	<	Less than (<)
=	Equals sign	>	Greater than (>)
?	Question mark	@	Commercial at
A-Z	Letters A-Z	[	Left square bracket
\	Reverse solidus (\)	]	Right square bracket
^	Caret	_	Underscore
`	Acute accent	a-z	Letters a-z
{	Left curly brace			Vertical bar
}	Right curly brace	~	Tilde

-Ÿ	Unused
	Non-breaking Space ( )
¡	Inverted exclamation, ¡ (¡)
¢	Cent sign ¢ (¢)
£	Pound sterling £ (£)
¤	General currency sign, ¤ (¤)
¥	Yen sign, ¥ (¥)
¦	Broken vertical bar, ¦ (¦)
§	Section sign, § (§)
¨	Umlaut, ¨ (¨)
©	Copyright, © (©)
ª	Feminine ordinal, ª (ª)
«	Left angle quote, « («)
¬	Not sign, ¬ (¬)
	Soft hyphen, - ()
®	Registered trademark, ® (®)
¯	Macron accent, ¯ (¯)
°	Degree sign, ° (°)
±	Plus or minus, ± (±)
²	Superscript two, ^{2} (²)
³	Superscript three, ^{3} (³)
´	Acute accent, ´ (´)
µ	Micro sign, µ (µ)
¶	Paragraph sign, ¶ (¶)
·	Middle dot, · (·)
¸	Cedilla, ¸ (¸)
¹	Superscript one, ^{1} (¹)
º	Masculine ordinal, º (º)
»	Right angle quote, » (»)
¼	Fraction one-fourth, ¼ (¼)
½	Fraction one-half, ½ (½)
¾	Fraction three-fourths, ¾ (¾)
¿	Inverted question mark, ¿ (¿)
À	Capital A, grave accent, À (À)
Á	Capital A, acute accent, Á (Á)
Â	Capital A, circumflex accent, Â (Â)
Ã	Capital A, tilde, Ã (Ã)

Ä Capital A, dieresis, Ä (Ä)
Å Capital A, ring, Å (Å)
Æ Capital AE dipthong, Æ (Æ)
Ç Capital C, cedilla, Ç (Ç)
È Capital E, grave accent, È (È)
É Capital E, acute accent, É (É)
Ê Capital E, circumflex accent, Ê (Ê)
Ë Capital E, dieresis, Ë (Ë)
Ì Capital I, grave accent, Ì (Ì)
Í Capital I, acute accent, Í (Í)
Î Capital I, circumflex accent, Î (Î)
Ï Capital I, dieresis, Ï (Ï)
Ð Capital Eth, Icelandic, Ð (Ð)
Ñ Capital N, tilde, Ñ (Ñ)
Ò Capital O, grave accent, Ò (Ò)
Ó Capital O, acute accent, Ó (Ó)
Ô Capital O, circumflex accent, Ô (Ô)
Õ Capital O, tilde, Õ (Õ)
Ö Capital O, dieresis, Ö (Ö)
× Multiply sign, × (×)
Ø Capital O, slash, Ø (Ø)
Ù Capital U, grave accent, Ù (Ù)
Ú Capital U, acute accent, Ú (Ú)
Û Capital U, circumflex accent, Û (Û)
Ü Capital U, dieresis or umlaut mark, Ü (Ü)
Ý Capital Y, acute accent, Ý (Ý)
Þ Capital THORN, Icelandic, Þ (Þ)
ß Small sharp s, German, ß (ß)
à Small a, grave accent, à (à)
á Small a, acute accent, á (á)
â Small a, circumflex accent, â (â)
ã Small a, tilde, ã (ã)
ä Small a, dieresis or umlaut mark, ä (ä)
å Small a, ring, å (å)
æ Small ae dipthong, æ (æ)
ç Small c, cedilla, ç (ç)
è Small e, grave accent, è (è)
é Small e, acute accent, é (é)
ê Small e, circumflex accent, ê (ê)
ë Small e, dieresis or umlaut mark, ë (ë)
ì Small i, grave accent, ì (ì)
í Small i, acute accent, í (í)
î Small i, circumflex accent, î (î)
ï Small i, dieresis or umlaut mark, ï (ï)
ð Small eth, Icelandic, ð (ð)
ñ Small n, tilde, ñ (ñ)
ò Small o, grave accent, ò (ò)

ó Small o, acute accent, ó (ó)
ô Small o, circumflex accent, ô (ô)
õ Small o, tilde, õ (õ)
ö Small o, dieresis or umlaut mark, ö (ö)
÷ Division sign, ÷ (÷)
ø Small o, slash, ø (ø)
ù Small u, grave accent, ù (ù)
ú Small u, acute accent, ú (ú)
û Small u, circumflex accent, û (û)
ü Small u, dieresis or umlaut mark, ü (ü)
ý Small y, acute accent, ý (ý)
þ Small thorn, Icelandic, þ (þ),
ÿ Small y, dieresis or umlaut mark, ÿ (ÿ)

K.1.2 Tags

Tags are used to delimit elements such as headings, paragraphs, lists, character highlighting and links. Normally an HTML element consists of a start-tag, which gives the element name and attributes, followed by the content and then the end tag. A start-tags is defined between a '<' and '>', and end tags between a '</' and '>'. For example to display text as bold:

```
<B>Header Level 1</B>
```

Some of the HTML only require a single start tag, these include:

</BR>. Line break.
</P>. Paragraph.
</LI>. List Item.
</DT>. Definition term.
</DD>. Definition Description.

Element content is a sequence of data character strings and nested elements. Some elements, such as anchors, cannot be nested.

K.1.3 Names

Names consist of a letter followed by letters, digits, periods or hyphens (normally limited to 72 characters). Entity names are case sensitive, but element and attribute names are not. For example:

'<FONT>', '<Font>', and '<font>'

are equivalent, but

'<' is different from '<'.

Start-tags always begin directly after the opening delimiter ('<').

K.1.4 Attributes

In a start-tag, white space and attributes are allowed between the element name and the closing delimiter. Attributes typically consist of an attribute name, an equal sign, and a value, which can be:

- A string literal, delimited by single quotes or double quotes and not containing any occurrences of the delimiting character.
- A name token (a sequence of letters, digits, periods, or hyphens). Name tokens are not case sensitive.

K.1.5 Comments

Comments are defined with a ‘<!’ and ends with a ‘>’. Each comment starts with `--' and includes all text up to and including the next occurrence of ‘--’. When defining a comment, white space is allowed after each comment, but not before the first comment. The entire comment declaration is ignored.

```
<!DOCTYPE HTML PUBLIC "-//IETF//DTD HTML 2.0//EN">
<HEAD>
<TITLE>Comment Document</TITLE>
<!-- Comment field 1 -->
<!-- Comment field 2 -->
<!>
</HEAD> <BODY>
```

K.1.6 HTML Public Text Identifiers

Documents that conform to the HTML 2.0 specification can include the following line at the start of the document:

```
<!DOCTYPE HTML PUBLIC "-//IETF//DTD HTML 2.0//EN">
```

K.2 Document structure and block structuring

An HTML document is a tree of elements, including a head and body, headings, paragraphs, lists, and so on. These include:

<HTML>	Document element. Consists of a head and a body. The head contains the title and optional elements, and the body is the main text consisting of paragraphs, lists and other elements.
<HEAD>	Head element. An unordered collection of information about the document.
<TITLE>	Title. Identifies the contents of the document in a global context.
<BASE>	Base address. Provides a base address for interpreting relative URLs when the document is read out of context.
<ISINDEX>	Keyword index. Indicates that the user agent should allow the user to search an index by giving keywords.
<LINK>	Link. Represents a hyperlink (*see* Hyperlinks) and has the same attributes as the <A> element.
<META>	Associated meta-information. A container for identifying specialized document meta-information.
<BODY>	Body element. Contains the text flow of the document, including headings, paragraphs, lists, etc.
<H1>...<H6>	Headings. The six heading elements, <H1> to <H6> identify section headings. Typical renderings are:

H1 Bold, very-large centered font. One or two blank lines above and below.
H2 Bold, large flush-left font. One or two blank lines above and below.
H3 Italic, large font, slightly indented from the left margin. One or two blank lines above and below.
H4 Bold, normal font, indented more than H3. One blank line above and below.
H5 Italic, normal font, indented as H4. One blank line above.
H6 Bold, indented same as normal text, more than H5. One blank line above.

<P> Paragraph. Indicates a paragraph. Typically, paragraphs are surrounded by a vertical space of one line or half a line. The first line in a paragraph is indented in some cases.

<PRE> Preformatted text. Represents a character cell block of text and can be used to define monospaced font. It may be used with the optional WIDTH attribute, which specifies the maximum number of characters for a line.

<ADDRESS> Address. Contains information such as address, signature and authorship. It is often used at the beginning or end of the body of a document.

<BLOCKQUOTE>
Block quote. Contains text quoted from another source. A typical rendering is a slight extra left and right indent, and/or italic font, and typically provides space above and below the quote.

<UL>, <LI> Unordered List. <UL> represents a list of items and is typically rendered as a bulleted list. The content of a <UL> element is a sequence of <LI> elements.

<OL> Ordered List. <OL> represents an ordered list of items which are sorted by sequence or order of importance and is typically rendered as a numbered list. The content of a <OL> element is a sequence of <LI> elements.

<DIR> Directory List. <DIR> is similar to the <UL> element and represents a list of short items. The content of a <DIR> element is a sequence of <LI> elements.

<MENU> Menu List. <MENU> is a list of items with typically one line per item. It is typically a more compact than an unordered list. The content of a <MENU> element is a sequence of <LI> elements.

<DL>, <DT>, <DD>
Definition list. Lists terms and corresponding definitions. Definition lists are typically formatted with the term flush-left and the definition, formatted paragraph style, indented after the term. The content of a <DL> element is a sequence of <DT> elements and/or <DD> elements, usually in pairs.

<CITE> Citation. <CITE> is used to indicate the title of a book or other citation. It is typically rendered as italics.

<CODE> Code. <CODE> indicates an example of code and is typically rendered in a mono-spaced font. It is intended for short words or phrases of code.

<EM> Emphasis. <EM> indicates an emphasized phrase and is typically rendered as italics.

<KBD> Typed text. <KBD> indicates text typed by a user and is typically rendered in a mono-spaced font.

<SAMP>	Literal characters. <SAMP> indicates a sequence of literal characters and is typically rendered in a mono-spaced font.
<STRONG>	Strong emphasis. <STRONG> indicates strong emphasis and is typically rendered in bold.
<VAR>	Placeholder variable. <VAR> indicates a placeholder variable and is typically rendered as italic.
<B>	Bold. <B> indicates bold text.
<I>	Italic. <I> indicates italic text.
<TT>	Teletype. <TT> indicates teletype (monospaced) text.
<A>	Anchor. The <A> element indicates a hyperlink . Attributes of the <A> element are: HREF URI of the head anchor of a hyperlink. NAME Name of the anchor. TITLE Advisory title of the destination resource. REL The REL attribute gives the relationship(s) described by the hyperlink. REV Same as the REL attribute, but the semantics of the relationship are in the reverse direction. URN Specifies a preferred, more persistent identifier for the head anchor of the hyperlink. METHODS Specifies methods to be used in accessing the destination, as a whitespace-separated list of names.
 	Line Break. specifies a line break between words.
<HR>	Horizontal Rule. <HR> is a divider between sections of text and is typically a full width horizontal rule.
<IMG>	Image. <IMG> refers to an image or icon. Attributes are: ALIGN alignment of the image with respect to the text baseline: • 'TOP' specifies that the top of the image aligns with the tallest item on the line containing the image. • 'MIDDLE' specifies that the center of the image aligns with the baseline of the line containing the image. • 'BOTTOM' specifies that the bottom of the image aligns with the baseline of the line containing the image. ALT Text to use in place of the referenced image resource. ISMAP Indicates an image map. SRC Specifies the URI of the image resource.

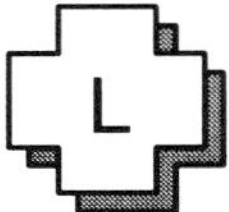

Java Reference

L.1 Package java.applet

L.1.1 Class ***java.applet.Applet***

The `Applet` class is a superclass of any applet. It provides a standard interface between applets and their environment. The following are defined:

```
// Constructors
public Applet();

// Methods
public void destroy();
public AppletContext getAppletContext();
public String getAppletInfo();
public AudioClip getAudioClip(URL url);
public AudioClip getAudioClip(URL url, String name);
public URL getCodeBase();
public URL getDocumentBase();
public Image getImage(URL url);
public Image getImage(URL url, String name);
public String getLocale();                                    // Java 1.1
public String getParameter(String name);
public String[][] getParameterInfo();
public void init();
public boolean isActive();
public void play(URL url);
public void play(URL url, String name);
public void resize(Dimension d);
public void resize(int width, int height);
public final void setStub(AppletStub stub);
public void showStatus(String msg);
public void start();
public void stop();
```

L.1.2 Interface ***java.applet.AppletContext***

The `AppletContext` interface corresponds to the applet's environment. The following are defined:

```
// Methods
public abstract Applet getApplet(String name);
public abstract Enumeration getApplets();
public abstract AudioClip getAudioClip(URL url);
public abstract Image getImage(URL url);
public abstract void showDocument(URL url);
public abstract void showDocument(URL url, String target);
public abstract void showStatus(String status);
```

*L.1.3 Interface **java.applet.AppletStub***

The `AppletStub` interface acts as the interface between the applet and the browser environment or applet viewer environment. The following are defined:

```
// Methods
public abstract void appletResize(int width, int height);
public abstract AppletContext getAppletContext();
public abstract URL getCodeBase();
public abstract URL getDocumentBase();
public abstract String getParameter(String name);
public abstract boolean isActive();
```

*L.1.4 Interface **java.applet.AudioClip***

The `AudioClip` interface is a simple abstraction for playing a sound clip. Multiple `AudioClip` items can be playing at the same time, and the resulting sound is mixed together to produce a composite. The following are defined:

```
// Methods
public abstract void loop();
public abstract void play();
public abstract void stop();
```

L.2 Package **java.awt**

*L.2.1 Class **java.awt.BorderLayout***

The `BorderLayout` class contains members named "North", "South", "East", "West", and "Center". These are laid out with a given size and constraints. The "North" and "South" components can be stretched horizontally and the "East" and "West" components can be stretched vertically. The "Center" component can be stretched horizontally and vertically. The following are defined:

```
// Constructors
public BorderLayout();
public BorderLayout(int hgap, int vgap);
// Constants
public static final String CENTER;                                   // Java 1.1
public static final String EAST;                                     // Java 1.1
public static final String NORTH;                                    // Java 1.1
public static final String SOUTH;                                    // Java 1.1
public static final String WEST;                                     // Java 1.1
// Methods
public void addLayoutComponent(Component comp, Object obj);          // Java 1.1
public void addLayoutComponent(String name, Component comp);         // Java 1.0
public int getHgap();                                                // Java 1.1
public float getLayoutAlignmentX(Container parent);                  // Java 1.1
public float getLayoutAlignmentY(Container parent);                  // Java 1.1
public int getVgap();                                                // Java 1.1
public void invalidateLayout(Container target);                      // Java 1.1
public void layoutContainer(Container target);
public Dimension maximumLayoutSize(Container target);                // Java 1.1
public Dimension minimumLayoutSize(Container target);
```

```
public Dimension preferredLayoutSize(Container target);
public void removeLayoutComponent(Component comp);
public int setHgap();                                                 // Java 1.1
public int setVgap();                                                 // Java 1.1
public String toString();
```

*L.2.2 Class **java.awt.Button***

The `Button` class creates labelled buttons, which can have an associated action when pushed. Three typical actions are: normal, when it has the input focus (the darkening of the outline lets the user know that this is an active object) and when the user clicks the mouse over the button. The following are defined:

```
// Constructors
public Button();
public Button(String label);

// Methods
public synchronized void addActionListener(ActionListern l);          // Java 1.1
public void addNotify();
public String getActionCommand();                                     // Java 1.1
public String getLabel();
protected String paramString();
public synchronized void removeActionListener(ActionListener l);
                                                                      // Java 1.1
public setActionCommand(String command);                              // Java 1.1
public void setLabel(String label);
```

*L.2.3 Class **java.awt.Checkbox***

The `Checkbox` class contains a checkbox which has an on/off state. The following are defined:

```
// Constructors
public Checkbox();
public Checkbox(String label);
public Checkbox(String label, boolean state);                         // Java 1.1
public Checkbox(String label, boolean state, Checkbox group);         // Java 1.1
public Checkbox(String label, CheckboxGroup group, boolean state);

// Methods
public synchronized void addItemListener(ItemListener l);             // Java 1.1
public void addNotify();
public CheckboxGroup getCheckboxGroup();
public String getLabel();
public boolean getState();
public Object[] getSelectedObject();                                  // Java 1.1
protected String paramString();
public synchronized void removeItemListener(ItemListener l);          // Java 1.1
public void setCheckboxGroup(CheckboxGroup g);
public void setLabel(String label);
public void setState(boolean state);
```

*L.2.4 Class **java.awt.CheckboxGroup***

The `CheckGroup` class groups a number of checkbox buttons. Only one of the checkboxes can be true (on) at a time. When one button is made true (on) the others will become false (off). The following are defined:

```
// Constructors
```

```
public CheckboxGroup();

// Methods
public Checkbox getCurrent();                                  // Java 1.0
public Checkbox getSelectedCurrent();                          // Java 1.1
public void setCurrent(Checkbox box);                          // Java 1.0
public void setSelectedCheckbox(Checkbox box);                 // Java 1.1
public String toString();
```

*L.2.5 Class **java.awt.CheckboxMenuItem***

The `CheckboxMenuItem` class allows for a checkbox that can be included in a menu. The following are defined:

```
// Constructors
public CheckboxMenuItem();                                     // Java 1.1
public CheckboxMenuItem(String label);
public CheckboxMenuItem(String label, boolean state);          // Java 1.1

// Methods
public synchronized void addItemListener(ItemListener l);      // Java 1.1
public void addNotify();
public boolean getState();
public synchronized Object[] getSelectObjects();               // Java 1.1
public String paramString();
public synchronized void removeItemListener(ItemListener l);   // Java 1.1
public void setState(boolean t);
```

*L.2.6 Class **java.awt.Choice***

The `Choice` class allows for a pop-up menu. The following are defined:

```
// Constructors
public Choice();

// Methods
public synchronized add(String item);                          // Java 1.1
public void addItem(String item);
public synchronized void addItemListener(ItemListener l);      // Java 1.1
public void addNotify();
public int countItems();                                       // Java 1.0
public String getItem(int index);
public String getItemCount();                                  // Java 1.1
public int getSelectedIndex();
public String getSelectedItem();
protected String paramString();
public synchronized Object[] getSelectedObjects();             // Java 1.1
public synchronized void insert(String item, int index);       // Java 1.1
public synchronized void remove(String item);                  // Java 1.1
public synchronized void remove(int position);                 // Java 1.1
public synchronized void removeAll();                          // Java 1.1
public synchronized void removeItemListener(ItemListener l);   // Java 1.1
public void select(int pos);
public void select(String str);
```

*L.2.7 Class **java.awt.Color***

This `Color` class supports the RGB colour format. A colour is represented by a 24-bit value of which the red, green and blue components are represented by an 8-bit value (0 to 255). The minimum intensity is 0, and the maximum is 255. The following are defined:

```
// Constants
public final static Color black, blue, cyan, darkGray, gray, green;
public final static Color lightGray, magenta, orange, pink, red;
public final static Color white, yellow;

// Constructors
public Color(float r, float g, float b);
public Color(int rgb);
public Color(int r, int g, int b);

// Methods
public Color brighter();
public Color darker();
public static Color decode (Strimg nm);                          // Java 1.1
public boolean equals(Object obj);
public int getBlue();
public static Color getColor(String nm);
public static Color getColor(String nm, Color v);
public static Color getColor(String nm, int v);
public int getGreen();
public static Color getHSBColor(float h, float s, float b);
public int getRed();
public int getRGB();
public int hashCode();
public static int HSBtoRGB(float hue, float saturation, float brightness);
public static float[] RGBtoHSB(int r, int g, int b, float hsbvals[]);
public String toString();
```

*L.2.8 Class **java.awt.Component***

The `Component` class is the abstract superclass for many of the Abstract Window Toolkit classes. The following are defined:

```
// Constants
public static final float BOTTOM_ALIGNMENT, CENTER_ALIGNMENT;
public static final float LEFT_ALIGNMENT, RIGHT_ALIGNMENT;
public static final float TOP_ALIGNMENT;

// Methods
public boolean action(Event evt, Object what);                   // Java 1.0
public synchronized void add(PopupMenu popup);                   // Java 1.1
public synchronized void addComponentListener(ComponentListener l);

                                                                 // Java 1.1
public synchronized void addFocusListener(FocusListener l);      // Java 1.1
public synchronized void addKeyListener(KeyListener l);          // Java 1.1
public synchronized void addMouseListener(MouseListener l);      // Java 1.1
public synchronized void addMouseMotionListener(MouseMotionListener l);
                                                                 // Java 1.1
public void addNotify();
public Rectangle bounds();                                       // Java 1.0
public int checkImage(Image image, ImageObserver observer);
public int checkImage(Image image, int width, int height,
        ImageObserver observer);
public boolean contains(int x, int y);                           // Java 1.1
public boolean contains(Point p);                                // Java 1.1
public Image createImage(ImageProducer producer);
public Image createImage(int width, int height);
public void deliverEvent(Event evt);                             // Java 1.0
public void disable();                                           // Java 1.0
public final void displayEvent(AWTEvent e);                      // Java 1.1
```

```
public void doLayout();                                        // Java 1.1
public void enable();                                          // Java 1.0
public void enable(boolean cond);                              // Java 1.0
public float getAlignmentX();                                  // Java 1.1
public float getAlignmentY();                                  // Java 1.1
public Color getBackground();
public Rectangle getBounds();                                  // Java 1.1
public ColorModel getColorModel();
public Component getComponentAt(int x, int y);                 // Java 1.1
public Component getComponentAt(Point p);                      // Java 1.1
public Cursor getCursor();                                     // Java 1.1
public Font getFont();
public FontMetrics getFontMetrics(Font font);
public Color getForeground();
public Graphics getGraphics();
public Locale getLocale();                                     // Java 1.1
public Point getLocation();                                    // Java 1.1
public Point getLocationOnScreen();                            // Java 1.1
public Dimension getMaximumSize();                             // Java 1.1
public Dimension getMinimumSize();                             // Java 1.1
public Container getParent();
public ComponentPeer getPeer();                                // Java 1.0
public Dimension getPreferredSize();                           // Java 1.1
public Dimension getSize();                                    // Java 1.1
public Toolkit getToolkit();
public final Object getTreeLock();                             // Java 1.1
public boolean gotFocus(Event evt, Object what);               // Java 1.0
public boolean handleEvent(Event evt);                         // Java 1.0
public void hide();                                            // Java 1.0
public boolean imageUpdate(Image img, int flags, int x, int y, int w, int h);
public boolean inside(int x, int y);                           // Java 1.0
public void invalidate();
public boolean isEnabled();
public boolean isFocusTransversable();                         // Java 1.1
public boolean isShowing();
public boolean isValid();
public boolean isVisible();
public boolean keyDown(Event evt, int key);                    // Java 1.0
public boolean keyUp(Event evt, int key);                      // Java 1.0
public void layout();                                          // Java 1.0
public void list();
public void list(PrintStream out);
public void list(PrintStream out, int indent);
public void list(PrintStream out);                             // Java 1.1
public Component locate(int x, int y);                         // Java 1.0
public Point location();                                       // Java 1.0
public boolean lostFocus(Event evt, Object what);              // Java 1.0
public Dimension minimumSize();                                // Java 1.0
public boolean mouseDown(Event evt, int x, int y);             // Java 1.0
public boolean mouseDrag(Event evt, int x, int y);             // Java 1.0
public boolean mouseEnter(Event evt, int x, int y);            // Java 1.0
public boolean mouseExit(Event evt, int x, int y);             // Java 1.0
public boolean mouseMove(Event evt, int x, int y);             // Java 1.0
public boolean mouseUp(Event evt, int x, int y);               // Java 1.0
public void move(int x, int y);                                // Java 1.0
public void nextFocus();                                       // Java 1.0
public void paint(Graphics g);
public void paintAll(Graphics g);
protected String paramString();
public boolean postEvent(Event evt);                           // Java 1.0
public Dimension preferredSize();                              // Java 1.0
```

```
public boolean prepareImage(Image image, ImageObserver observer);
public prepareImage(Image image, int width, int height, ImageObserver observer);
public void print(Graphics g);
public void printAll(Graphics g);
public synchronized void remove(MenuComponent popup);           // Java 1.1
public synchronized void removeComponentListener(ComponentListener l);
                                                                // Java 1.1
public synchronized void removeFocusListener(FocusListener l);  // Java 1.1
public synchronized void removeKeyListener(KeyListener l);      // Java 1.1
public synchronized void removeMouseListener(MouseListener l);  // Java 1.1
public synchronized void removeMouseMotionListener(MouseMotionListener l);
                                                                // Java 1.1
public void removeNotify();
public void repaint();
public void repaint(int x, int y, int width, int height);
public void repaint(long tm);
public void repaint(long tm, int x, int y, int width, int height);
public void requestFocus();
public void reshape(int x, int y, int width, int height);       // Java 1.0
public void resize(Dimension d);                                // Java 1.0
public void resize(int width, int height);                      // Java 1.0
public void setBackground(Color c);
public void setBounds(int x, int y, int width, int height);     // Java 1.1
public void setBounds(Rectangle r);                             // Java 1.1
public synchronized void setCursor(Cursor cursor);              // Java 1.1
public void setEnabled(boolean b);                              // Java 1.1
public void setFont(Font f);
public void setForeground(Color c);
public void setLocale(Locale l);                                // Java 1.1
public void setLocation(int x, int y);                          // Java 1.1
public void setLocation(Point p);                               // Java 1.1
public void setName(String name);                               // Java 1.1
public void setSize(int width, int height);                     // Java 1.1
public void setSize(Dimension d);                               // Java 1.1
public void setVisible(boolean b);                              // Java 1.1
public void show();                                             // Java 1.0
public void show(boolean cond);                                 // Java 1.0
public Dimension size();                                        // Java 1.0
public String toString();
public void transferFocus();                                    // Java 1.1
public void update(Graphics g);
public void validate();
```

*L.2.9 Class **java.awt.Container***

The `Container` class is the abstract superclass representing all components that can hold other components. The following are defined:

```
// Methods
public Component add(Component comp);
public Component add(Component comp, int pos);
public Component add(String name, Component comp);
public void add(Component comp, Object constraints);            // Java 1.1
public void add(Component comp, Object constraints, int index); // Java 1.1
public void addContainerListener(ContainerListener l);          // Java 1.1
public void addNotify();
public int countComponents();                                   // Java 1.0
public void deliverEvent(Event evt);                            // Java 1.0
public void doLayout();                                         // Java 1.1
public void getAlignmentX();                                    // Java 1.1
public void getAlignmentY();                                    // Java 1.1
```

```
public Component getComponent(int n);
public Component getComponentAt(int x, int y);                    // Java 1.1
public Component getComponentAt(Point p);                         // Java 1.1
public int getComponentCount();                                   // Java 1.1
public Component[] getComponents();
public getInsets();                                               // Java 1.1
public LayoutManager getLayout();
public Dimension getMaximumSize();                                // Java 1.1
public Dimension getMinimumSize();                                // Java 1.1
public Dimension getPreferredSize();                              // Java 1.1
public Insets insets();                                           // Java 1.0
public void invalidate();                                         // Java 1.1
pubic boolean isAncestorOf(Component c);                          // Java 1.1
public void layout();                                             // Java 1.0
public void list(PrintStream out, int indent);
public void list(PrintWriter out, int indent);                    // Java 1.1
public Component locate(int x, int y);                            // Java 1.0
public Dimension minimumSize();                                   // Java 1.0
public void paintComponents(Graphics g);
protected String paramString();
public Dimension preferredSize();                                 // Java 1.0
public void print(Graphics g);                                    // Java 1.1
public void printComponents(Graphics g);
public void remove(int index);                                    // Java 1.1
public void remove(Component comp);
public void removeAll();
public void removeContainerListener(ContainerListener l);         // Java 1.1
public void removeNotify();
public void setLayout(LayoutManager mgr);
public void validate();
```

*L.2.10 Class **java.awt.Cursor***

The `Cursor` class represents a mouse cursor. The following are defined:

```
// Constructors
public Cursor(int type);                                          // Java 1.1

// Constants
public static final int DEFAULT_CURSOR;                           // Java 1.1
public static final int CROSSHAIR_CURSOR, HAND_CURSOR;            // Java 1.1
public static final int MOVE_CURSOR;                              // Java 1.1
public static final int TEXT_CURSOR, WAIT_CURSOR;                 // Java 1.1
public static final int N_RESIZE_CURSOR, S_RESIZE_CURSOR;         // Java 1.1
public static final int E_RESIZE_CURSOR, W_RESIZE_CURSOR;         // Java 1.1
public static final int NE_RESIZE_CURSOR, NW_RESIZE_CURSOR;       // Java 1.1
public static final int SE_RESIZE_CURSOR, SW_RESIZE_CURSOR;       // Java 1.1

// Methods
public static Cursor getDefaultCursor();                          // Java 1.1
public static Cursor getPredefinedCursor();                       // Java 1.1
```

*L.2.11 Class **java.awt.Dialog***

The `Dialog` class supports a dialog window, in which a user can enter data. Dialog windows are invisible until the show method is used. The following are defined:

```
// Constructors
public Dialog(Frame parent);                                      // Java 1.1
public Dialog(Frame parent, boolean modal);
```

```
public Dialog(Frame parent, String title);                           // Java 1.1
public Dialog(Frame parent, String title, boolean modal);

// Methods
public void addNotify();
public String getTitle();
public boolean isModal();
public boolean isResizable();
public void setModal(boolean b);                                     // Java 1.1
protected String paramString();
public void setResizable(boolean resizable);
public void setTitle(String title);
public void show();                                                  // Java 1.1
```

L.2.12 Class *java.awt.Dimension*

The `Dimension` class contains the width and height of a component in an object. The following are defined:

```
// Fields
public int height;
public int width;

// Constructors
public Dimension();
public Dimension(Dimension d);
public Dimension(int width, int height);

// Methods
public boolean equals(Object obj);                                   // Java 1.1
public Dimension getSize();                                          // Java 1.1
public void setSize(Dimension d);                                    // Java 1.1
public void setSize(int width, int height);                          // Java 1.1
public String toString();
```

L.2.13 Class *java.awt.Event*

The `Event` class encapsulates user events from the GUI. The following are defined:

```
// Fields
public Object arg;
public int clickCount;
public Event evt;
public int id;
public int key;
public int modifiers;
public Object target;
public long when;
public int x;
public int y;

// possible values for the id field
public final static int ACTION_EVENT, GOT_FOCUS;
public final static int KEY_ACTION, KEY_ACTION_RELEASE;
public final static int KEY_PRESS, KEY_RELEASE;
public final static int LIST_DESELECT, LIST_SELECT;
public final static int LOAD_FILE, LOST_FOCUS;
public final static int MOUSE_DOWN, MOUSE_DRAG;
public final static int MOUSE_ENTER, MOUSE_EXIT;
public final static int MOUSE_MOVE, MOUSE_UP;
public final static int SAVE_FILE, SCROLL_ABSOLUTE;
```

```
public final static int SCROLL_BEGIN, SCROLL_END;                    // Java 1.1

public final static int SCROLL_LINE_DOWN, SCROLL_LINE_UP;
public final static int SCROLL_PAGE_DOWN, SCROLL_PAGE_UP;
public final static int WINDOW_DEICONIFY, WINDOW_DESTROY;
public final static int WINDOW_EXPOSE, WINDOW_ICONIFY;
public final static int WINDOW_MOVED;

// possible values for the key field when the
// action is KEY_ACTION or KEY_ACTION_RELEASE
public final static int DOWN, END;
public final static int F1, F2, F3, F4, F5, F6, F7, F8, F9, F10, F11, F12
public final static int HOME, LEFT, PGDN, PGUP, RIGHT, UP;
public final static int INSERT, DELETE;                              // Java 1.1
public final static int BACK_SPACE, ENTER;                           // Java 1.1
public final static int TAB, ESCAPE;                                 // Java 1.1
public final static int CAPS_LOCK, NUM_LOCK;                         // Java 1.1
public final static int SCROLL_LOCK, PAUSE;                          // Java 1.1
public final static int PRINT_SCREEN;                                // Java 1.1

// possible masks for the modifiers field
public final static int ALT_MASK;
public final static int CTRL_MASK;
public final static int META_MASK;
public final static int SHIFT_MASK;

// Constructors
public Event(Object target, int id, Object arg);
public Event(Object target, long when, int id,
                                int x, int y, int key, int modifiers);
public Event(Object target, long when, int id,
                  int x, int y, int key, int modifiers, Object arg);

// Methods
public boolean controlDown();
public boolean metaDown();
protected String paramString();
public boolean shiftDown();
public String toString();
public void translate(int dX, int dY);
```

*L.2.14 Class **java.awt.FileDialog***

The `FileDialog` class displays a dialog window. The following are defined:

```
// Fields
public final static int LOAD, SAVE;

// Constructors
public FileDialog(Frame parent);                                     // Java 1.1
public FileDialog(Frame parent, String title);
public FileDialog(Frame parent, String title, int mode);
// Methods
public void addNotify();
public String getDirectory();
public String getFile();
public FilenameFilter getFilenameFilter();
public int getMode();
protected String paramString();
public void setDirectory(String dir);
public void setFile(String file);
public void setFilenameFilter(FilenameFilter filter);
```

L.2.15 Class *java.awt.FlowLayout*

The `FlowLayout` class arranges components from left to right. The following are defined:

```
// Fields
public final static int CENTER, LEFT, RIGHT;

// Constructors
public FlowLayout();
public FlowLayout(int align);
public FlowLayout(int align, int hgap, int vgap);

// Methods
public void addLayoutComponent(String name, Component comp);
public int getAlignment();                                          // Java 1.1
public int getHgap();                                               // Java 1.1
public int getVgap();                                               // Java 1.1
public void layoutContainer(Container target);
public Dimension minimumLayoutSize(Container target);
public Dimension preferredLayoutSize(Container target);
public void removeLayoutComponent(Component comp);
public void setAlignment(int align);                                // Java 1.1
public void setHgap(int hgap);                                      // Java 1.1
public void setVgap(int vgap);                                      // Java 1.1
public String toString();
```

L.2.16 Class *java.awt.Font*

The `Font` class represents fonts. The following are defined:

```
// Fields
protected String name;
protected int size;
protected int style;

// style has the following bit masks
public final static int BOLD, ITALIC, PLAIN;

// Constructors
public Font(String name, int style, int size);

// Methods
public static Font decode(String str);                              // Java 1.1
public boolean equals(Object obj);
public String getFamily();
public static Font getFont(String nm);
public static Font getFont(String nm, Font font);
public String getName();
public int getSize();
public int getStyle();
public FontPeer getPeer();                                          // Java 1.1
public int hashCode();
public boolean isBold();
public boolean isItalic();
public boolean isPlain();
public String toString();
```

L.2.17 Class *java.awt.FontMetrics*

The `FontMetrics` class provides information about the rendering of a particular font. The following are defined:

```
// Fields
protected Font font;

// Constructors
protected FontMetrics(Font font);

// Methods
public int bytesWidth(byte data[], int off, int len);
public int charsWidth(char data[], int off, int len);
public int charWidth(char ch);
public int charWidth(int ch);
public int getAscent();
public int getDescent();
public Font getFont();
public int getHeight();
public int getLeading();
public int getMaxAdvance();
public int getMaxAscent();
public int getMaxDescent();                              // Java 1.0
public int[] getWidths();
public int stringWidth(String str);
public String toString();
```

*L.2.18 Class **java.awt.Frame***

The `Frame` class contains information on the top-level window. The following are defined:

```
// possible cursor types for the setCursor method
public final static int CROSSHAIR_CURSOR, DEFAULT_CURSOR;
public final static int E_RESIZE_CURSOR, HAND_CURSOR;
public final static int MOVE_CURSOR, N_RESIZE_CURSOR;
public final static int NE_RESIZE_CURSOR, NW_RESIZE_CURSOR;
public final static int S_RESIZE_CURSOR, SE_RESIZE_CURSOR;
public final static int SW_RESIZE_CURSOR, TEXT_CURSOR;
public final static int W_RESIZE_CURSOR, WAIT_CURSOR;
// Constructors
public Frame();
public Frame(String title);
// Methods
public void addNotify();
public void dispose();
public int getCursorType();                              // Java 1.0
public Image getIconImage();
public MenuBar getMenuBar();
public String getTitle();
public boolean isResizable();
protected String paramString();
public void remove(MenuComponent m);
public void setCursor(int cursorType);                   // Java 1.0
public void setIconImage(Image image);
public void setMenuBar(MenuBar mb);
public void setResizable(boolean resizable);
public void setTitle(String title);
```

*L.2.19 Class **java.awt.Graphics***

The `Graphics` class is an abstract class for all graphics contexts. This allows an application to draw onto components or onto off-screen images. The following are defined:

```
// Constructors
```

```
protected Graphics();

// Methods
public abstract void clearRect(int x, int y, int width, int height);
public abstract void clipRect(int x, int y, int width, int height);
public abstract void copyArea(int x, int y, int width, int height,
       int dx, int dy);
public abstract Graphics create();
public Graphics create(int x, int y, int width, int height);
public abstract void dispose();
public void draw3DRect(int x, int y, int width, int height, boolean raised);
public abstract void drawArc(int x, int y, int width, int height,
       int startAngle, int arcAngle);
public void drawBytes(byte data[], int offset, int length, int x, int y);
public void drawChars(char data[], int offset,      int length, int x, int y);
public abstract boolean      drawImage(Image img, int x, int y, Color bgcolor,
       ImageObserver observer);
public abstract boolean drawImage(Image img, int x, int y,
       ImageObserver observer);
public abstract boolean drawImage(Image img, int x, int y, int width,
       int height, Color bgcolor, ImageObserver observer);
public abstract boolean drawImage(Image img, int x, int y, int width,
       int height, ImageObserver observer);
public abstract boolean drawImage(Image img, int x, int y, int width,
       int height, Color bgcolor, ImageObserver observer);        // Java 1.1
public abstract void drawLine(int x1, int y1, int x2, int y2);
public abstract void drawOval(int x, int y,int width, int height);
public abstract void drawPolygon(int xPoints[], int yPoints[], int nPoints);
public void drawPolygon(Polygon p);
public abstract void drawPolyline(int xPoints[], int yPoints[], int nPoints);
                                                                  // Java 1.1
public void drawRect(int x, int y, int width, int height);
public abstract void drawRoundRect(int x, int y, int width,
       int height, int arcWidth, int arcHeight);
public abstract void drawString(String str, int x, int y);
public void fill3DRect(int x, int y, int width, int height, boolean raised);
public abstract void fillArc(int x, int y, int width, int height,
       int startAngle int arcAngle);
public abstract void fillOval(int x, int y, int width, int height);
public abstract void fillPolygon(int xPoints[], int yPoints[], int nPoints);
public void fillPolygon(Polygon p);
public abstract void fillRect(int x, int y, int width, int height);
public abstract void fillRoundRect(int x, int y, int width, int height,
       int arcWidth, int arcHeight);
public void finalize();
public abstract Shape getClip();                                  // Java 1.1
public abstract Rectangle getClipBounds();                        // Java 1.1
public abstract Rectangle getClipRect();                          // Java 1.0
public abstract Color getColor();
public abstract Font getFont();
public FontMetrics getFontMetrics();
public abstract FontMetrics getFontMetrics(Font f);
public abstract void setClip(int x, int y, int width, int height); // Java 1.1
public abstract void setClip(Shape clip);                         // Java 1.1
public abstract void setColor(Color c);
public abstract void setFont(Font font);
public abstract void setPaintMode();
public abstract void setXORMode(Color c1);
public String toString();
public abstract void translate(int x, int y);
```

L.2.20 Class *java.awt.Image*

The `Image` abstract class is the superclass of all classes that represents graphical images.

```
// Constants
public static final int SCALE_AREA_AVERAGING, SCALE_DEFAULT;
public static final int SCALE_FAST, SCALE_REPLICATE;
public static final int SCALE_SMOOTH;

// Fields
public final static Object UndefinedProperty;

// Constructors
public Image();

// Methods
public abstract void flush();
public abstract Graphics getGraphics();
public abstract int getHeight(ImageObserver observer);
public abstract Object getProperty(String name, ImageObserver observer);
public Image getScaledInstance(int width, int height, int hints);   // Java 1.1
public abstract ImageProducer getSource();
public abstract int getWidth(ImageObserver observer)
```

L.2.21 Class *java.awt.Insets*

The `Insets` object represents borders of a container and specifies the space that should be left around the edges of a container. The following are defined:

```
// Fields
public int bottom, left;
public int right, top;
// Constructors
public Insets(int top, int left, int bottom, int right);

// Methods
public Object clone();
public boolean equals(Object obj);                              // Java 1.1
public String toString();
```

L.2.22 Class *java.awt.Label*

The `label` class is a component for placing text in a container. The following are defined:

```
// Fields
public final static int CENTER, LEFT, RIGHT;

// Constructors
public Label();
public Label(String label);
public Label(String label, int alignment);

// Methods
public void addNotify();
public int getAlignment();
public String getText();
protected String paramString();
public void setAlignment(int alignment);
public void setText(String label);
```

L.2.23 Class *java.awt.List*

The `List` object can be used to produce a scrolling list of text items. It can be set up so that the user can either pick one or many items. The following are defined:

```
// Constructors
public List();
public List(int rows);                                              // Java 1.1
public List(int rows, boolean multipleSelections);

// Methods
public void add(String item);                                       // Java 1.1
public void addActionListener(ActionListener l);                    // Java 1.1
public void addItem(String item);
public void addItem(String item, int index);
public synchronized void addItemListener(ItemListener l);           // Java 1.1
public void addNotify();
public boolean allowsMultipleSelections();                          // Java 1.0
public void clear();                                                // Java 1.0
public int countItems();                                            // Java 1.0
public void delItem(int position);
public void delItems(int start, int end);                           // Java 1.0
public void deselect(int index);
public String getItem(int index);
public int getItemCount();                                          // Java 1.1
public synchronized String[] getItems();                            // Java 1.1
public Dimension getMinimumSize(int rows);                          // Java 1.1
public Dimension getMinimumSize();                                  // Java 1.1
public Dimension getPreferredSize(int rows);                        // Java 1.1
public Dimension getPreferredSize();                                // Java 1.1
public int getRows();
public int getSelectedIndex();
public int[] getSelectedIndexes();
public String getSelectedItem();
public String[] getSelectedItems();
public Object[] getSelectedObjects();                               // Java 1.1
public int getVisibleIndex();
public boolean isIndexSelected(int index);                          // Java 1.1
public MultipleMode();                                              // Java 1.1
public boolean isSelected(int index);                               // Java 1.0
public void makeVisible(int index);
public Dimension minimumSize();                                     // Java 1.0
public Dimension minimumSize(int rows);                             // Java 1.0
protected String paramString();
public Dimension preferredSize();                                   // Java 1.0
public Dimension preferredSize(int rows);                           // Java 1.0
public synchronized void remove(String item);                       // Java 1.1
public synchronized void remove(int position);                      // Java 1.1
public synchronized void removeActionListener(ActionListener l);    // Java 1.1
public synchronized void removeAll();                               // Java 1.1
public synchronized void removeItemListener(ItemListener l);        // Java 1.1
public void removeNotify();
public void replaceItem(String newValue, int index);
public void select(int index);
public synchronized void setMultipleMode(boolean b);                // Java 1.1
public void setMultipleSelections(boolean v);
```

L.2.24 Class *java.awt.MediaTracker*

The `MediaTracker` class contains a number of media objects, such as images and audio. The following are defined:

```
// Fields
public final static int ABORTED, COMPLETE;
```

```
public final static int ERRORED, LOADING;

// Constructors
public MediaTracker(Component comp);

// Methods
public void addImage(Image image, int id);
public void addImage(Image image, int id, int w, int h);
public boolean checkAll();
public boolean checkAll(boolean load);
public boolean checkID(int id);
public boolean checkID(int id, boolean load);
public Object[] getErrorsAny();
public Object[] getErrorsID(int id);
public boolean isErrorAny();
public boolean isErrorID(int id);
public synchronized removeImage(Image image);                  // Java 1.1
public synchronized removeImage(Image image, int id);          // Java 1.1
public synchronized removeImage(Image image, int id, int width, int height);
                                                                // Java 1.1
public int statusAll(boolean load);
public int statusID(int id, boolean load);
public void waitForAll();
public boolean waitForAll(long ms);
public void waitForID(int id);
public boolean waitForID(int id, long ms);
```

L.2.25 Class *java.awt.Menu*

The Menu object contains a pull-down component for a menu bar. The following are defined:

```
// Constructors
public Menu();                                                  // Java 1.1
public Menu(String label);
public Menu(String label, boolean tearOff);

// Methods
public MenuItem add(MenuItem mi);
public void add(String label);
public void addNotify();
public void addSeparator();
public int countItems();
public MenuItem getItem(int index);
public int getItemCount();                                      // Java 1.1
public synchronized void Insert(MenuItem menuitem, int index);  // Java 1.1
public void InsertSepatator(int index);                         // Java 1.1
public boolean isTearOff();
public void remove(int index);
public void remove(MenuComponent item);
public synchronized void removeAll();                           // Java 1.1
public void removeNotify();
```

L.2.26 Class *java.awt.MenuBar*

The MenuBar object contains a menu bar which is bound to a frame. The following are defined:

```
// Constructors
public MenuBar();

// Methods
```

```
public Menu add(Menu m);
public void addNotify();
public int countMenus();
public void deleteShortCut(MenuShortCut s);                    // Java 1.1
public Menu getHelpMenu();
public Menu getMenu(int i);
public int getMenuCount();                                     // Java 1.1
public MenuItem getShortcutMenuItem(MenuShortcut s);           // Java 1.1
public void remove(int index);
public void remove(MenuComponent m);
public void removeNotify();
public void setHelpMenu(Menu m);
public synchronized Enumeration shortcuts();                   // Java 1.1
```

L.2.27 Class *java.awt.MenuComponent*

The `MenuComponent` abstract class is the superclass of all menu-related components. The following are defined:

```
// Constructors
public MenuComponent();

// Methods
public final void dispatchEvent(AWTEvent e);                   // Java 1.1
public Font getFont();
public String getName();                                       // Java 1.1
public MenuContainer getParent();
public MenuComponentPeer getPeer();                            // Java 1.0
protected String paramString();
public boolean postEvent(Event evt);
public void removeNotify();
public void setFont(Font f);
public void setName(String name);                              // Java 1.1
public String toString();
```

L.2.28 Class *java.awt.MenuItem*

The `MenuItem` class contains all menu items. The following are defined:

```
// Constructors
public MenuItem();                                             // Java 1.1
public MenuItem(String label);
public MenuItem(String label, MenuShortcut s);                 // Java 1.1

// Methods
public void addActionListener(ActionListener l);               // Java 1.1
public void addNotify();
public void deleteShortcut();                                  // Java 1.1
public void disable();                                         // Java 1.0
public void enable();                                          // Java 1.0
public void enable(boolean cond);                              // Java 1.0
public String getLabel();
public MenuShortcut getShortcut();                             // Java 1.1
public boolean isEnabled();
public String paramString();
public synchronized void removeActionListener(ActionListener l);

                                                               // Java 1.1
public void setActionCommand(String command);                  // Java 1.1
public synchronized void setEnabled(boolean b);                // Java 1.1
public void setLabel(String label);
```

```
public void setShortcut(MenuShortcut s);                          // Java 1.1
```

L.2.29 Class *java.awt.MenuShortcut*

The `MenuShortcut` class has been added with Java 1.1. It represents a keystroke used to select a `MenuItem`. The following are defined:

```
// Constructors
public MenuShortcut(int key);                                     // Java 1.1
public MenuShortcut(int key, boolean useShiftModifier);           // Java 1.1

// Methods
public boolean equals(MenuShortcut s);                            // Java 1.1
public int getKey();                                              // Java 1.1
public String toString();                                         // Java 1.1
public boolean usesShiftModifier();                               // Java 1.1
```

L.2.30 Class *java.awt.Panel*

The `Panel` class provides space into which an application can attach a component. The following are defined:

```
// Constructors
public Panel();
public Panel(LayoutManger layout);                                // Java 1.1

// Methods
public void addNotify();
```

L.2.31 Class *java.awt.Point*

The `Point` class represents an (x, y) co-ordinate. The following are defined:

```
// Fields
public int x;
public int y;

// Constructors
public Point();                                                   // Java 1.1
public Point(Point p);                                            // Java 1.1
public Point(int x, int y);

// Methods
public boolean equals(Object obj);
public Point getLocation();                                       // Java 1.1
public int hashCode();
public void move(int x, int y);
public void setLocation(Point p);                                 // Java 1.1
public void setLocation(int x, int y);                            // Java 1.1
public String toString();
public void translate(int dx, int dy);
```

L.2.32 Class *java.awt.Polygon*

The `Polygon` class consists of an array of (x, y), which define the sides of a polygon. The following are defined:

```
// Fields
public int npoints, xpoints[],ypoints[];

// Constructors
```

```
public Polygon();
public Polygon(int xpoints[], int ypoints[], int npoints);

// Methods
public void addPoint(int x, int y);
public boolean contains(Point p);                          // Java 1.1
public boolean contains(int x, int y);                     // Java 1.1
public Rectangle getBoundingBox();                         // Java 1.0
public Rectangle getBounds();                              // Java 1.1
public boolean inside(int x, int y);                       // Java 1.0
```

L.2.33 Class *java.awt.PopupMenu*

The `PopupMenu` class has been added with Java 1.1. It represetns a pop-up menu rather than a pull-down menu. The following are defined:

```
// Constructors
public PopupMenu();                                        // Java 1.1
public PopupMenu(String label);                            // Java 1.1

// Methods
public synchronized void addNotify();                      // Java 1.1
public void show(Component origin, int x, int y);          // Java 1.1
```

L.2.34 Class *java.awt.Rectangle*

The `Rectangle` class defines an area defined by its top-left (x, y) co-ordinate, its width and its height. The following are defined:

```
// Fields
public int height, width, x, y;

// Constructors
public Rectangle();
public Rectangle(Rectangle r);                             // Java 1.1
public Rectangle(Dimension d);
public Rectangle(int width, int height);
public Rectangle(int x, int y, int width, int height);
public Rectangle(Point p);
public Rectangle(Point p, Dimension d);
// Methods
public void add(int newx, int newy);
public void add(Point pt);
public void add(Rectangle r);
public boolean contains(Point p);                          // Java 1.1
public boolean contains(int x, int y);                     // Java 1.1
public boolean equals(Object obj);
public Rectangle getBounds();                              // Java 1.1
public Point getLocation();                                // Java 1.1
public Dimension getSize();                                // Java 1.1
public void grow(int h, int v);
public int hashCode();
public boolean inside(int x, int y);                       // Java 1.0
public Rectangle intersection(Rectangle r);
public boolean intersects(Rectangle r);
public boolean isEmpty();
public void move(int x, int y);                            // Java 1.0
public void reshape(int x, int y, int width, int height);  // Java 1.0
public void resize(int width, int height);                 // Java 1.0
public void setBounds(Rectangle r);                        // Java 1.1
public void setBounds(int x, int y, int width, int height); // Java 1.1
```

```
public void setLocation(Point p);                               // Java 1.1
public void setLocation(int x, int y);                          // Java 1.1
public void setSize(Dimension d);                               // Java 1.1
public void setSize(int x, int y);                              // Java 1.1
public String toString();
public void translate(int dx, int dy);
public Rectangle union(Rectangle r);
```

L.2.35 Class *java.awt.Scrollbar*

The `Scrollbar` class is a convenient means of allowing a user to select from a range of values. The following are defined:

```
// Fields
public final static int HORIZONTAL, VERTICAL;

// Constructors
public Scrollbar();
public Scrollbar(int orientation);
public Scrollbar(int orientation, int value, int visible, int minimum,
       int maximum);

// Methods
public synchronized void addAdjustmenuListener(AdjustmentListener l);
                                                                // Java 1.1
public void addNotify();
public int getBlockIncrement();                                 // Java 1.1
public int getLineIncrement();                                  // Java 1.0
public int getMaximum();
public int getMinimum();
public int getOrientation();
public int getPageIncrement();                                  // Java 1.0
public int getUnitIncrement();                                  // Java 1.1
public int getValue();
public int getVisible();                                        // Java 1.0
protected String paramString();
public void setLineIncrement(int l);                            // Java 1.0
public synchronized void setMaximum(int max);                   // Java 1.1
public synchronized void setMinimum(int min);                   // Java 1.1
public synchronized void setOrientation(int orien);             // Java 1.1
public void setPageIncrement(int l);                            // Java 1.0
public void setValue(int value);
public void setValues(int value, int visible, int minimum, int maximum);
public void setVisibleAmount(int am);                           // Java 1.1
```

L.2.36 Class *java.awt.TextArea*

The `TextArea` class allows for a multi-line area for displaying text. The following are defined:

```
// Constructors
public TextArea();
public TextArea(int rows, int cols);
public TextArea(String text);
public TextArea(String text, int rows, int cols);
public TextArea(String text, int rows, int cols, int scrollbars);

                                                                // Java 1.1
// Constants
public static final int SCROLLBARS_BOTH;                        // Java 1.1
public static final int SCROLLBARS_HORIZONTAL_ONLY;             // Java 1.1
```

```
public static final int SCROLLBARS_NONE;                              // Java 1.1
public static final int SCROLLBARS_VERTICAL_ONLY;                     // Java 1.1

// Methods
public void addNotify();
public synchronized void append(String str);                          // Java 1.1
public void appendText(String str);                                   // Java 1.0
public int getColumns();
public Dimension getMinimumSize(int rows, int cols);                  // Java 1.1
public Dimension getMinimumSize();                                    // Java 1.1
public Dimension getPreferredSize(int rows, int cols);                // Java 1.1
public Dimension getPreferredSize();                                  // Java 1.1
public int getRows();
public int getScrollbarVisibility();                                  // Java 1.1
public void insertText(String str, int pos);                          // Java 1.1
public Dimension minimumSize();                                       // Java 1.0
public Dimension minimumSize(int rows, int cols);                     // Java 1.0
protected String paramString();
public Dimension preferredSize();                                     // Java 1.0
public Dimension preferredSize(int rows, int cols);                   // Java 1.0
public void replaceText(String str, int start, int end);              // Java 1.0
public void setColumns(int cols);                                     // Java 1.1
public void setRows(int rows);                                        // Java 1.1
```

L.2.37 Class *java.awt.TextComponent*

The `TextComponent` class is the superclass of any component that allows the editing of some text. The following are defined:

```
// Methods
public void addTextListener(TextListener l);                          // Java 1.1
public int getCaretPosition();                                        // Java 1.1
public String getSelectedText();
public int getSelectionEnd();
public int getSelectionStart();
public String getText();
public boolean isEditable();
protected String paramString();
public void removeNotify();
public void removeTextListener(TextListener l);                       // Java 1.1
public void select(int selStart, int selEnd);
public void selectAll();
public void setCaretPosition(int position);                           // Java 1.1
public void setEditable(boolean t);
public synchronized setSelectionEnd(int selectionEnd);                // Java 1.1
public synchronized setSelectionStart(int selectionStart);            // Java 1.1
public void setText(String t);
```

L.2.38 Class *java.awt.TextField*

The `TextField` class is a component that presents the user with a single editable line of text. The following are defined:

```
// Constructors
public TextField();
public TextField(int cols);
public TextField(String text);
public TextField(String text, int cols);

// Methods
public synchronized void addActionListener(ActionListener l);         // Java 1.1
```

```
public void addNotify();
public boolean echoCharIsSet();
public int getColumns();
public char getEchoChar();
public Dimension getMinimumSize(int cols);                    // Java 1.1
public Dimension getMinimumSize();                            // Java 1.1
public Dimension getPreferredSize(int cols);                  // Java 1.1
public Dimension getPreferredSize();                          // Java 1.1
public Dimension minimumSize();                               // Java 1.0

public Dimension minimumSize(int cols);                       // Java 1.0
protected String paramString();
public Dimension preferredSize();                             // Java 1.0
public Dimension preferredSize(int cols);                     // Java 1.0
public void setColumns(int cols);                             // Java 1.1
public void setEchoChar(char c);                              // Java 1.1
public void setEchoCharacter(char c);                         // Java 1.0
```

L.2.39 Class *java.awt.Toolkit*

The `Toolkit` class is the abstract superclass of all actual implementations of the Abstract Window Toolkit. The following are defined:

```
// Constructors
public Toolkit();

// Methods
public abstract int beep();                                   // Java 1.1
public abstract int checkImage(Image image, int width,
                               int height, ImageObserver observer);
public abstract Image createImage(ImageProducer producer);
public Image createImage(byte[] imagedatea);                  // Java 1.1
public Image createImage(byte[] imagedata, int imageoffset,
       int imagelength);                                      // Java 1.1
public abstract ColorModel getColorModel();
public static Toolkit getDefaultToolkit();
public abstract String[] getFontList();
public abstract FontMetrics getFontMetrics(Font font);
public abstract Image getImage(String filename);
public abstract Image getImage(URL url);
public int getMenuShortcutKeyMask();                          // Java 1.1
public abstract PrintJob getPrintJob(Frame frame, String jobtitle,
       Properties props);                                     // Java 1.1
public abstract int getScreenResolution();
public abstract Dimension getScreenSize();
public abstract Clipboard getSystemClipbaord();               // Java 1.1
public abstract EventQueue getSystemEventQueue();             // Java 1.1
public abstract boolean prepareImage(Image image, int width,
       int height, ImageObserver observer);
public abstract void sync();
```

L.2.40 Class *java.awt.Window*

The `Window` class is the top-level window; it has no borders and no menu bar. The following are defined:

```
// Constructors
public Window(Frame parent);
// Methods
public void addNotify();
public synchronized void addWindowListener(WindowListener l);       // Java 1.1
```

```
public void dispose();
public Component getFocusOwner();                          // Java 1.1
public Locale getLocale();                                 // Java 1.1
public Toolkit getToolkit();
public final String getWarningString();
public boolean isShowing();                                // Java 1.1
public void pack();
public postEvent(Event e);                                 // Java 1.1
public synchronized void removeWindowListener(WindowListener l);
public void show();
public void toBack();
public void toFront();
```

L.3 Package **java.awt.datatransfer**

L.3.1 Class ***java.awt.datatransfer.Clipboard***

The `Clipboard` class has been added with Java 1.1. It represents a clipboard onto which data can be transferred using cut-and-paste techniques. The following are defined:

```
// Constructors
public Clipboard(String name);                             // Java 1.1

// Methods
public synchronized Transferable getContents(Object requestor);
                                                           // Java 1.1
public String getName();                                   // Java 1.1
public synchronized void setContents(Transferable contents,
       Clipboard owner);                                   // Java 1.1
```

L.4 Package **java.awt.event**

L.4.1 Class ***java.awt.event.ActionEvent***

The `ActionEvent` class has been added with Java 1.1. It occurs when a event happens for a `Button`, `List`, `MenuItem` or `TextField`. The following are defined:

```
// Constructors
public ActionEvent(Object src, String cmd);                // Java 1.1

// Methods
public String getActionCommand();                          // Java 1.1
public int getModifiers();                                 // Java 1.1
public int paramString();                                  // Java 1.1
```

L.4.2 Interface ***java.awt.event.ActionListener***

The `ActionListener` interface has been added with Java 1.1. It defines the method which is called by an `ActionEvent`. The following is defined:

```
public void actionPerformed(ActionEvent e);                // Java 1.1
```

L.4.3 Class ***java.awt.event.AdjustmentEvent***

The `AdjustmentEvent` class has been added with Java 1.1. It occurs when a event happens for a `Scrollbar`. The following are defined:

```
// Constructors
public AdjustmentEvent(Object src, int id, int type, int value);    // Java 1.1

// Methods
public Adjustable getAdjustable();                                  // Java 1.1
public int getAdjustmentType();                                     // Java 1.1
public int getValue();                                              // Java 1.1
public String paramString();                                        // Java 1.1
```

L.4.4 Class ***java.awt.event.AdjustmentListener***

The `AdjustmentListener` interface has been added with Java 1.1. It defines the method which is called by an `AdjustmentEvent`. The following is defined:

```
public void adjustmentValueChanged(AdjustmentEvent e);              // Java 1.1
```

L.4.5 Class ***java.awt.event.ComponentEvent***

The `ComponentEvent` class has been added with Java 1.1. It occurs when a event happens for a `Component`. The following are defined:

```
// Constructors
public ComponentEvent(Object src, int id, int type, int value);

// Methods
public Component getComponent();                                    // Java 1.1
public String paramString();                                        // Java 1.1
```

L.4.6 Class ***java.awt.event.ComponentListener***

The `ComponentListener` interface has been added with Java 1.1. It defines the method which is called by a `ComponentEvent`. The following are defined:

```
public void componentHidden(ComponentEvent e);                      // Java 1.1
public void componentMoved(ComponentEvent e);                       // Java 1.1
public void componentResized(ComponentEvent e);                     // Java 1.1
public void componentShown(ComponentEvent e);                       // Java 1.1
```

L.4.7 Class ***java.awt.event.ContainerEvent***

The `ComponentEvent` class has been added with Java 1.1. It occurs when a event happens for a `Container`. The following are defined:

```
// Constructors
public ContainerEvent(Component src, int id, Compoent child);

// Methods
public Component getChild();                                        // Java 1.1
public Component getContainer();                                    // Java 1.1
public String paramString();                                        // Java 1.1
```

L.4.8 Class ***java.awt.event.ContainerListener***

The `ContainerListener` interface has been added with Java 1.1. It defines the method which is called by a ContainerEvent. The following are defined:

```
public void componentAdded(ComponentEvent e);                // Java 1.1
public void componentRemoved(ComponentEvent e);              // Java 1.1
```

L.4.9 Class *java.awt.event.ItemEvent*

The ItemEvent class has been added with Java 1.1. It occurs when a event happens for a Container. The following are defined:

```
// Constructors
public ItemEvent(ItemSelectable src, int id, Object item, int stateChanged);
                                                             // Java 1.1

// Methods
public Object getItem();                                     // Java 1.1
public ItemSelectable getItemSelectable();                   // Java 1.1
public int getStateChange();                                 // Java 1.1
public String paramString();                                 // Java 1.1
```

L.4.10 Class *java.awt.event.ItemListener*

The ItemListener interface has been added with Java 1.1. It defines the method which is called by an ItemEvent. The following is defined:

```
public void itemStateChanged(ItemEvent e);                   // Java 1.1
```

L.4.11 Class *java.awt.event.KeyEvent*

The KeyEvent class has been added with Java 1.1. It occurs when a event happens for a keypress. The following are defined:

```
// Constructors
public KeyEvent(Component src, int id, long when, int modifiers,
      int keyCode, char keyChar);                            // Java 1.1

// Constants
public static final int KEY_LAST, KEY_PRESSED, KEY_RELEASED, KEY_TYPED;
                              // Undefined Key and Character (Java 1.1)
public static final int VK_UNDEFINED, CHAR_UNDEFINED;
                                     // Alphanumeric keys (Java 1.1)
public static final int VK_A, VK_B, VK_C, VK_D, VK_E, VK_F, VK_G, VK_H;
public static final int VK_I, VK_J, VK_K, VK_L, VK_M, VK_N, VK_O, VK_P;
public static final int VK_Q, VK_R, VK_S, VK_T, VK_U, VK_V, VK_W, VK_X;
public static final int VK_Y, VK_Z;
public static final int VK_SPACE;
public static final int VK_0, VK_1, VK_2, VK_3, VK_4, VK_5, VK_6, VK_7;
public static final int VK_8, VK_9;
public static final int VK_NUMPAD0, VK_NUMPAD1, VK_NUMPAD2, VK_NUMPAD3;
public static final int VK_NUMPAD4, VK_NUMPAD5, VK_NUMPAD6, VK_NUMPAD7;
public static final int VK_NUMPAD8, VK_NUMPAD9;
                                         // Control keys (Java 1.1)
public static final int VK_BACK_SPACE, VK_ENTER, VK_ESCAPE, VK_TAB;
                                        // Modifier keys (Java 1.1)
public static final int VK_ALT, VK_CAPS_LOCK, VK_CONTROL, VK_META, VK_SHIFT;
                                        // Function keys (Java 1.1)
public static final int VK_F0, VK_F1, VK_F2, VK_F3, VK_F4, VK_F5, VK_F6;
public static final int VK_F7, VK_F8, VK_F9;
public static final int VK_PRINTSCREEN, VK_SCROLL_LOCK, VK_PAUSE;
public static final int VK_PAGE_DOWN, VK_PAGE_UP;
```

```
public static final int VK_DOWN, VK_UP, VK_RIGHT, VK_LEFT;
public static final int VK_END, VK_HOME, VK_ACCEPT, VK_NUM_LOCK, VK_CANCEL;
public static final int VK_CLEAR, VK_CONVERT, VK_FINAL, VK_HELP;
public static final int VK_KANA, VK_KANJI, VK_MODECHANGE, VK_NONCONVERT;
                                          // Punctuation keys (Java 1.1)
public static final int VK_ADD, VK_BACK_QUOTE, VK_BACK_SLASH;
public static final int VK_CLOSE_BRACKET, VK_COMMA, VK_DECIMAL;
public static final int VK_DIVIDE, VK_EQUALS, VK_MULTIPLY;
public static final int VK_OPEN_BRACKET, VK_PERIOD, VK_QUOTE;
public static final int VK_SEMICOLON, VK_SEPARATER, VK_SLASH;
public static final int VK_SUBTRACT;

// Methods
public void getKeyChar();                                     // Java 1.1

public int getKeyCode();                                      // Java 1.1
public boolean isActionKey();                                 // Java 1.1
public String paramString();                                  // Java 1.1
public void setKeyChar(char keyChar);                         // Java 1.1
public void setKeyCode(int keyCode);                          // Java 1.1
public void setModifiers(int modifiers);                      // Java 1.1
```

L.4.12 Class ***java.awt.event.KeyListener***

The `KeyListener` interface has been added with Java 1.1. It defines the method which is called by a `KeyEvent`. The following is defined:

```
public void keyPressed(KeyEvent e);                           // Java 1.1
public void keyReleased(KeyEvent e);                          // Java 1.1
public void keyTyped(KeyEvent e);                             // Java 1.1
```

L.4.13 Class ***java.awt.event.MouseEvent***

The `MouseEvent` class has been added with Java 1.1. It occurs when a event happens for a `MouseEvent`. The following are defined:

```
// Constructors
public MouseEvent(Component src, int id, long when, int modifiers, int x,
       int y, intclickCount, boolean popupTrigger);           // Java 1.1

// Constants
public static final int MOUSE_CLICKED, MOUSE_DRAGGED;
public static final int MOUSE_ENTERED, MOUSE_EXITED;
public static final int MOUSE_FIRST, MOUSE_LAST;
public static final int MOUSE_MOVED, MOUSE_PRESSED;
public static final int MOUSE_RELEASED;

// Methods
public int getClickCount();                                   // Java 1.1

public Point getPoint();                                      // Java 1.1
public int getX();                                            // Java 1.1
public int getY();                                            // Java 1.1
public boolean isPopupTrigger();                              // Java 1.1
public String paramString();                                  // Java 1.1
public synchronized void translatePoint(int x, int y);        // Java 1.1
```

L.4.14 Class ***java.awt.event.MouseListener***

The `MouseListener` interface has been added with Java 1.1. It defines the method which is called by a mouse click event. The following are defined:

```
public void mouseClicked(MouseEvent e);                    // Java 1.1
public void mouseEntered(MouseEvent e);                    // Java 1.1
public void mouseExited(MouseEvent e);                     // Java 1.1
public void mousePressed(MouseEvent e);                    // Java 1.1
public void mouseReleased(MouseEvent e);                   // Java 1.1
```

L.4.15 Class *java.awt.event.MouseMouseListener*

The `MouseMouseListener` interface has been added with Java 1.1. It defines the method which is called by a mouse drag or move event. The following are defined:

```
public void mouseDragged(MouseEvent e);                    // Java 1.1
public void mouseMoved(MouseEvent e);                      // Java 1.1
```

L.4.16 Class *java.awt.eventTextEvent*

The `TextEvent` class has been added with Java 1.1. It occurs when a event happens for an event within `TextField`, `TextArea` or other `TextComponent`. The following are defined:

```
// Constructors
public TextEvent(Object src, int id);                      // Java 1.1

// Constants
public static final int TEXT_FIRST, TEXT_LAST;
public static final int TEXT_VALUE_CHANGED;

// Methods
public String paramString();                               // Java 1.1
```

L.4.17 Class *java.awt.event.TextListener*

The `TextListener` interface has been added with Java 1.1. It defines the method which is called by a `TextEvent`. The following is defined:

```
public void textValueChanged(TextEvent e);                 // Java 1.1
```

L.4.18 Class *java.awt.eventWindowEvent*

The `WindowEvent` class has been added with Java 1.1. It occurs when an event happens within a `Window` object. The following are defined:

```
// Constructors
public WindowEvent(Window src, int id);                    // Java 1.1

// Constants
public static final int WINDOW_ACTIVATED, WINDOW_CLOSED;
public static final int WINDOW_CLOSING, WINDOW_DEACTIVATED;
public static final int WINDOW_DEICONIFIED, WINDOW_FIRST;
public static final int WINDOW_ICONIFIED, WINDOW_LAST;
public static final int WINDOW_OPENED;

// Methods
public Window getWindow();                                 // Java 1.1
public String paramString();                               // Java 1.1
```

L.4.19 Class *java.awt.event.WindowListener*

The `WindowListener` interface has been added with Java 1.1. It defines the method which is called by an `WindowEvent`. The following are defined:

```
public void windowActivated(WindowEvent e);                // Java 1.1
public void windowClosed(WindowEvent e);                   // Java 1.1
public void windowDeactivated(WindowEvent e);              // Java 1.1
public void windowDeiconified(WindowEvent e);              // Java 1.1
public void windowIconified(WindowEvent e);                // Java 1.1
public void windowOpened(WindowEvent e);                   // Java 1.1
```

L.5 Package **java.awt.image**

This package has been added with Java 1.1 and supports image processing classes.

L.6 Package **java.io**

L.6.1 Class *java.io.BufferedOutputStream*

The `BufferedOutputStream` implements a buffered output stream. These streams allow the program to write to an input device without having to worry about the interfacing method. The following are defined:

```
// Fields
protected byte buf[];
protected int count;

// Constructors
public BufferedOutputStream(OutputStream out);
public BufferedOutputStream(OutputStream out, int size);

// Methods
public void flush();
public void write(byte b[], int off, int len);
public void write(int b);
```

L.6.2 Class *java.io.BufferedReader*

The `BufferReader` class has been added with Java 1.1. It represents a buffered character input stream. The following are defined:

```
// Constructors
public BufferedReader(Reader in, int sz);                  // Java 1.1
public BufferedReader(Reader in);                          // Java 1.1

// Methods
public void close() throws IOException;                    // Java 1.1
public void mark(int readAheadLimit) throws IOException;   // Java 1.1
public boolean markSupported() throws IOException;         // Java 1.1
public int read() throws IOException;                      // Java 1.1
public int read(char [] cbuf, int off, int len) throws IOException; // Java 1.1
public String readLine() throws IOException;               // Java 1.1
public boolean ready() throws IOException;                 // Java 1.1
public void reset() throws IOException;                    // Java 1.1
```

```
public long skip(long n) throws IOException;                    // Java 1.1
```

L.6.3 Class *java.io.BufferedWriter*

The `BufferWriter` class has been added with Java 1.1. It represents a buffered character output stream. The following are defined:

```
// Constrsuctors
public BufferedWriter(Writer out, int sz);                      // Java 1.1
public BufferedWriter(Writer in);                               // Java 1.1

// Methods
public void close() throws IOException;                         // Java 1.1
public void flush() throws IOException;                         // Java 1.1
public void newLine() throws IOException;                       // Java 1.1
public void write(int c) throws IOException;                    // Java 1.1
public void write(char [] cbuf, int off, int len) throws IOException;
                                                                // Java 1.1
```

L.6.4 Class *java.io.ByteArrayInputStream*

The `ByteArrayInputStream` class supports input from a byte array. The following are defined:

```
// Fields
protected byte buf[];
protected int count;
protected int mark;                                             // Java 1.1
protected int pos;

// Constructors
public ByteArrayInputStream(byte buf[]);
public ByteArrayInputStream(byte buf[], int offset, int length);

// Methods
public int available();
public void mark(int markpos);                                  // Java 1.1
public boolean markSupported();                                 // Java 1.1
public int read();
public int read(byte b[], int off, int len);
public void reset();
public long skip(long n);
```

L.6.5 Class *java.io.ByteArrayOutputStream*

The `ByteArrayOutputStream` class allows supports output to a byte array. The following are defined:

```
// Fields
protected byte buf[];
protected int count;

// Constructors
public ByteArrayOutputStream();
public ByteArrayOutputStream(int size);

// Methods
public void reset();
public int size();
public byte[] toByteArray();
```

```
public String toString();
public String toString(int hibyte);                          // Java 1.0
public String toString(String enc);                          // Java 1.1
public void write(byte b[], int off, int len);
public void write(int b);
public void writeTo(OutputStream out);
```

L.6.6 Interface *java.io.DataInput*

The `DataInput` interface gives support for streams to read in a machine-independent way. The following are defined:

```
// Methods
public abstract boolean readBoolean();
public abstract byte readByte();
public abstract char readChar();
public abstract double readDouble();
public abstract float readFloat();
public abstract void readFully(byte b[]);
public abstract void readFully(byte b[], int off, int len);
public abstract int readInt();
public abstract String readLine();
public abstract long readLong();
public abstract short readShort();
public abstract int readUnsignedByte();
public abstract int readUnsignedShort();
public abstract String readUTF();
public abstract int skipBytes(int n);
```

L.6.7 Class *java.io.DataInputStream*

The `DataInputStream` class allows an application to read data in a machine-independent way. It uses standard Unicode strings which conforms to the UTF-81 specification. The following are defined:

```
// Constructors
public DataInputStream(InputStream in);

// Methods
public final int read(byte b[]);
public final int read(byte b[], int off, int len);
public final boolean readBoolean();
public final byte readByte();
public final char readChar();
public final double readDouble();
public final float readFloat();
public final void readFully(byte b[]);
public final void readFully(byte b[], int off, int len);
public final int readInt();
public final String readLine();                              // Java 1.0
public final long readLong();
public final short readShort();
public final int readUnsignedByte();
public final int readUnsignedShort();
public final String readUTF();
public final static String readUTF(DataInput in);
public final int skipBytes(int n);
```

L.6.8 Interface *java.io.DataOutput*

The `DataOutput` interface gives support for streams to write in a machine-independent way.

The following are defined:

```
// Methods
public abstract void write(byte b[]);
public abstract void write(byte b[], int off, int len);
public abstract void write(int b);
public abstract void writeBoolean(boolean v);
public abstract void writeByte(int v);
public abstract void writeBytes(String s);
public abstract void writeChar(int v);
public abstract void writeChars(String s);
public abstract void writeDouble(double v);
public abstract void writeFloat(float v);
public abstract void writeInt(int v);
public abstract void writeLong(long v);
public abstract void writeShort(int v);
public abstract void writeUTF(String str);
```

L.6.9 Class *java.io.DataOutputStream*

The `DataOutputStream` class allows an application to write data in a machine-independent way. It uses standard Unicode strings which conforms to the UTF-81 specification. The following are defined:

```
// Fields
protected int written;
// Constructors
public DataOutputStream(OutputStream out);

// Methods
public void flush();
public final int size();
public void write(byte b[], int off, int len);
public void write(int b);
public final void writeBoolean(boolean v);
public final void writeByte(int v);
public final void writeBytes(String s);
public final void writeChar(int v);
public final void writeChars(String s);
public final void writeDouble(double v);
public final void writeFloat(float v);
public final void writeInt(int v);
public final void writeLong(long v);
public final void writeShort(int v);
public final void writeUTF(String str);
```

L.6.10 Class *java.io.EOFException*

Exception that identifies that the end-of-file has been reached unexpectedly during input. The following are defined:

```
// Constructors
public EOFException();
public EOFException(String s);
```

L.6.11 Class *java.io.File*

The `File` class implements the file manipulation operations in an operating system independent way. The following are defined:

```
// Fields
public final static String pathSeparator;
public final static char pathSeparatorChar;
public final static String separator;
public final static char separatorChar;

// Constructors
public File(File dir, String name);
public File(String path);
public File(String path, String name);

// Methods
public boolean canRead();
public boolean canWrite();
public boolean delete();
public boolean equals(Object obj);
public boolean exists();
public String getAbsolutePath();
public String getCanonicalPath();                         // Java 1.1
public String getName();
public String getParent();
public String getPath();
public int hashCode();
public boolean isAbsolute();
public boolean isDirectory();
public boolean isFile();
public long lastModified();
public long length();
public String[] list();
public String[] list(FilenameFilter filter);
public boolean mkdir();
public boolean mkdirs();
public boolean renameTo(File dest);
public String toString();
```

L.6.12 Class *java.io.FileDescriptor*

The `FileDescriptor` class provides a way to cope with opening files or sockets. The following are defined:

```
// Fields
public final static FileDescriptor err, in, out;

// Constructors
public FileDescriptor();

// Methods
public void sync();                                       // Java 1.1
public boolean valid();
```

L.6.13 Class *java.io.FileInputStream*

The `FileInputStream` class provides supports for an input file. The following are defined:

```
// Constructors
public FileInputStream(File file);
public FileInputStream(FileDescriptor fdObj);
public FileInputStream(String name);

// Methods
```

```
public int available();
public void close();
protected void finalize();
public final FileDescriptor getFD();
public int read();
public int read(byte b[]);
public int read(byte b[], int off, int len);
public long skip(long n);
```

L.6.14 Interface *java.io.FilenameFilter*

The `FilenameFile` interface is used to filter filenames. The following is defined:

```
// Methods
public abstract boolean accept(File dir, String name);
```

L.6.15 Class *java.io.FileNotFoundException*

Exception that identifies that a file could not be found. The following are defined:

```
// Constructors
public FileNotFoundException();
public FileNotFoundException(String s);
```

L.6.16 Class *java.io.FileOutputStream*

The `FileOutputStream` class provides supports for an output file. The following are defined:

```
// Constructors
public FileOutputStream(File file);
public FileOutputStream(String name, boolean append);            // Java 1.1
public FileOutputStream(FileDescriptor fdObj);
public FileOutputStream(String name);

// Methods
public void close();
protected void finalize();
public final FileDescriptor getFD();
public void write(byte b[]);
public void write(byte b[], int off, int len);
public void write(int b);
```

L.6.17 Class *java.io.FilterInputStream*

The `FilterInputStream` class is the superclass of all classes that filter input streams. The following are defined:

```
// Fields
protected InputStream in;

// Constructors
protected FilterInputStream(InputStream in);

// Methods
public int available();
public void close();
public void mark(int readlimit);
public boolean markSupported();
public int read();
public int read(byte b[]);
```

```
public int read(byte b[], int off, int len);
public void reset();
public long skip(long n);
```

L.6.18 Class *java.io.FilterOutputStream*

The `FilterOutputStream` class is the superclass of all classes that filter output streams. The following are defined:

```
// Fields
protected OutputStream out;

// Constructors
public FilterOutputStream(OutputStream out);

// Methods
public void close();
public void flush();
public void write(byte b[]);
public void write(byte b[], int off, int len);
public void write(int b);
```

L.6.19 Class *java.io.InputStream*

The `InputStream` class is the superclass of all classes representing an input stream of bytes. The following are defined:

```
// Constructors
public InputStream();

// Methods
public int available();
public void close();
public void mark(int readlimit);
public boolean markSupported();
public abstract int read();
public int read(byte b[]);
public int read(byte b[], int off, int len);
public void reset();
public long skip(long n);
```

L.6.20 Class *java.io.InterruptedIOException*

Exception that identifies that an I/O operation has been interrupted. The following are defined:

```
// Fields
public int bytesTransferred;

// Constructors
public InterruptedIOException();
public InterruptedIOException(String s);
```

L.6.21 Class *java.io.IOException*

Exception that identifies that an I/O exception has occurred. The following are defined:

```
// Constructors
public IOException();
public IOException(String s);
```

L.6.22 Class *java.io.LineNumberInputStream*

The `LineNumberInputStream` class provides support for the current line number in an input stream. Each line is delimited by either a carriage return character ('\r'), new-line character ('\n') or both together. The following are defined:

```
// Constructors
public LineNumberInputStream(InputStream in);

// Methods
public int available();
public int getLineNumber();
public void mark(int readlimit);
public int read();
public int read(byte b[], int off, int len);
public void reset();
public void setLineNumber(int lineNumber);
public long skip(long n);
```

L.6.23 Class *java.io.OutputStream*

The `InputStream` class is the superclass of all classes representing an output stream of bytes. The following are defined:

```
// Constructors
public OutputStream();

// Methods
public void close();
public void flush();
public void write(byte b[]);
public void write(byte b[], int off, int len);
public abstract void write(int b);
```

L.6.24 Class *java.io.PipedInputStream*

The `PipedInputStream` class provides support for pipelined input communications. The following are defined:

```
// Constructors
public PipedInputStream();
public PipedInputStream(PipedOutputStream src);

// Methods
public void close();
public void connect(PipedOutputStream src);
public int read();
public int read(byte b[], int off, int len);
```

L.6.25 Class *java.io.PipedOutputStream*

The `PipedOutputStream` class provides support for pipelined output communications. The following are defined:

```
// Constructors
public PipedOutputStream();
public PipedOutputStream(PipedInputStream snk);

// Methods
public void close();
```

```
public void connect(PipedInputStream snk);
public void write(byte b[], int off, int len);
public void write(int b);
```

L.6.26 Class *java.io.PrintStream*

The `PrintStream` class provides support for output print streams. The following are defined:

```
// Constructors
public PrintStream(OutputStream out);                          // Java 1.0
public PrintStream(OutputStream out, boolean autoflush);       // Java 1.0

// Methods
public boolean checkError();
public void close();
public void flush();
public void print(boolean b);
public void print(char c);
public void print(char s[]);
public void print(double d);
public void print(float f);
public void print(int i);
public void print(long l);
public void print(Object obj);
public void print(String s);
public void println();
public void println(boolean b);
public void println(char c);
public void println(char s[]);
public void println(double d);
public void println(float f);
public void println(int i);
public void println(long l);
public void println(Object obj);
public void println(String s);
public void write(byte b[], int off, int len);
public void write(int b);
```

L.6.27 Class *java.io.PushbackInputStream*

The `PushbackInputStream` class provides support to put bytes back into an input stream. The following are defined:

```
// Fields
protected int pushBack;

// Constructors
public PushbackInputStream(InputStream in);

// Methods
public int available();
public boolean markSupported();
public int read();
public int read(byte bytes[], int offset, int length);
public void unread(int ch);
```

L.6.28 Class *java.io.RandomAccessFile*

The `RandomAccessFile` class support reading and writing from a random access file. The following are defined:

```
// Constructors
public RandomAccessFile(File file, String mode);
public RandomAccessFile(String name, String mode);

// Methods
public void close();
public final FileDescriptor getFD();
public long getFilePointer();
public long length();
public int read();
public int read(byte b[]);
public int read(byte b[], int off, int len);
public final boolean readBoolean();
public final byte readByte();
public final char readChar();
public final double readDouble();
public final float readFloat();
public final void readFully(byte b[]);
public final void readFully(byte b[], int off, int len);
public final int readInt();
public final String readLine();
public final long readLong();
public final short readShort();
public final int readUnsignedByte();
public final int readUnsignedShort();
public final String readUTF();
public void seek(long pos);
public int skipBytes(int n);
public void write(byte b[]);
public void write(byte b[], int off, int len);
public void write(int b);
public final void writeBoolean(boolean v);
public final void writeByte(int v);
public final void writeBytes(String s);
public final void writeChar(int v);
public final void writeChars(String s);
public final void writeDouble(double v);
public final void writeFloat(float v);
public final void writeInt(int v);
public final void writeLong(long v);
public final void writeShort(int v);
public final void writeUTF(String str);
```

L.6.29 Class *java.io.SequenceInputStream*

The `SequenceInputStream` supports the combination of several input streams into a single input stream. The following are defined:

```
// Constructors
public SequenceInputStream(Enumeration e);
public SequenceInputStream(InputStream s1, InputStream s2);

// Methods
public void avialable();                                    // Java 1.1
public void close();
public int read();
public int read(byte buf[], int pos, int len);
```

L.6.30 Class *java.io.StreamTokenizer*

The `StreamTokenizer` class splits an input stream into tokens. These tokens can be defined by number, quotes strings or comment styles. The following are defined:

```
// Fields
public double nval;
public String sval;
public int ttype;

// possible values for the ttype field
public final static int TT_EOF, TT_EOL, TT_NUMBER, TT_WORD;

// Constructors
public StreamTokenizer(InputStream I);

// Methods
public void commentChar(int ch);
public void eolIsSignificant(boolean flag);
public int lineno();
public void lowerCaseMode(boolean fl);
public int nextToken();
public void ordinaryChar(int ch);
public void ordinaryChars(int low, int hi);
public void parseNumbers();
public void pushBack();
public void quoteChar(int ch);
public void resetSyntax();
public void whitespaceChars(int low, int hi);
public void slashStarComments(boolean flag);
public String toString();
public void whitespaceChars(int low, int hi);
public void wordChars(int low, int hi);
```

L.6.31 Class *java.io.StringBufferInputStream*

The `StringBufferInputStream` class supports stream input buffers. The following are defined:

```
// Fields
protected String buffer;
protected int count, pos;

// Constructors
public StringBufferInputStream(String s);

// Methods
public int available();
public int read();
public int read(byte b[], int off, int len);
public void reset();
public long skip(long n);
```

L.6.32 Class *java.io.UTFDataFormatException*

Exception that identifies that a malformed UTF-8 string has been read in a data input stream. The following are defined:

```
// Constructors
public UTFDataFormatException();
public UTFDataFormatException(String s);
```

L.7 Package **java.lang**

L.7.1 Class ***java.lang.ArithmeticException***

Exception that is thrown when an exceptional arithmetic condition has occurred, such as a division-by-zero or a square root of a negative number. The following are defined:

```
// Constructors
public ArithmeticException();
public ArithmeticException(String s);
```

L.7.2 Class ***java.lang.ArrayIndexOutOfBoundsException***

Exception that is thrown when an illegal index term in an array has been accessed. The following are defined:

```
// Constructors
public ArrayIndexOutOfBoundsException();
public ArrayIndexOutOfBoundsException(int index);
public ArrayIndexOutOfBoundsException(String s);
```

L.7.3 Class ***java.lang.ArrayStoreException***

Exception that is thrown when the wrong type of object is stored in an array of objects. The following are defined:

```
// Constructors
public ArrayStoreException();
public ArrayStoreException(String s);
```

L.7.4 Class ***java.lang.Boolean***

The `Boolean` class implements the primitive type boolean of an object. Other methods are included for a converting a boolean to a String and vice versa. The following are defined:

```
public final static Boolean FALSE, TRUE;
public final static Boolean TYPE;                          // Java 1.1

// Constructors
public Boolean(boolean value);
public Boolean(String s);

// Methods
public boolean booleanValue();
public boolean equals(Object obj);
public static boolean getBoolean(String name);
public int hashCode();
public String toString();
public static Boolean valueOf(String s);
```

L.7.5 Class ***java.lang.Character***

The `Character` class implements the primitive type character of an object. Other methods are defined for determining the type of a character, and converting characters from uppercase to lowercase and vice versa. The following are defined:

```
// Constants
public final static int MAX_RADIX, MAX_VALUE;
public final static int MIN_RADIX, MIN_VALUE;
public final static int TYPE;                                  // Java 1.1
// Character type constants
public final static byte COMBINING_SPACE_MARK;                 // Java 1.1
public final static byte CONNECTOR_PUNCUATION, CONTROL;        // Java 1.1
public final static byte CURRENCY_SYMBOL, DASH_PUNCTUATION;    // Java 1.1
public final static byte DIGIT_NUMBER, ENCLOSING_MARK;         // Java 1.1
public final static byte END_PUNCTUATION, FORMAT;              // Java 1.1
public final static byte LETTER_NUMBER, LINE_SEPERATOR;        // Java 1.1
public final static byte LOWERCASE_LETTER, MATH_SYMBOL;        // Java 1.1
public final static byte MODIFIER_LETTER, MODIFIER_SYMBOL;     // Java 1.1

public final static byte NON_SPACING_MARK, OTHER_LETTER;       // Java 1.1
public final static byte OTHER_NUMBER, OTHER_PUNCTUATION;      // Java 1.1
public final static byte OTHER_SYMBOL, PARAGRAPH_SEPARATOR;    // Java 1.1
public final static byte PRIVATE_USE, SPACE_SEPARATOR;         // Java 1.1
public final static byte START_PUNCTUATION, SURROGATE;         // Java 1.1
public final static byte TITLECASE_LETTER, UNASSIGNED;         // Java 1.1
public final static byte UPPERCASE_LETTER;                     // Java 1.1

// Constructors
public Character(char value);

// Methods
public char charValue();
public static int digit(char ch, int radix);
public boolean equals(Object obj);
public static char forDigit(int digit, int radix);
public static char getNumericValue(char ch);                   // Java 1.1
public static char getType(char ch);                           // Java 1.1
public static boolean isDefined(char ch);
public static boolean isDigit(char ch);
public static boolean isISOControl(char ch);                   // Java 1.1
public static boolean isIdentifierIgnoreable(char ch);         // Java 1.1
public static boolean isJavaIndentierPart(char ch);            // Java 1.1
public static boolean isJavaIndentierStart(char ch);           // Java 1.1
public static boolean isJavaLetter(char ch);                   // Java 1.0
public static boolean isJavaLetterOrDigit(char ch);            // Java 1.0
public static boolean isLetter(char ch);
public static boolean isLetterOrDigit(char ch);
public static boolean isLowerCase(char ch);
public static boolean isSpace(char ch);                        // Java 1.0
public static boolean isSpaceChar(char ch);                    // Java 1.0
public static boolean isTitleCase(char ch);
public static boolean isUnicodeIdentifierPart(char ch);        // Java 1.1
public static boolean isUnicodeIdentifierStart(char ch);       // Java 1.1
public static boolean isUpperCase(char ch);
public static boolean isWhitespace(char ch);                   // Java 1.1
public static char toLowerCase(char ch);
public String toString();
public static char toTitleCase(char ch);
public static char toUpperCase(char ch);
```

L.7.6 Class *java.lang.*Class

The `Class` class implements the class Class and interfaces in a running Java application. The following are defined:

```
// Methods
public static Class forName(String className);
```

```
public ClassLoader getClassLoader();
public Class[] getInterfaces();
public String getName();
public Class getSuperclass();
public boolean isInterface();
public Object newInstance();
public String toString();
```

L.7.7 Class *java.lang.*Class*CastException*

Exception that is thrown when an object is casted to a subclass which it is not an instance. The following are defined:

```
// Constructors
public ClassCastException();
public ClassCastException(String s);
```

L.7.8 Class *java.lang.Compiler*

The Compiler class supports Java-to-native-code compilers and related services. The following are defined:

```
// Methods
public static Object command(Object any);
public static boolean compileClass(Class clazz);
public static boolean compileClasses(String string);
public static void disable();
public static void enable();
```

L.7.9 Class *java.lang.Double*

The `Double` class implements the primitive type double of an object. Other methods are included for a converting a double to a String and vice versa. The following are defined:

```
// Fields
public final static double MAX_VALUE, MIN_VALUE;
public final static double NaN, NEGATIVE_INFINITY, POSITIVE_INFINITY;
public final static double TYPE;                          // Java 1.1

// Constructors
public Double(double value);
public Double(String s);

// Methods
public static long doubleToLongBits(double value);
public double doubleValue();
public boolean equals(Object obj);
public float floatValue();
public int hashCode();
public int intValue();
public boolean isInfinite();
public static boolean isInfinite(double v);
public boolean isNaN();
public static boolean isNaN(double v);
public static double longBitsToDouble(long bits);
public long longValue();
public String toString();
public static String toString(double d);
public static Double valueOf(String s);
```

L.7.10 Class *java.lang.Error*

Exception that is thrown when there are serious problems that a reasonable application should not try to catch. The following are defined:

```
// Constructors
public Error();
public Error(String s);
```

L.7.11 Class *java.lang.Exception*

Exception that is thrown that indicates conditions that a reasonable application might want to catch.

```
// Constructors
public Exception();
public Exception(String s);
```

L.7.12 Class *java.lang.Float*

The `Float` class implements the primitive type float of an object. Other methods are included for a converting a float to a String and vice versa. The following are defined:

```
// Fields
public final static float MAX_VALUE MIN_VALUE;
public final static float NaN, NEGATIVE_INFINITY, POSITIVE_INFINITY;
public final static float TYPE;                                    // Java 1.1

// Constructors
public Float(double value);
public Float(float value);
public Float(String s);

// Methods
public double doubleValue();
public boolean equals(Object obj);
public static int floatToIntBits(float value);
public float floatValue();
public int hashCode();
public static float intBitsToFloat(int bits);
public int intValue();
public boolean isInfinite();
public static boolean isInfinite(float v);
public boolean isNaN();
public static boolean isNaN(float v);
public long longValue();
public String toString();
public static String toString(float f);
public static Float valueOf(String s);
```

L.7.13 Class *java.lang.IllegalAccessError*

Exception that is thrown when an application attempts to access or modify a field, or to call a method that it does not have access to. The following are defined:

```
// Constructors
public IllegalAccessError();
public IllegalAccessError(String s);
```

L.7.14 Class *java.lang.IllegalArgumentException*

Exception that is thrown when a method has been passed an illegal or inappropriate argument. The following are defined:

```
// Constructors
public IllegalArgumentException();
public IllegalArgumentException(String s);
```

L.7.15 Class *java.lang.IllegalThreadStateException*

Exception that is thrown to indicate that a thread is not in an appropriate state for the requested operation. The following are defined:

```
// Constructors
public IllegalThreadStateException();
public IllegalThreadStateException(String s);
```

L.7.16 Class *java.lang.IndexOutOfBoundsException*

Exception that is thrown to indicate that an index term is out of range. The following are defined:

```
// Constructors
public IndexOutOfBoundsException();
public IndexOutOfBoundsException(String s);
```

L.7.17 Class *java.lang.Integer*

The `Integer` class implements the primitive type integer of an object. Other methods are included for a converting a integer to a String and vice versa. The following are defined:

```
// Fields
public final static int MAX_VALUE, MIN_VALUE;
public final static int TYPE;                                // Java 1.1

// Constructors
public Integer(int value);
public Integer(String s);

// Methods
public Integer decode(String nm);                            // Java 1.1
public double doubleValue();
public boolean equals(Object obj);
public float floatValue();
public static Integer getInteger(String nm);
public static Integer getInteger(String nm, int val);
public static Integer getInteger(String nm, Integer val);
public int hashCode();
public int intValue();
public long longValue();
public static int parseInt(String s);
public static int parseInt(String s, int radix);
public static String toBinaryString(int i);
public static String toHexString(int i);
public static String toOctalString(int i);
public String toString();
public static String toString(int i);
public static String toString(int i, int radix);
public static Integer valueOf(String s);
```

```
public static Integer valueOf(String s, int radix);
```

L.7.18 Class *java.lang.InternalError*

Exception that is thrown when an unexpected internal error has occurs. The following are defined:

```
// Constructors
public InternalError();
public InternalError(String s);
```

L.7.19 Class *java.lang.InterruptedException*

Exception that is thrown when a thread is waiting, sleeping, or otherwise paused for a long time and another thread interrupts it using the interrupt method in class Thread. The following are defined:

```
// Constructors
public InterruptedException();
public InterruptedException(String s);
```

L.7.20 Class *java.lang.Long*

The `Long` class implements the primitive type long of an object. Other methods are included for a converting a long to a String and vice versa. The following are defined:

```
// Fields
public final static long MAX_VALUE, MIN_VALUE;
public final static long TYPE;                                // Java 1.1

// Constructors
public Long(long value);
public Long(String s);

// Methods
public double doubleValue();
public boolean equals(Object obj);
public float floatValue();
public static Long getLong(String nm);
public static Long getLong(String nm, long val);
public static Long getLong(String nm, Long val);
public int hashCode();
public int intValue();
public long longValue();
public static long parseLong(String s);
public static long parseLong(String s, int radix);
public static String toBinaryString(long i);
public static String toHexString(long i);
public static String toOctalString(long i);
public String toString();
public static String toString(long i);
public static String toString(long i, int radix);
public static Long valueOf(String s);
public static Long valueOf(String s, int radix);
```

L.7.21 Class *java.lang.Math*

The `Math` class contains methods to perform basic mathematical operations. The following are defined:

```
// Fields
public final static double E;
public final static double PI;

// Methods
public static double abs(double a);
public static float abs(float a);
public static int abs(int a);
public static long abs(long a);
public static double acos(double a);
public static double asin(double a);
public static double atan(double a);
public static double atan2(double a, double b);
public static double ceil(double a);
public static double cos(double a);
public static double exp(double a);
public static double floor(double a);
public static double IEEEremainder(double f1, double f2);
public static double log(double a);
public static double max(double a, double b);
public static float max(float a, float b);
public static int max(int a, int b);
public static long max(long a, long b);
public static double min(double a, double b);
public static float min(float a, float b);
public static int min(int a, int b);
public static long min(long a, long b);
public static double pow(double a, double b);
public static double random();
public static double rint(double a);
public static long round(double a);
public static int round(float a);
public static double sin(double a);
public static double sqrt(double a);
public static double tan(double a);
```

L.7.22 Class *java.lang.NegativeArraySizeException*

Exception that is thrown when an array is created with a negative size.

```
// Constructors
public NegativeArraySizeException();
public NegativeArraySizeException(String s);
```

L.7.23 Class *java.lang.NullPointerException*

Exception that is thrown when an application attempts to use a null pointer. The following are defined:

```
// Constructors
public NullPointerException();
public NullPointerException(String s);
```

L.7.24 Class *java.lang.Number*

The `Number` class contains the superclass of classes for float, double, integer and long. It can be used to convert values into int, long, float or double. The following are defined:

```
// Methods
public abstract double doubleValue();
public abstract float floatValue();
```

```
public abstract int intValue();
public abstract long longValue();
```

L.7.25 Class *java.lang.NumberFormatException*

Exception that is thrown when an application attempts to convert a string to one of the numeric types, but that the string does not have the appropriate format.

```
// Constructors
public NumberFormatException();
public NumberFormatException(String s);
```

L.7.26 Class *java.lang.Object*

The `Object` class contains the root of the class hierarchy. The following are defined:

```
// Constructors
public Object();

// Methods
protected Object clone();
public boolean equals(Object obj);
protected void finalize();
public final Class getClass();
public int hashCode();
public final void notify();
public final void notifyAll();
public String toString();
public final void wait();
public final void wait(long timeout);
public final void wait(long timeout, int nanos);
```

L.7.27 Class *java.lang.OutOfMemoryError*

Exception that is thrown when an application runs out of memory. The following are defined:

```
// Constructors
public OutOfMemoryError();
public OutOfMemoryError(String s);
```

L.7.28 Class *java.lang.Process*

The `Process` class contains methods which are used to control the process. The following are defined:

```
// Constructors
public Process();
// Methods
public abstract void destroy();
public abstract int exitValue();
public abstract InputStream getErrorStream();
public abstract InputStream getInputStream();
public abstract OutputStream getOutputStream();
public abstract int waitFor();
```

L.7.29 Class *java.lang.Runtime*

The `Runtime` class allows the application to interface with the environment in which it is running. The following are defined:

```
// Methods
public Process exec(String command);
public Process exec(String command, String envp[]);
public Process exec(String cmdarray[]);
public Process exec(String cmdarray[], String envp[]);
public void exit(int status);
public long freeMemory();
public void gc();
public InputStream getLocalizedInputStream(InputStream in);        // Java 1.0
public OutputStream getLocalizedOutputStream(OutputStream out);    // Java 1.0
public static Runtime getRuntime();
public void load(String filename);
public void loadLibrary(String libname);
public void runFinalization();
public long totalMemory();
public void traceInstructions(boolean on);
public void traceMethodCalls(boolean on);
```

L.7.30 Class *java.lang.SecurityManager*

The `SecurityManager` class is an abstract class that allows applications to determine if it is safe to execute a given operation. The following are defined:

```
// Fields
 protected boolean inCheck;

// Constructors
 protected SecurityManager();

// Methods
public void checkAccept(String host, int port);
public void checkAccess(Thread g);
public void checkAccess(ThreadGroup g);
public void checkConnect(String host, int port);
public void checkConnect(String host, int port, Object context);
public void checkCreateClassLoader();
public void checkDelete(String file);
public void checkExec(String cmd);
public void checkExit(int status);
public void checkLink(String lib);
public void checkListen(int port);
public void checkPackageAccess(String pkg);
public void checkPackageDefinition(String pkg);
public void checkPropertiesAccess();
public void checkPropertyAccess(String key);
public void checkRead(FileDescriptor fd);
public void checkRead(String file);
public void checkRead(String file, Object context);
public void checkSetFactory();
public boolean checkTopLevelWindow(Object window);
public void checkWrite(FileDescriptor fd);
public void checkWrite(String file);
protected int classDepth(String name);
protected int classLoaderDepth();
protected ClassLoader currentClassLoader();
protected Class[] getClassContext();
public boolean getInCheck();
public Object getSecurityContext();
protected boolean inClass(String name);
protected boolean inClassLoader();
```

L.7.31 Class *java.lang.StackOverflowError*

Exception that is thrown when a stack overflow occurs. The following are defined:

```
// Constructors
public StackOverflowError();
public StackOverflowError(String s);
```

L.7.32 Class *java.lang.String*

The `String` class represents character strings. As in C, a string is delimted by inverted commas. It contains string manipulation methods, such as `concat` (string concatenation), `equals` (if string is equal to), `toLowCase` (to convert a string to lowercase), and so on. The following are defined:

```
// Constructors
public String();
public String(byte ascii[], int hibyte);                        // Java 1.0
public String(byte ascii[], int hibyte, int offset, int count);     // Java 1.0
public String(char value[]);
public String(char value[], int offset, int count);
public String(String value);
public String(StringBuffer buffer);
public String(byte ascii[], int offset, int length, String enc);    // Java 1.1

// Methods
public char charAt(int index);
public int compareTo(String anotherString);
public String concat(String str);
public static String copyValueOf(char data[]);
public static String copyValueOf(char data[], int offset, unt count);
public boolean endsWith(String suffix);
public boolean equals(Object anObject);
public boolean equalsIgnoreCase(String anotherString);
public void getBytes(int srcBegin, int srcEnd, byte dst[], int dstBegin);
public void getChars(int srcBegin, int srcEnd,      char dst[], int dstBegin);
public int hashCode();
public int indexOf(int ch);
public int indexOf(int ch, int fromIndex);
public int indexOf(String str);
public int indexOf(String str, int fromIndex);
public String intern();
public int lastIndexOf(int ch);
public int lastIndexOf(int ch, int fromIndex);
public int lastIndexOf(String str);
public int lastIndexOf(String str, int fromIndex);
public int length();
public boolean regionMatches(boolean ignoreCase, int toffset,
                                    String other, int ooffset, int len);
public boolean regionMatches(int toffset, String other,  int offset, int len);
public String replace(char oldChar, char newChar);
public boolean startsWith(String prefix);
public boolean startsWith(String prefix, int toffset);
public String substring(int beginIndex);
public String substring(int beginIndex, int endIndex);
public char[] toCharArray();
public String toLowerCase();
public String toLowerCase(Locale locale);                       // Java 1.1
```

```
public String toString();
public String toUpperCase();
public String toUpperCase(Locale locale);                          // Java 1.1

public String trim();
public static String valueOf(boolean b);
public static String valueOf(char c);
public static String valueOf(char data[]);
public static String valueOf(char data[], int offset, int count);
public static String valueOf(double d);
public static String valueOf(float f);
public static String valueOf(int i);
public static String valueOf(long l);
public static String valueOf(Object obj);
```

L.7.33 Class *java.lang.StringBuffer*

The `StringBuffer` class implements a string buffer. The following are defined:

```
// Constructors
public StringBuffer();
public StringBuffer(int length);
public StringBuffer(String str);
// Methods
public StringBuffer append(boolean b);
public StringBuffer append(char c);
public StringBuffer append(char str[]);
public StringBuffer append(char str[], int offset, int len);
public StringBuffer append(double d);
public StringBuffer append(float f);
public StringBuffer append(int i);
public StringBuffer append(long l);
public StringBuffer append(Object obj);
public StringBuffer append(String str);
public int capacity();
public char charAt(int index);
public void ensureCapacity(int minimumCapacity);
public void getChars(int srcBegin, int srcEnd, char dst[], int dstBegin);
public StringBuffer insert(int offset, boolean b);
public StringBuffer insert(int offset, char c);
public StringBuffer insert(int offset, char str[]);
public StringBuffer insert(int offset, double d);
public StringBuffer insert(int offset, float f);
public StringBuffer insert(int offset, int i);
public StringBuffer insert(int offset, long l);
public StringBuffer insert(int offset, Object obj);
public StringBuffer insert(int offset, String str);
public int length();
public StringBuffer reverse();
public void setCharAt(int index, char ch);
public void setLength(int newLength);
public String toString();
```

L.7.34 Class *java.lang.StringIndexOutOfBoundsException*

Exception that is thrown when a string is indexed with a negative value or a value which is greater than or equal to the size of the string. The following are defined:

```
// Constructors
public StringIndexOutOfBoundsException();
public StringIndexOutOfBoundsException(int index);
```

```
public StringIndexOutOfBoundsException(String s)
```

L.7.35 Class *java.lang.System*

The `System` class implements a number of system methods. The following are defined:

```
// Fields
public static PrintStream err, in, out;
// Methods
public static void arraycopy(Object src, int src_position,
       Object dst, int dst_position, int length);
public static long currentTimeMillis();
public static void exit(int status);
public static void gc();
public static Properties getProperties();
public static String getProperty(String key);
public static String getProperty(String key, String def);
public static SecurityManager getSecurityManager();
public static void load(String filename);
public static void loadLibrary(String libname);
public static void runFinalization();
public static void setProperties(Properties props);
public static void setSecurityManager(SecurityManager s);
```

L.7.36 Class *java.lang.Thread*

The `Thread` class implements one or more threads. The following are defined:

```
// Fields
public final static int MAX_PRIORITY, MIN_PRIORITY, NORM_PRIORITY;

// Constructors
public Thread();
public Thread(Runnable target);
public Thread(Runnable target, String name);
public Thread(String name);
public Thread(ThreadGroup group, Runnable target);
public Thread(ThreadGroup group,Runnable target, String name);
public Thread(ThreadGroup group, String name);

// Methods
public static int activeCount();
public void checkAccess();
public int countStackFrames();
public static Thread currentThread();
public void destroy();
public static void dumpStack();
public static int enumerate(Thread tarray[]);
public final String getName();
public final int getPriority();
public final ThreadGroup getThreadGroup();
public void interrupt();
public static boolean interrupted();
public final boolean isAlive();
public final boolean isDaemon();
public boolean isInterrupted();
public final void join();
public final void join(long millis);
public final void join(long millis, int nanos)
public final void resume();
public void run();
public final void setDaemon(boolean on);
```

```
public final void setName(String name);
public final void setPriority(int newPriority);
public static void sleep(long millis);
public static void sleep(long millis, int nanos)
public void start();
public final void stop();
public final void stop(Throwable obj);
public final void suspend();
public String toString();
public static void yield();
```

L.7.37 Class *java.lang.ThreadGroup*

The `ThreadGroup` class implements a set of threads. The following are defined:

```
// Constructors
public ThreadGroup(String name);
public ThreadGroup(ThreadGroup parent, String name);

// Methods
public int activeCount();
public int activeGroupCount();
public final void checkAccess();
public final void destroy();
public int enumerate(Thread list[]);
public int enumerate(Thread list[], boolean recurse);
public int enumerate(ThreadGroup list[]);
public int enumerate(ThreadGroup list[], boolean recurse);
public final int getMaxPriority();
public final String getName();
public final ThreadGroup getParent();
public final boolean isDaemon();
public void list();
public final boolean parentOf(ThreadGroup g);
public final void resume();
public final void setDaemon(boolean daemon);
public final void setMaxPriority(int pri);
public final void stop();
public final void suspend();
public String toString();
public void uncaughtException(Thread t, Throwable e);
```

L.7.38 Class *java.lang.Throwable*

The `Throwable` class is the superclass of all errors and exceptions in the Java language. The following are defined:

```
// Constructors
public Throwable();
public Throwable(String message);

// Methods
public Throwable fillInStackTrace();
public String getMessage();
public void printStackTrace();
public void printStackTrace(PrintStream s);
public String toString();
```

L.7.39 Class *java.lang.UnknownError*

Exception that is thrown when an unknown error occurs. The following are defined:

```
// Constructors
public UnknownError();
public UnknownError(String s);
```

L.8 Package java.net

L.8.1 Class *java.net.DatagramPacket*

The `DatagramPacket` class implements datagram packets. The following are defined:

```
// Constructors
public DatagramPacket(byte[] ibuf, int ilength);
public DatagraamPacket(byte[] ibuf, int ilength, inetAddress iadd, int iport);

// Methods
public synchronized InetAddress getAddress();
public synchronized byte[] getData();
public synchronized int getLength();
public synchronized int getPort();
public synchronized void setAddress(InetAddress iaddr);          // Java 1.1
public synchronized void setDate(byte[] ibuf);                   // Java 1.1
public synchronized void setLength(int ilength);                 // Java 1.1
public synchronized void setPort(int iport);                     // Java 1.1
```

L.8.2 Class *java.net.InetAddress*

The `InetAddress` class represents Internet addresses. The following are defined:

```
// Methods
public InetAddress[] getAllByName(String host);
public InetAddress getByName(String host);
public InetAddress getLocalHost(String host);
public boolean equals(Object obj);
public byte[] getAddress();
public String getHostAddress();
public String getHostName();
public int hashCode();
public boolean isMulticastAddress();                             // Java 1.1
public String toString();
```

L.8.3 Class *java.net.ServerSocket*

The `ServerSocket` class represents servers which listen for a connection from clients. The following are defined:

```
// Constructors
public ServerSocket(int port);
public ServerSocket(int port, int backlog);
public ServerSocket(int port, int backlog, InetAddress bindAddr);
                                                                 // Java 1.1
// Methods
public Socket accept();
public void close();
public InetAddress getInetAddress();
```

```
public synchronized int getSoTimeout();                    // Java 1.1
public String toString();
```

L.8.4 Class *java.net.Socket*

The `Socket` class represents socket connections over a network. The following are defined:

```
// Constructors
public Socket(String host, int port);
public Socket(InetAddress addr, int port);
public Socket(InetAddress addr, int port, boolean stream);   // Java 1.0
public Socket(String host, int port, InetAddress addr, int localport);
                                                             // Java 1.1
public Socket(InetAddress addr, int port, InetAddress localAddress,
       int localport);                                       // Java 1.1

// Methods
public synchronized void close();
public InetAddress getInetAddress();
public InputStream getInputStream();
public InetAddress getLocalAddress();                        // Java 1.1
public int getLocalPort();
public OutputStream getOutputStream();
public int getPort();
public int getSoLinger();                                    // Java 1.1
public synchronized int getSoTimed();                        // Java 1.1
public boolean getTcpNoDelay();                              // Java 1.1
public void setSoLinger(boolean on, int val);                // Java 1.1
public synchronized void setSoTimed(int timeout);            // Java 1.1
public void setTcpNoDelay(boolean on);                       // Java 1.1
public String toString();
```

L.8.5 Class *java.net.SocketImpl*

The `SocketImpl` class represents socket connections over a network. The following are defined:

```
// Methods
public abstract void accept(SocketImpl s);
public abstract int available();
public abstract void bind(InetAddress host, int port);
public abstract void close();
public abstract void connect(String host, int port);
public abstract void connect(InetAddress addr, int port);
public abstract void create(boolean stream);
public FileDescriptor getFileDescriptor();
public InetAddress getInetAddress();
public abstract InetAddress getInputStream();
```

L.8.6 Class *java.net.URL*

The `URL` class represents Uniform Resource Locators. The following are defined:

```
// Constructors
public URL(String protocol, String host, int port, String file);
public URL(String protocol, String host, String file);
public URL(String spec);
public URL(URL context, String spec);

// Methods
public boolean equals(Object obj);
```

```
public final Object getContent();
public String getFile();
public String getHost();
public int getPort();
public String getProtocol();
public String getRef();
public int hashcode();
public URLConnection openConnection();
public final InputStream openStream();
public boolean sameFile(URL other);
public String toExternalForm();
public String toString();
```

L.9 Package **java.utils**

L.9.1 Class ***java.utils.BitSet***

The `BitSet` class implements boolean operations. The following are defined:

```
// Constructors
public BitSet();
public BitSet(int nbits);

// Methods
public void and(BitSet set);
public void clear(int bit);
public Object clone();
public boolean equals(Object obj);
public boolean get(int bit);
public int hashCode();
public void or(BitSet set);
public void set(int bit);
public int size();
public String toString();
public void xor(BitSet set);
```

L.9.2 Class ***java.utils.Calender***

The `Calender` class has been added with Java 1.1. It supports dates and times.

L.9.3 Class ***java.utils.Date***

The `Date` class supports dates and times. The following are defined:

```
// Constructors
public Date();
public Date(int year, int month, int date);                     // Java 1.0
public Date(int year, int month, int date, int hrs, int min);   // Java 1.0
public Date(int year, int month, int date, int hrs, int min, int sec);
                                                                // Java 1.0
public Date(long date);                                         // Java 1.0
public Date(String s);                                          // Java 1.0

// Methods
public boolean after(Date when);
public boolean before(Date when);
public boolean equals(Object obj);
```

```
public int getDate();                                                    // Java 1.0
public int getDay();                                                     // Java 1.0
public int getHours();                                                   // Java 1.0
public int getMinutes();                                                 // Java 1.0
public int getMonth();                                                   // Java 1.0
public int getSeconds();                                                 // Java 1.0
public long getTime();
public int getTimezoneOffset();                                          // Java 1.0
public int getYear();                                                    // Java 1.0
public int hashCode();
public static long parse(String s);
public void setDate(int date);                                           // Java 1.0
public void setHours(int hours);                                         // Java 1.0
public void setMinutes(int minutes);                                     // Java 1.0
public void setMonth(int month);                                         // Java 1.0
public void setSeconds(int seconds);                                     // Java 1.0
public void setTime(long time);
public void setYear(int year);                                           // Java 1.0
public String toGMTString();                                             // Java 1.0
public String toLocaleString();                                          // Java 1.0
public String toString();
public static long UTC(int year, int month, int date, int hrs, int min,
       int sec);                                                         // Java 1.0
```

L.9.4 Class *java.utils.Dictionary*

The `Dictionary` class is the abstract parent of any class which maps keys to values. The following are defined:

```
// Constructors
public Dictionary();

// Methods
public abstract Enumeration elements();
public abstract Object get(Object key);
public abstract boolean isEmpty();
public abstract Enumeration keys();
public abstract Object put(Object key, Object value);
public abstract Object remove(Object key);
public abstract int size();
```

L.9.5 Class *java.utils.EmptyStackException*

The `EmptyStackException` is thrown when the stack is empty. The following is defined:

```
// Constructors
public EmptyStackException();
```

L.9.6 Class *java.utils.Hashtable*

This `Hashtable` class supports a hashtable which maps keys to values. The following are defined:

```
// Constructors
public Hashtable();
public Hashtable(int initialCapacity);
public Hashtable(int initialCapacity, float loadFactor);
// Methods
public void clear();
public Object clone();
```

```
public boolean contains(Object value);
public boolean containsKey(Object key);
public Enumeration elements();
public Object get(Object key);
public boolean isEmpty();
public Enumeration keys();
public Object put(Object key, Object value);
protected void rehash();
public Object remove(Object key);
public int size();
public String toString();
```

L.9.7 Class *java.utils.NoSuchElementException*

The `NoSuchElementException` is thrown when there are no more elements in the enumeration. The following are defined:

```
// Constructors
public NoSuchElementException();
public NoSuchElementException(String s);
```

L.9.8 Class *java.utils.Observable*

The `Observable` class represents an observable object. The following are defined:

```
// Constructors
public Observable();
// Methods
public void addObserver(Observer o);
protected void clearChanged();
public int countObservers();
public void deleteObserver(Observer o);
public void deleteObservers();
public boolean hasChanged();
public void notifyObservers();
public void notifyObservers(Object arg);
protected void setChanged();
```

L.9.9 Class *java.utils.Properties*

The `Properties` class represents a persistent set of properties. The following are defined:

```
// Fields
protected Properties defaults;

// Constructors
public Properties();
public Properties(Properties defaults);

// Methods
public String getProperty(String key);
public String getProperty(String key, String defaultValue);
public void list(PrintStream out);
public void load(InputStream in);
public Enumeration propertyNames();
public void save(OutputStream out, String header);
```

L.9.10 Class *java.utils.Random*

The `Random` class implements pseudo-random generator functions. The following are defined:

```
// Constructors
public Random();
public Random(long seed);

// Methods
public double nextDouble();
public float nextFloat();
public double nextGaussian();
public int nextInt();
public long nextLong();
public void setSeed(long seed);
```

L.9.11 Class *java.utils.Stack*

The `Stack` class implements a last-in-first-out (LIFO) stack.

```
// Constructors
public Stack();

// Methods
public boolean empty();
public Object peek();
public Object pop();
public Object push(Object item);
public int search(Object o);
```

L.9.12 Class *java.utils.StringTokenizer*

The `StringTokenizer` class allows strings to be split into tokens. The following are defined:

```
// Constructors
public StringTokenizer(String str);
public StringTokenizer(String str, String delim);
public StringTokenizer(String str, String delim, boolean returnTokens);

// Methods
public int countTokens();
public boolean hasMoreElements();
public boolean hasMoreTokens();
public Object nextElement();
public String nextToken();
public String nextToken(String delim);
```

L.9.13 Class *java.utils.Vector*

The `Vector` class implements a growable array of objects. The following are defined:

```
// Fields
protected int capacityIncrement;
protected int elementCount;
protected Object elementData[];

// Constructors
public Vector();
public Vector(int initialCapacity);
public Vector(int initialCapacity, int capacityIncrement);

// Methods
public final void addElement(Object obj);
public final int capacity();
```

```
public Object clone();
public final boolean contains(Object elem);
public final void copyInto(Object anArray[]);
public final Object elementAt(int index);
public final Enumeration elements();
public final void ensureCapacity(int minCapacity)
public final Object firstElement();
public final int indexOf(Object elem);
public final int indexOf(Object elem, int index);
public final void insertElementAt(Object obj, int index);
public final boolean isEmpty();
public final Object lastElement();
public final int lastIndexOf(Object elem);
public final int lastIndexOf(Object elem, int index);
public final void removeAllElements();
public final boolean removeElement(Object obj);
public final void removeElementAt(int index);
public final void setElementAt(Object obj, int index);
public final void setSize(int newSize);
public final int size();
public final String toString();
public final void trimToSize();
```

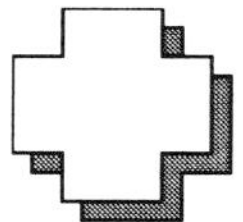

Glossary

100Base-FX IEEE-defined standard for 100 Mbps Ethernet using multimode fiber-optic cable.

100Base-TX (802.3u) IEEE-defined standard for 100Mbps Ethernet using two pairs of Cat-5 twisted-pair cable.

100VG-AnyLAN HP-derived network architecture based on the IEEE 802.12 standard that uses 100Mbps transmission rates. It uses a centrally controlled access method referred to as the Demand Priority Protocol (DPP), where the end node requests permission to transmit and the hub determines which node may do so, depending on the priority of the traffic.

10Base-T IEEE-defined standard for 10Mbps Ethernet using twisted-pair cables.

802.10 IEEE-defined standard for LAN security. It is sometimes used by network switches as a VLAN protocol and uses a technique where frames on any LAN carry a virtual LAN identification. For large networks this can be modified to provided security over the Internet.

802.12 Demand Priority Protocol IEEE-defined standard of transmitting 100Mbps over voice grade (telephone) twisted-pair cabling. *See* 100VG-AnyLAN.

802.1d IEEE-defined bridging standard for Spanning Tree protocol that is used to determine factors on how bridges (or switches) forward packets and avoid networking loops. Networks, which use redundant loops (for alternative routes), need to implement the IEEE 802.1d standard to stop packets from looping forever.

802.2 A set of IEEE-defined specifications for Logical Link Control (LLC) layer. It provides some network functions and interfaces the IEEE 802.5, or IEEE 802.3, standards to the transport layer.

802.3 IEEE-defined standard for CSMA/CD networks. IEEE 802.3 is the most popular implement of Ethernet.

802.3u IEEE-defined standard for 100Mbps Fast Ethernet. It also covers a technique called autosensing which allows 100 Mbps devices to connecting to 10 Mbps devices.

802.4 IEEE-defined token bus specifications.

802.5 IEEE-defined standard for token ring networks.

Adapter Device which usually connects a node onto a network, normally called a network interface adapter (NIC).

Adaptive cut-through switching
A forwarding technique on a switch which determines when the error count on frames received has exceeded the pre-configured limits. When this count is exceeded, it modifies its own operating state so that it no longer performs cut-through switching and goes into a store-and-forward mode. The cut-through method is extremely fast but suffers from the inability to check the CRC field. Thus if incorrect frames are transmitted they could have severe effects on the network segment. This is overcome with an adaptive cut-through switch by checking the CRC as the frame moves through the switch. When errors become too great the switch implements a store-and-forward method.

Adaptive delta modulation PCM
Similar to delta modulation PCM, but uses a number of bits to code the slope of the signal.

Adaptive Huffman coding
Uses a variable Huffman coding technique which responds to local changes in probabilities.

Address A unique label for the location of data or the identity of a communications device. This address can either be numeric or alphanumeric.

Address aging The time that a dynamic address stays in the address routing table of a bridge or switch.

Address Resolution Protocol (ARP)
A TCP/IP process which dynamically binds an IP address to a hardware address (such as an Ethernet MAC address). It can only operate across a single network segment.

Address tables These are used routers, switches and hubs to store either physical (such as MAC addresses) or higher-level addresses (such as IP addresses). The tables map node addresses to network addresses or physical domains. These address tables are dynamic and change due to nodes moving around the network.

Agent A program which allows users to configure or fault-find nodes on a network.

Aging The removing of address in the address table of a router or switch that no longer are referenced to forward a packet.

American National Standards Institute (ANSI)
ANSI is a non-profit organization which is made up of expert committees that publish standards for national industries.

American Standard Code for Information Interchange (ASCII)
An ANSI-defined character alphabet which has since been adopted as a standard international alphabet for the interchange of characters.

Amplitude modulation (AM)
Information is contained in the amplitude of a carrier.

Amplitude-Shift Keying (ASK)
Uses two, or more, amplitudes to represent binary digits. Typically used to transmit binary over speech-limited channels.

Application layer The highest layer of the OSI model.

Asynchronous Communication which does not depend on a clock.

Asynchronous transmission
Transmission where individual characters are sent one-by-one. Normally each character is delimited by a start and a stop bit. With asynchronous communications the transmitter and receiver only have to be roughly synchronized.

ATM (Asynchronous Transfer Mode)
Networking technology which involves sending 53-byte fast packets (ATM cell), as specified by the ANSI T1S1 subcommittee. The first 5 bytes are the header and the remaining bytes are the information field which can hold 48 bytes of data. Optionally the data can contain a 4-byte ATM adaptation layer and 44 bytes of actual data. The ATM adaptation layer field allows for fragmentation and reassembly of cells into larger packets at the source and destination respectively. The control field also contains bits which specify whether this is a flow control cell or an ordinary data cell, a bit to indicate whether this packet can be deleted in a congested network, and so on.

AUI Connection between the network adapter and an external transceiver.

Automatic broadcast control
Technique which minimizes broadcast and multicast traffic flooding through a switch. A switch acts as a proxy server and screens previously resolved ARP. This eliminates broadcasts associated with them.

Autonegotiation Technique used by a IEEE 802.3u node which determines whether a device that it is receiving or transmitting data in one of a number of Ethernet modes (100Base-TX, 100Base-TX Full Duplex, 10Base-T, 10Base-T Full Duplex or 100Base-T4). When the mode is learned, the device then adjusts to either transmission mode.

Autosensing Used by a 100Base-TX device to determine if the incoming data is transmitted at 10Mbps or 100Mbps.

Back pressure Technique which slows the incoming data rate into the buffer of a 802.3 port preventing it from receiving too much data. Switches which implement back pressure will transmit a jam signal to stop data input.

Backbone network The portion of a communications facility that connects primary nodes. A primary shared communications path that serves multiple users at designated jumping-off points.

Bandwidth In an analogue system it is defined as the range of frequencies contained in a signal. As an approximation it is the difference between the highest and lowest frequency in the signal. In a digital transmission system it is normally quoted at the bit per second.

Bandwidth allocation control protocol (BACP)
Protocol which monitor network traffic and allows or disallows access to users, depending on their needs. It is awaiting approval by the IETF.

Baseband Data transmission using unmodulated signals.

Basic rate interface (BRI)
Connection between ISDN and the user. It has three separate channels, one D-channel (which carries control information) and two B channels (which carry data).

Baud rate The number signaling elements sent per second with a RS-232, or modem, communications. In RS-232 the baud rate is equal to the bit-rate. With modems, two or more bits can be encoded as a single signaling element, such as two bits being represented by four different phase shifts (or one signaling element). The signaling element could change its amplitude, frequency or phase-shift to increase the bit-rate. Thus the bit-rate is a better measure of information transfer.

Bit stuffing The insertion of extra bits to stop the appearance of a defined sequence. In HDLC the bit sequence 01111110 delimits the start and end of a frame. Bit stuffing stops this bit sequence from occurring anywhere in the frame by the receiver inserting a 0 whenever their are five consecutive 1's transmitted. At the receive if five consecutive 1's are followed by a 0 then the 0 is deleted.

BNC A commonly used connector for coaxial cable.

BOOTP A standard TCP/IP protocol which allows nodes to be dynamically allocated an IP address.

Bridge A device which physically links two or more networks using the same communications protocols, such as Ethernet/ Ethernet or token ring/ token ting. It allows for the filtering of data between network segments.

Broadband Data transmission using multiplexed data using an analogue signal or high-frequency electromagnetic waves.

Broadcast Message sent to all users on the network.

Broadcast domain Network where broadcasts can be reported to all nodes on the network bounded by routers. A broadcast packet cannot traverse a router.

Broadcast storm Flood of broadcast packets generated by a broadcast transmission where high numbers of receivers are targeted for a long period of time.

Buffer A temporary-storage space in memory.

Bus A network topology where all nodes share a common transmission medium.

Byte A group of eight bits, see octet.

Capacity The maximum data rate in Mbps.

Carrier Sense Multiple Access/ Carrier Detect (CSMA/CD)
A network where all node share a common bus. Nodes must contend for the bus and if a collision occurs then all colliding nodes back-off for a random time period.

Cat-3 cable An EIA/TIA-568 wiring standard for unshield or shielded twisted pair cables.

Cat-5 cable An EIA/TIA-568 wiring standard for unshield or shielded twisted pair cables for the transmission of over 100 Mbps.

CHAP (challenge-handshake authentication protocol)
Identification method used by PPP to determine the originator of a connection.

Checksum An error-detection scheme in which bits are grouped to form integer values and then each of the value is summated. Normally, the negative of this value is then added as a checksum. At the receiver, all the grouped values and the checksum are summated and, in the absence of errors, the result should be zero.

Client Node or program that connects to a server node or program.

Coaxial cable A transmission medium consisting of one or more central wire conductors, surrounded by a insulating layer and encased in either a wire mesh or extruded metal sheathing. It supports RF frequencies from 50 to about 500 MHz. It comes in either a 10-mm diameter (thick coax) or a 5-mm diameter (thin coax).

Collision Occurs when one or more devices try t6 transmit over an Ethernet network.

Copper distributed data interface (CDDI)
FDDI over copper.

Cost An arbitrary value used by routers to compare different routes. Typically it is measure by hop counts, typical time delays or bandwidth.

CRC Cyclic Redundancy Check. An error-detection scheme. Used in most HDLC-related data link applications.

cross-talk Interference noise caused by conductors radiating electromagnetic radiation to couple into other conductors.

Cut-through switching
Technique where a switching device directs a packet to the destination port(s) as soon as it receives the destination and source address scanned from the packet header.

Data Communications Equipment (DCE)
Devices which establish, maintain and terminate a data communications conversation.

Data link layer Second layer of the OSI model which is responsible for link, error and flow control. It normally covers the framing of data packets, error control and physical addressing. Typical data link layers include Ethernet and FDDI.

Data Terminal Equipment (DTE)
Device at the end of the data communications connection.

Delta modulation PCM
Uses a single-bit code to represent the analogue signal. A 1 is transmission when the current sample increases its level, else a 0 is transmitted.

Delta modulation PCM requires a higher sampling rate that the Nyquist rate, but the actual bit rate is normally lower.

Destination MAC address
A 6-byte data unique of the destination MAC address. It is normally quoted as a 12-digit hexadecimal number (such as A5:B2:10:64:01:44).

Destination network address
A unique Internet Protocol (IP) or Internet Packet Exchange (IPX) address of the destination node.

Differential encoding
Source coding method which is used to code the difference between two samples. Typically used in real-time signals where there is limited change between one sample and the next, such as in audio and speech.

Dynamic host control protocol (DHCP)
It manages a pool of IP addresses for computers without a known IP address. This allows a finite number of IP addresses to be reused quickly and efficiently by many clients.

Electronic Industries Association (EIA)
Organization that have defined many of the serial communications standards.

Entropy coding Coding scheme which does not take into account the characteristics of the data and treats all the bits in the same way. It produces lossless coding. Typical methods used are statistical encoding and suppressing repetitive sequences.

Ethernet A local area network which uses coaxial, twisted-pair or fiber optic cable as a communication medium. It transmits at a rate of 10 Mbps and was developed by DEC, Intel and Xerox Corporation. The IEEE 802.3 network standard is based upon Ethernet.

Ethernet address A 48-bit number that identifies a node on an Ethernet network. Ethernet addresses are assigned by the Xerox Corporation.

Even parity An error-detection scheme where defined bit-grouping have an even number of 1's.

Extended Binary Coded Decimal Interchange Code (EBCDIC)
An 8-bit code alphabet developed by IBM allowing 256 different bit patterns for character definitions.

Fast Ethernet See IEEE 802.3u standard.

Fat pipe Term used to indicate a high level of bandwidth for the defined port.

Fiber Distributed Data Interface (FDDI)
A standard network technology that uses a dual counter-rotating token-passing fiber ring. It operates at 100 Mbps and provides for reliable backbone connections.

File server Computer that allows the sharing of file over a network.

File transfer protocol (FTP)
A protocol for transmitting files between host computers using the TCP/IP protocol.

Firewall Device which filters incoming and outgoing traffic.

Flow control Procedure to regulate the flow of data between two nodes.

Forward adaptive bit allocation
This technique, used in audio compression, makes bit allocation decisions adaptively, depending on signal content.

Fragment free cut-through switching
A modified cut-through switching technique where a switch or switch module waits until it has received a large enough packet to determine if it is error free.

Frame Normally associated a packet which has layer 2 information added to it. Packets are thus contained within frames. Frames and packets have variable lengths as opposed to cells which have fixed length.

Frame check sequence (FCS)
Standard error detection scheme.

Frequency-shift Keying (FSK)
Uses two, or more, frequencies to represent binary digits. Typically used to transmit binary data over speech-limited channels.

Full duplex Simultaneous, two-way communications.

Gateway A device that connects networks using different communications protocols, such as between Ethernet and FDDI. It provides protocol translation, in contrast to a bridge which connects two networks that are of the same protocol.

GIF Standard image compression technique which is copyrighted by CompuServe Incorporated. It uses LZW compression and supports a palette of 256 24-bit colors (16.7M colors). GIF support local and global color tables and animated images.

Half-duplex (HDX)

Two-way communications, one at a time.

Handshaking A reliable method for two devices to pass data.

HDLC ISO standard for the data link layer.

Hello packet Message transmitted from a root bridge to all other bridges in the network to constantly verify the Spanning Tree setup.

Hop The number of gateways and routers in a transmission path.

Hop count Used by the RIP routing protocol to measure the distance between a source and a destination.

Host A computer that communicates over a network. A host can both initiate communications and respond to communications that are addressed to it.

Huffman coding Uses a variable length code for each of the elements within the data. It normally analyzes the probability of element in the data and codes the most probable with fewer bits than the least probable.

Hub A hub is a concentration point for data and repeats data from one node to all other connected nodes.

Hypertext markup language (HTML)
Standard language that allows the integration of text and images over a distributed network.

Integrated systems digital network (ISDN)
Communication technology that contains two data channels (2B) and a control channel (H). It supports two 64 kbps data channels and sets up a circuit-switched connection.

International Telegraph Union Telecommunications Standards Sector (ITU-TSS)
Organization which has replaced the CCITT.

Internet Connection of nodes on a global network which use a DARPA-defined Internet address.

internet Two or more connected networks that may, or may not, use the same communication protocol.

Internet address An address that conforms to the DARPA-defined Internet protocol. A unique, four-byte number identifies a host or gateway on the Internet. This consists of a network number followed by a host number. The host number can be further divide into a subnet number.

Internet Engineering Task Force (IETF)
A committee that reviews and supports Internet protocol proposals.

IP (Internet Protocol)
Part of the TCP/IP which provides for node addressing.

IP address An address which is used to identify a node on the Internet.

IP multicast Addressing technique that allows IP traffic to be propagated from one source to a group of destinations.

IPX (Internet Packet Exchange)
Novell NetWare communications protocol which is similar to the IP protocol. The packets include network addresses and can be routed from one network to another.

IPX address Station address on a Novell NetWare network. It consists of two fields: a network number field and a node number field. The node number is the station address of the device and the network number is assigned to the network when the network is started-up. It is written in the form: NNNNNNNN:XXXXXX-XXXXXX, where N's represent the network number and X's represent the station address. An example of an IPX address is: DC105333:542C10-FF1432.

ISO International Standards Organization

ITU-T The Consultative Committee for International Telephone and Telegraph (now known at the ITU-TSS) is an advisory committee established by the United Nations. They attempt to establish standards for inter-country data transmission on a worldwide basis.

Jabber Occurs when the transmission of network signals exceeds the maximum allowable transmission time (20 ms to 150 ms). The medium becomes overrunned with data packets caused by a faulty node or wiring connection.

Jitter Movement of the edges of pulse over time, that may introduce error and loss of synchronization.

JPEG Image compression technique defined by the Joint Photographic Expert Group (JPEG), a subcommittee of the ISO/IEC. It uses a DCT, quantization, run-length and Huffman coding.

Latency Defines the amount of time between a device receiving data and it being forwarded on. Hubs have the lowest latency (less than 10μs), switches the next lowest (between 40μs and 60μs), then bridges (200μs to 300μs) and routers have the highest latency (around 1000 μs).

Learning bridge Bridge which learns the connected nodes to it. It uses this information to forward or drop frames.

Leased line A permanent telephone line connection reserved exclusively by the leased customer. There is no need for any connection and disconnection procedures.

Lempel-Ziv coding Coding method which takes into account repetition in phases, words or parts of words. It uses pointers to refer to previously defined sequences.

Lempel-Ziv Welsh (LZW) coding Coding method which takes into account repetition in phases, words or parts of words. It builds up a dictionary of previously send (or stored) sequences.

Line driver A device which converts an electrical signal to a form that is transmittable over a transmission line. Typically, it provides the required power, current and timing characteristics.

Link layer Layer 2 of the OSI model.

Link segment A point-to-point link terminated on either side by a repeater. Nodes can not be attached to a link segment.

Lossless compression Where information, once uncompressed is identical to the original uncompressed data.

Lossy compression Where information, once uncompressed, cannot be fully recovered.

MAC address A 6-byte data unique data-link layer address. It is normally quoted as a 12-digit hexadecimal number (such as A5:B2:10:64:01:44).

Masking effect Where noise is only heard by a person when there are no other sounds to mask it.

MDI (Medium Dependent Interface) The IEEE standard for the twisted-pair interface to 10Base-T (or 100Base-TX).

Media Access Control (MAC) Media-specific access-control for Token Ring and Ethernet.

Media Interface Controller (MIC) Media-specific access-control for token ring and Ethernet.

Medium Attachment Unit (MAU)
Method of converting digital data into a form which can be transmitted over a band-limited channel. Methods use either ASK, FSK, PSK or a mixture of ASK, FSK and PSK.

Modem (Modulator- Demodulator)
A device which converts binary digits into a form which can be transmitted over a speech-limited transmission channel.

MTU (Maximum Transmission Unit)
The largest packet that the IP protocol will send through the selected interface or segment.

Multicast Packets which are sent to all nodes on a subnet of a group within a network. This differs from a broadcast which forwards packet to all users on the network.

Multimode fiber Fiber-optic cable that has the ability to carry more than one frequency (mode) of light at a time.

N-series connectors Connector used with thick coaxial cable.

Network driver interface specification (NDIS)
Software specification for network adapter drivers. It support multiple protocols and multiple adapters, and is used in many operating systems, such as Windows 95/88/NT.

Network layer Third layer of the OSI model, which is responsible for ensuring that data passed to it from the transport layer is routed and delivered through the network. It provides end-to-end addressing and routing, and has support for a number of protocols, including IP, IPX, CLNP, X.25, or DDP.

Network termination (NT1)
Network termination for ISDN.

Node Any point in a network which provides communications services or where devices interconnect.

Intranet A company specific network which has additional security against external users.

Octet Same as a byte, a group of eight bits (typically used in communications terminology).

Odd parity An error-detection scheme where a defined bit-grouping has an odd number of 1's.

Open Data-Link Interface (ODLI)
Software specification for network adapter drivers using in NetWare and Apple networks. It supports multiple protocols and multiple adapters.

Optical Repeater A device that receives, restores, and re-times signals from one optical fiber segment to another.

Packet A sequence of binary digits that is transmitted as a unit in a computer network. A packet usually contains control information and data. They normally are contain with data link frames.

Packet switching Network switching in which data is processed in units of whole packets rather than attempting process data by dividing packets into fixed-length cells.

Password authentication protocol (PAP)
Protocol which checks a users password.

Phase-Locked Loop (PLL)
Tunes into a small range of frequencies in a signal and follows any variations in them.

Phase-Shift Keying (PSK)
Uses two, or more, phase-shifts to represent binary digits. Typically used to transmit binary data over speech-limited channels.

Physical layer Lowest layer of the OSI model which is responsible for the electrical, mechanical, and handshaking procedures over the interface that connects a device to a transmission medium

Ping Standard protocol used to determine if TCP/IP node are alive. Initially a node sends an ICMP (Internet Control Message Protocol) echo request packet to the remote node with the specified IP address and waits for echo response packets to return.

Point of presence (POP)
Physical access point to a long distance carrier interchange.

Point-to-point protocol (PPP)
Standard protocol to transfer data over the Internet asynchronously or synchronously.

Port Physical connection on a bridge or hub that connects to a network, node or other device.

Protocol A specification for coding of messages exchanged between two communications processes.

Quadrature modulation
Technique used in PAL and NSTC where the U and V information are added to the carrier with a 90° phase difference between them.

Quantization Involves converting an analogue level into a discrete quantized level. The number of bits used in the quantization process determines the number of quantization levels.

Quartet signaling Signaling technique used in 100VG-AnyLAN networks that allows data transmission at 100 Mbps over frame pairs of UTP cabling.

Repeater A device that receives, restores, and re-times signals from one segment of a network and passes them on to another. Both segments must have the same type of transmission medium and share the same set of protocols. A repeater cannot translate protocols.

Reverse address resolution protocol (RARP)
The opposite of ARP which maps an IP address to a MAC address.

RJ-45 Connector used with US telephones and with twisted-pair cables. It is also used in ISDN networks, hubs and switches

RMON An SNMP MIB that specifies the types of information listed in a number of special MIB groups that are commonly used for traffic management. Some of the popular groups used are Statistics, History, Alarms, Hosts, Hosts Top N, Matrix, Filters, Events, and Packet Capture.

Routing node A node that transmits packets between similar networks. A node that transmits packets between dissimilar networks is called a gateway.

RS-232C EIA-defined standard for serial communications.

RS-422, 423 EIA-defined standard which uses a higher transmission rates and cable lengths than RS-232.

RS-449 EIA-defined standard for the interface between a DTE and DCE for 9- and 37-way D-type connectors.

RS-485 EIA-defined standard which is similar to RS-422 but uses a balanced connection.

Run-length encoding (RLE)
Coding technique which represents long runs of a certain bit sequence with a special character.

SAP Service Access Point. Field defined by the IEEE 802.2 specification that is part of the address specification.

SAP Service Advertisement Protocol, Used by the IPX protocol to provide a means of informing network clients, via routers and servers of available network resources and services.

Segment A segment is any length of LAN cable terminated at both ends. In a bus network, segments are electrically continuous pieces of the bus, connected at by repeaters. It can also be bounded by bridges and routers.

Serial line internet protocol (SLIP)
A standard used for the point-to-point serial connections running TCP/IP.

Simplex One-way communication.

SNMP (Simple Network Management Protocol)
Standard protocol for managing network devices, such as hubs, bridges, and switches.

Source encoding Coding method which takes into account the characteristics of the information. Typically used in motion video and still image compression.

Statistical encoding Coding method analyses the statistical pattern of the data. Commonly occurring data is coded with a few bits and uncommon data by a large number of bits.

Subsampling Reduces digitized bit rate by sampling the luminance and chromaniance at different rates.

Suppressing repetitive sequences
Compression technique where long sequences of the same data is compressed with a short code.

Switch A very fast, low-latency, multiport bridge that is used to segment local area networks.

Synchronous Data which is synchronized by a clock.

TCP Part of the TCP/IP protocol and provides an error-free connection between two cooperating programs.

TCP/IP Internet An Internet is made up of networks of nodes that can communicate with each other using TCP/IP protocols.

Telnet Standard program which allows remote users to log into a station using the TCP/IP protocol.

TIFF Graphics format that supports many different types of images in a number of modes. It is supported by most packages and, in one mode provides for enhanced high-resolution images with 48-bit color.

Time to live A field in the IP header which defines the number of routers that a packet is allowed to traverse before being discarded.

Token A token transmits data around a token ring network.

Topology The physical and logical geometry governing placement of nodes on a network.

Transceiver A device that transmits and receives signals.

Transform encoding
Source-encoding scheme where the data is transformed by a mathematical transform in order to reduce the transmitted (or stored) data. A typical technique is the discrete cosine transform (DCT) and the fast fourier transform (FFT).

Transport layer Fourth layer of the OSI model. It allows end-to-end control of transmitted data and the optimized use of network resources.

Universal asynchronous receiver transmitter (UART)
Device which converts parallel data into a serial form, which can be transmitted over a serial line, and vice-versa.

V.24 ITU-T-defined specification, similar to RS-232C.

V.25 bis ITU-T specification describing procedures for call setup and disconnection over the DTE-DCE interface in a PSDN.

V.32 ITU-T standard serial communication for bi-directional data transmissions at speeds of 4.8 or 9.6 Kbps.

V.34 Improved v.32 specification with higher transmission rates (28.8 Kbps) and enhanced data compression.

Variable-length-code LZW (VLC-LZW) code
Uses a variation of LZW coding where variable-length codes are used to replace patterns detected in the original data.

Virtual circuit Logical circuit which connects two networked devices together.

X-ON/ X-OFF The Transmitter On/ Transmitter Off characters are used to control the flow of information between two nodes.

X.21 ITU-T-defined specification for the interconnection of DTEs and DCEs for synchronous communications.

X.25 ITU-T-defined specification for packet-switched network connections.

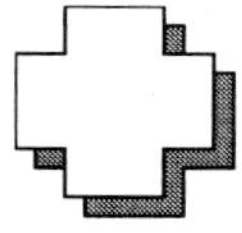

Abbreviations

AA	auto-answer
AAN	autonomously attached network
ABM	asynchronous balanced mode
AC	access control
ACK	acknowledge
ACL	access control list
ADC	analogue-to-digital converter
ADPCM	adaptive delta pulse code modulation
AES	audio engineering society
AFI	authority and format identifier
AM	amplitude modulation
AMI	alternative mark inversion
ANSI	American National Standard Institute
APCM	adaptive pulse code modulation
API	application program interface
ARM	asynchronous response mode
ARP	address resolution protocol
ASCII	American standard code for information exchange
ASK	amplitude-shift keying
AT	attention
ATM	asynchronous transfer mode
AUI	attachment unit interface
BCC	blind carbon copy
BCD	binary coded decimal
BIOS	basic input/output system
B-ISDN	broadband ISDN
BMP	bitmapped
BNC	British Naval Connector
BOOTP	boot protocol
BPDU	bridge protocol data units
bps	bits per second
CAD	computer-aided design
CAN	concentrated area network
CASE	common applications service elements
CATNIP	common architecture for the Internet
CCITT	International Telegraph and Telephone Consultative
CD	carrier detect
CDE	common desktop environment
CD-R	CD-recordable
CD-ROM	compact disk - read-only memory

CGI	common gateway interface
CGM	computer graphics metafile
CIF	common interface format
CMC	common mail call
CMOS	complementary MOS
CPU	central processing unit
CRC	cyclic redundancy
CRLF	carriage return, line feed
CRT	cathode ray tube
CSDN	circuit-switched data network
CSMA	carrier sense multiple access
CSMA/CA	CSMA with collision avoidance
CSMA/CD	CSMA with collision detection
CS-MUX	circuit-switched multiplexer
CSPDN	circuit-switched public data network
CTS	clear to send
DA	destination address
DAC	digital-to-analogue converter
DAC	dual attachment concentrator
DARPA	Defense Advanced Research Projects Agency
DAS	dual attachment station
DAT	digital audio tape
dB	decibel
DBF	NetBEUI frame
DC	direct current
DCC	digital compact cassette
DCD	data carrier detect
DCE	data circuit-terminating equipment
DCT	discrete cosine transform
DD	double density
DDE	dynamic data exchange
DES	data encryption standard
DHCP	dynamic host configuration program
DIB	directory information base
DISC	disconnect
DLC	data link control
DLL	dynamic link library
DM	disconnect mode
DNS	domain name server
DOS	disk operating system
DPCM	differential PCM
DPSK	differential phase-shift keying
DQDB	distributed queue dual bus
DR	dynamic range
DRAM	dynamic RAM
DSP	domain specific part
DSS	digital signature standard

DTE	data terminal equipment
DTR	data terminal ready
EaStMAN	Edinburgh/Stirling MAN
EBCDIC	extended binary coded decimal interchange code
EBU	European broadcast union
EEPROM	electrically erasable PROM
EF	empty flag
EFM	eight-to-fourteen modulation
EGP	exterior gateway protocol
EIA	Electrical Industries Association
EISA	extended international standard interface
ENQ	inquiry
EOT	end of transmission
EPROM	erasable PROM
EPS	encapsulated postscript
ETB	end of transmitted block
ETX	end of text
FAT	file allocation table
FAX	facsimile
FC	frame control
FCS	frame check sequence
FDDI	fiber distributed data interface
FDM	frequency division multiplexing
FDX	full duplex
FEC	forward error correction
FM	frequency modulation
FRMR	frame reject
FSK	frequency-shift keying
FTP	file transfer protocol
GFI	group format identifier
GGP	gateway-gateway protocol
GIF	graphics interface format
GUI	graphical user interface
HD	high density
HDB3	high-density bipolar code no. 3
HDLC	high-level data link control
HDTV	high-definition television
HDX	half duplex
HF	high frequency
HMUX	hybrid multiplexer
HPFS	high performance file system
HTML	hypertext mark-up language
HTTP	hypertext transfer protocol
Hz	Hertz
I/O	input/output
IA5	international alphabet no. 5
IAB	internet advisory board

IAP	internet access provider
ICMP	internet control message protocol
ICP	internet connectivity provider
IDEA	international data encryption algorithm
IDI	initial domain identifier
IDP	initial domain part
IEEE	Institute of Electrical and Electronic Engineers
IEFF	internet engineering task force
IGP	interior gateway protocol
ILD	injector laser diode
IMAC	isochronous MAC
IP	internet protocol
IPP	internet presence provider
IPX	internet packet exchange
ISA	international standard interface
ISDN	integrated services digital network
IS-IS	immediate system to intermediate system
ISO	International Standards Organization
ISP	internet service provider
ITU	International Telecommunications Union
JANET	joint academic network
JFIF	jpeg file interchange format
JISC	Joint Information Systems Committee
JPEG	Joint Photographic Expert Group
LAN	local area network
LAPB	link access procedure balanced
LAPD	link access procedure
LCN	logical channel number
LD-CELP	low-delay code excited linear prediction
LED	light emitting diode
LGN	logical group number
LIP	large IPX packets
LLC	logical link control
LRC	longitudinal redundancy check
LSL	link support level
LSP	link state protocol
LZ	Lempel-Ziv
LZW	LZ-Welsh
MAC	media access control
MAN	metropolitan area network
MAU	multi-station access unit
MDCT	modified discrete cosine transform
MDI	media dependent interface
MHS	message handling service
MIC	media interface connector
MIME	multi-purpose internet mail extension
MODEM	modulation/demodulator

MOS	metal oxide semiconductor
MPEG	motion picture experts group
NAK	negative acknowledge
NCP	netware control protocols
NCSA	National Center for Supercomputer Applications
NDIS	network device interface standard
NETBEUI	NetBIOS extended user interface
NIC	network interface card
NIS	network information system
NLSP	netware link-state routing protocol
NRZI	non-return to zero with inversion
NSAP	network service access point
NSCA	National Center for Supercomputer Applications
NTE	network terminal equipment
NTFS	NT file system
NTP	network time protocol
NTSC	National Television Standards Committee
ODI	open data-link interface
OH	off-hook
OSI	open systems interconnection
OSPF	open shortest path first
OUI	originators unique identifier
PA	point of attachment
PAL	phase alternation line
PC	personal computer
PCM	pulse code modulation
PDN	public data network
PHY	physical layer protocol
PING	packet internet gopher
PISO	parallel-in-serial-out
PKP	public key partners
PLL	phase-locked loop
PLS	physical signaling
PMA	physical medium attachment
PMD	physical medium dependent
PPP	point-to-point protocol
PPSDN	public packet-switched data network
PS	postscript
PSDN	packet-switched data network
PSE	packet switched exchange
PSK	phase-shift keying
PSTN	public-switched telephone network
QAM	quadrature amplitude modulation
QCIF	quarter common interface format
QIC	quarter inch cartridge
QT	quicktime
RAID	redundant array of inexpensive disks

RAM	random-access memory
RD	receive data
REJ	reject
RFC	request for comment
RGB	red, green and blue
RI	ring in
RIP	routing information protocol
RLE	run-length encoding
RNR	receiver not ready
RO	ring out
ROM	read-only memory
RPC	remote procedure call
RR	receiver ready
RSA	Rivest, Shamir and Adleman
RTF	rich text format
RTMP	routing table maintenance protocol
S/PDIF	Sony/Philips digital interface format
SABME	set asynchronous balanced mode extended
SAC	single attachment concentrator
SAP	service advertising protocol
SAPI	service access point identifier
SAS	single attachment station
SB-ADCMP	sub-band ADPCM
SCMS	serial copy management system
SCSI	small computer systems interface
SD	sending data
SDH	synchronous digital hierarchy
SDIF	Sony digital interface
SDLC	synchronous data link control
SECAM	séquential couleur à mémoire
SEL	selector/extension local address
SHEFC	Scottish Higher Education Funding Council
SIPO	serial-in parallel-out
SIPP	simple internet protocol plus
SMDS	switched multi-bit data stream
SMP	symmetrical multiprocessing
SMT	station management
SMTP	simple message transport protocol
SNA	systems network architecture (IBM)
SNMP	simple network management protocol
SNR	signal-to-noise ratio
SONET	synchronous optical network
SPX	sequenced packet exchange
QTV	studio-quality television
SRAM	static RAM
STA	spanning-tree architecture
STM	synchronous transfer mode

STP	shielded twisted-pair
SVGA	super VGA
TCP	transmission control protocol
TDAC	time-division aliasing cancellation
TDM	time-division multiplexing
TEI	terminal equipment identifier
TIFF	tagged input file format
TR	transmit data
TUBA	TCP and UDP with bigger addresses
UDP	user datagram protocol
UI	unnumbered information
UNI	universal network interface
UPS	uninterruptable power supplies
URI	universal resource identifier
URL	uniform resource locator
UTP	unshielded twisted pair
UV	ultra violet
VCI	virtual circuit identifier
VCR	video cassette recorder
VGA	variable graphics adapter
VIM	vendor-independent messaging
VLC-LZW	variable-length-code LZW
VLM	virtual loadable modules
VRC	vertical redundancy check
WAIS	wide area information servers
WAN	wide area network
WIMPs	windows, icons, menus and pointers
WINS	windows internet name service
WINSOCK	windows sockets
WORM	write-once read many
WWW	World Wide Web
XDR	external data representation
XOR	exclusive-OR

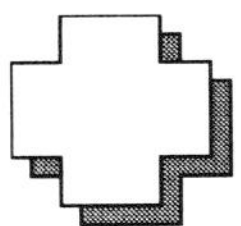

Miscellaneous and Quick Reference

NetBIOS name types

Microsoft networks identify computers by their NetBIOS name. Each is 16 characters long, and the 16th character represents the purpose of the name. An example list of a WINS database is:

```
Name                    Type     Status
FRED          <00>      UNIQUE   Registered
BERT          <00>      UNIQUE   Registered
STAFF         <1C>      GROUP    Registered
STAFF         <1E>      GROUP    Registered
```

The values for the 16th byte are:

00	Workstation	03	Message service
06	RAS server service	1B	Domain master browser
1C	Domain group name	1D	Master browers name
1E	Normal group name (workgroup)	1F	NetDDE service
20	Server service	21	RAS client
BE	Network Monitor Agent	BF	Network Monitor Utility

On Windows NT, the names in the WINS database can be shown with the `nbstat` command.

Windows NT TCP/IP setup

Windows NT uses the files LMHOSTS, HOSTS and NETWORKS to map TCP/IP names and network addresses. These are stored in the *<winNT root>*`\SYSTEM32\DRIVERS\ETC`. `LMHOSTS` maps IP addresses to a computer name. An example format is:

```
#IP-address          host-name
146.176.1.3          bills_pc
146.176.144.10       fred_pc      #DOM:STAFF
```

where comments have a preceding '#' symbol. To preserve compatibility with previous version of Microsoft LAN Manager, special commands have been included after the comment symbol. These include:

`#PRE`
`#DOM:`*domain*
`#INCLUDE` *fname*
`#BEGIN_ALTERNATE`
`#END_ALTERNATE`

where

`#PRE` specifies the name is preloaded into the memory of the computer and no further references to the `LMHOSTS` file will be made.
`#DOM:`*domain* specifies the name of the domain that the node belongs to.
`#BEGIN_ALTERNATE` and `#END_ALTERNATE` are used to group multiple `#include`'s
`#include` *fname* specifies other `LMHOST` files to include.

The `HOSTS` file format is IP address followed by the fully qualified name (FQDN) and then any aliases. Comments have a preceding '#' symbol. For example:

```
#IP Address         FQDN           Aliases
146.176.1.3         superjanet     janet
146.176.144.10      hp
146.176.145.21      mimas
146.176.144.11      mwave
146.176.144.13      vax
146.176.146.23      oberon
146.176.145.23      oberon
```

Windows NT TCP/IP commands (quick reference)

Command	*Description*	*Examples*
`arp`	Modifies Address Resolution Protocol tables. `-s` *IP-address* [*MAC-address*] ; manually modify `-a` [*IP-address*] ; display ARP entry `-d` *IP-address* ; delete entry	`arp -s 146.176.151.10 FF-AA-10-3F-A1-3F`
`finger`	Queries users on a remote computer. `@`*hostname* ; name of remote computer `-l` ; extend list	`finger -l fred@miranda` `finger @moon`

`ftp`	Remote file transfer. After connected the following commands can be used:
	
	`ascii binary bye cd`
	`dir get hash help`
	`lcd ls mget mput`
	`open prompt pwd quit`
	`remote help user`
		`ftp intel.com`

`hostname`	Displays the TCP/IP hostname of the local node.	`hostname`
`ipconfig`	Displays the TCP/IP settings on the local computer. `/all` ; show all settings	`ipconfig /all`
`lpq`	Sends a query to a TCP/IP host or printer. `-S` *print_server* `-P` *printer*	`lpq -p lp_laser` `lpq -s mirands -p dot_matrix`
`lpr`	Prints to a TCP/IP-based printer. `-S` *print_server* `-P` *printer*	`lpr -p lp_laser file.ps`

`nbstat`	Displays mapping of NetBIOS names to IP addresses. `-a` *NetBIOS-name* ; display name table for ; computer `-A` *IP-address* ; display name table for ; computer `-n` ; display NetBIOS table of ; local computer	`nbstat -A freds`
`netstat`	Displays status of TCP/IP connections. `-p` *protocol* ; display for given protocol `-r` ; show routing tables `-s` ; display statistics `-R` ; reload `HMHOSTS` `-S` ; display NetBIOS sessions by ; NetBIOS names `-s` ; display NetBIOS sessions by IP ; addresses	*See Section 19.4.5*

`nslookup`	Queries DNS servers. After connected the following commands can be used:
	
	`help`
	`finger [`*username*`]`
	`port=`*port*
	`querytype=`*type*
	; *type* can be A (address),
	; CNAME (canonical name which is an alias for ; another host), MX (mail exchanger which
	; handles mail for a given host), NS (name server
	; for the domain), PTR (pointer record which
	; maps an IP address to a hostname), SOA (start
	; of authority record) or ANY.
		See Section 19.4.4

`ping`	Test TCP/IP connectivity. `-a` ; resolve IP addresses to hostnames `-n` *count* ; set number of echo packets `-l` *size* ; specify packet size `-t` ; continuously ping `-i` *ttl* ; set time-to-live field `-w` *timeout* ; specify timeout in ms	*See Section 19.4.1*
`rcp`	Remote copy. [*hostname*[`.`*username*]] `-a` ; ASCII copy `-b` ; binary copy `-h` ; also hidden files `-r` ; recursively copy	`rcp -r \*.txt miranda.bill/home`
`rexec`	Execute remote command.	`rexec miranda -l bill "ls -l"`

`route`	Manipulates TCP/IP routing table. `-f` ; delete all routes `-p` ; make a permanent route `add` ; add a route `change` ; modify an existing route `delete` ; delete a route `gateway` ; specifies gateway `mask` *netmask* ; define subnet mask `print` ; print current table	`route gateway 146.151.176.12`
`rsh`	Executes remote shell. `-l` *username* ; user name `command` ; command to execute	`rsh -l bill "ls -l"`
`telnet`	Remote login.	`telnet www.intel.com`
`tftp`	Trivial FTP (uses UDP).	
`tracert`	Trace route. `-d` ; do not resolve IP addresses `-h` *max_hops* ; maximum number of hops `-w` *timeout* ; specify timeout	*See Section 19.4.6*

Windows NT system administration commands (quick reference)

Command	*Description*	*Examples*

`at`	Runs commands at a specified time. Options include:
	
	`\\computer-name`
	`time`
	`/every:date` ; such as day of the week such ; as M/T/W/Th/F/S/Su or day ; of the month
		`at 14:00 \\freds "cmd ping miranda > log"`
		
		`at 00:00 /every:M/W/F "cmd lpr log.txt"`

`attrib`	Displays or changes file attributes. Attributes include: `+r, -r,` (read) `+a, -a,` (archive) `+s, -s,` (system) `+h, -h,` (hidden) `/s` (include sub-directories)	`attrib +h test.txt`
`backup`	Backup program.	
`cacls`	Command-line Access Control Lists (ACLs). `/g` *username*`:`*right* ; grant user the following ; rights: r (read), c (change), ; f (full control). `/p` *username* ; replace rights, these are as ; above, but n (none) is added `/r` *username* ; delete all rights `/t` ; recursive change	`calcs list.txt /g fred:cf` `calcs *.* /r bill /t`

`chkdsk`	Checks disk. Options include: `/f` ; automatically fix errors	`chkdsk c: /f`
`cmd`	Run command-line shell.	
`convert`	Converts drive partition from FAT to NTFS,	`convert d: /fs:ntfs`
`convlog`	Converts files from Microsoft Information Server, FTP server and Gopher servers, and produces log files in NSCA or EMWAC format. `-t` [*emwave* | *ncsa*] ; specify EMWAC or NCSA `-s` [*f* | *w* | *g*] ; specify FTP (f), WWW (w) or ; Gopher (g) `-o` *outdir* ; specify output directory	`convlog -sg -ncsa -o c:\temp *.log`
`diskperf`	Toggles the disk performance counter.	
`ipxroute`	IPX routing. `servers` ; list NetWare servers	`ipxsroute ervers`

`jetpack`	Compacts WINS databases.	`net stop wins`
		`jetpack win.mdb tmp.mdb`
		`net start wins`

`netmon`	Network monitoring tool.	
`ntbackup`	Backup file system.	
`rasadmin`	Remote Access Server (RAS) administration.	
`rasautou`	Remote Access Server (RAS) debugging.	
`rasdial`	Remote Access Server (RAS) dial-up. `/phone` *tel*; telephone number	`rasdial miranda /phone:1112222`
`rasphone`	Edit RAS phonebook.	
`rdisk`	Create emergency repair disk.	
`regedit`	Edit registry.	
`restore`	Restores files after a backup.	
`start`	Starts applications from the command line.	
`winnt`	16-bit Window NT installation program. `/r` *dir* ; specify install directory `/s` *dir* ; installation source files	
`winnt32`	32-bit Window NT installation program.	

NT control services commands (quick reference)

Command	*Description*	*Example*
`net accounts`	Controls account settings `/domain` *dom* ; specify default domain	
`net computer`	Adds or deletes computers from current domain. `\\`*computer-name* `/add`; add computer `/del`; delete computer	`net computer \\freds /add` `net computer \\bills /del`
`net config server`	Configure server.	
`net config workstation`	Configure workstation.	
`net continue`	Unpauses a command that was paused with `net pause`.	
`net file`	Closes an opened file. When used on its own without arguments it gives the ID of all opened files. The `/close` option is used with the ID number to close a given file. `/close` ; close file	

`net group`	Creates, edits or deletes groups.
	
	`/add`
	; add new group or users to the named group
	; specified group
	`/delete`
	; delete group or users to the named group
	; specified group
		`net group "Staff" /add`
		`net group "Staff" /add fred`

`net help`	Help messages for net.	
`net helpmsg`	Detailed help for a given error message.	
`net localgroup`	Create or deletes local groups or local users.	
`net name`	Administers list of names for the Messenger service.	
`net pause`	Pauses a service.	`net pause lpdsvc` *; pause print service*
`net print`	Administers print queues. `\\`*computer-name* `/delete` ; delete job	
`net send`	Sends a text message to users or computers.	`net send bill "Hello"`
`net session`	Displays information of a current session.	
`net share`	Administers networks shares.	

`net start`	Starts a service.	`net start snmp` ; *start SNMP*
`net statistics`	Displays service statistics.	
`net stop`	Stops a service.	`net stop lpdsvc` ; *stop print server*
`net time`	Sets or queries time on a remote computer.	
`net use`	Administers networked resources.	
`net user`	Administers user accounts. `password` ; prompts for password `/active:[y/n]` ; active status `/add` ; add user `/delete` ; delete user `/expires:[`*date*`|NEVER]` ; expire time `/fullname: "`*name*`"` ; full name `/homedir:` *homedirpath* ; home directory `/passwdchg: [y |n ]` ; password change `/times: [`*times* `| ALL]` ; login times	`net user bill_c /add` `net user bill_c /active:y`
`net view`	Displays networked resources. `\\`*computer-name* `/domain [`*domain*`]` ; list of domain or ; computers within the ; specified domain	`net view \\freds` `net view /domain`

Example Internet domain name server files

In the following example setup, there are two name servers (`ees99` and `eepc02`) within the `eece.napier.ac.uk` domain. The Internet domain naming process is run with the `named` program and reads from the `named.boot` file (to use a file other than `/etc/named.boot` the `-b` option is used). Its contents are given next and it lists six main subnet (`146.176.144.`*x* to `146.176.151.`*x*). The files `net/net144` to `net/net151` contain the definition of the hosts that connect to these subnets.

```
;  @(#)named.boot.slave 1.13    (Berkeley)  87/07/21
;  boot file for secondary name server
; Note that there should be one primary entry for each SOA record.
;
directory       /usr/local/adm/named

; type    domain                          source host/file        backup file

primary   eece.napier.ac.uk               eece.napier.ac.uk
primary   144.176.146.in-addr.arpa        net/net144
primary   145.176.146.in-addr.arpa        net/net145
primary   146.176.146.in-addr.arpa        net/net146
primary   147.176.146.in-addr.arpa        net/net147
primary   150.176.146.in-addr.arpa        net/net150
primary   151.176.146.in-addr.arpa        net/net151
primary   0.0.127.IN-ADDR.ARPA            named.local
primary   0.0.127.IN-ADDR.ARPA            named.local
cache     .                               root.cache
```

The first line (after the comments, which begin with a semi-colon) defines that the master file (`eece.napier.ac.uk`) contains authoritative data for the `eece.napier.ac.uk` domain. All domain names are then relative to this domain. For example, a computer within this domain which has name `pc444` will have the full domain name of:

```
pc444.eece.napier.ac.uk
```

The second line of the file defines that the file `net/net144` contains the authoritative data on the `144.176.146.in-addr.arpa` domain, and so on. The `cache` line specifies that data in the `root.cache` file is to be place in the backup cache.

The following shows the contents of the `eece.napier.ac.uk` file on the `ees99` computer. Each master zone should begin with an `SOA` record for the zone. The `A` entry defines an address, `NS` defines a name server, `CNAME` defines an alias and `MX` a mail server. It can be seen that `ees99` has the `mw` alias, thus `ees99.eece.napier.ac.uk` is the same as `mw.eece.napier.ac.uk`.

```
@           IN    SOA   ees99.eece.napier.ac.uk.   mike.ees99.eece.napier.ac.uk. (
                        199806171          ; Serial
                        10800              ; Refresh every 3 hrs
                        1800               ; Retry every 1/2 hr
                        604800             ; Expire (seconds)
                        259200 )           ; Minimum time-to-live
            IN    NS    ees99.eece.napier.ac.uk.
            IN    NS    eepc02.eece.napier.ac.uk.
localhost   IN    A     127.0.0.1
@           IN    MX    11       146.176.151.139.
hp350       IN    A     146.176.144.10
mwave       IN    A     146.176.144.11
hplb69      IN    CNAME mwave
vax         IN    A     146.176.144.13
miranda     IN    A     146.176.144.14
triton      IN    A     146.176.144.20
mimas       IN    A     146.176.146.21
ees99       IN    A     146.176.151.99
mw          IN    CNAME ees99
```

The `SOA` lists a serial number, which has to be increased each time the master file is changed. This is because secondary servers check the serial number at an interval specified by the refresh time (`Refresh`). If the serial number increases then a zone transfer is done to load the new data. If the master server cannot be contacted when a refresh is due then the retry time (`Retry`) specifies the interval at which refreshes occur. If the master server cannot be contacted within the expire time interval (`Expire`) then all data from the zone is discard by secondary servers.

The following file lists some of the contents of the `net/net151` file. The `NS` entry defines the two name servers (`ees99` and `eepc02`) and the `PTR` entry maps an IP address to a domain name.

```
@         IN    SOA   ees99.eece.napier.ac.uk.  mike.ees99.eece.napier.ac.uk. (
                         199706091          ; Serial
                         10800              ; Refresh
                         1800               ; Retry
                         604800             ; Expire
                         259200 )           ; Minimum
          IN    NS    ees99.eece.napier.ac.uk.
          IN    NS    eepc02.eece.napier.ac.uk.
```

```
50       IN    PTR   ee50.eece.napier.ac.uk.
61       IN    PTR   eepc01.eece.napier.ac.uk.
62       IN    PTR   eepc02.eece.napier.ac.uk.
222      IN    PTR   pctest.eece.napier.ac.uk.
2        IN    PTR   pc345.eece.napier.ac.uk.
3        IN    PTR   pc307.eece.napier.ac.uk.
4        IN    PTR   pc320.eece.napier.ac.uk.
5        IN    PTR   pc331.eece.napier.ac.uk.
6        IN    PTR   pc401.eece.napier.ac.uk.
7        IN    PTR   pc404.eece.napier.ac.uk.
```

and for the net/net151 file:

```
@     IN    SOA   ees99.eece.napier.ac.uk.  mike.ees99.eece.napier.ac.uk. (
                        199806171       ; Serial
                        10800           ; Refresh
                        1800            ; Retry
                        604800          ; Expire
                        259200 )        ; Minimum
      IN    NS    ees99.eece.napier.ac.uk.
      IN    NS    eepc02.eece.napier.ac.uk.
10    IN    PTR   ees10.eece.napier.ac.uk.
11    IN    PTR   ees11.eece.napier.ac.uk.
12    IN    PTR   ees12.eece.napier.ac.uk.
13    IN    PTR   ees13.eece.napier.ac.uk.
14    IN    PTR   ees14.eece.napier.ac.uk.
15    IN    PTR   ees15.eece.napier.ac.uk.
```

The entries take the form:

<domain> *<op_ttl>* *<opt_class>* *<type>* *<resource_record_data>*

where

domain either a . which defines the root domain, a @ which defines the current origin, or a standard domain name. If it is a standard domain and it does not end with a ., then current origin is appended to the domain, else the domain names unmodified.

op_ttl is an optional number for the time to live.

opt_class is the object address type. This can be IN (for DARPA Internet) or HS (for Hesiod class).

resource_record This contains one of the following definitions:

- A (address).
- CNAME (canonical name which is an alias for another host).
- GID (group ID).
- HINFO (host information)
- MB (mailbox domain name).
- MG (mailbox domain name).
- MIFO (mailbox or mail list information).
- MR (mail rename domain name).
- MX (mail exchanger which handles mail for a given host).
- NS (name server for the domain).
- PTR (pointer record which maps an IP address to a hostname).
- SOA (start of authority record). The domain of originating host,

domain address of maintainer, a serial and the other parameters (refresh, retry time, expire time and minimum TTL) are also defined.

- TXT (text string).
- UID (user information).
- WKS (well known service). This defines an IP address followed by a list of services.

The contents of the `sobasefile` is given next. This defines the two domain name servers.

```
@      IN       SOA   ees99.eece.napier.ac.uk.   mike.ees99.eece.napier.ac.uk. (
                          19970307         ; Serial
                          10800            ; Refresh
                          1800             ; Retry
                          604800           ; Expire
                          259200 )         ; Minimum
       IN       NS    ees99.eece.napier.ac.uk.
       IN       NS    eepc02.eece.napier.ac.uk.
```

TCP/IP services

Port	*Protocol*	*Service*	*Comment*
1	TCP	TCPmux	
7	TCP/UDP	echo	
9	TCP/UDP	discard	Null
11	TCP	systat	Users
13	TCP/UDP	daytime	
15	TCP	netstat	
17	TCP	qotd	Quote
18	TCP/UDP	msp	Message send protocol
19	TCP/UDP	chargen	ttytst source
21	TCP	ftp	
23	TCP	telnet	
25	TCP	smtp	Mail
37	TCP/UDP	time	Timserver
39	UDP	rlp	Resource location
42	TCP	nameserver	IEN 116
43	TCP	whois	Nicname
53	TCP/UDP	domain	Domain name server
57	TCP	mtp	Deprecated
67	TCP	bootps	BOOTP server
67	UDP	bootps	
68	TCP/UDP	bootpc	BOOTP client
69	UDP	tftp	
70	TCP/UDP	gopher	Internet Gopher
77	TCP	rje	Netrjs
79	TCP	finger	
80	TCP/UDP	www	WWW HTTP
87	TCP	link	Ttylink
88	TCP/UDP	kerberos	Kerberos v5
95	TCP	supdup	
101	TCP	hostnames	
102	TCP	iso-tsap	ISODE.
105	TCP/UDP	csnet-ns	CSO name server

107	TCP/UDP	rtelnet	Remote Telnet
109	TCP/UDP	pop2	POP version 2
110	TCP/UDP	pop3	POP version 3
111	TCP/UDP	sunrpc	
113	TCP	auth	Rap identity authentication
115	TCP	sftp	
117	TCP	uucp-path	
119	TCP	nntp	USENET News Transfer Protocol
123	TCP/UDP	ntp	Network Time Protocol
137	TCP/UDP	netbios-ns	NETBIOS Name Service
138	TCP/UDP	netbios-dgm	NETBIOS Datagram Service
139	TCP/UDP	netbios-ssn	NETBIOS session service
143	TCP/UDP	imap2	Interim Mail Access Protocol Ver2
161	UDP	snmp	Simple Net Management Protocol
162	UDP	snmp-trap	SNMP trap
163	TCP/UDP	cmip-man	ISO management over IP (CMOT)
164	TCP/UDP	cmip-agent	
177	TCP/UDP	xdmcp	X Display Manager
178	TCP/UDP	nextstep	NeXTStep NextStep
179	TCP/UDP	bgp	BGP
191	TCP/UDP	prospero	
194	TCP/UDP	irc	Internet Relay Chat
199	TCP/UDP	smux	SNMP Unix Multiplexer
201	TCP/UDP	at-rtmp	AppleTalk routing
202	TCP/UDP	at-nbp	AppleTalk name binding
204	TCP/UDP	at-echo	AppleTalk echo
206	TCP/UDP	at-zis	AppleTalk zone information
210	TCP/UDP	z3950	NISO Z39.50 database
213	TCP/UDP	ipx	IPX
220	TCP/UDP	imap3	Interactive Mail Access
372	TCP/UDP	ulistserv	UNIX Listserv
512	TCP/UDP	exec	Comsat
513	TCP	login	
513	UDP	who	Whod
514	TCP	shell	No passwords used
514	UDP	syslog	
515	TCP	printer	Line printer spooler
517	UDP	talk	
518	UDP	ntalk	
520	UDP	route	RIP
525	UDP	timed	Timeserver
526	TCP	tempo	Newdate
530	TCP	courier	Rpc
531	TCP	conference	Chat
532	TCP	netnews	Readnews
533	UDP	netwall	Emergency broadcasts
540	TCP	uucp	Uucp daemon
543	TCP	klogin	Kerberized 'rlogin' (v5)
544	TCP	kshell	Kerberized 'rsh' (v5)
556	TCP	remotefs	Brunhoff remote filesystem
749	TCP	kerberos-adm	Kerberos 'kadmin' (v5)
750	UDP	#kerberos	Kerberos (server) UDP
750	TCP	#kerberos	Kerberos (server) TCP

760	TCP	krbupdate	Kerberos registration
761	TCP	kpasswd	Kerberos "passwd"
765	TCP	webster	Network dictionary
871	TCP	supfilesrv	SUP server
1127	TCP	supfiledbg	SUP debugging
1524	TCP	ingreslock	
1524	UDP	ingreslock	
1525	TCP	prospero-np	Prospero non-privileged
1525	UDP	prospero-np	
2105	TCP	eklogin	Kerberos encrypted rlogin
5002	TCP	rfe	Radio Free Ethernet

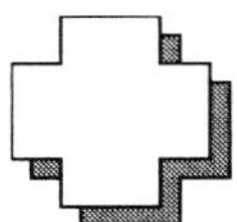

Additional Material

P.1 PGP

PGP (Pretty Good Privacy) uses the RSA algorithm with a 128-bit key. It was developed by Phil Zimmermann and gives encryption, authentication, digital signatures and compression. Its source code is freely available over the Internet and its usage is also free of charge, but it has encountered two main problems:

- The source code is freely available on the Internet causing the US government to claim that it violates laws which relate to the export of munitions. Current versions have since been produced outside of the US to overcome this problem.
- It uses algorithms which have patents, such as RSA, IDEA and MD5.

Figure P.1 shows the basic encryption process.

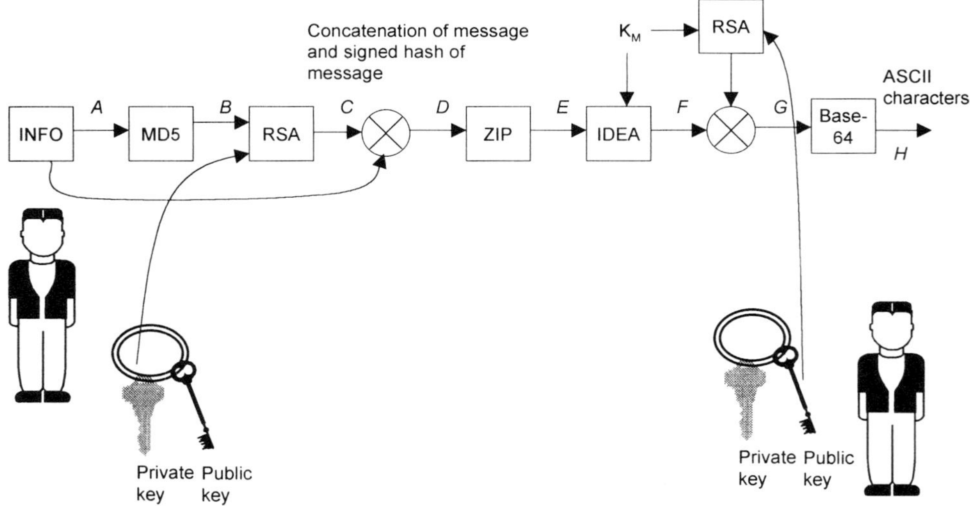

Figure P.1 PGP encryption.

The steps taken are:

A. Sender hashes the information using the MD5 algorithm.
B. Hashed message is the encrypted using RSA with the sender's private key (this is used to authenticate the sender as the senders public key will be used to decrypt the message).

C. Encrypted message is then concatenated with the original message.
D. Message is compressed using LZ compression.
E. A 128-bit IDEA key (K_M) is generated by some random input, such as the content of the message and the typing speed.
F. K_M is then used with the IDEA encryption. K_M is also encrypted with the receiver's public key.
G. Output from IDEA encryption and the encrypted K_M key are concatenated together.
H. Output is encoded as ASCII characters using Base-64 (See Section 21.7.4 on Electronic Mail).

To decrypt the message the receiver goes through the following steps:

A. Receiver reverses the Base-64 conversion.
B. The receiver decrypts the K_M key using their own private RSA key.
C. The K_M key is then used with the IDEA algorithm to decode the message.
D. The message is then decompressed using an UNZIP program.
E. The two fragments produced after the uncompression will be the plaintext message and an MD5/RSA encrypted message. The plaintext message is the original message, where the MD5/RSA encrypted message can be used to authenticate the sender. This is done by applying the sender's public key to the uncompressed encrypted part of the message. This should produce the original plaintext message.

PGP allows for three RSA key sizes. There are:

- 384 bits. This is intended for the casual user and can be cracked by serious crackers.
- 512 bits. This is intended for the commercial user and can only be cracked by organizations with a large budget and extensive computing facilities.
- 1024 bits. This is intended for military uses and, at present, cannot be cracked by anyone. This is the recommend key size for most users of reasonably powerful computers (386/486/Pentium/*etc*). In the future, a 2048-bit code may be used.

P.2 NetWare 4 and NDS

P.2.1 Introduction

The main disadvantages of NetWare 3.x are:

- It uses SPX/IPX which is incompatible with TCP/IP traffic.
- It is difficult to synchronize servers with user information.
- The file structure is local to individual servers.
- Server architecture is flat and cannot be organized into a hierarchical structure.

These have been address with NetWare 4.1, in which the bindery has been replace by Novell Directory Services (NDS). NDS is a combination of features from OSI X.500 and Banyan StreeTalk. Its main characteristics are:

- Hierarchical server structure.
- Network-wide users and groups.
- Global objects. NDS integrates users, groups, printers, servers, volumes and other physical resources into a hierarchical tree structure.
- System-wide login with a single password. This allows users to access resources which are connected to remote servers.
- NDS processes logins between NetWare 3.1 and NetWare 4.1 servers, if the login names and passwords are the same.
- Supports distributed file system.
- Synchronization services. NDS allows for directory synchronization, which allows directories to be mirrored on different partitions or different servers. This provides increased reliability in that if a server develops a fault then the files on that server can be replicated by another server.
- Standardized organizational structure for applications, printers, servers and services. This provides a common structure across different organizations.
- It integrates most of the administrative task in Windows-based NWADMIN.EXE program.
- It is a truly distributed system where the directory information can be distributed around the tree.
- Unlimited number of licenses per server. NetWare 3.1 limits the number of licenses to 250 per server.
- Support for NFS server for UNIX resources.
- Multiple login scripts, as opposed to system and user login scripts in NetWare 3.1.
- Windows NT support.

NDS is basically a common, distributed Directory database of logical and physical resources into a single information system. Many other applications have used Directory databases, such as electronic mail and network management. NDS servers within a network access the Directory database for the connected resources and how they are accessed. Thus application programs do not need to know the physical location and on which server is it connected, only its logical name.

The main reason to upgrade to NDS is that it better reflects the organizational structure of networked equipment within the organization. NetWare 3.1 is a server-based approach where resources are grouped around servers. This leads to increased maintenance around these servers, thus updates to one server may have to be updated on other servers. NDS allows for a central administration with a structure that reflects organizational structures.

P.2.2 NetWare directory services (NDS)

One of the major changes between NetWare 3.x and NetWare 4.1 is NDS. A major drawback of the NetWare 3.x bindery files is that they were independently maintained on each server. NDS addresses this by setting up a single logical database, which contains information on all network-attached resources. It is logically a single database, but may be physically located on different servers over the network. As the database is global to the network, a user can log in to all authorized network-attached resources, rather than requiring to login into each separate server. Thus, administration is focused on the single database.

As with NetWare 3.x bindery services, NDS organizes network resources by objects, properties, and values. NDS differs from the bindery services in that it defines two types of object:

- Leaf objects – which are network resources such as disk volumes, printers, printer queues, and so on.
- Container objects – which are cascadable organization units that contain leaf objects. A typical organizational unit might be company, department or group.

NDS organizes networked resources in a hierarchical or tree structure (as most organizations are structured in this way). The top of the tree is the root object, to which there is only a single root for an entire global NDS database. Servers then use container objects to connect to branches coming off the root object. This structure is similar to the organization of a directory file structure and can be used to represent the hierarchical structure of an organization. Figure P.2 illustrates a sample NDS database with root, container and leaf objects. In this case, the organization splits into four main containers: Electrical, Mechanical, Production and Administration. Each of these containers has associated leaf objects, such as disk volumes, printer queues, and so on. This is a similar approach to Workgroups in Microsoft Windows.

To improve fault tolerance, NDS allows branches of the tree (or partitions) to be stored on multiple file servers. These mirrors are then synchronized to keep them up-to-date. Another advantage of replicating partitions is that local copies of files can be stored so that network traffic is reduced.

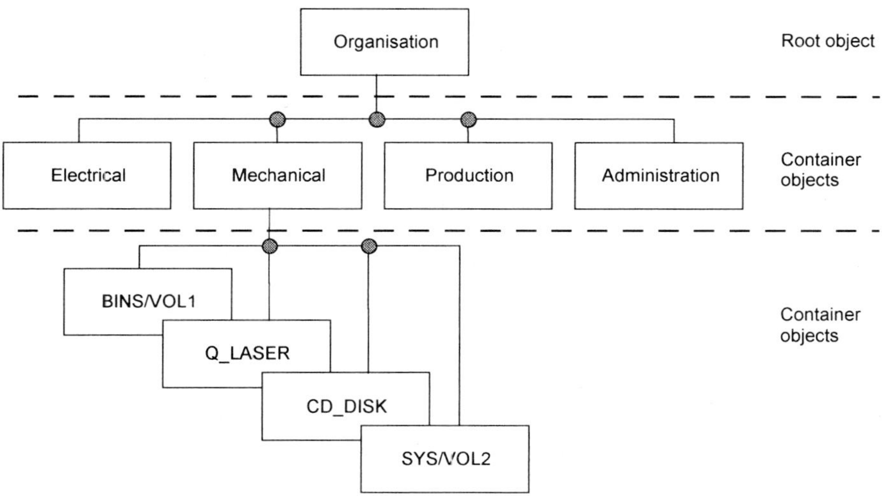

Figure P.2 NDS structure.

The container objects are:

[ROOT]. This is the top level of the inverted tree and contains all the objects within the organizational structure.

Organization. This object class defines the organizational name (such as FRED_ AND_CO). It is normally the next level after [ROOT] (or below the C=Country object).

User. This object defines an individual user. The first user created in a NetWare 4 system is the ADMIN user, which is typically the only user with rights to add and delete objects on the whole of the NDS structure.

NCP (NetWare Control Protocol) Server. This appears for all NetWare 4 servers.

Volume. This identifies the mounted volume for file services. A network file system data links to the Directory tree through Volume objects.

The most commonly uses objects are:

Bindery. These allow compatibility with existing Bindery-based NetWare 3, NetWare 3 clients and NetWare 4 servers which do not completely implement NDS. They display objects that isn't a user, group, queue, profile or print server, which was created using the bindery services.

Organizational unit. This object represents the OU part of the NDS tree. These divide the NDS tree into subdivisions, which can represent different geographical site, different divisions or workgroups. Different division might be PRODUCTION, ACCOUNT, RESEARCH, and so on. Each Organizational Unit has its own login script.

Organization role. This object represents a defined role within an organization object. It is thus easy to identifier users who have an administrative role with the organization.

Group. This object represents a grouping of users. All users within a group inherit the same access rights.

Directory map. This object points to a file system directory on a mounted volume. It is typically used to create a global file system which has physically separate parts.

Alias. This identifies an object with another name. For example, a print queue which is called NET_PRINT1 might an alias name of HP _LASER_JET_6.

Printer. This can either be connected to the printer port of a PC, or connected to a NetWare server.

Print queue. This object represents the queue of print jobs.

Profile. This object defines a special scripting file. This can be a global login script, a location login script or a special login script.

Print server. This object allows print jobs to be queued, waiting to be serviced by the associated printer.

P.2.3 NDS tree

Figure P.3 shows the top levels of the NDS tree. These are:

- [ROOT]. This is the top level of the tree. The top of the NDS tree is the [ROOT] object.
- C=Country. This object can be used, or not, to represent different counties, typically where an organization is distributed over two or more countries. If it is used then it must be placed below the [ROOT] object. NDS normally does not use the Country object and uses the Organization Unit to define the geographically located sites, such as SALES_UK.[ROOT], SALES_USA.[ROOT], and so on.
- L=Locality. This object defines locations within other objects, and identifies network portions. The Country and Locality objects are included in the X.500 specification, but they are not normally used, because many NetWare 4 utilities do not recognized it. When used, it must be place between below the [Root] object, Country object, Organization object, or Organizational Unit object.
- LP=Licensed Product. This object is automatically created when a license certificate is installed. When used, it must be place between below the [Root] object, Country object, Organization object, or Organizational Unit object.
- O=Organization. This object represents the name of the organization, a company division or a department. Each NDS Directory tree has at least one Organization object, and it must be placed below the [Root] object (unless the tree uses the Country or Locality object).
- OU=Organization Unit. This object normally represents the name of the organizational unit within the organization, such as Production, Accounts, and so on. At this level, User objects can be added and a system level login script is created. It is normally placed below the Organizational object.

The structure of the NDS should reflect the organization of the company, for its organizational structure, its locations and the organization of its networks. Normally there is only one Organization object as this makes it easier to merge the NDS tree with other organizations. With every Organization object, there are normally several Organization Units.

Apart from the container objects (C, O, OU, and so on) there are leaf objects. These are assigned a CN (for Common Name). They include:

CN=AFP Server
CN=Bindery
CN=Bindery Queue
CN=Computer
CN=Directory Map
CN=Group
CN=Organizational Role
CN=Print Queue
CN=Print Server
CN=Printer
CN=Profile
CN=Server
CN=User
CN=Volume

If possible, the NDS tree depth should have between four and eight levels. This makes management easier and allows resources to be easily accessed.

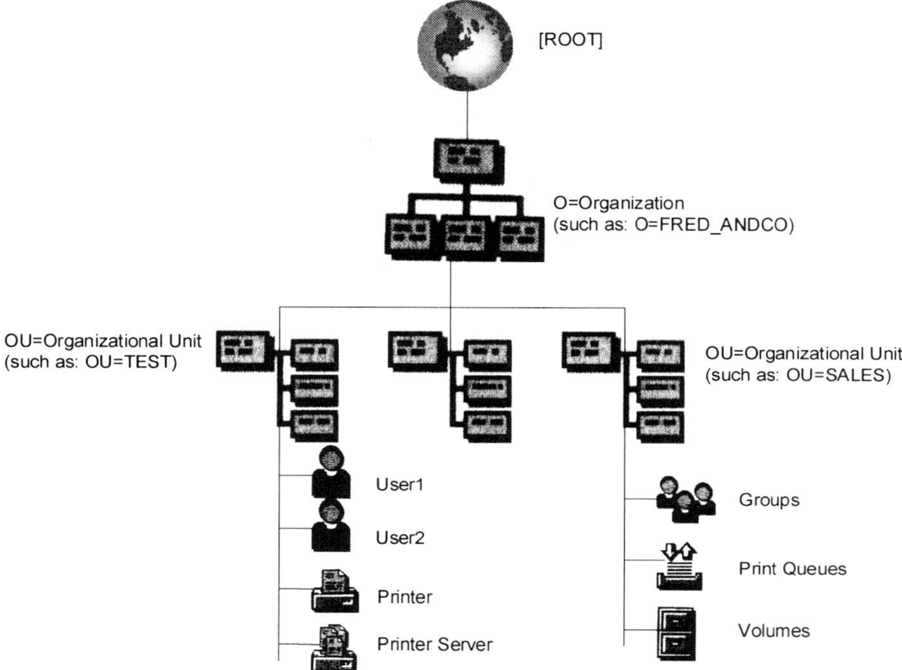

Figure P.3 NDS structure.

P.2.4 Typical naming syntax

The NDS tree can use many different naming formats, but a standardized naming structure has been developed. These are:

	Syntax	*Example*
[ROOT]	*company*_TREE	FRED_TREE
Organization	*company_name*	O=FRED
Organization Units	*location* (or *department*)	OU=SALES
Servers	*location-department*-SRV#	SALES-SRV1
Printer Servers	*location-department*-PS#	SALES-LZ5-PS3
Printers	*printer*-P#	HPLJ5-P2
Print Queues	*type*-P#	HPLJ5-P2
Volumes	*server_volume*	SALES-SRV1_DATA

P.2.5 Object names

The place at which an object is placed is called its context. Two objects which are placed in the same container have the same context. For example, if the user FRED_B works for the Fred & Co. (O=FRED_AND_CO), within the Test Department (OU=TEST), which is within the Engineering Unit (OU=ENGINEERING) then his context will be:

```
OU=TEST.OU=ENGINEERING.O=FRED_AND_CO
```

An object is either identified by its distinguishing name (such as `LP_LASER5`) or by its complete name (CN). In the name, periods separate the objects (these periods are similar to back slashes or forward slashes, which is common in many operating systems). For a complete name, which is referred to from the [ROOT] object, a leading period is used. Whereas, a relative name does not have a leading period. For example, a complete name for a User object `FRED_B` could be:

```
.CN=FRED_B.OU=TEST.OU=ENGINEERING.O=FRED_AND_CO
```

This defines a User, which has an Organization of `FRED_AND_CO`, which has an Organization Unit called `ENGINEERING`, there is then a subdivision below this called `TEST`. It is also possible to define a relative distinguishing name (RDN) which defines the relative path with respect to the current context.

Periods can be added to the start or the end of the context. They have the following definitions:

- Leading period. NDS ignores the current context of the object and resolves the name at the [ROOT] object.
- Trailing period. NDS selects a new context when resolving an object's complete name at the [ROOT] object.

For example, the partial name for the User object `FRED_B` relative to other objects in `OU=TEST` would be:

```
.CN=FRED_B.
```

The partial name of the User object `FRED_B` that has a complete name of:

```
.CN=FRED_B.OU=TEST.OU=ENGINEERING.O=FRED_AND_CO
▲
'------------------------------------------------- Distinguishing
                                                   name. Resolve from
                                                   [ROOT]
```

relative to a server object with a complete name of:

```
.CN=OU=SALES-SRV1.OU=SALES.O=FRED_AND_CO
```

is:

```
                                   ┌-------------- Move up one
                                   |               level
                                   ▼
CN=FRED_B.OU=TEST.OU=ENGINEERING.
▲
'------------------------------------------------- Relative name.
```

The `HPLJ5-P2` printer object which has the complete name of:

```
.CN=HPLJ5-P2.OU=TEST.OU=ENGINEERING.O=FRED_AND_CO
```

would be referred, within the OU=TEST.OU=ENGINEERING.O=FRED_AND_CO container, as:

```
CN=HPLJ5-P2
```

Typeless name

Notice that a relative name has a trailing period to identify that it is a partial name. It is also possible not to include the object types (such as CN for common name, OU for Organizational Unit and O for Organization). This is called a typeless name, and NDS makes a guess as to the object types. For example:

```
FRED_B.TEST.ENGINEERING.FRED_AND_CO
```

is the same as one of the previous examples. When guessing NDS uses the following rules:

- The object which is furthest to the left is assumed to be a common name (leaf object).
- The object which is furthest to the right is assumed to be the organization (container object).
- All other objects are assumed to be Organizational Units (container objects).

P.2.6 CX

The CX (Change conteXt) command is used to display or modify the context, or to view containers and leaf objects in the Directory tree. In a Command Prompt window, the following can be used:

Command	*Description*
CX	displays current context
CX /?	display help manual
CX /CONT	display all containers in the current context
CX /T	display all containers at and below the current context
CX .	move up one level
CX ..	move up two levels
CX /CONT	display containers in the [ROOT]
CX *content*	display context for *context*
CX /R	change current context to [ROOT]
CX /A	display all containers and objects in the current context.
CX /R /A /T	display all containers and objects, from the [ROOT] down.

For example to set the current context to the TEST.ENGINEERING.FRED_AND_CO container:

```
CX TEST.ENGINEERING.FRED_AND_CO
```

Then to change the context to ENGINEERING:

```
CX ENGINEERING.FRED_AND_CO
```

or

```
CX .
```

P.2.7 Startup files and scripts

Much of the initialization of a client is done with start-up files and scripts. The main startup files are:

- `CONFIG.SYS` and `AUTOEXEC.BAT`. These are standard start-up files for the PC and normally setup the environment of the computer. The `AUTOEXEC.BAT` file should include the `STARTNET.BAT` file.
- `STARTNET.BAT`. Provides a network connection.
- `NET.CFG`. Customizes the NetWare set up, such as setting ODI and VLM settings.

The login scripts are:

- Container Login Scripts. These set up the Organization and Organizational Unit properties, and generally replace System login scripts.
- Profile Login Scripts. These set up the environment of User groups.
- User Login Scripts. These customize the User environment. If no User login script exists then a Default Login Script is executed.

The NET.CFG file is similar to NetWare 3 but has extra lines to define the NetWare 4 options. An example file is:

```
Link Driver NE2000
         Int #1 11
         Port #1 320
         Frame Ethernet_II
         Frame Ethernet_802.3
         Protocol IPX 0 Ethernet_802.3

NetWare DOS Requester
      NAME CONTEXT= "OU=electrical.OU=engineering.O=napier"
      PREFERRED SERVER = EEE-SRV1
      FIRST NETWORK DRIVE = G
      NETWARE PROTOCOL = NDS, BIND
```

This defines that the name context for the user (with `NAME CONTEXT`) and that the preferred server is `EEE-SRV1`. The first network drive will be G: and the NetWare protocol is NDS and Bindery.

Drive disks can be mounted by adding lines to the Login Script (such as NETSTART.BAT, which is started from the AUTOEXEC.BAT file). For example, to mount the F:, G: and M: drives then the following could be added:

```
MAP ROOT F:= .EEE-SRV1.ENGINEERING.NAPIER\SYS:APPS
MAP ROOT G:= .CRAIGLOCKHART_1.MAJOR.NAPIER.AC.UK\SYS:APPS
MAP ROOT M:= .CRAIGLOCKHART_3.MAJOR.NAPIER.AC.UK\SYS:MAIL
```

P.2.8 Volume mapping

Volumes can be mounted as drives using the syntax:

MAP *drive_letter*:=CN=*servername_volumename.context*:

For example, to map the DATA volume of the TEST server to drive letter F: then the following is used:

```
MAP F:=CN=TEST_DATA.OU=TEST.:
```

P.2.9 Country Object

The country object is commonly not used as it fixes the geographical location of objects. It has the advantage, though, is that is fits into a common Internet naming structure (such as, `www.eece.napier.ac.uk`) or X.500 names. Most network though have the Organizational Unit following the [ROOT] level. For example and an educational organization in the UK will have a country object of UK and the organization object of AC (as defined in the Internet name). The Organization Unit would then be the name of the academic organization (in this case, Napier). Next, the facilities and departments are defined, as follows:

```
[Root]
   c=uk
      o=ac
         ou=napier
            ou=Arts
            ou=Business
            ou=Engineering
               ou=electrical
               ou=mechanical
               ou=computing
```

Thus, the context name for a printer (`LJET5`) in the Electrical department would be:

```
CN=LJET5.OU=ELECTRICAL.OU=ENGINEERING.O=NAPIER.C=UK
```

This is obviously similar to the Internet name for the device, which would be:

```
ljet5.electrical.engineering.napier.ac.uk
```

P.2.10 User class

NDS has an object-oriented database, where each object (such as Users, Printers, and so on) has associated properties. The User object has the following properties:

- Login Name. Normally the first character of the first name followed by the last name, such as BBuchanan (for Bill Buchanan). [Required]
- Given Name. Users first name. [Required]
- Last Name. Users last name. [Required]
- Full Name. Users full name. [Required]
- Generational Qualification. [Optional]
- Middle Initial. [Required]
- Other Name. [Optional]
- Title. Job title. [Required]

- Description. [Optional]
- Location. City or location. [Required]
- Department. [Optional]
- Telephone Number. Full telephone number [Required]
- Fax Number. Full Fax number [Required]
- Language. Spoken language [Optional]
- Network address. [System adds this]
- Default server. Server that the user initially logs into [Optional]
- Home Directory. Volume:subdirectory\user, such as DATA:HOME\BILL_B [Required]
- Required Password. Force password, or not. [Required]
- Account Balance. [Optional]
- Login Script. [Optional]
- Print Job Configuration. [Optional]
- Post Office Box. [Optional]
- Street. [Optional]
- City. [Optional]
- Start or Province. [Optional]
- Zip Code. [Optional]
- See also. [Optional]

P.2.11 Bindery services

In NetWare 4.1 the bindery services have been replaced by NetWare Directory Services (NDS). Many networks take some time to be fully upgrade from bindery to NDS, thus NetWare 4.1 supports bindery. This allows a NetWare 3 server to be upgraded to NetWare 4.1, but still run a bindery. It also allows NetWare 4 to integrate with NetWare 3 servers (typically, which run print services).

NetWare 4 servers support the following bindery-based resources:

- Bindery objects class.
- Bindery programs.
- Bindery-based NetWare client software.
- Groups.
- Print servers.
- Profiles.
- Queues.
- Users.

On a bindery service, NDS supports a flat structure for leaf objects in an Organization or Organizational Unit object. All objects within the specified container can then be accessed by NDS objects and by bindery-based servers and client workstations. On a bindery enabled server, a client gets it login script from that server. The login and script are not automatically transmitted to other servers (as they would with NDS).

Figure P.4 shows an example of a bindery-enabled server which is as an Organizational Unit object (OU).

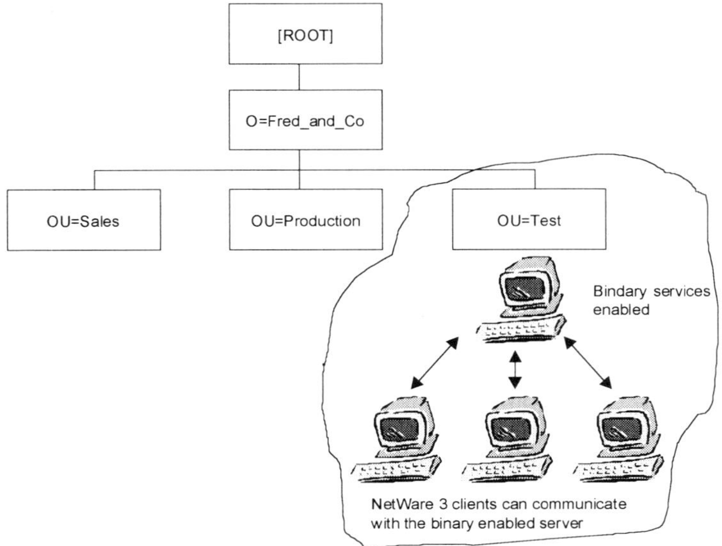

Figure P.4 Bindery services.

P.2.12 Time synchronization

Time synchronization between servers is important as it allows NDS events and modifications to be accurately time stamped. There are two main options:

- Single reference configuration. This provides a single source of time reference and is typically used with networks with less than thirty servers.
- Time provider group configuration. This provides a single reference primary server and, at least, two other primary time providers. It is typically used when there are more that thirty servers connected to the network. The other primary time providers allow for a system failure of the reference primary server.

The time synchronization is provided with:

- SAP. With SAP the SAP type is 0000 0010 0110 1011 (026Bh). The main disadvantage with this method is that SAP generally adds to network traffic and, as SAP is self configuring, an incorrectly set up can cause incorrect time to be transmitted.
- Configured List Communication. This method uses allows each server to keep a list of server which it can communicate with. This will generally lead to less network traffic as it does not use an SAP broadcast. It also stops incorrectly set up servers from transmitting incorrect time information (as the server will only communicate with preferred time servers).

NetWare uses the TIMESYNC.NLM (NetWare Loadable Module) module to synchronize their local time. The server then calculates Universal Coordinated Time (UTC) which provides a world standard time. UTC (Co-ordinated Universal Time) is a machine-independent time standard. It assumes that there are 86 400 seconds each day (24×60×60) and once every year or two an extra second is added (a "leap" second). This is normally added on 31st December or 30th June.

Most computer systems defined time with GMT (Greenwich Mean Time), which is UT (Universal Time). UTC is based on an atomic clock, whereas GMT is based on astronomical observations. Unfortunately, because the earth rotation GMT is not uniform, and is not as accurate as UTC. UTC is calculated by:

UTC = LOCAL TIME + *timezone_offset* + *current_daylight_adjustment*

With TIMESYNC.NLM the system time is not actually changed, the local clock is either speeded up (if the time is behind) or slowed down (if it is ahead). This makes for gradual changes in the system time. This is especially important for server synchronization where directories and files are kept up-to-date between servers. An incorrectly set time on one server could cause an older file to replace a newer file. Every NDS object and their associated properties have a timestamp associated with then.

NDS timestamp

Particular problems caused with time on computer systems are the Year 2000 bug (where dates are referenced to just the last two digits of the year) and where there is a roll-over in the counter value which stores the system time. The Year 2000 bug can be easily eradicated by making sure that all references to time take into account the full year format.

The PC contains a 32-bit counter which is updated every second and is reference to the 1st January 1970 (the starting date for the PC). This provides for 4,294,967,296 seconds (715,827,882 minutes, 11,930,465 hours, 497,103 days and 1361 years). The format of the NDS timestamp uses this format and adds other fields to define the place the event occurred and an indication of events that occur within a single second. It uses 64 bits and its format is:

- Seconds (32 bits). This stores the number of seconds since 1/1/1970. This allows for 4 billion seconds, which is approximately 1371 years, before a roll-over occurs.
- Replica Number (16 bits). This is a unique number which defines where the event occurred and the timestamp issued.
- Event ID (16 bits). Defines each event that occurs within a second a different Event ID. This is requires as many events can occur within a single second. This value is reset on every second, and thus allows up to 65,536 events each second.

NDS always uses the most recently time stamped object or properties for any updates. When an object is deleted its property is marked as "not present". It will only be deleted once the replica synchronization process propagates the change to all other replicas.

Time server types

NetWare 4.1 servers are set up as time servers when they are installed. They can either be:

- Primary time servers. A primary time server provides time information to others, but must contact at least primary (or reference) server for their own time.
- Secondary time servers. These are time consumers, which receive their time from other

servers (such as from a primary, reference or single reference time server).

- Reference time servers. These servers do not need to contact any other servers, and provide a time source for other primary time servers. This is a good option where there is a large network, as the primary time servers can provide local time information (this is called a time provider group).
- Single reference time servers. These servers do not need to contact other time servers to get their own time and are used as a single source of time. This is normally used in a small network, where there is a single reference time server with one or more secondary time servers. The single reference time server and reference time server normally get their local time information from another source, such as Internet time, radio or satellite time. This is the default condition for installation.

P.3 Windows NT

P.3.1 Administrative tools

Windows NT has a whole host of administrative tools. Figure P.5 shows the available tools. These include:

- User Manager for Domains. This tool supports the management of security for a domain.
- System Policy Manager. This tool manages the rights granted to groups and user accounts.
- Remote Access Admin. This is a tool which allows users to remotely login into a network server.
- Event Viewer. This is a tool which monitors system, security and application events.
- Disk Administrator. This is a tool which gives a graphical interface for managing disks.
- License Manager. This tool helps to manage and track licenses within an organization. Licensing can be done per seat or per sever.

P.3.2 User and group accounts

Each user within a domain has a user account and is assigned to one or more groups. Each group is granted permissions for the file system, accessing printers, and so on. Group accounts are useful because they simplify an organization into a single administrative unit. They also provide a convenient method of controlling access for several users who will be using Windows NT to perform similar tasks. By placing multiple users in a group, the administrator can assign rights and/or permissions to the group.

Each user on a Windows NT system has the following:

- A user name (such as `fred_bloggs`).
- A password (assigned by the administrator then changed by the user).
- The groups in which the user account is a member (for example, `staff`).
- Any user rights for using the assigned computer.

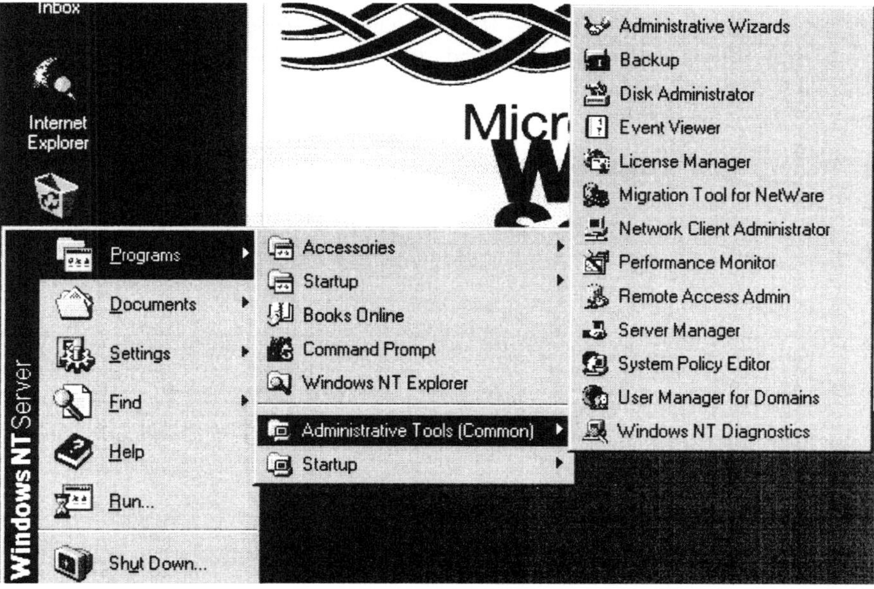

Figure P.5 Windows NT administrative tools.

Each time a user attempts to perform a particular action on a computer, Windows NT checks the user account to determine whether the user has the authority to perform that action (such as read the file, write to the file, delete the file, and so on).

Normally there are three main default user accounts: Administrator, Guest and an 'Initial User' account. The system manager uses the Administrator account to perform such tasks as installing software, adding/deleting user accounts, setting up network peripherals, installing hardware, and so on.

Guest accounts allow occasional users to log on and be granted limited rights on the local computer. The system manager must be sure that the access rights are limited so that hackers or inexperienced users cannot do damage to the local system.

The 'Initial User' account is created during installation of the Windows NT workstation. This account, assigned a name during installation, is a member of the Administrator's group and therefore has all the Administrator's rights and privileges.

After the system has been installed, the Administrator can allocate new user accounts, either by creating new user accounts, or by copying existing accounts.

P.3.3 User Manager

The User Manager allows the creation and management user accounts, creation and management of groups, and the management the security policies. Figure P.6 shows the user addition window (which is selected with the User → New User option). The title bar defines the domain or computer that is displayed for administration. Figure P.7 shows a window with several added users. It can be seen that initially that there are two users: Administrator (the System Manager) and the Guest. It can be seen that there are also some default groups:

- Administrators, Members can fully administer the computer/domain.
- Backup Operators. Members can bypass file security to backup files.

- Guests. Users granted guest logins.
- Power Users. Members can share files and directories.
- Replicators. Supports file replication in a domain.
- Users. Ordinary users.

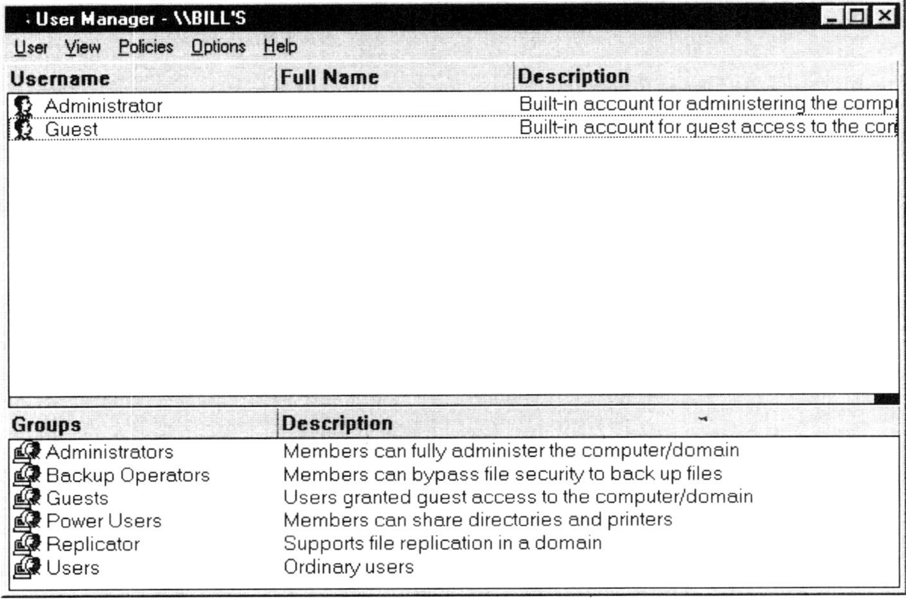

Figure P.6 User Manager.

When administering a computer running Windows NT Workstation or Windows NT Server that is not a domain controller, the list contains only local groups. These computers do not maintain global groups.

Users either have a global account icon or a local account icon. These are defined as:

- Global account. Typically, users have a global account which is a normal user account in a user's home domain. To define trust relationships for another domain, the user must have a global account on one domain to be granted access to another domain.
- Local account. A local account accommodates a user whose global account is in a domain which is not trusted by the current domain.

Typically, the system manager creates new accounts by copying existing users accounts. The items copied directly from an existing user account to a new user account are as follows:

- The description of the user (such as `Fred Bloggs, Ext 4444`).
- Group account membership (such as `Production`).
- Profile settings (such as home directory).
- If set, the attribute to stop the user from changing their password (sometimes the manager does not want the user to change the default password).
- If set, the attribute that causes the password to remain unexpired (sometimes the system manager forces users to change their passwords from time to time).

The items which are cleared and completed by the system manager are:

- The username and full name.
- The attribute that prompts users to change their passwords when they next login (normally a default password is initially set up and is changed when a user initially logs in).
- The attribute which disables the account (the manager must reset this before a user can log in).

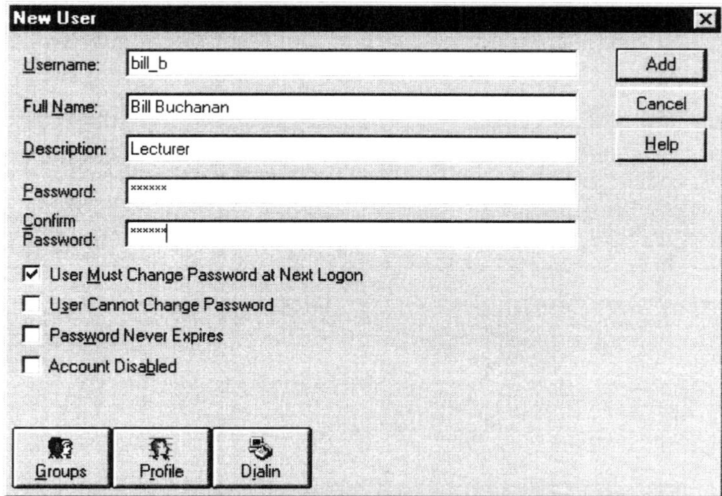

Figure P.7 Adding a user.

Figure P.8 shows the addition of several users and Figure P.9 shows the members of the Users group (showing the full name of the users). The groups that a user belongs to can be defined by selecting User → Properties then User Properties → Groups. Figure P.10 shows an example for the user bill_b. It can be seen that this user is a member of Users, but not of the other groups (such as Administers). A user can be added to a group by selecting the group in the right-hand side window and selecting the Add button. A user can be removed from a group by selecting the group on the left-hand window and selecting the Remove button.

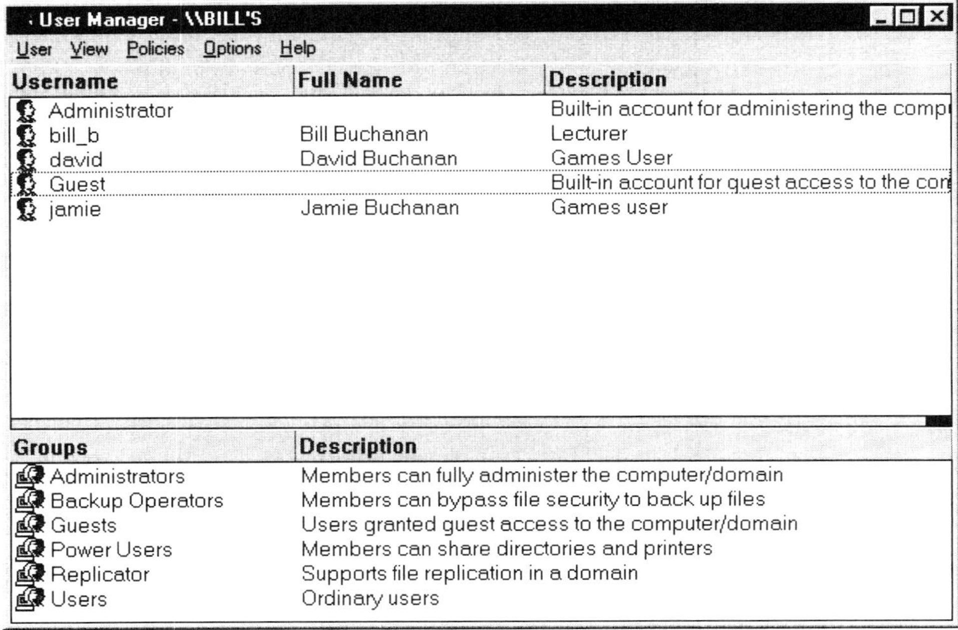

Figure P.8 Multiple users.

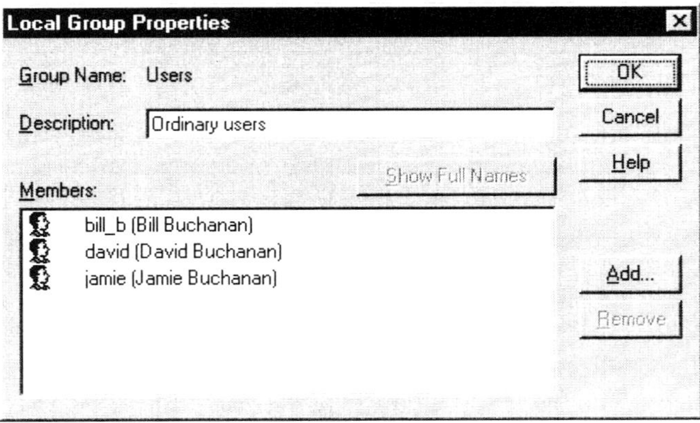

Figure P.9 Local group properties.

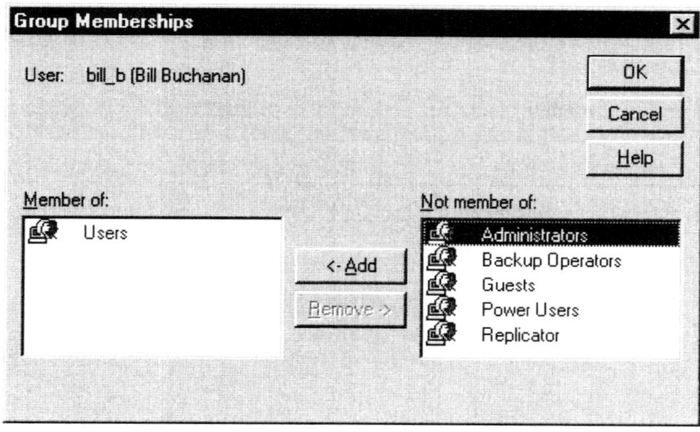

Figure P.10 Group membership.

P.3.4 System Policies Editor

The Systems Policy Editor provides policies for the following:

- Account policy. This policy defines the operation of the user password, such as minimum password size, number of acceptable bad passwords, whether there is a lock-out on the login, and so on.
- User Rights policy. This policy manages the rights of groups and user accounts. These rights allow users to access certain resources within the domain.
- Audit policy. This policy tracks auditing security events. These might be successful and/or unsuccessful events. The Event Viewer program can be used to view these events.
- Trust Relationships. Trust relationships create a link between two Windows NT Server domains and they allow users to be granted rights on other servers.

Account policy

This policy defines the operation of all user passwords, such as minimum password size, number of acceptable bad passwords, whether there is a lock-out on the login, and so on. Figure P.11 shows a sample window. It can be seen that, in this case, all passwords expire in 42 days. There is also no limit in the minimum time for a password change. Figure P.11 also shows that blank passwords are accepted and that there is no account lockout on a bad password entry. It can be seen that the number of acceptable bad passwords can also be set. This lockout can be reset after a number of minutes, if required.

User rights

The user rights policy manages the rights of groups and user accounts. These rights allow users to access certain resources within the domain. Normally these rights are granted to groups rather that to individual user accounts, as it is easier to administer. Figure P.12 shows an example.

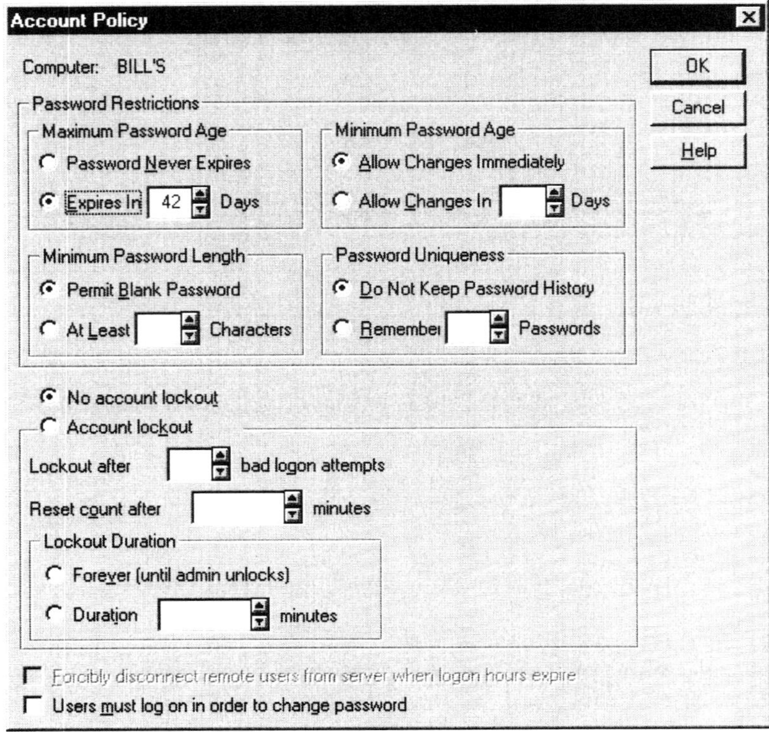

Figure P.11 Account policy window.

The rights include (with typical user groups are):

- Access this computer from network (Administrator, Everyone, Power Users).
- Backup files and directories (Administrator, Backup Operators)
- Change system time (Administrator, Power Users).
- Force shutdown from a remote computer (Administrator, Power Users).
- Load and unload device drivers (Administrator).
- Log on locally (Administrators, Backup Operators, Guests, Power Users, Users).
- Managing audits and security log (Administrator).
- Restore files and directories (Administrator, Backup Operators).
- Shut down the system (Administrators, Backup Operators, Power Users, Users).
- Take ownership of files or other objects (Administrator).

Initially the administrator selects the right and then chooses to Add… or Remove the group or user from the right.

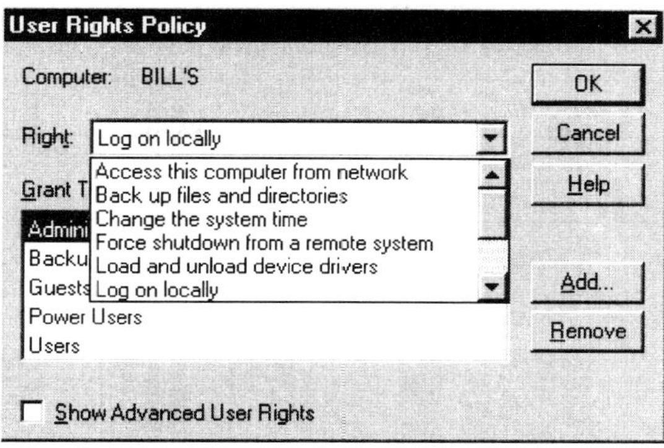

Figure P.12 User rights policy.

Audit

This policy tracks auditing security events. These might be successful and/or unsuccessful events. The Event Viewer program can be used to view these events. Figure P.13 shows an example of an audit policy window. In this case, all successful and unsuccessful login and logoffs will be logged. In addition, successful file and object accesses will be logged.

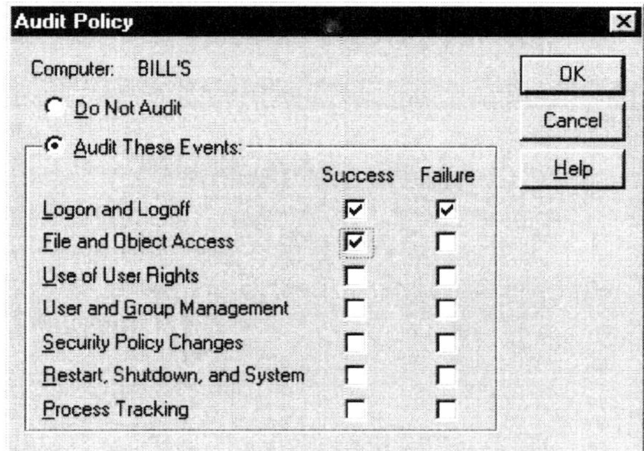

Figure P.13 Audit Policy.

Trust Relationships

Trust relationships create a link between two Windows NT Server domains and they allow users to be granted rights on other servers.

P.3.5 Control Panel and Windows NT Diagnostics

As with Windows 95 and Windows 98, Windows NT has a Control Panel from which users can modify or view system settings. Figure P.14 shows the Control Panel. The Windows NT Diagnostics program allows the user to view system settings. Figure P.15 shows the initial window.

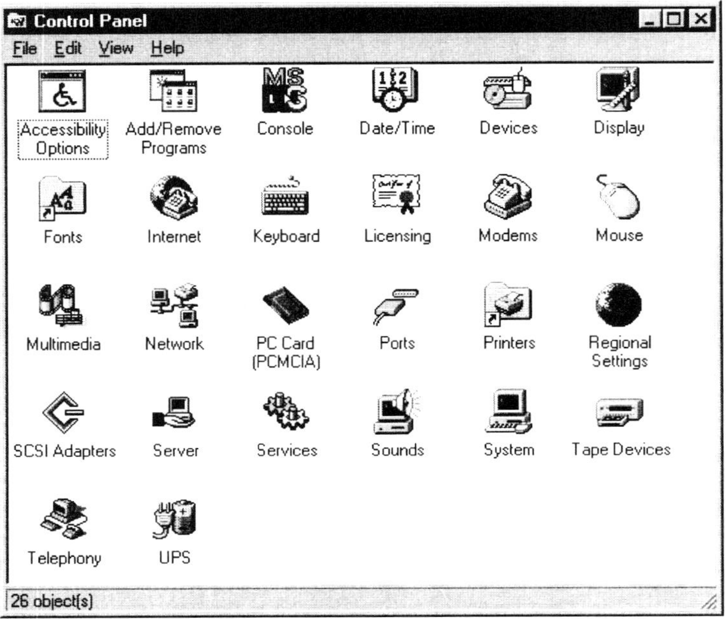

Figure P.14 Control Panel.

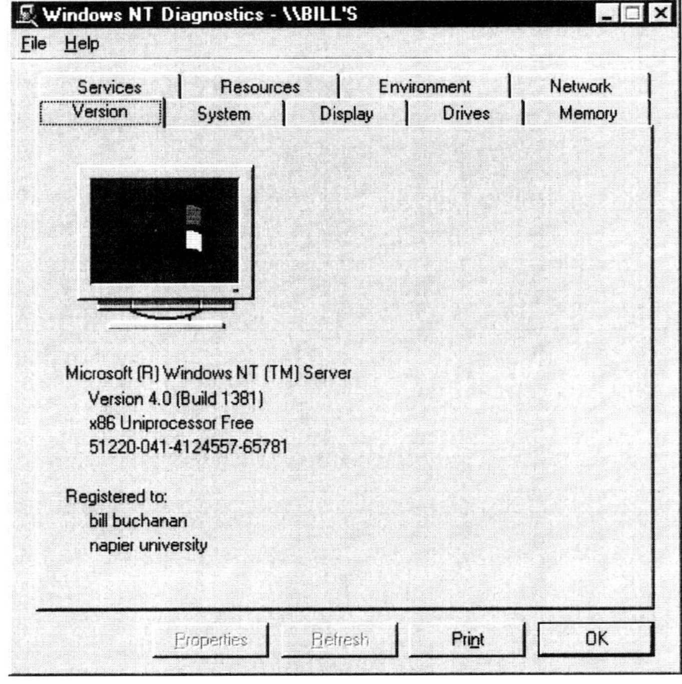

Figure P.15 Windows NT Diagnostics.

From the system selection (called the Windows NT Diagnostics window). The user can select to view the Resources, Services, Environment, Network, Memory, Drives, Display, System and Version. Figure P.16 shows an example Resources window. It can be seen that the currently assigned IRQs are displayed (for example, the serial port is using IRQ3 and IRQ4). The user can select to view the I/O memory map (I/O Port), DMA channels, memory and devices.

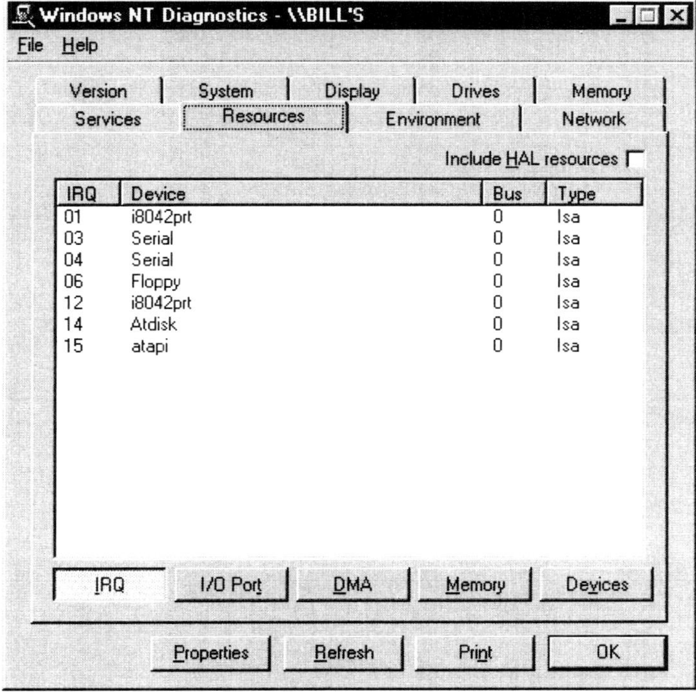

Figure P.16 Resources window.

Figure P.17 shows the system devices. In this case the devices connected are:

- PC Compatiable EISA/ISA (Device for handling I/O ports, system timer and interrupts).
- i8042ptr. PS/2 style mouse/ keyboard (both use the Intel 8042 IC).
- Parport. Parallel port (LPT1: 378h-37Ah).
- Serial. Serial port (COM1: IRQ4, 3F8h-3FEh; COM2: IRQ3, 2F8h-2FEh). See Figure P.19 for an example setting.
- Atdisk (Hard disk drive: IRQ14, 1F0h-1F7h, 3F6h).
- Floppy (floppy disk drive: IRQ6, 3F0h-3F5h, 3F7h, DMA2).
- Atapi (CDROM: IRQ15, 170h-177h).
- VgaSave (VGA: 3B0h-3BBh, 3C0h-3DFh, 1CEh-1CFh).

Figure P.19 shows the usage of the I/O ports.

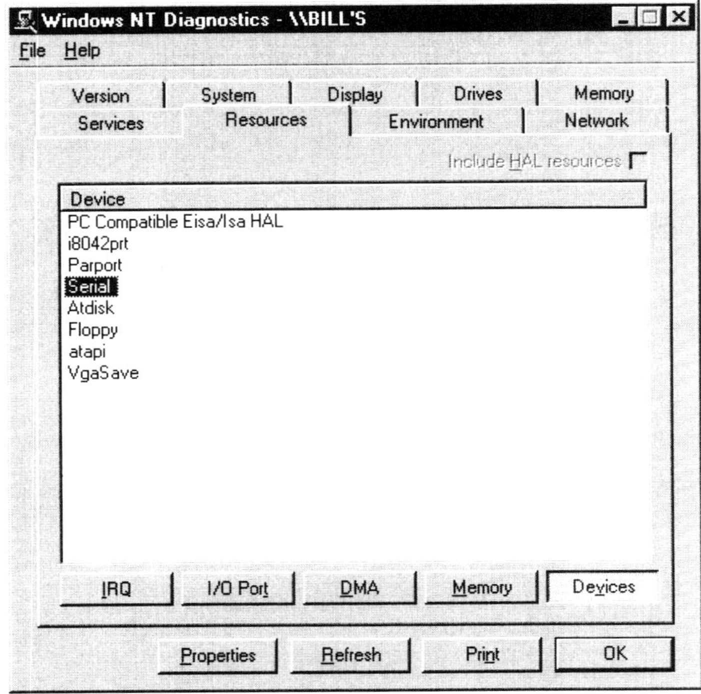

Figure P.17 Devices used.

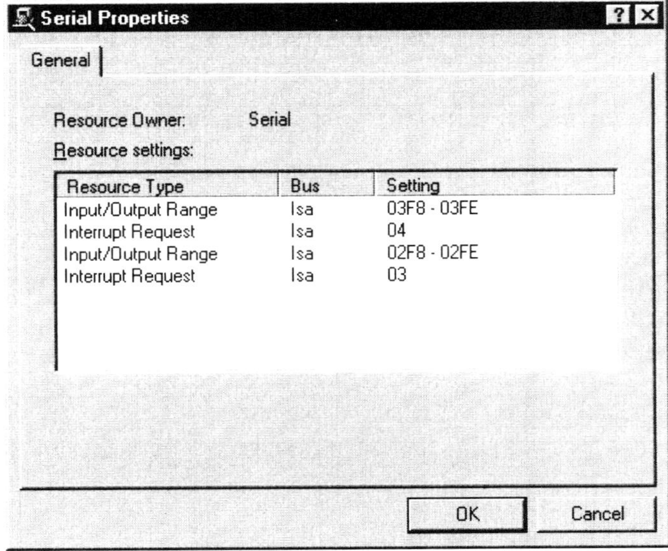

Figure P.18 Serial port properties.

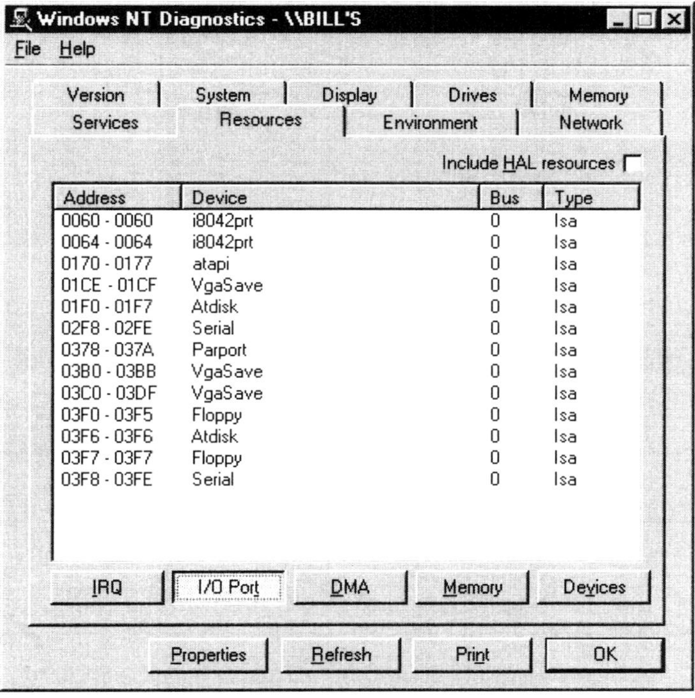

Figure P.19 I/O memory usage.

P.3.6 Server Manager

The Server Manager allows the Administrator to manage domains and computers. Figure P.20 shows the initial window. The computers within the domain are displayed within the window and Figure P.21 shows the window which appears after the Computer is selected. With computers in the domain the user can view information on the selected computer, such as:

- View connected users (see Figure P.22) and send messages to them.
- Open and share resources. Figure P.23 shows an example of a system with shared resources (C$, ADMINS$ and IPC$). An icon of a folder with a hand holding it represents a shared directory, the icon of a connector represents a named pipe, a printer icon represents a shared printer and a question mark represents an unrecognized resource. The IPC$ resources allows programs to intercommunicate (inter processes control) and must be running for shared resources. The ADMIN$ name is used during the remote administration of the computer. This name always points to the directory in which NT is installed (in Figure P.23 this is C:\WINNT). The C$ name represents the C: drive (other drives can be mounted using D$, E$, and so on). Other names are: NETLOGON$ (Domain Net Login), PRINT$ (print queues) and REPL$ (replication export server).
- View and manage shared files and directories. Figure P.24 shows an example window. From this window the user can add or delete shared directories (New Share and Stop Share).

- Manage directory replication.
- Manage the list of administrative alert recipients.
- Manage services of any computer in the domain (Server Manager manages only the local computer).

For domains the following can be administered:

- Promote backup domain controllers to the primary domain controller.
- Synchronize servers with the primary domain controller.
- Add and/or remove computers to/from a domain.

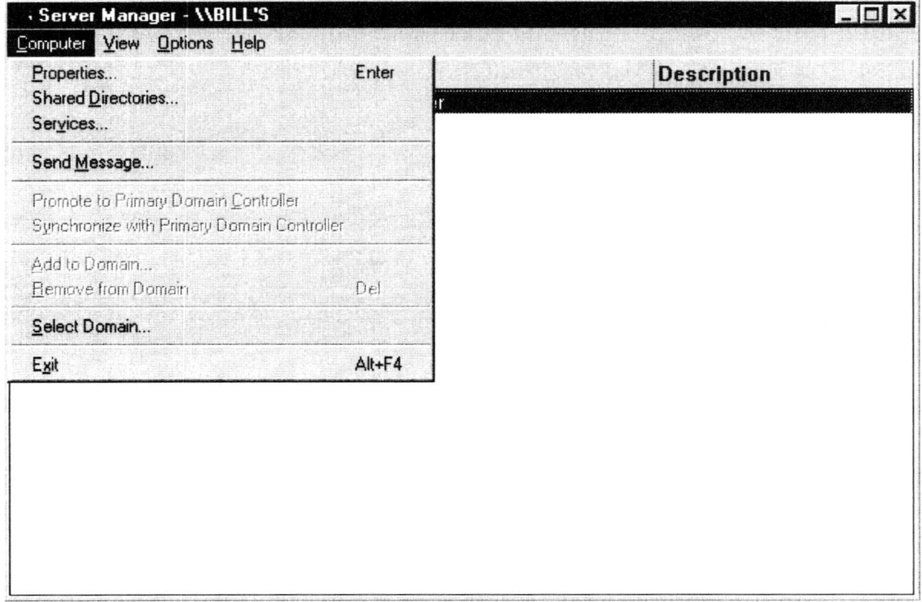

Figure P.20 Server Manager.

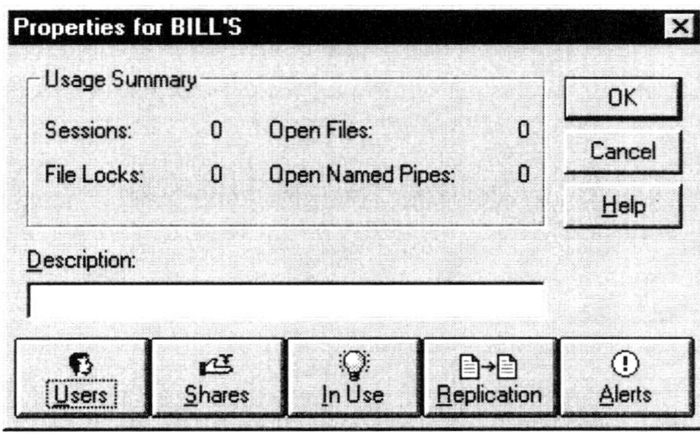

Figure P.21 Properties.

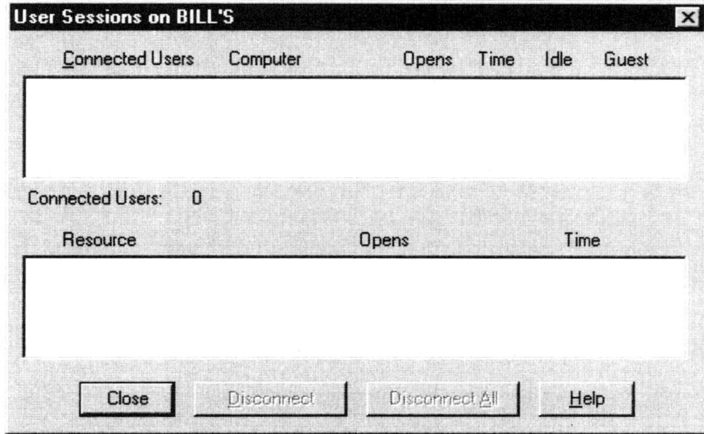

Figure P.22 User Session.

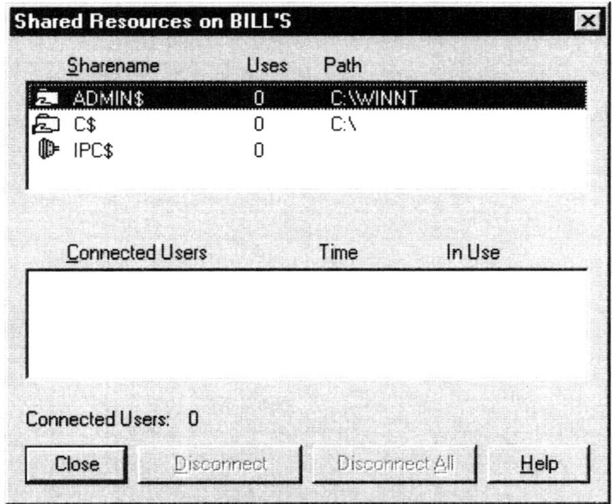

Figure P.23 Shared resources.

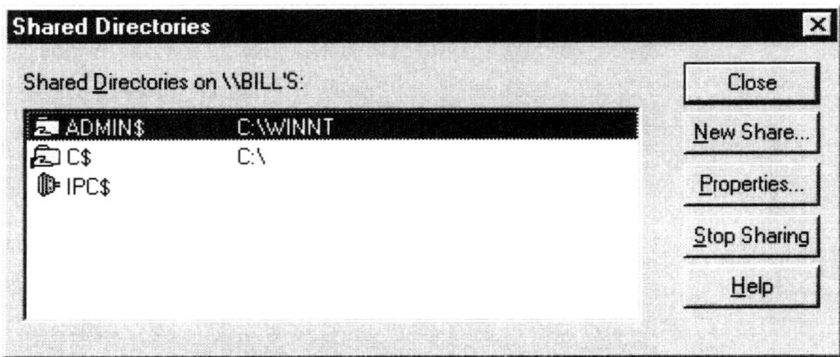

Figure P.24 Shared directories.

P.3.7 Services

Services are processes which perform certain functions within the computer, but will generally slow the computer down. They can be run either from the Server Manager (for local and remote computers) or from Control Panel, then Services. Figure P.25 shows a sample window. A service can either be started, stopped, paused or continued, this is shown in the Status column. The Startup column identifies how the service is started. If it is Manual then the user or a dependent service must start it, if it is Disabled then the service is not started, else if it is automatic then the service started every time the computer starts. Table P.1 gives a list of typical services, their status and their startup condition.

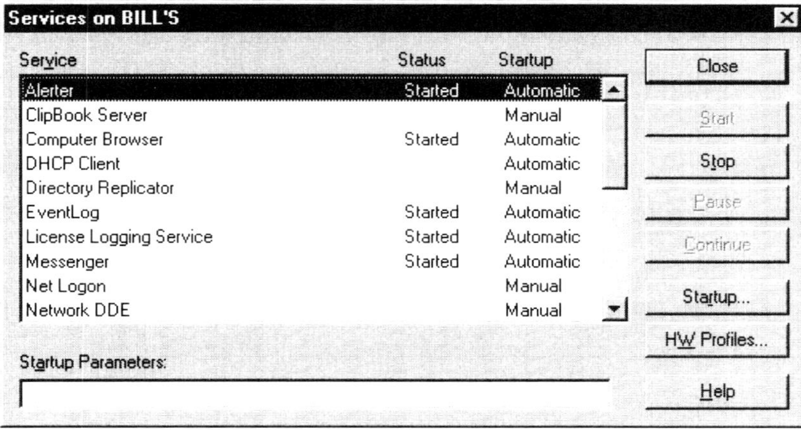

Figure P.25 Services.

Table P.1 Typical services.

Service	*Status*	*Startup*	*Description*
Alerter	Started	Automatic	Notifies selected users and computers of administrative alerts that occur on the selected computer
ClipBook Server		Manual	Allow the Clipbook to be viewed on other remote computers
Computer Browser	Started	Automatic	Builds-up an up-to-date list of computers. For example, Server Manager uses this to display connected computers with a domain
DHCP Client	Started	Automatic	
Directory Replicator		Manual	Replicates directories between computers
EventLog	Started	Automatic	
License Logging Service	Started	Automatic	

Messenger	Started	Automatic	Sends and receives messages sent by administrators or by the Alerter service.
Net Logon		Manual	Network login for the primary and backup domain controllers.
Network DDE		Manual	Network Dynamic Data Exchange (DDE) between applications
Network DDE DSDM		Manual	Network Dynamic Data Exchange Share Database Manager
NT LM Security Support Provider		Manual	
Plug and Play	Started	Automatic	
Remote Procedure Call (RPC) Locator		Manual	Interprocess communication over a network
RPC Service	Started	Automatic	
Schedule		Manual	Runs scheduler
Server	Started	Automatic	File, print sharing and RCP
Spooler	Started	Automatic	Provides print spooler services
TCP/IP NetBIOS Helper	Started	Manual	
Telephony Service		Manual	
UPS		Manual	Manages an interruptible power support (UPS)
Workstation	Started	Automatic	Provides network connections and communications

P.3.8 Event Viewer

The Event Viewer allows the user (normally the Administrator) to view events on the system. It is normally started by default when NT is started, but can be stopped by selecting Services from Control Panel, and then stopping event logging. Figure P.26 shows a security log and Figure P.27 shows an application log. Each window has certain columns, such as: Source (software that logged the event), User (user name), Category (classification of the event, as defined by the source, such as Logon and Logoff, Policy Change, Privilege Use, System Event, Object Access, Detailed Tracking, and Account Management), Computer (name of computer), Event (event number) and Type (such as Error, Warning, Information, Success Audit, or Failure Audit)

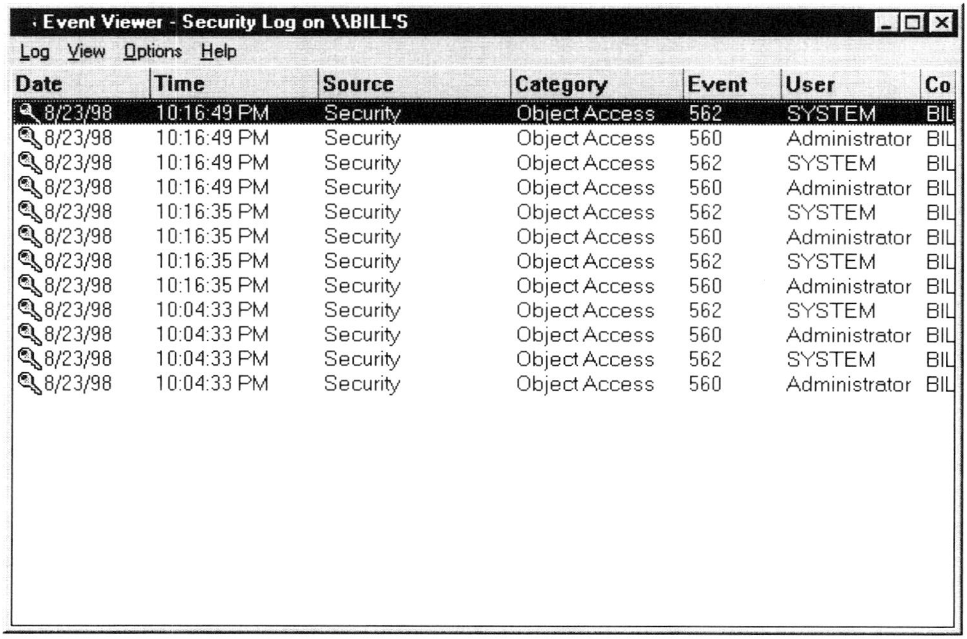

Event Viewer - Security Log on \\BILL'S

Log View Options Help

Date	Time	Source	Category	Event	User	Co
8/23/98	10:16:49 PM	Security	Object Access	562	SYSTEM	BIL
8/23/98	10:16:49 PM	Security	Object Access	560	Administrator	BIL
8/23/98	10:16:49 PM	Security	Object Access	562	SYSTEM	BIL
8/23/98	10:16:49 PM	Security	Object Access	560	Administrator	BIL
8/23/98	10:16:35 PM	Security	Object Access	562	SYSTEM	BIL
8/23/98	10:16:35 PM	Security	Object Access	560	Administrator	BIL
8/23/98	10:16:35 PM	Security	Object Access	562	SYSTEM	BIL
8/23/98	10:16:35 PM	Security	Object Access	560	Administrator	BIL
8/23/98	10:04:33 PM	Security	Object Access	562	SYSTEM	BIL
8/23/98	10:04:33 PM	Security	Object Access	560	Administrator	BIL
8/23/98	10:04:33 PM	Security	Object Access	562	SYSTEM	BIL
8/23/98	10:04:33 PM	Security	Object Access	560	Administrator	BIL

Figure P.26 Security log.

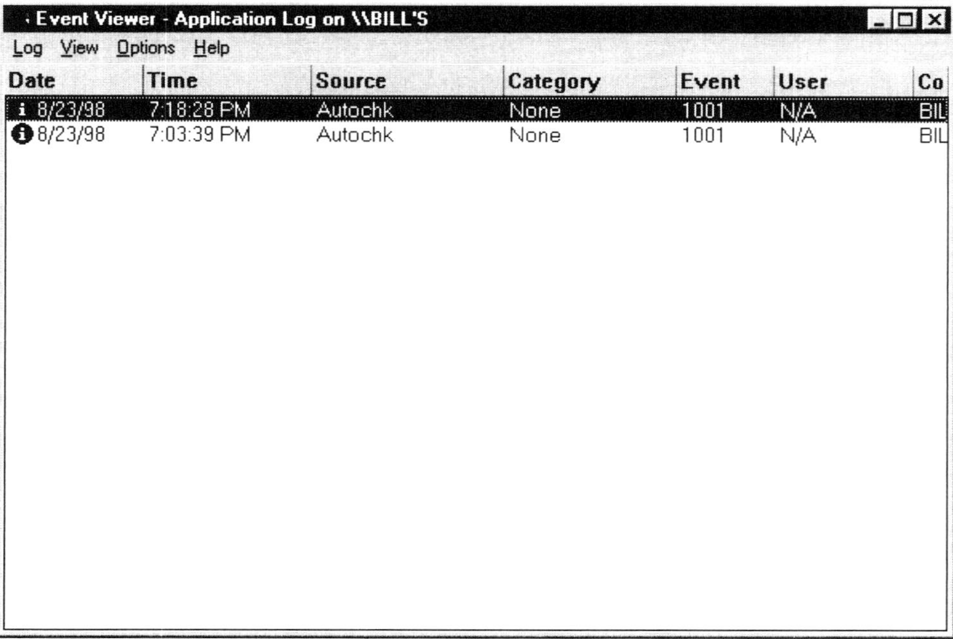

Figure P.27 Application log.

P.3.9 Master server

A master server holds all the licensing data for an organization. Licenses can be:

- Per Seat. In this mode, each computer must have a Client Access License to access a Windows NT Server for basic network services, such as file and print services. This type of licensing is typically used when clients connect to more than one server.
- Per Server. In this mode, the Client Access Licenses are assigned to a particular server, which allows one connection to that server for basic network services (such as file and print services). There must thus be enough licenses for the number of clients which connect to the server. This type of licensing is typical when computers only connect to a single server.

P.3.10 Boot process

Normally a computer runs a POST (Power On Self Test), then the MBR (Master Boot Record) is loaded. It then loads the PBR (Primary Boot Record) from the primary partition (normally the C: drive, A: drive, or, even a bootable CD-ROM drive). Next, the boot processes continues with the following files (and their location):

`Boot.ini.`	Root directory.
`Bootsect.dos.`	Root directory.
`Device drivers.`	*<Ntroot>*`\System32\Drivers`
`Hal.dll.`	*<Ntroot>*`\System32`
`Ntboodd.sys.`	Root directory.
`Ntdetect.com.`	Root directory.
`Ntldr.`	Root directory.
`Ntoskrnl.exe.`	*<Ntroot>*`\System32`
`System.`	*<Ntroot>*`\System32\Config`

The boot sequence is then as follows:

1. `Ntldr` is loaded and switches the processor to operate as a 32-bit processor with a flat memory structure.
2. `Ntldr` starts the appropriate file system mini-drivers (such as NTFS or FAT).
3. `Ntldr` reads the `Boot.ini` file and displays the bootable operating systems, such as Windows NT, Windows NT (VGA mode) and Windows 98. The user can then select the operating system to boot. Normally, there is a default operating system which is chosen within a given time.
4. `Ntldr` loads the required operating system. For NT, it runs `Ntdetect.com` which automatically scans the hardware and reports the hardware list to `Ntldr`. For non-NT, `Bootsect.dos` is loaded.
5. `Ntldr` loads `Ntoskrnl.exe` and `Hal.dll`.
6. `Ntldr` load `Ntoskrnl.exe`. This starts the system loader.

An example `boot.ini` file is given next:

```
[boot loader]
timeout=30
default=C:\="Microsoft Windows"
```

```
[operating systems]
multi(0)disk(0)rdisk(0)partition(1)\WINNT="Windows NT Server
                    Version 4.00"
C:\="Microsoft Windows"
multi(0)disk(0)rdisk(0)partition(1)\WINNT="Windows NT Server
                    Version 4.00 [VGA mode]" /basevideo /sos
```

The timeout parameter is defined in the `[boot loader]` section. It contains the number of seconds that boot options are displayed, before the default operating system is loaded. The `[operating systems]` option defines each of the bootable systems. The `basevideo` operation uses a standard VGA driver. It can be seen, in this case, that there are three bootable file systems, these are:

```
Windows NT Server Version 4.00
Microsoft Windows
Windows NT Server Version 4.00 [VGA mode]
```

If the system has MS-DOS (such as Windows 3.x) then the line:

```
C:\= "MS-DOS"
```

is included.

P.3.11 Task Manager

The Task Manager shows the currently running processes, and, if required, they can be stopped. It can be called by pressing Ctrl-Alt-Del and then selecting Task Manager. Figure P.28 shows a example of some processes running. The window icon with gray indicates a program, and a window icon with white indicates a status window. The open file icon indicates an open folder.

Figure P.28 Applications.

The processes window (as shown in Figure P.29) gives an indication of:

- Image name. Name of the process.
- PID. Process Identification.
- CPU Time. Total time that the process has used the processor.
- Memory usage. Total memory usage.

Figure P.30 shows the performance window. It shows memory and processor usage.

Windows NT Task Manager

File Options View Help

Applications | Processes | Performance

Image Name	PID	CPU	CPU Time	Mem Usage
System Idle Process	0	00	0:17:45	16 K
System	2	00	0:00:17	80 K
smss.exe	18	00	0:00:00	80 K
csrss.exe	26	00	0:00:07	100 K
winlogon.exe	32	00	0:00:01	0 K
services.exe	38	00	0:00:04	16 K
lsass.exe	41	00	0:00:00	80 K
Explorer.exe	42	01	0:00:16	896 K
spoolss.exe	63	00	0:00:00	80 K
llssrv.exe	73	00	0:00:00	16 K
RpcSs.exe	77	00	0:00:00	80 K
nddeagnt.exe	89	00	0:00:00	0 K
taskmgr.exe	99	04	0:00:08	944 K
ntvdm.exe	107	00	0:00:00	1604 K
wowexec.exe		00	0:00:04	
WINWORD.EXE	117	95	0:00:13	2816 K

End Process

Processes: 15 | CPU Usage: 100% | Mem Usage: 19640K / 30800K

Figure P.29 Processes.

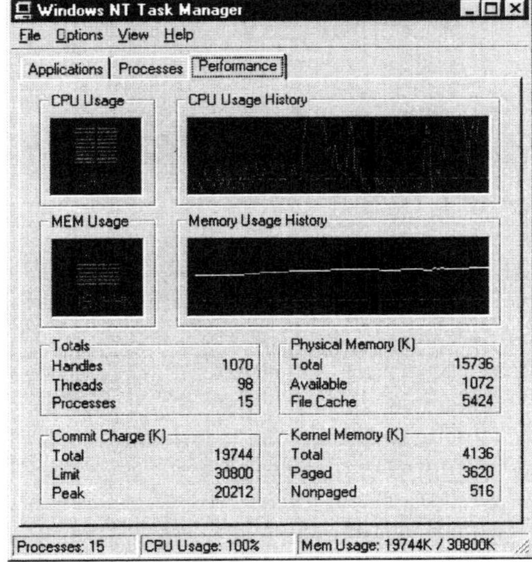

Figure P.30 Performance.

P.3.12 Enabling/disabling drivers

Device drivers can be enabled and disabled from Control Panel→Devices. Figure P.31 shows a sample window. The devices (with their method of starting) include:

Atapi (Boot)	Atdisk (Boot)	Beep (Boot)
BusMouse (System)	Cdaudio (System)	Cdfs (System)
Cdrom (System)	Disk (Boot)	Floppy (System)
i8042 (System)	Keyboard (System)	NDIS (System)
Modem (Manual)	Mouse (System)	NetBEUI Protocol (Automatic)
NetBEUI Interface (Manual)		NetDetect (Manual)
NTFS (System)	Null (System)	NWLInk IPX/SPX (Automatic)
NWLink	NetBIOS (Automatic)	Parallel (Automatic)
Parport (Automatic)	Pcmcia (Boot)	PnP ISA (System)
Scsiprnt (System)	TCP/IP Service (Automatic)	
VgaSave (System)	VgaStart (System)	WINS Client (Automatic)

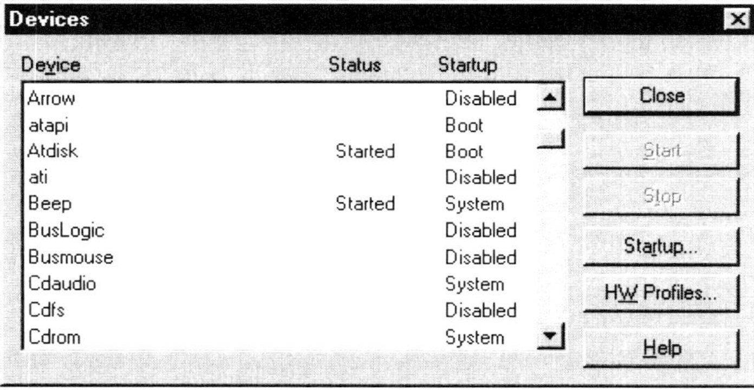

Figure P.31 Devices.

P.4 Viruses

Computer systems, especially PCs, are susceptible to computer viruses. A computer virus is a program which infects other programs by modifying them and making a direct, or a modified, copy of itself. Most viruses are annoying and do little damage to the computer system, apart from replicating themselves.

The PC has always been plagued with viruses for many reasons, including:

- Ease of access to system files. These include boot files (IO.SYS and MSDOS. SYS) and start-up files.
- Ease of access to hardware drivers and interrupts. This danger has been reduced with Windows 95/98 and NT, which do not allow user programs to access the hardware directly and the operating system.

- The lack of file attributes. Most PC systems have simple file attributes for read, write and execute. These attributes can also be easily changed, as the files are not owned by anyone. UNIX makes this difficult as users have limited access to files in other user directories and also to system files.
- The challenge. There are more PCs on the planet than all the other types of computer systems put together. Thus, for some people, there is a thrill in the thought of infecting computers around the world.
- Ease to transfer. Most users use floppy disk drives to transfer files. These are a perfect mechanism for transporting virus. Many companies have even banned the use of floppy disk drives to try and reduce the spread of viruses.
- Boot disks. Until recently, many floppy disks were formatted so that they could be used as a boot disk. Unfortunately, a boot disk allows viruses to be transferred through the boot sector and remain in memory after the computer is booted.
- Accidental release. Many viruses started life as a prototype program or an experimental program, which were accidentally released in the 'wild'. These are typically developed by undergraduate students who had just learned machine code and also PC system architecture. Most virus are not meant to damage to the computer system, but they typically have bugs in them which causes them to do some damage, such as to slow the computer down, or halt the computer.
- Thrill of event-driven programming. Many programmers love to write programs which are triggered by an event. Typical virus events are date (such as Friday the 13th and April 1st), time, and hardware or software interrupts.
- Criminal activity. This is actually not very common, as there are very few criminals who are also computer programmers. Virus programmers typically write virus programs for the thrill of it.
- Jokes. These are harmless programs that do 'amusing' things (although many users think they have been infected by a serious virus).

The damage caused by viruses is normally graded into five main definitions:

- Trivial. No physical damage and all the user has to do is to run an anti-virus program to get rid of the virus. Typically, the virus will occasionally display a message.
- Minor. Some files require to be re-installed from a backup or from a disk.
- Moderate. Virus trashes the hard disk, scrambles the FAT, or low-level formats the drive. This is typically recovered from a recent backup.
- Major. Virus gradually corrupts data files, of which the user is unaware of. This is more severe than moderate damage as earlier backups are likely to contain the virus. Many weeks or months worth or data may be lost.
- Severe. Virus gradually corrupts data files, but the user continues to use the computer. There is thus no simple way of knowing whether the data files, or the backup files, are good or bad).
- Unlimited. Virus gives a hacker access to your network, by stealing the supervisor password. The damage is then done by the hacker, who controls the network.

P.4.1 Virus types

Partition Viruses

When starting PCs, the system automatically reads the partition sector and executes the code it finds there. The partition sector (or Master Boot Record) is the first sector on a hard disk and it contains system start-up information, such as:

- Number of sectors in each partition.
- The start of the DOS partition starts.
- Small programs.

Viruses, which attach themselves to the partition sector, modify the code associated with the partition section. The system must be booted from a clean boot virus to eradicate this type of virus. Partition viruses only attack hard disks, as floppy disks do not have partition sectors. A partition is created with the FDISK program. Hard disks cannot be accessed unless they have a valid partition. Typical partitions are FAT-16, FAT-32 and NTFS.

Boot sector viruses

The boot sector resides on the first sector of a partition on a hard disk, or the first sector on a floppy disk. On starting, the PC reads from the active partition on the hard disk (identified by C:) or tries to read from the boot sector of the floppy disk. Boot sector viruses replace the boot sector with new code and moves, or deletes the original code.

Non-bootable floppy disks have executable code in their boot sector, which displays the message "Not bootable disk" when the computer is booted from it. Thus any floppy disk, whether it is bootable or non-bootable, can contain a virus, and can thus infect the PC.

File viruses

File viruses append or insert themselves into/onto executable files, such as .COM and .EXE programs. An indirect-action file virus installs itself into memory when the infected file is run, and infects other files when they are subsequently accessed.

Overwriting viruses

Overwriting viruses overwrite all, or part, of the original program. They are easy to detect as missing files are easily detected.

P.4.2 Anti-virus programs

Viruses use two main methods to introduce themselves to a computer system. These are through:

- Boot records on floppy and hard disks.
- Files. These are contained in either binary executable files or are macro viruses, which require another program to run them. A typical virus is the Word Macro virus, which effects the macros in Word 6.0. It cannot run itself and needs to hide within a Word document.

Virus scanning programs use a number of techniques to detect virus, these include:

- Scanning. Most viruses have a digital signature within a file or boot sector which identifies the virus. For example, the Murphy virus has the ASCII characters of Murphy added to the end of a file. The virus-scanning program thus has a table of known digital signatures which it compares with a byte-by-byte read of the disk drive. Care must be taken, though, that the file does not normally contain a genuine sequence of bytes which are mistaken for the virus. The scanner will thus report a virus when there is none.
- Change detection. This allows the virus checker to keep a list of the date and time that files were last changed. Any changes to the files can then be checked for viruses. Typically, the change detector keeps a track of changes to the main boot files. Unfortunately, virus programmers can change the date of a file so that it has the date of the original 'uninfected' file.
- Heuristic analysis. This technique attempts to detect viruses by watching for appearance or behavior that is characteristic of some class of known viruses. This allows for anti-virus program to detect new viruses which are not defined in the scanning technique.
- Verification. Scanning, change detection and heuristics can only identify a possible virus, only a verification program can prove that it really exists. Verifying program operates on the identified virus file and proves if it has been infected.

Disinfection

Disinfection involves removing a virus from the system, if possible it should be used with the minimum of damage. The two main methods are:

- Specific knowledge disinfection. With this method the disinfection program reverses the actions of the virus infection. This requires some knowledge of how the virus infects the system in the first place.
- Generic disinfection. This method has information about what the original file or boot record. This typically involves methods, such as storing system files, re-installing system files, and rebuilding the boot record.

It is sometimes difficult to disinfect system which has a memory-resident virus, as the disinfection program could be tampered with as it is being run. Thus, it is often advisable to boot the anti-virus program from a clean boot disk.

An disinfector is intelligent in that it will not damage an infected file, and will generally prompt the user to delete the infected file rather than damage it (which could cause more problems).

P.4.3 Trojan horses

A Trojan horse virus hides itself as a tempting-looking executable file or in a compressed file. Until recently virus checkers did not test compressed files for viruses, but most currently available virus checkers checker test the uncompressed files within a compressed file. Many also test files as they are executed, and the virus would thus be detected as the compressed file was uncompressed. A Trojan horse virus attached to an executable file will typically have an interesting name, such as:

```
hello.exe
nice.exe
```

```
mail.exe
```

One Trojan horse virus, available from a WWW page, contains a single executable which is said to contain over 100 virus. This, the owner of the WWW page, states that it can be used against someone's worst enemy. The WWW page quotes that it can be spread by sending the user an email with the attached executable file or to leave it on a targeted users computer. Writing a Trojan horse virus is extremely simple and can simply involve a batch file which has:

```
cd \                    ; or 'cd /' on a UNIX system
rm -r *.*
```

which, on a UNIX system or a PC system with the `rm` utility installed, will change the directory to the top-level and then try to delete all subdirectories below the top-level directory. It is also extremely easy for a user to write a program which scans all the subdirectories below the current directory, and delete their contents.

P.4.4 Polymorphic viruses

A polymorphic virus encrypts itself so that its signature is invisible to a virus scanner. It propagates itself by first decrypting itself using a decryption routine. This decrypted program can then do damage, such as spreading itself or deleting files on the computer. A non-polymorphic encrypted virus uses a decryption routine that does not change. Thus is it easy to detect as the signature is unchanging. A polymorphic virus uses a changeable decryption routine (known as a mutation engine), in which the signature is different each time. The mutation engine normally uses a random number and a simple mathematical algorithm to change the virus's signature.

Typical COM/EXE virus infectors are:

- Yankee Doodle. It infects EXE and COM files (and adds 2772 bytes). When an infected file is run, the virus automatically loads into memory and infects any program which is run. At 5:00 pm the virus causes the system to sometimes play "Yankee Doodle".
- SatanBug. It infects EXE and COM files (and adds between 3500 to 5000 bytes). When an infected file is executed, the virus installs itself in memory and infects any program which is executed. A SatanBug infected system generally runs slow and some programs may fail, but does no long term damage to the computer.
- Frodo. It infects EXE and COM files (and adds between 4096 bytes). As above it infects all programs which are executed. When the date is between September 22 and December 31, the virus typically hangs the computer. This is because there are bugs in the virus code) which intends to overwrite the boot record with a program to display the message "Frodo Lives" when the machine boots.

P.4.5 Stealth viruses

Stealth viruses hide themselves in system files, or in boot records. They are then called when a certain action occurs on the computer. Typically, the system files which are used to start the computer are infected, such as IO.SYS and MSDOS.SYS (which are called when the computer starts). Once the virus is in memory, it can hook itself onto a given action, such as interrupt routine. For example, by hooking on interrupt 21h the virus can interrupt all accesses to DOS services, such as reading and writing to the hard disk.

A stealth size virus attaches itself to a file, when the file is accessed the virus copies itself, which increases the size of the file. A read stealth virus intercepts a request to an infected file and presents uninfected contents. Thus to the user, files seem to be operating as normal. Unfortunately, this stealth rapidly spreads through the system.

A stealth virus is relatively easy to detect and erase, as they do not change their digital signature. It is important, though, when eradicating a stealth virus that the system is booted with a clean boot disk. Typical stealth viruses include:

- Stoned. The Stoned virus infects floppy and hard disk boot sectors. When the computer starts from an infected diskette, the virus infects the master boot record of the first physical hard disk. It installs itself in memory and occasionally displays the message "Your PC is now Stoned!". It then infects the boot sectors of any inserted disk. Luckily, it does not cause damage to computer files and there is no loss of data.
- Ping Pong. This virus infects diskettes and the hard disk partition (non-master) boot record. It sometimes produces a bouncing dot on the screen after booting and adds approximately 975 bytes to the size of files.

P.4.6 Slow viruses

A slow virus operates by targeting the copying or modify of a file, and leave the original file untouched. Thus, a virus scanner may not pick-up that the file has been modified by the virus as it looks as if it was modified by some other normal operation. An example slow virus is:

- Dark Avenger. This is a resident COM and EXE file virus and adds 1800 bytes onto infected COM files. When an infected program is run, the virus installs itself in memory. It then only infects EXE or COM files which are run, opened, renamed, or operated on. Roughly every 16 times an infected program is run, it overwrites a random sector of the disk and displays the message: "Eddie lives...somewhere in time!".

These viruses can be defeated by memory-resident virus checkers who test the creation of new files (an integrity shell).

P.4.7 Retro viruses

Retro viruses (or anti-anti viruses) try to disrupt the operation of an anti-virus program. They are designed by analyzing publicly available anti-virus programs (such as Dr Solomon's or Mcafee's) and identifying potential weaknesses. Typical methods include:

- Modifying operation. In this method the virus program changes the operation of the virus scanner, such as stopping it from scanning certain files, or blocking any system messages, or even, mimicking the output of the anti-virus program.
- Modify database. In this method the virus program modifies the virus signature database file so that the retro virus has the wrong digital signature (or even modify it when it is loaded into the computer's memory).
- Detect and run. This method involves the virus program detecting when the anti-virus program is run, and then hiding from it, either by deleting itself from memory or hiding in a file for a short time.

P.4.8 Worms

Before the advent of LANs and the Internet, the most common mechanism for spreading virus was through floppy disks and CD-ROM disks. Anti-virus programs can easily keep up-to-date with the latest virus, and modify their databases. This is a relatively slow method of spreading a virus and will take many months, if not years, to spread a virus over a large geographical area. Figure P.32 illustrates the spread of viruses.

LANs and the Internet have changed all this. A virus can now be transmitted over a LAN in a fraction of a second, and around the world in less than a second. Thus a virus can be created and transmitted around the world before an anti-virus program can even detect that it is available.

A worm is a program which runs on a computer and creates two threads. A thread in a program is a unit of code that can get a time slice from the operating system to run concurrently with other code units. Each process consists of one or more execution threads that identify the code path flow as it is run on the operating system. This enhances the running of an application by improving throughput and responsiveness. With the worm, the first thread searches for a network connection and when it finds a connection it copies itself to that computer. Next, the worm makes a copy of itself, and runs it on the system. Thus, a single copy will become two, then four, eight, and so on. This continues until the system, and the other connected systems, will be shutdown. The only way to stop the worm is to shutdown all the effected computers at the same time and then restart them. Figure P.33 illustrates a worm virus.

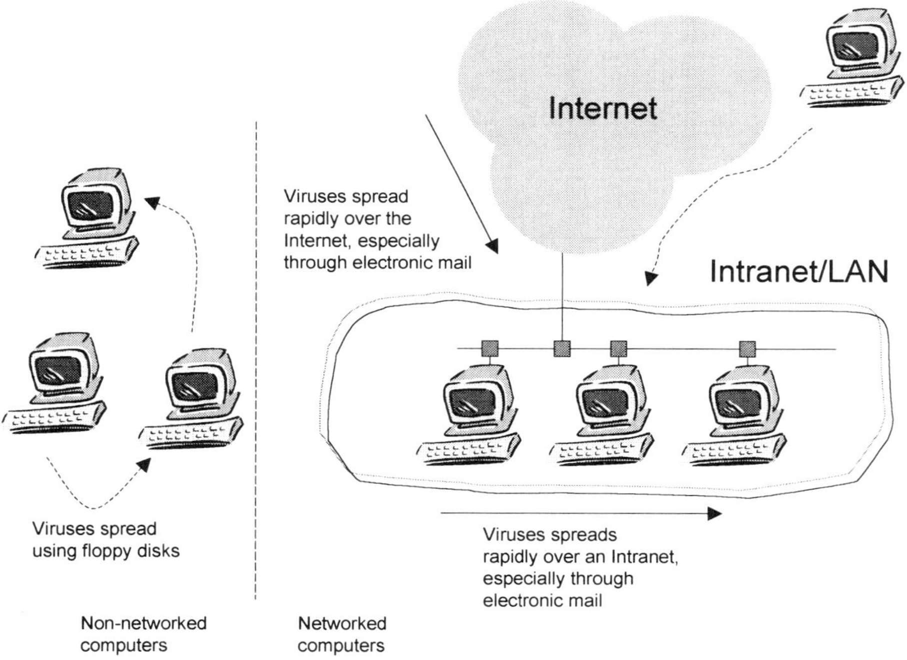

Figure P.32 Spread of viruses.

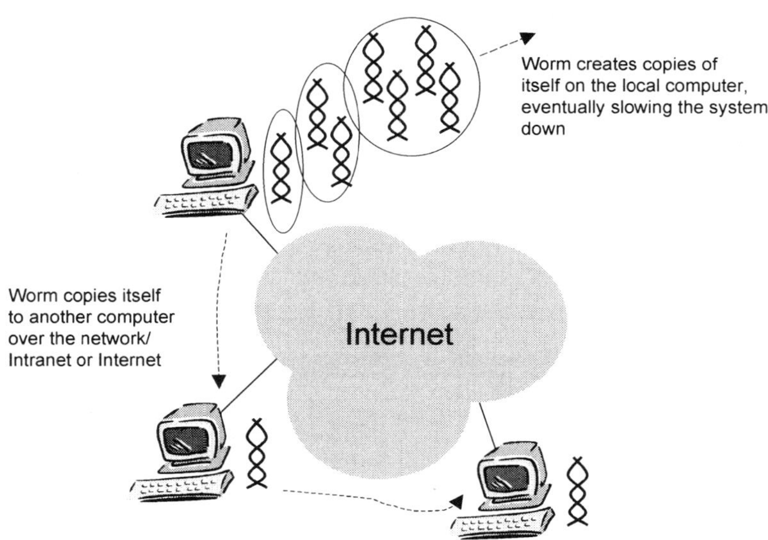

Figure P.33 Worm viruses.

P.4.9 Macro viruses

One of the most common viruses is the macro virus, which attacks macro or scripting facilities which are available in word processors (such as Microsoft Word), spread-sheets (such as Microsoft Excel), and remote transfer programs. A typical virus is the WM/CAP virus which modifies macros within Microsoft Word Version 6.0. When a macro is executed it can cause considerable damage, such as: deleting files, corrupting files, and so on.

The greatest increase in macro viruses is the number of viruses which use Microsoft Visual Basic for Applications (VBA) as this integrates with Microsoft Office (although recent releases have guarded against macro viruses). A macro virus in Word 6.0 is spread by:

- The file is either transmitted over email, over a LAN or from floppy disk.
- The infected file is opened and the `normal.dot` main template file is modified so that it contains the modified macros. Any files that are opened or created will now have the modified macros. The WM/CAP macro virus does not do much damage and simply overwrites existing macros.
- VBA is made to be event-driven, so operations, such as File Open or File Close, can have an attached macro. This makes it easy for virus programmers to write new macros.

Figure P.34 shows an example of a macro created by Visual Basic programming. The developed macro (`Macro1()`) simply loads a file called AUTHOR.doc, selects all the text, converts the text to bold, and then saves the file as AUTHOR.rtf. It can be seen that this macro is associated with `normal.dot`.

P.4.10 Other viruses

Other viruses include:

- Armored. This type of virus uses special code which makes them difficult to detect, or even understand.
- Companion. This type of virus creates an executable program which contains the virus, which then calls the mimicked program. For example, a virus could create a program called EXCEL.EXE which would contain the virus, and this could call the standard Microsoft Excel program.
- Phage. This type of virus modifies the code of a program, and unlike the previously discussed viruses they do not attach themselves onto a file. This type of virus is extremely dangerous as it is very difficult to recovery the original program code, without having some knowledge of the changes that the phage virus is likely to do.

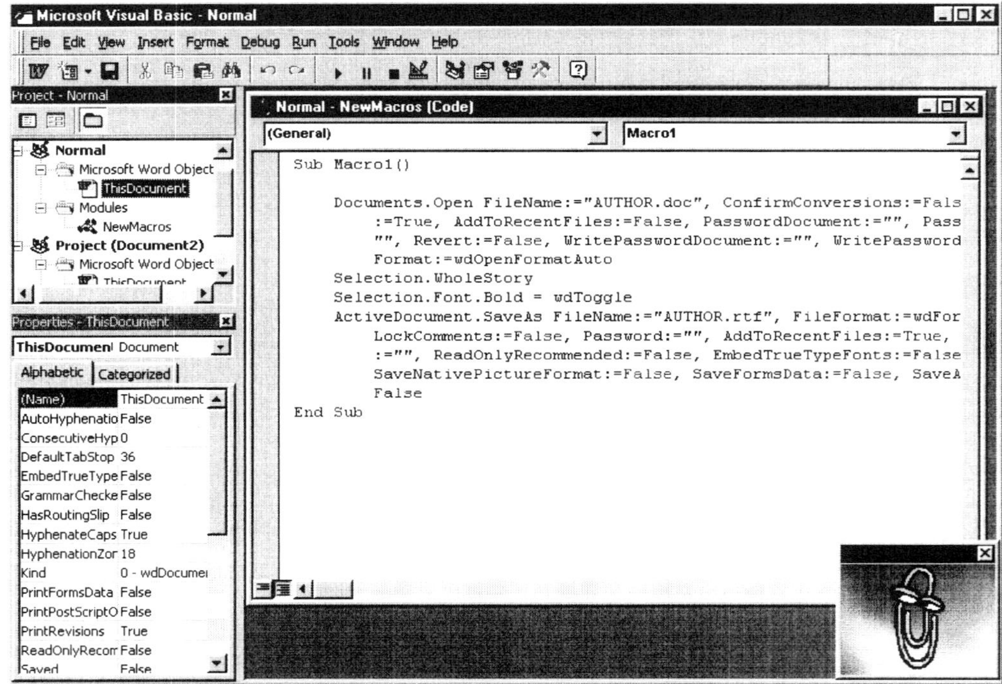

Figure P.34 Sample macro using VB programming.

P.5 Security

P.5.1 Introduction

Security involves protecting the hardware and software on a system from both internal attack and from external attack (hackers). An internal attack normally involves uneducated users causing damage, such as deleting important files or crashing systems. This effect can be minimized if the system manager properly protects the system. Typical actions are to limit the files that certain users can access and also the actions they can perform on the system.

Most system managers have seen the following:

- Sending a file of the wrong format to the system printer (such as sending a binary file). Another typical one is where there is a problem on a networked printer (such as lack of paper), but the user keeps re-sending the same print job.
- Deleting the contents of sub-directories, or moving files from one place to another (typically, these days, with the dragging of a mouse cursor). This problem can be reduced by regular backups.
- Deleting important system files (in a PC, these are normally AUTOEXEC.BAT and CONFIG.SYS). This can be overcome by the system administrator protecting important system files, such as making them read-only or hidden.
- Telling other people their user passwords or not changing a password from the initial default one. This can be overcome by the system administrator forcing the user to change their password at given time periods.

Security takes many forms, such as:

- Data protection. This is typically where sensitive or commercially important information is kept. It might include information databases, design files or software code files. One method of reducing this risk to encrypt important files with a password and/or some form of data encryption.
- Software protection. This involves protecting all the software packages from damage or from being misconfigured. A misconfigured software package can cause as much damage as a physical attack on a system, because it can take a long time to find the problem.
- Physical system protection. This involves protecting systems from intruders who might physically attack the systems. Normally, important systems are locked in rooms and then within locked rack-mounted cabinets.
- Transmission protection. This involves a hacker tampering with a transmission connection. It might involve tapping into a network connection or total disconnection. Tapping can be avoided by many methods, including using optical fibers which are almost impossible to tap into (as it would typically involve sawing through a cable with hundreds of fiber cables, which would each have to be connected back as they were connected initially). Underground cables can avoid total disconnection, or its damage can be reduced by having redundant paths (such as different connections to the Internet).

P.5.2 Hacking methods

The best form of protection is to disallow hackers into the network in the first place. Organizational networks are hacked for a number of reasons and in a number of ways. The most common methods are:

- IP spoofing attacks. This is where the hacker steals an authorized IP address, as illustrated in Figure P.35. Typically, it is done by determining the IP address of a computer and waiting until there is no-one using that computer, then using the unused IP address. Several users have been accused of accessing unauthorized material because other users have used their IP address. A login system which monitors IP addresses and the files that they are accessing over the Internet cannot be used as evidence against the user, as it is easy to steal IP addresses.

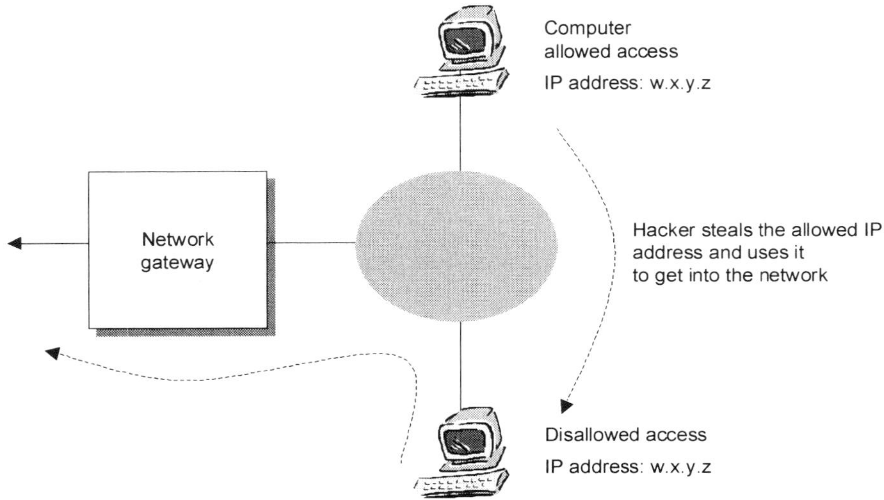

Figure P.35 IP spoofing.

- Packet-sniffing. This is where the hacker listens to TCP/IP packets which come out of the network and steals the information in them. Typical information includes user logins, e-mail messages, credit card number, and so on. This method is typically used to steal an IP address, before an IP spoofing attack. Figure P.36 shows an example where a hacker listens to a conversation between a server and a client. Most TELNET and FTP program actually transmit the user name and password as text values, these can be easily viewed by a hacker, as illustrated in Figure P.37.

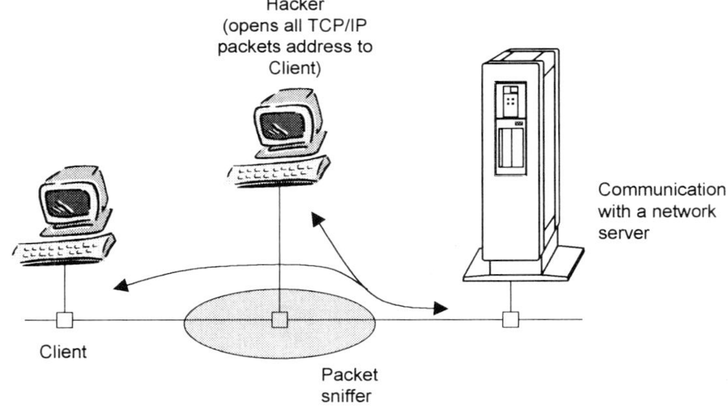

Figure P.36 Packet sniffing.

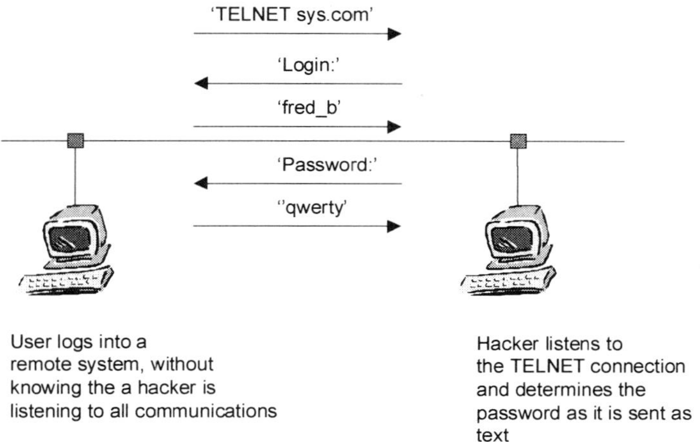
'TELNET sys.com'
'Login:'
'fred_b'
'Password:'
''qwerty'
User logs into a remote system, without knowing the a hacker is listening to all communications
Hacker listens to the TELNET connection and determines the password as it is sent as text

Figure P.37 Packet sniffing on a TELNET connection.

- Passwords attacks. This is a common weak-point in any system, and hackers will generally either find a user with a easy password (especially users which have the same password as their login name) or will use a special program which cycles through a range of passwords. This type of attack is normally easy to detect. The worst nightmare of this type of attack is when a hacker determines the system administrator password (or a user who has system privileges). This allows the hacker to change system set-ups, delete files, and even change user passwords.
- Sequence number prediction attacks. Initially, in a TCP/IP connection, the two computers exchange a start-up packet which contains sequence numbers. These sequence numbers are based on the computer's system clock and then run in a predictable manner, which can be determined by the hacker.
- Session hi-jacking attacks. In this method, the hacker taps into a connection between two computers, typically between a client and a server. The hacker then simulates the connection by using its IP address.
- Shared library attacks. Many systems have an area of shared library files. These are called by applications when they are required (for input/output, networking, graphics, and so on). A hacker may replace standard libraries for ones that have been tampered with, which allows the hacker to access system files and to change file privileges. Figure P.38 illustrates how a hacker might tamper with dynamic libraries (which are called as a program runs), or with static libraries (which are used when compiling a program). This would allow the hacker to possibly do damage to the local computer, send all communications to a remote computer, or even view everything that is viewed on the user screen. The hacker could also introduce viruses and cause unpredictable damage to the computer (such as remotely rebooting it, or crashing it at given times).
- Social engineering attacks. This type of attack is aimed at users who have little understanding of their computer system. A typical attack is where the hacker sends an email message to a user, asking for their password. Many unknowing users are tricked by this attack. A few examples are illustrated in Figure P.39. From the initial user login, the hacker can then access the system and further invade the system.

- Technological vulnerability attack. This normally involves attacking some part of the system (typically the operating system) which allows a hacker to be access to the system. A typical one is for the user to gain access to a system and then run a program which re-boots the system or slows it down by running a processor intensive program. This can be overcome in operating systems such as NT and UNIX by granting re-boot rights only to the system administrator.

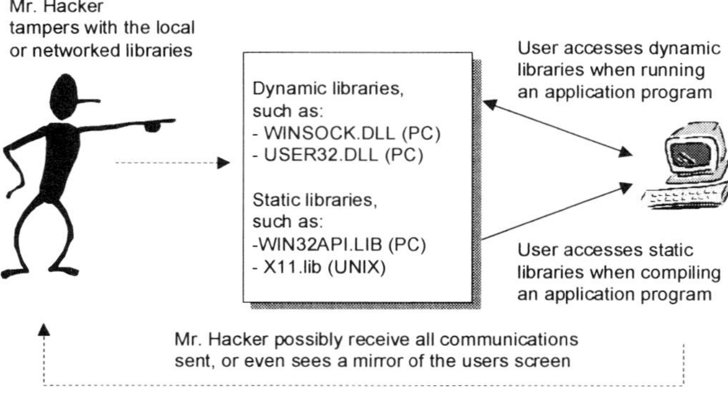

Figure P.38 Shared library attack.

Figure P.39 Social Engineering attack.

- Trust-access attacks. This allows a hacker to add their system to the list of systems which are allowed to log into the system without a user password. In UNIX this file is the *.rhosts* (trusted hosts) which is contained in the user's home directory. A major problem is when the trusted hosts file is contained in the root directory, as this allows a user to log in as the system administrator.

P.5.3 Security policies

A well protected system depends mainly on the system manager. It is up to the manager to define security policies which define how users can operate the system. A good set of policies would be:

- Restrictions on users who can use a given account. The system administrator needs to define the users who can login on a certain account.
- Password requirements and prohibitions. This defines the parameters of the password, such as minimum password size, time between password changes, and so on.
- Internet access/restrictions. This limits whether or not a user is allowed access to the Internet.
- User account deletion. The system administrator automatically deletes user accounts which are either not in use or moved to another system.
- Application program rules. This defines the programs which a user is allowed to run (typically games can be barred from some users).
- Monitoring consent. Users should be informed about how the system monitors their activities. It is important, for example, to tell users that their Internet accesses are being monitored. This gives the user no excuse when they are found to be accessing restricted sites.

P.5.4 Passwords

Passwords are normally an important part of any secure network. They can be easily hacked with the use of a program which continually tries different passwords within a given range (normally called directory-based attacks). These can be easily overcome by only allowing a user three bad logins before the system locks the user out for a defined time. NetWare and Windows NT both use this method, but UNIX does not. The system manager, though, can determine if an attack has occurred with the BADLOG file. This file stores a list of all the bad logins for user and the location of the user.

Passwords are a basic system for provide security on a network, and they are only as secure as the user makes them. Good rules for passwords are:

- Use slightly usual names, such as *vinegarwine*, *dancertop* or *helpcuddle*. Do not use names of a wife, husband, child or pet. Many users, especially ones who know the user can easily guess the user's password.
- Use numbers after the name, such as *vinedrink55* and *applefox32*. This makes the password difficult to crack as users are normally only allowed a few chances to login correctly before they are logged out (and a bad login event written to a bad login file).
- Have a few passwords which are changed at regular intervals. This is especially important for system managers. Every so often, these passwords should be changed to new ones.
- Make the password at least six characters long. This stops 'hackers' from watching the movement of the users fingers when they login, or from running a program which tries every permutation of characters. Every character added, multiplies the number of combinations by a great factor (for example, if just the characters from 'a' to 'z' and '0' to '9'

are taken then every character added increases the number of combinations by a factor of 36).

- Change some letters for numbers, or special characters. Typically, 'o' becomes a 0 (zero), 'i' becomes 1 (one), 's' becomes 5 (five), spaces become '$', 'b' becomes '6', and so on. So a password of 'silly password' might become '5illy$pa55w0rd' (the user makes a rule for 's' and 'o'). The user must obviously remember the rule that has been used for changing the letters to other characters. This method overcomes the technique of hackers and hacker programs, where combinations of words from a dictionary are hashed to try and make the hashed password.

The two main protocols used are:

- Password Authentication Protocol (PAP). This provides for a list of encrypted passwords.
- Challenge Handshake Authentication Protocol (CHAP). This is a challenge-response system which requires a list of unencrypted passwords. When a user logs into the system a random key is generated and sent to the user for encrypting the password. The user then uses this key to encrypt the password, and the encrypted password is sent back to the system. If it matches its copy of the encrypted password then it lets the user log into the system. The CHAP system then continues to challenge the user for encrypted data. If the user gets these wrong then the system disconnects the login.

Administrator user name

The administrator account has the highest privileges and will thus be the most highly prized hackable account. If it is hacked into then the whole system is vulnerable to attack until it can be proved that there are no hidden traps. One method of reducing the risk of being hacked is to change the name of the administrator from time to time. The administrator password should be totally secure with a relatively large password size.

Password security

The US Government defines certain security levels: D, C1, C2, B1, B2, B3 and A1, which are published in the *Trusted Computer Security Evaluation Criteria* books (each which have different colored cover to define there function), these include:

- Orange book. Describes system security.
- Red book. Interpretation of the Orange book in a network context.
- Blue book. Application of Orange book to systems not covered in the original book.

Windows NT uses the C2 security level It has the following features:

- Object control. Users own certain objects and they have control over how they are accessed.
- User names and passwords.
- No object reuse. Once a user or a group has been deleted, the user and group numerical IDs are not used again. New users or groups are granted a new ID number.
- Security auditing system. This allows the system administrator to trace security aspects,

such as user login, bad logins, program access, file access, and so on.
- Defined keystroke for system access. In Windows NT, the CNTRL-ALT-DEL keystroke is used by a user to log into the system.

Password hacking programs

There are many password cracking hacking programs, especially for UNIX and Windows NT, these include:

- Crack. This is a freely available program which can be used on Windows NT or UNIX systems.
- PwDump. This utility allows the user to view the encrypted password file for a local or a remote Windows NT system. These encoded passwords can then be used to hack into the system.
- ScanNT. This is a commercial package for Windows NT.

With PwDump the encoded (hashed) passwords are sent back to user, such as:

```
Administrator:500:EF10EDD3421010325A3B2178::Sys Admin::
Fred:501:32FAB36412032188:...:Fred Bloggs::
```

These encoded passwords can be used to hack into the system because the login process is as follows:

- Server contacts the clients and sends a random nonsense (nonce) message.
- Client then encrypts the nonce and user account name using the encoded password.
- Client sends the result and the text username to the server.
- Server validates this response using the encoded user password.

Many systems are prone to the man in the middle attack, where a hacker sends a forged challenge to the server. Next, the hacker waits for a real request by same user that is to be used to hack into the system. The hacker then sends a challenge to the user, as if it was the server, and incepts the response. This response can then be sent onto the server, from which the hacker can log in. If the hacker already knows the encoded password then there is no need to wait on a response from the hacked user as it can be impersonated immediately with the encoded password.

P.5.5 Hardware security

Passwords are a simple method of securing a system. A better method is to use a hardware-restricted system which either bar users from a specific area or even restrict users from login into a system. Typical methods are:

- Smart cards. With this method a user can only gain access to the system after they have inserted their personal smart card into the computer and then enter their PIN code.
- Biometrics. This is a better method than a smart card where a physical feature of the user is scanned. The scanned parameter requires to be unchanging, such as fingerprints or a retina images.

P.5.6 Hacker problems

Once a hacker has entered into a system, there are many methods which can be used to further penetrate into the system, such as:

- Modifying search paths. All systems set up a search path in which the system looks into to find the required executable. For example, in a UNIX system, a typical search path is /bin, /usr/bin, and so on. A hacker can change the search paths for a user and then replace standard programs with ones that have been modified. For example, the hacker could replace the email program for one that sends emails directly to the hacker or any directory listings could be sent to the hacker's screen.
- Modifying shared libraries. As discussed previously.
- Running processor intensive task which slows the system down, this task will be run in the background and will generally not be seen by the user. The hacker can further attack the system by adding the processor intensive task to the system start-up file (such as the `rc` file on a UNIX system).
- Running network intensive tasks which will slow the network down, and typically slow down all the connected computers. As with the processor intensive task, the networking intensive task can be added to the system start-up file.
- Infecting the system with a virus or worm.

Most PCs have now virus scanners which test the memory and files for viruses and thus virus are easy to detect. A more sinister virus is spread over the Internet, such as the Internet worm which was released on November 1988. This is a program which runs on a computer and creates two threads. A thread in a program is a unit of code that can get a time slice from the operating system to run concurrently with other code units. Each process consists of one or more execution threads that identify the code path flow as it is run on the operating system. This enhances the running of an application by improving throughput and responsiveness. With the worm, the first thread searches for a network connection and when it finds a connection it copies itself to that computer. Next, the worm makes a copy of itself and runs it on the system. Thus a single copy will become two, then four, eight, and so on. This will then continue until the system, and the other connected systems, will be shutdown. The only way to stop the worm is to shutdown all the effected computers at the same time and then restart them.

P.6 Firewalls

P.6.1 Firewalls

A firewall (or security gateway) protects a network against intrusion from outside sources. They tend to differ in their approach but can be characterized as follows:

- Firewalls which block traffic.
- Firewalls which permit traffic.

They can be split into three main types:

- Network-level firewalls (packet filters). This type of firewalls examines the parameters of the TCP/IP packet to determine if it should dropped or not. This can be done by examining the destination address, the source address, the destination port, the source port, and so on. The firewall must thus contain a list of barred IP addresses or allowable IP addresses. Typically a system manager will determine IP addresses of sites which are barred and add them to the table. Certain port numbers will also be barred, typically TELNET and FTP ports are barred and SMTP is allowed, as this allows mail to be routed into and out of the network, but no remote connections.
- Application-level firewalls. This type of firewall uses an intermediate system (a proxy server) to isolate the local computer from the external network. The local computer communicates with the proxy server, which in turns communicates with the external network, the external computer then communicates with the proxy which in turn communicates with the local computer. The external network never actually communicates directly with the local computer. The proxy server can then be set-up to be limited to certain types of data transfer, such as allowing HTTP (for WWW access), SMTP (for electronic mail), outgoing FTP, but blocking incoming FTP.
- Circuit-level firewalls. A circuit-level firewall is similar to an application-level firewall but it does not bother about the transferred protocol.

Network-level firewalls

The network-level firewall (or packet filter) is the simplest form of firewall and are also known as screen routers. It basically keeps a record of allowable source and destination IP addresses, and deletes all packets which do not have them. This technique is known as address filtering. The packet filter keeps a separate source and destination table for both directions, that is, into and out of the intranet. This type of method is useful for companies which have geographically spread sites, as the packet filter allows incoming traffic from other friendly sites, but blocks other non-friendly traffic. This is illustrated by Figure P.40.

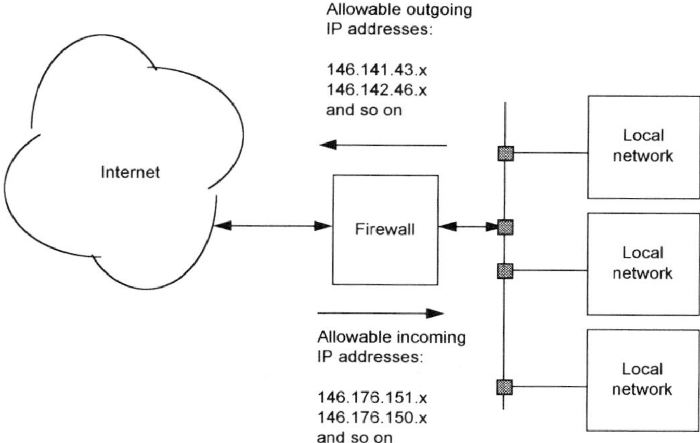

Figure P.40 Packet filter firewalls.

Unfortunately, this method suffers from the fact that IP addresses can be easily forged. For example, a hacker might determine the list of good source addresses and then add one of them to any packets which are addressed to the intranet. This type of attack is known as address spoofing and is the most common method of attacking a network.

Application-level firewall

The application-level firewall uses a proxy server to act as an intermediate system between the external network and the local computer. Normally the proxy only supports a given number of protocols, such as HTTP (for WWW access) or FTP. It is thus possible to block certain types of protocols, typically outgoing FTP (Figure P.41).

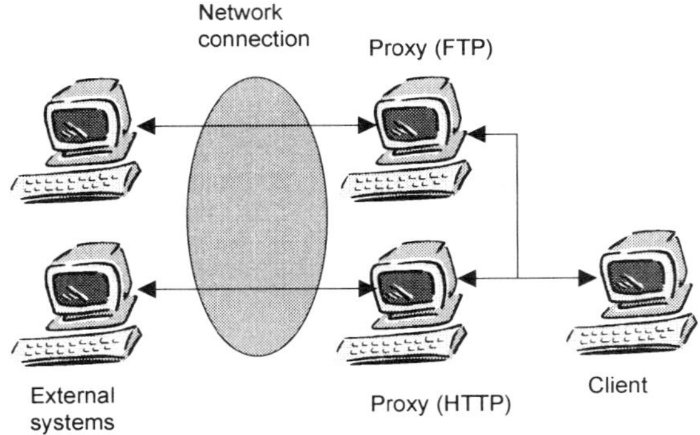

Figure P.41 Application-level firewall.

The proxy server thus isolates the local computer from the external network. The local computer communicates with the proxy server, which in turn communicates with the external network, the external computer then communicates with the proxy, which in turn, communicates with the local computer. The external network never actually communicates directly with the local computer. Figure P.42 shows a WWW browser is set up to communicate with a proxy server to get its access. In the advanced options (Figure P.43) different proxy servers can be specified. In this case for HTTP (WWW access), FTP, Gopher, Secure and Socks (Windows Sockets). It can also be seen that a proxy server can be bypassed by specify a number of IP addresses (or DNS).

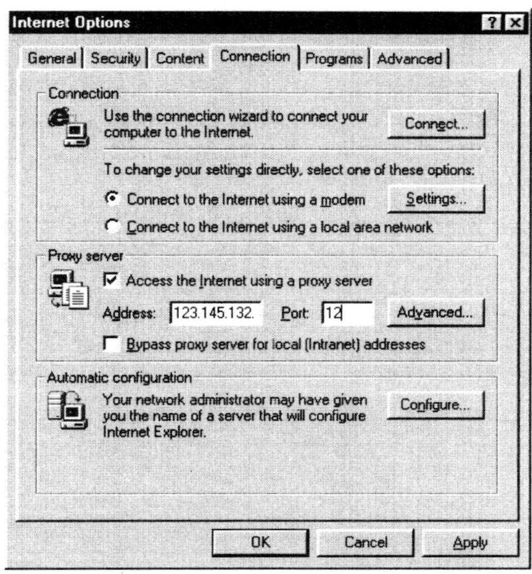

Figure P.42 Internet options showing proxy server selection.

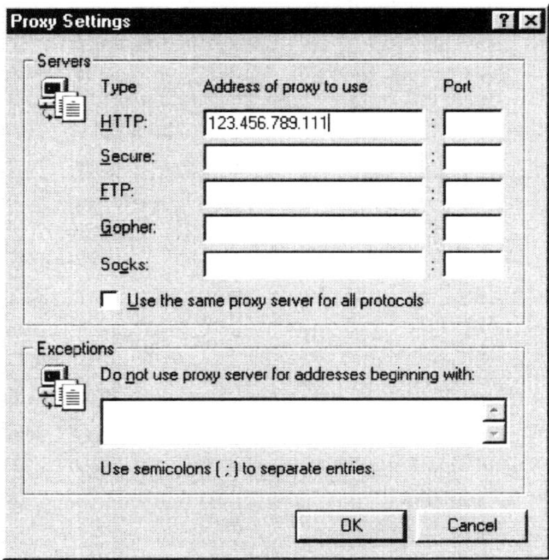

Figure P.43 Proxy settings.

P.6.2 Firewall architectures

The three main types of firewall are shown in Figure P.44, these are:

- Dual-homed host firewall. With this type of firewall a dedicated computer isolates the local network from the external network (typically, the Internet). It contains two network cards to connect to the two networks and should not run any routing software, and must rely on application-layer routing. This type of firewall is fairly secure and easy to maintain.

- Screen-host firewall. With this method, a router is placed between the external network and the firewall. Users on the local network connect to the firewall and the firewall only communicates with the router. There is thus no direct connection between the external network and the firewall (as all communications must go through the router). The security of this method is better than that of the dual-homed type, as there is no direct electrical connection to the external network.
- Screened-subnet firewall. With this method, a router is place on either side of the firewall. The firewall is thus isolated from the two connected networks. One router filters traffic between the local network and the firewall, and the other filters traffic between the external network and firewall. This provides the greatest level of security for both incoming and outgoing traffic.

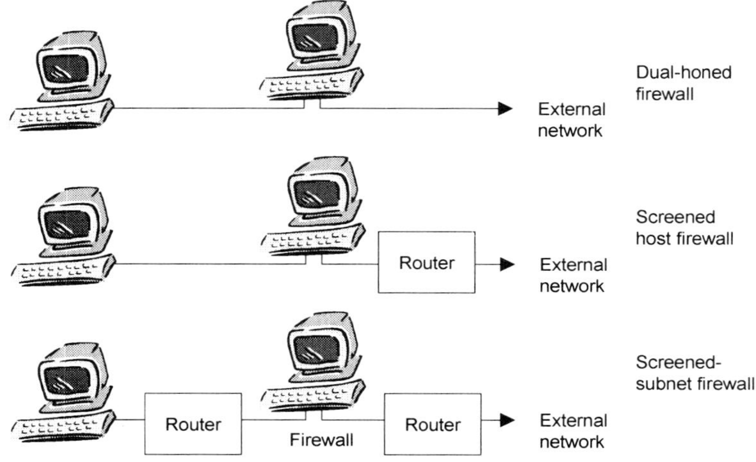

Figure P.44 Firewall types.

P.6.3 Security ratings

The Orange Book produced by the US Department of Defense (DOD) defines levels of security for systems. It is an important measure of a firewall's security. There are four main division, which split into seven main security ratings. Division D is the lowest security level and Division A is the highest. The rating are:

- Division D. This rating provides no protection on files or for users. For example, a DOS-based computer has no real security on files and users, thus it has a Division D rating.
- Division C. This rating splits into two groups: C1 rating and C2 rating. C1 contains a trust computing base (TCB) which separates users and data. It suffers from the fact that all the data on the system has the same security level. Thus, users cannot make distinctions between highly secure data and not-so secure data. A C1 system has user names and passwords, as well as some form of control of users and objects. C2 has a higher level of security and provide for some form of accountability and audit. This allows events to be logged and traced, for example, it might contain a list of user logins, network address logins, resource accesses, bad logins, and so on.

- Division B. This rating splits into three groups: B1, B2 and B3. Division B rated systems have all the security of a C2 rating, but have more security because they have a different level of security for all system accesses. For example, each computer can have a different security level, each printer can also have different security levels, and so on. Each object (such as a computer, printer, and so on) has a label associated with it. It is with this label that the security is set by. Non-labeled resources cannot be connected to the system. In a B2 rated system, users are notified of any changes of an object that they are using. The TCB also includes separate operator and administrator functions. In a B3 rated system the TCB excludes information which is not related to security. The system should also be designed to be simple to trace, but also well tested to prevent external hackers. It should also have a full-time administrator, audit trails and system recovery methods.
- Division A. This is the highest level of security. It is similar to B3, but has formal methods for the systems security policy. The system should also have a security manager, who should document the installation of the system, and any changes to the security model.

P.6.4 Application level gateways

Application-level gateways provide an extra layer of security when connecting an intranet to the Internet. They have three main components:

- A gateway node.
- Two firewalls which connect on either side of the gateway and only transmit packets which are destined for or to the gateway.

Figure P.45 shows the operation of an application level gateway. In this case, Firewall A discards anything that is not addressed to the gateway node, and discards anything that is not sent by the gateway node. Firewall B, similarly discards anything from the local network that is not addressed to the gateway node, and discards anything that is not sent by the gateway node. Thus, to transfer files from the local network into the global network, the user must do the following:

- Log onto the gateway node.
- Transfer the file onto the gateway.
- Transfer the file from the gateway onto the global network.

To copy a file from the network, an external user must:

- Log onto the gateway node.
- Transfer from the global network onto the gateway.
- Transfer the file from the gateway onto the local network.

A common strategy in organizations is to allow only electronic mail to pass from the Internet to the local network. This specifically disallows file transfer and remote login. Unfortunately, electronic mail can be used to transfer files. To overcome this problem the firewall can be designed specifically to disallow very large electronic mail messages, so it will limit the ability to transfer files. This tends not to be a good method as large files can be split up into small parts then sent individually.

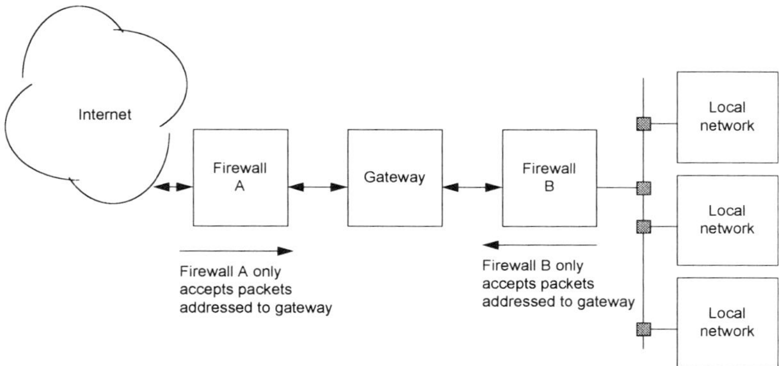

Figure P.45 Application level gateway.

P.6.5 Encrypted tunnels

Packet filters and application level gateways suffer from insecurity, which can allow non-friendly users into the local network. Packet filters can be tricked with fake IP addresses and application level gateways can be hacked into by determining the password of certain users of the gateway then transferring the files from the network to the firewall, on to the gateway, on to the next firewall and out. The best form of protection for this type of attack is to allow only a limited number of people to transfer files onto the gateway.

The best method of protection is to encrypt the data leaving the network then to decrypt it on the remote site. Only friendly sites will have the required encryption key to receive and send data. This has the extra advantage that the information cannot be easily tapped-into.

Only the routers which connect to the Internet require to encrypt and decrypt, as illustrated in Figure P.46. Typically, remote users connect to a corporation intranet by connecting over a modem which is connected to the corporation intranet, and using a standard Internet connection protocol, such as Point-to-Point Protocol (PPP). This can be expensive in both phone calls or in providing enough modem for all connected users. These costs can be drastically reduced if the user connects to an ISP, as they provide local rate charges. For this a new protocol, called Point-to-Point Tunneling Protocol (PPTP) has been developed to allow remote users connections to intranets from a remote connection (such as from a modem or ISDN). It operates as follows:

- Users connect to an ISP, using a protocol such as Point-to-Point Protocol (PPP) and requests that the information is sent to an intranet. The ISP has special software and hardware to handle PPTP.
- The data send to the ISP, using PPTP, is encrypted before it is sent into the Internet.
- The ISP sends the encrypted data (wrapped in an IP packet) to the Intranet.
- Data is passed through the firewall, which has the software and hardware to process PPTP packets.
- Next, the user logs in using Password Authentication Protocol (PAP) and Challenge Handshake Authentication (CHAP).
- Finally, the intranet server reads the IP packet and decrypts the data.

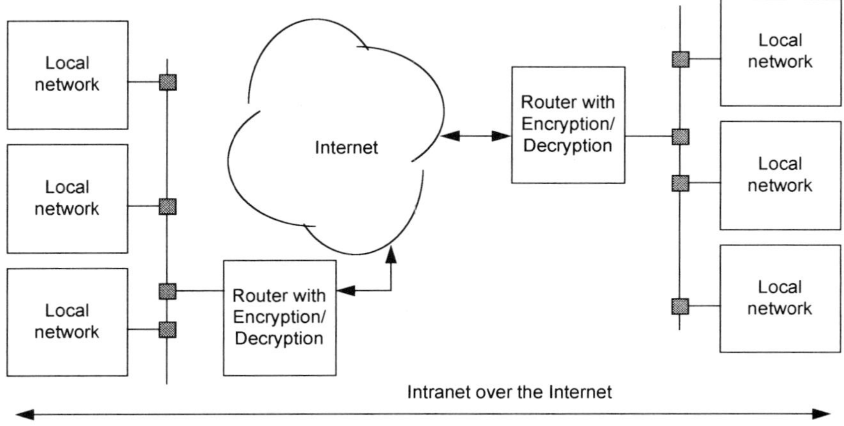

Figure P.46 Encryption tunnels.

P.6.6 Filtering routers

Filtering routers run software which allows many different parameters of the incoming and outgoing packets to be examined, such as:

- Source IP address. The router will have a table of acceptable source IP addresses. This will limit the access to the external network as only authorized users will be granted IP addresses. This unfortunately is prone to IP spoofing, where a local user can steal an authorized IP address. Typically, it is done by determining the IP address of a computer and waiting until there is no-one using the computer, then using the unused IP address. Several users have been accused of accessing unauthorized material because other users have used their IP address. A login system which monitors IP addresses and the files that they are accessing over the Internet cannot be used as evidence against the user, as it is easy to steal IP addresses.
- Destination IP address. The router will have a table of acceptable outgoing destination IP addresses, addresses which are not in the table are blocked. Typically, this will be used to limit the range of destination addresses to the connected organizational intranet, or to block certain addresses (such as pornography sites).
- Protocol. The router holds a table of acceptable protocols, such as TCP and/or UDP.
- Source port. The router will have a table of acceptable TCP ports. For example, electronic mail (SMTP) on port 25 could be acceptable, but remote login on port 543 will be blocked.
- Destination port. The router will have a table of acceptable TCP ports. For example, ftp on port 20 could be acceptable, but telnet connections on port 23 will be blocked.
- Rules. Other rules can be added to the system which define a mixture of the above. For example, a range of IP addresses can be allowed to transfer on a certain port, but another range can be blocked for this transfer.

Filter routers are either tightly bound when they are installed and then relaxed, or are relaxed and then bound. The type depends on the type of organization. For example, a financial institution will have a very strict router which will allow very little traffic, apart from the authorized traffic. The router can be opened-up when the systems have been proved to be secure (they can also be closed quickly when problems occur).

An open organization, such as an education institution will typically have an open system, where users are allowed to access any location on any port, and external users are allowed any access to the internal network. This can then be closed slowly when internal or external users breach the security or access unauthorized information. For example, if a student is access a pornographic site consistently then the IP address for that site could be blocked (this method is basically closing the door after the horse has bolted).

To most users the filtering router is an excellent method of limited traffic access, but to the determined hacker it can be easily breached, as the hacker can fake both IP addresses and also port addresses. It is extremely easy for a hacker to write their own TCP/IP driver software to address whichever IP address, and port numbers that they want.

P.7 Authentication

P.7.1 Introduction

It is obviously important to encrypt a transmitted message, but how can it be proved that the message was sent by the user who encrypted the message. This is achieved with message authentication. The two users who are communicating are sometimes known as the principals. It should be assumed that an intruder (hacker) can intercept and listen to messages at any part of the communications, whether it be the initial communication between the two parties and their encryption keys or when the encrypted messages are sent. The intruder could thus playback any communications between the parties and pretend to be the other.

P.7.2 Shared secret-key authentication

With this approach a secret key, K_{12} (between Fred and Bert) is used by both users. This would be transmitted through a secure channel, such as a telephone call, personal contact, mail message, and so on. The conversation will then be:

- The initiator (Fred) sends a challenge to the responder (Bert) which is a random number.
- The responder transmits it back using a special algorithm and the secret key. If the initiator receives back the correctly encrypted value then it knows that the responder is allowed to communicate with the user.

The random number should be large enough so that it is not possible for an intruder to listen to the communication and repeat it. There is little chance of the same 128-bit random number occurring within days, months or even years.

This method has validated Bert to Fred, but not Fred to Bert. Thus, Bert needs to know that the person receiving his communications is Fred. Thus Bert initiates the same procedure as before, sending a random number to Fred, who then encrypts it and sends it back. After this has been successfully received by Bert, encrypted communications can begin.

P.7.3 Diffie-Hellman key exchange

In the previous section, a private key was passed over a secure line. The Diffie-Hellman method allows for keys to be passed electronically. For this, Fred and Bert pick two large prime numbers:

a=Prime Number 1
b=Prime Number 2

where:

$(a-1)/2$ is also prime. The values of a and b are public keys. Next, Fred picks a private key (c) and Bert picks a private key (d). Fred sends the values of:

$(a, b, b^c \bmod a)$

Bert then responds by sending:

$(b^d \bmod a)$

For example:

a=43 (first prime number), b=7 (second prime number), c=9 (Fred's private key), d=8 (Bert's private key). Note, that the value of a (43) would not be used as (a–1)/2 is not prime (21).

Thus, the values sent by Fred will be:

(43, 7, 42)

The last value is 42 as 7^9 is 40,353,607, and 40,353,607 mod 43, is 42. Bert will respond back with:

(6)

as 7^8 is 5,764,801, and 5,764,801 mod 43 is 6.

Next Fred and Bert will calculate:

$b^{cd} \bmod a$

and both use this as their secret key. Figure P.47 shows an example of the interchange. It is difficult for an intruder to determine the values of c and d, when the values of a, b, c and d are large.

Unfortunately, this method suffers from the man-in-the-middle attack, where the intruder incepts the communications between Fred and Bert. Figure P.48 shows an interceptor (Bob) who has chosen a private key of e. Thus, Fred thinks he is talking to Bert, and vice-versa, but Bob is attacking as the man-in-the-middle. Bob then uses two different keys when talking with Fred and Bert.

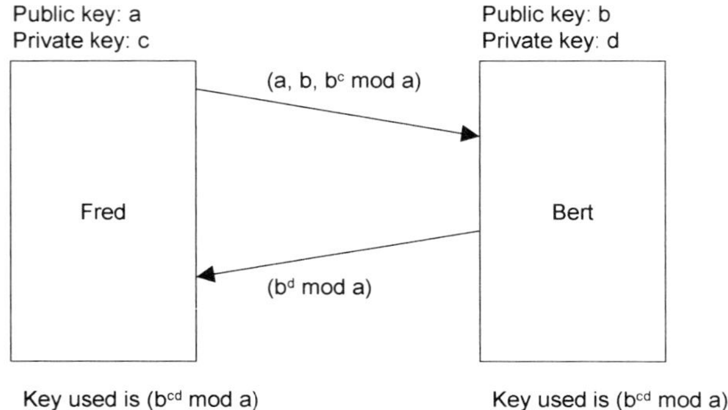

Figure P.47 Dillie-Hellman key exchange.

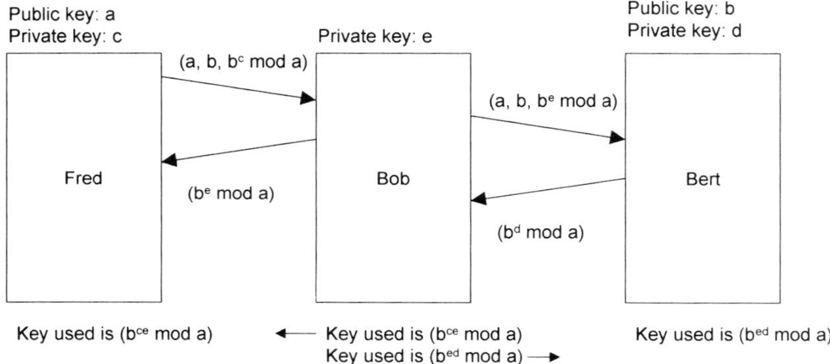

Figure P.48 Man-in-the-middle attack.

P.7.4 Key distribution center

The Dillie-Hellman method suffers from the man-in-the-middle attack, it also requires a separate key for each communication channel. A KDC (key distribution center) overcomes these problems with a single key and a secure channel for authentication. In a KDC, the authentication and session keys are managed through the KDC. One method is the wide-mouth protocol, which does the following:

- Fred selects a session key (K_{SESS}).
- Fred sends an encrypted message which contains the session key. The message is encrypted with K_{KDC1}, which is the key that Fred uses to pass messages to and from the KDC.
- The KDC decrypts this encrypted message using the K_{KDC1} key. It also extracts the session key (K_{SESS}). This session key is added to the encrypted message and then encrypted with K_{KDC2}, which is the key that Bert uses to pass messages to and from the KDC.

This method is relatively secure as there is a separate key used between the transmission between Fred and the KDC, and between Bert and the KDC. These keys are secret to Fred and the KDC, and between Bert and the KDC. The drawback with the method is that if the intruder determines secret key used for Fred to communicate with the KDC then it is possible to trick the KDC that it is communicating with Fred. As the key is unchanging, the theft of a key may take some time to discover and the possible damage widespread. The intruder can simply choose a new session key each time there is a new session.

P.7.5 Digital signatures

Digital signatures provide a way of validating an electronic document, in the same way as a hand-written signature does on a document. It must provide:

- Authentication of the sender. This is important as the recipient can verify the sender of the message.
- Authentication of the contents of the message. This is important, as the recipient knows that a third party has not modified the original contents of the message. Normally, this is also time-stamped.
- Authentication that the contents have not been changed by the recipient. This is important in legal cases where the recipient can prove that the message was as the original.

Secret-key signatures

The secret-key signature involves a user selecting a secret-key which is passed to a central authority, who keeps the key private. When Fred wants to communicate with Bert, he passes the plaintext to the central authority and encrypts it with a secret-key and the time-stamp. The central authority then passes an encrypted message to Bert using the required secret-key. A time-stamp is added to the message that is sent to Bert. This provides for a legal verification of the time the message was sent, and also stops intruders from replaying a transmitted message. The main problem with this method is that the central authority (typically, banks, government departmental or legal professionals) must be trust-worthy and reliable. They can also read all of the transmitted messages.

Message digests

Public and private-key signatures provide for both authentication and secrecy, but in many cases all that is required is that a text message is sent with the required authentication. A method of producing authentication is message digest, which generates a unique message digest for every message. The most common form of message digest is MD5 (RFC1321, R.Rivest). It is designed to be relatively fast to compute and does not require any large substitution tables. In summary, its operation is:

- It takes as input a message of arbitrary length.
- Produces an 128-bit "fingerprint" (or message digest) of the input.
- It is not possible to produce two message which have the same message digest, or to produce any message from a prespecified target message digest.

MD5 algorithm

Initially, the message with *b* bits is arranged as follows:

$$m_0\ m_1\ m_2\ m_3\ m_4\ m_5\ m_6\ \ldots\ m_{b-1}$$

Next five steps are performed:

- Adding padding bits. The message is padded so that its length is 64 bits less than being a multiple of 512 bits. For example, if the message is 900 bits long, then an extra 60 bits will be added so that it is 64 bits short of 1024 bits. The padded bits are a single '1' bit followed by '0' bits. At least one bit must be added and, at the most, 512 bits are added.

- Append Length. A 64-bit representation of b (the length of the message before the padding bits were added) is appended to the result of the previous step. The resulting message will thus be a multiple of 512 bits, or:

  $m_0\ m_1\ m_2\ m_3\ m_4\ m_5\ m_6\ \ldots\ m_{n-1}$

  where n is a multiple of 512.

- MD Buffer initialized. A four-word buffer (A, B, C, D) is used to compute the message digest. These are initialized to the following hexadecimal values (low-order bytes first):

  A: 01 23 45 67h (0000 0001 0010 ... 0111) B: 89 ab cd efh
  C: fe dc ba 98h (1111 1110 1101 ... 1000) D: 76 54 32 10h

- Message processed in 16-word blocks. Next four auxiliary functions are defined which operate on three 32-bit words and produce a single 32-bit word. These are:

$$F(x,y,z) = X.Y + \overline{X}.Z$$
$$G(x,y,z) = X.Z + Y.\overline{Z}$$
$$H(x,y,z) = X \oplus Y \oplus Z$$
$$I(x,y,z) = Y \oplus \left(X + \overline{Z}\right)$$

This step also involves a 64-element table T[1 ... 64] which is made up of the function. T[i] is equal to the integer part of 4,294,967,296 times abs(sin(i)), where i is in radians.
The algorithm is as follows:

```
/* Process each 16-word block. */
For i = 0 to N/16-1 do
    /* Copy block i into X. */
    For j = 0 to 15 do
        Set X[j] to M[i*16+j].
    end /* of loop on j */

    /* Save A as AA, B as BB, C as CC, and D as DD. */
    AA = A   BB = B
    CC = C   DD = D

    /* Round 1. */
    /* Let [abcd k s i] denote the operation  a = b + ((a + F(b,c,d) + X[k] + T[i]) <<< s). */
    /* Do the following 16 operations. */
    [ABCD  0  7  1] [DABC  1 12  2] [CDAB  2 17  3] [BCDA  3 22  4]
```

```
[ABCD  4  7  5] [DABC  5 12  6] [CDAB  6 17  7] [BCDA  7 22  8]
[ABCD  8  7  9] [DABC  9 12 10] [CDAB 10 17 11] [BCDA 11 22 12]
[ABCD 12  7 13] [DABC 13 12 14] [CDAB 14 17 15] [BCDA 15 22 16]

/* Round 2. */
/* Let [abcd k s i] denote the operation   a = b + ((a + G(b,c,d) + X[k] + T[i]) <<< s). */

/* Do the following 16 operations. */
[ABCD  1  5 17] [DABC  6  9 18] [CDAB 11 14 19] [BCDA  0 20 20]
[ABCD  5  5 21] [DABC 10  9 22] [CDAB 15 14 23] [BCDA  4 20 24]
[ABCD  9  5 25] [DABC 14  9 26] [CDAB  3 14 27] [BCDA  8 20 28]
[ABCD 13  5 29] [DABC  2  9 30] [CDAB  7 14 31] [BCDA 12 20 32]

/* Round 3. */
/* Let [abcd k s t] denote the operation   a = b + ((a + H(b,c,d) + X[k] + T[i]) <<< s). */

/* Do the following 16 operations. */
[ABCD  5  4 33] [DABC  8 11 34] [CDAB 11 16 35] [BCDA 14 23 36]
[ABCD  1  4 37] [DABC  4 11 38] [CDAB  7 16 39] [BCDA 10 23 40]
[ABCD 13  4 41] [DABC  0 11 42] [CDAB  3 16 43] [BCDA  6 23 44]
[ABCD  9  4 45] [DABC 12 11 46] [CDAB 15 16 47] [BCDA  2 23 48]

/* Round 4. */
/* Let [abcd k s t] denote the operation   a = b + ((a + I(b,c,d) + X[k] + T[i]) <<< s). */
/* Do the following 16 operations. */
[ABCD  0  6 49] [DABC  7 10 50] [CDAB 14 15 51] [BCDA  5 21 52]
[ABCD 12  6 53] [DABC  3 10 54] [CDAB 10 15 55] [BCDA  1 21 56]
[ABCD  8  6 57] [DABC 15 10 58] [CDAB  6 15 59] [BCDA 13 21 60]
[ABCD  4  6 61] [DABC 11 10 62] [CDAB  2 15 63] [BCDA  9 21 64]

/* Then perform the following additions */

A = A + AA   B = B + BB   C = C + CC   D = D + DD
end
/* of loop on i */
```

Note that the <<< symbol represent the rotate left operation, where the bits are rotated to the left.

- Output. The message digest is produced from A, B, C and D, where A is the low-order byte and D the high-order byte.

Standard test results give the following message digests:

Message	*Message digest*
""	d41d8cd98f00b204e9800998ecf8427e
"a"	0cc175b9c0f1b6a831c399e269772661
"abc"	900150983cd24fb0d6963f7d28e17f72
"abcdefghijklmnopqrstuvwxyz"	f96b697d7cb7938d525a2f31aaf161d0
"ABCDEFGHIJKLMNOPQRSTUVWXYZabcdefghijklmnopqrstuvwxyz0123456789"	c3fcd3d76192e4007dfb496cca67e13b
"12345678901234567890123456789012345678901234567890123456789012345678901234567890"	57edf4a22be3c955ac49da2e2107b67a

P.8 ATM

P.8.1 ATM signaling and call set-up

ATM, as with most telecommunications systems, uses a single-pass approach to setting up a connection. Initially, the source connection (the source end-system) communicates a connection request to the destination connection (the destination end-point). The routing protocol manages the routing of the connection request and all subsequent data flow. The call is established with:

- A set-up message. This is initially sent, across the UNI, to the first ATM switch. It contains:

  - Destination end-system address.
  - Desired traffic.
  - Quality of service.
  - Information Elements (IE) defining particular desired higher layer protocol bindings and so on.

- The initial ATM switch sends back a local call proceeding acknowledgement to the source end-system.
- The initial ATM switch invokes an ATM routing protocol, and propagates a signaling request across the network, it finally reaches the ATM switch connected to the destination end-system.
- The destination ATM switch connected to the destination end-system forwards the set-up message to the end-system, across its UNI.
- The destination end-system either accepts or rejects the connection. If necessary, the destination can negotiate the connection parameters. If the destination end-system rejects the connection request, it returns a release message. This is also sent back to the source end-system and clears the connection, such as clearing any allocated VCI labels. A release message can also be used by any of the end-systems, or by the network, to clear an established connection.
- If the destination end-system accepts the call then the ATM switch, which connects to it, returns a connect message through the network, along the same path.
- When the source end-system receives and acknowledges the connect message, either node can then start transmitting data on the connection.

P.8.2 ATM addressing scheme

ATM, like any other network, requires a network address scheme which identifies the source and destination addresses. The ITU-T have developed a standardized, telephone-like, numbering system called E.164 for addressing public ATM networks. Unfortunately, E.164 addresses are public addresses and cannot typically be used within private networks. The ATM Forum have since extended ATM addressing to include private networks. For a private networking scheme in UNI 3.0/3.1 they evaluated two different models. These basically differ in the way that the ATM protocol layer is viewed in relation to existing protocol layers, such as IP and IPX layers, and are:

- Peer model. This model treats the ATM layer as a peer of existing network layers and uses the same addressing schemes within the ATM networks. Thus, ATM endpoints would be identified by their existing network layer address (such as an IP or an IPX address). The ATM signaling requests would then carry these addresses for the source and destination ATM switch. Also network layer routing protocols, such as RIP, and so on, can be used to route ATM signaling requests using existing network layer addresses. The peer model allows for simplified addressing.
- Overlay model. This model decouples the ATM layer from any existing network protocol and defines a new addressing structure and a new routing protocol. This, as with Ethernet, FDDI and Token Ring which use MAC addresses, allow all existing protocols to operate over an ATM network. For this reason, the model is known as the subnetwork or overlay model. Thus, all ATM switches need an ATM address, and possibly, also a network layer address (such as an IP or IPX address).

The disadvantage with the overlay model is that there needs to be an ATM address resolution protocol which maps network addresses (IP or IPX). The peer model does not need address resolution protocols, and, because it uses existing routing protocols.

The ATM Forum decided to implement the overlay model for UNI 3.0/3.1 signaling. This is mainly the peer model would be difficult to implement as they must essentially act as multiprotocol routers and support address tables for all current protocols, as well as all of their existing routing protocols. In addition, currently available routing protocols for LANs and WANs do not map well into the QoS parameter.

The ATM Forum chose a private network addressing scheme based on the OSI Network Service Access Point (NSAP) address. They are not true NSAP addresses, and are either ATM private network addresses or ATM end-point identifiers. They basically are subnetwork points of attachment.

An NSAP ATM address for a private network has 20 bytes, while a public network uses a E.164 address, as defined by the ITU-T. NSAP-based addresses have three main fields:

- Authority and Format Identifier (AFI). This defines type and format of the Initial Domain Identifier (IDI).
- Initial Domain Identifier (IDI). This defines the address allocation and administration authority.
- Domain Specific Part (DSP). This defines the actual routing information.

There are three formats that private ATM addressing use for different definitions for the AFI and IDI parts. These are:

- NSAP Encoded E.164 format. The E.164 number is contained in the IDI.
- DCC Format. The IDI is a Data Country Code (DCC) which identities the country, as specified in ISO 3166. These addresses are administered, in each country, by the ISO National Member Body.
- ICD Format. The IDI contains the International Code Designator (ICD). The ICD is allocated by the ISO 6523 registration authority. They identify particular international organizations.

NSAP

NSAP is defined in ISO/IEC 8348. It divides the address into two main parts: Initial Domain Part (IDP), which splits into the Authority and Format Identifier (AFI), and Initial Domain Identifier (IDI). The format is thus:

```
| IDP          | DSP
| AFI  | IDI  |
```

In the ISO/IEC 10589 specification, the DSP address includes an ID and SEL (1 byte selector) field which are used by level 1 routing. Typically, the ID part is taken from the ISO/IEC 8802 48-bit MAC address. The format is thus:

```
| IDP          | DSP                              |
| AFI  | IDI  |                 | ID  | SEL   |
```

In the UNI-3.1 specification, the ID field is six bytes. It is also defines that the NSAP address format uses a maximum length of 20 bytes.

ICD Format

The format of an ICD scheme is:

```
AFI                    47                             1 byte
ICI                    xxxx                           2 bytes
Version                xx                             1 byte
Network                xxxxxx                         3 bytes
Tele traffic area      xx                             1 byte
Member identifier      xxxx                           2 byte
Member access point    xx                             1 byte
Area                   xx                             1 byte
Switch                 xx                             1 byte
MAC-address            xxxxxxxxxxxx                   6 bytes
Nselector              xx                             1 byte
```

An example format is:

```
+--+--+--+--+--+--+--+--+--+--+--+--+--+--+--+--+--+--+--+--+
|47|00 23|00|00 00 03|xx|xx xx|xx|xx|xx| ESI MAC address |xx|
+--+--+--+--+--+--+--+--+--+--+--+--+--+--+--+--+--+--+--+--+
```

DCC Format

The DCC format is defined by the National Standards Organization 39528+1100. Its fields include:

```
AFI (39 or 38)        ISO DCC format                  (1 byte)
IDI                                                   (2 bytes)
CFI                   Country Format Identifier       (4 bits)
CDI                   Country Domain Identifier       (12 bits)
SFI                   SURFNet Format Identifier       (4 bits)
          0=CLN S in case of an organizational SDI
          1=in case of a network SDI
          2=ATM in case of an organization SDI
SDI                   SURFNet Domain Identifier       (2 bytes)
                      decimal encoded administrative numbers
ASDI                  Additional Domain Identifier    (4 bits)
```

```
NYU                 Not yet used                (5 bytes)
ESI                 End System Identifier       (6 bytes)
SEL                 Selector                    (1 byte)
```

The EaStMAN network uses a 13 byte prefix, in the form:

`39.826f.1107.16.7000.00.nnmm.ee.ff`

where 39 is the AFI – ISO DCC, 826f is the IDI (indicating the UK), 1 is the CFI, 107 is the CDI, 1 for the SFI, 107 the SDI (country and domain), 16 for ASNI (for region), 00 (not yet used), `nnmm` for site code, `ee` for campus number and `ff` for switch number.

The `nn` part represents the institution, these are:

01	University of Edinburgh	02	Moray House
03	Queen Margaret College	04	Napier University
05	Heriot-Watt University	06	Edinburgh College of Art
07	University of Stirling		

and `mm` is the ring access point. These are assigned in a clockwise direction on the ring, starting from Kings Building (University of Edinburgh). In summary, the `nnmm` codes are assigned as follows:

University of Edinburgh:

Kings Buildings	0101	Pollock Halls	0102
Old College	0103	New College	0106

Moray House:

MH-H	0204	MH-Cramond	0209

Queen Margaret College:

QMC-Leith	0305	QMC-Corstorphine	030a

Napier University:

Merchiston	0407	Sighthill	040b

Heriot-Watt:

Riccarton campus	0508

Edinburgh College of Art:

ECA-L	060c	ECA-G	060d

University of Stirling:

Stirling 070e

The last two byte values (`ee` and `ff`) are allocated by the local institution. Typically, they can be used to identify the campus number (`ee`) and the switch number within the campus (`ff`).

E.164 Format (ATM Forum/95-0427R1)

The E.164 format provides a geographical scheme, but, as it is derived from the telephone system, the addresses are in short supply. Therefore, the E.164 NSAP format is recommned, which is extensible to ISDN. Its format is:

```
+--+--+--+--+--+--+--+--+--+--+--+--+--+--+--+--+--+--+--+--+
|45| Internat E.164 number |  HO-DSP   |xx xx xx xx xx xx|xx|
+--+--+--+--+--+--+--+--+--+--+--+--+--+--+--+--+--+--+--+--+
```

Where:

```
45        AFI for E.164 binary syntax
IDI       International E.164 number
HO-DSP    Extends the E.164 address to logically identify
          many devices in a single geographical location.
ESI       Similar to HO-DSP, extends the E.164 address.
SEL       Selector
```

P.9 Gigabit Ethernet

The IEEE 802.3 working group initiated the 802.3z Gigabit Ethernet task force to create the Gigabit Ethernet standard (which was finally defined in 1998). The Gigabit Ethernet Alliance (GEA) was founded in May 1996 and promotes Gigabit Ethernet collaboration between organizations. Companies, which initially were involved in the GEA, include: 3Com, Bay Networks, Cisco Systems, Compaq, Intel, LSI Logic, Sun and VLSI.

The amount of available bandwidth for a single segment is massive. For example, almost 125 million characters (125MB) can be sent in a single second. For example, a large reference book with over 1000 pages could be send over a network segment, ten times in a single second. Compare it also with a ×24, CD-ROM drive which transmits at a maximum rate of 3.6MB/s (24×150kB/sec). Gigabit Ethernet operates almost 35 times faster than this drive. With network switches, this bandwidth can be multiplied a given factor, as they allow multiple simultaneous connections.

Gigabit Ethernet is excellent challenger for network backbones as it interconnects 10/100BASE-T switches, and also provides a high-bandwidth to high-performance servers. Initial aims were:

- Half/full-duplex operation at 1000Mbps.
- Standard 802.3 Ethernet frame format. Gigabit Ethernet uses the same variable-length frame (64- to 1514-byte packets), and thus allows for easy upgrades.
- Standard CSMA/CD access method.
- Computability with existing 10BASE-T and 100BASE-T technologies.
- Development of an optional Gigabit Media Independent Interface (GMII).

The compatibility with existing 10/100BASE standards make the upgrading to Gigabit Ethernet much easier, and considerably less risky than changing to other networking types, such as FDDI and ATM. It will happily interconnect with, and autosense, existing slower rated Ethernet devices. Figure P.49 illustrates the functional elements of Gigabit Ethernet, its main characteristics are:

- Full-duplex communication. As defined by the IEEE 802.3x specification, two nodes connected via a full-duplex, switched path can simultaneously send and receive frames. Gigabit Ethernet supports new full-duplex operating modes for switch-to-switch and switch-to-end-station connections, and half-duplex operating modes for shared connections using repeaters and the CSMA/CD access method.
- Standard flow control. Gigabit Ethernet uses standard Ethernet flow control to avoid congestion and overloading. When operating in half-duplex mode, Gigabit Ethernet adopts the same fundamental CSMA/CD access method to resolve contention for the shared media.
- Enhanced CSMA/CD method. This maintains a 200m collision diameter at gigabit speeds. Without this, small Ethernet packets could complete their transmission before the transmitting node could sense a collision, thereby violating the CSMA/CD method. To resolve this issue, both the minimum CSMA/CD carrier time and the Ethernet slot time (the time, measured in bits, required for a node to detect a collision) have been extended from 64 bytes (which is 51.2μs for 10BASE and 5.12μs for 100BASE) to 512 bytes (which is 4.1μs for 1000BASE). The minimum frame length is still 64 bytes. Thus, frames smaller than 512 bytes have a new carrier extension field following the CRC field. Packets larger than 512 bytes are not extended.
- Packet bursting. The slot time changes effect the small-packet performance, but this has been offset by a new enhancement to the CSMA/CD algorithm, called packet bursting. This allows servers, switches and other devices to send bursts of small packets in order to fully utilize the bandwidth.

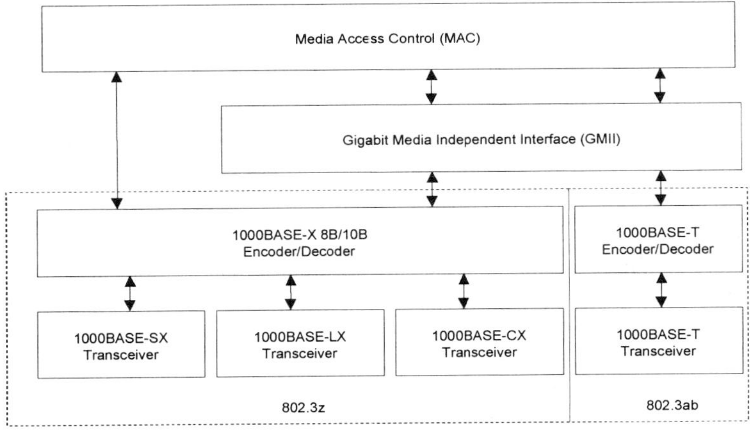

Figure P.49 Gigabit Ethernet functional elements.

Devices operating in full-duplex mode (such as switches and buffered distributors) are not subject to the carrier extension, slot time extension or packet bursting changes. Full-duplex devices use the regular Ethernet 96-bit interframe gap (IFG) and 64-byte minimum frame size.

P.9.1 Ethernet transceiver

The IEEE 802.3z task force spent much of their time defining the Gigabit Ethernet standard for the transceiver (physical layer), which is responsible for the mechanical, electrical and procedural characteristics for establishing, maintaining and deactivating the physical link between network devices. The physical layers are:

- 1000BASE-SX (Low cost, multi-mode fiber cables). These can be used for short interconnections and short backbone networks. The IEEE 802.3z task force have tried to integrate the new standard with existing cabling, whether it be twisted-pair cable, coaxial cable or fiber optic cable. These tests involved firing lasers in long lengths of multi-mode fiber cables. Through these test it was found that a jitter component results which is caused by a phenomenon known as differential mode delay (DMD). The 1000BASE-SX standard has resolved this by defining the launch of the laser signal, and enhanced conformance tests. Typical lengths are: 62.5 μm, multi-mode fiber (up to 220m). 50μm, multi-mode fiber (550m).
- 1000BASE-LX (Multi-mode/single mode-mode fiber cables). These can be used for longer runs, such as on backbones and campus networks. Single-mode fibers are covered by the long-wavelength standard, and provide for greater distances. External patch cords are used to reduce DMD. Typical lengths are: 62.5 μm, multi-mode fiber (up to 550m), 50μm, multi-mode fiber (up to 550m). 50μm, single-mode fiber (up to 5km).
- 1000BASE-CX (Shielded Balanced Copper). This standard supports interconnection of equipment using a copper-based cable, typically up to 25m. As with the upper two standard, it uses the Fiber Channel-based 8B/10B coding at the serial line rate of 1.25Gbps. The 1000BASE-T is likely to supersede this standard, but it has been relatively easy to define, and to implement.
- 1000BASE-T (UTP). This is a useful standard for connecting directly to workstations. The 802.3ab Task Force has been assigned the task of defining the 1000BASE-T physical layer standard for Gigabit Ethernet over four pairs of Cat-5 UTP cable, for cable distances of up to 100m, or networks with a diameter of 200m. As it can be used with existing cabling, it allows easy upgrades. Unfortunately, it requires new technology and new coding schemes in order to meet the potentially difficult and demanding parameters set by the previous Ethernet and Fast Ethernet standards.

P.9.2 Fiber Channel Components

The IEEE 802.3 committee based much of the physical layer technology on the ANSI-backed X3.230 Fiber Channel project. This allowed many manufacturers to re-use physical-layer Fiber Channel components for new Gigabit Ethernet designs, and has allowed a faster development time than is normal, and increased the volume production of the components. These include optical components and high-speed 8B/10B encoders.

The 1000BASE-T standard uses enhanced DSP (Digital Signal Processing) and enhanced silicon technology to enable Gigabit Ethernet over UTP cabling. As Figure P.49 shows, it does not use the 8B/10B encoding.

P.9.3 Buffered distributors

Along with repeaters, bridges and switches, a new device, called a buffered distributor (or full-duplex repeater), has been developed for Gigabit Ethernet. It is a full-duplex, multiport, hub-like device that connects two or more Gigabit Ethernet segments. Unlike a bridge, and like a repeater, it forwards all the Ethernet frames from one segment to the others, but unlike a standard repeater, a buffered distributor buffers one, or more, incoming frames on each link before forwarding them. This reduces collisions on connected segments. The maximum bandwidth for a buffered distributor will still only be 1Gbps, as opposed to Gigabit switches which allow multi-gigabit bandwidths.

P.9.4 Quality of Service

Many, real-time, networked applications require a given Quality of Server (QoS), which might related to bandwidth requirements, latency (network delays) and jitter. Unfortunately, there is nothing built-into Ethernet that allows for a QoS, thus new techniques have been developed to overcome this. These include:

- RSVP. Allows nodes to request and guarantee a QoS, and works at a higher-level to Ethernet. For this, each network component in the chain must support RSVP and communicate appropriately. Unfortunately, this may require an extensive investment to totally support RSVP, thus many vendors have responded in implementing proprietary schemes, which may make parts of the network vendor-specific.
- IEEE 802.1p and IEEE 802.1Q. Allows a QoS over Ethernet by "tagging" packets with an indication of the priority or class of service desired for the frames. These tags allow applications to communicate the priority of frames to internetworking devices. RSVP support can be achieved by mapping RSVP sessions into 802.1p service classes.
- Routing. Implemented at a higher layer.

P.9.5 Gigabit Ethernet migration

The greatest advantage of Gigabit Ethernet is that it is easy to upgrade existing Ethernet-based networks to higher bit rates. Typical migration might be:

- Switch-to-switch links. Involves upgrading the connections between switches to 1Gbps. As 1000BASE switches support both 100BASE and 1000BASE then not all the switches require to be upgraded at the same time, this allows for gradual migration.
- Switch-to-Server Links. Involves upgrading the connection between a switch and the server to 1Gbps. The server requires an upgraded Gigabit Ethernet interface card.
- Switched Fast Ethernet Backbone. Involves upgrading a Fast Ethernet backbone switch to a 100/1000BASE switch. It this supports both 100BASE and 1000BASE switching, using existing cabling.
- Shared FDDI Backbone. Involves replacing FDDI attachments on the ring with Gigabit Ethernet switches or repeaters. The Gigabit uses the existing fiber-optic cable, and provides a greatly increased segment bandwidth.
- Upgrade NICs on nodes to 1Gbps. It is unlikely that users will require 1Gbps connections, but this facility is possible.

P.9.6 1000BASE-T

One of the biggest challenges of Gigabit Ethernet is to use existing Cat-5 cables, as this will allow fast upgrades. Two critical parameters, which are negligible at 10BASE speeds, are:

- Return loss. Defines the amount of signal energy that is reflected back towards the transmitter due to impedance mismatches in the link (typically from connector and cable bends).
- Far-End Crosstalk. Noise that is leaked from another cable pair.

The 1000BASE-T Task Force estimates that less than 10% of the existing Cat-5 cable was improperly installed (as defined in ANSI/TIA/EIA568-A in 1995) and might not support 1000BASE-T (or even, 100BASE-TX). 100BASE-T uses two pairs, one for transmit and one for receive, and transmits at a symbol rate of 125Mbaud with a 3-level code. 1000BASE-T uses:

- All four pairs with a symbol rate of 125 Mbaud (symbols/sec). One symbol contains two bits of information.
- Each transmitted pulse uses a 5-level PAM (Pulse Amplitude Modulation) line code, which allows two bits to be transmitted at a time.
- Simultaneous sends and receives on each pair. Each connection uses a hybrid circuit to split the send and receive signals.
- Pulse shaping. Matches the characteristics of the transmitted signal to the channel so that the signal-to-noise ratio is minimized. It effectively reduces low frequency terms (which contain little data information, can cause distortion and cannot be passed over the transformer-coupled hybrid circuit), reduces high-frequency terms (which increases crosstalk) and rejects any external high-frequency noise. It is thought that the transmitted signal spectrum for 1000BASE will be similar to 100BASE.
- Forward Error Correction (FEC). This provides a second level of coding that helps to recover the transmitted symbols in the presence of high noise and crosstalk. The FEC bit uses the fifth level of the 5-level PAM.

A 5-level code (–2, –1, 0, +1, +2) allows two bits to be sent at a time, if all four pairs are used then 8 bits are sent at a time. If each pair transmits at a rate of 125Mbaud (symbols/sec), the resulting bit rate will be 1Gbps.

P.10 UDP

TCP allows for a reliable connection-based transfer of data. The User Datagram Protocol (UDP) is an unreliable connection-less approach, where datagrams are sent into the network without an acknowledgements or connections. It is defined in RFC 768 and uses IP as its underlying protocol. It has the advantage of TCP in that it has a minimal protocol mechanism, but does not guarantee delivery of any of the data. Figure P.50 shows its format. The fields are:

- Source port. This is an optional field is set to a zero if not used. It identifies the local port number which should be used when the destination host requires to contact the originator
- Destination Port to connect to on the destination.

- Length. Number of bytes in the datagram, including the UDP header and the data.
- Checksum. The 16-bit 1's complement of the 1's complement sum of the IP header, the UDP header, the data (which, if necessary, is padded with zero bytes at the end, to make an even number of bytes).

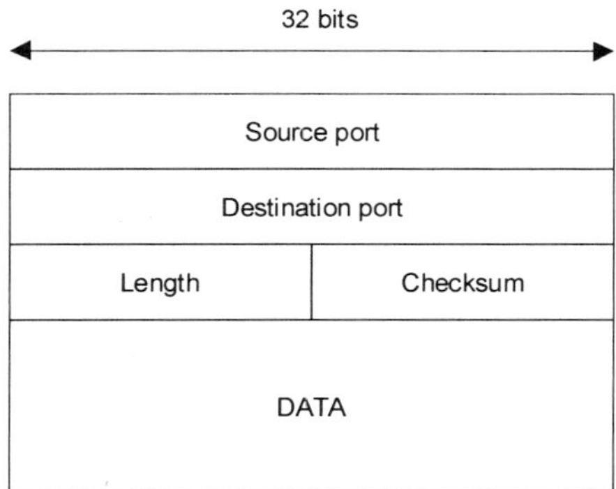

Figure P.50 UDP header format.

When used with IP the UDP/IP header is shown in Figure P.51. The Protocol field is set to 17 to identify UDP.

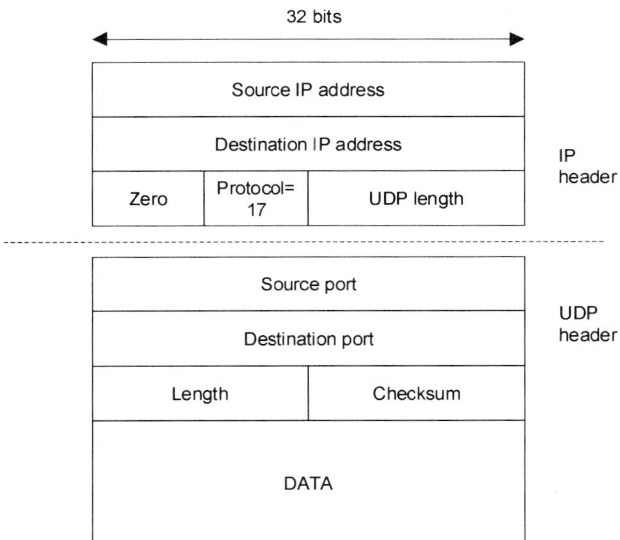

Figure P.51 UDP/IP header format.

P.11 TCP specification

TCP is made reliable with the following:

- Sequence numbers. Each TCP packet is sent with a sequence number. Theoretically, each data byte is assigned a sequence number. This sequence number of the first data byte in the segment is transmitted with that segment and is called the segment sequence number (SSN).
- Acknowledgements. Packets contain an acknowledgement number, which is the sequence number of the next expected transmitted data byte in the reverse direction. On sending, a host puts stores the transmitted data in a storage buffer, and starts a time. If the packet is acknowledged then this data is deleted, else, if no acknowledgement is received before the timer runs out, the packet is retransmitted.
- Window. With this, a host sends a window value which specifies the number of bytes, starting with the acknowledgement number, that the host can receive.

P.11.1 Connection establishment, clearing and data transmission

The main interfaces in TCP are shown in Figure P.52. The calls from the application program to TCP include:

- OPEN and CLOSE. To open and close a connection.
- SEND and RECEIVE. To send and receive.
- STATUS. To receive status information.

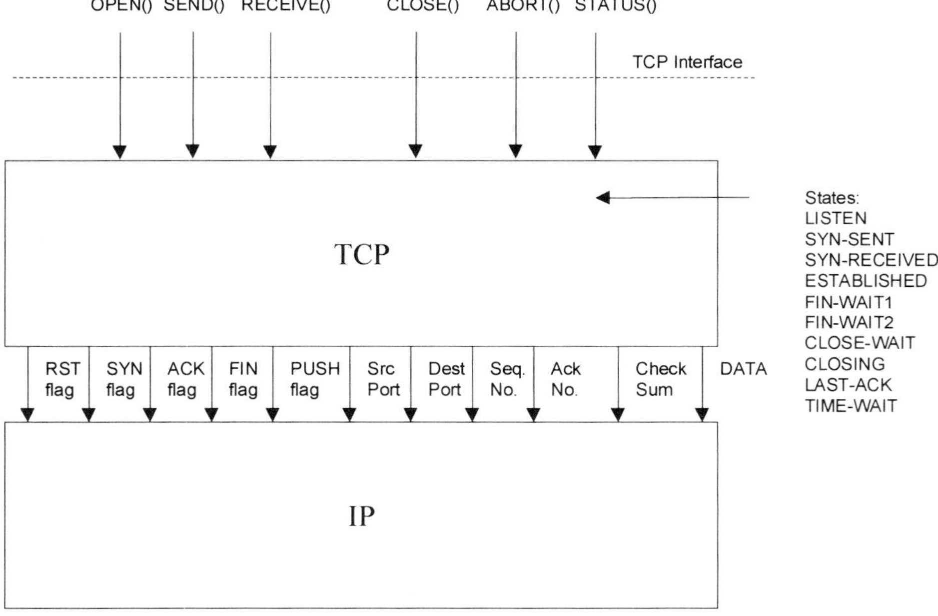

Figure P.52 TCP interface.

The OPEN call initiates a connection with a local port and foreign socket arguments. A Transmission Control Block (TCB) stores the information on the connection. After a successful connection, TCP adds a local connection name by which the application program refers to the connection in subsequent calls.

The OPEN call supports two different types of call, as illustrated in Figure P.53, these are:

- Passive OPEN. TCP waits for a connection from a foreign host, such as from an active OPEN. In this case, the foreign socket is defined by a zero. This is typically used by servers, such as TELNET and FTP servers. The connection can either be from a fully specified or an unspecified socket.
- Active OPEN. TCP actively connects to a foreign host, typically a server (which is opened with a passive OPEN). Two processes which issue active OPENs to each other, at the same time, will also be connected.

A connect is established with the transmission of TCP packets with the SYN control flag set and uses a three-way handshake. A connection is cleared by the exchange of packets with the FIN control flag set. Data flows in a stream using the SEND call to send data and RECEIVE to receive data.

The PUSH flag is used to send data in the SEND immediately to the recipient. This is required as a sending TCP is allowed to collect data from the sending application program and sends the data in segments when convenient. Thus, the push flag forces it to be sent. When the receiving TCP sees the PUSH flag, it does not wait for any more data from the sending TCP before passing the data to the receiving process.

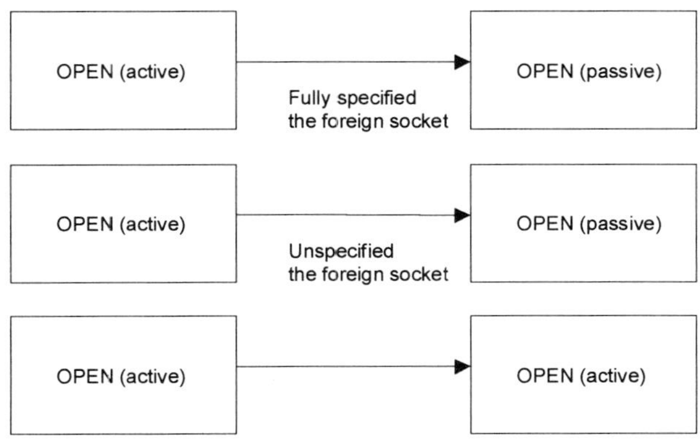

Figure P.53 TCP connections.

P.11.2 TCB parameters

Table P.2 outlines the send and receive packet parameters, as well as the current segment parameter, which are stored in the TCB. Along with this, the local and remote port number require to be stored.

Table P.1 TCB parameters.

Send Sequence Variables	Receive Sequence Variables	Current Packet Variable
SND.UNA Send unacknowledged	RCV.NXT Receive next	SEG.SEQ segment sequence number
SND.NXT Send next	RCV.WND Receive window	SEG.ACK segment acknowledgement number
SND.WND Send window	RCV.UP Receive urgent pointer	SEG.LEN segment length
SND.UP Send urgent pointer	IRS Initial receive sequence number	SEG.WND segment window
SND.WL1 Segment sequence number used for last window update		SEG.UP segment urgent pointer
SND.WL2 Segment acknow-ledgement number used for last window update		SEG.PRC segment precedence value
ISS Initial send sequence number		

P.11.3 Connection states

Figure P.54 outlines the states in which the connection goes into, and the events which cause them. The events from applications programs are: OPEN, SEND, RECEIVE, CLOSE, ABORT, and STATUS, and the events from the incoming TCP packets include the SYN, ACK, RST and FIN flags. The definition of each of the connection states are:

- LISTEN. This is the state in which TCP is waiting for a remote connection on a given port.
- SYN-SENT. This is the state where TCP is waiting for a matching connection request after it has sent a connection request.
- SYN-RECEIVED. This is the state where TCP is waiting for a confirming connection request acknowledgement after having both received and sent a connection request.
- ESTABLISHED. This is the state that represents an open connection. Any data received can be delivered to the application program. This is the normal state after for data to be transmitted.
- FIN-WAIT-1. This is the state in which TCP is waiting for a connection termination request, or an acknowledgement of a connection termination, from the remote TCP.
- FIN-WAIT-2. This is the state in which TCP is waiting for a connection termination request from the remote TCP.
- CLOSE-WAIT. This is the state where TCP is waiting for a connection termination request from the local application.
- CLOSING. This is the state where TCP is waiting for a connection termination request acknowledgement from the remote TCP.
- LAST-ACK. This is the state where TCP is waiting for an acknowledgement of the connection termination request previously sent to the remote TCP.
- TIME-WAIT. This is the state in which TCP is waiting for enough time to pass to be sure the remote TCP received the acknowledgement of its connection termination request.
- CLOSED. This is the fictional state, which occurs after the connection has been closed.

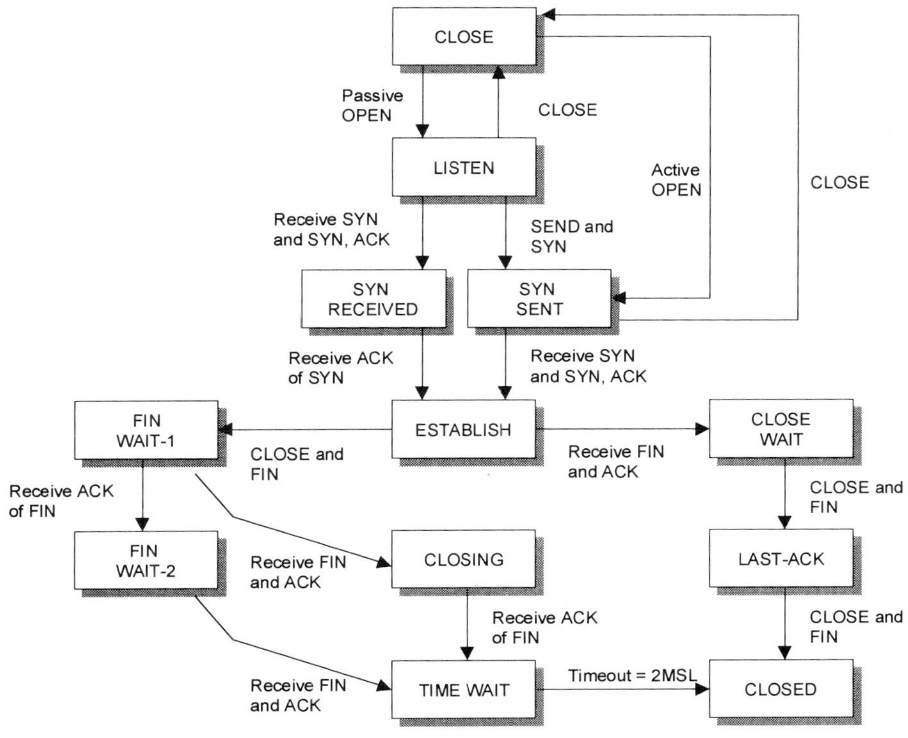

Figure P.54 TCP connection states.

Sequence Numbers

TCP packets contain a 32-bit sequence number (0 to 4,294,967,295), which related to every byte sent. It uses a cumulative acknowledgement scheme, where an acknowledgement with a value of VAL, validates all bytes up to, but not including, byte VAL. The number of bytes which the packet starts at the first data byte and are then numbered consecutively.

When sending data, TCP should receive acknowledgements for the transmitted data. The required TCB parameters will be:

SND.UNA	Oldest unacknowledged sequence number.
SND.NXT	Next sequence number to send.
SEG.ACK	Acknowledgement from the receiving TCP (next sequence number expected by the receiving TCP).
SEG.SEQ	First sequence number of a segment.
SEG.LEN	Number of bytes in the TCP packet.
SEG.SEQ+SEG.LEN–1	Last sequence number of a segment.

One receiving data, the following TCB parameters are required:

RCV.NXT	Next sequence number expected on an incoming segments, and is the left or lower edge of the receive window.

RCV.NXT+RCV.WND–1	Last sequence number expected on an incoming segment, and is the right or upper edge of the receive window.
SEG.SEQ	First sequence number occupied by the incoming segment.
SEG.SEQ+SEG.LEN–1	Last sequence number occupied by the incoming segment.

ISN selection

The Initial Sequence Number (ISN) is selected so that previous sockets are not confused with new sockets. Typically this can happen when a host application crashes and then quickly re-establishes the connection before the other side can time-out the connection. To avoid this a 32-bit initial sequence number (ISN) generator is created when the connection is made. It is generated by a 32-bit clock, which is incremented approximately every 4μs (giving a ISN cycle of 4.55 hours). Thus within 4.55 hours, each ISN will be unique.

As each connection has a send and receive sequence number, these is an initial send sequence number (ISS) and an initial receive sequence number (IRS). When establishing a connection, the two TCPs synchronize their initial sequence numbers. This is done by exchanging connection establishing packets, with the SYN bit set and with the initial sequence numbers (these packets are typically called SYNs). Thus four packets must be initially exchanged, these are:

- A sends to B. SYN with A_{SEQ}.
- B sends to A. ACK of the sequence number (A_{SEQ}).
- B sends to A. SYN with B_{SEQ}.
- A sends to B. ACK of the sequence number (B_{SEQ}).

Note that the two intermediate steps can be combined into a single message. This is sometimes knows as a three-way handshake. This handshake is necessary as the sequence numbers are not tied to a global clock, only to local clocks, and has many advantages, including the fact that old packets will be discarded as they occurred in a previous time.

To makes sure that a sequence number is not duplicated, a host must wait for a maximum segment lifetime (MSL) before starting to retransmit packets (segments) after start-up or when recovering from a crash. An example MSL is 2 minutes. Although, if it is recovering, and it has a memory of the previous sequence numbers, it may not need to wait for the MSL, as it can use sequence numbers which are much greater than the previously used sequence numbers.

P.11.4 Opening and closing a connection

Figure P.55 shows a basic three-way handshake. The steps are:

1. The initial state on the initiator is CLOSED and, on the recipient, it is LISTEN (the recipient is waiting for a connection).
2. The initiator goes into the SYN-SENT state and sends a packet with the SYN bit set and the indicates that the starting sequence number will be 999 (the current sequence number, thus the next number sent will be 1000). When this is received the recipient goes into the SYN-RECEIVED state.
3. The recipient sends back a TCP packet with the SYN and ACK bits set (which identifies that it is a SYN packet and also that it is acknowledging the previous SYN packet). In this case, the recipient tells the originator that it will start transmitting at a sequence

number of 100. The acknowledgement number is 1000, which is the sequence number that the recipient expects to receive next. When this is received, the originator goes into the ESTABLISHED state.

4. The originator sends back a TCP packet with the SYN and ACK bits set and the acknowledgement number is 101, which is the sequence number it expects to see next.
5. The originator transmits data with the sequence number of 1000.

Originator					**Recipient**
1. CLOSED					LISTEN
2. SYN-SENT	→	<SEQ=999><CTL=SYN>			SYN-RECEIVED
3. ESTABLISHED		<SEQ=100><ACK=1000>	<CTL=SYN,ACK	←	SYN-RECEIVED
4. ESTABLISHED	→	<SEQ=1000><ACK=101>	<CTL=ACK>		ESTABLISHED
5. ESTABLISHED	→	<SEQ=1000><ACK=101>	<CTL=ACK><DATA>		ESTABLISHED

Figure P.55 TCP connection.

Note that the acknowledgement number acknowledges every sequence number up to, but not including the acknowledgement number.

Figure P.56 shows how the three-way handshake prevents old duplicate connection initiations from causing confusion. In state 3, a duplicate SYN has received, which is from a previous connection. The Recipient sends back an acknowledgement for this (4), but when this is received by the Originator, the Originator sends back a RST (reset) packet. This causes the Recipient to go back into a LISTEN state. It will then receive the SYN packet send in 2, and after acknowledging it, a connection is made.

TCP connections are half-open if one of the TCPs has closed or aborted, and the other end is still connected. It can also occur if the two connections have become desynchronized because of a system crash. This connection is automatically reset if data is sent in either direction. This is because the sequence numbers will be incorrect, otherwise the connection will time-out.

A connection is normally closed with the CLOSE call. A host who has closed cannot continue to send, but can continue to RECEIVE until it is told to close by the other side. Figure P.57 shows a typical sequence for closing a connection. Normally the application program sends a CLOSE call for the given connection. Next, a TCP packet is sent with the FIN bit set, the originator enters into the FIN-WAIT-1 state. When the other TCP has acknowledged the FIN and sent a FIN of its own, the first TCP can ACK this FIN.

Originator				**Recipient**
1. CLOSED				LISTEN
2. SYN-SENT	→	<SEQ=999><CTL=SYN>		
3. (duplicate)	→	<SEQ=900><CTL=SYN>		
4. SYN-SENT		<SEQ=100><ACK=901> <CTL=SYN,ACK>	←	SYN-RECEIVED
5. SYN-SENT	→	<SEQ=901><CTL=RST>		LISTEN
6.		(packet 2 received)	→	
7. SYN-SENT		<SEQ=100><ACK=1000><CTL=SYN,ACK	←	SYN-RECEIVED
8. ESTABLISHED	→	<SEQ=1000><ACK=101><CTL=ACK><DATA>		ESTABLISHED

Figure P.56 TCP connection with duplicate connections.

	Originator			**Recipient**
1.	ESTABLISHED (*CLOSE call*)			ESTABLISHED
2.	FIN-WAIT-1	→ <SEQ=1000><ACK=99> <CTL=SFIN,ACK>		CLOSE-WAIT
3.	FIN-WAIT-2	<SEQ=99><ACK=1001> <CTL=ACK>	←	CLOSE-WAIT
4.	TIME-WAIT	<SEQ=99><ACK=101><CTL=FIN,ACK>	←	LAST-ACK
5.	TIME-WAIT	→ <SEQ=1001><ACK=102><CTL=ACK>		CLOSED

Figure P.57 TCP close connection.

P.12 Example PGP encryption

The Pretty Good Privacy (PGP) program developed by Philip Zimmermann is widely available over the Internet. It runs as a stand-alone application, and uses various options to use the package. Table P.2 outlines some of the options.

Table P.2 PGP options

Option	*Description*
pgp -e textfile her_userid [other userids]	Encrypts a plaintext file with the recipent's public key. In this case, it produces a file named textfile.pgp.
pgp -s textfile [-u your_userid]	Sign a plaintext file with a secret key. In this case, it produces a file named textfile.pgp.
pgp -es textfile her_userid [other userids] [-u your_userid]	Signs a plaintext file with the senders secret key, and then encrypt it with recipent's public key. In this case, it produces a file named textfile.pgp.
pgp -c textfile	Encrypt with conventional encryption only.
pgp ciphertextfile [-o plaintextfile]	Decrypt or check a signature for a ciphertext (.pgp) file.

To produce output in ASCII for email or to publish over the Internet, the –a option is used with other options. Table P.3 shows the key management functions.

Table P.3 PGP key management options

Option	*Description*
pgp –kg	Generate a unique public and private key.
pgp -ka keyfile [keyring]	Adds key file's contents to the user's public or secret key ring.

pgp -kr userid [keyring]	Removes a key or a user ID from the user's public or secret key ring.
pgp -ke your_userid [keyring]	Edit user ID or pass phrase.
pgp -kx userid keyfile [keyring]	Extract a key from the public or secret key ring.
pgp -kv[v] [userid] [keyring]	View the contents of the public key ring.
pgp -kc [userid] [keyring]	Check signatures on the public key ring.
pgp -ks her_userid [-u your_userid] [keyring]	Sign someone else's public key on your public key ring.
pgp -krs userid [keyring]	Remove selected signatures from a userid on a keyring.

P.12.1 RSA Key Generation

Both the public and the private keys are generated with:

```
pgp -kg
```

Initially, the user is asked about the key sizes. The larger the key the more secure it is. A 1024 bit key is very secure.

```
C:\pgp> pgp -kg
Pretty Good Privacy(tm) 2.6.3i - Public-key encryption for the masses.
(c) 1990-96 Philip Zimmermann, Phil's Pretty Good Software. 1996-01-18
International version - not for use in the USA. Does not use RSAREF.
Current time: 1998/12/29 23:13 GMT

Pick your RSA key size:
    1)   512 bits- Low commercial grade, fast but less secure
    2)   768 bits- High commercial grade, medium speed, good security
    3)  1024 bits- "Military" grade, slow, highest security
Choose 1, 2, or 3, or enter desired number of bits: 3

Generating an RSA key with a 1024-bit modulus.
```

Next, the program asks for a user ID, which is normally the users name and his/her password. This ID helps other users to find the required public key.

```
You need a user ID for your public key.  The desired form for this
user ID is your name, followed by your E-mail address enclosed in
<angle brackets>, if you have an E-mail address.
For example:  John Q. Smith <12345.6789@compuserve.com>
Enter a user ID for your public key:
Fred Bloggs <fred_b@myserver.com>
```

Next PGP also asks for a pass phrase, which is used to protect the private key if another person gets hold of it. No person can use the secret key file, unless they know the pass phrase. Thus the pass phase is like a password but is typically much longer. The phase is also required when the user is encrypting a message with his/her private key.

```
You need a pass phrase to protect your RSA secret key.
Your pass phrase can be any sentence or phrase and may have many
words, spaces, punctuation, or any other printable characters.
Enter pass phrase: fred bloggs
Enter same pass phrase again: fred bloggs
Note that key generation is a lengthy process.
```

The public and private keys are randomly derived from measuring the intervals between keystrokes. For this the software asks for the user to type a number of keys.

```
We need to generate 384 random bits.  This is done by measuring the
time intervals between your keystrokes.  Please enter some random text
on your keyboard until you hear the beep:

We need to generate 384 random bits.  This is done by measuring the
time intervals between your keystrokes.  Please enter some random text
on your keyboard until you hear the beep:
<keyboard typing>

   0 * -Enough, thank you.
..................................****
...................................****

Pass phrase is good.  Just a moment....
Key signature certificate added.
Key generation completed.
```

This has successfully generated the public and private keys. The public key is placed on the public key ring (PUBRING.PGP) and the private key is place on the users secret key ring (SECRING.PGP).

```
C:\pgp> dir *.pgp
SECRING  PGP          518  12-29-98 11:20p secring.pgp
PUBRING  PGP          340  12-29-98 11:20p pubring.pgp
```

The -kx option can be used to extract the new public key from the public key ring and place it in a separate public key file, which can be send to people who want to send an encrypted message to the user.

```
C:\pgp> pgp -kx fred_b
Extracting from key ring: 'pubring.pgp', userid "fred_b".
Key for user ID: Fred Bloggs <fred_b@myserver.com>
1024-bit key, key ID CD5AE745, created 1998/12/29

Extract the above key into which file? mykey

Key extracted to file 'mykey.pgp'.
```

The public key file (mykey.pgp) can be sent to other users, and can be added to their public key rings. Care must be taken never to send anyone a private key, but even if it is sent then it is still protected by the pass phase.

Often a user wants to publish their public key on their WWW page or transmit it by email. Thus, it requires to be converted into an ASCII format. For this the –kxa options can be used, such as:

```
C:\pgp> pgp -kxa fred_b

Extracting from key ring: 'pubring.pgp', userid "fred_b".
Key for user ID: Fred Bloggs <fred_b@myserver.com>
1024-bit key, key ID CD5AE745, created 1998/12/29

Extract the above key into which file? mykey
Transport armor file: mykey.asc
Key extracted to file 'mykey.asc'.

Extract the above key into which file? mykey

Transport armor file: mykey.asc

Key extracted to file 'mykey.asc'.
```

The file mykey.asc now contains an ASCII form of the key, such as:

```
Type Bits/KeyID     Date       User ID
pub  1024/CD5AE745 1998/12/29 Fred Bloggs <fred_b@myserver.com>

-----BEGIN PGP PUBLIC KEY BLOCK-----
Version: 2.6.3i

mQCNAzaJY84AAAEEAK0nvnuYcwGEaNdeqcDGXD6IrMFwX3iKtdGkZgyPyiENLb+C
bGX7P2zSG0z1d8c4f5OKYR/RgxzN4ILsAKthGaweGD0FJRgeIvn6FHJxEzmdBWIh
ME/8h2HZfegSXta8hFAMc8o9ASamolk5KBL0YWfsQ1DNbR+dMJpPqQ7NWudFAAUT
tCFGcmVkIEJsb2dncyA8ZnJlZF9iQG15c2VydmVyLmNvbT6JAJUDBRA2iWPOmk+p
Ds1a50UBAfkoA/4gO5DllYko4DfjPnq4ItDtN55SgoE3upPWL52R5RQZF1BoJEF6
eLT/kejD5b7gli/yP1S456bh/k8ifi9RwSPUFN/zFUsVVYrSjZKD3kzClV1/QgTy
YmlDHHHgou6rYFXk7mGEtWc4g4D1rzds+ppc/UjN8uNp5KQUg1FsVatvPA==
=X5Xx
-----END PGP PUBLIC KEY BLOCK-----
```

Now, someone's public key can be added to the Fred's public key ring. In this case, Fred Bloggs wants to send a message to Bert Smith. Bert's public key, in an ASCII form, is:

```
Type Bits/KeyID     Date       User ID
pub  1024/770CA60D 1998/12/30 Bert Smith <Bert_s.otherserver.com>

-----BEGIN PGP PUBLIC KEY BLOCK-----
Version: 2.6.3i

mQCNAzaKE5AAAAEEAN+5td9acGlPcTKp5J42UpwbDqz6mHOaxcO1lp6CoPE3+AXT
jfREEQ+TC0ZxMP6cCcwtEMnjVqu2M7F6li3v/AVqQIRZZkFsEOZ+8hlseHB0FR8Y
f8FDpmgld6wNpp8ocOyVul/sBQl549u0C/KnVQ6LtXo7UlsBtnbua9J3DKYNAAUR
tCNCZXJ0IFNtaXRoIDxCZXJ0X3Mub3RoZXJzZXJ2ZXIuY29tPokAlQMFEDaKE5B2
7mvSdwymDQEB2xkEANLMEDncVrFjR71abUIWHqquEFK+sqnOHPbHyIBni18x03UM
jeQJM1WA9/uIPqzeABJdD6anX4oK3yiByQjI5CT5+OdmU0y4e2+k1ab5mxxUWs7S
Tib3K5LLvPGxsOInOdunjFKaBLkrfU/L+zid3iW9FV6Zy8P07yDL2SmobRbh
=6rTj
-----END PGP PUBLIC KEY BLOCK-----
```

Fred can add Bert's key onto his public key ring with the –ka option:

```
C:\pgp> pgp -ka bert.pgp

Looking for new keys...
pub  1024/770CA60D 1998/12/30  Bert Smith <Bert_s.otherserver.com>

Checking signatures...
pub  1024/770CA60D 1998/12/30 Bert Smith <Bert_s.otherserver.com>
sig!      770CA60D 1998/12/30  Bert Smith <Bert_s.otherserver.com>


Keyfile contains:
   1 new key(s)

One or more of the new keys are not fully certified.
Do you want to certify any of these keys yourself (y/N)?
```

Bert's key has been added to Fred's public key ring. This ring can be listed with the –kv, as given next:

```
C:\pgp> pgp -kv

Key ring: 'pubring.pgp'
Type Bits/KeyID    Date       User ID
pub  1024/770CA60D 1998/12/30 Bert Smith <Bert_s.otherserver.com>
pub  1024/CD5AE745 1998/12/29 Fred Bloggs <fred_b@myserver.com>
2 matching keys found.
```

Next, a message can be send to Bert, using his public key.

```
C:\pgp>edit message.txt
Bert,

This is a secret message. Please
delete it after you have read it!

Fred.

C:\pgp>pgp -e message.txt

Recipients' public key(s) will be used to encrypt.
A user ID is required to select the recipient's public key.
Enter the recipient's user ID: bert smith

Key for user ID: Bert Smith <Bert_s.otherserver.com>
1024-bit key, key ID 770CA60D, created 1998/12/30

WARNING:  Because this public key is not certified with a trusted
signature, it is not known with high confidence that this public key
actually belongs to: "Bert Smith <Bert_s.otherserver.com>".

Are you sure you want to use this public key (y/N)? y
.
Ciphertext file: message.pgp
```

If the message needs to be transmitted by electronic mail or via a WWW page, it can be converted into text format with the –ea option, as given next:

```
C:\pgp>pgp -ea message.txt
Recipients' public key(s) will be used to encrypt.
A user ID is required to select the recipient's public key.
Enter the recipient's user ID: bert smith

Key for user ID: Bert Smith <Bert_s.otherserver.com>
1024-bit key, key ID 770CA60D, created 1998/12/30

WARNING:  Because this public key is not certified with a trusted
signature, it is not known with high confidence that this public key
actually belongs to: "Bert Smith <Bert_s.otherserver.com>".
But you previously approved using this public key anyway.
.
Transport armor file: message.asc
```

The message.asc file is now in a form which can be transmitted in ASCII characters. In this case, it is:


```
-----BEGIN PGP MESSAGE-----
Version: 2.6.3i

hIwDdu5r0ncMpg0BBAC7jOUx74vLb701lOCO0/5Fkc6pDJinqpA7isJH+JYbFkDj
wSv6vF/jAEonEPL8RVtqWncNDwjjwwV9OVPEZeaZ0qgZTWdbdSUilfqxZsaBo8Uz
dmmbzxd7CDTpnSYEyFWosPyzdxJqlsICig79Loh71lBdJXEhKnMy+1VMieNYtKYA
AABrB8LTMj2lkk9t6JfS2yOc1t9EfpVMLX+rxtPZ+Tq1aCOwfid4E77FyiKN260N
APzF8J6elXhBgNM3zesA8fR8KdEnrI2BYC2XsBzTxOiKnpqoLMwWl0A7TTyhv24L
1PhwFi/YQ2SPhemdpqY=
=ooNT
-----END PGP MESSAGE-----
```


Bert can now simply decrypt the received message.

```
C:\pgp\bert> pgp message.pgp

File is encrypted.  Secret key is required to read it.
Key for user ID: Bert Smith <Bert_s.otherserver.com>
1024-bit key, key ID 770CA60D, created 1998/12/30

You need a pass phrase to unlock your RSA secret key.
Enter pass phrase: Bert Smith
Pass phrase is good.  Just a moment......
Plaintext filename: message
```

Or Bert can convert the ASCII form into a binary format with the –da option, and the decrypt the message as before.

```
C:\pgp\bert> pgp -da message.asc
Stripped transport armor from 'message.asc', producing 'message.pgp'.
```

P.13 Visual Basic implementation of sockets

Visual Basic support a WinSock control which allows the connection of hosts over a network. It supports both UDP and TCP. Figure P.58 shows a sample Visual Basic screen with a

WinSock object (in this case, it is named Winsock1). To set the Procotol used then either select the Properties window on the WinSock object, click Protocol and select either sckTCPProtocol, or sckUDPProtocol. Otherwise, within the code it can be set to TCP with:

```
Winsock1.Protocol = sckTCPProtocol
```

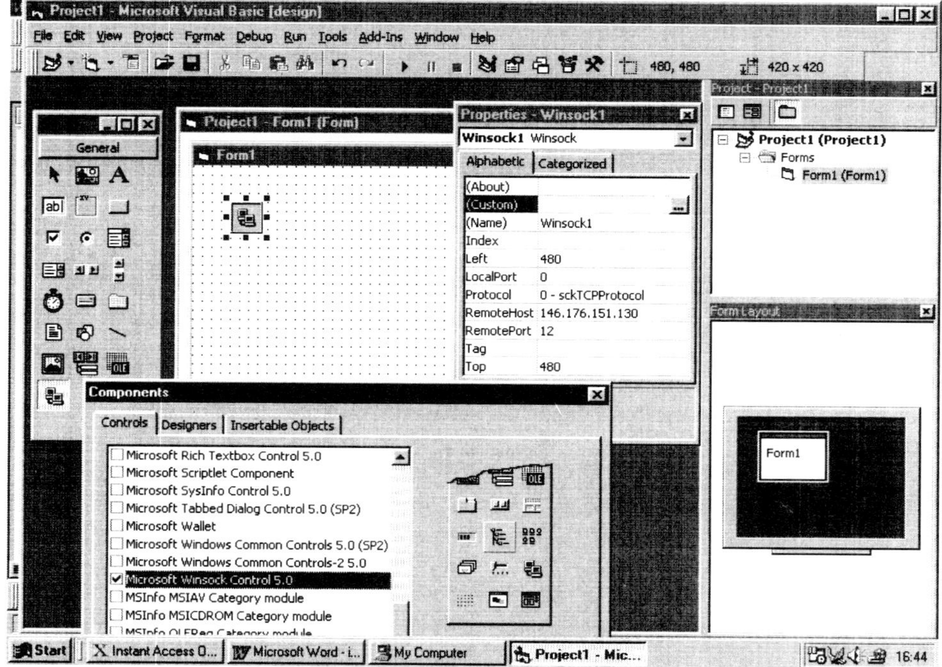

Figure P.58 WinSock object.

The WinSock object has various properties, such as:

obj.`RemoteHost`	Defines the IP address or DNS of the remote host.
obj.`LocalPort`	Defines the local port number

The methods that are used with the WinSock object are:

obj.`Connect`	Connects to a remote host (client invoked).
obj.`Listen`	Listens for a connection (server invoked).
obj.`GetData`	Reads data from the input steam.
obj.`SendData`	Sends data to an output stream.

The main events are:

`ConnectionRequest`	Occurs when a remote host wants to make a connection with a server.

`DataArrival`	Occurs when data has arrived from a connection (data is then read with GetData).

P.13.1 Creating a server

A server must listen for connection. To do this, do the following:

1. Create a new Standard EXE project.
2. Change the name of the default form to myServer.
3. Change the caption of the form to "Server Application" (see Figure P.59).
4. Put a Winsock control on the main format and change its name to myTCPServer.
5. Add two TextBox controls to the form. Name the first SendTextData, and the second ShowText.
6. Add the code given below to the form.

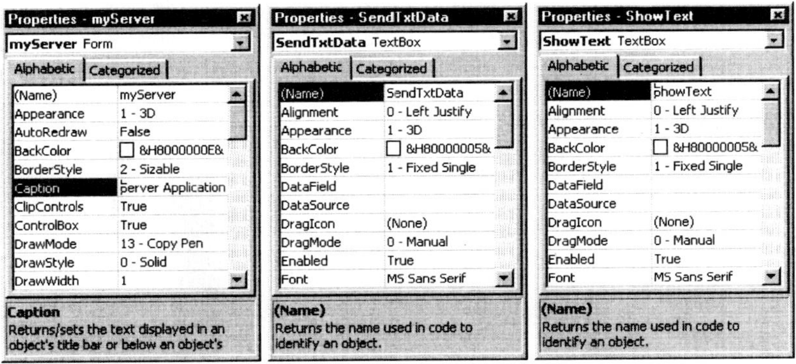

Figure P.59 Server setups.

```
Private Sub Form_Load()
   ' Set the local port to 1001 and listen for a connection
   myTCPServer.LocalPort = 1001
   myTCPServer.Listen
   myClient.Show
End Sub

Private Sub myTCPServer_ConnectionRequest (ByVal requestID As Long)
   ' Check state of socket, if it is not closed then close it.
   If myTCPServer.State <> sckClosed Then myTCPServer.Close
   ' Accept the request with the requestID parameter.
   myTCPServer.Accept  requestID
End Sub

Private Sub SendTextData_Change()
   ' SendTextData contains the data to be sent.
   ' This data is setn using the SendData method
   myTCPServer.SendData = SendTextData.Text
End Sub

Private Sub myTCPServer_DataArrival (ByVal bytesTotal As Long)
   ' Read incoming data into the str variable,
   ' then display it to ShowText
   Dim str As String
   myTCPServer.GetData = str
```

```
   ShowText.Text = str
End Sub
```

Figure P.60 shows the server setup.

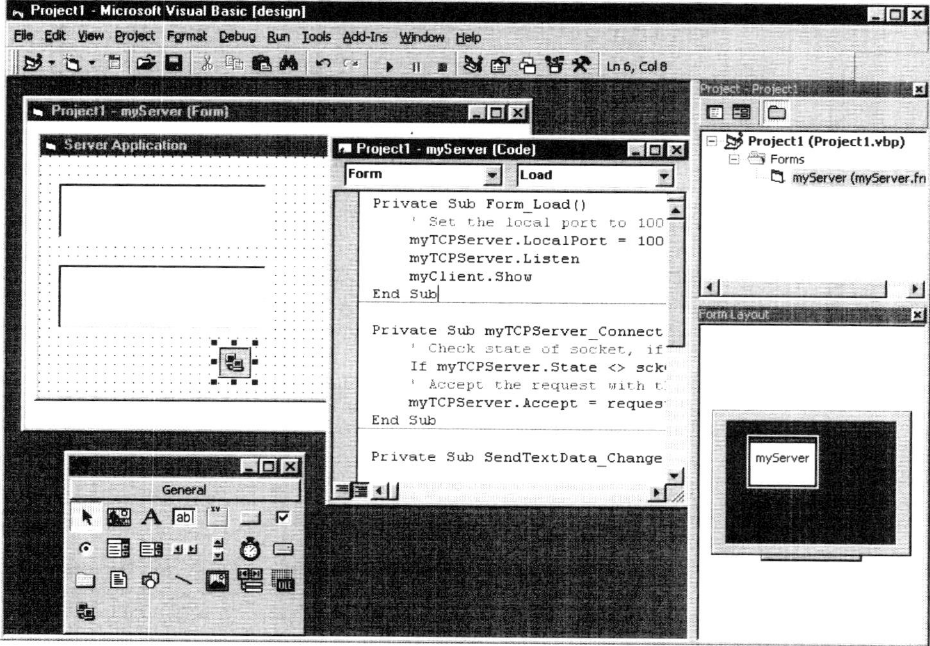

Figure P.60 Server form.

P.13.2 Creating a client

The client must actively seek a connection. To create a client, do the following:

1. A new form to the project, and name it myClient.
2. Change the caption of the form to "Client Application".
3. Add a Winsock control to the form and name it myTCPClient.
4. Add two TextBox controls to the form. Name the first SendTextData, and the second ShowText.
5. Draw a CommandButton control on the form and name it cmdConnect.
6. Change the caption of the CommandButton control to Connect (see Figure P.61).
7. Add the code given below to the form.

```
Private Sub Form_Load()
   ' In this case it will connect to 146.176.151.130
   ' change this to the local IP address or DNS of the local computer
   myTCPClient.RemoteHost = "146.176.151.130"
   myTCPClient.RemotePort = 1001
End Sub

Private Sub cmdConnect_Click()
```

```
   ' Connect to the server
   myTCPClient.Connect
End Sub

Private Sub SendTextData_Change()
   tcpClient.SendData txtSend.Text
End Sub

Private Sub tcpClient_DataArrival (ByVal bytesTotal As Long)
   Dim str As String
   myTCPClient.GetData str
   ShowText.Text = str
End Sub
```

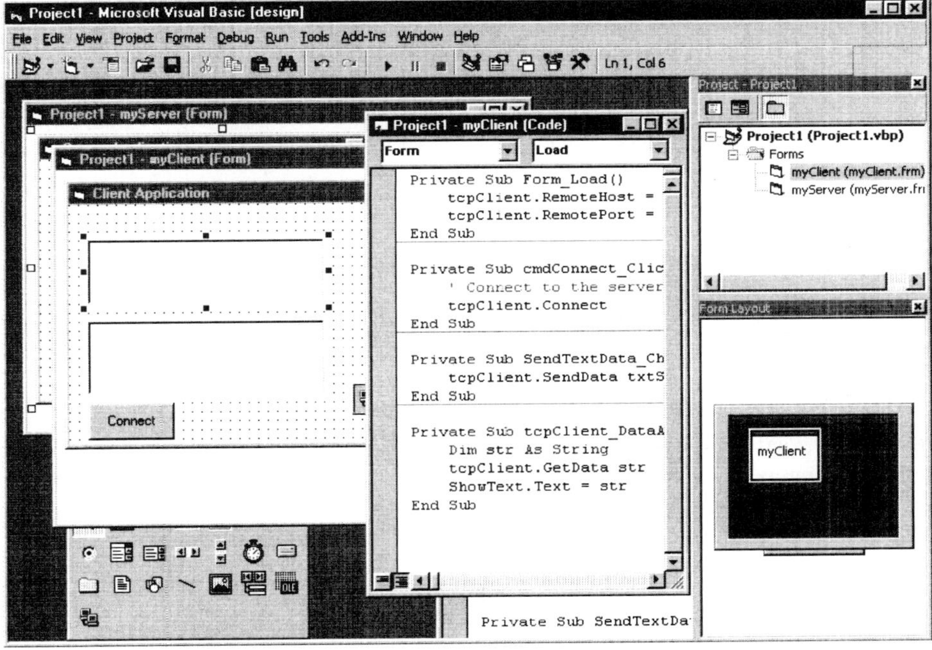

Figure P.61 Client form.

The program, when it is run, will act as a client and a server. Any text typed in the SendTxtData TextBox will be sent to the ShowText TextBox on the other form.

P.13.3 Multiple connections

In Visual Basic, it is also possible to create multiple connections to a server. This is done by creating multiple occurances of the server object. A new one is created every time there is a new connection (with the Connection_Request event). Each new server accepts the incoming connection. The following code, which has a Winsock control on a form called multServer, is given below.

```
Private ConnectNo As Long

Private Sub Form_Load()
```

```
   ConnectNo = 0
   multServer(0).LocalPort = 1001
   multServer(0).Listen
End Sub

Private Sub multServer_ConnectionRequest _
                     (Index As Integer, ByVal requestID As Long)
   If Index = 0 Then
      ConnectNo = ConnectNo + 1
      Load multServer(ConnectNo)
      multServer(ConnectNo).LocalPort = 0
      multServer(ConnectNo).Accept requestID
      Load txtData(ConnectNo)
   End If
End Sub
```

P.13.4 Connect event

The Connect event connects to a server. If an error occurs then a flag (ErrorOccurred) is set to True, else it is False. Its syntax is:

`Private Sub` *object*.`Connect(ErrorOccurred As Boolean)`

P.13.5 Close event

The Close event occurs when the remote computer closes the connection. Applications should use the Close method to correctly close their connection. Its syntax is:

object`_Close()`

P.13.6 DataArrival event

The DataArrival event occurs when new data arrives, and returns the number of bytes read (bytesTotal). Its syntax is:

object`_DataArrival (bytesTotal As Long)`

P.13.7 Bind method

The Bind method specifies the Local port (LocalPort) and the Local IP address (LocalIP) to be used for TCP connections. Its syntax is:

object.`Bind` *LocalPort*, *LocalIP*

P.13.8 Listen method

The Listen method creates a socket and goes into listen mode (for server applications). Its stays in this mode until a ConnectionRequest event occurs, which indicates an incoming connection. After this, the Accept method should be used to accept the connection. Its syntax is:

object.`Listen`

P.13.9 Accept method

The Accept method accepts incoming connections after a ConnectionRequest event. Its syntax is:

object.`Accept requestID`

The requestID parameter is passed into the ConnectionRequest event and is used with the Accept method.

P.13.10 Close method

The Close method closes a TCP connection. Its syntax is:

object.`Close`

P.13.11 SendData method

The SendData methods sends data (`Data`) to a remote computer. Its syntax is:

object.`SendData` *Data*

P.13.12 GetData method

The GetData method gets data (Data) from an object. Its syntax is:

object.`GetData` *data*, [*type*,] [*maxLen*]

P.14 ARP

ARP (Address Resolution Protocol) translates IP addresses to Ethernet addresses. This is used when IP packets are send from a computer, and the Ethernet address is added to the Ethernet frame. A table look-up, called the ARP table, is used to translate the addresses. One column has the IP address and the other has the Ethernet address. The following is an example ARP table:

IP address	Ethernet address
146.176.150.2	00-80-C8-22-6BE2
146.176.150.3	00-80-C8-22-CD4E
146.176.150.4	00-80-C8-23-114C

A typical conversation is as follows:

1. Application sends an application message to TCP.
2. TCP sends the corresponding TCP message to the IP module. The destination IP address is known by the application, the TCP module, and the IP module.
3. At this point the IP packet has been constructed and is ready to be given to the Ethernet driver, but first the destination Ethernet address must be determined.
4. The ARP table is used to look-up the destination Ethernet address.

The sequence is as follows:

1. An ARP request packet with a broadcast Ethernet address (`FF-FF-FF-FF-FF-FF`) is sent out on the network to every computer. Other typical Ethernet broadcast addresses are

given in Section P.14.1.
2. The outgoing IP packet is queued.
3. All the computers on the network segment read the broadcast Ethernet frame, and examine the Type field to determine if it is an ARP packet. If it is then it is passed to the ARP module.
4. If the IP address of a receiving station matches the IP address in the IP packet then it sends a response directly to the source Ethernet address.
5. The originator then receives the Ethernet frame and checks the Type field to determine if it an ARP packet. If it is then it adds the sender's IP address and Ethernet address to its ARP table.
6. The IP packet can now be send with the correct Ethernet address.

Each computer has a separate ARP table for each of its Ethernet interfaces.

P.14.1 Ethernet Multicast/Broadcast Addresses

The following is a list of typical Ethernet Multicast addresses:

Ethernet address	*Type field*	*Usage*
01-00-5E-00-00-00	0800	Internet Multicast (RFC-1112)
01-80-C2-00-00-00	0802	Spanning tree (for bridges)
09-00-09-00-00-01	8005	HP Probe
09-00-09-00-00-04	8005	HP DTC
09-00-1E-00-00-00	8019	Apollo DOMAIN
09-00-2B-00-00-03	8038	DEC Lanbridge Traffic Monitor (LTM)
09-00-4E-00-00-02	8137	Novell IPX
0D-1E-15-BA-DD-06	????	HP
CF-00-00-00-00-00	9000	Ethernet Configuration Test protocol

The following is a list of typical Ethernet Broadcast addresses:

Ethernet address	*Type field*	*Usage*
FF-FF-FF-FF-FF-FF	0600	XNS packets, Hello or gateway search.
FF-FF-FF-FF-FF-FF	0800	IP (such as RWHOD with UDP)
FF-FF-FF-FF-FF-FF	0804	CHAOS
FF-FF-FF-FF-FF-FF	0806	ARP (for IP and CHAOS) as needed
FF-FF-FF-FF-FF-FF	0BAD	Banyan
FF-FF-FF-FF-FF-FF	1600	VALID packets. Hello or gateway search.
FF-FF-FF-FF-FF-FF	8035	Reverse ARP
FF-FF-FF-FF-FF-FF	807C	Merit Internodal (INP)
FF-FF-FF-FF-FF-FF	809B	EtherTalk

P.15 IP multicasting

Many applications of modern communications require the transmission of IP datagrams to multiple hosts. Typical applications are video conferencing, remote teaching, and so on. This is supported by IP multicasting, where a host group is identified by a single IP address. The main parameters of IP multicasting are:

1. The group membership is dynamic.

2. Hosts may join and leave the group at any time.
3. There is also no limit to the location or number of members in a host group.
4. A host may be a member of more than one group at a time.
5. A host group may be permanent or transient. Permanent groups are well-known and are administratively assigned a permanent IP address. The group is then dynamically associated with this IP address. IP multicast addresses that are not reserved to permanent groups are available for dynamic assignment to transient groups.
6. Multicast routers forward IP multicast datagrams into the Internet.

P.15.1 Group addresses

A special group of addresses are assigned to multicasting. These are known as Class D address, an the begin with 1110 as their starting 4 bits (Class E addresses with the upper bits of 1111 are reserved for future uses). The Class D addresses thus range from:

224.0.0.0 (11100000 00000000 00000000 00000000)

239.255.255.255 (11101111 11111111 11111111 11111111)

The address 224.0.0.0 is reserved. 224.0.0.1 is also assigned to the permanent group of all IP hosts (including gateways), and is used to address all multicast hosts on the directly connected network. Reserved and allocated addresses are:

224.0.0.0	Reserved
224.0.0.1	All Systems on current subnet
224.0.0.2	All Routers on current subnet
224.0.0.3	Unassigned
224.0.0.4	DVMRP Routers
224.0.0.5	OSPFIGP All Routers
224.0.0.6	OSPFIGP Designated Routers
224.0.0.7	ST Routers
224.0.0.8	ST Hosts
224.0.0.9	RIP2 Routers
224.0.0.10-224.0.0.255	Unassigned
224.0.1.0	VMTP Managers Group
224.0.1.1	NTP Network Time Protocol
224.0.1.2	SGI-Dogfight
224.0.1.3	Rwhod
224.0.1.4	VNP
224.0.1.5	Artificial Horizons - Aviator
224.0.1.6	NSS - Name Service Server
224.0.1.7	AUDIONEWS - Audio News Multicast
224.0.1.8	SUN NIS+ Information Service
224.0.1.9	MTP Multicast Transport Protocol
224.0.1.10-224.0.1.255	Unassigned
224.0.2.1	rwho Group (BSD) (unofficial)
224.0.2.2	SUN RPC PMAPPROC_CALLIT
224.0.3.0-224.0.3.255	RFE Generic Service

224.0.4.0-224.0.4.255	RFE Individual Conferences
224.1.0.0-224.1.255.255	ST Multicast Groups
224.2.0.0-224.2.255.255	Multimedia Conference Calls
232.*x*.*x*.*x*	VMTP transient groups

All the above addresses are listing in the Domain Name Service under MCAST.NET and 224.IN-ADDR.ARPA. On an Ethernet or IEEE 802 network, the 23 low-order bits of the IP Multicast address are placed in the low-order 23 bits of the Ethernet or IEEE 802 net multicast address.

P.15.2 Conformance

There are three levels of conformance:

1. Level 0. No IP multicasting support. In this, a Level 0 host ignores, or deletes, all Class D addressed datagrams.
2. Level 1. Sending support, but no receiving. In this, a Level 1 host can send multicast datagrams, but cannot receive them.
3. Level 2: Full multicasting support. In this, a Level 2 host can send and receive IP multicasting. It also requires the implementation of the Internet Group Management Protocol (IGMP).

P.16 Assigned Internet Protocol numbers

This section contains information extracted from RFC1700 [Reynolds and Postel] on assigned number values.

IP Special addresses

The main forms of IP addresses are:

```
IP-address ::=  { <Network-number>, <Host-number> }
```

and

```
IP-address ::=  { <Network-number>, <Subnet-number>, <Host-number> }
```

Special addresses are:

{0, 0}.	Host on this network. This address can only be used as a source address
{0, <Host-number>}	Host on this network
{ –1, –1}.	Limited broadcast. This address can only be used as a destination address, and should not be forwarded outside the current subnet.
{<Network-number>, –1}	Directed broadcast to specified network. This address can only be used as a destination address.

{<Network-number>, <Subnet-number>, –1}

	Directed broadcast to specified subnet. This address can only be used as a destination address.
{<Network-number>, -1, -1}	Directed broadcast to all subnets of specified subnetted network. This address can only be used as a destination address.
{127, <any>}	Internal host loopback address. This address should never appear outside a host.

where –1 represents an address of all 1's.

IP Versions

Decimal	Keyword	Version
0		Reserved
1-3		Unassigned
4	IP	Internet Protocol
5	ST	ST Datagram Mode
6	SIP	Simple Internet Protocol
7	TP/IX	TP/IX: The Next Internet
8	PIP	The P Internet Protocol
9	TUBA	TUBA
10-14		Unassigned
15		Reserved

IP protocol numbers

Decimal	Keyword	Protocol
0		Reserved
1	ICMP	Internet Control Message
2	IGMP	Internet Group Management
3	GGP	Gateway-to-Gateway
4	IP	IP in IP (encasulation)
5	ST	Stream
6	TCP	Transmission Control
7	UCL	UCL
8	EGP	Exterior Gateway Protocol
9	IGP	any private interior gateway
10	BBN-RCC-MON	BBN RCC Monitoring
11	NVP-II	Network Voice Protocol
12	PUP	PUP
13	ARGUS	ARGUS
14	EMCON	EMCON
15	XNET	Cross Net Debugger
16	CHAOS	Chaos
17	UDP	User Datagram
18	MUX	Multiplexing
19	DCN-MEAS	DCN Measurement Subsystems
20	HMP	Host Monitoring
21	PRM	Packet Radio Measurement
22	XNS-IDP	XEROX NS IDP
23	TRUNK-1	Trunk-1
24	TRUNK-2	Trunk-2
25	LEAF-1	Leaf-1
26	LEAF-2	Leaf-2
27	RDP	Reliable Data Protocol
28	IRTP	Internet Reliable Transaction
29	ISO-TP4	ISO Transport Protocol Class 4
30	NETBLT	Bulk Data Transfer Protocol
31	MFE-NSP	MFE Network Services Protocol
32	MERIT-INP	MERIT Internodal Protocol
33	SEP	Sequential Exchange Protocol
34	3PC	Third Party Connect Protocol
35	IDPR	Inter-Domain Policy Routing Protocol
36	XTP	XTP
37	DDP	Datagram Delivery Protocol
38	IDPR-CMTP	IDPR Control Message Transport Proto
39	TP++	TP++ Transport Protocol
40	IL	IL Transport Protocol
41	SIP	Simple Internet Protocol
42	SDRP	Source Demand Routing Protocol

43	SIP-SR	SIP Source Route
44	SIP-FRAG	SIP Fragment
45	IDRP	Inter-Domain Routing Protocol
46	RSVP	Reservation Protocol
47	GRE	General Routing Encapsulation
48	MHRP	Mobile Host Routing Protocol
49	BNA	BNA
50	SIPP-ESP	SIPP Encap Security Payload
51	SIPP-AH	SIPP Authentication Header
52	I-NLSP	Integrated Net Layer Security TUBA
53	SWIPE	IP with Encryption
54	NHRP	NBMA Next Hop Resolution Protocol
61		any host internal protocol
62	CFTP	CFTP
63		any local network
64	SAT-EXPAK	SATNET and Backroom EXPAK
65	KRYPTOLAN	Kryptolan
66	RVD	MIT Remote Virtual Disk Protocol
67	IPPC	Internet Pluribus Packet Core
68		any distributed file system
69	SAT-MON	SATNET Monitoring
70	VISA	VISA Protocol
71	IPCV	Internet Packet Core Utility
72	CPNX	Computer Protocol Network Executive
73	CPHB	Computer Protocol Heart Beat
74	WSN	Wang Span Network
75	PVP	Packet Video Protocol
76	BR-SAT-MON	Backroom SATNET Monitoring
77	SUN-ND	SUN ND PROTOCOL-Temporary
78	WB-MON	WIDEBAND Monitoring
79	WB-EXPAK	WIDEBAND EXPAK
80	ISO-IP	ISO Internet Protocol
81	VMTP	VMTP
82	SECURE-VMTP	SECURE-VMTP
83	VINES	VINES
84	TTP	TTP
85	NSFNET-IGP	NSFNET-IGP
86	DGP	Dissimilar Gateway Protocol
87	TCF	TCF
88	IGRP	IGRP
89	OSPFIGP	OSPFIGP
90	Sprite-RPC	Sprite RPC Protocol
91	LARP	Locus Address Resolution Protocol
92	MTP	Multicast Transport Protocol
93	AX.25	AX.25 Frames
94	IPIP	IP-within-IP Encapsulation Protocol
95	MICP	Mobile Internetworking Control Pro.
96	SCC-SP	Semaphore Communications Sec. Pro.
97	ETHERIP	Ethernet-within-IP Encapsulation
98	ENCAP	Encapsulation Header
99		any private encryption scheme
100	GMTP	GMTP

Ports

0		Reserved
1	tcpmux	TCP Port Service Multiplexer
2	compressnet	Management Utility
3	compressnet	Compression Process
5	rje	Remote Job Entry
7	echo	Echo
11	discard	Discard
13	systat	Active Users
15	daytime	Daytime
17	qotd	Quote of the Day
18	msp	Message Send Protocol
19	chargen	Character Generator
20	ftp-data	File Transfer [Default Data]
21	ftp	File Transfer [Control]
23	telnet	Telnet
25	smtp	Simple Mail Transfer
27	nsw-fe	NSW User System FE
29	msg-icp	MSG ICP
31	msg-auth	MSG Authentication

```
33      dsp                Display Support Protocol
37      time               Time
38      rap                Route Access Protocol
39      rlp                Resource Location Protocol
41      graphics           Graphics
42      nameserver         Host Name Server
43      nicname            Who Is
44      mpm-flags          MPM FLAGS Protocol
45      mpm                Message Processing Module
46      mpm-snd            MPM [default send]
47      ni-ftp             NI FTP
48      auditd             Digital Audit Daemon
49      login              Login Host Protocol
50      re-mail-ck         Remote Mail Checking Protocol
51      la-maint           IMP Logical Address Maintenance
52      xns-time           XNS Time Protocol
53      domain             Domain Name Server
54      xns-ch             XNS Clearinghouse
55      isi-gl             ISI Graphics Language
56      xns-auth           XNS Authentication
58      xns-mail           XNS Mail
61      ni-mail            NI MAIL
62      acas               ACA Services
64      covia              Communications Integrator (CI)
65      tacacs-ds          TACACS-Database Service
66      sql*net            Oracle SQL*NET
67      bootps             Bootstrap Protocol Server
68      bootpc             Bootstrap Protocol Client
69      tftp Trivial File Transfer
70      gopher             Gopher
71      netrjs-1           Remote Job Service
72      netrjs-2           Remote Job Service
73      netrjs-3           Remote Job Service
74      netrjs-4           Remote Job Service
76      deos               Distributed External Object St
78      vettcp             vettcp
79      finger             Finger
80      www-http           World Wide Web HTTP
81      hosts2-ns          HOSTS2 Name Server
82      xfer               XFER Utility
83      mit-ml-dev         MIT ML Device
84      ctf                Common Trace Facility
85      mit-ml-dev         MIT ML Device
86      mfcobol            Micro Focus Cobol
88      kerberos           Kerberos
89      su-mit-tg          SU/MIT Telnet Gateway
90      dnsix              DNSIX Securit Attribute Token
91      mit-dov            MIT Dover Spooler
92      npp                Network Printing Protocol
93      dcp                Device Control Protocol
94      objcall            Tivoli Object Dispatcher
95      supdup             SUPDUP
96      ixie               DIXIE Protocol Specification
97      swift-rvf          Swift Remote Vitural File Prot
98      tacnews            TAC News
99      metagram           Metagram Relay
101     hostname           NIC Host Name Server
102     iso-tsap           ISO-TSAP
103     gppitnp            Genesis Point-to-Point Trans N
104     acr-nema           ACR-NEMA Digital Imag. & Comm.
105     csnet-ns           Mailbox Name Nameserver
106     3com-tsmux         3COM-TSMUX
107     rtelnet            Remote Telnet Service
108     snagas             SNA Gateway Access Server
109     pop2               Post Office Protocol - Version 2
110     pop3               Post Office Protocol - Version 3
111     sunrpc             SUN Remote Procedure Call
112     mcidas             McIDAS Data Transmission Proto
113     auth               Authentication Service
114     audionews          Audio News Multicast
115     sftp               Simple File Transfer Protocol
116     ansanotify         ANSA REX Notify
117     uucp-path          UUCP Path Service
118     sqlserv            SQL Services
119     nntp               Network News Transfer Protocol
```

```
120    cfdptkt          CFDPTKT
121    erpc             Encore Expedited Remote Pro.Ca
122    smakynet         SMAKYNET
123    ntp              Network Time Protocol
124    ansatrader       ANSA REX Trader
125    locus-map        Locus PC-Interface Net Map Ser
126    unitary          Unisys Unitary Login
127    locus-con        Locus PC-Interface Conn Server
128    gss-xlicen       GSS X License Verification
129    pwdgen           Password Generator Protocol
130    cisco-fna        cisco FNATIVE
131    cisco-tna        cisco TNATIVE
132    cisco-sys        cisco SYSMAINT
133    statsrv          Statistics Service
134    ingres-net       INGRES-NET Service
135    loc-srv          Location Service
136    profile          PROFILE Naming System
137    netbios-ns       NETBIOS Name Service
138    netbios-dgm      NETBIOS Datagram Service
139    netbios-ssn      NETBIOS Session Service
140    emfis-data       EMFIS Data Service
141    emfis-cntl       EMFIS Control Service
142    bl-idm           Britton-Lee IDM
143    imap2            Interim Mail Access Protocol
144    news             NewS
145    uaac             UAAC Protocol
146    iso-tp0          ISO-IP0
147    iso-ip           ISO-IP
148    cronus           CRONUS-SUPPORT
149    aed-512          AED 512 Emulation Service
150    sql-net          SQL-NET
151    hems             HEMS
152    bftp             Background File Transfer Progr
153    sgmp SGMP
154    netsc-prod       NETSC
155    netsc-dev        NETSC
156    sqlsrv           SQL Service
157    knet-cmp         KNET/VM Command/Message Protoc
158    pcmail-srv       PCMail Server
159    nss-routing      NSS-Routing
160    sgmp-traps       SGMP-TRAPS
161    snmp             SNMP
162    snmptrap         SNMPTRAP
163    cmip-man         CMIP Manager
164    cmip-agent       CMIP Agent
165    xns-courier      Xerox
166    s-net            Sirius Systems
167    namp             NAMP
168    rsvd             RSVD
169    send             SEND
170    print-srv        Network PostScript
171    multiplex        Network Innovations Multiplex
172    cl/1             Network Innovations CL/1
173    xyplex-mux       Xyplex
174    mailq            MAILQ
175    vmnet            VMNET
176    genrad-mux       GENRAD-MUX
177    xdmcpX           Display Manager Control Prot
178    nextstep         NextStep Window Server
179    bgp              Border Gateway Protocol
180    ris              Intergraph
181    unify            Unify
182    audit            Unisys Audit SITP
183    ocbinder         OCBinder
184    ocserver         OCServer
185    remote-kis       Remote-KIS
186    kis              KIS Protocol
187    aci              Application Communication Inte
188    mumps            Plus Five's MUMPS
189    qft              Queued File Transport
190    gacp             Gateway Access Control Protoco
191    prospero         Prospero Directory Service
192    osu-nms          OSU Network Monitoring System
193    srmp             Spider Remote Monitoring Proto
194    irc              Internet Relay Chat Protocol
```

195	dn6-nlm-aud	DNSIX Network Level Module Aud
196	dn6-smm-red	DNSIX Session Mgt Module Audit
197	dls	Directory Location Service
198	dls-mon	Directory Location Service Mon
199	smux	SMUX
200	src	IBM System Resource Controller
201	at-rtmp	AppleTalk Routing Maintenance
202	at-nbp	AppleTalk Name Binding
203	at-3	AppleTalk Unused
204	at-echo	AppleTalk Echo
205	at-5	AppleTalk Unused
206	at-zis	AppleTalk Zone Information
207	at-7	AppleTalk Unused
208	at-8	AppleTalk Unused
209	tam	Trivial Authenticated Mail Pro
210	z39.50	ANSI Z39.50
211	914c/g	Texas Instruments 914C/G Termi
212	anet	ATEXSSTR
213	ipx	IPX
214	vmpwscs	VM PWSCS
215	softpc	Insignia Solutions
216	atls	Access Technology License Serv
217	dbase	dBASE Unix
218	mpp	Netix Message Posting Protocol
219	uarps	Unisys ARPs
220	imap3	Interactive Mail Access Protoc
221	fln-spx	Berkeley rlogind with SPX auth
222	rsh-spx	Berkeley rshd with SPX auth
223	cdc	Certificate Distribution Cente
243	sur-meas	Survey Measurement
245	link	LINK
246	dsp3270	Display Systems Protocol
344	pdap	Prospero Data Access Protocol
345	wserv	Perf Analysis Workbench
346	zserv	Zebra server
347	fatserv	Fatmen Server
348	csi-sgwp	Cabletron Management Protocol
371	clearcase	Clearcase
372	ulistserv	Unix Listserv
373	legent-1	Legent Corporation
374	legent-2	Legent Corporation
375	hassle	Hassle
376	nip	Amiga Envoy Network Inquiry Pr
377	tnETOS	NEC Corporation
378	dsETOS	NEC Corporation
379	is99c	TIA/EIA/IS-99 modem client
380	is99s	TIA/EIA/IS-99 modem server
381	hp-collector	hp performance data collector
382	hp-managed-node	hp performance data managed no
383	hp-alarm-mgr	hp performance data alarm mana
384	arns	A Remote Network Server System
385	ibm-app	IBM Application
386	asa	ASA Message Router Object Def.
387	aurp	Appletalk Update-Based Routing
388	unidata-ldm	Unidata LDM Version 4
389	ldap	Lightweight Directory Access P
390	uis	UIS
391	synotics-relay	SynOptics SNMP Relay Port
392	synotics-broker	SynOptics Port Broker Port
393	dis	Data Interpretation System
394	embl-ndt	EMBL Nucleic Data Transfer
395	netcp	NETscout Control Protocol
396	netware-ip	Novell Netware over IP
397	mptn	Multi Protocol Trans. Net.
398	kryptolan	Kryptolan
400	work-sol	Workstation Solutions
401	ups	Uninterruptible Power Supply
402	genie	Genie Protocol
403	decap decap	
404	nced nced	
405	ncld ncld	
406	imsp Interactive	Mail Support Proto
407	timbuktu	Timbuktu
408	prm-sm	Prospero Resource Manager Sys.
409	prm-nm	Prospero Resource Manager Node

410 decladebug DECLadebug Remote Debug Protoc
411 rmt Remote MT Protocol
412 synoptics-trap Trap Convention Port
413 smsp SMSP
414 infoseek InfoSeek
415 bnet BNet
416 silverplatter Silverplatter
417 onmux Onmux
418 hyper-g Hyper-G
419 ariel1 Ariel
420 smpte SMPTE
421 ariel2 Ariel
422 ariel3 Ariel
423 opc-job-start IBM Operations Planning and Co
424 opc-job-track IBM Operations Planning and Co
425 icad-el ICAD
426 smartsdp smartsdp
427 svrloc Server Location
428 ocs_cmu OCS_CMU
429 ocs_amu OCS_AMU
430 utmpsd UTMPSD
431 utmpcd UTMPCD
432 iasd IASD
433 nnsp NNSP
434 mobileip-agent MobileIP-Agent
435 mobilip-mn MobilIP-MN
436 dna-cml DNA-CML
437 comscm comscm
438 dsfgw dsfgw
439 dasp dasp
440 sgcp sgcp
441 decvms-sysmgt decvms-sysmgt
442 cvc_hostd cvc_hostd
443 https https MCom
444 snpp Simple Network Paging Protocol
445 microsoft-ds Microsoft-DS
446 ddm-rdb DDM-RDB
447 ddm-dfm DDM-RFM
448 ddm-byte DDM-BYTE
449 as-servermap AS Server Mapper
450 tserver TServer
512 exec remote process execution;
513 login remote login
514 cmd like exec, but automatic
515 printer spooler
517 talk like tenex link, but across
518 ntalk
519 utime unixtime
520 efs extended file name server
525 timed timeserver
526 tempo newdate
530 courier rpc
531 conference chat
532 netnews readnews
533 netwall for emergency broadcasts
539 apertus-ldp Apertus Technologies Load Dete
540 uucp uucpd
541 uucp-rlogin uucp-rlogin Stuart Lynne
543 klogin
544 kshell krcmd
550 new-rwho new-who
555 dsf
556 remotefs rfs server
560 rmonitor rmonitord
561 monitor
562 chshell chcmd
564 9pfs plan 9 file service
565 whoami whoami
570 meter demon
571 meter udemon
600 ipcserver Sun IPC server
607 nqs nqs
606 urm Cray Unified Resource Manager
608 sift-uft Sender-Initiated/Unsolicited F
609 npmp-trap npmp-trap

610	npmp-local	npmp-local
611	npmp-gui	npmp-gui
634	ginadginad	
666	mdqs	
666	doom	doom Id Software
704	elcsd	errlog copy/server daemon
709	entrustmanager	EntrustManager
729	netviewdm1	IBM NetView DM/6000 Server/Cli
730	netviewdm2	IBM NetView DM/6000 send
731	netviewdm3	IBM NetView DM/6000 receive/tc
741	netgwnetGW	
742	netrcs	Network based Rev. Cont. Sys.
744	flexlm	Flexible License Manager
747	fujitsu-dev	Fujitsu Device Control
748	ris-cm	Russell Info Sci Calendar Mana
749	kerberos-adm	kerberos administration
750	rfile	
751	pump	
752	qrh	
753	rrh	
754	tell send	
758	nlogin	
759	con	
760	ns	
761	rxe	
762	quotad	
763	cycleserv	
764	omserv	
765	webster	
767	phonebook	phcne
769	vid	
770	cadlock	
771	rtip	
772	cycleserv2	
773	submit	
774	rpasswd	
775	entomb	
776	wpages	
780	wpgs	
786	concert	Concert
800	mdbs_daemon	
801	device	
996	xtreelic	Central Point Software
997	maitrd	
998	busboy	
999	garcon	
1000	cadlock	

Multicast

224.0.0.0	Base Address (Reserved)
224.0.0.1	All Systems on this Subnet
224.0.0.2	All Routers on this Subnet
224.0.0.3	Unassigned
224.0.0.4	DVMRP Routers
224.0.0.5	OSPFIGP OSPFIGP All Routers
224.0.0.6	OSPFIGP OSPFIGP Designated Routers
224.0.0.7	ST Routers
224.0.0.8	ST Hosts
224.0.0.9	RIP2 Routers
224.0.0.10	IGRP Routers
224.0.0.11	Mobile-Agents
224.0.0.12-224.0.0.255	Unassigned
224.0.1.0	VMTP Managers Group
224.0.1.1	NTP Network Time Protocol
224.0.1.2	SGI-Dogfight
224.0.1.3	Rwhod
224.0.1.4	VNP
224.0.1.5	Artificial Horizons - Aviator
224.0.1.6	NSS - Name Service Server
224.0.1.7	AUDIONEWS - Audio News Multicast
224.0.1.8	SUN NIS+ Information Service
224.0.1.9	MTP Multicast Transport Protocol
224.0.1.10	IETF-1-LOW-AUDIO

224.0.1.11	IETF-1-AUDIO
224.0.1.12	IETF-1-VIDEO
224.0.1.13	IETF-2-LOW-AUDIO
224.0.1.14	IETF-2-AUDIO
224.0.1.15	IETF-2-VIDEO
224.0.1.16	MUSIC-SERVICE
224.0.1.17	SEANET-TELEMETRY
224.0.1.18	SEANET-IMAGE
224.0.1.19	MLOADD
224.0.1.20	any private experiment
224.0.1.21	DVMRP on MOSPF
224.0.1.22	SVRLOC
224.0.1.23	XINGTV
224.0.1.24	microsoft-ds
224.0.1.25	nbc-pro
224.0.1.26	nbc-pfn
224.0.1.27-224.0.1.255	Unassigned
224.0.2.1	"rwho" Group (BSD) (unofficial)
224.0.2.2	SUN RPC PMAPPROC_CALLIT
224.0.3.000-224.0.3.255	RFE Generic Service
224.0.4.000-224.0.4.255	RFE Individual Conferences
224.0.5.000-224.0.5.127	CDPD Groups
224.0.5.128-224.0.5.255	Unassigned
224.0.6.000-224.0.6.127	Cornell ISIS Project
224.0.6.128-224.0.6.255	Unassigned
224.1.0.0-224.1.255.255	ST Multicast Groups
224.2.0.0-224.2.255.255	Multimedia Conference Calls
224.252.0.0-224.255.255.255	DIS transient groups
232.0.0.0-232.255.255.255	VMTP transient groups

IP type of service

TOS Value	Description
0000	Default
0001	Minimize Monetary Cost
0010	Maximize Reliability
0100	Maximize Throughput
1000	Minimize Delay
1111	Maximize Security

Type of Service recommended values:

Protocol	TOS Value	
TELNET (1)	1000	(minimize delay)
FTP		
Control	1000	(minimize delay)
Data (2)	0100	(maximize throughput)
TFTP	1000	(minimize delay)
SMTP (3)		
Command phase	1000	(minimize delay)
DATA phase	0100	(maximize throughput)
Domain Name Service		
UDP Query	1000	(minimize delay)
TCP Query	0000	
Zone Transfer	0100	(maximize throughput)
NNTP	0001	(minimize monetary cost)
ICMP		
Errors	0000	
Requests	0000 (4)	
Responses	<same as request> (4)	
Any IGP	0010	(maximize reliability)
EGP	0000	
SNMP	0010	(maximize reliability)
BOOTP	0000	

ICMP type numbers

Type	Name
0	Echo Reply
1	Unassigned
2	Unassigned
3	Destination Unreachable
4	Source Quench

```
   5     Redirect
   6     Alternate Host Address
   7     Unassigned
   8     Echo
   9     Router Advertisement
  10     Router Selection
  11     Time Exceeded
  12     Parameter Problem
  13     Timestamp
  14     Timestamp Reply
  15     Information Request
  16     Information Reply
  17     Address Mask Request
  18     Address Mask Reply
  19     Reserved (for Security)
 20-29   Reserved (for Robustness Experiment)
 30      Traceroute
 31      Datagram Conversion Error
 32      Mobile Host Redirect
 33      IPv6 Where-Are-You
 34      IPv6 I-Am-Here
 35      Mobile Registration Request
 36      Mobile Registration Reply
 37-255  Reserved

Type     Name
  0      Echo Reply
         Codes
             0  No Code
  3      Destination Unreachable
         Codes
             0  Net Unreachable
             1  Host Unreachable
             2  Protocol Unreachable
             3  Port Unreachable
             4  Fragmentation Needed and Don't Fragment was Set
             5  Source Route Failed
             6  Destination Network Unknown
             7  Destination Host Unknown
             8  Source Host Isolated
             9  Communication with Destination Network is
                Administratively Prohibited
            10  Communication with Destination Host is
                Administratively Prohibited
            11  Destination Network Unreachable for Type of Service
            12  Destination Host Unreachable for Type of Service
  4      Source Quench
         Codes
             0  No Code
  5      Redirect
         Codes
             0  Redirect Datagram for the Network (or subnet)
             1  Redirect Datagram for the Host
             2  Redirect Datagram for the Type of Service and Network
             3  Redirect Datagram for the Type of Service and Host
  6      Alternate Host Address
         Codes
             0  Alternate Address for Host
  7      Unassigned
  8      Echo
         Codes
             0  No Code
  9      Router Advertisement
         Codes
                  0  No Code
 10      Router Selection
         Codes
             0  No Code
 11      Time Exceeded
         Codes
             0  Time to Live exceeded in Transit
             1  Fragment Reassembly Time Exceeded
 12      Parameter Problem
         Codes
             0  Pointer indicates the error
```

```
         1  Missing a Required Option
         2  Bad Length
 13     Timestamp
        Codes
         0  No Code
 14     Timestamp Reply
        Codes
         0  No Code
 15     Information Request
        Codes
         0  No Code
 16     Information Reply
        Codes
         0  No Code
 17     Address Mask Request
        Codes
         0  No Code
 18     Address Mask Reply
        Codes
         0  No Code
 30     Traceroute
 31     Datagram Conversion Error
 32     Mobile Host Redirect
 33     IPv6 Where-Are-You
 34     IPv6 I-Am-Here
 35     Mobile Registration Request
 36     Mobile Registration Reply
```

TCP options

Type	Length	Description
0	-	End of Option List
1	-	No-Operation
2	4	Maximum Segment Lifetime
3	3	WSOPT - Window Scale
4	2	SACK Permitted
5	N	SACK
6	6	Echo (obsoleted by option 8)
7	6	Echo Reply (obsoleted by option by 8)
8	10	TSOPT - Time Stamp Option
9	2	Partial Order Connection Permited
10	5	Partial Order Service Profile
11		CC
12		CC.NEW
13		CC.ECHO
14	3	TCP Alternate Checksum Request
15	N	TCP Alternate Checksum Data
16		Skeeter
17		Bubba
18	3	Trailer Checksum Option

Domain Names

Decimal	Name
0	Reserved
1	Internet (IN)
2	Unassigned
3	Chaos (CH)
4	Hessoid (HS)
5-65534	Unassigned
65535	Reserved

For the Internet (IN) class the following are defined:

TYPE	Value	Description
A	1	Host address
NS	2	Authoritative name server
MD	3	Mail destination (Obsolete - use MX)
MF	4	Mail forwarder (Obsolete - use MX)
CNAME	5	Canonical name for an alias
SOA	6	Start of a zone of authority
MB	7	Mailbox domain name
MG	8	Mail group member

MR	9	Mail rename domain name
NULL	10	Null RR
WKS	11	Well-known service description
PTR	12	Domain name pointer
HINFO	13	Host information
MINFO	14	Mailbox or mail list information
MX	15	Mail exchange
TXT	16	Text strings
RP	17	For Responsible Person
AFSDB	18	For AFS Data Base location

Mail encoder header types

Keyword	Description
EDIFACT	EDIFACT format
EDI-X12	EDI X12 format
EVFU	FORTRAN format
FS	File System format
Hex	Hex binary format
LZJU90	LZJU90 format
LZW	LZW format
Message	Encapsulated Message
PEM	Privacy Enhanced Mail
PGP	Pretty Good Privacy
Postscript	Postscript format
Shar	Shell Archive format
Signature	Signature
Tar	Tar format
Text	Text
uuencode	uuencode format
URL	external URL-reference

BOOTP and DHCP parameters

Tag	Name	Data Length	Meaning
0	Pad	0	None
1	Subnet Mask	4	Subnet Mask Value
2	Time Offset	4	Time Offset in Seconds from UTC
3	Gateways	N	N/4 Gateway addresses
4	Time Server	N	N/4 Timeserver addresses
5	Name Server	N	N/4 IEN-116 Server addresses
6	Domain Server	N	N/4 DNS Server addresses
7	Log Server	N	N/4 Logging Server addresses
8	Quotes Server	N	N/4 Quotes Server addresses
9	LPR Server	N	N/4 Printer Server addresses
10	Impress Server	N	N/4 Impress Server addresses
11	RLP Server	N	N/4 RLP Server addresses
12	Hostname	N	Hostname string
13	Boot File Size	2	Size of boot file in 512 byte chunks
14	Merit Dump File		Client to dump and name the file to dump it to
15	Domain Name	N	The DNS domain name of the client
16	Swap Server	N	Swap Server addeess
17	Root Path	N	Path name for root disk
18	Extension File	N	Path name for more BOOTP info
19	Forward On/Off	1	Enable/Disable IP Forwarding
20	SrcRte On/Off	1	Enable/Disable Source Routing
21	Policy Filter	N	Routing Policy Filters
22	Max DG Assembly	2	Max Datagram Reassembly Size
23	Default IP TTL	1	Default IP Time to Live
24	MTU Timeout	4	Path MTU Aging Timeout
25	MTU Plateau	N	Path MTU Plateau Table
26	MTU Interface	2	Interface MTU Size
27	MTU Subnet	1	All Subnets are Local
28	Broadcast Address	4	Broadcast Address
29	Mask Discovery	1	Perform Mask Discovery
30	Mask Supplier	1	Provide Mask to Others
31	Router Discovery	1	Perform Router Discovery
32	Router Request	4	Router Solicitation Address
33	Static Route	N	Static Routing Table
34	Trailers	1	Trailer Encapsulation
35	ARP Timeout	4	ARP Cache Timeout
36	Ethernet	1	Ethernet Encapsulation
37	Default TCP TTL	1	Default TCP Time to Live

38	Keepalive Time	4	TCP Keepalive Interval
39	Keepalive Data	1	TCP Keepalive Garbage
40	NIS Domain	N	NIS Domain Name
41	NIS Servers	N	NIS Server Addresses
42	NTP Servers	N	NTP Server Addresses
43	Vendor Specific	N	Vendor Specific Information
44	NETBIOS Name Srv	N	NETBIOS Name Servers
45	NETBIOS Dist Srv	N	NETBIOS Datagram Distribution
46	NETBIOS Note Type	1	NETBIOS Note Type
47	NETBIOS Scope	N	NETBIOS Scope
48	X Window Font	N	X Window Font Server
49	X Window Manmager	N	X Window Display Manager
50	Address Request	4	Requested IP Address
51	Address Time	4	IP Address Lease Time
52	Overload	1	Overloaf "sname" or "file"
53	DHCP Msg Type	1	DHCP Message Type
54	DHCP Server Id	4	DHCP Server Identification
55	Parameter List	N	Parameter Request List
56	DHCP Message	N	DHCP Error Message
57	DHCP Max Msg Size	2	DHCP Maximum Message Size
58	Renewal Time	4	DHCP Renewal (T1) Time
59	Rebinding Time	4	DHCP Rebinding (T2) Time
60	Class Id	N	Class Identifier
61	Client Id	N	Client Identifier
62	Netware/IP Domain	N	Netware/IP Domain Name
63	Netware/IP Option	N	Netware/IP sub Options
128-154	Reserved		
255	End	0	None

Directory system names

Keyword	Attribute (X.520 keys)
CN	CommonName
L	LocalityName
ST	StateOrProvinceName
O	OrganizationName
OU	OrganizationalUnitName
C	CountryName

Content types and subtypes

Type	Subtype
text	plain
	richtext
	tab-separated-val
multipart	mixed
	alternative
	digest
	parallel
	appledouble
	header-set
message	rfc822
	partial
	external-body
	news
application	octet-stream
	postscript
	oda
	atomicmail
	andrew-inset
	slate
	wita
	dec-dx
	dca-rft
	activemessage
	rtf
	applefile
	mac-binhex40
	news-message-id
	news-transmission
	wordperfect5.1
	pdf
	zip

	macwriteii
	msword
	remote-printing
image	jpeg
	gif
	ief
	tiff
audio	basic
video	mpeg
	quicktime

Character Sets

US-ASCII	ISO-8859-1	ISO-8859-2	ISO-8859-3
ISO-8859-4	ISO-8859-5	ISO-8859-6	ISO-8859-7
ISO-8859-8	ISO-8859-9		

Access Types

FTP	ANON-FTP	TFTP	AFS
LOCAL-FILE	MAIL-SERVER		

Conversion Values

7BIT	8BIT	BASE64	BINARY
QUOTED-PRINTABLE			

MIME / X.400 mapping table

MIME content-type	X.400 Body Part
text/plain	
charset=us-ascii	ia5-text
charset=iso-8859-x	Extended Body Part - GeneralText
text/richtext	no mapping defined
application/oda	Extended Body Part - ODA
application/octet-stream	bilaterally-defined
application/postscript	Extended Body Part - mime-postscript-body
image/g3fax	g3-facsimile
image/jpeg	Extended Body Part - mime-jpeg-body
image/gif	Extended Body Part - mime-gif-body
audio/basic	no mapping defined
video/mpeg	no mapping defined

X.400 to MIME Table

X.400 Basic Body Part	MIME content-type
ia5-text	text/plain;charset=us-ascii
voice	No Mapping Defined
g3-facsimile	image/g3fax
g4-class1	no mapping defined
teletex	no mapping defined
videotex	no mapping defined
encrypted	no mapping defined
bilaterally-defined	application/octet-stream
nationally-defined	no mapping defined
externally-defined	See Extended Body Parts

X.400 Extended body part conversion

X.400 Extended Body Part	MIME content-type
GeneralText	text/plain;charset=iso-8859-x
ODA	application/oda
mime-postscript-body	application/postscript
mime-jpeg-body	image/jpeg
mime-gif-body	image/gif

Inverse ARP

Number	Operation Code (op)
1	REQUEST
2	REPLY
3	request Reverse
4	reply Reverse
5	DRARP-Request
6	DRARP-Reply
7	DRARP-Error
8	InARP-Request
9	InARP-Reply
10	ARP-NAK

Number	Hardware Type (hrd)
1	Ethernet (10Mb)
2	Experimental Ethernet (3Mb)
3	Amateur Radio AX.25
4	Proteon ProNET Token Ring
5	Chaos
6	IEEE 802 Networks
7	ARCNET
8	Hyperchannel
9	Lanstar
10	Autonet Short Address
11	LocalTalk
12	LocalNET
13	Ultra link
14	SMDS
15	Frame Relay
16	Asynchronous Transmission Mode
17	HDLC
18	Fibre Channel
19	Asynchronous Transmission Mode
20	Serial Line
21	Asynchronous Transmission Mode

IEEE 802 numbers of interest

Link Service Access Point IEEE binary	Internet binary	decimal	Description
00000000	00000000	0	Null LSAP
01000000	00000010	2	Indiv LLC Sublayer Mgt
11000000	00000011	3	Group LLC Sublayer Mgt
00100000	00000100	4	SNA Path Control
01100000	00000110	6	Reserved (DOD IP)
01110000	00001110	14	PROWAY-LAN
01110010	01001110	78	EIA-RS 511
01111010	01011110	94	ISI IP
01110001	10001110	142	PROWAY-LAN
01010101	10101010	170	SNAP
01111111	11111110	254	ISO CLNS IS 8473
11111111	11111111	255	Global DSAP

IANA Ethernet address block

The Internet Assigning Numbers Authority (IANA) owns the starting Ethernet address of:

0000 0000 0000 0000 0111 1010 (which is 00-00-5E)

This address can be used with the multicast bit (which is the first bit to the address) to create an Internet Multicast. It has the form:

```
1000 0000 0000 0000 0111 1010 xxxx xxx0 xxxx xxxx xxxx xxxx
|                                      |
Multicast Bit                          0 = Internet Multicast
                                       1 = Assigned by IANA for
                                           other uses
```

This gives an address range from 01-00-5E-00-00-00 to 01-00-5E-7F-FF-FF .

P.16.2 Ethernet vendor address

An Ethernet address is 48 bits. The first 24 bits identifies the manufacturer and the next 24 bits identifies the serial number. The manufacturer codes include:

```
00000C  Cisco
00000E  Fujitsu
00000F  NeXT
000010  Sytek
00001D  Cabletron
000020  DIAB
000022  Visual Technology
00002A  TRW
000032  GEC Computers Ltd
00005A  S & Koch
00005E  IANA
000065  Network General
00006B  MIPS
000077  MIPS
00007A  Ardent
000089  Cayman Systems
000093  Proteon
00009F  Ameristar Technology
0000A2  Wellfleet
0000A3  NAT
0000A6  Network General
0000A7  NCD
0000A9  Network Systems
0000AA  Xerox
0000B3  CIMLinc
0000B7  Dove
0000BC  Allen-Bradley
0000C0  Western Digital
0000C5  Farallon phone net card
0000C6  HP INO
0000C8  Altos
0000C9  Emulex
0000D7  Dartmouth College
0000DD  Gould
0000DE  Unigraph
0000E2  Acer Ccunterpoint
0000EF  Alantec
0000FD  High Level Hardvare
000102  BBN
001700  Kabel
008064  Wyse Technology
00802D  Xylogics, Inc.
00808C  Frontier Software Development
0080C2  IEEE 802.1 Committee
0080D3  Shiva
00AA00  Intel
00DD00  Ungermann-Bass
00DD01  Ungermann-Bass
020701  Racal InterLan
020406  BBN
026086  Satelcom MegaPac
02608C  3Com
02CF1F  CMC
080002  3Com
080003  ACC
080005  Symbolics
080008  BBN
080009  Hewlett-Packard
08000A  Nestar Systems
08000B  Unisys
080011  Tektronix, Inc.
080014  Excelan
080017  NSC
08001A  Data General
08001B  Data General
```

08001E	Apollo
080020	Sun
080022	NBI
080025	CDC
080026	Norsk Data (Nord)
080027	PCS Computer Systems
080028	TI
08002B	DEC
08002E	Metaphor
08002F	Prime Computer
080036	Intergraph
080037	Fujitsu-Xerox
080038	Bull
080039	Spider Systems
080041	DCA
080046	Sony
080047	Sequent
080049	Univation
08004C	Encore
08004E	BICC
080056	Stanford University
08005A	IBM
080067	Comdesign
080068	Ridge
080069	Silicon Graphics
08006E	Concurrent
080075	DDE
08007C	Vitalink
080080	XIOS
080086	Imagen/QMS
080087	Xyplex
080089	Kinetics
08008B	Pyramid
08008D	XyVision
080090	Retix Inc
800010	AT&T

Ethernet multicast addresses

An Ethernet multicast address has a multicast bit, a 23-bit vendor identifier part and a 24-bit vendor assigned part.

Ethernet Address	Type Field	Usage
01-00-5E-00-00-00- 01-00-5E-7F-FF-FF	0800	Internet Multicast
01-00-5E-80-00-00- 01-00-5E-FF-FF-FF	????	Internet reserved by IANA
01-80-C2-00-00-00	-802-	Spanning tree (for bridges)
09-00-02-04-00-01?	8080?	Vitalink printer
09-00-02-04-00-02?	8080?	Vitalink management
09-00-09-00-00-01	8005	HP Probe
09-00-09-00-00-01	-802-	HP Probe
09-00-09-00-00-04	8005?	HP DTC
09-00-1E-00-00-00	8019?	Apollo DOMAIN
09-00-2B-00-00-00	6009?	DEC MUMPS?
09-00-2B-00-00-01	8039?	DEC DSM/DTP?
09-00-2B-00-00-02	803B?	DEC VAXELN?
09-00-2B-00-00-03	8038	DEC Lanbridge Traffic Monitor (LTM)
09-00-2B-00-00-04	????	DEC MAP End System Hello
09-00-2B-00-00-05	????	DEC MAP Intermediate System Hello
09-00-2B-00-00-06	803D?	DEC CSMA/CD Encryption?
09-00-2B-00-00-07	8040?	DEC NetBios Emulator?
09-00-2B-00-00-0F	6004	DEC Local Area Transport (LAT)
09-00-2B-00-00-1x	????	DEC Experimental
09-00-2B-01-00-00	8038	DEC LanBridge Copy packets
09-00-2B-01-00-01	8038	DEC LanBridge Hello packets
09-00-4E-00-00-02?	8137?	Novell IPX
09-00-56-00-00-00- 09-00-56-FE-FF-FF	????	Stanford reserved
09-00-56-FF-00-00- 09-00-56-FF-FF-FF	805C	Stanford V Kernel, version 6.0
09-00-77-00-00-01	????	Retix spanning tree bridges

09-00-7C-02-00-05	8080?	Vitalink diagnostics
0D-1E-15-BA-DD-06	????	HP
AB-00-00-01-00-00	6001	DEC Maintenance Operation Protocol
AB-00-00-02-00-00	6002	DEC Maintenance Operation Protocol
AB-00-00-03-00-00	6003	DECNET Phase IV end node Hello
AB-00-00-04-00-00	6003	DECNET Phase IV Router Hello packets
AB-00-00-05-00-00	????	Reserved DEC through
AB-00-03-FF-FF-FF		
AB-00-03-00-00-00	6004	DEC Local Area Transport (LAT) - old
AB-00-04-00-xx-xx	????	Reserved DEC customer private use
AB-00-04-01-xx-yy	6007	DEC Local Area VAX Cluster groups Sys. Communication Architecture (SCA)
CF-00-00-00-00-00	9000	Ethernet Configuration Test protocol (Loopback)

Ethernet broadcast address

FF-FF-FF-FF-FF-FF	0600	XNS packets, Hello or gateway search?
FF-FF-FF-FF-FF-FF	0800	IP (e.g. RWHCD via UDP) as needed
FF-FF-FF-FF-FF-FF	0804	CHAOS
FF-FF-FF-FF-FF-FF	0806	ARP (for IP and CHAOS) as needed
FF-FF-FF-FF-FF-FF	0BAD	Banyan
FF-FF-FF-FF-FF-FF	1600	VALID packets, Hello or gateway search?
FF-FF-FF-FF-FF-FF	8035	Reverse ARP
FF-FF-FF-FF-FF-FF	807C	Merit Internodal (INP)
FF-FF-FF-FF-FF-FF	809B	EtherTalk

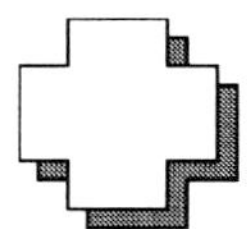

Index

1 CD information

Title: Handbook of Data Communications and Networks
Author: Dr William Buchanan.
Email: w.buchanan@napier.ac.uk
or bill@mailhost.dcs.napier.ac.uk
WWW: http://www.eece.napier.ac.uk/~bill_b
or http://www.dcs.napier.ac.uk/~bill

The following is a listing of the CD-ROM:

```
HANDBOOK PDF     17,229,201  09/08/98   23:48 handbook.pdf
INDEX    PDX            974  09/08/98   23:56 index.pdx
INDEX    LOG            832  09/08/98   23:56 INDEX.log
INDEX          <DIR>         09/08/98   23:52 INDEX
AR32E301 EXE      4,018,104  18/06/97   12:00 AR32E301.EXE
AR16E301 EXE      3,916,243  17/06/97   16:27 AR16E301.EXE
MISC     PDF      1,662,054  09/01/99   7:11p misc.PDF
README   RTF          6,459  01/01/99   4:33p readme.rtf
XLS            <DIR>         09/08/98   23:26 xls
IMAGES         <DIR>         09/08/98   23:26 images
OTHERS         <DIR>         09/08/98   23:26 others
BMP            <DIR>         09/08/98   23:26 bmp
PROGRAMS       <DIR>         09/08/98   23:26 programs
RFC            <DIR>         09/08/98   23:26 rfc
HTML           <DIR>         09/08/98   23:26 html
32BIT          <DIR>         09/08/98   23:26 32BIT
16BIT          <DIR>         09/08/98   23:26 16BIT
QUICKTIM       <DIR>         09/08/98   23:26 QUICKTIM
PDF            <DIR>         09/08/98   23:26 pdf
MAIN           <DIR>         09/08/98   23:26 main
PPT            <DIR>         09/08/98   23:26 ppt
SETUP          <DIR>         09/08/98   23:27 setup
AVI            <DIR>         09/08/98   23:27 avi
SOUNDS         <DIR>         09/08/98   23:27 sounds
PS             <DIR>         09/08/98   23:27 ps
```

The main file is HANDBOOK.PDF and the Abode Acrobat Reader must be installed before it can be read. To do this either use AE32E301.EXE for Windows 95/98/NT/2000 to AE16E301.EXE for Windows 3.x. After this is installed just double-click on the HANDBOOK.PDF file. Note, Acrobat Reader can also be installed from the SETUP.EXE file in the 32bit or 16bit directories. The MISC.PDF document has extra material which could not be included in the final version of HANDBOOK.PDF. Updates to this file can be found on one of the WWW page links given above.

2 Directory contents

XLS	Microsoft Excel Spreadsheets.
IMAGES	Image files.
OTHERS	Text files.
BMP	Bit-mapped files.
PROGRAMS	C programs.
RFC	Contains a current listing of RFC files (INDEX.RFC contains an index).
HTML	HTML source files.
32BIT	Adobe Acrobat Reader v. 3.01 for Windows 95/98/Win32/NT/2000.
16BIT	Adobe Acrobat Reader v. 3.01 for Windows 3.1 (16-bit).
QUICKTIM	Apple QuickTime 2.12 for Windows (QT32.EXE for 32-bit version, QT16.EXE for 16-bit version).
PDF	Additional PDF files.
MAIN	Miscellaneous graphics files.
PPT	Microsoft PowerPoint slides.

SETUP	Adobe Acrobat installation files.
AVI	AVI files.
SOUNDS	Sound files.
PS	Postscript files.

3 RFC files

A complete listing of the RFC files from RFC0000.TXT to RFC2468.TXT is given in the RFC subdirectory. These can be viewed by any text editor.

4 Handbook screens

After installing Adobe Acrobat Reader, the main document can be loaded by double-clicking on the HANDBOOK.PDF file. Figure 1 shows the main screen. On the left-hand side the main contents are shown as bookmarks. These can be selected by clicking on them, or can be expanded by clicking to the pointer triangle, as shown in Figure 2.

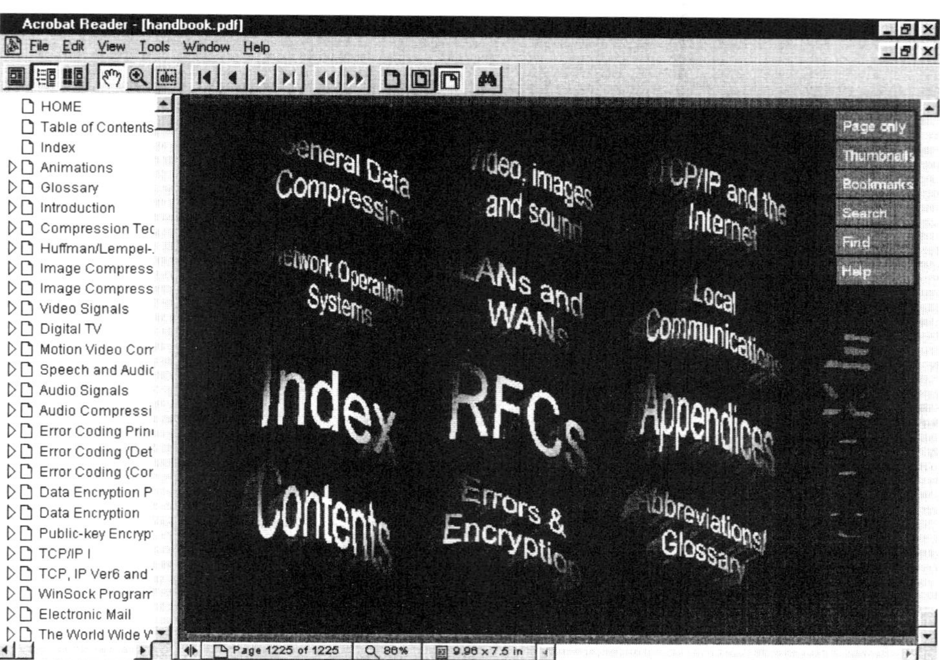

Figure 1: Main screen.

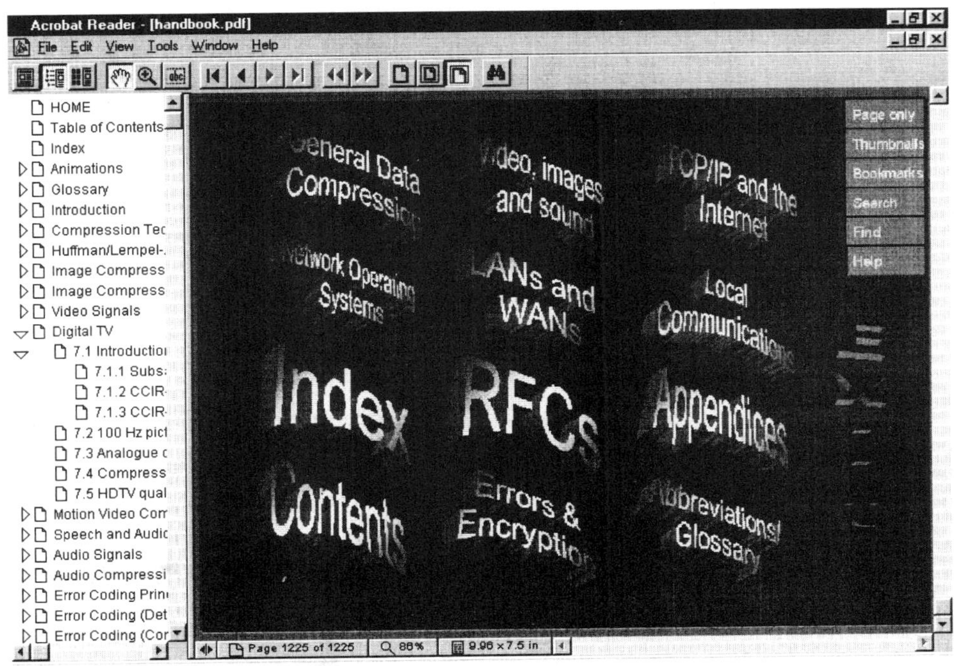

Figure 2: Expanding bookmarks.

It is also possible to view pages as thumbnails, as shown in Figure 3, and a full page view is shown in Figure 4.

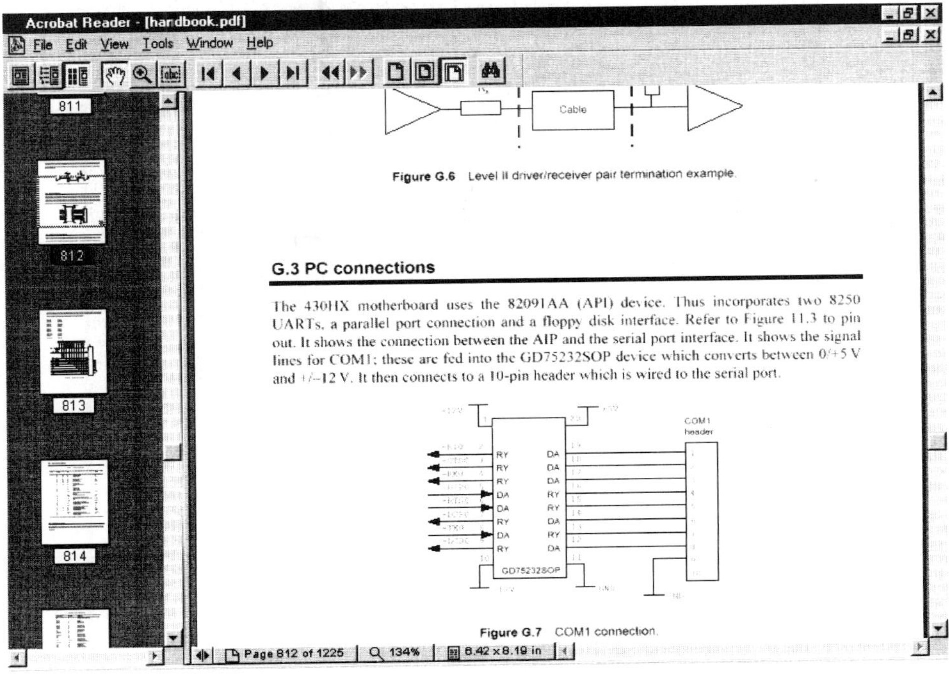

Figure 3: Showing thumbnails.

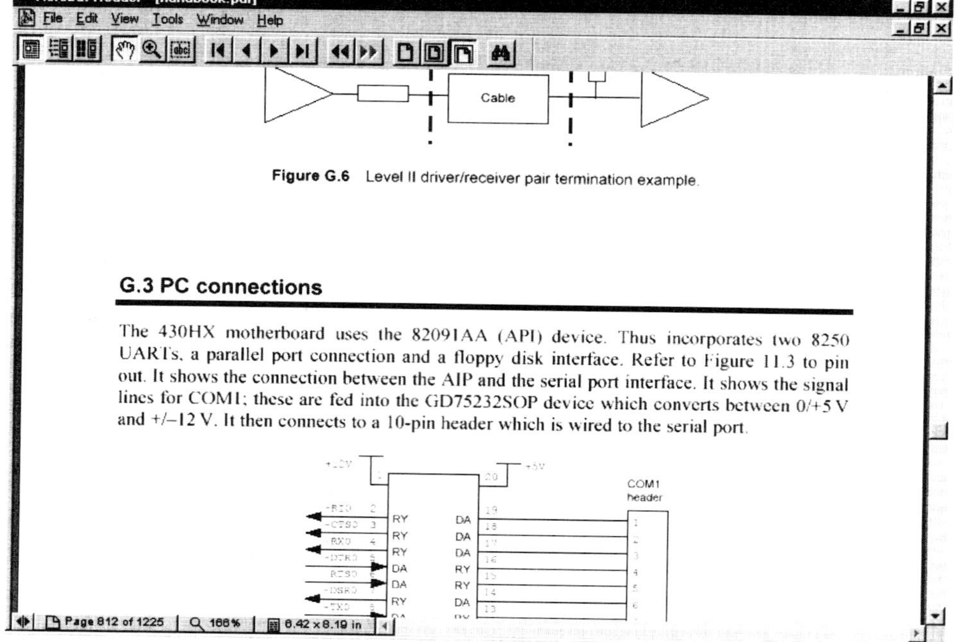

Figure 4: Displaying page only.

The index can be viewed by selecting the Index bookmark. A term in the index can be selected by clicking on the page number on the index window, as shown in Figure 5.

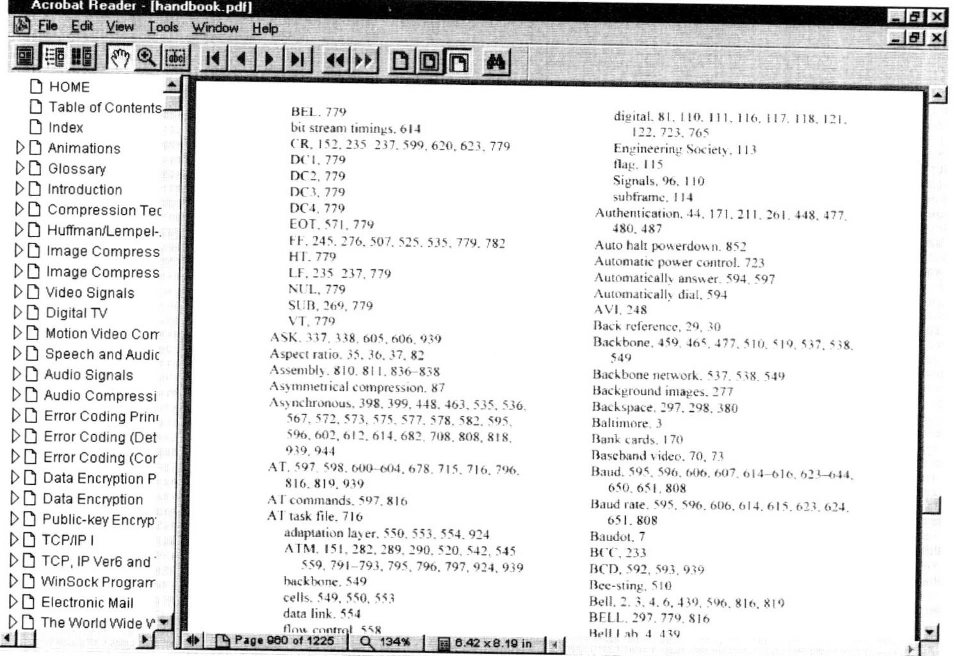

Figure 5: Selecting from the index.

Figure 6 shows how the Table of Content page can be viewed. From this window a section can be selected by single-clicking on the required section name. Figure 7 shows how a spreadsheet is selected.

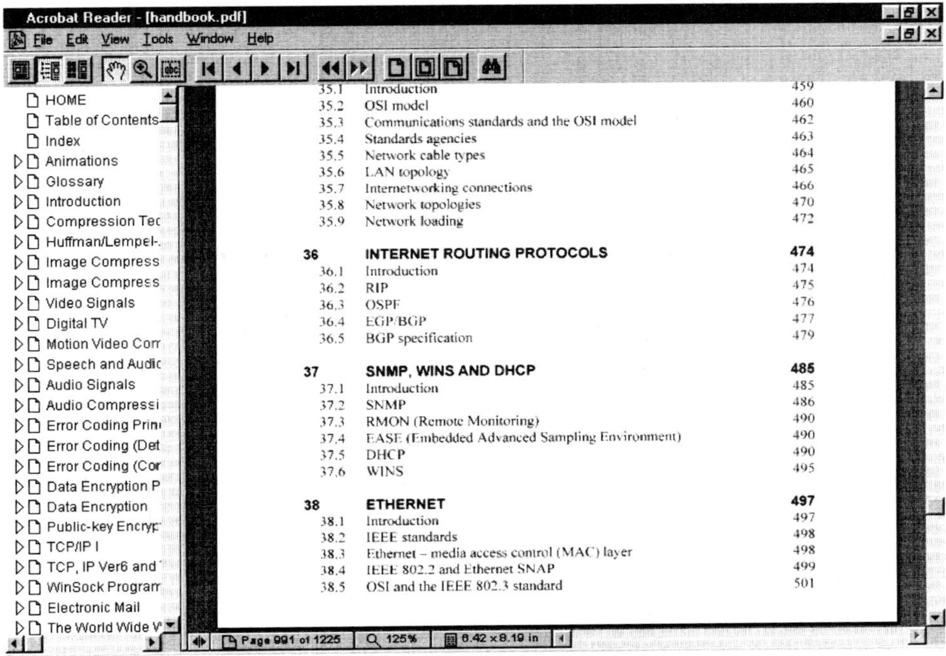

Figure 6: Selecting from the index.

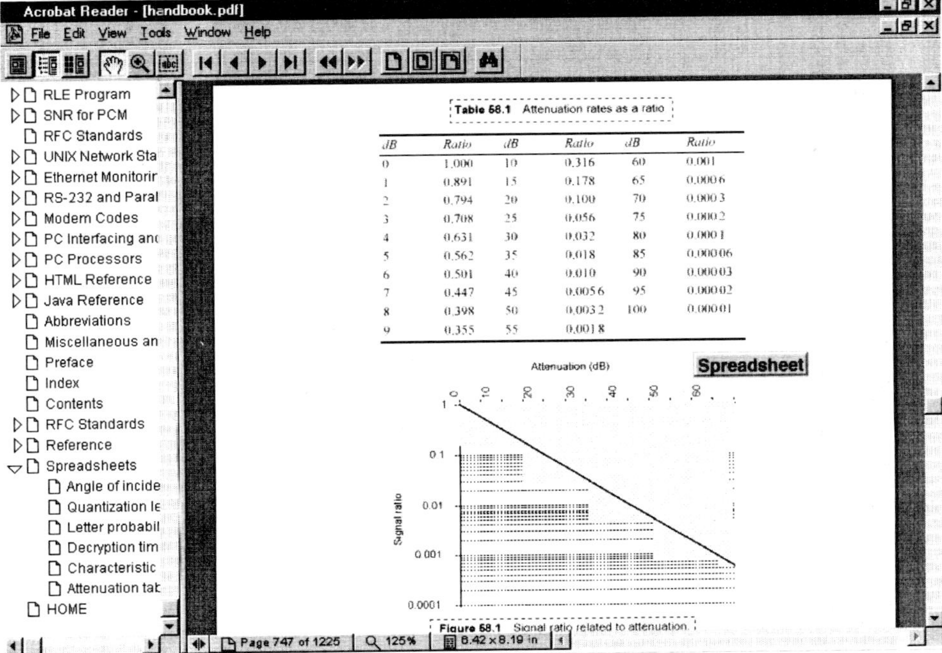

Figure 7: Spreadsheet view button.

Figure 8 shows an examples of the Contents window. In this, a green link indicates a link to a point in the document. Figure 9 shows an example of links within the RFC chapter.

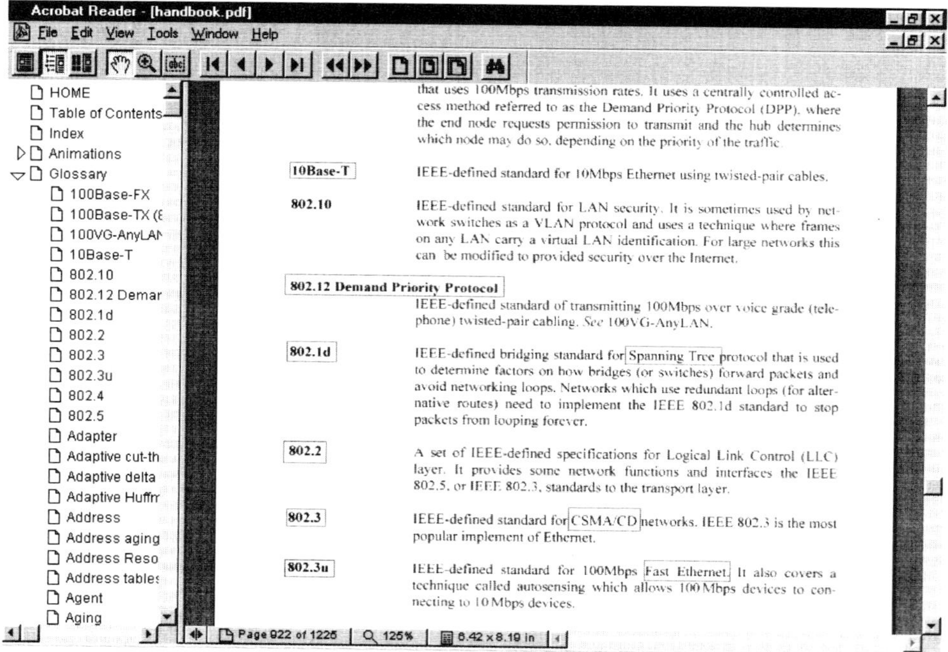

Figure 8: Glossary terms.

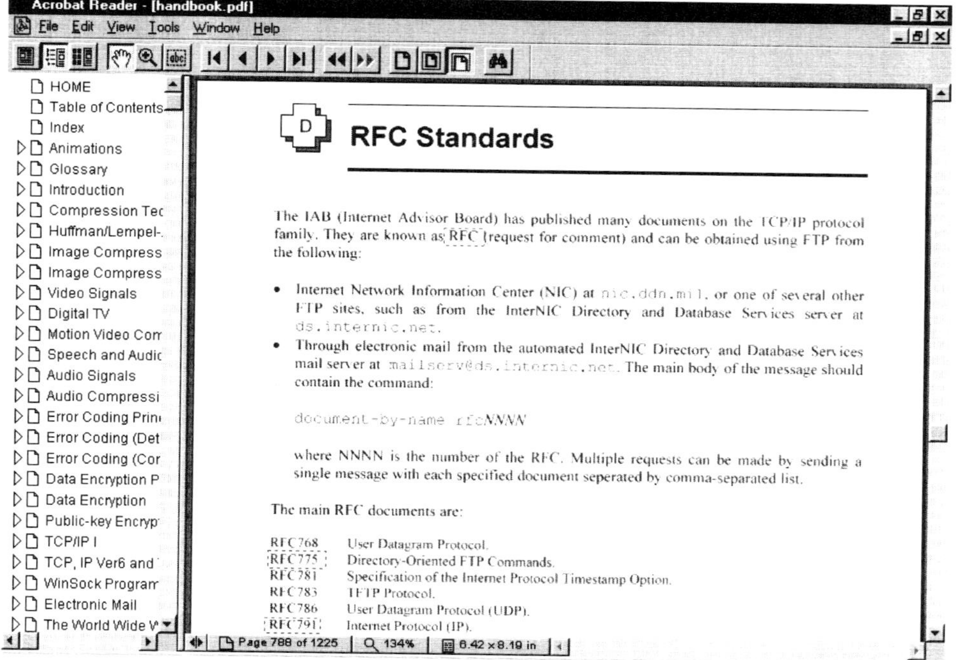

Figure 9: RFC chapter.

There are various animations on the CD. These are identified within a solid red box. Figure 10 and Figure 11 show an example.

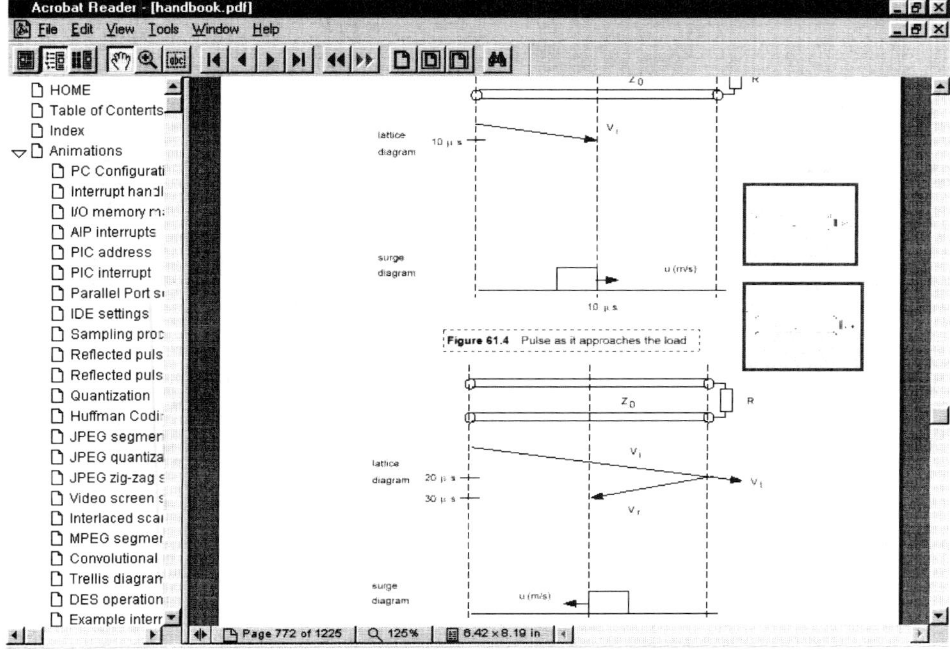

Figure 10: Sample animation.

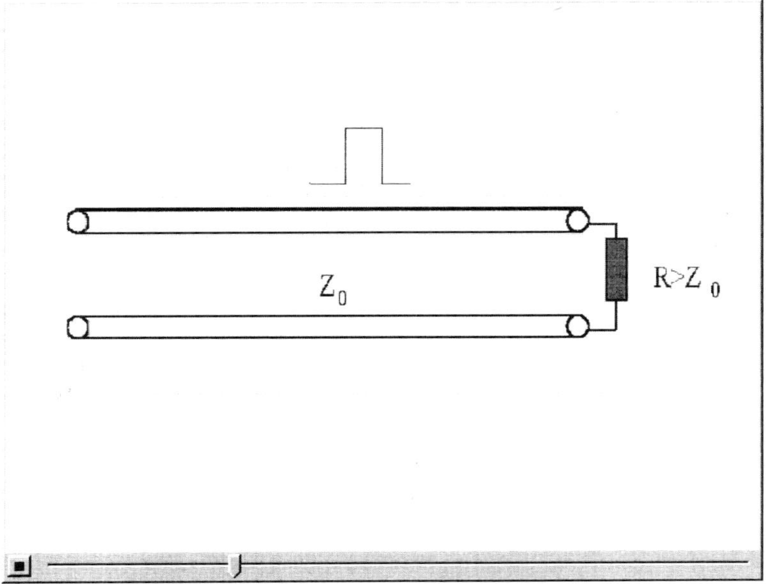

Figure 11: Sample animation.